Lecture Notes in Computer Science 8704

Commenced Publication in 1973
Founding and Former Series Editors:
Gerhard Goos, Juris Hartmanis, and Jan van Leeuwen

Advanced Research in Computing and Software Science

Subline of Lectures Notes in Computer Science

Paolo Baldan   Daniele Gorla (Eds.)

CONCUR 2014 – Concurrency Theory

25th International Conference, CONCUR 2014
Rome, Italy, September 2-5, 2014
Proceedings

Springer

Volume Editors

Paolo Baldan
University of Padova
Department of Mathematics
Via Trieste, 63
35121 Padova, Italy
E-mail: baldan@math.unipd.it

Daniele Gorla
University of Rome "La Sapienza"
Department of Computer Science
Via Salaria, 113
00198 Rome, Italy
E-mail: gorla@di.uniroma1.it

ISSN 0302-9743
e-ISSN 1611-3349
ISBN 978-3-662-44583-9
e-ISBN 978-3-662-44584-6
DOI 10.1007/978-3-662-44584-6
Springer Heidelberg New York Dordrecht London

Library of Congress Control Number: 2014946233

LNCS Sublibrary: SL 1 – Theoretical Computer Science and General Issues

Typesetting: Camera-ready by author, data conversion by Scientific Publishing Services, Chennai, India

Printed on acid-free paper

Springer is part of Springer Science+Business Media (www.springer.com)

Preface

This volume contains the proceedings of the 25th Conference on Concurrency Theory (CONCUR 2014), held in Rome, at the University of Rome "La Sapienza" and organized by University of Rome "La Sapienza" and the University of Padova. The purpose of the CONCUR conference is to bring together researchers, developers and students in order to advance the theory of concurrency and promote its applications. The principal topics include basic models of concurrency such as abstract machines, domain theoretic models, game theoretic models, categorical models, process algebras, graph transformation systems, coalgebraic models and Petri nets; logics for concurrency such as modal logics, probabilistic and stochastic logics, temporal logics, and resource logics; models of specialized systems such as biology-inspired systems, circuits, hybrid systems, mobile and collaborative systems, multi-core processors, probabilistic systems, real-time systems, service-oriented computing, and synchronous systems; verification and analysis techniques for concurrent systems such as abstract interpretation, atomicity checking, model checking, race detection, pre-order and equivalence checking, run-time verification, state-space exploration, static analysis, synthesis, testing, theorem proving, and type systems; related programing models such as distributed, component-based, object-oriented, web services and security issues in concurrent systems. This was the 25th edition of CONCUR. To mark this special occasion, the conference program included an invited presentation by Tony Hoare on the history of the theory of concurrency, the way it intertwines with the origin and development of the CONCUR conferences and the future perspectives in the field of concurrency theory. The conference program was further greatly enhanced by the enlightening invited talks by

- Javier Esparza (Technische Universität München, Germany)
- Jane Hillston (University of Edinburgh, UK),
- Catuscia Palamidessi (Inria Saclay and LIX, France),
- Vasco Vasconcelos (Universidade de Lisboa, Portugal).

This edition of the conference attracted 124 submissions. We wish to thank all the authors for their interest in CONCUR 2014. After careful discussions, the Program Committee selected 35 papers for presentation at the conference. Each submission was refereed by at least three reviewers, who delivered detailed and insightful comments and suggestions. The conference chairs warmly thank all the members of the Program Committee and all the referees for their hard and professional work, as well as for the friendly and constructive discussions. We would also like to thank the authors for having done their best to revise their papers taking into account the comments and suggestions by the referees.

The conference this year was co-located with two other conferences: TGC 2014 (9th International Symposium on Trustworthy Global Computing) and IFIP-TCS (8th International IFIP Conference on Theoretical Computer Science). Additionally, CONCUR 2014 included the following satellite workshops:

- EXPRESS/SOS (Combined 21st International Workshop on Expressiveness in Concurrency and 11th Workshop on Structured Operational Semantics), organized by Johannes Borgström and Silvia Crafa;
- YR-CONCUR (5th Young Researchers Workshop on Concurrency Theory), organized by Matteo Cimini;
- BEAT (3rd International Workshop on Behavioural Types), organized by Simon Gay, on behalf of COST Action IC1201 (BETTY);
- FOCLASA (13th International Workshop on the Foundations of Coordination Languages and Self-Adaptation), organized by Javier Cámara and José Proença;
- PV 2014 (Workshop on Parameterized Verification), organized by Giorgio Delzanno and Parosh A. Abdulla;
- TRENDS (Trends in Concurrency Theory), organized by Ilaria Castellani and Mohammad Mousavi, under the auspices of IFIP WG 1.8.

We would like to thank all the people who contributed to the success of CONCUR 2014, in particular the workshop organization chair Silvia Crafa, the Organizing Committee and the administrative staff. Furthermore, we thank the University of Rome "La Sapienza" and the Department of Mathematics of the University of Padova for their financial support. We are also grateful to Andrei Voronkov for his excellent EasyChair conference management system, which was extremely helpful for the electronic submission of papers, the Program Committee discussions and the production of the proceedings.

June 2014

Paolo Baldan
Daniele Gorla

Organization

Program Committee

Luca Aceto	University of Reykjavik, Iceland
Christel Baier	TU Dresden, Germany
Paolo Baldan	University of Padova, Italy
Luís Caires	Universidade Nova Lisboa, Portugal
David de Frutos Escrig	Universidad Complutense Madrid, Spain
Stéphanie Delaune	ENS Cachan, France
Wan Fokkink	Vrije Universiteit Amsterdam and CWI, The Netherlands
Yuxi Fu	Shanghai Jiaotong University, China
Fabio Gadducci	University of Pisa, Italy
Daniele Gorla	University of Rome "La Sapienza", Italy
Rachid Guerraoui	EPFL Zurich, Switzerland
Joshua Guttman	Worcester Polytechnic Institute, USA
Keijo Heljanko	Aalto University, Finland
Bartek Klin	University of Warsaw, Poland
K. Narayan Kumar	Chennai Mathematical Institute, India
Antonín Kučera	Masaryk University, Czech Republic
Barbara König	University of Duisburg-Essen, Germany
Radu Mardare	Aalborg University, Denmark
Andrzej Murawski	University of Warwick, UK
Madhusudan Parthasarathy	University of Illinois, USA
Anna Philippou	University of Cyprus, Cyprus
Shaz Qadeer	Microsoft Research, USA
Arend Rensink	University of Twente, The Netherlands
Peter Selinger	Dalhousie University, Canada
Alwen Tiu	Nanyang Technological University, Singapore
Daniele Varacca	Paris Diderot, France
Björn Victor	Uppsala University, Sweden
James Worrell	Oxford University, UK

Additional Reviewers

Abbes, Samy
Almagor, Shaull
Åman Pohjola, Johannes
Aminof, Benjamin
André, Étienne
Arcaini, Paolo
Atig, Mohamed Faouzi
Bacci, Giorgio
Bacci, Giovanni
Baez, John

Bakhshi, Rena
Barbanera, Franco
Barbosa, Luis
Barnat, Jiri
Bartoletti, Massimo
Basilico, Nicola
Basu, Samik
Bengtson, Jesper
Bergenthum, Robin
Bernardi, Giovanni
Bernardo, Marco
Berwanger, Dietmar
Bloem, Roderick
Blom, Stefan
Bollig, Benedikt
Boker, Udi
Bono, Viviana
Bonsangue, Marcello
Borgström, Johannes
Bortolussi, Luca
Brazdil, Tomas
Brenguier, Romain
Burckhardt, Sebastian
Caillaud, Benoît
Capecchi, Sara
Carbone, Marco
Cassel, Sofia
Cassez, Franck
Chen, Taolue
Ciancia, Vincenzo
Cimini, Matteo
Czerwiński, Wojciech
Delahaye, Benoit
Della Monica, Dario
Delzanno, Giorgio
Demangeon, Romain
Deng, Yuxin
Dinsdale-Young, Thomas
Dodds, Mike
Donaldson, Alastair
Doyen, Laurent
Emmi, Michael
Enea, Constantin
Escobar, Santiago
Faella, Marco
Feng, Xinyu
Ferreira, Carla
Filiot, Emmanuel
Forejt, Vojtech
Fossati, Luca
Fábregas, Ignacio
Gastin, Paul
Gebler, Daniel
Ghica, Dan
Giunti, Marco
Gogacz, Tomasz
Gregorio-Rodríguez, Carlos
Guenot, Nicolas
Göller, Stefan
Haase, Christoph
Hague, Matthew
Hawblitzel, Chris
He, Chaodong
Heckel, Reiko
Heindel, Tobias
Hendriks, Dimitri
Hildebrandt, Thomas
Hofman, Piotr
Horn, Florian
Horsman, Clare
Hunter, Paul
Jancar, Petr
Jansen, David N.
Jurdzinski, Marcin
Kakutani, Yoshihiko
Kerstan, Henning
Kiefer, Stefan
Klein, Joachim
Klueppelholz, Sascha
Knowles, Kenneth
Korenciak, Lubos
Kouzapas, Dimitrios
Kraehmann, Daniel
Krcal, Jan
Kretinsky, Jan
Krivine, Jean
Kähkönen, Kari
Küpper, Sebastian
Laarman, Alfons
Lanese, Ivan

Laneve, Cosimo
Lange, Martin
Langerak, Rom
Lazic, Ranko
Le, Duy-Khanh
Lin, Anthony
Lin, Anthony Widjaja
Lodaya, Kamal
Long, Huan
Malkis, Alexander
Markey, Nicolas
Marti-Oliet, Narciso
Martins, Francisco
Mazza, Damiano
McIver, Annabelle
Mcneile, Ashley
Melatti, Igor
Miculan, Marino
Mio, Matteo
Mio, Matteo
Montesi, Fabrizio
Mousavi, Mohammadreza
Mukhopadhyay, Partha
Mukund, Madhavan
Möller, Bernhard
Namjoshi, Kedar
Neatherway, Robin
Nigam, Vivek
Norman, Gethin
Novotný, Petr
Obdrzalek, Jan
Orejas, Fernando
Ortega-Mallén, Yolanda
Ouaknine, Joel
Ouederni, Meriem
Padovani, Luca
Palomino, Miguel
Parlato, Gennaro
Parthasarathy, Madhusudan
Perdrix, Simon
Perevoshchikov, Vitaly
Perez, Jorge A.
Peron, Adriano
Piazza, Carla
Popeea, Corneliu
Praveen, M.
Preoteasa, Viorel
Pérez, Jorge A.
Qiu, Daowen
Quaas, Karin
Ramanujam, R.
Randour, Mickael
Rehak, Vojtech
Riveros, Cristian
Rodriguez, Cesar
Rodriguez, Ismael
Romero-Hernández, David
Rosa-Velardo, Fernando
Rubin, Sasha
Rujiters, Enno
Saarikivi, Olli
Sack, Joshua
Saivasan, Prakash
Salvo, Ivano
Sanguinetti, Guido
Sankur, Ocan
Saurin, Alexis
Schmidt, Benedikt
Schmitz, Sylvain
Schnoebelen, Philippe
Sezgin, Ali
Siirtola, Antti Tapani
Sistla, A. Prasad
Smith, Geoffrey
Sobocinski, Pawel
Sokolova, Ana
Srba, Jiri
Srivathsan, B
Staton, Sam
Stirling, Colin
Strejcek, Jan
Stückrath, Jan
Sumii, Eijiro
Suresh, S.P.
T. Vasconcelos, Vasco
Thomas, Ehrhard
Toninho, Bernardo
Torres Vieira, Hugo
Trivedi, Ashutosh
Turrini, Andrea

Tzevelekos, Nikos
van Breugel, Franck
Van Eijck, Jan
Van Glabbeek, Rob
Vogler, Walter
Wahl, Thomas
Yildiz, Bugra
Yin, Qiang
Yoshida, Nobuko
Zappa Nardelli, Francesco
Zavattaro, Gianluigi
Zhang, Chihao
Zhang, Lijun
Zhu, Yunyun

Invited Talks

Laws of Programming: The Algebraic Unification of Theories of Concurrency

Tony Hoare

Microsoft Research (Cambridge) Ltd.

Abstract. I began my academic research career in 1968, when I moved from industrial employment as a programmer to the Chair of Computing at the Queens University in Belfast. My chosen research goal was to discover an axiomatic basis for computer programming. Originally I wanted to express the axioms as algebraic equations, like those which provide the basis of arithmetic or group theory. But I did not know how. After many intellectual vicissitudes, I have now discovered the simple secret. I would be proud of this discovery, if I were not equally ashamed at taking so long to discover it.

The Benefits of Sometimes Not Being Discrete

Jane Hillston

LFCS, School of Informatics, University of Edinburgh
jane.hillston@ed.ac.uk
http://www.quanticol.eu

Abstract. Discrete representations of systems are usual in theoretical computer science and they have many benefits. Unfortunately they also suffer from the problem of state space explosion, sometimes termed the *curse of dimensionality.* In recent years, research has shown that there are cases in which we can reap the benefits of discrete representation during system description but then gain from more efficient analysis by approximating the discrete system by a continuous one. This paper will motivate this approach, explaining the theoretical foundations and their practical benefits.

Deterministic Negotiations: Concurrency for Free

Javier Esparza

Fakultät für Informatik, Technische Universität München, Germany

Abstract. We give an overview of recent results and work in progress on deterministic negotiations, a concurrency model with atomic multi-party negotiations as primitive actions.

The Progress of Session Types

Vasco Thudichum Vasconcelos

LaSIGE, Faculty of Sciences, University of Lisbon

The session types can be traced back to 1993, when Kohei Honda presented "Types for dyadic interaction" in the seventh edition of this conference [5]. This seminal work, introducing basic type constructors and type composition operators for "dyadic interaction", was followed by two other papers, the first introducing a channel-based programming language [11], and later extending these ideas to a more general setting where channels may carry channels, while integrating recursive types [6].

Session types aim at modelling generic, meaningful structures of interaction. The first versions, those prevalent until 2008, encompassed exactly two interacting partners, as in term "dyadic". We have seen applications to concurrent, message passing systems, including the pi calculus (or a mild variation of it) as in [6] or in functional languages equipped with channel operations [12], but also to object-oriented systems, where session types mediate access to object's methods [3, 4].

By the turn of the millennium, communication had become a central concern in computational systems. Structures limited to describing binary interactions fall short of capturing the big picture of complex systems, even if they manage to represent all the individual binary interactions, necessarily in an unrelated manner. Different proposals address this matter, for example, by extending binary session types to scenarios of multiple participants [7] or by starting from new type constructs to describe multiparty interactions [2].

Types that capture the interaction patterns of a collection of participants find multiple applications these days, including the conventional verification of source code conformance against types, or, when the above deems not possible, the monitoring of running code against types, signalling divergences or providing for adaptation measures. They may as well be used for code generation, mechanically laying down the whole communication code, to be manually completed with the "computational" code, or for testing code against communication traces extracted from types.

The success of session types is due in part for its simplicity. With a suitable syntax, types become intuitive descriptions of protocols [10]. Furthermore, session types interact easily with programming languages; in fact they were developed to be integrated in programming languages. What is more surprising is that they can equip programming languages that were not designed with session types in mind, as for example conventional object-oriented languages [3, 4] or concurrent functional languages with channel based communication [12]. Also, recent developments revealed deep connections between session types and linear logic [1].

What lies ahead? There is strong sense of linearity associated to session types. Session types make (possibly long) series of interaction look like atomic, free from interference from other computations. This is usually achieved via a tight control on who keeps a reference to the interaction medium (the channel or the object reference, for example). At times, more flexible mechanisms would be welcome, but there is fine balance between flexibility and the kind of properties session types ensure. There are also important models of computation that pose difficulties to session types as we know them. These are systems whose assumptions lie outside those that underlay session types. I recall distributed systems with nodes that may die, and the actor system of computation [8, 9].

References

1. Caires, L., Pfenning, F.: Session types as intuitionistic linear propositions. In: Gastin, P., Laroussinie, F. (eds.) CONCUR 2010. LNCS, vol. 6269, pp. 222–236. Springer, Heidelberg (2010)
2. Caires, L., Vieira, H.T.: Conversation types. Theoretical Computer Science 411(51-52), 4399–4440 (2010)
3. Fähndrich, M., Aiken, M., Hawblitzel, C., Hodson, O., Hunt, G., Larus, J.R., Levi, S.: Language support for fast and reliable message-based communication in Singularity OS. SIGOPS Operating Systems Review 40(4), 177–190 (2006)
4. Gay, S., Vasconcelos, V.T., Ravara, A., Gesbert, N., Caldeira, A.Z.: Modular session types for distributed object-oriented programming. In: Proceedings of POPL 2010, pp. 299–312. ACM (2010)
5. Honda, K.: Types for dyadic interaction. In: Best, E. (ed.) CONCUR 1993. LNCS, vol. 715, pp. 509–523. Springer, Heidelberg (1993)
6. Honda, K., Vasconcelos, V.T., Kubo, M.: Language primitives and type discipline for structured communication-based programming. In: Hankin, C. (ed.) ESOP 1998. LNCS, vol. 1381, pp. 122–138. Springer, Heidelberg (1998)
7. Honda, K., Yoshida, N., Carbone, M.: Multiparty asynchronous session types. In: Proceedings of POPL 2008, pp. 273–284. ACM (2008)
8. Mostrous, D., Vasconcelos, V.T.: Session typing for a featherweight erlang. In: De Meuter, W., Roman, G.-C. (eds.) COORDINATION 2011. LNCS, vol. 6721, pp. 95–109. Springer, Heidelberg (2011)
9. Neykova, R., Yoshida, N.: Multiparty session actors. In: Kühn, E., Pugliese, R. (eds.) COORDINATION 2014. LNCS, vol. 8459, pp. 131–146. Springer, Heidelberg (2014)
10. Scribble project home page, http://www.scribble.org
11. Takeuchi, K., Honda, K., Kubo, M.: An interaction-based language and its typing system. In: Halatsis, C., Philokyprou, G., Maritsas, D., Theodoridis, S. (eds.) PARLE 1994. LNCS, vol. 817, pp. 398–413. Springer, Heidelberg (1994)
12. Vasconcelos, V.T., Gay, S.J., Ravara, A.: Typechecking a multithreaded functional language with session types. Theoretical Computer Science 368(1-2), 64–87 (2006)

Generalized Bisimulation Metrics*

Konstantinos Chatzikokolakis [1,2], Daniel Gebler [3],
Catuscia Palamidessi [4,2], and Lili Xu [2,5]

[1] CNRS
[2] LIX, Ecole Polytechnique
[3] VU University Amsterdam
[4] INRIA
[5] Institute of Software, Chinese Academy of Science

Abstract. The bisimilarity pseudometric based on the Kantorovich lifting is one of the most popular metrics for probabilistic processes proposed in the literature. However, its application in verification is limited to linear properties. We propose a generalization of this metric which allows to deal with a wider class of properties, such as those used in security and privacy. More precisely, we propose a family of metrics, parametrized on a notion of distance which depends on the property we want to verify. Furthermore, we show that the members of this family still characterize bisimilarity in terms of their kernel, and provide a bound on the corresponding metrics on traces. Finally, we study the case of a metric corresponding to differential privacy. We show that in this case it is possible to have a dual form, easier to compute, and we prove that the typical constructs of process algebra are non-expansive with respect to this metrics, thus paving the way to a modular approach to verification.

* This work has been partially supported by the project ANR-12-IS02-001 PACE, the project ANR-11-IS02-0002 LOCALI, the INRIA Equipe Associée PRINCESS, the INRIA Large Scale Initiative CAPPRIS, and the EU grant 295261 MEALS.

Table of Contents

25th Anniversary Talk

Invited Talks

Process Calculi

Model Checking and Abstraction

Synthesis

Quantitative Models

Automata and Multithreading

Complexity

Process Calculi and Types

Categories, Graphs and Quantum Systems

Automata and Time

Games

Laws of Programming: The Algebraic Unification of Theories of Concurrency

Tony Hoare

Microsoft Research (Cambridge) Ltd.

Abstract. I began my academic research career in 1968, when I moved from industrial employment as a programmer to the Chair of Computing at the Queens University in Belfast. My chosen research goal was to discover an axiomatic basis for computer programming. Originally I wanted to express the axioms as algebraic equations, like those which provide the basis of arithmetic or group theory. But I did not know how. After many intellectual vicissitudes, I have now discovered the simple secret. I would be proud of this discovery, if I were not equally ashamed at taking so long to discover it.

1 Historical Background

In 1969 [6], I reformulated Bob Floyd's assertional method of assigning meanings to programs [4] as a formal logic for conducting verification proofs. The basic judgment of the logic was expressed as a triple, often written

$$\{p\}q\{r\}.$$

The first operand of the triple (its precondition p) is an assertion, i.e., a description of the state of the computer memory before the program is executed. The middle operand (q) is the program itself, and the third operand (its post-condition r) is also an assertion, describing the state of memory after execution.

I now realise that there is no need to confine the precondition and the post-condition to be simply assertions. They can be arbitrary programs. The validity of the logic in application to programming is not affected. However the restrictions are fully justified by the resulting simplification in application of the logic to program verification.

The logic itself was specified as a collection of proof rules, similar to the system of natural deduction for reasoning in propositional logic. I illustrated the rules by the proof of correctness of a single small and over-simplified program, a very long division. This method of verification has since been used by experts in the proof of many more programs of much greater significance.

In the 1970s, my interests turned to concurrent programming, which I had failed to understand when I was manager of an operating system project for my industrial employer (the project failed [10]). To develop and confirm my understanding, I hoped to find simple proof rules for verification of concurrent programs. In fact, I regarded simplicity of proof as an objective criterion of

P. Baldan and D. Gorla (Eds.): CONCUR 2014, LNCS 8704, pp. 1–6, 2014.

the quality of any feature proposed for inclusion in a high level programming language - just as important as a possibly conflicting criterion, efficiency of implementation. As a by-product of the search for the relevant proof rules, I developed two features for shared-memory multiprogramming: a proposal for the conditional critical region [7], and later for a program structure known as the monitor [8].

At this time, the microprocessor revolution was offering an abundance of cheap computer power for programs running on multiple processors. They were connected by simple wires, and did not share a common memory. It was therefore important that communication on the wires entailed a minimum of software or hardware overhead. The criterion of efficiency led me to the proposal of a programming language structure known as Communicating Sequential Processes (CSP) [9]. The results of this research were exploited in the design of a comparatively successful microprocessor series, the INMOS transputer, and its less widely used programming language *occam* [14]. However, I was worried by the absence of a formal verification system for CSP.

In 1977 I moved to Oxford University, where Dana Scott had developed a tradition of denotational semantics for the formal definition of programming languages [20]. This tradition defines the meaning of a program in terms of all its possible behaviours when executed. I exploited the research capabilities of my Doctoral students at Oxford, Steve Brookes and Bill Roscoe; they developed a trace-based denotational semantics of CSP and proved that it satisfies a powerful and elegant set of algebraic laws [3].

Roscoe exploited the trace-based semantics in a model checking tool called *FDR* [19]. Its purpose was to explore the risk of deadlocks, non-termination and other errors in a program. On discovery of a potential failure, the trace of events leading up to the error helps the programmer to explore its possible causes, and choose which of them to correct.

2 The Origins of CONCUR

In the 1990s, I was a co-investigator on the basic research action CONCUR, funded by the European Community as an ESPRIT project. Its goal was a unification of three rival theories of concurrent programming: CSP (described above), a Calculus of Communicating Systems (CCS, due to Robin Milner) [17], and an Algebra of Communicating Processes (ACP, due to Jan Bergstra) [1,2]. These three designs differed not only in the details of their syntax, but also in the way that their semantic foundations were formalised.

Milners CCS was defined by an operational semantics. Its basic judgment is also a triple, called a transition

$$r \xrightarrow{q} p.$$

But in this case, the first operand (r) is the program being executed, the second operand (q) is a possible initial action of the program, and the third operand (p) is the program which remains to be executed after the first action has been performed. The operational semantics is given as a set of rules for deriving transitions, similar to those for deriving triples by verification logic.

Such an operational semantics is most directly useful in the design and implementation of interpreters and compilers for the language. In fact, the restriction of q to a single atomic action is motivated by this application. Relaxation of the restriction leads to a 'big step' semantics, which is equally valid for describing concurrent programs, but less useful for describing implementations.

The semantics of ACP was expressed as a set of algebraic equations and inequations, of just the kind I originally wanted in 1969. Equations between pairs of operands are inherently simpler and more comprehensible than triples, and algebraic substitution is a simpler and more powerful method of reasoning than that described by proof rules. Thus algebra is directly useful in all forms of reasoning about programs, including their optimisation for efficient execution on available hardware.

Unfortunately, we did not exploit this power of algebra to achieve the unification between theories that was the goal of the CONCUR project. In spite of the excellent research of the participants, this goal eluded us. I explained the failure as ultimately due to the three different methods of describing the semantics. I saw them as rivals, rather than complementary methods, useful for different purposes.

Inspired by this failure, in the 1990s I worked with my close colleague He Jifeng on a book entitled *"Unifying Theories of Programming"* (published in 1998) [5]. It was based on a model in which programs are relations between the initial and final states of their execution. To represent errors like non-termination, the relations were required to satisfy certain 'healthiness' constraints. Unfortunately, we could not find a simple and realistic model for concurrency and communication in a relational framework.

3 The Laws of Programming [11]

In the 1980s, the members of the Programming Research Group at Oxford were pursuing several lines of research in the theory of programming. There were many discussions of our apparently competing approaches. However, we all agreed on a set of algebraic laws covering sequential programming. The laws stated that the operator of sequential composition (;) is associative, has a unit (**skip**), and distributes through non-deterministic choice ($\sqcup$). This choice operator is associative, commutative and idempotent. I now recommend introduction of concurrent composition as a new and independent operator ($\|$). It shares all the algebraic properties of sequential composition and in addition it is commutative.

These algebraic properties are very familiar. They are widely taught in secondary schools. They are satisfied by many different number systems in arithmetic. And their application to computer programs commands almost immediate assent from experienced programmers.

A less familiar idea in the algebra of programming is a fundamental refinement ordering ($p < q$), which holds between similar or comparable programs.

It means that q can do everything that p can do, but maybe more. Thus p is a more determinate program than q; in all circumstances, it is therefore a valid implementation of q. Furthermore, if q has no errors, then neither has p. The algebraic principle of substitution of sub-terms within a term is strengthened to state that replacement of any sub-term of p by a sub-term that refines it will lead to a refinement of the original term p. This property is often formalised by requiring all the operators of the algebra to be monotonic with respect to the refinement ordering. Equality, and the substitution of equals, is just an extreme special case of refinement.

The most important new law governing concurrency is called the exchange law [12,13]. I happened upon it in 2007, and explored and developed it in collaboration with Ian Wehrman, then an intern with me at Microsoft Research. The law has the form of a refinement, expressing a sort of mutual distribution between sequential composition and concurrent composition. It is modelled after the interchange law, which is part of the mathematical definition of a two-category [16]. Although the law has four operands, it is similar in shape to other familiar laws of arithmetic:

$$(p \parallel q); (p \parallel q) < (p; p) \parallel (q; q)$$

The exchange law can be interpreted as expressing the validity of interleaving of threads as an implementation their concurrent composition. Such an interleaving is still widely used in time-sharing a limited number of processing units among a larger number of threads. But the law does not exclude the possibility of true concurrency, whereby actions from different threads occur simultaneously. As a result, the law applies both to shared-memory concurrency with conditional critical regions, as well as to communicating process concurrency, with either synchronous or buffered communication. Such a combination of programming idioms occurs widely in practical applications of concurrent systems.

4 Unification of Theories

This small collection of algebraic laws also plays a central role in the unification of other theories of concurrency, and other methods of presenting its semantics. For example, the deductive rules of Hoare logic can themselves be proved from the laws by elementary algebraic reasoning, just as the rules of natural deduction are proved from the Boolean Algebra of propositions. The proofs are based on a simple algebraic definition of the Hoare triple:

$$\{p\}q\{r\} \triangleq p\,; q < r$$

Hoare logic has more recently been extended by John Reynolds and Peter O'Hearn to include separation logic [15,18], which provides methods for reasoning about object orientation as well as concurrency. It thereby fills two serious gaps in the power of the original Hoare logic. The two new rules of concurrent separation logic can be simply proved from the single exchange law. And vice-versa: the exchange law can be proved from the rules of separation logic.

The concurrency rules for the transitions of Milners CCS can be similarly derived from the exchange law. Again, the proof is reversible. The definition of the Milner transition is remarkably similar to that of the Hoare triple:

$$r \xrightarrow{q} p \triangleq q\,;p < r$$

As a consequence, every theorem of Hoare logic can be translated to a theorem of Milner semantics by changing the order of the operands. And vice versa.

Additional operational rules that govern transitions for sequential composition can also be proved algebraically. The derivation from the same algebraic laws of two distinct (and even rival) systems for reasoning about programs is good evidence for the validity of the laws, and for their usefulness in application to programs.

Finally, a denotational semantics in terms of traces has an important role in defining a mathematical model for the laws. The model is realistic to the actual internal behaviour of a program when it is executed. It therefore provides an effective way of describing the events leading up to an error in the program, and in helping its diagnosis and correction.

5 Prospects

The main initial value of the unification of theories in the natural sciences is to enable experts to agree on the foundation, and collaborate in development of different aspects and different applications of it. To persuade a sceptical engineer (or manager) to adopt a theory for application on their next project, agreement among experts is an essential prerequisite. It is far too risky to apply a theory on which experts disagree.

In the longer term, the full value of a theory of programming will only be realised when their use by programmers is supported by a modern program development toolset. Such a toolset will contain a variety of sophisticated tools, based on different presentations of the same underlying theory. For example, program analysers and verification aids are based on deductive logic. Programming language interpreters and compilers are based on operational semantics. Program generators and optimisers are based on algebraic transformations. Finally, debugging aids will be based on a denotational model of program behaviour.

The question then arises: how do we know that all these different tools are fully consistent with each other? This is established by proof of the consistency of the theories on which the separate tools have been based. Mutual derivation of the theories is the strongest and simplest form of consistency: it establishes in principle the mutual consistency of tools that are based on the separate theories. What is more, the consistency is established by a proof that can be given even in advance of the detailed design of the toolset.

Acknowledgements. My sincere thanks are due to Daniele Gorla and Paolo Baldan for their encouragement and assistance in the preparation of this contribution.

References

1. Bergstra, J., Klop, J.: Fixed point semantics in process algebra. Technical Report IW 208, Mathematical Centre, Amsterdam (1982)
2. Bergstra, J.A., Klop, J.W.: Process algebra for synchronous communication. Information and Control 60(1-3), 109–137 (1984)
3. Brookes, S.D., Hoare, C.A.R., Roscoe, A.W.: A theory of communicating sequential processes. Journal of the ACM 31(3), 560–599 (1984)
4. Floyd, R.: Assigning meanings to programs. In: Proceedings of Symposium on Applied Mathematics, vol. 19, pp. 19–32 (1967)
5. Hoare, C.A.R., He, J.: Unifying Theories of Programming. Prentice Hall International Series in Computer Science (1998)
6. Hoare, C.A.R.: An axiomatic basis for computer programming. Communications of the ACM 12(10), 576–580 (1969)
7. Hoare, C.A.R.: Towards a theory of parrallel programming. In: Hoare, C.A.R., Perrott, R.H. (eds.) Operating Systems Techniques, Proceedings of Seminar at Queen's University, Belfast, Northern Ireland, pp. 61–71. Academic Press (1972)
8. Hoare, C.A.R.: Monitors: An operating system structuring concept. Communications of the ACM 17(10), 549–557 (1974)
9. Hoare, C.A.R.: Communicating sequential processes. Communications of the ACM 21(8), 666–677 (1978)
10. Hoare, C.A.R.: The emperor's old clothes. Communications of the ACM 24(2), 75–83 (1981)
11. Hoare, C.A.R., Hayes, I.J., He, J., Morgan, C., Roscoe, A.W., Sanders, J.W., Sørensen, I.H., Spivey, J.M., Sufrin, B.: Laws of programming. Communications of the ACM 30(8), 672–686 (1987)
12. Hoare, C.A.R., Möller, B., Struth, G., Wehrman, I.: Concurrent kleene algebra. In: Bravetti, M., Zavattaro, G. (eds.) CONCUR 2009. LNCS, vol. 5710, pp. 399–414. Springer, Heidelberg (2009)
13. Hoare, C.A.R., Wehrman, I., O'Hearn, P.W.: Graphical models of separation logic. In: Engineering Methods and Tools for Software Safety and Security. IOS Press (2009)
14. INMOS. occam Programming Manual. Prentice Hall (1984)
15. Ishtiaq, S.S., O'Hearn, P.W.: BI as an assertion language for mutable data structures. In: Proc. of POPL, pp. 14–26 (2001)
16. Mac Lane, S.: Categories for the working mathematician, 2nd edn. Springer, Heidelberg (1998)
17. Milner, R.: A Calculus of Communication Systems. LNCS, vol. 92. Springer, Heidelberg (1980)
18. Reynolds, J.C.: Separation logic: A logic for shared mutable data structures. In: Proc. of LICS, pp. 55–74 (2002)
19. Roscoe, A.W.: Model-checking CSP. In: A Classical Mind: Essays in Honour of C.A.R. Hoare. Prentice Hall International (UK) Ltd. (1994)
20. Scott, D., Strachey, C.: Toward a mathematical semantics for computer languages. Oxford Programming Research Group Technical Monograph, PRG-6 (1971)

The Benefits of Sometimes Not Being Discrete

Jane Hillston

LFCS, School of Informatics, University of Edinburgh, UK
`jane.hillston@ed.ac.uk`
`http://www.quanticol.eu`

Abstract. Discrete representations of systems are usual in theoretical computer science and they have many benefits. Unfortunately they also suffer from the problem of state space explosion, sometimes termed the *curse of dimensionality*. In recent years, research has shown that there are cases in which we can reap the benefits of discrete representation during system description but then gain from more efficient analysis by approximating the discrete system by a continuous one. This paper will motivate this approach, explaining the theoretical foundations and their practical benefits.

1 Introduction

Over the last twenty to thirty years, areas of quantitative modelling and analysis, such as performance, dependability and reliability modelling have embraced formal models [37]. This trend has been motivated by the increasing concurrency of the systems under consideration and the difficulties of constructing the underlying mathematical models, which are used for analysis, by hand. In particular concurrent modelling formalisms such as stochastic Petri nets and stochastic process algebras have been widely adopted as high-level modelling languages for generating underlying Markovian models. Moreover, there has been much work exploring how the properties of the high-level languages can be exploited to assist in the analysis of the underlying model through a variety of techniques (e.g. decomposition [23, 39], aggregation based on bisimulation [38], etc).

However, a combination of improved model construction techniques, and the increasing scale and complexity of the systems being developed, has led to ever larger models; and these models now frequently defy analysis even after model reduction techniques such as those mentioned above. The problem is the well-known curse of dimensionality: the state space of a discrete event system can grow exponentially with the number of components in the system.

Fortunately, over the last decade a new approach has emerged which offers a way to avoid this state space explosion problem, at least for one class of models. When the system under consideration can be presented as a *population* model and the populations involved are known to be *large*, then a good approximation of the discrete behaviour can be achieved through a continuous or fluid approximation. Moreover, this model is scale-free in the sense that the computational effort to solve it remains the same even as the populations involved grow larger. Of course, there is a cost, in the sense that some information is lost and it is no longer possible to analyse the system in terms of

P. Baldan and D. Gorla (Eds.): CONCUR 2014, LNCS 8704, pp. 7–22, 2014.

individual behaviours. But when average behaviours or expectations are required, for example in situations of collective behaviour, the fluid approach has substantial benefits.

The rest of this paper is organised as follows. Section 2 gives an intuitive explanation of how the fluid approximation approach has been widely used in biological modelling for many years, before presenting the mathematical foundations for the approach as provided by Kurtz's Theorem in Section 3. The attraction of combining the technique with the compositional models generated by process algebras is explained in Section 4, with discussion of how the mapping has been developed for a variety of process algebras. In Section 5 we give an overview of extending these results into the model checking arena, and in Section 6 we briefly summarise and conclude.

2 Biologists Just Do It!

In several disciplines fluid approximations have long been used, often without concern for formal foundations. The most noticeable example of this is in biological modelling of intra-cellular processes. These processes result from the collisions of molecules within the cell, an inherently discrete process. Yet, the most common form of mathematical model for these processes is a system of ordinary differential equations (ODEs) which captures the collective behaviour in terms of concentrations of different molecular states, rather than the states of individual molecules. At heart, this is a fluid approximation, as highlighted by Kurtz [46] and Gillespie [32]. But it has been so widely adopted that many biologists no longer recognise that there is a fundamental shift in representation taking place.

That there was an implicit transformation taking place during model construction became more obvious when formal representations started to be used to describe intracellular biological processes [58]. In the early 2000s researchers recognised that the intracellular processes were highly concurrent systems, amenable to description formalisms used to describe concurrency in computer systems. This led to a plethora of adopted and developed process algebras for describing cellular processes e.g. [18, 56, 57, 24]. Whilst most focussed on the discrete representation and subsequent discrete event simulation of an underlying continuous time Markov chain (CTMC) using Gillespie's algorithm [32], work such as [17, 20] established that it was also possible to derive the systems of ODEs more familiar to biologists from process algebra descriptions.

3 Kurtz's Theorem

At the foundations of fluid approximation is a fundamental result by Kurtz, dating back to the 1970s [45], which establishes that a sequence of CTMCs which satisfy some conditions and represent essentially the same system under growing populations, converges to a set of ODEs. At convergence the behaviour of the CTMC is indistinguishable from the behaviour of the set of ODEs. However, this theoretical limit is at an infinite population. Nevertheless in many practical cases we find empirically that sufficient convergence is often achieved at much lower populations, as illustrated in Fig. 1.

In order to explain this result in more detail we introduce a simple representation of Markov models of populations of interacting agents. Such models may be readily

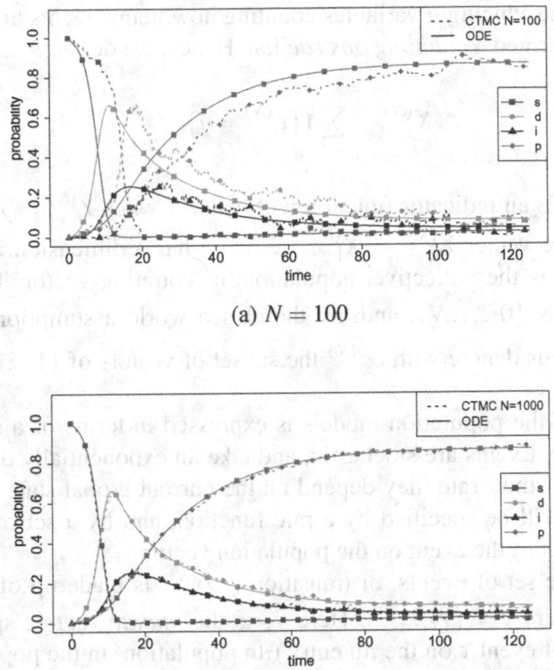

(a) $N = 100$

(b) $N = 1000$

Fig. 1. Comparison between the limit fluid ODE and a single stochastic trajectory of a network epidemic example, for total populations $N = 100$ and $N = 1000$. This demonstrates how the accuracy of the approximation of behaviour captured by the fluid ODE improves as the population size grows.

derived from stochastic process algebras such as PEPA or EMPA [38, 7]. We consider the case of models of processes evolving in continuous time, although a similar theory can be considered for discrete-time models (see, for instance, [13]). In principle, we can have different classes of agents, and many agents for each class in the system. To keep notation simple, we assume here that the number of agents is constant and equal to N (making a closed world assumption) but analogous results can be derived for systems which include the birth and death of agents.

In particular, let us assume that each agent is a finite state machine, with internal states taken from a finite set S, and labelled by integers: $S = \{1, 2, \ldots, n\}$. We have a population of N agents, and denote the state of agent i at time t, for $i = 1, \ldots, N$, by $Y_i^{(N)}(t) \in S$. Note that we have made explicit the dependence on N, the total population size.

A configuration of a system is thus represented by the tuple $(Y_1^{(N)}, \ldots, Y_N^{(N)})$. This representation is based on treating each agent as a distinct individual with identity conferred by the position in the vector. However, when dealing with population models, it is customary to assume that single agents in the same internal state cannot be distinguished, hence we can move from the *individual representation* to the *collective representation* by introducing n variables counting how many agents are in each state. This is sometimes termed a *counting abstraction*. Hence, we define

$$X_j^{(N)} = \sum_{i=1}^{N} \mathbf{1}\{Y_i^{(N)} = j\}, \tag{1}$$

where $\mathbf{1}\{Y_i^{(N)} = j\}$ is an indicator function with value 1 when $Y_i^{(N)} = j$ and zero, otherwise. Note that the vector $\mathbf{X}^{(N)} = (X_1^{(N)}, \ldots, X_n^{(N)})$ has a dimension independent of N; it is referred to as the collective, population, or counting vector. The domain of each variable $X_j^{(N)}$ is $\{0, \ldots, N\}$, and, by the closed world assumption, it holds that $\sum_{j=1}^{n} X_j^{(N)} = N$. Let us denote with $\mathscr{S}^{(N)}$ the subset of vectors of $\{1, \ldots, N\}^n$ that satisfy this constraint.

The dynamics of the population models is expressed in terms of a set of possible *events* or *transitions*. Events are stochastic, and take an exponentially distributed time to happen. Moreover their rate may depend on the current global state of the system. Hence, each event will be specified by a rate function, and by a set of update rules, specifying the impact of the event on the population vector.

In this model, the set of events, or transitions, $\mathscr{T}^{(N)}$, is made up of elements $\tau \in \mathscr{T}^{(N)}$, which are pairs $\tau = (\mathbf{v}_\tau, r_\tau^{(N)})$. Here $\mathbf{v}_\tau$ is the *update vector*; specifically $\mathbf{v}_{\tau,i}$ records the impact of event τ on the ith entry (ith population) in the population vector. The rate function, $r_\tau^{(N)} : \mathscr{S}^{(N)} \to \mathbb{R}_{\geq 0}$, depends on the current state of the system, and specifies the speed of the corresponding transition. It is assumed to be equal to zero if there are not enough agents available to perform a τ transition, and it is required to be *Lipschitz continuous* (when interpreted as a function on real numbers).

Thus we define a population model $\mathscr{X}^{(N)} = (\mathbf{X}^{(N)}, \mathscr{T}^{(N)}, \mathbf{x}_0^{(N)})$, where $\mathbf{x}_0^{(N)}$ is the initial state. Given such a model, it is straightforward to construct the CTMC $\mathbf{X}^{(N)}(t)$ associated with it; its state space is $\mathscr{S}^{(N)}$, while its infinitesimal generator matrix $Q^{(N)}$ is the $|\mathscr{S}^{(N)}| \times |\mathscr{S}^{(N)}|$ matrix defined by

$$q_{\mathbf{x},\mathbf{x}'} = \sum \{r_\tau(\mathbf{x}) \mid \tau \in \mathscr{T},\ \mathbf{x}' = \mathbf{x} + \mathbf{v}_\tau\}.$$

As explained above, fluid approximation approximates a CTMC by a set of ODEs. These differential equations can be interpreted in two different ways: they can be seen as an approximation of the average of the system (usually a first order approximation, see [9, 68]). This is often termed a *mean field* approximation. Alternatively, they can be interpreted as an approximate description of system trajectories for large populations. We will focus on this second interpretation, which corresponds to a functional version of the law of large numbers. In this interpretation, instead of having a sequence of random variables, like the sample mean, converging to a deterministic value, like the true mean, in this case we have a sequence of CTMCs (which can be seen as random trajectories in $\mathbb{R}^n$) for increasing population size, which converge to a deterministic trajectory, the solution of the fluid ODE.

In order to consider the convergence, we must formally define the sequence of CTMCs to be considered. To allow models of different population sizes to be compared we normalise the populations by dividing each variable by the total population N. In this way, the normalised population variables $\hat{\mathbf{X}}^{(N)} = \frac{\mathbf{X}^{(N)}}{N}$, or population densities, will always range between 0 and 1 (for the closed world models we consider here), and so the behaviour for different population sizes can be compared. In the case of a constant population, normalised variables are usually referred to as the *occupancy measures*, as they represent the fraction of agents which occupy each state.

After normalisation we must appropriately scale the update vectors, initial conditions, and rate functions [13]. Let $\mathscr{X}^{(N)} = (\mathbf{X}^{(N)}, \mathscr{T}^{(N)}, \mathbf{X_0}^{(N)})$ be the non-normalised model with total population N and $\hat{\mathscr{X}}^{(N)} = (\hat{\mathbf{X}}^{(N)}, \hat{\mathscr{T}}^{(N)}, \hat{\mathbf{X}}_\mathbf{0}^{(N)})$ the corresponding normalised model. We require that:

- initial conditions scale appropriately: $\hat{\mathbf{X}}_\mathbf{0}^{(N)} = \frac{\mathbf{X_0}^{(N)}}{N}$;
- for each transition $(\mathbf{v}_\tau, r_\tau^{(N)}(\mathbf{X}))$ of the non-normalised model, define $\hat{r}_\tau^{(N)}(\hat{\mathbf{X}})$ to be the rate function expressed in the normalised variables (obtained from $r_\tau^{(N)}$ by a change of variables). The corresponding transition in the normalised model is $(\mathbf{v}_\tau, \hat{r}_\tau^{(N)}(\hat{\mathbf{X}}))$, with update vector equal to $\frac{1}{N}\mathbf{v}_\tau$.

We further assume, for each transition τ, that there exists a bounded and Lipschitz continuous function $f_\tau(\hat{\mathbf{X}}) : E \to \mathbb{R}^n$ on normalised variables (where E contains all domains of all $\hat{\mathscr{X}}^{(N)}$), independent of N, such that $\frac{1}{N}\hat{r}_\tau^{(N)}(\mathbf{x}) \to f_\tau(\mathbf{x})$ *uniformly* on E. We denote the state of the CTMC of the N-th non-normalised (resp. normalised) model at time t as $\mathbf{X}^{(N)}(t)$ (resp. $\hat{\mathbf{X}}^{(N)}(t)$).

3.1 Deterministic Limit Theorem

In order to present the "classic" deterministic limit theorem, consider a sequence of normalised models $\hat{\mathscr{X}}^{(N)}$ and let $\mathbf{v}_\tau$ be the (non-normalised) update vectors. The *drift* $F^{(N)}(\hat{\mathbf{X}})$ of $\hat{\mathscr{X}}$, which is formally the mean instantaneous increment of model variables in state $\hat{\mathbf{X}}$, is defined as

$$F^{(N)}(\hat{\mathbf{X}}) = \sum_{\tau \in \hat{\mathscr{T}}} \frac{1}{N} \mathbf{v}_\tau \hat{r}_\tau^{(N)}(\hat{\mathbf{X}}) \tag{2}$$

Furthermore, let $f_\tau : E \to \mathbb{R}^n$, $\tau \in \hat{\mathscr{T}}$ be the limit rate functions of transitions of $\hat{\mathscr{X}}^{(N)}$. The *limit drift* of the model $\hat{\mathscr{X}}^{(N)}$ is therefore

$$F(\hat{\mathbf{X}}) = \sum_{\tau \in \hat{\mathscr{T}}} \mathbf{v}_\tau f_\tau(\hat{\mathbf{X}}), \tag{3}$$

and $F^{(N)}(\mathbf{x}) \to F(\mathbf{x})$ uniformly as $N \longrightarrow \infty$, as easily checked. The fluid ODE is

$$\frac{d\mathbf{x}}{dt} = F(\mathbf{x}), \quad with\ \mathbf{x}(0) = \mathbf{x_0} \in S.$$

Given that F is Lipschitz in E (since all f_τ are), this ODE has a unique solution $\mathbf{x}(t)$ in E starting from $\mathbf{x_0}$. Then, one can prove the following theorem:

Theorem 1 (Deterministic approximation [45, 27]). *Let the sequence $\hat{\mathbf{X}}^{(N)}(t)$ of Markov processes and $\mathbf{x}(t)$ be defined as above, and assume that there is some point*

$\mathbf{x_0} \in S$ *such that* $\hat{\mathbf{X}}^{(N)}(0) \rightarrow \mathbf{x_0}$ *in probability. Then, for any* finite *time horizon* $T < \infty$, *it holds that as* $N \longrightarrow \infty$:

$$\mathbb{P}\left\{ \sup_{0 \le t \le T} ||\hat{\mathbf{X}}^{(N)}(t) - \mathbf{x}(t)|| > \varepsilon \right\} \rightarrow 0.$$

Notice that the Theorem makes assertions about the trajectories of the population counts at all finite times, but nothing about what happens at steady state, i.e. when time goes to infinity.

3.2 Fast Simulation

Based on the Deterministic Approximation Theorem, we can consider the implications for a single individual in the population when the population size goes to infinity. Even as the collective behaviour tends to a deterministic process, each individual agent will still behave randomly. However, the Deterministic Approximation Theorem implies that the dynamics of a single agent, in the limit, becomes independent of other agents, and it will sense them only through the collective system state, or *mean field*, described by the fluid limit. This asymptotic decoupling allows us to find a simple, *time-inhomogenous*, Markov chain for the evolution of the single agent, a result often known as *fast simulation* [28, 30].

To see this decoupling we focus on a single individual $Y_h^{(N)}(t)$, which is a (Markov) process on the state space $S = \{1, \ldots, n\}$, conditional on the global state of the population $\hat{\mathbf{X}}^{(N)}(t)$. The evolution of this agent can be obtained by computing the rates q_{ij} at which its state changes from i to j, by projecting on a single agent the rate of global transitions that induce a change of state of at least one agent from i to j. Such a rate $q_{ij}(\hat{\mathbf{X}})$ still depends on the global system state, hence to track the evolution of agent $Y_h^{(N)}(t)$ we still need to know the global state of the system $\hat{\mathbf{X}}^{(N)}(t)$: e.g. solving any model checking problem on $Y_h^{(N)}(t)$ would requires us to work with the full Markov model $\hat{\mathbf{X}}^{(N)}(t)$.

However, as the size of the system increases, the deterministic limit theorem tells us the stochastic fluctuations of $\hat{\mathbf{X}}^{(N)}(t)$ tend to vanish, and this effect propagates to the stochastic behaviour of $Y_h^{(N)}(t)$, which can be approximated by making it dependent only on the fluid limit $\mathbf{x}(t)$. More precisely, we need to construct the time-inhomogeneous CTMC $z(t)$ with state space S and rates $q_{ij}(\mathbf{x}(t))$, computed along the fluid trajectory. The following theorem [28] guarantees that $z(t)$ is a good approximation of $Y_h^{(N)}(t)$:

Theorem 2 (Fast simulation theorem). *For any finite time horizon* $T < \infty$, $\mathbb{P}\{Y_h^{(N)}(t) \neq z(t),$ *for some* $t \le T\} \rightarrow 0$, *as* $N \rightarrow \infty$.

This theorem states that, in the limit of an infinite population, each agent will behave independently from all the others, sensing only the mean state of the global system, described by the fluid limit $\mathbf{x}(t)$. This *asymptotic decoupling* of the system, which can be generalised to any subset of $k \geq 1$ agents, is also known in the literature under the name of *propagation of chaos* [5].

Remark 1. For simplicity here we have considered a single class of agents without births or deaths. Nevertheless the same results hold for a model consisting of multiple classes of agents. In this case we construct a single agent class but partition the state space S into subsets, each of which represents the states of a distinct agent, and such that there are no transitions between subsets. The agents whose initial state is in each subset corresponds to agents of that class. Furthermore, events that capture birth and death can easily be included by allowing update vectors which are unbalanced in the sense that the total positive update is greater than or less than the total negative update. Such *open* systems can be handled in the same theory, see [12] for further details, but for clarity we will restrict to closed world models in this paper.

4 Stochastic Process Algebra with Fluid Interpretation

Kurtz's Theorem, or the Deterministic Approximation Theorem, has been established for many years. It has been widely used but when it is used directly from a CTMC model, it is the modeller's responsibility to prove that the model satisfies the necessary conditions for application of the theory, and moreover, to derive the corresponding ODEs. This must be done on a model-by-model basis. In recent years, the approach has been used for several performance and dependability models e.g. [3–5, 30].

This situation made it attractive to incorporate mean field or fluid approximation into the formal high-level language approaches which have developed over the last two decades for constructing CTMC models for quantitative analysis. From the perspective of the formal modelling community, this gives access to a scalable analysis technique which is immune to the problem of state space explosion; indeed, a technique which increases in accuracy as the size of the model grows. From the perspective of modellers already familiar with the mean field approach, it offers the possibility to establish the conditions for convergence at the language level via the semantics, once and for all, removing the need to fulfil the proof obligation on a model-by-model basis. Moreover the derivation of the ODEs can be automated in the implementation of the language.

Work has developed in both stochastic Petri nets, e.g. [66, 60, 61] and stochastic process algebras, e.g. [43, 40, 16]. Here we focus on the work in the process algebra context as it is more readily related to the agent-based CTMC model presented in the previous section. It is straightforward to see that components or agents within the process algebra description can be regarded as agents within the CTMC model, typically occupying different partitions within the notional complete state space for agents, as explained at the end of Section 3. When multiple instances of a component are present in the same context within the model, these constitute a *population*. In terms of the language the dynamic combinators are associated with the description of the behaviour of individual agents, essentially finite state machines, whereas static combinators, principally parallel composition, specify the structure of the system, which is now interpreted as the formation and interaction of populations.

The fluid approximation approach is only applicable to models where we have interactions of large populations (parallel compositions of large numbers of components with the same behaviour) within which each component has relatively simple behaviour rather than interactions between individuals each with complex behaviour. When this is the case we need to make the shift from a state representation based on individuals, to one based on counting (analogous to the shift represented by equation (1)). How this is handled depends on the process algebra but is generally straightforward. For example, in PEPA models there is a simple procedure to reduce the syntactic representation to a state vector [40, 65], but in languages such as Bio-PEPA the mapping is more straightforward because the language was designed to support fluid approximation [24]. The actions of the algebra correspond to the events in the CTMC model, and the definition of the process and its continuation via an action is the basis for the definition of the update vector.

The first work relating process algebra and mean field models can be found in the thesis of Sumpter [62]. Sumpter developed models of social insects in the discrete synchronous process algebra WSCCS [63]. He then heuristically derived difference equations to capture the mean field representation of the model. This work inspired the work of Norman and Shankland [54], in which WSCCS is used to build models of the spread of infectious diseases and difference equation representations are derived. This led on to further work with ever more rigour introduced into the relationship between the difference equation/ODE models and the process algebra descriptions from which they were derived [52, 53, 51], but in later work the authors switched from using WSCCS to using PEPA and Bio-PEPA for their modelling of epidemics.

As previously mentioned, work in systems biology stimulated more widespread interest in the relationship between process algebra description and ODE models. The first work here was the mapping given from PEPA models constructed in a particular style, representing a reagent-centric view of biological signal transductions pathways, to equivalent ODE models, by Calder *et al.* [17]. This was subsequently generalised to more arbitrary PEPA models with large populations, where the mapping to the ODE was made completely systematic, based on an intermediate structure termed the *activity matrix* [40]. In the work of Bortolussi and Policriti the authors consider a different style of process algebra, stochastic Concurrent Constraint Programming (sCCP), and demonstrate a mapping, both from process algebra to ODEs and from ODEs to process algebra descriptions [16]. At around the same time Cardelli also constructed a systematic mapping from process algebra (in this case a variant of CCS) to ODEs, using a Chemical Parametric Form as an intermediary in this case [20]. The relationship between this interpretation of the process algebra model and the discrete-state stochastic semantics is explored in [19].

After these initial explorations of the possibilities to relate the inherently discrete representation of a process algebra model with a fluid approximation of the underlying Markov process, there came a sequence of papers establishing the mapping on a firmer foundation and considering the convergence properties which can be inferred from Kurtz's Theorem. For example in [31], Geisweiller *et al.*, working with a generalised form of PEPA models which allow two forms of synchronisation — both the usual PEPA synchronisation based on the bounded capacity, and the biological notion of mass action — show that the syntactically derived ODE models are indeed those which are obtained by the application of Kurtz's Theoreom, guaranteeing convergence in the limit. In [65], Tribastone *et al.* show how it is possible to fully formalise the derivation of the ODEs for PEPA models, via a structured operational semantics. In [16] Bortolussi and Policriti construct a process algebra that matches a given set of ODEs in the limit. An alternative approach to the derivation of the fluid approximation model is taken in the work on Kappa [26], where the ODEs are derived as an abstract interpretation.

Some authors also considered how to make the derivation of ODEs from process algebra descriptions easier. As previously mentioned, the PEPA variant, Bio-PEPA [24] was explicitly constructed to maintain a counting abstraction, initially making the derivation of the activity matrix easier and later supporting a semantics in the style of [65]. Hayden and Bradley developed another variant of PEPA, termed Grouped PEPA, which makes clearer the population structures within models [34].

The system ODEs derived from a process algebra model are generally not amenable to algebraic solution, but instead are analysed by numerical simulation. This solution generates a trajectory, tracking the population counts of each local state over time, which can be interpreted as the expected population value over time. Such expected population counts are rarely the objective of quantitative modelling in computer science, although they are often the focus in biological systems. In computer systems derived measures such as throughput, response times, or first passage times are of more interest. In [64], Tribastone *et al.* establish when performance measures such as throughput and response time may legitimately be derived from a fluid approximation. Hayden *et al.* develop an approach to derive the more sophisticated first passage time distributions [36]. When the "passage" of interest relates to an individual component within the model the approach taken relies on the use of the fast simulation result. In further work [35], Hayden *et al.* show how response-time measures specified by stochastic probes can be readily calculated via the mean field approach.

5 Fluid Model Checking

Stochastic process algebra models have long been also analysed using quantitative model checking. In the case of stochastic model checking, there are some consolidated approaches, principally based on checking Continuous Stochastic Logic (CSL) formulae [2, 1, 59], and these are supported by software tools which are in widespread use such as PRISM [47, 48] and MRMC [41]. However these methods often depend on an explicit representation of the state space and consequently suffer from the state space explosion problem, which limits their applicability, particularly for population models. Even when statistical model checking is used, and the state space is only built on-the-fly, the size of population models may make adequate statistical sampling costly or even unattainable.

Thus it is natural to ask the question, to what extent can the fluid approximation techniques presented earlier in this paper be exploited to mitigate the problem of quantitative model checking of population CTMC-based models. The first work in this direction was presented in [11, 12], in which fluid approximation is used to carry out approximate model checking of behaviours of individual agents in large population models, specified as CSL formulae. This work builds on the Fast Simulation Theorem [30, 28], which characterises the limit behaviour of a single agent in terms of the solution of the fluid equation. Recall that the Fast Simulation Theorem states that a single agent senses the rest of the population only through its "average" evolution, as given by the fluid equation. Thus if the modeller wishes to verify a property of an individual agent within a population of many interacting agents (possibly with a small set of different capabilities) the approach is to check the property in a limit model which consists of the discrete representation of the individual agent taking into account the average evolution of the rest of the system. In practice, for CTMC models, the discrete representation of the individual agent is a *time-inhomogeneous* CTMC (ICTMC), where the rates of transitions between states are determined by the fluid approximation of the rest of the system. Model checking of ICTMCs is far more complex than the homogeneous-time case, but this is compensated because only the local states of one agent need to be considered, so the state space is typically small. The authors termed this approach Fluid Model Checking. Preliminary ideas on using fluid approximation in continuous time for model checking population models, and in particular for an extension of the logic CSL, were informally sketched in [43], but no model checking algorithms were presented. Subsequently the work was more fully developed in [44], which relies substantially on [11].

In the Fluid Model Checking approach the technicalities come from the time-inhomogeneous nature of the process being checked. As in the CTMC case, model checking CSL formulas of ICTMC can be expressed in terms of reachability calculations on an ICTMC, typically with modified structure that makes some states absorbing. However, these calculations are more complex as rates are not constant, but changing over time as the state of the whole system evolves and influences the considered agent. This introduces discontinuities in the satisfaction probabilities as, for example, states in the ICTMC may change from being in the goal set to not, as time progresses. Thus the solution of the Kolmogorov equations to calculate the reachability must be conducted in a piecewise manner, between the time points at which the sets of goal states and unsafe states change over time. Convergence and quasi-decidability results are presented that guarantee the asymptotic consistency of the model checking [12].

Like all results from Kurtz's theorem, the Fluid Model Checking result pertains to models within a finite time horizon. However useful properties in CSL are sometimes expressed in terms of the steady state operator $\mathscr{S}$. Subsequently, Bortolussi and Hillston consolidated the Fluid Model Checking approach by incorporating the next state operator and the steady state operator [14]. This latter involved establishing when Kurtz's result can safely be extended to the infinite time horizon in this context.

A limitation of the Fluid Model Checking approach is that only properties of a single individual agent (or small set of agents) within a population can be checked. But for population models it is natural to wish to evaluate more global properties such as if a proportion of agents within a population have reached a particular state within a given time period. In [15], Bortolussi and Lanciani present an alternative approach which is able to deal with such properties. Their work is based on a second-order fluid approximation known as Linear Noise Approximation [68]. This can be regarded as a functional version of the Central Limit Approximation [45].

The basic idea of [15] is to lift local specifications to collective ones by means of the Central Limit Theorem. Thus the properties that they consider are first expressed as a property of an individual agent, specified by a deterministic timed automaton with a single clock. This clock is taken to be global — it is never reset and keeps track of global passing of time. For an individual this will be a linear-time property. Such an individual property $\varphi(t)$ is then lifted to the population level to estimate the probability that a given number of agents within the system which satisfy $\varphi(t)$.

The method presented in [15] allows us to quickly estimate this probability by exploiting the Central Limit or Linear Noise Approximation (LNA). The key idea is to keep some estimation of the variability in the system. Rather than solely using the fluid approximation of average behaviour of the normalised behaviour $x(t)$, fluctuations in the form of Gaussian processes of the order of $\sqrt{N}$, where N is the population size, are included.

$$\mathbf{X}^{(N)}(t) \approx Nx(t) + \sqrt{N}Z(t),$$

where $Z(t)$ is a Gaussian stochastic process, i.e. a process whose finite dimensional projection (marginal distributions at any fixed and finite set of times) are Gaussian. $Z(t)$ has zero mean, and a covariance given by the solution of an additional set of $\mathcal{O}(N^2)$ ODEs. More details can be found in [15, 68].

For the purposes of model checking the authors combine the automaton-based property specification with the model of an individual agent, using a product construction (taking into account the clock constraints). This produces a population model with more variables, counting pairs of state-property configurations. The LNA is applied to this new model. The authors show that for a large class of individual properties, it is possible to introduce a variable $X_\varphi(t)$ in the extended model that counts how many individual agents satisfy the local property up to time t. From the Gaussian approximation of $X_\varphi(t)$, then one can easily compute the probabilities of interest. In [15], the authors discuss preliminary results, which are quite accurate and computationally efficient.

A further use of mean field approximation in model checking has recently been developed for discrete time, synchronous-clock population processes by Loreti *et al.* [49]. Although also derived from Kurtz's Theorem, this work takes a different approach as it is an *on-the-fly* model checker, only examining states as they are required for checking the property, rather than constructing the whole state space initially [25, 8, 33]. Similarly to Fluid Model Checking [11], in [49] the authors focus on a single individual or small set of individuals, with properties expressed in PCTL, and consider their evolution in the mean field created by the rest of the system. Again fast simulation provides the foundation for the approach, but for the discrete case, Loreti *et al.* follow the approach of [50] in which the behaviour of each agent is captured by a finite state discrete time Markov chain (DTMC).

As previously, the authors consider a system comprised of N agents, each with some initial state. A system *global state* $\mathbf{C}^{(N)} = \langle c_1, \ldots, c_N \rangle$ is the N-tuple of the current local states of its object instances. The dynamics of the system arise from all agents proceeding in discrete time, synchronously. A transition matrix $\mathbf{K}^{(N)}$ defines the state transitions of the object and their probabilities, and this may depend on the distribution of states of *all* agents in the system. More specifically, $\mathbf{K}^{(N)}$ is a function $\mathbf{K}^{(N)}(\mathbf{m})$ of the *occupancy measure* vector $\mathbf{m}$ of the current global state $\mathbf{C}^{(N)}$ (switching to the counting abstraction and normalising). State labels are associated with the states of an agent in its specification, and a global state is taken to assume the labels of the first component in the N-tuple. Further global system atomic properties can be expressed.

In [49] the authors develop a model checking algorithm which can applied in both the exact probabilistic case, and for the approximate mean-field semantics of the models. Here we focus on the latter approach. In this discrete case, for N large, the overall behaviour of the system in terms of its occupancy measure can be approximated by the (deterministic) solution of a mean-field *difference equation*. Loreti *et al.* show that the deterministic iterative procedure developed in [50] to compute the average overall behaviour of the system and behaviour of individual agents in that context, combines well with on-the-fly probabilistic model checking for bounded PCTL formulas on the selected agents. Just as in Fluid Model Checking [11], since the transition probabilities of individual agents may depend on the occupancy measure at a given time, the truth values of formulas may vary with time. The asymptotic correctness of the model checking procedure has been proven and a prototype implementation of the model checker, FlyFast, which has been applied to a variety of models [49].

One drawback of mean-field or fluid approximation is that the convergence results apply to infinite populations and currently there are not useful bounds on the errors introduced when smaller populations are considered. Some promising work in this direction was recently published by Bortolussi and Hayden [10]. In this paper the authors consider the transient dynamics and the steady state of certain classes of discrete-time population Markov processes. They combine stochastic bounds in terms of martingale inequalities and Chernoff inequalities, with control-theoretic methods to study the stability of a system perturbed by non-deterministic noise terms, and with algorithms to over-approximate the set of reachable states. The key idea is to abstract stochastic noise non-deterministically and apply techniques from control theory to examine the phase space of the mean field limit. This gives a more refined view of the dynamic behaviour allowing tighter bounds than the previously proposed bounds of Darling and Norris [28] which expand exponentially with time.

6 Conclusions and Future Perspectives

The fluid approximation technique is suitable for models comprised of interactions of populations of components, each component having relatively simple behaviour (few or moderate numbers of local states) but many components within the population. Moreover, in these cases the accuracy of the approximation increases as the size of the population grows. Building such models with a discrete formal description technique supports careful specification of the interactions between the components. This is in contrast to when mean field or fluid approximation is applied in fields such as epidemiology where predefined sets of ODEs are used, without consideration for the implicit assumptions about the interactions of individuals.

However, the population models amenable to fluid approximation are not the only systems which suffer from state space explosion and the technique is not suitable for models comprised of a small number of individual components, each of which has very complex behaviour resulting in a large number of local states. Moreover, recent work by Tschaikowski and Tribastone has shown that if the mapping to ODEs is carried out naively, there can be a problem of *fluid state space explosion* [67]. Nevertheless, the approach offers new possibilities for model analysis, tackling systems which would previously have been completely intractable and opening new arenas of research.

Acknowledgement. This work is partially supported by the EU project QUANTICOL, 600708. The author is very grateful to Luca Bortolussi and Stephen Gilmore for help in the preparation of this paper.

References

1. Aziz, A., Singhal, V., Balarin, F., Brayton, R., Sangiovanni-Vincentelli, A.: Verifying continuous time Markov chains. In: Alur, R., Henzinger, T.A. (eds.) CAV 1996. LNCS, vol. 1102, pp. 269–276. Springer, Heidelberg (1996)
2. Baier, C., Haverkort, B., Hermanns, H., Katoen, J.P.: Model checking continuous-time Markov chains by transient analysis. In: Emerson, E.A., Sistla, A.P. (eds.) CAV 2000. LNCS, vol. 1855, pp. 358–372. Springer, Heidelberg (2000)
3. Bakhshi, R., Cloth, L., Fokkink, W., Haverkort, B.R.: Mean-field analysis for the evaluation of gossip protocols. In: Proceedings of 6th Int. Conference on the Quantitative Evaluation of Systems (QEST 2009), pp. 247–256 (2009)
4. Bakhshi, R., Cloth, L., Fokkink, W., Haverkort, B.R.: Mean-field framework for performance evaluation of push-pull gossip protocols. Perform. Eval. 68(2), 157–179 (2011)
5. Benaïm, M., Le Boudec, J.: A class of mean field interaction models for computer and communication systems. In: Performance Evaluation (2008)
6. Benaïm, M., Le Boudec, J.Y.: On mean field convergence and stationary regime. CoRR, abs/1111.5710 (2011)
7. Bernardo, M., Gorrieri, R.: A Tutorial on EMPA: A Theory of Concurrent Processes with Nondeterminism, Priorities, Probabilities and Time. Theor. Comput. Sci. 202(1-2), 1–54 (1998)
8. Bhat, G., Cleaveland, R., Grumberg, O.: Efficient On-the-Fly Model Checking for CTL*. In: Logic in Computer Science (LICS 1995), pp. 388–397 (1995)
9. Bortolussi, L.: On the approximation of stochastic concurrent constraint programming by master equation, vol. 220, pp. 163–180 (2008)
10. Bortolussi, L., Hayden, R.A.: Bounds on the deviation of discrete-time Markov chains from their mean-field model. Perform. Eval. 70(10), 736–749 (2013)
11. Bortolussi, L., Hillston, J.: Fluid model checking. In: Koutny, M., Ulidowski, I. (eds.) CONCUR 2012. LNCS, vol. 7454, pp. 333–347. Springer, Heidelberg (2012)
12. Bortolussi, L., Hillston, J.: Fluid model checking. CoRR, 1203.0920 (2012)
13. Bortolussi, L., Hillston, J., Latella, D., Massink, M.: Continuous approximation of collective systems behaviour: A tutorial. In: Performance Evaluation (2013)
14. Bortolussi, L., Hillston, J.: Checking Individual Agent Behaviours in Markov Population Models by Fluid Approximation. In: Bernardo, M., de Vink, E., Di Pierro, A., Wiklicky, H. (eds.) SFM 2013. LNCS, vol. 7938, pp. 113–149. Springer, Heidelberg (2013)
15. Bortolussi, L., Lanciani, R.: Model Checking Markov Population Models by Central Limit Approximation. In: Joshi, K., Siegle, M., Stoelinga, M., D'Argenio, P.R. (eds.) QEST 2013. LNCS, vol. 8054, pp. 123–138. Springer, Heidelberg (2013)
16. Bortolussi, L., Policriti, A.: Dynamical systems and stochastic programming: To ordinary differential equations and back. In: Priami, C., Back, R.-J., Petre, I. (eds.) Transactions on Computational Systems Biology XI. LNCS, vol. 5750, pp. 216–267. Springer, Heidelberg (2009)
17. Calder, M., Gilmore, S., Hillston, J.: Automatically deriving ODEs from process algebra models of signalling pathways. In: Proceedings of Computational Methods in Systems Biology (CMSB 2005), pp. 204–215 (2005)

18. Cardelli, L.: Brane Calculi. In: Danos, V., Schachter, V. (eds.) CMSB 2004. LNCS (LNBI), vol. 3082, pp. 257–278. Springer, Heidelberg (2005)
19. Cardelli, L.: On process rate semantics. Theor. Comput. Sci. 391(3), 190–215 (2008)
20. Cardelli, L.: From Processes to ODEs by Chemistry. In: Ausiello, G., Karhumäki, J., Mauri, G., Ong, L. (eds.) 5th IFIP International Conference On Theoretical Computer Science - TCS 2008. IFIP, vol. 273, pp. 261–281. Springer, Boston (2008)
21. Chen, T., Han, T., Katoen, J.-P., Mereacre, A.: LTL model checking of time-inhomogeneous markov chains. In: Liu, Z., Ravn, A.P. (eds.) ATVA 2009. LNCS, vol. 5799, pp. 104–119. Springer, Heidelberg (2009)
22. Chen, T., Han, T., Katoen, J.P., Mereacre, A.: Model checking of continuous-time Markov chains against timed automata specifications. Logical Methods in Computer Science 7(1) (2011)
23. Ciardo, G., Trivedi, K.S.: A Decomposition Approach for Stochastic Reward Net Models. Perform. Eval. 18(1), 37–59 (1993)
24. Ciocchetta, F., Hillston, J.: Bio-PEPA: A framework for the modelling and analysis of biological systems. Theor. Comput. Sci. 410(33-34), 3065–3085 (2009)
25. Courcoubetis, C., Vardi, M., Wolper, P., Yannakakis, M.: Memory-efficient algorithms for the verification of temporal properties Form. Methods Syst. Des. 1(2-3), 275–288 (1992)
26. Danos, V., Feret, J., Fontana, W., Harmer, R., Krivine, J.: Abstracting the Differential Semantics of Rule-Based Models: Exact and Automated Model Reduction. In: Proceedings of Logic in Computer Science (LICS 2010), pp. 362–381 (2010)
27. Darling, R.W.R.: Fluid limits of pure jump Markov processes: A practical guide (2002), `http://arXiv.org`
28. Darling, R.W.R., Norris, J.R.: Differential equation approximations for Markov chains. Probability Surveys 5 (2008)
29. Donatelli, S., Haddad, S., Sproston, J.: Model checking timed and stochastic properties with CSL^{TA}. IEEE Trans. Software Eng. 35(2), 224–240 (2009)
30. Gast, N., Gaujal, B.: A mean field model of work stealing in large-scale systems. In: Proceedings of ACM SIGMETRICS 2010, pp. 13–24 (2010)
31. Geisweiller, N., Hillston, J., Stenico, M.: Relating continuous and discrete PEPA models of signalling pathways. Theor. Comput. Sci. 404(1-2), 97–111 (2008)
32. Gillespie, D., Petzold, L.: Numerical simulation for biochemical kinetics. In: System Modeling in Cellular Biology, pp. 331–353. MIT Press (2006)
33. Gnesi, S., Mazzanti, F.: An Abstract, on the Fly Framework for the Verification of Service-Oriented Systems. In: Results of the SENSORIA Project, pp. 390–407 (2011)
34. Hayden, R.A., Bradley, J.T.: A fluid analysis framework for a Markovian process algebra. Theor. Comput. Sci. 411(22-24), 2260–2297 (2010)
35. Hayden, R.A., Bradley, J.T., Clark, A.D.: Performance specification and evaluation with unified stochastic probes and fluid analysis. IEEE Trans. Software Eng. 39(1), 97–118 (2013)
36. Hayden, R.A., Stefanek, A., Bradley, J.T.: Fluid computation of passage-time distributions in large Markov models. Theor. Comput. Sci. 413(1), 106–141 (2012)
37. Hermanns, H., Herzog, U., Katoen, J.-P.: Process algebra for performance evaluation. Theor. Comput. Sci. 274(1-2), 43–87 (2002)
38. Hillston, J.: A Compositional Approach to Performance Modelling. Cambridge University Press (1995)
39. Hillston, J.: Exploiting Structure in Solution: Decomposing Compositional Models. In: Brinksma, E., Hermanns, H., Katoen, J.-P. (eds.) EEF School 2000 and FMPA 2000. LNCS, vol. 2090, pp. 278–314. Springer, Heidelberg (2001)
40. Hillston, J.: Fluid flow approximation of PEPA models. In: Proceedings of the Second International Conference on the Quantitative Evaluation of SysTems, QEST 2005, pp. 33 – 42 (2005)

41. Katoen, J.-P., Khattri, M., Zapreev, I.S.: A Markov Reward Model Checker. In: Proceedings of Quantitative Evaluation of Systems, QEST 2005, pp. 243–244 (2005)
42. Katoen, J.-P., Mereacre, A.: Model checking HML on piecewise-constant inhomogeneous markov chains. In: Cassez, F., Jard, C. (eds.) FORMATS 2008. LNCS, vol. 5215, pp. 203–217. Springer, Heidelberg (2008)
43. Kolesnichenko, A., Remke, A., de Boer, P.T., Haverkort, B.R.: Comparison of the mean-field approach and simulation in a peer-to-peer botnet case study. In: Thomas, N. (ed.) EPEW 2011. LNCS, vol. 6977, pp. 133–147. Springer, Heidelberg (2011)
44. Kolesnichenko, A., Remke, A., de Boer, P.-T., Haverkort, B.R.: A logic for model-checking of mean-field models. In: Proceedings of 43rd Int. Conference on Dependable Systems and Networks, DSN 2013 (2013)
45. Kurtz, T.G.: Solutions of ordinary differential equations as limits of pure jump Markov processes. Journal of Applied Probability 7, 49–58 (1970)
46. Kurtz, T.G.: Approximation of population processes. SIAM (1981)
47. Kwiatkowska, M., Norman, G., Parker, D.: Probabilistic symbolic model checking with PRISM: A hybrid approach. International Journal on Software Tools for Technology Transfer 6(2), 128–142 (2004)
48. Kwiatkowska, M., Norman, G., Parker, D.: PRISM 4.0: Verification of probabilistic real-time systems. In: Gopalakrishnan, G., Qadeer, S. (eds.) CAV 2011. LNCS, vol. 6806, pp. 585–591. Springer, Heidelberg (2011)
49. Latella, D., Loreti, M., Massink, M.: On-the-fly Fast Mean-Field Model-Checking. In: Abadi, M., Lluch Lafuente, A. (eds.) TGC 2013. LNCS, vol. 8358, pp. 297–314. Springer, Heidelberg (2014)
50. Le Boudec, J.-Y., McDonald, D., Mundinger, J.: A Generic Mean Field Convergence Result for Systems of Interacting Objects. In: Proceedings of Quantitative Evaluation of Systems (QEST 2007), pp. 3–18 (2007)
51. McCaig, C.: From individuals to populations: changing scale in process algebra models of biological systems. PhD thesis, University of Stirling (2008)
52. McCaig, C., Norman, R., Shankland, C.: Process Algebra Models of Population Dynamics. In: Horimoto, K., Regensburger, G., Rosenkranz, M., Yoshida, H. (eds.) AB 2008. LNCS, vol. 5147, pp. 139–155. Springer, Heidelberg (2008)
53. McCaig, C., Norman, R., Shankland, C.: From Individuals to Populations: A Symbolic Process Algebra Approach to Epidemiology. Mathematics in Computer Science 2(3), 535–556 (2009)
54. Norman, R., Shankland, C.: Developing the Use of Process Algebra in the Derivation and Analysis of Mathematical Models of Infectious Disease. In: Moreno-Díaz Jr., R., Pichler, F. (eds.) EUROCAST 2003. LNCS, vol. 2809, pp. 404–414. Springer, Heidelberg (2003)
55. Norris, J.R.: Markov Chains. Cambridge University Press (1997)
56. Priami, C.: Stochastic pi-Calculus. Comput. J. 38(7), 578–589 (1995)
57. Regev, A., Panina, E.M., Silverman, W., Cardelli, L., Shapiro, E.Y.: BioAmbients: an abstraction for biological compartments. Theor. Comput. Sci. 325(1), 141–167 (2004)
58. Regev, A., Shapiro, E.: Cellular Abstractions: Cells as computation. Nature 419(6905), 343–343 (2002)
59. Rutten, J., Kwiatkowska, M., Norman, G., Parker, D.: Mathematical Techniques for Analyzing Concurrent and Probabilistic Systems. CRM Monograph Series, vol. 23. American Mathematical Society (2004)
60. Silva, M., Recalde, L.: Petri nets and integrality relaxations: A view of continuous Petri net models. IEEE Trans. on Systems, Man, and Cybernetics, Part C 32(4), 314–327 (2002)
61. Silva, M., Recalde, L.: On fluidification of Petri Nets: from discrete to hybrid and continuous models. Annual Reviews in Control 28(2), 253–266 (2004)

62. Sumpter, D.T.J.: From Bee to Society: An Agent-based Investigation of Honey Bee Colonies. PhD thesis, University of Manchester (2000)
63. Tofts, C.M.N.: A Synchronous Calculus of Relative Frequency. In: Baeten, J.C.M., Klop, J.W. (eds.) CONCUR 1990. LNCS, vol. 458, pp. 467–480. Springer, Heidelberg (1990)
64. Tribastone, M., Ding, J., Gilmore, S., Hillston, J.: Fluid rewards for a stochastic process algebra. IEEE Trans. Software Eng. 38(4), 861–874 (2012)
65. Tribastone, M., Gilmore, S., Hillston, J.: Scalable differential analysis of process algebra models. IEEE Trans. Software Eng. 38(1), 205–219 (2012)
66. Trivedi, K.S., Kulkarni, V.G.: FSPNs: Fluid Stochastic Petri Nets. In: Ajmone Marsan, M. (ed.) ICATPN 1993. LNCS, vol. 691, pp. 24–31. Springer, Heidelberg (1993)
67. Tschaikowski, M., Tribastone, M.: Tackling continuous state-space explosion in a Markovian process algebra. Theor. Comput. Sci. 517, 1–33 (2014)
68. Van Kampen, N.G.: Stochastic Processes in Physics and Chemistry. Elsevier (1992)

Deterministic Negotiations: Concurrency for Free

Javier Esparza

Fakultät für Informatik, Technische Universität München, Germany

Abstract. We give an overview of recent results and work in progress on deterministic negotiations, a concurrency model with atomic multi-party negotiations as primitive actions.

Concurrency theory has introduced and investigated system models based on a variety of communication primitives: shared variables with semaphores, monitors, or locks; rendez-vous; message-passing with point-to-point channels; coordination (message-passing with tuple space); or broadcast. Recently, we have started the study of a new primitive: *negotiation*. Perhaps surprisingly, while negotiation has long been identified as an interaction paradigm by the artificial intelligence community [7,21,4,16], its theoretical study as communication primitive has not been yet undertaken.

From a concurrency theory point of view, an atomic negotiation is a synchronized choice: a set of agents meet to choose one out of a set of possible outcomes. In [10], Jörg Desel and I have presented a model of concurrency model with atomic multi-party negotiations, called *atoms*, as primitive actions. The model is close to Petri nets, and it uses much of its terminology. For an intuitive introduction, consider Figure 1, which shows a negotiation between four *agents*, numbered 1 to 4. Atoms are represented by black bars. A bar has a white circle or *port* for each participanting agent. For instance, the initial atom n_0 has four parties, while atoms n_1 and n_2 have only two. The local state of an agent is the set of atoms it is currently ready to engage in. Initially, all agents are only ready to engage in the initial atom n_0. A *marking* is a tuple of local states, one for each agent. An atom is enabled at a marking if all its parties are ready to engage in it. Enabled atoms can occur, meaning that their parties agree on one of the possible outcomes. After choosing an outcome, the edges leaving the atom and labelled with the outcome determine the negotiations that each of the parties is ready to engage in next. For example, in Figure 1, at the initial atom the agents decide whether, say, to accept a proposal for discussion (outcome y(es)) or not (outcome n(o)). If the agents agree on n, then after that they are ready to engage in the final atom n_f, i.e, n_f is the only enabled atom, and after n_f occurs the negotiation terminates. If they agree on y, then after that agents 1 and 2 are ready to engage in n_1, while 3 and 4 are ready to engage in n_2, and so both atoms are enabled. After n_1 and n_2 occur, n_3 becomes enabled, and the four agents decide in n_3 whether to accept (outcome a) or reject (outcome r) the proposal; in case of rejection, the atoms n_1 and n_2 become enabled again, modeling that the two teams of agents make revisions to the proposal and discuss it again.

P. Baldan and D. Gorla (Eds.): CONCUR 2014, LNCS 8704, pp. 23–31, 2014.

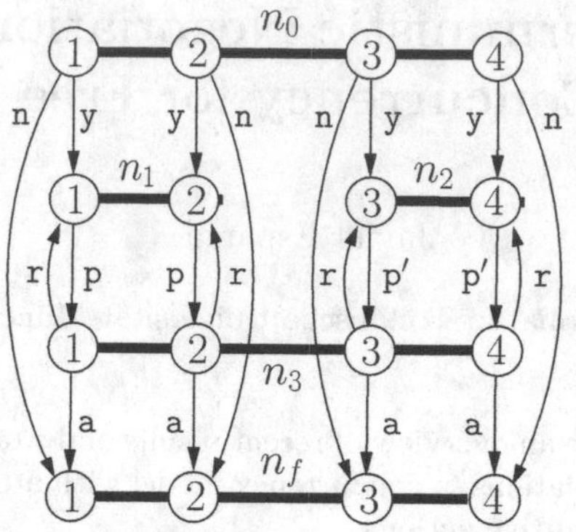

Fig. 1. A negotiation between four agents

Negotiations can handle data, although this is not part of the graphical representation. Agents also have an *internal* local state, typically determined by the current values of a set of variables local to the agent. For example, assume the goal of the negotiation of Figure 1 is to agree on the price of a commodity. The internal state of the i-th agent could then be given by a variable x_i, holding the agent's current proposal for the price. Each outcome of an atom has an associated state transformer, which only acts on the internal states of the agents that take part in the atom. For instance, the state transformer for the outcome p in atom n_1 would be a function that takes as input the current values of x_1 and x_2, and sets them to a new common value, the new price proposed by the two agents. In general, a state transformer is a relation between global states.

While every negotiation diagram can be translated into an equivalent Petri net, negotiations allow one to express some common situations more succintly [10]. For instance, consider a system in which k agents decide, independently of each other, whether they wish to accept or reject a proposal, and then conduct a negotiation requiring unanimity, that is, the proposal can only be approved if all agents support it. It is not difficult to see that the size of the Petri net modelling such a behaviour grows exponentially with k (essentially, the net needs a different transition for each subset of agents to cover the case in which exactly that subset rejects the proposal), while the size of the negotiation diagram grows only linearly in k.

The main merit of the negotiation model, however, is not succinctness, but the fact that it draws our attention to classes of systems which have a very natural definition within the model, but look contrived—and uninterestingly so—in others. In particular, in [10] we have defined *deterministic* negotiations[1]. A negotiation is deterministic if at every reachable state every agent is ready to participate in at most one atom. The negotiation of Figure 1 is deterministic: in fact, this follows directly from the fact that after choosing an outcome at an atom, the edges labelled with it direct the parties to one atom. But consider

[1] We also introduce *weakly deterministic* negotiations, but we don't discuss them in this note.

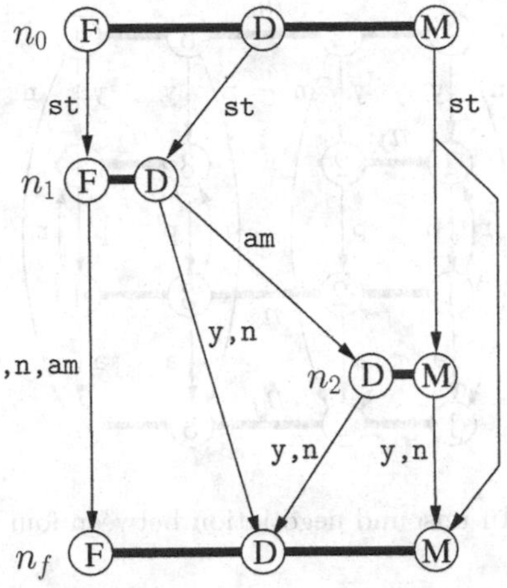

Fig. 2. A nondeterministic negotiation

now the negotiation of Figure 2. It is a negotiation between three agents, called Father, Daughter, and Mother, whose goal is to decide whether Daughter can go to a party. After the initial atom, Daughter and Father are ready to engage in n_1, while Mother is ready to engage in *both* n_2 and n_f (graphically denoted by the hyperedge connecting Mother's port at atom n_0 to her ports at atoms n_2 and n_f). Negotiations with proper hyperedges of this kind are called nondeterministic. Observe that after the initial atom occurs, the only atom enabled is n_1, and so Father and Daughter negotiate first, with possible outcomes `yes` (`y`), `no` (`n`), and `ask_mother` (`am`). Whether Mother participates in n_2 or in n_f is decided by the outcome of n_1: If Father and Daughter choose `am`, then atom n_2 becomes enabled, and Daughter and Mother negotiate with possible outcomes `yes`, `no`. If they choose `y` or `no`, then atom n_f becomes enabled, and the negotiation terminates.

The results of [10,11] and some recent work [12,9] show that deterministic negotiations are an exception to the "concurrency curse": the rule of thumb stating that all analysis problems for an interesting class of concurrent systems (where the input is the concurrent system itself, not its state space) will be at least NP- or PSPACE-hard. While the state space of a deterministic negotiation can grow exponentially in its size, we have derived algorithms for important analysis and synthesis problems that work directly on the negotiation diagram, without constructing its state space, and have polynomial complexity. This is our rationale for the title of this paper: in deterministic negotiations concurrency is "for free", in the sense that the capacity of the model to describe concurrent interaction does not require one to pay the usual "exponential fee". In the rest of this note we present a brief overview of our results.

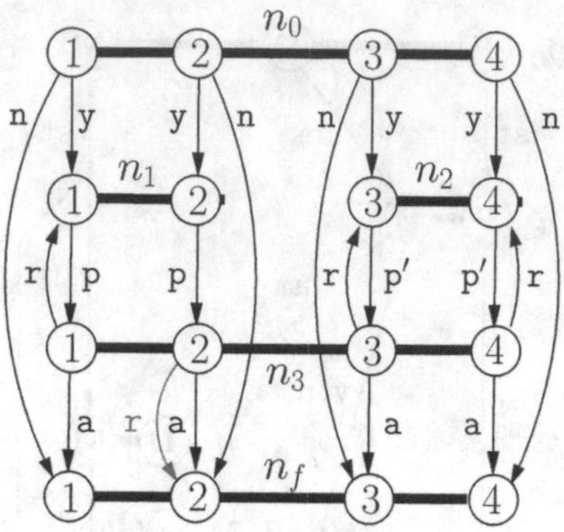

Fig. 3. An unsound negotiation between four agents

Soundness and Summarization. Like any other model of concurrency, negotiations can deadlock. Consider for instance the slight modification of the negotiation of Figure 1 shown in Figure 3. The only difference is that after outcome r agent 2 is now only ready to engage in the final atom. The negotiation reaches a deadlock after the sequence y p p′ r p′ of outcomes. Loosely speaking, a negotiation is *sound* if, whatever its current state, it can always finish, i.e., execute the final atom. (Readers familiar with workflow models will recognize this notion as the one defined in [1].) In particular, soundness implies deadlock-freedom. The negotiations of Figure 1 and Figure 2 are sound. The *soundness problem* consists of determining whether a given negotiation is sound, and it constitutes a first fundamental problem in the analysis of negotiations. A second problem comes from the fact that negotiations are expected to terminate. In particular, all negotiation diagrams have an initial and a final atom, and so an associated input/output relation on global states, where a global state, as usual, is a tuple of local states of the agents. The relation contains the pairs (q, q') of global states such that, if the agents start in state q, then the negotiation can terminate in state q'. Under the fairness assumption that it terminates, a sound negotiation is equivalent to a single atom whose state transformer determines the possible final internal states of all parties as a function of their initial internal states. The *summarization problem* consists of computing such an atomic negotiation, called a *summary*.

In [11] we have shown that the soundness and summarization problems for deterministic negotiations can be solved in polynomial time. The algorithm for the summarization problem takes the form of a reduction procedure in which the original negotiation is progressively reduced to a simpler one by means of three reduction rules. Each rule preserves soundness and summaries (i.e., the negotiation before the application of the rule is sound iff the negotiation after the application is sound, and both have the same summary). The rules are graphically described in Figure 4; for each rule, the figure at the top shows a fragment of a negotiation to which the rule can be applied, and the figure at the bottom the result of applying it. The rules generalize to a concurrent setting

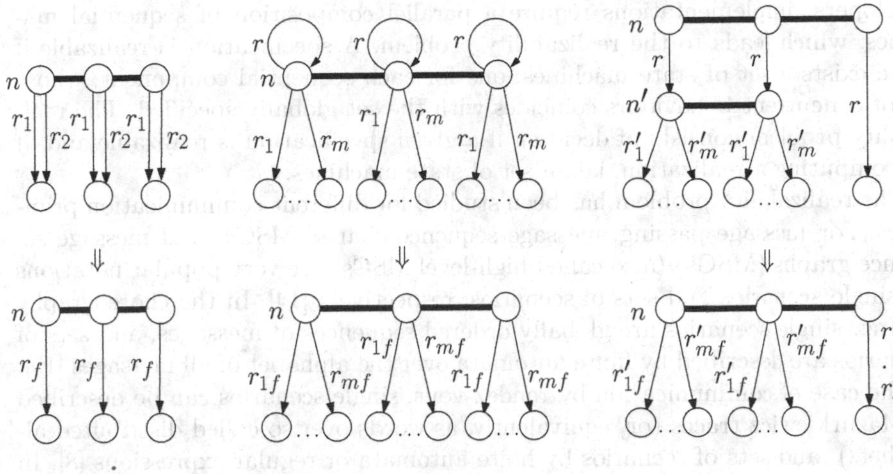

Fig. 4. The reduction rules

those commonly used to transform a finite automaton into an equivalent regular expression. Intuitively, the merge rule merges two outcomes that "move" each participating agent to the same next atom (outcomes r_1 and r_2 in the figure) into one single outcome with a fresh label. The iteration rule replaces the iteration of an outcome r followed by an outcome r_i by an outcome r_{if} with the same overall effect[2], and the shortcut rule merges the outcomes of two atoms n and n' that can occur one after the other into one single outcome with the same effect. Notice that the figure has to be complemented with the description of the transformers associated to the outcomes. For instance, the transformer of the outcome r_{1f} in the iteration rule is the composition of the Kleene star of the transformer of r and the transformer of r_1.

In [11] we show that a deterministic negotiation can be reduced to a single atom by repeated application of the rules if and only if it is sound. Moreover, the rules only have to be applied a polynomial number of times. The algorithm for checking soundness is a byproduct of this reduction algorithm: the negotiation is sound if and only if the reduction algorithm reduces it to a single atom.

Realizability. In [9] we are studying the *realizability problem* for deterministic negotiations. Design requirements for distributed systems are often captured with the help of scenarios specifying the interactions among agents during a run of the system. Formal notations for scenarios allow one to specify multiple scenarios by means of operations like choice, concatenation, and repetition. A set of scenarios specified in this way can be viewed as an early, global view of the desired system behaviours. While this view is usually more intuitive for

[2] This rule also reduces a negotiation with two atoms, one initial and one final, to one single atom, which is then both initial and final. In this case we have $n := n_0$, $n' := n_f$, and $m = 0$.

developers, implementations require a parallel composition of sequential machines, which leads to the realizability problem. A specification is realizable if there exists a set of state machines, one for each sequential component, whose set of concurrent behaviours coincides with the set globally specified. The realizablity problem consists of deciding if a given specification is realizable and, if so, computing a realization, i.e., a set of state machines.

The realizability problem has been studied for different communication primitives. For message passing, message sequence charts (MSCs) and message sequence graphs (MSGs) (also called high-level MSCs) are very popular notations for single scenarios and sets of scenarios, respectively [14]. In the choreography setting, single scenarios are globally ordered sequences of messages, and sets of scenarios are described by finite automata over the alphabet of all messages [19]. In the case of communication by rendez-vous, single scenarios can be described as Mazurkiewicz traces (or, equivalently, as words over so called distributed alphabets), and sets of scenarios by finite automata or regular expressions [8]. In all these settings, the complexity of the problem is high, ranging from PSPACE-hard to undecidable [3,5,13].

From a realizability point of view, negotiation diagrams are an implementation formalism. Indeed, the negotiation on the left of Figure 1 is the composition of the four state machines shown in Figure 5.

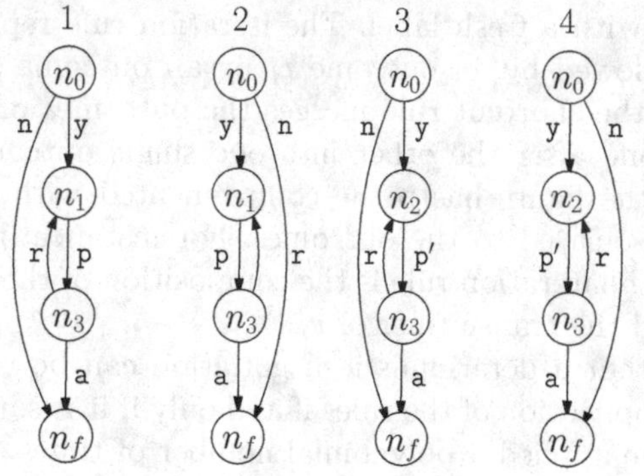

Fig. 5. Distributed view of the negotiation of Figure 1

We have designed a negotiation language for the global description of the runs of a deterministic negotiation, whose terms we call negotiation programs. For instance, the program corresponding to the negotiation of Figure 1 is shown in Figure 6.

The first two lines specify the agents of the system, and, for each outcome, the agents that have to agree to choose the outcome. The outer **do**-block corresponds to the atom n_0. The block offers a choice betwen the outcomes y and n; in the language, outcomes play the role of guards, and are prefixed by the operator "[]". After outcome y, the two outcomes p and p′ can be taken in parallel. The operator ∘ is the *layer composition* operator of Zwiers (see e.g. [22,15]). Loosely speaking,

```
agent A1, A2, A3, A4
outcome y, n, a, r : {a1, ..., a4}
        p : {a1, a2}
        p' : {a3, a4}
do [] y: (p || p') ∘
        do [] a: end
           [] r: (p || p') loop
        od end
   [] n: end
od
```

Fig. 6. Program equivalent to the negotiation of Figure 1

in every execution σ of $P_1 \circ P_2$, all actions of P_1 in which an agent a participates take place before all actions of P_2 in which the same agent participates. The other actions can occur in arbitrary order; in particular, if the sets of agents of P_1 and P_2 are disjoint, then $P_1 \circ P_2$ and $P_2 \circ P_1$ are equivalent programs, and in this case we write $P_1 \parallel P_2$. (Our language has only layer composition as primitive, and parallel composition is just a special case.) Finally, the inner **do**-block offers a choice between two alternatives, corresponding to the outcomes a and r. Alternatives are labeled with the keywords **loop** and **end**, indicating what happens after the chosen alternative has been executed: in the case of a loop-alternative, the block restarts, and for an end-altenative it terminates.

Our results shows that the realizability problem for deterministic negotiations has excellent semantic and computational properties:

(a) Every negotiation program has a sound realization.
(b) Negotiation programs are expressively complete: every sound deterministic negotiation diagram has an equivalent negotiation program[3].
(c) Negotiation programs can be distributed in linear time. We provide an algorithm to derive a negotiation from a program that generalizes classical constructions to derive an automaton from a regular expression. The negotiation diagram can then be projected onto its components.

Observe that (a) and (b) provide a *syntactic* characterization of soundness, which is a semantic property. Further, (a) and (b) can also be interpreted as a sort of Structure Theorem.. Just as flowcharts (or if-goto programs) model unstructured sequential programs (see Figure 5), negotiations can be viewed as a model of unstructured parallel programs. Therefore, (a) and (b) show that every *sound* unstructured program has an equivalent structured program. In other words, the soundness requirement, which is a desirable requirement for any well designed negotiation, turns out to have an unexpected beneficial side-effect: it *forces* the negotiation to be well structured.

Negotiation Games. In recent work with Philipp Hoffmann we have started the study of games on negotiations [12]. As for games played on pushdown

[3] Where equivalent means exhibiting the same completed Mazurkiewicz traces.

automata [20], vector addition systems with states (VASS) [6], counter machines [17], or asynchronous automata [18], games on negotiations can be translated into games played on the (reachable part of the) state space. Since the number of states of a negotiation may grow exponentially in the number of agents, the game arena can be exponentially larger than the negotiation. We explore the complexity of solving games *in the size of the negotiation*, not on the size of the arena. We study games formalizing the two most interesting questions related to a negotiation. Can a given *coalition* (i.e., a given subset of agents) force termination of the negotiation, or force that a given atom occurs? More precisely, a given coalition divides the atoms into those controlled by the coalition—the atoms in which a majority of the parties belong to the coalition—and those controlled by the environment. This induces a concurrent games with three players, called Coalition, Environment, and Scheduler. At each move, Scheduler selects a subset of the atoms enabled at the current marking, ensuring that they can occur in parallel (more precisely, Scheduler ensures that the sets of parties of the selected atoms are pairwise disjoint). Then, Coalition and Environment, independently of each other, choose an outcome for each of the selected atoms they control. This leads to a new marking, after which the next move can be played. Coalition wins if the play eventually executes the final atom, or the given goal atom.

Our results show that these two problems are EXPTIME-complete in the size of the negotiation, even if it is deterministic. So, at first sight, it seems as if, after all, the "concurrency curse" would also apply to deterministic negotiations. But then we are able to show that, surprisingly, the problems are polynomial for *sound* deterministic negotiations. The algorithm to decide the winner of a game is obtained by lifting the well-known attractor construction for reachability games from the arena of the state space to the negotiation diagram: While the usual attractor construction iteratively computes increasingly larger sets of *markings* from which Coalition can force termination, the lifted constructions computes increasingly larger sets of *atoms*.

This result is very satisfactory. Since unsound negotiations are ill-designed, we are not very interested in them, and the restriction to sound negotiations has as collateral effect a strong improvement in the complexity of the problem. Moreover, the restriction comes "at no cost", because deciding soundness of deterministic negotiations is also decidable in polynomial time.

A Short Conclusion. Concurrency theorists learn to live with the ubiquitous state-explosion problem, and to accept as a seemingly unavoidable consequence that all analysis problems for an interesting class of concurrent systems (where the input is the concurrent system itself, not its state space) will be at least NP- or PSPACE-hard. Our results show that deterministic negotiations escape this "concurrency curse". In future work we wish to investigate extensions of the model that increase its expressive power and retain at least some of its good analyzability properties. The next step is the study of weak deterministic negotiations, already outlined in [10].

References

1. van der Aalst, W.M.P.: The application of Petri nets to workflow management. J. Circuits, Syst. and Comput. 08(01), 21–66 (1998)
2. Abramsky, S., Gavoille, C., Kirchner, C., Meyer auf der Heide, F., Spirakis, P.G. (eds.): ICALP 2010. LNCS, vol. 6199. Springer, Heidelberg (2010)
3. Alur, R., Etessami, K., Yannakakis, M.: Realizability and verification of msc graphs. In: Orejas, F., Spirakis, P.G., van Leeuwen, J. (eds.) ICALP 2001. LNCS, vol. 2076, pp. 797–808. Springer, Heidelberg (2001)
4. Atdelzater, T.F., Atkins, E.M., Shin, K.G.: Qos negotiation in real-time systems and its application to automated flight control. IEEE Transactions on Computers 49(11), 1170–1183 (2000)
5. Basu, S., Bultan, T., Ouederni, M.: Deciding choreography realizability. In: POPL, pp. 191–202. ACM (2012)
6. Brázdil, T., Jancar, P., Kucera, A.: Reachability games on extended vector addition systems with states. In: Abramsky, et al. (eds.) [2], pp. 478–489
7. Davis, R., Smith, R.G.: Negotiation as a metaphor for distributed problem solving. Artificial Intelligence 20(1), 63–109 (1983)
8. Diekert, V., Rozenberg, G., Rozenburg, G.: The book of traces, vol. 15. World Scientific (1995)
9. Esparza, J., Desel, J.: Realizability of deterministic negotiations (in preparation)
10. Esparza, J., Desel, J.: On negotiation as concurrency primitive. In: D'Argenio, P.R., Melgratti, H. (eds.) CONCUR 2013 – Concurrency Theory. LNCS, vol. 8052, pp. 440–454. Springer, Heidelberg (2013), Extended version in CoRR abs/1307.2145
11. Esparza, J., Desel, J.: On negotiation as concurrency primitive II: Deterministic cyclic negotiations. In: Muscholl, A. (ed.) FOSSACS 2014 (ETAPS). LNCS, vol. 8412, pp. 258–273. Springer, Heidelberg (2014), Extended version in CoRR abs/1403.4958
12. Esparza, J., Hoffmann, P.: Negotiation games. CoRR, abs/1405.6820 (2014)
13. Genest, B., Gimbert, H., Muscholl, A., Walukiewicz, I.: Optimal Zielonka-type construction of deterministic asynchronous automata. In: Abramsky, et al. (eds.) [2], pp. 52–63
14. Genest, B., Muscholl, A., Peled, D.: Message sequence charts. In: Desel, J., Reisig, W., Rozenberg, G. (eds.) Lectures on Concurrency and Petri Nets. LNCS, vol. 3098, pp. 537–558. Springer, Heidelberg (2004)
15. Janssen, W., Poel, M., Zwiers, J.: Action systems and action refinement in the development of parallel systems - an algebraic approach. In: Groote, J.F., Baeten, J.C.M. (eds.) CONCUR 1991. LNCS, vol. 527, pp. 298–316. Springer, Heidelberg (1991)
16. Jennings, N.R., Faratin, P., Lomuscio, A.R., Parsons, S., Wooldridge, M.J., Sierra, C.: Automated negotiation: prospects, methods and challenges. Group Decision and Negotiation 10(2), 199–215 (2001)
17. Kučera, A.: Playing games with counter automata. In: Finkel, A., Leroux, J., Potapov, I. (eds.) RP 2012. LNCS, vol. 7550, pp. 29–41. Springer, Heidelberg (2012)
18. Mohalik, S., Walukiewicz, I.: Distributed games. In: Pandya, P.K., Radhakrishnan, J. (eds.) FSTTCS 2003. LNCS, vol. 2914, pp. 338–351. Springer, Heidelberg (2003)
19. Salaün, G., Bultan, T., Roohi, N.: Realizability of choreographies using process algebra encodings. IEEE T. Services Computing 5(3), 290–304 (2012)
20. Walukiewicz, I.: Pushdown processes: Games and model-checking. Inf. Comput. 164(2), 234–263 (2001)
21. Winsborough, W.H., Seamons, K.E., Jones, V.E.: Automated trust negotiation. In: Proceedings of DARPA Information Survivability Conference and Exposition, DISCEX 2000, vol. 1, pp. 88–102. IEEE (2000)
22. Zwiers, J.: Compositionality, Concurrency, and Partial Correctness. LNCS, vol. 321. Springer, Heidelberg (1989)

Generalized Bisimulation Metrics*

Konstantinos Chatzikokolakis[1,2], Daniel Gebler[3],
Catuscia Palamidessi[4,2], and Lili Xu[2,5]

[1] CNRS
[2] LIX, Ecole Polytechnique
[3] VU University Amsterdam
[4] INRIA
[5] Institute of Software, Chinese Academy of Science

Abstract. The bisimilarity pseudometric based on the Kantorovich lifting is one of the most popular metrics for probabilistic processes proposed in the literature. However, its application in verification is limited to linear properties. We propose a generalization of this metric which allows to deal with a wider class of properties, such as those used in security and privacy. More precisely, we propose a family of metrics, parametrized on a notion of distance which depends on the property we want to verify. Furthermore, we show that the members of this family still characterize bisimilarity in terms of their kernel, and provide a bound on the corresponding metrics on traces. Finally, we study the case of a metric corresponding to differential privacy. We show that in this case it is possible to have a dual form, easier to compute, and we prove that the typical constructs of process algebra are non-expansive with respect to this metrics, thus paving the way to a modular approach to verification.

1 Introduction

Originally proposed in the seminal works of Desharnais et al. [17,18], the bisimilarity pseudometric based on the Kantorovich lifting has become very popular in the process algebra community. One reason of this success is that, when dealing with probabilistic processes, metrics are more suitable than equivalences, since the latter are not robust wrt small variation of probabilities. Another important reason is that, thanks to the dual presentation of the Kantorovich lifting in terms of the mass transportation problem, the metric can be computed using linear programming algorithms [4,7,2]. Furthermore, this metric is an extension of probabilistic bisimilarity, in the sense that two states have distance distance 0 if and only if they are bisimilar. In fact, the metric also shares with bisimilarity the fact of being based on a similar coinductive definition. More precisely, it is defined as the greatest fixpoint of a transformation that has the same structure

* This work has been partially supported by the project ANR-12-IS02-001 PACE, the project ANR-11-IS02-0002 LOCALI, the INRIA Equipe Associée PRINCESS, the INRIA Large Scale Initiative CAPPRIS, and the EU grant 295261 MEALS.

P. Baldan and D. Gorla (Eds.): CONCUR 2014, LNCS 8704, pp. 32–46, 2014.

as the one used for bisimilarity.[1] This allows to transfer some of the concepts and methods that have been extensively explored in process algebra, and to use lines of reasoning which the process algebra community is familiar with. Along the same lines, a nice property of the Kantorovich bisimilarity pseudometric is that the standard operators of process algebra are not expansive wrt it. This can be seen as a generalization of the result that bisimulation is a congruence, and can be used in a similar way, for compositional reasoning and verification.

Last but not least, the Kantorovich bisimilarity metric provides a bound on the corresponding distance on probabilistic traces [12] (corresponding in the sense that the definition is based on the same Kantorovich lifting). This means that it can be used to verify certain probabilistic properties on traces. More specifically, it can be used to verify properties that are expressed in terms of difference between probabilities of sets of traces. These properties are linear, in the sense that the difference increases linearly wrt variations on the distributions.

Many properties, however, such as several privacy and security ones, are not linear. This is the case of the popular property of differential privacy [19], which is expressed in terms of ratios of probabilities. In fact, there are processes that have small Kantorovich distance, and which are not ϵ-differentially private for any finite ϵ. Another example are the properties used in quantitative information flow, which involve logarithmic functions on probabilities.

The purpose of this work is to generalize the Kantorovich lifting to obtain a family of metrics suitable for the verification of a wide class of properties, following the principles that:

i. the metrics of this family should depend on a parameter related to the class of properties (on traces) that we wish to verify,
ii. each metric should provide a bound on the corresponding metric on traces,
iii. the kernel of these metric should correspond to probabilistic bisimilarity,
iv. the general construction should be coinductive,
v. the typical process-algebra operators should be non-expansive,
vi. it should be feasible to compute these metrics.

In this paper we have achieved the first four desiderata. For the last two, so far we have studied the particular case of the multiplicative variant of the Kantorovich metric, which is based on the notion of distance used in the definition of differential privacy. We were able to find a dual form of the lifting, which allows to reduce the problem of its computation to a linear optimization problem solvable with standard algorithms. We have also proved that several typical process-algebra operators are non-expansive, and we have given explicitly the expression of the bound. For some of them we were able to prove this result in a general form, i.e., non-expansiveness wrt all the metrics of the family, and with the bound represented by the same expression.

[1] In the original definition the Kantorovich bisimilarity pseudometric was defined as the greatest fixpoint, but such definition requires using the reverse order on metrics. More recently, authors tend to use the natural order, and define the bisimilarity metric as the least fixpoint, see [12,1,2]. Here we follow the latter approach.

As an example of application of our method, we show of to instantiate our construction to obtain the multiplicative variant of the Kantorovich metric, and how to use it to verify the property of differential privacy.

All proofs are given in the report version of this paper [11].

Related Work. Bisimulation metrics based on the standard Kantorovich distance have been used in various applications, such as systems biology [25], games [9], planning [13] and security [8]. We consider in this paper discrete state spaces. Bisimulation metrics on uncountable state spaces have been explored in [18]. We define bisimulation metrics as fixed point of an appropriate functor. Alternative characterizations were provided in terms of coalgebras [6] and real-valued modal logics [18]. The formulation of the Kantorovich lifting as primal and dual linear program is due to [5].

Verification of differential privacy has been itself an active area of research. Prominent approaches based on formal methods are those based on type systems [22] and logical formulations [3]. Earlier papers [26,27] define a bisimulation distance, which however suffered from the fact that the respective kernel relation (states in distance 0) does not fully characterize probabilistic bisimilarity.

2 Preliminaries

2.1 Labelled Concurrent Markov Chains

Given a set X, we denote by $Prob(X), Disc(X)$ the set of all and discrete probability measures over X respectively; the support of a measure μ is defined as $supp(\mu) = \{x \in X | \mu(x) > 0\}$. A *labelled concurrent Markov chain* (henceforth LCMC) $\mathcal{A}$ is a tuple (S, A, D) where S is a countable set of *states*, A is a countable set of action *labels*, and $D \subseteq S \times A \times Disc(S)$ is a *transition relation*. We write $s \xrightarrow{a} \mu$ for $(s, a, \mu) \in D$.

An *execution* α is a (possibly infinite) sequence $s_0 a_1 s_1 a_2 s_2 \ldots$ of alternating states and labels, such that for each $i : s_i \xrightarrow{a_{i+1}} \mu_{i+1}$ and $\mu_{i+1}(s_{i+1}) > 0$. We use $lstate(\alpha)$ to denote the last state of a finite execution α. We use $Exec^*(\mathcal{A})$ and $Exec(\mathcal{A})$ to represent the set of finite executions and of all executions of $\mathcal{A}$, respectively. A *trace* is a sequence of labels in $A^* \cup A^\omega$ obtained from executions by removing the states. We use $[\,]$ to represent the empty trace, and $\frown$ to concatenate two traces.

A *labelled Markov chain* (henceforth LMC) $\mathcal{A}$ is a *fully probabilistic* LCMC, namely a LCMC where from each state of $\mathcal{A}$ there is at most one transition available. We denote by $L(s)$ and $\pi(s)$ the label and distribution of the unique transition starting from s (if any).

In a LMC $\mathcal{A}$, a state s of $\mathcal{A}$ induces a probability measure over traces as follows. The basic measurable events are the cones of finite traces, where the cone of a finite trace $\boldsymbol{t}$, denoted by $C_{\boldsymbol{t}}$, is the set $\{\boldsymbol{t}' \in A^* \cup A^\omega | \boldsymbol{t} \leq \boldsymbol{t}'\}$, where $\leq$ is the standard prefix preorder on sequences. The probability induced by s on a cone $C_{\boldsymbol{t}}$, denoted by $\Pr[s \triangleright C_{\boldsymbol{t}}]$, is defined recursively as follows:

$$\Pr[s \rhd C_t] = \begin{cases} 1 & \text{if } t = [\,] \\ 0 & \text{if } t = a^\frown t' \text{ and } a \neq L(s) \\ \sum_{s_i} \mu(s_i)\Pr[s_i \rhd C_{t'}] & \text{if } t = a^\frown t' \text{ and } s \xrightarrow{a} \mu \end{cases} \quad (1)$$

This probability measure is extended to arbitrary measurable sets in the σ-algebra of traces in the standard way. We write $\Pr[s \rhd \sigma]$ to represent the probability induced by s on the set of traces σ.

2.2 Pseudometrics

A pseudometric is a relaxed notion of a normal metric in which distinct elements can have distance zero. We consider here a generalized notion where the distance can also be infinite, and we use $[0, +\infty]$ to denote the non-negative fragment of the real numbers $\mathbb{R}$ enriched with $+\infty$. Formally, an (extended) pseudometric on a set X is a function $m : X^2 \to [0, +\infty]$ with the following properties: $m(x, x) = 0$ (reflexivity), $m(x, y) = m(y, x)$ (symmetry), and $m(x, y) \leq m(x, z) + m(z, y)$ (triangle inequality). A metric has the extra condition that $m(x, y) = 0$ implies $x = y$. Let $\mathcal{M}_X$ denote the set of all pseudo-metrics on X with the ordering $m_1 \preceq m_2$ iff $\forall x, y. m_1(x, y) \leq m_2(x, y)$. It can be shown that $(\mathcal{M}_X, \preceq)$ is a complete lattice with bottom element $\bot$ such that $\forall x, y. \bot(x, y) = 0$ and top element $\top$ such that $\forall x, y. \top(x, y) = \infty$.

The *ball* (wrt m) of radius r centered at $x \in X$ is defined as $B_r^m(x) = \{x' \in X : m(x, x') \leq r\}$. A point $x \in X$ is called *isolated* iff there exists $r > 0$ such that $B_r^m(x) = \{x\}$. The *diameter* (wrt m) of $A \subseteq X$ is defined as $\text{diam}_m(A) = \sup_{x,x' \in A} m(x, x')$. The *kernel* $\ker(m)$ is an equivalence relation on X defined as

$$(x, x') \in \ker(m) \quad \text{iff } m(x, x') = 0$$

3 A General Family of Kantorovich Liftings

We introduce here a family of liftings from pseudometrics on a set X to pseudometrics on $Prob(X)$. This family is obtained as a generalization of the Kantorovich lifting, in which the Lipschitz condition plays a central role.

Given two pseudometric spaces (X, d_X), (Y, d_Y), we say that $f : X \to Y$ is 1-Lipschitz wrt d_X, d_Y iff $d_Y(f(x), f(x')) \leq d_X(x, x')$ for all $x, x' \in X$. We denote by 1-Lip$[(X, d_X), (Y, d_Y)]$ the set of all such functions.

A function $f : X \to \mathbb{R}$ can be lifted to a function $\hat{f} : Prob(X) \to \mathbb{R}$ by taking its expected value. For discrete distributions (countable X) it can be written as:

$$\hat{f}(\mu) = \textstyle\sum_{x \in X} \mu(x) f(x) \quad (2)$$

while for continuous distributions we need to restrict f to be measurable wrt the corresponding σ-algebra on X, and take $\hat{f}(\mu) = \int f d\mu$.

Given a pseudometric m on X, the *standard Kantorovich lifting* of m is a pseudometric $K(m)$ on $Prob(X)$, defined as:

$$K(m)(\mu, \mu') = \sup\{|\hat{f}(\mu) - \hat{f}(\mu')| : f \in 1\text{-Lip}[(X, m), (\mathbb{R}, d_{\mathbb{R}})]\}$$

where $d_{\mathbb{R}}$ denotes the standard metric on reals. For continuous distributions we implicitly take the sup to range over measurable functions.

Generalization. A generalization of the Kantorovich lifting can be naturally obtained by extending the range of f from $(\mathbb{R}, d_{\mathbb{R}})$ to a generic metric space (V, d_V), where $V \subseteq \mathbb{R}$ is a convex subset of the reals[2], and d_V is a metric on V. A function $f : X \to V$ can be lifted to a function $\hat{f} : \mathit{Prob}(X) \to V$ in the same way as before (cfr. (2)); the requirement that V is convex ensures that $\hat{f}(\mu) \in V$.

Then, similarly to the standard case, given a pseudometric space (X, m) we can define a lifted pseudometric $K_V(m)$ on $\mathit{Prob}(X)$ as:

$$K_V(m)(\mu, \mu') = \sup\{d_V(\hat{f}(\mu), \hat{f}(\mu')) : f \in 1\text{-Lip}[(X, m)(V, d_V)]\} \quad (3)$$

The subscript V in K_V is to emphasize the fact that for each choice of (V, d_V) we may get a different lifting. We should also point out the difference between m, the pseudometric on X being lifted, and d_V, the metric (not pseudo) on V which parametrizes the lifting.

The constructed $K_V(m)$ can be shown to be an extended pseudometric for any choice of (V, d_V), i.e. it is non-negative, symmetric, identical elements have distance zero, and it satisfies the triangle inequality. However, without extra conditions, it is not guaranteed to be bounded (even if m itself is bounded). For the purposes of this paper this is not an issue. In the report version [11] we show that under the condition that d_V is *ball-convex* (i.e. all its balls are convex sets, which holds for all metrics in this paper), the following bound can be obtained:

$$K_V(m)(\mu, \mu') \le \text{diam}_m(\mathit{supp}(\mu) \cup \mathit{supp}(\mu'))$$

Examples. The standard Kantorovich lifting is obtained by taking $(V, d_V) = (\mathbb{R}, d_{\mathbb{R}})$. When 1-bounded pseudometrics are used, like in the construction of the standard bisimilarity metric, then we can equivalently take $V = [0, 1]$.

Moreover, a multiplicative variant of the Kantorovich lifting can be obtained by taking $(V, d_V) = ([0, 1], d_{\otimes})$ (or equivalently $([0, \infty), d_{\otimes})$) where $d_{\otimes}(x, y) = |\ln x - \ln y|$. The resulting lifting is discussed in detail in Section 5 and its relation to differential privacy is shown in Section 5.1.

4 A General Family of Bisimilarity Pseudometrics

In this section we define a general family of pseudometrics on the states of an LCMC which have the property of extending probabilistic bisimilarity in the usual sense. Following standard lines, we define a transformation on state pseudometrics by first lifting a state pseudometric to a pseudometric on distributions

[2] V could be further generalized to be a convex subset of a vector space. It is unclear whether such a generalization would be useful, hence it is left as future work.

(over states), using the generalized Kantorovich lifting defined in previous section. Then we apply the standard Hausdorff lifting to obtain a pseudometric on sets of distributions. This last step is to take into account the nondeterminism of the LCMC, i.e., the fact that in general, from a state, we can make transitions to different distributions. The resulting pseudometric naturally corresponds to a state pseudometric, obtained by associating each set of distributions to the states which originate them. Finally, we define the intended bisimilarity pseudometric as the least fixpoint of this transformation wrt the ordering $\preceq$ on the state pseudometrics (or equivalently, as the greatest fixpoint wrt the reverse of $\preceq$). We recall that $m \preceq m'$ means that $m(s,s') \leq m'(s,s')$ for all $s,s' \in S$.

Let $\mathcal{A} = (S, A, D)$ be a LCMC, let (V, d_V) be a metric space (for some convex $V \subseteq \mathbb{R}$), and let $\mathcal{M}$ be the set of pseudometrics m on S such that $\mathrm{diam}_m(S) \leq \mathrm{diam}_{d_V}(V)$. Recall that $\inf \emptyset = \mathrm{diam}_{d_V}(V)$ and $\sup \emptyset = 0$.

Definition 1. *The transformation $F_V : \mathcal{M} \to \mathcal{M}$ is defined as follows.*

$$F_V(m)(s,t) = \max\{ \sup_{s \xrightarrow{a} \mu} \inf_{t \xrightarrow{a} \nu} K_V(m)(\mu,\nu), \sup_{t \xrightarrow{a} \nu} \inf_{s \xrightarrow{a} \mu} K_V(m)(\nu,\mu)\}$$

We can also characterize F_V in terms of the following zigzag formulation:

Proposition 1. *For any $\epsilon \geq 0$, $F_V(m)(s,t) \leq \epsilon$ if and only if:*

- *if $s \xrightarrow{a} \mu$, then there exists ν such that $t \xrightarrow{a} \nu$ and $K_V(m)(\mu,\nu) \leq \epsilon$,*
- *if $t \xrightarrow{a} \nu$, then there exists μ such that $s \xrightarrow{a} \mu$ and $K_V(m)(\nu,\mu) \leq \epsilon$.*

The following result states that K_V and F_V are monotonic wrt $(\mathcal{M}, \preceq)$.

Proposition 2. *Let $m, m' \in \mathcal{M}$. If $m \preceq m'$ then:*

$$F_V(m)(s,s') \leq F_V(m')(s,s') \quad \textit{for all states } s,s'$$
$$K_V(m)(\mu,\mu') \leq K_V(m')(\mu,\mu') \quad \textit{for all distributions } \mu,\mu'$$

Since $(\mathcal{M}, \preceq)$ is a complete lattice and F_V is monotone on $\mathcal{M}$, by Tarski's theorem [24] F_V has a least fixpoint, which coincides with the least pre-fixpoint. We define the *bisimilarity pseudometric* bm_V as this least fixpoint:

Definition 2. *The bisimilarity pseudometric bm_V is defined as:*

$$bm_V = \min\{m \in \mathcal{M} \mid F_V(m) = m\} = \min\{m \in \mathcal{M} \mid F_V(m) \preceq m\}$$

In addition, if the states of $\mathcal{A}$ are finite, then the closure ordinal of F_V is ω (cf: [17], Lemma 3.10). Hence we can approximate bm_V by iterating the function F_V from the bottom element:

Proposition 3. *Assume S is finite. Let $m_0 = \bot$ and $m_{i+1} = F_V(m_i)$. Then $bm_V = \sup_i m_i$.*

Next section shows that bm_V is indeed a bisimilarity metric, in the sense that its kernel coincides with probabilistic bisimilarity.

4.1 Bisimilarity as 0-distance

We now show that under certain conditions, the pseudometric constructed by $K_V(m)$ characterizes bisimilarity at its kernel. Recall that the kernel $\ker(m)$ of m is an equivalence relation relating states at distance 0.

Given an equivalence relation R on S, its lifting $\mathcal{L}(R)$ is an equivalence relation on $Disc(S)$, defined as

$$(\mu, \mu') \in \mathcal{L}(R) \quad \text{iff} \quad \forall s \in S : \mu([s]_R) = \mu'([s]_R)$$

where $[s]_R$ denotes the equivalence class of s wrt R.

To obtain the characterization result we assume that (a) the set of states is finite, and (b) the distance d_V is non-discrete. Under these conditions, the kernel operator and the lifting operator commute (cfr. [15] for the analogous property for the standard Kantorovich lifting).

Lemma 1. *If S is finite and d_V is non-discrete, then* $\mathcal{L}(\ker(m)) = \ker(K_V(m))$.

We recall the notions of probabilistic bisimulation and bisimilarity, following the formulation in terms of post-fixpoints of a transformation on state relations:

Definition 3.

- *The transformation $B : S \times S \to S \times S$ is defined as: $(s, s') \in B(R)$ iff*
  - *if $s \xrightarrow{a} \mu$, then there exists μ' such that $t \xrightarrow{a} \mu'$ and $(\mu, \mu') \in \mathcal{L}(R)$,*
  - *if $s' \xrightarrow{a} \mu'$, then there exists μ such that $s \xrightarrow{a} \mu$ and $(\mu', \mu) \in \mathcal{L}(R)$.*
- *A relation $R \subseteq S \times S$ is called a bisimulation if it is a post-fixpoint of R, i.e. $R \subseteq B(R)$.*

It is easy to see that B is monotonic on $(2^{S\times S}, \subseteq)$ and that the latter is a complete lattice, hence by Tarski's theorem there exists the greatest fixpoint of B, and it coincides with the greatest bisimulation:

Definition 4. *The bisimilarity relation $\sim \subseteq S \times S$ is defined as:*

$$\sim \;=\; \max\{R \mid R = B(R)\} \;=\; \max\{R \mid R \subseteq B(R)\} \;=\; \bigcup\{R \mid R \subseteq B(R)\}$$

We are now ready to show the correspondence between pre-fixpoint metrics and bisimulations. Using Lemma 1, we can see that the definition of B corresponds to the characterization of F_V in Proposition 1, for $\epsilon = 0$. Hence we have:

Proposition 4. *For every $m \in \mathcal{M}$, if $F_V(m) \preceq m$ then* $\ker(m) \subseteq B(\ker(m))$, *i.e.,* $\ker(m)$ *is a bisimulation.*

As a consequence, $\ker(bm_V) \subseteq \sim$. The converse of Proposition 4 does not hold, because the fact that $\ker(m) \subseteq B(\ker(m))$ does not say anything about the effect of F_V on the distance between elements that are not on the kernel. However, in the case of bisimilarity we can make a connection: consider the greatest metric $m_\sim$ whose kernel coincides with bisimilarity, namely, $m_\sim(s,s') = 0$ if $s \sim s'$ and $m_\sim(s,s') = \mathrm{diam}_{d_V}(V)$ otherwise. We have that $F_V(m_\sim) \preceq m_\sim$, and therefore $\sim = \ker(m_\sim) \subseteq bm_V$. Therefore we can conclude that the kernel of the bisimilarity pseudometrics coincides with bisimilarity.

Theorem 1. $\ker(bm_V) = \sim$ *for every* (V, d_V),

4.2 Relation with Trace Distributions

In this section, we show the relation between the bisimilarity metric bm_V and the corresponding metric on traces, in the case of LMCs (labeled Markov chains). Note that we restrict to the fully probabilistic case here, where probabilities on traces can defined in the way shown in the preliminaries. The full case of LCMCs can be treated by using schedulers, but a proper treatment involves imposing scheduler restrictions which complicate the formalism. Since these problems are orthogonal to the goals of this paper, we keep the discussion simple by restricting to the fully probabilistic case.

The distance between trace distributions (i.e. distributions over A^ω) will be measured by the Kantorovich lifting of the *discrete metric*. Given (V, d_V), let $\delta_V = \mathrm{diam}_{d_V}(V)$. Then let dm_{δ_V} be the δ_V-valued discrete metric on A^ω, defined as $dm_{\delta_V}(\boldsymbol{t},\boldsymbol{t}') = 0$ if $\boldsymbol{t} = \boldsymbol{t}'$, and $dm_{\delta_V}(\boldsymbol{t},\boldsymbol{t}') = \delta_V$ otherwise.

Then $K_V(dm_{\delta_V})(\mu,\mu')$ is a pseudometric on $Prob(A^\omega)$, whose kernel coincides with probabilistic trace equivalence.

Proposition 5. $K_V(dm_{\delta_V})(\mu,\mu') = 0$ *iff* $\mu(\sigma) = \mu'(\sigma)$ *for all measurable* $\sigma \subseteq A^\omega$.

The following theorem expresses that our bisimilarity metric bm_V is a bound on the distance on traces, which extends the standard relation between probabilistic bisimilarity and probabilistic trace equivalence.

Theorem 2. *Let* $\mu = \Pr[s \rhd \cdot]$ *and* $\mu' = \Pr[s' \rhd \cdot]$. *Then* $K_V(dm_{\delta_V})(\mu,\mu') \leq bm_V(s,s')$

It should be noted that, although the choice of $K_V(dm_{\delta_V})$ as our trace distribution metric might seem arbitrary, this metric is in fact of great interest. In the case of the standard bisimilarity pseudometric, i.e. when $(V, d_V) = ([0,1], d_{\mathbb{R}})$, this metric is equal to the well-known *total variation* distance (also known as *statistical distance*), defined as $tv(\mu,\mu') = \sup_\sigma |\mu(\sigma) - \mu'(\sigma)|$:

$$K(dm_{\delta_V}) = tv \tag{4}$$

Theorem 2 reduces to the result of [12] relating the total variation distance to the bisimilarity pseudometric. Moreover, in the case of the multiplicative pseudometric, discussed in the next section, $K_V(dm_{\delta_V})$ is the same as the multiplicative distance between distributions, discussed in Section 5.1, which plays a central role in differential privacy.

Table 1. The standard Kantorovich metric and its multiplicative variant

	Standard $K(m)(\mu,\mu')$	Multiplicative $K_\otimes(m)(\mu,\mu')$
Primal	$\max_f \lvert\hat{f}(\mu)-\hat{f}(\mu')\rvert$ subject to $\forall s,s'.\,\lvert f(s)-f(s')\rvert \le m(s,s')$	$\max_f \lvert\ln\hat{f}(\mu)-\ln\hat{f}(\mu')\rvert$ subject to $\forall s,s'.\,\lvert\ln f(s)-\ln f(s')\rvert \le m(s,s')$
Dual	$\min_\ell \sum_{i,j}\ell_{ij}m(s_i,s_j)$ subject to $\forall i,j.\ \ell_{ij}\ge 0$ $\forall i.\ \sum_j \ell_{ij}=\mu(s_i)$ $\forall j.\ \sum_i \ell_{ij}=\mu'(s_j)$	$\min \ln z$ subject to $\forall i,j.\ \ell_{ij},r_i\ge 0$ $\forall i.\ \sum_j \ell_{ij}-r_i=\mu(s_i)$ $\forall j.\ \sum_i \ell_{ij}e^{m(s_i,s_j)}-r_j\le z\cdot\mu'(s_j)$

5 The Multiplicative Variant

In this section we investigate the multiplicative variant of the Kantorovich pseudometric, obtained by considering as distance d_V the ratio between two numbers instead than their difference. This is the distance used to define differential privacy. We show that this variant has a dual form, which can be used to compute the metric by using linear programming techniques. In the next section, we will show how to use it to verify differential privacy.

Definition 5. *The multiplicative variant $K_\otimes$ of the Kantorovich lifting is defined as the instantiation of K_V with $([0,1], d_\otimes)$ where $d_\otimes(x,y) = \lvert\ln x - \ln y\rvert$.*

It is well known that the standard Kantorovich metric has a dual form which can be interpreted in terms of *the Transportation Problem*, namely, the lowest total cost of transporting the mass of one distribution μ to the other distribution μ' given the distance m between locations (in our case, states). The dual form is shown in Table 1. Note that both the primal and the dual forms are linear optimization problems. The dual form is particularly suitable for computation, via standard linear programming techniques.

For our multiplicative variant, the objective function of the primal form is not a linear expression, hence the linear programming techniques cannot be applied directly. However, since $\ln\hat{f}(\mu)-\ln\hat{f}(\mu') = \ln\frac{\hat{f}(\mu)}{\hat{f}(\mu')}$ and ln is a monotonically increasing function, the primal problem is actually a linear-fractional program. It is known that such kind of program can be converted to an equivalent linear programming problem and then to a dual program. The dual form of the multiplicative variant obtained in this way is shown in Table 1. (For the sake of simplicity, the table shows only the dual form of $\ln\hat{f}(\mu)-\ln\hat{f}(\mu')$. The dual form of $\ln\hat{f}(\mu')-\ln\hat{f}(\mu)$ can be obtained by simply switching the roles of μ and

μ'.) Hence, the multiplicative pseudometric can be computed by using linear programming techniques.

5.1 Application to Differential Privacy

Differential privacy [19] is a notion of privacy originating from the area of statistical databases, which however has been recently applied to several other areas. The standard context is that of an analyst who wants to perform a statistical query to a database. Although obtaining statistical information is permitted, privacy issues arise when this information can be linked to that of an individual in the database. In order to hide this link, differentially private mechanisms add noise to the outcome of the query, in a way such that databases differing in a single individual have similar probability of producing the same observation.

More concretely, let $\mathcal{X}$ be the set of all databases; two databases $x, x' \in \mathcal{X}$ are *adjacent*, written $x \smile x'$, if they differ in the value of a single individual. A *mechanism* is a function $M : \mathcal{X} \to \mathit{Prob}(\mathcal{Z})$ where $\mathcal{Z}$ is some set of reported values. Intuitively, $M(x)$ gives the outcome of the query when applied to database x, which is a probability distribution since noise is added.

Let $tv_\otimes$ be a multiplicative variant of the total variation distance on $\mathit{Prob}(\mathcal{Z})$ (simply called "multiplicative distance" in [23]), defined as:

$$tv_\otimes(\mu, \mu') = \sup_Z |\ln \frac{\mu(Z)}{\mu'(Z)}|$$

Then differential privacy can be defined as follows.[3]

Definition 6. *A mechanism* $M : \mathcal{X} \to \mathit{Prob}(\mathcal{Z})$ *is* ϵ*-differentially private iff*

$$tv_\otimes(M(x), M(x')) \le \epsilon \qquad \forall x \smile x'$$

Intuitively, the definition requires that, when run on adjacent databases, the mechanism should produce similar results, since the distance between the corresponding distributions should be bounded by ϵ (a privacy parameter).

In our setting, we assume that the mechanism M is modelled by a LMC, and the result of the mechanism running on x is the trace produced by the execution of the LMC starting from some corresponding state s_x. That is, $\mathcal{Z} = A^\omega$ and

$$M(x) = \Pr[s_x \rhd \cdot] \tag{5}$$

The relation between differential privacy and the multiplicative bisimilarity metric comes from the fact that $tv_\otimes$ can be obtained as the $K_\otimes$ lifting of the discrete metric on A^ω.

Lemma 2. *Let* $\delta_V = \mathrm{diam}_{d_\otimes}([0,1]) = +\infty$ *and let* dm_{δ_V} *be the discrete metric on* A^ω*. Then* $tv_\otimes = K_\otimes(dm_{\delta_V})$*.*

[3] The definition can be generalized to an arbitrary set of secrets $\mathcal{X}$ equipped with a "distinguishability metric" $d_\mathcal{X}$ [10]. The results of this section extend to this setting.

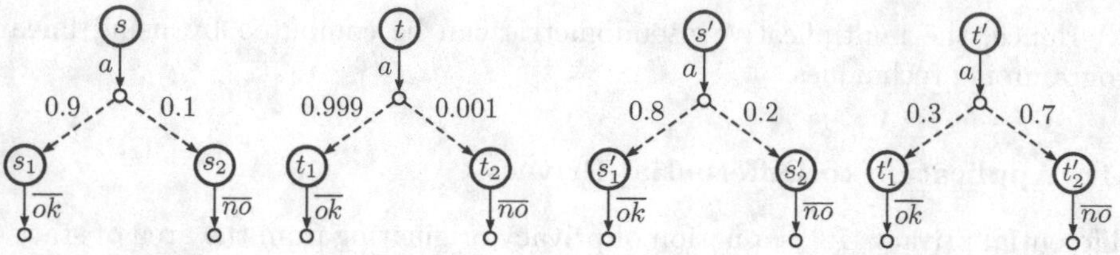

(a) $bm(s, t) = 0.099$ while $\epsilon = \ln 100$. **(b)** $bm(s', t') = 0.5$ while $\epsilon' = \ln 3.5$.

Fig. 1. The bisimilarity pseudometric bm does not imply differential privacy

Let $bm_{\otimes}$ be the instantiation of the bisimilarity metric bm_V with $K_{\otimes}$. The above Lemma, together with Theorem 2, imply the following result, which makes $bm_{\otimes}$ useful to verify differential privacy:

Theorem 3. *Let M be the mechanism defined by (5), and assume that*

$$bm_{\otimes}(s_x, s_{x'}) \le \epsilon \qquad \text{for all } x \smile x'$$

Then M satisfies ϵ-differential privacy.

Note that the use of the multiplicative $bm_{\otimes}$ is crucial in the above result. The following example shows that the standard bisimilarity metric bm (generated by the original Kantorovich lifting) may be very different from the level of differential privacy, which is expected, since bm bounds the additive total variation metric (Theorem 2 and (4)) instead of the multiplicative $tv_{\otimes}$.

Example 1. Consider the processes s, t shown in Fig. 1 (a). We have that $bm(s, t) = 0.1 - 0.001 = 0.099$ while their level of differential privacy is $\epsilon = \ln \frac{0.1}{0.001} = \ln 100$. Moreover, for the processes s', t' shown in Fig. 1 (b) we have $bm(s', t') = 0.7 - 0.2 = 0.5$ while their level of differential privacy is $\epsilon' = \ln \frac{0.7}{0.2} = \ln 3.5$. Using the original Kantorovich metric, s and t are considered more indistinguishable than s' and t', in sharp contrast to the corresponding differential privacy levels.

Approximate differential privacy. An approximate, also known as (ϵ, δ) version of differential privacy is also widely used [20], relaxing the definition by an additive factor δ. It requires that:

$$M(x)(Z) \le e^{\epsilon} M(x')(Z) + \delta \qquad \forall x \smile x', Z \subseteq \mathcal{Z}$$

The α-distance on distributions is proposed in [3] to capture (ϵ, δ)-differential privacy. For two real numbers a, b and a skew parameter $\alpha \ge 1$, the α-distance between a and b is $\max\{a - \alpha b, b - \alpha a, 0\}$. An instantiation of the Kantorovich lifting based on the α-distance seems promising for extending Theorem 3 to the approximate case; we leave this extension as future work.

Weak probabilistic anonymity. Weak probabilistic anonymity was proposed in [16] as a measure of the degree of protection of user's identities. It is defined in a way similar to differential privacy, with the crucial difference (apart from the lack of an adjacency relation) that it uses the (additive) total variation instead of the multiplicative one. Formally, let $\mathcal{X}$ contain the users' identities, and let $M : \mathcal{X} \to Prob(\mathcal{Z})$ be the system in which users operate. We say that M is *ϵ-weakly probabilistically anonymous* iff $tv(M(x), M(x')) \leq \epsilon$ for all $x, x' \in \mathcal{X}$.

For systems modelled by LMCs, by (4) and Theorem 2, we have that if $bm(s_x, s_{x'}) \leq \epsilon$ for all $x, x' \in \mathcal{X}$, then M satisfies ϵ-weak probabilistic anonymity. Hence bm can be used to verify this anonymity property.

6 Process Algebra

Process algebras allow to syntactically describe probabilistic processes in terms of a small set of well-understood operators. The operational semantics of a process term is a LCMC with transitions derived from SOS rules.

In order to specify and verify systems in a compositional manner, it is necessary that the behavioral semantics is compatible with all operators of the language that describe these systems. For behavioral equivalence semantics there is the common agreement that compositional reasoning requires that the considered behavioral equivalence is a congruence wrt all operators. On the other hand, for behavioral metric semantics there are several proposals of properties that operators should satisfy in order to facilitate compositional reasoning [18,1]. In this section we will show that the standard non-recursive process algebra operators are non-expansiveness [18] (as most prominent compositionality property) with respect to the bisimilarity metric.

We introduce a simple probabilistic process algebra that comprises the following operators i) constants 0 (stop process) and ϵ (skip process); ii) a family of n-ary prefix operators $a.([p_1]\_\oplus\ldots\oplus[p_n]\_)$ with $a \in Act$, $n \geq 1, p_1, \ldots, p_n \in (0,1]$ and $\sum_{i=1}^n p_i = 1$; iii) binary operators $\_;\_$ (sequential composition), $\_+\_$ (alternative composition), $\_+_p\_$ (probabilistic alternative composition), $\_\mid\_$ (synchronous parallel composition), $\_\parallel\_$ (asynchronous parallel composition), and $\_\parallel_p\_$ (probabilistic parallel composition). We assume a set of actions Act with the distinguished action $\surd \in A$ to denote successful termination. The operational semantics of all operators is specified by the rules in Table 2.

We use distribution terms in the target of rules (right hand side of the conclusion of the rules) in order to describe distributions. We briefly recall the semantics of distribution terms of [21,14]. The expression $\delta(x)$ denotes a Dirac distribution on x. The expression $\mu; \delta(y)$ denotes a distribution such that $(\mu; \delta(y))(x; y) = \mu(x)$, the expression $\mu \oplus_p \nu$ denotes a distribution such that $(\mu \oplus_p \nu)(x) = p\mu(x) + (1-p)\nu(x)$, and $(\mu \parallel \nu)(s \parallel t) = \mu(s)\nu(t)$.

The probabilistic prefix operator expresses that the process $a.([p_1]t_1 \oplus \ldots \oplus_p lus[p_n]t_n)$ can perform action a and evolves to process t_i with probability p_i. The sequential composition and the alternative composition are as usual. The synchronous parallel composition $s \mid t$ describes the simultaneous evolution of processes s and t, while the asynchronous parallel composition $t \parallel t$ describes the interleaving of s and t where both processes can progress by alternating at any rate the execution of their actions. The probabilistic alternative and

Table 2. Probabilistic process algebra operators

$$\frac{}{\varepsilon \xrightarrow{\surd} \delta(0)} \qquad \frac{}{a.\bigoplus_{i=1}^{n}[p_i]x_i \xrightarrow{a} \bigoplus_{i=1}^{n} p_i\delta(x_i)}$$

$$\frac{x \xrightarrow{a} \mu \quad a \neq \surd}{x;y \xrightarrow{a} \mu;\delta(y)} \qquad \frac{x \xrightarrow{\surd} \mu \quad y \xrightarrow{a} \nu}{x;y \xrightarrow{a} \nu} \qquad \frac{x \xrightarrow{a} \mu}{x+y \xrightarrow{a} \mu} \qquad \frac{y \xrightarrow{a} \nu}{x+y \xrightarrow{a} \nu}$$

$$\frac{x \xrightarrow{a} \mu \quad y \xrightarrow{a} \nu}{x \mid y \xrightarrow{a} \mu \mid \nu} \qquad \frac{x \xrightarrow{a} \mu}{x \parallel y \xrightarrow{a} \mu \parallel \delta(y)} \qquad \frac{y \xrightarrow{a} \nu}{x \parallel y \xrightarrow{a} \delta(x) \parallel \nu}$$

$$\frac{x \xrightarrow{a} \mu \quad y \not\xrightarrow{a}}{x +_p y \xrightarrow{a} \mu} \qquad \frac{x \not\xrightarrow{a} \quad y \xrightarrow{a} \nu}{x +_p y \xrightarrow{a} \nu} \qquad \frac{x \xrightarrow{a} \mu \quad y \xrightarrow{a} \nu}{x +_p y \xrightarrow{a} \mu \oplus_p \nu}$$

$$\frac{x \xrightarrow{a} \mu \quad y \not\xrightarrow{a}}{x \parallel_p y \xrightarrow{a} \mu \parallel_p \delta(y)} \qquad \frac{x \not\xrightarrow{a} \quad y \xrightarrow{a} \nu}{x \parallel_p y \xrightarrow{a} \delta(x) \parallel_p \nu} \qquad \frac{x \xrightarrow{a} \mu \quad y \xrightarrow{a} \nu}{x \parallel_p y \xrightarrow{a} \mu \parallel_p \delta(y) \oplus_p \delta(x) \parallel_p \nu}$$

probabilistic parallel composition replaces the nondeterministic choice of their non-probabilistic variants by a probabilistic choice. The probabilistic alternative composition $s +_p t$ evolves to the probabilistic choice between a distribution reached by s (with probability p) and a distribution reached by t (with probability $1-p$) for actions which can be performed by both processes. For actions that can be performed by either only s or only t, the probabilistic alternative composition $s +_p t$ behaves just like the nondeterministic alternative composition $s + t$. Similarly, the probabilistic parallel composition $s \parallel_p t$ evolves to a probabilistic choice between the nondeterministic choices of asynchronous parallel composition of s and t.

We start by showing an important auxiliary property how the distance between convex combinations of probability distributions relates to the distance between the combined probability distributions.

Proposition 6. *Let* $\mu_1, \mu_2, \mu_1', \mu_2' \in Disc(X)$ *and* $p \in [0,1]$. *Then* $K_\otimes(bm_\otimes)(p\mu_1+(1-p)\mu_2, p\mu_1'+(1-p)\mu_2') \leq \max(K_\otimes(bm_\otimes)(\mu_1,\mu_2), K_\otimes(bm_\otimes)(\mu_1',\mu_2'))$

Non-expansiveness is the most wildly studied compositionality property stating that the distance between composed processes is at most the sum of the distance between its parts.

Definition 7. *A n-ary operator* f *is non-expansive wrt a pseudometric* m *if*

$$m(f(s_1,\ldots,s_n), f(t_1,\ldots,t_n)) \leq \sum_{i=1}^{n} m(s_i,t_i)$$

Now we can show that all (non-recursive) operators of the probabilistic process algebra introduced above are non-expansive. In fact, we will provide upper bounds on distance between the composed processes which are in case of the (nondeterministic and probabilistic) alternative composition even stricter than the non-expansiveness condition.

Theorem 4. *Let s, t, s', t' be probabilistic processes. Then*

1. $bm_{\otimes}(s; t, s'; t') \leq bm_{\otimes}(s, s') + bm_{\otimes}(t, t')$
2. $bm_{\otimes}(s + t, s' + t') \leq \max(bm_{\otimes}(s, s'), bm_{\otimes}(t, t'))$
3. $bm_{\otimes}(s +_p t, s' +_p t') \leq \max(bm_{\otimes}(s, s'), bm_{\otimes}(t, t'))$
4. $bm_{\otimes}(s \mid t, s' \parallel t') \leq bm_{\otimes}(s, s') + bm_{\otimes}(t, t')$
5. $bm_{\otimes}(s \parallel t, s' \parallel t') \leq bm_{\otimes}(s, s') + bm_{\otimes}(t, t')$
6. $bm_{\otimes}(s \parallel_p t, s \parallel_p t') \leq bm_{\otimes}(s, s') + bm_{\otimes}(t, t')$

A similar result can be gained for the bisimilarity metric *bm* based on the standard Kantorovich lifting. This generalizes a similar result of [18] which considered only PTSs without nondeterministic branching and only a small set of process combinators.

For the generalized bisimilarity metric bm_V we can formulate a similar result for the nondeterministic alternative composition.

Theorem 5. *Let s, t, s', t' be probabilistic processes. Then*

$$bm_V(s + t, s' + t') \leq \max(bm_V(s, s'), bm_V(t, t'))$$

7 Conclusion and Future Work

We have proposed a family of Kantorovich pseudometrics depending on the notion of distance used to specify properties over traces. We have developed the theory of this notion, and showed how we can use it to verify the corresponding kind of properties. We have also showed that for the multiplicative variant, which is an interesting case because it corresponds to differential privacy, it is possible to give a dual form that makes the metric computable by standard techniques.

Future work include the investigation of methods to compute other members of this family, and of conditions that make possible a general dual form.

References

1. Bacci, G., Bacci, G., Larsen, K.G., Mardare, R.: Computing Behavioral Distances, Compositionally. In: Chatterjee, K., Sgall, J. (eds.) MFCS 2013. LNCS, vol. 8087, pp. 74–85. Springer, Heidelberg (2013)
2. Bacci, G., Bacci, G., Larsen, K.G., Mardare, R.: On-the-fly exact computation of bisimilarity distances. In: Piterman, N., Smolka, S.A. (eds.) TACAS 2013. LNCS, vol. 7795, pp. 1–15. Springer, Heidelberg (2013)
3. Barthe, G., Köpf, B., Olmedo, F., Béguelin, S.Z.: Probabilistic relational reasoning for differential privacy. In: Proc. of POPL. ACM (2012)
4. van Breugel, F., Worrell, J.B.: An algorithm for quantitative verification of probabilistic transition systems. In: Larsen, K.G., Nielsen, M. (eds.) CONCUR 2001. LNCS, vol. 2154, pp. 336–350. Springer, Heidelberg (2001)
5. van Breugel, F., Worrell, J.B.: Towards quantitative verification of probabilistic transition systems. In: Orejas, F., Spirakis, P.G., van Leeuwen, J. (eds.) ICALP 2001. LNCS, vol. 2076, pp. 421–432. Springer, Heidelberg (2001)
6. van Breugel, F., Worrell, J.: A behavioural pseudometric for probabilistic transition systems. Theor. Comp. Sci. 331(1), 115–142 (2005)

7. van Breugel, F., Worrell, J.: Approximating and computing behavioural distances in probabilistic transition systems. Theor. Comp. Sci. 360(1-3), 373–385 (2006)
8. Cai, X., Gu, Y.: Measuring anonymity. In: Bao, F., Li, H., Wang, G. (eds.) ISPEC 2009. LNCS, vol. 5451, pp. 183–194. Springer, Heidelberg (2009)
9. Chatterjee, K., de Alfaro, L., Majumdar, R., Raman, V.: Algorithms for Game Metrics. In: FSTTCS, vol. 2, pp. 107–118. Leibniz-Zentrum fuer Informatik (2008)
10. Chatzikokolakis, K., Andrés, M.E., Bordenabe, N.E., Palamidessi, C.: Broadening the scope of Differential Privacy using metrics. In: De Cristofaro, E., Wright, M. (eds.) PETS 2013. LNCS, vol. 7981, pp. 82–102. Springer, Heidelberg (2013)
11. Chatzikokolakis, K., Gebler, D., Palamidessi, C., Xu, L.: Generalized bisimulation metrics. Tech. rep., INRIA (2014)
12. Chen, D., van Breugel, F., Worrell, J.: On the complexity of computing probabilistic bisimilarity. In: Birkedal, L. (ed.) FOSSACS 2012. LNCS, vol. 7213, pp. 437–451. Springer, Heidelberg (2012)
13. Comanici, G., Precup, D.: Basis function discovery using spectral clustering and bisimulation metrics. In: Vrancx, P., Knudson, M., Grześ, M. (eds.) ALA 2011. LNCS, vol. 7113, pp. 85–99. Springer, Heidelberg (2012)
14. D'Argenio, P.R., Gebler, D., Lee, M.D.: Axiomatizing Bisimulation Equivalences and Metrics from Probabilistic SOS Rules. In: Muscholl, A. (ed.) FOSSACS 2014. LNCS, vol. 8412, pp. 289–303. Springer, Heidelberg (2014)
15. Deng, Y., Du, W.: The kantorovich metric in computer science: A brief survey. ENTCS 253(3), 73–82 (2009)
16. Deng, Y., Palamidessi, C., Pang, J.: Weak probabilistic anonymity. In: Proc. of SecCo. ENTCS, vol. 180 (1), pp. 55–76. Elsevier (2007)
17. Desharnais, J., Jagadeesan, R., Gupta, V., Panangaden, P.: The metric analogue of weak bisimulation for probabilistic processes. In: Proc. of LICS, pp. 413–422. IEEE (2002)
18. Desharnais, J., Jagadeesan, R., Gupta, V., Panangaden, P.: Metrics for labelled Markov processes. Theor. Comp. Sci. 318(3), 323–354 (2004)
19. Dwork, C.: Differential privacy. In: Bugliesi, M., Preneel, B., Sassone, V., Wegener, I. (eds.) ICALP 2006. LNCS, vol. 4052, pp. 1–12. Springer, Heidelberg (2006)
20. Dwork, C., Kenthapadi, K., McSherry, F., Mironov, I., Naor, M.: Our data, ourselves: Privacy via distributed noise generation. In: Vaudenay, S. (ed.) EUROCRYPT 2006. LNCS, vol. 4004, pp. 486–503. Springer, Heidelberg (2006)
21. Lee, M.D., Gebler, D., D'Argenio, P.R.: Tree Rules in Probabilistic Transition System Specifications with Negative and Quantitative Premises. In: Proc. EXPRESS/SOS 2012. EPTCS, vol. 89, pp. 115–130 (2012)
22. Reed, J., Pierce, B.C.: Distance makes the types grow stronger: A calculus for differential privacy. In: Proc. of ICFP, pp. 157–168. ACM (2010)
23. Smith, A.: Efficient, differentially private point estimators. arXiv preprint arXiv:0809.4794 (2008)
24. Tarski, A.: A lattice-theoretical fixpoint theorem and its applications. Pacific Journal of Mathematics 5(2), 285–309 (1955)
25. Thorsley, D., Klavins, E.: Approximating stochastic biochemical processes with wasserstein pseudometrics. Systems Biology, IET 4(3), 193–211 (2010)
26. Tschantz, M.C., Kaynar, D., Datta, A.: Formal verification of differential privacy for interactive systems (extended abstract). ENTCS 276, 61–79 (2011)
27. Xu, L., Chatzikokolakis, K., Lin, H.: Metrics for differential privacy in concurrent systems. In: Ábrahám, E., Palamidessi, C. (eds.) FORTE 2014. LNCS, vol. 8461, pp. 199–215. Springer, Heidelberg (2014)

Choreographies, Logically

Marco Carbone[1], Fabrizio Montesi[2,⋆], and Carsten Schürmann[1]

[1] IT University of Copenhagen, Copenhagen, Denmark
[2] University of Southern Denmark, Odense, Denmark

Abstract. In Choreographic Programming, a distributed system is programmed by giving a *choreography*, a global description of its interactions, instead of separately specifying the behaviour of each of its processes. Process implementations in terms of a distributed language can then be automatically projected from a choreography.

We present Linear Compositional Choreographies (LCC), a proof theory for reasoning about programs that modularly combine choreographies with processes. Using LCC, we logically reconstruct a semantics and a projection procedure for programs. For the first time, we also obtain a procedure for extracting choreographies from process terms.

1 Introduction

Choreographic Programming is a programming paradigm for distributed systems inspired by the "Alice and Bob" notation, where programs, called *choreographies*, are global descriptions of how endpoint processes interact during execution [14,21,1]. The typical set of programs defining the actions performed by each process is then generated by *endpoint projection* (EPP) [17,12,8,5,9,15].

The key aspect of choreography languages is that process interactions are treated linearly, i.e., they are executed exactly once. Previous work [8,9,15] developed correct notions of EPP by using session types [11], linear types for communications inspired by linear logic [10]. Despite the deep connections between choreographies and linearity, the following question remains unanswered:

Is there a formal connection between choreographies and linear logic?

Finding such a connection would contribute to a more precise understanding of choreographies, and possibly lead to answering open questions about them.

A good starting point for answering our question is a recent line of work on a Curry-Howard correspondence between the internal π-calculus [18] and linear logic [7,22]. In particular, proofs in Intuitionistic Linear Logic (ILL) correspond to π-calculus terms (proofs-as-programs) and ILL propositions correspond to session types [7]. An ILL judgement describes the interface of a process, for example:

$$P \triangleright x{:}A, y{:}B \vdash z : C$$

Above, process P needs to be composed with other processes that provide the behaviours (represented as types) A on channel x and B on channel y, in order to provide behaviour C on channel z. The focus is on how the process can be

⋆ Work performed while the author was employed at the IT University of Copenhagen.

P. Baldan and D. Gorla (Eds.): CONCUR 2014, LNCS 8704, pp. 47–62, 2014.

composed with other *external* processes, abstracting from the *internal* communications enacted inside the process itself (which may contain communicating sub-processes). On the contrary, choreographies are descriptions of the internal interactions among the processes inside a system, and therefore type systems for choreographies focus on checking such internal interactions [8,9]. It is thus unclear how the linear typing of ILL can be related to choreographies.

In this paper, we present Linear Compositional Choreographies (LCC), a proof theory inspired by linear logic for typing programs that modularly combine choreographies with processes in the internal π-calculus. The key aspect of LCC is to extend ILL judgements to describe interactions among internal processes in a system. Thanks to LCC, not only do we obtain a logical understanding of choreographic programming, but we also provide the foundations for tackling the open problem of extracting a choreography from a system of processes.

Main Contributions. We summarise our main contributions:

Linear Compositional Choreographies (LCC). We present LCC, a generalisation of ILL where judgements can also describe the internal interactions of a system (§ 3). LCC proofs are equipped with unique proof terms, called LCC programs, following the Curry-Howard interpretation of proofs-as-programs. LCC programs are in a language where choreographies and processes are modularly combined by following protocols given in the type language of LCC (à la session types [11]).

Logically-derived semantics. We derive a semantics for LCC programs from our proof theory (§ 4): (i) some rule applications in LCC proofs can be permuted (commuting conversions), defining equivalences (structural congruence) on LCC programs (§ 4.1); (ii) some proofs can be safely reduced to smaller proofs, corresponding to executing communications (§ 4.2). By following our semantics, we prove that all internal communications in a system can be reduced (proof normalisation), i.e., LCC programs are deadlock-free by construction (§ 4.3).

Choreography Extraction and Endpoint Projection. LCC consists of two fragments: the *action fragment*, which manipulates the external interfaces of processes, and the *interaction fragment*, which handles internal communications. We derive automatic transformations from proofs in either fragment to proofs in the other, yielding procedures of endpoint projection and choreography extraction (§ 5) that preserve the semantics of LCC programs. This is the first work addressing extraction for a fragment of the π-calculus, providing the foundations for a new development methodology where programmers can compose choreographies with existing process code (e.g., software libraries) and then obtain a choreography that describes the overall behaviour of the entire composition.

2 From ILL to LCC

In this section, we informally introduce processes and choreographies, and revisit the Curry-Howard correspondence between the internal π-calculus and ILL [7]. Building on ILL, we introduce the intuition behind the proof theory of LCC.

Processes and Choreographies. Consider the following processes:

$$\underbrace{\overline{x}(tea);\ x(tr);\ \overline{tr}(p)}_{P_{\mathsf{client}}} \qquad \underbrace{x(tea);\ \overline{x}(tr);\ tr(p);\ \overline{b}(m)}_{P_{\mathsf{server}}} \qquad \underbrace{b(m)}_{P_{\mathsf{bank}}} \tag{1}$$

The three processes above, given as internal π-calculus terms [18], denote a system composed by three endpoints (client, server, and bank). Their parallel execution is such that: client sends to server a request for tea on a channel x; then, server replies to client on the same channel x with a new channel tr (for transaction); client uses tr for sending to server the payment p; after receiving the payment, server deposits some money m by sending it over channel b to bank.

Programming with processes is error-prone, since they do not give an explicit description of how endpoints interact [14]. By contrast, choreographies specify how messages flow during execution [21]. For example, the choreography

$$\begin{array}{l} 1.\ \mathsf{client} \to \mathsf{server} : x(tea);\ \mathsf{server} \to \mathsf{client} : x(tr); \\ 2.\ \mathsf{client} \to \mathsf{server} : tr(p);\ \mathsf{server} \to \mathsf{bank} : b(m) \end{array} \tag{2}$$

defines the communications that occur in (1). We read $\mathsf{client} \to \mathsf{server} : x(tea)$ as "process client sends tea to process server through channel x".

ILL and the π-calculus. The processes in (1) can be typed by ILL, using propositions as session types that describe the usage of channels. For example, channel x in P_{client} has type $\mathbf{string} \otimes (\mathbf{string} \multimap \mathsf{end}) \multimap \mathsf{end}$, meaning: send a string; then, receive a channel of type $\mathbf{string} \multimap \mathsf{end}$ and, finally, stop (end). Concretely, in P_{client}, the channel of type $\mathbf{string} \multimap \mathsf{end}$ received through x is channel tr. The type of tr says that the process sending tr, i.e., P_{server}, will use it to receive a **string**; therefore, process P_{client} must implement the dual operation of that implemented by P_{server}, i.e., the output $\overline{tr}(p)$. Similarly, channel b has type $\mathbf{int} \otimes \mathsf{end}$ in P_{server}. We can formalise this intuition with the following three ILL judgements, where $A = \mathbf{string} \otimes (\mathbf{string} \multimap \mathsf{end}) \multimap \mathsf{end}$ and $B = \mathbf{int} \otimes \mathsf{end}$:

$$P_{\mathsf{client}} \triangleright \cdot \vdash x : A \qquad P_{\mathsf{server}} \triangleright x : A \vdash b : B \qquad P_{\mathsf{bank}} \triangleright b : B \vdash z : \mathsf{end}$$

Recall that $P_{\mathsf{server}} \triangleright x : A \vdash b : B$ reads as "given a context that implements channel x with type A, process P_{server} implements channel b with type B". Given these judgements, we compose P_{client}, P_{server}, and P_{bank} using channels x and b as:

$$(\boldsymbol{\nu} x)\,(\ P_{\mathsf{client}}\ \ |_x\ \ (\boldsymbol{\nu} b)\,(\ P_{\mathsf{server}}\ \ |_b\ \ P_{\mathsf{bank}}\ )\ ) \tag{3}$$

The compositions in (3) can be typed using the Cut rule of ILL:

$$\frac{P \triangleright \Delta_1 \vdash x : A \qquad Q \triangleright \Delta_2, x : A \vdash y : B}{(\boldsymbol{\nu} x)\,(P \mid Q) \triangleright \Delta_1, \Delta_2 \vdash y : B}\ \mathsf{Cut} \tag{4}$$

Above, Δ_1 and Δ_2 are sets of typing assignments, e.g., $z : D$. We interpret rule Cut as "If a process provides A on channel x, and another requires A on channel x to provide B on channel y, their parallel execution provides B on channel y".

Proofs in ILL correspond to process terms in the internal π-calculus [7], and applications of rule Cut can always be eliminated, a proof normalisation procedure known as cut elimination. This procedure provides a model of computation for processes. We illustrate a *cut reduction*, a step of cut elimination, in the following (we omit process terms for readability):

$$\dfrac{\dfrac{C_1 \vdash A \quad C_2 \vdash B}{C_1, C_2 \vdash A \otimes B}\otimes\mathsf{R} \qquad \dfrac{A, B \vdash D}{A \otimes B \vdash D}\otimes\mathsf{L}}{C_1, C_2 \vdash D}\,\mathsf{Cut} \quad \Longrightarrow \quad \dfrac{C_1 \vdash A \qquad \dfrac{C_2 \vdash B \quad A, B \vdash D}{C_2, A \vdash D}\,\mathsf{Cut}}{C_1, C_2 \vdash D}\,\mathsf{Cut}$$

The proof on the left-hand side applies a cut to two proofs, one providing $A \otimes B$, and the other providing D when provided with $A \otimes B$. The cut-reduction above ($\Rightarrow$) shows how this proof can be simplified to a proof where the cut on $A \otimes B$ is reduced to two cuts on the smaller formulas A and B. A cut-reduction corresponds to executing a communication between two processes, one outputting on a channel of type $A \otimes B$, and another inputting from the same channel [7]. Executing the communication yields a new system corresponding to the proof on the right-hand side. Cut-free proofs correspond to systems that have successfully completed all their internal communications.

Towards LCC. Cut reductions in ILL model the interactions between the internal processes in a system, which is exactly what choreographies describe syntactically. Therefore, in order to capture choreographies, we wish our proof theory to reason about transformations such as the cut reduction above.

ILL judgements give us no information on the applications of rule Cut in a proof. In contrast, standard type systems for choreographies [8,9,15] have different judgements: instead of interfaces for later composition, they contain information about internal processes and their interactions. Following this observation, we make two important additions to ILL judgements. First, we extend them to describe multiple processes by using *hypersequents*, i.e., collections of multiple ILL sequents [2]. Second, we represent the *connections* between sequents in a hypersequent, since two processes need to share a common connection for interacting. The following is an LCC judgement:

$$P \rhd \Delta_1 \vdash x{:}\bullet A \mid \Delta_2, x{:}\bullet A \vdash y{:}B$$

Above, we composed two ILL sequents with the operator $|$, which captures the parallel composition of processes. The two sequents are *connected* through channel x, denoted by the marking $\bullet$. We will use hypersequents and marking let us reason about interactions by handling both ends of a connection.

LCC judgements can express cut elimination as a proof. For example,

$$Q \rhd z_1{:}C_1, z_2{:}C_2 \vdash x{:}\bullet A \otimes B \mid x{:}\bullet A \otimes B \vdash w{:}D$$

represents the left-hand side of the cut reduction seen previously, where a process requires C_1 and C_2 to perform an interaction of type $A \otimes B$ with another process that can then provide D. Importantly, the connection of type $A \otimes B$ between the two sequents cannot be composed with external systems since it is used for internal interactions. Using our judgements, we can capture cut reductions:

$$Q' \rhd z_1{:}C_1 \vdash y{:}\bullet A \mid z_2{:}C_2 \vdash x{:}\bullet B \mid y{:}\bullet A, x{:}\bullet B \vdash w{:}D$$

The new judgement describes a system that still requires C_1 and C_2 in order to provide D, but now with three processes: one providing A from C_1, one providing B from C_2 and, finally, one using A and B for providing D. Also, the first two sequents are connected to the third one. This corresponds to the right-hand side of the cut reduction seen previously, where process Q reduces to process Q'.

We can now express the different internal states of a system before and after a cut reduction, by the structure of its connections in our judgements. This intuition is behind the new rules for typing choreographies presented in § 3.

3 Linear Compositional Choreographies

We present Linear Compositional Choreographies (LCC), a proof theory for typing programs that can modularly combine choreographies and processes.

Types. LCC propositions, or types, are defined as:

$$\textit{(Propositions)} \quad A, B ::= \mathbf{1} \quad | \quad A \otimes B \quad | \quad A \multimap B \quad | \quad A \oplus B \quad | \quad A \& B$$

LCC propositions are the same as in ILL: $\otimes$ and $\multimap$ are the multiplicative connectives, while $\oplus$ and $\&$ are additives. $\mathbf{1}$ is the atomic proposition. $A \otimes B$ is interpreted as "output a channel of type A and then behave as specified by type B". On the other hand, $A \multimap B$, the linear implication, reads "receive a channel of type A and then continue as B". Proposition $A \oplus B$ selects a branch of type A or B, while $A\&B$ offers the choice of A or B.

Hypersequents. Elements are types identified by variables, possibly marked by $\bullet$. Contexts are sets of elements, while hypersequents are sets of ILL sequents:

$$\textit{(Element)} \quad T ::= x{:}A \mid x{:}\bullet A \qquad \textit{(Contexts)} \quad \Delta, \Theta ::= \cdot \mid \Delta, T$$
$$\textit{(Hypersequents)} \quad \Psi ::= \Delta \vdash T \mid \Psi | \Psi$$

Contexts Δ and hypersequents Ψ are equivalent modulo associativity and commutativity. A hypersequent Ψ is the parallel composition of sequents. Given a sequent $\Delta \vdash T$, we call Δ its hypotheses and T its conclusion.

We make the standard assumption that a variable can appear at most once in any hypersequent, *unless* it is marked with $\bullet$. In LCC, bulleted variables appear exactly twice in a hypersequent, once as a hypothesis and once as a conclusion of two respective sequents which we say are then "connected". A provable hypersequent always has exactly one sequent with a non-bulleted conclusion, which we call the conclusion of the hypersequent. Similarly, we call non-bulleted hypotheses the hypotheses of the hypersequent. Intuitively, a provable hypersequent is a tree of sequents, whose root is the only sequent with a non-bulleted conclusion, and whose sequents have exactly one child for each of their bulleted hypotheses.

Processes and Choreographies. We give the syntax of our proof terms, or LCC programs, in Fig. 1. The syntax is an extension of that of the internal π-calculus with choreographic primitives. The internal π-calculus allows us to focus on a simple, yet very expressive fragment of the π-calculus [19], as in [7]. Terms can be *processes* performing I/O actions or *choreographies* of interactions.

Processes. An *(output)* $\overline{x}(y); (P|Q)$ sends a *fresh* name y over channel x and then proceeds with the parallel composition $P|Q$, whereas an *(input)* $x(y); P$ receives y over x and proceeds as P. In a *(left sel)* $x.\mathsf{inl}; P$, we send over channel x our choice of the left branch offered by the receiver. The term *(right sel)* $x.\mathsf{inr}; P$ selects the right branch instead. Selections communicate with the term *(case)*

$$
\begin{array}{llllll}
P, Q, R ::= & \overline{x}(y);(P|Q) & \textit{(output)} & \mid x(y);P & \textit{(input)} & \\
& \mid x.\mathsf{inl};P & \textit{(left sel)} & \mid x.\mathsf{inr};P & \textit{(right sel)} & \\
& \mid x.\mathsf{case}(P,Q) & \textit{(case)} & \mid P \mid_x Q & \textit{(par)} & \textit{Processes} \\
& \mid \mathsf{close}[x] & \textit{(close)} & \mid \mathsf{wait}[x];P & \textit{(wait)} & \\
& \mid (\boldsymbol{\nu} x)\, P & \textit{(res)} & & & \\
\textit{Choreographies} & \mid \overrightarrow{x(y)};P & \textit{(global com)} & \mid \overrightarrow{\mathsf{close}[x]}\;;P & \textit{(global close)} & \\
& \mid \overrightarrow{x.\mathsf{l}}(P,Q) & \textit{(global left sel)} & \mid \overrightarrow{x.\mathsf{r}}(P,Q) & \textit{(global right sel)} &
\end{array}
$$

Fig. 1. LCC programs

$x.\mathsf{case}(P,Q)$, which offers a left branch P and a right branch Q. The term *(par)* $P \mid_x P$ models parallel composition; here, differently from the output case, the two composed processes are not independent, but share the communication channel x. The term *(res)* is the standard restriction. Terms *(close)* and *(wait)* model, respectively, the request and acceptance for closing a channel, following real-world closing handshakes in communication protocols such as TCP.

Choreographies. The term *(res)*for name restriction is the same as for processes. A *(global com)* $\overrightarrow{x(y)};P$ describes a system where a fresh name y is communicated over a channel x, and then continues as P, where y is bound in P. The terms *(global left sel)* and *(global right sel)* model systems where, respectively, a left branch or a right branch is selected on channel x. Unused branches in global selections, e.g., Q in $\overrightarrow{x.\mathsf{l}}(P,Q)$, are unnecessary in our setting since they are never executed; however, their specification will be convenient for our technical development of endpoint projection, which will follow our concretisation transformation in LCC. Finally, term *(global close)* models the closure of a channel.

Note that, differently from § 2, we omit process identifiers in choreographies since our typing will make them redundant (cf. § 6).

Judgements. An LCC judgement has the form $P \triangleright \Psi$ where Ψ is a hypersequent and P is a proof term. If we regard LCC as a type theory for our term language, we say that the hypersequent Ψ types the term P.

3.1 Rules

The proof theory of LCC consists of the *action fragment* and the *interaction fragment*, which reason respectively about processes and choreographies.

Action Fragment. The action fragment includes ILL-style left and right rules, reported in Fig. 2, and the structural rules Conn and Scope, described separately.

Unit. The rules for unit are standard. Rule 1R types a process that requests to close channel x and terminates. Symmetrically, rule 1L types a process that waits for a request to close x, making sure that x does not occur in P.

$$\frac{P \triangleright \Psi_1|\Delta_1 \vdash y{:}A \quad Q \triangleright \Psi_2|\Delta_2 \vdash x{:}B}{\overline{x}(y);(P|Q) \triangleright \Psi_1|\Psi_2|\Delta_1,\Delta_2 \vdash x{:}A \otimes B}\otimes\mathsf{R} \qquad \frac{P \triangleright \Psi|\Delta, y{:}A, x{:}B \vdash T}{x(y);P \triangleright \Psi|\Delta, x{:}A \otimes B \vdash T}\otimes\mathsf{L}$$

$$\frac{P \triangleright \Psi|\Delta, y{:}A \vdash x{:}B}{x(y);P \triangleright \Psi|\Delta \vdash x{:}A \multimap B}\multimap\mathsf{R} \qquad \frac{P \triangleright \Psi_1|\Delta_1 \vdash y{:}A \quad Q \triangleright \Psi_2|\Delta_2, x{:}B \vdash T}{\overline{x}(y);(P|Q) \triangleright \Psi_1|\Psi_2|\Delta_1,\Delta_2, x{:}A \multimap B \vdash T}\multimap\mathsf{L}$$

$$\frac{}{\mathsf{close}[x] \triangleright \cdot \vdash x{:}\mathbf{1}}\mathsf{1R} \qquad \frac{P \triangleright \Psi|\Delta, x{:}A \vdash T}{x.\mathsf{inl};P \triangleright \Psi|\Delta, x{:}A\&B \vdash T}\&\mathsf{L}_1 \qquad \frac{Q \triangleright \Psi|\Delta, x{:}B \vdash T}{x.\mathsf{inr};Q \triangleright \Psi|\Delta, x{:}A\&B \vdash T}\&\mathsf{L}_2$$

$$\frac{P \triangleright \Psi|\Delta \vdash T}{\mathsf{wait}[x];P \triangleright \Psi|\Delta, x{:}\mathbf{1} \vdash T}\mathsf{1L} \qquad \frac{P \triangleright \Psi|\Delta \vdash x{:}A}{x.\mathsf{inl};P \triangleright \Psi|\Delta \vdash x{:}A \oplus B}\oplus\mathsf{R}_1 \qquad \frac{Q \triangleright \Psi|\Delta \vdash x{:}B}{x.\mathsf{inr};Q \triangleright \Psi|\Delta \vdash x{:}A \oplus B}\oplus\mathsf{R}_2$$

$$\frac{P \triangleright \Psi|\Delta \vdash x{:}A \quad Q \triangleright \Psi|\Delta \vdash x{:}B}{x.\mathsf{case}(P,Q) \triangleright \Psi|\Delta \vdash x{:}A\&B}\&\mathsf{R} \qquad \frac{P \triangleright \Psi|\Delta, x{:}A \vdash T \quad Q \triangleright \Psi|\Delta, x{:}B \vdash T}{x.\mathsf{case}(P,Q) \triangleright \Psi|\Delta, x{:}A \oplus B \vdash T}\oplus\mathsf{L}$$

Fig. 2. Left and Right Rules of the Action Fragment

Tensor. Rule $\otimes$R types the output $\overline{x}(y);(P|Q)$, combining the conclusions of the hypersequents of P and Q. The continuations P and Q will handle, respectively, the transmitted channel y and channel x. Ensuring that channels y and x are handled by different parallel processes avoids potential deadlocks caused by their interleaving [7,22]. Dually, rule $\otimes$L types an input $x(y);P$, by requiring the continuation to use channels y and x following their respective types.

Linear Implication. The proof term for rule $\multimap$ R is an input $x(y);P$, meaning that the process needs to receive a name of type A before offering behaviour B on channel x. Rule $\multimap$ L types the dual term $\overline{x}(y);(P|Q)$. Note that the prefixes in the proof terms are the same as for the tensor rules. This does not introduce ambiguity, since continuations are typed differently and thus it is never the case that both connectives could be used for typing the same term [7].

Additives. The rules for the additive connectives are standard. In a left selection $x.\mathsf{inl};P$, we send over x our choice of the left branch offered by the receiver. The term $x.\mathsf{inr};P$, is similar, but selects the right branch instead. Selections communicate with the term $x.\mathsf{case}(P,Q)$, which offers a left branch P and a right branch Q. In LCC, for example, rule &R states that $x.\mathsf{case}(P,Q)$ provides x with type $A\&B$ whenever P and Q provide x with type A and B respectively.

Connection and Scoping. We pull apart the standard Cut rule of ILL, as (4) in § 2, and obtain two rules that depend on hypersequents as an interim place to store information. The first rule, Conn, merges two hypersequents by forming a connection:

$$\frac{P \triangleright \Psi_1|\Delta_1 \vdash x{:}A \qquad Q \triangleright \Psi_2|\Delta_2, x{:}A \vdash T}{P \mid_x Q \triangleright \Psi_1|\Psi_2|\Delta_1 \vdash x{:}\bullet A|\Delta_2, x{:}\bullet A \vdash T}\mathsf{Conn}$$

The proof term for Conn is parallel composition: in the conclusion, the two terms P and Q are composed in parallel and share channel x.

The second rule, called Scope, delimits the scope of a connection:

$$\frac{P \triangleright \Psi \mid \Delta_1 \vdash x{:}\bullet A \mid \Delta_2, x{:}\bullet A \vdash T}{(\nu x)\, P \triangleright \Psi \mid \Delta_1, \Delta_2 \vdash T}\mathsf{Scope}$$

The proof term for Scope is a restriction of the scoped channel.

Interaction Fragment. Connections are first-class citizens in LCC and are object of logical reasoning. We give rules for composing connections, one for each connective, which correspond to choreographies. Such rules form, together with rule Scope, the interaction fragment of LCC.

Unit. A connection of type **1** between two sequents can always be introduced:

$$\frac{P \triangleright \Psi | \Delta \vdash T}{\overrightarrow{\mathsf{close}[x]}; P \triangleright \Psi | \cdot \vdash x : \bullet \mathbf{1} | \Delta, x : \bullet \mathbf{1} \vdash T} \; \mathbf{1}\mathsf{C}$$

Observe that the choreography term $\overrightarrow{\mathsf{close}[x]}; P$ describes the same behaviour as the process term $\mathsf{close}[x] \mid_x \mathsf{wait}[x]; P$, and indeed their typing is the same. In general, in LCC the typing of process terms and choreographic terms describing equivalent behaviour is the same. We will formalise this intuition in § 5.

Tensor. The connection rule for $\otimes$ combines two connections between three sequents. Technically, when two sequents $\Delta_1 \vdash y : \bullet A$ and $\Delta_2 \vdash x : \bullet B$ are connected to a third sequent $\Delta_3, y : \bullet A, x : \bullet B \vdash T$, we can merge the two connections into a single one, obtaining the sequents $\Delta_1, \Delta_2 \vdash x : \bullet A \otimes B$ and $\Delta_3, x : \bullet A \otimes B \vdash T$:

$$\frac{P \triangleright \Psi | \Delta_1 \vdash y : \bullet A | \Delta_2 \vdash x : \bullet B | \Delta_3, y : \bullet A, x : \bullet B \vdash T}{\overrightarrow{x(y)}; P \triangleright \Psi | \Delta_1, \Delta_2 \vdash x : \bullet A \otimes B | \Delta_3, x : \bullet A \otimes B \vdash T} \; \otimes\mathsf{C}$$

Rule $\otimes$C corresponds to typing a choreographic communication $\overrightarrow{x(y)}; P$. This rule is the formalisation in LCC of the cut reduction discussed in § 2. Term P will then perform communications on channel y with type A and x with type B.

Linear Implication. The rule for $\multimap$ manipulates connections with a causal dependency: if $\Delta_1 \vdash y : \bullet A$ is connected to $\Delta_2, y : \bullet A \vdash x : \bullet B$, which is connected to $\Delta_3, x : \bullet B \vdash T$, then $\Delta_2 \vdash x : \bullet A \multimap B$ is connected to $\Delta_1, \Delta_3, x : \bullet A \multimap B \vdash T$.

$$\frac{P \triangleright \Psi | \Delta_1 \vdash y : \bullet A | \Delta_2, y : \bullet A \vdash x : \bullet B | \Delta_3, x : \bullet B \vdash T}{\overrightarrow{x(y)}; P \triangleright \Psi | \Delta_2 \vdash x : \bullet A \multimap B | \Delta_1, \Delta_3, x : \bullet A \multimap B \vdash T} \; \multimap\mathsf{C}$$

Rule $\multimap$ C types a communication $\overrightarrow{x(y)}; P$. The prefix $\overrightarrow{x(y)}$ is the same as that of rule $\otimes$C, similarly to the action fragment for the connectives $\otimes$ and $\multimap$. Differently from rule $\otimes$C, the usage of channel x in the continuation P has a causal dependency on y whereas in $\otimes$C the two channels proceed separately.

Additives. The rules for the additive connectives follow similar reasoning:

$$\frac{P \triangleright \Psi | \Psi' | \Delta_1 \vdash x : \bullet A | \Delta_2, x : \bullet A \vdash T \qquad Q \triangleright \Psi' | \Delta_1 \vdash x : B}{\overrightarrow{x.\mathsf{l}}(P, Q) \triangleright \Psi | \Psi' | \Delta_1 \vdash x : \bullet A \& B | \Delta_2, x : \bullet A \& B \vdash T} \; \&\mathsf{C}_1$$

$$\frac{P \triangleright \Psi | \Delta_1 \vdash x : A \qquad Q \triangleright \Psi | \Psi' | \Delta_1 \vdash x : \bullet B | \Delta_2, x : \bullet B \vdash T}{\overrightarrow{x.\mathsf{r}}(P, Q) \triangleright \Psi | \Psi' | \Delta_1 \vdash x : \bullet A \& B | \Delta_2, x : \bullet A \& B \vdash T} \; \&\mathsf{C}_2$$

$$\frac{P \triangleright \Psi|\Psi'|\Delta_1 \vdash x:\bullet A|\Delta_2, x:\bullet A \vdash T \qquad Q \triangleright \Psi'|\Delta_2, x:B \vdash T}{\overrightarrow{x}.\mathsf{l}(P,Q) \triangleright \Psi|\Psi'|\Delta_1 \vdash x:\bullet A \oplus B|\Delta_2, x:\bullet A \oplus B \vdash T}\ \oplus\mathsf{C}_1$$

$$\frac{P \triangleright \Psi|\Delta_2, x:A \vdash T \qquad Q \triangleright \Psi|\Psi'|\Delta_1 \vdash x:\bullet B|\Delta_2, x:\bullet B \vdash T}{\overrightarrow{x}.\mathsf{r}(P,Q) \triangleright \Psi|\Psi'|\Delta_1 \vdash x:\bullet A \oplus B|\Delta_2, x:\bullet A \oplus B \vdash T}\ \oplus\mathsf{C}_2$$

Rule $\&\mathsf{C}_1$ types a choreography that selects the left branch on x and then proceeds P, provided that x is not used in Q since the latter is unused.

We call C-rules the interaction rules for manipulating connections. C-rules represent of cut reductions in ILL, following the intuition presented in § 2.

Example 1. We formalise and extend our example from § 2 as follows:

$$P_{\mathsf{client}'} = x.\mathsf{inr};\ \overline{x}(tea);\ \Big(\mathsf{close}[tea] \mid x(tr);\ \overline{tr}(p);\ (\mathsf{close}[p]|\mathsf{wait}[tr];\ \mathsf{close}[x]\,)\Big)$$

$$P_{\mathsf{server}'} = x.\mathsf{case}\left(\begin{array}{l} x(water);\ b.\mathsf{inl};\ \mathsf{wait}[water];\ \mathsf{wait}[x];\ \mathsf{close}[b], \\ x(tea);\ \overline{x}(tr);\ \left(\begin{array}{l} tr(p);\ \mathsf{wait}[tea];\ \mathsf{wait}[p];\ \mathsf{close}[tr] \mid \\ b.\mathsf{inr};\ \overline{b}(m);\ \big(\mathsf{close}[m] \mid \mathsf{wait}[x];\ \mathsf{close}[b]\big)\end{array}\right)\end{array}\right)$$

$$P_{\mathsf{bank}'} = b.\mathsf{case}(\ \ \mathsf{wait}[b];\ \mathsf{close}[z],\ \ \ b(m);\ \mathsf{wait}[m];\ \mathsf{wait}[b];\ \mathsf{close}[z]\ \ )$$

$$P = (\nu x)\,(P_{\mathsf{client}'} \mid_x (\nu b)\,(P_{\mathsf{server}'} \mid_b P_{\mathsf{bank}'}))$$

$$C = (\nu x)\,(\nu b)\ \overrightarrow{x}.\mathsf{r}\left(\begin{array}{l} x(water);\ b.\mathsf{inl};\ \mathsf{wait}[water];\ \mathsf{wait}[x];\ \mathsf{close}[b], \\ \overrightarrow{x(tea)};\ \overrightarrow{x(tr)};\ \overrightarrow{tr(p)};\ \overrightarrow{b}.\mathsf{r}\left(\begin{array}{l}\mathsf{wait}[b];\ \mathsf{close}[z] \\ \\ \overrightarrow{b(m)};\ \overrightarrow{\mathsf{close}[tea,p,tr,m,x,b]}\end{array}\right)\end{array}\right)$$

Process $P_{\mathsf{client}'}$ implements a new version of the client, which selects the right choice of a branching on channel x and then asks for some tea; then, it proceeds as P_{client} from § 2. Note that we have enhanced the processes with all expected closing of channels. The server $P_{\mathsf{server}'}$, instead, now offers to the client a choice between buying a tea (as in § 2) and getting a free glass of water. Since the water is free, the payment to the bank is not performed in this case. In either case, the bank is notified of whether a payment will occur or not, respectively right and left branch in P_{bank}. The processes are composed as a system in P.

Term C is the equivalent choreographic representation of P. We can type channel x as $(\mathbf{string} \otimes \mathsf{end}) \oplus (\mathbf{string} \otimes (\mathbf{string} \multimap \mathsf{end}) \multimap \mathsf{end})$ in both C and P. The type of channel b is: $\mathsf{end} \oplus (\mathbf{string} \otimes \mathsf{end})$. For clarity, we have used concrete data types instead of the abstract basic type $\mathbf{1}$. □

4 Semantics

We now derive an operational semantics for LCC programs from our proof theory, by obtaining the standard relations of structural equivalence $\equiv$ and reduction $\rightarrow$ as theorems of LCC. For example, the π-calculus rule $(\nu w)\,(P \mid_x Q) \equiv (\nu w)\,P \mid_x Q$ (for $w \notin \mathsf{fn}(Q)$) can be derived as a proof transformation, since:

$$\dfrac{\dfrac{P \triangleright \Psi|\ \Delta_1 \vdash y:\bullet D|\ \Delta, y:\bullet D \vdash x:A \qquad Q \triangleright \Psi'|\ \Delta', x:A \vdash T}{P \mid_x Q \triangleright \Psi|\ \Psi'|\ \Delta_1 \vdash y:\bullet D|\ \Delta, y:\bullet D \vdash x:\bullet A|\ \Delta', x:\bullet A \vdash T}\ \mathsf{Conn}}{(\nu y)\,(P \mid_x Q) \triangleright \Psi|\ \Psi'|\ \Delta_1, \Delta \vdash x:\bullet A|\ \Delta', x:\bullet A \vdash T}\ \mathsf{Scope}$$

[Scope/Conn/L]	$(\nu y)\,(P \mid_x Q) \equiv (\nu y)\,P \mid_x Q$	$\big(y \notin \mathsf{fn}(Q)\big)$
[Scope/Conn/R]	$(\nu y)\,(P \mid_x Q) \equiv P \mid_x (\nu y)\,Q$	$\big(y \notin \mathsf{fn}(P)\big)$
[Scope/Scope]	$(\nu y)\,(\nu x)\,P \equiv (\nu x)\,(\nu y)\,P$	
[Scope/1L]	$(\nu x)\,\mathsf{wait}[y];P \equiv \mathsf{wait}[y];(\nu x)\,P$	
[Scope/ ⊗ R/L], [Scope/ ⊸ L/L]	$(\nu w)\,\overline{x}(y);(P \mid Q) \equiv \overline{x}(y);((\nu w)\,P \mid Q)$	$\big(w \notin \mathsf{fn}(Q)\big)$
[Scope/ ⊗ R/R], [Scope/ ⊸ L/R]	$(\nu w)\,\overline{x}(y);(P \mid Q) \equiv \overline{x}(y);(P \mid (\nu w)\,Q)$	$\big(w \notin \mathsf{fn}(P)\big)$
[Scope/ ⊗ L], [Scope/ ⊸ R]	$(\nu w)\,x(y);P \equiv x(y);(\nu w)\,P$	
[Scope/ ⊕ R$_1$], [Scope/&L$_1$]	$(\nu w)\,x.\mathsf{inl};P \equiv x.\mathsf{inl};(\nu w)\,P$	
[Scope/ ⊕ R$_2$], [Scope/&L$_2$]	$(\nu w)\,x.\mathsf{inr};P \equiv x.\mathsf{inr};(\nu w)\,P$	
[Scope/ ⊕ L], [Scope/&R]	$(\nu w)\,x.\mathsf{case}(P,Q) \equiv x.\mathsf{case}((\nu w)\,P\,,\,(\nu w)\,Q)$	
[Scope/1C]	$(\nu w)\ \overrightarrow{\mathsf{close}[x]};P \equiv \overrightarrow{\mathsf{close}[x]};(\nu w)\,P$	
[Scope/ ⊗ C], [Scope/ ⊸ C]	$(\nu w)\ \overrightarrow{x(y)};P \equiv \overrightarrow{x(y)};(\nu w)\,P$	
[Scope/ ⊕ C$_1$/L], [Scope/&C$_1$/L]	$(\nu w)\ \overrightarrow{x.\mathsf{l}}(P,Q) \equiv \overrightarrow{x.\mathsf{l}}((\nu w)\,P,Q)$	$\big(w \notin \mathsf{fn}(Q)\big)$
[Scope/ ⊕ C$_1$/L/R], [Scope/&C$_1$/L/R]	$(\nu w)\ \overrightarrow{x.\mathsf{l}}(P,Q) \equiv \overrightarrow{x.\mathsf{l}}((\nu w)\,P,(\nu w)\,Q)$	$\big(w \in \mathsf{fn}(Q)\big)$
[Scope/ ⊕ C$_2$/R], [Scope/&C$_2$/R]	$(\nu w)\ \overrightarrow{x.\mathsf{r}}(P,Q) \equiv \overrightarrow{x.\mathsf{r}}(P,(\nu w)\,Q)$	$\big(w \notin \mathsf{fn}(P)\big)$
[Scope/ ⊕ C$_2$/L/R], [Scope/&C$_2$/L/R]	$(\nu w)\ \overrightarrow{x.\mathsf{r}}(P,Q) \equiv \overrightarrow{x.\mathsf{r}}((\nu w)\,P,(\nu w)\,Q)$	$\big(w \in \mathsf{fn}(P)\big)$

Fig. 3. Commuting Conversions (≡) for Scope (Restriction)

is equivalent to (≡)

$$\dfrac{\dfrac{P \triangleright \Psi \mid \Delta_1 \vdash y : \bullet D \mid \Delta, y : \bullet D \vdash x : A}{(\nu y)\,P \triangleright \Psi \mid \Delta_1, \Delta \vdash x : A}\,\mathsf{Scope} \qquad Q \triangleright \Psi' \mid \Delta', x : A \vdash T}{(\nu y)\,P \mid_x Q \triangleright \Psi \mid \Psi' \mid \Delta_1, \Delta \vdash x : \bullet A \mid \Delta', x : \bullet A \vdash T}\,\mathsf{Conn}$$

4.1 Commuting Conversions (≡)

The structural equivalence of LCC (≡) is defined in terms of *commuting conversions*, i.e., admissible permutations of rule applications in proofs. In ILL, commuting conversions concern the cut rule. However, since in LCC the cut rule has been split into Scope and Conn, we need to introduce two sets of commuting conversions, one for rule Scope, and one for rule Conn. In the sequel, we report commuting conversions between proofs by giving the corresponding process and choreography terms (cf. [14] for the complete LCC proofs).

Commuting Conversions for Scope**.** Commuting conversions for Scope correspond to permuting restriction with other operators in LCC programs. We report them in Fig. 3, where we assume variables to be distinct. For example, [Scope/ ⊗ R/L] says that an application of rule Scope to the conclusion of rule ⊗R can be commuted so that we can apply ⊗R to the conclusion of Scope. Note that the top-level LCC terms of some cases are identical, e.g., [Scope/⊗R/L] and [Scope/ ⊸ L/L], but the subterms are different since they have different typing.

Commuting Conversions for Conn**.** The commuting conversions for rule Conn, reported in Fig. 4, correspond to commuting the parallel operator with

[Conn/Conn]	$(P \mid_y Q) \mid_x R \equiv P \mid_y (Q \mid_x R)$	
[Conn/1L/L]	$\mathsf{wait}[x]; P \mid_y Q \equiv \mathsf{wait}[x]; (P \mid_y Q)$	
[Conn/1L/R]	$P \mid_y \mathsf{wait}[x]; Q \equiv \mathsf{wait}[x]; (P \mid_y Q)$	
[Conn/ ⊗R/R/L], [Conn/⊸L/R/L]	$P \mid_w \overline{x}(y); (Q \mid R) \equiv \overline{x}(y); ((P \mid_w Q) \mid R)$	
[Conn/ ⊗R/R/R], [Conn/⊸L/R/R]	$P \mid_w \overline{x}(y); (Q \mid R) \equiv \overline{x}(y); (Q \mid (P \mid_w R))$	
[Conn/ ⊗ L/L]	$x(y); P \mid_w Q \equiv x(y); (P \mid_w Q)$	
[Conn/ ⊗ L/R], [Conn/ ⊸ R/R]	$P \mid_w x(y); Q \equiv x(y); (P \mid_w Q)$	
[Conn/ ⊸ L/L/R]	$\overline{x}(y); (P \mid Q) \mid_w R \equiv \overline{x}(y); (P \mid (Q \mid_w R))$	
[Conn/ ⊕ R₁/R], [Conn/&L₁/R]	$P \mid_w x.\mathsf{inl}; Q \equiv x.\mathsf{inl}; (P \mid_w Q)$	
[Conn/ ⊕ R₂/R], [Conn/&L₂/R]	$P \mid_w x.\mathsf{inr}; Q \equiv x.\mathsf{inr}; (P \mid_w Q)$	
[Conn/ ⊕ L/L]	$x.\mathsf{case}(P, Q) \mid_w R \equiv x.\mathsf{case}((P \mid_w R), (Q \mid_w R))$	
[Conn/ ⊕ L/R], [Conn/&R/R]	$P \mid_w x.\mathsf{case}(Q, R) \equiv x.\mathsf{case}((P \mid_w Q), (P \mid_w R))$	
[Conn/&L₁/L]	$x.\mathsf{inl}; P \mid_w Q \equiv x.\mathsf{inl}; (P \mid_w Q)$	
[Conn/&L₂/L]	$x.\mathsf{inr}; P \mid_w Q \equiv x.\mathsf{inr}; (P \mid_w Q)$	
[Conn/1C/L]	$\overrightarrow{\mathsf{close}[x]}; P \mid_w Q \equiv \overrightarrow{\mathsf{close}[x]}; (P \mid_w Q)$	
[Conn/1C/R]	$P \mid_w \overrightarrow{\mathsf{close}[x]}; Q \equiv \overrightarrow{\mathsf{close}[x]}; (P \mid_w Q)$	
[Conn/⊗ C/L], [Conn/ ⊸C/L]	$\overrightarrow{x(y)}; P \mid_w Q \equiv \overrightarrow{x(y)}; (P \mid_w Q)$	$(y \notin \mathsf{fn}(Q))$
[Conn/⊗C/R], [Conn/ ⊸ C/R]	$P \mid_w \overrightarrow{x(y)}; Q \equiv \overrightarrow{x(y)}; (P \mid_w Q)$	$(y \notin \mathsf{fn}(P))$
[Conn/ ⊕ C₁/L]	$\overrightarrow{x.\mathsf{l}}(P, Q) \mid_w R \equiv \overrightarrow{x.\mathsf{l}}((P \mid_w R), (Q \mid_w R))$	$(w \in \mathsf{fn}(P) \cap \mathsf{fn}(Q))$
[Conn/ ⊕ C₁/R], [Conn/&C₁/R]	$P \mid_w \overrightarrow{x.\mathsf{l}}(Q, R) \equiv \overrightarrow{x.\mathsf{l}}((P \mid_w Q), (P \mid_w R))$	$(w \in \mathsf{fn}(Q) \cap \mathsf{fn}(R))$
[Conn/⊕C₁/R/L],[Conn/&C₁/R/L]	$P \mid_w \overrightarrow{x.\mathsf{l}}(Q, R) \equiv \overrightarrow{x.\mathsf{l}}((P \mid_w Q), R)$	$(w \in \mathsf{fn}(Q), w \notin \mathsf{fn}(R))$
[Conn/ ⊕ C₂/L]	$\overrightarrow{x.\mathsf{r}}(P, Q) \mid_w R \equiv \overrightarrow{x.\mathsf{r}}((P \mid_w R), (Q \mid_w R))$	$(w \in \mathsf{fn}(P) \cap \mathsf{fn}(Q))$
[Conn/ ⊕ C₂/R], [Conn/&C₂/R]	$P \mid_w \overrightarrow{x.\mathsf{r}}(Q, R) \equiv \overrightarrow{x.\mathsf{r}}((P \mid_w Q), (P \mid_w R))$	$(w \in \mathsf{fn}(Q) \cap \mathsf{fn}(R))$
[Conn/⊕C₂/R/R],[Conn/&C₂/R/L]	$P \mid_w \overrightarrow{x.\mathsf{r}}(Q, R) \equiv \overrightarrow{x.\mathsf{r}}(Q, (P \mid_w R))$	$(w \notin \mathsf{fn}(Q), w \in \mathsf{fn}(R))$
[Conn/&C₁/L]	$\overrightarrow{x.\mathsf{l}}(P, Q) \mid_w R \equiv \overrightarrow{x.\mathsf{l}}((P \mid_w R), Q)$	$(w \in \mathsf{fn}(P), w \notin \mathsf{fn}(Q))$
[Conn/&C₂/L]	$\overrightarrow{x.\mathsf{r}}(P, Q) \mid_w R \equiv \overrightarrow{x.\mathsf{r}}(P, (Q \mid_w R))$	$(w \notin \mathsf{fn}(P), w \in \mathsf{fn}(Q))$

Fig. 4. Commuting Conversions (≡) for Conn (Parallel Composition)

other terms. For example, rule [Conn/Conn] is the standard associativity of parallel in the π-calculus. Also, [Conn/ ⊗ C/L] says that $\overrightarrow{x(y)}$ in $\overrightarrow{x(y)}; P \mid_w Q$ can always be executed before Q as far as x and y do not occur in Q. This captures the concurrent behaviour of choreographies in [9]. Note that some of the rules are not standard for the π-calculus, e.g., [Conn/ ⊸ R/R], but this does not alter the intended semantics of parallel (cf. § 6, Semantics).

Since conversions preserve the concluding judgement of a proof, we have that:

Theorem 1 (Subject Congruence). *$P \triangleright \Psi$ and $P \equiv Q$ implies that $Q \triangleright \Psi$.*

4.2 Reductions (→)

As for structural equivalence, we derive the reduction semantics for LCC programs from proof transformations. The obtained rules, reported in Fig. 5, are standard for both processes and choreographies (cf. [19,9]): processes are reduced when they are the parallel composition of compatible actions, while choreographies can always be reduced. With an abuse of notation, we labelled each reduction with the channel it uses. Choreography reductions are also annotated with

$$
\begin{array}{ll}
[\beta_{\mathbf{1}}] & (\boldsymbol{\nu} x)\,(\mathsf{close}[x] \mid_x \mathsf{wait}[x]; Q) \xrightarrow{x} Q \\
[\beta_{\otimes}] & (\boldsymbol{\nu} x)\,(\overline{x}(y); (P|Q) \mid_x x(y); R) \xrightarrow{x} (\boldsymbol{\nu} y)\,(\boldsymbol{\nu} x)\,(P \mid_y (Q \mid_x R)) \\
[\beta_{\multimap}] & (\boldsymbol{\nu} x)\,(x(y); P \mid_x \overline{x}(y); (Q|R)) \xrightarrow{x} (\boldsymbol{\nu} x)\,(\boldsymbol{\nu} y)\,((Q \mid_y P) \mid_x R) \\
[\beta_{\oplus_1}] & (\boldsymbol{\nu} x)\,(x.\mathsf{inl}; P \mid_x x.\mathsf{case}(Q, R)) \xrightarrow{x} (\boldsymbol{\nu} x)\,(P \mid_w Q) \\
[\beta_{\oplus_2}] & (\boldsymbol{\nu} x)\,(x.\mathsf{inr}; P \mid_x x.\mathsf{case}(Q, R)) \xrightarrow{x} (\boldsymbol{\nu} x)\,(P \mid_x R) \\
[\beta_{\&_1}] & (\boldsymbol{\nu} x)\,(x.\mathsf{case}(P, Q) \mid_x x.\mathsf{inl}; R) \xrightarrow{x} (\boldsymbol{\nu} x)\,(P \mid_x R) \\
[\beta_{\&_2}] & (\boldsymbol{\nu} x)\,(x.\mathsf{case}(P, Q) \mid_x x.\mathsf{inr}; R) \xrightarrow{x} (\boldsymbol{\nu} x)\,(Q \mid_x R)
\end{array}
$$

$$
\begin{array}{ll}
[\beta_{\mathbf{1}\mathsf{C}}]\ (\boldsymbol{\nu} x)\ \overrightarrow{\mathsf{close}[x]}; P \xrightarrow{\bullet x} P & [\beta_{\otimes\mathsf{C}}], [\beta_{\multimap\mathsf{C}}]\ (\boldsymbol{\nu} x)\ \overrightarrow{x(y)}; P \xrightarrow{\bullet x} (\boldsymbol{\nu} y)\,(\boldsymbol{\nu} x)\,P \\
[\beta_{\&\mathsf{C}_1}], [\beta_{\oplus\mathsf{C}_1}]\ (\boldsymbol{\nu} x)\ \overrightarrow{x.\mathsf{l}}(P, Q) \xrightarrow{\bullet x} (\boldsymbol{\nu} x)\,P & [\beta_{\&\mathsf{C}_2}], [\beta_{\oplus\mathsf{C}_2}]\ (\boldsymbol{\nu} x)\ \overrightarrow{x.\mathsf{r}}(P, Q) \xrightarrow{\bullet x} (\boldsymbol{\nu} x)\,Q
\end{array}
$$

Fig. 5. Reductions

$\bullet$. We use t to range over labels of the form x or $\bullet x$, and $\tilde{t}$ to denote a sequence of such labels. As for commuting conversions, reductions preserve judgements:

Theorem 2 (Subject Reduction). *$P \triangleright \Psi$ and $P \xrightarrow{t} Q$ implies that $Q \triangleright \Psi$.*

4.3 Scope Elimination (Normalisation)

We can use commuting conversions and reductions to permute and reduce all applications of Scope in a proof until the proof is Scope-free. Since applications of Scope correspond to restrictions in LCC programs, the latter can always progress until all communications on restricted channels are executed. We denote the reflexive and transitive closure of $\xrightarrow{t}$ up to $\equiv$ with $\xrightarrow{\tilde{t}}\!\!\!\!\rightarrow$.

Theorem 3 (Deadlock-freedom). *$P \triangleright \Psi$ implies there exist Q restriction-free and $\tilde{t}$ such that $P \xrightarrow{\tilde{t}}\!\!\!\!\rightarrow Q$ and $Q \triangleright \Psi$.*

5 Choreography Extraction and Endpoint Projection

In LCC, a judgement containing connections can be derived by either (i) using the action fragment, corresponding to processes, or (ii) using the interaction fragment, corresponding to choreographies. Consider the two following proofs:

$$
\dfrac{\dfrac{\dfrac{}{\mathsf{close}[x] \triangleright \cdot \vdash x : \mathbf{1}}\,\mathsf{1R} \quad \dfrac{\dfrac{}{\mathsf{close}[y] \triangleright \cdot \vdash y : \mathbf{1}}\,\mathsf{1R}}{\mathsf{wait}[x]; \mathsf{close}[y] \triangleright x : \mathbf{1} \vdash y : \mathbf{1}}\,\mathsf{1L}}{\mathsf{close}[x] \mid_x \mathsf{wait}[x]; \mathsf{close}[y] \triangleright \cdot \vdash x : \bullet\mathbf{1} | x : \bullet\mathbf{1} \vdash y : \mathbf{1}}\,\mathsf{Conn}}{(\boldsymbol{\nu} x)\,(\mathsf{close}[x] \mid_x \mathsf{wait}[x]; \mathsf{close}[y]) \triangleright \cdot \vdash y : \mathbf{1}}\,\mathsf{Scope}
\qquad
\dfrac{\dfrac{\dfrac{}{\mathsf{close}[y] \triangleright \cdot \vdash y : \mathbf{1}}\,\mathsf{1R}}{\overrightarrow{\mathsf{close}[x]}\ ; \mathsf{close}[y] \triangleright \cdot \vdash x : \bullet\mathbf{1} | x : \bullet\mathbf{1} \vdash y : \mathbf{1}}\,\mathsf{1C}}{(\boldsymbol{\nu} x)\,(\overrightarrow{\mathsf{close}[x]}\ ; \mathsf{close}[y]) \triangleright \cdot \vdash y : \mathbf{1}}\,\mathsf{Scope}
$$

The two proofs above reach the same hypersequent following, respectively, methodologies (i) and (ii). In this section, we formally relate the two methodologies, deriving procedures of choreography extraction and endpoint projection from proof equivalences. As an example, consider the following equivalence, $[\alpha\gamma_{\otimes}]$:

$$
\begin{array}{llcl}
[\alpha\gamma_1] & \mathsf{close}[x] \mid_x \mathsf{wait}[x]; P & \overset{x}{\underset{\leftarrow\cdots}{\dashrightarrow}} & \overrightarrow{\mathsf{close}[x]}; P \\
[\alpha\gamma_\otimes] & \overline{x}(y); (P|Q) \mid_x x(y); R & \overset{x}{\underset{\leftarrow\cdots}{\dashrightarrow}} & \overrightarrow{x(y)}; (P \mid_y (Q \mid_x R)) \\
[\alpha\gamma_\multimap] & x(y); P \mid_x \overline{x}(y); (Q|R) & \overset{x}{\underset{\leftarrow\cdots}{\dashrightarrow}} & \overrightarrow{x(y)}; ((Q \mid_y P) \mid_x R) \\
[\alpha\gamma_{\&_1}] & x.\mathsf{case}(P, Q) \mid_x x.\mathsf{inl}; R & \overset{x}{\underset{\leftarrow\cdots}{\dashrightarrow}} & \overrightarrow{x.\mathsf{l}}((P \mid_x R), Q) \\
[\alpha\gamma_{\&_2}] & x.\mathsf{case}(P, Q) \mid_x x.\mathsf{inr}; R & \overset{x}{\underset{\leftarrow\cdots}{\dashrightarrow}} & \overrightarrow{x.\mathsf{r}}(P\ ,\ Q \mid_x R) \\
[\alpha\gamma_{\oplus_1}] & x.\mathsf{inl}; P \mid_x x.\mathsf{case}(Q, R) & \overset{x}{\underset{\leftarrow\cdots}{\dashrightarrow}} & \overrightarrow{x.\mathsf{l}}((P \mid_x Q)\ ,\ R) \\
[\alpha\gamma_{\oplus_2}] & x.\mathsf{inr}; P \mid_x x.\mathsf{case}(Q, R) & \overset{x}{\underset{\leftarrow\cdots}{\dashrightarrow}} & \overrightarrow{x.\mathsf{r}}(Q\ ,\ (P \mid_x R))
\end{array}
$$

Fig. 6. Extraction and Projection

$$
\dfrac{\dfrac{P \rhd \Psi_1|\Delta_1 \vdash y{:}A \qquad Q \rhd \Psi_2|\Delta_2 \vdash x{:}B}{\overline{x}(y); (P|Q) \rhd \Psi_1|\Psi_2|\Delta_1, \Delta_2 \vdash x{:}A \otimes B}\otimes\mathsf{R} \qquad \dfrac{R \rhd \Psi_3|\Delta_3, y{:}A, x{:}B \vdash T}{x(y); R \rhd \Delta_3, x{:}A \otimes B \vdash T}\otimes\mathsf{L}}{\overline{x}(y); (P|Q) \mid_x x(y); R \rhd \Psi_1|\Psi_2|\Psi_3|\Delta_1, \Delta_2 \vdash x{:}\bullet A \otimes B|\Delta_3, x{:}\bullet A \otimes B \vdash T}\mathsf{Conn}
$$

can be extracted to ($\overset{x}{\dashrightarrow}$), can be projected from ($\overset{x}{\leftarrow\cdots}$)

$$
\dfrac{\dfrac{P \rhd \Psi_1|\Delta_1 \vdash y{:}A \qquad \dfrac{Q \rhd \Psi_2|\Delta_2 \vdash x{:}B \qquad R \rhd \Psi_3 \mid \Delta_3, y{:}A, x{:}B \vdash T}{Q \mid_x R \rhd \Psi_2|\Psi_3|\Delta_2 \vdash x{:}\bullet B|\Delta_3, y{:}A, x{:}\bullet B \vdash T}\mathsf{Conn}^x}{P \mid_y (Q \mid_x R) \rhd \Psi_1 \mid \Psi_2 \mid \Psi_3 \mid \Delta_1 \vdash y{:}\bullet A \mid \Delta_2 \vdash x{:}\bullet B \mid \Delta_3, y{:}\bullet A, x{:}\bullet B \vdash T}\mathsf{Conn}^y}{\overrightarrow{x(y)}; (P \mid_y (Q \mid_x R)) \rhd \Psi_1 \mid \Psi_2 \mid \Psi_3 \mid \Delta_1, \Delta_2 \vdash x{:}\bullet A \otimes B \mid \Delta_3, x{:}\bullet A \otimes B \vdash T}\otimes\mathsf{C}^x
$$

The equivalence $[\alpha\gamma_\otimes]$ allows to transform a proof of a connection of type $A \otimes B$ from the action fragment into an equivalent proof in the interaction fragment, and vice versa. We report the equivalences for extraction and projection in Fig. 6, presenting their proof terms. We read these equivalences from left to right for extraction, denoted by $\dashrightarrow$, and from right to left for projection, denoted by $\cdots\rightarrow$. Note how a choreography term corresponds to the parallel composition of two processes with the same behaviour. It is also clear why the unselected process Q in $\overrightarrow{x.\mathsf{l}}(P, Q)$ is necessary for projecting the corresponding case process.

Using commuting conversions, extraction can always be applied to proofs containing instances of Conn, i.e., programs containing subterms of the form $P \mid_x Q$. Similarly, projection can always be applied to proofs with instances of a C-rule, i.e., programs with choreography interactions. We denote the reflexive and transitive closure of $\overset{x}{\dashrightarrow}$ up to $\equiv$ with $\overset{\tilde{x}}{\dashrightarrow\!\!\!\rightarrow}$ (resp. $\overset{\tilde{x}}{\cdots\cdots\twoheadrightarrow}$ for $\overset{x}{\cdots\rightarrow}$).

Theorem 4 (Extraction and Projection). *Let $P \rhd \Psi$. Then:*

(choreography extraction) $P \overset{\tilde{x}}{\dashrightarrow\!\!\!\rightarrow} Q$ for some $\tilde{x}$ and Q such that $Q \rhd \Psi$ and Q does not contain subterms of the form $R \mid_x R'$;

(endpoint projection) $P \overset{\tilde{x}}{\cdots\cdots\twoheadrightarrow} Q$ for some $\tilde{x}$ and Q such that $Q \rhd \Psi$ and Q does not contain choreography terms.

Example 2. Using the equivalences in Fig. 6 and $\equiv$, we can transform the processes to the choreography in Example 1 and vice versa. □

We now present the main property guaranteed by LCC: the extraction and projection procedures preserve the semantics of the transformed programs.

Theorem 5 (Correspondence). *Let $P \triangleright \Psi$ and P' be restriction-free. Then:* *(choreography extraction)* $P \xrightarrow{\tilde{x}}\!\!\!\rightarrow P'$ *implies* $P \overset{\tilde{x}}{\dashrightarrow}\!\!\!\rightarrow Q$ *such that* $Q \xrightarrow{\bullet\tilde{x}}\!\!\!\rightarrow P'$. *(endpoint projection)* $P \xrightarrow{\bullet\tilde{x}}\!\!\!\rightarrow P'$ *implies* $P \overset{\tilde{x}}{\cdots\cdots}\!\!\!\twoheadrightarrow Q$ *such that* $Q \xrightarrow{\tilde{x}}\!\!\!\rightarrow P'$.

6 Related Work and Discussion

Related Work. Our action fragment is inspired by π-DILL [7]. The key difference is that we split rule Cut into Conn and Scope, which allows us to (i) reason about choreographies and (ii) type processes where restriction and parallel are used separately. Extra typable processes are always convertible to those where a Conn is immediately followed by a Scope, hence equivalent to those in [7]. Wadler [22] introduces a calculus where processes correspond to proofs in classical linear logic. We conjecture that LCC can be adapted to the classical setting.

Our commuting conversions can be seen as a logical characterisation of *swapping* [9], which permutes independent communications in a choreography. Previous works [12,8,9,15] have formally addressed choreographies and EPP but without providing choreography extraction. Choreography extraction is a known hard problem [4], and our work is the first to address it for a language supporting channel passing. Probably, the work closest to ours wrt extraction is [13], where global types are extracted from session types; choreographies are more expressive than global types, since they capture the interleaving of different sessions. In the future, we plan to address standard features supported by [8,9,15] such as multiparty sessions, asynchrony, replicated services and nondeterminism.

Our mixing of choreographies with processes is similar to that found in [3] for global protocols and [15] for choreographies. The work [3] deals with the simpler setting of protocols, whereas we handle programs supporting name passing and session interleaving, both nontrivial problems [6,9,15]. The type system in [15] does not keep information on where the endpoints of connections are actually located as in our hypersequents, which enables extraction in our setting.

Process identifiers. In standard choreography calculi, the processes involved in a communication are usually identified explicitly as in the choreography (2) in § 2 [12,8,9,15]. In LCC, processes are implicitly identified in judgements by using separate sequents in a hypersequent. Omitting process identifiers is thus just a matter of presentational convenience: a way of retaining them would be to annotate each sequent in a hypersequent with a process identifier (cf. [14]).

Exponentials and Infinite Behaviour. Our work focuses on the multiplicative and additive fragments of linear logic, but we conjecture that the known cut rule for exponentials can be split into a connection rule and a scope rule such as the ones for the linear case. We believe that the results in this paper can be generalised to exponentials without altering its foundations. A logical characterisation of infinite behaviour for ICC may similarly be added to our framework, following the developments in [20]. We leave both extensions as future work.

ILL. LCC is a generalisation of ILL, since we can represent any instance of the Cut rule in ILL with consecutive applications of rules Conn and Scope.

Semantics. LCC includes more term equivalences than the π-calculus, e.g., [Conn / $\multimap$ R/R/2] in Fig. 4. We inherit this from linear logic [22]. However, the extra equivalences do not produce any new reductions in well-typed systems (cf. [16]).

Acknowledgments. This work was partially funded by the Demtech grant number 10-092309 from the Danish Council for Strategic Research.

References

1. Business Process Model and Notation, `http://www.omg.org/spec/BPMN/2.0/`
2. Avron, A.: Hypersequents, logical consequence and intermediate logics for concurrency. Ann. Math. Artif. Intell. 4, 225–248 (1991)
3. Baltazar, P., Caires, L., Vasconcelos, V.T., Vieira, H.T.: A Type System for Flexible Role Assignment in Multiparty Communicating Systems. In: Palamidessi, C., Ryan, M.D. (eds.) TGC 2012. LNCS, vol. 8191, pp. 82–96. Springer, Heidelberg (2013)
4. Basu, S., Bultan, T.: Choreography conformance via synchronizability. In: WWW, pp. 795–804 (2011)
5. Basu, S., Bultan, T., Ouederni, M.: Deciding choreography realizability. In: POPL, pp. 191–202 (2012)
6. Bettini, L., Coppo, M., D'Antoni, L., De Luca, M., Dezani-Ciancaglini, M., Yoshida, N.: Global progress in dynamically interleaved multiparty sessions. In: van Breugel, F., Chechik, M. (eds.) CONCUR 2008. LNCS, vol. 5201, pp. 418–433. Springer, Heidelberg (2008)
7. Caires, L., Pfenning, F.: Session types as intuitionistic linear propositions. In: Gastin, P., Laroussinie, F. (eds.) CONCUR 2010. LNCS, vol. 6269, pp. 222–236. Springer, Heidelberg (2010)
8. Carbone, M., Honda, K., Yoshida, N.: Structured communication-centered programming for web services. ACM TOPLAS 34(2), 8 (2012)
9. Carbone, M., Montesi, F.: Deadlock-freedom-by-design: multiparty asynchronous global programming. In: POPL, pp. 263–274 (2013)
10. Girard, J.-Y.: Linear logic. Theor. Comput. Sci. 50, 1–102 (1987)
11. Honda, K., Vasconcelos, V.T., Kubo, M.: Language primitives and type discipline for structured communication-based programming. In: Hankin, C. (ed.) ESOP 1998. LNCS, vol. 1381, pp. 122–138. Springer, Heidelberg (1998)
12. Lanese, I., Guidi, C., Montesi, F., Zavattaro, G.: Bridging the gap between interaction- and process-oriented choreographies. In: Proc. of SEFM, pp. 323–332. IEEE (2008)
13. Lange, J., Tuosto, E.: Synthesising choreographies from local session types. In: Koutny, M., Ulidowski, I. (eds.) CONCUR 2012. LNCS, vol. 7454, pp. 225–239. Springer, Heidelberg (2012)
14. Montesi, F.: Choreographic Programming. Ph.D. thesis, IT University of Copenhagen (2013), `http://www.itu.dk/people/fabr/papers/phd/thesis.pdf`

15. Montesi, F., Yoshida, N.: Compositional choreographies. In: D'Argenio, P.R., Melgratti, H. (eds.) CONCUR 2013 – Concurrency Theory. LNCS, vol. 8052, pp. 425–439. Springer, Heidelberg (2013)
16. Pérez, J.A., Caires, L.; Pfenning, F., Toninho, B.: Linear logical relations for session-based concurrency. In: Seidl, H. (ed.) ESOP 2012. LNCS, vol. 7211, pp. 539–558. Springer, Heidelberg (2012)
17. Qiu, Z., Zhao, X., Cai, C., Yang, H.: Towards the theoretical foundation of choreography. In: WWW, pp. 973–982. IEEE (2007)
18. Sangiorgi, D.: pi-calculus, internal mobility, and agent-passing calculi. Theor. Comput. Sci. 167(1&2), 235–274 (1996)
19. Sangiorgi, D., Walker, D.: The π-calculus: a Theory of Mobile Processes. Cambridge University Press (2001)
20. Toninho, B., Caires, L., Pfenning, F.: Higher-order processes, functions, and sessions: A monadic integration. In: Felleisen, M., Gardner, P. (eds.) Programming Languages and Systems. LNCS, vol. 7792, pp. 350–369. Springer, Heidelberg (2013)
21. W3C WS-CDL Working Group. Web services choreography description language version 1.0 (2004), `http://www.w3.org/TR/2004/WD-ws-cdl-10-20040427/`
22. Wadler, P.: Propositions as sessions. In: ICFP, pp. 273–286 (2012)

Deadlock Analysis of Unbounded Process Networks

Elena Giachino[1], Naoki Kobayashi[2], and Cosimo Laneve[1]

[1] Dept. of Computer Science and Egineering, University of Bologna – INRIA FOCUS, Italy
[2] Dept. of Computer Science, University of Tokyo, Japan

Abstract. Deadlock detection in concurrent programs that create networks with arbitrary numbers of nodes is extremely complex and solutions either give imprecise answers or do not scale. To enable the analysis of such programs, (1) we define an algorithm for detecting deadlocks of a basic model featuring recursion and fresh name generation: the *lam programs*, and (2) we design a type system for value passing CCS that returns lam programs. As a byproduct of these two techniques, we have an algorithm that is more powerful than previous ones and that can be easily integrated in the current release of `TyPiCal`, a type-based analyser for pi-calculus.

1 Introduction

Deadlock-freedom of concurrent programs has been largely investigated in the literature [2,4,1,11,18,19]. The proposed algorithms automatically detect deadlocks by building graphs of dependencies (a, b) between resources, meaning that the release of a resource referenced by a depends on the release of the resource referenced by b. The absence of cycles in the graphs entails deadlock freedom. When programs have infinite states, in order to ensure termination, current algorithms use finite approximate models that are excerpted from the dependency graphs. The cases that are particularly critical are those of programs that create networks with an arbitrary number of nodes.

To illustrate the issue, consider the following pi-calculus-like process that computes the factorial:

```
Fact(n,r,s) =  if n=0 then r?m. s!m
               else  new t in (r?m. t!(m*n)) | Fact(n-1,t,s)
```

Here, `r?m` waits to receive a value for `m` on `r`, and `s!m` sends the value `m` on `s`. The expression **new** `t in P` creates a fresh communication channel `t` and executes `P`. If the above code is invoked with `r!1 | Fact(n,r,s)`, then there will be a synchronisation between `r!1` and the input `r?m` in the body of `Fact(n,r,s)`. In turn, this may delegate the computation of the factorial to another process in parallel by means of a subsequent synchronisation on a new channel `t`. That is, in order to compute the factorial of `n`, `Fact` builds a network of $n+1$ nodes, where node `i` takes as input a value `m` and outputs `m*i`. Due to the inability of statically reasoning about unbounded structures, the current analysers usually return false positives when fed with `Fact`. For example, this is the case of `TyPiCal` [12,11], a tool developed for pi-calculus. (In particular, `TyPiCal` fails to recognise that there is no circularity in the dependencies among r, s, and t.)

In this paper we develop a technique to enable the deadlock analysis of processes with arbitrary networks of nodes. Instead of reasoning on finite approximations of such

P. Baldan and D. Gorla (Eds.): CONCUR 2014, LNCS 8704, pp. 63–77, 2014.

processes, we associate them with terms of a basic recursive model, called *lam* – for *deadLock Analysis Model* –, which collects dependencies and features recursion and dynamic name creation [5,6]. For example, the lam function corresponding to `Fact` is

$$\texttt{fact}(a_1,a_2,a_3,a_4) = (a_2,a_3) + (\nu\, a_5,a_6)((a_2,a_6)\ \&\ \texttt{fact}(a_5,a_6,a_3,a_4))$$

where (a_2,a_3) displays the dependency between the actions `r?m` and `s!m` and (a_2,a_5) the one between `r?m` and `t!(m*n)`. The function `fact` is defined operationally by unfolding the recursive invocations; see Section 3. The unfolding of $\texttt{fact}(a_1,a_2,a_3,a_4)$ yields the following sequence of abstract states (bound names in the definition of `fact` are replaced by fresh ones in the unfoldings).

$$\begin{aligned}
\texttt{fact}(a_1,a_2,a_3,a_4) &\longrightarrow (a_2,a_3) + ((a_2,a_6)\ \&\ \texttt{fact}(a_5,a_6,a_3,a_4))\\
&\longrightarrow (a_2,a_3) + (a_2,a_6)\ \&\ (a_6,a_3) + (a_2,a_6)\ \&\ (a_6,a_8)\ \&\ \texttt{fact}(a_7,a_8,a_3,a_4)\\
&\longrightarrow (a_2,a_3) + (a_2,a_6)\ \&\ (a_6,a_3) + (a_2,a_6)\ \&\ (a_6,a_8)\ \&\ (a_8,a_3)\\
&\qquad +(a_2,a_6)\ \&\ (a_6,a_8)\ \&\ (a_8,a_{10})\ \&\ \texttt{fact}(a_9,a_{10},a_3,a_4)\\
&\longrightarrow \cdots
\end{aligned}$$

While the model of `fact` is not finite-state, in Section 4 we demonstrate that it is decidable whether the computations of a lam program will ever produce a circular dependency. In our previous work [5,6], the decidability was established only for a restricted subset of lams.

We then define a type system that associates lams to processes. Using the type system, for example, the lam program `fact` can be extracted from the factorial process `Fact`. For the sake of simplicity, we address the *(asynchronous) value passing CCS* [15], a simpler calculus than pi-calculus, because it is already adequate to demonstrate the power of our lam-based approach. The syntax, semantics, and examples of value passing CCS are in Section 5; the type system is defined in Section 6. As a byproduct of the above techniques, our system is powerful enough to detect deadlocks of programs that create networks with arbitrary numbers of processes. It is also worth to notice that our system admits type inference and can be easily extended to pi-calculus. We discuss the differences of our techniques with respect to the other ones in the literature in Section 7 where we also deliver some concluding remark.

2 Preliminaries

We use an infinite set $\mathscr{A}$ of *(level) names*, ranged over by $a, b, c, \cdots$. A relation on a set A of names, denoted $\mathrm{R}, \mathrm{R}', \cdots$, is an element of $\mathscr{P}(A \times A)$, where $\mathscr{P}(\cdot)$ is the standard powerset operator and $\cdot \times \cdot$ is the cartesian product. Let

– R^+ be the *transitive closure* of R.
– $\{\mathrm{R}_1, \cdots, \mathrm{R}_m\} \Subset \{\mathrm{R}'_1, \cdots, \mathrm{R}'_n\}$ if and only if, for all R_i, there is R'_j such that $\mathrm{R}_i \subseteq \mathrm{R}'^{+}_j$.
– $(a_0,a_1), \cdots, (a_{n-1},a_n) \in \{\mathrm{R}_1, \cdots, \mathrm{R}_m\}$ if and only if there is R_i such that $(a_0,a_1), \cdots, (a_{n-1},a_n) \in \mathrm{R}_i$.
– $\{\mathrm{R}_1, \cdots, \mathrm{R}_m\}\ \&\ \{\mathrm{R}'_1, \cdots, \mathrm{R}'_n\} \stackrel{def}{=} \{\mathrm{R}_i \cup \mathrm{R}'_j \mid 1 \le i \le m \text{ and } 1 \le j \le n\}$.

We use $\mathscr{R}, \mathscr{R}', \cdots$ to range over $\{\mathrm{R}_1, \cdots, \mathrm{R}_m\}$ and $\{\mathrm{R}'_1, \cdots, \mathrm{R}'_n\}$, which are elements of $\mathscr{P}(\mathscr{P}(A \times A))$.

Definition 1. *A relation* R ***has a circularity*** *if* $(a,a) \in \mathrm{R}^+$ *for some* a. *A set of relations* $\mathscr{R}$ ***has a circularity*** *if there is* $\mathrm{R} \in \mathscr{R}$ *that has a circularity.*

For instance $\{\{(a,b),(b,c)\},\ \{(a,b),(c,b),(d,b),(b,c)\},\ \{(e,d),(d,c)\},\ \{(e,d)\}\}$ has a circularity because the second element of the set does.

3 The Language of Lams

In addition to the set of (*level*) *names*, we will also use *function names*, ranged over by $\mathtt{f}, \mathtt{g}, \mathtt{h}, \cdots$. A sequence of names is denoted by $\widetilde{a}$ and, with an abuse of notation, we also use $\widetilde{a}$ to address the *set of names* in the sequence.

A *lam program* is a pair $(\mathscr{L}, \mathtt{L})$, where $\mathscr{L}$ is a *finite set* of *function definitions* $\mathtt{f}(\widetilde{a}) = \mathtt{L}_\mathtt{f}$, with $\widetilde{a}$ and $\mathtt{L}_\mathtt{f}$ being the *formal parameters* and the *body* of $\mathtt{f}$, and $\mathtt{L}$ is the *main lam*. The syntax of the function bodies and the main lam is

$$\mathtt{L} \ ::= \ \mathtt{0} \ \mid \ (a,b) \ \mid \ \mathtt{f}(\widetilde{a}) \ \mid \ \mathtt{L} \,\&\, \mathtt{L} \ \mid \ \mathtt{L} + \mathtt{L} \ \mid \ (\nu\, a)\mathtt{L}$$

The lam $\mathtt{0}$ enforces no dependency, the lam (a,b) enforces the dependency (a,b), while $\mathtt{f}(\widetilde{a})$ represents a function invocation. The composite lam $\mathtt{L} \,\&\, \mathtt{L}'$ enforces the dependencies of $\mathtt{L}$ *and* of $\mathtt{L}'$, while $\mathtt{L} + \mathtt{L}'$ nondeterministically enforces the dependencies of $\mathtt{L}$ *or* of $\mathtt{L}'$, $(\nu\, a)\mathtt{L}$ creates a fresh name a and enforces the dependencies of $\mathtt{L}$ that may use a. Whenever parentheses are omitted, the operation "&" has precedence over "+". We will shorten $\mathtt{L}_1 \,\&\cdots\&\, \mathtt{L}_n$ into $\&_{i\in 1..n}\mathtt{L}_i$ and $(\nu\, a_1)\cdots(\nu\, a_n)\mathtt{L}$ into $(\nu\, a_1 \cdots a_n)\mathtt{L}$. Function definitions $\mathtt{f}(\widetilde{a}) = \mathtt{L}_\mathtt{f}$ and $(\nu\, a)\mathtt{L}$ are *binders* of $\widetilde{a}$ in $\mathtt{L}_\mathtt{f}$ and of a in $\mathtt{L}$, respectively, and the corresponding occurrences of $\widetilde{a}$ in $\mathtt{L}_\mathtt{f}$ and of a in $\mathtt{L}$ are called *bound*. A name x in $\mathtt{L}$ is *free* if it is not underneath a $(\nu\, a)$ (similarly for function definitions). Let $var(\mathtt{L})$ be the set of free names in $\mathtt{L}$.

In the syntax of $\mathtt{L}$, the operations "&" and "+" are associative, commutative with $\mathtt{0}$ being the identity on &, and definitions and lams are equal up-to alpha renaming of bound names. Namely, if $a \notin var(\mathtt{L})$, the following axioms hold:

$$(\nu\, a)\mathtt{L} = \mathtt{L} \qquad (\nu\, a)\mathtt{L}' \,\&\, \mathtt{L} = (\nu\, a)(\mathtt{L}' \,\&\, \mathtt{L}) \qquad (\nu\, a)\mathtt{L}' + \mathtt{L} = (\nu\, a)(\mathtt{L}' + \mathtt{L})$$

Additionally, when $\mathtt{V}$ ranges over lams that do not contain function invocations, the following axioms hold:

$$\mathtt{V} \,\&\, \mathtt{V} = \mathtt{V} \qquad \mathtt{V} + \mathtt{V} = \mathtt{V} \qquad \mathtt{V} \,\&\, (\mathtt{L}' + \mathtt{L}'') = \mathtt{V} \,\&\, \mathtt{L}' + \mathtt{V} \,\&\, \mathtt{L}''$$

These axioms permit to rewrite a lam without function invocations as a *collection* (operation +) *of relations* (elements of a relation are gathered by the operation &). Let $\equiv$ be the least congruence containing the above axioms. They are restricted to terms $\mathtt{V}$ that do not contain function invocations. In fact, $\mathtt{f}(\widetilde{d})\,\&\,((a,b)+(b,c)) \neq (\mathtt{f}(\widetilde{d})\,\&\,(a,b)) + (\mathtt{f}(\widetilde{d})\,\&\,(b,c))$ because the evaluation of the two lams (see below) may produce terms with different names.

Definition 2. *A lam* $\mathtt{V}$ *is in* ***normal form****, denoted* $\mathtt{nf}(\mathtt{V})$*, if* $\mathtt{V} = (\nu\widetilde{a})(\mathtt{V}_1 + \cdots + \mathtt{V}_n)$*, where* $\mathtt{V}_1, \cdots, \mathtt{V}_n$ *are dependencies only composed with* &.

Proposition 1. *For every* $\mathtt{V}$*, there is* $\mathtt{nf}(\mathtt{V})$ *such that* $\mathtt{V} \equiv \mathtt{nf}(\mathtt{V})$.

In the rest of the paper, we always assume lam programs $(\mathscr{L}, \mathtt{L})$ to be *well formed*.

Definition 3. *A lam program* $(\mathscr{L}, \mathtt{L})$ *is* ***well formed*** *if* (1) *function definitions in* $\mathscr{L}$ *have pairwise different function names and all function names occurring in the function bodies and* L *are defined;* (2) *the arity of function invocations occurring anywhere in the program matches the arity of the corresponding function definition;* (3) *every function definition in* $\mathscr{L}$ *has shape* $\mathtt{f}(\widetilde{a}) = (\nu\widetilde{c})\mathtt{L}_\mathtt{f}$, *where* $\mathtt{L}_\mathtt{f}$ *does not contain any* ν*-binder and* $var(\mathtt{L}_\mathtt{f}) \subseteq \widetilde{a} \cup \widetilde{c}$.

Operational semantics. Let a *lam context*, noted $\mathfrak{L}[\,]$, be a term derived by the following syntax:

$$\mathfrak{L}[\,] \quad ::= \quad [\,] \quad | \quad \mathtt{L} \,\&\, \mathfrak{L}[\,] \quad | \quad \mathtt{L} + \mathfrak{L}[\,]$$

As usual $\mathfrak{L}[\mathtt{L}]$ is the lam where the hole of $\mathfrak{L}[\,]$ is replaced by L. According to the syntax, lam contexts have no ν-binder; that is, the hassle of name captures is avoided. The operational semantics of a program $(\mathscr{L}, \mathtt{L})$ is a transition system where *states* are lams, the *transition relation* is the least one satisfying the rule

$$\text{(RED)}\quad \frac{\mathtt{f}(\widetilde{a}) = (\nu\widetilde{c})\mathtt{L}_\mathtt{f} \in \mathscr{L} \qquad \widetilde{c'} \text{ are fresh} \qquad \mathtt{L}_\mathtt{f}[\widetilde{c'}/\widetilde{c}][\widetilde{a'}/\widetilde{a}] = \mathtt{L}'_\mathtt{f}}{\mathfrak{L}[\mathtt{f}(\widetilde{a'})] \longrightarrow \mathfrak{L}[\mathtt{L}'_\mathtt{f}]}$$

and the initial state is the lam $\mathtt{L}'$ such that $\mathtt{L} \equiv (\nu\widetilde{c})\mathtt{L}'$ and $\mathtt{L}'$ does not contain any ν-binder. We write $\longrightarrow^*$ for the reflexive and transitive closure of $\longrightarrow$.

By (RED), a lam L is evaluated by successively replacing function invocations with the corresponding lam instances. Name creation is handled by replacing bound names of function bodies with fresh names. For example, if $\mathtt{f}(a) = (\nu c)((a,c) \,\&\, \mathtt{f}(c))$ and $\mathtt{f}(a')$ occurs in the main lam, then $\mathtt{f}(a')$ is replaced by $(a',c') \,\&\, \mathtt{f}(c')$, where c' is a fresh name.

Let us discuss some examples.

1. $(\{\mathtt{f}(a,b,c) = (a,b) \,\&\, \mathtt{g}(b,c) + (b,c),\ \mathtt{g}(d,e) = (d,e) + (e,d)\},\ \mathtt{f}(a,b,c))$. Then

$$\begin{aligned}\mathtt{f}(a,b,c) &\longrightarrow (a,b) \,\&\, \mathtt{g}(b,c) + (b,c) \longrightarrow (a,b) \,\&\, ((b,c) + (c,b)) + (b,c)\\ &\longrightarrow (a,b) \,\&\, (b,c) + (a,b) \,\&\, (c,b) + (b,c)\end{aligned}$$

   The lam in the final state *does not contain function invocations.* This is because the above program is not recursive. Additionally, the evaluation of $\mathtt{f}(a,b,c)$ *has not created names.* This is because the bodies of f and g do not contain ν-binders.

2. $(\{\mathtt{f}'(a) = (\nu b)(a,b) \,\&\, \mathtt{f}'(b)\},\ \mathtt{f}'(a_0))$. Then

$$\begin{aligned}\mathtt{f}'(a_0) &\longrightarrow (a_0,a_1) \,\&\, \mathtt{f}'(a_1) \longrightarrow (a_0,a_1) \,\&\, (a_1,a_2) \,\&\, \mathtt{f}'(a_2)\\ &\longrightarrow^n (a_0,a_1) \,\&\, \cdots \,\&\, (a_{n+1},a_{n+2}) \,\&\, \mathtt{f}'(a_{n+2})\end{aligned}$$

   In this case, *because of the* (νb) *binder*, the lam grows in the number of dependencies as the evaluation progresses.

3. $(\{\mathtt{f}''(a) = (\nu b)(a,b) + (a,b) \,\&\, \mathtt{f}''(b)\},\ \mathtt{f}''(a_0))$. Then

$$\begin{aligned}\mathtt{f}''(a_0) &\longrightarrow (a_0,a_1) + (a_0,a_1) \,\&\, \mathtt{f}''(a_1)\\ &\longrightarrow (a_0,a_1) + (a_0,a_1) \,\&\, (a_1,a_2) + (a_0,a_1) \,\&\, (a_1,a_2) \,\&\, \mathtt{f}''(a_2)\\ &\longrightarrow^n (a_0,a_1) + \cdots + (a_0,a_1) \,\&\, \cdots \,\&\, (a_{n+1},a_{n+2}) \,\&\, \mathtt{f}''(a_{n+2})\end{aligned}$$

   In this case, the lam grows in the number of "+"-terms, which in turn become larger and larger as the evaluation progresses.

Flattening and circularities. Lams represent elements of the set $\mathscr{P}(\mathscr{P}(\mathscr{A} \times \mathscr{A}))$. This property is displayed by the following flattening function.

Let $\mathscr{L}$ be a set of function definitions and let $I(\cdot)$, called *flattening*, be a function on lams that (i) maps function name $\mathtt{f}$ defined in $\mathscr{L}$ to elements of $\mathscr{P}(\mathscr{P}(A \times A))$ and (ii) is defined on lams as follows

$$I(\mathbf{0}) = \{\varnothing\}, \qquad I((a,b)) = \{\{(a,b)\}\}, \qquad I(\mathtt{L} \,\&\, \mathtt{L}') = I(\mathtt{L}) \,\&\, I(\mathtt{L}'),$$
$$I(\mathtt{L} + \mathtt{L}') = I(\mathtt{L}) \cup I(\mathtt{L}'), \qquad I((\nu\, a)\mathtt{L}) = I(\mathtt{L})[^{a'}/_{a}] \text{ with } a' \text{ fresh},$$
$$I(\mathtt{f}(\widetilde{c})) = I(\mathtt{f})[^{\widetilde{c}}/_{\widetilde{a}}] \text{ (where } \widetilde{a} \text{ are the formal parameters of } \mathtt{f}).$$

Note that $I(\mathtt{L})$ is unique up to a renaming of names that do not occur free in $\mathtt{L}$. Let $I^{\perp}$ be the map such that, for every $\mathtt{f}$ defined in $\mathscr{L}$, $I^{\perp}(\mathtt{f}) = \{\varnothing\}$. For example, let $\mathscr{L}$ defines $\mathtt{f}$ and $\mathtt{g}$ and let

$$I(\mathtt{f}) = \{\{(a,b),(b,c)\}\} \qquad I(\mathtt{g}) = \{\{(b,a)\}\}$$
$$\mathtt{L}'' = \mathtt{f}(a,b,c) + (a,b) \,\&\, \mathtt{g}(b,c) \,\&\, \mathtt{f}(d,b,c) + \mathtt{g}(d,e) \,\&\, (d,c) + (e,d).$$

Then

$$I(\mathtt{L}'') = \{\{(a,b),(b,c)\},\ \{(a,b),(c,b),(d,b),(b,c)\},\ \{(e,d),(d,c)\},\ \{(e,d)\}\}$$
$$I^{\perp}(\mathtt{L}'') = \{\varnothing, \{(a,b)\}, \{(d,c)\}, \{(e,d)\}\}.$$

Definition 4. *A lam* $\mathtt{L}$ ***has a circularity*** *if* $I^{\perp}(\mathtt{L})$ *has a circularity. A lam program* $(\mathscr{L}, \mathtt{L})$ ***has a circularity*** *if there is* $\mathtt{L} \longrightarrow^{*} \mathtt{L}'$ *and* $\mathtt{L}'$ *has a circularity.*

The property of "having a circularity" is preserved by $\equiv$ while the "absence of circularities" of a composite lam can be carried to its components.

Proposition 2. *1. if* $\mathtt{L} \equiv \mathtt{L}'$ *then* $\mathtt{L}$ *has a circularity if and only if* $\mathtt{L}'$ *has a circularity;*
2. $\mathtt{L} \,\&\, \mathtt{L}'$ *has no circularity implies both* $\mathtt{L}$ *and* $\mathtt{L}'$ *have no circularity (similarly for* $\mathtt{L} + \mathtt{L}'$ *and for* $(\nu\, a)\mathtt{L}$*).*

4 The Decision Algorithm for Detecting Circularities

In this section we assume a lam program $(\mathscr{L}, \mathtt{L})$ such that pairwise different function definitions in $\mathscr{L}$ have disjoint formal parameters. Without loss of generality, we assume that $\mathtt{L}$ does not contain any ν-binder.

Fixpoint definition of the interpretation function. The basic item of our algorithm is the computation of lam functions' interpretation. This computation is performed by means of a standard fixpoint technique that is detailed below.

Let A be the set of formal parameters of definitions in $\mathscr{L}$ and let $\varkappa$ be a special name that does not occur in $(\mathscr{L}, \mathtt{L})$. We use the domain $\left(\mathscr{P}(\mathscr{P}(A \cup \{\varkappa\} \times A \cup \{\varkappa\})),\ \subseteq\right)$ which is a *finite* lattice [3].

Definition 5. *Let* $\mathtt{f}_i(\widetilde{a_i}) = (\nu\, \widetilde{c_i})\mathtt{L}_i$*, with* $i \in 1..n$*, be the function definitions in* $\mathscr{L}$*. The family of flattening functions* $I^{(k)}_{\mathscr{L}} : \{\mathtt{f}_1, \cdots, \mathtt{f}_n\} \to \mathscr{P}(\mathscr{P}(A \cup \{\varkappa\} \times A \cup \{\varkappa\}))$ *is defined as follows*

$$I^{(0)}_{\mathscr{L}}(\mathtt{f}_i) = \{\varnothing\} \qquad I^{(k+1)}_{\mathscr{L}}(\mathtt{f}_i) = \{\mathtt{proj}_{\widetilde{a_i}}(R^{+}) \mid R \in I^{(k)}_{\mathscr{L}}(\mathtt{L}_i)\}$$

where $\mathtt{proj}_{\widetilde{a}}(R) \stackrel{def}{=} \{(a,b) \mid (a,b) \in R \text{ and } a, b \in \widetilde{a}\} \cup \{(\varkappa,\varkappa) \mid (c,c) \in R \text{ and } c \notin \widetilde{a}\}$.

We notice that $I^{(0)}_{\mathscr{L}}$ is the function $I^{\perp}$ of the previous section.

Proposition 3. *Let* $\mathtt{f}(\widetilde{a}) = (\nu \widetilde{c})\mathtt{L}_{\mathtt{f}} \in \mathscr{L}$. *(i) For every* k, $I^{(k)}_{\mathscr{L}}(\mathtt{f}) \in \mathscr{P}(\mathscr{P}((\widetilde{a} \cup \{\varkappa\}) \times (\widetilde{a} \cup \{\varkappa\})))$. *(ii) For every* k, $I^{(k)}_{\mathscr{L}}(\mathtt{f}) \Subset I^{(k+1)}_{\mathscr{L}}(\mathtt{f})$.

Since, for every k, $I^{(k)}_{\mathscr{L}}(\mathtt{f}_i)$ ranges over a finite lattice, by the fixpoint theory [3], there exists m such that $I^{(m)}_{\mathscr{L}}$ is a fixpoint, namely $I^{(m)}_{\mathscr{L}} \approx I^{(m+1)}_{\mathscr{L}}$ where $\approx$ is the equivalence relation induced by $\Subset$. In the following, we let $I_{\mathscr{L}}$, called the *interpretation function* (of a lam), be the least fixpoint $I^{(m)}_{\mathscr{L}}$.

Example 1. For example, let $\mathscr{L}$ be the factorial function in Section 1. Then

$$I^{(0)}_{\mathscr{L}}(\mathtt{fact}) = \{\varnothing\} \qquad I^{(1)}_{\mathscr{L}}(\mathtt{fact}) = \{\{(a_2, a_3)\}, \varnothing\} \qquad I^{(2)}_{\mathscr{L}}(\mathtt{fact}) = \{\{(a_2, a_3)\}, \varnothing\}$$

That is, in this case, $I_{\mathscr{L}} = I^{(1)}_{\mathscr{L}}$. □

Theorem 1. *A lam program* $(\mathscr{L}, \mathtt{L})$ *has a circularity if and only if* $I_{\mathscr{L}}(\mathtt{L})$ *has a circularity.*

For example, let $\mathscr{L}$ be the factorial function in Section 1 and let $\mathtt{L} = (a_3, a_2) \,\&\, \mathtt{fact}(a_1, a_2, a_3.a_4)$. From Example 1, we have $I_{\mathscr{L}}(\mathtt{fact}) = \{\{(a_2, a_3)\}, \varnothing\}$. Since $I_{\mathscr{L}}(\mathtt{L})$ has a circularity, by Theorem 1, there is $\mathtt{L} \longrightarrow^{*} \mathtt{L}'$ such that $I^{\perp}(\mathtt{L}')$ has a circularity. In fact it displays a circularity after the first transition:

$$\mathtt{L} \longrightarrow (a_3, a_2) \,\&\, ((a_2, a_3) + ((a_2, a_5) \,\&\, \mathtt{fact}(a_5, a_6, a_3, a_4))) \,.$$

5 Value-Passing CCS

In the present and next sections, we apply the foregoing theory of lams to refine Kobayashi's type system for deadlock-freedom of concurrent programs [11]. In his type system, the deadlock-freedom is guaranteed by a combination of *usages*, which are a kind of behavioral types capturing channel-wise communication behaviors, and *capability/obligation levels*, which are natural numbers capturing inter-channel dependencies (like "a message is output on x only if a message is received along y"). By replacing numbers with (lam) level names, we can achieve a more precise analysis of deadlock-freedom because of the algorithm in Section 4. The original type system in [11] is for the pi-calculus [16], but for the sake of simplicity, we consider a variant of the value-passing CCS [15], which is sufficient for demonstrating the power of our lam-based approach.

Our *value-passing CCS* uses several disjoint countable sets of names: in addition to level names, there are *integer and channel names*, ranged over by $x, y, z, \cdots$, *process names*, ranged over by $A, B, \cdots$, and *usage names*, ranged over by $\alpha, \beta, \cdots$. A *value-passing CCS program* is a pair $(\mathscr{D}, P)$, where $\mathscr{D}$ is a *finite set* of *process name definitions* $A(\widetilde{a}; \widetilde{x}) = P_A$, with $\widetilde{a}; \widetilde{x}$ and P_A respectively being the *formal parameters* and the *body* of A, and P is the *main process*.

The syntax of processes P_A and P is shown in Figure 1. A process can be the inert process $\mathbb{0}$, a message $x!e$ sent on a name x that carries (the value of) an expression e, an input $x?y.P$ that consumes a message $x!v$ and behaves like $P[^v/_y]$, a parallel composition of processes $P \mid Q$, a conditional **if** e **then** P **else** Q that evaluates e and behaves either

$$
\begin{array}{l}
P \text{ (processes) } ::= \mathbb{0} \mid x!e \mid x?y.P \mid (P \mid Q) \mid \textbf{if } e \textbf{ then } P \textbf{ else } Q \mid (\nu\widetilde{a}; x : \mathrm{T})P \mid A(\widetilde{a}; \widetilde{e}) \\
e \text{ (expressions) } ::= x \mid v \mid e_1 \textbf{ op } e_2 \\
\mathrm{T} \text{ (types) } ::= \textbf{int} \mid U \\
U \text{ (usages) } ::= \mathbf{0} \mid !^{a_1}_{a_2} \mid ?^{a_1}_{a_2}.U \mid (U_1|U_2) \mid \alpha \mid \mu\alpha.U
\end{array}
$$

Fig. 1. The Syntax of value-passing CCS

like P or like Q depending on whether the value is $\neq 0$ (*true*) or $= 0$ (*false*), a restriction $(\nu\widetilde{a}; x : \mathrm{T})P$ that behaves like P except that communications on x with the external environment are prohibited, an invocation $A(\widetilde{a}; \widetilde{e})$ of the process corresponding to A.

An expression e can be a name x, an integer value v, or a generic binary operation on integers v **op** v', where **op** ranges over a set including the usual operators like $+$, $\leq$, etc. Integer expressions without names (*constant expressions*) may be evaluated to an integer value (the definition of the evaluation of constant expressions is omitted). Let $[\![e]\!]$ be the evaluation of a constant expression e ($[\![e]\!]$ is undefined when the integer expression e contains integer names). Let also $[\![x]\!] = x$ when x is a non-integer name.

We defer the explanation of the meaning of types T (and usages U) until Section 6. It is just for the sake of simplicity that processes are annotated with types and level names. They do not affect the operational semantics of processes, and can be automatically inferred by using an inference algorithm similar to those in [11,10].

Similarly to lams, $A(\widetilde{a}; \widetilde{x}) = P_A$ and $(\nu\widetilde{a}; x : \mathrm{T})P$ are *binders* of $\widetilde{a}; \widetilde{x}$ in P_A and of $\widetilde{a}, x$ in P, respectively. We use the standard notions of alpha-equivalence, free and bound names of processes and, with an abuse of notation, we let $var(P)$ be the free names in P. In process name definitions $A(\widetilde{a}; \widetilde{x}) = P_A$, we always assume that $var(P_A) \subseteq \widetilde{a}, \widetilde{x}$.

Definition 6. *The **structural equivalence** $\equiv$ on processes is the least congruence containing alpha-conversion of bound names, commutativity and associativity of $\mid$ with identity $\mathbb{0}$, and closed under the following rule:*

$$((\nu\widetilde{a}; x : T)P) \mid Q \equiv (\nu\widetilde{a}; x : T)(P \mid Q) \qquad \widetilde{a}, x \notin var(Q)\,.$$

*The **operational semantics** of a program $(\mathscr{D}, P)$ is a transition system where the states are processes, the initial state is P, and the **transition relation** $\rightarrow_{\mathscr{D}}$ is the least one closed under the following rules:*

$$
\text{(R-COM)}\ \frac{[\![e]\!] = v}{x!e \mid x?y.P \rightarrow_{\mathscr{D}} P[^{v}/_{y}]}
\qquad
\text{(R-PAR)}\ \frac{P \rightarrow_{\mathscr{D}} P'}{P \mid Q \rightarrow_{\mathscr{D}} P' \mid Q}
\qquad
\text{(R-NEW)}\ \frac{P \rightarrow_{\mathscr{D}} Q}{(\nu\widetilde{a}; x : T)P \rightarrow_{\mathscr{D}} (\nu\widetilde{a}; x : T)Q}
$$

$$
\text{(R-IFT)}\ \frac{[\![e]\!] \neq 0}{\textbf{if } e \textbf{ then } P \textbf{ else } Q \rightarrow_{\mathscr{D}} P}
\qquad
\text{(R-IFF)}\ \frac{[\![e]\!] = 0}{\textbf{if } e \textbf{ then } P \textbf{ else } Q \rightarrow_{\mathscr{D}} Q}
\qquad
\text{(R-CALL)}\ \frac{[\![\widetilde{e}]\!] = \widetilde{v} \quad A(\widetilde{a}; \widetilde{x}) = P \in \mathscr{D}}{A(\widetilde{a'}; \widetilde{e}) \rightarrow_{\mathscr{D}} P[^{\widetilde{a'}}/_{\widetilde{a}}][^{\widetilde{v}}/_{\widetilde{x}}]}
$$

$$
\text{(R-CONG)}\ \frac{P \equiv P' \quad P' \rightarrow_{\mathscr{D}} Q' \quad Q' \equiv Q}{P \rightarrow_{\mathscr{D}} Q}
$$

We often omit the subscript of $\rightarrow_{\mathscr{D}}$ when it is clear from the context. We write $\rightarrow^*$ for the reflexive and transitive closure of $\rightarrow$.

The deadlock-freedom of a process P, which is the basic property that we will verify, means that P does not get stuck into a state where there is a message or an input. The formal definition is below.

Definition 7 (deadlock-freedom). *A program* $(\mathscr{D}, P)$ *is* ***deadlock-free*** *if the following condition holds: whenever* $P \rightarrow^* P'$ *and either (i)* $P' \equiv (\nu\widetilde{a}_1; x_1 : T_1)\cdots(\nu\widetilde{a}_k; x_k : T_k)(x!v \mid Q)$, *or (ii)* $P' \equiv (\nu\widetilde{a}_1; x_1 : T_1)\cdots(\nu\widetilde{a}_k; x_k : T_k)(x?y.Q_1 \mid Q_2)$, *then there exists* R *such that* $P' \rightarrow R$.

Example 2 (The dining philosophers). Consider the program consisting of the process definitions

$$
\begin{array}{l}
\mathit{Phils}(a_1, a_2, a_3, a_4; n : \mathbf{int}, \mathit{fork}_1 : U_1, \mathit{fork}_2 : U_2) = \\
\quad \mathbf{if}\ n = 1\ \mathbf{then}\ \mathit{Phil}(a_1, a_2, a_3, a_4; \mathit{fork}_1, \mathit{fork}_2)\ \mathbf{else} \\
\qquad (\nu a_5, a_6; \mathit{fork}_3 : U_3 \mid U_3 \mid !^{a_5}_{a_6})(\ \ \mathit{Phils}(a_1, a_2, a_5, a_6; n-1, \mathit{fork}_1, \mathit{fork}_3) \\
\qquad\qquad \mid \mathit{Phil}(a_5, a_6, a_3, a_4; \mathit{fork}_3, \mathit{fork}_2) \quad \mid \quad \mathit{fork}_3!1\ )
\end{array}
$$

$$
\begin{array}{l}
\mathit{Phil}(a_1, a_2, a_3, a_4; \mathit{fork}_1 : U_1, \mathit{fork}_2 : U_2) = \\
\quad \mathit{fork}_1?x_1.\mathit{fork}_2?x_2.(\mathit{fork}_1!x_1 \mid \mathit{fork}_2!x_2 \mid \mathit{Phil}(a_1, a_2, a_3, a_4; \mathit{fork}_1, \mathit{fork}_2)\ )
\end{array}
$$

and of the main process P:

$$
\begin{array}{l}
(\nu a_1, a_2; \mathit{fork}_1 : U_1 \mid U_1 \mid !^{a_1}_{a_2})(\nu a_3, a_4; \mathit{fork}_2 : U_2 \mid U_2 \mid !^{a_1}_{a_2}) \\
(\mathit{Phils}(a_1, a_2, a_3, a_4; m, \mathit{fork}_1, \mathit{fork}_2) \mid \mathit{Phil}(a_1, a_2, a_3, a_4; \mathit{fork}_1, \mathit{fork}_2) \mid \mathit{fork}_1!1 \mid \mathit{fork}_2!1\ )
\end{array}
$$

Here, $U_1 = \mu\alpha.?^{a_2}_{a_1}.(!^{a_1}_{a_2} \mid \alpha)$, $U_2 = \mu\alpha.?^{a_4}_{a_3}.(!^{a_3}_{a_4} \mid \alpha)$, and $U_3 = \mu\alpha.?^{a_6}_{a_5}.(!^{a_5}_{a_6} \mid \alpha)$, but please ignore the types for the moment. Every philosopher $\mathit{Phil}(a_1, a_2, a_3, a_4; \mathit{fork}_1, \mathit{fork}_2)$ grabs the two forks fork_1 and fork_2 in this order, releases the forks, and repeats the same behavior. The main process creates a ring consisting of $m + 1$ philosophers, where only one of the philosophers grabs the forks in the opposite order to avoid deadlock. This program is indeed deadlock-free in our definition. On the other hand, if we replace $\mathit{Phil}(a_1, a_2, a_3, a_4; \mathit{fork}_1, \mathit{fork}_2)$ with $\mathit{Phil}(a_1, a_2, a_3, a_4; \mathit{fork}_2, \mathit{fork}_1)$ in the main process, then the resulting process is not deadlock-free. □

The dining philosophers example is a paradigmatic case of the power of the analysis described in the next section. This example cannot be type-checked in Kobayashi's previous type system [11]: see Remark 1 in Section 6.

6 The Deadlock Freedom Analysis of Value-Passing CCS

We now explain the syntax of types in Figure 1. A type is either **int** or a usage. The former is used to type *integer* names; the latter is used to type *channel* names [11,9]. A usage describes how a channel can be used for input and output. The usage **0** describes a channel that cannot be used, $!^{a_1}_{a_2}$ describes a channel that is used for output, $?^{a_1}_{a_2}.U$ describes a channel that is first used for input and then used according to U, and $U \mid U'$ describes a channel that is used according to U and U', possibly in parallel. For example, in process $x!2 \mid x?z.y!z$, y has the usage $!^{a_1}_{a_2}$ (please, ignore the subscript and superscript for the moment), and x has the usage $!^{a_3}_{a_4} \mid ?^{a_5}_{a_6}.\mathbf{0}$. The usage $\mu\alpha.U$ describes a channel that is used recursively according to $U[^{\mu\alpha.U}/_{\alpha}]$. The operation $\mu\alpha.-$ is a binder and we use the standard notions of alpha-equivalence, free and bound usage names. For example, $\mu\alpha.!^{a_1}_{a_2}.\alpha$ describes a channel that can be sequentially used for output an arbitrary number of times; $\mu\alpha.?^{a_1}_{a_2}.!^{a_3}_{a_4}.\alpha$ describes a channel that should be used for input and output alternately. We often omit a trailing $\mathbf{0}$ and just write $?^{a_1}_{a_1}$ for $?^{a_1}_{a_1}.\mathbf{0}$.

The superscripts and subscripts of ? and ! are level names of lams (recall Section 3), and are used to control the causal dependencies between communications [11]. The superscript, called an *obligation level*, describes the degree of the obligation to use the channel for the specified operation. The subscript, called a *capability level*, describes the degree of the capability to use the channel for the specified operation (and successfully find a partner of the communication).

In order to detect deadlocks we consider the following two conditions:

1. If a process has an obligation of level a, then it can exercise only capabilities of level a' *less than* a before fulfilling the obligation. This corresponds to a dependency (a', a). For example, if x has type $?^{a_1}_{a_2}$ and y has type $!^{a_3}_{a_4}$, then the process $x?u.y!u$ has lam (a_2, a_3).

2. The whole usage of each channel must be consistent, in the sense that if there is a capability of level a to perform an input (respectively, a message), there must be a corresponding obligation of level a to perform a corresponding message (respectively, input). For example, the usage $!^{a_1}_{a_2} \mid ?^{a_2}_{a_1}$ is consistent, but neither $!^{a_1}_{a_2} \mid ?^{a_1}_{a_2}$ nor $!^{a_1}_{a_2}$ is.

To see how the constraints above guide our deadlock analysis, consider the (deadlocked) process: $x?u.y!u \mid y?u.x!u$. Because of condition 2 above, the usage of x and y must be of the form $?^{a_1}_{a_2} \mid !^{a_2}_{a_1}$ and $?^{a_3}_{a_4} \mid !^{a_4}_{a_3}$ respectively. Due to 1, we derive (a_2, a_4) for $x?u.y!u$, and (a_4, a_2) for $y?u.x!u$. Hence the process is deadlocked because the lam $(a_2, a_4)\&(a_4, a_2)$ has a circularity. On the other hand, for the process $x?u.y!u \mid y?u.\mathbf{0} \mid x!u$, we derive the lam (a_2, a_4), which has no circularity. Indeed, this last process is not deadlocked. While we use lams to detect deadlocks, Kobayashi [11] used natural numbers for obligation/capability levels.

As explained above, usages describe the channel-wise behavior of a process, and they form a tiny process calculus. The usage reduction relation $U \rightsquigarrow U'$ defined below means that if a channel of usage U is used for a communication, the channel may be used according to U' afterwards.

Definition 8. *Let $=$ be the least congruence on usages containing alpha-conversion of bound names, commutativity and associativity of $\mid$ with identity $\mathbf{0}$, and closed under the following rule:*

$$\text{(UC-MU)} \qquad \mu\alpha.U \;=\; U[^{\mu\alpha.U}/_{\alpha}]$$

The reduction relation $U \rightsquigarrow U'$ is the least relation closed under the rules:

$$\text{(UR-COM)}\ \ !^{a_1}_{a_2} \mid ?^{a_3}_{a_4}.U \rightsquigarrow U \qquad \text{(UR-PAR)}\ \ \frac{U_1 \rightsquigarrow U_1'}{U_1 \mid U_2 \rightsquigarrow U_1' \mid U_2} \qquad \text{(UR-CONG)}\ \ \frac{U_1 = U_1' \quad U_1' \rightsquigarrow U_2' \quad U_2' = U_2}{U_1 \rightsquigarrow U_2}$$

As usual, we let $\rightsquigarrow^*$ be the reflexive and transitive closure of $\rightsquigarrow$.

The following relation $rel(U)$ guarantees the condition 2 on capabilities and obligations above, that each capability must be accompanied by a corresponding obligation. This must hold during the whole computation, hence the definition below. The predicate $rel(U)$ is computable because it may be reduced to Petri Nets reachability (see [10] for the details about the encoding).

Definition 9. *U is **reliable**, written $rel(U)$, when the following conditions hold:*

1. *whenever $U \leadsto^* U'$ and $U' = !^{a_1}_{a_2} \mid U_1$, there are U_2 and U_3 such that $U_1 = ?^{a_2}_{a_3}.U_2 \mid U_3$ for some a_3; and*
2. *whenever $U \leadsto^* U'$ and $U' = ?^{a_1}_{a_2}.U_1 \mid U_2$, there is U_3 such that $U_2 = !^{a_2}_{a_3} \mid U_3$ for some a_3.*

The following type system uses *type environments*, ranged over $\Gamma, \Gamma', \cdots$, that map integer and channel names to types and process names to sequences $[\widetilde{a};\widetilde{T}]$. When $x \notin dom(\Gamma)$, we write $\Gamma, x{:}T$ for the environment such that $(\Gamma, x{:}T)(x) = T$ and $(\Gamma, x{:}T)(y) = \Gamma(y)$, otherwise. The operation $\Gamma_1 \mid \Gamma_2$ is defined by:

$$(\Gamma_1 \mid \Gamma_2)(x) = \begin{cases} \Gamma_1(x) & \text{if } x \in dom(\Gamma_1) \text{ and } x \notin dom(\Gamma_2) \\ \Gamma_2(x) & \text{if } x \in dom(\Gamma_2) \text{ and } x \notin dom(\Gamma_1) \\ [\widetilde{a};\widetilde{T}] & \text{if } \Gamma_1(x) = \Gamma_2(x) = [\widetilde{a};\widetilde{T}] \\ \mathbf{int} & \text{if } \Gamma_1(x) = \Gamma_2(x) = \mathbf{int} \\ U_1 \mid U_2 & \text{if } \Gamma_1(x) = U_1 \text{ and } \Gamma_2(x) = U_2 \end{cases}$$

The map $\Gamma_1 \mid \Gamma_2$ is undefined if, for some x, $(\Gamma_1 \mid \Gamma_2)(x)$ does not match any of the cases. Let $var(\Gamma) = \{a \mid \text{there is } x : \Gamma(x) = U \text{ and } a \in var(U)\}$.

There are three kinds of type judgments:

$\Gamma \vdash e : T$ – the expression e has type T in Γ;
$\Gamma \vdash P : \mathtt{L}$ – the process P has lam $\mathtt{L}$ in Γ;
$\Gamma \vdash (\mathscr{D}, P) : (\mathscr{L}, \mathtt{L})$ – the program $(\mathscr{D}, P)$ has lam program $(\mathscr{L}, \mathtt{L})$ in Γ.

As usual, $\Gamma \vdash e : T$ means that e evaluates to a value of type T under an environment that respects the type environment Γ. The judgment $\Gamma \vdash P : \mathtt{L}$ means that P uses each channel x according to $\Gamma(x)$, with the causal dependency as described by $\mathtt{L}$. For example, $x{:}?^{a_1}_{a_2}, y{:}!^{a_3}_{a_4} \vdash x?u.y!u : (a_2, a_3)$ should hold.

The typing rules of value-passing CCS are defined in Figure 2, where we use the predicate $noact(\Gamma)$ and the function $ob(U)$ defined as follows:

$noact(\Gamma)$ = *true* if and only if, for every channel name $x \in dom(\Gamma)$, $\Gamma(x) = \mathbf{0}$;
$ob(\Gamma) = \bigcup_{x \in dom(\Gamma), \Gamma(x)=U} ob(U)$ where

$$ob(\mathbf{0}) = \emptyset \qquad ob(!^{a_1}_{a_2}) = \{a_1\} \qquad ob(?^{a_1}_{a_2}.U) = \{a_1\}$$
$$ob(U \mid U') = ob(U) \cup ob(U') \qquad ob(\mu\alpha.U) = ob(U[^{\mathbf{0}}/_{\alpha}])$$

The predicate $noact(\Gamma)$ is used for controlling weakening (as in linear type systems). For example, if we did not require $noact(\Gamma)$ in rule T-ZERO, then we would obtain $x{:}?^{a_1}_{a_2}.\mathbf{0} \vdash \mathbf{0} : \mathbb{0}$. Then, by using T-IN and T-OUT, we would obtain: $x{:}?^{a_1}_{a_2}.\mathbf{0} \mid !^{a_2}_{a_1} \vdash \mathbf{0} \mid x!1 : \mathbb{0}$, and wrongly conclude that the output on x does not get stuck. It is worth to notice that, in the typing rules, we identify usages up to $=$.

Processes:

$$\text{(T-Zero)}\ \frac{noact(\Gamma)}{\Gamma \vdash \mathbf{0} : \mathbf{0}} \qquad \text{(T-Out)}\ \frac{\Gamma \vdash e : \mathbf{int}}{\Gamma, x{:}!^{a_1}_{a_2} \vdash x!e : \mathbf{0}} \qquad \text{(T-In)}\ \frac{\Gamma, x : U, y : \mathbf{int} \vdash P : \mathtt{L}}{\Gamma, x{:}?^{a_1}_{a_2}.U \vdash x?y.P : \mathtt{L}\ \&\ (\&_{a \in ob(\Gamma)}(a_2, a))}$$

$$\text{(T-Par)}\ \frac{\Gamma \vdash P : \mathtt{L} \qquad \Gamma' \vdash P' : \mathtt{L}'}{\Gamma \mid \Gamma' \vdash P \mid P' : \mathtt{L}\ \&\ \mathtt{L}'} \qquad \text{(T-New)}\ \frac{\Gamma, x : U \vdash P : \mathtt{L} \qquad rel(U) \qquad \widetilde{a} \cap var(\Gamma) = \emptyset}{\Gamma \vdash (\nu \widetilde{a}; x : U)P : (\nu \widetilde{a})\mathtt{L}}$$

$$\text{(T-If)}\ \frac{\Gamma \vdash e : \mathbf{int} \qquad \Gamma' \vdash P : \mathtt{L} \qquad \Gamma' \vdash P' : \mathtt{L}'}{\Gamma \mid \Gamma' \vdash \mathbf{if}\ e\ \mathbf{then}\ P\ \mathbf{else}\ P' : \mathtt{L} + \mathtt{L}'} \qquad \text{(T-Call)}\ \frac{\Gamma(A) = [\widetilde{a}; \widetilde{\mathtt{T}}] \qquad |\widetilde{a}| = |\widetilde{a}'| \qquad \Gamma \vdash \widetilde{e} : \widetilde{\mathtt{T}}}{\Gamma \vdash A(\widetilde{a}'; \widetilde{e}) : \mathtt{f}_A(\widetilde{a}')}$$

Expressions:

$$\text{(T-Int)}\ \frac{noact(\Gamma)}{\Gamma \vdash n : \mathbf{int}} \qquad \text{(T-Var)}\ \frac{noact(\Gamma)}{\Gamma, x : \mathtt{T} \vdash x : \mathtt{T}} \qquad \text{(T-Op)}\ \frac{\Gamma \vdash e : \mathbf{int}\ \Gamma \vdash e' : \mathbf{int}}{\Gamma \vdash e\ \mathbf{op}\ e' : \mathbf{int}} \qquad \text{(T-Seq)}\ \frac{(\Gamma_i \vdash e_i : T_i)^{i \in 1..n}}{\Gamma_1 \mid \cdots \mid \Gamma_n \vdash e_1, \ldots, e_n : T_1, \ldots, T_n}$$

Programs:

$$\text{(T-Prog)}\ \frac{\begin{array}{c}\mathscr{D} = \bigcup_{i \in 1..n}\{A_i(\widetilde{a}_i; \widetilde{x}_i : \widetilde{\mathtt{T}}_i) = P_i\} \qquad \Gamma = (A_i : [\widetilde{a}_i; \widetilde{\mathtt{T}}_i])^{i \in 1..n} \\ (\Gamma, \widetilde{x}_i : \widetilde{\mathtt{T}}_i \vdash P_i : \mathtt{L}_i)^{i \in 1..n} \qquad \Gamma' \vdash P : \mathtt{L} \qquad \mathscr{L} = \bigcup_{i \in 1..n}\{\mathtt{f}_{A_i}(\widetilde{a}_i) = \mathtt{L}_i\}\end{array}}{\Gamma \mid \Gamma' \vdash (\mathscr{D}, P) : (\mathscr{L}, \mathtt{L})}$$

Fig. 2. The type system of value-passing CCS (we assume a function name $\mathtt{f}_A$ for every process name A)

A few key rules are discussed. Rule (T-In) is the unique one that introduces dependency pairs. In particular, the process $x?u.P$ will be typed with a lam that contains pairs (a_2, a), where a_2 is the capability of x and a is the obligation of every channel in P (because they are all causally dependent from x). Rule (T-Out) just records in the type environment that x is used for output. Rule (T-Par) types a parallel composition of processes by collecting the environments – operation "$\mid$" – (like in other linear type systems [13,9]) and the lams of the components. Rule (T-Call) types a process name invocation in terms of a (lam) function invocation and constrains the sequences of level names in the two invocations to have equal lengths ($|\widetilde{a}| = |\widetilde{a}'|$) and the types of expressions to match with the types in the process declaration.

Example 3. We illustrate the type system in Figure 2 by typing two simple processes:

$$P = (\nu a_1, a_2; x{:}?^{a_1}_{a_2} \mid !^{a_2}_{a_1})(\nu a_3, a_4; y{:}?^{a_3}_{a_4} \mid !^{a_4}_{a_3})(x?z.y!z \mid y?z.x!z)$$
$$Q = (\nu a_1, a_2; x{:}?^{a_1}_{a_2} \mid !^{a_2}_{a_1})(\nu a_3, a_4; y{:}?^{a_3}_{a_4} \mid !^{a_4}_{a_3})(x?z.y!z \mid y?z.\mathbf{0} \mid x!1)$$

The proof tree of P is

$$\frac{\dfrac{\dfrac{y{:}!^{a_4}_{a_3},\ z : \mathbf{int} \vdash y!z : \mathbf{0}}{x{:}?^{a_1}_{a_2},\ y{:}!^{a_4}_{a_3} \vdash x?z.y!z : (a_2, a_4)} \qquad \dfrac{x{:}!^{a_2}_{a_1},\ z : \mathbf{int} \vdash x!z : \mathbf{0}}{x{:}!^{a_2}_{a_1},\ y{:}?^{a_3}_{a_4} \vdash y?z.x!z : (a_4, a_2)}}{x{:}?^{a_1}_{a_2} \mid !^{a_2}_{a_1},\ y{:}?^{a_3}_{a_4} \mid !^{a_4}_{a_3} \vdash x?z.y!z \mid y?z.x!z : (a_2, a_4)\ \&\ (a_4, a_2)}}{\emptyset \vdash P : (\nu a_1, a_2)(\nu a_3, a_4)((a_2, a_4)\ \&\ (a_4, a_2))}$$

and we notice that the lam in the conclusion has a circularity (in fact, P is deadlocked). The typing of Q is

$$\dfrac{\dfrac{\dfrac{z:\mathbf{int}\vdash z:\mathbf{int}}{y{:}!^{a_4}_{a_3},\ z:\mathbf{int}\vdash y!z:\mathbb{0}}}{x{:}?^{a_1}_{a_2},\ y{:}!^{a_4}_{a_3}\vdash x?z.y!z:(a_2,a_4)} \quad \dfrac{y:\mathbb{0},\ z:\mathbf{int}\vdash\mathbb{0}:\mathbb{0}}{y{:}?^{a_3}_{a_4}\vdash y?z.\mathbb{0}:\mathbb{0}} \quad \dfrac{\varnothing\vdash 1:\mathbf{int}}{x{:}!^{a_2}_{a_1}\vdash x!1:\mathbb{0}}}{\dfrac{x{:}?^{a_1}_{a_2}\mid !^{a_2}_{a_1},\ y{:}?^{a_3}_{a_4}\mid !^{a_4}_{a_3}\vdash x?z.y!z\mid y?z.\mathbb{0}\mid x!1:(a_2,a_4)}{\varnothing\vdash Q:(\nu\, a_1,a_2)(\nu\, a_3,a_4)(a_2,a_4)}}$$

The lam in the conclusion has no circularity. In fact, Q is not deadlocked. □

Example 3 also spots one difference between the type system in [11] and the one in Figure 2. Here the inter-channel dependencies check is performed *ex-post* by resorting to the lam algorithm in Section 4; in [11] this check is done *during* the type checking(/inference) and, for this reason, the process P is not typable in previous Kobayashi's type systems. In this case, the two analysers both recognize that P is deadlocked; Example 4 below discusses a case where the precision is different.

The following theorem states the soundness of our type system.

Theorem 2. *Let $\Gamma\vdash(\mathscr{D},P):(\mathscr{L},\mathtt{L})$ such that noact(Γ). If $(\mathscr{L},\mathtt{L})$ has no circularity then $(\mathscr{D},P)$ is deadlock-free.*

The following examples highlight the difference of the expressive power of the system in Figure 2 and the type system in [11].

Example 4. Let $(\mathscr{D},P)$ be the dining philosopher program in Example 2 and U_1 and U_2 be the usages defined therein. We have $\Gamma\vdash(\mathscr{D},P):(\mathscr{L},\mathtt{L})$ where

$$\begin{array}{l}
\Gamma = \mathit{Phils}:[a_1,a_2,a_3,a_4;\mathbf{int},U_1,U_2],\mathit{Phil}:[a_1,a_2,a_3,a_4;U_1,U_2]\\
\mathscr{L}=\{\ \mathtt{f}_{\mathit{Phils}}(a_1,a_2,a_3,a_4)=\mathtt{f}_{\mathit{Phil}}(a_1,a_2,a_3,a_4)\\
\qquad\qquad\qquad +(\nu a_5,a_6)(\mathtt{f}_{\mathit{Phils}}(a_1,a_2,a_5,a_6)\ \&\ \mathtt{f}_{\mathit{Phil}}(a_5,a_6,a_3,a_4)),\\
\qquad \mathtt{f}_{\mathit{Phil}}(a_1,a_2,a_3,a_4)=(a_1,a_4)\ \&\ (a_3,a_1)\ \&\ (a_3,a_2)\ \&\ \mathtt{f}_{\mathit{Phil}}(a_1,a_2,a_3,a_4)\ \}\\
\mathtt{L}=(\nu a_1,a_2,a_3,a_4)(\mathtt{f}_{\mathit{Phils}}(a_1,a_2,a_3,a_4)\ \&\ \mathtt{f}_{\mathit{Phil}}(a_1,a_2,a_3,a_4))
\end{array}$$

For example, let

$$\begin{array}{l}
P_1 = \mathit{fork}_1?x_1.\mathit{fork}_2?x_2.(\,\mathit{fork}_1!x_1\mid \mathit{fork}_2!x_2\mid \mathit{Phil}(a_1,a_2,a_3,a_4;\mathit{fork}_1,\mathit{fork}_2)\,)\\
P_2 = \mathit{fork}_2?x_2.(\,\mathit{fork}_1!x_1\mid \mathit{fork}_2!x_2\mid \mathit{Phil}(a_1,a_2,a_3,a_4;\mathit{fork}_1,\mathit{fork}_2)\,)\\
P_3 = \mathit{fork}_1!x_1\mid \mathit{fork}_2!x_2\mid \mathit{Phil}(a_1,a_2,a_3,a_4;\mathit{fork}_1,\mathit{fork}_2)
\end{array}$$

Then the body P_1 of Phil is typed as follows:

$$\dfrac{\dfrac{\dfrac{\Gamma_2,\mathit{fork}_1{:}!^{a_1}_{a_2}\vdash \mathit{fork}_1!x_1:\mathbb{0}\qquad \Gamma_2,\mathit{fork}_2{:}!^{a_3}_{a_4}\vdash \mathit{fork}_2!x_2:\mathbb{0}\qquad \Gamma_2,\mathit{fork}_1:U_1,\mathit{fork}_2:U_2\vdash \mathit{Phil}(a_1,a_2,a_3,a_4;\mathit{fork}_1,\mathit{fork}_2):\mathtt{f}_{\mathit{Phil}}(a_1,a_2,a_3,a_4)}{\Gamma_2,\mathit{fork}_1{:}!^{a_1}_{a_2}\mid U_1,\mathit{fork}_2{:}!^{a_3}_{a_4}\mid U_2\vdash P_3:\mathtt{f}_{\mathit{Phil}}(a_1,a_2,a_3,a_4)}}{\Gamma_1,\mathit{fork}_1{:}!^{a_1}_{a_2}\mid U_1,\mathit{fork}_2:U_2\vdash P_2:(a_3,a_1)\ \&\ (a_3,a_2)\ \&\ \mathtt{f}_{\mathit{Phil}}(a_1,a_2,a_3,a_4)}}{\Gamma,\mathit{fork}_1:U_1,\mathit{fork}_2:U_2\vdash P_1:(a_1,a_4)\ \&\ (a_3,a_1)\ \&\ (a_3,a_2)\ \&\ \mathtt{f}_{\mathit{Phil}}(a_1,a_2,a_3,a_4)}$$

where $\Gamma_1=\Gamma,x_1:\mathbf{int}$, $\Gamma_2=\Gamma,x_2:\mathbf{int}$, $U_1=\mu\alpha.?^{a_2}_{a_1}.(!^{a_1}_{a_2}\mid\alpha)$ and $U_2=\mu\alpha.?^{a_4}_{a_3}.(!^{a_3}_{a_4}\mid\alpha)$. Because $(\mathscr{L},\mathtt{L})$ has no circularity, by Theorem 2, we can conclude that $(\mathscr{D},P)$ is deadlock-free. □

Remark 1. The dining philosopher program cannot be typed in Kobayashi's type system [11]. That is because his type system assigns obligation/capability levels to each input/output *statically*. Thus only a fixed number of levels (represented as natural numbers) can be used to type a process in his type system. Since the above process can create a network consisting of an arbitrary number of dining philosophers, we need an unbounded number of levels to type the process. (Kobayashi [11] introduced a heuristic to partially mitigate the restriction on the number of levels being fixed, but the heuristic does not work here.) A variant of the dining philosopher example has been discussed in [8]. Since the variant is designed so that a finite number of levels are sufficient, it is typed both in [11] and in our new type system.

Similarly to the dining philosopher program, the system in [11] returns a false positive for the process Fact in Section 1, while it is deadlock-free according to our new system. We detail the arguments in the next example.

Example 5. Process Fact of Section 1 is written in the value passing CCS as follows.

$$\begin{array}{l}\texttt{Fact}(a_1,a_2,a_3,a_4;n:\textbf{int},r{:}?^{a_1}_{a_2},s{:}!^{a_3}_{a_4}) = \\ \quad \textbf{if } n=0 \textbf{ then } r?n.s!n \textbf{ else} \\ \qquad (\nu\, a_5,a_6;t{:}?^{a_5}_{a_6} \mid !^{a_6}_{a_5})(r?n.t!(m\times n) \mid \texttt{Fact}(a_5,a_6,a_3,a_4;n-1,t,s))\end{array}$$

Let $\Gamma = \texttt{Fact} : [a_1,a_2,a_3,a_4;\textbf{int},?^{a_1}_{a_2},!^{a_3}_{a_4}]$ and P be the body of the definition above. Then we have $\Gamma, n:\textbf{int}, r{:}?^{a_1}_{a_2}, s{:}!^{a_3}_{a_4} \vdash P : \mathtt{L}$ for $\mathtt{L} = (a_2,a_3) + (\nu\, a_5,a_6)((a_2,a_6)\ \&\ \mathtt{f}_{\mathtt{Fact}}(a_5,a_6,a_3,a_4))$. Thus, we have: $\Gamma \vdash (\mathscr{D},P'){:}(\mathscr{L},\mathtt{L}')$ for:

$$\begin{array}{l}P' = (\nu\, a_1,a_2;r{:}?^{a_1}_{a_2} \mid !^{a_2}_{a_1})(\nu\, a_3,a_4;s{:}?^{a_4}_{a_3} \mid !^{a_3}_{a_4})(r!1 \mid \texttt{Fact}(a_1,a_2,a_3,a_4;m,r,s) \mid s?x.\mathbf{0}) \\ \mathscr{L} = \{\mathtt{f}_{\mathtt{Fact}}(a_1,a_2,a_3,a_4) = \mathtt{L}\} \\ \mathtt{L}' = (\nu\, a_1,a_2,a_3,a_4)(\mathbb{0}\ \&\ \mathtt{f}_{\mathtt{Fact}}(a_1,a_2,a_3,a_4)\ \&\ \mathbb{0})\end{array}$$

where m is an integer constant. Since $(\mathscr{L},\mathtt{L}')$ does not have a circularity, we can conclude that $(\mathscr{D},P')$ is deadlock-free.

Type Inference. An *untyped* value-passing CCS program is a program where restrictions are $(\nu\, x)P$, process invocations are $A(\widetilde{e})$ and process definitions are $A(\widetilde{x}) = P$. Given an untyped value-passing CCS program $(\mathscr{D},P)$, with $var(P)=\varnothing$, there is an inference algorithm to decide whether there exists a program $(\mathscr{D}',P')$ that coincides with $(\mathscr{D},P)$, except for the type annotations, and such that $\Gamma \vdash (\mathscr{D}',P') : (\mathscr{L},\mathtt{L})$. The algorithm is almost the same as that of the type system in [10] and, therefore, we do not re-describe it here. The only extra work compared with the previous algorithm is the lam program extraction, which is done using the rules in Figure 2. Finally, it suffices to analyze the extracted lams by using the fixpoint technique in Section 4.

Synchronous Value Passing CCS and pi Calculus. The type system above can be easily extended to the pi-calculus, where channel names can be passed around through other channels. To that end, we extend the syntax of types as follows.

$$\mathtt{T} ::= \textbf{int} \mid \mathsf{ch}(\mathtt{T},U).$$

The type $\mathsf{ch}(\mathtt{T},U)$ describes a channel that is used according to the usage U, and $\mathtt{T}$ is the type of values passed along the channel. Only a slight change of the typing rules is sufficient, as summarized below.

(T-OUT')
$$\frac{\Gamma \vdash e : \mathrm{T}}{\Gamma, x : \mathsf{ch}(\mathrm{T}, !^{a_1}_{a_2}) \vdash x!e : \&_{a \in ob(\Gamma)}(a_2, a)}$$

(T-IN')
$$\frac{\Gamma, x : \mathsf{ch}(\mathrm{T}, U), y : \mathrm{T} \vdash P : \mathrm{L}}{\Gamma, x : \mathsf{ch}(\mathrm{T}, ?^{a_1}_{a_2}.U) \vdash x?y.P : \mathrm{L} \,\&\, (\&_{a \in ob(\Gamma)}(a_2, a))}$$

In particular, (T-OUT') introduces dependencies between an output channel and the values sent along the channel. We notice that, in case of *synchronous value passing CCS* (as well as pi-calculus), where messages have continuations, rule (T-OUT') also introduces dependency pairs between the capability of the channel and the obligations in the continuation.

7 Related Work and Conclusions

In this paper we have designed a new deadlock detection technique for the value-passing CCS (and for the pi-calculus) that enables the analysis of networks with arbitrary numbers of nodes. Our technique relies on a decidability result of a basic model featuring recursion and fresh name generation: the *lam programs*. This model has been introduced and studied in [5,6] for detecting deadlock of an object-oriented programming language [7], but the decidability was known only for a subset of lams where only linear recursion is allowed [6], and only approximate algorithms have been given for the full lam model.

The application of the lam model to deadlock-freedom of the value-passing CCS (and pi-calculus) is also new, and the resulting deadlock-freedom analysis significantly improves the previous deadlock-freedom analysis [11], as demonstrated through the dining philosopher example. In particular, Kobayashi's type system provides a mechanism for dealing with a limited form of unbounded dependency chains, but the mechanism is rather ad hoc and fragile with respect to a syntactic change. For example, while

```
Fib(n,r) =  if n<2 then r?n else new s in new t in
                  (Fib!(n-1,s) | s?x.(Fib!(n-2,t)|t?y.r!(x+y))
```

is typable, the variation obtained by swapping new s in and new t in is untypable. Neither Fact nor the dining philosopher example are typable in [11]. More recently, in [17], Padovani has introduced another type system for deadlock-freedom, which has a better support than Kobayashi's one for reasoning about unbounded dependency chains, by using a form of polymorphism on levels. However, since the levels in his type system are also integers, neither the Fact example nor the dining philosopher example are typable. In addition, Padovani's type system cannot deal with non-linear channels, like the fork channels in the dining philosopher example. That said, our type system does not subsume Padovani's one, as our system does not support recursive types.

Like other type-based analyses, our method cannot reason about value-dependent behaviors. For example, consider the following process:

$$(\textbf{if } b \textbf{ then } x?z.y!z \textbf{ else } y!1 \mid x?z.) \mid (\textbf{if } b \textbf{ then } x!1 \mid y?z. \textbf{ else } y?z.x!z).$$

It is deadlock-free, but our type system would extract the lam expression: $((a_x, a_y) + \mathbb{0}) \,\&\, (\mathbb{0} + (a_y, a_x))$ (where a_x and a_y are the capability levels of the inputs on x and y respectively), detecting a (false) circular dependency.

The integration of TyPiCal with the deadlock detection technique of this paper is left for future work. We expect that we can extend our analysis to cover lock-freedom [8,17], too. To that end, we can require that a lam is not only circularity-free but is also *well founded*, and/or combine the deadlock-freedom analysis with the termination analysis, following the technique in [14].

Acknowledgments. This work was partially supported by JSPS Kakenhi 23220001 and by the EU project FP7-610582 ENVISAGE: Engineering Virtualized Services.

References

1. Abadi, M., Flanagan, C., Freund, S.N.: Types for safe locking: Static race detection for Java. TOPLAS 28 (2006)
2. Boyapati, C., Lee, R., Rinard, M.: Ownership types for safe program.: preventing data races and deadlocks. In: OOPSLA, pp. 211–230. ACM (2002)
3. Davey, B.A., Priestley, H.A.: Introduction to Lattices and Order. Cambridge University Press (2002)
4. Flanagan, C., Qadeer, S.: A type and effect system for atomicity. In: PLDI, pp. 338–349. ACM (2003)
5. Giachino, E., Laneve, C.: A beginner's guide to the dead*l*ock *a*nalysis *m*odel. In: Palamidessi, C., Ryan, M.D. (eds.) TGC 2012. LNCS, vol. 8191, pp. 49–63. Springer, Heidelberg (2013)
6. Giachino, E., Laneve, C.: Deadlock detection in linear recursive programs. In: Bernardo, M., Damiani, F., Hähnle, R., Johnsen, E.B., Schaefer, I. (eds.) SFM 2014. LNCS, vol. 8483, pp. 26–64. Springer, Heidelberg (2014)
7. Giachino, E., Laneve, C., Lienhardt, M.: A framework for deadlock detection in ABS. In: Software and System Modeling (to appear, 2014)
8. Kobayashi, N.: A type system for lock-free processes. Information and Computation 177, 122–159 (2002)
9. Kobayashi, N.: Type systems for concurrent programs. In: Aichernig, B.K. (ed.) Formal Methods at the Crossroads. From Panacea to Foundational Support. LNCS, vol. 2757, pp. 439–453. Springer, Heidelberg (2003)
10. Kobayashi, N.: Type-based information flow analysis for the pi-calculus. Acta Informatica 42(4-5), 291–347 (2005)
11. Kobayashi, N.: A new type system for deadlock-free processes. In: Baier, C., Hermanns, H. (eds.) CONCUR 2006. LNCS, vol. 4137, pp. 233–247. Springer, Heidelberg (2006)
12. Kobayashi, N.: TyPiCal: Type-based static analyzer for the Pi-Calculus (2007), `http://www-kb.is.s.u-tokyo.ac.jp/~koba/typical/`
13. Kobayashi, N., Pierce, B.C., Turner, D.N.: Linearity and the pi-calculus. ACM Transactions on Programming Languages and Systems 21(5), 914–947 (1999)
14. Kobayashi, N., Sangiorgi, D.: A hybrid type system for lock-freedom of mobile processes. ACM TOPLAS 32(5) (2010)
15. Milner, R.: A Calculus of Communication Systems. LNCS, vol. 92. Springer, Heidelberg (1980)
16. Milner, R., Parrow, J., Walker, D.: A calculus of mobile processes, ii. Inf. and Comput. 100, 41–77 (1992)
17. Padovani, L.: Deadlock and Lock Freedom in the Linear π-Calculus. In: CSL-LICS 2014 (2014)
18. Suenaga, K.: Type-based deadlock-freedom verification for non-block-structured lock primitives and mutable references. In: Ramalingam, G. (ed.) APLAS 2008. LNCS, vol. 5356, pp. 155–170. Springer, Heidelberg (2008)
19. Vasconcelos, V.T., Martins, F., Cogumbreiro, T.: Type inference for deadlock detection in a multithreaded polymorphic typed assembly language. In: PLACES. EPTCS, vol. 17, pp. 95–109 (2009)

Trees from Functions as Processes

Davide Sangiorgi[1] and Xian Xu[2]

[1] University of Bologna/INRIA, Bologna, Italy
[2] East China University of Science and Technology, Shanghai, China

Abstract. Lévy-Longo Trees and Böhm Trees are the best known tree structures on the λ-calculus. We give general conditions under which an encoding of the λ-calculus into the π-calculus is sound and complete with respect to such trees. We apply these conditions to various encodings of the call-by-name λ-calculus, showing how the two kinds of tree can be obtained by varying the behavioural equivalence adopted in the π-calculus and/or the encoding. The conditions are presented in the π-calculus but can be adapted to other concurrency formalisms.

1 Introduction

The π-calculus is a well-known model of computation with processes. Since its introduction, its comparison with the λ-calculus has received a lot of attention. Indeed, a deep comparison between a process calculus and the λ-calculus is interesting for several reasons: it is a significant test of expressiveness, and helps in getting deeper insight into its theory. From the λ-calculus perspective, it provides the means to study λ-terms in contexts other than purely sequential ones, and with the instruments available in the process calculus. A more practical motivations for describing functions as processes is to provide a semantic foundation for languages which combine concurrent and functional programming and to develop parallel implementations of functional languages.

Beginning with Milner's seminal work [8], a number of λ-calculus strategies have been encoded into the π-calculus, including call-by-name, strong call-by-name (and call-by-need variants), call-by-value, parallel call-by-value (see [12, Chapter 15]). In each case, several variant encodings have appeared, by varying the target language or details of the encoding itself. Usually, when an encoding is given, a few basic results about its correctness are established, such as operational correctness and validity of reduction (i.e., the property that the encoding of a λ-term and the encoding of a derivative of it are behaviourally undistinguishable). Only in a few cases the question of the equality on λ-terms induced by the encoding has been tackled, e.g., [3–5,11,12]. In this paper, we refer to this question as the *full abstraction* issue: for an encoding $[\![\,]\!]$ of the λ-calculus into π-calculus, an equality $=_\lambda$ on the λ-terms, and an equality $=_\pi$ on the π-terms, full abstraction is achieved when for all λ-terms M, N we have $M =_\lambda N$ iff $[\![M]\!] =_\pi [\![N]\!]$. Full abstraction has two parts: soundness, which is the implication from right to left, and completeness, which is its converse.

The equality $=_\lambda$ usually is not the ordinary Morris-style contextual equivalence on the λ-terms: the π-calculus is richer — and hence more discriminating — than

P. Baldan and D. Gorla (Eds.): CONCUR 2014, LNCS 8704, pp. 78–92, 2014.

the λ-calculus; the latter is purely sequential, whereas the former can also express parallelism and non-determinism. (Exception to this are encodings into forms of π-calculus equipped with rigid constraints, e.g., typing constraints, which limit the set of legal π-calculus contexts.)

Indeed, the interesting question here is understanding what $=_\lambda$ is when $=_\pi$ is a well-known behavioural equivalence on π-terms. This question essentially amounts to using the encoding in order to build a λ-model, and then understanding the λ-model itself. While seldomly tackled, the outcomes of this study have been significant: for a few call-by-name encodings it has been shown that, taking (weak) bisimulation on the π-terms, then $=_\lambda$ corresponds to a well-known tree structure in the λ-calculus theory, namely the *Lévy-Longo Trees* (LTs) [12].

There is however another kind of tree structure in the λ-calculus, even more important: the *Böhm Trees* (BTs). BTs play a central role in the classical theory of the λ-calculus. The local structure of some of the most influential models of the λ-calculus, like Scott and Plotkin's P_ω [13], Plotkin's T^ω [10], is precisely the BT equality; and the local structure of Scott's D_∞ (historically the first mathematical, i.e., non-syntactical, model of the untyped λ-calculus) is the equality of the 'infinite η contraction' of BTs. The full abstraction results in the literature for encodings of λ-calculus into π-calculus, however, only concern LTs.

A major reason for the limited attention that the full abstraction issue for encodings of λ-calculus into π-calculus has received is that understanding what kind of the structure the encoding produces may be difficult, and the full abstraction proof itself is long and tedious. The contribution of this paper is twofold:

1. We present general conditions for soundness and completeness of an encoding of the λ-calculus with respect to both LTs *and* BTs. The conditions can be used both on coinductive equivalences such as bisimilarity, and on contextual equivalences such as may and must equivalences.
2. We show that by properly tuning the notion of observability and/or the details of the encoding it is possible to recover BTs in place of LTs.

Some conditions only concern the behavioural equivalence chosen for the π-calculus, and are independent of the encoding; a few conditions are purely syntactic (e.g., certain encoded contexts should be guarded); the only behavioural conditions are equality of β-convertible terms, equality among certain unsolvable terms, and existence of an inverse for certain contexts resulting from the encoding (i.e., the possibility of extracting their immediate subterms, up-to the behavioural equivalence chosen in the π-calculus). We use these properties to derive full abstraction results for BTs and LTs for various encodings and various behavioural equivalence of the π-calculus. For this we exploit a few basic properties of the encodings, making a large reuse of proofs.

In the paper we use the conditions with the π-calculus, but they could also be used in other concurrency formalisms.

Structure of the paper. Section 2 collects background material. Section 3 introduces the notion of encoding of the λ-calculus, and concepts related to this. Section 4 presents the conditions for soundness and completeness. Sections 5 and 6 apply the conditions on a few encodings of call-by-name and strong call-by-name from the literature, and for various behavioural equivalences on the π-calculus. Section 7 briefly discusses refinements of the π-calculus, notably with linear types. Some conclusions are reported in Section 8.

2 Background

The λ-calculus We use M, N to range over the set Λ of λ-terms, and x, y, z to range over variables. The standard syntax of λ-terms, and the rules for call-by-name and strong call-by-name (where reduction may continue underneath a λ-abstraction), can be recalled in [2]. We assume the standard concepts of free and bound variables and substitutions, and identify α-convertible terms. We write Ω for the divergent term $(\lambda x.\, xx)(\lambda x.\, xx)$. Intuitively, a term M has *order of unsolvability* n $(0 \leqslant n < \omega)$ if it behaves like Ω after n initial abstractions; M has *order of unsolvability* ∞ if it can reduce to an unbounded number of nested abstractions; M is *solvable* otherwise, with a head normal form of the shape $\lambda\widetilde{x}.\, yM_1 \ldots M_n$.

Definition 1 (Lévy-Longo trees and Böhm trees). *The* Lévy–Longo Tree *of $M \in \Lambda$ is the labelled tree, LT(M), defined coinductively as follows:*

1. *$LT(M) = \top$ if M is an unsolvable of order ∞;*
2. *$LT(M) = \lambda x_1 \ldots x_n.\, \bot$ if M is an unsolvable of order n;*
3. *$LT(M)$ = tree with $\lambda\widetilde{x}.\, y$ as the root and $LT(M_1)$,...,$LT(M_n)$ as the children, if M has head normal form $\lambda\widetilde{x}.\, yM_1 \ldots M_n$, $n \geqslant 0$.*

Two terms M, N have the same LT if $LT(M) = LT(N)$. The definition of Böhm trees (BTs) is obtained from that of LTs using BT in place of LT in the definition above, and demanding that $BT(M) = \bot$ whenever M is unsolvable (in place of clauses (1) and (2)). See [6] for a thorough tutorial on observational equivalences for such trees.

The (asynchronous) π-calculus We first consider encodings into the *asynchronous* π-calculus because its theory is simpler and because it is the usual target language for encodings of the λ-calculus. In all encodings we consider, the encoding of a λ-term is parametric on a name, that is, is a function from names to π-calculus processes. We call such expressions *abstractions*. For the purposes of this paper unary abstractions, i.e., with only one parameter, suffice. The actual instantiation of the parameter of an abstraction F is done via the *application* construct $F\langle a\rangle$. We use P, Q for process, F for abstractions. Processes and abstractions form the set of π-*agents* (or simply *agents*), ranged over by A. Small letters $a, b, \ldots, x, y, \ldots$ range over the infinite set of names. We use a tilde to indicate tuples; and given a tuple $\widetilde{t}$, we write t_i for the i-th component of the tuple. Substitutions are ranged over by σ. The grammar of the calculus is thus:

$$
\begin{array}{ll}
A := P \;\big|\; F & \text{(agents)}\\
P := \mathbf{0} \;\big|\; a(\widetilde{b}).\, P \;\big|\; \overline{a}\langle\widetilde{b}\rangle \;\big|\; P_1 \mid P_2 \;\big|\; \nu a\, P \;\big|\; !a(\widetilde{b}).\, P \;\big|\; F\langle a\rangle & \text{(processes)}\\
F := (a)\, P & \text{(abstractions)}
\end{array}
$$

Since the calculus is polyadic, we assume a *sorting system* [9] to avoid disagreements in the arities of the tuples of names carried by a given name. We will not present the sorting system because not essential. The reader should take for granted that all agents described obey a sorting. A *context* C of π is a π-agent in which some subterms have been replaced by the hole $[\cdot]$ or, if the context is polyadic, with indexed holes $[\cdot]_1, \ldots, [\cdot]_n$; then $C[A]$ or $C[\widetilde{A}]$ is the agent resulting from replacing the holes with the terms A or $\widetilde{A}$. If the initial expression was an abstraction, we call the context an *abstraction π-context*; otherwise it is a *process π-context*. (A hole itself may stand for an abstraction or a process.) A name is *fresh* if it does not occur in the objects under consideration.

The standard operational semantics of the asynchronous π-processes (as well as the one for synchronous π-processes) is recalled in [12]. Transitions are of the form $P \xrightarrow{a(\widetilde{b})} P'$ (an input, $\widetilde{b}$ are the bound names of the input prefix that has been fired), $P \xrightarrow{\nu \widetilde{d}\, \overline{a}\langle \widetilde{b}\rangle} P'$ (an output, where $\widetilde{d} \subseteq \widetilde{b}$ are private names extruded in the output), and $P \xrightarrow{\tau} P'$ (an internal action). We use μ to range over the labels of transitions. We write $\Longrightarrow$ for the reflexive transitive closure of $\xrightarrow{\tau}$, and $\stackrel{\mu}{\Longrightarrow}$ for $\Longrightarrow\xrightarrow{\mu}\Longrightarrow$; then $\stackrel{\widehat{\mu}}{\Longrightarrow}$ is $\stackrel{\mu}{\Longrightarrow}$ if μ is not τ, and $\Longrightarrow$ otherwise; finally $P \xrightarrow{\widehat{\mu}} P'$ holds if $P \xrightarrow{\mu} P'$ or ($\mu = \tau$ and $P = P'$). In bisimulations or similar coinductive relations for the asynchronous π-calculus, no name instantiation is required in the input clause or elsewhere (provided α-convertible processes are identified); i.e., the *ground* versions of the relations are congruences or precongruences [12].

Definition 2 (bisimilarity). Bisimilarity *is the largest symmetric relation* $\approx$ *on π-processes such that whenever* $P \approx Q$ *and* $P \xrightarrow{\mu} P'$ *then* $Q \stackrel{\widehat{\mu}}{\Longrightarrow} Q'$ *for some* Q' *and* $P \approx Q'$.

A key preorder in our work will be *expansion* [1,12]; this is a refinement of bisimulation that takes into account the number of internal actions in simulation. Intuitively, Q expands P if they are weak bisimilar and moreover Q has no fewer internal actions when simulating P.

Definition 3 (expansion relation). *A relation $\mathcal{R}$ on π-processes is an* expansion relation *if whenever* $P \mathcal{R} Q$*: (1) if* $P \xrightarrow{\mu} P'$ *then* $Q \stackrel{\mu}{\Longrightarrow} Q'$ *and* $P' \mathcal{R} Q'$*; (2) if* $Q \xrightarrow{\mu} Q'$ *then* $P \xrightarrow{\widehat{\mu}} P'$ *and* $P' \mathcal{R} Q'$.

We write $\preccurlyeq$ for the largest expansion relation, and simply call it *expansion*. We also need its 'divergence-sensitive' variant, written $\preccurlyeq^{\Uparrow}$, as an auxiliary relation when tackling must equivalences. Using $\Uparrow$ to indicate divergence (i.e., $P\Uparrow$ if P can undergo an infinite sequence of τ transitions), then $\preccurlyeq^{\Uparrow}$ is obtained by adding into Definition 3 the requirement that $Q\Uparrow$ implies $P\Uparrow$. We write $\succcurlyeq$ and ${}^{\Uparrow}\!\succcurlyeq$ for the inverse of $\preccurlyeq$ and $\preccurlyeq^{\Uparrow}$, respectively. The predicate $\Downarrow$ indicates barb-observability (i.e., $P\Downarrow$ if $P \Longrightarrow\xrightarrow{\mu}$ for some μ other than τ). As instance of a contextual divergence-sensitive equivalence, we consider *must-termination*, because of the simplicity of its definition — other choices would have been possible.

Definition 4 (may and must equivalences). *The* π*-processes* P *and* Q *are* may equivalent, *written* $P \sim_{\mathrm{may}} Q$, *if in all process contexts* C *we have* $C[P]{\Downarrow}$ *iff* $C[Q]{\Downarrow}$. *They are* must-termination equivalent *(briefly* must equivalent*), written* $P \sim_{\mathrm{must}} Q$, *if in all process contexts* C *we have* $C[P]{\Uparrow}$ *iff* $C[Q]{\Uparrow}$.

The behavioural relations defined above use the standard observables of π-calculus; they can be made coarser by using the observables of asynchronous calculi, where one takes into account that, since outputs are not blocking, only output transitions from tested processes are immediately detected by an observer. In our examples, the option of asynchronous observable will make a difference only in the case of may equivalence. In *asynchronous may equivalence*, $\sim^{\mathrm{asy}}_{\mathrm{may}}$, the barb-observability predicate $\Downarrow$ is replaced by the asynchronous barb-observability predicate $\Downarrow_{\mathrm{asy}}$, whereby $P{\Downarrow_{\mathrm{asy}}}$ holds if $P \Longrightarrow\xrightarrow{\mu}$ and μ is an output action. We have $\preccurlyeq \subseteq \approx \subseteq \sim_{\mathrm{may}} \subseteq \sim^{\mathrm{asy}}_{\mathrm{may}}$, and $\preccurlyeq^{\Uparrow} \subseteq \sim_{\mathrm{must}}$. The following results will be useful later. A process is *inactive* if it may never perform a visible action.

Lemma 1. *For all process contexts* C*, we have: (1) if* P *is inactive, then* $C[P]{\Downarrow}$ *implies* $C[Q]{\Downarrow}$ *for all* Q*,* $C[P]{\Downarrow_{\mathrm{asy}}}$ *implies* $C[Q]{\Downarrow_{\mathrm{asy}}}$ *for all* Q*, and* $C[a(\widetilde{x}).\,P]{\Downarrow_{\mathrm{asy}}}$ *implies* $C[P]{\Downarrow_{\mathrm{asy}}}$*; (2) if* $P{\Uparrow}$ *then for all* Q*,* $C[Q]{\Uparrow}$ *implies* $C[P]{\Uparrow}$.

Lemma 2. $\nu a\,(\overline{a}\langle\widetilde{b}\rangle \mid a(\widetilde{x}).\,P) \succcurlyeq^{\Uparrow} P\{\widetilde{b}/\widetilde{x}\}$.

3 Encodings of the λ-calculus and Full Abstraction

In this paper an 'encoding of the λ-calculus into π-calculus' is supposed to be *compositional* (a mapping to π-calculus agents defined structurally on λ-terms), and *uniform.* The 'uniformity' condition refers to the treatment of the free variables: if the λ-term M and M' are the same modulo a renaming of free variables, then also their encodings should be same modulo a renaming of free names; since, in our encodings, λ-variables are included in the set of π-calculus names, a way of ensuring uniformity is to require that the encoding commutes with (name) substitution, i.e., $[\![M\sigma]\!] \equiv [\![M]\!]\sigma$.

A compositional encoding can be extended to contexts. We sometimes use: (1) $C^x_\lambda \stackrel{\mathrm{def}}{=} [\![\lambda x.\,[\cdot]]\!]$, an *abstraction contexts of* $[\![\,]\!]$ (the hole represents the body of an abstraction); (2) $C^{x,n}_{\mathrm{var}} \stackrel{\mathrm{def}}{=} [\![x[\cdot]_1 \cdots [\cdot]_n]\!]$ (for $n \geqslant 0$), a *variable contexts of* $[\![\,]\!]$ (an application context in which the head is a variable and the holes represent the following sequence of terms). In the remainder of the paper, 'encoding' refers to a 'compositional and uniform encoding of the λ-calculus into the π-calculus'.

Definition 5 (soundness, completeness, full abstraction, validity of β rule). *An encoding* $[\![\,]\!]$ *and a relation* $\mathcal{R}$ *on* π*-agents are: (1)* sound for LTs *if* $[\![M]\!]\,\mathcal{R}\,[\![N]\!]$ *implies* $LT(M) = LT(N)$*, for all* $M, N \in \Lambda$*; (2)* complete for LTSs *if* $LT(M) = LT(N)$ *implies* $[\![M]\!]\,\mathcal{R}\,[\![N]\!]$*, for all* $M, N \in \Lambda$*; (3)* fully abstract for LTs *if they are both sound and complete for LTs.*

The same definitions will also be applied to BTs — just replace 'LT' with 'BT'. Moreover, $[\![\,]\!]$ *and* $\mathcal{R}$ validate rule β *if* $[\![(\lambda x.\,M)N]\!]\,\mathcal{R}\,[\![M\{N/x\}]\!]$*, for all* x, M, N.

4 Conditions for Completeness and Soundness

We first give the conditions for completeness of an encoding $[\![\,]\!]$ from the λ-calculus into π with respect to a relation $\asymp$ on π-agents; then those for soundness. In both cases, the conditions involve an auxiliary relation $\leq$ on π-agents.

Completeness conditions. In the conditions for completeness the auxiliary precongruence $\leq$ is required so to validate an 'up-to $\leq$ and contexts' technique. Such technique is inspired by the 'up-to expansion and contexts' technique for bisimulation [12], which allows us the following flexibility in the bisimulation game required on a candidate relation $\mathcal{R}$: given a pair of derivatives P and Q, it is not necessary that the pair (P, Q) itself be in $\mathcal{R}$, as in the ordinary definition of bisimulation; it is sufficient to find processes $\widetilde{P}, \widetilde{Q}$, and a context C such that $P \succcurlyeq C[\widetilde{P}]$, $Q \succcurlyeq C[\widetilde{Q}]$, and $\widetilde{P} \mathrel{\mathcal{R}} \widetilde{Q}$; that is, we can manipulate the original derivatives in terms of $\preccurlyeq$ so to isolate a common context C; this context is removed and only the resulting processes $\widetilde{P}, \widetilde{Q}$ need to be in $\mathcal{R}$. In the technique, the expansion relation is important: replacing it with bisimilarity breaks correctness. Also, some care is necessary when a hole of the contexts occurs underneath an input prefix, in which case a closure under name substitutions is required. Below, the technique is formulated in an abstract manner, using generic relations $\asymp$ and $\leq$. In the encodings we shall examine, $\asymp$ will be any of the congruence relations in Section 2, whereas $\leq$ will always be the expansion relation (or its divergence-sensitive variant, when $\asymp$ is must equivalence).

Definition 6 (up-to-$\leq$-and-contexts technique). *Relation* $\asymp$ validates the up-to-$\leq$-and-contexts technique *if for any symmetric relation $\mathcal{R}$ on π-processes we have $\mathcal{R} \subseteq \asymp$ whenever for any pair $(P, Q) \in \mathcal{R}$, if $P \xrightarrow{\mu} P'$ then $Q \stackrel{\widehat{\mu}}{\Longrightarrow} Q'$ and there are processes $\widetilde{P}, \widetilde{Q}$ and a context C such that $P' \geq C[\widetilde{P}]$, $Q' \geq C[\widetilde{Q}]$, and, if $n \geqslant 0$ is the length of the tuples $\widetilde{P}$ and $\widetilde{Q}$, at least one of the following two statements is true, for each $i \leqslant n$: (1) $P_i \asymp Q_i$; (2) $P_i \mathrel{\mathcal{R}} Q_i$ and, if $[\cdot]_i$ occurs under an input in C, also $P_i\sigma \mathrel{\mathcal{R}} Q_i\sigma$ for all substitutions σ.*

Below is the core of the completeness conditions. Some of these conditions ((1)-(3)) only concern the chosen behavioural equivalence $\asymp$ and its auxiliary relation $\leq$, and are independent of the encoding; the most important condition is the validity of the up-to-$\leq$-and-contexts technique. Other conditions (such as (4)) are purely syntactic; we use the standard concept of guarded context (in which the hole appears underneath some prefix) [12]. The only behavioural conditions on the encoding are (5), (6) (plus (ii) in Theorem 1). They concern validity of β rule and equality of certain unsolvables — very basic requirements for the operational correctness of an encoding.

Definition 7. *Let $\asymp$ and $\leq$ be relations on π-agents such that:*

1. *$\asymp$ is a congruence and $\asymp \supseteq \geq$;*
2. *$\leq$ is an expansion relation and is a precongruence;*
3. *$\asymp$ validates the up-to-$\leq$-and-contexts technique.*

Now, an encoding $[\![\,]\!]$ *of* λ*-calculus into* π*-calculus is* faithful for $\asymp$ under $\leq$ *if*

4. *the variable contexts of* $[\![\,]\!]$ *are guarded;*
5. $[\![\,]\!]$ *and* $\geq$ *validate rule* β*;*
6. *if* M *is an unsolvable of order* 0 *then* $[\![M]\!] \asymp [\![\Omega]\!]$.

Theorem 1 (completeness). *Let* $[\![\,]\!]$ *be an encoding of the* λ*-calculus into* π*-calculus, and* $\asymp$ *a relation on* π*-agents. Suppose there is a relation* $\leq$ *on* π*-agents such that* $[\![\,]\!]$ *is faithful for* $\asymp$ *under* $\leq$*. We have:*

(i) if the abstraction contexts of $[\![\,]\!]$ *are guarded, then* $[\![\,]\!]$ *and* $\asymp$ *are complete for LTs;*
(ii) if $[\![M]\!] \asymp [\![\Omega]\!]$ *whenever* M *is unsolvable of order* ∞*, then* $[\![\,]\!]$ *and* $\asymp$ *are complete for BTs.*

The proofs for LTs and BTs are similar. In the proof for LTs, for instance, we consider the relation $\mathcal{R} \stackrel{\text{def}}{=} \{([\![M]\!], [\![N]\!]) \text{ s.t. } LT(M) = LT(N)\}$ and show that for each $([\![M]\!], [\![N]\!]) \in \mathcal{R}$ one of the following conditions is true, for some abstraction context C^x_λ, variable context $C^{x,n}_{\text{var}}$, and terms M_i, N_i:

(a) $[\![M]\!] \asymp [\![\Omega]\!]$ and $[\![N]\!] \asymp [\![\Omega]\!]$;
(b) $[\![M]\!] \geq C^x_\lambda[[\![M_1]\!]]$, $[\![N]\!] \geq C^x_\lambda[[\![N_1]\!]]$ and $([\![M_1]\!], [\![N_1]\!]) \in \mathcal{R}$;
(c) $[\![M]\!] \geq C^{x,n}_{\text{var}}[[\![M_1]\!], \ldots, [\![M_n]\!]]$, $[\![N]\!] \geq C^{x,n}_{\text{var}}[[\![N_1]\!], \ldots, [\![N_n]\!]]$ and $([\![M_i]\!], [\![N_i]\!]) \in \mathcal{R}$ for all i.

Here, (a) is used when M and N are unsolvable of order 0, by appealing to clause (6) of Definition 7. In the remaining cases we obtain (b) or (c), depending on the shape of the LT for M and N, and appealing to clause (5) of Definition 7. The crux of the proof is exploiting the property that $\asymp$ validates the up-to-$\leq$-and-contexts technique so to derive $\mathcal{R} \subseteq \asymp$ (the continuations of $[\![M]\!]$ and $[\![N]\!]$ are somehow related via the expansion and common context). Intuitively, this is possible because the variable and abstraction contexts of $[\![\,]\!]$ are guarded, and therefore the first action from terms such as $C^x_\lambda[[\![M_1]\!]]$ and $C^{x,n}_{\text{var}}[[\![M_1]\!], \ldots, [\![M_n]\!]]$ only consumes the context, and because $\leq$ is an expansion relation (clause (2) of Definition 7). Note that condition (2) of Definition 6 requires closure under substitutions when a hole is underneath a prefix. In clause (c) above we can derive closure under substitutions from $([\![N_i]\!], [\![N_i]\!]) \in \mathcal{R}$ because the LT equality is preserved by variable renaming and because we assume an encoding to act uniformly on the free names (Section 3).

In the results for BTs, the condition on abstraction contexts being guarded is not needed because the condition can be proved redundant in presence of the condition in the assertion (ii) of the theorem. Intuitively, the reason is that, if in a term the head reduction never unveils a variable, then the term is unsolvable and can be equated to Ω using condition (ii); if it does unveil a variable, then in the encoding the subterms following the variable are underneath at least one prefix (because the variable contexts of the encoding are guarded, by condition (4)) and then we are able to apply a reasoning similar to that in clause (c) above. Also, we do not need to explicitly prove $[\![\lambda x.\,\Omega]\!] \asymp [\![\Omega]\!]$, this can be derived from condition (ii) and clauses (5) and (6) of Definition 7.

Soundness conditions. In the conditions for soundness, one of the key requirements will be that certain contexts have an *inverse.* This intuitively means that it is possible to extract any of the processes in the holes of the context, up to the chosen behavioural equivalence. To have some more flexibility, we allow the appearance of the process of a hole after a rendez-vous with the external observer. This allows us to: initially restrict some names that are used to consume the context; then export such names before revealing the process of the hole. The reason why the restriction followed by the export of these names is useful is that the names might occur in the process of the hole; initially restricting them allows us to hide the names to the external environment; exporting them allows to remove the restrictions once the inversion work on the context is completed. The drawback of this initial rendez-vous is that we have to require a prefix-cancellation property on the behavioural equivalence; however, the requirement is straightforward to check in common behavioural equivalences.

We give the definition of inversion only for abstraction π-contexts whose holes are themselves abstractions. We only need this form of contexts when reasoning on λ-calculus encodings.

Definition 8. *Let C be an abstraction π-context with n holes, each occurring exactly once, each hole itself standing for an abstraction. We say that C* has inverse with respect to a relation $\mathcal{R}$ *on π-agents, if for every $i = 1, \ldots, n$ and for every $\widetilde{A}$ there exists a process π-context D_i and fresh names a, z, b such that*

$$D_i[C[\widetilde{A}]]\ \mathcal{R}\ (\boldsymbol{\nu}\widetilde{b}\,)(\overline{a}\langle\widetilde{c}\rangle \mid b(z).\, A_i\langle z\rangle)\,, \qquad \textit{for } b \in \widetilde{b} \subseteq \widetilde{c}.$$

It is useful to establish inverse properties for contexts for the finest possible behavioural relation, so to export the result to coarser relations. In our work, the finest such relation is the divergence-sensitive expansion ($\preccurlyeq^{\Uparrow}$).

Example 1. We show examples of inversion using contexts that are similar to some abstraction and variable contexts in encodings of λ-calculus.

1. Consider a context $C \stackrel{\text{def}}{=} (p)\, p(x,q).\, ([\cdot]\langle q\rangle)$. If F fills the context, then an inverse for ${}^{\Uparrow}\!\succcurlyeq$ is the context

$$D \stackrel{\text{def}}{=} \boldsymbol{\nu} b\, (\overline{a}\langle b\rangle \mid b(r).\, \boldsymbol{\nu} p\, ([\cdot]\langle p\rangle \mid \overline{p}\langle x, r\rangle))$$

   where all names are fresh (i.e., not free in F). Indeed we have, using simple algebraic manipulations (such as the law of Lemma 2):

$$\begin{aligned} D[C[F]] &\ {}^{\Uparrow}\!\succcurlyeq \boldsymbol{\nu} b\, (\overline{a}\langle b\rangle \mid b(r).\, \boldsymbol{\nu} p\, (p(x,q).\, F\langle q\rangle \mid \overline{p}\langle x, r\rangle)) \\ &\ {}^{\Uparrow}\!\succcurlyeq \boldsymbol{\nu} b\, (\overline{a}\langle b\rangle \mid b(r).\, F\langle r\rangle) \end{aligned}$$

2. Consider now a context $C \stackrel{\text{def}}{=} (p)\, (\boldsymbol{\nu} r, y\,)(\overline{x}\langle r\rangle \mid \overline{r}\langle y, p\rangle \mid !y(q).\, [\cdot]\langle q\rangle)$. If F fills the hole, then an inverse context is

$$D \stackrel{\text{def}}{=} ((\boldsymbol{\nu} x, p, b\,)([\cdot]\langle p\rangle \mid x(r).\, r(y,z).\, (\overline{a}\langle x, b\rangle \mid b(u).\, \overline{y}\langle u\rangle))) \qquad (1)$$

where again all names are fresh with respect to F. We have:

$$\begin{aligned} D[C[F]] &= (\boldsymbol{\nu} x, p, b)((C[F])\langle p\rangle \mid x(r).r(y,z).(\overline{a}\langle x,b\rangle \mid b(u).\overline{y}\langle u\rangle)) \\ &\Uparrow\!\succcurlyeq (\boldsymbol{\nu} x, p, b)(\boldsymbol{\nu} r, y)(\overline{x}\langle r\rangle \mid \overline{r}\langle y,p\rangle \mid !y(q).F\langle q\rangle) \mid \\ &\qquad\qquad x(r).r(y,z).(\overline{a}\langle x,b\rangle \mid b(u).\overline{y}\langle u\rangle)) \\ &\Uparrow\!\succcurlyeq (\boldsymbol{\nu} x, b)(\boldsymbol{\nu} y\,(!y(q).F\langle q\rangle \mid (\overline{a}\langle x,b\rangle \mid b(u).\overline{y}\langle u\rangle))) \\ &\Uparrow\!\succcurlyeq (\boldsymbol{\nu} x, b)(\overline{a}\langle x,b\rangle \mid b(r).(\boldsymbol{\nu} y\,(!y(q).F\langle q\rangle \mid \overline{y}\langle r\rangle))) \\ &\Uparrow\!\succcurlyeq (\boldsymbol{\nu} x, b)(\overline{a}\langle x,b\rangle \mid b(r).F\langle r\rangle) \end{aligned}$$

Definition 9. *A relation $\mathcal{R}$ on π-agents* has the rendez-vous cancellation *property if whenever $\boldsymbol{\nu}\widetilde{b}\,(\overline{a}\langle\widetilde{c}\rangle \mid b(r).P)\ \mathcal{R}\ \boldsymbol{\nu}\widetilde{b}\,(\overline{a}\langle\widetilde{c}\rangle \mid b(r).Q)$ where $b \in \widetilde{b} \subseteq \widetilde{c}$ and a, b are fresh, then also $P\ \mathcal{R}\ Q$.*

The cancellation property is straightforward for a behavioural relation $\asymp$ because, in the initial processes, the output $\overline{a}\langle\widetilde{c}\rangle$ is the only possible initial action, after which the input at b must fire (the assumption 'a, b fresh' facilitates matters, though it is not essential).

As for completeness, so for soundness we isolate the common conditions for LTs and BTs. Besides the conditions on inverse of contexts, the other main requirement is about the inequality among some structurally different λ-terms (condition 6).

Definition 10. *Let $\asymp$ and $\leq$ be relations on π-agents where*

1. *$\asymp$ is a congruence, $\leq$ a precongruence,*
2. *$\asymp\ \supseteq\ \geq$;*
3. *$\asymp$ has the rendez-vous cancellation property.*

An encoding $[\![\,]\!]$ of the λ-calculus into π-calculus is respectful for $\asymp$ under $\leq$ *if*

4. *$[\![\,]\!]$ and $\geq$ validate rule β;*
5. *if M is an unsolvable of order 0, then $[\![M]\!] \asymp [\![\Omega]\!]$;*
6. *the terms $[\![\Omega]\!]$, $[\![x\widetilde{M}]\!]$, $[\![x\widetilde{M'}]\!]$, and $[\![y\widetilde{M''}]\!]$ are pairwise unrelated by $\asymp$, assuming that $x \neq y$ and that tuples $\widetilde{M}$ and $\widetilde{M'}$ have different lengths;*
7. *the abstraction and variable contexts of $[\![\,]\!]$ have inverse with respect to $\geq$.*

The condition on variable context having an inverse is the most delicate one. In the encodings of the π-calculus we have examined, however, the condition is simple to achieve.

Theorem 2 (soundness). *Let $[\![\,]\!]$ be an encoding of the λ-calculus into π-calculus, and $\asymp$ a relation on π-agents. Suppose there is a relation $\leq$ on π-agents such that $[\![\,]\!]$ is respectful for $\asymp$ under $\leq$. We have:*

1. *if, for any M, the term $[\![\lambda x.\,M]\!]$ is unrelated by $\asymp$ to $[\![\Omega]\!]$ and to any term of the form $[\![x\widetilde{M}]\!]$, then $[\![\,]\!]$ and $\asymp$ are sound for LTs;*
2. *if*
   (a) $[\![M]\!] \asymp [\![\Omega]\!]$ whenever M is unsolvable of order ∞,

(b) *M solvable implies that the term* $[\![\lambda x.\, M]\!]$ *is unrelated by* $\asymp$ *to* $[\![\Omega]\!]$ *and to any term of the form* $[\![x\widetilde{M}]\!]$,

then $[\![\,]\!]$ *and* $\asymp$ *are sound for BTs.*

For the proof of Theorem 2, we use a coinductive definition of LT and BT equality, as forms of bisimulation. Then we show that the relation $\{(M,N) \mid [\![M]\!] \asymp [\![N]\!]\}$ implies the corresponding tree equality. In the case of internal nodes of the trees, we exploit conditions such as (6) and (7) of Definition 10.

Full abstraction We put together Theorems 1 and 2.

Theorem 3. *Let* $[\![\,]\!]$ *be an encoding of the* λ*-calculus into* π*-calculus,* $\asymp$ *a congruence on* π*-agents. Suppose there is a precongruence* $\leq$ *on* π*-agents such that*

1. $\leq$ *is an expansion relation and* $\asymp \supseteq \geq$;
2. $\asymp$ *validates the up-to-*$\leq$*-and-contexts technique;*
3. *the variable contexts of* $[\![\,]\!]$ *are guarded;*
4. *the abstraction and variable contexts of* $[\![\,]\!]$ *have inverse with respect to* $\geq$;
5. $[\![\,]\!]$ *and* $\geq$ *validate rule* β;
6. *if* M *is an unsolvable of order 0 then* $[\![M]\!] \asymp [\![\Omega]\!]$;
7. *the terms* $[\![\Omega]\!]$, $[\![x\widetilde{M}]\!]$, $[\![x\widetilde{M'}]\!]$, *and* $[\![y\widetilde{M''}]\!]$ *are pairwise unrelated by* $\asymp$, *assuming that* $x \neq y$ *and that tuples* $\widetilde{M}$ *and* $\widetilde{M}'$ *have different lengths.*

We have:

(i) *if*
 (a) *the abstraction contexts of* $[\![\,]\!]$ *are guarded, and*
 (b) *for any* M *the term* $[\![\lambda x.\, M]\!]$ *is unrelated by* $\asymp$ *to* $[\![\Omega]\!]$ *and to any term of the form* $[\![x\widetilde{M}]\!]$,
 then $[\![\,]\!]$ *and* $\asymp$ *are fully abstract for LTs;*

(ii) *if*
 (a) *M solvable implies that the term* $[\![\lambda x.\, M]\!]$ *is unrelated by* $\asymp$ *to* $[\![\Omega]\!]$ *and to any term of the form* $[\![x\widetilde{M}]\!]$, *and*
 (b) $[\![M]\!] \asymp [\![\Omega]\!]$ *whenever* M *is unsolvable of order* ∞,
 then $[\![\,]\!]$ *and* $\asymp$ *are fully abstract for BTs.*

In Theorems 1(i) and 3(i) for LTs the abstraction contexts are required to be guarded. This is reasonable in encodings of strategies, such as call-by-name, where evaluation does not continue underneath a λ-abstraction, but it is too demanding when evaluation can go past a λ-abstraction, such as strong call-by-name. We therefore present also the following alternative condition:

$$M, N \text{ unsolvable of order } \infty \text{ implies } [\![M]\!] \asymp [\![N]\!]. \qquad (*)$$

Theorem 4. *Theorems 1(i) and 3(i) continue to hold when the condition that the abstraction contexts be guarded is replaced by* $(*)$ *above.*

5 Examples with call-by-name

In this section we apply the theorems on soundness and completeness in the previous section to two well-known encodings of call-by-name λ-calculus: the one in Figure 1.a is Milner's original encoding [8]. The one in Figure 1.b is a variant encoding in which a function communicates with its environment via a rendez-vous (request/answer) pattern. An advantage of this encoding is that it can be easily tuned to call-by-need, or even used in combination with call-by-value [12].

For each encoding we consider soundness and completeness with respect to four behavioural equivalences: bisimilarity ($\approx$), may ($\sim_{\text{may}}$), must ($\sim_{\text{must}}$), and asynchronous may ($\sim^{\text{asy}}_{\text{may}}$). The following lemma allows us to apply the up-to-$\leq$-and-contexts technique.

Lemma 3. *Relations $\approx$, $\sim_{\text{may}}$, and $\sim^{\text{asy}}_{\text{may}}$ validate the up-to-$\preccurlyeq$-and-contexts technique; relation $\sim_{\text{must}}$ validates the up-to-$\preccurlyeq^{\Uparrow}$-and-contexts technique.*

The result in Lemma 3 for bisimulation is from [12]. The proofs for the may equivalences follow the definitions of the equivalences, reasoning by induction on the number of steps required to bring out an observable. The proof for the must equivalence uses coinduction to reason on divergent paths. Both for the may and for the must equivalences, the role of expansion ($\preccurlyeq$) is similar to its role in the technique for bisimulation.

Theorem 5. *The encoding of Figure 1.a is fully abstract for LTs when the behavioural equivalence for π-calculus is $\approx$, $\sim_{\text{may}}$, or $\sim_{\text{must}}$; and fully abstract for BTs when the behavioural equivalence is $\sim^{\text{asy}}_{\text{may}}$.*

The encoding of Figure 1.b is fully abstract for LTs under any of the equivalences $\approx$, $\sim_{\text{may}}$, $\sim_{\text{must}}$, or $\sim^{\text{asy}}_{\text{may}}$.

As Lemma 3 brings up, in the proofs, the auxiliary relation for $\approx$, $\sim_{\text{may}}$, and $\sim^{\text{asy}}_{\text{may}}$ is $\preccurlyeq$; for $\sim_{\text{must}}$ it is $\preccurlyeq^{\Uparrow}$. With Lemma 3 at hand, the proofs for the soundness and completeness statements are simple. Moreover, there is a large reuse of proofs and results. For instance, in the completeness results for LTs, we only have to check that: the variable and abstraction contexts of the encoding are guarded; β rule is validated; all unsolvable of order 0 are equated. The first check is straightforward and is done only once. For the β rule, it suffices to establish its validity for $\preccurlyeq^{\Uparrow}$, which is the finest among the behavioural relations considered; this is done using distributivity laws for private replications [12], which are valid for strong bisimilarity and hence for $\preccurlyeq^{\Uparrow}$, and the law of Lemma 2. Similarly, for the unsolvable terms of order 0 it suffices to prove that they are all 'purely divergent' (i.e., divergent and unable to even perform some visible action), which follows from the validity of the β rule for $\preccurlyeq^{\Uparrow}$.

Having checked the conditions for completeness, the only two additional conditions needed for soundness for LTs are conditions (6) and (7) of Definition 10, where we have to prove that certain terms are unrelated and that certain contexts have an inverse. The non-equivalence of the terms in condition (6) can be established for the coarsest equivalences, namely $\sim^{\text{asy}}_{\text{may}}$ and $\sim_{\text{must}}$, and then

$$\begin{aligned}
[\![\lambda x.\, M]\!] &\stackrel{\text{def}}{=} (p)\, p(x,q).\, [\![M]\!]\langle q\rangle \\
[\![x]\!] &\stackrel{\text{def}}{=} (p)\, \overline{x}\langle p\rangle \\
[\![MN]\!] &\stackrel{\text{def}}{=} (p)\, (\boldsymbol{\nu} r, x\,)\Big([\![M]\!]\langle r\rangle \mid \overline{r}\langle x,p\rangle \mid \\
&\qquad !x(q).\, [\![N]\!]\langle q\rangle\Big) \quad \text{(for } x \text{ fresh)}
\end{aligned}$$

Fig. 1.a: Milner's encoding

$$\begin{aligned}
[\![\lambda x.\, M]\!] &\stackrel{\text{def}}{=} (p)\, \boldsymbol{\nu} v\, (\overline{p}\langle v\rangle \mid v(x,q).\, [\![M]\!]\langle q\rangle) \\
[\![x]\!] &\stackrel{\text{def}}{=} (p)\, \overline{x}\langle p\rangle \\
[\![MN]\!] &\stackrel{\text{def}}{=} (p)\, \boldsymbol{\nu} r\, \Big([\![M]\!]\langle r\rangle \mid \\
&\qquad r(v).\, \boldsymbol{\nu} x\, (\overline{v}\langle x,p\rangle \mid \\
&\qquad !x(q).\, [\![N]\!]\langle q\rangle)\Big) \quad \text{(for } x \text{ fresh)}
\end{aligned}$$

Fig. 1.b: a variant encoding

Fig. 1. The two encodings of call-by-name

exported to the other equivalences. It suffices to look at visible traces of length 1 at most, except for terms of the form $[\![x\widetilde{M}]\!]$ and $[\![x\widetilde{M'}]\!]$, when tuples $\widetilde{M}$ and $\widetilde{M'}$ have different lengths, in which case one reasons by induction on the shortest of the two tuples.

The most delicate point is the existence of an inverse for the abstraction and the variable contexts. This can be established for the finest equivalence ($\preccurlyeq^{\Uparrow}$), and then exported to coarser equivalences. The two constructions needed for this are similar to those examined in Example 1.

For Milner's encoding, in the case of $\sim^{\text{asy}}_{\text{may}}$, we actually obtain the BT equality. One may find this surprising at first: BTs are defined from weak head reduction, in which evaluation continues underneath a λ-abstraction; however Milner's encoding mimics the call-by-name strategy, where reduction stops when a λ-abstraction is uncovered. We obtain BTs with $\sim^{\text{asy}}_{\text{may}}$ by exploiting Lemma 1(1) as follows. The encoding of a term $\lambda x.\, M$ is $(p)\, p(x,q).\, [\![M]\!]\langle q\rangle$. In an asynchronous semantics, an input is not directly observable; with $\sim^{\text{asy}}_{\text{may}}$ an input prefix can actually be erased provided, intuitively, that an output is never liberated. We sketch the proof of $[\![M]\!] \sim^{\text{asy}}_{\text{may}} [\![\Omega]\!]$ whenever M is unsolvable of order ∞, as required in condition (ii) of Theorem 3. Consider a context C with $C[[\![M]\!]]\Downarrow$, and suppose the observable is reached after n internal reductions. Term M, as ∞-unsolvable, can be β-reduced to $M' \stackrel{\text{def}}{=} (\lambda x)^n.\, N$, for some N. By validity of β-rule for $\succcurlyeq$, also $C[[\![M']\!]]\Downarrow$ in at most n steps; hence the subterm $[\![N]\!]$ of $[\![M']\!]$ does not contribute to the observable, since the abstraction contexts of the encodings are guarded and M' has n initial abstractions. We thus derive $C[[\![(\lambda x)^n.\, \Omega]\!]]\Downarrow$ and then, by repeatedly applying the third statement of Lemma 1(1) (as Ω is inactive), also $C[[\![\Omega]\!]]\Downarrow$. (The converse implication is given by the first statement in Lemma 1(1).)

6 An Example with Strong call by name

In this section we consider a different λ-calculus strategy, strong call-by-name, where the evaluation of a term may continue underneath a λ-abstraction. The main reason is that we wish to see the impact of this difference on the equivalences induced by the encodings. Intuitively, evaluation underneath a λ-abstraction is fundamental in the definition of BTs and therefore we expect that obtaining the

$$[\![\lambda x.\, M]\!] \stackrel{\text{def}}{=} (p)\,(\boldsymbol{\nu} x, q)(\overline{p}\langle x, q\rangle \mid [\![M]\!]\langle q\rangle) \qquad [\![x]\!] \stackrel{\text{def}}{=} (p)\,(x(p').\,(p' \rhd p))$$

$$[\![MN]\!] \stackrel{\text{def}}{=} (p)\,(\boldsymbol{\nu} q, r)([\![M]\!]\langle q\rangle \mid q(x, p').\,(p' \rhd p \mid !\overline{x}\langle r\rangle.\,[\![N]\!]\langle r\rangle)) \qquad \text{(for } x \text{ fresh)}$$

$$\text{where } r \rhd q \stackrel{\text{def}}{=} r(y, h).\,\overline{q}\langle y, h\rangle$$

Fig. 2. Encoding of strong call-by-name

BT equality will be easier. However, the LT equality will still be predominant: in BTs a λ-abstraction is sometimes unobservable, whereas in an encoding into π-calculus a λ-abstraction always introduces a few prefixes, which are observable in the most common behavioural equivalences.

The encoding of strong call-by-name, from [7], is in Figure 2. The encoding behaves similarly to that in Figure 1.b; reduction underneath a 'λ' is implemented by exploting special wire processes (such as $q \rhd p$). They allow us to split the body M of an abstraction from its head λx; then the wires make the liaison between the head and the body. It actually uses the *synchronous* π-calculus, because some of the output prefixes have a continuation. Therefore the encoding also offers us the possibility of discussing the portability of our conditions to the synchronous π-calculus. For this, the only point in which some care is needed is that in the synchronous π-calculus, bisimilarity and expansion need some closure under name substitutions, in the input clause (on the placeholder name of the input), and the outermost level (i.e., before the bisimulation or expansion game is started) to become congruence or precongruence relations. Name substitutions may be applied following the early, late or open styles. The move from a style to another one does not affect the results in terms of BTs and LTs in the paper. We omit the definitions, see e.g., [12].

In short, for any of the standard behavioural congruences and expansion precongruences of the synchronous π-calculus, the conditions concerning $\asymp$ and $\leq$ of the theorems in Section 4 remain valid. In Theorem 6 below, we continue to use the symbols $\approx$ and $\preccurlyeq$ for bisimilarity and expansion, assuming that these are bisimulation congruences and expansion precongruences in any of the common π-calculus styles (early, late, open). (Again, in the case of must equivalence the expansion preorder should be divergence sensitive.) The proof of Theorem 6 is similar to that of Theorem 5. The main difference is that, since in strong call-by-name the abstraction contexts are not guarded, we have to adopt the modification in one of the conditions for LTs suggested in Theorem 4. Moreover, for the proof of validity of β rule for $\preccurlyeq$, we use the following law to reason about wire processes $r \rhd q$ (and similarly for $\preccurlyeq^{\Uparrow}$):

- $\boldsymbol{\nu} q\,(q \rhd p \mid P) \succcurlyeq P\{p/q\}$ provided p does not appear free in P, and q only appears free in P only once, in a subexpression of the form $\overline{q}\langle\widetilde{v}\rangle.\,\mathbf{0}$.

Theorem 6. *The encoding of Figure 2 is fully abstract for LTs when the behavioural equivalence for the π-calculus is $\approx$, $\sim_{\text{may}}$, or $\sim_{\text{may}}^{\text{asy}}$; and fully abstract for BTs when the behavioural equivalence is $\sim_{\text{must}}$.*

Thus we obtain the BT equality for the must equivalence. Indeed, under strong call-by-name, all unsolvable terms are divergent. In contrast with Milner's encoding of Figure 1.a, under asynchronous may equivalence we obtain LTs because in the encoding of strong call-by-name the first action of an abstraction is an output, rather than an input as in Milner's encoding, and outputs are observable in asynchronous equivalences.

7 Types and Asynchrony

We show, using Milner's encoding (Figure 1.a), that we can sometimes switch from LTs to BTs by taking into account some simple *type* information together with asynchronous forms of behavioural equivalences. The type information needed is the linearity of the parameter name of the encoding (names p, q, r in Figure 1.a). Linearity ensures us that the external environment can never cause interferences along these names: if the input capability is used by the process encoding a λ-term, then the external environment cannot exercise the same (competing) capability. In an asynchronous behavioural equivalence input prefixes are not directly observable (as discussed earlier for asynchronous may).

Linear types and asynchrony can easily be incorporate in a bisimulation congruence by using a contextual form of bisimulation such as *barbed congruence* [12]. In this case, barbs (the observables of barbed congruence) are only produced by output prefixes (as in asynchronous may equivalence); and the contexts in which processes may be tested should respect the type information ascribed to processes (in particular the linearity mentioned earlier). We write $\approx_{\mathrm{bc}}^{\mathrm{lin,asy}}$ for the resulting asynchronous typed barbed congruence. Using Theorem 3(ii) we obtain:

Theorem 7. *The encoding of Figure 1.a is fully abstract for BTs when the behavioural equivalence for the π-calculus is* $\approx_{\mathrm{bc}}^{\mathrm{lin,asy}}$.

The auxiliary relation is still $\preccurlyeq$; here asynchrony and linearity are not needed.

8 Conclusions and Future Work

In this paper we have studied soundness and completeness conditions with respect to BTs and LTs for encodings of λ-calculus into the π-calculus. While the conditions have been presented on the π-calculus, they can be adapted to some other concurrency formalisms. For instance, expansion, a key preorder in our conditions, can always be extracted from bisimilarity as its "efficiency" preorder. It might be difficult, in contrast, to adapt our conditions to sequential languages; a delicate condition, for instance, appears to be the one on inversion of variable contexts.

We have used the conditions to derive tree characterisations for various encodings and various behavioural equivalences, including bisimilarity, may and must equivalences, and asynchronous may equivalence. The proofs of the conditions can often be transported from a behavioural equivalence to another one, with little or no extra work (e.g., exploiting containments among equivalences and preorders). Overall, we found the conditions particularly useful when dealing with contextual equivalences, such as may and must equivalences. It is unclear to us how soundness and completeness could be proved for them by relying on, e.g., direct characterisations of the equivalences (such as trace equivalence or forms of acceptance trees) and standard proof techniques for them.

It would be interesting to examine additional conditions on the behavioural equivalences of the π-calculus capable to retrieve, as equivalence induced by an encoding, that of BTs under η contraction, or BTs under infinite η contractions [2]. Works on linearity in the π-calculus, such as [14] might be useful.

In the paper we have considered encodings of call-by-name or strong call-by-name. These strategies fit the definition of BTs and LTs, in which reduction always picks the leftmost redex. We do not know, in contrast, what kind of tree structures could be obtained from encodings of the call-by-value strategy.

Acknowledgements. We thank the anonymous reviewers for useful comments. This work has been supported by project ANR 12IS02001 PACE and NSF of China (61261130589), and partially supported by NSF of China (61202023 and 61173048).

References

1. Arun-Kumar, S., Hennessy, M.: An efficiency preorder for processes. Acta Informatica 29, 737–760 (1992)
2. Barendregt, H.P.: The Lambda Calculus: Syntax, semantics. North-Holland (1984)
3. Berger, M., Honda, K., Yoshida, N.: Sequentiality and the π-calculus. In: Abramsky, S. (ed.) TLCA 2001. LNCS, vol. 2044, pp. 29–45. Springer, Heidelberg (2001)
4. Berger, M., Honda, K., Yoshida, N.: Genericity and the pi-calculus. Acta Informatica 42(2-3), 83–141 (2005)
5. Demangeon, R., Honda, K.: Full abstraction in a subtyped pi-calculus with linear types. In: Katoen, J.-P., König, B. (eds.) CONCUR 2011. LNCS, vol. 6901, pp. 280–296. Springer, Heidelberg (2011)
6. Dezani-Ciancaglini, M., Giovannetti, E.: From Bohm's theorem to observational equivalences: an informal account. ENTCS 50(2), 83–116 (2001)
7. Hirschkoff, D., Madiot, J.M., Sangiorgi, D.: Duality and i/o-types in the π-calculus. In: Koutny, M., Ulidowski, I. (eds.) CONCUR 2012. LNCS, vol. 7454, pp. 302–316. Springer, Heidelberg (2012)
8. Milner, R.: Functions as processes. Mathematical Structures in Computer Science 2(2), 119–141 (1992)
9. Milner, R.: Communicating and Mobile Systems: The π-Calculus. CUP (1999)
10. Plotkin, G.D.: T^ω as a universal domain. Journal of Computer and System Sciences 17, 209–236 (1978)
11. Sangiorgi, D.: Lazy functions and mobile processes. In: Proof, Language and Interaction: Essays in Honour of Robin Milner, pp. 691–720. MIT Press (2000)
12. Sangiorgi, D., Walker, D.: The π-calculus: a Theory of Mobile Processes. CUP (2001)
13. Scott, D.: Data types as lattices. SIAM Journal on Computing 5(3), 522–587 (1976)
14. Yoshida, N., Honda, K., Berger, M.: Linearity and bisimulation. Journal of Logic and Algebraic Programming 72(2), 207–238 (2007)

Bisimulations Up-to: Beyond First-Order Transition Systems

Jean-Marie Madiot[1], Damien Pous[1], and Davide Sangiorgi[2]

[1] ENS Lyon, Université de Lyon, CNRS, INRIA, France
[2] Università di Bologna, INRIA, Italy

Abstract. The bisimulation proof method can be enhanced by employing *'bisimulations up-to'* techniques. A comprehensive theory of such enhancements has been developed for first-order (i.e., CCS-like) labelled transition systems (LTSs) and bisimilarity, based on the notion of compatible function for fixed-point theory.

We transport this theory onto languages whose bisimilarity and LTS go beyond those of first-order models. The approach consists in exhibiting fully abstract translations of the more sophisticated LTSs and bisimilarities onto the first-order ones. This allows us to reuse directly the large corpus of up-to techniques that are available on first-order LTSs. The only ingredient that has to be manually supplied is the compatibility of basic up-to techniques that are specific to the new languages. We investigate the method on the π-calculus, the λ-calculus, and a (call-by-value) λ-calculus with references.

1 Introduction

One of the keys for the success of bisimulation is its associated proof method, whereby to prove two terms equivalent, one exhibits a relation containing the pair and one proves it to be a bisimulation. The bisimulation proof method can be enhanced by employing relations called *'bisimulations up-to'* [14,19,20]. These need not be bisimulations; they are simply *contained in* a bisimulation. Such techniques have been widely used in languages for mobility such as π-calculus or higher-order languages such as the λ-calculus, or Ambients (e.g., [23,16,11]).

Several forms of bisimulation enhancements have been introduced: 'bisimulation up-to bisimilarity' [17] where the derivatives obtained when playing bisimulation games can be rewritten using bisimilarity itself; 'bisimulation up-to transitivity' where the derivatives may be rewritten using the up-to relation; 'bisimulation up-to-context' [21], where a common context may be removed from matching derivatives. Further enhancements may exploit the peculiarities of the definition of bisimilarity on certain classes of languages: e.g., the up-to-injective-substitution techniques of the π-calculus [7,23], techniques for shrinking or enlarging the environment in languages with information hiding mechanisms (e.g., existential types, encryption and decryption constructs [1,25,24]), frame equivalence in the psi-calculi [18], or higher-order languages [12,10]. Lastly, it is important to notice that one often wishes to use *combinations* of up-to techniques.

P. Baldan and D. Gorla (Eds.): CONCUR 2014, LNCS 8704, pp. 93–108, 2014.

For instance, up-to context alone does not appear to be very useful; its strength comes out in association with other techniques, such as up-to bisimilarity or up-to transitivity.

The main problem with up-to techniques is proving their soundness (i.e. ensuring that any 'bisimulation up-to' is contained in bisimilarity). In particular, the proofs of complex combinations of techniques can be difficult or, at best, long and tedious. And if one modifies the language or the up-to technique, the entire proof has to be redone from scratch. Indeed the soundness of some up-to techniques is quite fragile, and may break when such variations are made. For instance, in certain models up-to bisimilarity may fail for weak bisimilarity, and in certain languages up-to bisimilarity and context may fail even if bisimilarity is a congruence relation and is strong (treating internal moves as any other move).

This problem has been the motivation for the development of a theory of enhancements, summarised in [19]. Expressed in the general fixed-point theory on complete lattices, this theory has been fully developed for both strong and weak bisimilarity, in the case of first-order labelled transition systems (LTSs) where transitions represent pure synchronisations among processes. In this framework, up-to techniques are represented using *compatible* functions, whose class enjoys nice algebraic properties. This allows one to derive complex up-to techniques algebraically, by composing simpler techniques by means of a few operators.

Only a small part of the theory has been transported onto other forms of transition systems, on a case by case basis. Transferring the whole theory would be a substantial and non-trivial effort. Moreover it might have limited applicability, as this work would probably have to be based on specific shapes for transitions and bisimilarity (a wide range of variations exist, e.g., in higher-order languages).

Here we explore a different approach to the transport of the theory of bisimulation enhancements onto richer languages. The approach consists in exhibiting fully abstract translations of the more sophisticated LTSs and bisimilarities onto first-order LTSs and bisimilarity. This allows us to import directly the existing theory for first-order bisimulation enhancements onto the new languages. Most importantly, the schema allows us to combine up-to techniques for the richer languages. The only additional ingredient that has to be provided manually is the soundness of some up-to techniques that are specific to the new languages. This typically includes the up-to context techniques, since those contexts are not first-order.

Our hope is that the method proposed here will make it possible to obtain a single formalised library about up-to techniques, that can be reused for a wide range of calculi: currently, all existing formalisations of such techniques in a proof assistant are specific to a given calculus: π-calculus [5,4], the psi-calculi [18], or a miniML language [6].

We consider three languages: the π-calculus, the call-by-name λ-calculus, and an imperative call-by-value λ-calculus. Other calculi like the Higher-Order π-calculus can be handled in a similar way; we omit the details here for lack of space. We moreover focus on weak bisimilarity, since its theory is more delicate than that of strong bisimilarity. When we translate a transition system into a first-order one, the grammar for the labels can be complex (e.g. include terms, labels, or contexts). What makes the system 'first-order' is that labels are taken as syntactic atomic objects, that may only be checked for syntactic equality. Note that full abstraction of the translation does not imply that the up-to techniques come for free: further conditions must be ensured. We shall see this with the π-calculus, where early bisimilarity can be handled but not the late one.

Forms of up-to context have already been derived for the languages we consider in this paper [11,23,22]. The corresponding soundness proofs are difficult (especially in λ-calculi), and require a mix of induction (on contexts) and coinduction (to define bisimulations). Recasting up-to context within the theory of bisimulation enhancements has several advantages. First, this allows us to combine this technique with other techniques, directly. Second, substitutivity (or congruence) of bisimilarity becomes a corollary of the compatibility of the up-to-context function (in higher-order languages these two kinds of proofs are usually hard and very similar). And third, this allows us to decompose the up-to context function into smaller pieces, essentially one for each operator of the language, yielding more modular proofs, also allowing, if needed, to rule out those contexts that do not preserve bisimilarity (e.g., input prefix in the π-calculus).

The translation of the π-calculus LTS into a first-order LTS follows the schema of abstract machines for the π-calculus (e.g., [26]) in which the issue of the choice of fresh names is resolved by ordering the names. Various forms of bisimulation enhancements have appeared in papers on the π-calculus or dialects of it. A translation of higher-order π-calculi into first-order processes has been proposed by Koutavas et al [8]. While the shape of our translations of λ-calculi is similar, our LTSs differ since they are designed to recover the theory of bisimulation enhancements. In particular, using the LTSs from [8] would lead to technical problems similar to those discussed in Remark 2. In the λ-calculus, limited forms of up-to techniques have been developed for applicative bisimilarity, where the soundness of the up-to context has still open problems [12,11]. More powerful versions of up-to context exist for forms of bisimilarity on open terms (e.g., open bisimilarity or head-normal-form bisimilarity) [13]. Currently, the form of bisimilarity for closed higher-order terms that allows the richest range of up-to techniques is environmental bisimilarity [22,9]. However, even in this setting, the proofs of combinations of up-to techniques are usually long and non-trivial. Our translation of higher-order terms to first-order terms is designed to recover environmental bisimilarity.

In Section 6, we show an example of how the wide spectrum of up-to techniques made available via our translations allows us to simplify relations needed in bisimilarity proofs, facilitating their description and reducing their size.

2 First-Order Bisimulation and Up-to Techniques

A *first-order Labelled Transition System*, briefly LTS, is a triple $(Pr, Act, \longrightarrow)$ where Pr is a non-empty set of states (or processes), Act is the set of *actions* (or *labels*), and $\longrightarrow \subseteq Pr \times Act \times Pr$ is the *transition relation*. We use P, Q, R to range over the processes of the LTS, and μ to range over the labels in *Act*, and, as usual, write $P \xrightarrow{\mu} Q$ when $(P, \mu, Q) \in \longrightarrow$. We assume that *Act* includes a special action τ that represents an internal activity of the processes. We derive bisimulation from the notion of *progression* between relations.

Definition 1. *Suppose $\mathcal{R}, \mathcal{S}$ are relations on the processes of an LTS. Then $\mathcal{R}$* strongly progresses to *$\mathcal{S}$, written $\mathcal{R} \rightsquigarrow_{\mathbf{sp}} \mathcal{S}$, if $\mathcal{R} \subseteq \mathcal{S}$ and if $P \mathcal{R} Q$ implies:*

- *whenever $P \xrightarrow{\mu} P'$ there is Q' s.t. $Q \xrightarrow{\mu} Q'$ and $P' \mathcal{S} Q'$;*
- *whenever $Q \xrightarrow{\mu} Q'$ there is P' s.t. $P \xrightarrow{\mu} P'$ and $P' \mathcal{S} Q'$.*

A relation $\mathcal{R}$ is a strong bisimulation *if $\mathcal{R} \rightsquigarrow_{\mathbf{sp}} \mathcal{R}$; and* strong bisimilarity, $\sim$, *is the union of all strong bisimulations.*

To define weak progression we need weak transitions, defined as usual: first, $P \xrightarrow{\hat{\mu}} P'$ means $P \xrightarrow{\mu} P'$ or $\mu = \tau$ and $P = P'$; and $\xRightarrow{\hat{\mu}}$ is $\Longrightarrow\xrightarrow{\hat{\mu}}\Longrightarrow$ where $\Longrightarrow$ is the reflexive transitive closure of $\xrightarrow{\tau}$. *Weak progression*, $\mathcal{R} \rightsquigarrow_{\mathbf{wp}} \mathcal{S}$, and *weak bisimilarity*, $\approx$, are obtained from Definition 1 by allowing the processes to answer using $\xRightarrow{\hat{\mu}}$ rather than $\xrightarrow{\mu}$.

Below we summarise the ingredients of the theory of bisimulation enhancements for first-order LTSs from [19] that will be needed in the sequel. We use f and g to range over functions on relations over a fixed set of states. Each such function represents a potential up-to technique; only the *sound* functions, however, qualify as up-to techniques:

Definition 2. *A function f is* sound for $\sim$ *if $\mathcal{R} \rightsquigarrow_{\mathbf{sp}} f(\mathcal{R})$ implies $\mathcal{R} \subseteq \sim$, for all $\mathcal{R}$; similarly, f is* sound for $\approx$ *if $\mathcal{R} \rightsquigarrow_{\mathbf{wp}} f(\mathcal{R})$ implies $\mathcal{R} \subseteq \approx$, for all $\mathcal{R}$.*

Unfortunately, the class of sound functions does not enjoy good algebraic properties. As a remedy to this, the subset of *compatible* functions has been proposed. The concepts in the remainder of the section can be instantiated with both strong and weak bisimilarities; we thus use $\mathbf{p}$ to range over $\mathbf{sp}$ or $\mathbf{wp}$.

Definition 3. *We write $f \rightsquigarrow_{\mathbf{p}} g$ when $\mathcal{R} \rightsquigarrow_{\mathbf{p}} \mathcal{S}$ implies $f(\mathcal{R}) \rightsquigarrow_{\mathbf{p}} g(\mathcal{S})$ for all $\mathcal{R}$ and $\mathcal{S}$. A monotone function f on relations is* $\mathbf{p}$-compatible *if $f \rightsquigarrow_{\mathbf{p}} f$.*

In other terms [19], f is $\mathbf{p}$-compatible iff $f \circ \mathbf{p} \subseteq \mathbf{p} \circ f$ where $\mathbf{p}(\mathcal{S})$ is the union of all $\mathcal{R}$ such that $\mathcal{R} \rightsquigarrow_{\mathbf{p}} \mathcal{S}$ and $\circ$ denotes function composition. Note that $\mathcal{R} \rightsquigarrow_{\mathbf{p}} \mathcal{S}$ is equivalent to $\mathcal{R} \subseteq \mathbf{p}(\mathcal{S})$.

Lemma 1. *If f is* $\mathbf{sp}$*-compatible, then f is sound for $\sim$; if f is* $\mathbf{wp}$*-compatible, then f is sound for $\approx$.*

Simple examples of compatible functions are the identity function and the function mapping any relation onto bisimilarity (for the strong or weak case, respectively). The class of compatible functions is closed under function composition and union (where the union $\cup F$ of a set of functions F is the point-wise union mapping $\mathcal{R}$ to $\bigcup_{f \in F} f(\mathcal{R})$), and thus under omega-iteration (where the omega-iteration f^ω of a function f maps $\mathcal{R}$ to $\bigcup_{n \in \mathbb{N}} f^n(\mathcal{R})$).

Other examples of compatible functions are typically contextual closure functions, mapping a relation into its closure w.r.t. a given set of contexts. For such functions, the following lemma shows that the compatibility of up-to-context implies substitutivity of (strong or weak) bisimilarity.

Lemma 2. *If f is* **sp**-*compatible, then $f(\sim) \subseteq \sim$; similarly if f is* **wp**-*compatible, then $f(\approx) \subseteq \approx$.*

Certain closure properties for compatible functions however only hold in the strong case. The main example is the *chaining operator* $\frown$, which implements relational composition:

$$f \frown g\,(\mathcal{R}) \triangleq f(\mathcal{R})\, g(\mathcal{R})$$

where $f(\mathcal{R})\, g(\mathcal{R})$ indicates the composition of the two relations $f(\mathcal{R})$ and $g(\mathcal{R})$. Using chaining we can obtain the compatibility of the function 'up to transitivity' mapping any relation $\mathcal{R}$ onto its reflexive and transitive closure $\mathcal{R}^\star$. Another example of **sp**-compatible function is 'up to bisimilarity' ($\mathcal{R} \mapsto \sim\mathcal{R}\sim$).

In contrast, in the weak case bisimulation up to bisimilarity is unsound. This is a major drawback in up-to techniques for weak bisimilarity, which can be partially overcome by resorting to the *expansion* relation $\gtrsim$ [3]. Expansion is an asymmetric refinement of weak bisimilarity whereby $P \gtrsim Q$ holds if P and Q are bisimilar and, in addition, Q is at least as efficient as P, in the sense that Q is capable of producing the same activity as P without ever performing more internal activities (the τ-actions); see [15] for its definition. Up-to-expansion yields a function ($\mathcal{R} \mapsto \gtrsim\mathcal{R}\lesssim$) that is **wp**-compatible. As a consequence, the same holds for the 'up-to expansion and contexts' function. More sophisticated up-to techniques can be obtained by carefully adjusting the interplay between visible and internal transitions, and by taking into account termination hypotheses [19].

Some further compatible functions are the functions **sp** and **wp** themselves (indeed a function f is **p**-compatible if $f \circ \mathbf{p} \subseteq \mathbf{p} \circ f$, hence trivially f can be replaced by **p** itself). Intuitively, the use of **sp** and **wp** as up-to techniques means that, in a diagram-chasing argument, the two derivatives need not be related; it is sufficient that the derivatives of such derivatives be related. Accordingly, we sometimes call functions **sp** and **wp** *unfolding* functions. We will use **sp** in the example in Section 6 and **wp** in Sections 4 and 5, when proving the **wp**-compatibility of the up to context techniques.

Last, note that to use a function f in combinations of up-to techniques, it is actually not necessary that f be **p**-compatible: for example proving that f progresses to $f \cup g$ and g progresses to g is enough, as then $f \cup g$ would be compatible. Extending this reasoning allows us to make use of 'second-order up-to techniques' to reason about compatibility of functions. When F is a set of functions, we say that F is **p**-*compatible up to* if for all f in F, it holds that $f \rightsquigarrow_{\mathbf{p}} (g \cup (\cup F))^\omega$ for a function g that has already been proven compatible. (We sometimes say that F is **p**-*compatible up to* g, to specify which compatible function is employed.) Lemma 1 and 2 remain valid when 'f is compatible' is replaced by '$f \in F$ and F is compatible up to'.

Terminology We will simply say that a function is *compatible* to mean that it is both **sp**-compatible and **wp**-compatible; similarly for compatibility up to. In languages defined from a grammar, a context C is a term with numbered holes $[\cdot]_1, \ldots, [\cdot]_n$, and each hole $[\cdot]_i$ can appear any number of times in C.

3 The π-calculus

The syntax and operational semantics of the π-calculus are recalled in [15]. We consider the early transition system, in which transitions are of the forms

$$P \xmapsto{ab}_\pi P' \qquad P \xmapsto{\overline{a}b}_\pi P' \qquad P \xmapsto{\overline{a}(b)}_\pi P' .$$

In the third transition, called bound output transition, name b is a binder for the free occurrences of b in P' and, as such, it is subject to α-conversion. The definition of bisimilarity takes α-conversion into account. The clause for bound output of strong early bisimilarity says ($\mathrm{fn}(Q)$ indicates the names free in Q):

– if $P \xmapsto{\overline{a}(b)}_\pi P'$ and $b \notin \mathrm{fn}(Q)$ then $Q \xmapsto{\overline{a}(b)}_\pi Q'$ for some Q' such that $P' \sim Q'$.

(The complete definition of bisimilarity is recalled in [15]). When translating the π-calculus semantics to a first-order one, α-conversion and the condition $b \notin \mathrm{fn}(Q)$ have to be removed. To this end, one has to force an agreement between two bisimilar process on the choice of the bound names appearing in transitions. We obtain this by considering *named processes* (c, P) in which c is bigger or equal to all names in P. For this to make sense we assume an enumeration of the names and use $\leq$ as the underlying order, and $c+1$ for name following c in the enumeration; for a set of names N, we also write $c \geq N$ to mean $c \geq a$ for all $a \in N$.

The rules below define the translation of the π-calculus transition system to a first-order LTS. In the first-order LTS, the grammar for labels is the same as that of the original LTS; however, for a named process (c, P) the only name that may be exported in a bound output is $c+1$; similarly only names that are below or equal to $c+1$ may be imported in an input transition. (Indeed, testing for all fresh names $b > c$ is unnecessary, doing it only for one ($b = c+1$) is enough.) This makes it possible to use the ordinary definition of bisimilarity for first-order LTS, and thus recover the early bisimilarity on the source terms.

$$\frac{P \xmapsto{\tau}_\pi P'}{(c,P) \xrightarrow{\tau} (c,P')} \qquad \frac{P \xmapsto{ab}_\pi P'}{(c,P) \xrightarrow{ab} (c,P')}\ b \leq c \qquad \frac{P \xmapsto{\overline{a}b}_\pi P'}{(c,P) \xrightarrow{\overline{a}b} (c,P')}\ b \leq c$$

$$\frac{P \xmapsto{ab}_\pi P'}{(c,P) \xrightarrow{ab} (b,P')}\ b = c+1 \qquad \frac{P \xmapsto{\overline{a}(b)}_\pi P'}{(c,P) \xrightarrow{\overline{a}(b)} (b,P')}\ b = c+1$$

We write π^1 for the first-order LTS derived from the above translation of the π-calculus. Although the labels of the source and target transitions have a similar shape, the LTS in π^1 is first-order because labels are taken as purely syntactic objects (without α-conversion). We write $\sim^{\mathrm{e}}$ and $\approx^{\mathrm{e}}$ for strong and weak early bisimilarity of the π-calculus.

Theorem 1. *Assume $c \geq \mathrm{fn}(P) \cup \mathrm{fn}(Q)$. Then we have: $P \sim^{\mathrm{e}} Q$ iff $(c, P) \sim (c, Q)$, and $P \approx^{\mathrm{e}} Q$ iff $(c, P) \approx (c, Q)$.*

The above full abstraction result allows us to import the theory of up-to techniques for first-order LTSs and bisimilarity, both in the strong and the weak case. We have however to prove the soundness of up-to techniques that are specific to the π-calculus. Function isub implements 'up to injective name substitutions':

$$\begin{aligned}\mathsf{isub}(\mathcal{R}) \triangleq \{((d, P\sigma), (d, Q\sigma)) \ \text{ s.t. } (c, P)\ \mathcal{R}\ (c, Q),\ \mathrm{fn}(P\sigma) \cup \mathrm{fn}(Q\sigma) \leq d,\\ \text{and } \sigma \text{ is injective on } \mathrm{fn}(P) \cup \mathrm{fn}(Q)\ \ \}\ .\end{aligned}$$

A subtle drawback is the need of another function manipulating names, str, allowing us to replace the index c in a named process (c, P) with a lower one:

$$\mathsf{str}(\mathcal{R}) \triangleq \{((d, P), (d, Q)) \ \text{ s.t. } (c, P)\ \mathcal{R}\ (c, Q) \text{ and } \mathrm{fn}(P, Q) \leq d\ \}\ .$$

Lemma 3. *The set $\{\mathsf{isub}, \mathsf{str}\}$ is compatible up to.*

The up-to-context function is decomposed into a set of smaller context functions, called *initial* [19], one for each operator of the π-calculus. The only exception to this is the input prefix, since early bisimilarity in the π-calculus is not preserved by this operator. We write $\mathcal{C}_o$, $\mathcal{C}_\nu$, $\mathcal{C}_!$, $\mathcal{C}_|$, and $\mathcal{C}_+$ for these initial context functions, respectively returning the closure of a relation under the operators of output prefix, restriction, replication, parallel composition, and sum.

Definition 4. *If $\mathcal{R}$ is a relation on π^1, we define $\mathcal{C}_o(\mathcal{R})$, $\mathcal{C}_\nu(\mathcal{R})$, $\mathcal{C}_!(\mathcal{R})$, $\mathcal{C}_|(\mathcal{R})$ and $\mathcal{C}_+(\mathcal{R})$ by saying that whenever $(c, P)\ \mathcal{R}\ (c, Q)$,*

- *$(c, \overline{a}b.P)\ \ \mathcal{C}_o(\mathcal{R})\ \ (c, \overline{a}b.Q)$, for any a, b with $a, b \leq c$,*
- *$(c, \nu a.P)\ \ \mathcal{C}_\nu(\mathcal{R})\ \ (c, \nu a.Q)$,*
- *$(c, !P)\ \ \mathcal{C}_!(\mathcal{R})\ \ (c, !Q)$;*

and, whenever $(c, P_1)\ \mathcal{R}\ (c, Q_1)$ and $(c, P_2)\ \mathcal{R}\ (c, Q_2)$,

- *$(c, P_1 \mid Q_1)\ \ \mathcal{C}_|(\mathcal{R})\ \ (c, P_2 \mid Q_2)$,*
- *$(c, P_1 + Q_1)\ \ \mathcal{C}_+(\mathcal{R})\ \ (c, P_2 + Q_2)$.*

While bisimilarity in the π-calculus is not preserved by input prefix, a weaker rule holds (where $=$ can be $\sim^{\mathrm{e}}$ or $\approx^{\mathrm{e}}$):

$$\frac{P = Q \quad \text{and } P\{c/b\} = Q\{c/b\} \text{ for each } c \text{ free in } P, Q}{a(b).P = a(b).Q} \tag{1}$$

We define $\mathcal{C}_i$, the function for input prefix, accordingly: we have $(d, a(b).P)\ \mathcal{C}_i(\mathcal{R})\ (d, a(b).Q)$ if $a \leq d$ and $(d+1, P\{c/b\})\ \mathcal{R}\ (d+1, Q\{c/b\})$ for all $c \leq d+1$.

Theorem 2. *The set $\{\mathcal{C}_o, \mathcal{C}_i, \mathcal{C}_\nu, \mathcal{C}_!, \mathcal{C}_|, \mathcal{C}_+\}$ is* **sp**-*compatible up to* $\mathsf{isub} \cup \mathsf{str}$.

Weak bisimilarity is not preserved by sums, only by guarded sums, whose function is $\mathcal{C}_{g+} \triangleq \mathcal{C}_{+}^{\omega} \circ (\mathcal{C}_o \cup \mathcal{C}_i)$.

Theorem 3. *The set* $\{\mathcal{C}_o, \mathcal{C}_i, \mathcal{C}_\nu, \mathcal{C}_!, \mathcal{C}_|, \mathcal{C}_{g+}\}$ *is* **wp**-*compatible up to* $\mathsf{isub} \cup \mathsf{str} \cup \mathbf{b}$ *where* $\mathbf{b} = (\mathcal{R} \mapsto \sim\mathcal{R}\sim)$ *is 'up to bisimilarity'.*

The compatibility of these functions is not a logical consequence of the up to context results in the π-calculus; instead we prove them from scratch [15], with the benefit of having a separate proof for each initial context.

As a byproduct of the compatibility of these initial context functions, and using Lemma 2, we derive the standard substitutivity properties of strong and weak early bisimilarity, including the rule (1) for input prefix.

Corollary 1. *In the* π-*calculus, relations* $\sim^{\mathrm{e}}$ *and* $\approx^{\mathrm{e}}$ *are preserved by the operators of output prefix, replication, parallel composition, restriction;* $\sim^{\mathrm{e}}$ *is also preserved by sum, whereas* $\approx^{\mathrm{e}}$ *is only preserved by guarded sums. Moreover, rule (1) is valid both for* $\sim^{\mathrm{e}}$ *and* $\approx^{\mathrm{e}}$.

Remark 1. Late bisimilarity makes use of transitions $P \xmapsto{a(b)}_{\pi} P'$ where b is bound, the definition of bisimulation containing a quantification over names. To capture this bisimilarity in a first-order LTS we would need to have two transitions for the input $a(b)$: one to fire the input a, leaving b uninstantiated, and another to instantiate b. While such a translation does yield full abstraction for both strong and weak late bisimilarities, the decomposition of an input transition into two steps prevents us from obtaining the compatibility of up to context.

4 Call-by-name λ-calculus

To study the applicability of our approach to higher-order languages, we investigate the pure call-by-name λ-calculus, referred to as ΛN in the sequel.

We use M, N to range over the set Λ of λ-terms, and x, y, z to range over variables. The standard syntax of λ-terms, and the rules for call-by-name reduction, are recalled in [15]. We assume the familiar concepts of free and bound variables and substitutions, and identify α-convertible terms. The only values are the λ-abstractions $\lambda x.M$. In this section and in the following one, results and definitions are presented on closed terms; extension to open terms is made using closing abstractions (i.e., abstracting on all free variables). The reduction relation of ΛN is $\longmapsto_{\mathsf{n}}$, and $\Longmapsto_{\mathsf{n}}$ its reflexive and transitive closure.

As bisimilarity for the λ-calculus we consider *environmental bisimilarity* [22,9], which allows a set of up-to techniques richer than Abramsky's applicative bisimilarity [2], even if the two notions actually coincide, together with contextual equivalence. Environmental bisimilarity makes a clear distinction between the tested terms and the environment. An element of an environmental bisimulation has, in addition to the tested terms M and N, a further component $\mathcal{E}$, the environment, which expresses the observer's current knowledge. When an input from the observer is required, the arguments supplied are terms that the observer can build using the current knowledge; that is, terms obtained by composing the values in $\mathcal{E}$ using the operators of the calculus. An *environmental relation* is a set of elements each of which is of the form $(\mathcal{E}, M, N)$ or $\mathcal{E}$, and where M, N are closed terms and $\mathcal{E}$ is a relation on closed values. We use $\mathcal{X}, \mathcal{Y}$ to range over environmental relations. In a triple $(\mathcal{E}, M, N)$ the relation component $\mathcal{E}$ is the *environment*, and M, N are the *tested terms*. We write $M \; \mathcal{X}_{\mathcal{E}} \; N$ for $(\mathcal{E}, M, N) \in \mathcal{X}$. We write $\mathcal{E}^\star$ for the closure of $\mathcal{E}$ under contexts. We only define the weak version of the bisimilarity; its strong version is obtained in the expected way.

Definition 5. *An environmental relation $\mathcal{X}$ is an* environmental bisimulation *if*

1. *$M \; \mathcal{X}_{\mathcal{E}} \; N$ implies:*
   (a) if $M \longmapsto_{\mathsf{n}} M'$ then $N \Longmapsto_{\mathsf{n}} N'$ and $M' \; \mathcal{X}_{\mathcal{E}} \; N'$;
   (b) if $M = V$ then $N \Longmapsto_{\mathsf{n}} W$ and $\mathcal{E} \cup \{(V, W)\} \in \mathcal{X}$ (V and W are values);
   (c) the converse of the above two conditions, on N;
2. *if $\mathcal{E} \in \mathcal{X}$ then for all $(\lambda x.P, \lambda x.Q) \in \mathcal{E}$ and for all $(M, N) \in \mathcal{E}^\star$ it holds that $P\{M/x\} \; \mathcal{X}_{\mathcal{E}} \; Q\{N/x\}$.*

Environmental bisimilarity, $\approx^{\mathsf{env}}$, is the largest environmental bisimulation.

For the translation of environmental bisimilarity to first-order, a few issues have to be resolved. For instance, an environmental bisimilarity contains both triples $(\mathcal{E}, M, N)$, and pure environments $\mathcal{E}$, which shows up in the difference between clauses (1) and (2) of Definition 5. Moreover, the input supplied to tested terms may be constructed using arbitrary contexts.

We write ΛN^1 for the first-order LTS resulting from the translation of ΛN. The states of ΛN^1 are sequences of λ-terms in which only the last one need not be a value. We use Γ and Δ to range over sequences of values only; thus (Γ, M) indicates a sequence of λ-values followed by M; and Γ_i is the i-th element in Γ.

For an environment $\mathcal{E}$, we write $\mathcal{E}_1$ for an ordered projection of the pairs in $\mathcal{E}$ on the first component, and $\mathcal{E}_2$ is the corresponding projection on the second component. In the translation, intuitively, a triple $(\mathcal{E}, M, N)$ of an environmental bisimulation is split into the two components $(\mathcal{E}_1, M)$ and $(\mathcal{E}_2, N)$. Similarly, an environment $\mathcal{E}$ is split into $\mathcal{E}_1$ and $\mathcal{E}_2$. We write $C[\Gamma]$ for the term obtained by replacing each hole $[\cdot]_i$ in C with the value Γ_i. The rules for transitions in ΛN^1 are as follows:

$$\frac{M \longmapsto_{\mathsf{n}} M'}{(\Gamma, M) \xrightarrow{\tau} (\Gamma, M')} \qquad \frac{\Gamma_i(C[\Gamma]) \longmapsto_{\mathsf{n}} M'}{\Gamma \xrightarrow{i,C} (\Gamma, M')} \tag{2}$$

The first rule says that if M reduces to M' in ΛN then M can also reduce in ΛN^1, in any environment. The second rule implements the observations in clause (2) of Definition 5: in an environment Γ (only containing values), any component Γ_i can be tested by supplying, as input, a term obtained by filling a context C with values from Γ itself. The label of the transition records the position i and the context chosen. As the rules show, the labels of ΛN^1 include the special label τ, and can also be of the form i, C where i is a integer and C a context.

Theorem 4. *$M \approx^{env}_{\mathcal{E}} N$ iff $(\mathcal{E}_1, M) \approx (\mathcal{E}_2, N)$ and $\mathcal{E} \in \approx^{env}$ iff $\mathcal{E}_1 \approx \mathcal{E}_2$.*

(The theorem also holds for the strong versions of the bisimilarities.) Again, having established full abstraction with respect to a first-order transition system and ordinary bisimilarity, we can inherit the theory of bisimulation enhancements. We have however to check up-to techniques that are specific to environmental bisimilarity. A useful such technique is 'up to environment', which allows us to replace an environment with a larger one; $\mathsf{w}(\mathcal{R})$ is the smallest relation that includes $\mathcal{R}$ and such that, whenever $(V, \Gamma, M)\ \mathsf{w}(\mathcal{R})\ (W, \Delta, N)$ then also $(\Gamma, M)\ \mathsf{w}(\mathcal{R})\ (\Delta, N)$, where V and W are any values. (Here w stands for 'weakening' as, from Lemmas 2 and 4, if $(V, \Gamma, M) \approx (W, \Delta, N)$ then $(\Gamma, M) \approx (\Delta, N)$.)

Lemma 4. *Function* w *is compatible.*

Somehow dual to weakening is the strengthening of the environment, in which a component of an environment can be removed. However this is only possible if the component removed is 'redundant', that is, it can be obtained by gluing other pieces of the environment within a context; strengthening is captured by the following str function: $(\Gamma, C_v[\Gamma], M)\ \mathsf{str}(\mathcal{R})\ (\Delta, C_v[\Delta], N)$ whenever $(\Gamma, M)\ \mathcal{R}\ (\Delta, N)$ and C_v is a value context (i.e., the outermost operator is an abstraction). We derive the compatibility up to of str in Theorem 5.

For up-to context, we need to distinguish between arbitrary contexts and evaluation contexts. There are indeed substitutivity properties, and corresponding up-to techniques, that only hold for the latter contexts. A hole $[\cdot]_i$ of a context C is in a *redex position* if the context obtained by filling all the holes but $[\cdot]_i$ with values is an evaluation context. Below C ranges over arbitrary contexts, whereas E ranges over contexts whose first hole is in redex position.

$$\begin{aligned} \mathcal{C}(\mathcal{R}) &\triangleq \{((\Gamma, C[\Gamma]), (\Delta, C[\Delta])) &&\text{s.t. } \Gamma\ \mathcal{R}\ \Delta\} \\ \mathcal{C}_{\mathsf{e}}(\mathcal{R}) &\triangleq \{((\Gamma, E[M, \Gamma]), (\Delta, E[N, \Delta])) &&\text{s.t. } (\Gamma, M)\ \mathcal{R}\ (\Delta, N)\} \end{aligned}$$

Theorem 5. *The set $\{\mathsf{str}, \mathcal{C}, \mathcal{C}_{\mathsf{e}}\}$ is* **sp***-compatible up to the identity function, and* **wp***-compatible up to* $\mathbf{wp} \cup \mathbf{e}$ *where* $\mathbf{e} \triangleq (\mathcal{R} \mapsto \gtrsim\mathcal{R}\lesssim)$ *is 'up to expansion'.*

For the proof, we establish the progression property separately for each function in $\{\mathsf{str}, \mathcal{C}, \mathcal{C}_{\mathsf{e}}\}$, using simple diagram-chasing arguments (together with induction on the structure of a context). Once more, the compatibility of the up to context functions entails also the substitutivity properties of environmental bisimilarity. In [22] the two aspects (substitutivity and up-to context) had to be proved separately, with similar proofs. Moreover the two cases of contexts (arbitrary contexts and evaluation contexts) had to be considered at the same time, within the same proof. Here, in contrast, the machinery of compatible function allows us to split the proof into two simpler proofs.

Remark 2. A transition system ensuring full abstraction as in Theorem 4 does not guarantee the compatibility of the up-to techniques specific to the language

$$M ::= x \mid MM \mid \nu\ell M \mid V \qquad V ::= \lambda x.M \mid \mathsf{set}_\ell \mid \mathsf{get}_\ell \qquad E ::= [\cdot] \mid EV \mid ME$$

$$\frac{}{(s;(\lambda x.M)V) \longmapsto_{\mathsf{R}} (s;M\{V/x\})} \qquad \frac{\ell \notin \mathsf{dom}(s)}{(s;\nu\ell\, M) \longmapsto_{\mathsf{R}} (s[\ell \mapsto I];M)}$$

$$\frac{\ell \in \mathsf{dom}(s)}{(s;\mathsf{get}_\ell V) \longmapsto_{\mathsf{R}} (s;s[\ell])} \qquad \frac{\ell \in \mathsf{dom}(s)}{(s;\mathsf{set}_\ell V) \longmapsto_{\mathsf{R}} (s[\ell \mapsto V];I)}$$

$$\frac{(s;M) \longmapsto_{\mathsf{R}} (s';M')}{(s;E[M]) \longmapsto_{\mathsf{R}} (s';E[M'])}$$

Fig. 1. The imperative λ-calculus

in consideration. For instance, a simpler and maybe more natural alternative to the second transition in (2) is the following one:

$$\frac{}{\Gamma \xrightarrow{i,C} (\Gamma, \Gamma_i(C[\Gamma]))} \tag{3}$$

With this rule, full abstraction holds, but up-to context is unsound: for any Γ and Δ, the singleton relation $\{(\Gamma,\Delta)\}$ is a bisimulation up to $\mathcal{C}$: indeed, using rule (3), the derivatives of the pair Γ,Δ are of the shape $\Gamma_i(C[\Gamma])$, $\Delta_i(C[\Delta])$, and they can be discarded immediately, up to the context $[\cdot]_i C$. If up-to context were sound then we would deduce that any two terms are bisimilar. (The rule in (2) prevents such a behaviour since it ensures that the tested values are 'consumed' immediately.)

5 Imperative call-by-value λ-calculus

In this section we study the addition of imperative features (higher-order references, that we call locations), to a call-by-value λ-calculus. It is known that finding powerful reasoning techniques for imperative higher-order languages is a hard problem. The language, ΛR, is a simplified variant of that in [10,22]. The syntax of terms, values, and evaluation contexts, as well as the reduction semantics are given in Figure 1. A λ-term M is run in a *store*: a partial function from locations to closed values, whose domain includes all free locations of both M and its own co-domain. We use letters s,t to range over stores. New store locations may be created using the operator $\nu\ell\, M$; the content of a store location ℓ may be rewritten using $\mathsf{set}_\ell V$, or read using $\mathsf{get}_\ell V$ (the former instruction returns a value, namely the identity $I \triangleq \lambda x.x$, and the argument of the latter one is ignored). We denote the reflexive and transitive closure of $\longmapsto_{\mathsf{R}}$ by $\Longmapsto_{\mathsf{R}}$.

Note that in contrast with the languages in [10,22], locations are not directly first-class values; the expressive power is however the same: a first-class location ℓ can always be encoded as the pair $(\mathsf{get}_\ell, \mathsf{set}_\ell)$.

We present the first-order LTS for ΛR, and then we relate the resulting strong and weak bisimilarities directly with contextual equivalence (the reference equivalence in λ-calculi). Alternatively, we could have related the first-order bisimilarities to the environmental bisimilarities of ΛR, and then inferred the correspondence with contextual equivalence from known results about environmental bisimilarity, as we did for ΛN.

We write $(s; M) \downarrow$ when M is a value; and $(s; M) \Downarrow$ if $(s; M) \Longmapsto_{\mathsf{R}}\downarrow$. For the definition of contextual equivalence, we distinguish the cases of values and of arbitrary terms, because they have different substitutivity properties: values can be tested in arbitrary contexts, while arbitrary terms must be tested only in evaluation contexts. As in [22], we consider contexts that do not contain free locations (they can contain bound locations). We refer to [22] for more details on these aspects.

Definition 6. – *For values V, W, we write $(s; V) \equiv (t; W)$ when $(s; C[V])\Downarrow$ iff $(t; C[W])\Downarrow$, for all location-free context C.*
– *For terms M and N, we write $(s; M) \equiv (t; N)$ when $(s; E[M])\Downarrow$ iff $(t; E[N])\Downarrow$, for all location-free evaluation context E.*

We now define ΛR^1, the first-order LTS for ΛR. The states and the transitions for ΛR^1 are similar to those for the pure λ-calculus of Section 4, with the addition of a component for the store. The two transitions (2) of call-by-name λ-calculus become:

$$\frac{(s; M) \longmapsto_{\mathsf{R}} (s'; M')}{(s; \Gamma, M) \xrightarrow{\tau} (s'; \Gamma, M')} \qquad \frac{\Gamma' = \Gamma, \mathsf{getset}(r) \quad (s \uplus r[\Gamma']; \Gamma_i(C[\Gamma'])) \longmapsto_{\mathsf{R}} (s'; M')}{(s; \Gamma) \xrightarrow{i, C, \mathsf{cod}(r)} (s'; \Gamma', M')}$$

The first rule is the analogous of the first rule in (2). The important differences are on the second rule. First, since we are *call-by-value*, C now ranges over $\mathbb{C}_v$, the set of *value contexts* (i.e., contexts of the form $\lambda x.C'$) without free locations. Moreover, since we are now *imperative*, in a transition we must permit the creation of new locations, and a term supplied by the environment should be allowed to use them. In the rule, the new store is represented by r (whose domain has to be disjoint from that of s). Correspondingly, to allow manipulation of these locations from the observer, for each new location ℓ we make set_ℓ and get_ℓ available, as an extension of the environment; in the rule, these are collectively written $\mathsf{getset}(r)$, and Γ' is the extended environment. Finally, we must initialise the new store, using terms that are created out of the extended environment Γ'; that is, each new location ℓ is initialised with a term $D_\ell[\Gamma']$ (for $D_\ell \in \mathbb{C}_v$). Moreover, the contexts D_ℓ chosen must be made visible in the label of the transition. To take care of these aspects, we view r as a *store context*, a tuple of assignments $\ell \mapsto D_\ell$. Thus the initialisation of the new locations is written $r[\Gamma']$; and, denoting by $\mathsf{cod}(r)$ the tuple of the contexts D_ℓ in r, we add $\mathsf{cod}(r)$ to the label of the transition. Note also that, although C and D_ℓ are location-free, their holes may be instantiated with terms involving the set_ℓ and get_ℓ operators, and these allow manipulation of the store.

Once more, on the (strong and weak) bisimilarities that are derived from this first-order LTS we can import the theory of compatible functions and bisimulation enhancements. Concerning additional up-to functions, specific to ΛR, the functions w, str, $\mathcal{C}$ and $\mathcal{C}_{\mathsf{e}}$ are adapted from Section 4 in the expected manner—contexts C_v, C and E must be location-free. A further function for ΛR is store, which manipulates the store by removing locations that do not appear elsewhere (akin to garbage collection); thus, $\mathsf{store}(\mathcal{R})$ is the set of all pairs

$$((s \uplus r[\Gamma']; \Gamma', M), (t \uplus r[\Delta']; \Delta', N))$$

such that $(s; \Gamma, M) \mathrel{\mathcal{R}} (t; \Delta, N)$, and with $\Gamma' = \Gamma, \mathsf{getset}(r)$ and $\Delta' = \Delta, \mathsf{getset}(r)$.

Lemma 5. *The set* $\{\mathsf{w}, \mathsf{str}, \mathcal{C}_{\mathsf{e}}, \mathsf{store}, \mathcal{C}\}$ *is* **sp**-*compatible up to the identity function and is* **wp**-*compatible up to* $\mathbf{wp} \cup \mathbf{e}$.

The techniques $\mathcal{C}$ and $\mathcal{C}_{\mathsf{e}}$ allow substitutivity under location-free contexts, from which we can derive the soundness part of Theorem 6.

Theorem 6. $(s; M) \equiv (t; N)$ *iff* $(s; M) \approx (t; N)$.

Proof (sketch). Soundness ($\Leftarrow$) follows from congruence by $\mathcal{C}_{\mathsf{e}}$ (Lemmas 5 and 2) and completeness ($\Rightarrow$) is obtained by standard means. See [15] for details.

Note that substitutivity of bisimilarity is restricted either to values ($\mathcal{C}$), or to evaluation contexts ($\mathcal{C}_{\mathsf{e}}$). The following lemma provides a sufficient condition for a given law between arbitrary terms to be preserved by arbitrary contexts.

Lemma 6. *Let* $\asymp$ *be any of the relations* $\sim, \approx$, *and* $\gtrsim$. *Suppose* L, R *are* ΛR *terms with* $(s; \Gamma, L) \asymp (s; \Gamma, R)$ *for all environments* Γ *and stores* s. *Then also* $(s; \Gamma, C[L]) \asymp (s; \Gamma, C[R])$, *for any store* s, *environment* Γ *and context* C.

Proof (sketch). We first prove a simplified result in which C is an evaluation context, using techniques $\mathcal{C}_{\mathsf{e}}$ and store. We then exploit this partial result together with up-to expansion to derive the general result. See [15] for more details.

We use this lemma at various places in the example we cover in Section 6. For instance we use it to replace a term $N_1 \triangleq (\lambda x.E[x])M$ (with E an evaluation context) with $N_2 \triangleq E[M]$, under an arbitrary context. Such a property is delicate to prove, even for closed terms, because the evaluation of M could involve reading from a location of the store that itself could contain occurrences of N_1 and N_2.

6 An Example

We conclude by discussing an example from [10]. It consists in proving a law between terms of ΛR extended with integers, operators for integer addition and subtraction, and a conditional—those constructs are straightforward to accommodate in the presented framework. For readability, we also use the standard notation for store assignment, dereferencing and sequence: $(\ell := M) \triangleq \mathsf{set}_\ell M$, $!\ell \triangleq \mathsf{get}_\ell I$, and $M; N \triangleq (\lambda x.N)M$ where x does not appear in N. The two terms are the following ones:

- $M \triangleq \lambda g.\nu\ell\ \ell := 0; g(\mathsf{incr}_\ell);\ \mathtt{if}\ !\ell\ \mathtt{mod}\ 2 = 0\ \mathtt{then}\ I\ \mathtt{else}\ \Omega$
- $N \triangleq \lambda g.g(F); I$,

where $\mathsf{incr}_\ell \triangleq \lambda z.\ell := !\ell + 2$, and $F \triangleq \lambda z.I$. Intuitively, those two terms are weakly bisimilar because the location bound by ℓ in the first term will always contain an even number.

This example is also considered in [22] where it is however modified to fit the up-to techniques considered in that paper. The latter are less powerful than those available here thanks to the theory of up-to techniques for first-order LTSs (e.g., up to expansion is not considered in [22]—its addition to environmental bisimulations is non-trivial, having stores and environments as parameters).

We consider two proofs of the example. In comparison with the proof in [22]: (i) we handle the original example from [10], and (ii) the availability of a broader set of up-to techniques and the possibility of freely combining them allows us to work with smaller relations. In the first proof we work up to the store (through the function store) and up to expansion—two techniques that are not available in [22]. In the second proof we exploit the up-to-transitivity technique of Section 2, which is only sound for strong bisimilarity, to further reduce the size of the relation we work with.

First proof. We first employ Lemma 6 to reach a variant similar to that of [22]: we make a 'thunk' out of the test in M, and we make N look similar. More precisely, let $\mathsf{test}_\ell \triangleq \lambda z.\mathtt{if}\ !\ell\ \mathtt{mod}\ 2 = 0\ \mathtt{then}\ I\ \mathtt{else}\ \Omega$, we first prove that

- $M \approx M' \triangleq \lambda g.\nu\ell\ \ell := 0; g(\mathsf{incr}_\ell); \mathsf{test}_\ell I$, and
- $N \approx N' \triangleq \lambda g.g(F); FI$.

It then suffices to prove that $M' \approx N'$, which we do using the following relation:

$$\mathcal{R} \triangleq \left\{ \left(s, M', (\mathsf{incr}_\ell, \mathsf{test}_\ell)_{\ell \in \tilde{\ell}}\right), \left(\emptyset, N', (F, F)_{\ell \in \tilde{\ell}}\right) \ \text{ s.t. } \forall \ell \in \tilde{\ell},\ \ s(\ell) \text{ is even} \right\} .$$

The initial pair of terms is generalised by adding any number of private locations, since M' can use itself to create more of them. Relation $\mathcal{R}$ is a weak bisimulation up to store, $\mathcal{C}$ and expansion. More details can be found in [15].

Second proof. Here we also preprocess the terms using Lemma 6, to add a few artificial internal steps to N, so that we can carry out the reminder of the proof using strong bisimilarity, which enjoys more up-to techniques than weak bisimilarity:

- $M \approx M' \triangleq \lambda g.\nu\ell\ \ell := 0; g(\mathsf{incr}_\ell); \mathsf{test}_\ell I$,
- $N \approx N'' \triangleq \lambda g.I; I; g(\mathsf{incr}_0); \mathsf{test}_0 I$.

where incr_0 and test_0 just return I on any input, taking the same number of internal steps as incr_ℓ and test_ℓ. We show that $M' \sim N''$ by proving that the following relation $\mathcal{R}$ is a strong bisimulation *up to unfolding, store, weakening, strengthening, transitivity and context* (a technique unsound in the weak case):

$$\mathcal{S} \triangleq \{(M', N'')\} \cup \{(\ell \mapsto 2n, \mathsf{incr}_\ell, \mathsf{test}_\ell), (\emptyset, \mathsf{incr}_0, \mathsf{test}_0) \ \text{ s.t. } \ n \in \mathbb{N}\}$$

This relation uses a single location; there is one pair for each integer that can be stored in the location. In the diagram-chasing arguments for $\mathcal{S}$, essentially a pair of derivatives is proved to be related under the function

$$\mathbf{sp} \circ \mathbf{sp} \circ \mathsf{star} \circ (\mathsf{str} \cup \mathsf{store} \cup \mathcal{C} \cup \mathsf{w})^{\omega}$$

where $\mathsf{star} : \mathcal{R} \mapsto \mathcal{R}^{\star}$ is the reflexive-transitive closure function. (Again, we refer to [15] for more details.)

The difference between the relation $\mathcal{R}$ in the first proof and the proofs in [10,22] is that $\mathcal{R}$ only requires locations that appear free in the tested terms; in contrast, the relations in [10,22] need to be closed under all possible extensions of the store, including extensions in which related locations are mapped onto arbitrary context-closures of related values. We avoid this thanks to the up-to store function. The reason why, both in [10,22] and in the first proof above, several locations have to be considered is that, with bisimulations akin to environmental bisimulation, the input for a function is built using the values that occur in the candidate relation. In our example, this means that the input for a function can be a context-closure of M and N; hence uses of the input may cause several evaluations of M and N, each of which generates a new location. In this respect, it is surprising that our second proof avoids multiple allocations (the candidate relation $\mathcal{S}$ only mentions one location). This is due to the massive combination of up-to techniques whereby, whenever a new location is created, a double application of up to context (the 'double' is obtained from up-to transitivity) together with some administrative work (given by the other techniques) allows us to absorb the location.

Acknowledgement. The authors acknowledge support from the ANR projects 2010-BLAN-0305 PiCoq and 12IS02001 PACE.

References

1. Abadi, M., Gordon, A.D.: A bisimulation method for cryptographic protocols. In: Hankin, C. (ed.) ESOP 1998. LNCS, vol. 1381, pp. 12–26. Springer, Heidelberg (1998)
2. Abramsky, S.: The lazy lambda calculus. In: Turner, D. (ed.) Research Topics in Functional Programming, pp. 65–116. Addison-Wesley (1989)
3. Arun-Kumar, S., Hennessy, M.: An efficiency preorder for processes. Acta Informatica 29, 737–760 (1992)
4. Chaudhuri, K., Cimini, M., Miller, D.: Formalization of the bisimulation-up-to technique and its meta theory. Draft (2014)
5. Hirschkoff, D.: A full formalisation of pi-calculus theory in the calculus of constructions. In: Gunter, E.L., Felty, A.P. (eds.) TPHOLs 1997. LNCS, vol. 1275, pp. 153–169. Springer, Heidelberg (1997)
6. Hur, C.-K., Neis, G., Dreyer, D., Vafeiadis, V.: The power of parameterization in coinductive proof. In: POPL, pp. 193–206. ACM (2013)

7. Jeffrey, A., Rathke, J.: Towards a theory of bisimulation for local names. In: LICS, pp. 56–66 (1999)
8. Koutavas, V., Hennessy, M.: First-order reasoning for higher-order concurrency. Computer Languages, Systems & Structures 38(3), 242–277 (2012)
9. Koutavas, V., Levy, P.B., Sumii, E.: From applicative to environmental bisimulation. Electr. Notes Theor. Comput. Sci. 276, 215–235 (2011)
10. Koutavas, V., Wand, M.: Small bisimulations for reasoning about higher-order imperative programs. In: POPL 2006, pp. 141–152. ACM (2006)
11. Lassen, S.B.: Relational reasoning about contexts. In: Higher-order Operational Techniques in Semantics, pp. 91–135. Cambridge University Press (1998)
12. Lassen, S.B.: Relational Reasoning about Functions and Nondeterminism. PhD thesis, Department of Computer Science, University of Aarhus (1998)
13. Lassen, S.B.: Bisimulation in untyped lambda calculus: Böhm trees and bisimulation up to context. Electr. Notes Theor. Comput. Sci. 20, 346–374 (1999)
14. Lenisa, M.: Themes in Final Semantics. Ph.D. thesis, Universitá di Pisa (1998)
15. Madiot, J.-M., Pous, D., Sangiorgi, D.: Web appendix to this paper, `http://hal.inria.fr/hal-00990859`
16. Merro, M., Nardelli, F.Z.: Behavioral theory for mobile ambients. J. ACM 52(6), 961–1023 (2005)
17. Milner, R.: Communication and Concurrency. Prentice Hall (1989)
18. Pohjola, J.Å., Parrow, J.: Bisimulation up-to techniques for psi-calculi. Draft (2014)
19. Pous, D., Sangiorgi, D.: Enhancements of the bisimulation proof method. In: Advanced Topics in Bisimulation and Coinduction. Cambridge University Press (2012)
20. Rot, J., Bonsangue, M., Rutten, J.: Coalgebraic bisimulation-up-to. In: van Emde Boas, P., Groen, F.C.A., Italiano, G.F., Nawrocki, J., Sack, H. (eds.) SOFSEM 2013. LNCS, vol. 7741, pp. 369–381. Springer, Heidelberg (2013)
21. Sangiorgi, D.: On the bisimulation proof method. J. of MSCS 8, 447–479 (1998)
22. Sangiorgi, D., Kobayashi, N., Sumii, E.: Environmental bisimulations for higher-order languages. ACM Trans. Program. Lang. Syst. 33(1), 5 (2011)
23. Sangiorgi, D., Walker, D.: The Pi-Calculus: a theory of mobile processes. Cambridge University Press (2001)
24. Sumii, E., Pierce, B.C.: A bisimulation for dynamic sealing. Theor. Comput. Sci. 375(1-3), 169–192 (2007)
25. Sumii, E., Pierce, B.C.: A bisimulation for type abstraction and recursion. J. ACM 54(5) (2007)
26. Turner, N.D.: The polymorphic pi-calculus: Theory and Implementation. PhD thesis, Department of Computer Science, University of Edinburgh (1996)

Parameterized Model Checking of Rendezvous Systems*

Benjamin Aminof[1], Tomer Kotek[2], Sasha Rubin[2], Francesco Spegni[3], and Helmut Veith[2]

[1] IST Austria
[2] TU Wien, Austria
[3] UnivPM Ancona, Italy

Abstract. A standard technique for solving the parameterized model checking problem is to reduce it to the classic model checking problem of finitely many finite-state systems. This work considers some of the theoretical power and limitations of this technique. We focus on concurrent systems in which processes communicate via pairwise rendezvous, as well as the special cases of disjunctive guards and token passing; specifications are expressed in indexed temporal logic without the next operator; and the underlying network topologies are generated by suitable Monadic Second Order Logic formulas and graph operations. First, we settle the exact computational complexity of the parameterized model checking problem for some of our concurrent systems, and establish new decidability results for others. Second, we consider the cases that model checking the parameterized system can be reduced to model checking some fixed number of processes, the number is known as a cutoff. We provide many cases for when such cutoffs can be computed, establish lower bounds on the size of such cutoffs, and identify cases where no cutoff exists. Third, we consider cases for which the parameterized system is equivalent to a single finite-state system (more precisely a Büchi word automaton), and establish tight bounds on the sizes of such automata.

1 Introduction

Many concurrent systems consist of an arbitrary number of identical processes running in parallel, possibly in the presence of an environment or control process. The parameterized model checking problem (PMCP) for concurrent systems is to decide if a given temporal logic specification holds irrespective of the number of participating processes.

Although the PMCP is undecidable in general (see [12,6]) for some combinations of communication primitives, network topologies, and specification languages, it is often proved decidable by a reduction to model checking finitely

* The second, third, fourth and fifth authors were supported by the Austrian National Research Network S11403-N23 (RiSE) of the Austrian Science Fund (FWF) and by the Vienna Science and Technology Fund (WWTF) through grants PROSEED, ICT12-059, and VRG11-005.

P. Baldan and D. Gorla (Eds.): CONCUR 2014, LNCS 8704, pp. 109–124, 2014.

many finite-state systems [9,5,6,3,1]. In many of these cases it is even possible to reduce the problem of whether a parameterized system satisfies a temporal specification for any number of processes to the same problem for systems with at most c processes. The number c is known as a *cutoff* for the parameterized system. In other cases the reduction produces a single finite-state system, often in the form of an automaton such as a Büchi automaton, that represents the set of all execution traces of systems of all sizes.

The goal of this paper is to better understand the power and limitations of these techniques, and this is done by addressing three concrete questions.

Question 1: For which combinations of communication primitive, specification language, and network topologies is the PMCP decidable? In the decidable cases, what is the computational complexity of the PMCP?

In case a cutoff c exists, the PMCP is decidable by a reduction to model checking c many finite-state systems. The complexity of this procedure depends on the size of the cutoff. Thus we ask:

Question 2: When do cutoffs exist? In case a cutoff exists, can it be computed? And if so, what is a lower bound on the cutoff?

The set of execution traces of a parameterized system (for a given process type P) is defined as the projection onto the local states of P of all (infinite) runs of systems of all sizes.[1] In case this set is ω-regular, one can reduce the PMCP of certain specifications (including classic ones such as coverability) to the language containment problem for automata (this is the approach taken in [9, Section 4]). Thus we ask:

Question 3: Is the set of executions of the system ω-regular? If so, what is a lower bound on the sizes of the non-deterministic Büchi word automata recognizing the set of executions?

System Model. In order to model and verify a concurrent system we should specify three items: (i) the communication primitive, (ii) the specification language, and (iii) the set of topologies.

We focus on concurrent systems in which processes communicate via pairwise rendezvous [9], as well as two other communication primitives (expressible in terms of pairwise rendezvous), namely disjunctive guards [5] and token-passing systems [6,3,1]. Two special cases are systems with one process template U (in other words, all processes run the same code), and systems with two process templates C, U in which there is exactly one copy of C; in other words, all processes run the same code, except for one (which is called the controller).

Specifications of parameterised systems are typically expressed in indexed temporal logic [2] which allows one to quantify over processes (e.g., $\forall i \neq j.\ \mathsf{AG}\ (\neg(\text{critical}, i) \vee \neg(\text{critical}, j))$ says that no two processes are in their critical sections at the same time). We focus on a fragment of this logic where the process quantifiers only appear at the front of a temporal logic formula — allowing the

[1] Actually we consider the destuttering of this set, as explained in Section 2.5.

process quantifiers to appear in the scope of path quantifiers results in undecidability even with no communication between processes [10].

The sets of topologies we consider all have either bounded tree-width, or more generally bounded clique-width, and are expressible in one of three ways. (1) Using MSO, a powerful and general formalism for describing sets of topologies, which can express e.g. planarity, acyclicity and ℓ-connectivity. (2) As *iteratively constructible* sets of topologies, an intuitive formalism which creates graph sequences by iterating graph operations [8]. Many typical classes of topologies (e.g., all rings, all stars, all cliques) are iteratively constructible. (3) As *clique-like* sets of topologies, which includes the set of cliques and the set of stars, but excludes the set of rings. Iteratively constructible and clique-like sets of topologies are MSO-definable, the former in the presence of certain auxiliary relations.

Prior Work and Our Contributions. For each communication primitive (rendezvous, disjunctive guards, token passing) and each question (decidability and complexity, cutoffs, equivalent automata) we summarise the known answers and our contributions. Obviously, the breadth of questions along these axis is great, and we had to limit our choices as to what to address. Thus, this article is not meant to be a comprehensive taxonomy of PMCP. That is, it is not a mapping of the imaginary hypercube representing all possible choices along these axis. Instead, we started from the points in this hypercube that represent the most prominent known results and, guided by the three main questions mentioned earlier, have explored the unknown areas in each point's neighborhood.

Pairwise Rendezvous.

Decidability and Complexity: The PMCP for systems which communicate by pairwise rendezvous, on clique topologies, with a controller C, for 1-index $\mathsf{LTL}\backslash\mathsf{X}$ specifications is EXPSPACE-complete [9,7] (and PSPACE without a controller [9, Section 4]). We show the PMCP is undecidable if we allow the more general 1-index $\mathsf{CTL}^*\backslash\mathsf{X}$ specifications. Thus, for the results on pairwise rendezvous we fix the specification language to be 1-index $\mathsf{LTL}\backslash\mathsf{X}$. We introduce sets of topologies that naturally generalise cliques and stars, and exclude rings (the PMCP is already undecidable for uni-directional rings and 1-index safety specifications [12,6]), which we call *clique-like* sets of topologies, and show that the PMCP of 1-index $\mathsf{LTL}\backslash\mathsf{X}$ on clique-like topologies is EXPSPACE-complete (PSPACE-complete without a controller). We also prove that the program complexity is EXPSPACE-complete (respectively PTIME).

Cutoffs: We show that even for clique topologies there are not always cutoffs.

Equivalent automata: We prove that the set of (destuttered) executions of systems with a controller are not, in general, ω-regular, already for clique topologies. On the other hand, we extend the known result that the set of (destuttered) executions for systems with only user processes U (i.e., without a controller) is ω-regular for clique topologies [9] to clique-like topologies, and give an effective construction of the corresponding Büchi automaton.

Disjunctive Guards.

In this section we focus on clique topologies and 1-index LTL\X specifications. Though we sometimes consider more general cases (as in Theorem 10), we postpone these cases for future work.

Decidability and Complexity: We show the PMCP is undecidable if we allow 1-index CTL*\X specifications, already for clique topologies. We prove that for systems with a controller the complexity of the PMCP is PSPACE-complete and the program complexity is coNP-complete, whereas for systems without a controller the complexity is PSPACE-complete and the program complexity is in PTIME. We note that the PTIME and PSPACE upper bounds follow from [9,5], although we improve the time complexity for the case with a controller.

Cutoffs: Cutoffs exist for such systems and are of size $|U| + 2$ [5]. We prove these cutoffs are tight.

Equivalent automaton: We prove that the set of (destuttered) executions is accepted by an effectively constructible Büchi automaton of size $O(|C| \times 2^{|U|})$. It is very interesting to note that this size is smaller than the smallest system size one gets (in the worst-case) from the cutoff result, namely $|C| \times |U|^{|U|+2}$. Hence, the PMCP algorithm obtained from the cutoff is less efficient than the one obtained from going directly to a Büchi automaton. As far as we know, this is the first theoretical proof of the existence of this phenomenon. We also prove that, in general, our construction is optimal, i.e., that in some cases every automaton for the set of (destuttered) executions must be of size $2^{\Omega(|U|+|C|)}$.

Token Passing Systems.

In this section we focus on MSO-definable set of topologies of bounded tree-width or clique-width, as well as on iteratively-constructible sets of topologies.

Decidability and Complexity: We prove that the PMCP is decidable for indexed CTL*\Xon such topologies. This considerably generalises the results of [1], where decidability for this logic was shown for a few concrete topologies such as rings and cliques.

Cutoffs: For the considered topologies and indexed CTL*\X we prove that the PMCPs have *computable* cutoffs. From [1] we know that there is a (computable) set of topologies and a system template such that there is no algorithm that given an indexed CTL*\X formula can compute the associated cutoff (even though a cutoff for *the given formula* always exists). This justifies our search of sets of topologies for which the PMCP for CTL*\X has computable cutoffs. We also give a lower bound on cutoffs for iteratively-constructible sets and indexed LTL\X.

Equivalent automaton: Our ability to compute cutoffs for 1-index LTL\X formulas and the considered topologies implies that the (destuttered) sets of execution traces are ω-regular, and the construction of Büchi automata which compute these traces is effective.

Due to space limitations, in many cases proofs/sketches are not given, and only a statement of the basic technique used for the proof is given. The reader is referred to the full version of the article for more details.

2 Definitions and Preliminaries

A *labeled transition system (LTS)* is a tuple $(S, R, I, \Phi, \mathsf{AP}, \Sigma)$, where S is the set of *states*, $R \subseteq S \times \Sigma \times S$ is the *transition relation*, $I \subseteq S$ are the *initial states*, $\Phi : S \to 2^{\mathsf{AP}}$ is the *state-labeling*, AP is a set of *atomic propositions* or *atoms*, and Σ is the *transition-labels alphabet*. When AP and Σ are clear from the context we drop them. A *finite LTS* is an LTS in which S, R, Σ are finite and $\Phi(s)$ is finite for every $s \in S$. Transitions $(s, a, s') \in R$ may be written $s \xrightarrow{\mathsf{a}} s'$. A *transition system (TS)* (S, R, I, Σ) is an LTS without the labeling function and without the set of atomic propositions. A *run* is an infinite path that starts in an initial state. For a formal definition of path, state-labeled path, action-labeled path, refer to the full version of this paper.

2.1 Process Template, Topology, Pairwise Rendezvous System

We define how to (asynchronously) compose processes that communicate via pairwise rendezvous into a single system. We consider time as being discrete (i.e. not continuous). Processes are not necessarily identical, but we assume only a finite number of different process types. Roughly, at every vertex of a topology (a directed graph with vertices labeled by process types) there is a process of the given type running; at every time step either, and the choice is nondeterministic, exactly one process makes an internal transition, or exactly two processes with an edge between them in the topology instantaneously synchronize on a message (sometimes called an action) $\mathsf{m} \in \Sigma_{\mathsf{sync}}$. The sender of the message m performs an $\mathsf{m}!$ transition, and the receiver an $\mathsf{m}?$ transition. Note that the sender can not direct the message to a specific neighbouring process (nor can the receiver choose from where to receive it), but the pair is chosen non-deterministically. [2]

Fix a countable set of atoms (also called atomic propositions) $\mathsf{AP}_{\mathsf{pr}}$. Fix a finite synchronization alphabet Σ_{sync} (that does not include the symbol τ), and define the *communication alphabet* $\Sigma = \{\mathsf{m}!, \mathsf{m}? \mid \mathsf{m} \in \Sigma_{\mathsf{sync}}\}$.

Process Template, System Arity, System Template. A *process template* is a finite LTS $P = (S, R, \{\iota\}, \Phi, \mathsf{AP}_{\mathsf{pr}}, \Sigma \cup \{\tau\})$. Since $\mathsf{AP}_{\mathsf{pr}}$ and the transition-labels alphabet are typically fixed, we will omit them. The *system arity* is a natural number $r \in \mathbb{N}$. It refers to the number of different process types in the system. A *(r-ary) system template* is a tuple of process templates $\overline{P} = (P_1, \cdots, P_r)$ where r is the system arity. The process template $P_i = (S_i, R_i, \{\iota_i\}, \Phi_i)$ is called the *ith process template*.

Topology G. An *r-topology* is a finite structure $G = (V, E, T_1, \cdots, T_r)$ where $E \subseteq V \times V$, and the $T_i \subseteq V$ partition V. The *type* of $v \in V$ denoted *type*(v) is the unique j such that $v \in T_j$. We might write V_G, E_G and type_G to stress G.

We sometimes assume that $V := \{1, \cdots, n\}$ for some $n \in \mathbb{N}$. For instance, an *r-ary clique topology* with $V = \{1, \cdots, n\}$ has $E = \{(i, j) \in [n]^2 \mid i \neq j\}$ (and some partition of the nodes into sets $T_1, \cdots, T_r$); and the 1*-ary ring topology* with $V = \{1, \cdots, n\}$ has $E = \{(i, j) \in [n]^2 \mid j = i + 1 \mod n\}$ and $T_1 = V$.

[2] In models in which we allow processes to send in certain directions, e.g., send left and send right in a bi-directional ring, then PMCP is quickly undecidable [1].

(Pairwise-Rendezvous) System. Given system arity r, system template $\overline{P} = (P_1, \cdots, P_r)$ with $P_i = (S_i, R_i, \{\iota_i\}, \Phi_i)$, and r-topology $G = (V, E, \overline{T})$, define the *system* $\overline{P}^G$ as the LTS $(Q, \Delta, Q_0, \Lambda, \mathsf{AP}_{\mathsf{pr}} \times V, \Sigma_{\mathsf{sync}} \cup \{\tau\})$ where

- The set Q is the set of functions $f : V \to \cup_{i \leq r} S_i$ such that $f(v) \in S_i$ iff $type(v) = i$ (for all $v \in V, i \leq r$). Such functions (sometimes written as vectors) are called *configurations*.
- The set Q_0 consists of the unique *initial configuration* f_ι defined as $f_\iota(v) = \iota_{type(v)}$ (for all $v \in V$).
- The set of *global transitions* Δ are tuples $(f, \mathsf{m}, g) \in Q \times (\Sigma_{\mathsf{sync}} \cup \{\tau\}) \times Q$ where one of the following two conditions hold:
  - $\mathsf{m} = \tau$ and there exists $v \in V$ such that $f(v) \xrightarrow{\tau} g(v)$ is a transition of the process template $P_{type(v)}$, and for all $w \neq v$, $f(w) = g(w)$; this is called an *internal transition*,
  - $\mathsf{m} \in \Sigma_{\mathsf{sync}}$ and there exists $v \neq w \in V$ with $(v, w) \in E$ such that $f(v) \xrightarrow{\mathsf{m}!} g(v)$ and $f(w) \xrightarrow{\mathsf{m}?} g(w)$ and for all $z \notin \{v, w\}$, $f(z) = g(z)$; this is called a *synchronous transition*. We say that the process at v *sends the message* m and the process at w *receives the message* m.
- The labeling function $\Lambda : Q \to 2^{\mathsf{AP}_{\mathsf{pr}} \times V}$ is defined by $(p, v) \in \Lambda(f) \iff p \in \Phi_{type(v)}(f(v))$ (for all configurations f, atoms $p \in \mathsf{AP}_{\mathsf{pr}}$ and vertices $v \in V$).

In words then, a topology of size n specifies n-many processes, which processes have the same type, and how the processes are connected. In the internal transition above only the process at vertex v makes a transition, and in the synchronous transition above only the process at vertex v and its neighbour at w make a transition. Let $\pi = f_0 f_1 \cdots$ be a state-labeled path in $\overline{P}^G$. The *projection of π to vertex $v \in V$*, written $proj_v(\pi)$, is the sequence $f_0(v) f_1(v) \cdots$ of states of $P_{type(v)}$. If $type(v) = j$ we say that the *vertex v runs (a copy of) the process P_j*, or that *the process at v is P_j*.

2.2 Disjunctively-Guarded System, and Token Passing System

We define guarded protocols and token-passing systems as restricted forms of pairwise rendezvous systems. In fact, the restrictions are on the system template and the synchronization alphabet. Write $P_i = (S_i, R_i, \{\iota_i\}, \Phi_i, \mathsf{AP}_{\mathsf{pr}}, \Sigma \cup \{\tau\})$.

Disjunctively-Guarded System Template. A system $\overline{P}^G$ is *disjunctively-guarded* if $\overline{P}$ is. A system template $\overline{P}$ is *disjunctively-guarded* if **(i)** The state sets of the process templates are pairwise disjoint, i.e., $S_i \cap S_j = \emptyset$ for $1 \leq i < j \leq r$. **(ii)** The transition-labels alphabet Σ is $\{\tau\} \cup \{\mathsf{q}!, \mathsf{q}? \mid q \in \cup_{i \leq r} S_i\}$ **(iii)** For every state $s \in S$, there is a transition labeled $s \xrightarrow{\mathsf{s}?} s$. **(iv)** For every state $s \in S$, the only transitions labeled $\mathsf{s}?$ are of the form $s \xrightarrow{\mathsf{s}?} s$. Intuitively, in this kind of systems a process can decide to move depending on the local state of some neighbor process, but it cannot relate the state of any two processes at the same time, nor it can force another process to move from its local state. Our definition of disjunctively-guarded systems on a clique topology is a reformulation of the definition of concrete system in [5, Section 2].

Token Passing System. One can express a token passing system (TPS) as a special case of pairwise rendezvous. In this work we only consider the case of a single valueless token, whose formal definition can be found in [1,6]. A token passing system (TPS) $\overline{P}^G$ can be thought of the asynchronous parallel composition of the processes templates in $\overline{P}$ over topology G according to the types of vertices. At any time during the computation, exactly one vertex has the token. The token starts with the unique process in P_1. At each time step either exactly one process makes an internal transition, or exactly two processes synchronize when one process sends the token to another along an edge of G.

2.3 Indexed Temporal Logic

We assume the reader is familiar with the syntax and semantics of CTL^* and LTL. Indexed temporal logics were introduced by [2] to model specifications of certain distributed systems. They are obtained by adding *vertex quantifiers* to a given temporal logic over indexed atomic propositions. For example, in a system with two process templates, the formula $\forall i : type(i) = 1. \mathsf{AG}((good, i))$ states that every process of type 1 on all computations at all points of time satisfies the atom *good*. In a system with one process template, the formula $\forall i \neq j. \mathsf{AG}(\neg(critical, i) \vee \neg(critical, j))$ states that it is never the case that two processes both satisfy the atom *critical* at the same time.

Syntax. Fix an infinite set $\mathsf{Vars} = \{i, j, \ldots\}$ of vertex variables (called index variables for the clique topology). A *vertex quantifier* is an expression of the form $\exists x : type(x) = m$ or $\forall x : type(x) = m$ where $m \in \mathbb{N}$. An *indexed CTL* formula over vertex variables Vars and atomic propositions* AP_{pr} is a formula of the form $Q_1 i_1, \ldots, Q_k i_k : \varphi.$, where each $i_n \in \mathsf{Vars}$, each Q_{i_n} is an index quantifier, and φ is a CTL^* formula over atomic predicates $\mathsf{AP}_{\mathsf{pr}} \times \mathsf{Vars}$.

The semantics is fully described in the full version of this paper. For 1-ary systems we may write $\forall x$ instead of $\forall x : type(x) = 1$. In the syntax of indexed formulas we may write $type(x) = P_m$ instead of $type(x) = m$. i-CTL^* denotes the set of all indexed CTL^* sentences, and k-CTL^* for the set of all k-indexed formulas in i-CTL^*, i.e., formulas with k quantifiers. We similarly define indexed versions of various fragments of CTL^*, e.g., i-LTL, k-$\mathsf{LTL}\backslash\mathsf{X}$ and k-$\mathsf{CTL}^*_d\backslash\mathsf{X}$ (k-CTL^* formulas with nesting depth of path quantifiers at most d). We write $\overline{P}^G \equiv_{\text{k-}\mathsf{CTL}^*_d\backslash\mathsf{X}} \overline{P}^{G'}$, if $\overline{P}^G$ and $\overline{P}^{G'}$ agree on all k-$\mathsf{CTL}^*_d\backslash\mathsf{X}$ formulas.

Note. The index variables are bound *outside* of all the temporal path quantifiers (A and E). In particular, for an existentially quantified LTL formula to be satisfied there must exist a valuation of the index variables such that ϕ holds for all runs (and not one valuation for each run). Thus this logic is sometimes called prenex indexed temporal logic. Note that if one allows index quantifiers inside the scope of temporal path quantifiers then one quickly reaches undecidability even for systems with no communication [10].

For the remainder of this paper specifications only come from i-CTL*\X, i.e., without the next-time operators. It is usual in the context of parameterized systems to consider specification logics without the "next" (X) operator.

2.4 Parameterized Topology, Parameterized System, PMCP, Cutoff

Parameterized Topology $\mathcal{G}$. An *(r-ary) parameterized topology* $\mathcal{G}$ is a set of r-topologies. Moreover, we assume membership in $\mathcal{G}$ is decidable. Typical examples are the set of r-ary cliques or the set of 1-ary rings.

Parameterized Model Checking Problem. Fix an r-ary parameterized topology $\mathcal{G}$, a set of r-ary system templates $\mathcal{P}$, and a set of indexed temporal logic sentences $\mathcal{F}$. The *parameterized model checking problem (PMCP)* for this data, written $\mathsf{PMCP}_{\mathcal{G}}(\mathcal{P},\mathcal{F})$, is to decide, given a formula $\varphi \in \mathcal{F}$ and a system template $\overline{P} \in \mathcal{P}$, whether for all $G \in \mathcal{G}$, $\overline{P}^G \models \varphi$. The complexity of the $\mathsf{PMCP}_{\mathcal{G}}(\mathcal{P},\mathcal{F})$, where the formula $\varphi \in \mathcal{F}$ is fixed and only the system template is given as an input, is called the *program complexity.*

Cutoff. A *cutoff* for $\mathsf{PMCP}_{\mathcal{G}}(\mathcal{P},\mathcal{F})$ is a natural number c such that for every $\overline{P} \in \mathcal{P}$ and $\varphi \in \mathcal{F}$, the following are equivalent: **(i)** $\overline{P}^G \models \varphi$ for all $G \in \mathcal{G}$ with $|V_G| \leq c$; **(ii)** $\overline{P}^G \models \varphi$ for all $G \in \mathcal{G}$.

Lemma 1. *If $\mathsf{PMCP}_{\mathcal{G}}(\mathcal{P},\mathcal{F})$ has a cutoff, then $\mathsf{PMCP}_{\mathcal{G}}(\mathcal{P},\mathcal{F})$ is decidable*

Proof. If c is a cutoff, let $G_1,\ldots,G_n$ be all topologies G in $\mathcal{G}$ such that $|V_G| \leq c$. The algorithm that solves PMCP takes $\overline{P},\varphi$ as input and checks whether or not $\overline{P}^{G_i} \models \varphi$ for all $1 \leq i \leq n$. □

2.5 Destuttering and Process Executions

The *destuttering of an infinite word* $\alpha \in \Sigma^\omega$ is the infinite word $\alpha^\delta \in \Sigma^\omega$ defined by replacing every maximal finite consecutive sequence of repeated symbols in α by one copy of that symbol. Thus, the destuttering of $(aaba)^\omega$ is $(ab)^\omega$; and the destuttering of aab^ω is ab^ω. The *destuttering of set* $L \subseteq \Sigma^\omega$, written L^δ, is the set $\{\alpha^\delta \mid \alpha \in L\} \subseteq \Sigma^\omega$.

It is known that LTL\X can not distinguish between a word and its destuttering, which is the main motivation for the following definition.

Process Executions. For parameterized r-topology $\mathcal{G}$, r-ary system template $\overline{P} = (P_1,\cdots,P_r)$ and $t \leq r$, define the set of *(process) executions (with respect to $t,\overline{P},\mathcal{G}$)*, written $t\text{-EXEC}_{\mathcal{G},\overline{P}}$, as the destuttering of the following set:

$$\bigcup_{G\in\mathcal{G}}\{proj_v(\pi) \mid \pi \text{ is a state-labelled run of } \overline{P}^G \text{ and } v \in V_G \text{ is of type } t\}.$$

When $\mathcal{G}$ or $\overline{P}$ is clear from the context we may omit them.

The point is that for universal 1-index LTL\X we can reduce the PMCP to model checking a single system whose runs are $t\text{-EXEC}_{\mathcal{G},\overline{P}}$. This is explained in details in the full version of this paper.

2.6 Two Prominent Kinds of Pairwise-Rendezvous Systems

Identical Processes. Concurrent systems in which all processes are identical are modeled with system arity $r = 1$. In this case there is a single process template P, and a topology may be thought of as a directed graph $G = (V, E)$ (formally $G = (V, E, T_1)$ with $T_1 = V$). We write $\text{USER-EXEC}_{\mathcal{G}}(U)$ for the set of executions of the user processes in a 1-ary system, i.e., $1\text{-EXEC}_{\mathcal{G},U}$.

Identical Processes with a Controller. Concurrent systems in which all processes are identical except for one process (typically called a controller or the environment) are modeled with system arity $r = 2$, and system templates of the form (P_1, P_2), and we restrict the topologies so that exactly one vertex has type 1 (i.e., runs the controller). We call such topologies *controlled*. We often write (C, U) instead of (P_1, P_2), and $G = (V, E, v)$ instead of $(V, E, \{v\}, V \setminus \{v\})$. We write $\text{CONTROLLER-EXEC}_{\mathcal{G}}(C, U)$ for the set of executions of the controller process, i.e., $1\text{-EXEC}_{\mathcal{G},(C,U)}$. We write $\text{USER-EXEC}_{\mathcal{G}}(C, U)$ for the set of executions of the user processes in this 2-ary system, i.e., $2\text{-EXEC}_{\mathcal{G},(C,U)}$.

2.7 Classes of Parameterized Topologies

Here we define the classes of parameterized topologies which we will use in the sequel. The classes we define all have bounded clique-width.

w-terms and Clique-width. An *r-ary w-topology* $(V, E, T_1, \ldots, T_r, C_1, \ldots, C_w)$ extends $(V, E, T_1, \ldots, T_r)$ by a partition $(C_1, \ldots, C_w)$ of V. For every $u \in V$, if $u \in C_i$ then we say u has color i. We define the w-terms inductively. ϵ is a w-term. If x, y are w-terms, then $add_{i,t}(x)$, $recol_{i,j}(x)$, $edge_{i,j}(x)$ and $x \sqcup y$ are w-terms for $i, j \in [w]$, $t \in [r]$. Every w-term x has an associated w-topology $[[x]]$:

- $[[\epsilon]]$ has $V = E = \emptyset$ and empty labeling.
- $[[add_{i,t}(x)]]$ is formed by adding a new vertex of color i and type t to $[[x]]$.
- $[[recol_{i,j}(x)]]$ is formed by recoloring every vertex with color i of $[[x]]$ by j.
- $[[edge_{i,j}(x)]]$ is formed from $[[x]]$ by adding an edge from every vertex of color i to every vertex of color j.
- $[[x \sqcup y]]$ is the disjoint union of x and y and the union of the labelings.

A topology G has *clique-width at most w* if there is a w-term ρ such that G is isomorphic to $[[\rho(\epsilon)]]$ (forgetting the coloring $C_1, \ldots, C_w$). Every topology of size n has clique-width at most n. A class of topologies $\mathcal{G}$ has *bounded clique-width* if there exists w such that every graph in $\mathcal{G}$ has clique-width at most w. It is well-known if $\mathcal{G}$ has bounded tree-width, then it has bounded clique-width.

Monadic Second Order Logic MSO. MSO is a powerful logic for graphs and graph-like structures. It is the extension of First Order Logic with set quantification. MSO can define classic graph-theoretic concepts such as planarity, connectivity, c-regularity and c-colorability. We assume the reader is familiar with Monadic Second Order logic as described e.g. in [4]. A parameterized topology $\mathcal{G}$ is *MSO-definable* if there exists an MSO-formula Φ such that $G \in \mathcal{G}$ iff $G \models \Phi$. E.g., $\exists U \forall x \forall y (E(x, y) \rightarrow (U(x) \leftrightarrow \neg U(y)))$ defines the set of bipartite graphs.

We denote by $\equiv_q^{\mathsf{MSO}}$ the equivalence relation of topologies of being indistinguishable by MSO-formulas of quantifier rank q.

Theorem 1 (Courcelle's Theorem, see [4]). *Let $w \geq 1$ and let $\varphi \in MSO$. The MSO theory of r-topologies of clique-width w is decidable. I.e., there is an algorithm that on input $\varphi \in MSO$, decides whether there is an r-topology G of clique-width at most w such that $G \models \varphi$. Moreover, the number of equivalence classes in $\equiv_q^{MSO}$ is finite and computable, and a topology belonging to each class is computable.*

We now define a user-friendly and expressive formalism that can be used to generate natural parameterized topologies.

Iteratively Constructible Parameterized Topologies. A parameterized topology is *iteratively constructible* if it can be built from an initial labeled graph by means of a repeated fixed succession of elementary operations involving addition of vertices and edges, deletion of edges, and relabeling. More precisely, an r-ary parameterized topology $\mathcal{G}$ is *iteratively-constructible* if there are w-terms $\rho(x), \sigma(x)$ with one variable x and no use of disjoint union, and a w-graph H_0 such that (i) $G \in \mathcal{G}$ iff $G = \sigma(\rho^n(H_0))$ for some $n \in \mathbb{N}$, where $\rho^0(H) = H$, (ii) exactly one vertex of H_0 has type 1, and (iii) no vertex of type 1 is added in ρ or σ. For terms $\rho(\cdot)$ and $\rho'(\cdot)$ we write $\rho :: \rho'$ instead of $\rho(\rho'(\cdot))$. Intuitively, ρ "builds up" the topology, and σ puts on the "finishing touch" (see examples below). The unique vertex of type 1 can act as the controller if it is assigned a unique process template, and it is the initial token position in TPSs.

Example 1 (Cliques and rings). The set of cliques (irreflexive) is iteratively constructible: let H_0 consist of a single vertex v of color 1 and type 1, let $\rho(x)$ be $edge_{1,1} :: add_{1,2}(x)$, and $\sigma(x)$ be the identity.

The set of uni-directional rings is iteratively constructible: let H_0 consist of two vertices, one of color 1 and type 1 and one of color 2 and type 2 with an edge from 1 to 2. Let $\rho(x)$ be $recol_{4,2} :: recol_{2,3} :: edge_{2,4} :: add_{4,2}$ and $\sigma(x) = edge_{2,1}$.

Clique-Like (and Controllerless Clique-like) Parameterized Topologies. We now define other sets of topologies of bounded clique-width that generalise cliques and stars, but not rings.

Let H be an r-ary topology with vertex set V_H of size m in which each vertex has a distinct type. Let $\rho_2(x) = add_{1,type(1)} :: \cdots :: add_{m,type(m)}$. Let $\rho_1(x)$ be the m-term obtained by the composition of $edge_{i,j}$ for all $(i,j) \in E_H$ (in an arbitrary order). Let $\rho(x) = \rho_1(x) :: \rho_2(x)$. We have $[[\rho(\epsilon)]] = H$.

An r-ary parameterized topology $\mathcal{G}$ is *clique-like* if there is an r-ary topology H and a partition $B_{sng}, B_{clq}, B_{ind}$ of V_H such that $G \in \mathcal{G}$ iff there exists a function $num : B_{clq} \cup B_{ind} \to \mathbb{N}$ such that $[[\rho^{num}(\epsilon)]] = G$, and ρ^{num} is obtained from ρ by (i) repeating each $add_{i,type(i)}$ $num(i)$ times rather than once, and (ii) finally performing $edge_{i,i}$ for all $i \in B_{clq}$. Intuitively, G is obtained from H by substituting each vertex in B_{clq} with a clique, each vertex in B_{ind} with an independent set, and leaving every vertex in B_{sng} as a single vertex.

We say that $\mathcal{G}$ is *generated by H and $B_{sng}, B_{clq}, B_{ind}$*. The cardinality of B_{sng} is the *number of controllers* in $\mathcal{G}$. In case $B_{sng} = \emptyset$ we say that $\mathcal{G}$ is *controllerless*.

Example. Cliques, stars and complete bipartite graphs. Let H be the 2-topology with vertex set $V_H = \{1, 2\}$ and edge set $\{(1,2),(2,1)\}$ and $type(i) = i$ for $i \in [2]$. The set of 2-ary cliques in which exactly one index has type 1 is clique-like using H as defined, $B_{clq} = \{2\}$, $B_{ind} = \emptyset$ and $B_{sng} = \{1\}$. The set of stars in which exactly one index has type 1 is clique-like using H above, $B_{clq} = \emptyset$, $B_{ind} = \{2\}$ and $B_{sng} = \{1\}$. The set of topologies that are complete bipartite graphs is clique-like using H above, $B_{ind} = \{1, 2\}$, and $B_{clq} = B_{sng} = \emptyset$.

Example. Rings are not clique-like. Clique-like parameterized topologies have diameter at most $|V_H|$ unless their diameter is infinite. Rings have unbounded but finite diameter and are therefore not clique-like.

3 Results for Pairwise-Rendezvous Systems

The known decidability results for parameterized pairwise-rendezvous systems are for clique topologies and specifications from 1-indexed LTL\X. [9]. Thus we might hope to generalise this result in two directions: more general specification languages and more general topologies. We first show, by reducing the non-halting problem of two-counter machines (2CMs) to the PMCP, that allowing branching specifications results in undecidability:

Theorem 2. *$PMCP_{\mathcal{G}}(\mathcal{P}, \mathcal{F})$ is undecidable where $\mathcal{F}$ is the set of 1-indexed CTL_2^*\X formulas, $\mathcal{G}$ is the set of 1-ary clique topologies, and $\mathcal{P}$ is the set of 1-ary system templates.*

We conclude that we should restrict the specification logic if we want decidability. In the rest of this section we focus on 1-indexed LTL\X and parameterized clique-topologies with or without a controller (note that the PMCP for 1-indexed LTL\X is undecidable for topologies that contain uni-directional rings [12,6]).

Pairwise Rendezvous: Complexity of PMCP. The proof of the following theorem extends the technique used in [9, Theorem 3.6] for clique topologies:

Theorem 3. *Fix an r-ary clique-like parameterized topology $\mathcal{G}$, let $\mathcal{F}$ be the set of 1-index LTL\X formulas, and $\mathcal{P}$ the set of r-ary system templates. Then $PMCP_{\mathcal{G}}(\mathcal{P}, \mathcal{F})$ is decidable in EXPSPACE.*

Thus, using the fact that PMCP is EXPSPACE-hard already for clique topologies and the coverability problem [7], we get:

Theorem 4. *Fix an r-ary clique-like parameterized topology $\mathcal{G}$, let $\mathcal{F}$ be the set of 1-index LTL\X formulas, and $\mathcal{P}$ the set of r-ary system templates. Then $PMCP_{\mathcal{G}}(\mathcal{P}, \mathcal{F})$ is EXPSPACE-complete. The same holds for program complexity.*

It is known that PMCP for 1-ary cliques is PSPACE-complete (the upper bound is from [9, Section 4], and the lower bound holds already for LTL\X model checking a single finite state system P, with no communication). We extend the upper bound to clique-like topologies in which $B_{sng} = \emptyset$, i.e., controllerless clique-like parameterized topologies. The proof follows [9] and is via a reduction to emptiness of Büchi automata, see Theorem 8.

Theorem 5. *Fix an r-ary controllerless clique-like parameterized topology $\mathcal{G}$, let $\mathcal{F}$ be the set of 1-index LTL\X formulas, and $\mathcal{P}$ the set of r-ary system templates. Then $PMCP_{\mathcal{G}}(\mathcal{P}, \mathcal{F})$ is PSPACE-complete, and the program complexity is in PTIME.*

Pairwise Rendezvous: Cutoffs.

Theorem 6. *Let $\mathcal{G}$ be the 1-ary parameterized clique topology and let $\mathcal{F}$ be the set of 1-index LTL\X formulas. There exists a process template P such that $PMCP_{\mathcal{G}}(\{P\}, \mathcal{F})$ has no cutoff.*

Proof (Sketch). Define process template $P = (S, R, I, \Phi)$ by $S := \{1, 2, 3\}$, $I = \{1\}$, $R = \{(1, \tau, 1), (1, a!, 2), (2, \tau, 1), (1, a?, 3)\}$, and $\Phi(i) = \{i\}$. Thus in a system with $n + 1$ processes one possible behaviour is, up to stuttering, $(12)^n 1^\omega$. This run does not appear in any system with $\leq n$ processes. Thus take the formula ϕ_n stating that for every process and every path, the initial segment, up to stuttering, is not of the form $(12)^n$ (for instance $1 \wedge (1\ \mathsf{U}\ (2 \wedge (2\ \mathsf{U}\ 1)))$ states that there is an initial prefix of the form 11*22*11*). □

Pairwise Rendezvous: Equivalence to Finite-State Systems. The following theorem says that if there is a cutoff for the set of 1-indexed LTL\X formulas then the set of executions is ω-regular. The proof uses the fact that 1-indexed LTL\X is expressive enough to describe finite prefixes of infinite words, and deducing that since all finite executions of a system of any size must already appear in systems up to the cutoff size, so do the infinite executions. This holds for general topologies, not only for clique-like ones.

Theorem 7. *Fix r-ary parameterized topology $\mathcal{G}$, let $\mathcal{F}$ be the set of 1-index LTL\X formulas, and let $\overline{P}$ be an r-ary system template. If $PMCP_{\mathcal{G}}(\{\overline{P}\}, \mathcal{F})$ has a cutoff, then for every $t \leq r$, the set of executions t-$\text{EXEC}_{\mathcal{G},\overline{P}}$ is ω-regular.*

The following theorem states that the set of executions of each process in a controllerless parameterized clique-like topology is ω-regular, i.e., recognizable by a Non-deterministic Büchi Word automaton (NBW)(see [13] for a definition). This is done by a reduction to the case of a clique topology and using the corresponding result in [9, Section 4][3]

Theorem 8. *For every controllerless clique-like r-ary parameterized topology $\mathcal{G}$, every r-ary system template $\overline{P}$, and every $i \leq r$, there is a linearly sized NBW (computable in PTIME) that recognises the set i-$\text{EXEC}_{\mathcal{G},\overline{P}}$.*

---

[3] The relevant result in [9, Section 4] is correct. However, its proof has some bugs and some of the statements (e.g., Theorem 4.8) are wrong. In the full version of this paper we give a correct proof for the main result of [9, Section 4].

By constructing an appropriate system template, and using a pumping argument, we are able to show that the set of executions of systems with a controller is not, in general, ω-regular. More precisely:

Theorem 9. *Let $\mathcal{G}$ be the 2-ary parameterized clique topology. There exist a system template (C, U) for which* CONTROLLER-EXEC$_{\mathcal{G}}(C, U)$ *is not ω-regular.*

4 Results for Disjunctive Guards

In the following we will consider parameterized systems as described in Section 2.6, i.e., with an arbitrary number of copies of one template U, and possibly with a unique controller C, arranged in a clique.

The following theorem follows similar lines as Theorem 2, and uses a reduction from the non-halting problem of 2CMs. The main complication here is that, unlike the case of pairwise rendezvous, mutual exclusion is not easily obtainable using disjunctive guards, and thus more complicated gadgets are needed to ensure that the counter operations are simulated correctly.

Theorem 10. *$PMCP_{\mathcal{G}}(\mathcal{P}, \mathcal{F})$ is undecidable where $\mathcal{F}$ is the set of 1-indexed $CTL_2^*\backslash X$ formulas, $\mathcal{G}$ is the 1-ary parameterized clique topology, and $\mathcal{P}$ is the set of 1-ary disjunctively-guarded system templates.*

We conclude that we should restrict the specification logic if we want decidability, and in the rest of this section we focus on 1-indexed LTL\X.

Disjunctive Guards: Cutoffs. By [5], for the r-ary parameterized clique topology and k-indexed LTL\X formulae, there is a cutoff of size $|U| + 2$ (where U is the process template). The following proposition shows that this cutoff is tight.

Proposition 1. *Let $\mathcal{G}$ be the r-ary parameterized clique topology, let $\mathcal{F}$ be the set of 1-index $LTL\backslash X$ formulas, and let $k > 0$. There is a disjunctively-guarded system template $\mathcal{P}$ of size $\Theta(k)$ such that $\Theta(k)$ is the smallest cutoff for $PMCP_{\mathcal{G}}(\mathcal{P}, \mathcal{F})$*

Proof (sketch). We show the case of 1-ary cliques. Similar examples exist for r-ary systems, with or without a controller. Consider the process template: $U = (S_U, R_U, I_U, \Phi_U)$ where $S_U = \{s_1, \ldots, s_k\}$, $R_U = \{(s_i, s_i, s_{i+1}) \mid i < k\} \cup \{(s_k, s_k, s_1)\} \cup \{(s_i, \top, s_i) \mid i \leq k\}$, $I_U = \{s_1\}$, and $\Phi_U(s_i) = \{s_i\}$; and the formula $\phi_k = \forall x.AG((s_k, x) \rightarrow G(s_k, x))$. Evidently, ϕ_k holds in all systems with at most k processes, but false in systems with $k + 1$ or more processes.

Disjunctive Guards: Equivalence to Finite-State Systems. There are several techniques for solving the PMCP for 1-indexed LTL\X formulae for systems using disjunctive guards. One such technique consists in finding an NBW that model-checks the set of all possible executions of the system, for any number of copies of user processes U. We begin by showing that in general, such an automaton is necessarily big. We show the following lower bound by encoding the language of palindromes of length $2k$.

Proposition 2. *Let $\mathcal{G}$ be the 2-ary parameterized controlled clique topology. For every $k > 0$ there exist a disjunctively-guarded system template (C, U) where* *the sizes of C and U are $\Theta(k)$ such that the smallest NBW whose language is* CONTROLLER-EXEC$_{\mathcal{G}}(C, U)$ *has size at least* $2^{\Omega(k)}$.

On the other hand, the cutoff $|U| + 2$ yields an NBW of size $|C| \times |U|^{\Omega(|U|)}$, and since this cutoff is tight, this technique can not yield a smaller NBW. In the following theorem we prove, surprisingly, that there is a smaller NBW, of size $O(|C| \times 2^{|U|})$.

Theorem 11. *Let $\mathcal{G}$ be the 2-ary parameterized controlled clique topology. For every disjunctively-guarded system template (C, U) there is an NBW of size $O(|C| \times 2^{|U|})$ recognizing the set* CONTROLLER-EXEC$_{\mathcal{G}}(C, U)$. *The same is true for* USER-EXEC$_{\mathcal{G}}(C, U)$.

Intuitively, each state in the NBW pairs the current *controller state* together with a set of *reachable user states*, i.e. sets of states of U that can be reached in some system of finite size, given the actual state of the controller C. In this construction, a state $s \in S_U$ is considered reachable iff it is the target of a sequence of transitions in R_U that (a) are not guarded, or (b) are guarded by other reachable states, or (c) are guarded by the current controller state. The NBW has $O(|C| \times 2^{|U|})$ (abstract) configurations, and it is shown that every path in the NBW can be concretized in some system of some finite size.

Disjunctive Guards: Complexity of PMCP. We inherit the PSPACE-hardness of model-checking LTL\X on a single finite-state system. For the upper bound, the construction in Theorem 11 can be done 'on-the-fly'

Theorem 12. *Let $\mathcal{G}$ be the 2-ary parameterized controlled clique topology or the 1-ary parameterized clique topology. Let $\mathcal{F}$ be the set of 1-index LTL\X formulas, and let $\mathcal{P}$ be the set of disjunctively guarded system templates (of suitable arity). The complexity of PMCP$_{\mathcal{G}}(\mathcal{P}, \mathcal{F})$ is PSPACE-complete.*

We inherit the PTIME program complexity (without controller) from Theorem 8. With a controller, the coNP upper bound results from a fine analysis of Theorem 11, and the coNP-hardness by coding of unsatisfiability (the user processes store an assignment, and the controller verifies it is not satisfying).

Theorem 13. *Fix $\mathcal{F}$ to be the set of 1-index LTL\X formulas. If $\mathcal{P}$ is the set of 1-ary disjunctively guarded system templates, and $\mathcal{G}$ is the 1-ary parameterized clique topology, then the program complexity of PMCP$_{\mathcal{G}}(\mathcal{P}, \mathcal{F})$ is PTIME.*

If $\mathcal{P}$ is the set of 2-ary disjunctively guarded system templates, and $\mathcal{G}$ is the 2-ary parameterized controlled clique topology, then the program complexity of PMCP$_{\mathcal{G}}(\mathcal{P}, \mathcal{F})$ is coNP-complete.

5 Results for Token Passing Systems

Theorem 14. *Let $\mathcal{G}$ be a parameterized topology that is either iteratively-constructible, or MSO-definable and of bounded clique-width. Then (i) The problem PMCP$_{\mathcal{G}}(\mathcal{P}$, i-CTL*\X) is decidable; (ii) There is an algorithm that given k and d outputs a cutoff for k-CTL$^*_d$\X.*

Decidability. We use the *finiteness* and *reduction* properties of k-$\mathsf{CTL}_d^*\backslash\mathsf{X}$ from [1]. The reduction property essentially says that the process templates in $\overline{P}$ play no role, i.e. we can assume the processes in $\overline{P}_{topo}$ do nothing except send and receive the token. The only atoms are p_j which indicate that j has the token. In a k-$\mathsf{CTL}_d^*\backslash\mathsf{X}$ formula $Q_1 x_1 \ldots Q_k x_k.\ \varphi$, every valuation of the variables $x_1, \ldots, x_k$ designates k vertices of the underlying topology G, say $\bar{g} = g_1, \ldots, g_k$. The formula φ can only use the atoms p_{g_j} for $g_j \in \bar{g}$. We denote the structures of φ by $G|\bar{g}$ to indicate (1) that the process templates are $\overline{P}_{topo}$ and (2) that $\bar{g}$ have been assigned to $x_1, \ldots, x_k$ by quantification. The finiteness property says that there is a computable finite set $CON_{d,k}$ such that every $G|\bar{g}$ is $\equiv_{\mathsf{k\text{-}CTL}_d^*\backslash\mathsf{X}}$-equivalent to a member of $CON_{d,k}$. We use the details of the construction of $CON_{d,k}$ to show essentially that $\equiv_{\mathsf{k\text{-}CTL}_d^*\backslash\mathsf{X}}$ is MSO-definable by reducing the quantification on infinite paths in k-$\mathsf{CTL}_d^*\backslash\mathsf{X}$ to MSO quantification on finite simple paths and cycles. Decidability of PMCP is achieved using the decidability of MSO on classes of parameterized topologies of bounded clique-width (Theorem 1). The decidability of PMCP on iteratively constructible parameterized topologies can be shown by employing methods of similar to [8].

Cutoffs. Cuttoffs are derived as the maximal size of a representative topology belonging to a $\equiv_q^{\mathsf{MSO}}$-equivalence class as guaranteed in Theorem 14 and are non-elementary due to the number of equivalence classes. For iterateively-constructible parameterized topologies the cutoffs may be much lower, though there exists a system template $\overline{P}$, and, for all $k \in \mathbb{N}$, an iteratively constructible parameterized topology $\mathcal{G}_k$ of clique-width at most k and a k-indexed $\mathsf{LTL}\backslash\mathsf{X}$ formula φ such that the cutoff of $\mathsf{PMCP}_{\mathcal{G}}(\{\overline{P}\}, \{\varphi\})$ is $2^{\Omega(\sqrt{k})}$.

6 Discussion and Related Work

The applicability of the reduction of the PMCP to finitely many classical model checking problems as a technique for solving the PMCP depends on the communication primitive, the specification language, and the set of topologies of the system. The wide-ranging nature of our work along these axes gives us some insights which may be pertinent to system models different from our own:

Decidability But no Cutoffs. Theorems 3 and 6 show that it can be the case that, for certain sets of specifications formula, cutoffs do not exist yet the PMCP problem is decidable.

Cutoffs may not be Optimal. Proposition 1 and Theorem 11 imply that even in cases that cutoffs exist and are computable, they may not yield optimal algorithms for solving the PMCP.

Formalisms for Topologies are Useful. Many results in Sections 3 and 5 show that decidability and complexity of PMCP can be extended from concrete examples of sets of topologies such as rings and cliques to infinite classes of topologies given as user-friendly yet powerful formalisms. The formalisms we study may be useful for other system models.

In the context of cutoffs, it is worth noting that we only considered cutoffs with respect to sets of formulas and process templates. As Theorem 6 shows, there is a parameterized topology $\mathcal{G}$, and a system template $\overline{P}$, for which no cutoff exists for the set of 1-indexed $\mathsf{LTL}\backslash\mathsf{X}$ formulas. Note, however, that if the formula φ is also fixed then a cutoff always exists. Indeed, given $\mathcal{G}, \overline{P}, \varphi$, letting $c := |V_G|$ yields a (minimal) cutoff if we choose G to be the smallest for which $\overline{P}^G \not\models \varphi$, or simply the smallest topology in $\mathcal{G}$ if all topologies in $\mathcal{G}$ satisfy φ. We reserve the question of computing the cutoff in such cases to future work.

As previously discussed, this work draws on and generalises the work in [9] on pairwise rendezvous on cliques, the work in [5] on disjunctive guards on cliques, and the work in [1,3,6] on token-passing systems. There are very few published complexity lower-bounds for PMCP (notable exceptions are [7,11]), and to the best of our knowledge, our lower bounds on the sizes of cutoffs are the first proven non-trivial lower bounds for these types of systems.

Acknowledgments. The first author is supported by ERC Start grant (279307: Graph Games) and the RiSE network (S11407-N23).

References

1. Aminof, B., Jacobs, S., Khalimov, A., Rubin, S.: Parameterized model checking of token-passing systems. In: McMillan, K.L., Rival, X. (eds.) VMCAI 2014. LNCS, vol. 8318, pp. 262–281. Springer, Heidelberg (2014)
2. Browne, M.C., Clarke, E.M., Grumberg, O.: Reasoning about networks with many identical finite state processes. Inf. Comput. 81, 13–31 (1989)
3. Clarke, E., Talupur, M., Touili, T., Veith, H.: Verification by network decomposition. In: Gardner, P., Yoshida, N. (eds.) CONCUR 2004. LNCS, vol. 3170, pp. 276–291. Springer, Heidelberg (2004)
4. Courcelle, B., Engelfriet, J.: Graph Structure and Monadic Second-Order Logic - A Language-Theoretic Approach. Encyclopedia of mathematics and its applications, vol. 138. Cambridge University Press (2012)
5. Emerson, E.A., Kahlon, V.: Reducing model checking of the many to the few. In: McAllester, D. (ed.) CADE 2000. LNCS, vol. 1831, pp. 236–254. Springer, Heidelberg (2000)
6. Emerson, E.A., Namjoshi, K.S.: On reasoning about rings. Int. J. Found. Comput. Sci. 14(4), 527–550 (2003)
7. Esparza, J.: Keeping a crowd safe: On the complexity of parameterized verification. In: STACS (2014)
8. Fischer, E., Makowsky, J.A.: Linear recurrence relations for graph polynomials. In: Avron, A., Dershowitz, N., Rabinovich, A. (eds.) Trakhtenbrot/Festschrift. LNCS, vol. 4800, pp. 266–279. Springer, Heidelberg (2008)
9. German, S.M., Sistla, A.P.: Reasoning about systems with many processes. J. ACM 39(3), 675–735 (1992)
10. John, A., Konnov, I., Schmid, U., Veith, H., Widder, J.: Counter attack on byzantine generals: Parameterized model checking of fault-tolerant distributed algorithms. CoRR abs/1210.3846 (2012)
11. Schmitz, S., Schnoebelen, P.: The Power of Well-Structured Systems. In: D'Argenio, P.R., Melgratti, H. (eds.) CONCUR 2013. LNCS, vol. 8052, pp. 5–24. Springer, Heidelberg (2013)
12. Suzuki, I.: Proving properties of a ring of finite-state machines. Inf. Process. Lett. 28(4), 213–214 (1988)
13. Vardi, M., Wolper, P.: Automata-theoretic techniques for modal logics of programs. J. Comput. Syst. Sci. 32(2), 182–221 (1986)

On the Completeness of Bounded Model Checking for Threshold-Based Distributed Algorithms: Reachability

Igor Konnov, Helmut Veith, and Josef Widder

Vienna University of Technology (TU Wien)

Abstract. Counter abstraction is a powerful tool for parameterized model checking, if the number of local states of the concurrent processes is relatively small. In recent work, we introduced parametric interval counter abstraction that allowed us to verify the safety and liveness of threshold-based fault-tolerant distributed algorithms (FTDA). Due to state space explosion, applying this technique to distributed algorithms with hundreds of local states is challenging for state-of-the-art model checkers. In this paper, we demonstrate that reachability properties of FTDAs can be verified by bounded model checking. To ensure completeness, we need an upper bound on the diameter, i.e., on the longest distance between states. We show that the diameters of accelerated counter systems of FTDAs, and of their counter abstractions, have a quadratic upper bound in the number of local transitions. Our experiments show that the resulting bounds are sufficiently small to use bounded model checking for parameterized verification of reachability properties of several FTDAs, some of which have not been automatically verified before.

1 Introduction

A system that consists of concurrent anonymous (identical) processes can be modeled as a counter system: Instead of recording which process is in which local state, we record for each local state, how many processes are in this state. We have one counter per local state ℓ, denoted by $\kappa[\ell]$. Each counter is bounded by the number of processes. A step by a process that goes from local state ℓ to local state ℓ' is modeled by decrementing $\kappa[\ell]$ and incrementing $\kappa[\ell']$.

We consider a specific class of counter systems, namely those that are defined by *threshold automata*. The technical motivation to introduce threshold automata is to capture the relevant properties of fault-tolerant distributed algorithms (FTDAs). FTDAs are an important class of distributed algorithms that work even if a subset of the processes fail [20]. Typically, they are parameterized in the number of processes and the number of tolerated faulty processes. These numbers of processes are parameters of the verification problem. We show that accelerated counter systems defined by threshold automata have a diameter whose bound is independent of the bound on the counters, but depends only on characteristics of the threshold automaton. This bound can be used for parameterized model checking of FTDAs, as we confirm by experimental evaluation.

P. Baldan and D. Gorla (Eds.): CONCUR 2014, LNCS 8704, pp. 125–140, 2014.

Modeling FTDAs as counter systems defined by threshold automata. A threshold automaton consists of rules that define the conditions and effects of changes to the local state of a process of a distributed algorithm. Conditions are *threshold guards* that compare the value of a shared integer variable to a linear combination of parameters, e.g., $x \geq n - t$, where x is a shared variable and n and t are parameters. This captures counting arguments which are used in FTDAs, e.g., a process takes a certain step only if it has received a message from a majority of processes. To model this, we use the shared variable x as the number of processes that have sent a message, n as the number of processes in the system, and t as the assumed number of faulty processes. The condition $x \geq n - t$ then captures a majority under the resilience condition that $n > 2t$. Resilience conditions are standard assumptions for the correctness of an FTDA. Apart from changing the local state, applying a rule can increase a shared variable, which naturally captures that a process has sent a message. Thus we consider threshold automata where shared variables are never decreased and where rules that form cycles do not modify shared variables, which is natural for modeling FTDAs.

Bounding the Diameter. For reachability it is not relevant whether we "move" processes one by one from state ℓ to ℓ'. If several processes perform the same transition one after the other, we can model this as a single update on the counters: The sequence where b processes one after the other move from ℓ to ℓ' can be encoded as a transition where $\kappa[\ell]$ is decreased by b and $\kappa[\ell']$ is increased by b. Value b is called the acceleration factor and may vary in a run depending on how many repetitions of the same transition should be captured. We call such runs of a counter system *accelerated.* The lengths of accelerated runs are the ones relevant for the diameter of the counter system.

The main technical challenge comes from the interactions of shared variables and threshold guards. We address it with the following three ideas: (i) *Acceleration* as discussed above. (ii) *Sorting*, that is, given an arbitrary run of a counter system, we can shorten it by changing the order of transitions such that there are possibly many consecutive transitions that can be merged according to (i). However, as we have arithmetic threshold conditions, not all changes of the order result in allowed runs. (iii) *Segmentation*, that is, we partition a run into segments, inside of which we can reorder the transitions; cf. (ii). In combination, these three ideas enable us to prove the main theorem: *The diameter of a counter system is at most quadratic in the number of rules; more precisely, it is bounded by the product of the number of rules and the number of distinct threshold conditions.* In particular, the diameter is independent of the parameters.

Using the Bound for Parameterized Model Checking. Parameterized model checking is concerned with the verification of concurrent or distributed systems, where the number of processes is not a priori fixed, that is, a system is verified for all sizes. In our case, the counter systems for all values of n and t that satisfy the resilience condition should be verified. A well-known parameterized model checking technique is to map all these counter systems to a *counter abstraction*, where the counter values are not natural numbers, but range over an abstract finite domain, e.g. [29]. In [16] we developed a more general form of counter abstraction for expressions used in threshold guards, which leads, e.g., to the abstract domain of four values that capture the parametric intervals $[0, 1)$ and $[1, t+1)$ and $[t+1, n-t)$ and $[n-t, \infty)$. It is easy to see [16] that a counter abstraction simulates all counter systems for all parameter values that satisfy the resilience condition. The bound d on the diameter of counter systems implies a bound $\hat{d}$ on the diameter of the counter abstraction. From this and simulation follows that if an abstract state is not reachable in the counter abstraction within $\hat{d}$ steps, no concretization of this state is reachable in any of the concrete counter systems. This allows us to efficiently combine counter abstraction with *bounded model checking* [6]. Typically, bounded model checking is restricted to finding bugs that occur after a bounded number of steps of the systems. However, if one can show that within this bound every state is reachable from an initial state, bounded model checking is a complete method for verifying reachability.

2 Our Approach at a Glance

Figure 1 represents a threshold automaton: The circles depict the local states, and the arrows represent rules (r_1 to r_5) that define how the automaton makes transitions. Rounded corner labels correspond to conditional rules, so that the rule can only be executed if the threshold guard evaluates to true. In our example, x and y are shared variables, and n, t, and f are parameters that are assumed to satisfy the resilience condition $n \geq 2t \wedge f \leq t$. The number of processes (that each execute the automaton) depends on the parameters, in this example we assume that n processes run concurrently. Finally, rectangular labels on arrows correspond to rules that increment a shared variable. The transitions of the counter system are then defined using the rules, e.g., when rule r_2 is executed, then variable y is incremented and the counters $\kappa[\ell_3]$ and $\kappa[\ell_2]$ are updated.

Consider a counter system in which the parameter values are $n = 3$, and $t = f = 1$. Let σ_0 be the configuration where $x = y = 0$ and all counters are set to 0 except $\kappa[\ell_1] = 3$. This configuration corresponds to a concurrent system where all three processes are in ℓ_1. For illustration, we assume that in this concurrent system processes have the identifiers 1, 2, and 3, and we denote by $r_i(j)$ that process j executes rule r_i. Recall that we have anonymous (symmetric) systems, so we use the identifiers only for illustration: the transition of the counter system is solely defined by the rule being executed.

As we are interested in the diameter, we have to consider the distance between configurations in terms of length of runs. In this example, we consider the distance of σ_0 to a configuration where $\kappa[\ell_5] = 3$, that is, all three processes are in local state ℓ_5. First, observe that the rule r_5 is locked in σ_0 as $y = 0$ and $t = 1$. Hence, we require that rule r_2 is executed at least once so that the value of y increases. However, due to the precedence relation on the rules, before that, r_1 must be executed, which is also locked in σ_0. The sequence of transitions $\tau_1 = r_3(1), r_4(1), r_3(2), r_4(2)$ leads from σ_0 to the configuration where $\kappa[\ell_1] = 1$, $\kappa[\ell_4] = 2$, and $x = 2$; we denote it by σ_1. In σ_1, rule r_1 is unlocked,

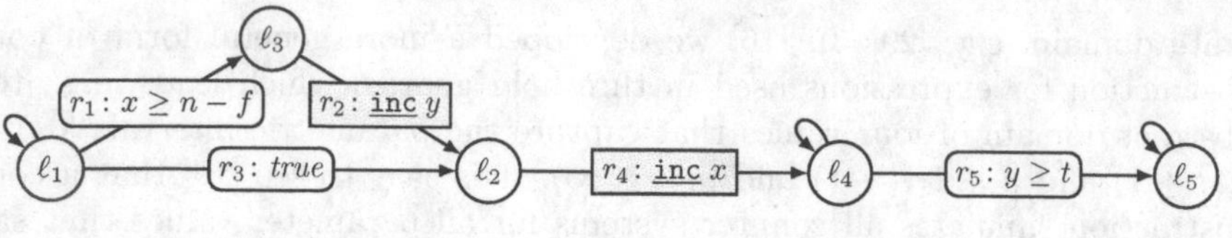

Fig. 1. Example of a Threshold Automaton

so we may apply $\tau_2 = r_1(3), r_2(3)$, to arrive at σ_2, where $y = 1$, and thus r_5 is unlocked. To σ_2 we may apply $\tau_3 = r_5(1), r_5(2), r_4(3), r_5(3)$ to arrive at the required configuration σ_3 with $\kappa[\ell_5] = 3$.

In order to exploit acceleration as much as possible, we would like to group together occurrences of the same rule. In τ_1, we can actually swap $r_4(1)$ and $r_3(2)$ as locally the precedence relation of each process is maintained, and both rules are unconditional. Similarly, in τ_3, we can move $r_4(3)$ to the beginning of the sequence τ_3. Concatenating these altered sequences, the resulting complete schedule is $\tau = r_3(1), r_3(2), r_4(1), r_4(2), r_1(3), r_2(3), r_4(3), r_5(1), r_5(2), r_5(3)$. We can group together the consecutive occurrences for the same rules r_i, and write the schedule using pairs consisting of rules and acceleration factors, that is, $(r_3, 2)$, $(r_4, 2)$, $(r_1, 1)$, $(r_2, 1)$, $(r_4, 1)$, $(r_5, 3)$.

In schedule τ, the occurrences of all rules are grouped together except for r_4. That is, in the accelerated schedule we have two occurrences for r_4, while for the other rules one occurrence is sufficient. Actually, there is no way around this: We cannot swap $r_2(3)$ with $r_4(3)$, as we have to maintain the local precedence relation of process 3. More precisely, in the counter system, r_4 would require us to decrease the counter $\kappa[\ell_2]$ at a point in the schedule where $\kappa[\ell_2] = 0$. We first have to increase the counter value by executing a transition according to rule r_2, before we can apply r_4. Moreover, we cannot move the subsequence $r_1(3), r_2(3), r_4(3)$ to the left, as $r_1(3)$ is locked in the prefix.

In this paper we characterize such cases. The issue here is that r_4 can unlock r_1 (we use the notation $r_4 \prec_U r_1$), while r_1 precedes r_4 in the control flow of the processes ($r_1 \prec_P r_4$). We coin the term *milestone* for transitions like $r_1(3)$ that cannot be moved, and show that the same issue arises if a rule r locks a threshold guard of rule r', where r precedes r' in the control flow. As processes do not decrease shared variables, we have at most one milestone per threshold guard. The sequence of transitions between milestones is called a segment. We prove that transitions inside a segment can be swapped, so that one can group transitions for the same rule in so-called batches. Each of these batches can then be replaced by a single accelerated transition that leads to the same configuration as the original batch. Hence, any segment can be replaced by an accelerated one whose length is at most the number of rules of a process. This and the number of milestones gives us the required bound on the diameter. This bound is independent of the parameters, and only depends on the number of threshold guards and the precedence relation between the rules of the processes.

Our main result is that the bound on the diameter is independent of the parameter values. In contrast, reachability of a specific local state depends on the parameter values: for a process to reach ℓ_5, at least $n - f$ processes must execute r_4 before at least t other processes must execute r_2. That is, the system must contain at least $(n-f)+t$ processes. In case of $t > f$, we obtain $(n-f)+t > n$, which is a contradiction, and ℓ_5 cannot be reached for such parameter values. The model checking problem we are interested in is whether a given state is unreachable for all parameter values that satisfy the resilience condition.

3 Parameterized Counter Systems

3.1 Threshold Automata

A threshold automaton describes a process in a concurrent system. It is defined by its local states, the shared variables, the parameters, and by rules that define the state changes and their conditions and effects on shared variables. Formally, a *threshold automaton* is a tuple $\mathsf{TA} = (\mathcal{L}, \mathcal{I}, \Gamma, \Pi, \mathcal{R}, RC)$ defined below.

States. The set $\mathcal{L}$ is the finite set of *local states*, and $\mathcal{I} \subseteq \mathcal{L}$ is the set of *initial local states*. The set Γ is the finite set of *shared variables* that range over $\mathbb{N}_0$. To simplify the presentation, we view the variables as vectors in $\mathbb{N}_0^{|\Gamma|}$. The finite set Π is a set of *parameter variables* that range over $\mathbb{N}_0$, and the *resilience condition* RC is a formula over parameter variables in linear integer arithmetic, e.g., $n > 3t \ \wedge \ t \geq f$. Then, we denote the set of *admissible parameters* by $\mathbf{P}_{RC} = \{\mathbf{p} \in \mathbb{N}_0^{|\Pi|} : \mathbf{p} \models RC\}$.

Rules. A rule defines a conditional transition between local states that may update the shared variables. Here we define the syntax and give only informal explanations of the semantics, which is defined via counter systems in Section 3.2.

Formally, a *rule* is a tuple $(\mathit{from}, \mathit{to}, \varphi^{\leq}, \varphi^{>}, \mathbf{u})$: The local states *from* and *to* are from $\mathcal{L}$. Intuitively, they capture from which local state to which a process moves, or, in terms of counter systems, which counters decrease and increase, respectively. A rule is only executed if the conditions $\varphi^{\leq}$ and $\varphi^{>}$ evaluate to true. Each condition consists of multiple guards. Each guard is defined using some shared variable $x \in \Gamma$, coefficients $a_0, \ldots, a_{|\Pi|} \in \mathbb{Z}$, and parameter variables $p_1, \ldots, p_{|\Pi|} \in \Pi$ so that

$$a_0 + \sum\nolimits_{i=1}^{|\Pi|} a_i \cdot p_i \ \leq \ x \qquad \text{and} \qquad a_0 + \sum\nolimits_{i=1}^{|\Pi|} a_i \cdot p_i \ > \ x$$

are a *lower guard* and *upper guard*, respectively (both, variables and coefficients, may differ for different guards). The *condition* $\varphi^{\leq}$ is a conjunction of lower guards, and the condition $\varphi^{>}$ is a conjunction of upper guards. Rules may increase shared variables. We model this using an update vector $\mathbf{u} \in \mathbb{N}_0^{|\Gamma|}$, which is added to the vector of shared variables, when the rule is executed. Then $\mathcal{R}$ is the finite set of rules.

Definition 1. *Given a threshold automaton* $(\mathcal{L}, \mathcal{I}, \Gamma, \Pi, \mathcal{R}, RC)$*, we define the* precedence relation $\prec_P$*, the* unlock relation $\prec_U$*, and the* lock relation $\prec_L$ *as subsets of* $\mathcal{R} \times \mathcal{R}$ *as follows:*

1. $r_1 \prec_P r_2$ *iff* $r_1.to = r_2.from$*. We denote by* $\prec_P^+$ *the transitive closure of* $\prec_P$*. If* $r_1 \prec_P r_2 \land r_2 \prec_P r_1$*, or if* $r_1 = r_2$*, we write* $r_1 \sim_P r_2$*.*
2. $r_1 \prec_U r_2$ *iff there is a* $\mathbf{g} \in \mathbb{N}_0^{|\Gamma|}$ *and* $\mathbf{p} \in \mathbf{P}_{RC}$ *satisfying* $(\mathbf{g}, \mathbf{p}) \models r_1.\varphi^{\leq} \land r_1.\varphi^{>}$ *and* $(\mathbf{g}, \mathbf{p}) \not\models r_2.\varphi^{\leq} \land r_2.\varphi^{>}$ *and* $(\mathbf{g} + r_1.\mathbf{u}, \mathbf{p}) \models r_2.\varphi^{\leq} \land r_2.\varphi^{>}$*.*
3. $r_1 \prec_L r_2$ *iff there is a* $\mathbf{g} \in \mathbb{N}_0^{|\Gamma|}$ *and* $\mathbf{p} \in \mathbf{P}_{RC}$ *satisfying* $(\mathbf{g}, \mathbf{p}) \models r_1.\varphi^{\leq} \land r_1.\varphi^{>}$ *and* $(\mathbf{g}, \mathbf{p}) \models r_2.\varphi^{\leq} \land r_2.\varphi^{>}$ *and* $(\mathbf{g} + r_1.\mathbf{u}, \mathbf{p}) \not\models r_2.\varphi^{\leq} \land r_2.\varphi^{>}$*.*

Definition 2. *Given a threshold automaton* $(\mathcal{L}, \mathcal{I}, \Gamma, \Pi, \mathcal{R}, RC)$*, we define the following quantities:* $\mathcal{C}^{\leq} = |\{r.\varphi^{\leq} : r \in \mathcal{R}, \exists r' \in \mathcal{R}.\ r' \not\prec_P^+ r \ \land\ r' \prec_U r\}|$*,* $\mathcal{C}^{>} = |\{r.\varphi^{>} : r \in \mathcal{R}, \exists r'' \in \mathcal{R}.\ r \not\prec_P^+ r'' \ \land\ r'' \prec_L r\}|$*. Finally,* $\mathcal{C} = \mathcal{C}^{\leq} + \mathcal{C}^{>}$*.*

We consider specific threshold automata, namely those that naturally capture FTDAs, where rules that form cycles do not increase shared variables.

Definition 3 (Canonical Threshold Automaton). *A threshold automaton* $(\mathcal{L}, \mathcal{I}, \Gamma, \Pi, \mathcal{R}, RC)$ *is canonical, if* $r.\mathbf{u} = \mathbf{0}$ *for all rules* $r \in \mathcal{R}$ *that satisfy* $r \prec_P^+ r$*.*

Order on rules. The relation $\sim_P$ defines equivalence classes of rules. For a given set of rules $\mathcal{R}$ let $\mathcal{R}/\!\sim$ be the set of equivalence classes defined by $\sim_P$. We denote by $[r]$ the equivalence class of rule r. For two classes c_1 and c_2 from $\mathcal{R}/\!\sim$ we write $c_1 \prec_C c_2$ iff there are two rules r_1 and r_2 in $\mathcal{R}$ satisfying $[r_1] = c_1$ and $[r_2] = c_2$ and $r_1 \prec_P^+ r_2$ and $r_1 \not\sim_P r_2$. Observe that the relation $\prec_C$ is a strict partial order (irreflexive and transitive). Hence, there are linear extensions of $\prec_C$. Below, we fix an *arbitrary* of these linear extensions to sort transitions in a schedule:

Notation. We denote by $\prec_C^{lin}$ a linear extension of $\prec_C$.

3.2 Counter Systems

Given a threshold automaton $\mathsf{TA} = (\mathcal{L}, \mathcal{I}, \Gamma, \Pi, \mathcal{R}, RC)$, a function $N \colon \mathbf{P}_{RC} \to \mathbb{N}_0$ that formalizes the number of processes to be modeled (e.g., n), and admissible parameter values $\mathbf{p} \in \mathbf{P}_{RC}$, we define a counter system as a transition system (Σ, I, R), that consists of the set of configurations Σ, which contain the counters and variables, the set of initial configurations I, and the transition relation R:

Configurations. A configuration $\sigma = (\boldsymbol{\kappa}, \mathbf{g}, \mathbf{p})$ consists of a vector of *counter values* $\sigma.\boldsymbol{\kappa} \in \mathbb{N}_0^{|\mathcal{L}|}$,[1] a vector of *shared variable values* $\sigma.\mathbf{g} \in \mathbb{N}_0^{|\Gamma|}$, and a vector of *parameter values* $\sigma.\mathbf{p} = \mathbf{p}$. The set Σ is the set of all configurations. The set of initial configurations I contains the configurations that satisfy $\sigma.\mathbf{g} = \mathbf{0}$, $\sum_{i \in \mathcal{I}} \sigma.\boldsymbol{\kappa}[i] = N(\mathbf{p})$, and $\sum_{i \notin \mathcal{I}} \sigma.\boldsymbol{\kappa}[i] = 0$.

---

[1] For simplicity we use the convention that $\mathcal{L} = \{1, \ldots, |\mathcal{L}|\}$.

Transition relation. A *transition* is a pair $t = (rule, factor)$ of a rule of the threshold automaton and a non-negative integer called the *acceleration factor*, or just factor for short. For a transition $t = (rule, factor)$ we refer by $t.\mathbf{u}$ to $rule.\mathbf{u}$, by $t.\varphi^{>}$ to $rule.\varphi^{>}$, etc. We say a transition t is *unlocked* in configuration σ if $\forall k \in \{0, \ldots, t.factor - 1\}.\ (\sigma.\boldsymbol{\kappa}, \sigma.\mathbf{g} + k \cdot t.\mathbf{u}, \sigma.\mathbf{p}) \models t.\varphi^{\leq} \wedge t.\varphi^{>}$. For transitions t_1 and t_2 we say that the two transitions are related iff $t_1.rule$ and $t_2.rule$ are related, e.g., for $\prec_P$ we write $t_1 \prec_P t_2$ iff $t_1.rule \prec_P t_2.rule$.

A transition t is *applicable (or enabled)* in configuration σ, if it is unlocked, and if $\sigma.\boldsymbol{\kappa}[t.from] \geq t.factor$. We say that σ' is the result of applying the (enabled) transition t to σ, and use the notation $\sigma' = t(\sigma)$, if

- t is enabled in σ
- $\sigma'.\mathbf{g} = \sigma.\mathbf{g} + t.factor \cdot t.\mathbf{u}$
- $\sigma'.\mathbf{p} = \sigma.\mathbf{p}$
- if $t.from \neq t.to$ then $\sigma'.\boldsymbol{\kappa}[t.from] = \sigma.\boldsymbol{\kappa}[t.from] - t.factor$ and $\sigma'.\boldsymbol{\kappa}[t.to] = \sigma.\boldsymbol{\kappa}[t.to] + t.factor$ and $\forall \ell \in \mathcal{L} \setminus \{t.from, t.to\}.\ \sigma'.\boldsymbol{\kappa}[\ell] = \sigma.\boldsymbol{\kappa}[\ell]$
- if $t.from = t.to$ then $\sigma'.\boldsymbol{\kappa} = \sigma.\boldsymbol{\kappa}$

The transition relation $R \subseteq \Sigma \times \Sigma$ of the counter system is defined as follows: $(\sigma, \sigma') \in R$ iff there is a $r \in \mathcal{R}$ and a $k \in \mathbb{N}_0$ such that $\sigma' = t(\sigma)$ for $t = (r, k)$. As updates to shared variables do not decrease their values, we obtain:

Proposition 1. *For all configurations σ, all rules r, and all transitions t applicable to σ, the following holds:*

1. *If $\sigma \models r.\varphi^{\leq}$ then $t(\sigma) \models r.\varphi^{\leq}$*
2. *If $t(\sigma) \not\models r.\varphi^{\leq}$ then $\sigma \not\models r.\varphi^{\leq}$*
3. *If $\sigma \not\models r.\varphi^{>}$ then $t(\sigma) \not\models r.\varphi^{>}$*
4. *If $t(\sigma) \models r.\varphi^{>}$ then $\sigma \models r.\varphi^{>}$*

Schedules. A *schedule* is a sequence of transitions. A schedule $\tau = t_1, \ldots, t_m$ is called *applicable* to configuration σ_0, if there is a sequence of configurations $\sigma_1, \ldots, \sigma_m$ such that $\sigma_i = t_i(\sigma_{i-1})$ for all i, $0 < i \leq m$. A schedule $t_1, \ldots, t_m$ where $t_i.factor = 1$ for $0 < i \leq m$ is a *conventional schedule*. If there is a $t_i.factor > 1$, then a schedule is called *accelerated*.

We write $\tau \cdot \tau'$ to denote the concatenation of two schedules τ and τ', and treat a transition t as schedule. If $\tau = \tau_1 \cdot t \cdot \tau_2 \cdot t' \cdot \tau_3$, for some τ_1, τ_2, and τ_3, we say that transition t precedes transition t' in τ, and denote this by $t \to_\tau t'$.

4 Diameter of Counter Systems

In this section, we will present the outline of the proof of our main theorem:

Theorem 1. *Given a canonical threshold automaton TA and a size function N, for each $\mathbf{p}$ in $\mathbf{P}_{RC}$ the diameter of the counter system is less than or equal to $d(\mathsf{TA}) = (\mathcal{C} + 1) \cdot |\mathcal{R}| + \mathcal{C}$, and thus independent of $\mathbf{p}$.*

From the theorem it follows that for all parameter values, reachability in the counter system can be verified by exploring runs of length at most $d(\mathsf{TA})$. However, the theorem alone is not sufficient to solve the parameterized model checking problem. For this, we combine the bound with the abstraction method in [16]. More precisely, the counter abstraction in [16] simulates the counter systems for *all* parameter values that satisfy the resilience condition. Consequently, the bound on the length of the run of the counter systems entails a bound for the counter abstraction. We exploit this in the experiments in Section 5.

4.1 Proof Idea

Given a rule r, a schedule τ and two transitions t_i and t_j, with $t_i \rightarrow_\tau t_j$, the subschedule $t_i \cdot \ldots \cdot t_j$ of τ is a *batch of rule* r if $t_\ell.rule = r$ for $i \le \ell \le j$, and if the subschedule is maximal, that is, $i = 1 \vee t_{i-1} \neq r$ and $j = m \vee t_{j+1} \neq r$. Similarly, we define a batch of a class c as a subschedule $t_i \cdot \ldots \cdot t_j$ where $[r_\ell] = c$ for $i \le \ell \le j$, and where the subschedule is maximal as before.

Definition 4 (Sorted schedule). *Given a schedule τ, and the relation $\prec_C^{lin}$, we define $sort(\tau)$ as the schedule that satisfies:*

1. *$sort(\tau)$ is a permutation of schedule τ.*
2. *two transitions from the same equivalence class maintain their relative order, that is, if $t \rightarrow_\tau t'$ and $t \sim_P t'$, then $t \rightarrow_{sort(\tau)} t'$.*
3. *for each equivalence class defined by $\sim_P$ there is at most one batch in $sort(\tau)$.*
4. *if $t \rightarrow_{sort(\tau)} t'$, then $t \sim_P t'$ or $[t] \prec_C^{lin} [t']$.*

The crucial observation is that if we have a schedule $\tau_1 = t \cdot t'$ applicable to configuration σ with $t.rule = t'.rule$, we can replace it with another applicable (one-transition) schedule $\tau_2 = t''$, with $t''.rule = t.rule$ and $t''.factor = t.factor + t'.factor$, such that $\tau_1(\sigma) = \tau_2(\sigma)$. Thus, we can reach the same configuration with a shorter schedule. More generally, we may replace a batch of a rule by a single accelerated transition whose factor is the sum of all factors in the batch.

In this section we give a bound on the diameter, i.e., the length of the shortest path between any two configurations σ and σ' for which there is a schedule τ applicable to σ satisfying $\sigma' = \tau(\sigma)$. A simple case is if $sort(\tau)$ is applicable to σ and each equivalence class defined by the precedence relation consists of a single rule (e.g., the control flow is a directed acyclic graph). Then by Definition 4 we have at most $|\mathcal{R}|$ batches in $sort(\tau)$, that is, one per rule. By the reasoning of above we can replace each batch by a single accelerated transition.

In general $sort(\tau)$ may not be applicable to σ, or there are equivalence classes containing multiple rules, i.e., rules form cycles in the precedence relation. The first issue comes from locking and unlocking. We identify milestone transitions, and show that two neighboring non-milestone transitions can be swapped according to *sort* in Section 4.3. We also deal with the issue of cycles in the precedence relation. It is ensured by *sort* that within a segment, all transitions that belong to a cycle form a batch. In Section 4.2, we replace such a batch by a batch where the remaining rules do not form a cycle. Removing cycles requires the assumption that shared variables are not incremented in cycles.

4.2 Removing Cycles

We consider the distance between two configurations σ and σ' that satisfy $\sigma.\mathbf{g} = \sigma'.\mathbf{g}$, i.e., along any schedule connecting these configurations, the values of shared variables are unchanged, and thus the evaluations of guards are also unchanged. By Definition 3, we can apply this section's result to batches of a class.

Definition 5. *Given a schedule* $\tau = t_1, t_2, \ldots$, *we denote by* $|\tau|$ *the length of the schedule. Further, we define the following vectors*

$$\mathbf{in}(\tau)[\ell] = \sum_{\substack{1 \le i \le |\tau| \\ t_i.to = \ell}} t_i.\mathit{factor}, \quad \mathbf{out}(\tau)[\ell] = \sum_{\substack{1 \le i \le |\tau| \\ t_i.\mathit{from} = \ell}} t_i.\mathit{factor}, \quad \mathbf{up}(\tau) = \sum_{1 \le i \le |\tau|} t_i.\mathbf{u}.$$

From the definition of a counter system, we directly obtain:

Proposition 2. *For all configurations* σ, *and all schedules* τ *applicable to* σ, *if* $\sigma' = \tau(\sigma)$, *then* $\sigma'.\boldsymbol{\kappa} = \sigma.\boldsymbol{\kappa} + \mathbf{in}(\tau) - \mathbf{out}(\tau)$, *and* $\sigma'.\mathbf{g} = \sigma.\mathbf{g} + \mathbf{up}(\tau)$.

Proposition 3. *For all configurations* σ, *and all schedules* τ *and* τ' *applicable to* σ, *if* $\mathbf{in}(\tau) = \mathbf{in}(\tau')$, $\mathbf{out}(\tau) = \mathbf{out}(\tau')$, *and* $\mathbf{up}(\tau) = \mathbf{up}(\tau')$, *then* $\tau(\sigma) = \tau'(\sigma)$.

Given a schedule $\tau = t_1, t_2, \ldots$ we say that the index set $I = \{i_1, \ldots, i_j\}$ forms a cycle in τ, if for all b, $1 \le b < j$, it holds that $t_{i_b}.to = t_{i_{b+1}}.\mathit{from}$, and $t_{i_j}.to = t_{i_1}.\mathit{from}$. Let $\mathcal{R}(\tau) = \{r \colon t_i \in \tau \wedge t_i.\mathit{rule} = r\}$.

Proposition 4. *For all schedules* τ, *if* τ *contains a cycle, then there is a schedule* τ' *satisfying* $|\tau'| < |\tau|$, $\mathbf{in}(\tau) = \mathbf{in}(\tau')$, $\mathbf{out}(\tau) = \mathbf{out}(\tau')$, *and* $\mathcal{R}(\tau') \subseteq \mathcal{R}(\tau)$.

Repeated application of the proposition leads to a cycle-free schedule (possibly the empty schedule), and we obtain:

Theorem 2. *For all schedules* τ, *there is a schedule* τ' *that contains no cycles,* $\mathbf{in}(\tau) = \mathbf{in}(\tau')$, $\mathbf{out}(\tau) = \mathbf{out}(\tau')$, *and* $\mathcal{R}(\tau') \subseteq \mathcal{R}(\tau)$.

The issue with this theorem is that τ' is not necessarily applicable to the same configurations as τ. In the following theorem, we prove that if a schedule satisfies a specific condition on the order of transitions, then it is applicable.

Theorem 3. *Let* σ *and* σ' *be two configurations with* $\sigma.\mathbf{g} = \sigma'.\mathbf{g}$, *and let* τ *be a schedule with* $\mathbf{up}(\tau) = \mathbf{0}$, *all transitions unlocked in* σ, *and where if* $t_i \to_\tau t_j$, *then* $t_j \not\prec_P t_i$. *If* $\sigma'.\boldsymbol{\kappa} - \sigma.\boldsymbol{\kappa} = \mathbf{in}(\tau) - \mathbf{out}(\tau)$, *then* τ *is applicable to* σ.

Corollary 1. *For all configurations* σ, *and all schedules* τ *applicable to* σ, *with* $\mathbf{up}(\tau) = \mathbf{0}$, *there is a schedule with at most one batch per rule applicable to* σ *satisfying that* τ' *contains no cycles,* $\tau'(\sigma) = \tau(\sigma)$, *and* $\mathcal{R}(\tau') \subseteq \mathcal{R}(\tau)$.

4.3 Identifying Milestones and Swapping Transitions

In this section we deal with locking and unlocking. To this end, we start by defining milestones. Then the central Theorem 4 establishes that two consequent non-milestone transitions can be swapped, if needed to sort the segment according to $\prec_C^{lin}$: the resulting schedule is still applicable, and leads to the same configuration as the original one.

Definition 6 (Left Milestone). *Given a configuration σ and a schedule $\tau = \tau' \cdot t \cdot \tau''$ applicable to σ, the transition t is a left milestone for σ and τ, if*

1. *there is a transition t' in τ' satisfying $t' \not\prec_P^+ t \wedge t' \prec_U t$,*
2. *$t.\varphi^{\le}$ is locked in σ, and*
3. *for all t' in τ', $t'.\varphi^{\le} \neq t.\varphi^{\le}$.*

Definition 7 (Right Milestone). *Given a configuration σ and a schedule $\tau = \tau' \cdot t \cdot \tau''$ applicable to σ, the transition t is a right milestone for σ and τ, if*

1. *there is a transition t'' in τ'' satisfying $t \not\prec_P^+ t'' \wedge t'' \prec_L t$,*
2. *$t.\varphi^{>}$ is locked in $\tau(\sigma)$, and*
3. *for all t'' in τ'', $t''.\varphi^{>} \neq t.\varphi^{>}$.*

Definition 8 (Segment). *Given a schedule τ and configuration σ, τ' is a segment if it is a subschedule of τ, and does not contain a milestone for σ and τ.*

Having defined milestones and segments, we arrive at our central result.

Theorem 4. *Let σ be a configuration, τ a schedule applicable to σ, and $\tau = \tau_1 \cdot t_1 \cdot t_2 \cdot \tau_2$. If transitions t_1 and t_2 are not milestones for σ and τ, and satisfy $[t_2] \prec_C^{lin} [t_1]$, then*

i. schedule $\tau' = \tau_1 \cdot t_2 \cdot t_1 \cdot \tau_2$ is applicable to σ,
ii. $\tau'(\sigma) = \tau(\sigma)$, and

Repeated application of the theorem leads to a schedule where milestones and sorted schedules alternate. By the definition of a milestone, there is at most one milestone per condition. Thus, the number of milestones is bounded by $\mathcal{C}$ (Definition 2). Together with Corollary 1, this is used to establish Theorem 1.

5 Experimental Evaluation

We have implemented the techniques in our tool ByMC [1]. Technical details about our approach to abstraction and refinement can be found in [13]. The input are the descriptions of our benchmarks in parametric Promela [17], which describe parameterized processes. Hence, as preliminary step ByMC computes the PIA data abstraction [16] to obtain finite state processes. Based on this, ByMC does preprocessing to compute threshold automata, the locking and unlocking relations, and to generate the inputs for our model checking back-ends.

Preprocessing. First, we compute the set of rules $\mathcal{R}$: Recall that a rule is a tuple $(\mathit{from}, \mathit{to}, \varphi^{\le}, \varphi^{>}, \mathbf{u})$. ByMC calls NuSMV to explore a single process system with unrestricted shared variables, in order to compute the (*from*, *to*) pairs. From this, ByMC computes the reachable local states. In the case of our benchmark CBC, e.g., that cuts the local states we have to consider from 2000 to 100, approximately. All our experiments — including the ones with FASTer [3] — are based on the reduced local state space. Then, for each pair (*from*, *to*), ByMC explores symbolic path to compute the guards and update vectors for the pair. This gives us the set of rules $\mathcal{R}$. Then, ByMC encodes Definition 1 in Yices, to construct the lock $\prec_L$ and unlock $\prec_U$ relations. Then, ByMC computes the relations $\{(r, r') : r' \not\preceq_P^+ r \ \wedge\ r' \prec_U r\}$ and $\{(r, r'') : r \not\preceq_P^+ r'' \ \wedge\ r'' \prec_L r\}$ as required by Definition 2. This provides the bounds.

Back-ends. ByMC generates the PIA counter abstraction [16] to be used by the following back-end model checkers. We have also implemented an automatic abstraction refinement loop for the counterexamples provided by NuSMV.
BMC. NuSMV 2.5.4 [10] (using MiniSAT) performs incremental bounded model checking with the bound $\hat{d}$. If a counterexample is reported, ByMC refines the system as explained in [16], if the counterexample is spurious.
BMCL. We combine NuSMV with the multi-core SAT solver Plingeling [5]: NuSMV does bounded model checking for 30 steps. Spurious counterexample are refined by ByMC. If there is no counterexample, NuSMV produces a single CNF formula with the bound $\hat{d}$, whose satisfiability is then checked with Plingeling.
BDD. NuSMV 2.5.4 performs BDD-based symbolic checking.
FAST. FASTer 2.1 [3] performs reachability analysis using plugin Mona-1.3.

5.1 Benchmarks

We encoded several asynchronous FTDAs in our parametric Promela, following the technique in [17]; they can be obtained from [1]. All models contain transitions with lower threshold guards. The benchmarks CBC also contain upper threshold guards. If we ignore self-loops, the precedence relation of all but NBAC and NBACC, which have non-trivial cycles, are partial orders.
Folklore Reliable Broadcast (FRB) [9]. In this algorithm, n processes have to agree on whether a process has broadcast a message, in the presence of $f \le n$ crashes. Our model of FRB has one shared variable and the abstract domain of two intervals $[0, 1)$ and $[1, \infty)$. In this paper, we are concerned with the safety property *unforgeability*: If no process is initialized with value 1 (message from the broadcaster), then no correct process ever accepts.
Consistent Broadcast (STRB) [31]. Here, we have $n - f$ correct processes and $f \ge 0$ Byzantine faulty ones. The resilience condition is $n > 3t \wedge t \ge f$. There is one shared variable and the abstract domain of four intervals $[0, 1)$, $[1, t + 1)$, $[t + 1, n - t)$, and $[n - t, \infty)$. Here, we check only unforgeability (see FRB), whereas in [16] we checked also liveness properties.
Byzantine Agreement (ABA) [8]. There are $n > 3t$ processes, $f \le t$ of them Byzantine faulty. The model has two shared variables. We have to consider two different cases for the abstract domain, namely, case ABA0 with the domain $[0, 1)$, $[1, t + 1)$, $[t + 1, \lceil \frac{n+t}{2} \rceil)$, and $[\lceil \frac{n+t}{2} \rceil, \infty)$ and case ABA1 with the domain $[0, 1)$, $[1, t + 1)$, $[t + 1, 2t + 1)$, $[2t + 1, \lceil \frac{n+t}{2} \rceil)$, and $[\lceil \frac{n+t}{2} \rceil, \infty)$. As for FRB, we check unforgeability. This case study, and all below, run out of memory when using Spin for model checking the counter abstraction [16].
Condition-Based Consensus (CBC) [27]. This is a restricted form of consensus solvable in asynchronous systems. We consider binary condition-based consensus in the presence of clean crashes, which requires four shared variables.

Table 1. Summary of experiments on AMD Opteron®Processor 6272 with 192 GB RAM and 32 CPU cores. Plingeling used up to 16 cores. "TO" denotes timeout of 24 hours; "OOM" denotes memory overrun of 64 GB; "ERR" denotes runtime error; "RTO" denotes that the refinement loop timed out.

Input	**Threshold A.**				**Bounds**			**Time, [HH:]MM:SS**				**Memory, GB**			
FTDA	$\lvert\mathcal{L}\rvert$	$\lvert\mathcal{R}\rvert$	$\mathcal{C}^{\leq}$	$\mathcal{C}^{>}$	d	$d^\star$	$\hat{d}$	BMCL	BMC	BDD	FAST	BMCL	BMC	BDD	FAST
FIG. 1	5	5	1	0	11	9	27	00:00:03	00:00:04	00:01	00:00:08	0.01	0.02	0.02	0.06
FRB	6	8	1	0	17	10	10	00:00:13	00:00:13	00:06	00:00:08	0.01	0.02	0.02	0.01
STRB	7	15	3	0	63	30	90	00:00:09	00:00:06	00:04	00:00:07	0.02	0.03	0.02	0.07
ABA0	37	180	6	0	1266	586	1758	00:21:26	02:20:10	00:15	00:08:40	6.37	1.49	0.07	3.56
ABA1	61	392	8	0	3536	1655	6620	TO 25%	TO 12%	00:33	02:36:25	TO	TO	0.08	15.65
CBC0	43	204	0	0	204	204	612	01:38:54	TO 57%	OOM	ERR	1.28	TO	OOM	ERR
CBC1	115	896	1	1	2690	2180	8720	TO 05%	TO 11%	TO	TO	TO	TO	TO	TO
NBACC	109	1724	6	0	12074	5500	16500	RTO	RTO	TO	TO	RTO	RTO	TO	TO
NBAC	77	1356	6	0	9498	4340	13020	RTO	RTO	TO	TO	RTO	RTO	TO	TO
WHEN A BUG IS INTRODUCED															
ABA0	32	139	6	0	979	469	1407	00:00:16	00:00:18	TO	00:05:57	0.04	0.04	TO	2.70
ABA1	54	299	8	0	2699	1305	5220	00:00:22	00:00:21	TO	ERR	0.06	0.06	TO	ERR

Under the resilience condition $n > 2t \wedge f \geq 0$, we have to consider two different cases depending on f: If $f = 0$ we have case CBC0 with the domain $[0, 1)$, $[1, \lceil\frac{n}{2}\rceil)$, $[\lceil\frac{n}{2}\rceil, n - t)$, and $[n - t, \infty)$. If $f \neq 0$, case CBC1 has the domain: $[0, 1)$, $[1, f)$, $[f, \lceil\frac{n}{2}\rceil)$, $[\lceil\frac{n}{2}\rceil, n - t)$, and $[n - t, \infty)$. We verified several properties, all of which resulted in experiments with similar characteristics. We only give $validity_0$ in the table, i.e., no process accepts value 0, if all processes initially have value 1.

Non-blocking Atomic Commitment (NBAC and NBACC) [30,15]. Here, n processes are initialized with YES or NO and decide on whether to commit a transaction. The transaction must be aborted if at least one process is initialized to NO. We consider the cases NBACC and NBAC of clean crashes and crashes, respectively. Both models contain four shared variables, and the abstract domain is $[0, 1)$ and $[1, n)$ and $[n-1, n)$, and $[n, \infty)$. The algorithm uses a failure detector, which is modeled as local variable that changes its value non-deterministically.

5.2 Evaluation

Table 1 summarizes the experiments. For the threshold automata, we give the number of local states $|\mathcal{L}|$, rules $|\mathcal{R}|$, and conditions according to Definition 2, i.e., $\mathcal{C}^{\leq}$ and $\mathcal{C}^{>}$. The column $\boldsymbol{d}$ provides the bound on the diameter as in Theorem 1, whereas the column $\boldsymbol{d}^\star$ provides an improved diameter: In the proof of Theorem 1, we bound the length of all segments by $|\mathcal{R}|$. However, by Definition 6, segments to the left of a left milestone cannot contain transitions for rules with the same condition as the milestone. The same is true for segments to the right of right milestones. BYMC explores all orders of milestones, an uses this observation about milestones to compute a more precise bound $\boldsymbol{d}^\star$ for the diameter. Our encoding of the counter abstraction only increments and decrements counters. If $|\hat{D}|$ is the size of the abstract domain, a transition in a counter system is simulated by at most $|\hat{D}| - 1$ steps in the counter abstraction; this leads to the diameter $\hat{d}$ for counter abstractions, which we use in our experiments.

As the experiments show, all techniques rapidly verify FRB, STRB, and FIG. 1. FRB and STRB had already been verified before using SPIN [16]. The more challenging examples are ABA0 and ABA1, where BDD clearly outperforms the other techniques. Bounded model checking is slower here, because the diameter bound does not exploit knowledge on the specification. FAST performs well on these benchmarks. We believe this is because many rules are always disabled, due to the initial states as given in the specification. To confirm this intuition, we introduced a bug into ABA0 and ABA1, which allows the processes to non-deterministically change their value to 1. This led to a dramatic slowdown of BDD and FAST, as reflected in the last two lines.

Using the bounds of this paper, BMCL verified CBC0, whereas all other techniques failed. BMCL did not reach the bounds for CBC1 with our experimental setup, but we believe that the bound is within the reach with a better hardware or an improved implementation. In this case, we report the percentages of the bounds we reached with bounded model checking.

In the experiments with NBAC and NBACC, the refinement loop timed out. We are convinced that we can address this issue by integrating the refinement loop with an incremental bounded model checker.

6 Related Work and Discussions

Specific forms of counter systems can be used to model parameterized systems of concurrent processes. Lubachevsky [25] discusses *compact* programs that reach each state in a bounded number of steps, where the bound is independent of the number of processes. In [25] he gives examples of compact programs, and in [24] he proves that specific semaphore programs are compact. We not only show compactness, but give a bound on the diameter. In our case, communication is not restricted to semaphores, but we have threshold guards. Counter abstraction [29] follows this line of research, but as discussed in [4], does not scale well for large numbers of local states.

Acceleration in infinite state systems (e.g., in flat counter automata [22]) is a technique that computes the transitive closure of a transition relation and applies it to the set of states. The tool FAST [2] uses acceleration to compute the set of reachable states in a symbolic procedure. This appears closely related to our acceleration factors. However, in [2] a transition is chosen and accelerated dynamically in the course of symbolic state space exploration, while we statically use acceleration factors and reordering of transitions.

One achieves completeness for reachability in bounded model checking by exploring all runs that are not longer than the diameter of the system [6]. The notion of *completeness threshold* [11] generalizes this idea to safety and liveness properties. As in general, computing the diameter is believed to be as hard as the model checking problem, one can use a coarser bound provided by the reoccurrence diameter [19]. In practice, the reoccurrence diameter of counter abstraction is prohibitively large, so that we give bounds on the diameter.

Partial orders are a useful concept for reasoning about distributed systems [20]. In model checking, *partial order reduction* [14,32,28] is used to reduce the search space. It is based on the idea that changing the order of steps of concurrent processes leads to "equivalent" behavior with respect to the specification. Typically, partial order reduction is used on-the-fly to prune runs that are equivalent to representative ones. In contrast, we bound the length of representative runs offline in order to ensure completeness of bounded model checking. A partial order reduction for threshold-guarded FTDAs was introduced in [7]. It can be used for model checking small instances, while we focus on parameterized model checking.

Our technique of determining which transitions can be swapped in a run reminds of *movers* as discussed by Lipton [23], or more generally the idea to show that certain actions can be grouped into larger atomic blocks to simplify proofs [12,21]. Movers address the issue of grouping many local transitions of a process together. In contrast, we conceptually group transitions of different processes together into one accelerated transition. Moreover, the definition of a mover by Lipton is independent of a specific run: a left mover (e.g., a "release" operation) is a transition that in *all runs* can "move to the left" with respect to transitions of other processes. In our work, we look at individual runs and identify which transitions (milestones) must not move in this run.

As next steps we will focus on liveness of fault-tolerant distributed algorithms. In fact the liveness specifications are in the fragment of linear temporal logic for which it is proven [18] that a formula can be translated into a cliquey Büchi automaton. For such automata, [18] provides a completeness threshold. Still, there are open questions related to applying our results to the idea of [18].

References

1. ByMC: Byzantine model checker (2013), `http://forsyte.tuwien.ac.at/software/bymc/` (accessed: June 2014)
2. Bardin, S., Finkel, A., Leroux, J., Petrucci, L.: Fast: acceleration from theory to practice. STTT 10(5), 401–424 (2008)
3. Bardin, S., Leroux, J., Point, G.: Fast extended release. In: Ball, T., Jones, R.B. (eds.) CAV 2006. LNCS, vol. 4144, pp. 63–66. Springer, Heidelberg (2006)
4. Basler, G., Mazzucchi, M., Wahl, T., Kroening, D.: Symbolic counter abstraction for concurrent software. In: Bouajjani, A., Maler, O. (eds.) CAV 2009. LNCS, vol. 5643, pp. 64–78. Springer, Heidelberg (2009)
5. Biere, A.: Lingeling, Plingeling and Treengeling entering the SAT competition 2013. In: Proceedings of SAT Competition 2013; Solver and p. 51 (2013)
6. Biere, A., Cimatti, A., Clarke, E.M., Zhu, Y.: Symbolic model checking without bdds. In: Cleaveland, W.R. (ed.) TACAS 1999. LNCS, vol. 1579, pp. 193–207. Springer, Heidelberg (1999)
7. Bokor, P., Kinder, J., Serafini, M., Suri, N.: Efficient model checking of fault-tolerant distributed protocols. In: DSN, pp. 73–84 (2011)
8. Bracha, G., Toueg, S.: Asynchronous consensus and broadcast protocols. J. ACM 32(4), 824–840 (1985)
9. Chandra, T.D., Toueg, S.: Unreliable failure detectors for reliable distributed systems. JACM 43(2), 225–267 (1996)

10. Cimatti, A., Clarke, E.M., Giunchiglia, E., Giunchiglia, F., Pistore, M., Roveri, M., Sebastiani, R., Tacchella, A.: Nusmv 2: An opensource tool for symbolic model checking. In: Brinksma, E., Larsen, K.G. (eds.) CAV 2002. LNCS, vol. 2404, pp. 359–364. Springer, Heidelberg (2002)
11. Clarke, E., Kroning, D., Ouaknine, J., Strichman, O.: Completeness and complexity of bounded model checking. In: Steffen, B., Levi, G. (eds.) VMCAI 2004. LNCS, vol. 2937, pp. 85–96. Springer, Heidelberg (2004)
12. Doeppner, T.W.: Parallel program correctness through refinement. In: POPL, pp. 155–169 (1977)
13. Gmeiner, A., Konnov, I., Schmid, U., Veith, H., Widder, J.: Tutorial on parameterized model checking of fault-tolerant distributed algorithms. In: Bernardo, M., Damiani, F., Hähnle, R., Johnsen, E.B., Schaefer, I. (eds.) SFM 2014. LNCS, vol. 8483, pp. 122–171. Springer, Heidelberg (2014)
14. Godefroid, P.: Using partial orders to improve automatic verification methods. In: Clarke, E., Kurshan, R.P. (eds.) CAV 1990. LNCS, vol. 531, pp. 176–185. Springer, Heidelberg (1991)
15. Guerraoui, R.: Non-blocking atomic commit in asynchronous distributed systems with failure detectors. Distributed Computing 15(1), 17–25 (2002)
16. John, A., Konnov, I., Schmid, U., Veith, H., Widder, J.: Parameterized model checking of fault-tolerant distributed algorithms by abstraction. In: FMCAD, pp. 201–209 (2013)
17. John, A., Konnov, I., Schmid, U., Veith, H., Widder, J.: Towards modeling and model checking fault-tolerant distributed algorithms. In: Bartocci, E., Ramakrishnan, C.R. (eds.) SPIN 2013. LNCS, vol. 7976, pp. 209–226. Springer, Heidelberg (2013)
18. Kroening, D., Ouaknine, J., Strichman, O., Wahl, T., Worrell, J.: Linear completeness thresholds for bounded model checking. In: Gopalakrishnan, G., Qadeer, S. (eds.) CAV 2011. LNCS, vol. 6806, pp. 557–572. Springer, Heidelberg (2011)
19. Kroning, D., Strichman, O.: Efficient computation of recurrence diameters. In: Zuck, L.D., Attie, P.C., Cortesi, A., Mukhopadhyay, S. (eds.) VMCAI 2003. LNCS, vol. 2575, pp. 298–309. Springer, Heidelberg (2002)
20. Lamport, L.: Time, clocks, and the ordering of events in a distributed system. Commun. ACM 21(7), 558–565 (1978)
21. Lamport, L., Schneider, F.B.: Pretending atomicity. Tech. Rep. 44, SRC (1989)
22. Leroux, J., Sutre, G.: Flat counter automata almost everywhere! In: Peled, D.A., Tsay, Y.-K. (eds.) ATVA 2005. LNCS, vol. 3707, pp. 489–503. Springer, Heidelberg (2005)
23. Lipton, R.J.: Reduction: A method of proving properties of parallel programs. Commun. ACM 18(12), 717–721 (1975)
24. Lubachevsky, B.D.: An approach to automating the verification of compact parallel coordination programs. II. Tech. Rep. 64, New York University. Computer Science Department (1983)
25. Lubachevsky, B.D.: An approach to automating the verification of compact parallel coordination programs. I. Acta Informatica 21(2), 125–169 (1984)
26. Lynch, N.: Distributed Algorithms. Morgan Kaufman (1996)
27. Mostéfaoui, A., Mourgaya, E., Parvédy, P.R., Raynal, M.: Evaluating the condition-based approach to solve consensus. In: DSN, pp. 541–550 (2003)
28. Peled, D.: All from one, one for all: on model checking using representatives. In: Courcoubetis, C. (ed.) CAV 1993. LNCS, vol. 697, pp. 409–423. Springer, Heidelberg (1993)

29. Pnueli, A., Xu, J., Zuck, L.D.: Liveness with $(0, 1, \infty)$- counter abstraction. In: Brinksma, E., Larsen, K.G. (eds.) CAV 2002. LNCS, vol. 2404, pp. 107–122. Springer, Heidelberg (2002)
30. Raynal, M.: A case study of agreement problems in distributed systems: Non-blocking atomic commitment. In: HASE, pp. 209–214 (1997)
31. Srikanth, T., Toueg, S.: Simulating authenticated broadcasts to derive simple fault-tolerant algorithms. Dist. Comp. 2, 80–94 (1987)
32. Valmari, A.: Stubborn sets for reduced state space generation. In: Rozenberg, G. (ed.) APN 1990. LNCS, vol. 483, pp. 491–515. Springer, Heidelberg (1991)

Lost in Abstraction: Monotonicity in Multi-threaded Programs*

Alexander Kaiser[1], Daniel Kroening[1], and Thomas Wahl[2]

[1] University of Oxford, United Kingdom
[2] Northeastern University, Boston, United States

Abstract. *Monotonicity* in concurrent systems stipulates that, in any global state, extant system actions remain executable when new processes are added to the state. This concept is not only natural and common in multi-threaded software, but also useful: if every thread's memory is finite, monotonicity often guarantees the decidability of safety property verification even when the number of running threads is unknown. In this paper, we show that the act of obtaining finite-data thread abstractions for model checking can be at odds with monotonicity: Predicate-abstracting certain widely used monotone software results in non-monotone multi-threaded Boolean programs — the monotonicity is *lost in the abstraction*. As a result, well-established sound and complete safety checking algorithms become inapplicable; in fact, safety checking turns out to be undecidable for the obtained class of unbounded-thread Boolean programs. We demonstrate how the abstract programs can be modified into monotone ones, without affecting safety properties of the non-monotone abstraction. This significantly improves earlier approaches of enforcing monotonicity via overapproximations.

1 Introduction

This paper addresses non-recursive procedures executed by multiple threads (e.g. dynamically generated, and possibly unbounded in number), which communicate via shared variables or higher-level mechanisms such as mutexes. OS-level code, including Windows, UNIX, and Mac OS device drivers, makes frequent use of such concurrency APIs, whose correct use is therefore critical to ensure a reliable programming environment.

The utility of *predicate abstraction* as a safety analysis method is known to depend critically on the choice of predicates: the consequences of a poor choice range from inferior performance to flat-out unprovability of certain properties. We propose in this paper an extension of predicate abstraction to multi-threaded programs that enables reasoning about intricate data relationships, namely

shared-variable: "shared variables `s` and `t` are equal",
single-thread: "local variable `l` of thread i is less than shared variable `s`", and
inter-thread: "local variable `l` of thread i is less than variable `l` *in all other threads*".

* This work is supported by the Toyota Motor Corporation, NSF grant no. 1253331 and ERC project 280053.

P. Baldan and D. Gorla (Eds.): CONCUR 2014, LNCS 8704, pp. 141–155, 2014.

Why such a rich predicate language? For certain concurrent algorithms such as the widely used *ticket* busy-wait lock algorithm [4] (the default locking mechanism in the Linux kernel since 2008; see Fig. 1), the verification of elementary safety properties **requires** single- and inter-thread relationships. They are needed to express, for instance, that a thread holds the minimum ticket value, an inter-thread relationship.

In the main part of the paper, we address the problem of full parameterized (unbounded-thread) program verification with respect to our rich predicate language. Such reasoning requires first that the n-thread abstract program $\hat{\mathcal{P}}^n$, obtained by existential inter-thread predicate abstraction of the n-thread concrete program $\mathcal{P}^n$, is rewritten into a single template program $\tilde{\mathcal{P}}$ to be executed by (any number of) multiple threads. In order to capture the semantics of these programs in the template $\tilde{\mathcal{P}}$, the template programming language must itself permit variables that refer to the currently executing or a generic passive thread; we call such programs *dual-reference (DR)*. We describe how to obtain $\tilde{\mathcal{P}}$, namely essentially as an overapproximation of $\hat{\mathcal{P}}^{\mathsf{b}}$, for a constant b that scales linearly with the number of inter-thread predicates used in the predicate abstraction.

Given the *Boolean* dual-reference program $\tilde{\mathcal{P}}$, we might now expect the unbounded-thread replicated program $\tilde{\mathcal{P}}^\infty$ to form a classical *well quasi-ordered transition system* [2], enabling the fully automated, algorithmic safety property verification in the abstract. This turns out not to be the case: the expressiveness of dual-reference programs renders parameterized program location reachability undecidable, despite the finite-domain variables. The root cause is the lack of *monotonicity* of the transition relation with respect to the standard partial order over the space of unbounded thread counters. That is, adding passive threads to the source state of a valid transition can invalidate this transition and in fact block the system. Since the input C programs are, by contrast, perfectly monotone, we say the monotonicity is *lost in the abstraction*. As a result, our abstract programs are in fact not well quasi-ordered.

Inspired by earlier work on *monotonic abstractions* [3], we address this problem by restoring the monotonicity using a simple *closure operator*, which enriches the transition relation of the abstract program $\tilde{\mathcal{P}}$ such that the obtained program $\tilde{\mathcal{P}}_m$ engenders a monotone (and thus well quasi-ordered) system. The closure operator essentially terminates passive threads that block transitions allowed by other passive threads. In contrast to those earlier approaches, which *enforce* (rather than restore) monotonicity in genuinely non-monotone systems, we exploit the fact that the input programs are monotone. As a result, the monotonicity closure $\tilde{\mathcal{P}}_m$ can be shown to be *safety-equivalent* to the intermediate program $\tilde{\mathcal{P}}$.

To summarize, the central contribution of this paper is a predicate abstraction strategy for unbounded-thread C programs, with respect to the rich language of inter-thread predicates. This language allows the abstraction to track properties that are essentially universally quantified over all passive threads. To this end, we first develop such a strategy for a fixed number of threads. Second, in preparation for extending it to the unbounded case, we describe how the abstract model, obtained by existential predicate abstraction for a given thread count n, can be expressed as a template program that can be multiply instantiated. Third, we show a sound and complete algorithm for reachability analysis for the obtained parameterized Boolean dual-reference programs.

```
struct Spinlock {
    natural s := 1; // ticket being served
    natural t := 1; }; // next free ticket

struct Spinlock lock; // shared

void spin_lock() {
    natural l := 0; // local
  ℓ1: l := fetch_and_add(lock.t);
  ℓ2: while (l ≠ lock.s)
          /* spin */; }

void spin_unlock() {
  ℓ3: lock.s++; }
```

The ticket algorithm: Shared variable *lock* has two integer components: `s` holds the ticket currently served (or, if none, the ticket served next), while `t` holds the ticket to be served after all waiting threads have had access. To request access to the locked region, a thread atomically retrieves the value of `t` and then increments `t`. The thread then busy-waits ("spins") until local variable `l` agrees with shared `s`. To unlock, a thread increments `s`.

See [21] for more intuition.

Fig. 1. Our goal is to verify "unbounded-thread mutual exclusion": no matter how many threads try to acquire and release the lock concurrently, no two of them should simultaneously be between the calls to functions spin_lock and spin_unlock

We overcome the undecidability of the problem by building a monotone closure that enjoys the same safety properties as the original abstract dual-reference program.

We omit in this paper practical aspects such as predicate discovery, the algorithmic construction of the abstract programs, and abstraction refinement. In our technical report [21], we provide, however, an extensive appendix, with proofs of all lemmas and theorems.

2 Inter-Thread Predicate Abstraction

In this section we introduce single- and inter-thread predicates, with respect to which we then formalize existential predicate abstraction. Except for the extended predicate language, these concepts are mostly standard and lay the technical foundations for the contributions of this paper.

2.1 Input Programs and Predicate Language

2.1.1 Asynchronous Programs. An *asynchronous program* $\mathcal{P}$ allows only one thread at a time to change its local state. We model $\mathcal{P}$, designed for execution by $n \geq 1$ concurrent threads, as follows. The variable set V of a program $\mathcal{P}$ is partitioned into sets S and L. The variables in S, called *shared*, are accessible jointly by all threads, and those in L, called *local*, are accessible by the individual thread that owns the variable. We assume the statements of $\mathcal{P}$ are given by a transition formula $\mathcal{R}$ over unprimed (current-state) and primed (next-state) variables, V and $V' = \{\mathrm{v}' : \mathrm{v} \in V\}$. Further, the initial states are characterized by the initial formula $\mathcal{I}$ over V. We assume $\mathcal{I}$ is expressible in a suitable logic for which existential quantification is computable (required later for the abstraction step).

As usual, the computation may be controlled by a local program counter pc, and involve non-recursive function calls. When executed by n threads, $\mathcal{P}$ gives rise to n-*thread program states* consisting of the valuations of the variables in $V_n = S \cup L_1 \cup \ldots L_n$, where $L_i = \{\mathtt{l}_i : \mathtt{l} \in L\}$. We call a variable set *uniformly indexed* if its variables either all have no index, or all have the same index. For a formula f and two uniformly-indexed variable sets X_1 and X_2, let $f\{X_1 \triangleright X_2\}$ denote f after replacing every occurrence of a variable in X_1 by the variable in X_2 with the same base name, if any; unreplaced if none. We write $f\{X_1 \blacktriangleright X_2\}$ short for $f\{X_1 \triangleright X_2\}\{X_1{}' \triangleright X_2{}'\}$. As an example, given $S = \{\mathtt{s}\}$ and $L = \{\mathtt{l}\}$, we have $(\mathtt{l}' = \mathtt{l} + \mathtt{s})\{L \blacktriangleright L_a\} = (\mathtt{l}'_a = \mathtt{l}_a + \mathtt{s})$. Finally, let $X \mathrel{\mathring{=}} X'$ stand for $\forall x \in X : x = x'$.

The n-*thread instantiation* $\mathcal{P}^n$ is defined for $n \geq 1$ as

$$\mathcal{P}^n = (\mathcal{R}^n, \mathcal{I}^n) = \left(\bigvee_{a=1}^{n} (\mathcal{R}_a)^n, \bigwedge_{a=1}^{n} \mathcal{I}\{L \triangleright L_a\}\right) \tag{1}$$

where
$$(\mathcal{R}_a)^n :: \mathcal{R}\{L \blacktriangleright L_a\} \wedge \bigwedge_{p:p \neq a} L_p \mathrel{\mathring{=}} L'_p \,. \tag{2}$$

Formula $(\mathcal{R}_a)^n$ asserts that the shared variables, and the variables of the *active* (executing) thread a are updated according to $\mathcal{R}$, while the local variables of passive threads $p \neq a$ are not modified (p ranges over $\{1, \ldots, n\}$). A state is *initial* if all threads are in a state satisfying $\mathcal{I}$. An n-*thread execution* is a sequence of n-thread program states whose first state satisfies $\mathcal{I}^n$ and whose consecutive states are related by $\mathcal{R}^n$. We assume the existence of an error location in $\mathcal{P}$; an *error state* is one where some thread resides in the error location. $\mathcal{P}$ is *safe* if no execution exists that ends in an error state. Mutex conditions can be checked using a ghost semaphore and redirecting threads to the error location if they try to access the critical section while the semaphore is set.

2.1.2 Predicate Language.

We extend the predicate language from [10] to allow the use of the *passive-thread variables* $L_P = \{\mathtt{l}_P : \mathtt{l} \in L\}$, each of which represents a local variable owned by a generic passive thread. The presence of variables of various categories gives rise to the following predicate classification.

Definition 1. *A predicate Q over S, L and L_P is* **shared** *if it contains variables from S only,* **local** *if it contains variables from L only,* **single-thread** *if it contains variables from L but not from L_P, and* **inter-thread** *if it contains variables from L and from L_P.*

Single- and inter-thread prediactes may contain variables from S. For example, in the ticket algorithm (Fig. 1), with $S = \{\mathtt{s}, \mathtt{t}\}$ and $L = \{\mathtt{l}\}$, examples of shared, local, single- and inter-thread predicates are: $\mathtt{s} = \mathtt{t}$, $\mathtt{l} = 5$, $\mathtt{s} = \mathtt{l}$ and $\mathtt{l} \neq \mathtt{l}_P$, respectively.

Semantics. Let $Q[1], \ldots, Q[m]$ be m predicates (any class). Predicate $Q[i]$ is evaluated in a given n-thread state v ($n \geq 2$) with respect to a choice of active thread a:

$$Q[i]_a \quad :: \quad \bigwedge_{p:p \neq a} Q[i]\{L \triangleright L_a\}\{L_P \triangleright L_p\} \,. \tag{3}$$

As special cases, for single-thread and shared predicates (no L_P variables), we have $Q[i]_a = Q[i]\{L \triangleright L_a\}$ and $Q[i]_a = Q[i]$, resp. We write $v \models Q[i]_a$ if $Q[i]_a$ holds in state v. Predicates $Q[i]$ give rise to an abstraction function α, mapping each n-thread program state v to an $m \times n$ bit matrix with entries

$$\alpha(v)_{i,a} = \begin{cases} \mathtt{T} & \text{if } v \models Q[i]_a \\ \mathtt{F} & \text{otherwise}. \end{cases} \tag{4}$$

Function α partitions the n-thread program state space via m predicates into $2^{m\times n}$ equivalence classes. As an example, consider the inter-thread predicates $\mathtt{l} \leq \mathtt{l}_P$, $\mathtt{l} > \mathtt{l}_P$, and $\mathtt{l} \neq \mathtt{l}_P$ for a local variable $\mathtt{l}$, $n = 4$ and the state $v :: (\mathtt{l}_1, \mathtt{l}_2, \mathtt{l}_3, \mathtt{l}_4) = (4, 4, 5, 6)$:

$$\alpha(v) = \begin{pmatrix} \mathtt{T\,T\,F\,F} \\ \mathtt{F\,F\,F\,T} \\ \mathtt{F\,F\,T\,T} \end{pmatrix}. \tag{5}$$

In the matrix, row $i \in \{1, 2, 3\}$ lists the truth of predicate $Q[i]$ for each of the four threads in the active role. Predicate $\mathtt{l} \leq \mathtt{l}_P$ captures whether a thread owns the minimum value for local variable $\mathtt{l}$ (true for $a = 1, 2$); $\mathtt{l} > \mathtt{l}_P$ tracks whether a thread owns the *unique* maximum value (true for $a = 4$) ; finally $\mathtt{l} \neq \mathtt{l}_P$ captures the uniqueness of a thread's copy of $\mathtt{l}$ (true for $a = 3, 4$).

Inter-thread predicates and abstraction. Predicates that reason universally about threads have been used successfully as targets in (inductive) invariant generation procedures [5,24]. In this paper we discuss their role in abstractions. The use of these fairly expressive and presumably expensive predicates is not by chance: automated methods that cannot reason about them [13,10,26] essentially fail for the ticket algorithm in Fig. 1: for a fixed number of threads that concurrently and repeatedly (e.g. in an infinite loop) request and release lock ownership, the inter-thread relationships need to be "simulated" via enumeration, incurring very high time and space requirements, even for a handful of threads. In the unbounded-thread case, they diverge. This is essentially due to known limits of thread-modular and Owicki-Gries style proof systems, which do not have access to inter-thread predicates [23]. In [21], we show that the number of *single-thread* predicates needed to prove correctness of the ticket algorithm depends on n, from which unprovability in the unbounded case follows.

2.2 Existential Inter-Thread Predicate Abstraction

Embedded into our formalism, the goal of *existential predicate abstraction* [8,18] is to derive an abstract program $\hat{\mathcal{P}}^n$ by treating the equivalence classes induced by Eq. (4) as abstract states. $\hat{\mathcal{P}}^n$ thus has $m \times n$ Boolean variables:

$$\hat{V}_n = \bigcup_{a=1}^{n} \hat{L}_a = \bigcup_{a=1}^{n} \{b[i]_a : 1 \leq i \leq m\}.$$

Variable $b[i]_a$ tracks the truth of predicate $Q[i]$ for active thread a. This is formalized in (6), relating concrete and abstract n-thread states (valuations of V_n and $\hat{V}_n$, resp.):

$$\mathcal{D}^n :: \bigwedge_{i=1}^{m} \bigwedge_{a=1}^{n} b[i]_a \Leftrightarrow Q[i]_a . \tag{6}$$

For a formula f, let f' denote f after replacing each variable by its primed version. We then have $\hat{\mathcal{P}}^n = (\hat{\mathcal{R}}^n, \hat{\mathcal{I}}^n) = \left(\bigvee_{a=1}^{n} (\hat{\mathcal{R}}_a)^n,\ \hat{\mathcal{I}}^n\right)$ where

$$(\hat{\mathcal{R}}_a)^n :: \exists V_n V_n' : (\mathcal{R}_a)^n \wedge \mathcal{D}^n \wedge (\mathcal{D}^n)', \tag{7}$$

$$\hat{\mathcal{I}}^n :: \exists V_n \quad : \mathcal{I}^n \wedge \mathcal{D}^n . \tag{8}$$

As an example, consider the decrement operation $\mathtt{l} := \mathtt{l} - 1$ on a local integer variable $\mathtt{l}$, and the inter-thread predicate $\mathtt{l} < \mathtt{l}_P$. Using Eq. (7) with $n = 2$, $a = 1$, we get 4 abstract transitions, which are listed in Table 1. The table shows that the abstraction is no longer asynchronous (treating $\mathtt{b}_1$ as belonging to thread 1, $\mathtt{b}_2$ to thread 2): in the highlighted transition, the executing thread 1 changes (its pc and hence) its local state, and so does thread 2. By contrast, on the right we have $\mathtt{l}_2 = \mathtt{l}_2'$ in all rows. The loss of asynchrony will become relevant in Sect. 3, where we define a suitable abstract Boolean programming language (which then necessarily must accommodate non-asynchronous programs).

Table 1. Abstraction $(\hat{\mathcal{R}}_1)^2$ for stmt. $\mathtt{l} := \mathtt{l} - 1$ against predicate $\mathtt{l} < \mathtt{l}_P$ (left); concrete witness transitions, i.e. elements of $(\mathcal{R}_1)^2$ (right). The highlighted row indicates asynchrony violations.

$\mathtt{b}_1$	$\mathtt{b}_2$	$\mathtt{b}_1'$	$\mathtt{b}_2'$	$\mathtt{l}_1$	$\mathtt{l}_2$	$\mathtt{l}_1'$	$\mathtt{l}_2'$
F	F	T	F	1	1	0	1
F	T	F	F	1	0	0	0
F	T	F	T	2	0	1	0
T	F	T	F	1	2	0	2

Proving the ticket algorithm (fixed-thread case). As in any existential abstraction, the abstract program $\hat{\mathcal{P}}^n$ overapproximates (the set of executions of) the concrete program $\mathcal{P}^n$; the former can therefore be used to verify safety of the latter. We illustrate this using the ticket algorithm (Fig. 1). Consider the predicates $Q[1] :: \mathtt{l} \neq \mathtt{l}_P$, $Q[2] :: \mathtt{t} > \max(\mathtt{l}, \mathtt{l}_P)$, and $Q[3] :: \mathtt{s} = \mathtt{l}$. The first two are inter-thread; the third is single-thread. The predicates assert the uniqueness of a ticket ($Q[1]$), that the next free ticket is larger than all tickets currently owned by threads ($Q[2]$), and that a thread's ticket is currently being served ($Q[3]$). The abstract reachability tree for $\hat{\mathcal{P}}^n$ and these predicates reveals that mutual exclusion is satisfied: there is no state with both threads in location ℓ_3. The tree grows exponentially with n.

3 From Existential to Parametric Abstraction

Classical existential abstraction as described in Sect. 2.2 obliterates the symmetry present in the concrete concurrent program, which is given as the n-thread instantiation of a single-thread template $\mathcal{P}$: the abstraction is instead formulated via predicates over the explicitly expanded n-thread program $\mathcal{R}^n$. As observed in previous work [10], such a "symmetry-oblivious" approach suffers from poor scalability for fixed-thread verification problems. Moreover, *parametric* reasoning over an unknown number of threads is impossible since the abstraction (7) directly depends on n.

To overcome these problems, we now derive an overapproximation of $\hat{\mathcal{P}}^n$ via a generic program template $\tilde{\mathcal{P}}$ that can be instantiated for any n. There is, however, one obstacle: instantiating a program (such as $\mathcal{P}$) formulated over shared variables and one copy of the thread-local variables naturally gives rise to asynchronous concurrency. The programs resulting from inter-thread predicate abstraction are, however, not asynchronous, as we have seen. As a result, we need a more powerful abstract programming language.

3.1 Dual-Reference Programs

In contrast to asynchronous programs, the variable set $\tilde{V}$ of a *dual-reference (DR)* program $\tilde{\mathcal{P}}$ is partitioned into two sets: $\tilde{L}$, the local variables of the active thread as before, and $\tilde{L}_P = \{\mathtt{l}_P : \mathtt{l} \in \tilde{L}\}$. The latter set contains passive-thread variables, which, intuitively, regulate the behavior of non-executing threads. To simplify reasoning about DR programs, we exclude classical shared variables from the description: they can be simulated using the active and passive flavors of local variables (see [21]).

The statements of $\tilde{\mathcal{P}}$ are given by a transition formula $\tilde{\mathcal{R}}$ over $\tilde{V}$ and $\tilde{V}'$, now potentially including passive-thread variables. Similarly, $\tilde{\mathcal{I}}$ may contain variables from $\tilde{L}_P$. The n-thread instantiation $\tilde{\mathcal{P}}^n$ of a DR program $\tilde{\mathcal{P}}$ is defined for $n \geq 2$ as

$$\tilde{\mathcal{P}}^n = (\tilde{\mathcal{R}}^n, \tilde{\mathcal{I}}^n) = \left(\bigvee\nolimits_{a=1}^{n} (\tilde{\mathcal{R}}_a)^n,\ \bigvee\nolimits_{a=1}^{n} (\tilde{\mathcal{I}}_a)^n\right) \tag{9}$$

where

$$(\tilde{\mathcal{R}}_a)^n :: \bigwedge_{p:p\neq a} \tilde{\mathcal{R}}\{\tilde{L} \triangleright\!\!\triangleright \tilde{L}_a\}\{\tilde{L}_P \triangleright\!\!\triangleright \tilde{L}_p\} \tag{10}$$

$$(\tilde{\mathcal{I}}_a)^n :: \bigwedge_{p:p\neq a} \tilde{\mathcal{I}}\{\tilde{L} \triangleright \tilde{L}_a\}\{\tilde{L}_P \triangleright \tilde{L}_p\} \tag{11}$$

Recall that $f\{X_1 \triangleright\!\!\triangleright X_2\}$ denotes index replacement of both current-state and next-state variables. Eq. (10) encodes the effect of a transition on the active thread a, and $n-1$ passive threads p. The conjunction ensures that the transition formula $\tilde{\mathcal{R}}$ holds no matter which thread $p \neq a$ takes the role of the passive thread: transitions that "work" only for select passive threads are rejected.

3.2 Computing an Abstract Dual-Reference Template

From the existential abstraction $\hat{\mathcal{P}}^n$ we derive a Boolean dual-reference template program $\tilde{\mathcal{P}}$ such that, for all n, the n-fold instantiation $\tilde{\mathcal{P}}^n$ overapproximates $\hat{\mathcal{P}}^n$. The variables of $\tilde{\mathcal{P}}$ are $\tilde{L} = \{b[i] : 1 \leq i \leq m\}$ and $\tilde{L}_P = \{b[i]_P : 1 \leq i \leq m\}$. Intuitively, the transitions of $\tilde{\mathcal{P}}$ are those that are feasible, for **some** n, in $\hat{\mathcal{P}}^n$, given active thread 1 and passive thread 2. We first compute the set $\tilde{\mathcal{R}}(n)$ of these transitions for fixed n. Formally, the components of $\tilde{\mathcal{P}}(n) = (\tilde{\mathcal{R}}(n), \tilde{\mathcal{I}}(n))$ are, for $n \geq 2$,

$$\tilde{\mathcal{R}}(n) :: \exists \hat{L}_3, \hat{L}'_3, \ldots, \hat{L}_n, \hat{L}'_n : (\hat{\mathcal{R}}_1)^n\{\hat{L}_1 \triangleright\!\!\triangleright \tilde{L}\}\{\hat{L}_2 \triangleright\!\!\triangleright \tilde{L}_P\} \tag{12}$$

$$\tilde{\mathcal{I}}(n) :: \exists \hat{L}_3, \ldots, \hat{L}_n : \quad \hat{\mathcal{I}}^n\ \{\hat{L}_1 \triangleright \tilde{L}\}\{\hat{L}_2 \triangleright \tilde{L}_P\} \tag{13}$$

We apply this strategy to the earlier example of the decrement statement $\mathtt{l} := \mathtt{l} - 1$. To compute Eq. (12) first with $n = 2$, we need $(\hat{\mathcal{R}}_1)^2$, which was enumerated previously in Table 1. Simplification results in a Boolean DR program with variables b and b_P and transition relation

$$\tilde{\mathcal{R}}(2) = (\neg\mathsf{b} \wedge \mathsf{b}_P \wedge \neg\mathsf{b}') \vee (\neg\mathsf{b}_P \wedge \mathsf{b}' \wedge \neg\mathsf{b}'_P). \quad (14)$$

Using (14) as the template $\tilde{\mathcal{R}}$ in (10) generates existential abstractions of many concrete decrement transitions; for instance, for $n = 2$ and $a = 1$ we get back the transition relation in Table 1. The question is now: does (14) suffice as a template, i.e. does $(\tilde{\mathcal{R}}(2))^n$ overapproximate $\hat{\mathcal{R}}^n$ for all n? The answer is no: the abstract 3-thread transitions shown in Table 2 are not permitted by $(\tilde{\mathcal{R}}(2))^n$ for any n, since neither $\neg\mathsf{b} \wedge \mathsf{b}_P$ nor $\mathsf{b}' \wedge \neg\mathsf{b}'_P$ are satisfied for all choices of passive threads (violations highlighted in the table).

We thus increase n to 3, recompute Eq. (12), and obtain

$$\tilde{\mathcal{R}}(3) :: \tilde{\mathcal{R}}(2) \vee (\neg\mathsf{b} \wedge \neg\mathsf{b}_P \wedge \neg\mathsf{b}' \wedge \neg\mathsf{b}'_P). \quad (15)$$

The new disjunct accommodates the abstract transitions highlighted in Table 2, which were missing before.

Table 2. Part of the abstraction $(\hat{\mathcal{R}}_1)^3$ for stmt. $\mathtt{l} := \mathtt{l} - 1$ against predicate $\mathtt{l} < \mathtt{l}_P$ (left); concrete witness transitions (right). The highlighted elements are inconsistent with (14) as a template.

b_1	b_2	b_3	b'_1	b'_2	b'_3	l_1	l_2	l_3	l'_1	l'_2	l'_3
F	F	F	F	F	F	1	0	0	0	0	0
F	F	T	F	F	F	1	1	0	0	1	0
F	F	T	F	F	T	2	1	0	1	1	0

Does $(\tilde{\mathcal{R}}(3))^n$ overapproximate $\hat{\mathcal{R}}^n$ for all n? When does the process of increasing n stop? To answer these questions, we first state the following diagonalization lemma, which helps us prove the overapproximation property for the template program.

Lemma 2. *$(\tilde{\mathcal{P}}(n))^n$ overapproximates $\hat{\mathcal{P}}^n$: For every $n \geq 2$ and every a, $(\hat{\mathcal{R}}_a)^n \Rightarrow (\tilde{\mathcal{R}}(n)_a)^n$ and $\hat{\mathcal{I}}^n \Rightarrow (\tilde{\mathcal{I}}(n)_a)^n$.*

We finally give a saturation bound for the sequence $(\tilde{\mathcal{P}}(n))$. Along with the diagonalization lemma, this allows us to obtain a template program $\tilde{\mathcal{P}}$ independent of n, and enable parametric reasoning in the abstract.

Theorem 3. *Let $\#_{IT}$ be the number of inter-thread predicates among the $Q[i]$. Then the sequence $(\tilde{\mathcal{P}}(n))$ stabilizes at $\mathsf{b} = 4 \times \#_{IT} + 2$, i.e. for $n \geq \mathsf{b}$, $\tilde{\mathcal{P}}(n) = \tilde{\mathcal{P}}(\mathsf{b})$.*

Corollary 4 (from L. 2,T. 3). *Let $\tilde{\mathcal{P}} := \tilde{\mathcal{P}}(\mathsf{b})$, for b as in Thm. 3. The components of $\tilde{\mathcal{P}}$ are thus $(\tilde{\mathcal{R}}, \tilde{\mathcal{I}}) = (\tilde{\mathcal{R}}(\mathsf{b}), \tilde{\mathcal{I}}(\mathsf{b}))$. Then, for $n \geq 2$, $\tilde{\mathcal{P}}^n$ overapproximates $\hat{\mathcal{P}}^n$.*

Building a template DR program thus requires instantiating the existentially abstracted transition relation for a number b of threads that is linear in the number of inter-thread predicates with respect to which to abstraction is built.

As a consequence of losing asynchrony in the abstraction, many existing model checkers for concurrent software become inapplicable [25,11,12]. For a fixed thread count n, the problem can be circumvented by forgoing the replicated nature of the concurrent programs, as done in [10] for boom tool: it proves the ticket algorithm correct up to $n = 3$, but takes a disappointing 30 minutes. The goal of the following section is to design an efficient and, more importantly, fully parametric solution.

4 Unbounded-Thread Dual-Reference Programs

The multi-threaded Boolean dual-reference programs $\tilde{\mathcal{P}}^n$ resulting from predicate-abstracting asynchronous programs against inter-thread predicates are symmetric and free of recursion. The symmetry can be exploited using classical methods that "counterize" the state space [17]: a global state is encoded as a vector of local-state counters, each of which records the number of threads currently occupying a particular local state.

These methods are applicable to unbounded thread numbers as well, in which case the local state counters range over unbounded natural numbers $[0, \infty[$. The fact that the abstract program executed by each thread is finite-state now might suggest that the resulting infinite-state counter systems can be modeled as vector addition systems (as done in [17]) or, more generally, as *well quasi-ordered transition systems* [15,1] (defined below). This would give rise to sound and complete algorithms for local-state reachability in such programs.

This strategy turns out to be wrong: the full class of Boolean DR programs is expressive enough to render safety checking for an unbounded number of threads undecidable, despite the finite-domain variables:

Theorem 5. *Program location reachability for Boolean DR programs run by an unbounded number of threads is undecidable.*

The proof reduces the halting problem for 2-counter machines to a reachability problem for a DR program $\tilde{\mathcal{P}}$. Counter values c_i are reduced to numbers of threads in program locations d_i of $\tilde{\mathcal{P}}$. A zero-test for counter c_i is reduced to testing the *absence of any thread* in location d_i. This condition can be expressed using passive-thread variables, but not using traditional single-thread local variables. (Details of the proof are in [21].)

Thm. 5 implies that the unbounded-counter systems obtained from asynchronous programs are in fact *not* well quasi-ordered. How come? Can this problem be fixed, in order to permit a complete verification method? If so, at what cost?

4.1 Monotonicity in Dual-Reference Programs

For a transition system $(\Sigma, \rightarrowtail)$ to be well-quasi ordered, we need two conditions to be in place [15,1,2]:

Well Quasi-Orderedness: There exists a reflexive and transitive binary relation $\preceq$ on Σ such that for every infinite sequence $v, w, \ldots$ of states in Σ there exist i, j with $i < j$ and $v_i \preceq v_j$.

Monotonicity: For any v, v', w with $v \rightarrowtail v'$ and $v \preceq w$ there exists w' such that $w \rightarrowtail w'$ and $v' \preceq w'$.

We apply this definition to the case of dual-reference programs. Representing global states of the abstract system $\tilde{\mathcal{P}}^n$ defined in Sect. 3 as counter tuples, we can define $\preceq$ as

$$(n_1, \ldots, n_k) \preceq (n'_1, \ldots, n'_k) \;::\; \forall i = 1..k : n_i \leq n'_i$$

where k is the number of thread-local states. We can now characterize monotonicity of DR programs as follows:

Lemma 6. *Let $\tilde{\mathcal{R}}$ be the transition relation of a DR program. Then the infinite-state transition system $\cup_{n=1}^{\infty} \tilde{\mathcal{R}}^n$ is monotone (with respect to $\preceq$) exactly if, for all $k \geq 2$:*

$$(v, v') \in \tilde{\mathcal{R}}^k \quad \Rightarrow \quad \forall l_{k+1}\, \exists l'_{k+1}, \pi : (\langle v, l_{k+1}\rangle, \pi(\langle v', l'_{k+1}\rangle)) \in \tilde{\mathcal{R}}^{k+1} \;. \qquad (16)$$

In (16), the expression $\forall l_{k+1} \exists l'_{k+1} \cdots$ quantifies over valuations of the local variables of thread $k+1$. The notation $\langle v, l_{k+1}\rangle$ denotes a $(k+1)$-thread state that agrees with v in the first k local states and whose last local state is l_{k+1}; similarly $\langle v', l'_{k+1}\rangle$. Symbol π denotes a permutation on $\{1, \ldots, k+1\}$ that acts on states by acting on thread indices, which effectively reorders thread local states.

Asynchronous programs are trivially monotone (and DR): Eq. (16) is satisfied by choosing $l'_{k+1} := l_{k+1}$ and π the identity. Table 3 shows instructions found in *non*-asynchronous programs that destroy monotonicity, and why. For example, the swap instruction in the first row gives rise to a DR program with a 2-thread transition $(0, 0, 0, 0) \in \tilde{\mathcal{R}}^2$. Choosing $l_3 = 1$ in (16) requires the existence of a transition in $\tilde{\mathcal{R}}^3$ of the form $(\mathtt{l}_1, \mathtt{l}_2, \mathtt{l}_3, \mathtt{l}'_1, \mathtt{l}'_2, \mathtt{l}'_3) = (0, 0, 1, \pi(0, 0, \mathtt{l}'_3))$, which is impossible: by equations (9) and (10), there must exist $a \in \{1, 2, 3\}$ such that for $\{p, q\} = \{1, 2, 3\} \setminus \{a\}$, both "$a$ swaps with p" and "a swaps with q" hold, i.e.

$$\mathtt{l}'_p = \mathtt{l}_a \wedge \mathtt{l}'_a = \mathtt{l}_p \quad \wedge \quad \mathtt{l}'_q = \mathtt{l}_a \wedge \mathtt{l}'_a = \mathtt{l}_q \,,$$

which is equivalent to $\mathtt{l}'_a = \mathtt{l}_p = \mathtt{l}_q \wedge \mathtt{l}_a = \mathtt{l}'_p = \mathtt{l}'_q$. It is easy to see that this formula is inconsistent with the partial assignment $(0, 0, 1, \pi(0, 0, \mathtt{l}'_3))$, no matter what $\mathtt{l}'_3$.

More interesting for us is the fact that asynchronous programs (= our input language) are monotone, while their parametric predicate abstractions may not be; this demonstrates that the monotonicity is in fact *lost in the abstraction*. Consider again the decrement instruction $\mathtt{l} := \mathtt{l} - 1$, but this time abstracted against the inter-thread predicate $Q :: \mathtt{l} = \mathtt{l}_P$. Parametric abstraction results in the two-thread and three-thread template instantiations

$$\tilde{\mathcal{R}}^2 = (\neg \mathtt{b}_1 \vee \neg \mathtt{b}'_1) \;\wedge\; \mathtt{b}_1 = \mathtt{b}_2 \;\wedge\; \mathtt{b}'_1 = \mathtt{b}'_2$$
$$\tilde{\mathcal{R}}^3 = (\neg \mathtt{b}_1 \vee \neg \mathtt{b}'_1) \;\wedge\; \mathtt{b}_1 = \mathtt{b}_2 = \mathtt{b}_3 \;\wedge\; \mathtt{b}'_1 = \mathtt{b}'_2 = \mathtt{b}'_3 \,.$$

Consider the transition $(0, 0) \rightarrow (1, 1) \in \tilde{\mathcal{R}}^2$ and the three-thread state $w = (0, 0, 1) \succ (0, 0)$: w clearly has no successor in $\tilde{\mathcal{R}}^3$ (it is in fact inconsistent), violating monotonicity. We discuss in Sect. 4.2 what happens to the decrement instruction with respect to predicate $\mathtt{l} < \mathtt{l}_P$.

Table 3. Each row shows a single-instruction program, whether the program gives rise to a monotone system and, if not, an assignment that violates Eq. (17). (Some of these programs are not finite-state.)

Dual-reference program		Monotonicity	
instruction	variables	mon.?	assgn. violating (17)
$\mathtt{l},\mathtt{l}_P := \mathtt{l}_P,\mathtt{l}$	$\mathtt{l} \in \mathbb{B}$	no	$\mathtt{l}=0, \mathtt{l}'=1$
$\mathtt{l},\mathtt{l}_P := \mathtt{l}+1,\mathtt{l}_P-1$	$\mathtt{l} \in \mathbb{N}$	yes	
$\mathtt{l}_P := \mathtt{l}_P+1$	$\mathtt{l} \in \mathbb{N}$	yes	
$\mathtt{l} := \mathtt{l}+\mathtt{l}_P$	$\mathtt{l} \in \mathbb{N}$	no	$\mathtt{l}=\mathtt{l}'=1$
$\mathtt{l}_P := c$	$\mathtt{l}, c \in \mathbb{N}$	yes	

4.2 Restoring Monotonicity in the Abstraction

Our goal is now to restore the monotonicity that was lost in the parametric abstraction. The standard covering relation $\preceq$ defined over local state counter tuples turns **monotone** and **Boolean** DR programs into instances of well quasi-ordered transition systems. Program location reachability is then decidable, even for unbounded threads.

In order to do so, we first derive a sufficient condition for monotonicity that can be checked **locally** over $\tilde{\mathcal{R}}$, as follows.

Theorem 7. *Let $\tilde{\mathcal{R}}$ be the transition relation of a DR program. Then the infinite-state transition system $\cup_{n=1}^{\infty}\tilde{\mathcal{R}}^n$ is monotone if the following formula over $\tilde{L} \times \tilde{L}'$ is valid:*

$$\exists \tilde{L}_P \tilde{L}'_P : \tilde{\mathcal{R}} \;\Rightarrow\; \forall \tilde{L}_P \exists \tilde{L}'_P : \tilde{\mathcal{R}}\,. \tag{17}$$

Unlike the monotonicity characterization given in Lemma 6, Eq. (17) is formulated only about the template program $\tilde{\mathcal{R}}$. It suggests that, if $\tilde{\mathcal{R}}$ holds for some valuation of its passive-thread variables, then no matter how we replace the current-state passive-thread variables $\tilde{L}_P$, we can find next-state passive-thread variables $\tilde{L}'_P$ such that $\tilde{\mathcal{R}}$ still holds. This is true for asynchronous programs, since here $\tilde{L}_P = \emptyset$. It fails for the swap instruction in the first row of Table 3: the instruction gives rise to the DR program $\tilde{\mathcal{R}} :: \mathtt{l}' = \mathtt{l}_P \wedge \mathtt{l}'_P = \mathtt{l}$. The assignment on the right in the table satisfies $\tilde{\mathcal{R}}$, but if $\mathtt{l}_P$ is changed to 0, $\tilde{\mathcal{R}}$ is violated no matter what value is assigned to $\mathtt{l}'_P$.

We are now ready to modify the possibly non-monotone abstract DR program $\tilde{\mathcal{P}}$ into a new, monotone abstraction $\tilde{\mathcal{P}}_m$. Our solution is similar in spirit to, but different in effect from, earlier work on *monotonic abstractions* [3], which proposes to delete processes that violate universal guards and thus block a transition. This results in an overappoximation of the original system and thus possibly spuriously reachable error states. By contrast, exploiting the monotonicity of the *concrete* program $\mathcal{P}$, we can build a monotone program $\tilde{\mathcal{P}}_m$ that is safe exactly when $\tilde{\mathcal{P}}$ is, thus fully preserving soundness and precision of the abstraction $\tilde{\mathcal{P}}$.

Definition 8. *The* ***non-monotone fragment*** ***(NMF)*** *of a DR program with transition relation $\tilde{\mathcal{R}}$ is the formula over $\tilde{L} \times \tilde{L}_P \times \tilde{L}'$:*

$$\mathcal{F}(\tilde{\mathcal{R}}) \quad :: \quad \neg\exists \tilde{L}'_P : \tilde{\mathcal{R}} \;\wedge\; \exists \tilde{L}_P \tilde{L}'_P : \tilde{\mathcal{R}}\,. \tag{18}$$

The NMF encodes partial assignments $(\mathtt{l}, \mathtt{l}_P, \mathtt{l}')$ that cannot be extended, via any $\mathtt{l}'_P$, to a full assignment satisfying $\tilde{\mathcal{R}}$, but can be extended for some valuation of $\tilde{L}_P$ other than $\mathtt{l}_P$. We revisit the two non-monotone instructions from Table 3. The NMF of $\mathtt{l}, \mathtt{l}_P := \mathtt{l}_P, \mathtt{l}$ is $\mathtt{l}' \neq \mathtt{l}_P$: this clearly cannot be extended to an assignment satisfying $\tilde{\mathcal{R}}$, but when $\mathtt{l}_P$ is changed to $\mathtt{l}'$, we can choose $\mathtt{l}'_P = \mathtt{l}$ to satisfy $\tilde{\mathcal{R}}$. The non-monotone fragment of $\mathtt{l} := \mathtt{l} + \mathtt{l}_P$ is $\mathtt{l}' \geq \mathtt{l} \wedge \mathtt{l}' \neq \mathtt{l} + \mathtt{l}_P$.

Eq. (18) is slightly stronger than the negation of (17): the NMF binds the values of the $\tilde{L}_P$ variables for which a violation of $\tilde{\mathcal{R}}$ is possible. It can be used to "repair" $\tilde{\mathcal{R}}$:

Lemma 9. *For a DR program with transition relation $\tilde{\mathcal{R}}$, the program with transition relation $\tilde{\mathcal{R}} \vee \mathcal{F}(\tilde{\mathcal{R}})$ is monotone.*

Lemma 9 suggests to add artificial transitions to $\tilde{\mathcal{P}}$ that allow arbitrary passive-thread changes in states of the non-monotone fragment, thus lifting the blockade previously caused by some passive threads. While this technique restores monotonicity, the problem is of course that such arbitrary changes will generally modify the program behavior; in particular, an added transition may lead a thread directly into an error state that used to be unreachable.

In order to instead obtain a *safety-equivalent* program, we prevent passive threads that block a transition in $\tilde{\mathcal{P}}^n$ from affecting the future execution. This can be realized by redirecting them to an auxiliary sink state. Let $\ell_\perp$ be a fresh program label.

Definition 10. *The **monotone closure** of DR program $\tilde{\mathcal{P}} = (\tilde{\mathcal{R}}, \tilde{\mathcal{I}})$ is the DR program $\tilde{\mathcal{P}}_m = (\tilde{\mathcal{R}}_m, \tilde{\mathcal{I}})$ with the transition relation $\tilde{\mathcal{R}}_m :: \tilde{\mathcal{R}} \vee (\mathcal{F}(\tilde{\mathcal{R}}) \wedge (\mathsf{pc}'_P = \ell_\perp))$.*

This extension of the transition relation has the following effects: (i) for any program state, if any passive thread can make a move, so can all, ensuring monotonicity, (ii) the added moves do not affect the safety of the program, and (iii) transitions that were previously possible are retained, so no behavior is removed. The following theorem summarizes these claims:

Theorem 11. *Let $\mathcal{P}$ be an asynchronous program, and $\tilde{\mathcal{P}}$ its parametric abstraction. The monotone closure $\tilde{\mathcal{P}}_m$ of $\tilde{\mathcal{P}}$ is monotone. Further, $(\tilde{\mathcal{P}}_m)^n$ is safe exactly if $\tilde{\mathcal{P}}^n$ is.*

Thm. 11 justifies our strategy for reachability analysis of an asynchronous program $\mathcal{P}$: form its parametric predicate abstraction $\tilde{\mathcal{P}}$ described in Sections 2 and 3, build the monotone closure $\tilde{\mathcal{P}}_m$, and analyze $(\tilde{\mathcal{P}}_m)^\infty$ using any technique for monotone systems.

Proving the parameterized ticket algorithm. Applying this strategy to the ticket algorithm yields a well quasi-ordered transition system for which the backward reachability method described in [1] returns "uncoverable", confirming that the ticket algorithm guarantees mutual exclusion, this time *for arbitrary thread counts*. We remind the reader that the ticket algorithm is challenging for existing techniques: cream [19], slab [11] and symmpa [10] handle only a fixed number of threads, and the resource requirements of these algorithms grow rapidly; none of them can handle even a handful of threads. The recent approach from [14] generates polynomial-size proofs, but again only for fixed thread counts.

5 Comparison with Related Work

Existing approaches for verifying asynchronous shared-memory programs typically do not exploit the monotone structure that source-level multi-threaded programs often naturally exhibit [20,7,9,26,19,10,12,14]. For example, the constraint-based approach in [19], implemented in cream, generates Owicki-Gries and rely-guarantee type proofs. It uses predicate abstraction in a CEGAR loop to generate environment invariants for fixed thread counts, whereas our approach directly checks the interleaved state space and exploits monotonicity. Whenever possible, cream generates thread-modular proofs by prioritizing predicates that do not refer to the local variables of other threads.

A CEGAR approach for fixed-thread symmetric concurrent programs has been implemented in symmpa [10]. It uses predicate abstraction to generate a Boolean Broadcast program (a special case of DR program). Their approach cannot reason about relationships between local variables across threads, which is crucial for verifying algorithms such as the ticket lock. Nevertheless, even the restricted predicate language of [10] can give rise to non-asynchronous programs. As a result, their technique cannot be extended to unbounded thread counts with well quasi-ordered systems technology.

Recent work on data flow graph representations of fixed-thread concurrent programs has been applied to safety property verification [14]. The inductive data flow graphs can serve as succinct correctness proofs for safety properties; for the ticket example they generate correctness proofs of size quadratic in n. Similar to [14], the technique in [12] uses data flow graphs to compute invariants of concurrent programs with unbounded threads (implemented in duet). In contrast to our approach, which uses an expressive predicate language, duet constructs proofs from relationships between either solely shared or solely local variables. These are insufficient for many benchmarks such as the parameterized ticket algorithm.

Predicates that, like our inter-thread predicates, reason over all participating processes/threads have been used extensively in invariant generation methods [5,16,22]. As a recent example, an approach that relies on abstract interpretation instead of model checking is [24]. Starting with a set of candidate invariants (assertions), the approach builds a *reflective abstraction*, from which invariants of the concrete system are obtained in a fixed point process. These approaches and ours share the insight that complex relationships over all threads may be required to prove easy-to-state properties such as mutual exclusion. They differ fundamentally in the way these relationships are used: abstraction with respect to a given set $\mathcal{Q}$ of quantified predicates determines the strongest invariant expressible as a Boolean formula over the set $\mathcal{Q}$; the result is unlikely to be expressible in the language that defines $\mathcal{Q}$. Future work will investigate how invariant generation procedures can be used towards *predicate discovery* in our technique.

The idea of "making" systems monotone, in order to enable wqo-based reasoning, was pioneered in earlier work [6,3]. Bingham and Hu deal with guards that require universal quantification over thread indices, by transforming such systems into Broadcast protocols. This is achieved by replacing conjunctively guarded actions by transitions that, instead of checking a universal condition, execute it assuming that any thread not satisfying it "resigns". This happens via a designated local state that isolates such threads from participation in future the computation. The same idea was further developed by Abdulla et al. in the context of *monotonic abstractions*. Our solution to the loss of monotonicity was in some way inspired by these works, but differs in two crucial aspects: first, our concrete input systems are asynchronous and thus monotone, so our incentive to *preserve* monotonicity in the abstract is strong. Second, exploiting the input monotonicity, we can achieve a monotonic abstraction that is safety-equivalent to the non-monotone abstraction and thus not merely an error-preserving approximation. This is essential, to avoid spurious counterexamples in addition to those unavoidably introduced by the predicate abstraction.

6 Concluding Remarks

We have presented in this paper a comprehensive verification method for arbitrarily-threaded asynchronous shared-variable programs. Our method is based on predicate abstraction and permits expressive universally quantified *inter-thread* predicates, which track relationships such as "my ticket number is the smallest, among all threads". Such predicates are required to verify, via predicate abstraction, some widely used algorithms like the ticket lock. We found that the abstractions with respect to these predicates result in non-monotone finite-data replicated programs, for which reachability is in fact undecidable. To fix this problem, we strengthened the earlier method of monotonic abstractions such that it does not introduce spurious errors into the abstraction.

We view the treatment of monotonicity as the major contribution of this work. Program design often naturally gives rise to "monotone concurrency", where adding components cannot disable existing actions, up to component symmetry. Abstractions that interfere with this feature are limited in usefulness. Our paper shows how the feature can be inexpensively restored, allowing such abstraction methods and powerful infinite-state verification methods to coexist peacefully.

References

1. Abdulla, P.A.: Well (and better) quasi-ordered transition systems. B. Symb. Log. (2010)
2. Abdulla, P.A., Cerans, K., Jonsson, B., Tsay, Y.-K.: General decidability theorems of infinite-state systems. In: LICS (1996)
3. Abdulla, P.A., Delzanno, G., Rezine, A.: Monotonic abstraction in parameterized verification. ENTCS (2008)
4. Andrews, G.R.: Concurrent programming: principles and practice. Benjamin-Cummings Publishing Co., Inc., Redwood City (1991)
5. Arons, T., Pnueli, A., Ruah, S., Xu, J., Zuck, L.D.: Parameterized verification with automatically computed inductive assertions. In: Berry, G., Comon, H., Finkel, A. (eds.) CAV 2001. LNCS, vol. 2102, pp. 221–234. Springer, Heidelberg (2001)
6. Bingham, J.D., Hu, A.J.: Empirically efficient verification for a class of infinite-state systems. In: Halbwachs, N., Zuck, L.D. (eds.) TACAS 2005. LNCS, vol. 3440, pp. 77–92. Springer, Heidelberg (2005)
7. Chaki, S., Clarke, E., Kidd, N., Reps, T., Touili, T.: Verifying concurrent message-passing C programs with recursive calls. In: Hermanns, H., Palsberg, J. (eds.) TACAS 2006. LNCS, vol. 3920, pp. 334–349. Springer, Heidelberg (2006)
8. Clarke, E.M., Grumberg, O., Long, D.E.: Model checking and abstraction. In: TOPLAS (1994)

9. Cook, B., Kroening, D., Sharygina, N.: Verification of Boolean programs with unbounded thread creation. Theoretical Comput. Sci. (2007)
10. Donaldson, A.F., Kaiser, A., Kroening, D., Tautschnig, M., Wahl, T.: Counterexample-guided abstraction refinement for symmetric concurrent programs. In: FMSD (2012)
11. Dräger, K., Kupriyanov, A., Finkbeiner, B., Wehrheim, H.: SLAB: A certifying model checker for infinite-state concurrent systems. In: Esparza, J., Majumdar, R. (eds.) TACAS 2010. LNCS, vol. 6015, pp. 271–274. Springer, Heidelberg (2010)
12. Farzan, A., Kincaid, Z.: Verification of parameterized concurrent programs by modular reasoning about data and control. In: POPL (2012)
13. Farzan, A., Kincaid, Z.: DUET: Static analysis for unbounded parallelism. In: Sharygina, N., Veith, H. (eds.) CAV 2013. LNCS, vol. 8044, pp. 191–196. Springer, Heidelberg (2013)
14. Farzan, A., Kincaid, Z., Podelski, A.: Inductive data flow graphs. In: POPL (2013)
15. Finkel, A., Schnoebelen, P.: Well-structured transition systems everywhere! Theoretical Comput. Sci. (2001)
16. Flanagan, C., Qadeer, S.: Predicate abstraction for software verification. In: POPL, pp. 191–202. ACM (2002)
17. German, S., Sistla, P.: Reasoning about systems with many processes. JACM (1992)
18. Graf, S., Saïdi, H.: Construction of abstract state graphs with PVS. In: Grumberg, O. (ed.) CAV 1997. LNCS, vol. 1254, pp. 72–83. Springer, Heidelberg (1997)
19. Gupta, A., Popeea, C., Rybalchenko, A.: Predicate abstraction and refinement for verifying multi-threaded programs. In: POPL (2011)
20. Henzinger, T., Jhala, R., Majumdar, R.: Race checking by context inference. In: PLDI (2004)
21. Kaiser, A., Kroening, D., Wahl, T.: Lost in abstraction: Monotonicity in multi-threaded programs (extended technical report). CoRR (2014)
22. Lahiri, S.K., Bryant, R.E.: Constructing quantified invariants via predicate abstraction. In: Steffen, B., Levi, G. (eds.) VMCAI 2004. LNCS, vol. 2937, pp. 267–281. Springer, Heidelberg (2004)
23. Malkis, A.: Cartesian Abstraction and Verification of Multithreaded Programs. PhD thesis, Albert-Ludwigs-Universität Freiburg (2010)
24. Sanchez, A., Sankaranarayanan, S., Sánchez, C., Chang, B.-Y.E.: Invariant generation for parametrized systems using self-reflection. In: Miné, A., Schmidt, D. (eds.) SAS 2012. LNCS, vol. 7460, pp. 146–163. Springer, Heidelberg (2012)
25. La Torre, S., Madhusudan, P., Parlato, G.: Model-checking parameterized concurrent programs using linear interfaces. In: Touili, T., Cook, B., Jackson, P. (eds.) CAV 2010. LNCS, vol. 6174, pp. 629–644. Springer, Heidelberg (2010)
26. Witkowski, T., Blanc, N., Kroening, D., Weissenbacher, G.: Model checking concurrent Linux device drivers. In: ASE (2007)

Synthesis from Component Libraries with Costs

Guy Avni and Orna Kupferman

School of Computer Science and Engineering, The Hebrew University, Israel

Abstract. *Synthesis* is the automated construction of a system from its specification. In real life, hardware and software systems are rarely constructed from scratch. Rather, a system is typically constructed from a library of components. Lustig and Vardi formalized this intuition and studied LTL synthesis from component libraries. In real life, designers seek optimal systems. In this paper we add optimality considerations to the setting. We distinguish between quality considerations (for example, size – the smaller a system is, the better it is), and pricing (for example, the payment to the company who manufactured the component). We study the problem of designing systems with minimal quality-cost and price. A key point is that while the quality cost is individual – the choices of a designer are independent of choices made by other designers that use the same library, pricing gives rise to a resource-allocation game – designers that use the same component share its price, with the share being proportional to the number of uses (a component can be used several times in a design). We study both closed and open settings, and in both we solve the problem of finding an optimal design. In a setting with multiple designers, we also study the game-theoretic problems of the induced resource-allocation game.

1 Introduction

Synthesis is the automated construction of a system from its specification. The classical approach to synthesis is to extract a system from a proof that the specification is satisfiable. In the late 1980s, researchers realized that the classical approach to synthesis is well suited to *closed* systems, but not to *open* (also called *reactive*) systems [1,24]. A reactive system interacts with its environment, and a correct system should have a *strategy* to satisfy the specification with respect to all environments. It turns out that the existence of such a strategy is stronger than satisfiability, and is termed *reliability*.

In spite of the rich theory developed for synthesis, in both the closed and open settings, little of this theory has been reduced to practice. This is in contrast with verification algorithms, which are extensively applied in practice. We distinguish between algorithmic and conceptual reasons for the little impact of synthesis in practice. The algorithmic reasons include the high complexity of the synthesis problem (PSPACE-complete in the closed setting [28] and 2EXPTIME-complete in the open setting [24], for specifications in LTL) as well as the intricacy of the algorithms in the open setting – the traditional approach involves determinization of automata on infinite words [27] and a solution of parity games [19].

We find the argument about the algorithmic challenge less compelling. First, experience with verification shows that even nonelementary algorithms can be practical,

P. Baldan and D. Gorla (Eds.): CONCUR 2014, LNCS 8704, pp. 156–172, 2014.

since the worst-case complexity does not arise often. For example, while the model-checking problem for specifications in second-order logic has nonelementary complexity, the model-checking tool MONA [14] successfully verifies many specifications given in second-order logic. Furthermore, in some sense, synthesis is not harder than verification: the complexity of synthesis is given with respect to the specification only, whereas the complexity of verification is given with respect to the specification and the system, which is typically much larger than the specification. About the intercity of the algorithms, in the last decade we have seen quite many alternatives to the traditional approach – Safraless algorithms that avoid determinization and parity games, and reduce synthesis to problems that are simpler and are amenable to optimizations and symbolic implementations [17,21,22].

The arguments about the conceptual and methodological reasons are more compelling. We see here three main challenges, relevant in both the closed and open settings. First, unlike verification, where a specification can be decomposed into sub-specifications, each can be checked independently, in synthesis the starting point is one comprehensive specification. This inability to decompose or evolve the specification is related to the second challenge. In practice, we rarely construct systems from scratch or from one comprehensive specification. Rather, systems are constructed from existing components. This is true for both hardware systems, where we see IP cores or design libraries, and software systems, where web APIs and libraries of functions and objects are common. Third, while in verification we only automate the check of the system, automating its design is by far more risky and unpredictable – there are typically many ways to satisfy a satisfiable or realizable specification, and designers will be willing to give up manual design only if they can count on the automated synthesis tool to construct systems of comparable quality. Traditional synthesis algorithms do not attempt to address the quality issue.

In this paper we continue earlier efforts to cope with the above conceptual challenges. Our contribution extends both the setting and the results of earlier work. The realization that design of systems proceeds by composition of underlying components is not new to the verification community. For example, [18] proposed a framework for component-based modelling that uses an abstract layered model of components, and [12] initiated a series of works on interface theories for component-based design, possibly with a reuse of components in a library [13]. The need to consider components is more evident in the context of software, where, for example, recursion is possible, so components have to be equipped with mechanisms for call and return [4]. The setting and technical details, however, are different from these in the synthesis problem we consider here. The closer to our work here is [23], which studied LTL synthesis from reusable component libraries. Lustig and Vardi studied two notions of component composition. In the first notion, termed data-flow composition, components are cascaded so that the outputs of one component are fed to other components. In the second notion, termed control-flow composition, the composition is flat and control flows among the different components. The second notion, which turns out to be the decidable one [23], is particularly suitable for modelling web-service orchestration, where users are typically offered services and interact with different parties [3].

Let us turn now to the quality issue. Traditional formal methods are based on a Boolean satisfaction notion: a system satisfies, or not, a given specification. The richness of today's systems, however, calls for specification formalisms that are *multi-valued*. The multi-valued setting arises directly in probabilistic and weighted systems and arises indirectly in applications where multi-valued satisfaction is used in order to model quantitative properties of the system like its size, security level, or quality. Reasoning about quantitative properties of systems is an active area of research in recent years, yielding quantitative specification formalisms and algorithms [11,16,10,2,9]. In quantitative reasoning, the Boolean satisfaction notion is refined and one can talk about the cost, or reward, of using a system, or, in our component-based setting, the cost of using a component from the library.

In order to capture a wide set of scenarios in practice, we associate with each component in the library two costs: a *quality cost* and a *construction cost*. The quality cost, as describes above, concerns the performance of the component and is paid each time the component is used. The construction cost is the cost of adding the component to the library. Thus, a design that uses a component pays its construction cost once. When several designs use the same component, they share its construction cost. This corresponds to real-life scenarios, where users pay, for example, for web-services, and indeed their price is influenced by the market demand.

We study synthesis from component libraries with costs in the closed and open settings. In both settings, the specification is given by means of a deterministic automaton $\mathcal{S}$ on finite words (DFA).[1] In the closed setting, the specification is a regular language over some alphabet Σ and the library consists of box-DFAs (that is, DFAs with exit states) over Σ. In the open setting, the specification $\mathcal{S}$ is over sets I and O of input and output signals, and the library consists of box-I/O-transducers. The boxes are black, in the sense that a design that uses components from the library does not see Σ (or $I \cup O$) nor it sees the behaviour inside the components. Rather, the mode of operation is as in the control-flow composition of [23]: the design gives control to one of the components in the library. It then sees only the exit state through which the component completes its computation and relinquishes control. Based on this information, the design decides which component gets control next, and so on.

In more technical details, the synthesis problem gets as input the specification $\mathcal{S}$ as well as a library $\mathcal{L}$ of components $\mathcal{B}_1, \ldots, \mathcal{B}_n$. The goal is to return a correct design – a transducer $\mathcal{D}$ that reads the exit states of the components and outputs the next component to gain control. In the closed setting, correctness means that the language over Σ that is generated by the composition defined by $\mathcal{D}$ is equal to the language of $\mathcal{S}$. In the open setting, correctness means that the interaction of the composition defined by $\mathcal{D}$ with all input sequences generates a computation over $I \cup O$ that is in the language of $\mathcal{S}$.

We first study the problem without cost and reduce it to the solution of a two-player safety game $\mathcal{G}_{\mathcal{L},\mathcal{S}}$. In the closed setting, the game is of full information and the problem

[1] It is possible to extend our results to specifications in LTL. We prefer to work with deterministic automata, as this setting isolates the complexity and technical challenges of the design problem and avoids the domination of the doubly-exponential complexity of going from LTL to deterministic automata.

can be solved in polynomial time. In the open setting, the flexibility that the design have in responding to different input sequences introduces partial information to the game, and the problem is EXPTIME-complete. We note that in [23], where the open setting was studied and the specification is given by means of an LTL formula, the complexity is 2EXPTIME-complete, thus one could have expected our complexity to be only polynomial. We prove, however, hardness in EXPTIME, showing that it is not just the need to transfer the LTL formula to a deterministic formalism that leads to the high complexity.

We then turn to integrate cost to the story. As explained above, there are two types of costs associated with each component $\mathcal{B}_i$ in $\mathcal{L}$. The first type, quality cost, can be studied for each design in isolation. We show that even there, the combinatorial setting is not simple. While for the closed setting an optimal design can be induced from a memoryless strategy of the designer in the game $\mathcal{G}_{\mathcal{L},\mathcal{S}}$, making the problem of finding an optimal design NP-complete, seeking designs of optimal cost may require sophisticated compositions in the open setting. In particular, we show that optimal designs may be exponentially larger than other correct designs[2], and that an optimal design may not be induced by a memoryless strategy in $\mathcal{G}_{\mathcal{L},\mathcal{S}}$. We are still able to bound the size of an optimal transducer by the size of $\mathcal{G}_{\mathcal{L},\mathcal{S}}$, and show that the optimal synthesis problem is NEXPTIME-complete.

The second type of cost, namely construction cost, depends not only on choices made by the designer, but also on choices made by designers of other specifications that use the library. Indeed, recall that the construction cost of a component is shared by designers that use this component, with the share being proportional to the number of uses (a component can be used several times in a design). Hence, the setting gives rise to a *resource-allocation game* [26,15]. Unlike traditional resource-allocation games, where players' strategies are sets of resources, here each strategy is a multiset – the components a designer needs. As has been the case in [7], the setting of multisets makes the game less stable. We show that the game is not guaranteed to have a *Nash Equilibrium* (NE), and that the problem of deciding whether an NE exists is Σ_2^P-complete. We then turn to the more algorithmic related problems and show that the problems of finding an optimal design given the choices of the other designers (a.k.a. the *best-response* problem, in algorithmic game theory) and of finding designs that minimize the total cost for all specifications (a.k.a. the *social optimum*) are both NP-complete.

Due to lack of space, many proofs and examples are missing in this version. They can be found in the full version, in the authors' URLs.

2 Preliminaries

Automata, Transducers, and Boxes. A *deterministic finite automaton* (DFA, for short) is a tuple $\mathcal{A} = \langle \Sigma, Q, \delta, q_0, F \rangle$, where Σ is an alphabet, Q is a set of states, $\delta : Q \times \Sigma \to Q$ is a partial transition function, $q_0 \in Q$ is an initial states, and $F \subseteq Q$ is a set of accepting states. We extend δ to words in an expected way, thus $\delta^* : Q \times \Sigma^* \to Q$ is such that for $q \in Q$, we have $\delta^*(q, \epsilon) = q$ and for $w \in \Sigma^*$ and $\sigma \in \Sigma$, we have $\delta^*(q, w \cdot \sigma) = \delta(\delta^*(q, w), \sigma)$. When $q = q_0$, we sometimes omit it, thus $\delta^*(w)$ is the

[2] Recall that "optimal" here refers to the quality-cost function.

state that $\mathcal{A}$ reaches after reading w. We assume that all states are reachable from q_0, thus for every $q \in Q$ there exists a word $w \in \Sigma^*$ such that $\delta^*(w) = q$. We refer to the *size* of $\mathcal{A}$, denoted $|\mathcal{A}|$, as the number of its states.

The *run* of $\mathcal{A}$ on a word $w = w_1, \ldots w_n \in \Sigma^*$ is the sequence of states $r = r_0, r_1, \ldots, r_n$ such that $r_0 = q_0$ and for every $0 \leq i \leq n-1$ we have $r_{i+1} = \delta(r_i, w_{i+1})$. The run r is accepting iff $r_n \in F$. The *language* of $\mathcal{A}$, denoted $L(\mathcal{A})$, is the set of words $w \in \Sigma^*$ such that the run of $\mathcal{A}$ on w is accepting, or, equivalently, $\delta^*(w) \in F$. For $q \in Q$, we denote by $L(\mathcal{A}^q)$ the language of the DFA that is the same as $\mathcal{A}$ only with initial state q.

A *transducer* models an interaction between a system and its environment. It is similar to a DFA except that in addition to Σ, which is referred to as the input alphabet, denoted Σ_I, there is an output alphabet, denoted Σ_O, and rather than being classified to accepting or rejecting, each state is labeled by a letter from Σ_O[3]. Formally, a transducer is a tuple $\mathcal{T} = \langle \Sigma_I, \Sigma_O, Q, q_0, \delta, \nu \rangle$, where Σ_I is an input alphabet, Σ_O is an output alphabet, Q, q_0, and $\delta : Q \times \Sigma_I \to Q$ are as in a DFA, and $\nu : Q \to \Sigma_O$ is an output function. We require $\mathcal{T}$ to be *receptive*. That is, δ is complete, so for every input word $w \in \Sigma_I^*$, there is a run of $\mathcal{T}$ on w. Consider an input word $w = w_1, \ldots, w_n \in \Sigma_I^*$. Let $r = r_0, \ldots, r_n$ be the run of $\mathcal{T}$ on w. The *computation of* $\mathcal{T}$ *in* w is then $\sigma_1, \ldots, \sigma_n \in (\Sigma_I \times \Sigma_O)^*$, where for $1 \leq i \leq n$, we have $\sigma_i = \langle w_i, \nu(r_{i-1}) \rangle$. We define the language of $\mathcal{T}$, denoted $L(\mathcal{T})$, as the set of all its computations. For a specification $L \subseteq (\Sigma_I \times \Sigma_O)^*$, we say that $\mathcal{T}$ *realizes* L iff $L(\mathcal{T}) \subseteq L$. Thus, no matter what the input sequence is, the interaction of $\mathcal{T}$ with the environment generates a computation that satisfies the specification.

By adding *exit states* to DFAs and transducers, we can view them as components from which we can compose systems. Formally, we consider two types of components. Closed components are modeled by *box-DFAs* and open components are modeled by *box-transducers*. A box-DFA augments a DFA by a set of exit states. Thus, a box-DFA is a tuple $\langle \Sigma, Q, \delta, q_0, F, E \rangle$, where $E \subseteq Q$ is a nonempty set of exit states. There are no outgoing transitions from an exit state. Also, the initial state cannot be an exit state and exit states are not accepting. Thus, $q_0 \notin E$ and $F \cap E = \emptyset$. Box-transducers are defined similarly, and their exit states are not labeled, thus $\nu : Q \setminus E \to \Sigma_O$.

Component Libraries. A *component library* is a collection of boxes $\mathcal{L} = \{\mathcal{B}_1, \ldots, \mathcal{B}_n\}$. We say that $\mathcal{L}$ is a *closed library* if the boxes are box-DFAs, and is an *open library* if the boxes are box-transducers. Let $[n] = \{1, \ldots, n\}$. In the first case, for $i \in [n]$, let $\mathcal{B}_i = \langle \Sigma, C_i, \delta_i, c_i^0, F_i, E_i \rangle$. In the second case, $\mathcal{B}_i = \langle \Sigma_I, \Sigma_O, C_i, \delta_i, c_i^0, \nu_i, E_i \rangle$. Note that all boxes in $\mathcal{L}$ share the same alphabet (input and output alphabet, in the case of transducers). We assume that the states of the components are disjoint, thus for every $i \neq j \in [n]$, we have $C_i \cap C_j = \emptyset$. We use the following abbreviations $\mathcal{C} = \bigcup_{i \in [n]} C_i$, $\mathcal{C}_0 = \bigcup_{i \in [n]} \{c_i^0\}$, $\mathcal{F} = \bigcup_{i \in [n]} F_i$, and $\mathcal{E} = \bigcup_{i \in [n]} E_i$. We define the *size* of $\mathcal{L}$ as $|\mathcal{C}|$.

We start by describing the intuition for composition of closed libraries. A *design* is a recipe to compose the components of a library $\mathcal{L}$ (allowing multiple uses) into a DFA. A run of the design on a word starts in an initial state of one of the components in $\mathcal{L}$. We say that this component has the initial *control*. When a component is in control, the

[3] These transducers are sometimes referred to as *Moore machines*.

run uses its states, follows its transition function, and if the run ends, it is accepting iff it ends in one of the components' accepting states. A component relinquishes control when the run reaches one of its exit states. It is then the design's duty to assign control to the next component, which gains control through its initial state.

Formally, a design is a transducer $\mathcal{D}$ with input alphabet $\mathcal{E}$ and output alphabet $[n]$. We can think of $\mathcal{D}$ as running beside the components. When a component reaches an exit state e, then $\mathcal{D}$ reads the input letter e, proceeds to its next state, and outputs the index of the component to gain control next. Note that $\mathcal{D}$ does not read the alphabet Σ and has no information about the states that the component visits. It only sees which exit state has been reached.

Consider a design $\mathcal{D} = \langle \mathcal{E}, [n], D, \delta, d^0, \nu \rangle$ and a closed library $\mathcal{L}$. We formalize the behavior of $\mathcal{D}$ by means of the *composition DFA* $\mathcal{A}_{\mathcal{L},\mathcal{D}}$ that simulates the run of $\mathcal{D}$ along with the runs of the box-DFAs. Formally, $\mathcal{A}_{\mathcal{L},\mathcal{D}} = \langle \Sigma, Q_{\mathcal{L},\mathcal{D}}, \delta_{\mathcal{L},\mathcal{D}}, q^0_{\mathcal{L},\mathcal{D}}, F_{\mathcal{L},\mathcal{D}} \rangle$ is defined as follows. The set of states $Q_{\mathcal{L},\mathcal{D}} \subseteq (\mathcal{C} \setminus \mathcal{E}) \times D$ consists of pairs of a *component state* from $\mathcal{C}$ and an *design state* from S. The component states are consistent with ν, thus $Q_{\mathcal{L},\mathcal{D}} = \bigcup_{i \in [n]} (C_i \setminus E_i) \times \{q : \nu(q) = i\}$. In exit states, the composition immediately moves to the initial state of the next component, which is why the component states of $\mathcal{A}_{\mathcal{L},\mathcal{D}}$ do not include $\mathcal{E}$. Consider a state $\langle c, q \rangle \in Q_{\mathcal{L},\mathcal{D}}$ and a letter $\sigma \in \Sigma$. Let $i \in [n]$ be such that $c \in C_i$. When a run of $\mathcal{A}_{\mathcal{L},\mathcal{D}}$ reaches the state $\langle c, q \rangle$, the component $\mathcal{B}_i$ is in control. Recall that c is not an exit state. Let $c' = \delta_i(c, \sigma)$. If $c' \notin E_i$, then $\mathcal{B}_i$ does not relinquish control after reading σ and $\delta_{\mathcal{L},\mathcal{D}}(\langle c, q \rangle, \sigma) = \langle c', q \rangle$. If $c' \in E_i$, then $\mathcal{B}_i$ relinquishes control through c', and it is the design's task to choose the next component to gain control. Let $q' = \delta(q, c')$ and let $j = \nu(q')$. Then, $\mathcal{B}_j$ is the next component to gain control (possibly $j = i$). Accordingly, we advance $\mathcal{D}$ to q' and continue to the initial state of $\mathcal{B}_j$. Formally, $\delta_{\mathcal{L},\mathcal{D}}(\langle c, q \rangle, \sigma) = \langle c^0_j, q' \rangle$. (Recall that $c^0_j \notin E_j$, so the new state is in $Q_{\mathcal{L},\mathcal{D}}$.) Note also that a visit in c' is skipped. The component that gains initial control is chosen according to $\nu(d^0)$. Thus, $q^0_{\mathcal{L},\mathcal{D}} = \langle c^0_j, d^0 \rangle$, where $j = \nu(d^0)$. Finally, the accepting states of $\mathcal{A}_{\mathcal{L},\mathcal{D}}$ are these in which the component state is accepting, thus $F_{\mathcal{L},\mathcal{D}} = \mathcal{F} \times D$.

The definition of a composition for an open library is similar. There, the composition is a transducer $\mathcal{T}_{\mathcal{L},\mathcal{D}} = \langle \Sigma_I, \Sigma_O, Q_{\mathcal{L},\mathcal{D}}, \delta_{\mathcal{L},\mathcal{D}}, q^0_{\mathcal{L},\mathcal{D}}, \nu_{\mathcal{L},\mathcal{D}} \rangle$, where $Q_{\mathcal{L},\mathcal{D}}$, $q^0_{\mathcal{L},\mathcal{D}}$, and $\delta_{\mathcal{L},\mathcal{D}}$ are as in the closed setting, except that $\delta_{\mathcal{L},\mathcal{D}}$ reads letters in Σ_I, and $\nu_{\mathcal{L},\mathcal{D}}(\langle c, q \rangle) = \nu_i(c)$, for $i \in [n]$ such that $c \in C_i$.

3 The Design Problem

The *design problem* gets as input a component library $\mathcal{L}$ and a specification that is given by means of a DFA $\mathcal{S}$. The problem is to decide whether there exists a correct design for $\mathcal{S}$ using the components in $\mathcal{L}$. In the closed setting, a design $\mathcal{D}$ is correct if $L(\mathcal{A}_{\mathcal{L},\mathcal{D}}) = L(\mathcal{S})$. In the open setting, $\mathcal{D}$ is correct if the transducer $\mathcal{T}_{\mathcal{L},\mathcal{D}}$ realizes $\mathcal{S}$. Our solution to the design problem reduces it to the problem of finding the winner in a turn-based two-player game, defined below.

A *turn-based two-player game* is played on an arena $\langle V, \Delta, V_0, \alpha \rangle$, where $V = V_1 \cup V_2$ is a set of vertices that are partitioned between Player 1 and Player 2, $\Delta \subseteq V \times V$ is a set of directed edges, $V_0 \subseteq V$ is a set of initial vertices, and α is an objective for Player 1, specifying a subset of V^ω. We consider here *safety games*, where $\alpha \subseteq V$ is a set of vertices that are *safe* for Player 1. The game is played as follows. Initially, Player 1 places a token on one of the vertices in V_0. Assume the token is placed on a vertex $v \in V$ at the beginning of a round. The player that owns v is the player that moves the token to the next vertex, where the legal vertices to continue to are $\{v' \in V : \langle v, v' \rangle \in \Delta\}$. The outcome of the game is a *play* $\pi \in V^\omega$. The play is winning for Player 1 if for every $i \geq 1$, we have $\pi_i \in \alpha$. Otherwise, Player 2 wins.

A *strategy* for Player i, for $i \in \{1, 2\}$, is a recipe that, given a prefix of a play, tells the player what his next move should be. Thus, it is a function $f_i : V^* \cdot V_i \to V$ such that for every play $\pi \cdot v \in V^*$ with $v \in V_i$, we have $\langle v, f_i(\pi \cdot v) \rangle \in \Delta$. Since Player 1 moves first, we require that $f_1(\epsilon)$ is defined and is in V_0. For strategies f_1 and f_2 for players 1 and 2 respectively, the play $out(f_1, f_2) \in V^\omega$ is the unique play that is the outcome the game when the players follow their strategies. A strategy f_i for Player i is *memoryless* if it depends only in the current vertex, thus it is a function $f_i : V_i \to V$.

A strategy is *winning* for a player if by using it he wins against every strategy of the other player. Formally, a strategy f_1 is winning for Player 1 iff for every strategy f_2 for Player 2, Player 1 wins the play $out(f_1, f_2)$. The definition for Player 2 is dual. It is well known that safety games are *determined*, namely, exactly one player has a winning strategy, and admits *memoryless* strategies, namely, Player i has a winning strategy iff he has a memoryless winning strategy. Deciding the winner of a safety game can done in linear time.

Solving the Design Problem. We describe the intuition of our solution for the design problems. Given a library $\mathcal{L}$ and a specification $\mathcal{S}$ we construct a safety game $\mathcal{G}_{\mathcal{L},\mathcal{S}}$ such that Player 1 wins $\mathcal{G}_{\mathcal{L},\mathcal{S}}$ iff there is a correct design for $\mathcal{S}$ using the components in $\mathcal{L}$. Intuitively, Player 1's goal is to construct a correct design, thus he chooses the components to gain control. Player 2 challenges the design that Player 1 chooses, thus he chooses a word (over Σ in the closed setting and over $\Sigma_I \times \Sigma_O$ in the open setting) and wins if his word is a witness for the incorrectness of Player 1's design.

Closed Designs. The input to the closed-design problem is a closed-library $\mathcal{L}$ and a DFA $\mathcal{S}$ over the alphabet Σ. The goal is to find a correct design $\mathcal{D}$. Recall that $\mathcal{D}$ is correct if the DFA $\mathcal{A}_{\mathcal{L},\mathcal{D}}$ that is constructed from $\mathcal{L}$ using $\mathcal{D}$ satisfies $L(\mathcal{A}_{\mathcal{L},\mathcal{D}}) = L(\mathcal{S})$. We assume that $\mathcal{S}$ is the minimal DFA for the language $L(\mathcal{S})$.

Theorem 1. *The closed-design problem can be solved in polynomial time.*

Proof: Given a closed-library $\mathcal{L}$ and a DFA $\mathcal{S} = \langle \Sigma, S, \delta_S, s^0, F_S \rangle$, we describe a safety game $\mathcal{G}_{\mathcal{L},\mathcal{S}}$ such that Player 1 wins $\mathcal{G}_{\mathcal{L},\mathcal{S}}$ iff there is a design of $\mathcal{S}$ using components from $\mathcal{L}$. Recall that $\mathcal{L}$ consists of box-DFAs $\mathcal{B}_i = \langle \Sigma, C_i, \delta_i, c_i^0, F_i, E_i \rangle$, for $i \in [n]$, and that we use $\mathcal{C}$, $\mathcal{C}_0$, $\mathcal{E}$, and $\mathcal{F}$ to denote the union of all states, initial states, exit states, and accepting states in all the components of $\mathcal{L}$. The number of vertices in $\mathcal{G}_{\mathcal{L},\mathcal{S}}$ is $|(\mathcal{C}_0 \cup \mathcal{E}) \times S|$ and it can be constructed in polynomial time. Since solving safety games can be done in linear time, the theorem follows.

We define $\mathcal{G}_{\mathcal{L},\mathcal{S}} = \langle V, E, V_0, \alpha \rangle$. First, $V = (\mathcal{C}_0 \cup \mathcal{E}) \times S$ and $V_0 = \mathcal{C}_0 \times \{s^0\}$. Recall that Player 1 moves when it is time to decide the next (or first) component to gain control. Accordingly, $V_1 = \mathcal{E} \times S$. Also, Player 2 challenges the design suggested by Player 1 and chooses the word that is processed in a component that gains control, so $V_2 = \mathcal{C}_0 \times S$.

Consider a vertex $\langle e, s\rangle \in V_1$. Player 1 selects the next component to gain control. This component gains control through its initial state. Accordingly, E contains edges $\langle\langle e, s\rangle, \langle c_i^0, s\rangle\rangle$, for every $i \in [n]$. Note that since no letter is read when control is passed, we do not advance the state in $\mathcal{S}$. Consider a vertex $v = \langle c_i^0, s\rangle \in V_2$. Player 2 selects the word that is read in the component $\mathcal{B}_i$, or equivalently, he selects the exit state from which $\mathcal{B}_i$ relinquishes control. Thus, E contains an edge $\langle\langle c_i^0, s\rangle, \langle e, s'\rangle\rangle$ iff there exists a word $u \in \Sigma^*$ such that $\delta_i^*(u) = e$ and $\delta_{\mathcal{S}}^*(s, u) = s'$.

We now turn to define the winning condition. All the vertices in V_1 are in α. A vertex $v \in V_2$ is not in α if it is possible to extend the word traversed for reaching v to a witness for the incorrectness of $\mathcal{D}$. Accordingly, a vertex $\langle c_i^0, s\rangle$ is not in α if one of the following holds. First ("the suffix witness"), there is a finite word that is read inside the current component and witnesses the incorrectness. Formally, there is $u \in \Sigma^*$ such that $\delta_i^*(u) \in F_i$ and $\delta_{\mathcal{S}}^*(s, u) \notin F_{\mathcal{S}}$, or $\delta_i^*(u) \in C_i \setminus (F_i \cup E_i)$ and $\delta_{\mathcal{S}}^*(s, u) \in F_{\mathcal{S}}$. Second ("the infix witness"), there are two words that reach the same exit state of the current component yet the behavior of $\mathcal{S}$ along them is different. Formally, there exist words $u, u' \in \Sigma^*$ such that $\delta_i^*(u) = \delta_i^*(u') \in E_i$ and $\delta_{\mathcal{S}}^*(s, u) \neq \delta_{\mathcal{S}}^*(s, u')$. Intuitively, the minimality of $\mathcal{S}$ enables us to extend either u or u' to an incorrectness witness. Given $\mathcal{L}$ and $\mathcal{S}$, the game $\mathcal{G}_{\mathcal{L},\mathcal{S}}$ can be constructed in polynomial time.

In the full version we prove that there is a correct design iff Player 1 wins $\mathcal{G}_{\mathcal{L},\mathcal{S}}$. ■

We continue to study the open setting. Recall that there, the input is a DFA $\mathcal{S}$ over the alphabet $\Sigma_I \times \Sigma_O$ and an open library $\mathcal{L}$. The goal is to find a correct design $\mathcal{D}$ or return that no such design exists, where $\mathcal{D}$ is correct if the composition transducer $\mathcal{T}_{\mathcal{L},\mathcal{D}}$ realizes $L(\mathcal{S})$.

Lustig and Vardi [23] studied the design problem in a setting in which the specification is given by means of an LTL formula. They showed that the problem is 2EXPTIME-complete. Given an LTL formula one can construct a deterministic parity automaton that recognizes the language of words that satisfy the formula. The size of the automaton is doubly-exponential in the size of the formula. Thus, one might guess that the design problem in a setting in which the specification is given by means of a DFA would be solvable in polynomial time. We show that this is not the case and that the problem is EXPTIME-complete. As in [23], our upper bound is based on the ability to "summarize" the activity inside the components. Starting with an LTL formula, the solution in [23] has to combine the complexity involved in the translation of the LTL formula into an automaton with the complexity of finding a design, which is done by going throughout a universal word automaton that is expanded to a tree automaton. Starting with a deterministic automaton, our solution directly uses games: Given an open-library $\mathcal{L}$ and a DFA $\mathcal{S}$, we describe a safety game $\mathcal{G}_{\mathcal{L},\mathcal{S}}$ such that Player 1 wins $\mathcal{G}_{\mathcal{L},\mathcal{S}}$ iff there is a design for $\mathcal{S}$ using components from $\mathcal{L}$. The number of vertices in $\mathcal{G}_{\mathcal{L},\mathcal{S}}$ is exponential in S and $\mathcal{C}$. Since solving safety games can be done in linear time, membership in EXPTIME follows. The interesting contribution, however, is the lower bound, showing that the problem is EXPTIME-hard even when the specification is given by means of a deterministic automaton. For that, we describe a reduction from the problem of deciding whether Player 1 has a winning strategy in a *partial-information safety game*, known to be EXPTIME-complete [8].

Partial-information games are a variant of the *full-information* games defined above in which Player 1 has imperfect information [25]. That is, Player 1 is unaware of the location on which the token is placed and is only aware of the observation it is in. Accordingly, in his turn, Player 1 cannot select the next location to move the token to. Instead, the edges in the game are labeled with actions, denoted Γ. In each round, Player 1 selects an action and Player 2 resolves nondeterminism and chooses the next location the token moves to. In our reduction, the library $\mathcal{L}$ consists of box-transducers $\mathcal{B}_a$, one for every action $a \in \Gamma$. The exit states of the components correspond to the observations in $\mathcal{O}$. That is, when a component exits through an observation $L_i \in \mathcal{O}$, the design decides which component $\mathcal{B}_a \in \mathcal{L}$ gains control, which corresponds to a Player 1 strategy that chooses the action $a \in \Gamma$ from the observation L_i. The full definition of partial-information games as well as the upper and lower bounds are described in the full version.

Theorem 2. *The open-design problem is EXPTIME-complete.*

4 Libraries with Costs

Given a library and a specification, there are possibly many, in fact infinitely many, designs that are solutions to the design problem. As a trivial example, assume that $L(\mathcal{S}) = a^*$ and that the library contains a component $\mathcal{B}$ that traverses the letter a (that is, $\mathcal{B}$ consists of an accepting initial state that has an a-transition to an exist state). An optimal design for $\mathcal{S}$ uses $\mathcal{B}$ once: it has a single state with a self loop in which $\mathcal{B}$ is called. Other designs can use $\mathcal{B}$ arbitrarily many numbers. When we wrote "optimal" above, we assumed that the smaller the design is, the better it is. In this section we would like to formalize the notion of optimality and add to the composition picture different costs that components in the libraries may have.

In order to capture a wide set of scenarios in practice, we associate with each component in $\mathcal{L}$ two costs: a *construction cost* and a *quality cost*. The costs are given by the functions *c-cost*, *q-cost* : $\mathcal{L} \to \mathbb{R}^+ \cup \{0\}$, respectively. The construction cost of a component is the cost of adding it to the library. Thus, a design that uses a component pays its construction cost once, and (as would be the case in Section 5), when several designs use a component, they share its construction cost. The quality cost measures the performance of the component, and involves, for example, its number of states or security level. Accordingly, a design pays the quality cost of a component every time it uses it, and the fact the component is used by other designs is not important.[4]

Formally, consider a library $\mathcal{L} = \{\mathcal{B}_1, \ldots, \mathcal{B}_n\}$ and a design $\mathcal{D} = \langle [n], E, D, d^0, \delta, \nu \rangle$. The number of times $\mathcal{D}$ uses a component B_i is $nused(\mathcal{D}, \mathcal{B}_i) = |\{d \in D : \nu(d) = i\}|$.

[4] One might consider a different quality-cost model, which takes into an account the cost of *computations*. The cost of a design is then the maximal or expected cost of its computations. Such a cost model is appropriate for measures like the running time or other complexity measures. We take here a global approach, which is appropriate for measures like the number of states or security level.

The set of components that are used in $\mathcal{D}$, is $used(\mathcal{D}) = \{\mathcal{B}_i : nused(\mathcal{D}, \mathcal{B}_i) \geq 1\}$. The cost of a design is then $cost(\mathcal{D}) = \sum_{\mathcal{B} \in used(\mathcal{D})} c\text{-}cost(\mathcal{B}) + nused(\mathcal{D}, \mathcal{B}) \cdot q\text{-}cost(\mathcal{B})$.

We state the problem of finding the cheapest design as a decision problem. For a specification DFA $\mathcal{S}$, a library $\mathcal{L}$, and a threshold μ, we say that an input $\langle \mathcal{S}, \mathcal{L}, \mu \rangle$ is in BCD (standing for "bounded cost design") iff there exists a correct design $\mathcal{D}$ such that $cost(\mathcal{D}) \leq \mu$. In this section we study the BCD problem in a setting with a single user. Thus, decisions are independent of other users of the library, which, recall, may influence the construction cost.

In section 3, we reduced the design problem to the problem of the solution of a safety game. In particular, we showed how a winning strategy in the game induces a correct design. Note that while we know that safety games admits memoryless strategies, there is no guarantee that memoryless strategies are guaranteed to lead to optimal designs. We first study this point and show that, surprisingly, while memoryless strategies are sufficient for obtaining an optimal design in the closed setting, this is not the case in the open setting. The source of the difference is the fact that the language of a design in the open setting may be strictly contained in the language of the specification. The approximation may enable the user to generate a design that is more complex and is still cheaper in terms of cost. This is related to the fact that over approximating the language of a DFA may result in exponentially bigger DFAs [5]. We are still able to bound the size of the cheapest design by the size of the game.

4.1 On the Optimality and Non-Optimality of Memoryless Strategies

Consider a closed library $\mathcal{L}$ and a DFA $\mathcal{S}$. Recall that a correct design for $\mathcal{S}$ from components in $\mathcal{L}$ is induced by a winning strategy of Player 1 in the game $\mathcal{G}_{\mathcal{L},\mathcal{S}}$ (see Theorem 1). If the winning strategy is not memoryless, we can trim it to a memoryless one and obtain a design whose state space is a subset of the design induced by the original strategy. Since the design has no flexibility with respect to the language of $\mathcal{S}$, we cannot do better. Hence the following lemma.

Lemma 1. *Consider a closed library $\mathcal{L}$ and a DFA $\mathcal{S}$. For every $\mu \geq 0$, if there is a correct design $\mathcal{D}$ with $cost(\mathcal{D}) \leq \mu$, then there is a correct design $\mathcal{D}'$ induced by a memoryless strategy for Player 1 in $\mathcal{G}_{\mathcal{L},\mathcal{S}}$ such that $cost(\mathcal{D}') \leq \mu$.*

While Lemma 1 seems intuitive, it does not hold in the setting of open systems. There, a design has the freedom to generate a language that is a subset of $L(\mathcal{S})$, as long as it stays receptive. This flexibility allows the design to generate a language that need not be related to the structure of the game $\mathcal{G}_{\mathcal{L},\mathcal{S}}$, which may significantly reduce its cost. Formally, we have the following.

Lemma 2. *There is an open library $\mathcal{L}$ and a family of DFAs $\mathcal{S}_n$ such that $\mathcal{S}_n$ has a correct design $\mathcal{D}_n$ with cost 1 but every correct design for $\mathcal{S}_n$ that is induced by a memoryless strategy for Player 1 in $\mathcal{G}_{\mathcal{L},\mathcal{S}_n}$ has cost n.*

Proof: We define $\mathcal{S}_n = \langle \Sigma_I \times \Sigma_O, S_n, \delta_{\mathcal{S}_n}, s_0^0, F_{\mathcal{S}_n} \rangle$, where $\Sigma_I = \{\tilde{0}, \tilde{1}, \#\}$, $\Sigma_O = \{0, 1, \text{-}\}$, and S_n, $\delta_{\mathcal{S}_n}$ and $F_{\mathcal{S}_n}$ are as follows. Essentially, after reading a prefix of i #'s, for $1 \leq i \leq n$, the design should arrange its outputs so that the i-th and $(n+i)$-th letters agree (they are $\tilde{0}$ and 0, or are $\tilde{1}$ and 1). One method to do it is to count the number of #'s and then check the corresponding indices. Another method is to keep track of all the first n input letters and make sure that they repeat. The key idea is that while in the second method we strengthen the specification (agreement is checked with respect to all i's, ignoring the length of the #-prefix), it is still receptive, which is exactly the flexibility that the open setting allows. We define $\mathcal{S}_n$ and the library $\mathcal{L}$ so that the structure of $\mathcal{G}_{\mathcal{L},\mathcal{S}_n}$ supports the first method, but counting each # has a cost of 1. Consequently, a memoryless strategy of Player 1 in $\mathcal{G}_{\mathcal{L},\mathcal{S}_n}$ induces a design that counts, and is therefore of cost n, whereas an optimal design follows the second method, and since it does not count the number of #'s, its cost is only 1.

We can now describe the DFA $\mathcal{S}_n$ in more detail. It consists of n chains, sharing an accepting sink s_{acc} and a rejecting sink s_{rej}. For $0 \leq i \leq n-1$, we describe the i-th chain, which is depicted in Figure 1. When describing $\delta_{\mathcal{S}_n}$, for ease of presentation, we sometimes omit the letter in Σ_I or Σ_O and we mean that every letter in the respective alphabet is allowed. For $0 \leq i < n-1$, we define $\delta_{\mathcal{S}_n}(s_0^i, \#) = s_0^{i+1}$ and $\delta_{\mathcal{S}_n}(s_0^n, \#) = s_0^n$. Note that words of the form $\#^i a_1 \tilde{1} a_2 b 0$ or $\#^i a_1 \tilde{0} a_2 b 1$ are not in $L(\mathcal{S}_n)$, where if $0 \leq i \leq n-1$, then $a_1 \in (\tilde{0}+\tilde{1})^i$, $a_2 \in (\tilde{0}+\tilde{1})^{n-i-1}$, and $b \in (0+1)^i$, and if $i > n-1$ then the lengths of a_1, a_2, and b are $n-1$, 0, and $n-1$, respectively. We require that after reading a word in $\#^*(\tilde{0}+\tilde{1})^n$ there is an output of n letters in $\{0,1\}$. Thus, for $n \leq j \leq n+i+1$, we define $\delta_{\mathcal{S}_n}(s_j^{i,0}, \text{-}) = \delta_{\mathcal{S}_n}(s_j^{i,1}, \text{-}) = s_{rej}$. Also, $\mathcal{S}_n$ accepts every word that has a # after the initial prefix of # letters. Thus, for $1 \leq j \leq i$, we define $\delta_{\mathcal{S}_n}(s_j^i, \#) = s_{acc}$, and for $i+1 \leq j \leq n+i+1$ and $t \in \{0,1\}$ we define $\delta_{\mathcal{S}_n}(s_j^{i,t}, \#) = s_{acc}$.

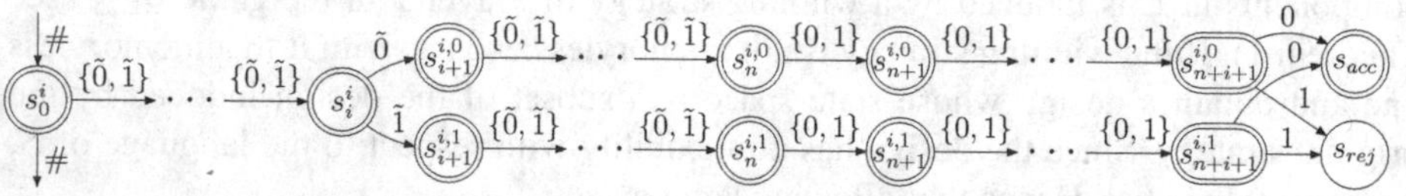

Fig. 1. A description of the i-th chain of the specification $\mathcal{S}_n$

The library $\mathcal{L}$ is depicted in Figure 2. The quality and construction costs of all the components is 0, except for $\mathcal{B}_1$ which has $q\text{-}cost(\mathcal{B}_1) = 1$ and $c\text{-}cost(\mathcal{B}_1) = 0$.

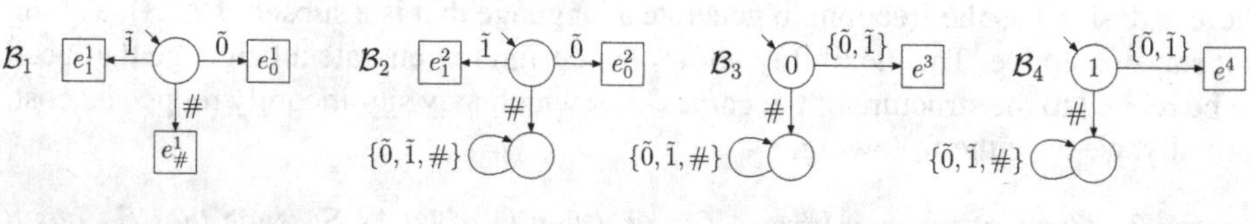

Fig. 2. The library $\mathcal{L}$. Exit states are square nodes and the output of a state is written in the node.

In the full version we prove that every correct design must cost at least 1 and describe such a design, which, as explained above, does not track the number of #'s that are read and can thus use $\mathcal{B}_1$ only once. On the other hand, a design that corresponds to a winning memoryless strategy in $\mathcal{G}_{\mathcal{L},\mathcal{S}}$ uses $\mathcal{B}_1$ n times, thus it costs n, and we are done. ■

4.2 Solving the BCD Problem

Theorem 3. *The BCD problem is NP-complete for closed designs.*

Proof: Consider an input $\langle \mathcal{S}, \mathcal{L}, \mu \rangle$ to the BCD problem. By Lemma 1, we can restrict the search for a correct design $\mathcal{D}$ with $cost(\mathcal{D}) \leq \mu$ to these induced by a memoryless strategy for Player 1 in $\mathcal{G}_{\mathcal{L},\mathcal{S}}$. By the definition of the game $\mathcal{G}_{\mathcal{L},\mathcal{S}}$, such a design has at most $|\mathcal{C}_0 \times S|$ states. Since checking if a design is correct and calculating its cost can be done in polynomial time, membership in NP follows. For the lower bound we show a reduction from SET-COVER, which we describe in the full version. ■

We turn to study the open setting, which is significantly harder than the closed one. For the upper bound, we first show that while we cannot restrict attention to designs induced by memoryless strategies, we can still bound the size of optimal designs:

Theorem 4. *For an open library $\mathcal{L}$ with ℓ components and a specification $\mathcal{S}$ with n states, a cheapest correct design $\mathcal{D}$ has at most $\binom{n}{n/2} \cdot \ell$ states.*

Proof: Given $\mathcal{S}$ and $\mathcal{L}$, assume towards contradiction that the cheapest smallest design $\mathcal{D}$ for $\mathcal{S}$ using the components in $\mathcal{L}$ has more than $\binom{n}{n/2} \cdot \ell$ states.

Consider a word $w \in L(\mathcal{T}_{\mathcal{L},\mathcal{D}})$. Let $\mathcal{B}_{i_1}, \ldots, \mathcal{B}_{i_m} \in \mathcal{L}$ be the components that are traversed in the run r of $\mathcal{T}_{\mathcal{L},\mathcal{D}}$ that induces w. Let $w = w_1 \cdot \ldots \cdot w_m$, where, for $1 \leq j \leq m$, the word w_j is induced in the component $\mathcal{B}_{i_j}$. We say that w is suffix-less if $w_m = \epsilon$, thus r ends in the initial state of the last component to gain control. We denote by $\pi_w(\mathcal{D}) = e_{i_1}, \ldots, e_{i_{m-1}} \in \mathcal{E}^*$ the sequence of exit states that r visits.

For a state $d \in D$, we define the set $S_d \subseteq S$ so that $s \in S_d$ iff there exists a suffix-less word $w \in (\Sigma_I \times \Sigma_O)^*$ such that $\delta^*_{\mathcal{S}}(w) = s$ and $\delta^*_{\mathcal{D}}(\pi_w(\mathcal{D})) = d$. Since $\mathcal{D}$ has more than $\binom{n}{n/2} \cdot \ell$ states, there is a component $\mathcal{B}_i \in \mathcal{L}$ such that the set $D' \subseteq \mathcal{L}$ of states that are labeled with $\mathcal{B}_i$ is larger than $\binom{n}{n/2}$. Thus, there must be two states $d, d' \in D'$ that have $S_{d'} \subseteq S_d$. Note that $\nu(d) = \nu(d') = i$.

In the full version we show that we can construct a new correct design $\mathcal{D}'$ by merging d' into d. Since for every component $\mathcal{B} \in \mathcal{L}$, we have $nused(\mathcal{D}, \mathcal{B}) \geq nused(\mathcal{D}', \mathcal{B})$, it follows that $cost(\mathcal{D}) \geq cost(\mathcal{D}')$. Moreover, $\mathcal{D}'$ has less states than $\mathcal{D}$, and we have reached a contradiction. ■

Before we turn to the lower bound, we argue that the exponential blow-up proven in Theorem 4 cannot be avoided:

Theorem 5. *For every $n \geq 1$, there is an open library $\mathcal{L}$ and specification $\mathcal{S}_n$ such that the size of $\mathcal{L}$ is constant, the size of $\mathcal{S}_n$ is $O(n^2)$, and every cheapest correct design for $\mathcal{S}_n$ that uses components from $\mathcal{L}$ has at least 2^n states.*

Proof: Consider the specification $\mathcal{S}_n$ and library $\mathcal{L}$ that are described in Lemma 2. As detailed in the full version, every correct design that costs 1 cannot count #'s and should thus remember vectors in $\{0, 1\}^n$. ■

Theorem 6. *The BCD problem for open libraries is NEXPTIME-complete.*

Proof: Membership in NEXPTIME follows from Theorem 4 and the fact we can check the correctness of a design and calculate its cost in polynomial time. For the lower bound, in the full version, we describe a reduction from the problem of exponential tiling to the BCD problem for open libraries. The idea behind the reduction is as follows. Consider an input $\langle T, V, H, n\rangle$ to EXP-TILING, where $T = \{t_1, \ldots, t_m\}$ is a set of tiles, $V, H \subseteq T \times T$ are *vertical* and *horizontal* relations, respectively, and $n \in \mathbb{N}$ is an index given in unary. We say that $\langle T, V, H, n\rangle \in$ EXP-TILING if it is possible to fill a $2^n \times 2^n$ square with the tiles in T that respects the two relations.

Given an input $\langle T, V, H, n\rangle$, we construct an input $\langle \mathcal{L}, \mathcal{S}, k\rangle$ to the open-BCD problem such that there is an exponential tiling iff there is a correct design $\mathcal{D}$ with $cost(\mathcal{D}) \leq 2^{2n+1} + 1$. The idea behind the reduction is similar to that of Lemma 2. We define $\Sigma_I = \{\tilde{0}, \tilde{1}, \#, c, v, h, \_\}$ and $\Sigma_O = \{0, 1, \_\} \cup T$. For $x \in \{0,1\}^n$, we use $\tilde{x}$ to refer to the $\{\tilde{0}, \tilde{1}\}$ copy of x. The library $\mathcal{L}$ has the same components as in Lemma 2 with an additional *tile component* $\mathcal{B}_t$ for every $t \in T$. The component $\mathcal{B}_t$ outputs t in its initial state, and when reading c, v, or h, it relinquishes control. When reading every other letter, it enters an accepting sink. The construction costs of the components in $\mathcal{L}$ is 0. We define $q\text{-}cost(\mathcal{B}_1) = 2^{2n} + 1$, and $q\text{-}cost(\mathcal{B}_t) = 1$ for all $t \in T$. The other components' quality cost is 0.

Consider a correct design $\mathcal{D}$ with $cost(\mathcal{D}) \leq 2^{2n+1} + 1$. We define $\mathcal{S}$ so that a correct design must use $\mathcal{B}_1$ at least once, thus $\mathcal{D}$ uses it exactly once. Intuitively, $a \cdot b$, for $a, b \in \{0,1\}^n$, can be thought of as two coordinates in a $2^n \times 2^n$ square. We define $\mathcal{S}$ so that after reading the word $\tilde{a} \cdot \tilde{b} \in \{\tilde{0}, \tilde{1}\}^{2n}$, a component is output, which can be thought of as the tile in the (a, b) coordinate in the square. The next letter that can be read is either c, v, or h. Then, $\mathcal{S}$ enforces that the output is $a \cdot b$, $(a+1) \cdot b$, and $a \cdot (b+1)$, respectively. Thus, we show that $\mathcal{D}$ uses exactly 2^{2n} tile components and the tiling that it induces is legal. ■

5 Libraries with Costs and Multiple Users

In this section we study the setting in which several designers, each with his own specification, use the library. The construction cost of a component is now shared by the designers that use it, with the share being proportional to the number of times the component is used. For example, if $c\text{-}cost(\mathcal{B}) = 8$ and there are two designers, one that uses $\mathcal{B}$ once and a second that uses $\mathcal{B}$ three times, then the construction costs of $\mathcal{B}$ of the two designers are 2 and 6, respectively. The quality cost of a component is not shared. Thus, the cost a designer pays for a design depends on the choices of the other users and he has an incentive to share the construction costs of components with other designers. We model this setting as a multi-player game, which we dub *component library games* (CLGs, for short). The game can be thought of as a one-round game in which each player (user) selects a design that is correct according to his specification. In this section we focus on closed designs.

Formally, a CLG is a tuple $\langle \mathcal{L}, \mathcal{S}_1, \ldots, \mathcal{S}_k\rangle$, where $\mathcal{L}$ is a closed component library and, for $1 \leq i \leq k$, the DFA $\mathcal{S}_i$ is a specification for Player i. A strategy of Player i is a design that is correct with respect to $\mathcal{S}_i$. We refer to a choice of designs for all the players as a *strategy profile*. Consider a profile $P = \langle \mathcal{D}_1, \ldots, \mathcal{D}_k\rangle$ and a component $\mathcal{B} \in \mathcal{L}$. The construction cost of $\mathcal{B}$ is split proportionally between the players that use it. Formally, for $1 \leq i \leq k$, recall that we use $nused(\mathcal{B}, \mathcal{D}_i)$ to denote the number of times $\mathcal{D}_i$ uses $\mathcal{B}$. For a profile P, let $nused(\mathcal{B}, P)$ denote the number of times $\mathcal{B}$ is used by all the designs in P. Thus, $nused(\mathcal{B}, P) = \sum_{1 \le i \le k} nused(\mathcal{B}, \mathcal{D}_i)$. Then, the construction cost that Player i pays in P for $\mathcal{B}$ is $c\text{-}cost_i(P, \mathcal{B}) = c\text{-}cost(\mathcal{B}) \cdot \frac{nused(\mathcal{B},\mathcal{D}_i)}{nused(\mathcal{B},P)}$. Since the quality costs of the components is not shared, it is calculated as in Section 4. Thus, the cost Player i pays in profile P, denoted $cost_i(P)$ is $\sum_{\mathcal{B} \in \mathcal{L}} c\text{-}cost_i(P, \mathcal{B}) + nused(\mathcal{B}, \mathcal{D}_i) \cdot q\text{-}cost(\mathcal{D}_i)$. We define the cost of a profile P, denoted $cost(P)$, as $\sum_{i \in [k]} cost_i(P)$.

For a profile P and a correct design $\mathcal{D}$ for Player i, let $P[i \leftarrow \mathcal{D}]$ denote the profile obtained from P by replacing the choice of design of Player i by $\mathcal{D}$. A profile P is a *Nash equilibrium* (NE) if no Player i can benefit by unilaterally deviating from his choice in P to a different design; i.e., for every Player i and every correct design $\mathcal{D}$ with respect to $\mathcal{S}_i$, it holds that $cost_i(P[i \leftarrow \mathcal{D}]) \ge cost_i(P)$.

Theorem 7. *There is a CLG with no NE.*

Proof: We adapt the example for multiset cost-sharing games from [6] to CLGs. Consider the two-player CLG over the alphabet $\Sigma = \{a, b, c\}$ in which Player 1 and 2's specifications are (the single word) languages $\{ab\}$ and $\{c\}$, respectively. The library is depicted in Figure 3, where the quality costs of all components is 0, $c\text{-}cost(\mathcal{B}_1) = 12$, $c\text{-}cost(\mathcal{B}_2) = 5$, $c\text{-}cost(\mathcal{B}_3) = 1$, and $c\text{-}cost(\mathcal{B}_4) = c\text{-}cost(\mathcal{B}_5) = 0$. Both players have two correct designs. For Player 1, the first design uses $\mathcal{B}_1$ twice and the second design uses $\mathcal{B}_1$ once and $\mathcal{B}_2$ once. There are also uses of $\mathcal{B}_4$ and $\mathcal{B}_5$, but since they can be used for free, we do not include them in the calculations. For Player 2, the first design uses $\mathcal{B}_2$ once, and the second design uses $\mathcal{B}_1$ once. The table in Figure 3 shows the players' costs in the four possible CLG's profiles, and indeed none of the profiles is a NE. ■

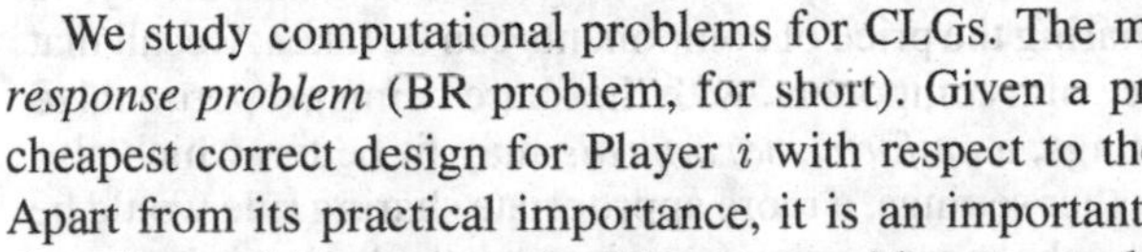

	$\{\mathcal{B}_1, \mathcal{B}_1\}$	$\{\mathcal{B}_1, \mathcal{B}_2\}$
$\{\mathcal{B}_3\}$	5, 12	5, 13
$\{\mathcal{B}_1\}$	4, 8	6, 7

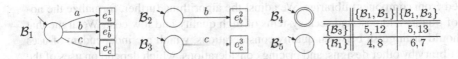

Fig. 3. The library of the CLG with no NE, and the costs of the players in its profiles

We study computational problems for CLGs. The most basic problem is the *best-response problem* (BR problem, for short). Given a profile P and $i \in [k]$, find the cheapest correct design for Player i with respect to the other players' choices in P. Apart from its practical importance, it is an important ingredient in the solutions to the other problems we study. The next problem we study is finding the *social optimum* (SO, for short), namely the profile that minimizes the total cost of all players; thus the one obtained when the players obey some centralized authority. For both the BR and SO problems, we study the decision (rather than search) variants, where the input includes a threshold μ. Finally, since CLGs are not guaranteed to have a NE, we study the problem of deciding whether a given CLG has a NE. We term this problem ∃NE.

Note that the BCD problem studied in Section 4 is a special case of BRP when there is only one player. Also, in a setting with a single player, the SO and BR problems coincide, thus the lower bound of Theorem 3 applies to them. In Lemma 1 we showed that if there is a correct design $\mathcal{D}$ with $cost(\mathcal{D}) \le \mu$, then there is also a correct design $\mathcal{D}'$, based on a memoryless strategy and hence having polynomially many states, such that for every component $\mathcal{B}$, we have $nused(\mathcal{D}', \mathcal{B}) \leq nused(\mathcal{D}, \mathcal{B})$. The arguments there apply in the more general case of CLGs. Thus, we have the following.

Theorem 8. *The BR and SO problems are NP-complete.*

We continue to study the ∃NE problem. We show that ∃NE is complete for Σ_2^P – the second level of the polynomial hierarchy. Namely, decision problems solvable in polynomial time by a nondeterministic Turing machine augmented by an *oracle* for an NP-complete problem.

Theorem 9. *The ∃NE problem is Σ_2^P-complete.*

Proof: The full proof can be found in the full version. We describe its idea in the following. For the upper bound, we describe a nondeterministic Turing machine with an oracle to SBR problem – the strict version of the BR problem, where we seek a design whose cost is strictly smaller than μ. Given a CLG $\mathcal{G} = \langle \mathcal{L}, \mathcal{S}_1, \ldots, \mathcal{S}_k \rangle$, we guess a profile $P = \langle \mathcal{D}_1, \ldots, \mathcal{D}_k \rangle$, where for $1 \leq i \leq k$, the design $\mathcal{D}_i$ has at most $|\mathcal{C}_0 \times S_i|$ states, where S_i are the states of $\mathcal{S}_i$. We check whether the designs are correct, and use the oracle to check whether there is a player that can benefit from deviating from P. For the lower bound, we show a reduction from the complement of the Π_2^P-complete problem *min-max vertex cover* [20]. ■

6 Discussion

Traditional synthesis algorithms assumed that the system is constructed from scratch. Previous work adjusted synthesis algorithms to a reality in which systems are constructed from component libraries. We adjust the algorithms further, formalize the notions of quality and cost and seek systems of high quality and low cost. We argue that one should distinguish between quality considerations, which are independent of uses of the library by other designs, and pricing considerations, which depend on uses of the library by other designs.

Once we add multiple library users to the story, synthesis is modeled by a resource-allocation game and involves ideas and techniques form algorithmic game theory. In particular, different models for sharing the price of components can be taken. Recall that in our model, users share the price of a component, with the share being proportional to the number of uses. In some settings, a *uniform sharing rule* may fit better, which also makes the game more stable. In other settings, a more appropriate sharing rule would be the one used in *congestion games* – the more a component is used, the higher is its price, reflecting, for example, a higher load. Moreover, synthesis of different specifications in different times gives rise to *dynamic allocation* of components, and synthesis of collections of specifications by different users gives rise to *coalitions* in the games. These notions are well studied in algorithmic game theory and enable an even better modeling of the rich settings in which traditional synthesis is applied.

References

1. Abadi, M., Lamport, L., Wolper, P.: Realizable and unrealizable concurrent program specifications. In: Ronchi Della Rocca, S., Ausiello, G., Dezani-Ciancaglini, M. (eds.) ICALP 1989. LNCS, vol. 372, pp. 1–17. Springer, Heidelberg (1989)

2. Almagor, S., Boker, U., Kupferman, O.: Formalizing and reasoning about quality. In: Fomin, F.V., Freivalds, R., Kwiatkowska, M., Peleg, D. (eds.) ICALP 2013, Part II. LNCS, vol. 7966, pp. 15–27. Springer, Heidelberg (2013)
3. Alonso, G., Casati, F., Kuno, H.A., Machiraju, V.: Web Services - Concepts, Architectures and Applications. In: Data-Centric Systems and Applications. Springer (2004)
4. Alur, R., Etessami, K., Madhusudan, P.: A temporal logic of nested calls and returns. In: Jensen, K., Podelski, A. (eds.) TACAS 2004. LNCS, vol. 2988, pp. 467–481. Springer, Heidelberg (2004)
5. Avni, G., Kupferman, O.: When does abstraction help? IPL 113, 901–905 (2013)
6. Avni, G., Kupferman, O., Tamir, T.: Congestion and cost-sharing games with multisets of resources (submitted, 2014)
7. Avni, G., Kupferman, O., Tamir, T.: Network-formation games with regular objectives. In: Muscholl, A. (ed.) FOSSACS 2014. LNCS, vol. 8412, pp. 119–133. Springer, Heidelberg (2014)
8. Berwanger, D., Doyen, L.: On the power of imperfect information. In: Proc. 28th TST& TCS, pp. 73–82 (2008)
9. Bohy, A., Bruyère, V., Filiot, E., Raskin, J.-F.: Synthesis from LTL specifications with mean-payoff objectives. In: Piterman, N., Smolka, S.A. (eds.) TACAS 2013. LNCS, vol. 7795, pp. 169–184. Springer, Heidelberg (2013)
10. Boker, U., Chatterjee, K., Henzinger, T.A., Kupferman, O.: Temporal specifications with accumulative values. In: Proc. 26th LICS, pp. 43–52 (2011)
11. de Alfaro, L., Faella, M., Henzinger, T.A., Majumdar, R., Stoelinga, M.: Model checking discounted temporal properties. TCS 345(1), 139–170 (2005)
12. de Alfaro, L., Henzinger, T.A.: Interface theories for component-based design. In: Henzinger, T.A., Kirsch, C.M. (eds.) EMSOFT 2001. LNCS, vol. 2211, pp. 148–165. Springer, Heidelberg (2001)
13. Doyen, L., Henzinger, T.A., Jobstmann, B., Petrov, T.: Interface theories with component reuse. In: Proc. 8th EMSOFT, pp. 79–88 (2008)
14. Elgaard, J., Klarlund, N., Möller, A.: Mona 1.x: New techniques for WS1S and WS2S. In: Hu, A.J., Vardi, M.Y. (eds.) CAV 1998. LNCS, vol. 1427, pp. 516–520. Springer, Heidelberg (1998)
15. Fabrikant, A., Papadimitriou, C., Talwar, K.: The complexity of pure nash equilibria. In: Proc. 36th STOC, pp. 604–612 (2004)
16. Faella, M., Legay, A., Stoelinga, M.: Model checking quantitative linear time logic. ENTCS 220(3), 61–77 (2008)
17. Filiot, E., Jin, N., Raskin, J.-F.: Antichains and compositional algorithms for LTL synthesis. FMSD 39(3), 261–296 (2011)
18. Gößler, G., Sifakis, J.: Composition for component-based modeling. Sci. Comput. Program. 55(1-3), 161–183 (2005)
19. Jurdziński, M.: Small progress measures for solving parity games. In: Reichel, H., Tison, S. (eds.) STACS 2000. LNCS, vol. 1770, pp. 290–301. Springer, Heidelberg (2000)
20. Ko, K.-I., Lin, C.-L.: On the complexity of min-max optimization problems and their approximation. In: Minimax and Applications. Nonconvex Optimization and Its Applications, vol. 4, pp. 219–239. Springer (1995)
21. Kupferman, O., Piterman, N., Vardi, M.Y.: Safraless compositional synthesis. In: Ball, T., Jones, R.B. (eds.) CAV 2006. LNCS, vol. 4144, pp. 31–44. Springer, Heidelberg (2006)
22. Kupferman, O., Vardi, M.Y.: Safraless decision procedures. In: Proc. 46th FOCS, pp. 531–540 (2005)
23. Lustig, Y., Vardi, M.Y.: Synthesis from component libraries. STTT 15, 603–618 (2013)

24. Pnueli, A., Rosner, R.: On the synthesis of a reactive module. In: Proc. 16th POPL, pp. 179–190 (1989)
25. Raskin, J.-F., Chatterjee, K., Doyen, L., Henzinger, T.: Algorithms for ω-regular games with imperfect information. LMCS 3(3) (2007)
26. Roughgarden, T., Tardos, E.: How bad is selfish routing? JACM 49(2), 236–259 (2002)
27. Safra, S.: On the complexity of ω-automata. In: Proc. 29th FOCS, pp. 319–327 (1988)
28. Sistla, A.P., Clarke, E.M.: The complexity of propositional linear temporal logic. JACM 32, 733–749 (1985)

Compositional Controller Synthesis for Stochastic Games

Nicolas Basset, Marta Kwiatkowska, and Clemens Wiltsche

Department of Computer Science, University of Oxford, United Kingdom

Abstract. Design of autonomous systems is facilitated by automatic synthesis of correct-by-construction controllers from formal models and specifications. We focus on stochastic games, which can model the interaction with an adverse environment, as well as probabilistic behaviour arising from uncertainties. We propose a synchronising parallel composition for stochastic games that enables a compositional approach to controller synthesis. We leverage rules for compositional assume-guarantee verification of probabilistic automata to synthesise controllers for games with multi-objective quantitative winning conditions. By composing winning strategies synthesised for the individual components, we can thus obtain a winning strategy for the composed game, achieving better scalability and efficiency at a cost of restricting the class of controllers.

1 Introduction

With increasing pervasiveness of technology in civilian and industrial applications, it has become paramount to provide formal guarantees of safety and reliability for autonomous systems. We consider the development of correct-by-construction controllers satisfying high-level specifications, based on formal system models. Automated synthesis of controllers has been advocated, for example, for autonomous driving [3] and distributed control systems [15].

Stochastic Games. When designing autonomous systems, often a critical element is the presence of an uncertain and adverse environment, which introduces stochasticity and requires the modelling of the non-cooperative aspect in a game-theoretical setting [7,14]. Hence, we model a system we wish to control as a two-player turn-based stochastic game [18], and consider automated synthesis of strategies that are winning against every environment (Player $\Box$), which we can then interpret as controllers of the system (Player $\Diamond$). In addition to probabilities, one can also annotate the model with rewards to evaluate various quantities, for example, profit or energy usage, by means of expectations.

Compositionality. We model systems as a composition of several smaller components. For controller synthesis for games, a compositional approach requires that we can derive a strategy for the composed system by synthesising only for the individual components. Probabilistic automata (PAs) are naturally suited to modelling multi-component probabilistic systems, where synchronising composition is well-studied [17]; see also [19] for a taxonomic discussion. While PAs

P. Baldan and D. Gorla (Eds.): CONCUR 2014, LNCS 8704, pp. 173–187, 2014.

can be viewed as stochastic games with strategies already applied, it is not immediately clear how to compose games in a natural way.

Our Composition. We formulate a composition of stochastic games where component games synchronise on shared actions and interleave otherwise. The composition is inspired by interface automata [9], which are well-suited to compositional assume-guarantee verification of component-based systems, where we preserve the identity of Player $\Diamond$ and Player $\Box$ in the composition by imposing similar compatibility conditions. Our composition is commutative and associative, and reduces to PA composition when only Player $\Box$ is present. The results we prove are independent of specific winning conditions, and thus provide a general framework for the development of compositional synthesis methods.

Compositional Synthesis. We show that any rule for compositional verification of PAs carries over as a synthesis rule for games. First, after applying winning strategies to a game, the resulting PAs still satisfy the winning condition for any environment. Then, compositional rules for PAs can be used, such as the assume-guarantee rules in [13] developed for winning conditions involving multi-objective total expected reward and probabilistic LTL queries. These first two steps, described also as "schedule-and-compose" [8], are applicable when strategies can only be implemented locally in practice, for instance, when wanting to control a set of physically separated electrical generators and loads in a microgrid, where no centralised strategy can be implemented. One key property of our game composition is that strategies applied to individual components can be composed to a strategy for the composed game, while preserving the probability measure over the traces. Hence, we obtain a winning strategy for the composed game, which alleviates the difficulty of having to deal with the product state space by trading off the expressiveness of the generated strategies.

Winning Conditions. Each player plays according to a *winning condition*, specifying the desirable behaviour of the game, for example "the probability of reaching a failure state is less than 0.01." We are interested in synthesis for zero-sum games for which several kinds of winning conditions are defined in the literature, including ω-regular [4], expected total and average reward [10], and multi-objective versions thereof [7]. Our game composition is independent of such winning conditions, since they are definable on the trace distributions.

Work We Build Upon. In this paper we extend the work of [13] by lifting the compositional verification rules for PAs to compositional synthesis rules for games. Typically, such rules involve multi-objective queries, and we extend the synthesis methods for such queries in [6,7] to compositional strategy synthesis.

Contributions. Several notions of (non-stochastic) game composition have recently been proposed [11,12], but they do not preserve player identity, i.e. which player controls which actions, and hence are not applicable to synthesising strategies for a specific player. In this paper, we make the following contributions.

- We define a composition for stochastic games, which, to our knowledge, is the first composition for competitive probabilistic systems that preserves the control authority of Player $\Diamond$, enabling compositional strategy synthesis.

- We show how to apply strategies synthesised for the individual components to the composition, such that the trace distribution is preserved.
- We lift compositional rules for PAs to the game framework for synthesis.
- We apply our theory to demonstrate how to compositionally synthesise controllers for games with respect to multi-objective total expected reward queries, and demonstrate the benefits on a prototype implementation.

Structure. In Section 2 we introduce stochastic games, their normal form, and their behaviour under strategies. In Section 3 we define our game composition, and show that strategies for the individual components can be applied to the composed game. We demonstrate in Section 4 how to use proof rules for PAs and previously developed synthesis methods to compositionally synthesise strategies.

2 Stochastic Games, Induced PAs and DTMCs

We introduce notation and main definitions for stochastic games and their behaviour under strategies.

Distributions. A *discrete probability distribution* (or *distribution*) over a (countable) set Q is a function $\mu : Q \to [0,1]$ such that $\sum_{q \in Q} \mu(q) = 1$; its *support* $\{q \in Q \,|\, \mu(q) > 0\}$ is the set of values where μ is nonzero. We denote by $\mathcal{D}(Q)$ the set of all distributions over Q with finite support. A distribution $\mu \in \mathcal{D}(Q)$ is *Dirac* if $\mu(q) = 1$ for some $q \in Q$, and if the context is clear we just write q to denote such a distribution μ. We denote by $\boldsymbol{\mu}$ the *product distribution* of $\mu^i \in \mathcal{D}(Q^i)$ for $1 \leq i \leq n$, defined on $Q^1 \times \cdots \times Q^n$ by $\boldsymbol{\mu}(q^1, \ldots, q^n) \stackrel{\text{def}}{=} \mu^1(q^1) \cdot \cdots \cdot \mu^n(q^n)$.

Stochastic Games. We consider turn-based action-labelled stochastic two-player games (henceforth simply called *games*), which distinguish two types of nondeterminism, each controlled by a separate player. Player $\Diamond$ represents the controllable part for which we want to synthesise a strategy, while Player $\Box$ represents the uncontrollable environment. Examples of games are shown in Figure 1.

Definition 1. *A* game *is a tuple* $\langle S, (S_\Diamond, S_\Box), \varsigma, \mathcal{A}, \longrightarrow \rangle$, *where S is a countable set of* states *partitioned into* Player $\Diamond$ states $S_\Diamond$ *and* Player $\Box$ states $S_\Box$; $\varsigma \in \mathcal{D}(S)$ *is an* initial distribution; $\mathcal{A}$ *is a countable set of* actions; *and* $\longrightarrow \subseteq S \times (\mathcal{A} \cup \{\tau\}) \times \mathcal{D}(S)$ *is a* transition relation, *such that, for all s, $\{(s, a, \mu) \in \longrightarrow\}$ is finite.*

We adopt the infix notation by writing $s \xrightarrow{a} \mu$ for a *transition* $(s, a, \mu) \in \longrightarrow$, and if $a = \tau$ we speak of a τ-transition. The action labels $\mathcal{A}$ on transitions model observable behaviours, whereas τ can be seen as internal: it cannot be used in winning conditions and is not synchronised in the composition.

We denote the set of *moves* by $S_\bigcirc \stackrel{\text{def}}{=} \{(a, \mu) \,|\, \exists s \in S \,.\, s \xrightarrow{a} \mu\}$. A move (a, μ) is *incoming* to a state s if $\mu(s) > 0$, and is *outgoing* from a state s if $s \xrightarrow{a} \mu$. Note that, as for PAs [17], there could be several moves associated to each action. We define the set of actions *enabled* in a state s by $\mathsf{En}(s) \stackrel{\text{def}}{=} \{a \in \mathcal{A} \,|\, \exists \mu \,.\, s \xrightarrow{a} \mu\}$.

A finite (infinite) *path* $\lambda = s_0(a_0, \mu_0)s_1(a_1, \mu_1)s_2 \ldots$ is a finite (infinite) sequence of alternating states and moves, such that $\varsigma(s_0) > 0$, and, for all $i \geq 0$,

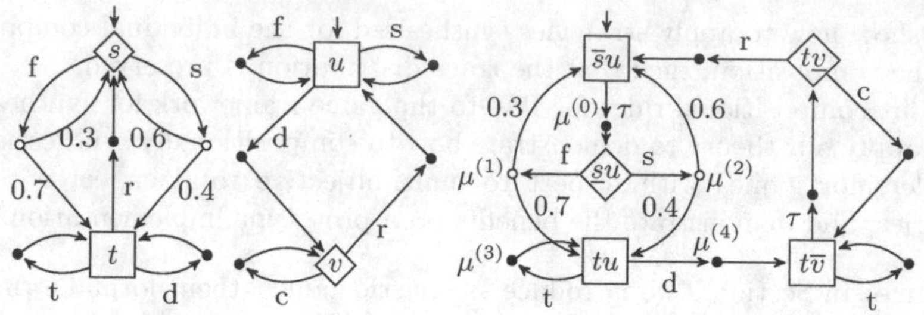

Fig. 1. Three games: the game on the right is the composition of the normal forms of the other two games. Dirac distributions are shown as filled circles.

$s_i \xrightarrow{a_i} \mu_i$ and $\mu_i(s_{i+1}) > 0$. A finite path λ ends in a state, and we write $\mathsf{last}(\lambda)$ for the last state of λ. We denote the set of finite (infinite) paths of a game G by Ω_G^+ (Ω_G), and by $\Omega_{G,\#}^+$ the set of paths ending in a Player $\#$ state, for $\# \in \{\Diamond, \Box\}$. A finite (infinite) *trace* is a finite (infinite) sequence of actions. Given a path, its trace is the sequence of actions along λ, with τ projected out. Formally, $\mathsf{trace}(\lambda) \stackrel{\text{def}}{=} \text{PROJ}_{\{\tau\}}(a_0 a_1 \ldots)$, where, for $\alpha \subseteq \mathcal{A} \cup \{\tau\}$, PROJ_α is the morphism defined by $\text{PROJ}_\alpha(a) = a$ if $a \notin \alpha$, and ϵ (the empty trace) otherwise.

Strategies. Nondeterminism for each player is resolved by a strategy. A *strategy* for Player $\#$, for $\# \in \{\Diamond, \Box\}$, is a function $\sigma_\# : \Omega_{G,\#}^+ \to \mathcal{D}(S_\bigcirc)$ such that $\sigma_\#(\lambda)(a, \mu) > 0$ only if $\mathsf{last}(\lambda) \xrightarrow{a} \mu$. The set of Player $\#$ strategies in game G is denoted by $\Sigma_\#^G$. A strategy is called *memoryless* if, for each path λ, the choice $\sigma_\#(\lambda)$ is uniquely determined by $\mathsf{last}(\lambda)$.

Normal Form of a Game. We can transform every game into its corresponding normal form, which does not affect the winning conditions. Transforming a game to normal form is the first step of our game composition.

Definition 2. *A game is in* normal form *if the following hold:*

- *Every τ-transition $s \xrightarrow{\tau} \mu$ is from a Player $\Box$ state s to a Player $\Diamond$ state s' with a Dirac distribution $\mu = s'$.*
- *Every Player $\Diamond$ state s can only be reached by an incoming move (τ, s). In particular, every distribution μ of a non-τ-transition, as well as the initial distribution, assigns probability zero to all Player $\Diamond$ states.*

Given a game G without τ-transitions, one can construct its normal form $\mathcal{N}(G)$ by splitting every state $s \in S_\Diamond$ into a Player $\Box$ state $\overline{s}$ and a Player $\Diamond$ state $\underline{s}$, s.t.

- the incoming (resp. outgoing) moves of $\overline{s}$ (resp. $\underline{s}$) are precisely the incoming (resp. outgoing) moves of s, with every Player $\Diamond$ state $t \in S_\Diamond$ replaced by $\overline{t}$;
- and the only outgoing (resp. incoming) move of $\overline{s}$ (resp. $\underline{s}$) is $(\tau, \underline{s})$.

Intuitively, at $\overline{s}$ the game is idle until Player $\Box$ allows Player $\Diamond$ to choose a move in $\underline{s}$. Hence, any strategy for G carries over naturally to $\mathcal{N}(G)$, and we can operate w.l.o.g. with normal-form games. Also, τ can be considered as a scheduling choice. In the transformation to normal form, at most one such scheduling choice is introduced for each Player $\Box$ state, but in the composition more choices can be added, so that Player $\Box$ resolves nondeterminism arising from concurrency.

Game Unfolding. Strategy application is defined on the unfolded game. The *unfolding* $\mathcal{U}(G) = \langle \Omega_G^+, (\Omega_{G,\Box}^+, \Omega_{G,\Diamond}^+), \varsigma, \mathcal{A}, \longrightarrow' \rangle$ of the game G is such that $\lambda \xrightarrow{a}' \mu_{\lambda,a}$ if and only if $\mathsf{last}(\lambda) \xrightarrow{a} \mu$ and $\mu_{\lambda,a}(\lambda(a,\mu)s) \stackrel{\text{def}}{=} \mu(s)$ for all $s \in S$.

An unfolded game is a set of trees (the roots are the support of the initial distribution), with potentially infinite depth, but finite branching. The entire history is stored in the states, so memoryless strategies suffice for unfolded games; formally, each strategy $\sigma_\Diamond \in \Sigma_\Diamond^G$ straightforwardly maps to a memoryless strategy $\mathcal{U}(\sigma_\Diamond) \in \Sigma_\Diamond^{\mathcal{U}(G)}$ by letting $\mathcal{U}(\sigma_\Diamond)(\lambda)(a, \mu_{\lambda,a}) = \sigma_\Diamond(\lambda)(a,\mu)$. We denote by $\mathcal{U}(G)_\bigcirc$ the set of moves of the unfolded form of a game G and by $\mathcal{U}(G)_{\#\bigcirc}$ the set of moves following a Player $\#$ state that is of the form $(a, \mu_{\lambda,a})$ with $\lambda \in \Omega_{G,\#}^+$. We remark that the unfolding of a normal form game is also in normal form.

2.1 Induced PA

When only one type of nondeterminism is present in a game, it is a *probabilistic automaton (PA)*. PAs are well-suited for compositional modelling [17], and can be used for verification, i.e. checking whether *all* behaviours satisfy a specification (when only Player $\Box$ is present), as well as strategy synthesis (when only Player $\Diamond$ is present) [14]. A PA is a game where $S_\Diamond = \emptyset$ and $S_\Box = S$, which we write here as $\langle S, \varsigma, \mathcal{A}, \longrightarrow \rangle$. This definition corresponds to modelling nondeterminism as an adverse, uncontrollable, environment, and so, by applying a Player $\Diamond$ strategy to a game to resolve the controllable nondeterminism, we are left with a PA where only uncontrollable nondeterminism for Player $\Box$ remains.

Definition 3. *Given an unfolded game* $\mathcal{U}(G) = \langle \Omega_G^+, (\Omega_{G,\Box}^+, \Omega_{G,\Diamond}^+), \varsigma, \mathcal{A}, \longrightarrow \rangle$ *in normal form and a strategy* $\sigma_\Diamond \in \Sigma_\Diamond^G$, *the* induced PA *is* $G^{\sigma_\Diamond} = \langle S', \varsigma, \mathcal{A}, \longrightarrow' \rangle$, *where* $S' \subseteq \Omega_{G,\Box}^+ \cup \mathcal{U}(G)_{\Diamond\bigcirc}$ *is defined inductively as the reachable states, and*

(I1) $\lambda \xrightarrow{\tau}' \mathcal{U}(\sigma_\Diamond)(\lambda')$ *iff* $\lambda \xrightarrow{\tau} \lambda'$ *(Player $\Diamond$ strategy chooses a move);*
(I2) $(a, \mu_{\lambda,a}) \xrightarrow{a}' \mu_{\lambda,a}$ *for* $(a, \mu_{\lambda,a}) \in \mathcal{U}(G)_{\Diamond\bigcirc}$ *(the chosen move is performed);*
(I3) $\lambda \xrightarrow{a}' \mu_{\lambda,a}$ *iff* $\lambda \xrightarrow{a} \mu_{\lambda,a}$ *and* $\lambda \in \Omega_{G,\Box}^+$ *(external transitions from older Player $\Box$ state remain unchanged).*

The unfolded form of the game in Figure 1(right) is shown in Figure 2(a), and strategy application is illustrated in Figure 2(b).

2.2 Induced DTMC

A discrete-time Markov chain (DTMC) is a model for systems with probabilistic behaviour only. When applying a Player $\Box$ strategy to an induced PA, all nondeterminism is resolved, and a DTMC is obtained. A (labelled) *DTMC D* is a PA such that, for each $s \in S$, there is at most one transition $s \xrightarrow{a} \mu$.

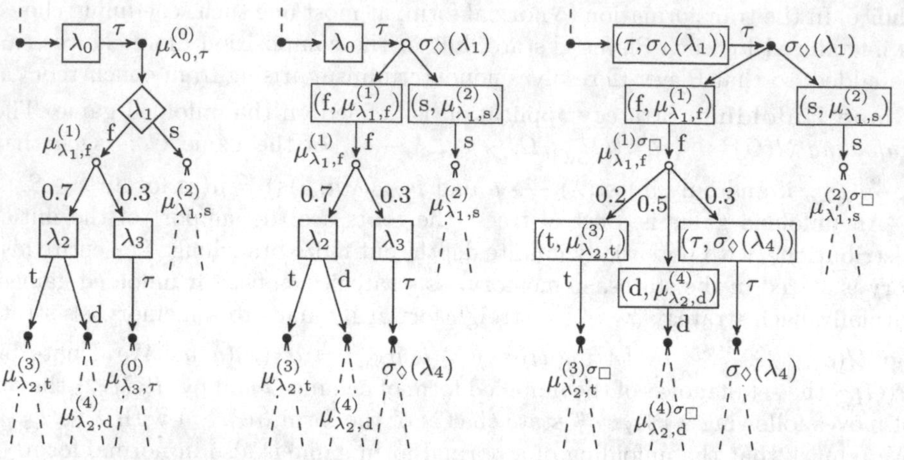

(a) Game in Figure 1 (right) unfolded. (b) Induced PA via the strategy $\sigma_\Diamond$ assigning uniform probabilities. (c) Induced DTMC via the strategy $\sigma_\Box = \mathcal{U}(\sigma'_\Box)$ assigning $\frac{2}{7}$ to t and $\frac{5}{7}$ to d.

Fig. 2. Game unfoldings. The dashed lines represent recursive application of unfolding and strategy application. Denote $\lambda_0 = \overline{s}u$, $\lambda_1 = \lambda_0(\tau, \mu^{(0)}_{\lambda_0,\tau})\underline{s}u$, $\lambda_2 = \lambda_1(\mathrm{f}, \mu^{(1)}_{\lambda_1,\mathrm{f}})\overline{s}u$, $\lambda_3 = \lambda_1(\mathrm{f}, \mu^{(1)}_{\lambda_1,\mathrm{f}})tu$, $\lambda_4 = \lambda_3(\tau, \mu^{(0)}_{\lambda_3,\tau})\underline{s}u$.

Definition 4. *Given an unfolded PA* $\mathcal{U}(M) = \langle S, \varsigma, \mathcal{A}, \longrightarrow\rangle$ *and a strategy* $\sigma_\Box \in \Sigma^M_\Box$, *the* induced DTMC $M^{\sigma_\Box} = \langle S', \varsigma', \mathcal{A}, \longrightarrow'\rangle$ *is such that* $S' \subseteq \mathcal{U}(M)_\bigcirc$ *is defined inductively as the states reachable via* ς' *and* $\longrightarrow'$, *where*

- $(a, \mu_{\lambda,a}) \xrightarrow{a}{}' \mu^{\sigma_\Box}_{\lambda,a}$, *such that, for all moves* $(b, \nu_{\lambda(a,\mu)t,b})$, *we let the distribution* $\mu^{\sigma_\Box}_{\lambda,a}(b, \nu_{\lambda(a,\mu)t,b}) \stackrel{\text{def}}{=} \mu(t)\sigma_\Box(\lambda(a,\mu)t)(b,\nu)$; *and*
- *for all moves* $(b, \nu_{t,b})$, $\varsigma'(b, \nu_{t,b}) \stackrel{\text{def}}{=} \varsigma(t)\sigma_\Box(t)(b,\nu)$.

Note that an induced PA is already unfolded, and does not need to be unfolded again. We illustrate in Figure 2(c) the application of a Player $\Box$ strategy.

Probability Measures. We define the probability measure $\Pr_D$ of a DTMC D in the usual way. The *cylinder set* of a path $\lambda \in \Omega^+_D$ (resp. trace $w \in \mathcal{A}^*$) is the set of infinite paths (resp. traces) with prefix λ (resp. w). For a finite path $\lambda = s_0(a_0, \mu_0)s_1(a_1, \mu_1)\ldots s_n$ we define $\Pr_D(\lambda)$, the measure of its cylinder set, by: $\Pr_D(\lambda) \stackrel{\text{def}}{=} \varsigma(s_0)\prod_{i=0}^{n-1}\mu_i(s_{i+1})$. We write $\Pr^{\sigma_\Diamond,\sigma_\Box}_G$ (resp. $\Pr^{\sigma_\Box}_M$) for the measure $\Pr_{G^{\sigma_\Diamond,\sigma_\Box}}$ (resp. $\Pr_{M^{\sigma_\Box}}$). The measures uniquely extend to infinite paths due to Carathéodory's extension theorem.

Given a finite trace w, $\mathsf{paths}(w)$ denotes the set of minimal finite paths with trace w, i.e. $\lambda \in \mathsf{paths}(w)$ if $\mathsf{trace}(\lambda) = w$ and there is no path $\lambda' \neq \lambda$ with $\mathsf{trace}(\lambda') = w$ and λ' being a prefix of λ. The measure of the cylinder set of w is $\Pr_D(w) \stackrel{\text{def}}{=} \sum_{\lambda\in\mathsf{paths}(w)} \Pr_D(\lambda)$, and we call Pr_D the *trace distribution* of D.

Winning Conditions. Providing strategies for both players resolves all nondeterminism in a game, resulting in a distribution over paths. A *specification* φ is a predicate on trace distributions, and for a DTMC D we write $D \models \varphi$ if $\varphi(P_D)$ holds. A specification φ is *defined on traces* if $\varphi(P_D) = \varphi(P_{D'})$ for all DTMCs D, D' such that $P_D(w) = P_{D'}(w)$ for all traces w.

3 Composing Stochastic Games

We now introduce composition operators for games and Player $\Diamond$ strategies, leading towards a framework for compositional synthesis in Section 4. Games can be composed of several component games, and our composition is inspired by interface automata [9], which have a natural interpretation as (concurrent) games.

3.1 Game Composition

We provide a synchronising composition of games so that controllability is preserved for Player $\Diamond$, that is, actions controlled by Player $\Diamond$ in the components are controlled by Player $\Diamond$ in the composition. We endow each component game with an *alphabet* of actions $\mathcal{A}$, where synchronisation on *shared actions* in $\mathcal{A}^1 \cap \mathcal{A}^2$ is viewed as a (blocking) communication over ports, as in interface automata, though for simplicity we do not distinguish inputs and outputs. Synchronisation is multi-way and we do not impose input-enabledness of IO automata [8].

Given n games G^i in normal form with respective state spaces S^i, for $i \in I$ (let $I = \{1, \ldots, n\}$ be the *index set* of the composed game), the state space of the composition is a subset of the Cartesian product $S^1 \times \ldots \times S^n$, whose states contain at most one Player $\Diamond$ component thanks to the normal form. We denote by s^i the ith component of $\boldsymbol{s} \in S^1 \times \ldots \times S^n$. Furthermore, every probability distribution in the composition is a product distribution. We say that a transition $\boldsymbol{s} \xrightarrow{a} \boldsymbol{\mu}$ *involves* the ith component if $s^i \xrightarrow{a} \mu^i$.

Definition 5. *Given n games in normal form* $G^i = \langle S^i, (S^i_\Diamond, S^i_\Box), \varsigma^i, \mathcal{A}^i, \longrightarrow^i \rangle$, $i \in I$, *their* composition *is the game* $\|_{i \in I} G^i \stackrel{\text{def}}{=} \langle \boldsymbol{S}, (\boldsymbol{S}_\Diamond, \boldsymbol{S}_\Box), \boldsymbol{\varsigma}, \mathcal{A}, \longrightarrow \rangle$, *where*

- $\boldsymbol{S} \subseteq \boldsymbol{S}_\Diamond \cup \boldsymbol{S}_\Box$, *with* $\boldsymbol{S}_\Box \subseteq S^1_\Box \times \cdots \times S^n_\Box$, *and* $\boldsymbol{S}_\Diamond \subseteq \{\boldsymbol{s} \in S^1 \times \cdots \times S^n \mid \exists! \iota .\ s^\iota \in S^\iota_\Diamond\}$ *inductively defined to contain the states reachable from the initial distribution through the transition relation;*
- $\boldsymbol{\varsigma} = \varsigma^1 \times \cdots \times \varsigma^n$; *(note that, due to the normal form,* $\boldsymbol{\varsigma}(\boldsymbol{s}) > 0$ *only if* $\boldsymbol{s} \in \boldsymbol{S}_\Box$*)*
- $\mathcal{A} = \bigcup_{i=1}^n \mathcal{A}^i$;
- *The transition relation* $\longrightarrow$ *is defined such that*
  - $\boldsymbol{s} \xrightarrow{a} \boldsymbol{\mu}$ *for* $a \neq \tau$ *if*
    - *(C1) at least one component is involved;*
    - *(C2) the components involved in the transition are exactly those having a in their action alphabet;*
    - *(C3) for the uninvolved components j (equivalently, that do not have a in their action alphabet), the state remains the same* ($\mu^j = s^j$);

(C4) if s is a Player $\Diamond$ state then its only Player $\Diamond$ component G^ι is involved in the transition; and

- $s \xrightarrow{\tau} t$ *if only one component G^i is involved, $s \in S_\Box$, and s.t. $\mathsf{En}(t) \neq \emptyset$.*

We take the view that identity of the players must be preserved through composition to facilitate synthesis, and thus Player $\Diamond$ actions of the individual components are controlled by a single Player $\Diamond$ in the composition. Player $\Box$ in the composition acts as a scheduler, controlling which component advances and, in Player $\Box$ states, selecting among available actions, whether synchronised or not. Synchronisation in Player $\Diamond$ states means that Player $\Diamond$ in one component may indirectly control some Player $\Box$ actions in another component. In particular, we can impose assume-guarantee contracts at the component level in the following sense. Player $\Diamond$ of different components can cooperate to achieve a common goal: in one component Player $\Diamond$ satisfies the goal B under an assumption A on its environment behaviour (i.e. $A \to B$), while Player $\Diamond$ in the other component ensures that the assumption is satisfied, against all Player $\Box$ strategies.

Our game composition is both associative and commutative, facilitating a modular model development, and is closely related to PA composition [17], with the condition *(C4)* added. As PAs are just games without Player $\Diamond$ states, the game composition restricted to PAs is the same as classical PA composition. The condition $\mathsf{En}(t) \neq \emptyset$ for τ-transitions ensures that a Player $\Diamond$ state is never entered if it were to result in deadlock introduced by the normal form transformation. Deadlocks that were present before the transformation are still present in the normal form. In the composition of normal form games, τ-transitions are only enabled in Player $\Box$ states, and Player $\Diamond$ states are only reached by such transitions; hence, composing normal form games yields a game in normal form.

In Figure 1, the game on the right is the composition of the normal forms of the two games on the left. The actions "f" and "s" are synchronised and controlled by Player $\Diamond$ in $\underline{s}u$. The "d" action is synchronised and controlled by Player $\Box$ in both components, and so it is controlled by Player $\Box$ in the composition, in tu. The action "t" is not synchronised, and thus available in $t\overline{v}$ and tu; it is, however, not available in $t\underline{v}$, as it is a Player $\Diamond$ state controlled by $\underline{v}$. The action "c" is also not synchronised, and is available in $t\underline{v}$. The "r" action is synchronised; it is available both in t and in $\underline{v}$, and hence also in $t\underline{v}$.

Constructing the composition of n components of size $|S|$ clearly requires time $\mathcal{O}(|S|^n)$. In strategy synthesis, the limiting factor is that applying the method on a large product game may be computationally infeasible. For example, the synthesis methods for multi-objective queries of [6] that we build upon are exponential in the number of objectives and polynomial in the size of the state space, and the theoretical bounds can be impractical even for small systems (see [7]). To alleviate such difficulties we focus on compositional strategy synthesis.

3.2 Strategy Composition

For compositional synthesis, we assume the following compatibility condition on component games: we require that actions controlled by Player $\Diamond$ in one game are enabled and fully controlled by Player $\Diamond$ in the composition.

Definition 6. *Games $G^1, \ldots, G^n$ are* compatible *if, for every Player $\Diamond$ state $\boldsymbol{s} \in S_\Diamond$ in the composition with $s^\iota \in S^\iota_\Diamond$, if $s^\iota \xrightarrow{a}{}^\iota \mu^\iota$ then there is exactly one distribution $\boldsymbol{\nu}$, denoted by $\langle \mu^\iota \rangle_{\boldsymbol{s},a}$, such that $\boldsymbol{s} \xrightarrow{a} \boldsymbol{\nu}$ and $\nu^\iota = \mu^\iota$. (That is, for $i \neq \iota$ such that $a \in \mathcal{A}^i$, there exists exactly one a-transition enabled in s^i.)*

Our compatibility condition is analogous to that for single-threaded interface automata [9]. It remains to be seen if this condition can be relaxed without affecting preservation properties of the winning conditions.

Composing Strategies. Given a path λ of a composed game $\mathcal{G} = \|_{i \in I} G^i$, for each individual component G^i one can retrieve the corresponding path $[\lambda]^i$ that contains precisely the transitions G^i is involved in. The *projection* $[\cdot]^i : \Omega^+_{\mathcal{G}} \to \Omega^+_{G^i}$ is defined inductively so that, for all states $\boldsymbol{t} \in S$ and paths $\lambda(a, \boldsymbol{\mu})\boldsymbol{t} \in \Omega^+_{\mathcal{G}}$ (with possibly $a = \tau$), we have $[\boldsymbol{s}]^i \stackrel{\text{def}}{=} s^i$; and inductively that $[\lambda(a, \boldsymbol{\mu})\boldsymbol{t}]^i \stackrel{\text{def}}{=} [\lambda]^i(a, \mu^i)t^i$ if $\mathsf{last}(\lambda)^i \xrightarrow{a}{}^i \mu^i$, and $[\lambda]^i$ otherwise.

Recall that a Player $\Diamond$ state $\boldsymbol{s}$ of the composed game has exactly one component s^ι that is a Player $\Diamond$ state in G^ι; we say that the ιth Player $\Diamond$ *controls* $\boldsymbol{s}$. Given a Player $\Diamond$ strategy for each component, the strategy for the composed game plays the strategy of the Player $\Diamond$ controlling the respective states.

Definition 7. *Let $\sigma^i_\Diamond$, $i \in I$, be Player $\Diamond$ strategies for compatible games. Their* composition, *$\sigma_\Diamond = \|_{i \in I} \sigma^i_\Diamond$, is defined such that $\sigma_\Diamond(\lambda)(a, \langle \mu^\iota \rangle_{\boldsymbol{s},a}) \stackrel{\text{def}}{=} \sigma^\iota_\Diamond([\lambda]^\iota)(a, \mu^\iota)$ for all $\lambda \in \Omega^+_{\mathcal{G}}$ with $\boldsymbol{s} = \mathsf{last}(\lambda) \in S_\Diamond$.*

From this definition, strategy composition is clearly associative. Note that, for each choice, the composed strategy takes into account the history of only one component, which is less general than using the history of the composed game.

3.3 Properties of the Composition

We now show that synthesising strategies for compatible individual components is sufficient to obtain a composed strategy for the composed game.

Functional Simulations. We introduce functional simulations, which are a special case of classical PA simulations [17], and show that they preserve winning conditions over traces. Intuitively, a PA M' functionally simulates a PA M if all behaviours of M are present in M', and if strategies translate from M to M'.

Given a distribution μ, and a partial function $f : S \to S'$ defined on the support of μ, we write $\overline{f}(\mu)$ for the distribution defined by $\overline{f}(\mu)(s') \stackrel{\text{def}}{=} \sum_{f(s)=s'} \mu(s)$. A *functional simulation* from a PA M to a PA M' is a partial function $f : S \to S'$ such that $\overline{f}(\varsigma) = \varsigma'$, and if $s \xrightarrow{a} \mu$ in M then $f(s) \xrightarrow{a}{}' \overline{f}(\mu)$ in M'.

Lemma 1. *Given a functional simulation from a PA M to a PA M' and a specification φ defined on traces, for every strategy $\sigma_\Box \in \Sigma^M_\Box$ there is a strategy $\sigma'_\Box \in \Sigma^{M'}_\Box$ such that $(M')^{\sigma'_\Box} \models \varphi \Leftrightarrow M^{\sigma_\Box} \models \varphi$.*

Key Lemma. The PA $\|_{i \in I} (G^i)^{\sigma^i_\Diamond}$ is constructed by first unfolding each component, applying the Player $\Diamond$ strategies, and then composing the resulting PAs, while the PA $(\|_{i\in I} G^i)^{\|_{i\in I}\sigma^i_\Diamond}$ is constructed by first composing the individual components, then unfolding, and applying the composed Player $\Diamond$ strategy. The following lemma justifies, via the existence of a functional simulation, that composing Player $\Diamond$ strategies preserves the trace distribution between such PAs, and hence yields a fully compositional approach.

Lemma 2. *Given compatible games G^i, $i \in I$, and respective Player $\Diamond$ strategies $\sigma^i_\Diamond$, there is a functional simulation from $(\|_{i\in I} G^i)^{\|_{i\in I}\sigma^i_\Diamond}$ to $\|_{i\in I} (G^i)^{\sigma^i_\Diamond}$.*

In general, there is no simulation in the other direction, as in the PA composition one can no longer distinguish states in the induced PA that were originally Player $\Diamond$ states, and so condition *(C4)* of the composition is never invoked.

4 Compositional Synthesis

Applying strategies synthesised for games to obtain induced PAs allows us to reuse compositional rules for PAs. Using Lemma 2, we can then lift the result back into the game domain. This process is justified in Theorem 1 below.

4.1 Composition Rules

We suppose the designer is supplying a game $\mathcal{G} = \|_{i\in I} G^i$ composed of *atomic games* G_i, together with specifications defined on traces φ_i, $i \in I$, and show how, using our framework, strategies $\sigma^i_\Diamond$ synthesised for G^i and φ_i can be composed to a strategy $\sigma_\Diamond = \|_{i\in I} \sigma^i_\Diamond$ for $\mathcal{G}$, satisfying a specification φ defined on traces.

Theorem 1. *Given a rule $\mathfrak{P}$ for PAs $\mathcal{M}_i$ and specifications φ^i_j and φ defined on traces, then the rule $\mathfrak{G}$ holds for all Player $\Diamond$ strategies $\sigma^i_\Diamond$ of compatible games G^i with the same action alphabets as the corresponding PAs, where*

$$\mathfrak{P} \equiv \frac{\mathcal{M}_i \models \varphi^i_j \quad 1 \le j \le m \quad i \in I}{\|_{i\in I} \mathcal{M}_i \models \varphi,} \quad \text{and} \quad \mathfrak{G} \equiv \frac{(G^i)^{\sigma^i_\Diamond} \models \bigwedge_{j=1}^m \varphi^i_j \quad i \in I}{(\|_{i\in I} G^i)^{\|_{i\in I}\sigma^i_\Diamond} \models \varphi.}$$

Theorem 1 enables the compositional synthesis of strategies in an automated way. First, synthesis is performed for atomic components G^i, $i \in I$, by obtaining for each i a Player $\Diamond$ strategy $\sigma^i_\Diamond$ for $G^i \models \bigwedge_{j=1}^m \varphi^i_j$. We apply $\mathfrak{P}$ with $\mathcal{M}_i \stackrel{\text{def}}{=} (G^i)^{\sigma^i_\Diamond}$ to deduce that φ holds in $\|_{i=1}^n (G^i)^{\sigma^i_\Diamond}$ and, using Lemma 1 and 2, that $\|_{i\in I} \sigma^i_\Diamond$ is a winning strategy for Player $\Diamond$ in $\|_{i=1}^n G^i$. The rules can be applied recursively, making use of associativity of the game and strategy composition.

4.2 Multi-objective Queries

In this section we leverage previous work on compositional verification for PAs in order to compositionally synthesise strategies for games.

Reward and LTL Objectives. The *expected value* of a function $\rho : \mathcal{A}^* \to \mathbb{R}_{\pm\infty}$ over traces in a DTMC D is $\mathbb{E}_D[\rho] \stackrel{\text{def}}{=} \lim_{n\to\infty} \sum_{w\in\mathcal{A}^n} \Pr_D(w)\rho(w)$, if the limit exists in $\mathbb{R}_{\pm\infty}$. We denote by $\mathbb{E}_G^{\sigma_\Diamond,\sigma_\Box}$ (resp. $\mathbb{E}_M^{\sigma_\Box}$) the expected value in a game G (resp. PA M) under the respective strategies. A *reward structure* of a game with actions $\mathcal{A}$ is a function $r : \mathcal{A}_r \to \mathbb{Q}$, where $\mathcal{A}_r \subseteq \mathcal{A}$. Given a reward structure r such that either $r(a) \leq 0$ or $r(a) \geq 0$ for all actions a occurring infinitely often on a path, the *total reward* for a trace $w = a_0a_1\ldots$ is $rew(r)(w) \stackrel{\text{def}}{=} \lim_{t\to\infty} \sum_{i=0}^{t} r(a_i)$, which is measurable thanks to the restrictions imposed on r. Given reward structures r_j, $j \in J$, for all $a \in \bigcup_{j\in J} \mathcal{A}_{r_j}$ we let $(\sum_{j\in J} r_j)(a)$ be the sum of the $r_j(a)$ that are defined for a.

To express LTL properties over traces, we use the standard LTL operators (cf. [16]); in particular, the operators F and G stand for *eventually* and *always*, respectively. For a DTMC D, and an LTL formula Ξ over actions $\mathcal{A}_\Xi$, define $\Pr_D(\Xi) \stackrel{\text{def}}{=} \Pr_D(\{\lambda \in \Omega_D \mid \text{PROJ}_{\mathcal{A}\setminus\mathcal{A}_\Xi}(\mathsf{trace}(\lambda)) \models \Xi\})$, that is, the measure of infinite paths with traces satisfying Ξ, where actions not in $\mathcal{A}_\Xi$ are disregarded.

A *reward* (resp. *LTL*) *objective* is of the form $r \blacktriangleright v$ (resp. $\Xi \blacktriangleright v$), where r is a reward structure, Ξ is an LTL formula, $v \in \mathbb{Q}$ is a bound, and $\blacktriangleright \in \{\geq, >\}$. A reward objective $r \blacktriangleright v$ (resp. LTL objective $\Xi \blacktriangleright v$) is true in a game G under a pair of strategies $(\sigma_\Diamond, \sigma_\Box)$ if and only if $\mathbb{E}_G^{\sigma_\Diamond,\sigma_\Box}[rew(r)] \blacktriangleright v$ (resp. $\Pr_G^{\sigma_\Diamond,\sigma_\Box}(\Xi) \blacktriangleright v$), and similarly for PAs and DTMCs. Minimisation of rewards can be expressed by reverting signs.

Multi-objective Queries. A *multi-objective query* (MQ) φ is a Boolean combination of reward and LTL objectives, and its truth value is defined inductively on its syntax. An MQ φ is a *conjunctive query* (CQ) if it is a conjunction of objectives. Given an MQ with bounds $v_1, v_2, \ldots$, call $\boldsymbol{v} = (v_1, v_2, \ldots)$ the *target*. Denote by $\varphi[\boldsymbol{x}]$ the MQ φ, where, for all i, $r_i \blacktriangleright v_i$ is replaced by $r_i \blacktriangleright x_i$, and $\Xi_i \blacktriangleright v_i$ is replaced by $\Xi_i \blacktriangleright x_i$. Given a game G (resp. PA M) we write $G^{\sigma_\Diamond,\sigma_\Box} \models \varphi$ (resp. $M^{\sigma_\Box} \models \varphi$), if the query φ evaluates to true under the respective strategies. We write $M \models \varphi$ if $M^{\sigma_\Box} \models \varphi$ for all $\sigma_\Box \in \Sigma_\Box^M$. We say that an MQ φ is *achievable* in a game G if there is a Player $\Diamond$ strategy $\sigma_\Diamond$ such that $G^{\sigma_\Diamond} \models \varphi$, that is, $\sigma_\Diamond$ is winning for φ against all possible Player $\Box$ strategies. We require that expected total rewards are bounded, that is, we ask for any reward structure r in an MQ φ that $G^{\sigma_\Diamond,\sigma_\Box} \models r < \infty \wedge r > -\infty$ for all $\sigma_\Diamond$ and $\sigma_\Box$.

Fairness. Since PA rules as used in Theorem 1 often include fairness conditions, we recall here the concept of unconditional process fairness based on [1]. Given a composed PA $\mathcal{M} = \|_{i\in I} M^i$, a strategy $\sigma_\Box$ is *unconditionally fair* if $\mathcal{M}^{\sigma_\Box} \models \mathsf{u}$, where $\mathsf{u} \stackrel{\text{def}}{=} \bigwedge_{i\in I} \mathsf{GF}\,\mathcal{A}_i \geq 1$, that is, each component makes progress infinitely often with probability 1. We write $\mathcal{M} \models^{\mathsf{u}} \varphi$ if, for all unconditionally fair strategies $\sigma_\Box \in \Sigma_\Box$, $\mathcal{M}^{\sigma_\Box} \models \varphi$; this is equivalent to $\mathcal{M} \models \mathsf{u} \to \varphi$ (the arrow $\to$ stands for the standard logical implication), and so MQs can incorporate fairness.

Applying Theorem 1. In particular, for the premises in Theorem 1 we can use the compositional rules for PAs developed in [13], which are stated for MQs. Thus, the specification φ for the composed game can, for example, be a CQ, or a summation of rewards, among others. Unconditional fairness corresponds precisely to the fairness conditions used in the PA rules of [13]. When the PA rules include fairness assumptions, note that, for a single component, unconditional fairness is equivalent to only requiring deadlock-freedom.

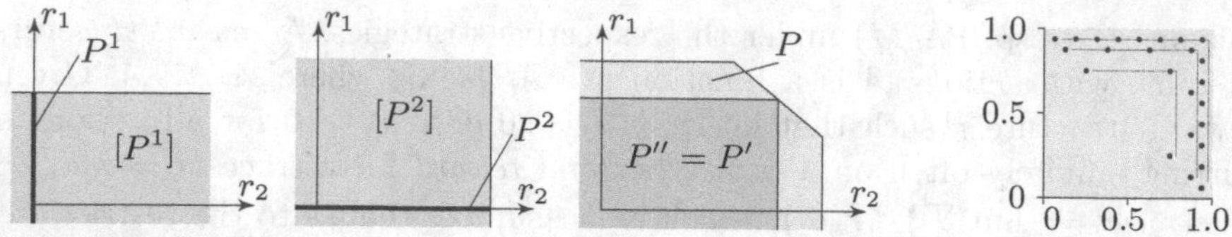

(a) MQs $\varphi^1 = r_1 \geq v_1$ (left), $\varphi^2 = r_2 \geq v_2$ (center), and $\varphi = r_1 \geq v_1 \wedge r_2 \geq v_2$ (right). Suppose the Pareto sets P^i for φ^i, $i \in \{1, 2\}$, consist of the thick lines, unbounded towards $-\infty$. Their liftings $[P^i]$ are shown in dark grey. P'' under-approximates the Pareto set P for φ.

(b) Each dot is a weight vector, where the accuracy a is one of $\{\frac{1}{2}, \frac{1}{4}, \frac{1}{8}\}$.

Fig. 3. Compositional Pareto set computation (a); weight vector selection (b)

4.3 Compositional Pareto Set Computation

We describe in this section how to pick the targets of the objectives φ_i in compositional rules, such as those in Theorem 1, so that φ is achievable.

Pareto Sets. Given an MQ φ with N objectives, vector $\boldsymbol{v} \in \mathbb{R}^N$ is a *Pareto vector* if and only if $\varphi[\boldsymbol{v} - \varepsilon]$ is achievable for all $\varepsilon > 0$, and $\varphi[\boldsymbol{v} + \varepsilon]$ is not achievable for any $\varepsilon > 0$. The downward closure of the set of all such vectors is called a *Pareto set*, where the downward closure of a set X is defined as $\mathsf{dwc}(X) \stackrel{\text{def}}{=} \{\boldsymbol{y} \in \mathbb{R}^N \mid \exists \boldsymbol{x} \in X \,.\, \boldsymbol{x} \geq \boldsymbol{y}\}$. Given $\varepsilon > 0$, an *ε-approximation of a Pareto set P* is a set of vectors Q satisfying that, for any $\boldsymbol{w} \in Q$, there is a vector $\boldsymbol{v} \in P$ such that $\|\boldsymbol{v} - \boldsymbol{w}\| \leq \varepsilon$, and for every $\boldsymbol{v} \in P$ there is a vector $\boldsymbol{w} \in Q$ such that $\|\boldsymbol{v} - \boldsymbol{w}\| \leq \varepsilon$, where $\|\cdot\|$ is the Manhattan norm.

Under-Approximating Pareto Sets. We can compositionally compute an under-approximation of the Pareto set for φ, which we illustrate in Figure 3.

Consider N reward structures, $r_1, \ldots, r_N$, and objectives φ^i, $i \in I$, over these reward structures for respective games G^i, as well as an objective φ, over the same reward structures, for the composed game $\mathcal{G} = \|_{i \in I} G^i$. Note that, for each $1 \leq j \leq N$, the reward structure r_j may be present in several objectives φ^i. Let P^i be the Pareto set for $G^i \models \varphi^i$, for $i \in I$, and so each point $\boldsymbol{v}^{(i)} \in P^i$ represents a target vector for the MQ $\varphi^i[\boldsymbol{v}^{(i)}]$ achievable in the game G^i.

For a Pareto set P^i, define the *lifting* $[P^i]$ to all N reward structures by $[P^i] \stackrel{\text{def}}{=} \{\boldsymbol{v} \in \mathbb{R}^N_{\pm\infty} \mid$ the coordinates of $\boldsymbol{v}$ appearing in φ^i are in $P^i\}$. The set $P' \stackrel{\text{def}}{=} \cap_{i \in I}[P^i]$ is the set of target vectors for all M reward structures, which are consistent with achievability of all objectives φ^i in the respective games. The projection[1] P'' of P' onto the space of reward structures appearing in φ then yields an under-approximation of the Pareto set P for φ in the composed game $\mathcal{G}$, that is, $P'' \subseteq P$. A point $\boldsymbol{v} \in P''$ can be achieved by instantiating the objectives φ^i with any targets $\boldsymbol{v}^{(i)}$ in P' that match $\boldsymbol{v}$.

---

[1] More generally, if φ contains items such as $r_i + r_j \blacktriangleright v_i + v_j$, as in the (SUM-REWARD) rule of [13], a new dimension is introduced combining the rewards as required.

4.4 Compositional MQ Synthesis

We now describe our compositional strategy synthesis method.

MQ Synthesis for Component Games. A game is *stopping* if, under any strategy pair, with probability 1 a part of the game is reached where the properties no longer change (see the technical report [2] for details.) A strategy is *ε-optimal* for an MQ φ with target $\boldsymbol{v}$ if it achieves $\varphi[\boldsymbol{v}-\varepsilon]$ for all $\varepsilon > 0$. From [6] we have that, for atomic stopping games with MQs, it is decidable whether an ε-optimal strategy exists (optimal strategies may not exist), and, ε-optimal strategies can be represented finitely using stochastic memory update [7].

We compute a strategy for an MQ in CNF $\bigwedge_{i=1}^{n} \bigvee_{j=1}^{m} r_{i,j} \geq v_{i,j}$ by implementing value iteration based on [6]. First, set an initial accuracy, $a \leftarrow \frac{1}{2}$. For each $0 \leq i < n$, select a corresponding $0 \leq j_i < m$. Then uniformly iterate over *weights* $\boldsymbol{x}_i \in [0, 1 - a/2]^m$ by gridding with accuracy a, keeping the j_ith dimension constant at $1 - a/2$. The pattern of selected vectors is shown for $m = 2$ dimensions in Figure 3(b). At each selection of $\boldsymbol{x}_i$, check, using the CQ algorithm of [7], if $\bigwedge_{i=1}^{n} (\sum_{j=1}^{m} \boldsymbol{x}_i^j \cdot r_{i,j} \geq \sum_{j=1}^{m} \boldsymbol{x}_i^j \cdot v_{i,j})$ is realisable, and, if so, return the winning strategy. Otherwise, if all options for selecting j_i are exhausted, refine the accuracy to $a \leftarrow \frac{a}{2}$ and repeat.

Every point $\boldsymbol{y} \in \mathbb{R}^n$ in a CQ Pareto set with weights $\boldsymbol{x}_1, \ldots, \boldsymbol{x}_n \in \mathbb{R}^m_{\geq 0}$ corresponds to intersection of half-spaces $\boldsymbol{x}_i \cdot \boldsymbol{z} \geq y^i$; the union over all choices of weight vectors is the ε-approximation of the corresponding MQ Pareto set.

MQ Synthesis for Composed Games. Our method for compositional strategy synthesis, based on synthesis for atomic games, is summarised as follows:

(S1) **User Input:** A composed game $\mathcal{G} = \|_{i \in I} G^i$, MQs φ^i, φ, and matching PA rules for use in Theorem 1.
(S2) **First Stage:** Obtain ε-approximate Pareto sets P^i for $G^i \models \varphi^i$, and compute P'' as in Section 4.3.
(S3) **User Feedback:** Pick targets $\boldsymbol{v}$ for φ from P'' and matching targets $\boldsymbol{v}^{(i)}$ for φ^i from P^i.
(S4) **Second Stage:** Synthesise strategies $\sigma^i_\Diamond$, for $G^i \models \varphi^i[\boldsymbol{v}^{(i)}]$, and compose them using Definition 7 (see the technical report [2] for composing strategies in the stochastic memory update representation.)
(S5) **Output:** A composed strategy $\|_{i \in I} \sigma^i_\Diamond$, winning for $\mathcal{G} \models \varphi[\boldsymbol{v}]$ by Theorem 1.

Steps (S1), (S4) and (S5) are sufficient if the targets are known, while (S2) and (S3) are an additional feature enabled by the Pareto set computation.

4.5 Case Study

We illustrate our approach with an example, briefly outlined here; see the technical report [2] for more detail. We model an Internet-enabled fridge that autonomously selects between different digital agents selling milk whenever restocking is needed. We compute the Pareto sets and strategies in a prototype implementation as an extension of PRISM-games [5].

Table 1. Run time comparison between compositional and monolithic strategy synthesis. For the CQ value iteration (cf. [7]) we use 60, 400 and 200 iterations for the fridge, trader and composed model, respectively. Computations were done on a 2.8 GHz Intel® Xeon® CPU, with 32 GB RAM, under Fedora 14.

Traders (n)	State Space Size			Running Time [s]				
	F	T_i	C	Composition	Compositional		Monolithic	
					Pareto Set	Strategies	Pareto Set	Strategy
1	11	7	17	0.006	31.0	0.2	10.9	0.26
2	23	7	119	0.1	400.0	0.37	6570.0	161.0
3	39	7	698	2.0	407.0	2.4	> 3h	–
4	59	7	3705	75.0	4870.0	1.4	> 5h	–

The fridge repeatedly invites offers from several traders, and decides whether to accept or decline the offers, based on the quality of each offer. The objective is for the traders to maximise the unit price, and for the fridge to maximise the amount of milk it purchases. For n traders T_i, $1 \leq i \leq n$, and a fridge F, denote the composition $C \stackrel{\text{def}}{=} (\|_{i=1}^n T_i) \parallel F$. We use the following reward objectives $O_i \equiv$ "offers made by T_i" $\geq v_{o_i}$, $A_i \equiv$ "offers of T_i accepted" $\geq v_{a_i}$, $Q_i \equiv$ "quality of offers made by T_i" $\geq v_{q_i}$, $\$_i \equiv$ "unit price of T_i" $\geq v_{\$_i}$, and $\# \equiv$ "amount of milk obtained by F" $\geq v_{\#}$, and synthesise strategies as explained in Section 4.1 according to the rule:

$$\frac{F \models \bigwedge_{j=1}^n (O_j \rightarrow A_j) \wedge (\bigwedge_{j=1}^n Q_j \rightarrow \#) \qquad T_i \models A_i \rightarrow (Q_i \wedge \$_i) \qquad 1 \leq i \leq n}{C \models \bigwedge_{j=1}^n (O_j \rightarrow \$_j) \wedge (\bigwedge_{j=1}^n O_j \rightarrow \#).}$$

The main advantages of compositional synthesis are a dramatic improvement in efficiency and the compactness of strategies, as indicated in Table 1. In general, the strategies are randomised and history dependent. For the case of two traders, with the target that we selected, we generate a strategy where the traders make an expensive offer in the first round with probability 0.91, but from then on consistently make less expensive bulk offers.

5 Conclusion

We have defined a synchronising composition for stochastic games, and formulated a compositional approach to controller synthesis by leveraging techniques for compositional verification of PAs [13] and multi-objective strategy synthesis of [6,7]. We have extended the implementation of [7] to synthesise ε-optimal strategies for two-player stochastic games for total expected reward objectives in conjunctive normal form. We intend to investigate relaxing the compatibility condition and consider notions of fairness weaker than unconditional fairness to broaden the applicability of our methods.

Acknowledgements. The authors thank Dave Parker and Vojtěch Forejt for helpful discussions. The authors are partially supported by ERC Advanced Grant VERIWARE.

References

1. Baier, C., Groesser, M., Ciesinski, F.: Quantitative analysis under fairness constraints. In: Liu, Z., Ravn, A.P. (eds.) ATVA 2009. LNCS, vol. 5799, pp. 135–150. Springer, Heidelberg (2009)
2. Basset, N., Kwiatkowska, M., Wiltsche, C.: Compositional controller synthesis for stochastic games. University of Oxford, Technical Report CS-RR-14-05 (2014)
3. Campbell, M., Egerstedt, M., How, J.P., Murray, R.M.: Autonomous driving in urban environments: approaches, lessons and challenges. Phil. Trans. R. Soc. A 368, 4649–4672 (1928)
4. Chatterjee, K., Jurdziński, M., Henzinger, T.A.: Simple stochastic parity games. In: Baaz, M., Makowsky, J.A. (eds.) CSL 2003. LNCS, vol. 2803, pp. 100–113. Springer, Heidelberg (2003)
5. Chen, T., Forejt, V., Kwiatkowska, M., Parker, D., Simaitis, A.: PRISM-games: A model checker for stochastic multi-player games. In: Piterman, N., Smolka, S.A. (eds.) TACAS 2013. LNCS, vol. 7795, pp. 185–191. Springer, Heidelberg (2013)
6. Chen, T., Forejt, V., Kwiatkowska, M., Simaitis, A., Wiltsche, C.: On stochastic games with multiple objectives. In: Chatterjee, K., Sgall, J. (eds.) MFCS 2013. LNCS, vol. 8087, pp. 266–277. Springer, Heidelberg (2013)
7. Chen, T., Kwiatkowska, M., Simaitis, A., Wiltsche, C.: Synthesis for multi-objective stochastic games: An application to autonomous urban driving. In: Joshi, K., Siegle, M., Stoelinga, M., D'Argenio, P.R. (eds.) QEST 2013. LNCS, vol. 8054, pp. 322–337. Springer, Heidelberg (2013)
8. Cheung, L., Lynch, N., Segala, R., Vaandrager, F.: Switched PIOA: Parallel composition via distributed scheduling. TCS 365(1-2), 83–108 (2006)
9. de Alfaro, L., Henzinger, T.A.: Interface automata. SIGSOFT Softw. Eng. Notes 26(5), 109–120 (2001)
10. Filar, J., Vrieze, K.: Competitive Markov decision processes. Springer (1996)
11. Gelderie, M.: Strategy composition in compositional games. In: Fomin, F.V., Freivalds, R., Kwiatkowska, M., Peleg, D. (eds.) ICALP 2013, Part II. LNCS, vol. 7966, pp. 263–274. Springer, Heidelberg (2013)
12. Ghosh, S., Ramanujam, R., Simon, S.: Playing extensive form games in parallel. In: Dix, J., Leite, J., Governatori, G., Jamroga, W. (eds.) CLIMA XI. LNCS, vol. 6245, pp. 153–170. Springer, Heidelberg (2010)
13. Kwiatkowska, M., Norman, G., Parker, D., Qu, H.: Compositional probabilistic verification through multi-objective model checking. Information and Computation 232, 38–65 (2013)
14. Kwiatkowska, M., Parker, D.: Automated verification and strategy synthesis for probabilistic systems. In: Van Hung, D., Ogawa, M. (eds.) ATVA 2013. LNCS, vol. 8172, pp. 5–22. Springer, Heidelberg (2013)
15. Ozay, N., Topcu, U., Murray, R.M., Wongpiromsarn, T.: Distributed synthesis of control protocols for smart camera networks. In: ICCPS 2011, pp. 45–54. IEEE (2011)
16. Pnueli, A.: The temporal logic of programs. In: FOCS 1977, pp. 46–57. IEEE (1977)
17. Segala, R.: Modelling and Verification of Randomized Distributed Real Time Systems. PhD thesis, Massachusetts Institute of Technology (1995)
18. Shapley, L.S.: Stochastic games. Proc. Natl. Acad. Sci. USA 39(10), 1095 (1953)
19. Sokolova, A., de Vink, E.P.: Probabilistic automata: system types, parallel composition and comparison. In: Baier, C., Haverkort, B.R., Hermanns, H., Katoen, J.-P., Siegle, M. (eds.) Validation of Stochastic Systems. LNCS, vol. 2925, pp. 1–43. Springer, Heidelberg (2004)

Synchronizing Strategies under Partial Observability

Kim Guldstrand Larsen, Simon Laursen, and Jiří Srba

Aalborg University, Department of Computer Science
Selma Lagerlöfs Vej 300, 9220 Aalborg East, Denmark
{kgl,simlau,srba}@cs.aau.dk

Abstract. Embedded devices usually share only partial information about their current configurations as the communication bandwidth can be restricted. Despite this, we may wish to bring a failed device into a given predetermined configuration. This problem, also known as resetting or synchronizing words, has been intensively studied for systems that do not provide any information about their configurations. In order to capture more general scenarios, we extend the existing theory of synchronizing words to synchronizing strategies, and study the synchronization, short-synchronization and subset-to-subset synchronization problems under partial observability. We provide a comprehensive complexity analysis of these problems, concluding that for deterministic systems the complexity of the problems under partial observability remains the same as for the classical synchronization problems, whereas for nondeterministic systems the complexity increases already for systems with just two observations, as we can now encode alternation.

1 Introduction

In February last year (2013), Aalborg University launched an experimental satellite [3] designed by students. There was a failure during the initialization phase executed by the satellite at the orbit, resulting in unknown orientation of the solar panel. This caused significant problems with energy supply and very limited communication capabilities of the satellite, especially when transmitting information that is energetically more expensive than receiving it. The task was to command the satellite from the Earth so that it returned to some predefined well-known position.

A simplified model of the problem is depicted in Figure 1a. In the example, we assume for illustration purposes that there are only eight possible rotation positions of a single solar panel, numbered by 1 to 8 in the figure. The thin lines with a dashed surface indicate the direction the panel is facing in a given position. This determines whether the panel is active and produces energy (facing towards light) or inactive and does not produce any energy. The thick line at position 5 indicates the current (unknown) position of the solar panel. The satellite cannot communicate the exact position of the solar panel, instead it is only capable of transmitting information as to whether the current position produces energy

P. Baldan and D. Gorla (Eds.): CONCUR 2014, LNCS 8704, pp. 188–202, 2014.

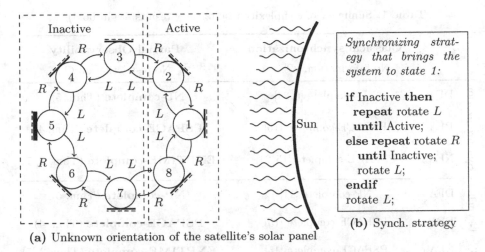

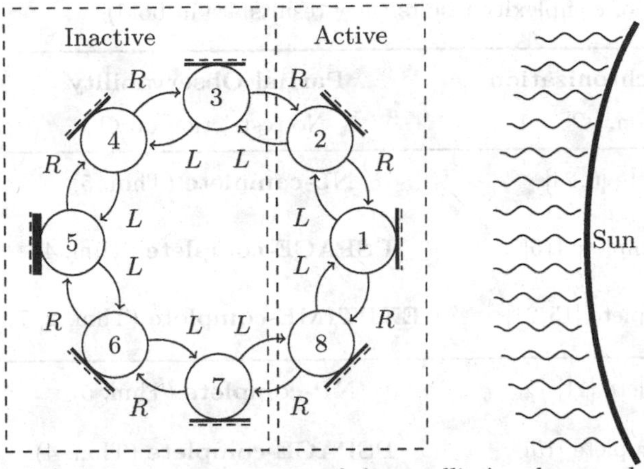

(a) Unknown orientation of the satellite's solar panel

```
Synchronizing strategy that brings the system to state 1:

if Inactive then
  repeat rotate L
  until Active;
else repeat rotate R
  until Inactive;
  rotate L;
endif
rotate L;
```

(b) Synch. strategy

Fig. 1. Satellite with partial observability and its synchronizing strategy

(observation Active) or not (observation Inactive). The panel can be commanded to rotate one position to the left (action L) or to the right (action R) and our task is to bring it from any possible (unknown) position into the position 1 where it produces most energy. As we cannot observe the actual position of the panel, we need to find a strategy that relies only on the fact whether the panel is Active or Inactive. Such a strategy indeed exists as shown in Figure 1b.

The classical concept of synchronizing words [6] for deterministic finite automata dates back more than 50 years and it concerns the existence of a word that brings a given automaton from any of its states to a single state (see [22,24] for recent survey papers). However, for our example in Figure 1a it is clear that there is no single synchronizing word—in this classical setting—over $\{L, R\}$ that can bring the panel into the same position. Instead, we need to design a strategy that relies on a partial information about the system, in our case on whether the panel is Active or Inactive.

1.1 Our Contribution

We introduce a general synthesis problem for synchronizing *strategies* of systems with *partial observability.* We deal with this problem in the setting of finite-state automata where each state has a single observation from a finite set of observations; we call the model *labelled transition system with partial observability* (LTSP). The task is to suggest a strategy for adaptive generation of actions based on the so-far seen observations. Such a strategy should bring the system from any possible initial state into a single synchronizing state. We also consider two other variants of the synchronization synthesis problem (i) with a bound on the maximal length of (runs of) the strategy (short-synchronization) and (ii) synchronization from a given set of states to a given set of states (subset-to-subset synchronization). We provide a comprehensive complexity study of these problems in the setting of total deterministic (DFA), partial deterministic (PFA) and

Table 1. Summary of complexity results (new results are in bold)

		Classical synchronization No information, $\lvert\mathcal{O}\rvert = 1$	Partial Observability No restriction on $\mathcal{O}$
Synchronization	DFA	NL-complete [6,24]	**NL-complete** (Thm. 5)
	PFA	PSPACE-complete [15]	**PSPACE-complete** (Thm. 4)
	NFA	PSPACE-complete [15,21]	**EXPTIME-complete** (Thm. 2, 7)
Short-synch.	DFA	NP-complete [11]	**NP-complete** (Thm. 5)
	PFA	PSPACE-complete [15]	**PSPACE-complete** (Thm. 4)
	NFA	PSPACE-complete [15]	**EXPTIME-complete** (Thm. 2, 7)
Subset-to-subset	DFA	PSPACE-complete [20]	**PSPACE-complete** (Thm. 4)
	PFA	PSPACE-complete ([20], on-the-fly)	**PSPACE-complete** (Thm. 4)
	NFA	PSPACE-complete ([20], on-the-fly)	**EXPTIME-complete** (Thm. 2, 6)

nondeterministic (NFA) finite automata. Our results, compared to the classical synchronization problems, are summarized in the right column of Figure 1.

Our first technical contribution is a translation from the synthesis of history-dependent synchronizing strategies on LTSP to the synthesis of memoryless winning reachability strategies for a larger two-player knowledge game. This allows us to argue for the EXPTIME containment of the synchronization problem on NFA. However, for DFA and PFA the knowledge game is insufficient to obtain tight complexity upper-bounds. For this reason, and as our second contribution, we define a notion of aggregated knowledge graph allowing us to derive a PSPACE containment for PFA and NL containment for DFA, despite the double-exponential size of the aggregated knowledge graph in the general nondeterministic case.

In order to complement the complexity upper-bounds with matching lower-bounds, we provide as our third contribution a novel polynomial-time reduction from alternating linear bounded automata into the synchronization problems for NFA with partial observability. This is achieved by showing that a combination of the partial observability and nondeterminism can capture alternation. This technique provides matching lower-bounds for all our three synchronization problems on NFA. The lower-bounds for DFA and PFA are derived from the classical problem setting.

In addition, we describe a polynomial-time reduction from a setting with an arbitrary number of observations to an equivalent setting with just two observations, causing only a logarithmic overhead as a factor in the size of the system. Thus all the lower-bound results mentioned earlier remain valid even when restricting the synchronizing strategy synthesis problem to only two observations.

1.2 Related Work

The study of synchronizing words initiated by Černý [6] is a variant of our more general strategy synthesis problem where all states return the same observation, and the existence of synchronizing words, short synchronizing words and subset-to-subset synchronizing words have been in different contexts studied up to now; see [22,24] for recent surveys. The computational complexities of word synchronization problems for DFA, PFA and NFA are summarized in left column of Table 1. Note that the NL-completeness for the classical synchronization problem on DFA (not explicitly mentioned in the literature) follows directly from the fact that the problem of synchronizing all states is equivalent to checking that any given pair of states can be synchronized [6, 24]. The PSPACE containment of subset-to-subset word synchronization for NFA and PFA follows from [20] by running the algorithm for DFA in an on-the-fly manner, while guessing step-by-step the synchronizing path.

Through the last years there has been an increasing interest in novel settings of the synchronization problem. Volkov et al. [12] study the problem for deterministic automata with positive cost on transitions, and constrain the cost of the synchronizing word. They also study a synchronization game, where the player who wants to synchronize the system proposes only every second character in the synchronizing word. Doyen et al. [9, 10] study the existence of infinite synchronizing words in a probabilistic setting. The theory of synchronizing words have also seen several practical applications, for instance in biocomputing [2], model-based testing [4], and robotics [1].

The notion of homing sequences [13, 16] is related to the study of synchronizing words and to our study of synchronizing strategies. A homing sequence is a sequence of input actions that makes it possible to determine the current state of the system by looking at the outputs from the system. Homing sequences are studied on the model of Mealy machine, essentially a DFA where each transition produces an output from a given finite alphabet (see [22] for a recent survey). Homing sequences have been, among others, studied in an adaptive variant where the next input symbol is determined by the knowledge of the previous outputs. This is related to our synchronizing strategies that depend on the history of observations, however, there are no complexity results for adaptive homing sequence on nondeterministic systems.

Pomeranz and Reddy [17] suggest to combine synchronizing words and adaptive homing sequences. They first apply a homing sequence and then find a word that brings the machine to one particular state. The theory is applied to sequential circuit testing for deterministic systems and their adaptive synchronization problem can be seen as a subclass of our systems with partial observability (the output actions of a Mealy machine can be encoded as observations).

The idea of gathering the knowledge of possible states where the system can be after a sequence of observable actions, formalized in the notion of knowledge game, is inspired by a similar technique from [5, 7]. Our aggregated knowledge graph technique is related to the super graph construction used in [14]. The complexity of the conditional planning problem from artificial intelligence have also recently been studied under different observability assumptions [19].

Finally, regarding our EXPTIME lower bound, similar complexity results are available for reachability on finite games with partial observability. In [18] the authors study reachability games where both players have only a partial information about the current configuration of the game and show 2-EXPTIME-hardness of deciding whether the first player has a winning strategy. Our synchronization problem for NFA can be also seen as a game, however, here the second player (nondeterminism) has a full information. This variant, called semiperfect-information game, was studied in [8] for a parity objective (including reachability) and the authors show that the problem is both in NP and coNP. Our synchronization problem for NFA is similar to the semiperfect-information game, however, with a very different objective of synchronizing from any given state. This is documented by the fact that the synchronization problem under partial observability for NFA becomes EXPTIME-complete.

2 Definitions

We shall now formally rephrase our problem. We define labelled transition systems with partial observability, introduce synchronizing strategies and formulate the three decision problems we are interested in.

Definition 1. *A labelled transition system with partial observability (LTSP) is a quintuple $T = (S, Act, \rightarrow, \mathcal{O}, \gamma)$ where S is a set of states, Act is an action alphabet, $\rightarrow \subseteq S \times Act \times S$ is the transition relation, written $s \xrightarrow{a} s'$ whenever $(s, a, s') \in \rightarrow$, $\mathcal{O}$ is a nonempty set of observations, and $\gamma : S \rightarrow \mathcal{O}$ is a function mapping each state to an observation.*

We shall study the synchronization problems for three natural subclasses of LTSP, namely DFA (deterministic finite automata), PFA (partial finite automata) and NFA (nondeterministic finite automata). An LTSP is called NFA if S, Act and $\mathcal{O}$ are all finite sets. If the transition relation is also deterministic, i.e. for every $s \in S$ and $a \in Act$ there is at most one $s' \in S$ such that $s \xrightarrow{a} s'$, then we call it PFA. If the transition relation is moreover complete, i.e. for all $s \in S$ and $a \in Act$ there is exactly one $s' \in S$ such that $s \xrightarrow{a} s'$, then we have a DFA. In the rest of the paper we focus on the NFA class and its PFA and DFA subclasses (implicitly assuming partial observability).

For the rest of this section, let $T = (S, Act, \rightarrow, \mathcal{O}, \gamma)$ be a fixed LTSP. A *path* in T is a finite sequence $\pi = s_1 a_1 s_2 a_2 \ldots a_{n-1} s_n$ where $s_i \xrightarrow{a_i} s_{i+1}$ for all i, $1 \leq i < n$. The length of π is the number of transitions, denoted as $|\pi| = n - 1$.

The last state s_n in such a path π is referred to as $last(\pi)$. The set of all finite paths in T is denoted by $paths(T)$. The observation sequence of π is the unique sequence of state observations $\gamma(\pi) = \gamma(s_1)\gamma(s_2)\ldots\gamma(s_n)$.

A *strategy* on T is a function from finite sequences of observations to next actions to be taken, formally

$$\delta : \mathcal{O}^+ \to Act \cup \{done\}$$

where $done \notin Act$ is a special symbol signalling that the achieved path is maximal. In the rest of the paper we consider only strategies that are feasible and terminating. A strategy δ is *feasible* if the action proposed by the strategy is executable from the last state of the path; formally we require that for every $\pi = s_1a_1s_2a_2\ldots a_{n-1}s_n \in paths(T)$ that *follows* the strategy, meaning that $\delta(\gamma(s_1a_1s_2a_2\ldots s_i)) = a_i$ for all i, $1 \le i < n$, either $\delta(\gamma(\pi)) = done$ or there is at least one $s' \in S$ such that $last(\pi) \xrightarrow{\delta(\gamma(\pi))} s'$. A strategy δ is *terminating* if it does not generate any infinite path, in other words there is no infinite sequence $\pi = s_1a_1s_2a_2\ldots$ such that $s_i \xrightarrow{a_i} s_{i+1}$ and $\delta(\gamma(s_1a_1s_2a_2\ldots s_i)) = a_i$ for all $i \ge 1$.

Given a subset of states $X \subseteq S$ and a feasible, terminating strategy δ, the set of all maximal paths that follow the strategy δ in T and start from some state in X, denoted by $\delta[X]$, is defined as follows:

$$\delta[X] = \{\pi = s_1a_1s_2a_2\ldots a_{n-1}s_n \in paths(T) \mid s_1 \in X \text{ and } \delta(\gamma(\pi)) = done \text{ and } \delta(\gamma(s_1a_1s_2a_2\ldots s_i)) = a_i \text{ for all } i, 1 \le i < n\} .$$

The set of final states reached when following δ starting from X is defined as $last(\delta[X]) = \{last(\pi) \mid \pi \in \delta[X]\}$ and the length of δ from X is defined as $length(\delta[X]) = \max\{|\pi| \mid \pi \in \delta[X]\}$. By $length(\delta)$ we understand $length(\delta[S])$.

We now define a synchronizing strategy that guarantees to bring the system from any of its states into a single state.

Definition 2 (Synchronizing strategy). *A strategy δ for an LTSP $T = (S, Act, \rightarrow, \mathcal{O}, \gamma)$ is* synchronizing *if δ is feasible, terminating and $last(\delta[S])$ is a singleton set.*

Note that synchronizing strategy for NFA means that any execution of the system (for all possible nondeterministic choices) will synchronize into the same singleton set. It is clear that a synchronizing strategy can be arbitrarily long as it relies on the full history of observable actions. We will now show that this is in fact not needed as we can find strategies that do not perform unnecessary steps.

Let $T = (S, Act, \rightarrow, \mathcal{O}, \gamma)$ be an LTSP and let $\omega \in \mathcal{O}^+$ be a sequence of observations. We define the set of possible states (called belief) where the system can be after observing the sequence ω by

$$belief(\omega) = \{last(\pi) \mid \pi \in paths(T),\ \gamma(\pi) = \omega\} .$$

A strategy δ for T is *belief-compatible* if for all $\omega_1, \omega_2 \in \mathcal{O}^+$ with $belief(\omega_1) = belief(\omega_2)$ we have $\delta(\omega_1) = \delta(\omega_2)$.

Lemma 1. *If there is a synchronizing strategy δ for a finite LTSP $T = (S, Act, \rightarrow, \mathcal{O}, \gamma)$ then T has also a belief-compatible synchronizing strategy δ' such that $length(\delta') \leq 2^{|S|}$ and $length(\delta') \leq length(\delta)$.*

We shall now define three versions of the synchronization problem studied in this paper. The first problem simply asks about the existence of a synchronizing strategy.

Problem 1 (Synchronization). Given an LTSP T, is there a synchronizing strategy for T?

The second problem of short-synchronization moreover asks about the existence of a strategy shorter than a given length bound. This can be, for instance, used for finding the shortest synchronizing strategy via the bisection method.

Problem 2 (Short-Synchronization). Given an LTSP T and a bound $k \in \mathbb{N}$, is there a synchronizing strategy δ for T such that $length(\delta) \leq k$?

Finally, the general subset-to-subset synchronization problem asks to synchronize only a subset of states, reaching not necessarily a single synchronizing state but any state from a given set of final states.

Problem 3 (Subset-to-Subset Synchronization). Given an LTSP T and subsets $S_{from}, S_{to} \subseteq S$, is there a feasible and terminating strategy δ for T such that $last(\delta[S_{from}]) \subseteq S_{to}$?

If we restrict the set of observations to a singleton set (hence the γ function does not provide any useful information about the current state apart from the length of the sequence), we recover the well-known decision problems studied in the body of literature related to the classical word synchronization (see e.g. [22, 24]). Note that in this classical case the strategy is now simply a fixed finite sequence of actions.

3 Complexity Upper-Bounds

In this section we shall introduce the concept of knowledge game and aggregated knowledge graph so that we can conclude with the complexity upper-bounds for the various synchronization problems with partial observability.

3.1 Knowledge Game

Let $T = (S, Act, \rightarrow, \mathcal{O}, \gamma)$ be a fixed LTSP. We define the set of successors from a given state $s \in S$ under the action $a \in Act$ as $succ(s, a) = \{s' \mid s \xrightarrow{a} s'\}$. For $X \subseteq S$ we define

$$succ(X, a) = \begin{cases} \{s' \in succ(s, a) \mid s \in X\} & \text{if } succ(s, a) \neq \emptyset \text{ for all } s \in X \\ \emptyset & \text{otherwise} \end{cases}$$

such that $succ(X, a)$ is nonempty iff every state from X enables the action a. We also define a function $split : 2^S \to 2^{(2^S)}$

$$split(X) = \{\{s \in X \mid \gamma(s) = o\} \mid o \in \mathcal{O}\} \setminus \emptyset$$

that partitions a given set of states X into the equivalence classes according to the observations that can be made.

We can now define the knowledge game, a two-player game played on a graph where each node represents a belief (a set of states where the players can end up by following a sequence of transitions). Given a current belief, *Player 1* plays by proposing a possible action that all states in the belief can perform. *Player 2* then determines which of the possible next beliefs (partitionings) the play continues from. *Player 1* wins the knowledge game if there is a strategy so that any play from the given initial belief reaches the same singleton belief $\{s\}$. Formally, we define the knowledge game as follows.

Definition 3. *Given an LTSP* $T = (S, Act, \to, \mathcal{O}, \gamma)$, *the corresponding* knowledge game *is a quadruple* $G(T) = (\mathcal{V}, \mathcal{I}, Act, \Rightarrow)$ *where*

- $\mathcal{V} = \{V \in 2^S \setminus \emptyset \mid \{V\} = split(V)\}$ *is the set of all unsplittable beliefs,*
- $\mathcal{I} = split(S)$ *is the set of initial beliefs, and*
- $\Rightarrow \subseteq \mathcal{V} \times Act \times \mathcal{V}$ *is the transition relation, written* $V_1 \overset{a}{\Rightarrow} V_2$ *for* $(V_1, a, V_2) \in \Rightarrow$, *such that* $V_1 \overset{a}{\Rightarrow} V_2$ *iff* $V_2 \in split(succ(V_1, a))$.

Example 1. In Figure 2a we show the knowledge game graph for our running example from Figure 1a. We only display the part of the graph reachable from the initial belief consisting of states $\{3, 4, 5, 6, 7\}$ where the solar panel is inactive. Assume that we want to synchronize from any of these states into the state 8. This can be understood as a two-player game where from the current belief *Player 1* proposes an action and *Player 2* picks a new belief reachable in one step under the selected action. The question is whether *Player 1* can guarantee that any play of the game reaches the belief $\{8\}$. This is indeed the case and the strategy of *Player 1* is, for example, to repeatedly propose the action L until the belief $\{8\}$ is eventually reached.

We shall now formalize the rules of the knowledge game. A *play* in a knowledge game $G(T) = (\mathcal{V}, \mathcal{I}, Act, \Rightarrow)$ is a sequence of beliefs $\mu = V_1 V_2 V_3 \ldots$ where $V_1 \in \mathcal{I}$ and for all $i \geq 1$ there is $a_i \in Act$ such that $V_i \overset{a_i}{\Rightarrow} V_{i+1}$. The set of all plays in $G(T)$ is denoted $plays(G(T))$.

A *strategy* (for *Player 1*) is a function $\rho : \mathcal{V} \to Act$. A play $\mu = V_1 V_2 V_3 \ldots$ *follows* the strategy ρ if $V_i \overset{\rho(V_i)}{\Longrightarrow} V_{i+1}$ for all $i \geq 1$. Note that the strategy is memoryless as it depends only on the current belief.

Player 1 wins the game $G(T)$ if there is $s \in S$ and a strategy ρ such that for every play $\mu = V_1 V_2 V_3 \ldots \in plays(G(T))$ that follows ρ there exists an $i \geq 1$ such that $V_i = \{s\}$.

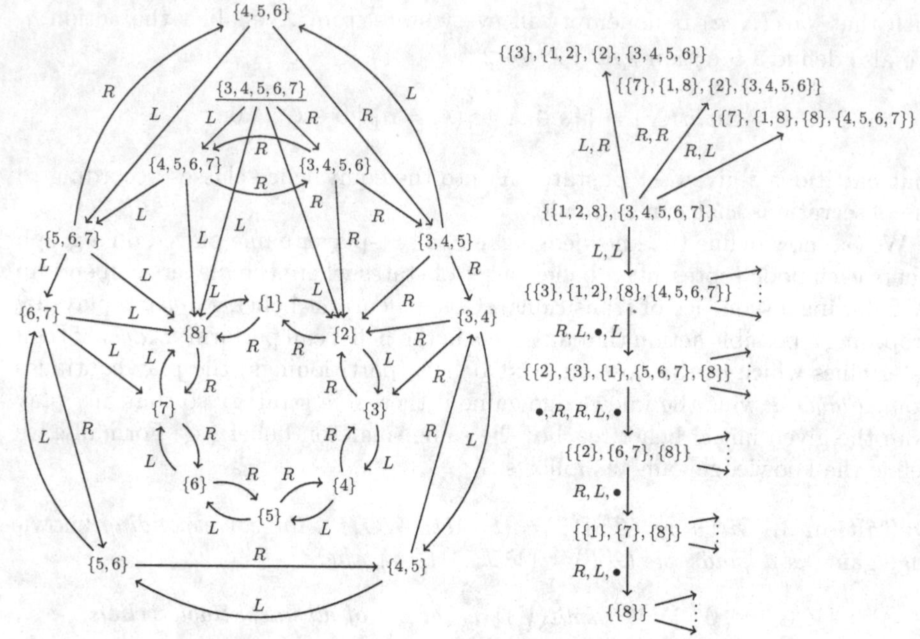

(a) A knowledge game example rooted at $\{3, 4, 5, 6, 7\}$

(b) A fragment of the aggregated knowledge graph

Fig. 2. Examples of a knowledge game and an aggregated knowledge graph

The length of a play $\mu = V_1V_2V_3\dots$ for reaching a singleton belief $\{s\}$ is $\mathit{length}(\mu, s) = i - 1$ where i is the smallest i such that $V_i = \{s\}$. The length of a winning strategy ρ in the game $G(T)$ that reaches the singleton belief $\{s\}$ is

$$\mathit{length}(\rho) = \max_{\mu \in \mathit{plays}(G(T)),\ \mu \text{ follows } \rho} \mathit{length}(\mu, s)\,.$$

Theorem 1. *Let $T = (S, Act, \rightarrow, \mathcal{O}, \gamma)$ and let $G(T) = (\mathcal{V}, \mathcal{I}, Act, \Rightarrow)$ be the corresponding knowledge game where $\mathcal{I} = \mathit{split}(S)$. Then* Player 1 *wins the knowledge game $G(T)$ iff there is a synchronizing strategy for T. Moreover for any winning strategy ρ in the game $G(T)$ there is a synchronizing strategy δ for T such that $\mathit{length}(\rho) = \mathit{length}(\delta)$, and for any synchronizing strategy δ for T there is a winning strategy ρ in the game $G(T)$ such that $\mathit{length}(\rho) \leq \mathit{length}(\delta)$.*

Proof. Assume that *Player 1* wins the knowledge game $G(T)$ with the strategy ρ so that all plays reach the belief $\{s\}$. We want to find a synchronizing strategy δ for T. Let the initial observation be $o_1 \in \mathcal{O}$; this gives the initial belief $V_1 = \{t \in S \mid \gamma(t) = o_1\}$. We can now use the winning strategy ρ to determine the first action of our synchronizing strategy $\delta(o_1) = \rho(V_1)$. By executing the action $\rho(V_1)$, we get the next observation o_2. Now assume that we have a sequence of observations $o_1o_2\dots o_{i-1}o_i$. We can inductively determine the current belief V_i

as

$$V_i = \{t \in succ(V_{i-1}, \rho(V_{i-1})) \mid \gamma(t) = o_i\}$$

for all $i > 1$. This gives us the synchronizing strategy

$$\delta(o_1 o_2 \ldots o_{i-1} o_i) = \begin{cases} done & \text{if} \quad V_i = \{s\} \\ \rho(V_i) & \text{otherwise} \end{cases}$$

that guarantees that all plays follow the winning strategy ρ. Hence in any play there exists an $i \geq 1$ such that $V_i = \{s\}$. By this construction it is clear that $length(\rho) = length(\delta)$.

For the other direction, assume that there is a synchronizing strategy δ for T. Then we know from Lemma 1 that there exists also a belief-compatible synchronizing strategy δ' where $length(\delta') \leq length(\delta)$. We want to find a winning strategy ρ for *Player 1* in $G(T)$. As we know by construction that all states in a belief V have the same observation, we use the notation $\gamma(V) = o$ if $\gamma(t) = o$ for all $t \in V$. Let the initial belief be $V_1 \in \mathcal{I}$. We use the synchronizing strategy δ' to determine the first action that *Player 1* winning strategy should propose by $\rho(V_1) = \delta'(\gamma(V_1))$. Now *Player 2* determines the next belief V_2 such that $V_1 \xrightarrow{\delta'(\gamma(V_1))} V_2$. In general, assume inductively that we reached a belief V_i along the play $\mu = V_1 V_2 \ldots V_i$. The winning strategy from V_i is given by

$$\rho(V_i) = \delta'(\gamma(V_1)\gamma(V_2) \ldots \gamma(V_i)) \ .$$

Note that this definition makes sense because δ' is belief-compatible (and hence different plays in the knowledge game that lead to the same belief will propose the same action). From the construction of the strategy and by Lemma 1 it is also clear that $length(\rho) = length(\delta') \leq length(\delta)$. □

We conclude with a theorem proving EXPTIME-containment of the three synchronization problems for NFA (and hence clearly also for PFA and DFA).

Theorem 2. *The synchronization, short-synchronization and subset-to-subset synchronization problems for NFA are in EXPTIME.*

The proof is done by exploring in polynomial time the underlying, exponentially large, graph of the knowledge game.

3.2 Aggregated Knowledge Graph

Knowledge games allowed us to prove EXPTIME upper-bounds for the three synchronization problems on NFA, however, it is in general not possible to guess winning strategies for *Player 1* in polynomial space. Hence we introduce the so-called aggregated knowledge graph where we ask a simple reachability question (one player game). This will provide better complexity upper-bounds for deterministic systems, despite the fact that the aggregated knowledge graph can be exponentially larger than the knowledge game graph.

Definition 4. *Let $G(T) = (\mathcal{V}, \mathcal{I}, Act, \Rightarrow)$ be a knowledge game. The aggregated knowledge graph is a tuple $AG(G(T)) = (\mathscr{C}, \mathcal{C}_0, \rightrightarrows)$ where*

- $\mathscr{C} = 2^{\mathcal{V}} \setminus \emptyset$ *is the set of configurations (set of aggregated beliefs),*
- $\mathcal{C}_0 = \mathcal{I}$ *is the initial configuration (set of all initial beliefs), and*
- $\rightrightarrows \subseteq \mathscr{C} \times \mathscr{C}$ *is the transition relation such that $\mathcal{C}_1 \rightrightarrows \mathcal{C}_2$, standing for $(\mathcal{C}_1, \mathcal{C}_2) \in \rightrightarrows$, is possible if for every $V \in \mathcal{C}_1$ there is an action $a_V \in Act \cup \{\bullet\}$ such that $V \overset{a_V}{\Longrightarrow} V'$ for at least one V' (by definition $V \overset{\bullet}{\Rightarrow} V$ if and only if $|V| = 1$), ending in $\mathcal{C}_2 = \{V' \mid V \in \mathcal{C}_1 \text{ and } V \overset{a_V}{\Longrightarrow} V'\}$.*

Example 2. Figure 2b shows a fragment of the aggregated knowledge graph for our running example from Figure 1a. The initial configuration is the aggregation of the initial beliefs and each transition is labelled with a sequence of actions for each belief in the aggregated configuration. The suggested path shows how to synchronize into the state 8. Note that the action $\bullet$, allowed only on singleton beliefs, stands for the situation where the belief is not participating in the given step.

Theorem 3. *Let $G(T) = (\mathcal{V}, \mathcal{I}, Act, \Rightarrow)$ be a knowledge game and let $AG(G(T)) = (\mathscr{C}, \mathcal{C}_0, \rightrightarrows)$ be the corresponding aggregated knowledge graph. Then $\mathcal{C}_0 \rightrightarrows^* \{\{s\}\}$ for some state s if and only if* Player 1 *wins the knowledge game $G(T)$. Moreover, for any winning strategy ρ in $G(T)$ that reaches the singleton belief $\{s\}$ we have $\mathcal{C}_0 \rightrightarrows^{length(\rho)} \{\{s\}\}$, and whenever $\mathcal{C}_0 \rightrightarrows^n \{\{s\}\}$ then there is a winning strategy ρ in $G(T)$ such that $length(\rho) \leq n$.*

The proof is done by translating the path in the aggregated knowledge graph into a winning strategy for *Player 1* in the knowledge game, and vice versa.

The aggregated knowledge graph can in general be exponentially larger than its corresponding knowledge game as the nodes are now subsets of beliefs (that are subsets of states). Nevertheless, we can observe that for DFA and PFA, the size of configurations in $AG(G(T))$ cannot grow.

Lemma 2. *Let T be an LTSP generated by DFA or PFA. Let $AG(G(T)) = (\mathscr{C}, \mathcal{C}_0, \rightrightarrows)$ be the corresponding aggregated knowledge graph. Whenever $C \rightrightarrows C'$ then $\sum_{V \in C} |V| \geq \sum_{V' \in C'} |V'|$.*

Theorem 4. *The synchronization, short-synchronization and subset-to-subset synchronization problems for DFA and PFA are decidable in PSPACE.*

Proof. By Theorem 3 and Theorem 1 we get that we can reach the configuration $\{\{s\}\}$ for some $s \in S$ in the aggregated graph $AG(G(T))$ if and only if there is a synchronizing strategy for the given LTSP T. From Lemma 2 we know that for DFA and PFA the size of each aggregated configuration reachable during any computation is bounded by the size of the set S and therefore can be stored in polynomial space. As PSPACE is closed under nondeterminism, the path to the configuration $\{\{s\}\}$ for some $s \in S$ can be guessed, resulting in a polynomial-space algorithm for the synchronizing problem. Theorem 3 also implies that the length of the shortest synchronizing strategy in T is the same as the length of the shortest path to the configuration $\{\{s\}\}$ for some s, giving us that the short-synchronization problems for DFA and PFA are also in PSPACE. Regarding the subset-to-subset synchronization problem from the set S_{from} to the set S_{to}, we can in a straightforward manner modify the aggregated knowledge graph so that the initial configuration is produced by splitting S_{from} according to the observations and we end in any configuration consisting solely of beliefs V that satisfy $V \subseteq S_{to}$ (while allowing the action $\bullet$ from any such belief to itself). □

Finally, for the synchronization and short-synchronization problems on DFA, we can derive even better complexity upper-bounds by using the aggregated knowledge graph.

Theorem 5. *The synchronization problem on DFA is in NL and the short-synchronization problem on DFA is in NP.*

The first claim is proved using our aggregated knowledge graph together with a generalization of the result from [6, 24] saying that all pairs of states in the system can synchronize iff all states can synchronize simultaneously. For the second claim we show that the shortest synchronizing strategy in DFA has length at most $(n-1)n^2$ where n is the number of states. The strategy can be guessed (in the aggregated knowledge graph) and verified in nondeterministic polynomial time.

4 Complexity Lower-Bounds

We shall now describe a technique that will allow us to argue about EXPTIME-hardness of the synchronization problems for NFA.

Theorem 6. *The subset-to-subset synchronization problem is EXPTIME-hard for NFA.*

Proof (Sketch). By a polynomial time reduction from the EXPTIME-complete [23] acceptance problem for alternating linear bounded automaton over the binary alphabet $\{a, b\}$. W.l.o.g. we assume that the existential and universal choices do not change the current head position and the tape content and we have special deterministic states for tape manipulation. We shall construct an LTSP over three observations $\{default, choice_1, choice_2\}$.

Each tape cell at position k is encoded as in Figure 3a. The actions t_a^k and t_b^k can reveal the current content of the cell, while the actions u_a^k and u_b^k are used to update the stored letter. The current control state q and the head position k are remembered via newly added states of the form (q, k). If (q, k) corresponds to a deterministic state, it will (by a sequence of two actions t_x^k and $u_{x'}^k$ as depicted in Figure 3b) test whether the k'th cell stores the required letter x and then it will update it to x'. For the pair (q, k) where q is an existential state, we add the transitions as in Figure 3c. Clearly, the strategy can select the action 1 or 2 in order to commit to one of the choices and all tape cells just mimic the selected

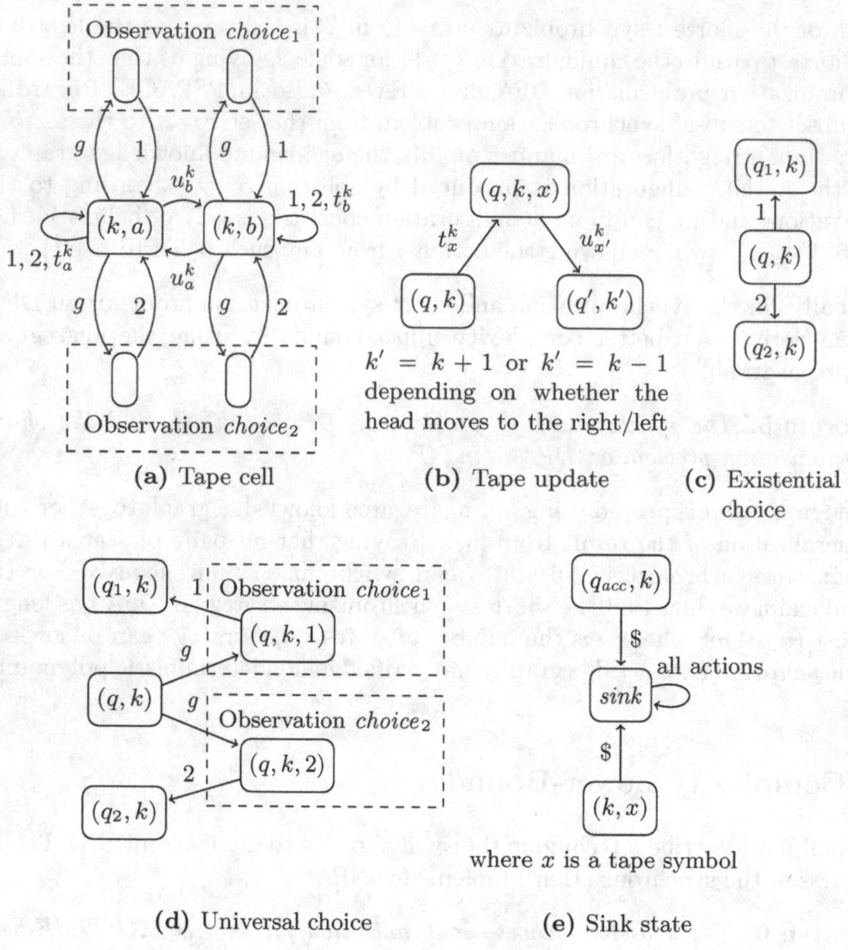

Fig. 3. Encoding idea

action via self-loops. So far we did not need any observations as the introduced states all belong to *default*.

The tricky part is regarding the transitions from the pair (q, k) where q is a universal state. The situation is depicted in Figure 3d. Here the strategy can propose only the action g while the nondeterminism is in control of whether we end up in $(q, k, 1)$ or $(q, k, 2)$. However, this choice is revealed by the observation $choice_1$ or $choice_2$, respectively. Notice that the nondeterminism in the cell encoding does not have to follow the same observation as in the control part. Nevertheless, if this happens, the strategy is allowed to "split" into two separate continuations.

Finally, if the accepting control state q_{acc} is reached, we allow to enter a new state *sink* under a new action \$, not only from (q_{acc}, k) but also from any cell state (k, a) and (k, b) as depiced in Figure 3e. This is the only way in which the LTSP can synchronize, assuming that we only start from the states $(q_0, 1)$ where

q_0 is the initial state and the cell positions that correspond to the initial content of the tape. This assumption is valid as we only consider the subset-to-subset synchronization problem in this theorem. □

Theorem 7. *The synchronization and short-synchronization problems are EXPTIME-hard for NFA.*

Proof (Sketch). Given the construction for the subset-to-subset synchronization problem, we need to guarantee that the execution starts from the predefined states also in the general synchronization problem. Hence we introduce additional transitions together with a new state *init* having the observation *default*. These transitions add a new action # that bring us from any state into one of the initial states of the subset-to-subset problem. There is also a new #-labelled transition from *init* into the initial control state and *init* has no other transitions. This implies that any synchronizing strategy must start by performing the action #. Note that for the short-synchronization case, we use Lemma 1 giving us an exponential upper-bound on the length of the shortest synchronizing strategy. □

The reader may wonder whether three different observations are necessary for proving EXPTIME-hardness of the synchronizing problems or whether one can show the hardness only with two. By analysis of the construction, we can observe that two observations are in fact sufficient. Moreover, there is a general polynomial-time reduction from a given synchronization problem with an arbitrary number of observations to just two observations, while increasing the size of the system by only a logarithmic factor.

Theorem 8. *The synchronization, short-synchronization and subset-to-subset synchronization problems on DFA, PFA and NFA are polynomial-time reducible to the equivalent problems with only two observations.*

Proof (Sketch). Let $T = (S, Act, \rightarrow, \mathcal{O}, \gamma)$ be a given finite LTSP and let $\ell = \lceil \log |\mathcal{O}| \rceil$. The idea is to encode every observation in binary, so that we need only ℓ bits for each observation. Now instead of entering a state s in the original system, we enter instead a chain of newly added states of length $\ell - 1$ that reveal via the binary observations 0/1 the actual observation of the state s (where s reveals the last bit). □

References

[1] Ananichev, D.S., Volkov, M.V.: Synchronizing monotonic automata. Theoretical Computer Science 327(3), 225–239 (2004)

[2] Benenson, Y., Adar, R., Paz-Elizur, T., Livneh, Z., Shapiro, E.: Dna molecule provides a computing machine with both data and fuel. Proceedings of the National Academy of Sciences of the USA 100(5), 2191–2196 (2003)

[3] Boel, T.: Årsag til fejl på aalborg-satellit: Solcellerne vendte væk fra solen. Ingeniøren (Weekly national news magazine about engineering) (March 8, 2013), http://ing.dk/artikel/aarsag-til-fejl-paa-aalborg-satellit-solcellerne-vendte-vaek-fra-solen-156828
[4] Broy, M., Jonsson, B., Katoen, J.-P., Leucker, M., Pretschner, A. (eds.): Model-Based Testing of Reactive Systems. LNCS, vol. 3472. Springer, Heidelberg (2005)
[5] Cassez, F., David, A., Larsen, K.G., Lime, D., Raskin, J.-F.: Timed control with observation based and stuttering invariant strategies. In: Namjoshi, K.S., Yoneda, T., Higashino, T., Okamura, Y. (eds.) ATVA 2007. LNCS, vol. 4762, pp. 192–206. Springer, Heidelberg (2007)
[6] Černý, J.: Poznámka k. homogénnym experimentom s konecnými automatmi. Mat. Fyz. Čas SAV 14, 208–215 (1964)
[7] Chatterjee, K., Doyen, L., Henzinger, T.A., Raskin, J.-F.: Algorithms for omega-regular games with imperfect information. In: Ésik, Z. (ed.) CSL 2006. LNCS, vol. 4207, pp. 287–302. Springer, Heidelberg (2006)
[8] Chatterjee, K., Henzinger, T.A.: Semiperfect-information games. In: Sarukkai, S., Sen, S. (eds.) FSTTCS 2005. LNCS, vol. 3821, pp. 1–18. Springer, Heidelberg (2005)
[9] Doyen, L., Massart, T., Shirmohammadi, M.: Infinite synchronizing words for probabilistic automata. In: Murlak, F., Sankowski, P. (eds.) MFCS 2011. LNCS, vol. 6907, pp. 278–289. Springer, Heidelberg (2011)
[10] Doyen, L., Massart, T., Shirmohammadi, M.: Synchronizing objectives for markov decision processes. In: iWIGP. EPTCS, pp. 61–75 (2011)
[11] Eppstein, D.: Reset sequences for monotonic automata. SIAM Journal on Computing 19(3), 500–510 (1990)
[12] Fominykh, F., Volkov, M.: P(l)aying for synchronization. In: Moreira, N., Reis, R. (eds.) CIAA 2012. LNCS, vol. 7381, pp. 159–170. Springer, Heidelberg (2012)
[13] Gill, A.: State-identification experiments in finite automata. Information and Control 4(2-3), 132–154 (1961)
[14] Krichen, M.: State identification. In: Broy (ed.) [4], pp. 35–67
[15] Martyugin, P.: Computational complexity of certain problems related to carefully synchronizing words for partial automata and directing words for nondeterministic automata. Theory of Com. Systems, 1–12 (2013)
[16] Moore, E.F.: Gedanken Experiments on Sequential Machines. In: Automata Studies, pp. 129–153. Princeton U. (1956)
[17] Pomeranz, I., Reddy, S.M.: Application of homing sequences to synchronous sequential circuit testing. IEEE Trans. Computers 43(5) (1994)
[18] Reif, J.H.: The complexity of two-player games of incomplete information. Journal of Computer and System Sciences 29(2), 274–301 (1984)
[19] Rintanen, J.: Complexity of conditional planning under partial observability and infinite executions. In: ECAI, pp. 678–683 (2012)
[20] Rystsov, I.K.: Polynomial complete problems in automata theory. Information Processing Letters 16(3), 147–151 (1983)
[21] Rystsov, I.K.: Rank of a finite automaton. Cybernetics and Systems Analysis 28(3), 323–328 (1992)
[22] Sandberg, S.: Homing and synchronizing sequences. In: Broy (ed.) [4], pp. 5–33
[23] Sipser, M.: Introduction to the Theory of Computation. Course Technology (2006)
[24] Volkov, M.V.: Synchronizing automata and the Černý conjecture. In: Martín-Vide, C., Otto, F., Fernau, H. (eds.) LATA 2008. LNCS, vol. 5196, pp. 11–27. Springer, Heidelberg (2008)

Probabilistic Robust Timed Games*

Youssouf Oualhadj[1], Pierre-Alain Reynier[2], and Ocan Sankur[3]

[1] Université de Mons (UMONS), Belgium
[2] LIF, Université d'Aix-Marseille and CNRS, France
[3] Université Libre de Bruxelles, Belgium

Abstract. Solving games played on timed automata is a well-known problem and has led to tools and industrial case studies. In these games, the first player (Controller) chooses delays and actions and the second player (Perturbator) resolves the non-determinism of actions. However, the model of timed automata suffers from mathematical idealizations such as infinite precision of clocks and instantaneous synchronization of actions. To address this issue, we extend the theory of timed games in two directions. First, we study the synthesis of *robust* strategies for Controller which should be tolerant to adversarially chosen clock imprecisions. Second, we address the case of a stochastic perturbation model where both clock imprecisions and the non-determinism are resolved randomly. These notions of robustness guarantee the implementability of synthesized controllers. We provide characterizations of the resulting games for Büchi conditions, and prove the EXPTIME-completeness of the corresponding decision problems.

1 Introduction

For real-time systems, timed games are a standard mathematical formalism which can model control synthesis problems under timing constraints. These consist in two-players games played on arenas, defined by timed automata, whose state space consists in discrete locations and continuous clock values. The two players represent the control law and the environment. Since the first theoretical works [2], symbolic algorithms have been studied [10], tools have been developed and successfully applied to several case studies.

Robustness. Because model-based techniques rely on abstract mathematical models, an important question is whether systems synthesized in a formalism are implementable in practice. In timed automata, the abstract mathematical semantics offers arbitrarily precise clocks and time delays, while real-world digital systems have response times that may not be negligible, and control software cannot ensure timing constraints exactly, but only up to some error, caused by clock imprecisions, measurement errors, and communication delays. A major challenge is thus to ensure that the synthesized control software is *robust*, *i.e.* ensures the specification even in presence of imprecisions [15].

* This work was partly supported by ANR projects ECSPER (ANR-2009-JCJC-0069) and ImpRo (ANR-2010-BLAN-0317), European project Cassting (FP7-ICT-601148), and ERC starting grant inVEST (FP7-279499), and ARC Project : Game Theory for the Automatic Synthesis of Computer Systems.

P. Baldan and D. Gorla (Eds.): CONCUR 2014, LNCS 8704, pp. 203–217, 2014.

Following these observations there has been a growing interest in lifting the theory of verification and synthesis to take robustness into account. Model-checking problems were re-visited by considering an unknown perturbation parameter to be synthesized for several kinds of properties [19,12,7], see also [9]. Robustness is also a critical issue in controller synthesis problems. In fact, due to the infinite precision of the semantics, synthesized strategies may not be realizable in a finite-precision environment; the controlled systems synthesized using timed games technology may not satisfy the proven properties at all. In particular, due to perturbations in timings, some infinite behaviors may disappear completely. A first goal of our work is to develop algorithms for *robust* controller synthesis: we consider this problem by studying robust strategies in timed games, namely, those guaranteeing winning despite imprecisions bounded by a parameter.

Adversarial or Stochastic Environments. We consider controller synthesis problems under two types of environments. In order to synthesize correct controllers for critical systems, one often considers an *adversarial* (or worst-case) environment, so as to ensure that *all* behaviors of the system are correct. However, in some cases, one is rather interested considering a *stochastic environment* which determines the resolution of non-determinism, and the choice of clock perturbations following probability distributions. We are then interested in satisfying a property *almost-surely*, that is, with probability 1, or *limit-surely*, that is, for every $\varepsilon > 0$, there should exist a strategy for Controller under which the property is satisfied with probability at least $1 - \varepsilon$.

Contributions. We formalize the robust controller synthesis problem against an adversarial environment as a (non-stochastic) game played on timed automata with an unknown imprecision parameter δ, between players Controller and Perturbator. The game proceeds by Controller suggesting an action and a delay, and Perturbator perturbing each delay by at most δ and resolving the non-determinism by choosing an edge with the given action. Thus, the environment's behaviors model both uncontrollable moves and the limited precision Controller has. We prove the EXPTIME-completeness of deciding whether there exists a positive δ for which Controller has a winning strategy for a Büchi objective, matching the complexity of timed games in the classical sense. Our algorithm also allows one to compute $\delta > 0$ and a witness strategy on positive instances.

For stochastic environments, we study two probabilistic variants of the semantics: we first consider the case of adversarially resolved non-determinism and independently and randomly chosen perturbations, and then the case where both the non-determinism and perturbations are randomly resolved and chosen. In each case, we are interested in the existence of $\delta > 0$ such that Controller wins almost-surely (resp. limit-surely). We give decidable characterizations based on finite abstractions, and EXPTIME algorithms. All problems are formulated in a parametric setting: the parameter δ is unknown and is to be computed by our algorithms. This is one of the technical challenges in this paper.

Our results on stochastic perturbations can also be seen as a new interpretation of robustness phenomena in timed automata. In fact, in the literature on robustness in timed automata, non-robust behaviors are due to the accumulation of the imprecisions δ along long runs, and in the proofs, one exhibits non-robustness by artificially constructing such runs (e.g. [12,22]). In contrast, in the present setting, we show that non-robust behaviors either occur almost surely, or can be avoided surely (Theorem 13).

Related Work. While several works have studied robustness issues for model-checking, there are very few works on robust controller synthesis in timed systems:
- The (non-stochastic) semantics we consider was studied for *fixed* δ in [11]; but the parameterized version of the problem was not considered.
- The restriction of the parameterized problem to (non-stochastic) *deterministic* timed automata was considered in [22]. Here, the power of Perturbator is restricted as it only modifies the delays suggested by Controller, but has no non-determinism to resolve. Therefore, the results consist in a robust Büchi acceptance condition for timed automata, but they do not generalize to timed games. Technically, the algorithm consists in finding an *aperiodic* cycle, which are cycles that are "stable" against perturbations. This notion was defined in [3] to study entropy in timed languages. We will also use aperiodic cycles in the present paper.
- A variant of the semantics we consider was studied in [8] for (deterministic) timed automata and shown to be EXPTIME-complete already for reachability due to an implicit presence of alternation. Timed games, Büchi conditions, or stochastic environments were not considered.
- Probabilistic timed automata where the non-determinism is resolved following probability distributions have been studied [16,4,17]. Our results consist in deciding almost-sure and limit-sure Büchi objectives in PTAs subject to random perturbations in the delays. Note that PTAs are equipped with a possibly different probability distribution for each action. Although we only consider uniform distributions, the two settings are equivalent for almost-sure and limit-sure objectives. Games played on PTA were considered in [14] for minimizing expected time to reachability with NEXPTIME ∩ co-NEXPTIME algorithms.

To the best of our knowledge, this work is the first to study a stochastic model of perturbations for synthesis in timed automata.

Due to space limitations, the proofs are omitted, but they are available in [18].

2 Robust Timed Games

Timed Automata. Given a finite set of clocks $\mathcal{C}$, we call *valuations* the elements of $\mathbb{R}_{\geq 0}^{\mathcal{C}}$. For a subset $R \subseteq \mathcal{C}$ and a valuation ν, $\nu[R \leftarrow 0]$ is the valuation defined by $\nu[R \leftarrow 0](x) = 0$ if $x \in R$ and $\nu[R \leftarrow 0](x) = \nu(x)$ otherwise. Given $d \in \mathbb{R}_{\geq 0}$ and a valuation ν, the valuation $\nu + d$ is defined by $(\nu + d)(x) = \nu(x) + d$ for all $x \in \mathcal{C}$. We extend these operations to sets of valuations in the obvious way. We write $\mathbf{0}$ for the valuation that assigns 0 to every clock.

An atomic clock constraint is a formula of the form $k \preceq x \preceq' l$ or $k \preceq x - y \preceq' l$ where $x, y \in \mathcal{C}$, $k, l \in \mathbb{Z} \cup \{-\infty, \infty\}$ and $\preceq, \preceq' \in \{<, \leq\}$. A *guard* is a conjunction of atomic clock constraints. A valuation ν satisfies a guard g, denoted $\nu \models g$, if all constraints are satisfied when each $x \in \mathcal{C}$ is replaced with $\nu(x)$. We write $\Phi_{\mathcal{C}}$ for the set of guards built on $\mathcal{C}$. A *zone* is a subset of $\mathbb{R}_{\geq 0}^{\mathcal{C}}$ defined by a guard.

A *timed automaton* $\mathcal{A}$ over a finite alphabet of actions Act is a tuple $(\mathcal{L}, \mathcal{C}, \ell_0, \mathsf{Act}, E)$, where $\mathcal{L}$ is a finite set of locations, $\mathcal{C}$ is a finite set of clocks, $E \subseteq \mathcal{L} \times \Phi_{\mathcal{C}} \times \mathsf{Act} \times 2^{\mathcal{C}} \times \mathcal{L}$ is a set of edges, and $\ell_0 \in \mathcal{L}$ is the initial location. An edge $e = (\ell, g, a, R, \ell')$ is also written as $\ell \xrightarrow{g,a,R} \ell'$. A state is a pair $q = (\ell, \nu) \in \mathcal{L} \times \mathbb{R}_{\geq 0}^{\mathcal{C}}$. An edge $e = (\ell, g, a, R, \ell')$ is enabled in a state (ℓ, ν) if ν satisfies the guard g.

The set of possible behaviors of a timed automaton can be described by the set of its runs, as follows. A *run* of $\mathcal{A}$ is a sequence $q_1e_1q_2e_2\ldots$ where $q_i \in \mathcal{L} \times \mathbb{R}_{\geq 0}^{\mathcal{C}}$, and writing $q_i = (\ell, \nu)$, either $e_i \in \mathbb{R}_{>0}$, in which case $q_{i+1} = (\ell, \nu + e_i)$, or $e_i = (\ell, g, a, R, \ell') \in E$, in which case $\nu \models g$ and $q_{i+1} = (\ell', \nu[R \leftarrow 0])$. The set of runs of $\mathcal{A}$ starting in q is denoted $\mathsf{Runs}(\mathcal{A}, q)$.

Parameterized Timed Game. In order to define perturbations, and to capture the reactivity of a controller to these, we define the following parameterized timed game semantics. Intuitively, the parameterized timed game semantics of a timed automaton is a two-player game parameterized by $\delta > 0$, where Player 1, also called Controller chooses a delay $d > \delta$ and an action $a \in \mathsf{Act}$ such that every a-labeled enabled edge is such that its guard is satisfied after any delay in the set $d+[-\delta, \delta]$ (and there exists at least one such edge). Then, Player 2, also called Perturbator chooses an actual delay $d' \in d+[-\delta, \delta]$ after which the edge is taken, and chooses one of the enabled a-labeled edges. Hence, Controller is required to always suggest delays that satisfy the guards whatever the perturbations are.

Formally, given a timed automaton $\mathcal{A} = (\mathcal{L}, \mathcal{C}, \ell_0, \mathsf{Act}, E)$ and $\delta > 0$, we define the *parameterized timed game* of $\mathcal{A}$ w.r.t. δ as a two-player turn-based game $\mathcal{G}_\delta(\mathcal{A})$ between players Controller and Perturbator. The state space of $\mathcal{G}_\delta(\mathcal{A})$ is partitioned into $V_C \cup V_P$ where $V_C = \mathcal{L} \times \mathbb{R}_{\geq 0}^{\mathcal{C}}$ belong to Controller, and $V_P = \mathcal{L} \times \mathbb{R}_{\geq 0}^{\mathcal{C}} \times \mathbb{R}_{\geq 0} \times \mathsf{Act}$ belong to Perturbator. The initial state is $(\ell_0, \mathbf{0}) \in V_C$. The transitions are defined as follows: from any state $(\ell, \nu) \in V_C$, there is a transition to $(\ell, \nu, d, a) \in V_P$ whenever $d > \delta$, for every edge $e = (\ell, g, a, R, \ell')$ such that $\nu + d \models g$, we have $\nu + d + \varepsilon \models g$ for all $\varepsilon \in [-\delta, \delta]$, and there exists at least one such edge e. Then, from any such state $(\ell, \nu, d, a) \in V_P$, there is a transition to $(\ell', \nu') \in V_C$ iff there exists an edge $e = (\ell, g, a, R, \ell')$ as before, and $\varepsilon \in [-\delta, \delta]$ such that $\nu' = (\nu + d + \varepsilon)[R \leftarrow 0])$. A *play* of $\mathcal{G}_\delta(\mathcal{A})$ is a finite or infinite sequence $q_1e_1q_2e_2\ldots$ of states and transitions of $\mathcal{G}_\delta(\mathcal{A})$, with $q_1 = (\ell_0, \mathbf{0})$, where e_i is a transition from q_i to q_{i+1}. It is said to be *maximal* if it is infinite or cannot be extended. A *strategy* for Controller is a function that assigns to every non-maximal play ending in some $(\ell, \nu) \in V_C$, a pair (d, a) where $d > \delta$ and a is an action such that there is a transition from (ℓ, ν) to (ℓ, ν, d, a). A strategy for Perturbator is a function that assigns, to every play ending in (ℓ, ν, d, a), a state (ℓ', ν') such that there is a transition from the former to the latter state. A play ρ is *compatible* with a strategy f for Controller if for every prefix ρ' of ρ ending in V_C, the next transition along ρ after ρ' is given by f. We define similarly compatibility for Perturbator's strategies. A play naturally gives rise to a unique run, where the states are in V_C, and the delays and the edges are those chosen by Perturbator.

Robust Timed Game Problem. Given $\delta > 0$, and a pair of strategies f, g, respectively for Controller and Perturbator, we denote ρ the unique maximal run that is compatible with both f and g. A *Büchi objective* is a subset of the locations of $\mathcal{A}$. A Controller's strategy f is winning for a Büchi objective B if for any Perturbator's strategy g the run ρ that is compatible with f and g is infinite and visits infinitely often a location of B. The *robust timed game problem* asks, for a timed automaton $\mathcal{A}$ and a Büchi objective B, if there exists $\delta > 0$ such that Controller has a winning strategy in $\mathcal{G}_\delta(\mathcal{A})$ for the objective B. When this holds, we say that Controller wins the robust timed game for $\mathcal{A}$, and otherwise that

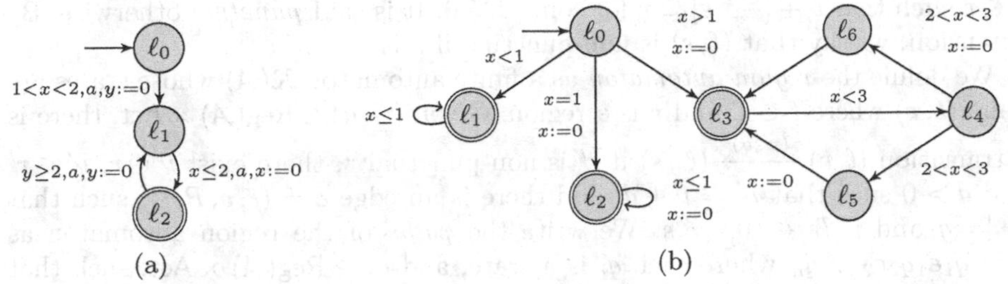

Fig. 1. On the left, a timed automaton from [19] that is not robustly controllable for the Büchi objective $\{\ell_2\}$. In fact, Perturbator can enforce that the value of x be increased by δ at each arrival at ℓ_1, thus blocking the run eventually (see [22]). On the right, a timed automaton that is robustly controllable for the Büchi objective $\{\ell_1, \ell_2, \ell_3\}$. We assume that all transitions have the same label. The cycle around ℓ_1 cannot be taken forever, as value of x increases due to perturbations. The cycle around ℓ_2 can be taken forever, but Controller cannot reach ℓ_2 due to the equality $x = 1$. Controller's strategy is thus to loop forever around ℓ_3. This is possible as for both choices of Perturbator in location ℓ_4, clock x will be reset, and thus perturbations do not accumulate. If one of the two resets were absent, Perturbator could force the run to always take that branch, and would win the game.

Perturbator does. Note that these games are determined since for each $\delta > 0$, the semantics is a timed game.

Figure 1 shows examples of timed automata where either Controller or Perturbator wins the robust timed game according to our definitions. The main result of this paper for this non-stochastic setting is the following.

Theorem 1. *The robust timed game problem is EXPTIME-complete.*

We focus on presenting the EXPTIME membership in Sections 4 and 5. The algorithm relies on a characterization of winning strategies in a refinement of the region automaton construction.

In order to formally introduce the appropriate notions for this characterization, we need definitions given in the following section.

3 Regions, Orbit Graphs and Shrunk DBMs

Regions and Region Automata. Following [12,19,3], we assume that the clocks are bounded above by a known constant [1] in all timed automata we consider. Fix a timed automaton $\mathcal{A} = (\mathcal{L}, \mathcal{C}, \ell_0, \mathsf{Act}, E)$. We define *regions* as in [1], as subsets of $\mathbb{R}^{\mathcal{C}}_{\geq 0}$. Any region r is defined by fixing the integer parts of the clocks, and giving a partition $X_0, X_1, \ldots, X_m$ of the clocks, ordered according to their fractional values: for any $\nu \in r$, $0 = \mathsf{frac}(\nu(x_0)) < \mathsf{frac}(\nu(x_1)) < \ldots < \mathsf{frac}(\nu(x_m))$ for any $x_0 \in X_0, \ldots, x_m \in X_m$, and $\mathsf{frac}(\nu(x)) = \mathsf{frac}(\nu(y))$ for any $x, y \in X_i$. Here, $X_i \neq \emptyset$ for all $1 \leq i \leq m$ but X_0 might be empty. For any valuation ν, let $[\nu]$ denote the region to which ν belongs. $\mathsf{Reg}(\mathcal{A})$ denotes the set of regions of $\mathcal{A}$. A region r is said to be *non-punctual* if it contains some

[1] Any timed automaton can be transformed to satisfy this property.

$\nu \in r$ such that $\nu + [-\varepsilon, \varepsilon] \subseteq r$ for some $\varepsilon > 0$. It is said *punctual* otherwise. By extension, we say that (ℓ, r) is non-punctual if r is.

We define the *region automaton* as a finite automaton $\mathcal{R}(\mathcal{A})$ whose states are pairs (ℓ, r) where $\ell \in \mathcal{L}$ and r is a region. Given $(r', a) \in \mathsf{Reg}(\mathcal{A}) \times \mathsf{Act}$, there is a transition $(\ell, r) \xrightarrow{(r',a)} (\ell', s)$ if r' is non-punctual [2], there exist $\nu \in r$, $\nu' \in r'$ and $d > 0$ such that $\nu' = \nu + d$, and there is an edge $e = (\ell, g, R, \ell')$ such that $r' \models g$ and $r'[R \leftarrow 0] = s$. We write the *paths* of the region automaton as $\pi = q_1 e_1 q_2 e_2 \ldots q_n$ where each q_i is a state, and $e_i \in \mathsf{Reg}(\mathcal{A}) \times \mathsf{Act}$, such that $q_i \xrightarrow{e_i} q_{i+1}$ for all $1 \leq i \leq n-1$. The *length* of the path is n, and is denoted by $|\pi|$. If a Büchi condition B is given, a cycle of the region automaton is *winning* if it contains an occurrence of a state (ℓ, r) with $\ell \in B$.

Vertices and Orbit Graphs. A *vertex* of a region r is any point of $\bar{r} \cap \mathbb{N}^{\mathcal{C}}$, where $\bar{r}$ denotes the topological closure of r. Let $\mathcal{V}(r)$ denote the set of vertices of r. We also extend this definition to $\mathcal{V}((\ell, r)) = \mathcal{V}(r)$.

With any path π of the region automaton, we associate a labeled bipartite graph $\Gamma(\pi)$ called the *folded orbit graph of* π [19] (FOG for short). Intuitively, the FOG of a path gives the reachability relation between the vertices of the first and the last regions, assuming the guards are closed. For any path π from state q to q', the node set of the graph $\Gamma(\pi)$ is defined as the disjoint union of $\mathcal{V}(q)$ and $\mathcal{V}(q')$. There is an edge from $v \in \mathcal{V}(q)$ to $v' \in \mathcal{V}(q')$, if, and only if v' is reachable from v along the path π when all guards are replaced by their closed counterparts. It was shown that any run along π can be written as a convex combination of runs along vertices; using this observation orbit graphs can be used to characterize runs along given paths [19]. An important property that we will use is that there is a monoid morphism from paths to orbit graphs. In fact, the orbit graph of a path can be obtained by combining the orbit graphs of a factorization of the path.

When the path π is a cycle around q, then $\Gamma(\pi)$ is defined on the node set $\mathcal{V}(q)$, by merging the nodes of the bipartite graph corresponding to the same vertex. A cycle π is *aperiodic* if for all $k \geq 1$, $\Gamma(\pi^k)$ is strongly connected. Aperiodic cycles are closely related to cycles whose FOG is a complete graph since a long enough iteration of the former gives a complete FOG and conversely, any cycle that has some power with a complete FOG is aperiodic. In the timed automaton of Fig. 1(b), the cycles around locations ℓ_2 and ℓ_3 are aperiodic while that of ℓ_1 is not. Complete FOG are of particular interest to us as they exactly correspond to paths whose reachability relation (between valuations of the initial and last region) is complete [3]. This means that there is no convergence phenomena along the path.

DBMs and Shrunk DBMs. We assume the reader is familiar with the data structure of *difference-bound matrix (DBM)* which are square matrices over $(\mathbb{R} \times \{<, \leq\}) \cup \{(\infty, <)\}$ used to represent zones. DBMs were introduced in [6,13] for analyzing timed automata; see also [5]. Standard operations used to explore the state space of timed automata have been defined on DBMs: intersection is written $M \cap N$, $\mathsf{Pre}\,(M)$ is the set of time predecessors of M, $\mathsf{Unreset}_R(M)$ is the set of valuations that end in M when the clocks in R are reset. We also consider $\mathsf{Pre}_{>\delta}(M)$, time predecessors with a delay of more than δ.

[2] Note this slight modification in the definition of the region automaton.

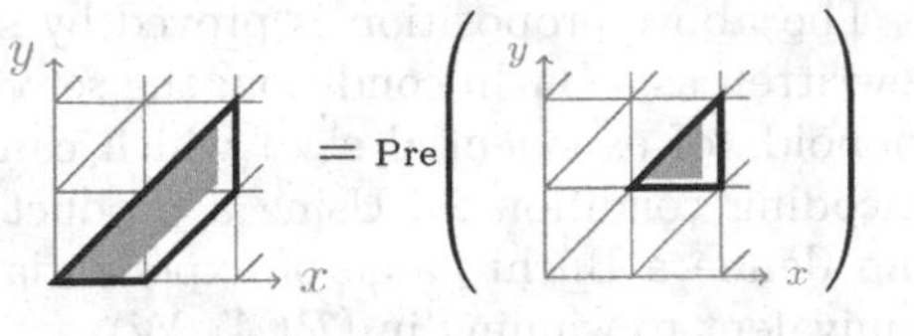

Fig. 2. Time-predecessors operation $(M, P) = \mathsf{Pre}(N, Q)$ applied on a shrunk DBM. Here, the shaded area on the left represents the set $M - \delta P$, while the zone with the thick contour represents M.

To analyze the parametric game $\mathcal{G}_\delta(\mathcal{A})$, we need to express *shrinkings* of zones, *e.g.* sets of states satisfying constraints of the form $g = 1 + \delta < x < 2 - \delta \wedge 2\delta < y$, where δ is a parameter. Formally, a shrunk DBM is a pair (M, P), where M is a DBM, and P is a nonnegative integer matrix called a *shrinking matrix* (SM). This pair represents the set of valuations defined by the DBM $M - \delta P$, for any $\delta > 0$. For instance, M is the guard g obtained by setting $\delta = 0$, and P is made of the integer multipliers of δ.

We adopt the following notation: when we write a statement involving a shrunk DBM (M, P), we mean that for some $\delta_0 > 0$, the statement holds for $(M - \delta P)$ for all $\delta \in [0, \delta_0]$. For instance, $(M, P) = \mathsf{Pre}_{>\delta}(N, Q)$ means that $M - \delta P = \mathsf{Pre}_{>\delta}(N - \delta Q)$ for all small enough $\delta > 0$. Additional operations are defined for shrunk DBMs: for any (M, P), we define $\mathsf{shrink}_{[-\delta,\delta]}(M, P)$ as the set of valuations ν with $\nu + [-\delta, \delta] \subseteq M - \delta P$, for small enough $\delta > 0$. Figure 2 shows an example of a shrunk DBM and an operation applied on it. Standard operations on zones can also be performed on shrunk DBMs in poly-time [21,8].

4 Playing in the Region Automaton

In this section, we will define an appropriate abstraction based on region automata in order to characterize winning in the robust timed game. We note that the usual region automaton does not carry enough information for our purpose; for instance, the blocking behavior in Fig.1(a) cannot be detected in the region automaton (which does contain infinite runs). We therefore define, on top of the usual region construction, a complex winning condition $\mathcal{W}$ characterizing accepting runs *along aperiodic cycles.* In order to be able to transfer the condition $\mathcal{W}$ to the continuous semantics, we study the properties of $\mathcal{W}$ on the abstract region game, and derive two necessary and sufficient conditions ($\mathcal{C}_C$ and $\mathcal{C}_P$) for winning which will be used in the next section to derive the algorithm.

Abstract Arena and Strategies. We fix a timed automaton $\mathcal{A} = (\mathcal{L}, \mathcal{C}, \ell_0, \mathsf{Act}, E)$ and a Büchi condition ϕ. We define a two-player turn-based game played on the region automaton $\mathcal{R}(\mathcal{A})$. In this game, Controller's strategy consists in choosing actions, while Perturbator's strategy consists in resolving non-determinism.

We consider standard notions of (finite-memory, memoryless) strategies in this game and, given a finite-memory strategy σ, we denote by $\mathcal{R}(\mathcal{A})[\sigma]$ the automaton obtained under strategy σ.

Winning Condition on $\mathcal{R}(\mathcal{A})$. We define set $\mathcal{W}$ of winning plays in the game $\mathcal{R}(\mathcal{A})$: an infinite play is winning iff the following two conditions are satisfied: 1) an accepting state in ϕ is visited infinitely often 2) disjoint finite factors with complete folded orbit graphs are visited infinitely often.

Proposition 2. *The game $(\mathcal{R}(\mathcal{A}), \mathcal{W})$ is determined, admits finite-memory strategies for both players, and wining strategies can be computed in EXPTIME.*

The above proposition is proved by showing that condition 2) of $\mathcal{W}$ can be rewritten as a Büchi condition: the set of folded orbit graphs constitute a finite monoid (of exponential size) which can be used to build a Büchi automaton encoding condition 2). Using a product construction for Büchi automata, one can define a Büchi game of exponential size where winning for any player is equivalent to winning in $(\mathcal{R}(\mathcal{A}), \mathcal{W})$.

From the Abstract Game to the Robust Timed Game. We introduce two conditions for Perturbator and Controller which are used in Section 5 to build concrete strategies in the robust timed game.

$\mathcal{C}_P$: there exists a finite memory strategy τ for Perturbator such that no cycle in $\mathcal{R}(\mathcal{A})[\tau]$ reachable from the initial state is winning aperiodic.
$\mathcal{C}_C$: there exists a finite-memory strategy σ for Controller such that every cycle in $\mathcal{R}(\mathcal{A})[\sigma]$ reachable from the initial state is winning aperiodic.

Intuitively, determinacy allows us to write that either all cycles are aperiodic, or none is, respectively under each player's winning strategies. We prove that these properties are sufficient and necessary for respective players to win $(\mathcal{R}(\mathcal{A}), \mathcal{W})$:

Lemma 3. *The winning condition $\mathcal{W}$ is equivalent to $\mathcal{C}_P$ and $\mathcal{C}_C$: 1.* Perturbator *wins the game $(\mathcal{R}(\mathcal{A}), \mathcal{W})$ iff property $\mathcal{C}_P$ holds. 2.* Controller *wins the game $(\mathcal{R}(\mathcal{A}), \mathcal{W})$ iff property $\mathcal{C}_C$ holds. In both cases, a winning strategy for $\mathcal{W}$ is also a witness for $\mathcal{C}_C$ (resp. $\mathcal{C}_P$).*

The proof is obtained by the following facts: finite-memory strategies are sufficient to win the game $(\mathcal{R}(\mathcal{A}), \mathcal{W})$, thanks to the previous proposition; given a folded orbit graph γ, there exists n such that γ^n is complete iff γ is aperiodic; last, the concatenation of a complete FOG with an arbitrary FOG is complete.

5 Solving the Robust Timed Game

In this section, we show that condition $\mathcal{C}_P$ (resp. $\mathcal{C}_C$) is sufficient to witness the existence of a winning strategy in the robust timed game for Perturbator (resp. Controller). By Lemma 3, the robust timed game problem is then reduced to $(\mathcal{R}(\mathcal{A}), \mathcal{W})$ and we obtain:

Theorem 4. *Let $\mathcal{A}$ be a timed automaton with a Büchi condition. Then,* Controller *wins the robust timed game for $\mathcal{A}$ iff he wins the game $(\mathcal{R}(\mathcal{A}), \mathcal{W})$.*

By Proposition 2, the robust timed game can be solved in EXPTIME. In addition, when Controller wins the robust timed game, one can also compute $\delta > 0$ and an actual strategy in $\mathcal{G}_\delta(\mathcal{A})$: Lemma 3 gives an effective strategy σ satisfying $\mathcal{C}_C$ and the proof of Lemma 6 will effectively derive a strategy (given as shrunk DBMs).

5.1 Sufficient Condition for Perturbator

We first prove that under condition $\mathcal{C}_P$, Perturbator wins the robust timed game. We use the following observations. Once one fixes a strategy for Perturbator satisfying $\mathcal{C}_P$, intuitively, one obtains a timed automaton (where there is no more non-determinism in actions), such that all accepting cycles are non-aperiodic. As Perturbator has no more non-determinism to resolve, results of [22] apply and the next lemma follows.

Lemma 5. *If $\mathcal{C}_P$ holds, then* Perturbator *wins the robust timed game.*

5.2 Sufficient Condition for Controller

Proving that $\mathcal{C}_C$ is a sufficient condition for Controller is the main difficulty in the paper; the proof for the next lemma is given in this section.

Lemma 6. *If $\mathcal{C}_C$ holds, then* Controller *wins the robust timed game.*

Proof Outline. We consider the non-deterministic automaton $\mathcal{B} = \mathcal{R}(\mathcal{A})[\sigma]$ which represents the behavior of game $\mathcal{R}(\mathcal{A})$ when Controller plays according to σ, given by condition $\mathcal{C}_C$. Without loss of generality, we assume that σ is a *memoryless strategy* played on the game $\mathcal{R}(\mathcal{A})[\sigma]$ (states of $\mathcal{R}(\mathcal{A})$ can be augmented with memory) and that $\mathcal{B}$ is trim. Given an edge $e = q \to q'$ of $\mathcal{B}$, we denote by $\mathsf{edge}(e)$ the underlying transition in $\mathcal{A}$.

Given a state p of $\mathcal{B}$, we denote by $\mathsf{Unfold}(\mathcal{B}, p)$ the infinite labeled tree obtained as the unfolding of $\mathcal{B}$, rooted in state p. Formally, nodes are labeled by states of $\mathcal{B}$ and given a node x labeled by q, $\sigma(q)$ is defined and there exists q' such that $q \xrightarrow{\sigma(q)} q'$ in $\mathcal{B}$. Then x has one child node for each such q'. We may abusively use nodes to refer to labels to simplify notations.

We first choose states $q_1, \ldots, q_n$ such that every cycle of $\mathcal{B}$ contains one of the q_i's. Let us denote by q_0 the initial state of $\mathcal{B}$, for $i = 0..n$, one can choose a finite prefix t_i of $\mathsf{Unfold}(\mathcal{B}, q_i)$ such that every leaf of t_i is labeled by some q_j, $j = 1..n$. Indeed, as $\mathcal{B}$ is trim and σ is a winning strategy for Controller in the game $(\mathcal{R}(\mathcal{A}), \mathcal{W})$, every branch of $\mathsf{Unfold}(\mathcal{B}, q_i)$ is infinite.

Strategies for standard timed games can be described by means of regions. In our robustness setting, we use shrinkings of regions. Let (ℓ_i, r_i) be the label of state q_i. To build a strategy for Controller, we will identify $\delta > 0$ and zones s_i, $i = 1..n$, obtained as shrinking of regions r_i. These zones satisfy that the controllable predecessors through the tree t_i computed with zones $(s_j)_j$ at leafs contains the zone s_i: this means that from any configuration in (ℓ_i, s_i), Controller has a strategy to ensure reaching a configuration in one of the (ℓ_j, s_j)'s, when following the tree t_i. These strategies can thus be repeated, yielding infinite outcomes. This idea is illustrated in Fig. 3 where the computations along some prefix t are depicted: the shrunk zone at a node represents the controllable predecessors of the shrunk zones of its children. Here, from the shrunk set of the root node, one can ensure reaching a shrinking of each leaf which is included in the shrinking of the starting state of its cycle, yielding a kind of fixpoint. We have in fact $(r, Q_i) \subseteq (r, Q_1)$ for $i = 3, 4$, and $(s, Q_5) \subseteq (s, Q_2)$.

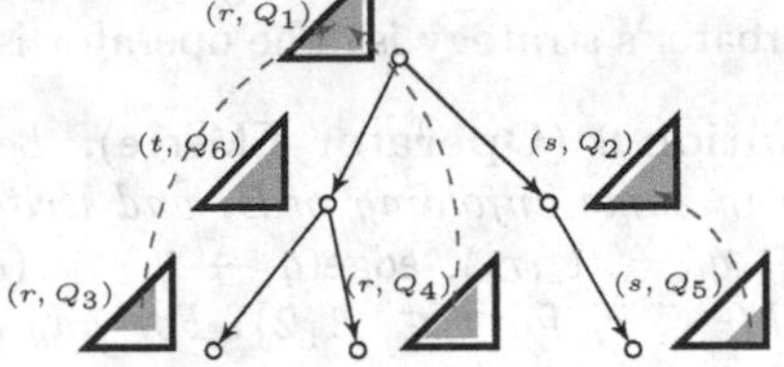

Fig. 3. Proof idea of Lemma 6. Dashed arrows represent cycles.

To identify these zones s_i, we will successively prove the three following facts:
1) Prefixes t_i's can be chosen such that every branch has a complete FOG
2) Controllable predecessors through t_i's of non-empty shrunk zones are non-empty shrunk zones
3) Controllable predecessors through t_i's can be faithfully over-approximated by the intersection of controllable predecessors through branches of t_i

Ensuring Branches with Complete FOGs. To prove property 1) of the Proof outline, we use condition $\mathcal{C}_C$ and the fact that long enough repetitions of aperiodic cycles yield complete FOGs. We obtain:

Lemma 7. *Under condition $\mathcal{C}_C$, there exists an integer N such that every path ρ in $\mathcal{B}$ of length at least N has a complete folded orbit graph.*

Controllable Predecessors and Merge. In order to compute winning states in $\mathcal{G}_\delta(\mathcal{A})$ through unfoldings, we define two operators. CPre is the standard set of controllable predecessors along a single edge:

Definition 8 (Operator CPre). *Let $e = q \to q_1$ be an edge in some unfolding of $\mathcal{B}$. Let us write $q = (\ell, r)$, $q_1 = (\ell_1, r_1)$, $\sigma(q) = (r', a)$ and $\mathsf{edge}(q \to q_1) = (\ell, g_1, a, R_1, \ell_1)$. Let M_1 be a DBM such that $M_1 \subseteq r_1$ and $\delta \geq 0$. We define the set of δ-controllable predecessors of M_1 through edge e as* $\mathsf{CPre}_e^\delta(M_1) = r \cap \mathsf{Pre}_{>\delta}\left(\mathsf{Shrink}_{[-\delta,\delta]}\left(r' \cap \mathsf{Unreset}_{R_1}(M_1)\right)\right)$.

The above definition is extended to paths. Intuitively, $\mathsf{CPre}_e^\delta(M_1)$ are the valuations in region r from which M_1 can be surely reached through a delay in r' and the edge e despite perturbations up to δ.

We now consider the case of branching paths, where Perturbator resolves non-determinism. In this case, in order for Controller to ensure reaching given subsets in all branches, one needs a stronger operator, which we call CMerge. Intuitively, $\mathsf{CMerge}_{e_1,e_2}^\delta(M_1, M_2)$ is the set of valuations in the region starting r from which Controller can ensure reaching either M_1 or M_2 by a *single* strategy, whatever Perturbator's strategy is. The operator is illustrated in Fig. 4.

Definition 9 (Operator CMerge). *Let $e_1 = q \to q_1$ and $e_2 = q \to q_2$ be edges in some unfolding of $\mathcal{B}$, and write $q = (\ell, r)$, $\sigma(q) = (r', a)$ and for $i \in \{1,2\}$, $q_i = (\ell_i, r_i)$, $\mathsf{edge}(q \to q_i) = (\ell, g_i, a, R_i, \ell_i)$. Let M_i be a DBM such that $M_i \subseteq r_i$ for $i \in \{1,2\}$. For any $\delta \geq 0$, define the set of δ-controllable predecessors of M_1, M_2 through edges e_1, e_2 as* $\mathsf{CMerge}_{e_1,e_2}^\delta(M_1, M_2) = r \cap \mathsf{Pre}_{>\delta}$ $\left(\mathsf{Shrink}_{[-\delta,\delta]}\left(r' \cap \bigcap_{i\in\{1,2\}} \mathsf{Unreset}_{R_i}(M_i)\right)\right)$.

We extend CMerge by induction to finite prefixes of unfoldings of $\mathcal{B}$. Consider a tree t and shrunk DBMs $(M_i, P_i)_i$ for its leaves, $\mathsf{CMerge}_t^\delta((M_i, P_i)_i)$ is the set of states for which there is a strategy ensuring reaching one of (M_i, P_i). Because $\mathsf{CMerge}_{e_1,e_2}^\delta$ is more restrictive than $\mathsf{CPre}_{e_1}^\delta \cap \mathsf{CPre}_{e_2}^\delta$, we always have $\mathsf{CMerge}_t^\delta \subseteq \bigcap_\beta \mathsf{CPre}_\beta^\delta$, where β ranges over all branches of t (See Fig. 4).

The following lemma states property 2) of the proof outline. Existence of the SM Q follows from standard results on shrunk DBMs. Non-emptiness of (M, Q) follows from the fact that every delay edge leads to a *non-punctual* region. Define a *full-dimensional subset of a set* $r \subseteq \mathbb{R}^n$ is a subset $r' \subseteq r$ such that there is

$\nu \in r'$ and $\varepsilon > 0$ satisfying $\mathsf{Ball}_{d_\infty}(\nu, \varepsilon) \cap r \subseteq r'$, where $\mathsf{Ball}_{d_\infty}(\nu, \varepsilon)$ is the standard ball of radius ε for the infinity norm on $\mathbb{R}^n$.

Lemma 10. *Let t be a finite prefix of* $\mathsf{Unfold}(\mathcal{B}, q)$, *$r$ the region labeling the root, and $r_1, \ldots, r_k$ those of the leafs. $M, N_1, \ldots, N_k$ be non-empty DBMs that are full dimensional subsets of $r, r_1, \ldots, r_k$ satisfying $M = \mathsf{CMerge}_t^0((N_j)_j)$. We consider shrinking matrices P_j, $1 \leq j \leq k$, of DBM N_j such that $(N_j, P_j) \neq \emptyset$. Then, there exists a SM Q such that $(M, Q) = \mathsf{CMerge}_t^\delta((N_j, P_j)_j)$, and $(M, Q) \neq \emptyset$.*

Over-Approximation of CMerge. Given a prefix t where each branch β_i ends in a leaf labeled with (r_i, P_i), we see $\cap_{\beta_i} \mathsf{CPre}_{\beta_i}^0((r_i, P_i))$ as an over-approximation of $\mathsf{CMerge}_t^0((r_i, P_i)_i)$. We will show that both sets have the same "shape", *i.e.* any facet that is not shrunk in one set, is not shrunk in the other one. This is illustrated in Fig. 4.

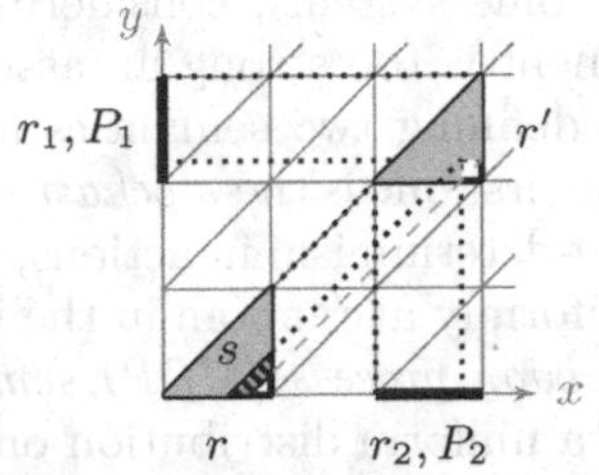

Fig. 4. We have $s = \mathsf{CMerge}^0((r_i, P_i)_i)$, which is strictly included in $\cap_i \mathsf{CPre}^0(r_i, P_i)$ but has the same shape

We introduce the notion of *0-dominance* as follows: for a pair of SMs P, Q, *Q 0-dominates P*, written $P \preceq_0 Q$, if $Q[i,j] \leq P[i,j]$, and $Q[i,j] = 0$ implies $P[i,j] = 0$ for all i, j. Informally, a set shrunk by P is smaller than that shrunk by Q, but yields the same shape. The 0-dominance is the notion we use for a "precise" over-approximation of CMerge:

Lemma 11. *Let t be a finite prefix of* $\mathsf{Unfold}(\mathcal{B}, q)$, *with $q = (\ell, r)$, and let $(\ell_i, r_i), 1 \leq i \leq k$ denote the (labels of) leafs of t. We denote by β_i, $1 \leq i \leq k$, the branches of t. Consider SMs P_i, $1 \leq i \leq k$, for regions r_i. Let us denote $(r, P) = \mathsf{CMerge}_t^0((r_i, P_i)_{1 \leq i \leq k})$ and $(r, Q) = \bigcap_{i=1}^k \mathsf{CPre}_{\beta_i}^0(r_i, P_i)$, then $P \preceq_0 Q$.*

Putting Everything Together. In order to complete the proof of Lemma 6, we first recall the following simple lemma:

Lemma 12 ([22]). *For any DBM M, there is a SM P_0 s.t. $(M, P_0) \neq \emptyset$, and is fully dimensional, and for any SM P and $\varepsilon > 0$ with $(M, P) \neq \emptyset$ and $M - \varepsilon P_0 \neq \emptyset$, we have $M - \varepsilon P_0 \subseteq (M, P)$.*

Remember we have identified states q_i and trees t_i, $i = 0..n$. Denote (ℓ_i, r_i) the label of q_i. For each $i = 1..n$, we denote by P_i the SM obtained by Lemma 12 for r_i. Consider now some tree t_i, $i = 0..n$, with $((r_j, P_j)_j)$ at leafs. Let β_j be a branch of t_i and denote by (r_j, P_j) its leaf. By Lemma 7, the FOG of β_j is complete, and thus from *any* valuation in r_i, one can reach every valuation in the target region r_j along β_j (see [3]), and thus $r_i = \mathsf{CPre}_{\beta_j}^0(r_j, P_j)$. This holds for every branch and we obtain $r_i = \bigcap_j \mathsf{CPre}_{\beta_j}^0(r_j, P_j)$. By Lemma 11, this entails $r_i = \mathsf{CMerge}_{t_i}^0((r_j, P_j)_j)$. We can choose $\varepsilon > 0$ small enough such that the zone $s_i = r_i - \varepsilon P_i$ is non-empty for every $i = 1..n$ and we obtain $r_i = \mathsf{CMerge}_{t_i}^0((s_j)_j)$. We can then apply Lemma 10, yielding some SM Q_i of r_i such that $\emptyset \neq (r_i, Q_i) = \mathsf{CMerge}_{t_i}^\delta((s_j)_j)$. There are two cases:

- $i = 0$: as r_0 is the singleton $\{\mathbf{0}\}$, we have $(r_0, Q_0) = r_0$, and thus $r_0 = \mathsf{CMerge}_{t_0}^\delta((s_j)_j)$. In other terms, for small enough δ's, Controller has a strategy in $\mathcal{G}_\delta(\mathcal{A})$ along t_0 to reach one of the (ℓ_j, s_j)'s starting from the initial configuration $(\ell_0, \mathbf{0})$.

– $i \geq 1$: Lemma 12 entails $s_i \subseteq \mathsf{CMerge}^{\delta}_{t_i}((s_j)_j)$, which precisely states that for small enough δ's, Controller has a strategy in $\mathcal{G}_\delta(\mathcal{A})$ along t_i, starting in (ℓ_i, s_i), to reach one of the (ℓ_j, s_j)'s.

These strategies can thus be combined and repeated, yielding the result.

6 Probabilistic Semantics

In some systems, considering the environment as a completely adversarial opponent is too strong an assumption. We thus consider stochastic environments by defining two semantics as probabilistic variants of the robust timed games. The first one is the *stochastic game semantics* where Perturbator only resolves the non-determinism in actions, but the perturbations are chosen independently and uniformly at random in the interval $[-\delta, \delta]$. The second semantics is the *Markov decision process (MDP) semantics*, where the non-determinism is also resolved by a uniform distribution on the edges, and there is no player Perturbator.

6.1 Stochastic Game Semantics

Formally, given $\delta > 0$, the state space is partitioned into $V_C \cup V_P$ as previously. At each step, Controller picks a delay $d \geq \delta$, and an action a such that for every edge $e = (\ell, g, a, R, \ell')$ such that $\nu + d \models g$, we have $\nu + d + \varepsilon \models g$ for all $\varepsilon \in [-\delta, \delta]$, and there exists at least one such edge e. Perturbator then chooses an edge e with label a, and a perturbation $\varepsilon \in [-\delta, \delta]$ is chosen independently and uniformly at random. The next state is determined by delaying $d + \varepsilon$ and taking the edge e. To ensure that probability measures exist, we restrict to *measurable strategies*.

In this semantics, we are interested in deciding whether Controller can ensure a given Büchi objective almost surely, for some $\delta > 0$. It turns out that the same characterization as in Theorem 4 holds in the probabilistic case.

Theorem 13. *It is EXPTIME-complete to decide whether for some $\delta > 0$,* Controller *has a strategy achieving a given Büchi objective almost surely in the stochastic game semantics. Moreover, if $\mathcal{C}_C$ holds then* Controller *wins almost-surely; if $\mathcal{C}_P$ holds then* Perturbator *wins almost-surely.*

This is a powerful result showing a strong distinction between robust and non-robust timed games: in the first case, a controller that ensures the specification almost surely can be computed, while in non-robust timed games, *any* controller will fail *almost surely*. Thus, while in previous works on robustness in timed automata (e.g. [19]) the emphasis was on additional behaviors that might appear in the worst-case due to the accumulation of perturbations, we show that in our setting, this is inevitable. Note that this also shows that limit-sure winning (see next section) is equivalent to almost-sure winning.

6.2 Markov Decision Process Semantics

The *Markov decision process semantics* consists in choosing both the perturbations, and the edges uniformly at random (and independently). Formally, it consists in restricting Perturbator to choose all possible edges uniformly at random in the stochastic game semantics. We denote by $\mathcal{G}_\delta^{\mathsf{MDP}}(\mathcal{A})$ the resulting game, and $\mathbb{P}^\sigma_{\mathcal{G}_\delta^{\mathsf{MDP}}(\mathcal{A}),s}$ the probability measure on $\mathsf{Runs}(\mathcal{A}, s)$ under strategy σ.

For a given timed Büchi automaton, denote ϕ the set of accepting runs. We are interested in the two following problems: (we let $s_0 = (\ell_0, \mathbf{0})$)

Almost-sure winning: does there exist $\delta > 0$ and a strategy σ for Controller such that $\mathbb{P}^\sigma_{\mathcal{G}_\delta^{\mathsf{MDP}}(\mathcal{A}),s_0}(\phi) = 1$?

Limit-sure winning: does there exist, for every $0 \leq \varepsilon \leq 1$, a perturbation upper bound δ, and a strategy σ for Controller such that $\mathbb{P}^\sigma_{\mathcal{G}_\delta^{\mathsf{MDP}}(\mathcal{A}),s_0}(\phi) \geq 1 - \varepsilon$?

Observe that if almost-sure winning cannot be ensured, then limit-sure winning still has a concrete interpretation in terms of controller synthesis: given a quantitative constraint on the quality of the controller, what should be the precision on clocks measurements to be able to synthesize a correct controller? Consider the timed automaton depicted on the right. It is easy to see that Controller loses the (non-stochastic) robust game, the stochastic game and in the MDP semantics with almost-sure condition, but he wins in the MDP semantics with limit-sure condition.

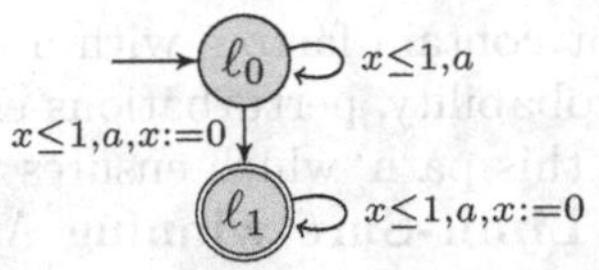

Fig. 5. This automaton is losing in the MDP semantics for the almost-sure winning but winning under the same semantics for the limit-sure winning. In fact, a blocking state (ℓ_0, x) with $x > 1 - \delta$ is reachable with positive probability for any δ.

Theorem 14. *It is EXPTIME-complete to decide whether* Controller *wins almost-surely (resp. limit-surely) in the MDP semantics of a timed Büchi automaton.*

To prove this theorem, we will define decidable characterizations on $\mathcal{R}(\mathcal{A})$ which we will see as a finite Markov decision process. In this MDP, the nondeterminism of actions is resolved according to a uniform distribution. Given a strategy $\hat{\sigma}$ for Controller and a state v, we denote by $\mathbb{P}^{\hat{\sigma}}_{\mathcal{R}(\mathcal{A}),v}$ the resulting measure on $\mathsf{Runs}(\mathcal{R}(\mathcal{A}), v)$. The initial state of $\mathcal{R}(\mathcal{A})$ is v_0. We will use well-known notions about finite MDPs; we refer to [20].

Almost-Sure winning We introduce the winning condition $\mathcal{W}'$: Controller's strategy $\hat{\sigma}$ in $\mathcal{R}(\mathcal{A})$ is winning in state v iff $\mathbb{P}^{\hat{\sigma}}_{\mathcal{R}(\mathcal{A}),v}(\phi) = 1$ and all runs in $\mathsf{Runs}(\mathcal{R}(\mathcal{A})[\hat{\sigma}], v)$ contain infinitely many disjoint factors whose FOGs are complete. Observe that this combines an almost-sure requirement with a sure requirement. This winning condition is our characterization for almost-sure winning:

Proposition 15. Controller *wins almost-surely in the MDP semantics of a timed Büchi automaton $\mathcal{A}$ iff* Controller *wins the game $(\mathcal{R}(\mathcal{A}), \mathcal{W}')$ in v_0.*

Intuitively, the first condition is required to ensure winning almost-surely, and the second condition allows to forbid blocking behaviors. Notice the resemblance with condition $\mathcal{W}$; the difference is that ϕ only needs to be ensured almost-surely rather than surely. We prove the decidability of this condition.

Lemma 16. *The game $(\mathcal{R}(\mathcal{A}), \mathcal{W}')$ admits finite-memory strategies, and winning strategies can be computed in EXPTIME.*

The proof of Prop. 15 uses the following ideas. We first assume that Controller wins the abstract game using some strategy $\hat{\sigma}$. We derive from $\hat{\sigma}$ a strategy σ in the MDP semantics by concretizing the delays chosen by σ. To do so, we consider the automaton $\mathcal{R}(\mathcal{A})[\hat{\sigma}]$ and proceed as in Section 5, which results in a strategy defined by means of shrinking matrices. Using the results of Section 5, we can prove that the outcomes of σ are never blocked, and thus the probabilities of paths in $\mathcal{R}(\mathcal{A})$ under $\hat{\sigma}$ are preserved by σ. As a consequence, σ wins almost-surely.

Conversely, by contradiction, we assume that Controller does not satisfy $\mathcal{W}'$ in $\mathcal{R}(\mathcal{A})$ while there exists an almost-sure strategy σ for the MDP semantics. We build from σ a strategy $\hat{\sigma}$ in $\mathcal{R}(\mathcal{A})$, and prove that it satisfies ϕ almost-surely. This entails the existence of a run ρ in $\mathcal{R}(\mathcal{A})[\hat{\sigma}]$ such that ρ eventually does not contain factors with a complete FOG. We finally show that, with positive probability, perturbations ensure that the run gets blocked along a finite prefix of this path, which ensures that σ is not almost-surely winning.

Limit-Sure winning As illustrated in Fig. 5, it is possible, for any $\varepsilon > 0$, to choose the parameter $\delta > 0$ small enough to ensure a winning probability of at least $1 - \varepsilon$. The idea is that in such cases one can ensure reaching the set of almost-sure winning states with arbitrarily high probability, although the run can still be blocked with small probability before reaching this set.

To characterize limit-sure winning, we define condition $\mathcal{W}''$ as follows. If WIN' denotes the set of winning states for Controller in the game $(\mathcal{R}(\mathcal{A}), \mathcal{W}')$, then $\mathcal{W}''$ is defined as the set of states from which one can almost surely reach WIN'.

Proposition 17. Controller *wins limit-surely in the MDP semantics of a timed Büchi automaton* $\mathcal{A}$ *from* s_0 *iff* Controller *wins the game* $(\mathcal{R}(\mathcal{A}), \mathcal{W}'')$ *in* v_0.

The proof of this proposition relies on the following lemma, and uses techniques similar as those introduced to prove Proposition 15.

Lemma 18. *The game* $(\mathcal{R}(\mathcal{A}), \mathcal{W}'')$ *admits finite-memory strategies, and winning strategies can be computed in EXPTIME.*

7 Conclusion

In this paper, we defined robust timed games with Büchi conditions and unknown imprecision parameters. Our formalism allows one to solve robust controller synthesis problems both against an adversarial (or worst-case) environment, and two variants of probabilistic environments. The procedures we have developed allow, when they exist, to effectively build a bound $\delta > 0$ on the perturbation and a winning strategy for Controller. Some questions remain open including the generalization of these results to concurrent timed games with parity conditions considered in [11]. We believe it is possible to derive symbolic algorithms but this will require extending the theory to detect aperiodic cycles in zone graphs rather than in the region graph.

References

1. Alur, R., Dill, D.L.: A theory of timed automata. Theoretical Computer Science 126(2), 183–235 (1994)
2. Asarin, E., Maler, O., Pnueli, A., Sifakis, J.: Controller synthesis for timed automata. In: SSC 1998, pp. 469–474. Elsevier Science (1998)

3. Basset, N., Asarin, E.: Thin and thick timed regular languages. In: Fahrenberg, U., Tripakis, S. (eds.) FORMATS 2011. LNCS, vol. 6919, pp. 113–128. Springer, Heidelberg (2011)
4. Beauquier, D.: On probabilistic timed automata. Theor. Comput. Sci. 292(1), 65–84 (2003)
5. Bengtsson, J., Yi, W.: Timed automata: Semantics, algorithms and tools. In: Desel, J., Reisig, W., Rozenberg, G. (eds.) ACPN 2003. LNCS, vol. 3098, pp. 87–124. Springer, Heidelberg (2004)
6. Berthomieu, B., Menasche, M.: An enumerative approach for analyzing time Petri nets. In: WCC 1983, pp. 41–46. North-Holland/IFIP (September 1983)
7. Bouyer, P., Markey, N., Reynier, P.-A.: Robust analysis of timed automata via channel machines. In: Amadio, R.M. (ed.) FOSSACS 2008. LNCS, vol. 4962, pp. 157–171. Springer, Heidelberg (2008)
8. Bouyer, P., Markey, N., Sankur, O.: Robust reachability in timed automata: A game-based approach. In: Czumaj, A., Mehlhorn, K., Pitts, A., Wattenhofer, R. (eds.) ICALP 2012, Part II. LNCS, vol. 7392, pp. 128–140. Springer, Heidelberg (2012)
9. Bouyer, P., Markey, N., Sankur, O.: Robustness in timed automata. In: Abdulla, P.A., Potapov, I. (eds.) RP 2013. LNCS, vol. 8169, pp. 1–18. Springer, Heidelberg (2013)
10. Cassez, F., David, A., Fleury, E., Larsen, K.G., Lime, D.: Efficient on-the-fly algorithms for the analysis of timed games. In: Abadi, M., de Alfaro, L. (eds.) CONCUR 2005. LNCS, vol. 3653, pp. 66–80. Springer, Heidelberg (2005)
11. Chatterjee, K., Henzinger, T.A., Prabhu, V.S.: Timed parity games: Complexity and robustness. Logical Methods in Computer Science 7(4) (2011)
12. De Wulf, M., Doyen, L., Markey, N., Raskin, J.-F.: Robust safety of timed automata. Formal Methods in System Design 33(1-3), 45–84 (2008)
13. Dill, D.L.: Timing assumptions and verification of finite-state concurrent systems. In: Sifakis, J. (ed.) AVMFSS 1989. LNCS, vol. 407, pp. 197–212. Springer, Heidelberg (1990)
14. Forejt, V., Kwiatkowska, M., Norman, G., Trivedi, A.: Expected reachability-time games. In: Chatterjee, K., Henzinger, T.A. (eds.) FORMATS 2010. LNCS, vol. 6246, pp. 122–136. Springer, Heidelberg (2010)
15. Henzinger, T.A., Sifakis, J.: The embedded systems design challenge. In: Misra, J., Nipkow, T., Sekerinski, E. (eds.) FM 2006. LNCS, vol. 4085, pp. 1–15. Springer, Heidelberg (2006)
16. Jensen, H.E.: Model checking probabilistic real time systems. In: Proc. 7th Nordic Workshop on Programming Theory, pp. 247–261. Citeseer (1996)
17. Kwiatkowska, M., Norman, G., Segala, R., Sproston, J.: Automatic verification of real-time systems with discrete probability distributions. Theor. Comput. Sci. 282(1), 101–150 (2002)
18. Oualhadj, Y., Reynier, P.-A., Sankur, O.: Probabilistic Robust Timed Games. Technical report (June 2014), `http://hal.archives-ouvertes.fr/hal-01010813`
19. Puri, A.: Dynamical properties of timed automata. Discrete Event Dynamic Systems 10(1-2), 87–113 (2000)
20. Putterman, M.L.: Markov Decision Processes: Discrete Stochastic Dynamic Programming. John Wiley and Sons, New York (1994)
21. Sankur, O., Bouyer, P., Markey, N.: Shrinking timed automata. In: FSTTCS 2011. LIPIcs, vol. 13, pp. 375–386. Leibniz-Zentrum für Informatik (2011)
22. Sankur, O., Bouyer, P., Markey, N., Reynier, P.-A.: Robust controller synthesis in timed automata. In: D'Argenio, P.R., Melgratti, H. (eds.) CONCUR 2013 – Concurrency Theory. LNCS, vol. 8052, pp. 546–560. Springer, Heidelberg (2013)

Perturbation Analysis in Verification of Discrete-Time Markov Chains*

Taolue Chen[1], Yuan Feng[2], David S. Rosenblum[3], and Guoxin Su[3]

[1] Department of Computer Science, Middlesex University London
[2] Centre for Quantum Computation and Intelligent Systems, UK
University of Technology, Sydney, Australia
[3] Department of Computer Science, School of Computing,
National University of Singapore, Singapore

Abstract. Perturbation analysis in probabilistic verification addresses the *robustness* and *sensitivity* problem for verification of stochastic models against qualitative and quantitative properties. We identify two types of perturbation bounds, namely *non-asymptotic bounds* and *asymptotic bounds*. Non-asymptotic bounds are exact, pointwise bounds that quantify the upper and lower bounds of the verification result subject to a given perturbation of the model, whereas asymptotic bounds are closed-form bounds that approximate non-asymptotic bounds by assuming that the given perturbation is sufficiently small. We perform perturbation analysis in the setting of Discrete-time Markov Chains. We consider three basic matrix norms to capture the perturbation distance, and focus on the computational aspect. Our main contributions include algorithms and tight complexity bounds for calculating both non-asymptotic bounds and asymptotic bounds with respect to the three perturbation distances.

1 Introduction

Probabilistic verification techniques, and in particular probabilistic model checking, have been successfully applied to a variety of domains ranging from wireless communication protocols to dynamic power management schemes, and to systems biology and quantum cryptography. Mature probabilistic model checking tools such as PRISM [22] support verification of most existing stochastic models, e.g., Discrete-time Markov Chains (DTMCs), Markov Decision Processes (MDPs), and stochastic games, against a wide range of qualitative and quantitative properties.

When modelling real-life systems with stochastic models, one usually has to face the issue that these systems are governed by empirical or unknown distributions, such as the failure rate of some system component. As a result, measurements or experiments are employed to determine, for instance, transition probabilities (for discrete-time systems) or transition rates (for continuous-time systems). Those statistical quantities

* The work is partially supported by the Australian Research Council (Grant Nos. DP130102764 and FT100100218) and the Singapore Ministry of Education (Grant R-252-000-458-133). Y. Feng is also supported by the Overseas Team Program of the Academy of Mathematics and Systems Science, CAS and the CAS/SAFEA International Partnership Program for Creative Research Team. T. Chen is partially supported by the EU Erasmus Grant.

P. Baldan and D. Gorla (Eds.): CONCUR 2014, LNCS 8704, pp. 218–233, 2014.

are imprecise. In the worst case, a tiny but non-trivial change to some quantities in the model might lead to a misleading or even invalid verification result.

These issues motivate the following important problem for probabilistic verification: *If some of the quantities in the stochastic model are perturbed, what is the influence on verification of the model?* In other words, given a stochastic model, we need to measure the *robustness* and *sensitivity* of verification results. The purpose of perturbation analysis, as the central topic of the current paper, is to shed light on this problem.

A straightforward approach is to modify the model manually for each set of values of each perturbed quantity, and then perform model checking multiple times with a model checker such as PRISM. Such a solution is simple but unsatisfactory: It is resource-consuming while providing little information about the impact of model perturbations on verification. Instead, in this paper, we present a sound and rigorous approach to characterising the *maximal/minimal* variation that might occur to the verification outcome, with respect to a given perturbation of the model. This yields a measure for the sensitivity and robustness of these verification results. Such an analysis also potentially reduces the overall time of verifying a large number of similar stochastic models if only approximated results are required. More specifically, we pursue two types of perturbation bounds, namely *non-asymptotic bounds*[1] and *asymptotic bounds*.

- Non-asymptotic bounds are pointwise bounds that quantify the maximum and the minimum of the verification result subject to a given perturbation of the model.
- Asymptotic bounds are closed-form, lightweight approximations of non-asymptotic bounds when the model perturbation is sufficiently small (i.e., close to 0).

The main task of perturbation analysis in the current paper is to compute these bounds, the formal definitions of which are presented in Section 2.

Contributions. In this paper, we focus on the computational aspect of perturbation analysis in DTMC verification. We consider three different perturbation distances for DTMCs based on three norms over stochastic matrices—the "entrywise" ∞-norm, the induced ∞-norm, and the "entrywise" 1-norm, which quantify the perturbation distance of DTMCs. These norms are widely adopted in literature (e.g. [8,15,31]), and somehow are easy to compute as they are "linear". Henceforth, we refer to the three distances as Type I, II and III distances, respectively. Our key contributions, summarised in Table 1, include two aspects:

- We present algorithms to compute non-asymptotic bounds under Type I, II and III distances, respectively, and identify tight computational complexity bounds. For Type I and II distances, we present polynomial-time algorithms, while for Type III distance, we show that the computation (technically, the aligned decision problem) is in PSPACE and is SQUARE-ROOT-SUM hard.
- We provide a unified treatment for the asymptotic bounds of an arbitrary degree for distances of all three types. In particular, we show how to compute the *linear* and the *quadratic* asymptotic bounds. This subsumes the resuls reported previously [31] regarding linear asymptotic bounds.

[1] The term is adopted from *non-asymptotic* analysis of random matrices and *non-asymptotic* information theory.

Table 1. Complexity of Computing Perturbation Bounds

Distance	Non-Asymptotic	Asymptotic	
		linear	quadratic
I & II	P	PL	NP
III	in PSPACE, SRS-hard		

The computation of non-asymptotic bounds is related to verification of stochastic models with uncertainty (e.g. Interval Markov Chains (IMCs)). Typically, two different semantics for IMCs are studied in the literature, namely *Uncertain Markov Chains* (UMCs) and *Interval Markov Decision Processes* (IMDPs) [28]. The non-asymptotic bounds adopt the UMC semantics. For Type I and II distances, since the UMC and IMDP semantics coincide, we apply a technique similar to the one by Puggelli *et al.* [24] to obtain polynomial-time algorithms for non-asymptotic bounds. However, for Type III distance, we can only obtain a PSPACE algorithm—we show a slightly better complexity upper bound, namely the complexity of the existential theory of reals. For the lower bound, we give a reduction from the well-known SQUARE-ROOT-SUM (SRS) problem. The exact complexity of the SRS problem, i.e., whether it is in P or even in NP, is open since 1976. This suggests that our PSPACE upper bound cannot be substantially improved without a breakthrough concerning this long-standing open problem.

The study of asymptotic bounds in probabilistic verification was initiated by Su *et al.* [30,31], where linear and quadratic asymptotic bounds with respect to a single perturbation distance function are studied. Apart from giving a unified formulation of general asymptotic bounds under the three types of distances, the current paper also improves the complexity results reported in the previous work. Our main techniques for this are from multivariate calculus: We resort the problem to optimisation problems of multivariate polynomials under (virtually) linear constraints. For linear asymptotic bounds, this enables us to derive an analytical expression, whereas for quadratic asymptotic bounds, we exploit quadratic programming. We also identify complexity upper bounds for the two cases.

For simplicity, we focus on reachability in this paper. However, the presented techniques can be generalised for ω-regular properties and various performance properties, such as expected rewards and long-run average rewards, without substantial difficulty (and see our previous work [31] for ω-regular properties).

Related Work. In general, perturbation theory for applied mathematics investigates solutions for mathematically formulated problems that involve parameters subject to perturbations [23]. There exists a line of research on perturbation analysis of DTMCs, the common goal of which is to find a suitable condition number for the distance of steady states and the distance of transition matrices between two DTMCs [27,13,29]. In these works, the condition number is defined as the supremum of the quotient of the deviation of the perturbed DTMCs and the allowed perturbation. The deviation is hence bounded universally for all DTMCs with respect to chosen norms of distance metrics. In the formal verification setting, the closest work is by Chatterjee [8], who studied the continuity and robustness of the value function in stochastic parity games with respect to imprecision in the transition probabilities. This can be regarded as a (rough) perturbation analysis for stochastic games. However, non-asymptotic and asymptotic bounds are not considered there, nor is their computational aspect. It is an interesting direction for future work to extend our results to the game setting. Moreover, Desharnaisa *et al.* [15] addressed perturbation analysis for labelled Markov processes (LMPs). The authors defined a distance akin to the Type II distance and gave a bound on the difference between LMPs measured by a behaviour pseudo-metric with respect to the perturbation.

Most available verification results are on IMCs, (arguably) the simplest variant models of DTMCs with uncertainty. In particular, Sen *et al.* [28] proved that model checking IMC against probabilistic computational tree logic (PCTL) is NP-hard. More general results on IMCs against ω-regular properties are reported in [9]. Chen *et al.* [12] presented thorough results on the complexity of model checking IMCs against reachability and PCTL properties, under both the UMC and the IMDP semantics. Benedikt *et al.* [5] considered the LTL model checking problem for IMCs.

Other related work includes *parameter synthesis* for stochastic models [14,21,19,20] and *model repair* [16,4,11]. In general, these studies attempt to identify some (or all) parameter configuration(s) in a parametric model such that a given property is satisfied. Hence, the approaches there are considerably different from ours.

Structure of the Paper. The rest of the paper is structured as follows: Section 2 presents definitions of models, model distances, and non-asymptotic and asymptotic bounds. Section 3 presents results on computation of non-asymptotic bounds with respect to three types of distances. Section 4 presents results on computation of asymptotic bounds. Section 5 concludes the paper and outlines future work. An extended version of the paper contains proofs and more details [10].

2 Models, Distances and Perturbation Bounds

Given a finite set S, we use $\Delta(S)$ to denote the set of (discrete) *probability distributions* over S, i.e., functions $\mu : S \to [0,1]$ with $\sum_{s\in S}\mu(s) = 1$.

Definition 1. *A* Discrete-time Markov Chain *(DTMC) is a tuple $\mathcal{D} = (S, \alpha, \mathbf{P})$, where*

- *S is a finite set of states,*
- *$\alpha \in \Delta(S)$ is the initial distribution, and*
- *$\mathbf{P} : S \times S \to [0,1]$ is a* transition probability matrix *such that for any state $s \in S$, $\sum_{s'\in S}\mathbf{P}(s,s') = 1$, i.e., $\mathbf{P}(s,\cdot) \in \Delta(S)$.*

An (infinite) *path* in $\mathcal{D}$ is a sequence $\pi = s_0 s_1 \cdots$ such that $s_i \in S$ and $\mathbf{P}(s_i, s_{i+1}) > 0$ for each $i \geq 0$. Denote the i-th state of π (i.e., s_i) as $\pi[i]$, and the set of paths in $\mathcal{D}$ as $Paths^{\mathcal{D}}$. The *probability distribution* $\Pr^{\mathcal{D}}$ over $Paths^{\mathcal{D}}$ is defined in a standard way [3, Chapter 10].

Definition 2. *A* Markov Decision Process *(MDP) is a tuple $\mathcal{M} = (S, \alpha, \mathcal{T})$, where*

- *S and α are defined the same as in Definition 1, and*
- *$\mathcal{T} : S \to \wp(\Delta(S))$ is the transition function s.t. $\mathcal{T}(s)$ is* finite *for each $s \in S$.*

Without loss of generality, we assume that $\mathcal{T}(s) \neq \emptyset$ for each $s \in S$. At each state s of $\mathcal{M}$, a probability distribution μ (over S) is chosen *nondeterministically* from the set $\mathcal{T}(s)$. A successor state s' is then chosen according to μ with probability $\mu(s')$. An (infinite) path π in $\mathcal{M}$ is a sequence of the form $s_0 \xrightarrow{\mu_1} s_1 \xrightarrow{\mu_2} \cdots$ where $s_i \in S$, $\mu_{i+1} \in \mathcal{T}(s_i)$ and $\mu_{i+1}(s_{i+1}) > 0$ for each $i \geq 0$. A *finite* path is a prefix of an infinite path ending in a state. Let $Paths^{\mathcal{M}}$ be the set of finite paths. A *scheduler* σ : $Paths^{\mathcal{M}} \to \Delta(S)$ maps a finite path (the *history*) to a distribution over S such that for any finite path $\pi = s_0 \xrightarrow{\mu_1} \cdots \xrightarrow{\mu_n} s_n$, $\sigma(\pi) \in \mathcal{T}(s_n)$. In particular, a *simple* scheduler σ chooses a distribution only based on the current state, and thus for each finite path π ending in s, $\sigma(\pi) = \sigma(s) \in \mathcal{T}(s)$. Note that we obtain a (possibly infinite-state) DTMC by fixing a scheduler in an MDP [3,25]. In the sequel, we write $\mathcal{M}_\sigma$ for such a DTMC given an MDP $\mathcal{M}$ and a scheduler σ.

We often relax the definition of MDPs by allowing $\mathcal{T}(s)$ to be *infinite*. As long as $\mathcal{T}(s)$ is compact (for instance in the paper, $\mathcal{T}(s) \subseteq \mathbb{R}^{|S|}$ with respect to the Euclidean topology), most interesting properties for MDPs are carried over. This feature is made use of by existing work on IMDPs mentioned in the Introduction.

For the convenience of perturbation analysis, we also define a parametric variant of DTMCs [31]. When performing perturbation analysis for a DTMC in practice, it is usually required that some of transitions remain unchanged. To accommodate this, we specify a set of transitions $C \subseteq S \times S$ for a DTMC $\mathcal{D}$ with state space S. The intuition behind this requirement is that only probabilities of transitions in C can be perturbed. The perturbed quantities are captured by a sequence of pair-wise distinct variables $\boldsymbol{x} = (x_1, \ldots, x_k)$ with $k = |C|$.

Definition 3. *The* parametric DTMC *of $\mathcal{D}$ on $\boldsymbol{x}$ is a tuple $\mathcal{D}(\boldsymbol{x}) = (S, \alpha, \mathbf{P}, F)$ where F is a one-to-one mapping from C to the variable set $\{x_i\}_{1 \leq i \leq k}$.*

For simplicity, we denote by $\mathbf{P}(\boldsymbol{x})$ the parametric variation of $\mathbf{P}$ with the (s,t)-entry being $\mathbf{P}(s,t) + F(s,t)$ if $(s,t) \in C$, and $\mathbf{P}(s,t)$ otherwise. We defer the specification of domains for variables from $\{x_i\}_{1 \leq i \leq k}$ in Section 2.2.

Reachability. For a given DTMC $\mathcal{D} = (S, \alpha, \mathbf{P})$, let $G \subseteq S$ be a set of *target* states. We consider the probability of reaching G. Formally, let $\Diamond G = \{\pi \in Paths^{\mathcal{D}} \mid \pi[i] \in G$ for some $i \geq 0\}$. We are interested in $\Pr^{\mathcal{D}}(\Diamond G)$. Let $S_0 = \{s \in S \mid \Pr^{\mathcal{D}}(s \models \Diamond G) = 0\}$, and $S_? = S \backslash (S_0 \cup G)$. Let $\widetilde{\mathbf{P}}$ be the matrix obtained by restricting $\mathbf{P}$ on $S_?$. Then, $I - \widetilde{\mathbf{P}}$ is invertible. Let $\boldsymbol{b}$ be a vector on $S_?$ such that $\boldsymbol{b}[s] = \sum_{t \in G} \mathbf{P}(s,t)$ for each $s \in S_?$. Let $\tilde{\alpha}$ be the restriction of α on $S_?$. We have that $\Pr^{\mathcal{D}}(\Diamond G) = \tilde{\alpha}^{\mathrm{T}}(I - \widetilde{\mathbf{P}})^{-1}\boldsymbol{b}$ [3, Chapter 10].For an MDP $\mathcal{M}$ with state space $S \supseteq G$, we can also define the *maximum* reachability probability $\sup_\sigma \Pr^{\mathcal{M}_\sigma}(\Diamond G)$, which can be calculated efficiently by linear programming [25].

2.1 Distance between DTMCs

A DTMC $(S, \alpha, \mathbf{P})$ induces a digraph in a standard way: The set of vertices of the digraph is S, and there is an edge from s to t *iff* $\mathbf{P}(s,t) > 0$. Given two DTMCs $\mathcal{D}_1 = (S, \alpha_1, \mathbf{P}_1)$ and $\mathcal{D}_2 = (S, \alpha_2, \mathbf{P}_2)$, we say that $\mathcal{D}_1$ and $\mathcal{D}_2$ are *structurally equivalent*, denoted as $\mathcal{D}_1 \equiv \mathcal{D}_2$, if for each pair of states $s, t \in S$, $\mathbf{P}_1(s,t) > 0$ iff $\mathbf{P}_2(s,t) > 0$. Namely, $\mathcal{D}_1$ and $\mathcal{D}_2$ have the same underlying digraphs. We now identify three distances for two (structurally equivalent) DTMCs.

Definition 4 (Distance of DTMCs). *Given two DTMCs $\mathcal{D}_1$ and $\mathcal{D}_2$ such that $\mathcal{D}_1 \equiv \mathcal{D}_2$, we define the distances d_{I}, d_{II}, and d_{III} as*

(1) $d_{\mathrm{I}}(\mathcal{D}_1, \mathcal{D}_2) = \max_{s,t \in S} |\mathbf{P}_1(s,t) - \mathbf{P}_2(s,t)|$,
(2) $d_{\mathrm{II}}(\mathcal{D}_1, \mathcal{D}_2) = \max_{s \in S}\{\sum_{t \in S} |\mathbf{P}_1(s,t) - \mathbf{P}_2(s,t)|\}$,
(3) $d_{\mathrm{III}}(\mathcal{D}_1, \mathcal{D}_2) = \sum_{s,t \in S} |\mathbf{P}_1(s,t) - \mathbf{P}_2(s,t)|$.

We call d_{I}, d_{II}, and d_{III} as Type I, Type II, and Type III distances respectively. We use $\star$ to range over $\{\mathrm{I}, \mathrm{II}, \mathrm{III}\}$, and $d_\star$ to denote a generic distance definition.

Remark 1. In matrix theory, Type I is the distance induced from the "entrywise" ∞-norm, Type II is induced from the ∞-norm (which is an induced norm of the ∞-norm for vectors), and Type III is induced from the "entrywise" 1-norm.

Let $C_s = \{t \in S \mid (s,t) \in C\}$ (C_s may be empty), and so $C = \biguplus_{s \in S}\{s\} \times C_s$. *To simplify notations in the remainder of the paper, when given a DTMC, we fix an associated C.* Accordingly, the distance definitions can be formulated as norms of variable vectors for parametric DTMCs. Recall that $\boldsymbol{x} = \{x_i\}_{1 \le i \le k}$ with $k = |C|$. Let $\boldsymbol{x}_s = (x_{s,t})_{t \in C_s}$ and so $\boldsymbol{x}$ is a concatenation of $\boldsymbol{x}_s$ for $s \in S$. Note that the distance between $\mathcal{D}(\boldsymbol{x})$ and $\mathcal{D}$ is exactly the corresponding norm of $\boldsymbol{x}$ which is defined as follows:

(1) $\|\boldsymbol{x}\|_{\mathrm{I}} = \max_{1 \le i \le k} |x_i|$,
(2) $\|\boldsymbol{x}\|_{\mathrm{II}} = \max_{s \in S}\{\sum_{t \in C_s} |x_{s,t}|\}$,
(3) $\|\boldsymbol{x}\|_{\mathrm{III}} = \sum_{1 \le i \le k} |x_i|$.

2.2 Perturbation Bounds

For the purpose of perturbation analysis, we define two types of perturbation bounds, namely *non-asymptotic bounds* and *asymptotic bounds*, which are the main research object of this paper.

We write $\mathcal{D} \sim_C \mathcal{D}'$ if $\mathcal{D}$ and $\mathcal{D}'$ differ only for transitions in C, and let

$$[\mathcal{D}]_{\star,\delta} = \{\mathcal{D}' \text{ is a DTMC} \mid \mathcal{D} \equiv \mathcal{D}', d_\star(\mathcal{D}, \mathcal{D}') \le \delta \text{ and } \mathcal{D} \sim_C \mathcal{D}'\}.$$

Definition 5 (Non-Asymptotic bound). *The upper and lower non-asymptotic bounds of a DTMC $\mathcal{D}$ are defined as follows:*

$$\rho_\star^+(\delta) = \sup\left\{\Pr^{\mathcal{D}'}(\Diamond G) \mid \mathcal{D}' \in [\mathcal{D}]_{\star,\delta}\right\} \text{ and } \rho_\star^-(\delta) = \inf\left\{\Pr^{\mathcal{D}'}(\Diamond G) \mid \mathcal{D}' \in [\mathcal{D}]_{\star,\delta}\right\}.$$

We use the subscript $\star$ to emphasise the fact that $\rho_\star^+(\delta)$ and $\rho_\star^-(\delta)$ are dependent on the distance $d_\star$. We can present an alternative characterisation for non-asymptotic bounds with parametric DTMCs. Let

$$\mathbf{U}_{\star,\delta} = \{\boldsymbol{x} \in \mathbf{R}^k \mid \mathcal{D}(\boldsymbol{x}) \in [\mathcal{D}]_{\star,\delta}\}.$$

Here, and in the sequel, we abuse the notation slightly to denote by $\mathcal{D}(\boldsymbol{x})$ the DTMC obtained by instantiating the variables in the corresponding parametric DTMC with the real vector $\boldsymbol{x}$. In particular, we have $\mathcal{D}(\mathbf{0}) = \mathcal{D}$ with $\mathbf{P}(\mathbf{0}) = \mathbf{P}$.

Note that, for each $\boldsymbol{x} \in \mathbf{U}_{\star,\delta}$, $\mathcal{D}(\boldsymbol{x})$ and $\mathcal{D}$ are structurally equivalent and $I - \widetilde{\mathbf{P}}(\boldsymbol{x})$ is invertible. We then write

$$p(\boldsymbol{x}) := \Pr^{\mathcal{D}(\boldsymbol{x})}(\Diamond G) = \alpha^{\mathrm{T}}(I - \widetilde{\mathbf{P}}(\boldsymbol{x}))^{-1}\boldsymbol{b}(\boldsymbol{x}). \tag{1}$$

There are alternative ways of generating or expressing $p(\boldsymbol{x})$ reported in [14,19,18]. Obviously, $p(\cdot)$ is a multivariate rational function on $\mathbf{U}_{\star,\delta}$ and thus is infinitely differentiable. It is then straightforward to observe that

$$\rho^+_\star(\delta) = \sup_{\boldsymbol{x}\in\mathbf{U}_{\star,\delta}} p(\boldsymbol{x}) \text{ and } \rho^-_\star(\delta) = \inf_{\boldsymbol{x}\in\mathbf{U}_{\star,\delta}} p(\boldsymbol{x}) \tag{2}$$

Also note that $\mathbf{U}_{\star,\delta}$ is convex and thus connected. Hence, by the continuity of $p(\cdot)$ and the Intermediate Value Theorem, for any value $y \in (\rho^-_\star(\delta), \rho^+_\star(\delta))$, there exists $\boldsymbol{x} \in \mathbf{U}_{\star,\delta}$ such that $p(\boldsymbol{x}) = y$.

Asymptotic bounds provide reasonably accurate approximations for $\rho^+(\delta)$ and $\rho^-(\delta)$ when $\delta > 0$ is close to 0. Let $r = \min_{(s,t)\in C}\{\mathbf{P}(s,t), 1 - \mathbf{P}(s,t)\} > 0$.

Definition 6 (Asymptotic bound). *An* asymptotic bound *of degree n for $\rho^+_\star(\cdot)$ (resp. $\rho^-_\star(\cdot)$) is a function $f^+_n : (0,r) \to \mathbf{R}$ (resp. $f^-_n : (0,r) \to \mathbf{R}$) such that*

$$f^+_n(\delta) - \rho^+_\star(\delta) = o(\delta^n) \text{ and } f^-_n(\delta) - \rho^-_\star(\delta) = o(\delta^n);$$

in other words,

$$\lim_{\delta\to 0} \frac{|f^+_n(\delta) - \rho^+_\star(\delta)|}{\delta^n} = 0 \text{ and } \lim_{\delta\to 0} \frac{|f^-_n(\delta) - \rho^-_\star(\delta)|}{\delta^n} = 0.$$

In words, Definition 6 states that, as δ tends to 0, the convergent rate of f^+_n (resp. f^-_n) to $\rho^+_\star$ (resp. $\rho^-_\star$) is *at least* of order n or, equivalently, f^+_n (resp. f^-_n) approaches to $\rho^+_\star(\delta)$ (resp. $\rho^-_\star$) at least as fast as any polynomial function on δ of degree n. We note that asymptotic perturbation bounds can be non-unique. We refer to asymptotic bounds of degree one as *linear asymptotic bounds* (*linear bounds* for short), and asymptotic bound of degree two as *quadratic asymptotic bounds* (*quadratic bounds* for short).

3 Computing Non-asymptotic Bounds

In this section, we present algorithms for computing non-asymptotic bounds and analyse the complexity. An obvious fact about non-asymptotic bounds is given by the following proposition:

Proposition 1. *Given a DTMC $\mathcal{D}$ and $\star \in \{\mathrm{I}, \mathrm{II}, \mathrm{III}\}$, $\rho^+_\star(\cdot)$ and $\rho^-_\star(\cdot)$ are continuous functions in $(0, r)$.*

3.1 Type I and Type II Distances

In this section, we deal with non-asymptotic bounds under Type I and II distances given in Definition 4. In particular, we focus on Type I distance, while Type II distance can be dealt with in a similar way.

In general terms, our strategy is to reduce the computation of $\rho_{\mathrm{I}}^{+}(\cdot)$ and $\rho_{\mathrm{I}}^{-}(\cdot)$ to linear programming via MDPs. Let $\mathcal{D} = (S, \alpha, \mathbf{P})$. Consider an MDP $\mathcal{M} = (S, \alpha, \mathcal{T})$, where for each state $s \in S$,

$$\mathcal{T}(s) = \{\mu \in \Delta(S) \mid |\mu(s') - \mathbf{P}(s, s')| \leq \delta \text{ for any } s' \in S\}$$

We have the following proposition:

Proposition 2. $\rho_{\mathrm{I}}^{+}(\delta) = \sup_{\sigma} \Pr^{\mathcal{M}_{\sigma}}(\Diamond G)$ *and* $\rho_{\mathrm{I}}^{-}(\delta) = \inf_{\sigma} \Pr^{\mathcal{M}_{\sigma}}(\Diamond G)$.

Proposition 2 allows us to reduce the problem of computing $\rho_{\mathrm{I}}^{+}(\delta)$ (resp. $\rho_{\mathrm{I}}^{-}(\delta)$) to computing the maximum (resp. minimum) reachability probability for the MDP $\mathcal{M}$, and the latter is resorted to the standard linear programming technique. Note that here the MDP $\mathcal{M}$ is merely a tool which can simplify the technical development, and that we are not considering verification of "perturbed" MDPs. *Below we only present an algorithm for* $\rho_{\mathrm{I}}^{+}(\delta)$*, since an algorithm for* $\rho_{\mathrm{I}}^{-}(\delta)$ *can be obtained in a dual manner.*

Recall that $\boldsymbol{x} = (x_{s,t})_{(s,t)\in C}$ is a concatenation of vectors $\boldsymbol{x}_s = (x_{s,t})_{t\in C_s}$ for each $s \in S$. Intuitively, $x_{s,t}$ captures the perturbed quantity at the (s,t)-entry of the transition probability matrix $\mathbf{P}$. We introduce a new vector of variables $\boldsymbol{y} = (y_s)_{s\in S}$. Intuitively, y_s captures the probability to reach G from state s. For each state s, $\Omega(s)$ is a set of vectors defined as:

$$\boldsymbol{x}_s \in \Omega(s) \text{ iff } \begin{cases} \sum_{t\in C_s} x_{s,t} = 0 \\ 0 \leq \mathbf{P}(s,t) + x_{s,t} \leq 1, \text{ for each } t \in C_s \\ -\delta \leq x_{s,t} \leq \delta, \text{ for each } t \in C_s \end{cases} \tag{3}$$

For simplicity, we also write

$$\Gamma(\boldsymbol{x}_s, \boldsymbol{y}) = \sum_{t\in C_s} (\mathbf{P}(s,t) + x_{s,t}) \cdot y_t + \sum_{t\notin C_s} \mathbf{P}(s,t) \cdot y_t.$$

Then, we consider the following (pseudo-) linear program (which can be derived directly from the MDP formulation [25,6]):

$$\begin{aligned} \text{minimise} \quad & \sum_{s\in S} \alpha(s) y_s \\ \text{subject to} \quad & y_s \geq \max_{\boldsymbol{x}_s\in\Omega(s)} \Gamma(\boldsymbol{x}_s, \boldsymbol{y}) \quad s \notin G \\ & y_s = 1 \quad\quad\quad\quad\quad\quad\quad\;\; s \in G \end{aligned} \tag{4}$$

Note that, for a fixed $\boldsymbol{y}$, $\max_{\boldsymbol{x}_s\in\Omega(s)} \Gamma(\boldsymbol{x}_s, \boldsymbol{y})$ in Problem (4) is itself a linear program where the constraint is given in (3). (It also follows that $\max_{\boldsymbol{x}_s\in\Omega(s)} \Gamma(\boldsymbol{x}_s, \boldsymbol{y})$ does exist although $\Omega(s)$ is infinite.) We denote its (Lagrange) *dual function* as $\min_{\boldsymbol{\lambda}_s} \Gamma'(\boldsymbol{\lambda}_s, \boldsymbol{y})$, where $\boldsymbol{\lambda}_s$ is the Lagrange multiplier vectors (dural variables) for the linear program. Strong duality implies that

$$\max_{\boldsymbol{x}_s} \Gamma(\boldsymbol{x}_s, \boldsymbol{y}) = \min_{\boldsymbol{\lambda}_s} \Gamma'(\boldsymbol{\lambda}_s, \boldsymbol{y}) \tag{5}$$

Hence, Problem (4) becomes the following problem:

$$\begin{aligned} &\text{minimise} && \sum_{s \in S} \alpha(s) y_s \\ &\text{subject to} && y_s \geq \min_{\boldsymbol{\lambda}_s} \Gamma'(\boldsymbol{\lambda}_s, \boldsymbol{y}) \quad s \notin G \\ & && y_s = 1 \qquad\qquad\qquad\quad s \in G \end{aligned} \tag{6}$$

It is not hard to observe that Problem (6) is equivalent to the following problem:

$$\begin{aligned} &\text{minimise} && \sum_{s \in S} \alpha(s) y_s \\ &\text{subject to} && y_s \geq \Gamma'(\boldsymbol{\lambda}_s, \boldsymbol{y}) \quad s \notin G \\ & && y_s = 1 \qquad\qquad\quad s \in G \end{aligned} \tag{7}$$

Note that Problem (7) is a linear program and is solvable in polynomial time.

By a similar argument (detailed in our extended paper [10]) we can demonstrate that ρ_{II}^{+} can be computed in polynomial time. We conclude our results in this subsection by the following theorem:

Theorem 1. *Given any DTMC, $\rho_{\mathrm{I}}^{+}(\cdot)$, $\rho_{\mathrm{I}}^{-}(\cdot)$, $\rho_{\mathrm{II}}^{+}(\cdot)$, $\rho_{\mathrm{II}}^{-}(\cdot)$ can be computed in polynomial time.*

Remark 2. It is worth mentioning that Chen *et al.* [12] gave a thorough answer on the complexity of model checking IMCs against PCTL under both the UMC and the IMDP semantics. The main technique there is (a generalised version of) the ellipsoid algorithm for linear programming. Their approach can also be adopted here to tackle the problem for Type I and II distances. However, the technique exploited here (and by Perggelli *et al.* [24]) allows us to use off-the-shelf linear program solvers (e.g., Matlab), while the approach by Chen *et al.* requires more efforts in implementation. Furthermore, our extended paper [10] presents more practical algorithms based on a "value iteration" scheme from MDPs, which underpin the tool support.

3.2 Type III Distance

In this section, we focus on Type III distance given in Definition 4. We note that the technique employed in the previous subsection for Type I and II distances *cannot* be used here. Nevertheless, we still formulate the problem as an optimisation problem (with the same optimisation variables as in Problem (4) in the previous subsection):

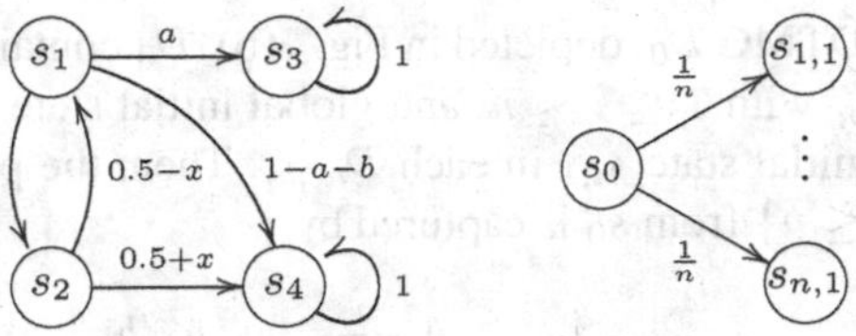

Fig. 1. Examples of DTMCs: (a) $\mathcal{D}_{a,b}$ and (b) $\mathcal{D}_0$

$$
\begin{aligned}
\text{maximise} \quad & \sum_{s\in S} \alpha(s) y_s \\
\text{subject to} \quad & y_s = 1 && \text{for } s \in G \\
& y_s = \sum_{t\in S} (\mathbf{P}(s,t) + x_{s,t}) \cdot y_t && \text{for } s \notin G \\
& \sum_{(s,t)\in C} |x_{s,t}| \le \delta \\
& \sum_{t\in C_s} x_{s,t} = 0 && \text{for all } s \in S \\
& 0 \le \mathbf{P}(s,t) + x_{s,t} \le 1 && \text{for each } (s,t) \in C
\end{aligned}
\tag{8}
$$

Clearly, Problem (8) is *not* a convex programming problem owing to the bilinear form $\sum_{t\in S}(\mathbf{P}(s,t) + x_{s,t}) \cdot y_t$. However, an obvious observation is that the corresponding decision problem, that is, deciding whether $\rho^+_{\text{III}}(\delta) \ge \theta$ for some given rational $\theta \in [0,1]$, can be formulated in the existential theory of reals. Since the decision problem of the existential theory of reals is in PSPACE [7], a PSPACE complexity upper bound follows.

Proposition 3. *The problem to decide whether* $\rho^+_{\text{III}}(\delta) \ge \theta$ *for a given rational* $\theta \in [0,1]$ *is in* PSPACE.

Below, we show that the computation of $\rho^+_{\text{III}}(\delta)$ is unlikely to admit a polynomial-time algorithm. Moreover, even the achievement of an NP upper bound (for its corresponding decision problem) is difficult. We illustrate this by an example.

Example 1. Consider a DTMC $\mathcal{D}_{a,b}$ depicted in Fig. 1(a). The transition matrix of $\mathcal{D}_{a,b}$ and its perturbed matrix are respectively as follows:

$$
\begin{bmatrix} 0 & b & a & 1-a-b \\ 0.5 & 0 & 0 & 0.5 \\ 0 & 0 & 1 & 0 \\ 0 & 0 & 0 & 1 \end{bmatrix} \quad \text{and} \quad \begin{bmatrix} 0 & b & a & 1-a-b \\ 0.5-x & 0 & 0 & 0.5+x \\ 0 & 0 & 1 & 0 \\ 0 & 0 & 0 & 1 \end{bmatrix}
$$

where $0 < a, b < a+b < 1$ and x ranges over $(-0.5, 0.5)$. Basic calculation reveals that the probability of reaching s_4 from s_1 is

$$
1 - \frac{a}{1 - b(0.5 - x)} = 1 - \frac{a/b}{1/b - 0.5 + x}.
$$

We construct another DTMC $\mathcal{D}_0$, depicted in Fig. 1(b). $\mathcal{D}_0$ contains n "copies" of $\mathcal{D}_{a,b}$, each denoted by $\mathcal{D}_{a_i,b_i}$ with $1 \le i \le n$, and global initial state s_0 that has $1/n$ probability to reach each initial state $s_{i,1}$ in each $\mathcal{D}_{a_i,b_i}$. Then, the probability of reaching states in $\{s_{i,4} \mid 1 \le i \le n\}$ from s_0 is captured by

$$p(x_1, \ldots, x_n) = 1 - \frac{1}{n} \sum_{i=1}^{n} \frac{a_i/b_i}{1/b_i - 0.5 + x_i}.$$

Let $\rho^+_{\text{III}}(2\delta)$ be the *maximum* of $p(\cdot)$ subject to $\sum_{i=1}^n |x_i| \le \delta$ and $x_i \in (-0.5, 0.5)$ for each $1 \le i \le n$. Essentially, to compute $\rho^+_{\text{III}}(2\delta)$, we need to *minimise*

$$1 - p(x_1, \ldots, x_n) = \frac{1}{n} \sum_{i=1}^{n} \frac{a_i/b_i}{1/b_i - 0.5 + x_i}$$

subject to the same constraints. Since $1 - p(\cdot)$ is a decreasing function, it is not hard to see that, in order to minimise it, each x_i must be non-negative and thus $\sum_{i=1}^n x_i = \delta$.

By the Cauchy–Schwarz inequality,

$$\sum_{i=1}^{n} (1/b_i - 0.5 + x_i) \cdot \sum_{i=1}^{n} \frac{a_i/b_i}{1/b_i - 0.5 + x_i} \ge \left(\sum_{i=1}^{n} \sqrt{\frac{a_i}{b_i}} \right)^2.$$

Namely, $\sum_{i=1}^n \frac{a_i/b_i}{1/b_i - 0.5 + x_i} \ge \left(\sum_{i=1}^n \sqrt{\frac{a_i}{b_i}} \right)^2 / (\sum_{i=1}^n 1/b_i - 0.5n + \delta)$. The equality in the above inequality holds if and only if

$$x_i = \frac{\sqrt{a_i/b_i}}{\sum_{i=1}^n \sqrt{a_i/b_i}} \cdot \left(\sum_{i=1}^{n} (1/b_i - 0.5) + \delta \right) - (1/b_i - 0.5) \ge 0.$$

Clearly, in general $\rho^+_{\text{III}}(2\delta)$ is *not* a rational number, and neither are x_i's. This example suggests the insight that in general one would not be able to use linear programming to compute $\rho^+_{\text{III}}(\cdot)$, which is in a sharp contrast to $\rho^+_{\text{I}}(\cdot)$ and $\rho^+_{\text{II}}(\cdot)$.

With a generalisation of Example 1, we can show that the SQUARE-ROOT-SUM (SRS) problem can be reduced to deciding whether $\rho^+_{\text{III}}(\delta) \ge \theta$ for some given rational $\theta \in [0, 1]$. An instance of the SRS problem is the decision of $\sum_{i=1}^n \sqrt{t_i} \le y$ for a given tuple $(t_1, \cdots, t_n, y)$ of natural numbers (greater than 1). The reduction is involved and is detailed in our extended paper [10].

Proposition 4. *Given a DTMC, deciding* $\rho^+_{\text{III}}(\delta) \ge \theta$ *for given* δ *and* θ *is* SQUARE-ROOT-SUM *hard.*

Using a similar construction, one can also show that computing $\rho^-_{\text{III}}(\cdot)$ is SRS hard. The SRS problem has been studied extensively, especially in computational geometry where the square root sum represents the sum of Euclidean distances between given pairs of points with integer/rational coordinates.[2] Allender *et al.* [1] showed that this

[2] For example, determining whether the length of a TSP tour of a set of points on the plane is bounded by a given threshold can be easily encoded as the SRS problem.

problem is decidable in the 4-th level of the Counting Hierarchy (an analogue of the polynomial-time hierarchy for counting classes); hence it is unlikely to be PSPACE-hard. But it remains open whether the problem can be decided in P or even in NP. Interesting examples that are related to formal verification can be studied by Etessami and Yannakakis [17], among others.

4 Computing Asymptotic Bounds

In this section, we consider the computation of asymptotic bounds. Recall that the reachability probability $p(\boldsymbol{x})$ (cf. (1)) is smooth, namely, infinitely differentiable. We present a unified characterisation for $\rho_\star^+$ and $\rho_\star^-$ with $\star \in \{\mathrm{I}, \mathrm{II}, \mathrm{III}\}$ using the Taylor expansion of $p(\boldsymbol{x})$. Define the following multi-variate index notations: Let $\iota = (\iota_1, \ldots, \iota_k)$ be a vector of integers. Let

$$|\iota| = \iota_1 + \ldots + \iota_k, \quad \iota! = \iota_1! \ldots \iota_k!, \quad \boldsymbol{x}^\iota = x_1^{\iota_1} \ldots x_k^{\iota_k}$$

$$\text{and} \quad \nabla^\iota p(\boldsymbol{x}) = \frac{\partial^{|\iota|} p(\boldsymbol{x})}{\partial x_1^{\iota_1} \cdots \partial x_k^{\iota_k}}.$$

Recall that $r = \min_{(s,t)\in C}\{\mathbf{P}(s,t), 1 - \mathbf{P}(s,t)\}$. For $n \in \mathbb{N}$, let $g_{\star,n}^+ : (0, r) \to \mathbf{R}$ such that $g_{\star,n}^+(\delta)$ is the solution of the following optimisation problem:

$$\begin{aligned} \text{maximise} \quad & \sum_{1\le|\iota|\le n} \frac{\nabla^\iota p(\mathbf{0})}{\iota!} \boldsymbol{x}^\iota \\ \text{subject to} \quad & \boldsymbol{x} \in \mathbf{U}_{\star,\delta} \end{aligned} \tag{9}$$

Theorem 2. *For each* $\star \in \{\mathrm{I}, \mathrm{II}, \mathrm{III}\}$, $g_{\star,n}^+(\cdot)$ *is an asymptotic bound of degree* n *for* $\rho_\star^+$.

An asymptotic bound of degree n for $\rho_\star^-$ can be obtained in a similar way as in Problem (9) by replacing *maximise* by *minimise*. We hence focus on the maximum case. The remainder of this section presents a method for computing $g_{\star,1}^+(\cdot)$ and $g_{\star,2}^+(\cdot)$, namely, the linear and quadratic bounds.

4.1 Linear Bounds

The linear bound of $\rho_\star^+$, $g_{\star,1}^+(\cdot)$, can be obtained by instantiating Problem (9) with $n = 1$. We show that $g_{\star,1}^+(\delta) = \kappa_\star\delta$, where $\kappa_\star$ is a solution of the following optimisation problem:

$$\begin{aligned} \text{maximise} \quad & \nabla p(\mathbf{0}) \cdot \boldsymbol{x} \\ \text{subject to} \quad & \sum_{t\in C_s} x_{s,t} = 0 \text{ for each } s \in S \\ & \|\boldsymbol{x}\|_\star = 1 \end{aligned} \tag{10}$$

We write $\nabla p(\mathbf{0})$ as $\boldsymbol{h} = (h_1, \cdots, h_k)$, which is of dimension $k = |C|$. Then $\boldsymbol{h}$ can be computed according to the following proposition.

Proposition 5. *For each* $1 \leq i \leq k$,

$$h_i = \alpha^T [I - \widetilde{\mathbf{P}}]^{-1} \widetilde{\mathbf{P}}_i [I - \widetilde{\mathbf{P}}]^{-1} b + \alpha^T [I - \widetilde{\mathbf{P}}]^{-1} \boldsymbol{b}_i$$

where $\widetilde{\mathbf{P}}_i$ *is the matrix on* $S_?$ *such that* $\widetilde{\mathbf{P}}_i(s,t) = 1$ *if* $\widetilde{\mathbf{P}}(\boldsymbol{x})(s,t)$ *contains* x_i *and 0 otherwise, and* $\mathbf{b}_i$ *is the vector on* $S_?$ *such that* $\mathbf{b}_i(s) = 1$ *if* $\boldsymbol{b}(\boldsymbol{x})(s)$ *contains* x_i *and 0 otherwise.*

When instantiated with the three types of norms (corresponding to the three types of distances, respectively), we obtain analytical solutions of Problem (10) for each $\star \in \{\mathrm{I}, \mathrm{II}, \mathrm{III}\}$.

Proposition 6. *The following statements hold:*

- *Let* $\boldsymbol{h}_s = (h_{s,t})_{t \in C_s}$ *(i.e.,* $\boldsymbol{h}$ *is a concatenation of* $\boldsymbol{h}_s$ *for each* $s \in S$*). We sort each* $\boldsymbol{h}_s$ *in non-decreasing order to get* $\boldsymbol{h}'_s = (h'_{s,1}, \ldots, h'_{s,k_s})$ *where* $k_s = |C_s|$*. Then*

$$\kappa_{\mathrm{I}} = \sum_{s \in S} \left(\sum_{1 \leq i \leq \lfloor k_s/2 \rfloor} h'_{s,i} - \sum_{\lceil k_s/2 \rceil + 1 \leq i \leq k_s} h'_{s,i} \right);$$

- $\kappa_{\mathrm{II}} = \sum_{s \in S} \frac{1}{2} (\max_{t \in C_s} \{h_{s,t}\} - \min_{t \in C_s} \{h_{s,t}\})$*; and*
- $\kappa_{\mathrm{III}} = \max_{s \in S} \frac{1}{2} (\max_{t \in C_s} \{h_{s,t}\} - \min_{t \in C_s} \{h_{s,t}\})$

As $\kappa_\star$ for each $\star \in \{\mathrm{I}, \mathrm{II}, \mathrm{III}\}$ is nonnegative, Theorem 2 immediately implies the following theorem.

Theorem 3. *For each* $\star \in \{\mathrm{I}, \mathrm{II}, \mathrm{III}\}$, $\kappa_\star \delta$ *is a linear bound for* $\rho_\star^+$.

Essentially, computing $\kappa_\star$ boils down to computing an inverse matrix, which can be done by Gaussian elimination. Hence we have

Proposition 7. *The problem of computing linear bounds is in* $\mathcal{O}(|\mathcal{D}|^3)$.

Remark 3. We can show that computing $\kappa_\star$ can be done in GapL, which concerns logspace-bounded computation. In a nutshell, $\sharp L$ is defined, in analogy to $\sharp P$, to be the set of functions that count the number of accepting computation paths of a nondeterministic logspace-bounded Turing machine. The class GapL is defined by Allender and Ogihara [2] to be the closure of $\sharp L$ under subtraction. Furthermore, the decision version of computing $\kappa_\star$ is in PL (probabilistic logspace), and the technical details can be found in the extended version of our paper [10].

4.2 Quadratic Bounds

Similar to the linear case, we can instantiate Problem (9) to obtain the quadratic bound $g_{\star,2}^+$. However, it is usually inefficient to solve Problem (9) for every given δ. Instead, we show that there exists a uniform direction vector for $\boldsymbol{x}$ for all sufficiently small δ, along which the quadratic bound is obtained. For this purpose, we consider an alternative optimisation problem:

$$\begin{aligned} \text{maximise} \quad & \sum_{|\iota|=2} \frac{\nabla^{\iota} p(\mathbf{0})}{\iota!} \boldsymbol{x}^{\iota} \\ \text{subject to} \quad & \sum_{t \in C_s} x_{s,t} = 0 \text{ for each } s \in S \\ & \|\boldsymbol{x}\|_{\star} = 1 \text{ and } \boldsymbol{h} \cdot \boldsymbol{x} = \kappa_{\star} \end{aligned} \tag{11}$$

For each $\star \in \{\text{I}, \text{II}, \text{III}\}$, let $v_\star$ be the solution of Problem (11). The following technical result states that the coefficient of the linear term of $g^{+}_{\star,2}$ is exactly $\kappa_\star$, and the coefficient of the quadratic term of $g^{+}_{\star,2}$ is $v_\star$ (obtained by solving Problem (11)).

Theorem 4. *For each $\star \in \{\text{I}, \text{II}, \text{III}\}$, the function $v_\star \delta^2 + \kappa_\star \delta$ is a quadratic bound for $\rho^{+}_{\star}$.*

To compute the quadratic bounds, we must solve a quadratic program, which is known to be NP-complete [26]. The following result is rather straightforward.

Proposition 8. *The problem of computing quadratic bounds is in* TFNP *(namely Total Function* NP*).*

5 Conclusion

In this paper, we have performed an in-depth study on perturbation analysis in the setting of DTMC verification. We defined non-asymptotic and asymptotic perturbation bounds and focused on their computation. In particular, we considered three fundamental matrix norms for stochastic matrices to quantify perturbations of DTMCs. With respect to these distances, we presented algorithms and complexity analysis for computing the non-asymptotic and asymptotic perturbation bounds.

An ongoing work is to generalise the results for continuous-time models, MDPs, stochastic games, etc. Moreover, in the current paper, we only consider structurally equivalent perturbations, but it would be interesting to see how to relax this assumption.

References

1. Allender, E., Bürgisser, P., Kjeldgaard-Pedersen, J., Miltersen, P.B.: On the complexity of numerical analysis. SIAM J. Comput. 38(5), 1987–2006 (2009)
2. Allender, E., Ogihara, M.: Relationships among PL, #L, and the determinant. ITA 30(1), 1–21 (1996)
3. Baier, C., Katoen, J.-P.: Principles of Model Checking. MIT Press (2008)
4. Bartocci, E., Grosu, R., Katsaros, P., Ramakrishnan, C.R., Smolka, S.A.: Model repair for probabilistic systems. In: Abdulla, P.A., Leino, K.R.M. (eds.) TACAS 2011. LNCS, vol. 6605, pp. 326–340. Springer, Heidelberg (2011)
5. Benedikt, M., Lenhardt, R., Worrell, J.: LTL model checking of Interval Markov Chains. In: Piterman, N., Smolka, S.A. (eds.) TACAS 2013. LNCS, vol. 7795, pp. 32–46. Springer, Heidelberg (2013)

6. Bianco, A., de Alfaro, L.: Model checking of probabalistic and nondeterministic systems. In: Thiagarajan, P.S. (ed.) FSTTCS 1995. LNCS, vol. 1026, pp. 499–513. Springer, Heidelberg (1995)
7. Canny, J.F.: Some algebraic and geometric computations in PSPACE. In: Simon, J. (ed.) STOC, pp. 460–467. ACM (1988)
8. Chatterjee, K.: Robustness of structurally equivalent concurrent parity games. In: Birkedal, L. (ed.) FOSSACS 2012. LNCS, vol. 7213, pp. 270–285. Springer, Heidelberg (2012)
9. Chatterjee, K., Sen, K., Henzinger, T.A.: Model-checking ω-regular properties of Interval Markov Chains. In: Amadio, R. (ed.) FOSSACS 2008. LNCS, vol. 4962, pp. 302–317. Springer, Heidelberg (2008)
10. Chen, T., Feng, Y., Rosenblum, D.S., Su, G.: Perturbation analysis in verification of discrete-time Markov chains. Technical report, Middlesex University London (2014), `http://www.cs.mdx.ac.uk/staffpages/taoluechen/pub-papers/concur14-full.pdf`
11. Chen, T., Hahn, E., Han, T., Kwiatkowska, M., Qu, H., Zhang, L.: Model repair for Markov decision processes. In: TASE 2013, pp. 85–92 (July 2013)
12. Chen, T., Han, T., Kwiatkowska, M.Z.: On the complexity of model checking interval-valued discrete time Markov chains. Inf. Process. Lett. 113(7), 210–216 (2013)
13. Cho, G.E., Meyer, C.D.: Comparison of perturbation bounds for the stationary distribution of a Markov chain. Linear Algebra Appl. 335, 137–150 (2000)
14. Daws, C.: Symbolic and parametric model checking of discrete-time Markov chains. In: Liu, Z., Araki, K. (eds.) ICTAC 2004. LNCS, vol. 3407, pp. 280–294. Springer, Heidelberg (2005)
15. Desharnais, J., Gupta, V., Jagadeesan, R., Panangaden, P.: Metrics for labelled Markov processes. Theor. Comput. Sci. 318(3), 323–354 (2004)
16. Donaldson, R., Gilbert, D.: A model checking approach to the parameter estimation of biochemical pathways. In: Heiner, M., Uhrmacher, A.M. (eds.) CMSB 2008. LNCS (LNBI), vol. 5307, pp. 269–287. Springer, Heidelberg (2008)
17. Etessami, K., Yannakakis, M.: Recursive Markov chains, stochastic grammars, and monotone systems of nonlinear equations. J. ACM 56(1) (2009)
18. Filieri, A., Ghezzi, C., Tamburrelli, G.: Run-time efficient probabilistic model checking. In: ICSE 2011, New York, NY, USA, pp. 341–350 (2011)
19. Hahn, E., Hermanns, H., Zhang, L.: Probabilistic reachability for parametric Markov models. International Journal on Software Tools for Technology Transfer 13(1), 3–19 (2011)
20. Hahn, E.M., Han, T., Zhang, L.: Synthesis for PCTL in parametric Markov decision processes. In: Bobaru, M., Havelund, K., Holzmann, G.J., Joshi, R. (eds.) NFM 2011. LNCS, vol. 6617, pp. 146–161. Springer, Heidelberg (2011)
21. Han, T., Katoen, J.-P., Mereacre, A.: Approximate parameter synthesis for probabilistic time-bounded reachability. In: RTSS, pp. 173–182. IEEE Computer Society (2008)
22. Kwiatkowska, M., Norman, G., Parker, D.: PRISM 4.0: Verification of probabilistic real-time systems. In: Gopalakrishnan, G., Qadeer, S. (eds.) CAV 2011. LNCS, vol. 6806, pp. 585–591. Springer, Heidelberg (2011)
23. Murdock, J.A.: Perturbation: Theory and Method. John Wiley & Sons, Inc. (1991)
24. Puggelli, A., Li, W., Sangiovanni-Vincentelli, A.L., Seshia, S.A.: Polynomial-time verification of PCTL properties of MDPs with convex uncertainties. In: Sharygina, N., Veith, H. (eds.) CAV 2013. LNCS, vol. 8044, pp. 527–542. Springer, Heidelberg (2013)
25. Puterman, M.L.: Markov Decision Processes: Discrete Stochastic Dynamic Programming. Wiley, New York (1994)
26. Sahni, S.: Computationally related problems. SIAM Journal on Computing 3, 262–279 (1974)

27. Schweitzer, P.J.: Perturbation theory and finite Markov chains. Journal of Applied Probability 5(2), 401–413 (1968)
28. Sen, K., Viswanathan, M., Agha, G.: Model-checking Markov chains in the presence of uncertainties. In: Hermanns, H., Palsberg, J. (eds.) TACAS 2006. LNCS, vol. 3920, pp. 394–410. Springer, Heidelberg (2006)
29. Solan, E., Vieille, N.: Perturbed Markov chains. J. Applied Prob. 40(1), 107–122 (2003)
30. Su, G., Rosenblum, D.S.: Asymptotic bounds for quantitative verification of perturbed probabilistic systems. In: Groves, L., Sun, J. (eds.) ICFEM 2013. LNCS, vol. 8144, pp. 297–312. Springer, Heidelberg (2013)
31. Su, G., Rosenblum, D.S.: Perturbation analysis of stochastic systems with empirical distribution parameters. In: ICSE, pp. 311–321. ACM (2014)

Robust Synchronization in Markov Decision Processes*,**

Laurent Doyen[1], Thierry Massart[2], and Mahsa Shirmohammadi[1,2]

[1] LSV, ENS Cachan & CNRS, France
[2] Université Libre de Bruxelles, Belgium

Abstract. We consider synchronizing properties of Markov decision processes (MDP), viewed as generators of sequences of probability distributions over states. A probability distribution is p-synchronizing if the probability mass is at least p in some state, and a sequence of probability distributions is weakly p-synchronizing, or strongly p-synchronizing if respectively infinitely many, or all but finitely many distributions in the sequence are p-synchronizing.

For each synchronizing mode, an MDP can be (*i*) *sure* winning if there is a strategy that produces a 1-synchronizing sequence; (*ii*) *almost-sure* winning if there is a strategy that produces a sequence that is, for all $\varepsilon > 0$, a $(1\text{-}\varepsilon)$-synchronizing sequence; (*iii*) *limit-sure* winning if for all $\varepsilon > 0$, there is a strategy that produces a $(1\text{-}\varepsilon)$-synchronizing sequence.

For each synchronizing and winning mode, we consider the problem of deciding whether an MDP is winning, and we establish matching upper and lower complexity bounds of the problems, as well as the optimal memory requirement for winning strategies: (a) for all winning modes, we show that the problems are PSPACE-complete for weak synchronization, and PTIME-complete for strong synchronization; (b) we show that for weak synchronization, exponential memory is sufficient and may be necessary for sure winning, and infinite memory is necessary for almost-sure winning; for strong synchronization, linear-size memory is sufficient and may be necessary in all modes; (c) we show a robustness result that the almost-sure and limit-sure winning modes coincide for both weak and strong synchronization.

1 Introduction

Markov Decision Processes (MDPs) are studied in theoretical computer science in many problems related to system design and verification [22,15,10]. MDPs are a model of reactive systems with both stochastic and nondeterministic behavior, used in the control problem for reactive systems: the nondeterminism represents the possible choices of the controller, and the stochasticity represents

* This work was partially supported by the Belgian Fonds National de la Recherche Scientifique (FNRS), and by the PICS project *Quaverif* funded by the French Centre National de la Recherche Scientifique (CNRS).

** Fuller version: [1].

P. Baldan and D. Gorla (Eds.): CONCUR 2014, LNCS 8704, pp. 234–248, 2014.

the uncertainties about the system response. The controller synthesis problem is to compute a control strategy that ensures correct behaviors of the system with probability 1. Traditional well-studied specifications describe correct behaviors as infinite sequences of states, such as reachability, Büchi, and co-Büchi, which require the system to visit a target state once, infinitely often, and ultimately always, respectively [3,4].

In contrast, we consider symbolic specifications of the behaviors of MDPs as sequences of probability distributions $X_i : Q \to [0,1]$ over the finite state space Q of the system, where $X_i(q)$ is the probability that the MDP is in state $q \in Q$ after i steps. The symbolic specification of stochastic systems is relevant in applications such as system biology and robot planning [6,14,17], and recently it has been used in several works on design and verification of reactive systems [2,9,20]. While the verification of MDPs may yield undecidability, both with traditional specifications [5,16], and symbolic specifications [20,13], decidability results are obtained for *eventually synchronizing* conditions under general control strategies that depend on the full history of the system execution [14]. Intuitively, a sequence of probability distributions is eventually synchronizing if the probability mass tends to accumulate in a given set of target states along the sequence. This is an analogue, for sequences of probability distributions, of the reachability condition.

In this paper, we consider an analogue of the Büchi and coBüchi conditions for sequences of distributions [12,11]: the probability mass should get synchronized infinitely often, or ultimately at every step. More precisely, for $0 \leq p \leq 1$ let a probability distribution $X : Q \to [0,1]$ be p-synchronized if it assigns probability at least p to some state. A sequence $\bar{X} = X_0 X_1 \dots$ of probability distributions is (*a*) *eventually p-synchronizing* if X_i is p-synchronized for some i; (*b*) *weakly p-synchronizing* if X_i is p-synchronized for infinitely many i's; (*c*) *strongly p-synchronizing* if X_i is p-synchronized for all but finitely many i's. It is easy to see that strongly p-synchronizing implies weakly p-synchronizing, which implies eventually p-synchronizing. The qualitative synchronizing properties, corresponding to the case where either $p = 1$, or p tends to 1, are analogous to the traditional reachability, Büchi, and coBüchi conditions.

We consider the following qualitative (winning) modes, summarized in Table 1: (*i*) *sure* winning, if there is a strategy that generates a {eventually, weakly, strongly} 1-synchronizing sequence; (*ii*) *almost-sure* winning, if there is a strategy that generates a sequence that is, for all $\varepsilon > 0$, {eventually, weakly, strongly} $(1-\varepsilon)$-synchronizing; (*iii*) *limit-sure* winning, if for all $\varepsilon > 0$, there is a strategy that generates a {eventually, weakly, strongly} $(1-\varepsilon)$-synchronizing sequence.

For eventually synchronizing deciding if a given MDP is winning is PSPACE-complete, and the three winning modes form a strict hierarchy [14]. In particular, there are limit-sure winning MDPs that are not almost-sure winning. An important and difficult result in this paper is that the new synchronizing modes are more robust: for weak and strong synchronization, we show that the almost-sure and limit-sure modes coincide. Moreover we establish the complexity of deciding if a given MDP is winning by providing tight (matching) upper and lower bounds:

Table 1. Winning modes and synchronizing objectives (where $\mathcal{M}_n^\alpha(T)$ denotes the probability that under strategy α, after n steps the MDP $\mathcal{M}$ is in a state of T)

	Eventually	Weakly	Strongly
Sure	$\exists\alpha \quad \exists n \ \mathcal{M}_n^\alpha(T) = 1$	$\exists\alpha \quad \forall N \, \exists n \geq N \ \mathcal{M}_n^\alpha(T) = 1$	$\exists\alpha \quad \exists N \, \forall n \geq N \ \mathcal{M}_n^\alpha(T) = 1$
Almost-sure	$\exists\alpha \quad \sup_n \mathcal{M}_n^\alpha(T) = 1$	$\exists\alpha \quad \limsup_{n\to\infty} \mathcal{M}_n^\alpha(T) = 1$	$\exists\alpha \quad \liminf_{n\to\infty} \mathcal{M}_n^\alpha(T) = 1$
Limit-sure	$\sup_\alpha \sup_n \mathcal{M}_n^\alpha(T) = 1$	$\sup_\alpha \limsup_{n\to\infty} \mathcal{M}_n^\alpha(T) = 1$	$\sup_\alpha \liminf_{n\to\infty} \mathcal{M}_n^\alpha(T) = 1$

for each winning mode we show that the problems are PSPACE-complete for weak synchronization, and PTIME-complete for strong synchronization.

Thus the weakly and strongly synchronizing properties provide conservative approximations of eventually synchronizing, they are robust (limit-sure and almost-sure coincide), and they are of the same (or even lower) complexity as compared to eventually synchronizing.

We also provide optimal memory bounds for winning strategies: exponential memory is sufficient and may be necessary for sure winning in weak synchronization, infinite memory is necessary for almost-sure winning in weak synchronization, and linear memory is sufficient for strong synchronization in all winning modes. We present a variant of strong synchronization for which memoryless strategies are sufficient.

Related works and applications. Synchronization problems were first considered for deterministic finite automata (DFA) where a *synchronizing word* is a finite sequence of control actions that can be executed from any state of an automaton and leads to the same state (see [23] for a survey of results and applications). While the existence of a synchronizing word can be decided in polynomial time for DFA, extensive research efforts are devoted to establishing a tight bound on the length of the shortest synchronizing word, which is conjectured to be $(n-1)^2$ for automata with n states [8]. Various extensions of the notion of synchronizing word have been proposed for non-deterministic and probabilistic automata [7,18,19,12], leading to results of PSPACE-completeness [21], or even undecidability [19,13].

For probabilistic systems, a natural extension of words is the notion of strategy that reacts and chooses actions according to the sequence of states visited along the system execution. In this context, an input word corresponds to the special case of a blind strategy that chooses the control actions in advance. In particular, almost-sure weak and strong synchronization with blind strategies has been studied [12] and the main result is the undecidability of deciding the existence of a blind almost-sure winning strategy for weak synchronization, and the PSPACE-completeness of the emptiness problem for strong synchronization [11,13]. In contrast, for general strategies (which also correspond to input trees), we establish the PSPACE-completeness and PTIME-completeness of deciding almost-sure weak and strong synchronization respectively.

A typical application scenario is the design of a control program for a group of mobile robots running in a stochastic environment. The possible behaviors of the robots and the stochastic response of the environment (such as obstacle encounters) are represented by an MDP, and a synchronizing strategy corresponds to a control program that can be embedded in every robot to ensure that they meet (or synchronize) eventually once, infinitely often, or eventually forever.

2 Markov Decision Processes and Synchronization

We closely follow the definitions of [14]. A *probability distribution* over a finite set S is a function $d : S \to [0,1]$ such that $\sum_{s \in S} d(s) = 1$. The *support* of d is the set $\mathsf{Supp}(d) = \{s \in S \mid d(s) > 0\}$. We denote by $\mathcal{D}(S)$ the set of all probability distributions over S. Given a set $T \subseteq S$, let $d(T) = \sum_{s \in T} d(s)$ and $\|d\|_T = max_{s \in T} d(s)$. For $T \neq \emptyset$, the *uniform distribution* on T assigns probability $\frac{1}{|T|}$ to every state in T. Given $s \in S$, the *Dirac distribution* on s assigns probability 1 to s, and by a slight abuse of notation, we denote it simply by s.

A *Markov decision process* (MDP) is a tuple $\mathcal{M} = \langle Q, \mathsf{A}, \delta \rangle$ where Q is a finite set of states, A is a finite set of actions, and $\delta : Q \times \mathsf{A} \to \mathcal{D}(Q)$ is a probabilistic transition function. A state q is *absorbing* if $\delta(q,a)$ is the Dirac distribution on q for all actions $a \in \mathsf{A}$.

Given state $q \in Q$ and action $a \in \mathsf{A}$, the successor state of q under action a is q' with probability $\delta(q,a)(q')$. Denote by $\mathsf{post}(q,a)$ the set $\mathsf{Supp}(\delta(q,a))$, and given $T \subseteq Q$ let $\mathsf{Pre}(T) = \{q \in Q \mid \exists a \in \mathsf{A} : \mathsf{post}(q,a) \subseteq T\}$ be the set of states from which there is an action to ensure that the successor state is in T. For $k > 0$, let $\mathsf{Pre}^k(T) = \mathsf{Pre}(\mathsf{Pre}^{k-1}(T))$ with $\mathsf{Pre}^0(T) = T$.

A *path* in $\mathcal{M}$ is an infinite sequence $\pi = q_0 a_0 q_1 a_1 \ldots$ such that $q_{i+1} \in \mathsf{post}(q_i, a_i)$ for all $i \geq 0$. A finite prefix $\rho = q_0 a_0 q_1 a_1 \ldots q_n$ of a path (or simply a finite path) has length $|\rho| = n$ and last state $\mathsf{Last}(\rho) = q_n$. We denote by $\mathsf{Play}(\mathcal{M})$ and $\mathsf{Pref}(\mathcal{M})$ the set of all paths and finite paths in $\mathcal{M}$ respectively.

Strategies. A *randomized strategy* for $\mathcal{M}$ (or simply a strategy) is a function $\alpha : \mathsf{Pref}(\mathcal{M}) \to \mathcal{D}(\mathsf{A})$ that, given a finite path ρ, returns a probability distribution $\alpha(\rho)$ over the action set, used to select a successor state q' of ρ with probability $\sum_{a \in \mathsf{A}} \alpha(\rho)(a) \cdot \delta(q,a)(q')$ where $q = \mathsf{Last}(\rho)$.

A strategy α is *pure* if for all $\rho \in \mathsf{Pref}(\mathcal{M})$, there exists an action $a \in \mathsf{A}$ such that $\alpha(\rho)(a) = 1$; and *memoryless* if $\alpha(\rho) = \alpha(\rho')$ for all ρ, ρ' such that $\mathsf{Last}(\rho) = \mathsf{Last}(\rho')$. Finally, a strategy α uses *finite-memory* if there exists a right congruence $\approx$ over $\mathsf{Pref}(\mathcal{M})$ (i.e., if $\rho \approx \rho'$, then $\rho \cdot a \cdot q \approx \rho' \cdot a \cdot q$ for all $\rho, \rho' \in \mathsf{Pref}(\mathcal{M})$ and $a \in \mathsf{A}$, $q \in Q$) of finite index such that $\rho \approx \rho'$ implies $\alpha(\rho) = \alpha(\rho')$. The index of $\approx$ is the *memory size* of the strategy.

Outcomes and Winning Modes. Given an initial distribution $d_0 \in \mathcal{D}(Q)$ and a strategy α in an MDP $\mathcal{M}$, a *path-outcome* is a path $\pi = q_0 a_0 q_1 a_1 \ldots$ in $\mathcal{M}$ such that $q_0 \in \mathsf{Supp}(d_0)$ and $a_i \in \mathsf{Supp}(\alpha(q_0 a_0 \ldots q_i))$ for all $i \geq 0$. The probability of a finite prefix $\rho = q_0 a_0 q_1 a_1 \ldots q_n$ of π is $d_0(q_0) \cdot \prod_{j=0}^{n-1} \alpha(q_0 a_0 \ldots q_j)(a_j) \cdot \delta(q_j, a_j)(q_{j+1})$. We denote by $Outcomes(d_0, \alpha)$ the set of all path-outcomes from

d_0 under strategy α. An *event* $\Omega \subseteq \mathsf{Play}(\mathcal{M})$ is a measurable set of paths, and given an initial distribution d_0 and a strategy α, the probability $Pr^{\alpha}(\Omega)$ of Ω is uniquely defined [22]. We consider the following classical winning modes. Given an initial distribution d_0 and an event Ω, we say that $\mathcal{M}$ is: *sure winning* if there exists a strategy α such that $Outcomes(d_0, \alpha) \subseteq \Omega$; *almost-sure winning* if there exists a strategy α such that $\Pr^{\alpha}(\Omega) = 1$; and *limit-sure winning* if $\sup_{\alpha} \Pr^{\alpha}(\Omega) = 1$.

For example, given a set $T \subseteq Q$ of target states, and $k \in \mathbb{N}$, we denote by $\Box T = \{q_0 a_0 q_1 \cdots \in \mathsf{Play}(\mathcal{M}) \mid \forall i : q_i \in T\}$ the safety event of always staying in T, by $\Diamond T = \{q_0 a_0 q_1 \cdots \in \mathsf{Play}(\mathcal{M}) \mid \exists i : q_i \in T\}$ the event of reaching T, and by $\Diamond^k T = \{q_0 a_0 q_1 \cdots \in \mathsf{Play}(\mathcal{M}) \mid q_k \in T\}$ the event of reaching T after exactly k steps. Hence, if $\Pr^{\alpha}(\Diamond T) = 1$ then almost-surely a state in T is reached under strategy α.

We consider a symbolic outcome of MDPs viewed as generators of sequences of probability distributions over states [20]. Given an initial distribution $d_0 \in \mathcal{D}(Q)$ and a strategy α in $\mathcal{M}$, the *symbolic outcome* of $\mathcal{M}$ from d_0 is the sequence $(\mathcal{M}_n^{\alpha})_{n \in \mathbb{N}}$ of probability distributions defined by $\mathcal{M}_k^{\alpha}(q) = Pr^{\alpha}(\Diamond^k \{q\})$ for all $k \geq 0$ and $q \in Q$. Hence, $\mathcal{M}_k^{\alpha}$ is the probability distribution over states after k steps under strategy α. Note that $\mathcal{M}_0^{\alpha} = d_0$ and the symbolic outcome is a deterministic sequence of distributions: each distribution $\mathcal{M}_k^{\alpha}$ has a unique (deterministic) successor.

Informally, synchronizing objectives require that the probability of a given state (or some group of states) tends to 1 in the sequence $(\mathcal{M}_n^{\alpha})_{n \in \mathbb{N}}$, either once, infinitely often, or always after some point. Given a set $T \subseteq Q$, consider the functions $sum_T : \mathcal{D}(Q) \to [0, 1]$ and $max_T : \mathcal{D}(Q) \to [0, 1]$ that compute $sum_T(X) = \sum_{q \in T} X(q)$ and $max_T(X) = \max_{q \in T} X(q)$. For $f \in \{sum_T, max_T\}$ and $p \in [0, 1]$, we say that a probability distribution X is p-synchronized according to f if $f(X) \geq p$, and that a sequence $\bar{X} = X_0 X_1 \dots$ of probability distributions is [12,11,14]:

(a) *event* (or *eventually*) *p-synchronizing* if X_i is p-synchronized for some $i \geq 0$;
(b) *weakly p-synchronizing* if X_i is p-synchronized for infinitely many i's;
(c) *strongly p-synchronizing* if X_i is p-synchronized for all but finitely many i's.

For $p = 1$, these definitions are analogous to the traditional reachability, Büchi, and coBüchi conditions [3], and the following winning modes can be considered [14]: given an initial distribution d_0 and a function $f \in \{sum_T, max_T\}$, we say that for the objective of {eventually, weak, strong} synchronization from d_0, $\mathcal{M}$ is:

- *sure winning* if there exists a strategy α such that the symbolic outcome of α from d_0 is {eventually, weakly, strongly} 1-synchronizing according to f;
- *almost-sure winning* if there exists a strategy α such that for all $\varepsilon > 0$ the symbolic outcome of α from d_0 is {eventually, weakly, strongly} $(1-\varepsilon)$-synchronizing according to f;
- *limit-sure winning* if for all $\varepsilon > 0$, there exists a strategy α such that the symbolic outcome of α from d_0 is {eventually, weakly, strongly} $(1-\varepsilon)$-synchronizing according to f;

Table 2. Computational complexity of the membership problem (new results in boldface)

	Eventually	Weakly	Strongly
Sure	PSPACE-C [14]	**PSPACE-C**	**PTIME-C**
Almost-sure	PSPACE-C [14]	**PSPACE-C**	**PTIME-C**
Limit-sure	PSPACE-C [14]		

Note that the winning modes for synchronization objectives differ from the classical winning modes in MDPs: they can be viewed as a specification of the set of sequences of distributions that are winning in a non-stochastic system (since the symbolic outcome is deterministic), while the traditional almost-sure and limit-sure winning modes for path-outcomes consider a probability measure over paths and specify the probability of a specific event (i.e., a set of paths). Thus for instance a strategy is almost-sure synchronizing if the (single) symbolic outcome it produces belongs to the corresponding winning set, whereas traditional almost-sure winning requires a certain event to occur with probability 1.

We often write $\|X\|_T$ instead of $max_T(X)$ (and we omit the subscript when $T = Q$) and $X(T)$ instead of $sum_T(X)$, as in Table 1 where the definitions of the various winning modes and synchronizing objectives for $f = sum_T$ are summarized.

Decision problems. For $f \in \{sum_T, max_T\}$ and $\lambda \in \{event, weakly, strongly\}$, the *winning region* $\langle\langle 1 \rangle\rangle^{\lambda}_{sure}(f)$ is the set of initial distributions such that $\mathcal{M}$ is sure winning for λ-synchronizing (we assume that $\mathcal{M}$ is clear from the context). We define analogously the sets $\langle\langle 1 \rangle\rangle^{\lambda}_{almost}(f)$ and $\langle\langle 1 \rangle\rangle^{\lambda}_{limit}(f)$. For a singleton $T = \{q\}$ we have $sum_T = max_T$, and we simply write $\langle\langle 1 \rangle\rangle^{\lambda}_{\mu}(q)$ (where $\mu \in \{sure, almost, limit\}$). It follows from the definitions that $\langle\langle 1 \rangle\rangle^{strongly}_{\mu}(f) \subseteq \langle\langle 1 \rangle\rangle^{weakly}_{\mu}(f) \subseteq \langle\langle 1 \rangle\rangle^{event}_{\mu}(f)$ and thus strong and weak synchronization are conservative approximations of eventually synchronization. It is easy to see that $\langle\langle 1 \rangle\rangle^{\lambda}_{sure}(f) \subseteq \langle\langle 1 \rangle\rangle^{\lambda}_{almost}(f) \subseteq \langle\langle 1 \rangle\rangle^{\lambda}_{limit}(f)$, and for $\lambda = event$ the inclusions are strict [14]. In contrast, weak and strong synchronization are more robust as we show in this paper that the almost-sure and limit-sure winning modes coincide.

Lemma 1. *There exists an MDP $\mathcal{M}$ and state q such that $\langle\langle 1 \rangle\rangle^{\lambda}_{sure}(q) \subsetneq \langle\langle 1 \rangle\rangle^{\lambda}_{almost}(q)$ for $\lambda \in \{weakly, strongly\}$.*

The *membership problem* is to decide, given an initial probability distribution d_0, whether $d_0 \in \langle\langle 1 \rangle\rangle^{\lambda}_{\mu}(f)$. It is sufficient to consider Dirac initial distributions (i.e., assuming that MDPs have a single initial state) because the answer to the general membership problem for an MDP $\mathcal{M}$ with initial distribution d_0 can be obtained by solving the membership problem for a copy of $\mathcal{M}$ with a new initial state from which the successor distribution on all actions is d_0.

For eventually synchronizing, the membership problem is PSPACE-complete for all winning modes [14]. In this paper, we show that the complexity of the

Table 3. Memory requirement (new results in boldface)

	Eventually	Weakly	Strongly	
			sum_T	max_T
Sure	exponential [14]	**exponential**	**memoryless**	**linear**
Almost-sure	infinite [14]	**infinite**	**memoryless**	**linear**
Limit-sure	unbounded [14]			

membership problem is PSPACE-complete for weak synchronization, and even PTIME-complete for strong synchronization. The complexity results are summarized in Table 2, and we present the memory requirement for winning strategies in Table 3.

3 Weak Synchronization

We establish the complexity and memory requirement for weakly synchronizing objectives. We show that the membership problem is PSPACE-complete for sure and almost-sure winning, that exponential memory is necessary and sufficient for sure winning while infinite memory is necessary for almost-sure winning, and we show that limit-sure and almost-sure winning coincide.

3.1 Sure Weak Synchronization

The PSPACE upper bound of the membership problem for sure weak synchronization is obtained by the following characterization.

Lemma 2. *Let $\mathcal{M}$ be an MDP and T be a target set. For all states q_{init}, we have $q_{\mathsf{init}} \in \langle\langle 1 \rangle\rangle^{weakly}_{sure}(sum_T)$ if and only if there exists a set $S \subseteq T$ such that $q_{\mathsf{init}} \in \mathsf{Pre}^m(S)$ for some $m \geq 0$ and $S \subseteq \mathsf{Pre}^n(S)$ for some $n \geq 1$.*

The PSPACE upper bound follows from the characterization in Lemma 2. A (N)PSPACE algorithm is to guess the set $S \subseteq T$, and the numbers m, n (with $m, n \leq 2^{|Q|}$ since the sequence $\mathsf{Pre}^n(S)$ of predecessors is ultimately periodic), and check that $q_{\mathsf{init}} \in \mathsf{Pre}^m(S)$ and $S \subseteq \mathsf{Pre}^n(S)$. The PSPACE lower bound follows from the PSPACE-completeness of the membership problem for sure eventually synchronization [14, Theorem 2].

Lemma 3. *The membership problem for $\langle\langle 1 \rangle\rangle^{weakly}_{sure}(sum_T)$ is PSPACE-hard even if T is a singleton.*

The proof of Lemma 2 suggests an exponential-memory strategy for sure weak synchronization that in $q \in \mathsf{Pre}^n(S)$ plays an action a such that $\mathsf{post}(q, a) \subseteq \mathsf{Pre}^{n-1}(S)$, which can be realized with exponential memory since $n \leq 2^{|Q|}$. It can be shown that exponential memory is necessary in general.

Theorem 1. *For sure weak synchronization in MDPs:*

1. *(Complexity). The membership problem is PSPACE-complete.*
2. *(Memory). Exponential memory is necessary and sufficient for both pure and randomized strategies, and pure strategies are sufficient.*

3.2 Almost-Sure Weak Synchronization

We present a characterization of almost-sure weak synchronization that gives a PSPACE upper bound for the membership problem. Our characterization uses the limit-sure eventually synchronizing objectives *with exact support* [14]. This objective requires that the probability mass tends to 1 in a target set T, and moreover that after the same number of steps the support of the probability distribution is contained in a given set U. Formally, given an MDP $\mathcal{M}$, let $\langle\langle 1\rangle\rangle^{event}_{limit}(sum_T, U)$ for $T \subseteq U$ be the set of all initial distributions such that for all $\varepsilon > 0$ there exists a strategy α and $n \in \mathbb{N}$ such that $\mathcal{M}^{\alpha}_{n}(T) \geq 1 - \varepsilon$ and $\mathcal{M}^{\alpha}_{n}(U) = 1$.

We show that an MDP is almost-sure weakly synchronizing in target T if (and only if), for some set U, there is a sure eventually synchronizing strategy in target U, and from the probability distributions with support U there is a limit-sure winning strategy for eventually synchronizing in $\mathsf{Pre}(T)$ with support in $\mathsf{Pre}(U)$. This ensures that from the initial state we can have the whole probability mass in U, and from U have probability $1 - \varepsilon$ in $\mathsf{Pre}(T)$ (and in T in the next step), while the whole probability mass is back in $\mathsf{Pre}(U)$ (and in U in the next step), allowing to repeat the strategy for $\varepsilon \to 0$, thus ensuring infinitely often probability at least $1 - \varepsilon$ in T (for all $\varepsilon > 0$).

Lemma 4. *Let $\mathcal{M}$ be an MDP and T be a target set. For all states q_{init}, we have $q_{\mathsf{init}} \in \langle\langle 1\rangle\rangle^{weakly}_{almost}(sum_T)$ if and only if there exists a set U such that*

- *$q_{\mathsf{init}} \in \langle\langle 1\rangle\rangle^{event}_{sure}(sum_U)$, and*
- *$d_U \in \langle\langle 1\rangle\rangle^{event}_{limit}(sum_{\mathsf{Pre}(T)}, \mathsf{Pre}(U))$ where d_U is the uniform distribution over U.*

Since the membership problems for sure eventually synchronizing and for limit-sure eventually synchronizing with exact support are PSPACE-complete ([14, Theorem 2 and Theorem 4]), the membership problem for almost-sure weak synchronization is in PSPACE by guessing the set U, and checking that $q_{\mathsf{init}} \in \langle\langle 1\rangle\rangle^{event}_{sure}(sum_U)$, and that $d_U \in \langle\langle 1\rangle\rangle^{event}_{limit}(sum_{\mathsf{Pre}(T)}, \mathsf{Pre}(U))$. We establish a matching PSPACE lower bound.

Lemma 5. *The membership problem for $\langle\langle 1\rangle\rangle^{weakly}_{almost}(sum_T)$ is PSPACE-hard even if T is a singleton.*

Simple examples show that winning strategies require infinite memory for almost-sure weak synchronization.

Theorem 2. *For almost-sure weak synchronization in MDPs:*

1. *(Complexity). The membership problem is PSPACE-complete.*
2. *(Memory). Infinite memory is necessary in general for both pure and randomized strategies, and pure strategies are sufficient.*

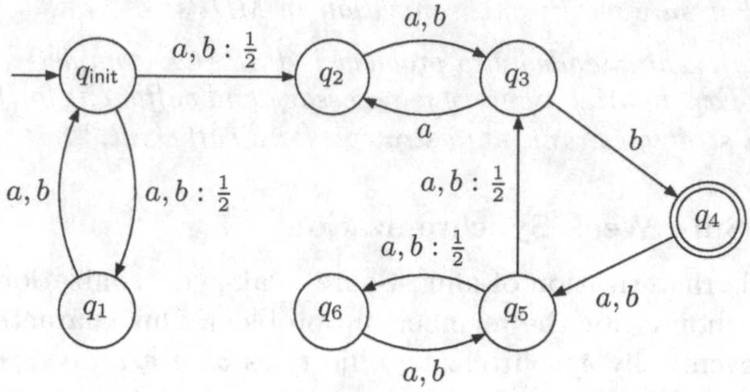

Fig. 1. An example to show $q_{\mathsf{init}} \in \langle\langle 1 \rangle\rangle_{limit}^{weakly}(q_4)$ implies $q_{\mathsf{init}} \in \langle\langle 1 \rangle\rangle_{almost}^{weakly}(q_4)$

3.3 Limit-Sure Weak Synchronization

We show that the winning regions for almost-sure and limit-sure weak synchronization coincide. The result is not intuitively obvious (recall that it does not hold for eventually synchronizing) and requires a careful analysis of the structure of limit-sure winning strategies to show that they always induce the existence of an almost-sure winning strategy. The construction of an almost-sure winning strategy from a family of limit-sure winning strategies is illustrated in the following example.

Consider the MDP $\mathcal{M}$ in Fig. 1 with initial state q_{init} and target set $T = \{q_4\}$. Note that there is a relevant strategic choice only in q_3, and that q_{init} is limit-sure winning for eventually synchronization in $\{q_4\}$ since we can inject a probability mass arbitrarily close to 1 in q_3 (by always playing a in q_3), and then switching to playing b in q_3 gets probability $1 - \varepsilon$ in T (for arbitrarily small ε). Moreover, the same holds from state q_4. These two facts are sufficient to show that q_{init} is limit-sure winning for weak synchronization in $\{q_4\}$: given $\varepsilon > 0$, play from q_{init} a strategy to ensure probability at least $p_1 = 1 - \frac{\varepsilon}{2}$ in q_4 (in finitely many steps), and then play according to a strategy that ensures from q_4 probability $p_2 = p_1 - \frac{\varepsilon}{4}$ in q_4 (in finitely many, and at least one step), and repeat this process using strategies that ensure, if the probability mass in q_4 is at least p_i, that the probability in q_4 is at least $p_{i+1} = p_i - \frac{\varepsilon}{2^{i+1}}$ (in at least one step). It follows that $p_i = 1 - \frac{\varepsilon}{2} - \frac{\varepsilon}{4} - \cdots - \frac{\varepsilon}{2^i} > 1 - \varepsilon$ for all $i \geq 1$, and thus $\limsup_{i \to \infty} p_i \geq 1 - \varepsilon$ showing that q_{init} is limit-sure weakly synchronizing in target $\{q_4\}$.

It follows from the result that we establish in this section (Theorem 3) that q_{init} is actually almost-sure weakly synchronizing in target $\{q_4\}$. To see this, consider the sequence $\mathsf{Pre}^i(T)$ for $i \geq 0$: $\{q_4\}, \{q_3\}, \{q_2\}, \{q_3\}, \ldots$ is ultimately periodic with period $r = 2$ and $R = \{q_3\} = \mathsf{Pre}(T)$ is such that $R = \mathsf{Pre}^2(R)$. The period corresponds to the loop $q_2 q_3$ in the MDP. It turns out that *limit-sure* eventually synchronizing in T implies *almost-sure* eventually synchronizing in R (by the proof of [14, Lemma 9]), thus from q_{init} a *single* strategy ensures that the probability mass in R is 1, either in the limit or after finitely many steps.

Note that in both cases since $R = \mathsf{Pre}^r(R)$ this even implies almost-sure weakly synchronizing in R. The same holds from state q_4.

Moreover, note that all distributions produced by an almost-sure weakly synchronizing strategy are themselves almost-sure weakly synchronizing. An almost-sure winning strategy for weak synchronization in $\{q_4\}$ consists in playing from q_{init} an *almost-sure* eventually synchronizing strategy in target $R = \{q_3\}$, and considering a decreasing sequence ε_i such that $\lim_{i\to\infty} \varepsilon_i = 0$, when the probability mass in R is at least $1 - \varepsilon_i$, inject it in $T = \{q_4\}$. Then the remaining probability mass defines a distribution (with support $\{q_1, q_2\}$ in the example) that is still almost-sure eventually synchronizing in R, as well as the states in T. Note that in the example, (almost all) the probability mass in $T = \{q_4\}$ can move to q_3 in an even number of steps, while from $\{q_1, q_2\}$ an odd number of steps is required, resulting in a *shift* of the probability mass. However, by repeating the strategy two times from q_4 (injecting large probability mass in q_3, moving to q_4, and injecting in q_3 again), we can make up for the shift and reach q_3 from q_4 in an even number of steps, thus in synchronization with the probability mass from $\{q_1, q_2\}$. This idea is formalized in the rest of this section, and we prove that we can always make up for the shifts, which requires a carefully analysis of the allowed amounts of shifting.

The result is easier to prove when the target T is a singleton, as in the example. For an arbitrary target set T, we need to get rid of the states in T that do not contribute a significant (i.e., bounded away from 0) probability mass in the limit, that we call the 'vanishing' states. We show that they can be removed from T without changing the winning region for limit-sure winning. When the target set has no vanishing state, we can construct an almost-sure winning strategy as in the case of a singleton target set.

Given an MDP $\mathcal{M}$ with initial state $q_{\mathsf{init}} \in \langle\!\langle 1 \rangle\!\rangle^{weakly}_{limit}(sum_T)$ that is limit-sure winning for the weakly synchronizing objective in target set T, let $(\alpha_i)_{i\in\mathbb{N}}$ be a family of limit-sure winning strategies such that $\limsup_{n\to\infty} \mathcal{M}^{\alpha_i}_n(T) \geq 1 - \varepsilon_i$ where $\lim_{i\to\infty} \varepsilon_i = 0$. Hence by definition of lim sup, for all $i \geq 0$ there exists a strictly increasing sequence $k_{i,0} < k_{i,1} < \cdots$ of positions such that $\mathcal{M}^{\alpha_i}_{k_{i,j}}(T) \geq 1 - 2\varepsilon_i$ for all $j \geq 0$. A state $q \in T$ is *vanishing* if $\liminf_{i\to\infty} \liminf_{j\to\infty} \mathcal{M}^{\alpha_i}_{k_{i,j}}(q) = 0$ for some family of limit-sure weakly synchronizing strategies $(\alpha_i)_{i\in\mathbb{N}}$. Intuitively, the contribution of a vanishing state q to the probability in T tends to 0 and therefore $\mathcal{M}$ is also limit-sure winning for the weakly synchronizing objective in target set $T \setminus \{q\}$.

Lemma 6. *If an MDP $\mathcal{M}$ is limit-sure weakly synchronizing in target set T, then there exists a set $T' \subseteq T$ such that $\mathcal{M}$ is limit-sure weakly synchronizing in T' without vanishing states.*

For a limit-sure weakly synchronizing MDP in target set T (without vanishing states), we show that from a probability distribution with support T, a probability mass arbitrarily close to 1 can be injected synchronously back in T (in at least one step), that is $d_T \in \langle\!\langle 1 \rangle\!\rangle^{event}_{limit}(sum_{\mathsf{Pre}(T)})$. The same holds from the initial state q_{init} of the MDP. This property is the key to construct an almost-sure weakly synchronizing strategy.

Lemma 7. *If an MDP $\mathcal{M}$ with initial state q_{init} is limit-sure weakly synchronizing in a target set T without vanishing states, then $q_{\mathsf{init}} \in \langle\langle 1 \rangle\rangle^{event}_{limit}(sum_{\mathsf{Pre}(T)})$ and $d_T \in \langle\langle 1 \rangle\rangle^{event}_{limit}(sum_{\mathsf{Pre}(T)})$ where d_T is the uniform distribution over T.*

To show that limit-sure and almost-sure winning coincide for weakly synchronizing objectives, from a family of limit-sure winning strategies we construct an almost-sure winning strategy that uses the eventually synchronizing strategies of Lemma 7. The construction consists in using successively strategies that ensure probability mass $1 - \varepsilon_i$ in the target T, for a decreasing sequence $\varepsilon_i \to 0$. Such strategies exist by Lemma 7, both from the initial state and from the set T. However, the mass of probability that can be guaranteed to be synchronized in T by the successive strategies is always smaller than 1, and therefore we need to argue that the remaining masses of probability (of size ε_i) can also get synchronized in T, and despite their possible shift with the main mass of probability.

Two main key arguments are needed to establish the correctness of the construction: (1) eventually synchronizing implies that a finite number of steps is sufficient to obtain a probability mass of $1 - \varepsilon_i$ in T, and thus the construction of the strategy is well defined, and (2) by the finiteness of the period r (such that $R = \mathsf{Pre}^r(R)$ where $R = \mathsf{Pre}^k(T)$ for some k) we can ensure to eventually make up for the shifts, and every piece of the probability mass can contribute (synchronously) to the target infinitely often.

Theorem 3. $\langle\langle 1 \rangle\rangle^{weakly}_{limit}(sum_T) = \langle\langle 1 \rangle\rangle^{weakly}_{almost}(sum_T)$ *for all MDPs and target sets T.*

Finally, we note that the complexity results of Theorem 1 and Theorem 2 hold for the membership problem with functions *max* and max_T by the lemma below. First, for $\mu \in \{sure, almost, limit\}$, we have $\langle\langle 1 \rangle\rangle^{weakly}_{\mu}(max_T) = \bigcup_{q \in T} \langle\langle 1 \rangle\rangle^{weakly}_{\mu}(q)$, showing that the membership problems for *max* are polynomial-time reducible to the corresponding membership problem for sum_T with singleton T. The reverse reduction is as follows. Given an MDP $\mathcal{M}$, a state q and an initial distribution d_0, we can construct an MDP $\mathcal{M}'$ and initial distribution d'_0 such that $d_0 \in \langle\langle 1 \rangle\rangle^{weakly}_{\mu}(q)$ iff $d'_0 \in \langle\langle 1 \rangle\rangle^{weakly}_{\mu}(max_{Q'})$ where Q' is the state space of $\mathcal{M}'$ (thus $max_{Q'}$ is simply the function max). The idea is to construct $\mathcal{M}'$ and d'_0 as a copy of $\mathcal{M}$ and d_0 where all states except q are duplicated, and the initial and transition probabilities are equally distributed between the copies. Therefore if the probability mass tends to 1 in some state, it has to be in q.

Lemma 8. *For weak synchronization and each winning mode, the membership problems with functions max and max_T are polynomial-time equivalent to the membership problem with function $sum_{T'}$ with a singleton T'.*

4 Strong Synchronization

In this section, we show that the membership problem for strongly synchronizing objectives can be solved in polynomial time, for all winning modes, and both with function max_T and function sum_T. We show that linear-size memory is necessary in general for max_T, and memoryless strategies are sufficient for sum_T.

It follows from our results that the limit-sure and almost-sure winning modes coincide for strong synchronization.

4.1 Strong Synchronization with Function *max*

First, note that for strong synchronization the membership problem with function max_T reduces to the membership problem with function max_Q where Q is the entire state space, by a construction similar to the proof of Lemma 8: states in $Q \setminus T$ are duplicated, ensuring that only states in T are used to accumulate probability.

The strongly synchronizing objective with function *max* requires that from some point on, almost all the probability mass is at every step in a single state. The sequence of states that contain almost all the probability corresponds to a sequence of deterministic transitions in the MDP, and thus eventually to a cycle of deterministic transitions.

The *graph of deterministic transitions* of an MDP $\mathcal{M} = \langle Q, \mathsf{A}, \delta \rangle$ is the directed graph $G = \langle Q, E \rangle$ where $E = \{\langle q_1, q_2 \rangle \mid \exists a \in \mathsf{A} : \delta(q_1, a)(q_2) = 1\}$. For $\ell \geq 1$, a *deterministic cycle in* $\mathcal{M}$ of length ℓ is a finite path $\hat{q}_\ell \hat{q}_{\ell-1} \cdots \hat{q}_0$ in G (that is, $\langle \hat{q}_i, \hat{q}_{i-1} \rangle \in E$ for all $1 \leq i \leq \ell$) such that $\hat{q}_0 = \hat{q}_\ell$. The cycle is *simple* if $\hat{q}_i \neq \hat{q}_j$ for all $1 \leq i < j \leq \ell$.

We show that sure (resp., almost-sure and limit-sure) strong synchronization is equivalent to sure (resp., almost-sure and limit-sure) reachability to a state in such a cycle, with the requirement that it can be reached in a synchronized way, that is by finite paths whose lengths are congruent modulo the length ℓ of the cycle. To check this, we keep track of a modulo-ℓ counter along the play.

Define the MDP $\mathcal{M} \times [\ell] = \langle Q', \mathsf{A}, \delta' \rangle$ where $Q' = Q \times \{0, 1, \cdots, \ell - 1\}$ and $\delta'(\langle q, i \rangle, a)(\langle q', i - 1 \rangle) = \delta(q, a)(q')$ (where $i - 1$ is $\ell - 1$ for $i = 0$) for all states $q, q' \in Q$, actions $a \in \mathsf{A}$, and $0 \leq i \leq \ell - 1$.

Lemma 9. *Let η be the smallest positive probability in the transitions of $\mathcal{M}$, and let $\frac{1}{1+\eta} < p \leq 1$. There exists a strategy α such that $\liminf_{n\to\infty} \|\mathcal{M}_n^\alpha\| \geq p$ from an initial state q_{init} if and only if there exists a simple deterministic cycle $\hat{q}_\ell \hat{q}_{\ell-1} \cdots \hat{q}_0$ in $\mathcal{M}$ and a strategy β in $\mathcal{M} \times [\ell]$ such that $\Pr^\beta(\Diamond\{\langle \hat{q}_0, 0 \rangle\}) \geq p$ from $\langle q_{\mathsf{init}}, 0 \rangle$.*

It follows directly from Lemma 9 with $p = 1$ that almost-sure strong synchronization is equivalent to almost-sure reachability to a deterministic cycle in $\mathcal{M} \times [\ell]$. The same equivalence holds for the sure and limit-sure winning modes.

Lemma 10. *A state q_{init} is sure (resp., almost-sure or limit-sure) winning for the strongly synchronizing objective (according to max_Q) if and only if there exists a simple deterministic cycle $\hat{q}_\ell \hat{q}_{\ell-1} \cdots \hat{q}_0$ such that $\langle q_{\mathsf{init}}, 0 \rangle$ is sure (resp., almost-sure or limit-sure) winning for the reachability objective $\Diamond\{\langle \hat{q}_0, 0 \rangle\}$ in $\mathcal{M} \times [\ell]$.*

Since the winning regions of almost-sure and limit-sure winning coincide for reachability objectives in MDPs [4], the next corollary follows from Lemma 10.

Corollary 1. *$\langle\langle 1\rangle\rangle^{strongly}_{limit}(max_T) = \langle\langle 1\rangle\rangle^{strongly}_{almost}(max_T)$ for all target sets T.*

If there exists a cycle c satisfying the condition in Lemma 10, then all cycles reachable from c in the graph G of deterministic transitions also satisfy the condition. Hence it is sufficient to check the condition for an arbitrary simple cycle in each strongly connected component (SCC) of G. It follows that strong synchronization can be decided in polynomial time (SCC decomposition can be computed in polynomial time, as well as sure, limit-sure, and almost-sure reachability in MDPs). The length of the cycle gives a linear bound on the memory needed to win, and the bound is tight.

Theorem 4. *For the three winning modes of strong synchronization according to max_T in MDPs:*

1. *(Complexity). The membership problem is PTIME-complete.*
2. *(Memory). Linear memory is necessary and sufficient for both pure and randomized strategies, and pure strategies are sufficient.*

4.2 Strong Synchronization with Function *sum*

The strongly synchronizing objective with function sum_T requires that eventually all the probability mass remains in T. We show that this is equivalent to a traditional reachability objective with target defined by the set of sure winning initial distributions for the safety objective $\Box T$.

It follows that almost-sure (and limit-sure) winning for strong synchronization is equivalent to almost-sure (or equivalently limit-sure) winning for the coBüchi objective $\Diamond\Box T = \{q_0 a_0 q_1 \cdots \in \mathsf{Play}(\mathcal{M}) \mid \exists j \cdot \forall i > j : q_i \in T\}$. However, sure strong synchronization is not equivalent to sure winning for the coBüchi objective: the MDP in Fig. 2 is sure winning for the coBüchi objective $\Diamond\Box\{q_{\mathsf{init}}, q_2\}$ from q_{init}, but not sure winning for the reachability objective $\Diamond S$ where $S = \{q_2\}$ is the winning region for the safety objective $\Box\{q_{\mathsf{init}}, q_2\}$ (and thus not sure strongly synchronizing). Note that this MDP is almost-sure strongly synchronizing in target $T = \{q_{\mathsf{init}}, q_2\}$ from q_{init}, and almost-sure winning for the coBüchi objective $\Diamond\Box T$, as well as almost-sure winning for the reachability objective $\Diamond S$.

Lemma 11. *Given a target set T, an MDP $\mathcal{M}$ is sure (resp., almost-sure or limit-sure) winning for the strongly synchronizing objective according to sum_T if and only if $\mathcal{M}$ is sure (resp., almost-sure or limit-sure) winning for the reachability objective $\Diamond S$ where S is the sure winning region for the safety objective $\Box T$.*

Corollary 2. *$\langle\langle 1\rangle\rangle^{strongly}_{limit}(sum_T) = \langle\langle 1\rangle\rangle^{strongly}_{almost}(sum_T)$ for all target sets T.*

The following result follows from Lemma 11 and the fact that the sure winning region for safety and reachability, and the almost-sure winning region for reachability can be computed in polynomial time for MDPs [4]. Moreover, memoryless strategies are sufficient for these objectives.

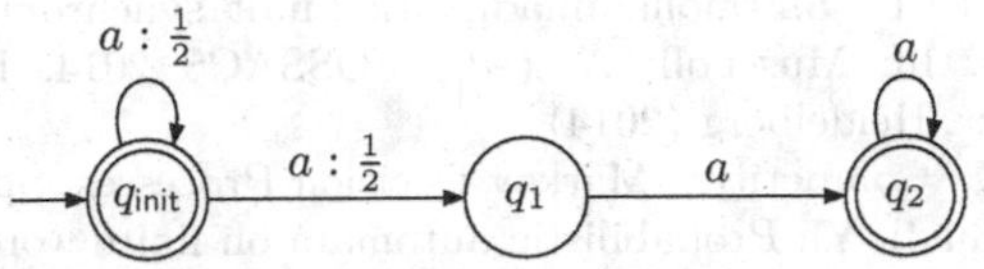

Fig. 2. An MDP such that q_{init} is sure-winning for coBüchi objective in $T = \{q_{\mathsf{init}}, q_2\}$ but not for strong synchronization according to sum_T

Theorem 5. *For the three winning modes of strong synchronization according to sum_T in MDPs:*

1. *(Complexity). The membership problem is PTIME-complete.*
2. *(Memory). Pure memoryless strategies are sufficient.*

References

1. ArXiv CoRR (2014), `http://arxiv.org/abs/1402.2840` (Full version)
2. Agrawal, M., Akshay, S., Genest, B., Thiagarajan, P.S.: Approximate verification of the symbolic dynamics of Markov chains. In: LICS, pp. 55–64. IEEE (2012)
3. de Alfaro, L., Henzinger, T.A.: Concurrent omega-regular games. In: Proc. of LICS, pp. 141–154 (2000)
4. de Alfaro, L., Henzinger, T.A., Kupferman, O.: Concurrent reachability games. Theor. Comput. Sci. 386(3), 188–217 (2007)
5. Baier, C., Bertrand, N., Größer, M.: On decision problems for probabilistic Büchi automata. In: Amadio, R.M. (ed.) FOSSACS 2008. LNCS, vol. 4962, pp. 287–301. Springer, Heidelberg (2008)
6. Baldoni, R., Bonnet, F., Milani, A., Raynal, M.: On the solvability of anonymous partial grids exploration by mobile robots. In: Baker, T.P., Bui, A., Tixeuil, S. (eds.) OPODIS 2008. LNCS, vol. 5401, pp. 428–445. Springer, Heidelberg (2008)
7. Burkhard, H.D.: Zum längenproblem homogener experimente an determinierten und nicht-deterministischen automaten. Elektronische Informationsverarbeitung und Kybernetik 12(6), 301–306 (1976)
8. Cerný, J.: Poznámka k. homogénnym experimentom s konecnymi automatmi. Matematicko-fyzikálny Časopis 14(3), 208–216 (1964)
9. Chadha, R., Korthikanti, V.A., Viswanathan, M., Agha, G., Kwon, Y.: Model checking MDPs with a unique compact invariant set of distributions. In: QEST, pp. 121–130. IEEE (2011)
10. Courcoubetis, C., Yannakakis, M.: The complexity of probabilistic verification. J. ACM 42(4), 857–907 (1995)
11. Doyen, L., Massart, T., Shirmohammadi, M.: Infinite synchronizing words for probabilistic automata. In: Murlak, F., Sankowski, P. (eds.) MFCS 2011. LNCS, vol. 6907, pp. 278–289. Springer, Heidelberg (2011)
12. Doyen, L., Massart, T., Shirmohammadi, M.: Synchronizing objectives for Markov decision processes. In: Proc. of iWIGP. EPTCS, vol. 50, pp. 61–75 (2011)
13. Doyen, L., Massart, T., Shirmohammadi, M.: Infinite synchronizing words for probabilistic automata (Erratum). CoRR abs/1206.0995 (2012)

14. Doyen, L., Massart, T., Shirmohammadi, M.: Limit synchronization in Markov decision processes. In: Muscholl, A. (ed.) FOSSACS 2014. LNCS, vol. 8412, pp. 58–72. Springer, Heidelberg (2014)
15. Filar, J., Vrieze, K.: Competitive Markov Decision Processes. Springer (1997)
16. Gimbert, H., Oualhadj, Y.: Probabilistic automata on finite words: Decidable and undecidable problems. In: Abramsky, S., Gavoille, C., Kirchner, C., Meyer auf der Heide, F., Spirakis, P.G. (eds.) ICALP 2010, Part II. LNCS, vol. 6199, pp. 527–538. Springer, Heidelberg (2010)
17. Henzinger, T.A., Mateescu, M., Wolf, V.: Sliding window abstraction for infinite Markov chains. In: Bouajjani, A., Maler, O. (eds.) CAV 2009. LNCS, vol. 5643, pp. 337–352. Springer, Heidelberg (2009)
18. Imreh, B., Steinby, M.: Directable nondeterministic automata. Acta Cybern. 14(1), 105–115 (1999)
19. Kfoury, D.: Synchronizing sequences for probabilistic automata. Studies in Applied Mathematics 29, 101–103 (1970)
20. Korthikanti, V.A., Viswanathan, M., Agha, G., Kwon, Y.: Reasoning about MDPs as transformers of probability distributions. In: QEST, pp. 199–208. IEEE (2010)
21. Martyugin, P.: Computational complexity of certain problems related to carefully synchronizing words for partial automata and directing words for nondeterministic automata. Theory Comput. Syst. 54(2), 293–304 (2014)
22. Vardi, M.Y.: Automatic verification of probabilistic concurrent finite-state programs. In: Proc. of FOCS, pp. 327–338. IEEE Computer Society (1985)
23. Volkov, M.V.: Synchronizing automata and the Cerny conjecture. In: Martín-Vide, C., Otto, F., Fernau, H. (eds.) LATA 2008. LNCS, vol. 5196, pp. 11–27. Springer, Heidelberg (2008)

Probabilistic Bisimulation: Naturally on Distributions*

Holger Hermanns[1], Jan Krčál[1], and Jan Křetínský[2]

[1] Saarland University – Computer Science, Saarbrücken, Germany
{hermanns,krcal}@cs.uni-saarland.de
[2] IST Austria
jan.kretinsky@ist.ac.at

Abstract. In contrast to the usual understanding of probabilistic systems as stochastic processes, recently these systems have also been regarded as transformers of probabilities. In this paper, we give a *natural* definition of strong bisimulation for probabilistic systems corresponding to this view that treats probability *distributions* as first-class citizens. Our definition applies in the same way to discrete systems as well as to systems with uncountable state and action spaces. Several examples demonstrate that our definition refines the understanding of behavioural equivalences of probabilistic systems. In particular, it solves a longstanding open problem concerning the representation of memoryless continuous time by memoryfull continuous time. Finally, we give algorithms for computing this bisimulation not only for finite but also for classes of uncountably infinite systems.

1 Introduction

Continuous time concurrency phenomena can be addressed in two principal manners: On the one hand, *timed automata* (TA) extend interleaving concurrency with real-valued clocks [2]. On the other hand, time can be represented by memoryless stochastic time, as in *continuous time Markov chains* (CTMC) and extensions, where time is represented in the form of exponentially distributed random delays [33,30,6,23]. TA and CTMC variations have both been applied to very many intriguing cases, and are supported by powerful real-time, respectively stochastic time model checkers [3,37] with growing user bases. The models are incomparable in expressiveness, but if one extends timed automata with the possibility to sample from exponential distributions [5,10,28], there appears to be

* This work is supported by the EU 7th Framework Programme under grant agreements 295261 (MEALS) and 318490 (SENSATION), European Research Council (ERC) under grant agreement 267989 (QUAREM), Austrian Science Fund (FWF) project S11402-N23 (RiSE), Czech Science Foundation under grant agreement P202/12/G061, the DFG Transregional Collaborative Research Centre SFB/TR 14 AVACS, and by the CAS/SAFEA International Partnership Program for Creative Research Teams. Jan Křetínský is currently on leave from Faculty of Informatics, Masaryk University, Czech Republic.

P. Baldan and D. Gorla (Eds.): CONCUR 2014, LNCS 8704, pp. 249–265, 2014.

a natural bridge from CTMC to TA. This kind of stochastic semantics of timed automata has recently gained considerable popularity by the statistical model checking approach to TA analysis [14,13].

Still there is a disturbing difference, and this difference is the original motivation [12] of the work presented in this paper. The obvious translation of an exponentially distributed delay into a clock expiration sampled from the very same exponential probability distribution fails in the presence of concurrency. This is because the translation is not fully compatible with the natural interleaving concurrency semantics for TA respectively CTMC. This is illustrated by the following example, which in the middle displays two small CTMC, which are supposed to run independently and concurrently.

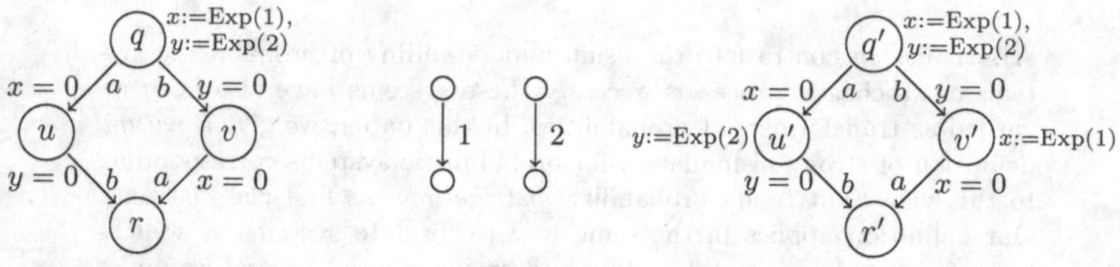

On the left and right we see two stochastic automata (a variation of timed automata formally defined in Section 3). They have clocks x and y which are initialized by sampling from exponential distributions, and then each run down to 0. The first one reaching 0 triggers a transition and the other clock keeps on running unless resampled, which happens on the right, but not on the left. The left model is obtained by first translating the respective CTMC, and then applying the natural TA interleaving semantics, while the right model is obtained by first applying the equally natural CTMC interleaving semantics prior to translation.

The two models have subtly different semantics in terms of their underlying dense probabilistic timed transition systems. This can superficially be linked to the memoryless property of exponential distributions, yet there is no formal basis for proving equivalence. Our paper closes this gap, which has been open for at least 15 years, by introducing a natural *continuous-space distribution-based* bisimulation. Our result is embedded in several further intriguing application contexts and algorithmic achievements for this novel bisimulation.

The theory of bisimulations is a well-established and elegant framework to describe equivalence between processes based on their behaviour. In the standard semantics of probabilistic systems [38,44], when a probabilistic step from a state to a distribution is taken, the random choice is resolved and we instead continue from one of the successor states. Recently, there has been considerable interest in instead regarding probabilistic systems as deterministic transformers of probability *distributions* [36,1,20], where the choice is not resolved and we continue from the distribution over successors. Thus, instead of the current state the transition changes the current distribution over the states. Although the distribution semantics is very natural in many contexts [29], it has been only partially reflected in the study of bisimulations [29,19,24,23].

Our definition arises as an unusual, but very simple instantiation of the standard coalgebraic framework for bisimulations [42]. (No knowledge of coalgebra is required from the reader though.) Despite its simplicity, the resulting notion is surprisingly fruitful, not only because it indeed solves the longstanding correspondence problem between CTMC and TA with stochastic semantics.

Firstly, it is more adequate than other equivalences when applied to systems with distribution semantics, including large-population models where different parts of the population act differently [39]. Indeed, as argued in [26], some equivalent states are not identified in the standard probabilistic bisimulations and too many are identified in the recent distribution based bisimulations [19,24]. Our approach allows for a bisimulation identifying precisely the desired states [26].

Secondly, our bisimulation over distributions induces an equivalence *on states*, and this relation equates behaviourally indistinguishable states which in many settings are unnecessarily distinguished by standard bisimulations. We shall discuss this phenomenon in the context of several applications. Nevertheless, the key idea to work with distributions instead of single states also bears disadvantages. The main difficulty is that even for finite systems the space of distributions is uncountable, thus bisimulation is difficult to compute. However, we show that it admits a concise representation using methods of linear algebra and we provide an algorithm for computing it. Further, in order to cover e.g. continuous-time systems, we need to handle both uncountably many states (that store the sampled time) and labels (real time durations). Fortunately, there is an elegant way to do so using the standard coalgebra framework. Moreover, it can easily be further generalized, e.g. adding rewards to the generic definition is a trivial task.

Our contribution is the following:

- We give a natural definition of bisimulation from the distribution perspective for systems with generally uncountable spaces of states and labels.
- We argue by means of several applications that the definition can be considered more useful than the classical notions of probabilistic bisimulation.
- We provide an algorithm to compute this distributional bisimulation on finite non-deterministic probabilistic systems, and present a decision algorithm for uncountable continuous-time systems induced by the stochastic automata mentioned above.

A full version of this paper is available [31].

2 Probabilistic Bisimulation on Distributions

A (potentially uncountable) set S is a *measurable space* if it is equipped with a σ-algebra, denoted by $\Sigma(S)$. The elements of $\Sigma(S)$ are called *measurable sets*. For a measurable space S, let $\mathcal{D}(S)$ denote the set of *probability measures* (or *probability distributions*) over S. The following definition is similar to the treatment of [51].

Definition 1. *A* non-deterministic labelled Markov process *(NLMP) is a tuple* $\mathbf{P} = (S, L, \{\tau_a \mid a \in L\})$ *where S is a measurable space of* states, *L is a measurable* *space of* labels, *and* $\tau_a : S \to \Sigma(\mathcal{D}(S))$ *assigns to each state* s *a measurable set of probability measures* $\tau_a(s)$ *available in* s *under* a.[(1)]

When in a state $s \in S$, NLMP reads a label $a \in L$ and *non-deterministically* chooses a successor distribution $\mu \in \mathcal{D}(S)$ that is in the set of convex combinations[(2)] over $\tau_a(s)$, denoted by $s \xrightarrow{a} \mu$. If there is no such distribution, the process halts. Otherwise, it moves into a successor state according to μ. Considering convex combinations is necessary as it gives more power than pure resolution of non-determinism [43].

Example 1. If all sets are finite, we obtain *probabilistic automata (PA)* defined [43] as a triple $(S, L, \longrightarrow)$ where $\longrightarrow \subseteq S \times L \times \mathcal{D}(S)$ is a probabilistic transition relation with $(s, a, \mu) \in \longrightarrow$ if $\mu \in \tau_a(s)$.

Example 2. In the continuous setting, consider a random number generator that also remembers the previous number. We set $L = [0,1]$, $S = [0,1] \times [0,1]$ and $\tau_x(\langle new, last \rangle) = \{\mu_x\}$ for $x = new$ and $\emptyset$ otherwise, where μ_x is the uniform distribution on $[0,1] \times \{x\}$. If we start with a uniform distribution over S, the measure of successors under any $x \in L$ is 0. Thus in order to get any information of the system we have to consider successors under sets of labels, e.g. intervals.

For a measurable set $A \subseteq L$ of labels, we write $s \xrightarrow{A} \mu$ if $s \xrightarrow{a} \mu$ for some $a \in A$, and denote by $S_A := \{s \mid \exists \mu : s \xrightarrow{A} \mu\}$ the set of states having some outgoing label from A. Further, we can lift this to probability distributions by setting $\mu \xrightarrow{A} \nu$ if $\mu(S_A) > 0$ and $\nu = \frac{1}{\mu(S_A)} \int_{s \in S_A} \nu_s \, \mu(d\,s)$ for some measurable function assigning to each state $s \in S_A$ a measure ν_s such that $s \xrightarrow{A} \nu_s$. Intuitively, in μ we restrict to states that do not halt under A and consider all possible combinations of their transitions; we scale up by $\frac{1}{\mu(S_A)}$ to obtain a distribution again.

Example 3. In the previous example, let υ be the uniform distribution. Due to the independence of the random generator on previous values, we get $\upsilon \xrightarrow{[0,1]} \upsilon$. Similarly, $\upsilon \xrightarrow{[0.1,0.2]} \upsilon_{[0.1,0.2]}$ where $\upsilon_{[0.1,0.2]}$ is uniform on $[0,1]$ in the first component and uniform on $[0.1, 0.2]$ in the second component, with no correlation.

Using this notation, a non-deterministic and probabilistic system such as NLMP can be regarded as a non-probabilistic, thus solely non-deterministic, labelled transition system over the uncountable space of probability distributions. The natural bisimulation from this distribution perspective is as follows.

Definition 2. *Let* $(S, L, \{\tau_a \mid a \in L\})$ *be a NLMP and* $R \subseteq \mathcal{D}(S) \times \mathcal{D}(S)$ *be a symmetric relation. We say that* R *is a (strong)* probabilistic bisimulation *if for each* $\mu \, R \, \nu$ *and measurable* $A \subseteq L$

---

(1) We further require that for each $s \in S$ we have $\{(a, \mu) | \mu \in \tau_a(s)\} \in \Sigma(L) \otimes \Sigma(\mathcal{D}(S))$ and for each $A \in \Sigma(L)$ and $Y \in \Sigma(\mathcal{D}(S))$ we have $\{s \in S \mid \exists a \in A. \tau_a(s) \cap Y \neq \emptyset\} \in \Sigma(S)$. Here $\Sigma(\mathcal{D}(S))$ is the Giry σ-algebra [27] over $\mathcal{D}(X)$.

(2) A distribution $\mu \in \mathcal{D}(S)$ is a *convex combination* of a set $M \in \Sigma(\mathcal{D}(S))$ of distributions if there is a measure ν on $\mathcal{D}(S)$ such that $\nu(M) = 1$ and $\mu = \int_{\mu' \in \mathcal{D}(S)} \mu' \nu(d\mu')$.

1. $\mu(S_A) = \nu(S_A)$*, and*
2. for each $\mu \xrightarrow{A} \mu'$ *there is a* $\nu \xrightarrow{A} \nu'$ *such that* $\mu' R \nu'$*.*

We set $\mu \sim \nu$ *if there is a probabilistic bisimulation* R *such that* $\mu R \nu$*.*

Example 4. Considering Example 2, states $\{x\} \times [0,1]$ form a class of $\sim$ for each $x \in [0,1]$ as the old value does not affect the behaviour. More precisely, $\mu \sim \nu$ iff marginals of their first component are the same.

Naturalness. Our definition of bisimulation is not created ad-hoc as it often appears for relational definitions, but is actually an instantiation of the standard bisimulation for a particular *coalgebra.* Although this aspect is not necessary for understanding the paper, it is another argument for naturalness of our definition. For reader's convenience, we present a short introduction to coalgebras and the formal definitions in [31]. Here we only provide an intuitive explanation by example.

Non-deterministic labelled transition systems are essentially given by the transition function $S \to \mathcal{P}(S)^L$; given a state $s \in S$ and a label $a \in L$, we can obtain the set of the successors $\{s' \in S \mid s \xrightarrow{a} s'\}$. The transition function corresponds to a coalgebra, which induces a bisimulation coinciding with the classical one of Park and Milner [40]. Similarly, PA are given by the transition function $S \to \mathcal{P}(\mathcal{D}(S))^L$; instead of successors there are distributions over successors. Again, the corresponding coalgebraic bisimulation coincides with the classical ones of Larsen and Skou [38] and Segala and Lynch [44].

In contrast, our definition can be obtained by considering states S' to be distributions in $\mathcal{D}(S)$ over the original state space and defining the transition function to be $S' \to ([0,1] \times \mathcal{P}(S'))^{\Sigma(L)}$. The difference to the standard non-probabilistic case is twofold: firstly, we consider all measurable sets of labels, i.e. all elements of $\Sigma(L)$; secondly, for each label set we consider the mass, i.e. element of $[0,1]$, of the current state distribution that does not deadlock, i.e. can perform some of the labels. These two aspects form the crux of our approach and distinguish it from other approaches.

3 Applications

We now argue by some concrete application domains that the distribution view on bisimulation yields a fruitful notion.

Memoryless vs. Memoryfull Continuous Time. First, we reconsider the motivating discussion from Section 1 revolving around the difference between continuous time represented by real-valued clocks, respectively memoryless stochastic time. For this we introduce a simple model of *stochastic automata* [10].

Definition 3. *A* stochastic automaton (SA) *is a tuple* $\mathcal{S} = (\mathcal{Q}, \mathcal{C}, \mathcal{A}, \to, \kappa, F)$ *where* $\mathcal{Q}$ *is a set of locations,* $\mathcal{C}$ *is a set of clocks,* $\mathcal{A}$ *is a set of actions,* $\to \subseteq \mathcal{Q} \times \mathcal{A} \times 2^{\mathcal{C}} \times \mathcal{Q}$ *is a set of edges,* $\kappa : \mathcal{Q} \to 2^{\mathcal{C}}$ *is a clock setting function, and* F *assigns to each clock its distribution over* $\mathbb{R}_{\geq 0}$*.*

Avoiding technical details, $\mathcal{S}$ has the following NLMP semantics $\mathbf{P}_{\mathcal{S}}$ with state space $S = \mathcal{Q} \times \mathbb{R}^{\mathcal{C}}$, assuming it is initialized in some location q_0: When a location q is entered, for each clock $c \in \kappa(q)$ a positive *value* is chosen randomly according to the distribution $F(c)$ and stored in the state space. Intuitively, the automaton idles in location q with all clock values decreasing at the same speed until some edge (q, a, X, q') becomes *enabled*, i.e. all clocks from X have value ≤ 0. After this *idling time* t, the action a is taken and the automaton enters the next location q'. If an edge is enabled on entering a location, it is taken immediately, i.e. $t = 0$. If more than one edge become enabled simultaneously, one of them is chosen non-deterministically. Its formal definition is given in [31]. We now are in the position to harvest Definition 2, to arrive at the novel bisimulation for stochastic automata.

Definition 4. *We say that locations q_1, q_2 of an SA $\mathcal{S}$ are* probabilistic bisimilar, *denoted $q_1 \sim q_2$, if $\mu_{q_1} \sim \mu_{q_2}$ in $\mathbf{P}_{\mathcal{S}}$ where μ_q denotes a distribution over the state space of $\mathbf{P}_{\mathcal{S}}$ given by the location being q, every $c \notin \kappa(q)$ being* 0, *and every $c \in \kappa(q)$ being independently set to a random value according to $F(c)$.*

This bisimulation identifies q and q' from Section 1 unlike any previous bisimulation on SA [10]. In Section 4 we discuss how to compute this bisimulation, despite being continuous-space. Recall that the model initialized by q is obtained by first translating two simple CTMC, and then applying the natural interleaving semantics, while the model, of q' is obtained by first applying the equally natural CTMC interleaving semantics prior to translation. The bisimilarity of these two models generalizes to the whole universe of CTMC and SA:

Theorem 1. *Let $SA(\mathcal{C})$ denote the stochastic automaton corresponding to a CTMC $\mathcal{C}$. For any CTMC $\mathcal{C}_1, \mathcal{C}_2$, we have*

$$SA(\mathcal{C}_1) \parallel_{SA} SA(\mathcal{C}_1) \quad \sim \quad SA(\mathcal{C}_1 \parallel_{CT} \mathcal{C}_1).$$

Here, $\parallel_{CT}$ and $\parallel_{SA}$ denotes the interleaving parallel composition of SA [11] (echoing TA parallel composition) and CTMC [33,30] (Kronecker sum of their matrix representations), respectively.

Bisimulation for Partial-Observation MDP (POMDP). A POMDP is a quadruple $\mathcal{M} = (S, \mathcal{A}, \delta, \mathcal{O})$ where (as in an MDP) S is a set of states, $\mathcal{A}$ is a set of actions, and $\delta : S \times \mathcal{A} \to \mathcal{D}(S)$ is a transition function. Furthermore, $\mathcal{O} \subseteq 2^S$ partitions the state space. The choice of actions is resolved by a policy yielding a Markov chain. Unlike in an MDP, such choice is not based on the knowledge of the current state, only on knowing that the current state belongs into an *observation* $o \in \mathcal{O}$. POMDPs have a wide range of applications in robotic control, automated planning, dialogue systems, medical diagnosis, and many other areas [45].

In the analysis of POMDP, the distributions over states, called *beliefs*, arise naturally and bisimulations over beliefs have already been considered [7,34]. However, to the best of our knowledge, no algorithms for computing belief bisimilation for POMDP exist. We fill this gap by our algorithm for computing distribution bisimulation for PA in Section 4. Indeed, two beliefs μ, ν in POMDP

$\mathcal{M}$ are belief bisimilar in the spirit of [7] iff μ and ν are distribution bisimilar in the induced PA $D_{\mathcal{M}} = (S, \mathcal{O} \times \mathcal{A}, \longrightarrow)$ where $(s, (o, a), \mu) \in \longrightarrow$ if $s \in o$ and $\delta(s, a) = \mu$.[(3)]

Further Applications. Probabilistic automata are especially apt for compositional modelling of *distributed systems.* The only information a component in a distributed system has about the current state of another component stems from their mutual communication. Therefore, each component can be also viewed from the outside as a partial-observation system. Thus, also in this context, distribution bisimulation is a natural concept. While $\sim$ is not a congruence w.r.t. standard parallel composition, it is apt for compositional modelling of distributed systems where only *distributed schedulers* are considered. For details, see [31,48].

Furthermore we can understand a PA as a description, in the sense of [25,39], of a representative *agent* in a large homogeneous *population.* The distribution view then naturally represents the ratios of agents being currently in the individual states and labels given to this large population of PAs correspond to global control actions [25]. For more details on applications, see [31].

4 Algorithms

In this section, we discuss computational aspects of deciding our bisimulation. Since $\sim$ is a relation over distributions over the system's state space, it is uncountably infinite even for simple finite systems, which makes it in principle intricate to decide. Fortunately, the bisimulation relation has a linear structure, and this allows us to employ methods of linear algebra to work with it effectively. Moreover, important classes of continuous-space systems can be dealt with, since their structure can be exploited. We exemplify this on a subset of deterministic stochastic automata, for which we are able to provide an algorithm to decide bisimilarity.

Finite Systems – Greatest Fixpoints. Let us fix a PA $(S, L, \longrightarrow)$. We apply the standard approach by starting with $\mathcal{D}(S) \times \mathcal{D}(S)$ and pruning the relation until we reach the fixpoint $\sim$. In order to represent $\sim$ using linear algebra, we identify a distribution μ with a vector $(\mu(s_1), \ldots, \mu(s_{|S|})) \in \mathbb{R}^{|S|}$.

Although the space of distributions is uncountable, we construct an implicit representation of $\sim$ by a system of equations written as columns in a matrix E.

Definition 5. *A matrix E with $|S|$ rows is a* bisimulation matrix *if for some bisimulation R, for any distributions μ, ν*

$$\mu \, R \, \nu \quad \textit{iff} \quad (\mu - \nu)E = \mathbf{0}.$$

For a bisimulation matrix E, an equivalence class of μ is then the set $(\mu + \{\rho \mid \rho E = \mathbf{0}\}) \cap \mathcal{D}(S)$, the set of distributions that are equal modulo E.

Example 5. The bisimulation matrix E below encodes that several conditions must hold for two distributions μ, ν to be bisimilar. Among others, if we multiply

(3) Note that [7] also considers rewards that can be easily added to $\sim$ and our algorithm.

$\mu - \nu$ with e.g. the second column, we must get 0. This translates to $(\mu(v) - \nu(v)) \cdot \mathbf{1} = 0$, i.e. $\mu(v) = \nu(v)$. Hence for bisimilar distributions, the measure of v has to be the same. This proves that $u \not\sim v$ (here we identify states and their Dirac distributions). Similarly, we can prove that $t \sim \frac{1}{2}t' + \frac{1}{2}t''$. Indeed, if we multiply the corresponding difference vector $(0, 0, 1, -\frac{1}{2}, -\frac{1}{2}, 0, 0)$ with any column of the matrix, we obtain 0.

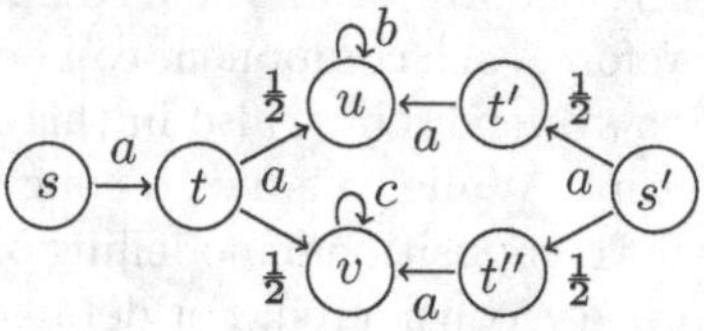

$$\begin{matrix} s: \\ s': \\ t: \\ t': \\ t'': \\ u: \\ v: \end{matrix} \begin{pmatrix} 1 & 0 & 0 & 0 & 0 \\ 1 & 0 & 0 & 0 & 0 \\ 1 & 0 & 0 & \frac{1}{2} & \frac{1}{2} \\ 1 & 0 & 0 & 0 & 1 \\ 1 & 0 & 0 & 1 & 0 \\ 1 & 0 & 1 & 0 & 0 \\ 1 & 1 & 0 & 0 & 0 \end{pmatrix}$$

Note that the unit matrix is always a bisimulation matrix, not relating anything with anything but itself. For which bisimulations do there exist bisimulation matrices? We say a relation R over distributions is *convex* if $\mu R \nu$ and $\mu' R \nu'$ imply $(p\mu + (1-p)\mu')\ R\ (p\nu + (1-p)\nu')$ for any $p \in [0, 1]$.

Lemma 1. *Every convex bisimulation has a corresponding bisimulation matrix.*

Since $\sim$ is convex (see [31]), there is a bisimulation matrix corresponding to $\sim$. It is a least restrictive bisimulation matrix E (note that all bisimulation matrices with the least possible dimension have identical solution space), we call it *minimal bisimulation matrix.* We show that the necessary and sufficient condition for E to be a bisimulation matrix is *stability* with respect to transitions.

Definition 6. *For a* $|S| \times |S|$ *matrix* P*, we say that a matrix* E *with* $|S|$ *rows is* P-stable *if for every* $\rho \in \mathbb{R}^{|S|}$,

$$\rho E = \mathbf{0} \implies \rho P E = \mathbf{0} \tag{1}$$

We first briefly explain the stability in a simpler setting.

Action-Deterministic Systems. Let us consider PA where in each state, there is at most one transition. For each $a \in L$, we let $P_a = (p_{ij})$ denote the transition matrix such that for all i, j, if there is (unique) transition $s_i \xrightarrow{a} \mu$ we set p_{ij} to $\mu(s_j)$, otherwise to 0. Then μ evolves under a into μP_a. Denote $\mathbf{1} = (1, \ldots, 1)^\top$.

Proposition 1. *In an action-deterministic PA,* E *containing* $\mathbf{1}$ *is a bisimulation matrix iff it is* P_a*-stable for all* $a \in L$.

To get a minimal bisimulation matrix E, we start with a single vector $\mathbf{1}$ which stands for an equation saying that the overall probability mass in bisimilar distributions is the same. Then we repetitively multiply all vectors we have by all the matrices P_a and add each resulting vector to the collection if it is linearly independent of the current collection, until there are no changes. In Example 5, the second column of E is obtained as $P_c\mathbf{1}$, the fourth one as $P_a(P_c\mathbf{1})$ and so on.

The set of all columns of E is thus given by the described iteration

$$\{P_a \mid a \in L\}^* \mathbf{1}$$

modulo linear dependency. Since P_a have $|S|$ rows, the fixpoint is reached within $|S|$ iterations yielding $1 \leq d \leq |S|$ equations. Each class then forms an $(|S| - d)$-dimensional affine subspace intersected with the set of probability distributions $\mathcal{D}(S)$. This is also the principle idea behind the algorithm of [50] and [19].

Non-Deterministic Systems. In general, for transitions under A, we have to consider c_i^A non-deterministic choices in each s_i among all the outgoing transitions under some $a \in A$. We use variables w_i^j denoting the probability that j-th transition, say (s_i, a_i^j, μ_i^j), is taken by the scheduler/player$^{(4)}$ in s_i. We sum up the choices into a "non-deterministic" transition matrix P_A^W with parameters W whose ith row equals $\sum_{j=1}^{c_i^A} w_i^j \mu_i^j$. It describes where the probability mass moves from s_i under A depending on the collection W of the probabilities the player gives each choice. By $\mathcal{W}_A$ we denote the set of all such W.

A simple generalization of the approach above would be to consider $\{P_A^W \mid A \subseteq L, W \in \mathcal{W}_A\}^*\mathbf{1}$. However, firstly, the set of these matrices is uncountable whenever there are at least two transitions to choose from. Secondly, not all P_A^W may be used as the following example shows.

Example 6. In each bisimulation class in the following example, the probabilities of $s_1 + s_2$, s_3, and s_4 are constant, as can also be seen from the bisimulation matrix E, similarly to Example 5. Further, E can be obtained as $(\mathbf{1}\ P_c\mathbf{1}\ P_b\mathbf{1})$. Observe that E is $P_{\{a\}}^W$-stable for W that maximizes the probability of going into the "class" s_3 (both s_1 and s_2 go to s_3, i.e. $w_1^1 = w_2^1 = 1$); similarly for the "class" s_4.

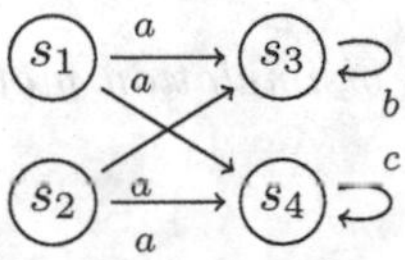

$$P_{\{a\}}^W = \begin{pmatrix} 0 & 0 & w_1^1 & w_1^2 \\ 0 & 0 & w_2^1 & w_2^2 \\ 0 & 0 & 0 & 0 \\ 0 & 0 & 0 & 0 \end{pmatrix} \qquad E = \begin{pmatrix} 1 & 0 & 0 \\ 1 & 0 & 0 \\ 1 & 0 & 1 \\ 1 & 1 & 0 \end{pmatrix}$$

However, for $\overline{W}$ with $w_1^1 \neq w_2^1$, e.g. s_1 goes to s_3 and s_2 goes with equal probability to s_3 and s_4 ($w_1^1 = 1, w_2^1 = w_2^2 = \frac{1}{2}$), we obtain from $P_{\{a\}}^{\overline{W}} E$ a new independent vector $(0, 0.5, 0, 0)^\top$ enforcing a partition finer than $\sim$. This does not mean that Spoiler wins the game when choosing such mixed $\overline{W}$ in some μ, it only means that Duplicator needs to choose a *different* W in a bisimilar ν in order to have $\mu P_A^{\overline{W}} \sim \nu P_A^W$ for the successors.

A fundamental observation is that we get the correct bisimulation when Spoiler is restricted to finitely many "extremal" choices and Duplicator is restricted for such extremal W to respond only with the very same W. (*)

To this end, consider $M_A^W = P_A^W E$ where E is the current matrix with each of e columns representing an equation. Intuitively, the ith row of M_A^W describes how

(4) We use the standard notion of Spoiler-Duplicator bisimulation game (see e.g. [42]) where in $\{\mu_0, \mu_1\}$ Spoiler chooses $i \in \{0, 1\}$, $A \subseteq L$, and $\mu_i \xrightarrow{A} \mu_i'$, Duplicator has to reply with $\mu_{1-i} \xrightarrow{A} \mu_{1-i}'$ such that $\mu_i(S_A) = \mu_{i-1}(S_A)$, and the game continues in $\{\mu_0', \mu_1'\}$. Spoiler wins iff at some point Duplicator cannot reply.

much of s_i is moved to various classes when a step is taken. Denote the linear forms in M_A^W over W by m_{ij}. Since the players can randomize and mix choices which transition to take, the set of vectors $\{(m_{i1}(w_i^1, \ldots, w_i^{c_i}), \ldots, m_{ib}(w_i^1, \ldots, w_i^{c_i})) \mid w_i^1, \ldots, w_i^{c_i} \geq 0, \sum_{j=1}^{c_i} w_i^j = 1\}$ forms a convex polytope denoted by C_i. Each vector in C_i is thus the ith row of the matrix M_A^W where some concrete weights w_i^j are "plugged in". This way C_i describes all the possible choices in s_i and their effect on where the probability mass is moved.

Denote vertices (extremal points) of a convex polytope P by $\mathcal{E}(P)$. Then $\mathcal{E}(C_i)$ correspond to pure (non-randomizing) choices that are "extremal" w.r.t. E. Note that now if $s_j \sim s_k$ then $C_j = C_k$, or equivalently $\mathcal{E}(C_j) = \mathcal{E}(C_k)$. Indeed, for every choice in s_j there needs to be a matching choice in s_k and vice versa. However, since we consider bisimulation between generally non-Dirac distributions, we need to combine these extremal choices. For an arbitrary distribution $\mu \in \mathcal{D}(S)$, we say that a tuple $c \in \prod_{i=1}^{|S|} \mathcal{E}(C_i)$ is *extremal in* μ if $\mu \cdot c^\top$ is a vertex of the polytope $\{\mu \cdot c'^\top \mid c' \in \prod_{i=1}^{|S|} C_i\}$. Note that each extremal c corresponds to particular pure choices, denoted by $W(c)$. Unfortunately, for choices $W(c)$ of Spoiler extremal in *some* distribution, Duplicator may in *another* distribution need to make different choices. Indeed, in Example 6 the tuple corresponding to $\overline{W}$ is extremal in the Dirac distribution of state s_1. Therefore, we define $\mathcal{E}(C)$ to be the set of tuples c extremal in the uniform distribution. Interestingly, tuples extremal in the uniform distribution are (1) extremal in all distributions and (2) reflect all extremal choices, i.e. for every c extremal in some μ, there is a c' extremal in the uniform distribution such that c' is also extremal in μ and $\mu \cdot c = \mu \cdot c'$. As a result, the fundamental property $(*)$ is guaranteed.

Proposition 2. *Let E be a matrix containing* $\mathbf{1}$. *It is a bisimulation matrix iff it is $P_A^{W(c)}$-stable for all $A \subseteq L$ and $c \in \mathcal{E}(C)$.*

Input : Probabilistic automaton $(S, L, \rightarrow)$
Output : A minimal bisimulation matrix E
foreach $A \subseteq L$ **do**
  compute P_A^W // non-deterministic transition matrix
$E \leftarrow (\mathbf{1})$
repeat
  foreach $A \subseteq L$ **do**
    $M_A^W \leftarrow P_A^W E$ // polytope of all choices
    compute $\mathcal{E}(C)$ from M_A^W // vertices, i.e. extremal choices
    foreach $c \in \mathcal{E}(C)$ **do**
      $M_A^{W(c)} \leftarrow M_A^W$ with values $W(c)$ plugged in
      $E_{new} \leftarrow$ columns of $M_A^{W(c)}$ linearly independent of columns of E
      $E \leftarrow (E\ E_{new})$
until E *does not change*

Algorithm 1: Bisimulation on probabilistic automata

Theorem 2. *Algorithm 1 computes a minimal bisimulation matrix.*

The running time is exponential. We leave the question whether linear programming or other methods [32] can yield E in polynomial time open. The algorithm can easily be turned into one computing other bisimulation notions from the literature, for which there were no algorithms so far, see Section 5.

Continuous-Time Systems - Least Fixpoints. Turning our attention to continuous systems, we finally sketch an algorithm for deciding bisimulation $\sim$ over a subclass of stochastic automata, this constitutes the first algorithm to compute a bisimulation on the uncountably large semantical object.

We need to adopt two restrictions. First, we consider only *deterministic* SA, where the probability that two edges become enabled at the same time is zero (when initiated in any location). Second, to simplify the exposition, we restrict all distributions occurring to exponential distributions. Notably, even for this class, our bisimulation is strictly coarser than the one induced by standard bisimulations [33,30,6] for continuous-time Markov chains. At the end of the section we discuss possibilities for extending the class of supported distributions. Both the restrictions can be effectively checked on SA.

Theorem 3. *Let $\mathcal{S} = (\mathcal{Q}, \mathcal{C}, \mathcal{A}, \rightarrow, \kappa, F)$ be a deterministic SA over exponential distributions. There is an algorithm to decide in time polynomial in $|\mathcal{S}|$ and exponential in $|\mathcal{C}|$ whether $q_1 \sim q_2$ for any locations q_1, q_2.*

The rest of the section deals with the proof. We fix $\mathcal{S} = (\mathcal{Q}, \mathcal{C}, \mathcal{A}, \rightarrow, \kappa, F)$ and $q_1, q_2 \in \mathcal{Q}$. First, we straightforwardly abstract the NLMP semantics $\mathbf{P}_{\mathcal{S}}$ over state space $S = \mathcal{Q} \times \mathbb{R}^{\mathcal{C}}$ by a NLMP $\hat{\mathbf{P}}$ over state space $\hat{S} = \mathcal{Q} \times (\mathbb{R}_{\geq 0} \cup \{-\})^{\mathcal{C}}$ where all negative values of clocks are expressed by $-$. Let ξ denote the obvious mapping of distributions $\mathcal{D}(S)$ onto $\mathcal{D}(\hat{S})$. Then ξ preserves bisimulation since two states s_1, s_2 that differ only in negative values satisfy $\xi(\tau_a(s_1)) = \xi(\tau_a(s_2))$ for all $a \in L$.

Lemma 2. *For any distributions μ, ν on S we have $\mu \sim \nu$ iff $\xi(\mu) \sim \xi(\nu)$.*

Second, similarly to an embedded Markov chain of a CTMC, we further abstract the NLMP $\hat{\mathbf{P}}$ by a *finite* deterministic PA $\bar{D} = (\bar{S}, \mathcal{A}, \longrightarrow)$ such that each state of $\bar{D}$ is a distribution over the uncountable state space $\hat{S}$.

- The set $\bar{S}$ is the set of states reachable via the transitions relation defined below from the distributions μ_{q_1}, μ_{q_2} corresponding to q_1, q_2 (see Definition 4).
- Let us fix a state $\mu \in \bar{S}$ (note that $\mu \in \mathcal{D}(\hat{S})$) and an action $a \in \mathcal{A}$ such that in the NLMP $\hat{\mathbf{P}}$ an a-transition occurs with positive probability, i.e. $\mu \xrightarrow{A_a} \nu$ for some ν and for $A_a = \{a\} \times \mathbb{R}_{\geq 0}$. Thanks to restricting to deterministic SA, $\hat{\mathbf{P}}$ is also deterministic and such a distribution ν is uniquely defined. We set $(\mu, a, M) \in \longrightarrow$ where M is the discrete distribution that assigns probability $p_{q,f}$ to state $\nu_{q,f}$ for each $q \in \mathcal{Q}$ and $f : \mathcal{C} \to \{-, +\}$ where $p_{q,f} = \nu(\hat{S}_{q,f})$, $\nu_{q,f}$ is the conditional distribution $\nu_q(X) := \nu(X \cap \hat{S}_{q,f})/\nu(\hat{S}_{q,f})$ for any measurable $X \subseteq \hat{S}$, and $\hat{S}_{q,f} = \{(q', v) \in \hat{S} \mid q' = q, v(c) \geq 0 \text{ iff } f(c) = + \text{ for each } c \in \mathcal{C}\}$ the set of states with location q and where the sign of clock values matches f.

For exponential distributions all the reachable states $\nu \in \bar{S}$ correspond to some location q where the subset $X \subseteq \mathcal{C}$ is newly sampled, hence we obtain:

Lemma 3. *For a deterministic SA over exponential distributions,* $|\bar{S}| \leq |\mathcal{Q}| \cdot 2^{|\mathcal{C}|}$.

Instead of a greatest fixpoint computation as employed for the discrete algorithm, we take a complementary approach and prove or disprove bisimilarity by a least fixpoint procedure. We start with the initial pair of distributions (states in $\bar{D}$) which generates further requirements that we impose on the relation and try to satisfy them. We work with a *tableau*, a rooted tree where each node is either an *inner node* with a pair of discrete probability distributions over states of $\bar{D}$ as a label, a *repeated node* with a label that already appears somewhere between the node and the root, or a *failure node* denoted by $\square$, and the children of each inner node are obtained by one *rule* from $\{\mathbf{Step}, \mathbf{Lin}\}$. A tableau not containing $\square$ is *successful.*

Step. For a node $\mu \sim \nu$ where μ and ν have *compatible timing*, we add for each label $a \in L$ one child node $\mu_a \sim \nu_a$ where μ_a and ν_a are the unique distributions such that $\mu \xrightarrow{a} \mu_a$ and $\nu \xrightarrow{a} \nu_a$. Otherwise, we add one failure node. We say that μ and ν have compatible timing if for all actions $a \in \mathcal{A}$ we have that $T_a[\mu]$ is equivalent to $T_a[\nu]$. Here $T_a[\rho]$ is a measure over $\mathbb{R}_{\geq 0}$ such that $T_a[\rho](I) := \rho(\hat{S}_{\{a\}\times I})$, i.e. the measure of states moving after time in I with action a.

Lin. For a node $\mu \sim \nu$ linearly dependent on the set of remaining nodes in the tableau, we add one child (repeat) node $\mu \sim \nu$. Here, we understand each node $\mu \sim \nu$ as a vector $\mu - \nu$ in the $|S_\mathcal{S}|$-dimensional vector space.

Note that compatibility of timing is easy to check. Furthermore, the set of rules is correct and complete w.r.t. bisimulation in $\hat{\mathbf{P}}$.

Lemma 4. *There is a successful tableau from* $\mu \sim \nu$ *iff* $\mu \sim \nu$ *in* $\hat{\mathbf{P}}$. *Moreover, the set of nodes of a successful tableau is a subset of a bisimulation.*

We get Theorem 3 since $q_1 \sim q_2$ iff $\xi(\mu_{q_1}) \sim \xi(\mu_{q_2})$ in $\hat{\mathbf{P}}$ and since, thanks to **Lin**:

Lemma 5. *There is a successful tableau from* $\mu \sim \nu$ *iff there is a finite successful tableau from* $\mu \sim \nu$ *of size polynomial in* $|\bar{S}|$.

Example 7. Let us demonstrate the rules by a simple example. Consider the following stochastic automaton $\mathcal{S}$ on the left.

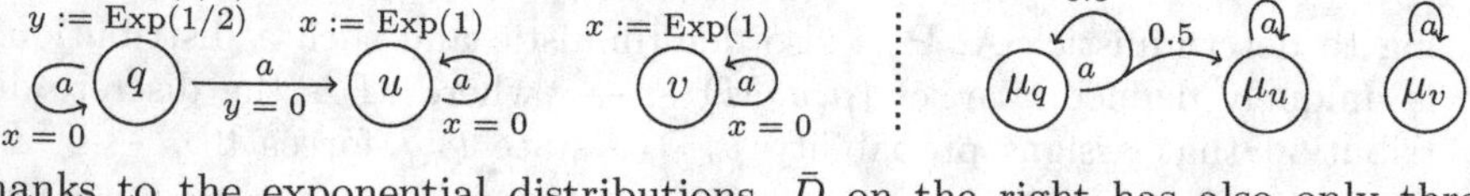

Thanks to the exponential distributions, $\bar{D}$ on the right has also only three states where $\mu_q = q \otimes Exp(1/2) \otimes Exp(1/2)$ is the product of two exponential distributions with rate $1/2$, $\mu_u = u \otimes Exp(1)$, and $\mu_v = v \otimes Exp(1)$. Note that for both clocks x and y, the probability of getting to zero first is 0.5.

$$\frac{1\cdot\mu_u \sim 1\cdot\mu_v}{1\cdot\mu_u \sim 1\cdot\mu_v}\,\textbf{Step} \qquad \cfrac{\cfrac{\cfrac{1\cdot\mu_q + 0\cdot\mu_u \sim 1\cdot\mu_v}{\frac{1}{2}\cdot\mu_q + \frac{1}{2}\cdot\mu_u \sim 1\cdot\mu_v}\,\textbf{Step}}{\frac{1}{4}\cdot\mu_q + \frac{3}{4}\cdot\mu_u \sim 1\cdot\mu_v}\,\textbf{Step}}{\cdots}\,\textbf{Step}$$

The finite tableau on the left is successful since it ends in a repeated node, thus it proves $u \sim v$. The infinite tableau on the right is also successful and proves $q \sim v$. When using only the rule **Step**, it is necessarily infinite as no node ever repeats. The rule **Lin** provides the means to truncate such infinite sequences. Observe that the third node in the tableau on the right above is linearly dependent on its ancestors.

Remark 1. Our approach can be turned into a complete proof system for bisimulation on models with *expolynomial* distributions [(5)]. For them, the states of the discrete transition system $\bar{D}$ can be expressed symbolically. In fact, we conjecture that the resulting semi-algorithm can be twisted to a decision algorithm for this expressive class of models. This is however out of the scope of this paper.

5 Related Work and Discussion

For an overview of coalgebraic work on probabilistic bisimulations we refer to a survey [46]. A considerable effort has been spent to extend this work to continuous-space systems: the solution of [15] (unfortunately not applicable to $\mathbb{R}$), the construction of [21] (described by [42] as "ingenious and intricate"), sophisticated measurable selection techniques in [18], and further approaches of [17] or [51]. In contrast to this standard setting where relations between states and their successor distributions must be handled, our work uses directly relations on distributions which simplifies the setting. The coalgebraic approach has also been applied to trace semantics of uncountable systems [35]. The topic is still very lively, e.g. in the recent [41] a different coalgebraic description of the classical probabilistic bisimulation is given.

Recently, distribution-based bisimulations have been studied. In [19], a bisimulation is defined in the context of language equivalence of Rabin's deterministic probabilistic automata and also an algorithm to compute the bisimulation on them. However, only finite systems with no non-determinism are considered. The most related to our notion are the very recent independently developed [24] and [48]. However, none of them is applicable in the continuous setting and for neither of the two any algorithm has previously been given. Nevertheless, since they are close to our definition, our algorithm with only small changes can actually compute them. Although the bisimulation of [24] in a rather complex way extends [19] to the non-deterministic case reusing their notions, it can be equivalently rephrased as our Definition 2 only considering singleton sets $A \subseteq L$.

(5) With density that is positive on an interval $[\ell, u)$ for $\ell \in \mathbb{N}_0$, $u \in \mathbb{N} \cup \{\infty\}$ given piecewise by expressions of the form $\sum_{i=0}^{I}\sum_{j=0}^{J} a_{ij} x^i e^{-\lambda_{ij} x}$ for $a_{ij}, \lambda_{ij} \in \mathbb{R} \cup \{\infty\}$. This class contains many important distributions such as exponential, or uniform, and enables efficient approximation of others.

Therefore, it is sufficient to only consider matrices P_A^W for singletons A in our algorithm. Apart from being a weak relation, the bisimulation of [48] differs in the definition of $\mu \xrightarrow{A} \nu$: instead of restricting to the states of the support that can perform *some* action of A, it considers those states that can perform *exactly* actions of A. Here each ith row of each transition matrix P_A^W needs to be set to zero if the set of labels from s_i is different from A.

There are also bisimulation relations over distributions defined over finite [9,29] or uncountable [8] state spaces. They, however, coincide with the classical [38] on Dirac distributions and are only directly lifted to non-Dirac distributions. Thus they fail to address the motivating correspondence problem from Section 1. Moreover, no algorithms were given. Further, weak bisimulations [23,22,16] (coarser than usual state based analogues) applied to models without internal transitions also coincide with lifting [29] of the classical bisimulation [38] while our bisimulation is coarser.

There are other bisimulations that identify more states than the classical [38] such as [47] and [4] designed to match a specific logic. Another approach to obtain coarser equivalences on probabilistic automata is via testing scenarios [49].

6 Conclusion

We have introduced a general and natural notion of a distribution-based probabilistic bisimulation, have shown its applications in different settings and have provide algorithms to compute it for finite and some classes of infinite systems. As to future work, the precise complexity of the finite case is certainly of interest. Further, the tableaux decision method opens the arena for investigating wider classes of continuous-time systems where the new bisimulation is decidable.

Acknowledgement. We would like to thank Pedro D'Argenio, Filippo Bonchi, Daniel Gebler, and Matteo Mio for valuable feedback and discussions.

References

1. Agrawal, M., Akshay, S., Genest, B., Thiagarajan, P.: Approximate verification of the symbolic dynamics of Markov chains. In: LICS (2012)
2. Alur, R., Dill, D.: A theory of timed automata. Theor. Comput. Sci. 126(2), 183–235 (1994)
3. Behrmann, G., David, A., Larsen, K.G., Pettersson, P., Yi, W.: Developing UPPAAL over 15 years. Softw., Pract. Exper. 41(2), 133–142 (2011)
4. Bernardo, M., Nicola, R.D., Loreti, M.: Revisiting bisimilarity and its modal logic for nondeterministic and probabilistic processes. Technical Report 06, IMT Lucca (2013)
5. Bravetti, M., D'Argenio, P.R.: Tutte le algebre insieme: Concepts, discussions and relations of stochastic process algebras with general distributions. In: Baier, C., Haverkort, B.R., Hermanns, H., Katoen, J.-P., Siegle, M. (eds.) Validation of Stochastic Systems. LNCS, vol. 2925, pp. 44–88. Springer, Heidelberg (2004)
6. Bravetti, M., Hermanns, H., Katoen, J.-P.: YMCA: Why Markov Chain Algebra? Electr. Notes Theor. Comput. Sci. 162, 107–112 (2006)

7. Castro, P., Panangaden, P., Precup, D.: Equivalence relations in fully and partially observable Markov decision processes. In: IJCAI (2009)
8. Cattani, S.: Trace-based Process Algebras for Real-Time Probabilistic Systems. PhD thesis, University of Birmingham (2005)
9. Crafa, S., Ranzato, F.: A spectrum of behavioral relations over lTSs on probability distributions. In: Katoen, J.-P., König, B. (eds.) CONCUR 2011. LNCS, vol. 6901, pp. 124–139. Springer, Heidelberg (2011)
10. D'Argenio, P., Katoen, J.-P.: A theory of stochastic systems, part I: Stochastic automata. Inf. Comput. 203(1), 1–38 (2005)
11. D'Argenio, P., Katoen, J.-P.: A theory of stochastic systems, part II: Process algebra. Inf. Comput. 203(1), 39–74 (2005)
12. D'Argenio, P.R., Baier, C.: What is the relation between CTMC and TA? Personal Communication (1999)
13. David, A., Larsen, K.G., Legay, A., Mikučionis, M., Poulsen, D.B., van Vliet, J., Wang, Z.: Statistical model checking for networks of priced timed automata. In: Fahrenberg, U., Tripakis, S. (eds.) FORMATS 2011. LNCS, vol. 6919, pp. 80–96. Springer, Heidelberg (2011)
14. David, A., Larsen, K.G., Legay, A., Mikučionis, M., Wang, Z.: Time for statistical model checking of real-time systems. In: Gopalakrishnan, G., Qadeer, S. (eds.) CAV 2011. LNCS, vol. 6806, pp. 349–355. Springer, Heidelberg (2011)
15. de Vink, E.P., Rutten, J.J.M.M.: Bisimulation for probabilistic transition systems: A coalgebraic approach. In: Degano, P., Gorrieri, R., Marchetti-Spaccamela, A. (eds.) ICALP 1997. LNCS, vol. 1256, pp. 460–470. Springer, Heidelberg (1997)
16. Deng, Y., Hennessy, M.: On the semantics of Markov automata. Inf. Comput. 222, 139–168 (2013)
17. Desharnais, J., Gupta, V., Jagadeesan, R., Panangaden, P.: Approximating labeled Markov processes. In: LICS (2000)
18. Doberkat, E.-E.: Semi-pullbacks and bisimulations in categories of stochastic relations. In: Baeten, J.C.M., Lenstra, J.K., Parrow, J., Woeginger, G.J. (eds.) ICALP 2003. LNCS, vol. 2719, pp. 996–1007. Springer, Heidelberg (2003)
19. Doyen, L., Henzinger, T., Raskin, J.-F.: Equivalence of labeled Markov chains. Int. J. Found. Comput. Sci. 19(3), 549–563 (2008)
20. Doyen, L., Massart, T., Shirmohammadi, M.: Limit synchronization in Markov decision processes. CoRR, abs/1310.2935 (2013)
21. Edalat, A.: Semi-pullbacks and bisimulation in categories of Markov processes. Mathematical Structures in Computer Science 9(5), 523–543 (1999)
22. Eisentraut, C., Hermanns, H., Krämer, J., Turrini, A., Zhang, L.: Deciding bisimilarities on distributions. In: QEST (2013)
23. Eisentraut, C., Hermanns, H., Zhang, L.: On probabilistic automata in continuous time. In: LICS (2010)
24. Feng, Y., Zhang, L.: When equivalence and bisimulation join forces in probabilistic automata. In: Jones, C., Pihlajasaari, P., Sun, J. (eds.) FM 2014. LNCS, vol. 8442, pp. 247–262. Springer, Heidelberg (2014)
25. Gast, N., Gaujal, B.: A mean field approach for optimization in discrete time. Discrete Event Dynamic Systems 21(1), 63–101 (2011)
26. Georgievska, S., Andova, S.: Probabilistic may/must testing: Retaining probabilities by restricted schedulers. Formal Asp. Comput. 24(4-6), 727–748 (2012)
27. Giry, M.: A categorical approach to probability theory. In: Banaschewski, B. (ed.) Categorical Aspects of Topology and Analysis. LNCS, vol. 915, pp. 68–85. Springer, Heidelberg (1982)

28. Harrison, P.G., Strulo, B.: Spades - a process algebra for discrete event simulation. J. Log. Comput. 10(1), 3–42 (2000)
29. Hennessy, M.: Exploring probabilistic bisimulations, part I. Formal Asp. Comput. (2012)
30. Hermanns, H., Herzog, U., Mertsiotakis, V.: Stochastic process algebras - between LOTOS and Markov chains. Computer Networks 30(9-10), 901–924 (1998)
31. Hermanns, H., Krčál, J., Křetínský, J.: Probabilistic bisimulation: Naturally on distributions. CoRR, abs/1404.5084 (2014)
32. Hermanns, H., Turrini, A.: Deciding probabilistic automata weak bisimulation in polynomial time. In: FSTTCS (2012)
33. Hillston, J.: A Compositional Approach to Performance Modelling. Cambridge University Press, New York (1996)
34. Jansen, D.N., Nielson, F., Zhang, L.: Belief bisimulation for hidden Markov models - logical characterisation and decision algorithm. In: Goodloe, A.E., Person, S. (eds.) NFM 2012. LNCS, vol. 7226, pp. 326–340. Springer, Heidelberg (2012)
35. Kerstan, H., König, B.: Coalgebraic trace semantics for probabilistic transition systems based on measure theory. In: Koutny, M., Ulidowski, I. (eds.) CONCUR 2012. LNCS, vol. 7454, pp. 410–424. Springer, Heidelberg (2012)
36. Korthikanti, V., Viswanathan, M., Agha, G., Kwon, Y.: Reasoning about MDPs as transformers of probability distributions. In: QEST (2010)
37. Kwiatkowska, M., Norman, G., Parker, D.: Prism 4.0: Verification of probabilistic real-time systems. In: Gopalakrishnan, G., Qadeer, S. (eds.) CAV 2011. LNCS, vol. 6806, pp. 585–591. Springer, Heidelberg (2011)
38. Larsen, K., Skou, A.: Bisimulation through probabilistic testing. In: POPL (1989)
39. May, R., et al.: Biological populations with nonoverlapping generations: stable points, stable cycles, and chaos. Science 186(4164), 645–647 (1974)
40. Milner, R.: Communication and concurrency. PHI Series in computer science. Prentice Hall (1989)
41. Mio, M.: Upper-expectation bisimilarity and Lukasiewicz μ-calculus. In: Muscholl, A. (ed.) FOSSACS 2014. LNCS, vol. 8412, pp. 335–350. Springer, Heidelberg (2014)
42. Sangiorgi, D., Rutten, J.: Advanced Topics in Bisimulation and Coinduction, 1st edn. Cambridge University Press, New York (2011)
43. Segala, R.: Modeling and Verification of Randomized Distributed Real-time Systems. PhD thesis, Massachusetts Institute of Technology, Cambridge, MA, USA (1995)
44. Segala, R., Lynch, N.: Probabilistic simulations for probabilistic processes. In: Jonsson, B., Parrow, J. (eds.) CONCUR 1994. LNCS, vol. 836, pp. 481–496. Springer, Heidelberg (1994)
45. Shani, G., Pineau, J., Kaplow, R.: A survey of point-based pomdp solvers. AAMAS 27(1), 1–51 (2013)
46. Sokolova, A.: Probabilistic systems coalgebraically: A survey. Theor. Comput. Sci. 412(38), 5095–5110 (2011)
47. Song, L., Zhang, L., Godskesen, J.C.: Bisimulations meet PCTL equivalences for probabilistic automata. In: Katoen, J.-P., König, B. (eds.) CONCUR 2011. LNCS, vol. 6901, pp. 108–123. Springer, Heidelberg (2011)
48. Song, L., Zhang, L., Godskesen, J.C.: Late weak bisimulation for Markov automata. CoRR, abs/1202.4116 (2012)
49. Stoelinga, M., Vaandrager, F.: A testing scenario for probabilistic automata. In: Baeten, J.C.M., Lenstra, J.K., Parrow, J., Woeginger, G.J. (eds.) ICALP 2003. LNCS, vol. 2719, pp. 464–477. Springer, Heidelberg (2003)

50. Tzeng, W.: A polynomial-time algorithm for the equivalence of probabilistic automata. SIAM J. Comput. 21(2), 216–227 (1992)
51. Wolovick, N.: Continuous probability and nondeterminism in labeled transaction systems. PhD thesis, Universidad Nacional de Córdoba (2012)

Averaging in LTL

Patricia Bouyer, Nicolas Markey, and Raj Mohan Matteplackel

LSV – CNRS and ENS Cachan – France

Abstract. For the accurate analysis of computerized systems, powerful quantitative formalisms have been designed, together with efficient verification algorithms. However, verification has mostly remained boolean—either a property is true, or it is false. We believe that this is too crude in a context where quantitative information and constraints are crucial: correctness should be quantified!

In a recent line of works, several authors have proposed quantitative semantics for temporal logics, using e.g. *discounting* modalities (which give less importance to distant events). In the present paper, we define and study a quantitative semantics of LTL with *averaging* modalities, either on the long run or within an until modality. This, in a way, relaxes the classical Boolean semantics of LTL, and provides a measure of certain properties of a model. We prove that computing and even approximating the value of a formula in this logic is undecidable.

1 Introduction

Formal verification of computerized systems is an important issue that aims at preventing bugs in the developed computerized systems. The model-checking approach to verification consists in automatically checking that the model of a system satisfies a correctness property. The standard approach is therefore a yes/no (that is, boolean) approach: either the system satisfies the specified property, or the system does not satisfy the property. Model-checking has been widely developed and spread over the last 35 years and is a real success story.

In many applications, quantitative information is crucial; quantities can already appear at the functional level of the system (such as timing constraints between events, or bounds on various quantities like the energy consumption, ...), and many quantitative models like timed automata [4] and their weighted extension [5,7] have therefore been proposed and studied. But quantities can even have more impact on the quality of the system: how good is a system w.r.t. a property? In that case the standard boolean approach might appear as too crude: among those systems that are incorrect (in a boolean sense), some might still be better than others. In order to take this into account, the model-checking approach to verification has to be lifted to a more quantitative perspective [18]. This would allow to *quantify* the quality of systems, and to investigate their tolerance to slight perturbations.

Partly supported by ERC Starting Grant EQualIS and EU FP7 project Cassting.

P. Baldan and D. Gorla (Eds.): CONCUR 2014, LNCS 8704, pp. 266–280, 2014.

There are three classical approaches for turning standard model checking to a quantitative perspective. A first approach, building on automata-based techniques to model checking, consists in defining quantitative semantics for finite state automata. This uses weighted automata [21,16], with different possible semantics. Quantitative decision problems for this setting are addressed in [13,15]. A second approach consists in defining distances between models, or between models and specifications, that can provide an accurateness measure of the model w.r.t. the specification. This approach has been developed e.g. in [12], and then extended into the *model measuring* problem [19]. A third approach is to define quantitative specification languages. For probabilistic systems, this approach is rather standard, and quantitative logics like CSL have been defined and used for model-checking [6]. More recently, this approach has been developed for quantitative but non-stochastic systems. We give more details on those approaches in the "related work" paragraph below.

Example 1 (Jobshop scheduling). Consider a finite set of machines, on which we want to schedule finitely many jobs with possibly dependencies between jobs. Standard analysis asks for the existence of a scheduler that satisfies some scheduling policy, or for optimal such schedulers. A more quality-oriented approach could consist in evaluating the average load along a schedule, or the least machine usage, or the average idle time of a given machine. Those cannot be expressed as a standard boolean model-checking question. ◁

Example 2 (Mobile-phone server). Consider a server that should acknowledge any request by some grant (representing the range of frequency—the bigger the range, the larger the grant). Then the quality of such a server could be expressed as the average over all requests of the range that is allocated in response. This cannot be expressed as a standard boolean model-checking question. ◁

In this paper, we propose quantitative measures of correctness based on the linear-time temporal logic LTL. More precisely, we propose a natural extension of LTL, called avgLTL, with two natural averaging modalities: a new average-until operator $\psi_1 \widetilde{\mathbf{U}} \psi_2$ that computes the average value of ψ_1 along the path until ψ_2 has a high value, and where the semantics of standard modalities are extended using a min-max approach; and a long-run average operator $\widetilde{\mathbf{G}} \psi$, which computes the limit of the values of ψ in the long run along the path. Developing the two examples above, we will show that this logic can express interesting properties.

We focus on the model-checking problem, which corresponds to computing the value of a run (or a Kripke structure) w.r.t. a given property, and on the corresponding decision (comparison with a threshold) and approximation problems. We show that all variants (*i.e.*, all kinds of thresholds, and both when the model is a single path and when it is a Kripke structure) of model-checking and approximation problems are undecidable. Such a robust undecidability is rather surprising (at least to us), given the positive results of [2] for a discounted semantics for LTL, of [22] for an extension of LTL with mean-payoff constraints. Despite the undecidability result for frequency-LTL (a boolean extension of LTL with frequency-constrained "until" modality) and for LTL with average assertions over weighted Kripke structures [8,10], we had hope that some variants of our problem would be decidable.

However we believe these undecidability results are interesting in several respects. (*i*) First, up to now (see related work below), quantitative specification languages based on LTL have always involved discounting factors, which allows to only consider a bounded horizon; this helps obtaining decidability results. In several papers though, averaging in LTL is mentioned, but left as open research directions. (*ii*) Also, we prove robust undecidability results, in the sense that undecidability is proven both for model-checking over a path and model-checking a Kripke structure, and for all thresholds; note that many cases require a specific proof. (*iii*) Finally, our proof techniques are non-trivial and may be interesting in other contexts; we were not able to get a direct encoding of two-counter machines for proving the undecidability of the model-checking problem over Kripke structures, and had to use a diagonal argument; this is due to convergence phenomena that arise in the context of quantitative model-checking, and which have mostly been omitted so far in the rest of the literature.

Related Work. Several recent papers have proposed quantitative-verification frameworks based on temporal logic. The authors of [14] were the first to suggest giving temporal logics a quantitative semantics: they extend CTL with various new modalities involving a discount on the future (the later the event, the smaller the impact on the value of the formula). In that framework, model-checking is proven decidable.

As regards linear-time temporal logics, a first attempt to define a quantitative semantics has been proposed in [17]. However, no modality is really quantitative, only the models are quantitative, yielding finitely non-boolean values. Still, the authors suggest discounting and long-run averaging as possible extensions of their work. Another approach is tackled in [1], where functions f are added to the syntax of LTL, with the value of $f(\psi_1, \ldots, \psi_k)$ on a path π being the result of applying f to the values of subformulas $\psi_1, \ldots, \psi_k$ on π. As explained in [1], this quantitative language is not that expressive: each formula only takes finitely many values. It follows that the verification problems are decidable.

Frequency-LTL , an extension of LTL with "frequency-until", has been studied in [9], and even though it has a boolean semantics, the frequency modality gives a quantitative taste to the logic: $\phi_1 \mathbf{U}^c \phi_2$ holds true along a path whenever there is a position along that path at which ϕ_2 holds, and the frequency of ϕ_1 along the prefix is at least c. This paper shows the undecidability of the satisfiability problem. We discuss this approach in more details in Section 8, since it shares some techniques with ours.

Finally the recent work [2] is the closest to ours. It studies LTL extended with a discounted until modality: roughly, the values of the subformulas are multiplied by a discount factor, which decreases and tends to zero with the distance to the evaluation point. This way, the further the witness, the lower the value. An automata-based algorithm is given to decide the threshold problem. Due to discounting, whether the value of a formula is larger than some threshold on a path can be checked on a bounded prefix of the path. On the other hand, adding local average (*i.e.*, the average of finitely many subformulas) yields undecidability (for the existence of a path with value 1/2). We will discuss with more details this paper in Section 8.

2 Average-LTL

Let $\mathcal{P}$ be a finite set of atomic propositions. A *quantitative Kripke structure* over $\mathcal{P}$ is a 4-tuple $\mathcal{K} = \langle V, v_0, E, \ell \rangle$ where V is a finite set of vertices, $v_0 \in V$ is the initial vertex, $E \subseteq V \times V$ is a set of transitions (which we assume total, meaning that for each $v \in V$, there exists $v' \in V$ s.t. $(v, v') \in E$) and $\ell \colon V \to ([0,1] \cap \mathbb{Q})^{\mathcal{P}}$ is a labelling function, associating with each state the value of each atomic proposition in that state. The Kripke structure $\mathcal{K}$ is said *qualitative* whenever for every $v \in V$ and $p \in \mathcal{P}$, $(\ell(v))(p) \in \{0,1\}$. A run or path in a Kripke structure $\mathcal{K}$ from $v \in V$ is a finite or infinite sequence $\pi = (v_i)_{i \in I}$ (where I is a (bounded or unbounded) interval of $\mathbb{N}$ containing 0) s.t. $v_0 = v$ and $(v_{i-1}, v_i) \in E$ for all relevant $i \in I \setminus \{0\}$. The size $|\pi|$ of π is the cardinality of I. In the sequel, we will be interested in the sequence $\ell(\pi) = (\ell(v_i))_{i \in I}$, and we will often identify a run with the sequence in $(([0,1] \cap \mathbb{Q})^{\mathcal{P}})^I$ it defines. Given a run $\pi = (v_i)_{i \in I}$ and an integer j, we write $\pi_{\geq j}$ for the run $(v_{i+j})_{i \geq 0, i+j \in I}$.

We now introduce the logic average-LTL (avgLTL for short) and its interpretation over infinite runs. The syntax of avgLTL over $\mathcal{P}$ is given by:

$$\varphi ::= p \mid \neg p \mid \varphi \vee \varphi \mid \varphi \wedge \varphi \mid \mathbf{X}\, \varphi \mid \varphi \,\mathbf{U}\, \varphi \mid \mathbf{G}\, \varphi \mid \varphi \,\widetilde{\mathbf{U}}\, \varphi \mid \widetilde{\mathbf{G}}\, \varphi.$$

where $p \in \mathcal{P}$. Notice that negation is only allowed on atomic propositions. We write LTL for the fragment where $\widetilde{\mathbf{U}}$ and $\widetilde{\mathbf{G}}$ are not allowed.

Let $\pi = (v_i)_{i \in \mathbb{N}}$ be an infinite run, and φ be an avgLTL formula. The valuation $[\![\pi, \varphi]\!]$ is then given as follows:

$$[\![\pi, p]\!] = (\ell(v_0))(p) \qquad [\![\pi, \neg p]\!] = 1 - (\ell(v_0))(p)$$

$$[\![\pi, \psi_1 \vee \psi_2]\!] = \max\{[\![\pi, \psi_1]\!], [\![\pi, \psi_2]\!]\} \qquad [\![\pi, \mathbf{X}\, \psi]\!] = [\![\pi_{\geq 1}, \psi]\!]$$

$$[\![\pi, \psi_1 \wedge \psi_2]\!] = \min\{[\![\pi, \psi_1]\!], [\![\pi, \psi_2]\!]\}$$

$$[\![\pi, \mathbf{G}\, \psi]\!] = \inf_{i \in \mathbb{N}} [\![\pi_{\geq i}, \psi]\!]$$

$$[\![\pi, \psi_1 \,\mathbf{U}\, \psi_2]\!] = \sup_{i \in \mathbb{N}} \min\{[\![\pi_{\geq i}, \psi_2]\!], \min_{0 \leq j < i}([\![\pi_{\geq j}, \psi_1]\!])\}$$

$$[\![\pi, \widetilde{\mathbf{G}}\, \psi]\!] = \liminf_{i \to \infty} \left(\textstyle\sum_{j=0}^{j<i} [\![\pi_{\geq j}, \psi]\!]\right)/i$$

$$[\![\pi, \psi_1 \,\widetilde{\mathbf{U}}\, \psi_2]\!] = \sup\left(\{[\![\pi, \psi_2]\!]\} \cup \left\{\min\{[\![\pi_{\geq i}, \psi_2]\!], (\textstyle\sum_{j=0}^{j<i} [\![\pi_{\geq j}, \psi_1]\!])/i\} \mid i > 0\right\}\right)$$

We recover the boolean semantics for the standard operators when all atomic propositions have either value 0 (false) or value 1 (true). Note that in that case we might abusively consider that $v_i \in 2^{\mathcal{P}}$, recording the set of atomic propositions with value 1 at each position. The first five rules are standard and natural in a quantitative setting. The semantics of the $\mathbf{U}$- and $\mathbf{G}$-modalities are also natural: they extends the standard equivalences $\psi_1 \,\mathbf{U}\, \psi_2 \equiv \psi_2 \vee (\psi_1 \wedge \mathbf{X}\,(\psi_1 \,\mathbf{U}\, \psi_2))$, and $\mathbf{G}\, \psi \equiv \psi \wedge \mathbf{X}\, \mathbf{G}\, \psi$ to a quantitative setting. The last two modalities are specific to our setting: formula $\psi_1 \widetilde{\mathbf{U}} \psi_2$ computes the average of formula ψ_1 for the i first steps, and then compares the value with that of ψ_2 at the $(i+1)$-st step. The best choice of i (if it exists) is then selected, and gives the value to the formula. Formula $\widetilde{\mathbf{G}} \psi$ computes the average of ψ in the long-run.

We come back to our two illustrative examples given in the introduction, to show how our logic can be used to express natural properties.

Example 3 (Jobshop scheduling). We come back to Example 1, assuming a set of n machines. Let load be an atomic proposition having value k/n at state s if k machines are in use in that state. Notice that we could equivalently use the local averaging operator $\oplus$ of [2] in order to have load defined as the average of the atomic propositions indicating which machines are in use. Then formula $\varphi_1 = \mathsf{load} \, \widetilde{\mathbf{U}} \, \mathsf{stop}$ evaluated on a schedule computes the average machine use along that schedule, if stop is a boolean atomic proposition which holds true when all jobs are finished. A schedule assigning value 1 to φ_1 could be seen as an optimal schedule, where no computation power is lost. A schedule assigning a small value to formula φ_1 is a schedule with a large loss of computation power.

On the other hand formula $\varphi_2 = \mathsf{load} \, \mathbf{U} \, \mathsf{stop}$ will evaluate to the smallest instantaneous machine use along a schedule. Note that syntactically it is a standard until, but it evaluates differently in our quantitative framework. ◁

Example 4 (Mobile phone server). The quality of the server of Example 2 can be expressed as the average over all requests of the frequency allocated in response. We can write such a property as $\varphi_3 = \widetilde{\mathbf{G}} \, (\neg \mathsf{req} \vee \mathsf{no\_grant} \, \mathbf{U} \, \mathsf{grant})$, where req and $\mathsf{no\_grant}$ are boolean atomic propositions with the obvious meaning, and grant is an atomic proposition with value in $[0,1]$ representing the quality of the allocated range of frequencies (the closer to 1, the better). Larger values of φ_3 then indicate better frequency allocation algorithms. ◁

We also evaluate formulas of avgLTL over Kripke structures. If v is a state of the Kripke structure $\mathcal{K}$ and $\varphi \in \mathsf{avgLTL}$, then we define: $[\![(\mathcal{K}, v), \varphi]\!] = \sup\{[\![\pi, \varphi]\!] \mid \pi$ is an infinite run of $\mathcal{K}$ from $v\}$. We simply write $[\![\mathcal{K}, \varphi]\!]$ when $v = v_0$ is the initial vertex of $\mathcal{K}$. Notice that considering the supremum here corresponds to the *existential* semantics of boolean LTL, where the aim is to find a path satisfying the formula.

Example 5. We develop a small toy example to illustrate how simple formulas can be evaluated in the (qualitative) Kripke structure depicted on Fig. 1.

Consider the avgLTL formulas $a \, \widetilde{\mathbf{U}} \, b$ and $c \, \widetilde{\mathbf{U}} \, b$. For the first formula we have $[\![a \cdot b \cdot c^\omega, a \, \widetilde{\mathbf{U}} \, b]\!] = 1$ (the supremum being reached at the second position along the run), and therefore $[\![\mathcal{K}, a \, \widetilde{\mathbf{U}} \, b]\!] = 1$.

Now, for the formula $c \, \widetilde{\mathbf{U}} \, b$ and the same run as above, we have $[\![a \cdot b \cdot c^\omega, c \, \widetilde{\mathbf{U}} \, b]\!] = 0$: indeed, the right-hand-side formula b has value zero everywhere except at position 1, but the average of c on the previous positions is zero. For the run $a \cdot (b \cdot c)^\omega$, considering all positions (but position 1) where b is non-zero, we get $[\![a \cdot (b \cdot c)^\omega, c \, \widetilde{\mathbf{U}} \, b]\!] = \sup\{n/(2n+1) \mid n \in \mathbb{N}_{>0}\} = 1/2$. Note that the value $1/2$ is not reached by any prefix. Now consider the run $\pi'_k =$

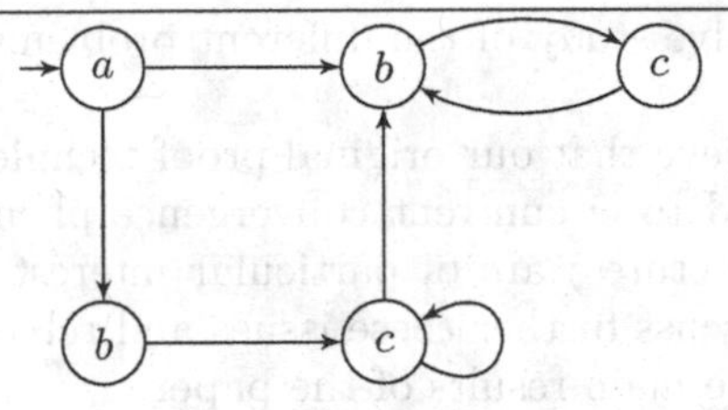

Fig. 1. A Kripke structure $\mathcal{K}$

$a \cdot b \cdot c^k \cdot (b \cdot c)^\omega$, for some positive integer k. Then we have $[\![\pi'_k, c \widetilde{\mathbf{U}} b]\!] = \sup\{(k+n)/(k+2n+2) \mid n \in \mathbb{N}\}$. When $k \geq 3$, the supremum is $k/(k+2)$, which is reached for $n = 0$ (*i.e.*, at the second occurrence of b). From this we get that $[\![\mathcal{K}, c \widetilde{\mathbf{U}} b]\!] = 1$. However no run witnesses that value. $\lhd$

3 The Problems We Consider

In this paper, we consider the following two problems:

Existence Problem: given a Kripke structure $\mathcal{K}$, an avgLTL formula φ, and a threshold $\bowtie c$ (with $\bowtie \in \{<, \leq, =, \geq, >\}$ and $c \in [0,1] \cap \mathbb{Q}$), is there a path π in $\mathcal{K}$ such that $[\![\pi, \varphi]\!] \bowtie c$?

Value Problem: given a Kripke structure $\mathcal{K}$, an avgLTL formula φ, and a threshold $\bowtie c$ (with $\bowtie \in \{<, \leq, =, \geq, >\}$ and $c \in [0,1] \cap \mathbb{Q}$), does $[\![\mathcal{K}, \varphi]\!] \bowtie c$?

Note that both problems are different since, as illustrated in Example 5, it can be the case that $[\![\mathcal{K}, \varphi]\!] = 1$ even though no path of $\mathcal{K}$ assigns value 1 to φ.

We also consider their approximation variants, defined as follows:

Approximate Existence Problem: given a Kripke structure $\mathcal{K}$, an avgLTL formula φ, a value $c \in [0,1] \cap \mathbb{Q}$ and $\varepsilon > 0$, is there a path π in $\mathcal{K}$ such that $c - \varepsilon < [\![\pi, \varphi]\!] < c + \varepsilon$?

Approximate Value Problem: given a Kripke structure $\mathcal{K}$, an avgLTL formula φ, a value $c \in [0,1] \cap \mathbb{Q}$ and $\varepsilon > 0$, does $c - \varepsilon < [\![\mathcal{K}, \varphi]\!] < c + \varepsilon$?

4 Model Checking avgLTL Is Undecidable

In the sequel, we prove that avgLTL model-checking is *robustly* undecidable, in the sense that all the problems above are undecidable, for all threshold conditions considered. We would like to emphasize that different kinds of threshold give rise to different problems, and could have led to different decidability results. For instance, given a Kripke structure $\mathcal{K}$ and an avgLTL formula φ, $[\![\mathcal{K}, \varphi]\!] > 1/2$ iff there exists an infinite run π in $\mathcal{K}$ such that $[\![\pi, \varphi]\!] > 1/2$. On the other hand, $[\![\mathcal{K}, \varphi]\!] = 1/2$ iff there exists a sequence of infinite runs $(\pi^n)_{n \in \mathbb{N}}$ such that $[\![\pi^n, \varphi]\!] \leq 1/2$ for every n, and $\lim_{n \to \infty} [\![\pi^n, \varphi]\!] = 1/2$. These remarks advocate for a clear and exhaustive study of the different problems with all the different thresholds.

Additionally, we believe that our original proof techniques (in particular the diagonal argument used to circumvent convergence phenomena for the model-checking of Kripke structures) are of particular interest and could be used in related settings. We discuss further these issues and related works in Section 8

We can now state the main results of the paper.

Theorem 6. *The existence problem is undecidable, for every threshold of the form* $\bowtie 1/2$, *with* $\bowtie \in \{<, \leq, =, \geq, >\}$.

Theorem 7. *The value problem is undecidable, for every threshold of the form* $\bowtie 1/2$, *with* $\bowtie \in \{<, \leq, =, \geq, >\}$.

We present these results as two distinct theorems, since proofs require very different techniques, even though a similar encoding is used.

Remark 8. Our proofs only involve qualitative Kripke structures. We present the results for $c = 1/2$, but our proofs could be adapted to handle any other rational value in $(0, 1)$ (e.g. by inserting fake actions in the encoding).

Now, if the approximate variants were decidable, then taking e.g. $c = 1$ and $\epsilon = 1/2$, we could decide e.g. whether a formula has value larger than $1/2$, contradicting the previous theorems. Hence:

Theorem 9. *The approximate existence and value problems are undecidable.*

The rest of the paper presents the main ideas of the proof. Due to lack of space, the full proofs could not be included here, but can be found in the research report [11] associated to this paper.

5 Proof of Theorem 6

We only give an explanation of the undecidability for the existence problem with threshold $\geq 1/2$ (the other types of thresholds require a twist in the construction, but no fundamental new argument).

The proof relies on an encoding of the halting problem for deterministic two-counter machines, which is well-known to be undecidable. A two-counter machine $\mathcal{M}$ is a finite-state machine, equiped with two kinds of transitions: *update*-transitions move from one state to another one while incrementing or decrementing one of the counters; *test*-transitions keep the counters unchanged, but may lead to two different states depending on the positiveness of one of the counters. The machine has a special state, called the *halting state*, from which no transitions is possible. We assume w.l.o.g. that all the other states have exactly one outgoing transition.

A configuration of $\mathcal{M}$ is given by the current state and the values of both counters. A run of $\mathcal{M}$ is a sequence of consecutive configurations *which might not properly update the counters.* It is said *valid* whenever the counters are properly updated along the run. There is a unique maximal valid run in $\mathcal{M}$ from the initial configuration: it is either halting or infinite.

The idea of our reduction is to build a Kripke structure which generates the encodings of all (including invalid) runs of $\mathcal{M}$: it has to take care of the discrete structure of $\mathcal{M}$, but does not check that counters are properly updated along the run. Correct update of counter values will be checked using an avgLTL formula.

Description of the Encoding. We first explain how we encode the runs of $\mathcal{M}$. We only give a simplified idea of the encoding. We write Q for the set of states of $\mathcal{M}$.

For $p \geq 2$, we write $\mathbb{B}_p$ for the set $\{0, 1, \ldots, p-1\}$. For $b \in \mathbb{B}_p$, we let $b^{+i} = b + i \mod p$. An element of $\mathbb{B}_p$ is abusively called a *bit*. These bits are used to distinguish between consecutive configurations. For the rest of this section, taking $p = 2$ would be sufficient, but the proof of Theorem 7 requires higher values for p. We encode configurations of $\mathcal{M}$ using the following finite set of atomic propositions: $\mathcal{P}_p = (Q \cup \{a_0, a_1\}) \times \mathbb{B}_p \cup \{\#\}$. The symbol $\#$ will be a marker for halting computations.

Exactly one atomic proposition from $\mathcal{P}_p$ will have value one at each position along the encoding (the other propositions having value zero). Given a bit b, a configuration $\gamma = (q, n_0, n_1)$ of $\mathcal{M}$ is encoded as the word $\mathsf{enc}_b(\gamma) = (q, b) \cdot (a_0, b)^{n_0} \cdot (a_1, b)^{n_1}$. For a halting configuration, we set $\mathsf{enc}_b(\gamma) = (q_{halt}, b)$.

The bit $b \in \mathbb{B}_p$ is incremented (modulo p) from one configuration to the next one. Let $\rho = \gamma_0 \cdot \gamma_1 \cdots$ be a (not necessary valid) run in $\mathcal{M}$. The p-encoding of ρ is then given by:

$$p\text{-}\mathsf{enc}(\rho) = \begin{cases} \mathsf{enc}_{b_0}(\gamma_0) \cdot \mathsf{enc}_{b_1}(\gamma_1) \cdot \mathsf{enc}_{b_2}(\gamma_2) \cdots & \text{if } \rho \text{ is infinite} \\ \mathsf{enc}_{b_0}(\gamma_0) \cdot \mathsf{enc}_{b_1}(\gamma_1) \cdots \mathsf{enc}_{b_{n-1}}(\gamma_{n-1}) \#^\omega & \text{if } \rho \text{ has length } n \end{cases}$$

with $b_j = j \mod p$ for every j. We write $\mathsf{enc}(\rho)$ if p is clear from the context.

We can easily construct a Kripke structure that generates the encodings of all possible (valid or invalid) runs of $\mathcal{M}$. For index p, we write $\mathcal{K}^p_{\mathcal{M}}$ for the corresponding Kripke structure. We now turn to the avgLTL formula, whose role is to check proper updates of the counters.

Definition of the Formulas. We will define a formula $\mathtt{consec}^p_{\mathcal{M}}$, which will be used to check that each single consecution in the run properly updates the counters. Then we define formula

$$\mathtt{halt}^p_M = \mathbf{F}\, q_{halt} \wedge \mathbf{G}\, \mathtt{consec}^p_{\mathcal{M}}.$$

It is rather clear that if we can build such a formula $\mathtt{consec}^p_{\mathcal{M}}$, then the above formula will check that the unique maximal valid run of $\mathcal{M}$ is halting. Unfortunately, things are not that easy, and formula $\mathbf{G}\, \mathtt{consec}^p_{\mathcal{M}}$ will only be able to check the validity of *finite* runs

We now focus on defining $\mathtt{consec}^p_{\mathcal{M}}$, using the average-until modality. We only give an intuition (the full definition requires the complete encoding). Consider a portion P of the p-encoding of a run ρ, which corresponds to a single-step of the computation of $\mathcal{M}$ where instruction q keeps both counter values unchanged:

$$\ldots (q, b) \cdot (a_0, b)^{n_0} \cdot (a_1, b)^{n_1} \cdot (q', b^{+1}) \cdot (a_0, b^{+1})^{n'_0} \cdot (a_1, b^{+1})^{n'_1} (q'', b^{+2}) \ldots$$

The formula has to enforce $n_0' = n_0$ and $n_1' = n_1$. This is the case if, and only if, for every $\alpha \in \{1 + n_0 + n_1, 1 + n_0 + n_1', 1 + n_0' + n_1, 1 + n_0' + n_1'\}$,

$$\frac{\alpha}{1 + n_0 + n_1 + 1 + n_0' + n_1'} = \frac{1}{2}.$$

The denominator is the length of the portion from (q, b) to the position just before (q'', b^{+2}), whereas the various values for α are the number of positions where some distinguished atomic proposition holds along this portion. For instance, $1 + n_0' + n_1$ is the number of positions where formula $\psi = (q', b^{+1}) \vee (a_0, b^{+1}) \vee (a_1, b)$ holds along P. Computing the above quotient will be done using an $\widetilde{\mathbf{U}}$-formula: $[\![P, \psi\, \widetilde{\mathbf{U}}\, (q'', b^{+2})]\!]$ precisely equals $\frac{\alpha}{1+n_0+n_1+1+n_0'+n_1'}$.

Using this idea, we are able to construct a formula $\texttt{consec}^p_{\mathcal{M}}$ (as a conjunction of several $\widetilde{\mathbf{U}}$-formulas) whose value is $1/2$ along a single step of the computation if, and only if, this step is valid (that is, it correctly updates the counters).

Correctness of the Reduction. Even though formula $\texttt{consec}^p_{\mathcal{M}}$ properly checks the validity of a single step of the computation, it might be the case that $[\![p\text{-enc}(\rho), \mathbf{G}\,\texttt{consec}^p_{\mathcal{M}}]\!] = 1/2$, even though the whole computation is not valid: this is due to the definition of the semantics of $\widetilde{\mathbf{U}}$ as the supremum over all positions of the average; in particular, a single error in the computation can be hidden in the rest of the run. Consider for instance the counter machine in Fig. 2. The unique initial and maximal valid run of $\mathcal{M}$ halts. However, if the first transition increments counter a_0 twice, and all further transitions are properly taken, then the resulting (invalid) run will assign value $1/2$ to formula $\mathbf{G}\,\texttt{consec}^p_{\mathcal{M}}$.

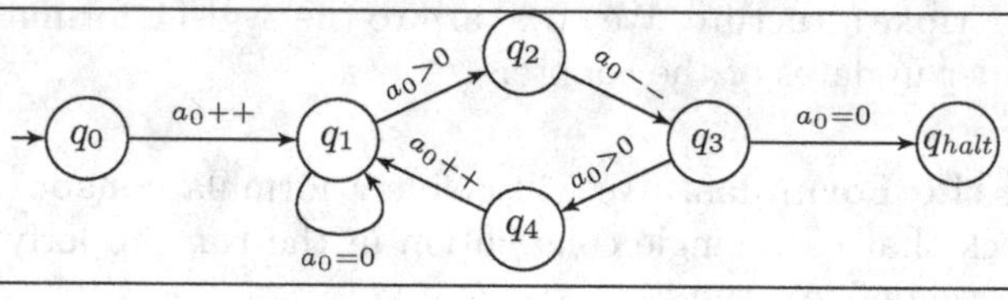

Fig. 2. There is an invalid infinite run ρ such that $[\![p\text{-enc}(\rho), \mathbf{G}\,\texttt{consec}^p_{\mathcal{M}}]\!] = 1/2$

Still, we are able to prove the following classification of runs of $\mathcal{M}$ in terms of the value of $\texttt{halt}^p_{\mathcal{M}}$. It proves the fact that formula $\texttt{consec}^p_{\mathcal{M}}$ properly checks the validity of a single step of the computation, provided the $\widetilde{\mathbf{U}}$-formulas cannot benefit from the supremum semantics. This is the case when the run in the Kripke structure ends with $\#^\omega$, which corresponds to finite runs of $\mathcal{M}$.

Classification 1. *Fix $p \geq 2$. Let ρ be a maximal run in $\mathcal{M}$.*

- *if ρ is infinite, then $[\![p\text{-enc}(\rho), \texttt{halt}^p_{\mathcal{M}}]\!] = 0$;*

– *if ρ is finite and valid, then* $[\![\text{p-enc}(\rho), \mathtt{halt}^p_{\mathcal{M}}]\!] = 1/2$;
– *if ρ is finite and invalid, then* $[\![\text{p-enc}(\rho), \mathtt{halt}^p_{\mathcal{M}}]\!] < 1/2$.

Corollary 10. *Fix $p \geq 2$. The following five statements are equivalent:*

1. *$\mathcal{M}$ halts;*
2. *the unique initial and maximal valid run $\rho_{\mathcal{M}}$ of $\mathcal{M}$ is such that* $[\![\text{p-enc}(\rho_{\mathcal{M}}), \mathtt{halt}^p_{\mathcal{M}}]\!] = 1/2$;
3. *there exists an initial maximal run ρ in $\mathcal{M}$ such that* $[\![\text{p-enc}(\rho), \mathtt{halt}^p_{\mathcal{M}}]\!] = 1/2$;
4. *there exists an initial maximal path π in $\mathcal{K}^p_{\mathcal{M}}$ such that* $[\![\pi, \mathtt{halt}^p_{\mathcal{M}}]\!] = 1/2$;
5. *there exists an initial maximal path π in $\mathcal{K}^p_{\mathcal{M}}$ such that* $[\![\pi, \mathtt{halt}^p_{\mathcal{M}}]\!] \geq 1/2$.

This corollary allows to conclude the undecidability proof of Theorem 6.

6 Proof of Theorem 7

As already mentioned, whether $[\![\mathcal{K}, \varphi]\!] > 1/2$ (and dually, $[\![\mathcal{K}, \varphi]\!] \leq 1/2$) is equivalent to the existence of a path whose value is strictly more than 1/2, which we just proved undecidable.

We now turn to the more interesting cases of = (the result for $\geq$ and $<$ directly follows, as we explain at the end of this proof). We were not able to write a direct proof as previously, because we could not distinguish between counter machines that have a halting computation (whose encoding has value 1/2 against formula $\mathtt{halt}^p_{\mathcal{M}}$ above) and counter machines that have sequences of computations whose encodings have values *converging* to 1/2.

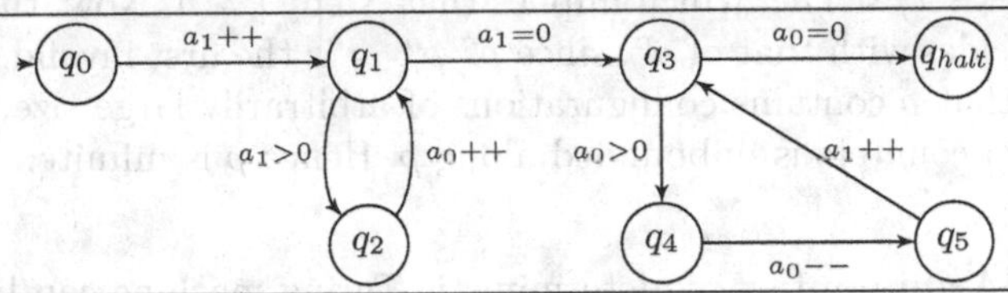

Fig. 3. A non-halting two-counter machine for which $[\![\mathcal{K}^p_{\mathcal{M}}, \mathtt{halt}^p_{\mathcal{M}}]\!] = 1/2$

Example 11. We consider the deterministic two-counter machine $\mathcal{M}$ of Fig. 3, having q_0 as its initial state. The unique initial and maximal valid run of $\mathcal{M}$ is infinite (it loops in $q_1 \leftrightarrows q_2$). A single error can make the transition from q_1 to q_3 available, from which valid consecutions lead to q_{halt}. The *weight* of this error can be arbitrarily small, as it can occur with an arbitrarily large value of a_0. It is not difficult to check that $[\![\mathcal{K}^p_{\mathcal{M}}, \mathtt{halt}^p_{\mathcal{M}}]\!] = 1/2$ (for any $p \geq 2$). ◁

Analysis of a Non-Halting Two-Counter Machine. We consider a deterministic accept/reject two-counter machine $\mathcal{M}$: such machines have two halting states, now named q_{accept} and q_{reject}. Their computations may still be infinite. We consider formula $\mathsf{consec}^p_{\mathcal{M}}$ again, and define $\mathsf{accept}^p_{\mathcal{M}} = \mathbf{F}\, q_{accept} \wedge \mathbf{G}\, \mathsf{consec}^p_{\mathcal{M}}$.

We first analyse the impact of the first error along a finite run ρ of $\mathcal{M}$ onto the value of $\mathbf{G}\, \mathsf{consec}^p_{\mathcal{M}}$, and we are able to show the following surprising but crucial lemma (remember the example of Fig. 2) whose proof requires long technical developments. The condition imposed on p is a sufficient condition for "detecting" invalid consecutions along finite runs. The computation leading to this value is explained in the long version [11] of this work.

Lemma 12. *Fix $p \geq 927$. Let ρ be a finite invalid run of $\mathcal{M}$. Assume $\rho_i \rho_{i+1}$ is the first invalid consecution along ρ, and write step_i for the portion of p-$\mathsf{enc}(\rho)$ corresponding to that consecution. Pick $n \geq 30$ such that $[\![\mathsf{step}_i, \mathsf{consec}^p_{\mathcal{M}}]\!] \leq 1/2 - 1/n$. Then $[\![p\text{-}\mathsf{enc}(\rho), \mathbf{G}\, \mathsf{consec}^p_{\mathcal{M}}]\!] \leq 1/2 - 1/n$.*

This allows to prove the next fundamental result:

Lemma 13. *Fix $p \geq 927$, and assume that $[\![\mathcal{K}^p_{\mathcal{M}}, \mathsf{accept}^p_{\mathcal{M}}]\!] = 1/2$, but that no run ρ of $\mathcal{M}$ has $[\![p\text{-}\mathsf{enc}(\rho), \mathsf{accept}^p_{\mathcal{M}}]\!] = 1/2$. Then the unique initial and maximal valid run of $\mathcal{M}$ is infinite.*

We sketch the proof of this lemma, since it contains an interesting argument.

Sketch of proof. Let ρ be the unique initial and maximal valid run of $\mathcal{K}^p_{\mathcal{M}}$. Let $(\rho^n)_{n\in\mathbb{N}}$ be a sequence of initial and maximal runs such that $[\![p\text{-}\mathsf{enc}(\rho^n), \mathsf{accept}^p_{\mathcal{M}}]\!] > 1/2 - 1/n$ (such a sequence exists by hypothesis, but runs ρ^n might be invalid). Pick $n \geq 30$, and let $\rho^n_{i_n} \rho^n_{i_n+1}$ be the first invalid consecution of ρ^n. Write step_{i_n} for the portion of p-$\mathsf{enc}(\rho^n)$ corresponding to that consecution. Applying Lemma 12, we get that $[\![\mathsf{step}_{i_n}, \mathsf{consec}^p_{\mathcal{M}}]\!] > 1/2 - 1/n$. Since $\rho^n_{i_n} \rho^n_{i_n+1}$ is an invalid consecution, we also have that $[\![\mathsf{step}_{i_n}, \mathsf{consec}^p_{\mathcal{M}}]\!] < 1/2$. It follows that $1/(|\mathsf{step}_{i_n}| - 1) < 1/n$, which implies that $|\mathsf{step}_{i_n}| > n$. Now, the prefix of ρ of size i_n coincides with that of ρ^n, since $\rho^n_{i_n} \rho^n_{i_n+1}$ is the first invalid consecution. We conclude that ρ contains configurations of arbitrarily large size, so that the sum of the two counters is unbounded along ρ. Hence ρ is infinite. □

A Diagonal Argument. Any deterministic Turing machine can be simulated by a deterministic two-counter machine [20]. In particular, given a deterministic Turing machine B, we can build a deterministic two-counter machine $\mathcal{M}(B)$ whose computation mimics the run of B on input B. Then $\mathcal{M}(B)$ accepts (resp. rejects, does not halt) if, and only if, B accepts (resp. rejects, does not halt on) input B.

We fix $p \geq 927$, and define the following function $\mathcal{H}$, which takes as input a deterministic Turing machine B:

$$\mathcal{H}(B) = \begin{cases} \textit{accept} & \text{if } [\![\mathcal{K}^p_{\mathcal{M}(B)}, \mathsf{accept}^p_{\mathcal{M}(B)}]\!] = 1/2 \\ \textit{reject} & \text{otherwise} \end{cases}$$

Proposition 14. *The function $\mathcal{H}$ is not computable.*

Proof. Towards a contradiction, assume $\mathcal{H}$ is computable. Let $\mathcal{T}_{\mathcal{H}}$ be a deterministic Turing machine that computes $\mathcal{H}$. Notice in particular that $\mathcal{T}_{\mathcal{H}}$ halts on all its inputs; we assume that it ends in its state $q^{\mathcal{T}}_{accept}$ when $\mathcal{H}$ accepts the input, and in $q^{\mathcal{T}}_{reject}$ when $\mathcal{H}$ returns *reject*.

We now define the following deterministic Turing machine $\mathcal{C}$, which takes as input a deterministic Turing machine B:

$\mathcal{C}(B)$: Simulate $\mathcal{T}_{\mathcal{H}}$ on B;
If the simulation ends in $q^{\mathcal{T}}_{accept}$ then goto $q^{\mathcal{C}}_{reject}$, otherwise goto $q^{\mathcal{C}}_{accept}$.

The Turing machine $\mathcal{C}$ terminates on all its inputs, since so does $\mathcal{T}_{\mathcal{H}}$; also, $\mathcal{C}$ is deterministic, and we can therefore run $\mathcal{C}$ on input $\mathcal{C}$ itself.

Assume $\mathcal{C}$ accepts input $\mathcal{C}$. This means that $\mathcal{H}(\mathcal{C})$ rejects, which means that $[\![\mathcal{K}^p_{\mathcal{M}(\mathcal{C})}, \texttt{accept}^p_{\mathcal{M}(\mathcal{C})}]\!] < 1/2$. This means that $\mathcal{M}(\mathcal{C})$ does not accept (by a straightforward extension of Corollary 10 to accept/reject two-counter machines), and therefore $\mathcal{C}$ does not accept $\mathcal{C}$, contradicting our hypothesis.

Hence $\mathcal{C}$ rejects input $\mathcal{C}$, so that $[\![\mathcal{K}^p_{\mathcal{M}(\mathcal{C})}, \texttt{accept}^p_{\mathcal{M}(\mathcal{C})}]\!] = 1/2$. However, since $\mathcal{C}$ does not accept $\mathcal{C}$, the unique initial and maximal valid run of $\mathcal{M}_{\mathcal{C}}$ is either infinite or rejecting. Applying Lemma 13 to $\mathcal{M}_{\mathcal{C}}$, we get that it is actually infinite. This means that the simulation of $\mathcal{T}_{\mathcal{H}}$ on input $\mathcal{C}$ does not terminate. This contradicts the fact that $\mathcal{T}_{\mathcal{H}}$ terminates on every input. Therefore $\mathcal{H}$ is not computable. □

Theorem 7 is a direct consequence of this lemma for threshold $= 1/2$. Now, using Classification 1, for a deterministic two-counter machine $\mathcal{M}$, it holds that $[\![\mathcal{K}^p_{\mathcal{M}}, \texttt{accept}^p_{\mathcal{M}}]\!] = 1/2$ iff $[\![\mathcal{K}^p_{\mathcal{M}}, \texttt{accept}^p_{\mathcal{M}}]\!] \geq 1/2$. Hence the above proof applies to threshold $\geq 1/2$ as well. The case of $< 1/2$ is the dual of $\geq 1/2$: if $\mathcal{K}$ is a Kripke structure and φ an avgLTL formula, $[\![\mathcal{K}, \varphi]\!] < 1/2$ iff it is not the case that $[\![\mathcal{K}, \varphi]\!] \geq 1/2$, which proves the result for threshold $< 1/2$ as well.

7 Proof of Theorem 9

We now discuss the undecidability of the approximate variants. It relies on the same encoding as that for the existence problem and threshold $> 1/2$. For that threshold, we have a classification of the runs similar to Classification 1, for formula $\texttt{halt}^{p,>}_{\mathcal{M}}$: for every maximal run ρ in $\mathcal{M}$:

- if ρ is infinite, then $[\![p\text{-enc}(\rho), \texttt{halt}^{p,>}_{\mathcal{M}}]\!] = 0$;
- if ρ is finite and valid, then $1/2 < [\![p\text{-enc}(\rho), \texttt{halt}^{p,>}_{\mathcal{M}}]\!] < 3/4$;
- if ρ is finite and invalid, then $[\![p\text{-enc}(\rho), \texttt{halt}^{p,>}_{\mathcal{M}}]\!] < 1/2$.

We deduce that $\mathcal{M}$ halts iff there exists an initial and maximal valid run π in $\mathcal{K}^p_{\mathcal{M}}$ with $1/2 < [\![\pi, \texttt{halt}^{p,>}_{\mathcal{M}}]\!] < 3/4$. This shows undecidability of the approximate existence problem.

Now, we also have in this case the equivalence with $1/2 < [\![\mathcal{K}^p_{\mathcal{M}}, \texttt{halt}^{p,>}_{\mathcal{M}}]\!] < 7/8$ (not 3/4 since there might be some convergence phenomenon towards value 3/4), which also shows the undecidability of the approximate value problem.

8 Discussion on Related Works

In this section, we would like to illustrate the difficulty of lifting temporal-logic model checking from the qualitative to the quantitative setting. As we saw in this paper, several new convergence phenomena do appear, which make the problem complex, but also make the proofs difficult. Our undecidability proofs in this paper involve difficult techniques to properly handle the convergence phenomena that appear in the semantics of the logic. This difficulty has led to several wrong arguments in the related litterature, as we now illustrate.

We first discuss the logic frequency-LTL of [9]. This logic has a boolean semantics, but extends LTL with a frequency-$\mathbf{U}$ modality, which gives it a quantitative taste: formula $\phi_1 \mathbf{U}^c \phi_2$ holds true along a path π whenever there is a position n along π at which ϕ_2 holds, and the number of previous positions where ϕ_1 holds is larger than or equal to $c \cdot n$ (hence c is a lower bound on the frequency of ϕ_1 on the prefix before ϕ_2 holds). Note that it need not be the case that the position n is the first position where ϕ_2 holds: for instance $abbcaaac$ satisfies formula $a \mathbf{U}^{\frac{1}{2}} c$, but at the first occurrence of c, the frequency of a on the prefix is $1/3$, which is less than $1/2$; the correct witness position for $a \mathbf{U}^{\frac{1}{2}} c$ is the second occurrence of c, where the frequency of a becomes $4/7$. In frequency-LTL, there is no convergence phenomena, but some possibly unbounded search for some witnessing position. Then evaluating $b \mathbf{U}^{\frac{1}{2}} c$ on a path π is not equivalent to comparing formula $b \widetilde{\mathbf{U}} c$ to value $1/2$ on path π: first because of convergence phenomena (as illustrated in Example 5), and because in our quantitative setting, the value of the right-hand-side subformula could be less than $1/2$.

It is shown in [9] that the validity problem for frequency-LTL is undecidable, and our reduction shares similarities with that reduction (but we believe that our reduction is simpler, and the result stronger, since it uses no nested $\widetilde{\mathbf{U}}$). However the undecidability proof (as written in [9]) has a flaw: it relies on the claim that "[t]he formula $b \mathbf{U}^{\frac{1}{2}} l \wedge \hat{b} \mathbf{U}^{\frac{1}{2}} l$ enforces the pattern $b^m \hat{b}^m l$..." (the order of b's and $\hat{b}$'s is enforced by another LTL formula). This claim is wrong in general, since the $\mathbf{U}^{\frac{1}{2}}$-formulas might not refer to the same occurrences of l. The proof can be patched[1], and one way is to restrict to paths that end with $\#^\omega$ for some marker #; in that way a backward argument can be used to check proper encoding of the execution of the two-counter machine (this is actually what we do in the proof of Theorem 6).

We now discuss the logic discounted-LTL of [2]. This logic gives a quantitative semantics to an extension of LTL, with a new discounted-$\mathbf{U}$ modality: given a discount function η, the value of formula $\phi_1 \mathbf{U}_\eta \phi_2$ along a path π is the supremum over all positions n along π of the minimum of the value of ϕ_2 at that position, discounted by $\eta(n)$, and of the values of ϕ_1 at every earlier position i, discounted by $\eta(i)$. Satisfiability is proven decidable; it is shown undecidable when adding the local average operator $\oplus$, which computes the average of two formulas.

[1] Personal communication with the authors.

Those results are then extended to the model-checking problem[2]. While the first result extends properly for threshold $< c$ (since the infimum over all paths is smaller than c if, and only if, there is a path that evaluates to a value smaller than c; hence convergence phenomena are avoided), it is not valid for Theorem 3 of [2] (which is stated with threshold $> c$). Also, undecidability of the model-checking problem with local-average operator (Theorem 6 of [2]) is not correct since it does not take convergence phenomena into account. A corrected version of the proof is available in [3]; while it does not use a diagonal argument as we do, the undecidability proof is not a direct encoding of a two-counter machine, but requires computing the value of two different formulas in order to encode the halting problem.

This all shows that extending temporal logics to a quantitative setting is more than a simple exercise: complex convergence phenomena come into play, which have to be understood and handled with extreme care. We hope that our work will provide new insights about these problems, and believe that our techniques can be useful for handling them.

9 Conclusion and Future Work

We believe that our logic avgLTL is a very relevant logic in many applications. It provides a way of *measuring* some properties, such as the average load of the CPUs in scheduling applications. We proved that the value of a formula can not be computed—and not even approximated. For the interesting case however (deciding whether $[\![\mathcal{K}, \phi]\!] \geq \eta$), we had to resort to an original diagonal argument to get around convergence phenomena.

Our negative results certainly echo back the fact, mentioned e.g. in [17], that averaging does not fit well with classical automata-based approaches for temporal logics. Indeed, averaging gives rise to new values that are not present in the original automaton. Discounting LTL instead of averaging has the same difficulty, but this is compensated by the fact that when discounting, the value of a formula can be approximated by considering only a finite prefix of a run [2].

We are currently investigating two directions in order to get decidability results: first by adding discounting on the right-hand-side formula (while keeping averaging on the left-hand-side); second, by considering the *qualitative* cases of avgLTL, namely whether a formula has value 0 or 1. One difficulty here is that in some cases the witnesses are a family of paths, instead of just a single path.

References

1. Almagor, S., Boker, U., Kupferman, O.: Formalizing and reasoning about quality. In: Fomin, F.V., Freivalds, R., Kwiatkowska, M., Peleg, D. (eds.) ICALP 2013, Part II. LNCS, vol. 7966, pp. 15–27. Springer, Heidelberg (2013)

[2] Note that the value of a Kripke structure is defined there as the infimum over all paths, and not as the supremum as we do here; this corresponds to the duality between existential *vs* universal quantification in standard LTL.

2. Almagor, S., Boker, U., Kupferman, O.: Discounting in LTL (To appear). In: Ábrahám, E., Havelund, K. (eds.) TACAS 2014 (ETAPS). LNCS, vol. 8413, pp. 424–439. Springer, Heidelberg (2014)
3. Almagor, S., Boker, U., Kupferman, O.: Discounting in LTL. Research Report 1406.4249, arXiv, 21 pages (2014)
4. Alur, R., Dill, D.L.: A theory of timed automata. Theoretical Computer Science 126(2), 183–235 (1994)
5. Alur, R., La Torre, S., Pappas, G.J.: Optimal paths in weighted timed automata. In: Di Benedetto, M.D., Sangiovanni-Vincentelli, A.L. (eds.) HSCC 2001. LNCS, vol. 2034, pp. 49–62. Springer, Heidelberg (2001)
6. Aziz, A., Sanwal, K., Singhal, V., Brayton, R.K.: Model-checking continuous-time Markov chains. ACM Transactions on Computational Logic 1(1), 162–170 (2000)
7. Behrmann, G., Fehnker, A., Hune, T., Larsen, K.G., Pettersson, P., Romijn, J., Vaandrager, F.: Minimum-cost reachability for priced timed automata. In: Di Benedetto, M.D., Sangiovanni-Vincentelli, A.L. (eds.) HSCC 2001. LNCS, vol. 2034, pp. 147–161. Springer, Heidelberg (2001)
8. Boker, U., Chatterjee, K., Henzinger, T.A., Kupferman, O.: Temporal specifications with accumulative values. In: LICS 2011, pp. 43–52. IEEE Comp. Soc. Press (2011)
9. Bollig, B., Decker, N., Leucker, M.: Frequency linear-time temporal logic. In: TASE 2012, pp. 85–92. IEEE Comp. Soc. Press (2012)
10. Bouyer, P., Gardy, P., Markey, N.: Quantitative verification of weighted kripke structures. Research Report LSV-14-08, Laboratoire Spécification et Vérification, ENS Cachan, France, 26 pages (2014)
11. Bouyer, P., Markey, N., Matteplackel, R.M.: Quantitative verification of weighted kripke structures. Research Report LSV-14-02, Laboratoire Spécification et Vérification, ENS Cachan, France, 35 pages (2014)
12. Černý, P., Henzinger, T.A., Radhakrishna, A.: Simulation distances. Theor. Computer Science 413(1), 21–35 (2012)
13. Chatterjee, K., Doyen, L., Henzinger, T.A.: Quantitative languages. ACM Transactions on Computational Logic 11(4) (2010)
14. de Alfaro, L., Faella, M., Henzinger, T.A., Majumdar, R., Stoelinga, M.: Model checking discounted temporal properties. Theor. Computer Science 345(1), 139–170 (2005)
15. Doyen, L.: Games and Automata: From Boolean to Quantitative Verification. Mémoire d'habilitation, ENS Cachan, France (2012)
16. Droste, M., Kuich, W., Vogler, W. (eds.): Handbook of Weighted Automata. Springer (2009)
17. Faella, M., Legay, A., Stoelinga, M.: Model checking quantitative linear time logic. In: QAPL 2008. ENTCS, vol. 220, pp. 61–77. Elsevier Science (2008)
18. Henzinger, T.A.: Quantitative reactive models. In: France, R.B., Kazmeier, J., Breu, R., Atkinson, C. (eds.) MODELS 2012. LNCS, vol. 7590, pp. 1–2. Springer, Heidelberg (2012)
19. Henzinger, T.A., Otop, J.: From model checking to model measuring. In: D'Argenio, P.R., Melgratti, H. (eds.) CONCUR 2013 – Concurrency Theory. LNCS, vol. 8052, pp. 273–287. Springer, Heidelberg (2013)
20. Minsky, M.L.: Computation: Finite and Infinite Machines. Prentice Hall, Inc. (1967)
21. Schützenberger, M.-P.: On the definition of a family of automata. Information and Control 4(2-3), 245–270 (1961)
22. Tomita, T., Hiura, S., Hagihara, S., Yonezaki, N.: A temporal logic with mean-payoff constraints. In: Aoki, T., Taguchi, K. (eds.) ICFEM 2012. LNCS, vol. 7635, pp. 249–265. Springer, Heidelberg (2012)

Decidable Topologies for Communicating Automata with FIFO and Bag Channels*

Lorenzo Clemente[1], Frédéric Herbreteau[2], and Grégoire Sutre[2]

[1] Université Libre de Bruxelles, Brussels, Belgium
[2] Univ. Bordeaux and CNRS, LaBRI, UMR 5800, Talence, France

Abstract. We study the reachability problem for networks of finite-state automata communicating over unbounded perfect channels. We consider communication topologies comprising both ordinary FIFO channels and *bag channels*, i.e., channels where messages can be freely re-ordered. It is well-known that when only FIFO channels are considered, the reachability problem is decidable if, and only if, there is no undirected cycle in the topology. On the other side, when only bag channels are allowed, the reachability problem is decidable for any topology by a simple reduction to Petri nets. In this paper, we study the more complex case where the topology contains *both* FIFO and bag channels, and we provide a complete characterisation of the decidable topologies in this generalised setting.

1 Introduction

Communicating finite-state automata (CFSA) are a fundamental model of computation where concurrent processes exchange messages over unbounded, reliable channels. Depending on the context, messages are delivered in the order they were sent (FIFO channel), or in any order (bag channel). On the one hand, FIFO channels can be used, e.g., to model communications through TCP sockets, as TCP preserves the order of messages. It is well-known that the reachability problem for CFSA with only FIFO channels is undecidable [5,18]. This problem becomes decidable when the communication topology is required to be acyclic [18,14]. Many other decidable subclasses and under/over-approximation techniques have been considered in the literature [3,2,7,6,14,8,4,11,10,1]. On the other hand, bag channels can be used, e.g., to model asynchronous procedure calls [19,12,9]. Indeed, libraries supporting asynchronous programming do not guarantee, in general, that procedures are executed in the order they were asynchronously called. The reachability problem for CFSA with only bag channels is decidable (without any further restriction), by an immediate reduction to reachability in Petri nets, which is known to be decidable [16,13,15].

Contributions. While the reachability problem is well-understood for communication topologies of just FIFO or bag channels, we go one step forward and we

* This work was partially supported by the ANR project Vacsim (ANR-11-INSE-004).

P. Baldan and D. Gorla (Eds.): CONCUR 2014, LNCS 8704, pp. 281–296, 2014.

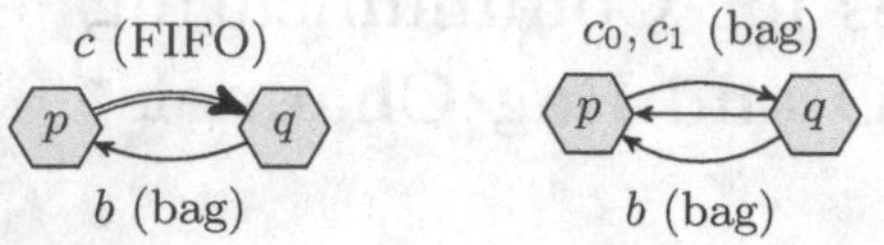

(a) Opposite channels **(b)** Synchronising c

Fig. 1. Opposite channels

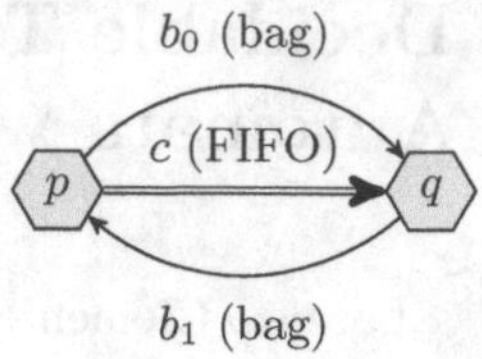

Fig. 2. Undecidable topology

study topologies comprising *both* FIFO *and* bag channels. Our main result is a complete characterisation of decidable topologies of FIFO and bag channels, and a detailed complexity analysis in the decidable case. As a consequence of our results, we show that certain non-trivial cycles comprising FIFO and bag channels can be allowed while preserving decidability.

In addition to being the right model in some contexts, bag channels also provide a non-trivial over-approximation of FIFO channels. Indeed, it is always possible to over-approximate the reachability set by turning *all* channels into bag channels. Thanks to our characterisation, a much finer analysis may be obtained, in practice, by selectively over-approximating only *some* of the FIFO channels.

Preview. Let us illustrate our main techniques with some example. While the topology in Fig. 1a is undecidable when all channels are FIFO, it becomes decidable when b is bag. Indeed, the FIFO channel c can be "made synchronous" by forcing receptions to occur right after transmissions. This, in turn, can be implemented by replacing c with two opposite bag channels c_0, c_1 implementing a simple rendezvous protocol. We thus obtain the topology in Fig. 1b, which is decidable since it contains only bag channels.

A more difficult case is the one in Fig. 3a. As above, reachability is undecidable if all channels are FIFO, but it becomes decidable when one channel, say b, is bag. However, the correctness argument is more involved here, since, unlike in the previous example, channel c cannot be made synchronous. The problem is that making c synchronous requires rescheduling the actions of the receiver q to occur earlier. But this is not possible since q might try to read on the other channel b, which could be empty. The crucial observation is that we can always schedule all actions of p to occur before all actions of q. Therefore, in this topology, the order between transmissions and receptions can be relaxed. The only property that matters is that the string of messages which is received is the same as the one which is sent. We can thus *split* the bag channel b into two bag channels b_0 and b_1 (see Fig. 3b), where q's potentially blocking receptions on b are replaced with non-blocking transmissions on b_1. The new process r just matches incoming messages on b_0 and b_1. In the new topology, c can be made synchronous, and we proceed as above to obtain the decidable topology in Fig. 3c.

Finally, we also have undecidable topologies which mix in a non-trivial way FIFO and bag channels. For example, consider the one in Fig. 2, where c is FIFO and b_0 and b_1 are bags. This topology is undecidable, even when b_0 and b_1 are *unary* bag channels (i.e., the message alphabet is a singleton). The idea is to use the two bag channels b_0 and b_1 to implement a synchronisation protocol

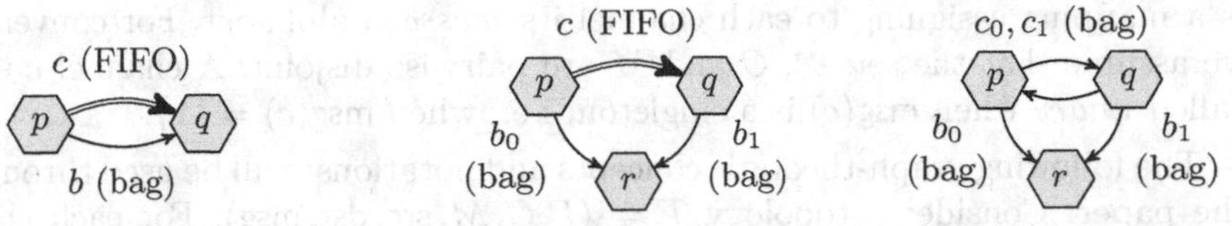

(a) Parallel channels b and c **(b)** Splitting b **(c)** Synchronising c

Fig. 3. Two parallel channels

between processes p and q. This protocol is then used by p to decide which message is to be received by q from c, thus simulating a channel machine (which has undecidable reachability).

Outline. The rest of the paper is organised as follows. We start with preliminaries in Sec. 2. In Sec. 3 we show techniques for synchronising and splitting channels (decidability). In Sec. 4 we explain how unary bag channels can be used to simulate rendezvous synchronisation (undecidability). In Sec. 5 we present our characterisation of decidable topologies and study the complexity of the decidable instances. Finally, in Sec. 6 we compare our techniques with the work of Chambart and Schnoebelen [8], and in Sec. 7 we end with directions for future work.

2 Preliminaries

A *labelled transition system* (LTS for short) is a tuple $\mathcal{A} = \langle S, S_I, S_F, A, \rightarrow \rangle$ where S is a set of *states* with *initial states* $S_I \subseteq S$ and *final states* $S_F \subseteq S$, A is a finite set of *actions*, and $\rightarrow \subseteq S \times A \times S$ is a *labelled transition relation.* For simplicity, we write $s \xrightarrow{a} s'$ in place of $(s, a, s') \in \rightarrow$. An LTS is called *finite* when its set of states is finite. A *run* in $\mathcal{A}$ is an alternating sequence $(s_0, a_1, s_1, \ldots, a_n, s_n)$ of states $s_i \in S$ and actions $a_i \in A$, with $n \geq 0$, such that $s_{i-1} \xrightarrow{a_i} s_i$ for all $1 \leq i \leq n$. The natural number n, which may be zero, is called the *length* of the run. An *accepting run* is a run starting in an initial state (i.e., with $s_0 \in S_I$) and ending in a final state (i.e., with $s_n \in S_F$). Given two runs $\sigma = (s_0, a_1, s_1, \ldots, a_m, s_m)$ and $\tau = (t_0, b_1, t_1, \ldots, b_n, t_n)$ such that $s_m = t_0$, their *join* is the run $\sigma \cdot \tau = (s_0, a_1, s_1, \ldots, a_m, s_m, b_1, t_1, \ldots, b_n, t_n)$.

Topologies. We consider systems that are composed of several processes communicating through the asynchronous exchange of messages. Communications rely on point-to-point FIFO channels between processes. In our setting, each channel is equipped with a message alphabet that specifies the set of messages that can be conveyed over the channel. To simplify the presentation, we assume a special message, written 1, that is always in the message alphabet. Formally, a communication *topology* is a tuple $\mathcal{T} = \langle P, C, M, \mathsf{src}, \mathsf{dst}, \mathsf{msg} \rangle$, where P is a finite set of *processes*, C is a finite set of *channels*, M is a finite set of *messages* containing the special message 1, $\mathsf{src} : C \rightarrow P$ and $\mathsf{dst} : C \rightarrow P$ are mappings assigning to each channel a *source* and a *destination* process, and $\mathsf{msg} : C \rightarrow \{N \subseteq M \mid 1 \in N\}$

is a mapping assigning to each channel its *message* alphabet. For convenience, we assume that the sets P, C and M are pairwise disjoint. A channel $c \in C$ is called *unary* when $\mathsf{msg}(c)$ is a singleton, i.e., when $\mathsf{msg}(c) = \{1\}$.

The following graph-theoretic concepts and notations[1] will be used throughout the paper. Consider a topology $\mathcal{T} = \langle P, C, M, \mathsf{src}, \mathsf{dst}, \mathsf{msg} \rangle$. For each channel $c \in C$, we let $\stackrel{c}{\Longrightarrow}$ denote the binary relation on P defined by $p \stackrel{c}{\Longrightarrow} q$ if $p = \mathsf{src}(c)$ and $q = \mathsf{dst}(c)$. The inverse of $\stackrel{c}{\Longrightarrow}$ is written $\stackrel{c}{\Longleftarrow}$. A *directed walk* in $\mathcal{T}$ is an alternating sequence $(p_0, c_1, p_1, \ldots, c_n, p_n)$ of processes $p_i \in P$ and channels $c_i \in C$, with $n \geq 0$, such that $p_{i-1} \stackrel{c_i}{\Longrightarrow} p_i$ for all $1 \leq i \leq n$. The natural number n, which may be zero, is called the *length* of the directed walk. To improve readability, directed walks will usually be written $p_0 \stackrel{c_1}{\Longrightarrow} p_1 \cdots \stackrel{c_n}{\Longrightarrow} p_n$. A directed walk is said to be *closed* when it starts and ends in the same process (i.e., when $p_0 = p_n$). A *directed path* is a directed walk in which all channels are pairwise distinct, and all processes—except, possibly, the first and last ones—are pairwise distinct. The notation $p \stackrel{*}{\Longrightarrow} q$ means that there is a directed walk—or, equivalently, there is a directed path—from p to q (i.e., with $p_0 = p$ and $p_n = q$). A *directed cycle* is a closed directed path of non-zero length.

We also need undirected variants of the above notions. For each channel $c \in C$, we let $\stackrel{c}{=\!=}$ denote the binary relation on P defined by $p \stackrel{c}{=\!=} q$ if $\{p, q\} = \{\mathsf{src}(c), \mathsf{dst}(c)\}$. Observe that $p \stackrel{c}{=\!=} q$ if, and only if, $p \stackrel{c}{\Longrightarrow} q$ or $p \stackrel{c}{\Longleftarrow} q$. The notions of *undirected walk*, *undirected path* and *undirected cycle* are defined as the directed ones, except that $\stackrel{c}{\Longrightarrow}$ is replaced by $\stackrel{c}{=\!=}$.

Communicating Processes. Given a topology $\mathcal{T} = \langle P, C, M, \mathsf{src}, \mathsf{dst}, \mathsf{msg} \rangle$, the set of possible *communication actions* for a process $p \in P$, written A^p_{com}, is the union of the set $\{c!m \mid c \in C \wedge \mathsf{src}(c) = p \wedge m \in \mathsf{msg}(c)\}$ of its *transmission actions* and of the set $\{c?m \mid c \in C \wedge \mathsf{dst}(c) = p \wedge m \in \mathsf{msg}(c)\}$ of its *reception actions*. The set of all communication actions is $A_{\mathrm{com}} = \bigcup_{p \in P} A^p_{\mathrm{com}}$. Actions not in A_{com} are called *internal actions*.

Definition 2.1. *A* system of communicating processes *is a pair* $\mathcal{S} = \langle \mathcal{T}, \{\mathcal{A}^p\}_{p \in P} \rangle$ *where* $\mathcal{T} = \langle P, \ldots \rangle$ *is a topology, and, for each* $p \in P$, $\mathcal{A}^p = \langle S^p, S^p_I, S^p_F, A^p, \rightarrow^p \rangle$ *is a finite labelled transition system such that* $A^p \cap A_{\mathrm{com}} = A^p_{\mathrm{com}}$. *For convenience, we assume*[2] *that the sets of actions* A^p, *with* $p \in P$, *are pairwise disjoint.*

We give the operational *semantics* of a system of communicating processes $\mathcal{S}$ as a global labelled transition system $[\![\mathcal{S}]\!] = \langle X, X_I, X_F, A, \rightarrow \rangle$. States of $\mathcal{S}$ are called *configurations* to prevent confusion with those of $\mathcal{A}^p$. A configuration of $[\![\mathcal{S}]\!]$ is pair $x = (\boldsymbol{s}, \boldsymbol{w})$ where $\boldsymbol{s}$ maps each process p to a state in S^p, and $\boldsymbol{w}$ maps each channel c to a word over its message alphabet $\mathsf{msg}(c)$. Formally, $X = (\prod_{p \in P} S^p) \times (\prod_{c \in C} \mathsf{msg}(c)^*)$. A configuration is initial (resp. final) when

[1] In this paper, we use contemporary graph terminology (see, e.g., [20]). For instance, the term *walk* is used for "paths" that may repeat channels and/or processes.

[2] This assumption is not restrictive as it only concerns internal actions. Indeed, the sets A^p_{com}, with $p \in P$, are already pairwise disjoint by definition.

each process is in its initial state (resp. final state) and all channels are empty. Formally, $X_I = (\prod_{p\in P} S^p_I) \times \{\varepsilon\}$ and $X_F = (\prod_{p\in P} S^p_F) \times \{\varepsilon\}$, where ε maps each channel $c \in C$ to the empty word ε. The set of actions A of $\mathcal{S}$ is given by $A = \bigcup_{p\in P} A^p$. Observe that $\{A^p\}_{p\in P}$ is a partition of A. We define the transition relation $\rightarrow$ of $[\![\mathcal{S}]\!]$ to be the set of all triples (x_1, a, x_2), where $x_1 = (\boldsymbol{s}_1, \boldsymbol{w}_1)$ and $x_2 = (\boldsymbol{s}_2, \boldsymbol{w}_2)$ are configurations, such that, for some process $p \in P$, the following conditions are satisfied:

- $s^p_1 \xrightarrow{a} s^p_2$ is a transition in $\mathcal{A}^p$, and $s^q_1 = s^q_2$ for all other processes $q \in P\setminus\{p\}$.
- If a is an internal action, then $\boldsymbol{w}_1 = \boldsymbol{w}_2$.
- If $a = c!m$, then $w^c_2 = w^c_1 \cdot m$ and $w^d_2 = w^d_1$ for all other channels $d \in C \setminus \{c\}$.
- If $a = c?m$, then $w^c_1 = m \cdot w^c_2$ and $w^d_2 = w^d_1$ for all other channels $d \in C \setminus \{c\}$.

So, in a transition $x_1 \xrightarrow{a} x_2$, exactly one process p moves (namely, the unique $p \in P$ such that $a \in A^p$), and the others stay put. The channels are updated according to the action a that is performed by the transition. Given a process $p \in P$, a *move* of p is any transition $x_1 \xrightarrow{a} x_2$ such that $a \in A^p$. Following [11], we define the *causal-equivalence* relation $\sim$ over runs as the least congruence, with respect to join, such that $(x_1, a, x_2, b, x_3) \sim (x_1, b, x'_2, a, x_3)$ whenever a, b are actions of distinct processes. Informally, two runs are causal-equivalent if they can be transformed one into the other by iteratively commuting adjacent moves that (i) are *not* from the same process and (ii) do *not* form a "matching send/receive pair". It is readily seen that causal-equivalent runs necessarily start in the same configuration and end in the same configuration.

Statement of the Problem. Given a topology $\mathcal{T}$, the *reachability problem* for systems of communicating processes with topology $\mathcal{T}$, denoted by REACH($\mathcal{T}$), is defined as follows:

Input: a system of communicating processes $\mathcal{S}$ with topology $\mathcal{T}$,
Output: whether there exists an accepting run in $[\![\mathcal{S}]\!]$.

Observe that we require all channels to be empty at the end of an accepting run. Also note that REACH($\mathcal{T}$) is parametrised by a topology $\mathcal{T}$. The main result of this paper is a characterisation of the topologies $\mathcal{T}$ for which REACH($\mathcal{T}$) is decidable. Our techniques are based on topological transformations that induce reductions between the associated reachability problems. We let $\preceq_m$ denote the *many-one reducibility* relation between decision problems. For example, REACH($\mathcal{T}$) $\preceq_m$ REACH($\mathcal{U}$) when $\mathcal{T}$ is obtained from $\mathcal{U}$ by removing channels.

When only unary channels are present, the reachability problem is decidable by an immediate reduction to reachability in Petri nets.

Theorem 2.2 ([16,13,15]). *If $\mathcal{T}$ is a topology with only unary channels, then* REACH($\mathcal{T}$) *is decidable.*

On the other hand, when the topology contains only non-unary channels the following characterisation for REACH($\mathcal{T}$) is well-known.

Theorem 2.3 ([18,14]). *Given a topology $\mathcal{T}$ with no unary channel,* $\text{REACH}(\mathcal{T})$ *is decidable if, and only if, $\mathcal{T}$ has no undirected cycle.*

In Sec. 5, we refine the latter condition to account for topologies with both unary and non-unary channels (see Theorem 5.3). We further generalise it in Sec. 6 to a more general setting comprising both FIFO and bag channels.

3 Synchronising and Splitting Channels

A useful technique in the analysis of communicating processes is to transform asynchronous communications into synchronous ones, without compromising the behaviour of the system. To this end, we use the notion of synchronous runs from [8,11]. Formally, a run $((\boldsymbol{s}_0, w_0), a_1, (\boldsymbol{s}_1, w_1), \ldots, a_n, (\boldsymbol{s}_n, w_n))$ is *synchronous* for a given channel c if $w_0^c = w_n^c = \varepsilon$ and $w_i^c = \varepsilon \vee w_{i+1}^c = \varepsilon$ for all $0 < i < n$. Intuitively, this means that each transmission on c is immediately followed by its matching reception, i.e., communication over c behaves like rendezvous synchronisation [17]. It is well-known that in a polyforest topology (i.e., with no undirected cycle), every run can be reordered to have *all* channels synchronous, and reachability can be solved by exploring the resulting finite transition system [18,14,8,11]. Since we are interested in analysing more complex topologies where not all channels can be simultaneously made synchronous in general, we need to consider channels individually rather than globally.

Synchronising Essential Channels. Whether a channel can always be made synchronous (by reordering moves in runs) is a semantic condition that depends on the complex behaviour of the whole system. In fact, this condition is an undecidable problem in general (by an easy reduction from the reachability problem). Therefore, we are interested in syntactic conditions that are sufficient for a channel to be made synchronous. One such condition is that of *essential channel* [8], which is a structural condition depending only on the topology.

Definition 3.1 ([8]). *A channel c is* essential *if all directed paths from* $\mathsf{src}(c)$ *to* $\mathsf{dst}(c)$ *contain c. In particular,* $\mathsf{src}(c) \neq \mathsf{dst}(c)$.

Lemma 3.2 ([8]). *If c is an essential channel, then every run that starts and ends with c empty is causal-equivalent to a run that is synchronous for c.*

Thus, we can replace asynchronous communications on an essential channel by synchronous ones. We loosely implement the latter through a topological transformation that replaces an essential channel c, with source p and destination q, by one unary channel c_m from p to q for each message $m \in \mathsf{msg}(c)$, as well as one unary channel c_{ack} back from q to p. These unary channels are enough to simulate synchronous communications over c. Each message m conveyed over c is placed in the corresponding unary channel c_m instead. After each transmission on c_m, the process p waits for an acknowledgement from c_{ack}. Conversely, q notifies p via c_{ack} after each reception from c_m. While it applies to arbitrary essential channels, this topological transformation is only useful for non-unary ones.

Definition 3.3. *Given a topology* $\mathcal{T} = \langle P, C, M, \mathsf{src}, \mathsf{dst}, \mathsf{msg}\rangle$ *and a channel* $c \in C$, *the* synchronisation *of* c *in* $\mathcal{T}$ *is the topology* $\mathcal{U} = \langle P, C', M, \mathsf{src}', \mathsf{dst}', \mathsf{msg}'\rangle$ *defined by* $C' = (C \setminus \{c\}) \cup \{c_m \mid m \in \mathsf{msg}(c)\} \cup \{c_{\mathsf{ack}}\}$ *and*

$$(\mathsf{src}'(d), \mathsf{dst}'(d), \mathsf{msg}'(d)) = \begin{cases} (\mathsf{src}(d), \mathsf{dst}(d), \mathsf{msg}(d)) & \text{if } d \in C \setminus \{c\} \\ (\mathsf{src}(c), \mathsf{dst}(c), \{1\}) & \text{if } d = c_m \\ (\mathsf{dst}(c), \mathsf{src}(c), \{1\}) & \text{if } d = c_{\mathsf{ack}} \end{cases}$$

where c_m, *for* $m \in \mathsf{msg}(c)$, *and* c_{ack} *are new channels that are not in* $P \cup C \cup M$.

Proposition 3.4 (Synchronisation). *If* c *is an essential channel in* $\mathcal{T}$, *then* $\text{REACH}(\mathcal{T}) \preceq_m \text{REACH}(\mathcal{U})$ *where* $\mathcal{U}$ *results from the synchronisation of* c *in* $\mathcal{T}$.

Remark 3.5. An essential channel could, alternatively, be removed by merging its endpoints (see [8]). Instead, our synchronisation construction replaces an essential channel by a collection of new unary channels. While either technique could be used for decidability (see Subsec. 5.2), synchronisation yields simpler proofs and avoids taking the product of LTSes at the process level, which cirvumvents an immediate exponential blow-up in our reduction to reachability in Petri nets (see Subsec. 5.3). From a practical viewpoint, the new unary channels are 1-bounded by construction; analyzers for Petri nets could take advantage of this fact.

Splitting Irreversible Channels. According to Proposition 3.4 above, a topology containing an essential non-unary channel c can always be simplified by synchronising c. The resulting topology is simpler in the sense that it contains one less non-unary channel. However, there are situations where a channel is not essential, and thus it cannot be synchronised in general, but it can be made essential after a small modification. We have shown one such simple case on Fig. 3a, where the channel c is not essential but it can be made so by *splitting*[3] the other channel $p \overset{b}{\Longrightarrow} q$ into two channels $p \overset{b_0}{\Longrightarrow} r$ and $q \overset{b_1}{\Longrightarrow} r$ for a new process r, see Fig. 3b. Here, r is a new process that simply matches messages received from b_0 and b_1; receptions on b become transmissions on b_1. Clearly, the new system with the split topology has at least the same runs as the original system. Moreover, in this case, the converse holds as well, since we can always schedule all actions of p before any action of q, thus causality between transmissions and receptions can be relaxed. Formally, the splitting operation is defined as follows.

Definition 3.6. *Given a topology* $\mathcal{T} = \langle P, C, M, \mathsf{src}, \mathsf{dst}, \mathsf{msg}\rangle$ *and a channel* $c \in C$, *the* split *of* c *in* $\mathcal{T}$ *is the topology* $\mathcal{U} = \langle P', C', M, \mathsf{src}', \mathsf{dst}', \mathsf{msg}'\rangle$ *defined by* $P' = P \cup \{r\}$, $C' = (C \setminus \{c\}) \cup \{c_0, c_1\}$, *and*

$$(\mathsf{src}'(d), \mathsf{dst}'(d), \mathsf{msg}'(d)) = \begin{cases} (\mathsf{src}(d), \mathsf{dst}(d), \mathsf{msg}(d)) & \text{if } d \in C \setminus \{c\} \\ (\mathsf{src}(c), r, \mathsf{msg}(c)) & \text{if } d = c_0 \\ (\mathsf{dst}(c), r, \mathsf{msg}(c)) & \text{if } d = c_1 \end{cases}$$

where r *is a new process and* c_0, c_1 *are new channels that are not in* $P \cup C \cup M$.

[3] Despite having similar names, this splitting notion and the splitting technique of [8] have little in common (see Sec. 6).

To justify splitting, we introduce the notion of causal run. Intuitively, in a run which is causal for a process p, only those processes that can "transitively" send messages to p may be scheduled before p.

Definition 3.7. *A run* $(x_0, a_1, x_1, \ldots, a_n, x_n)$ *is* causal *for a given process* p *if* $q \stackrel{*}{\Longrightarrow} p$ *for every process* q *such that* $a_i \in A^q$ *and* $a_j \in A^p$ *for some* $1 \leq i < j \leq n$.

Lemma 3.8. *Given a process* p*, every run is causal-equivalent to a run that is causal for* p.

Recall that the idea behind splitting is to relax the causality between transmissions and receptions. This does not introduce "spurious" runs provided that every run can be reordered to have all actions of the receiver after the last action of the sender. A sufficient condition is given by the notion of *irreversible channel.*

Definition 3.9. *A channel* c *is* reversible *if there is a directed path from its destination* $\mathsf{dst}(c)$ *to its source* $\mathsf{src}(c)$*. A channel is* irreversible *if it is not reversible.*

The following proposition states that the reachability problem for a given topology $\mathcal{T}$ can be reduced to the reachability problem for the topology obtained from $\mathcal{T}$ by splitting c. As we will see in Sec. 5, splitting will be the first of a series of reductions for decidable topologies.

Proposition 3.10 (Split). *If* c *is an irreversible channel in* $\mathcal{T}$*, then it holds that* $\text{REACH}(\mathcal{T}) \preceq_m \text{REACH}(\mathcal{U})$ *where* $\mathcal{U}$ *results from the split of* c *in* $\mathcal{T}$.

4 The Power of Unary Channels

Let u be a process in a topology $\mathcal{T}$, and let $\mathcal{U}$ be a topology which is the same as $\mathcal{T}$ except that u is expanded into a strongly-connected sub-topology. In this section, we show that the behaviour of u in $\mathcal{T}$ can be *distributed* over its expansion in $\mathcal{U}$. We achieve this by demonstrating how processes in a strongly-connected sub-topology can simulate global rendezvous synchronisation over a finite alphabet of shared actions, which allows them to synchronise with each others and to step-wise simulate the behaviour of u. This technique works even when distributing behaviour over unary channels only, and it will be used to show the undecidability part of our characterisation (see Subsec. 5.1).

To formally define how a process is expanded into a sub-topology, it is more convenient to describe how to *fuse* a set of channels D. Intuitively, the fusion of D in a given topology is the topology obtained by merging together, in a single process, all endpoints of channels in D, by removing D, and by redirecting the other channels in the natural way.

Definition 4.1. *Given a topology* $\mathcal{T} = \langle P, C, M, \mathsf{src}, \mathsf{dst}, \mathsf{msg} \rangle$ *and a set of channels* $D \subseteq C$*, the* fusion *of* D *in* $\mathcal{T}$ *is the topology* $\mathcal{U} = \langle P', C', M, \mathsf{src}', \mathsf{dst}', \mathsf{msg}' \rangle$ *defined by* $P' = (P \setminus \{\mathsf{src}(c), \mathsf{dst}(c) \mid c \in D\}) \cup \{u\}$ *where* u *is a new process that is not in* $P \cup C \cup M$*,* $C' = C \setminus D$*, and, for every* $c \in C'$*,* $\mathsf{msg}'(c) = \mathsf{msg}(c)$*,* $\mathsf{src}'(c) = \mathsf{src}(c)$ *if* $\mathsf{src}(c) \in P'$ *and* $\mathsf{src}'(c) = u$ *otherwise, and similarly for* dst'.

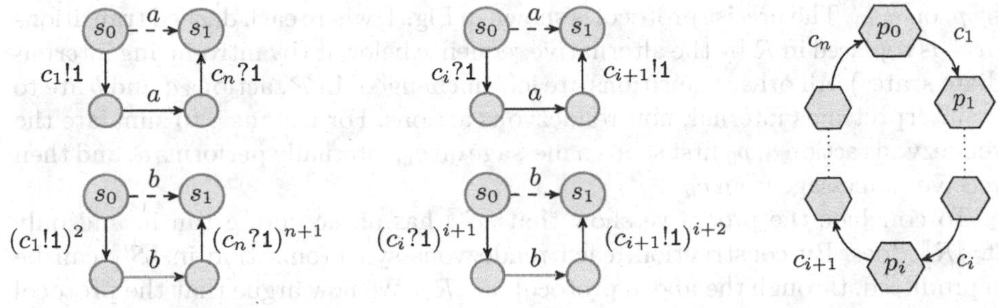

Fig. 4. Synchronisation protocol for a simple directed cycle

This section shows that fusing a strongly-connected sub-topology makes the reachability problem easier. We first deal with the simple case of directed cycles. The *support* of a directed walk/cycle is the set of channels that it visits.

Lemma 4.2. *If D is the support of a directed cycle in a topology $\mathcal{T}$ then* $\text{REACH}(\mathcal{U}) \preceq_m \text{REACH}(\mathcal{T})$ *where $\mathcal{U}$ results from the fusion of D in $\mathcal{T}$.*

Proof. Consider a directed cycle $p_0 \stackrel{c_1}{\Longrightarrow} p_1 \stackrel{c_2}{\Longrightarrow} \cdots \stackrel{c_n}{\Longrightarrow} p_n = p_0$ in $\mathcal{T}$ such that $D = \{c_1, \ldots, c_n\}$. This directed cycle is depicted in Fig. 4c. Denote by $\mathcal{U}$ the topology that results from the fusion of D in $\mathcal{T}$, and let u be the process in $\mathcal{U}$ that corresponds to the merging of all endpoints of D. Consider a system of communicating processes $\mathcal{S}$ with topology $\mathcal{U}$. We construct a new system $\mathcal{R}$ with topology $\mathcal{T}$ that simulates $\mathcal{S}$ by "distributing" the behaviour of the process u over $p_1, \ldots, p_n$. All other processes are left unchanged.

As a first step, let us assume that the processes $p_1, \ldots, p_n$ can perform multi-way rendezvous synchronisation over a finite alphabet of actions Σ. By *multi-way*, we mean that each time some process p_i performs an action a in Σ, then in fact *all* processes $p_1, \ldots, p_n$ perform the action a at the same time. For brevity, we will omit the "multi-way" qualifier from now on. It is easily shown that rendezvous synchronisation, even over a binary alphabet, allows $p_1, \ldots, p_n$ to coordinate in $\mathcal{T}$ and simulate the behaviour of u in $\mathcal{U}$. This way, we obtain from $\mathcal{S}$ a new system $\mathcal{S}'$, with topology $\mathcal{T}$, such that $[\![\mathcal{S}]\!]$ has an accepting run if, and only if, $[\![\mathcal{S}']\!]$ does. Moreover, $\mathcal{S}'$ does not use any channel in D.

As a second step, we explain how rendezvous synchronisation between $p_1, \ldots, p_n$ over a binary alphabet, say $\{a, b\}$, can be achieved through communications on the channels in D. In our simulation, rendezvous synchronisations are initiated by p_0, and then propagated along the directed cycle $p_0 \xrightarrow{c_1} p_1 \xrightarrow{c_2} \cdots \xrightarrow{c_n} p_n$ back to p_0. The latter then checks that the other processes correctly performed the desired rendezvous action. More precisely, whenever p_0 wants to handshake on some action (a or b), it sends some number of messages on c_1 to p_1 and waits for an acknowledgement on c_n before proceeding to its next move. In the meantime, the processes $p_1, \ldots, p_{n-1}$ do the same, but first receive and then transmit. The number of messages received by p_{i+1} from c_{i+1} is exactly the same as the number of messages sent by p_i on c_{i+1}. The precise protocol is shown in Fig. 4, where each dashed transitions in $\mathcal{S}'$ is replaced in $\mathcal{R}$ by the alternative sequence below it (by introducing intermediate states). All other transitions are left unchanged. In $\mathcal{R}$, actions a and b are to be interpreted as internal, non-rendezvous actions. For instance, to simulate the rendezvous action a, p_0 first sends a message on c_1, internally performs a, and then receives a message from c_n.

To conclude the proof, we show that $[\![\mathcal{S}']\!]$ has an accepting run if, and only if, $[\![\mathcal{R}]\!]$ does. By construction, each rendezvous synchronisation in $[\![\mathcal{S}']\!]$ can be reproduced, through the above protocol, in $[\![\mathcal{R}]\!]$. We now argue that the protocol does not introduce any spurious behaviour. Recall that the channels $c_1, \ldots, c_n$ are empty at the beginning. So, for each $1 \leq i \leq n-1$, the process p_i may only simulate a rendezvous action after p_0 has initiated a synchronisation round. Let us look at the first rendezvous action that is simulated by p_0. If this action is a, then p_0 sends one message on c_1 and receives one message from c_n. This entails that all the other processes $p_1, \ldots, p_{n-1}$ simulate the rendezvous action a. At the end of this synchronisation round, all the channels $c_1, \ldots, c_n$ are again empty. If the first rendezvous action that p_0 simulates is b, then p_0 sends two messages on c_1 and receives $n+1$ messages from c_n. Again, this entails that $p_1, \ldots, p_{n-1}$ simulate, each, the rendezvous action b. Indeed, by contradiction, if p_i simulates a, then it must continue simulating a since there are not enough messages in c_i anymore to simulate b. Therefore, it simply relays messages from c_i to c_{i+1}, and cannot produce on c_{i+1} the extra message that p_{i+1} expects to simulate b. By applying the same arguments to the remaining processes $p_{i+1}, \ldots, p_{n-1}$, we obtain that p_{n-1} is not able to produce on c_n the $n+1$ messages that p_0 expects to complete its simulation of b, a contradiction. Again, at the end of this synchronisation round, all the channels $c_1, \ldots, c_n$ are empty. By repeating this analysis for each synchronisation round, we obtain that every accepting run of $[\![\mathcal{R}]\!]$ can be mapped back to an accepting run of $[\![\mathcal{S}']\!]$. □

We now show that the previous lemma still holds for closed directed walks, i.e., where processes/channels can be repeated. The proof is by induction on the cardinality of D. As expected, the induction step follows from Lemma 4.2.

Proposition 4.3 (Fusion). *If D is the support of a closed directed walk in a topology $\mathcal{T}$, then* $\text{REACH}(\mathcal{U}) \preceq_m \text{REACH}(\mathcal{T})$ *where $\mathcal{U}$ results from the fusion of D in $\mathcal{T}$.*

5 Characterisation of Decidable Topologies

We are now ready to state and prove our characterisation of decidable topologies. The characterisation is expressed in the same vein as Theorem 2.3, and generalises it. We first introduce some additional definitions and notations. Consider a topology $\mathcal{T} = \langle P, C, M, \mathsf{src}, \mathsf{dst}, \mathsf{msg} \rangle$, and let $D \subseteq C$ be a subset of channels. Two processes p and q are *synchronizable over D*, written $p \approx_D q$, if there exists a directed path from p to q and a directed path from q to p, both using only channels in D. Note that $\approx_D$ is an equivalence relation on P.

A *jumping circuit* is a sequence $(p_0, c_1, q_1, p_1, \ldots, c_n, q_n, p_n)$ of processes $p_i, q_i \in P$ and channels $c_i \in C$, with $n \geq 1$, such that $c_1, \ldots, c_n$ are pairwise distinct non-unary channels, $p_0 = p_n$, and $p_{i-1} \stackrel{c_i}{=\!=} q_i \approx_D p_i$ for all $1 \leq i \leq n$, where $D = C \setminus \{c_1, \ldots, c_n\}$. Recall that the binary relation $\stackrel{c}{=\!=}$ is the union of $\stackrel{c}{\Longrightarrow}$ and $\stackrel{c}{\Longleftarrow}$. A *jumping cycle* is a jumping circuit $(p_0, c_1, q_1, p_1, \ldots, c_n, q_n, p_n)$ such that $q_i \not\approx_D q_j$ for all $1 \leq i < j \leq n$. To improve readability, jumping circuits and jumping cycles will often be written $p_0 \stackrel{c_1}{=\!=} q_1 \approx_D p_1 \cdots \stackrel{c_n}{=\!=} q_n \approx_D p_n$.

Remark 5.1. Every jumping circuit can be transformed into a jumping cycle.

Remark 5.2. For every jumping cycle $p_0 \stackrel{c_1}{=\!=} q_1 \approx_D p_1 \cdots \stackrel{c_n}{=\!=} q_n \approx_D p_n$, there exist n pairwise disjoint subsets $D_1, \ldots, D_n$ of the set $D = C \setminus \{c_1, \ldots, c_n\}$ such that $q_i \approx_{D_i} p_i$ for all $1 \leq i \leq n$.

Theorem 5.3. *Given a topology $\mathcal{T}$,* Reach($\mathcal{T}$) *is decidable if, and only if, $\mathcal{T}$ has no jumping cycle.*

The two directions of the theorem are proved in the two subsections below. To illustrate our characterisation, let us give some examples of decidable topologies. Certainly, polyforest topologies are decidable since they contain no undirected cycle, therefore no jumping cycle. Moreover, decidability is preserved if we add, for each channel of the polyforest, a unary channel in the opposite direction (see Lemma 5.4 below). Even further, we still get a decidable topology if each process is expanded into a sub-topology containing only unary channels. These operations introduce non-trivial cycles of unary and non-unary channels. Finally, adding unary channels looping on the same process always preserves decidability, as well as adding additional unary channels in parallel to already existing ones.

5.1 Undecidability

Consider a topology $\mathcal{T}$ containing a jumping cycle $\xi = (p_0, c_1, q_1, p_1, \ldots, c_n, q_n, p_n)$. By Remark 5.2, it holds that $p_0 \stackrel{c_1}{=\!=} q_1 \approx_{D_1} p_1 \cdots \stackrel{c_n}{=\!=} q_n \approx_{D_n} p_n$ for some pairwise disjoint subsets $D_1, \ldots, D_n$ of $C \setminus \{c_1, \ldots, c_n\}$. We may assume, w.l.o.g., that each D_i is the support of a closed directed walk in $\mathcal{T}$. To prove that Reach($\mathcal{T}$) is undecidable, we show that Reach($\mathcal{U}$) $\preceq_m$ Reach($\mathcal{T}$) for some topology $\mathcal{U}$ with an undirected cycle containing only non-unary channels, hence, for which reachability is undecidable by Theorem 2.3. To do so, we build a sequence of topologies $\mathcal{U}_0, \mathcal{U}_1, \ldots, \mathcal{U}_n$ by fusing together synchronizable processes, as follows: We define $\mathcal{U}_0 = \mathcal{T}$, and, for each $1 \leq i \leq n$, we let $\mathcal{U}_i$ result from the fusion of D_i in $\mathcal{U}_{i-1}$. Pairwise disjointness of $D_1, \ldots, D_n$ ensures that every channel in D_i is still a channel in $\mathcal{U}_{i-1}$. It is routinely checked that:

- D_i is still the support of a closed directed walk in $\mathcal{U}_{i-1}$, and
- ξ induces a closed undirected walk $u_0 \stackrel{c_1}{=\!=} u_1 \cdots \stackrel{c_n}{=\!=} u_n$ in $\mathcal{U}_n$.

The first item entails, by Proposition 4.3, that Reach($\mathcal{U}_i$) $\preceq_m$ Reach($\mathcal{U}_{i-1}$). By the transitivity of $\preceq_m$, we get that Reach($\mathcal{U}_n$) $\preceq_m$ Reach($\mathcal{U}_0$). Since $c_1, \ldots, c_n$

are pairwise distinct non-unary channels, we derive, from the second item, that $\mathcal{U}_n$ contains an undirected cycle with only non-unary channels. It follows from Theorem 2.3 that $\text{REACH}(\mathcal{U}_n)$ is undecidable. As $\mathcal{U}_0 = \mathcal{T}$, we conclude that $\text{REACH}(\mathcal{T})$ is undecidable.

5.2 Decidability

Starting from a topology $\mathcal{T}$ with no jumping cycle, we apply a sequence of topological transformations that produce a topology $\mathcal{U}$ with only unary channels, and such that $\text{REACH}(\mathcal{T}) \preceq_m \text{REACH}(\mathcal{U})$. Since the latter is decidable by Theorem 2.2, the former is decidable as well.

Given a topology $\mathcal{T}$ and a channel c in $\mathcal{T}$, an *acknowledgement channel* for c is a new unary channel, written $\overleftarrow{c}$, with source $\mathsf{dst}(c)$ and destination $\mathsf{src}(c)$. The following lemma says that adding an acknowledgement channel for an essential non-unary channel preserves the absence of jumping cycle. It immediately entails Corollary 5.5 below.

Lemma 5.4. *Consider a topology $\mathcal{T}$ and an essential non-unary channel c therein. Let $\mathcal{U}$ be the topology obtained from $\mathcal{T}$ by adding an acknowledgement channel for c. Then $\mathcal{T}$ contains a jumping cycle if $\mathcal{U}$ contains a jumping cycle.*

Proof (sketch). Assume that $\mathcal{T}$ has no jumping cycle, but adding $\overleftarrow{c}$ yields a topology $\mathcal{U}$ with a jumping cycle $p_0 \overset{c_1}{=\!=} q_1 \approx_{D_1} p_1 \cdots \overset{c_n}{=\!=} q_n \approx_{D_n} p_n = p_0$. Observe that $\overleftarrow{c}$ cannot be one of $c_1, \ldots, c_n$ since it is unary. If $\overleftarrow{c}$ does not appear in any D_i, then $\mathcal{T}$ has a jumping cycle. Hence, $\overleftarrow{c}$ appears on a closed directed walk π that synchronises two processes q_i and p_i. Since c is essential, it must appear on π too. We may assume, w.l.o.g., that $\overleftarrow{c}$ appears on the directed path from p_i to q_i on π, and that c appears on the directed path from q_i to p_i. We get that $q_i \overset{*}{\Longrightarrow} p \overset{c}{\Longrightarrow} q \overset{*}{\Longrightarrow} p_i \overset{*}{\Longrightarrow} q \overset{\overleftarrow{c}}{\Longrightarrow} p \overset{*}{\Longrightarrow} q_i$. Therefore, q_i can synchronise with p using only channel from $E_i = D_i \setminus \{c, \overleftarrow{c}\}$, and similarly p_i can synchronise with q via E_i. So we can build a jumping cycle in $\mathcal{T}$ from the jumping cycle in $\mathcal{U}$ by replacing $q_i \approx_{D_i} p_i$ by $q_i \approx_{E_i} p \overset{c}{\Longrightarrow} q \approx_{E_i} p_i$, a contradiction. □

Corollary 5.5. *Consider a topology $\mathcal{T}$ and an essential non-unary channel c therein. Let $\mathcal{U}$ be the topology resulting from the synchronisation of c in $\mathcal{T}$. Then $\mathcal{T}$ contains a jumping cycle if $\mathcal{U}$ contains a jumping cycle.*

Remark 5.6. The converse of Corollary 5.5 also holds, but it is not required for the proof of Theorem 5.3.

We say that a topology $\mathcal{T}$ is *divided* if the destination of every irreversible unary channel is a sink (i.e., is not the source of some channel) and is not the destination of some non-unary channel. The following two properties of divided topologies are crucial in the proof of Theorem 5.3.

Lemma 5.7. *Consider a topology $\mathcal{T}$ and a non-unary channel c therein. If $\mathcal{T}$ is divided, then so is the topology resulting from the synchronisation of c in $\mathcal{T}$.*

Lemma 5.8. *If $\mathcal{T}$ is a divided topology with no jumping cycle, then every non-unary channel in $\mathcal{T}$ is essential.*

We now prove the "if" direction of Theorem 5.3. Assume that $\mathcal{T}$ has no jumping cycle. Let $c_1, \ldots, c_n$ denote the non-unary channels of $\mathcal{T}$. We first build $\mathcal{U}_0$ from $\mathcal{T}$ by splitting all unary channels that are irreversible in $\mathcal{T}$. Note that $\mathcal{U}_0$ does not depend on the order in which the irreversible unary channels of $\mathcal{T}$ are split. It follows from Proposition 3.10 and the transitivity of $\preceq_m$ that $\text{Reach}(\mathcal{T}) \preceq_m \text{Reach}(\mathcal{U}_0)$. By construction, the topology $\mathcal{U}_0$ is divided, and it still has no jumping cycle. So, by Lemma 5.8, every non-unary channel in $\mathcal{U}_0$ is essential. Notice that $\mathcal{U}_0$ has the same non-unary channels as $\mathcal{T}$, namely $c_1, \ldots, c_n$. For each $1 \leq i \leq n$, let $\mathcal{U}_i$ be the topology resulting from the synchronisation of c_i in $\mathcal{U}_{i-1}$. By induction, it is immediate to prove that, for every $0 \leq i \leq n, \mathcal{U}_i$ has no jumping cycle, the induction step holding by Corollary 5.5, that $\mathcal{U}_i$ is divided, by Lemma 5.7, that $c_{i+1}, \ldots, c_n$ are still essential in $\mathcal{U}_i$, by Lemma 5.8, and that $\text{Reach}(\mathcal{U}_{i-1}) \preceq_m \text{Reach}(\mathcal{U}_i)$, by Proposition 3.4. By the transitivity of $\preceq_m$, we get that $\text{Reach}(\mathcal{T}) \preceq_m \text{Reach}(\mathcal{U}_n)$. Since $\mathcal{U}_n$ contains only unary channels, $\text{Reach}(\mathcal{U}_n)$ is decidable by Theorem 2.2. Thus $\text{Reach}(\mathcal{T})$ is decidable.

5.3 Complexity

We consider the reachability problem for systems of communicating processes whose topology has no jumping cycle. This problem, written ReachNJC, is the union of the problems $\text{Reach}(\mathcal{T})$ for topologies $\mathcal{T}$ with no jumping cycle. Note that deciding whether a topology has a jumping cycle is a simple graph-theoretic problem that can be solved in polynomial time. Hence, it can be checked efficiently whether a given system of communicating processes is an instance of ReachNJC or not. Here, we show that ReachNJC is equivalent to reachability in Petri nets.

The *size* of a labelled transition system $\mathcal{A} = \langle S, S_I, S_F, A, \rightarrow \rangle$ is defined as $|\mathcal{A}| = |S| + |\rightarrow|$. Similarly, the *size* of a topology $\mathcal{T} = \langle P, C, M, \mathsf{src}, \mathsf{dst}, \mathsf{msg} \rangle$ is $|\mathcal{T}| = |P| + |C| + \sum_{c \in C} |\mathsf{msg}(c)|$. Finally, the *size* of a system of communicating processes $\mathcal{S} = \langle \mathcal{T}, \{\mathcal{A}^p\}_{p \in P} \rangle$ is $|\mathcal{S}| = |\mathcal{T}| + \sum_{p \in P} |\mathcal{A}^p|$. The algorithm in Subsec. 5.2 transforms a system $\mathcal{S}$ over a topology with no jumping cycle, into a system $\mathcal{S}'$ with unary channels only. Since unary channels are essentially counters (over the natural numbers) that may only be incremented and decremented, $\mathcal{S}'$ can be naturally interpreted as a Petri net. Crucially, we show that $\mathcal{S}'$ (and thus the Petri net) has size polynomial in $|\mathcal{S}|$. This is possible since the synchronisation operation from Sec. 3 avoids taking the product of processes (at the cost of introducing 1-bounded unary channels/counters).

Theorem 5.9. ReachNJC *is equivalent to reachability in Petri nets under polynomial-time many-one reductions.*

6 Discussion

Unary vs. Bag Channels. A bag channel is a channel where messages can be freely reordered. Therefore, it suffices to *count* how many messages of each type are in the channel. So, a bag channel over a message alphabet of cardinality n can be implemented with n unary channels in parallel. A topology of bag and FIFO channels is a topology (as defined in Sec. 2) where, in addition, each channel has a flag indicating whether it is ordered (FIFO) or not (bag). Our characterisation from Sec. 5 immediately generalises to bag channels by modifying the definition of jumping cycle and requiring that the c_i's be non-unary FIFO channels (instead of just non-unary).

Unary/Bag vs. Lossy Channels. Another over-approximation incomparable with bag channels is provided by *lossy channels*, where messages might be lost at any moment [2,7]. A complete characterisation of decidable topologies mixing perfect and lossy channels has been presented in [8]. In order to reduce to basic decidable topologies, they consider two reduction rules. The first one is the *fusion* of essential channels. This is similar in spirit to our *synchronisation* (see Proposition 3.4), with the only difference that fusing channels requires to take the product of the underlying processes, while synchronising channels just replaces one channel with several 1-bounded unary channels. This allows us to obtain precise complexity bounds in Subsec. 5.3. The second reduction rule is *splitting* a complex topology $\mathcal{T}$ into $\mathcal{T}_1$ and $\mathcal{T}_2$ when all channels between $\mathcal{T}_1$ and $\mathcal{T}_2$ are unidirectional and lossy. Despite similar names, this is different to our splitting technique (see Proposition 3.10), because we split (irreversible) channels, and not topologies. However, while lossy channels cannot be split, the lossy channels involved in the splitting of $\mathcal{T}$ in the sense of [8] are *irreversible* in our terminology, and thus could be split *if they were perfect channels.* Moreover, if we replace lossy channels with perfect bag channels, it additionally holds that, if $\mathcal{T}_1, \mathcal{T}_2$ above are decidable in our characterisation (i.e., no jumping cycles), then the same holds for $\mathcal{T}$. Since also fusion/synchronisation preserves decidable topologies in both settings, we have that any decidable topology of perfect and lossy channels is still a decidable topology by replacing lossy channels with perfect bag channels.

Moreover, some topologies which are undecidable with lossy channels become decidable with bag channels. For example, the topology with three parallel channels c, d, e between two processes with c perfect FIFO and d, e lossy FIFO is undecidable, while if d, e are bag channels it becomes decidable.

Finally, while the topology in Fig. 2 is undecidable when channels b_0 and b_1 are either both bag channels or both lossy channels, our construction with unary bag channels is correct *even if those are unary and lossy.* Indeed, as soon as any message gets lost due to lossiness, our synchronization protocol gets irremediably stuck. However, the construction from [8] does not generalise to unary channels in this case. Thus, we extend their undecidability result to this topology.

7 Conclusions and Future Work

We have presented a complete characterisation of the decidable topologies for networks of finite-state automata communicating over FIFO and bag channels. Remarkably, every decidable topology can be solved using two simple techniques (synchronising essential channels and splitting irreversible channels), whereas every topology that cannot be solved with these two techniques is undecidable.

The same characterisation problem is solved in [8] but for networks mixing perfect and lossy FIFO channels. A direction for future research is to characterise decidable topologies of lossy/perfect FIFO/bag channels.

Relaxing FIFO channels to the bag type can be applied in other contexts as well. For example, the work [11] studies topologies of networks of *pushdown automata* communicating over FIFO channels, and it is natural to ask what happens when some channels and/or pushdown stores are bags instead of strings.

References

1. Abdulla, P.A., Atig, M.F., Cederberg, J.: Analysis of message passing programs using SMT-solvers. In: Van Hung, D., Ogawa, M. (eds.) ATVA 2013. LNCS, vol. 8172, pp. 272–286. Springer, Heidelberg (2013)
2. Abdulla, P., Jonsson, B.: Verifying programs with unreliable channels. Inf. Comput. 127(2), 91–101 (1996)
3. Boigelot, B., Godefroid, P.: Symbolic verification of communication protocols with infinite state spaces using QDDs. Form. Methods Sys. Des. 14, 237–255 (1999)
4. Bouajjani, A., Emmi, M.: Bounded phase analysis of message-passing programs. In: Flanagan, C., König, B. (eds.) TACAS 2012. LNCS, vol. 7214, pp. 451–465. Springer, Heidelberg (2012)
5. Brand, D., Zafiropulo, P.: On communicating finite-state machines. J. ACM 30(2), 323–342 (1983)
6. Cécé, G., Finkel, A.: Verification of programs with half-duplex communication. Inf. Comput. 202(2), 166–190 (2005)
7. Cécé, G., Finkel, A., Purushothaman Iyer, S.: Unreliable channels are easier to verify than perfect channels. Inf. Comput. 124(1), 20–31 (1996)
8. Chambart, P., Schnoebelen, P.: Mixing lossy and perfect fifo channels. In: van Breugel, F., Chechik, M. (eds.) CONCUR 2008. LNCS, vol. 5201, pp. 340–355. Springer, Heidelberg (2008)
9. Ganty, P., Majumdar, R.: Algorithmic verification of asynchronous programs. ACM Trans. Program. Lang. Syst. 34(1), 1–6 (2012)
10. Haase, C., Schmitz, S., Schnoebelen, P.: The power of priority channel systems. In: D'Argenio, P.R., Melgratti, H. (eds.) CONCUR 2013 – Concurrency Theory. LNCS, vol. 8052, pp. 319–333. Springer, Heidelberg (2013)
11. Heußner, A., Leroux, J., Muscholl, A., Sutre, G.: Reachability analysis of communicating pushdown systems. LMCS 8(3), 1–20 (2012)
12. Jhala, R., Majumdar, R.: Interprocedural analysis of asynchronous programs. SIGPLAN Not. 42(1), 339–350 (2007)
13. Kosaraju, S.R.: Decidability of reachability in vector addition systems. In: Proc. STOC 1982, pp. 267–281 (1982) (preliminary version)
14. La Torre, S., Madhusudan, P., Parlato, G.: Context-bounded analysis of concurrent queue systems. In: Ramakrishnan, C.R., Rehof, J. (eds.) TACAS 2008. LNCS, vol. 4963, pp. 299–314. Springer, Heidelberg (2008)
15. Leroux, J.: Vector addition system reachability problem: A short self-contained proof. In: Dediu, A.-H., Inenaga, S., Martín-Vide, C. (eds.) LATA 2011. LNCS, vol. 6638, pp. 41–64. Springer, Heidelberg (2011)

16. Mayr, E.: An algorithm for the general petri net reachability problem. In: Proc. STOC 1981, pp. 238–246 (1981)
17. Milner, R.: Communication and Concurrency. Prentice-Hall (1989)
18. Pachl, J.K.: Reachability problems for communicating finite state machines. Research Report CS-82-12, University of Waterloo (May 1982)
19. Sen, K., Viswanathan, M.: Model checking multithreaded programs with asynchronous atomic methods. In: Ball, T., Jones, R.B. (eds.) CAV 2006. LNCS, vol. 4144, pp. 300–314. Springer, Heidelberg (2006)
20. West, D.B.: Introduction to Graph Theory, 2nd edn. Prentice-Hall (2001)

Controllers for the Verification of Communicating Multi-pushdown Systems*

C. Aiswarya[1], Paul Gastin[2], and K. Narayan Kumar[3]

[1] Uppsala University, Sweden
aiswarya.cyriac@it.uu.se
[2] LSV, ENS Cachan, CNRS & INRIA, France
gastin@lsv.ens-cachan.fr
[3] Chennai Mathematical Institute, India
kumar@cmi.ac.in

Abstract. Multi-pushdowns communicating via queues are formal models of multi-threaded programs communicating via channels. They are turing powerful and much of the work on their verification has focussed on under-approximation techniques. Any error detected in the under-approximation implies an error in the system. However the successful verification of the under-approximation is not as useful if the system exhibits unverified behaviours. Our aim is to design *controllers* that observe/restrict the system so that it stays within the verified under-approximation. We identify some important properties that a good controller should satisfy. We consider an extensive under-approximation class, construct a distributed controller with the desired properties and also establish the decidability of verification problems for this class.

1 Introduction

Most of the critical hardware and software consists of several parallel computing units/components. Each of these may execute recursive procedures and may also have several unbounded data-structures to enhance its computing power. Several of such components may be running on the same processor giving rise to a multi-threaded system with many unbounded data-structures. Furthermore, such complex infinite state systems may communicate over a network and be physically distributed. The high computational power in combination with unconstrained interactions make the analysis of these systems very hard.

The verification of such systems is undecidable in general. Even the basic problem of control state reachability (or emptiness checking) is undecidable as soon a program has two stacks or a self queue. However, these systems are so important, that several under-approximation techniques have been invented for their verification. If the under-approximation fails to satisfy a requirement, that immediately indicates an error in the system. However, if the system is verified correct under such restrictions, the correctness is compromised if the system

* Supported by LIA InForMel, and DIGITEO LoCoReP.

P. Baldan and D. Gorla (Eds.): CONCUR 2014, LNCS 8704, pp. 297–311, 2014.

eventually exhibits behaviours outside the class. *Controlling* the system to only exhibit behaviours that have been verified to be correct is therefore crucial to positively use these under-approximation techniques. Alternately, we may use these controllers to raise a signal whenever the system behaviour departs from the verified class. For example, in the cruise control system of a car (or auto-pilot systems in trains/aircrafts), it will be useful to signal such a departure and switch from automatic to manual mode.

Our Contributions. We aim at obtaining a uniform controller for a class, which when run in parallel with the system, controls it so as to exhibit only those behaviours permitted by the class. Such a controller should possess nice properties like determinism, non-blocking, system independence etc. In Section 3, we identify and analyse such desirable features of a controller.

Our next contribution is to propose a very generous under-approximation class and to construct a controller satisfying all the desired properties. Our class bounds the number of phases – in a phase only one data-structure can be read in an unrestricted way though writes to all data-structures are allowed. But our notion of phases extends sensibly contexts of [14] and phases of [13]. In particular it permits *autonomous computations* within a phase instead of the well-queuing assumption. The latter corresponds to permitting reads from queues in the *main* program but not from any of the functions it calls. We permit recursive calls to be at any depth of recursion when reading from a queue. After such a read, however, returning from the function causes a phase change.

A concurrent system may be controlled in a global manner or in a distributed manner. If the concurrent processes are at a single location and communicate via shared variables, e.g., multi-threaded programs, a global controller is reasonable. We describe this sequential controller in Section 4. However, when these multi-threaded processes are physically distributed it is natural to demand a distributed controller. In Section 5, we illustrate the design of a controllable under-approximation class by extending our idea of phases to the distributed setting and constructing a distributed controller with all the desired properties.

Finally, we can prove using the split-width technique [5,6,8] that our generous under-approximation class can be model-checked against a wide variety of logics.

For lack of space, proofs of correctness and of decidability are omitted from this extended abstract and can be found in the full version [7].

Related Work: In the study of distributed automata a number of difficult synthesis theorems [9–11,19] have been proved. These theorems in conjunction with constructions for intersections yield controllers for these classes. Of particular interest is the theory of finite state machines communicating via queues, called message-passing automata (MPA). These have been well studied using labeled partial-orders (or graphs) called MSCs (Message-sequence charts) to represent behaviours. These systems are turing powerful and techniques restricting channel usage have been studied to obtain decidability. The most general class of this kind, called existentially k-bounded MSCs, consists of all behaviours (MSCs) that have at least one linearization in which the queue lengths are bounded by

k at every point. A deep result of [9] shows that for each k there is an MPA which accepts precisely the set of existentially k-bounded MSCs. Thus, if one uses such behaviours as an under-approximation class then this result implies the existence of a distributed controller. However, it is known that this controller cannot be made deterministic.

The bounding technique for verification has been extensively studied in the case of multi-pushdown systems (MPDS). For the restrictions studied in literature, bounded-context [18], bounded-phase [13], bounded-scope [16] and ordered stacks [2, 3], it is quite easy to construct deterministic controllers, though this question has not been addressed before. The context bounding technique is extended to pushdown systems communicating via queues under the restriction that queues may be read only when the stacks are empty (well-queuing) in [14], and under a dual restriction (on writes instead of reads) in [12]. Controllability is however not studied there. The k-Phase restriction we consider here is a natural joint generalization of these contexts (as well as the bounded-phase restriction for MPDS). In fact, for every bound k, there exist behaviours which are not captured by [13] and [14], but which are captured by our class with a bound of 3. (See Figure 1 for an example.)

2 Systems with Stacks and Queues

We provide a formal description of systems with data-structures and their behaviours. We restrict ourselves to systems with global states providing an (interleaved) sequential view. In Section 5 we extend this to the distributed case where there are a number of components each with their own collection of transitions. We consider a finite set $\mathbf{DS} = \mathbf{Stacks} \uplus \mathbf{Queues}$ of data-structures which are either stacks or queues and a finite set Σ of actions. Our systems have a finite set of control locations and use these (unbounded) stacks and queues. We obtain an interesting class of infinite state systems, providing an (interleaved) sequential view of multi-threaded recursive programs communicating via FIFO channels.

A stack-queue system (SQS) over data-structures **DS** and actions from Σ is a tuple $\mathcal{S} = (\mathsf{Locs}, \mathsf{Val}, \mathsf{Trans}, \mathsf{in}, \mathsf{Fin})$ where Locs is a finite set of locations, Val is a finite set of values that can be stored in the data-structures, $\mathsf{in} \in \mathsf{Locs}$ is the initial location, $\mathsf{Fin} \subseteq \mathsf{Locs}$ is the set of final locations, and Trans is the set of transitions which may write a value to, or read a value from, or do not involve a data-structure. For $\ell, \ell' \in \mathsf{Locs}$, $a \in \Sigma$, $d \in \mathbf{DS}$ and $v \in \mathsf{Val}$, we have

- internal transitions of the form $\ell \xrightarrow{\mathsf{a}} \ell'$,
- write transitions of the form $\ell \xrightarrow{a,d!v} \ell'$, and
- read transitions of the form $\ell \xrightarrow{a,d?v} \ell'$.

Intuitively, an SQS consists of a finite state system equipped with a collection of stacks and queues. In each step, it may use an internal transition to merely change its state, or use a write transition to append a value to the tail of a particular queue or stack or use a read transition to remove a value from the head (or tail) of a queue (of a stack respectively). The transition relation makes explicit the identity of the data-structure being accessed and the type of the operation. As observed in [1, 6, 13, 17] it is often convenient to describe the runs of such systems as a state-labeling of words decorated with a matching relation per data-structure instead of the traditional operational semantics using configurations and moves. This will prove all the more useful when we move to the distributed setting where traditionally semantics has always been given as state-labelings of appropriate partial orders [9, 11, 19].

A stack-queue Word (SQW) over **DS** and Σ is a tuple $\mathcal{W} = (w, (\rhd^d)_{d\in\mathbf{DS}})$ where $w = a_1 a_2 \cdots a_n \in \Sigma^+$ is the sequence of actions, and for each $d \in \mathbf{DS}$, the matching relation $\rhd^d \subseteq \{1, \ldots, n\}^2$ relates write events to data-structure d to their corresponding read events. The following conditions should be satisfied:
- write events should precede read events: $e \rhd^d f$ implies $e < f$,
- data-structure accesses are disjoint: if $e_1 \rhd^d e_2$ and $e_3 \rhd^{d'} e_4$ are distinct edges ($d \neq d'$ or $(e_1, e_2) \neq (e_3, e_4)$) then they are disjoint ($|\{e_1, e_2, e_3, e_4\}| = 4$),
- $\forall d \in$ **Stacks**, $\rhd^d$ conforms to LIFO: if $e_1 \rhd^d f_1$ and $e_2 \rhd^d f_2$ are different edges then we do not have $e_1 < e_2 < f_1 < f_2$.
- $\forall d \in$ **Queues**, $\rhd^d$ conforms to FIFO: if $e_1 \rhd^d f_1$ and $e_2 \rhd^d f_2$ are different edges then we do not have $e_1 < e_2$ and $f_2 < f_1$.

We let $\rhd = \bigcup_{d\in\mathbf{DS}} \rhd^d$ be the set of all matching edges and $\mathcal{E} = \{1, \ldots, n\}$ be the set of events of $\mathcal{W}$. The set of all stack-queue words is denoted by $\mathbb{SQW}$.

We say that an event e is a *read event* (on data-strucutre d) if there is an f such that $f \rhd^d e$. We define *write events* similarly and an event is *internal* if it is neither a read nor a write. To define the run of an SQS over a stack-queue word $\mathcal{W}$, we introduce two notations. For $e \in \mathcal{E}$, we denote by e^- the immediate predecessor of e if it exists, and we let $e^- = \bot \notin \mathcal{E}$ otherwise. We let $\max(\mathcal{W})$ be the maximal event of $\mathcal{W}$.

A run of an SQS $\mathcal{S}$ on a stack-queue word $\mathcal{W}$ is a mapping $\rho\colon \mathcal{E} \to \mathsf{Locs}$ satisfying the following consistency conditions (with $\rho(\bot) = \mathsf{in}$):

- if e is an internal event then $\rho(e^-) \xrightarrow{\lambda(e)} \rho(e) \in \mathrm{Trans}$,
- if $e \rhd^d f$ for some data-structure $d \in \mathbf{DS}$ then for some $v \in \mathrm{Val}$ we have both $\rho(e^-) \xrightarrow{\lambda(e), d!v} \rho(e) \in \mathrm{Trans}$ and $\rho(f^-) \xrightarrow{\lambda(f), d?v} \rho(f) \in \mathrm{Trans}$.

The run is accepting if $\rho(\max(\mathcal{W})) \in \mathsf{Fin}$. The *language* $\mathcal{L}(\mathcal{S})$ accepted by an SQS $\mathcal{S}$ is the set of stack-queue words on which it has an accepting run.

Notice that SQSs are closed under intersection, by means of the cartesian product. Let $\mathcal{S}_i = (\mathsf{Locs}_i, \mathrm{Val}_i, \mathrm{Trans}_i, \mathsf{in}_i, \mathsf{Fin}_i)$ for $i \in \{1, 2\}$ be two SQSs. The cartesian product is $\mathcal{S}_1 \times \mathcal{S}_2 = (\mathsf{Locs}_1 \times \mathsf{Locs}_2, \mathrm{Val}_1 \times \mathrm{Val}_2, \mathrm{Trans}, (\mathsf{in}_1, \mathsf{in}_2), \mathsf{Fin}_1 \times \mathsf{Fin}_2)$ where the set of transitions is defined by

- $(\ell_1, \ell_2) \xrightarrow{a} (\ell_1', \ell_2') \in \mathrm{Trans}$ if $\ell_i \xrightarrow{a} \ell_i' \in \mathrm{Trans}_i$ for $i \in \{1, 2\}$,
- $(\ell_1, \ell_2) \xrightarrow{a, d!(v_1, v_2)} (\ell_1', \ell_2') \in \mathrm{Trans}$ if $\ell_i \xrightarrow{a, d!v_i} \ell_i' \in \mathrm{Trans}_i$ for $i \in \{1, 2\}$,
- $(\ell_1, \ell_2) \xrightarrow{a, d?(v_1, v_2)} (\ell_1', \ell_2') \in \mathrm{Trans}$ if $\ell_i \xrightarrow{a, d?v_i} \ell_i' \in \mathrm{Trans}_i$ for $i \in \{1, 2\}$.

In fact, $\mathcal{S}_1 \times \mathcal{S}_2$ has an (accepting) run on a stack-queue word $\mathcal{W}$ iff both $\mathcal{S}_1$ and $\mathcal{S}_2$ have an (accepting) run on $\mathcal{W}$. Therefore, $\mathcal{L}(\mathcal{S}_1 \times \mathcal{S}_2) = \mathcal{L}(\mathcal{S}_1) \cap \mathcal{L}(\mathcal{S}_2)$.

3 Controllers and Controlled Systems

SQSs are turing powerful as soon as **DS** contains two stacks or a queue, and hence their verification is undecidable. However, since it is an important problem, various under-approximation techniques have been invented in the recent years [2,3,13,14,16,18], starting with the *bounded-context restriction* [18] for systems with only stacks. Here, the number of times the system switches from using one stack to another is bounded by a fixed number k. Reachability and many other properties become decidable when restricted to such behaviours.

A typical under-approximation technique describes a whole family of classes $\mathbb{C}_k$ parametrized by an integer k which is proportional to the coverage: the higher the parameter, the more behaviours are covered. For example, the bound on number of context switches k serves as this parameter for the context bounding technique. Ideally, the under-approximations defined by the classes $(\mathbb{C}_k)_k$ should be *universal*, i.e., should cover all behaviours: every stack-queue word $\mathcal{W}$ should be in $\mathbb{C}_k$ for some k. This is true for the context bounding technique.

Traditionally under-approximations yield decidability for verification problems such as reachability [18] and model checking against linear time properties expressed in various logics upto MSO [17]. For such properties, if the model-checking problem yields a negative answer then this immediately means that the full system fails the verification as well.

However, assume that a system $\mathcal{S}$ has been verified against some linear-time or reachability property (or properties) wrt. some under-approximation class $\mathbb{C}$. This give us little information on whether the full system satisfies these properties. Hence we need a mechanism, which we call a *controller*, to restrict the system so that it does not exhibit behaviours outside $\mathbb{C}$. Observe that w.r.t. linear-time properties restricting the system to even a proper subset of $\mathbb{C}$ would still be acceptable though not desirable. However, for reachability properties a proper restriction might lead to a system that no longer satisfies the property. Therefore, a controller should allow all and only the behaviours of $\mathbb{C}$.

We now describe formally our notion of a controller for a class and examine some key properties that make it interesting.

A controller for a class $\mathbb{C} \subseteq \mathsf{SQW}$ is an SQS $\mathcal{C}$ such that $\mathcal{L}(\mathcal{C}) = \mathbb{C}$. We say that a class $\mathbb{C}$ is *controllable* if it admits a controller.

Suppose the restriction of the behaviours of a system $\mathcal{S}$ to a class $\mathbb{C}$ has been verified against some linear-time or reachability property φ. Further suppose that $\mathbb{C}$ admits a controller $\mathcal{C}$. Then, the *controlled system* $\mathcal{S}' = \mathcal{S} \times \mathcal{C}$ is such that $\mathcal{L}(\mathcal{S}') = \mathcal{L}(\mathcal{S}) \cap \mathbb{C}$, and therefore satisfies φ. Thus, a controller for a class is independent of the system $\mathcal{S}$ as well as the property. Once we identify a controllable class with decidable verification we may verify and control any system in a completely generic and transparent manner without any additional work.

Notice that we could have introduced more general controllability. For instance, a class $\mathbb{C}$ is non-uniformly controllable if for each system $\mathcal{S}$, there exists another system $\mathcal{S}'$ such that $\mathcal{L}(\mathcal{S}') = \mathcal{L}(\mathcal{S}) \cap \mathbb{C}$. While this would allow more classes to be controllable, it would not be very useful since it does not yield an automatic way to build $\mathcal{S}'$ from $\mathcal{S}$.

Using the cartesian product makes the controller integrable into the system. The controller, by definition, does not have its own auxiliary data-structures, but only shares the data-structures of the system. Moreover, it does not access a data-structure out of sync with the system. We could also give more intrusive power to a controller by allowing its transitions to depend on the current state of the system and on the current value read/written by the system on data-structures. But again, such a system would not be generic, and also, by its strong observation power, would compromise the privacy of the system.

We now consider other properties that a good controller must satisfy and use that to arrive at a formal definition of such a controller.

The under-approximation classes are often defined based on the data-structure accesses, and do not depend on the action labels/internal actions. Hence an ideal controller should be definable independent of the action labels and must be oblivious to the internal moves. This can be done as follows.

We omit action labels from read/write transitions of $\mathcal{C}$: an abstract transition $\ell \xrightarrow{d!v} \ell'$ stands for transitions $\ell \xrightarrow{a,d!v} \ell'$ for all $a \in \Sigma$ and similarly for read transitions. Also, we do not describe internal transitions and assume instead that there are self-loops $\ell \xrightarrow{\mathrm{a}} \ell$ for all locations and actions.

This (abstract) controller should be deterministic and non-blocking, so that instantiating it with any alphabet will still be deterministic and non-blocking. Thus, the controller should have a unique run on any $\mathcal{W}$ and moreover this run does not depend on the internal events / action labels along the run, but depends only on the sequence of reads/writes on the different data-structures that appear along $\mathcal{W}$. The state of the controller at any point along this run unambiguously indicates whether the current prefix can be extended to a word that belongs to the class $\mathbb{C}$. With this we are ready to formalize our notion of a good controller.

A DS-controller is an SQS $\mathcal{C}$ which is oblivious to internal events and to action labels and which is deterministic and non-blocking. Formally, its (abstract) transitions should satisfy:

- for every $\ell \in \mathsf{Locs}$ and $d \in \mathbf{DS}$ there exists exactly one $\ell' \in \mathsf{Locs}$ and $v \in \mathrm{Val}$ such that $\ell \xrightarrow{d!v} \ell'$,
- for every $\ell \in \mathsf{Locs}$, $d \in \mathbf{DS}$ and $v \in \mathrm{Val}$ there exists exactly one $\ell' \in \mathsf{Locs}$ such that $\ell \xrightarrow{d?v} \ell'$.

All that we said so far suffices for a global (or seqeuntial) system. If the system to be verified and controlled is actually physically distributed, then a global sequential controller would not be integrable in the system. Instead we would need a distributed controller and this is much harder to achieve. We discuss this in Section 5.

Next we examine real examples of controllable under-approximations. While an under-approxiamation $\mathbb{C}_k$ is *nicely controllable* if it admits a controller with the above features, the class itself should satisfy some other properties for it to be useful. Firstly, $\mathbb{C}_k$ should have a wide coverage over the set of possible behaviours. A useful feature is that all behaviours fall in the class for an appropriately chosen parameter. Second, the definition of the class should be easy to describe. Finally, the verification problem for the class should be decidable. For instance, considering the collection of behaviours with clique/split/tree-width bounded by k satisfies the first and third properties but does not satisfy the second property. But more importantly, it is not clear that they have nice controllers of the form described above. We propose a meaningful class which has more coverage than bounded phase of [13], and is nicely controllable. We show the decidability of this class by demonstrating a bound on split-width.

4 Class and Controller: Sequential Case

We begin by identifying a class of behaviours, called k-Phase behaviours, which is verifiable and admits a **DS**-controller. Roughly speaking, a phase is a segment of the run where the reads are from a fixed data-structure. However, between successive reads, read-free recursive computations are permitted which may write to all data-structures, including their own call-stack. We formalize this below.

An autonomous computation involves a single recursive thread executing a recursive procedure without reading any other data structure. All read events are from a single stack while there is no restriction placed on the writes. We say that an edge $e \rhd f$ is *autonomous* if $e \rhd^s f$ for some $s \in \mathbf{Stacks}$ and all in-between read events are from the same stack s: if $e' \rhd^d f'$ with $e \leq f' \leq f$ then $d = s$. We shall write $\rhd_a$ for the subset of $\rhd$ consisting of the autonomous edges and $\rhd_{na}$ for $\rhd \setminus \rhd_a$ and refer to them as the non-autonomous edges. If $e \rhd_a f$ then e and f are called *autonomous* write and read events respectively.

A d-phase is a sequence of consecutive events in which all non-autonomous reads are from the data-structure $d \in \mathbf{DS}$. Writes to all data-structures are permitted. Moreover, a phase must not break an autonomous computation. Formally, a d-phase is identified by a pair of events $e \leq f$ (the first and the last events in the sequence) such that, if $e' \rhd_{na} f'$ with $e \leq f' \leq f$ then $e' \rhd^d f'$ and if $e' \rhd_a f'$ with $e \leq f' \leq f$ or $e \leq e' \leq f$ then $e \leq e' \leq f' \leq f$.

Example 1. Suppose $\mathbf{DS} = \{q, s_1, s_2\}$. A q-phase is depicted on the right. Straight lines (resp. curved lines) represent $\rhd^d$ edges from queues (resp. stacks). Autonomous computations are highlighted in white.

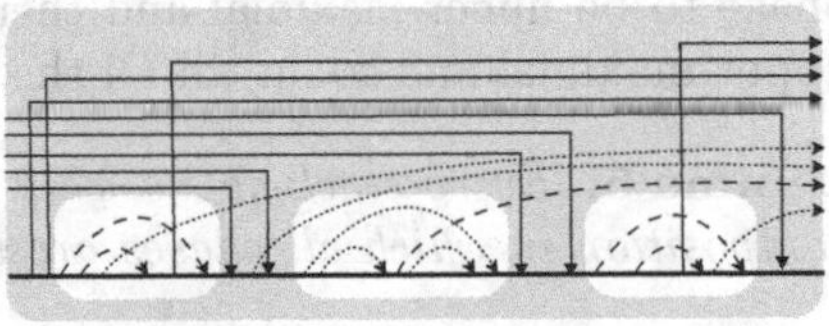

Remark 2. Permitting autonomous (recursive) computations during a phase is a natural generalization of *well-queueing* assumption of [14] where reads from

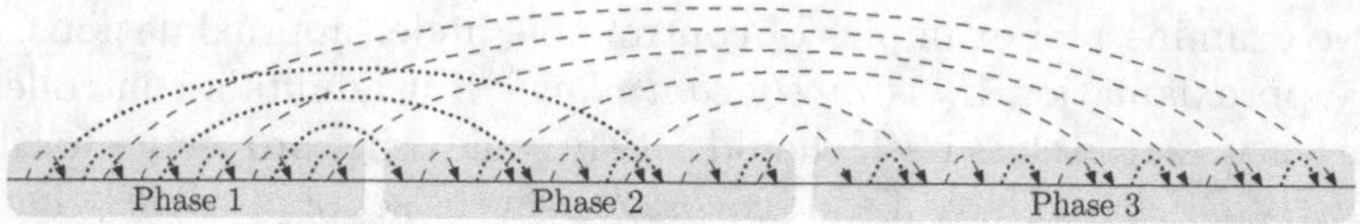

Fig. 1. A stack-queue word over two stacks and its maximal phase decomposition

queues are permitted only when the stack associated with a process is empty. The latter corresponds to permitting reads from queues in the *main* program but not from any of the functions it calls. We permit recursive calls to be at any depth of recursion when reading from the queue. After such a read, however, returning from the function causes a phase change.

Our aim is to obtain a decidable and controllable class by bounding the number of phases. In the presence of queues, reading and writing on a queue during a phase can be used to simulate a turing machine using just 1-phase computations. Allowing autonomous computations on one stack while reading and writing on another also results in the same effect. This motivates the following definition which rules out such *self-loops*.

A phase identified by a pair (e, f) has a *self-loop* if it contains a non-autonomous edge: $e \leq e' \rhd_{na} f' \leq f$.

A phase decomposition is a partition of the set of events into phases with no self-loops. A k-phase decomposition is a phase decomposition with at most k phases. We denote by k-Phase the class of stack-queue words that admit a k-phase decomposition.

Remark 3. Observe that by freely allowing autonomous computations (as opposed to well-queuing), every stack-queue word is in k-Phase for some k.

Remark 4. When restricted to systems with only stacks, k-Phase subsumes the *k bounded phase restriction* for multi-pushdown systems [13]. It also subsumes the *k bounded context restriction* for systems with stacks and queues [14]. In fact, for every bound k, there exist stack-queue words which are not captured by [13] and [14], but which are in 3-Phase. (See Figure 1.)

A phase with no self-loops identified by (e, f) is *upper-maximal* if it cannot be extended upwards in a phase with no self-loops: if (e, g) is a phase with no self-loops then $g \leq f$. Given any k phase decomposition, we may extend the first phase to be upper-maximal and then extend the next (remaining) phase to be upper-maximal and so on till all the phases are upper maximal.

Lemma 5. *Every stack-queue word in* k-Phase *admits a* maximal *k-phase decomposition in which all phases are upper-maximal.*

Now we take up the task of constructing a **DS**-controller for the class k-Phase. A crucial step towards this end is to identify autonomous reads. We show below that this can be achieved with a multi-pushdown automaton $\mathcal{B}$ observing the data-structure access. When the system $\mathcal{S}$ writes/reads some value on a stack s the automaton $\mathcal{B}$ will simultaneously write/read a bit on the same stack. $\mathcal{B}$ is obtained as a cartesian product of automata $\mathcal{B}_s$ ($s \in \textbf{Stacks}$) identifying the autonomous reads on stack s (described in Figure 2).

Here, $s!b$ (resp. $s?b$) means that the system $\mathcal{S}$ writes/reads on stack s and b is the tag bit that is simultaneously written/read by $\mathcal{B}_s$ on stack s. The other events do not change stack s. Moreover, $\bar{s}?$ is the observation of a read event of $\mathcal{S}$ which is *not* on stack s, and *else* means any event which is not explicitly specified.

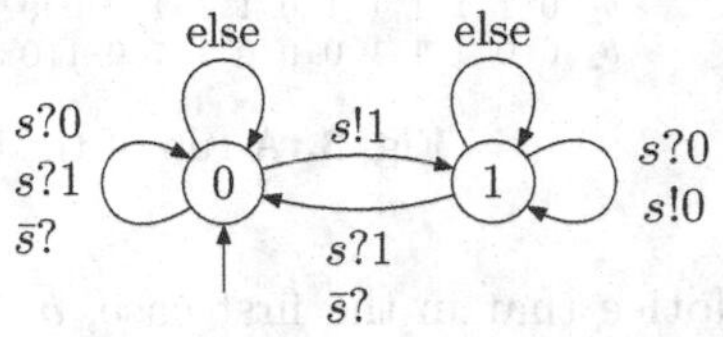

Fig. 2. The automaton $\mathcal{B}_s$

We say that e is a *possibly autonomous write* to stack s at event g if $e \rhd^s f$ and $e \le g < f$ and $e' \rhd^d f'$ with $e \le f' \le g$ implies $d = s$. Intuitively $\mathcal{B}_s$ will be in state 1 iff in the current prefix there is an unmatched write event e to stack s which is possibly autonomous. On a write to s the automaton moves from state 0 to 1 since this write is possibly autonomous, and pushes 1 on the stack to indicate that it is the first possibly autonomous write in the past. Then, as long as it does not read from a data-structure $d \neq s$, it stays in state 1, pushing 0 on the stack on a write to s and reading 0 from the stack on a read from s. If it reads 1 from the stack, then it has matched the first possibly autonomous write in the past, hence it goes back to state 0. On a read from $d \neq s$ it goes to state 0 since there cannot be any possibly autonomous write to s at this read event.

Lemma 6. *The automaton $\mathcal{B}_s$ is deterministic and non-blocking. Moreover, in the unique run of $\mathcal{B}_s$ on a word, the state b_s before a read from stack s determines whether this read is autonomous ($b_s = 1$) or not ($b_s = 0$).*

We now construct the deterministic **DS**-controller $\mathcal{C}^k$ for k-Phase. This controller computes the maximal phase decomposition of a behaviour and uses the automaton $\mathcal{B}$ to identify autonomous reads. We denote by $\boldsymbol{b} = (b_s)_{s \in \textbf{Stacks}}$ a state of $\mathcal{B}$. In addition, a state of $\mathcal{C}^k$ holds two other values:

- a counter $n \in \{1, \ldots, k, \infty\}$ which indicates the current phase number. The counter starts from value 1 and is non-decreasing along a run. The ∞ indicates that the number of phases has exceeded k. We follow the convention that $i+1$ has the usual meaning if $i < k$, $k+1 = \infty$ and $\infty + 1 = \infty$.
- a value $d \in \textbf{DS} \cup \{?\}$ which indicates that the current phase has non-autonomous reads from $d \in \textbf{DS}$ or that only autonomous reads have occurred so far ($d = ?$). Note that in the first phase all reads are autonomous (a non-autonomous read would create a self-loop). Hence, $d = ?$ iff $n = 1$.

The initial state of the controller is $(1, ?, \mathbf{0})$. On an internal event, the state remains unchanged. When the system writes to a data-structure the controller $\mathcal{C}^k$ writes its current phase number in addition to the bits written by $\mathcal{B}$.

$$(n, d, \boldsymbol{b}) \xrightarrow{d'!n} (n, d, \boldsymbol{b}) \qquad \text{if } d' \in \textbf{Queues} \tag{1}$$

$$(n, d, \boldsymbol{b}) \xrightarrow{d'!(n,c)} (n, d, \boldsymbol{b}') \quad \text{if } d' \in \textbf{Stacks} \wedge \boldsymbol{b} \xrightarrow{d'!c} \boldsymbol{b}' \text{ in } \mathcal{B} \tag{2}$$

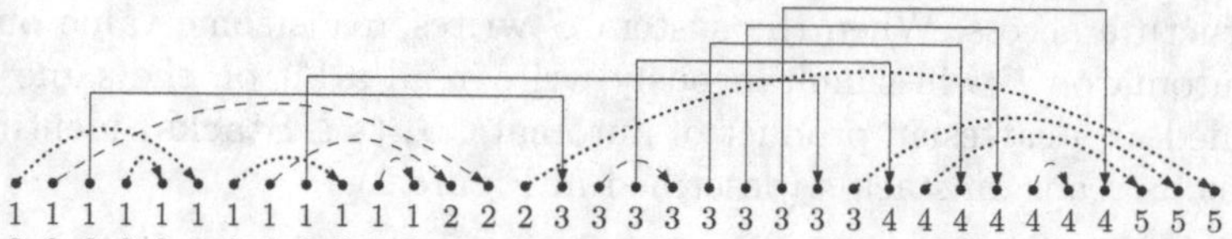

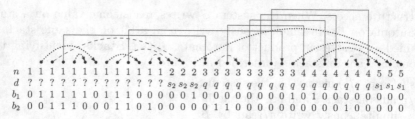

n	1	1	1	1	1	1	1	1	1	1	1	1	1	2	2	2	3	3	3	3	3	3	3	3	3	4	4	4	4	4	4	4	5	5	5
d	?	?	?	?	?	?	?	?	?	?	?	?	?	s_2	s_2	s_2	q	q	q	q	q	q	q	q	q	q	q	q	q	q	q	q	s_1	s_1	s_1
b_1	0	1	1	1	1	1	0	1	1	1	0	0	0	0	0	1	0	0	0	0	0	0	0	0	1	0	1	0	0	0	0	0	0	0	0
b_2	0	0	1	1	1	0	0	0	1	1	0	1	0	0	0	0	0	1	1	0	0	0	0	0	0	0	0	0	0	1	0	0	0	0	0

Fig. 3. A run of the deterministic sequential controller $\mathcal{C}^k$

Notice that in the first case, $\boldsymbol{b} \xrightarrow{d'!} \boldsymbol{b}$ is a transition in $\mathcal{B}$. A read event from a queue d' will stay in the same phase if d' is the current data-structure and the matching write comes from a previous phase (to avoid self-loops): if $d' \in$ **Queues** then we have the following transitions in $\mathcal{C}^k$

$$(n, d, \boldsymbol{b}) \xrightarrow{d'?m} (n, d', \boldsymbol{0}) \qquad \text{if } d' = d \wedge m < n \tag{3}$$
$$(n, d, \boldsymbol{b}) \xrightarrow{d'?m} (n+1, d', \boldsymbol{0}) \quad \text{otherwise} \tag{4}$$

Notice that in these cases, $\boldsymbol{b} \xrightarrow{d'?} \boldsymbol{0}$ is a transition in $\mathcal{B}$ since no stack can be on an autonomous computation at a read event from a queue. Further if $d = ?$, reading from a queue forces a phase change. This is needed, as otherwise there will be a self-loop on the first phase.

Finally, a read event from a stack s will stay in the same phase if it is an autonomous read ($b_s = 1$), or $s = d$ is the current data-structure and this read does not create a self-loop: if $s \in$ **Stacks** then in $\mathcal{C}^k$ we have the transitions

$$(n, d, \boldsymbol{b}) \xrightarrow{s?(m,c)} (n, d, \boldsymbol{b}') \qquad \text{if } (b_s = 1 \vee (s = d \wedge m < n)) \wedge \boldsymbol{b} \xrightarrow{s?c} \boldsymbol{b}' \text{ in } \mathcal{B} \tag{5}$$
$$(n, d, \boldsymbol{b}) \xrightarrow{s?(m,c)} (n+1, s, \boldsymbol{0}) \quad \text{otherwise} \tag{6}$$

Notice that in the last case, $\boldsymbol{b} \xrightarrow{s?c} \boldsymbol{0}$ is a transition in $\mathcal{B}$ and thus in all moves the third component stays consistent with moves of $\mathcal{B}$.

By construction the controller is deterministic and non-blocking. If the unique run of the controller on a $\mathcal{W}$ does not use a state of the form $(\infty, d, \boldsymbol{b})$ then $\mathcal{W}$ is in k-Phase. The set of positions labeled by states of the form $(i, d, \boldsymbol{b})$ identify the ith phase in a k phase decomposition. Conversely, let $\mathcal{W}$ be in k-Phase. Let $(\boldsymbol{b_e})$ be the state labeling position e in $\mathcal{W}$ in the unique run of $\mathcal{B}$ on $\mathcal{W}$. Let $(X_i)_{(i \leq l)}$ be the phases in the maximal decomposition of $\mathcal{W}$. It is easy to verify that the first position of X_i, $i \geq 2$ is a non-autonomous read and let d_i be the data-structure associated with this read. Then the labeling assigning $(1, ?, \boldsymbol{b_e})$ to any position $e \in X_1$ and $(i, d_i, \boldsymbol{b_e})$ to any event e in X_i, $2 \leq i \leq l$ is is an accepting run of the controller on $\mathcal{W}$.

Theorem 7. *The SQS $\mathcal{C}^k$ is a* **DS***-controller for the class k-*Phase *with* $(|\mathbf{DS}| \cdot (k+1)+1)2^{|\mathbf{Stacks}|}$ *states.*

5 Class and Controller: Distributed Case

In this section we describe a model intended to capture collections of SQS communicating via reliable FIFO channels (or queues). Such systems are called Stack-Queue Distributed System (SQDS). A behaviour of an SQDS is a tuple of stack-queue words with additional matching relations describing the inter-process communication via queues. Such behaviours extend Message Sequence Charts (MSCs) with matching relations for the internal stacks and queues. We call them stack-queue MSCs (SQMSC).

We then extend the notion of k-Phase to this distributed setting. We show that k-Phase enjoys a deterministic distributed controller with local acceptance conditions.

An architecture $\mathfrak{A}$ is a tuple $(\mathbf{Procs}, \mathbf{Stacks}, \mathbf{Queues}, \mathsf{Writer}, \mathsf{Reader})$ consisting of a set of processes **Procs**, a set of stacks **Stacks**, a set of queues **Queues** and functions Writer and Reader which assign to each stack/queue the process that will write (push/send) into it and the process that will read (pop/receive) from it respectively. We write **DS** for **Stacks** $\uplus$ **Queues**.

A stack d must be local to its process, so $\mathsf{Writer}(d) = \mathsf{Reader}(d)$. On the other hand, a queue d may be local to a process p if $\mathsf{Writer}(d) = p = \mathsf{Reader}(d)$, otherwise it provides a FIFO channel from $\mathsf{Writer}(d)$ to $\mathsf{Reader}(d)$.

A Stack-Queue Distributed System (SQDS) over an architecture $\mathfrak{A}$ and an alphabet Σ is a tuple $\mathcal{S} = (\mathsf{Locs}, \mathrm{Val}, (\mathrm{Trans}_p)_{p\in\mathbf{Procs}}, \mathsf{in}, \mathsf{Fin})$ where each $\mathcal{S}_p = (\mathsf{Locs}, \mathrm{Val}, \mathrm{Trans}_p, \mathsf{in}, \emptyset)$ is an SQS over **DS** and Σ in which the transitions are compatible with the architecture: Trans_p may have a write (resp. read) transitions on data-structure d only if $\mathsf{Writer}(d) = p$ (resp. $\mathsf{Reader}(p) = d$). Moreover, $\mathsf{Fin} \subseteq \mathsf{Locs}^{\mathbf{Procs}}$ is the global acceptance condition. We say that the acceptance condition is local if $\mathsf{Fin} = \prod_{p\in\mathbf{Procs}} \mathsf{Fin}_p$ where $\mathsf{Fin}_p \subseteq \mathsf{Locs}$ for all $p \in \mathbf{Procs}$.

A stack-queue MSC (SQMSC) over architecture $\mathfrak{A}$ and alphabet Σ is a tuple $\mathcal{M} = ((w_p)_{p\in\mathbf{Procs}}, (\rhd^d)_{d\in\mathbf{DS}})$ where $w_p \in \Sigma^*$ is the sequence of events on process p and $\rhd^d$ is the relation matching write events on data-structure d with their corresponding read events. We let $\mathcal{E}_p = \{(p,i) \mid 1 \le i \le |w_p|\}$ be the set of events on process $p \in \mathbf{Procs}$. For an event $e = (p,i) \in \mathcal{E}_p$, we set $\mathsf{pid}(e) = p$ and $\lambda(e)$ be the ith letter of w_p. We write $\rightarrow$ for the successor relation on processes: $(p,i) \rightarrow (p,i+1)$ if $1 \le i < |w_p|$ and we let $\rhd = \bigcup_{d\in\mathbf{DS}} \rhd^d$ be the set of all matching edges. We require the relation $< = (\rightarrow \cup \rhd)^+$ to be a strict partial order on the set of events. Finally, the matching relations should comply with the architecture: $\rhd^d \subseteq \mathcal{E}_{\mathsf{Writer}(d)} \times \mathcal{E}_{\mathsf{Reader}(d)}$. Moreover, data-structure accesses should be disjoint, stacks should conform to LIFO and queues should conform to FIFO (the formal definitions are taken verbatim from Section 2). An SQMSC is depicted in Figure 4.

As before, to define the run of an SQDS over a stack-queue MSC $\mathcal{M}$, we introduce two notations. For $p \in \mathbf{Procs}$ and $e \in \mathcal{E}_p$, we denote by e^- the unique event such that $e^- \rightarrow e$ if it exists, and we let $e^- = \bot_p \notin \mathcal{E}$ otherwise. We let $\max_p(\mathcal{M})$ be the maximal event of $\mathcal{E}_p$ if it exists and $\max_p(\mathcal{M}) = \bot_p$ otherwise.

A run of an SQDS $\mathcal{S}$ over a stack-queue MSC $\mathcal{M}$ is a mapping $\rho\colon \mathcal{E} \to \mathsf{Locs}$ satisfying the following consistency conditions (with $\rho(\bot_p) = \mathsf{in}$):

- if e is an internal event then $\rho(e^-) \xrightarrow{\lambda(e)} \rho(e) \in \mathsf{Trans}_{\mathsf{pid}(e)}$,
- if $e \rhd^d f$ for some data-structure $d \in \mathbf{DS}$ then for some $v \in \mathrm{Val}$ we have both $\rho(e^-) \xrightarrow{\lambda(e),d!v} \rho(e) \in \mathsf{Trans}_{\mathsf{pid}(e)}$ and $\rho(f^-) \xrightarrow{\lambda(f),d?v} \rho(f) \in \mathsf{Trans}_{\mathsf{pid}(f)}$.

The run is accepting if $(\rho(\max_p(\mathcal{M})))_{p\in\mathbf{Procs}} \in \mathsf{Fin}$. The *language* $\mathcal{L}(\mathcal{S})$ accepted by an SQDS $\mathcal{S}$ is the set of stack-queue MSCs on which it has an accepting run.

Notice that SQDSs are closed under intersection, by means of the cartesian product. The construction is similar to the one for SQSs in Section 2.

Bounded Acyclic Phase SQMSCs. We generalize the under-approximation class k-Phase to the distributed setting. We allow at most k phases per process. As in the sequential case, autonomous computations are freely allowed. However, cycles on phases can be caused be the richer structure of the SQMSC than simple self loops.

In the distributed setting, the definitions of **autonomous computations** and of d-**phases** are identical to the sequential case, cf. Section 4. Again, we write $\rhd_a$ for autonomous edges and $\rhd_{na}$ for non-autonomous edges. A phase, which is a sequence of consecutive events executed by a single process, is identified by a pair of events (e, f) such that $e \to^* f$.

A phase (e, f) **has a cycle** if there is a non-autonomous edge $e' \rhd_{na} f'$ with $e \leq e'$ and $f' \to^* f$. Notice that e' needs not be in the phase. So a cycle starts from the phase at e then follows the partial order to some non-autonomous write e' whose read f' is in the phase. A phase is *acyclic* if it has no cycles. Notice that a non-autonomous edge within a phase induces a cycle (self-loop) whereas autonomous edges are freely allowed within phases. As a matter of fact, when there is exactly one process, a phase has a cycle iff it has a self-loop.

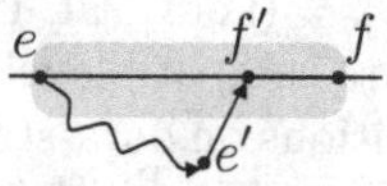

A phase decomposition of an SQMSC is a partition of its set of events into phases. A phase decomposition is *acyclic* if all phases are acyclic. It is a k-phase decomposition if there are at most k phases per process. We denote by k-Phase the set of SQMSCs that admits an acyclic k-phase decomposition.

An acyclic phase (e, f) is *upper-maximal* if extending it upwards would result in a cycle, i.e., for every other acyclic phase (e, f'), we have $f' \leq f$. See Figure 4 for an example. Lemma 5 easily lifts up to the distributed case as well.

Lemma 8. *Every SQMSC in* k-Phase *admits a* maximal *acyclic* k*-phase decomposition in which all phases are upper-maximal.*

Deterministic Distributed Controller. We extend the notion of nice controllers to the distributed setting. That means controllers should be *distributed* and have *local* acceptance conditions. A local controller for one process should

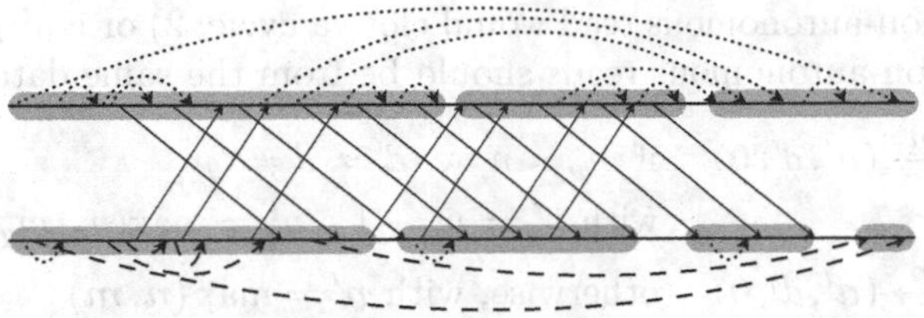

Fig. 4. A stack-queue MSC and its maximal phase decomposition

be able to control the behaviour of that process regardless of the states of the other local controllers. The communication between the local controllers is also only by means of overloading the actual messages sent between the processes. The local controllers are not allowed to send messages out of sync, as it would create new behaviours in the controlled system. Thus a controlled system should be again obtained as a cartesian product of the system with a controller where both are SQDS, but in addition the controller has local acceptance conditions.

Theorem 9. *The class* k-Phase *admits a deterministic distributed controller* $\mathcal{C}^k$ *with* $(|\mathbf{DS}| \cdot (k+2)^{|\mathbf{Procs}|}+1)2^{|\mathbf{Stacks}|}$ *states.*

The distributed controller is a generalisation of the sequential controller of Section 4. The main difference is that the local controller of process p remembers not only its current phase number, but a tuple $\boldsymbol{n} = (n_q)_{q\in\mathbf{Procs}}$ of phase numbers for each process. The intuition is that n_q is the largest phase of process q that is known to process p ($n_q = 0$ if no events of process q are in the past of the current event of process p).

For each stack s, we use the automaton $\mathcal{B}_s$ defined in Section 4 that identifies autonomous reads. For each process $p \in \mathbf{Procs}$, we let $\mathcal{B}_p$ be the product of the automata $\mathcal{B}_s$ where s is a stack of process p (i.e., $s \in \mathbf{Stacks}$ and $\mathsf{Writer}(s) = p$).

A state of the local controller $\mathcal{C}_p^k$ for process p is a tuple $(\boldsymbol{n}, d, \boldsymbol{b}_p)$ where $\boldsymbol{n} = (n_q)_{q\in\mathbf{Procs}}$ is the phase vector with $n_q \in \{0, 1, \ldots, k, \infty\}$, $d \in \mathbf{DS} \cup \{?\}$ with $\mathsf{Reader}(d) = p$ if $d \neq ?$, and $\boldsymbol{b}_p$ is a state of $\mathcal{B}_p$. The initial state of $\mathcal{C}_p^k$ is $\mathsf{in}_p = (\boldsymbol{n}, ?, \boldsymbol{0})$ with $n_p = 1$ and $n_q = 0$ for $q \neq p$. The local acceptance condition Fin_p is given by the set of states $(\boldsymbol{n}, d, \boldsymbol{b}_p)$ with $n_q \neq \infty$ for all $q \in \mathbf{Procs}$.

We describe now the local transitions of $\mathcal{C}_p^k$. They are similar to the transitions of the sequential controller given in Section 4. We start with write transitions, so let $d' \in \mathbf{DS}$ be such that $\mathsf{Writer}(d') = p$. On write events, the current phase vector is written on to the data-structure (in addition to the autonomous bit where needed).

$$(\boldsymbol{n}, d, \boldsymbol{b}_p) \xrightarrow{d'!\boldsymbol{n}} (\boldsymbol{n}, d, \boldsymbol{b}_p) \quad \text{if } d' \in \mathbf{Queues} \tag{7}$$

$$(\boldsymbol{n}, d, \boldsymbol{b}_p) \xrightarrow{d'!(\boldsymbol{n},c)} (\boldsymbol{n}, d, \boldsymbol{b}'_p) \quad \text{if } d' \in \mathbf{Stacks} \wedge \boldsymbol{b}_p \xrightarrow{d'!c} \boldsymbol{b}'_p \text{ in } \mathcal{B}_p \tag{8}$$

Let $d' \in \mathbf{Queues}$ be such that $\mathsf{Reader}(d') = p$. The transitions of $\mathcal{C}_p^k$ that read queue d' are given below. We should switch to the next phase 1) if $m_p = n_p$ since otherwise this non-autonomous read would close a cycle, 2) or if $d' \neq d \neq ?$ since in a phase all non-autonomous reads should be from the same data-structure.

$$(\boldsymbol{n}, d, \boldsymbol{b}_p) \xrightarrow{d'?\boldsymbol{m}} (\boldsymbol{n}', d', \boldsymbol{0}) \quad \text{if } m_p = n_p \vee (d' \neq d \neq ?) \tag{9}$$

$$\text{with } n'_p = n_p + 1 \wedge n'_q = \max(n_q, m_q) \text{ for } q \neq p$$

$$(\boldsymbol{n}, d, \boldsymbol{b}_p) \xrightarrow{d'?\boldsymbol{m}} (\boldsymbol{n}', d', \boldsymbol{0}) \quad \text{otherwise, with } \boldsymbol{n}' = \max(\boldsymbol{n}, \boldsymbol{m}) \tag{10}$$

Similarly, we give below read transitions from $d' \in$ **Stacks** with $\mathsf{Reader}(d') = p$. Here a switch of phase is required under the same conditions but only when the read is not autonomous.

$$(\boldsymbol{n}, d, \boldsymbol{b}_p) \xrightarrow{d'?(\boldsymbol{m},c)} (\boldsymbol{n}', d', \boldsymbol{0}) \quad \text{if } b_{d'} = 0 \wedge (m_p = n_p \vee (d' \neq d \neq ?)) \tag{11}$$

$$\text{with } n'_p = n_p + 1 \wedge n'_q = \max(n_q, m_q) \text{ for } q \neq p$$

$$(\boldsymbol{n}, d, \boldsymbol{b}_p) \xrightarrow{d'?(\boldsymbol{m},c)} (\boldsymbol{n}', d, \boldsymbol{b}'_p) \quad \text{otherwise,} \tag{12}$$

$$\text{with } \boldsymbol{n}' = \max(\boldsymbol{n}, \boldsymbol{m}) \wedge \boldsymbol{b}_p \xrightarrow{d'?c} \boldsymbol{b}'_p \text{ in } \mathcal{B}_p$$

One of the differences of a local controller from a sequential controller is that the first phase may also perform non-autonomous reads. However, in such case, it must be from a queue.

On read transitions (10 and 12) which stay in the same phase, the phase vector is updated by taking the maximum between the current phase vector and the read-phase vector ($\boldsymbol{n}' = \max(\boldsymbol{n}, \boldsymbol{m})$). On a phase switch, a similar update is performed but the current phase number of process p is incremented.

6 Decidability

In this section we explain briefly why k-Phase is a verifiable under-approximation for SQDS. Consider the reachability problem which is equivalent to asking if given an SQDS $\mathcal{S}$ and $k \in \mathbb{N}$ whether $\mathcal{S}$ accepts at least one $\mathcal{M}$ from k-Phase. A non-trivial extension of the technique of [14] allows to reduce the reachability problem of SQDS restricted to k-Phase to the reachability problem of multi-pushdown systems for bounded phase.

A more general question is to model-check properties expressed in linear time logics ranging from temporal logics to MSO($\rightarrow$,$\rhd^d$). Given a formula φ we have to determine whether every $\mathcal{M} \in k$-Phase that is accepted by $\mathcal{S}$ satisfies φ. Observe that we may equivalently ask whether *every* behaviour of the controlled system $\mathcal{S}'$ satisfies φ. Using a slightly different approach we can obtain decidability not only for reachability but also for the linear-time model-checking problems.

In this approach we show that every behaviour in k-Phase has *split-width* [5, 6, 8] or tree-width [17] or clique-width [4] (measures of the complexity of graphs that happen to be equivalent for our class of graphs) bounded by some function $f(k)$. Here, we show an exponential bound on the split-width. Then, results from [6, 8, 17] imply that MSO model-checking for $\mathcal{S}'$ is decidable and results from [5,8] imply that model-checking linear-time temporal logic formulas can be solved in double exponential time. This is optimal, since reachability of k-phase multi-pushdown systems is double exponential time hard [15].

References

1. Alur, R., Madhusudan, P.: Adding nesting structure to words. Journal of the ACM 56, 16:1–16:43 (2009)
2. Atig, M.F., Bollig, B., Habermehl, P.: Emptiness of multi-pushdown automata is 2ETIME-complete. In: Ito, M., Toyama, M. (eds.) DLT 2008. LNCS, vol. 5257, pp. 121–133. Springer, Heidelberg (2008)
3. Breveglieri, L., Cherubini, A., Citrini, C., Crespi-Reghizzi, S.: Multi-pushdown languages and grammars. Int. J. Found. Comput. Sci. 7(3), 253–292 (1996)
4. Courcelle, B.: The expression of graph properties and graph transformations in monadic second-order logic. In: Rozenberg, G. (ed.) Handbook of Graph Grammars, pp. 313–400. World Scientific (1997)
5. Cyriac, A.: Verification of Communicating Recursive Programs via Split-width. PhD thesis, ENS Cachan (2014), `http://www.lsv.ens-cachan.fr/~cyriac/download/Thesis_Aiswarya_Cyriac.pdf`
6. Cyriac, A., Gastin, P., Narayan Kumar, K.: MSO decidability of multi-pushdown systems via split-width. In: Koutny, M., Ulidowski, I. (eds.) CONCUR 2012. LNCS, vol. 7454, pp. 547–561. Springer, Heidelberg (2012)
7. Cyriac, A., Gastin, P., Narayan Kumar, K.: Controllers for the Verification of Communicating Multi-pushdown Systems. Technical report (2014), `http://hal.archives-ouvertes.fr/`
8. Cyriac, A., Gastin, P., Narayan Kumar, K.: Verifying Communicating Multi-pushdown Systems. Technical report (January 2014), `http://hal.archives-ouvertes.fr/hal-00943690`
9. Genest, B., Kuske, D., Muscholl, A.: A Kleene theorem and model checking algorithms for existentially bounded communicating automata. Information and Computation 204(6), 920–956 (2006)
10. Genest, B., Muscholl, A., Seidl, H., Zeitoun, M.: Infinite-state high-level MSCs: Model-checking and realizability. Journal of Computer and System Sciences 72(4), 617–647 (2006)
11. Henriksen, J.G., Mukund, M., Narayan Kumar, K., Sohoni, M.A., Thiagarajan, P.S.: A theory of regular MSC languages. Inf. Comput. 202(1), 1–38 (2005)
12. Heußner, A., Leroux, J., Muscholl, A., Sutre, G.: Reachability analysis of communicating pushdown systems. In: Ong, L. (ed.) FOSSACS 2010. LNCS, vol. 6014, pp. 267–281. Springer, Heidelberg (2010)
13. La Torre, S., Madhusudan, P., Parlato, G.: A robust class of context-sensitive languages. In: LICS, pp. 161–170. IEEE Computer Society Press (2007)
14. La Torre, S., Madhusudan, P., Parlato, G.: Context-bounded analysis of concurrent queue systems. In: Ramakrishnan, C.R., Rehof, J. (eds.) TACAS 2008. LNCS, vol. 4963, pp. 299–314. Springer, Heidelberg (2008)
15. La Torre, S., Madhusudan, P., Parlato, G.: An infinite automaton characterization of double exponential time. In: Kaminski, M., Martini, S. (eds.) CSL 2008. LNCS, vol. 5213, pp. 33–48. Springer, Heidelberg (2008)
16. La Torre, S., Napoli, M.: Reachability of multistack pushdown systems with scope-bounded matching relations. In: Katoen, J.-P., König, B. (eds.) CONCUR 2011. LNCS, vol. 6901, pp. 203–218. Springer, Heidelberg (2011)
17. Madhusudan, P., Parlato, G.: The tree width of auxiliary storage. In: Ball, T., Sagiv, M. (eds.) POPL, pp. 283–294. ACM (2011)
18. Qadeer, S., Rehof, J.: Context-bounded model checking of concurrent software. In: Halbwachs, N., Zuck, L.D. (eds.) TACAS 2005. LNCS, vol. 3440, pp. 93–107. Springer, Heidelberg (2005)
19. Zielonka, W.: Notes on finite asynchronous automata. R.A.I.R.O. — Informatique Théorique et Applications 21, 99–135 (1987)

Pairwise Reachability Analysis for Higher Order Concurrent Programs by Higher-Order Model Checking

Kazuhide Yasukata, Naoki Kobayashi, and Kazutaka Matsuda

The University of Tokyo, Hongo, Japan

Abstract. We propose a sound, complete, and automatic method for pairwise reachability analysis of higher-order concurrent programs with recursion, nested locks, joins, and dynamic thread creation. The method is based on a reduction to higher-order model checking (i.e., model checking of trees generated by higher-order recursion schemes). It can be considered an extension of Gawlitz et al.'s work on the join-lock-sensitive reachability analysis for dynamic pushdown networks (DPN) to higher-order programs. To our knowledge, this is the first application of higher-order model checking to sound and complete verification of (reasonably expressive models of) concurrent programs.

1 Introduction

Verification of concurrent programs is important but fundamentally difficult, especially in the presence of recursion. Ramalingam [19] has shown that the reachability problem for two pushdown systems with rendezvous-style synchronization primitives is undecidable. There are two major approaches to cope with this limitation. One is to give up the soundness, and underapproximate the actual behavior of a concurrent program by bounding the number of context switches [18], etc. The other approach is to restrict the synchronization primitives. Kahlon et al. [7] have shown that the pairwise reachability problem ("Given two pushdown systems P_1 and P_2 and control locations ℓ_1 and ℓ_2, is there a reachable global state where P_1 is at ℓ_1 and P_2 is at ℓ_2?") is decidable if the two pushdown systems synchronize only via nested locking. Lammich et al. [14] later extended it to dynamic pushdown networks (DPN), where processes may be dynamically created. Gawlitza et al. [4] have further extended the result to allow synchronization via joins in addition to nested locking.

In the present paper, we follow the latter line of work and extend it to deal with *higher-order* concurrent programs with recursion, dynamic process creation, joins, and nested locking. We consider the pairwise reachability problem: "Given a program P and two control locations ℓ_1 and ℓ_2, may the program reach a state where one process is at ℓ_1 and another is at ℓ_2?". We show that this problem can be reduced to *higher-order model checking* [15,9], hence it is decidable. The main idea is to transform a given program to a non-deterministic higher-order

P. Baldan and D. Gorla (Eds.): CONCUR 2014, LNCS 8704, pp. 312–326, 2014.

recursion scheme (a kind of tree grammar where non-terminals may take higher-order functions as arguments) $\mathcal{G}$ that generates a tree language $\mathcal{L}(\mathcal{G})$ consisting of all the possible execution histories (called *action trees* [14,4]) of the program ignoring the synchronization constraints imposed by joins and nested locking. Let L_1 be the set of action trees that respect the synchronization constraints imposed by joins and nested locking, and L_2 be the set of action trees that represent histories that end with a state where two of the processes are at ℓ_1 and ℓ_2. Then, ℓ_1 and ℓ_2 are pairwise reachable if and only if $\mathcal{L}(\mathcal{G}) \cap L_1 \cap L_2 \neq \emptyset$. Since both L_1 and L_2 are regular (where the regularity of L_1 is due to [4]), the latter condition can be decided by using higher-order model checking [15,9]. We formalize the reduction and prove its correctness. We also report preliminary experimental results, which confirm that the approach is feasible at least for small programs, despite the extremely high worst-case complexity of higher-order model checking (k-EXPTIME complete for order-k higher-order recursion schemes [15,12]).

To our knowledge, this is the first application of higher-order model checking to sound and *complete* verification of higher-order *concurrent* programs. The previous applications of higher-order model checking were mainly for higher-order *functional* programs [9,13,16]. For concurrent programs, the previous applications [10,5] were based on the underapproximation approach. One may think that higher-order functions are exotic features for multi-threaded programs. As demonstrated in [9,20], however, even if higher-order functions are not so often used explicitly in source programs, they are required for precisely modelling control/data structures such as exceptions and lists.

The rest of the paper is structured as follows. Section 2 introduces the target language and formally defines the pairwise reachability problem. After providing the necessary backgrounds (such as action trees [14,4] and higher-order model checking [15,9]) Section 3 provides the reduction of the pairwise reachability problem to higher-order model checking. Section 4 reports preliminary experiments. Section 5 discusses related work and Section 6 concludes the paper. Omitted proofs are found in the longer version, available from the authors' web page.

2 Target Language and Pairwise Reachability Problem

This section introduces a higher-order concurrent programming language and defines the pairwise reachability problem for it.

Definition 1. *A* ***program*** *is a finite set of* function definitions

$$\{F_1\,\tilde{x_1} = e_1,\ ...,\ F_n\,\tilde{x_n} = e_n\},$$

where F_i denotes a defined function symbol, and e ranges over the set Exp *of* expressions*, defined by:*

$$e ::= \$ \mid x \mid F \mid \mathbf{if}_{\_}\, e_1\, e_2 \mid e_1\, e_2 \mid \mathbf{join};\, e \mid \mathbf{acq}_i;\, e \mid \mathbf{rel}_i;\, e \mid \mathbf{spawn}(e_c);\, e \mid e^{\ell}$$

Here, i ranges over a finite set Lock of (non-reentrant) locks, and ℓ ranges over a finite set Label of program point labels. Note that the arity of each function may be 0. We require that the function symbols $F_1, ..., F_n$ are different from each other, and that any program p contains exactly one definition of a "main" function S, of the form $S = e$.

We explain the intuitive meaning of each expression; the formal operational semantics is given later. The expression \$ represents the termination of the current process. The expression $\mathbf{if}_{-}\, e_1\, e_2$ executes either e_1 or e_2 non-deterministically, and the expression $e_1 e_2$ applies the function e_1 to e_2. As defined later, function calls are based on the call-by-name semantics; call-by-value programs can be transformed to call-by-name programs by using the CPS transformation [17]. The expression $\mathbf{spawn}(e_c); e$ spawns a new child process that executes e_c, and continues to execute e without waiting for the child process. The expression $\mathbf{join}; e$ waits for the termination of all the processes that the current process has created, and then executes e. The expression $\mathbf{acq}_i; e$ waits to acquire the lock i, and then executes e. The expression $\mathbf{rel}_i; e$ releases the lock i and executes e. The label ℓ in the expression e^ℓ is used for specifying the pairwise reachability problem, and does not affect the operational semantics.

In this paper, we consider only "well-typed" programs, as defined below.

Definition 2 (types). *The set of **types** is inductively defined by:*

$$\tau ::= \mathbf{unit} \mid \tau_1 \to \tau_2$$

Here, **unit** *is the type of the unit value \$, and $\tau_1 \to \tau_2$ is the type of functions from τ_1 to τ_2. The **order** and the **arity** of types are inductively defined by:*

$$\begin{array}{ll} order(\mathbf{unit}) = 0 & order(\tau_1 \to \tau_2) = \max(order(\tau_1)+1, order(\tau_2)) \\ arity(\mathbf{unit}) = 0 & arity(\tau_1 \to \tau_2) = arity(\tau_2) + 1 \end{array}$$

*A **type environment** Γ is a map from a finite set of variables to types. The **type judgment relation** $\Gamma \vdash e : \tau$ for expressions is the least relation closed under the following rules:*

$$\frac{}{\emptyset \vdash \$: \mathbf{unit}} \qquad \frac{}{\Gamma\{x \mapsto \tau\} \vdash x : \tau} \qquad \frac{\Gamma \vdash e_1 : \mathbf{unit} \qquad \Gamma \vdash e_2 : \mathbf{unit}}{\Gamma \vdash \mathbf{if}_{-}\, e_1\, e_2 : \mathbf{unit}}$$

$$\frac{\Gamma \vdash e_1 : \tau_1 \to \tau_2 \qquad \Gamma \vdash e_2 : \tau_1}{\Gamma \vdash e_1\, e_2 : \tau_2} \qquad \frac{\Gamma \vdash e : \mathbf{unit}}{\Gamma \vdash \mathbf{join}; e : \mathbf{unit}} \qquad \frac{\Gamma \vdash e : \mathbf{unit}}{\Gamma \vdash \mathbf{acq}_i; e : \mathbf{unit}}$$

$$\frac{\Gamma \vdash e : \mathbf{unit}}{\Gamma \vdash \mathbf{rel}_i; e : \mathbf{unit}} \qquad \frac{\Gamma \vdash e : \mathbf{unit} \qquad \Gamma \vdash e_c : \mathbf{unit}}{\Gamma \vdash \mathbf{spawn}(e_c); e : \mathbf{unit}} \qquad \frac{\Gamma \vdash e : \mathbf{unit}}{\Gamma \vdash e^\ell : \mathbf{unit}}$$

*A program $p = \{F_1\, x_{11}, \ldots, x_{1k_1} = e_1, \ldots, F_n\, x_{n1}, \ldots, x_{nk_n} = e_n\}$ is **well-typed** under Γ if $\Gamma = \{\, F_i \mapsto \tau_{i1} \to \ldots \to \tau_{ik_i} \to \mathbf{unit} \mid i \in \{1, \ldots, n\}\}$ and $\Gamma \cup \{x_{j1} \mapsto \tau_{j1}, ..., x_{jk_j} \mapsto \tau_{jk_j}\} \vdash e_j : \mathbf{unit}$ holds for each $j \in \{1, \ldots, n\}$. The **order** of p (well-typed under Γ) is* $\max(\{\, order(\Gamma(F)) \mid F \in dom(\Gamma)\,\})$.

Remark 1. In the language above, we have only the unit-value as a base value. As we have higher-order functions, however, we can encode booleans and conditionals by using the Church encoding. Values in infinite data domains (such as unbounded integers) can be dealt with (soundly but incompletely) by using predicate abstraction [9,13].

Example 1. Here is an example of an order-2 program, well-typed under $\Gamma = \{S \mapsto \mathbf{unit}, F \mapsto (\mathbf{unit} \to \mathbf{unit}) \to \mathbf{unit} \to \mathbf{unit}, G \mapsto \mathbf{unit} \to \mathbf{unit}, H \mapsto (\mathbf{unit} \to \mathbf{unit}) \to \mathbf{unit} \to \mathbf{unit}\}$.

$$p = \left\{ \begin{array}{ll} S = F\,G\,\$ & F\,g\,t = \mathbf{if}_{-}(\mathbf{spawn}(H\,g\,\$);\, F\,g\,t)\,(\mathbf{join};\, t) \\ G\,t = t^{\ell} & H\,g\,t = \mathbf{acq}_1;\, (g\,(\mathbf{rel}_1;\, t)) \end{array} \right\}$$

This program is obtained by CPS-transforming the following C-like code:

```
main(){f(g);}                         g(){
f(g){                                   L: <critical section>;
  if(*){spawn{h(g);}; f(g);}            return;
  else{join(); }                      }
}                                     h(g){ acq(1); g(); rel(1);}
```

The root process spawns non-deterministically many child processes, and then waits for the child processes by the join operation. Each child acquires the lock 1 and then release 1 at the program point ℓ.

We now define the formal semantics of programs. A **configuration** c of a program is a map from a finite set consisting of sequences of natural numbers (where each sequence serves as a process identifier) to the set of triples (e, L, s) consisting of an expression e, a sequence L of locks, and a natural number s. Intuitively, $c(\pi) = (e, i_1 \cdots i_k, s)$ means that the process π is executing e, that it holds locks $i_1, \ldots, i_k$ that have been acquired in this order, and that it has created s child processes so far. The **reduction relation** $c \longrightarrow c'$ on configurations is defined by the following rules.

$$\frac{F\,x_1 \cdots x_k \to e \in p}{c \uplus \{\,\pi \mapsto (F\,e_1 \ldots e_k, L, s)\,\} \longrightarrow_p c \uplus \{\,\pi \mapsto ([e_1/x_1, \ldots, e_k/x_k]e, L, s)\,\}}$$

$$c \uplus \{\,\pi \mapsto (\mathbf{if}_{-}\,e_1\,e_2, L, s)\,\} \longrightarrow_p c \uplus \{\,\pi \mapsto (e_i, L, s)\,\} \;\; (i \in \{1,2\})$$

$$c \uplus \{\,\pi \mapsto (e^{\ell}, L, s)\,\} \longrightarrow_p c \uplus \{\,\pi \mapsto (e, L, s)\,\}$$

$$\frac{i \notin \mathbf{locked}(c \uplus \{\,\pi \mapsto (\mathbf{acq}_i;\, e, L, s)\,\})}{c \uplus \{\,\pi \mapsto (\mathbf{acq}_i;\, e, L, s)\,\} \longrightarrow_p c \uplus \{\,\pi \mapsto (e, L \cdot i, s)\,\}}$$

$$c \uplus \{\,\pi \mapsto (\mathbf{rel}_i;\, e, L \cdot i, s)\,\} \longrightarrow_p c \uplus \{\,\pi \mapsto (e, L, s)\,\}$$

$$c \uplus \{\,\pi \mapsto (\mathbf{spawn}(e_c);\, e, L, s)\,\} \longrightarrow_p c \uplus \{\,\pi \mapsto (e, L, s+1), \pi \cdot s \mapsto (e_c, \epsilon, 0)\,\}$$

$$\frac{\{\,j \mid \pi \cdot j \in dom(c)\,\} = \emptyset}{c \uplus \{\,\pi \mapsto (\mathbf{join};\, e, L, s)\,\} \longrightarrow_p c \uplus \{\,\pi \mapsto (e, L, s)\,\}}$$

$$c \uplus \{ \pi \mapsto (\$, \epsilon, s) \} \longrightarrow_p c$$

Here, $\mathbf{locked}(c)$ represents the set of all acquired locks, defined by:

$$\mathbf{locked}(c) = \bigcup_{c(\pi)=(e, i_1 \cdots i_k, s)} \{ i_1, \ldots, i_k \}.$$

Definition 3 (pairwise reachability). *Let p be a (well-typed) program and ℓ_1, ℓ_2 be labels. We say that (ℓ_1, ℓ_2) is **pairwise reachable** by p, written $p \models \ell_1 || \ell_2$, if there exist c, π_1, π_2 ($\pi_1 \neq \pi_2$) such that $\{ \epsilon \mapsto (S, \epsilon, 0) \} \longrightarrow_p^* c$ with $c(\pi_1) = (e_1^{\ell_1}, L_1, s_1)$ and $c(\pi_2) = (e_2^{\ell_2}, L_2, s_2)$ for some $e_2, e_2, L_1, L_2, s_1, s_2$. The **pairwise reachability problem** is the decision problem of checking whether $p \models \ell_1 || \ell_2$ holds.*

Example 2. Recall the example program showed in Example 1. The verification problem "can the program point ℓ (i.e., the critical section) be reached by multiple processes simultaneously?" can be reduced to the pairwise reachability for the program p and the pair (ℓ, ℓ). In this case, the answer to the problem is "no".

3 From Pairwise Reachability to Higher-Order Model Checking

In this section, we show that the pairwise reachability problem can be reduced to higher-order model checking [15], hence it is decidable. The basic idea of this reduction is to transform a program to a grammar called a higher-order recursion scheme [15], which generates *action trees* [4,14] that represent all the possible executions of the program. Since the set of action trees that represent valid executions (i.e., those that respect synchronization constraints on joins and nested locks) is regular, the pairwise reachability problem can be reduced to an inclusion problem between the tree language generated by the higher-order recursion scheme and the regular language, which can be further reduced to higher-order model checking. We first review action trees and higher-order model checking in Sections 3.1 and 3.2 respectively. We then present the reduction from the pairwise reachability analysis to higher-order model checking.

3.1 Action Trees

An action tree [4,14] is a finite tree that represents a history of executions of a program up to a certain state. It ignores how the executions of multiple processes are interleaved, and expresses only process-wise execution histories and the parent/child relationship between processes. Gawlitza et al. [4] originally introduced action trees to represent execution histories of dynamic pushdown networks, but the notion of action trees is independent of a particular computation model. Here we use them to represent execution histories of higher-order concurrent programs introduced in the previous section.

Definition 4 (action trees). *The set of* ***action trees****, ranged over by* γ*, is defined inductively by:*

$$\gamma ::= \bot \mid \langle \$ \rangle \mid \ell \mid \langle \mathsf{jo} \rangle\, \gamma \mid \langle \mathsf{sp} \rangle\, \gamma_1\, \gamma_2 \mid \langle \mathsf{Acq}_i \rangle\, \gamma \mid \langle \mathsf{Rel}_i \rangle\, \gamma.$$

Here, ℓ *ranges over (program) labels and* i *over locks.*

We have used the term representation of (labelled) trees above. Trees can also be considered as map from paths to labels, by: $(a\, \gamma_1 \cdots \gamma_n)^{\#} = \{\epsilon \mapsto a\} \cup \{ i \cdot u \mapsto b \mid \gamma_i^{\#}(u) = b \}$.

Each non-leaf node represents an action of each process. The tree $\langle \mathsf{jo} \rangle\, \gamma$ ($\langle \mathsf{Acq}_i \rangle\, \gamma$ and $\langle \mathsf{Rel}_i \rangle\, \gamma$ respectively) means that the process performed join (acquires and releases the lock i, respectively) and then behaved like γ. The tree $\langle \mathsf{sp} \rangle\, \gamma_1\, \gamma_2$ means that the process spawned a child process that behaved like γ_2, and the process itself behaved like γ_1. Thus, the leftmost path from the root node in an action tree represents a sequence of actions performed by the root process, and each leftmost path from the second child of a $\langle \mathsf{sp} \rangle$-node represents a sequence of actions performed by the spawned process. Each leaf node of an action tree represents the current state of each process: $\langle \$ \rangle$ means that the process has terminated, ℓ means that the process is at the program point ℓ, and $\bot$ means that the process is at a program point not labeled by any element of Label. Figure 1 shows an example of an action tree. It represents an execution history where the root process spawns a child process, waits for the child, and terminates (represented by $\langle \$ \rangle$), and the child process acquires and releases the lock i, and then terminates. It corresponds to the following execution of the program in Example 1:

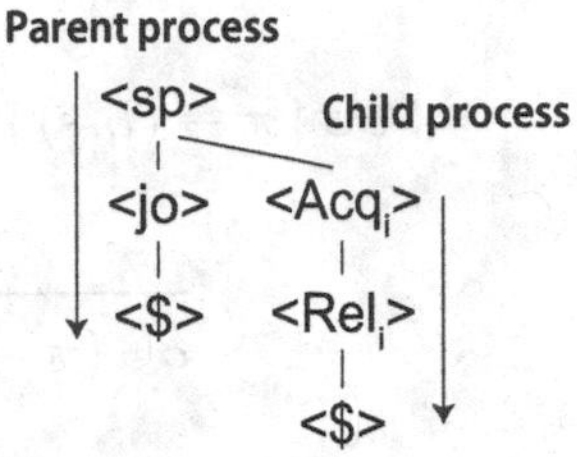

Fig. 1. Action tree

$$\begin{aligned}
&\{\epsilon \mapsto (S, \epsilon, 0)\} \longrightarrow \{\epsilon \mapsto (F\, G\, \$, \epsilon, 0)\} \longrightarrow^* \{\epsilon \mapsto (\mathbf{spawn}(H\, G\, \$); F\, G\, \$, \epsilon, 0)\} \\
&\longrightarrow \{\epsilon \mapsto (F\, G\, \$, \epsilon, 1), 0 \mapsto (H\, G\, \$, \epsilon, 0)\} \\
&\longrightarrow^* \{\epsilon \mapsto (\mathbf{join}; \$, \epsilon, 1), 0 \mapsto (\mathbf{acq}_1; G(\mathbf{rel}_1; \$), \epsilon, 0)\} \\
&\longrightarrow^* \{\epsilon \mapsto (\mathbf{join}; \$, \epsilon, 1), 0 \mapsto (G(\mathbf{rel}_1; \$), 1, 0)\} \\
&\longrightarrow^* \{\epsilon \mapsto (\mathbf{join}; \$, \epsilon, 1), 0 \mapsto ((\mathbf{rel}_1; \$)^{\ell}, 1, 0)\} \\
&\longrightarrow^* \{\epsilon \mapsto (\mathbf{join}; \$, \epsilon, 1), 0 \mapsto (\$, \epsilon, 0)\} \\
&\longrightarrow \{\epsilon \mapsto (\mathbf{join}; \$, \epsilon, 1)\} \longrightarrow \{\epsilon \mapsto (\$, \epsilon, 1)\} \longrightarrow \emptyset
\end{aligned}$$

Note that not every action tree represents a valid execution history. For example, consider the action tree. $\langle \mathsf{sp} \rangle\, (\langle \mathsf{Acq}_1 \rangle\, \ell_1)\, (\langle \mathsf{Acq}_1 \rangle\, \ell_2)$. It represents a state where the parent and child processes are at program points ℓ_1 and ℓ_2 respectively, after *both* having acquired the lock 1 (and not released it yet), which is obviously impossible. In order to exclude action trees that do not respect synchronization constraints, we introduce a transition system on *abstract* configurations, obtained by removing expressions from configurations introduced in the previous section.

Definition 5. *An* ***abstract configuration*** *is a map from a finite set consisting of sequences of natural numbers (where each sequence serves as a process identifier) to the set of pairs (L, s) consisting of a sequence L of locks, and a natural number s. The transition relation on abstract configurations is defined by:*

$$\frac{i \notin \mathbf{locked}(c \uplus \{\pi \mapsto (L,s)\})}{c \uplus \{\pi \mapsto (L,s)\} \xrightarrow{\pi,\langle \mathsf{Acq}_i\rangle} c \uplus \{\pi \mapsto (L \cdot i, s)\}}$$

$$c \uplus \{\pi \mapsto (L \cdot i, s)\} \xrightarrow{\pi,\langle \mathsf{Rel}_i\rangle} c \uplus \{\pi \mapsto (L, s)\}$$

$$c \uplus \{\pi \mapsto (L,s)\} \xrightarrow{\pi,\langle \mathsf{sp}\rangle} c \uplus \{\pi \mapsto (L, s+1), \pi \cdot s \mapsto (\epsilon, 0)\}$$

$$\frac{\{k \mid \pi \cdot k \in dom(c)\} = \emptyset}{c \uplus \{\pi \mapsto (L,s)\} \xrightarrow{\pi,\langle \mathsf{jo}\rangle} c \uplus \{\pi \mapsto (L, s)\}}$$

$$c \uplus \{\pi \mapsto (\epsilon,s)\} \xrightarrow{\pi,\langle \$\rangle} c$$

Here, $\mathbf{locked}(c)$ *is defined similarly to that for configurations, as*

$$\mathbf{locked}(c) = \bigcup_{c(\pi)=(i_1\cdots i_k, s)} \{i_1, \ldots, i_k\}.$$

Each action tree can be mapped to an abstract configuration as follows.

$$\begin{array}{ll}
\theta_{\pi,L,s}(\langle\$\rangle) = \emptyset & \theta_{\pi,L,s}(\langle\mathsf{sp}\rangle\,\gamma_1\,\gamma_2) = \theta_{\pi,L,s+1}(\gamma_1) \cup \theta_{\pi\cdot s,\epsilon,0}(\gamma_2) \\
\theta_{\pi,L,s}(\langle\mathsf{jo}\rangle\,\gamma) = \theta_{\pi,L,s}(\gamma) & \theta_{\pi,L,s}(\langle\mathsf{Acq}_i\rangle\,\gamma) = \theta_{\pi,L\cdot i,s}(\gamma) \\
\theta_{\pi,L\cdot i,s}(\langle\mathsf{Rel}_i\rangle\,\gamma) = \theta_{\pi,L,s}(\gamma) & \theta_{\pi,L,s}(\gamma) = \{\pi \mapsto (L,s)\}\ (\text{if } \gamma \in \{\bot\} \cup \mathsf{Label})
\end{array}$$

We write $\theta(t)$ for $\theta_{\epsilon,\epsilon,s}(t)$, and write $\gamma_1 \overset{u}{\rightsquigarrow} \gamma_2$ if $\gamma_1(u) = \bot$ and γ_2 is obtained from γ_1 by replacing $\bot$ at u with a tree of the form $a \bot \ldots \bot$ with $a = \langle\$\rangle, \langle\mathsf{sp}\rangle, \langle\mathsf{jo}\rangle, \langle\mathsf{Acq}_i\rangle, \langle\mathsf{Rel}_i\rangle$.

We can now define "valid" action trees as follows.

Definition 6. *An action tree γ is* ***join-lock sensitive*** *if there is a sequence $\bot = \gamma_0 \overset{u_1}{\rightsquigarrow} \gamma_1 \overset{u_2}{\rightsquigarrow} \ldots \overset{u_n}{\rightsquigarrow} \gamma_n = \gamma'$ such that $\theta(\gamma_0) \xrightarrow{\mathsf{pn}(u_1,\gamma),\gamma_1(u_1)} \theta(\gamma_1) \xrightarrow{\mathsf{pn}(u_2,\gamma),\gamma_2(u_2)} \ldots \xrightarrow{\mathsf{pn}(u_n,\gamma),\gamma_n} \theta(\gamma_n)$, where γ' is the action tree obtained from γ by replacing all $\ell \in \mathsf{Label}$ with $\bot$. Here, $\mathsf{pn}(u,\gamma)$ represents the identifier of the process that executes the action of $\gamma(u)$. It is defined by:*

$$\begin{array}{lll}
\mathsf{pn}(\epsilon,\gamma) = \epsilon & \mathsf{cn}(\epsilon,\gamma) = 0 & \\
\mathsf{pn}(u\cdot 1,\gamma) = \mathsf{pn}(u,\gamma) & \mathsf{cn}(u\cdot 1,\gamma) = \mathsf{cn}(u,\gamma) & \text{if } \gamma(u) \neq \langle\mathsf{sp}\rangle \\
\mathsf{pn}(u\cdot 2,\gamma) = \mathsf{pn}(u,\gamma)\cdot\mathsf{cn}(u,\gamma) & \mathsf{cn}(u\cdot 1,\gamma) = \mathsf{cn}(u,\gamma)+1 & \text{if } \gamma(u) = \langle\mathsf{sp}\rangle \\
 & \mathsf{cn}(u\cdot 2,\gamma) = 0 &
\end{array}$$

Note that $\mathsf{pn}(u_i, \gamma_i) = \mathsf{pn}(u_i, \gamma)$.

The following is the key property of action trees, which we use in our reduction from pairwise reachability analysis to higher-order model checking.

Lemma 1 (Gawlitza et al. 2011, Section 5 [4]). *The set $L_{\mathsf{sensitive}}$ of join-lock-sensitive action trees is a regular tree language.*

Remark 2. We have modified the original definition of join-lock sensitive (schedulable) action trees. Our notion of join-lock sensitive action trees corresponds to join-lock-well-formed, join-lock sensitive schedulable action trees [4].

3.2 Higher-Order Model Checking

In this subsection, we review *higher-order recursion schemes* and (a variation of) *higher-order model checking* [15].

The set of **sorts** is given by the grammar: $\kappa ::= \mathsf{o} \mid \kappa_1 \to \kappa_2$. Intuitively o describes trees, and $\kappa_1 \to \kappa_2$ describes functions from κ_1 to κ_2. The **order** of sorts is defined by: $\mathit{order}(\mathsf{o}) = 0$ and $\mathit{order}(\kappa_1 \to \kappa_2) = \max(\mathit{order}(\kappa_1)+1, \mathit{order}(\kappa_2))$.

Definition 7 (Higher-Order Recursion Scheme). *A (non-deterministic)* ***higher-order recursion scheme*** *(HORS, for short) is a quadruple: $\mathcal{G} = (\Sigma, \mathcal{N}, \mathcal{R}, S)$ where Σ is a ranked alphabet (i.e., a map from a finite set of symbols called* ***terminals*** *to their arities); $\mathcal{N}$ is a map from a finite set of symbols called* ***non-terminals*** *to sorts; $S \in \mathit{dom}(\mathcal{N})$ is the start symbol of sort o; and $\mathcal{R}$ is a finite set of* ***transition rules*** *of the form $A\,x_1 \cdots x_\ell \to t$, where t ranges over the set of applicative terms defined by $t ::= x \mid a \mid A \mid t_1 t_2$. Here, a ranges over $\mathit{dom}(\Sigma)$ and A ranges over $\mathit{dom}(\mathcal{N})$. If $A\,x_1 \cdots x_\ell \to t \in \mathcal{N}$, then $\mathcal{N}(A)$ must be of the form $\kappa_1 \to \cdots \to \kappa_\ell \to \mathsf{o}$ and $\mathcal{N} \cup \{x_1 : \kappa_1, \ldots, x_\ell : \kappa_\ell\} \vdash_\Sigma t : \mathsf{o}$ must be derivable by using the following rules (where non-terminals are treated as variables).*

$$\mathcal{K} \vdash_\Sigma a : \underbrace{\mathsf{o} \to \cdots \to \mathsf{o}}_{\Sigma(a)} \to \mathsf{o} \qquad\qquad \mathcal{K} \vdash_\Sigma x : \mathcal{K}(x)$$

$$\frac{\mathcal{K} \vdash_\Sigma t_1 : \kappa_1 \to \kappa_2 \qquad \mathcal{K} \vdash_\Sigma t_2 : \kappa_1}{\mathcal{K} \vdash_\Sigma t_1 t_2 : \kappa_2}$$

The order of a HORS $\mathcal{G}$, written $\mathit{order}(\mathcal{G})$ is $\max(\{\mathit{order}(A) \mid A \in \mathit{dom}(\mathcal{N})\})$.

Note that unlike *deterministic* HORS, there may be an arbitrary number of rewriting rules for each non-terminal. We omit the adjective 'non-deterministic' in the rest of this paper.

To define the rewriting relation, we define the notion of the reduction context.

Definition 8. *The set of* ***reduction contexts*** *is defined by:*

$$C ::= [\,] \mid a\,t_1 \ldots t_{i-1}\,C\,t_{i+1} \ldots t_n$$

For a reduction context C, we write $C[t]$ for the term obtained from C by replacing $[\,]$ with t.

Then, the rewriting relation $\longrightarrow_{\mathcal{G}}$ on terms is defined by:

$$C[A\,t_1 \ldots t_m] \longrightarrow_{\mathcal{G}} C[\,[t_1/x_1,\ldots,t_m/x_m]t\,] \quad (\text{if } A\,x_1 \ldots x_m \rightarrow t)$$

In this paper, we consider a HORS as a generator of a language of finite trees, rather than that of an infinite tree [15].

Definition 9 (Tree Languages of Recursion Schemes). *Let $\mathcal{G} = (\Sigma, \mathcal{N}, \mathcal{R}, S)$ be a HORS. The language generated by $\mathcal{G}$ is defined by:*

$$\mathcal{L}(\mathcal{G}) = \{t \mid t \text{ is a ranked } \Sigma\text{-labeled tree and } S \longrightarrow^*_{\mathcal{G}} t\}$$

Example 3. Consider an order-2 HORS $\mathcal{G} = (\Sigma, \mathcal{N}, \mathcal{R}, S)$ such that

$$\begin{aligned}
&\Sigma = \{\,a \mapsto 2, b \mapsto 1, c \mapsto 0\,\}\\
&\mathcal{N} = \{\,S \mapsto \mathtt{o}, F \mapsto (\mathtt{o} \rightarrow \mathtt{o}) \rightarrow \mathtt{o} \rightarrow \mathtt{o}, G \mapsto \mathtt{o} \rightarrow \mathtt{o}, T \mapsto (\mathtt{o} \rightarrow \mathtt{o}) \rightarrow \mathtt{o} \rightarrow \mathtt{o}\,\}\\
&\mathcal{R} = \{S \rightarrow F\,G\,c,\ F\,g\,x \rightarrow a\,(g\,x)\,(F\,(T\,g)\,x),\ F\,g\,x \rightarrow g\,x,\\
&\qquad G\,x \rightarrow b\,x,\ T\,g\,x \rightarrow g\,(g\,x)\}
\end{aligned}$$

Here is an example of reduction from S to $a\,(b\,c)\,(b^2\,c)$.

$$\begin{aligned}
&S \longrightarrow_{\mathcal{G}} F\,Gc \longrightarrow_{\mathcal{G}} a\,(G\,c)\,(F\,(T\,G)c) \longrightarrow_{\mathcal{G}} a\,(b\,c)(F\,(T\,G)\,c) \longrightarrow_{\mathcal{G}} a\,(b\,c)(T\,G\,c)\\
&\longrightarrow_{\mathcal{G}} a\,(b\,c)(G\,(G\,c)) \longrightarrow^*_{\mathcal{G}} a\,(b\,c)(b\,(b\,c))
\end{aligned}$$

The language $\mathcal{L}(\mathcal{G})$ is $\{\,a\,(b\,c)(a\,(b^2\,c)(a \ldots (a\,(b^{2^{n-1}}\,c)(b^{2^n}\,c))\ldots)) \mid n \in \mathbb{N}_+\,\}$.

The following theorem is an easy corollary of Ong's result on the model checking of (deterministic) HORS [15].

Theorem 1. *Given a HORS $\mathcal{G}$ and a regular tree language L, it is decidable whether $\mathcal{L}(\mathcal{G}) \subseteq L$.*

In the present paper, we call the inclusion problem above a higher-order model checking problem. The standard model checking problem for HORS [15] is the problem of deciding whether the infinite tree generated by a deterministic HORS satisfies a given property.

3.3 Reduction from Pairwise Reachability to Higher-Order Model Checking

Now we show how to reduce pairwise reachability to higher-order model checking. First, we define a transformation from a concurrent program to a HORS that generates action trees of the which join-lock sensitive subset represent all and only the possible reachable configurations of the program.

Definition 10. *Let $p = \{F_1\,x_{11},\ldots,x_{1k_1} = e_1,\ \ldots,\ F_n\,x_{n1},\ldots,x_{nk_n} = e_n\}$ be a well-typed higher-order concurrent program under Γ. A (non-deterministic)*

HORS $\mathcal{G}_p$ is defined by:

$$\Sigma = \{ \langle \mathsf{sp} \rangle \mapsto 2, \langle \mathsf{jo} \rangle \mapsto 1, \langle \$ \rangle \mapsto 0, \bot \mapsto 0 \} \\ \cup \{ \langle \mathsf{Acq}_i \rangle, \langle \mathsf{Rel}_i \rangle \mid i \in \mathsf{Lock} \} \cup \{ \ell \mapsto 0 \mid \ell \in \mathsf{Label} \}$$

$$\mathcal{G}_p = (\Sigma, \mathcal{N}_\Gamma \cup \mathcal{N}_0, \mathcal{R}_\Gamma \cup \mathcal{R}_0, S)$$

$$\mathcal{N}_\Gamma = \{F_1 \mapsto (\Gamma(F_1))^\sharp, \ldots, F_n \mapsto (\Gamma(F_n))^\sharp\}$$

$$\mathcal{N}_0 = \{E_\$ \mapsto \mathsf{o}, E_{\mathbf{if\_}} \mapsto (\mathsf{o} \to \mathsf{o} \to \mathsf{o}), E_{\mathbf{join}} \mapsto (\mathsf{o} \to \mathsf{o}), E_{\mathbf{spawn}} \mapsto (\mathsf{o} \to \mathsf{o} \to \mathsf{o})\} \\ \cup \{E_{\mathbf{acq}_i} \mapsto (\mathsf{o} \to \mathsf{o}) \mid i \in \mathsf{Lock}\} \cup \{E_{\mathbf{rel}_i} \mapsto (\mathsf{o} \to \mathsf{o}) \mid i \in \mathsf{Lock}\} \\ \cup \{E_\ell \mapsto (\mathsf{o} \to \mathsf{o}) \mid \ell \in \mathsf{Label}\}$$

$$\mathcal{R}_\Gamma = \{F_1 \tilde{x}_1 \to \mathcal{E}(e_1), \ldots, F_n \tilde{x}_n \to \mathcal{E}(e_n)\}$$

$$\mathcal{R}_0 = \{E_{\mathbf{if\_}}\, x\, y \to x,\ E_{\mathbf{if\_}}\, x\, y \to y,\ E_{\mathbf{join}}\, x \to \langle \mathsf{jo} \rangle\, x,\ E_{\mathbf{spawn}}\, x\, y \to \langle \mathsf{sp} \rangle\, x\, y\} \\ \cup \{E_{\mathbf{acq}_i}\, x \to \langle \mathsf{Acq}_i \rangle\, x \mid i \in \mathsf{Lock}\} \cup \{E_{\mathbf{rel}_i}\, x \to \langle \mathsf{Rel}_i \rangle\, x \mid i \in \mathsf{Lock}\} \\ \cup \{E_\ell\, x \to \ell \mid \ell \in \mathsf{Label}\} \cup \{E_\ell\, x \to x \mid \ell \in \mathsf{Label}\} \cup \{ E_\$ \to \langle \$ \rangle \} \\ \cup \{E\, \tilde{x} \to \bot \mid E \in \mathit{dom}(\mathcal{N}_\Gamma \cup \mathcal{N}_0) \setminus \{E_\ell \mid \ell \in \mathsf{Label}\}\}$$

Here, $(\cdot)^\sharp$ is a transformation from types of expressions to sorts, defined by: $\mathbf{unit}^\sharp = \mathsf{o}$ and $(\tau_1 \to \tau_2)^\sharp = \tau_1^\sharp \to \tau_2^\sharp$. The function $\mathcal{E}$ transforms an expression to an applicative term, defined inductively by:

$$\mathcal{E}(\$) = E_\$ \qquad \mathcal{E}(x) = x \qquad \mathcal{E}(F) = F \qquad \mathcal{E}(\mathbf{if\_}\, e_1\, e_2) = E_{\mathbf{if\_}}\, \mathcal{E}(e_1)\, \mathcal{E}(e_2)$$
$$\mathcal{E}(e_1\, e_2) = \mathcal{E}(e_1)\, \mathcal{E}(e_2) \qquad \mathcal{E}(\mathbf{join};\, e) = E_{\mathbf{join}}\, \mathcal{E}(e) \qquad \mathcal{E}(\mathbf{acq}_i;\, e) = E_{\mathbf{acq}_i}\, \mathcal{E}(e)$$
$$\mathcal{E}(\mathbf{rel}_i;\, e) = E_{\mathbf{rel}_i}\, \mathcal{E}(e) \quad \mathcal{E}(\mathbf{spawn}(e_c);\, e) = E_{\mathbf{spawn}}\, \mathcal{E}(e)\, \mathcal{E}(e_c) \quad \mathcal{E}(e^\ell) = E_\ell\, \mathcal{E}(e)$$

The idea of the transformation above is quite simple: just replace each synchronization primitive **op** with a non-terminal $E_{\mathbf{op}}$, which will generate a tree node $\langle \mathsf{op} \rangle$ indicating that the operation **op** has been performed. Additionally, in order to generate all the intermediate states of an execution, we allow each non-terminal to be reduced to $\bot$ or $\ell \in \mathsf{Label}$. Note that $\mathit{order}(\Gamma(F)) = \mathit{order}(\mathcal{N}(F))$ holds for every function symbol F of p. Thus, $\mathit{order}(\mathcal{G}_p) = \max(\mathit{order}(p), 1)$.

Example 4. The program p of Example 1 is transformed to the recursion scheme $\mathcal{G}_p = (\Sigma, \mathcal{N}_\Gamma \cup \mathcal{N}_0, \mathcal{R}_\Gamma \cup \mathcal{R}_0, S)$ where

$$\mathcal{N}_\Gamma = \{S \mapsto \mathsf{o},\ F \mapsto (\mathsf{o} \to \mathsf{o}) \to \mathsf{o} \to \mathsf{o},\ G \mapsto \mathsf{o} \to \mathsf{o},\ H \mapsto (\mathsf{o} \to \mathsf{o}) \to \mathsf{o} \to \mathsf{o}\}$$

$$\mathcal{R}_\Gamma = \left\{ \begin{array}{c} S \to F\, G\, E_\$ \\ F\, g\, t \to E_{\mathbf{if\_}}\, (E_{\mathbf{spawn}}\, (F\, g\, t)\, (H\, g\, E_\$))\, (E_{\mathbf{join}}\, t) \\ G\, t \to E_\ell\, t \\ H\, g\, t \to E_{\mathbf{acq}_i}\, (g\, (E_{\mathbf{rel}_i}\, t)) \end{array} \right\}.$$

The action tree in Figure 1 is generated by the following reduction sequence:

$$S \longrightarrow F\, G\, E_\$ \longrightarrow E_{\mathbf{if\_}}\, (E_{\mathbf{spawn}}\, (F\, G\, E_\$)\, (H\, G\, E_\$))\, (E_{\mathbf{join}}\, E_\$)$$
$$\longrightarrow E_{\mathbf{spawn}}\, (F\, G\, E_\$)\, (H\, G\, E_\$)) \longrightarrow \langle \mathsf{sp} \rangle\, (F\, G\, E_\$)\, (H\, G\, E_\$))$$
$$\longrightarrow^* \langle \mathsf{sp} \rangle\, (E_{\mathbf{if\_}}\, (E_{\mathbf{spawn}}\, (F\, G\, E_\$)\, (H\, G\, E_\$))\, (E_{\mathbf{join}}\, E_\$))\, (E_{\mathbf{acq}_i}\, (G\, (E_{\mathbf{rel}_i}\, E_\$)))$$
$$\longrightarrow^* \langle \mathsf{sp} \rangle\, (E_{\mathbf{join}}\, E_\$)\, (\langle \mathsf{Acq}_i \rangle\, (G\, (E_{\mathbf{rel}_i}\, E_\$))) \longrightarrow^* \langle \mathsf{sp} \rangle\, (\langle \mathsf{jo} \rangle\, E_\$)\, (\langle \mathsf{Acq}_i \rangle\, (E_{\mathbf{rel}_i}\, E_\$))$$
$$\longrightarrow^* \langle \mathsf{sp} \rangle\, (\langle \mathsf{jo} \rangle\, \langle \$ \rangle)\, (\langle \mathsf{Acq}_i \rangle\, (\langle \mathsf{Rel}_i \rangle\, E_\$)) \longrightarrow \langle \mathsf{sp} \rangle\, (\langle \mathsf{jo} \rangle\, \langle \$ \rangle)\, (\langle \mathsf{Acq}_i \rangle\, (\langle \mathsf{Rel}_i \rangle\, \langle \$ \rangle))$$

Now we show that the grammar generates all the action trees that corresponds to the reachable configurations of a program. We first prepare some definitions. To clarify the relationship between an applicative term (consisting of terminals and non-terminals of a HORS) and a configuration, we first define the terms that can appear at "run-time" and thus have unique corresponding configurations.

Definition 11. *An applicative term t is called a* **run-time term** *if (1) t has sort o, (2) t contains no labels nor $\bot$, and (3) no terminals occur in the arguments of non-terminals in t.*

Definition 12. *Let t be a run-time term of sort o. The action tree $t^\perp$ is defined by:*

$$(E_\ell\, t)^\perp = \ell \qquad (A\, t_1 \dots t_n)^\perp = \bot \text{ (if } A \notin \{ E_\ell \mid \ell \in \mathsf{Label} \}) \\ (a\, t_1 \dots t_n)^\perp = a\, t_1^\perp \dots t_n^\perp$$

We extend the map $\mathcal{X}_{\pi,L,S}(\cdot)$ to that on run-time trees, by:

$$\mathcal{X}_{\pi,L,s}(\langle \mathsf{sp} \rangle\, t_1\, t_2) = \mathcal{X}_{\pi,L,s+1}(t_1) \cup \mathcal{X}_{\pi\cdot s,\epsilon,0}(t_2) \quad \mathcal{X}_{\pi,L,s}(\langle \mathsf{jo} \rangle\, t) = \mathcal{X}_{\pi,L,s}(t) \\ \mathcal{X}_{\pi,L,s}(\langle \$ \rangle) = \emptyset \quad \mathcal{X}_{\pi,L,s}(\langle \mathsf{Acq}_i \rangle\, t) = \mathcal{X}_{\pi,L\cdot i,s}(t) \quad \mathcal{X}_{\pi,L\cdot i,s}(\langle \mathsf{Rel}_i \rangle\, t) = \mathcal{X}_{\pi,L,s}(t) \\ \mathcal{X}_{\pi,L,s}(t) = \{ \pi \mapsto (\mathcal{E}^{-1}(t), L, s) \} \text{ (for the other cases)}$$

Here, $\mathcal{E}^{-1}$ is the inverse of $\mathcal{E}$. We write $\mathcal{X}(t)$ for $\mathcal{X}_{\epsilon,\epsilon,0}(t)$.

The following lemmas establish the correspondence between p and $\mathcal{G}_p$.

Lemma 2. *Suppose $S \longrightarrow^*_{\mathcal{G}_p} t$ where t is a run-time term and $t^\perp$ is join-lock sensitive. Then, $\{ \epsilon \mapsto (S, \epsilon, 0) \} \longrightarrow^*_p \mathcal{X}(t)$ holds.*

Lemma 3. *If $\{ \epsilon \mapsto (S, \epsilon, 0) \} \longrightarrow^*_p c$, then there exists a run-time term t such that $S \longrightarrow^*_{\mathcal{G}_p} t$, $\mathcal{X}(t) = c$, and $t^\perp$ is join-lock-sensitive.*

By the above lemmas, the pairwise reachability problem on p is reduced to higher-order model checking on $\mathcal{G}_p$.

Theorem 2. *Let p be a program and (ℓ_1, ℓ_2) be a pair of labels. Let L_{ℓ_1,ℓ_2} be the set $\{ \gamma \mid \exists u_1, u_2 . \gamma(u_1) = \ell_1 \wedge \gamma(u_2) = \ell_2 \wedge u_1 \neq u_2 \}$ of action trees. Then, $p \models \ell_1 || \ell_2$ if and only if $\mathcal{L}(\mathcal{G}_p) \not\subseteq \overline{L_{\mathsf{sensitive}}} \cup \overline{L_{\ell_1,\ell_2}}$ holds.*

Proof. Suppose $p \models \ell_1 || \ell_2$. Then there exists c such that $\{ (S, \epsilon, 0) \} \longrightarrow^*_p c$ with $c(\pi_1) = (e_1^{\ell_1}, L_1, s_1)$, $c(\pi_2) = (e_2^{\ell_2}, L_2, s_2)$, and $\pi_1 \neq \pi_2$. By Lemma 3, there exists a run-time term t such that $S \longrightarrow^*_{\mathcal{G}_p} t$, $\mathcal{X}(t) = c$, and $t^\perp$ is join-lock-sensitive. By the conditions $\mathcal{X}(t) = c$ and $t^\perp$, $t^\perp \in L_{\ell_1,\ell_2}$. Since $S \longrightarrow^*_{\mathcal{G}_p} t \longrightarrow^*_{\mathcal{G}_p} t^\perp$, we have $t^\perp \in \mathcal{L}(\mathcal{G}_p) \cap L_{\mathsf{sensitive}} \cap L_{\ell_1,\ell_2}$, i.e., $\mathcal{L}(\mathcal{G}_p) \not\subseteq \overline{L_{\mathsf{sensitive}}} \cup \overline{L_{\ell_1,\ell_2}}$.

Conversely, suppose $\mathcal{L}(\mathcal{G}_p) \not\subseteq \overline{L_{\mathsf{sensitive}}} \cup \overline{L_{\ell_1,\ell_2}}$, i.e., $\gamma \in \mathcal{L}(\mathcal{G}_p) \cap L_{\mathsf{sensitive}} \cap L_{\ell_1,\ell_2}$ for some action tree γ. Then there exists a run-time term t such that $t^\perp = \gamma$ and $S \longrightarrow^*_{\mathcal{G}_p} t$. By Lemma 2, we have $\{ \epsilon \mapsto (S, \epsilon, 0) \} \longrightarrow^*_p \mathcal{X}(t)$. By the conditions $t^\perp = \gamma$ and $\gamma \in L_{\ell_1,\ell_2}$, t has two distinct sub-terms (at redex-positions) of the form $E_{\ell_1}\, t_1$ and $E_{\ell_2}\, t_2$. Thus, there exist π_1, π_2 such that $\mathcal{X}(t)(\pi_1) = (e_1^{\ell_1}, L_1, s_1)$ and $\mathcal{X}(t)(\pi_2) = (e_2^{\ell_2}, L_2, s_2)$ with $\pi_1 \neq \pi_2$. Therefore, we have $p \models \ell_1 || \ell_2$ as required. □

Because $\overline{L_{\mathsf{sensitive}}} \cup \overline{L_{\ell_1,\ell_2}}$ is a regular tree language, by Theorems 1 and 2, the pairwise reachability $p \models \ell_1 || \ell_2$ is decidable.

Complexity. Recall that the order of $\mathcal{G}_p$ is $\max(order(p), 1)$ and the model checking of order-k HORS is k-EXPTIME [15]. Therefore, the pairwise reachability analysis for order-k programs is k-EXPTIME for $k \geq 1$. As for the lower-bound, the problem of checking whether $\ell \in \mathcal{L}(\mathcal{G})$ (where ℓ is a singleton tree consisting of the leaf ℓ) is already $(k-1)$-EXPTIME-hard [12]. Since it can be easily reduced to a pairwise reachability analysis problem, the pairwise reachability is $(k-1)$-EXPTIME-hard. It should be noted, however, that if a regular tree language L is fixed, and also if both the largest arity and order of symbols are fixed, then $\mathcal{L}(\mathcal{G}) \subseteq L$ can be decided in time linear in the size of $\mathcal{G}$ [11]. In our method, the regular tree language $\overline{L_{\mathsf{sensitive}} \cup L_{\ell_1,\ell_2}}$ is determined by the sets Label and Lock. We can fix Label as $\{\ell_1, \ell_2\}$ by omitting the other labels from the input program. Therefore, if we fix (i) the largest arity and order of functions in a program and (ii) the number of locks used in the program (i.e., $|\mathsf{Lock}|$), then the pairwise reachability can also be decided in time linear in the size of the program.

4 Preliminary Experiments

We carried out preliminary experiments to check the feasibility of our verification method. As the underlying model checker, we used HorSat [1]. At the time of writing this paper, we have not yet fully automated the translation from programs to HORS, but doing so is not difficult.

Table 1 shows the experimental results. The column "Order" and "# of functions" indicate the order and the number of function definitions of each program. The rightmost column shows the times spent for higher-order model checking (excluding those for translations, which can be performed instantly once automated). Note that the Reachability column shows the answers for pairwise reachability, not for the original verification problems. The benchmark program `example` has been taken from Example 1. The program `example_wrong` is a variation of `example`, obtained by omitting **join** operation of F. The other benchmark programs have been obtained by encoding exceptions, Java-style "synchronized" constructs (but with non-reentrant locks), and lists. The encoding uses higher-order functions, so it demonstrates an advantage of being able to deal with higher-order programs. For example, `exception` models the following OCaml-like program:

```
let rec read_and_update x =
  let n = read_int(x) (* may raise an Eof exception *) in
    lock g; c := c+n; (* may raise an Overflow exception *)
    unlock g; read_and_update x;;
let rec f file =
  let x = open_in file in
  try read_and_update(x) with
    Eof -> close x | Overflow -> unlock g;
spawn(f("foo"));spawn(f("bar"));join();print c
```

Table 1. The experimental results

Program	Order	# of functions	Checked pair	Reachability	Elapsed time [sec.]
`example`	2	4	(ℓ, ℓ)	no	0.11
`example_wrong`	2	4	(ℓ, ℓ)	yes	0.08
`exception`	3	7	(ℓ, ℓ)	no	0.25
`exception_wrong`	3	7	(ℓ, ℓ)	yes	0.12
`synchronized`	3	7	(ℓ_1, ℓ_1)	no	0.28
			(ℓ_1, ℓ_2)	yes	1.01
`list`	4	8	(ℓ_1, ℓ_1)	no	0.81
			(ℓ_1, ℓ_2)	no	1.75

Here, the goal of verification is to check that no race occurs on the shared variable `c`. The above program is encoded into the following order-3 program of the language in Section 2;

$$R\ h\ k = \mathbf{if}_{\_}(h\ True)\ (\mathbf{acq}_g; (\mathbf{if}_{\_}\ (\mathbf{rel}_g; R\ h\ k)\ (h\ False))^{\ell})$$
$$F\ k = R\ H\ k \qquad H\ b = b\ \$\ (\mathbf{rel}_g; \$) \qquad P\ k = k^{\ell}$$
$$S = \mathbf{spawn}(F); (\mathbf{spawn}(F); (\mathbf{join}; (P\ \$))) \qquad True\ x\ y = x \qquad False\ x\ y = y$$

Here, the function R corresponds to `read_and_update`; we have abstracted away `x` and instead added an exception handler `h` and a continuation parameter `k` in order to precisely model exception primitives. The program `exception_wrong` is a variation of `exception` obtained by omitting $\mathbf{acq}_g$ and $\mathbf{rel}_g$ operations.

All the benchmark programs have been verified within a few seconds. Although the programs are very small, this is encouraging, considering the worst-case complexity of higher-order model checking. As discussed at the end of the previous section, the pairwise reachability can be decided in time linear *with respect to the size of HORS* under a certain assumption; thus, the results indicate that our method may scale for larger programs.

5 Related Work

As already mentioned in Section 1, the present work is based on the series of work on verification of pushdown systems with nested locking [7,14,4], and extends it to deal with higher-order programs. The target language of our verification is more expressive than dynamic pushdown networks and corresponds to an extension of collapsible pushdown systems [6] (which are higher-order pushdown systems extended with collapse operations) with concurrency primitives. Gawlitza et al. [4] used a clever encoding of a configuration of a dynamic pushdown network, so that the (forward) reachable set of configurations can be represented as a regular tree language and the reachability problem can be reduced to the inclusion between regular tree languages. One of our insights was that thanks to the decidability of higher-order model checking, actually we need not represent the reachable set as a regular language; a higher-order tree language (generated by a HORS) suffices. This has enabled not only the higher-order extension, but also a conceptual simplification of the verification method in our opinion.

Higher-order model checking has recently been applied to program verification, but most of them have been for sequential programs [9,13,16]. Kobayashi and Igarashi [10] have shown that the reachability problem for higher-order concurrent programs with a bounded number of context switches can be reduced to higher-order model checking. (They have also shown a reduction from verification of higher-order concurrent programs to an extension of higher-order model checking, but the latter is undecidable.) Hague [5] has shown the decidability of reachability of ordered, phase-bounded and scope-bounded concurrent collapsible pushdown systems. These methods underapproximate the reachable set of ordinary higher-order concurrent programs (without the "bound" conditions). To our knowledge, there is no realistic implementation of those methods. There are also overapproximation approaches to static analysis or verification of higher-order concurrent programs [2,3,8].

6 Conclusion

We have shown the decidability of pairwise reachability of higher-order concurrent programs with recursion, dynamic process creation, joins, and nested locking. To our knowledge, this is the first realistic application of higher-order model checking to verification of concurrent programs. Despite the extremely high worst-case complexity of higher-order model checking, preliminary experiments show that our approach is feasible at least for small programs.

Acknowledgement. We would like to thank Markus Müller-Olm for the discussion on the subject and information about dynamic pushdown networks, and anonymous referees for useful comments. We would also like to thank Taku Terao for providing his higher-order model checker for the experiments. This work was supported by JSPS Kakenhi 23220001.

References

1. Broadbent, C.H., Kobayashi, N.: Saturation-based model checking of higher-order recursion schemes. In: Rocca, S.R.D. (ed.) CSL. LIPIcs, vol. 23, pp. 129–148. Schloss Dagstuhl - Leibniz-Zentrum fuer Informatik (2013)
2. D'Osualdo, E., Kochems, J., Ong, C.-H.L.: Automatic verification of erlang-style concurrency. In: Logozzo, F., Fähndrich, M. (eds.) Static Analysis. LNCS, vol. 7935, pp. 454–476. Springer, Heidelberg (2013)
3. Feret, J.: Abstract interpretation of mobile systems. Journal of Logic and Algebraic Programming 63(1) (2005)
4. Gawlitza, T.M., Lammich, P., Müller-Olm, M., Seidl, H., Wenner, A.: Join-lock-sensitive forward reachability analysis for concurrent programs with dynamic process creation. In: Jhala, R., Schmidt, D. (eds.) VMCAI 2011. LNCS, vol. 6538, pp. 199–213. Springer, Heidelberg (2011)

5. Hague, M.: Saturation of concurrent collapsible pushdown systems. In: Proceedings of FSTTCS 2013. LIPIcs, vol. 24, pp. 313–325 (2013)
6. Hague, M., Murawski, A., Ong, C.-H.L., Serre, O.: Collapsible pushdown automata and recursion schemes. In: Proceedings of 23rd Annual IEEE Symposium on Logic in Computer Science, pp. 452–461. IEEE Computer Society (2008)
7. Kahlon, V., Ivančić, F., Gupta, A.: Reasoning about threads communicating via locks. In: Etessami, K., Rajamani, S.K. (eds.) CAV 2005. LNCS, vol. 3576, pp. 505–518. Springer, Heidelberg (2005)
8. Kobayashi, N.: Type systems for concurrent programs. In: Aichernig, B.K. (ed.) Formal Methods at the Crossroads. From Panacea to Foundational Support. LNCS, vol. 2757, pp. 439–453. Springer, Heidelberg (2003)
9. Kobayashi, N.: Model checking higher-order programs. J. ACM 60(3), 20 (2013)
10. Kobayashi, N., Igarashi, A.: Model-checking higher-order programs with recursive types. In: Felleisen, M., Gardner, P. (eds.) Programming Languages and Systems. LNCS, vol. 7792, pp. 431–450. Springer, Heidelberg (2013)
11. Kobayashi, N., Ong, C.-H.L.: A type system equivalent to the modal mu-calculus model checking of higher-order recursion schemes. In: Proceedings of LICS 2009, pp. 179–188 (2009)
12. Kobayashi, N., Ong, C.-H.L.: Complexity of model checking recursion schemes for fragments of the modal mu-calculus. Logical Methods in Computer Science 7(4) (2011)
13. Kobayashi, N., Sato, R., Unno, H.: Predicate abstraction and CEGAR for higher-order model checking. In: Proceedings of PLDI 2011, pp. 222–233 (2011)
14. Lammich, P., Müller-Olm, M., Wenner, A.: Predecessor sets of dynamic pushdown networks with tree-regular constraints. In: Bouajjani, A., Maler, O. (eds.) CAV 2009. LNCS, vol. 5643, pp. 525–539. Springer, Heidelberg (2009)
15. Ong, C.-H.L.: On model-checking trees generated by higher-order recursion schemes. In: LICS, pp. 81–90 (2006)
16. Ong, C.-H.L., Ramsay, S.: Verifying higher-order programs with pattern-matching algebraic data types. In: Proceedings of POPL 2011, pp. 587–598 (2011)
17. Plotkin, G.D.: Call-by-name, call-by-value and the lambda-calculus. Theor. Comput. Sci. 1(2), 125–159 (1975)
18. Qadeer, S., Rehof, J.: Context-bounded model checking of concurrent software. In: Halbwachs, N., Zuck, L.D. (eds.) TACAS 2005. LNCS, vol. 3440, pp. 93–107. Springer, Heidelberg (2005)
19. Ramalingam, G.: Context-sensitive synchronization-sensitive analysis is undecidable. ACM Trans. Program. Lang. Syst. 22(2), 416–430 (2000)
20. Sato, R., Unno, H., Kobayashi, N.: Towards a scalable software model checker for higher-order programs. In: Proceedings of PEPM 2013, pp. 53–62 (2013)

A Linear-Time Algorithm for the Orbit Problem over Cyclic Groups

Anthony Widjaja Lin[1] and Sanming Zhou[2]

[1] Yale-NUS College, Singapore

[2] Department of Mathematics and Statistics, The University of Melbourne, Australia

Abstract. The orbit problem is at the heart of symmetry reduction methods for model checking concurrent systems. It asks whether two given configurations in a concurrent system (represented as finite sequences over some finite alphabet) are in the same orbit with respect to a given finite permutation group (represented by their generators) acting on this set of configurations. It is known that the problem is in general as hard as the graph isomorphism problem, which is widely believed to be not solvable in polynomial time. In this paper, we consider the restriction of the orbit problem when the permutation group is cyclic (i.e. generated by a single permutation), an important restriction of the orbit problem. Our main result is a linear-time algorithm for this subproblem.

1 Introduction

Since the inception of model checking, a key challenge in verifying concurrent systems has always been how to circumvent the state explosion problem due to the growth in the number of processes. Among others, symmetry reduction [10, 14, 18] has emerged to be an effective technique in combatting the state explosion problem. The essence of symmetry reduction is to identify symmetries in the system and avoid exploring states that are "similar" (under these symmetries) to previously explored states, thereby speeding up model checking.

Every symmetry reduction method has to deal with the following problems: (1) how to identify symmetries in the given system, and (2) how to check that two configurations are similar under these symmetries. For concurrent systems with n processes, Problem 1 amounts to searching for a group G of permutations on $[n] := \{1, \ldots, n\}$ such that the system behaves in an identical way under the action of permuting the indices of the processes by any $\pi \in G$. For example, for a distributed protocol with a ring topology, the group G could be a *rotation group* generated by the "cyclical right shift" permutation RS that maps $i \mapsto i + 1$ mod n for each $i \in [n]$. The reader is referred to the recent survey [23] for more detailed discussions and techniques for handling Problem 1, a computationally difficult problem in general. Now the group G partitions the state space of the concurrent system (i.e. Γ^n for some finite set Γ) into equivalence classes called *(G-)orbits.* Problem 2 is essentially the *orbit problem (over finite permutation groups)*: given G and two configurations $\mathbf{v}, \mathbf{w} \in \Gamma^n$, determine whether $\mathbf{v}$ and $\mathbf{w}$

P. Baldan and D. Gorla (Eds.): CONCUR 2014, LNCS 8704, pp. 327–341, 2014.

are in the same G-orbit. For example, if G is generated by RS with $n = 4$, the two configurations $(1, 1, 0, 0)$ and $(0, 0, 1, 1)$ are in the same orbit.

The orbit problem (OP) was first studied in the context of model checking by Clarke *et al.* [10] in which it was shown to be in NP but is as hard as the graph isomorphism problem, which is widely believed to be not solvable in polynomial time. The difficulty of the problem is due to the fact that the input group G is represented by a set S of generators and that the size of G can be exponential in $|S|$ in the worst case. There is also a closely related variant of OP called the *constructive orbit problem (COP)*, which asks to compute the lexicographically smallest element $\mathbf{w} \in \Gamma^n$ in the orbit of a given configuration $\mathbf{v} \in \Gamma^n$ with respect to a given group G. OP is easily reducible to COP, though the reverse direction is by no means clear. COP was initially studied in the context of graph canonisation by Babai and Luks [3], in which COP was shown to be NP-hard (in contrast, whether OP is NP-hard is open). In the context of model checking, COP was first studied by Clarke *et al.* [9], in which a number of "easy groups" for which COP becomes solvable in P are given including polynomial-sized groups (e.g. rotation groups), the full symmetry group S_n (i.e. containing all permutations on $[n]$), and disjoint/wreath products of easy groups (cf. [13]).

In this paper, we consider the orbit problem over *cyclic groups* (i.e. generated by a single permutation $\pi \in S_n$), which is an important subproblem of OP. Firstly, an algorithm for this subproblem has immediate applications for OP in the general case. For example, given a permutation group G with generators $\pi_1, \ldots, \pi_k$, we can check if the two configurations $\mathbf{v}$ and $\mathbf{w}$ are in the same orbit of the cyclic subgroup generated by any one of π_j. [If yes, then $\mathbf{v}$ and $\mathbf{w}$ are also in the same G-orbit.] It is also possible to combine cyclic groups with other easy groups from [9] via disjoint/wreath product operators. Secondly, it subsumes a commonly occurring class of symmetries for concurrent systems: the rotation groups. Unlike the case of rotation groups however, the size of cyclic groups can be exponential in n (see Proposition 3 below), which rules out a naive enumeration of the group elements. Finally, OP over cyclic groups is intimately connected to the classical orbit problem over rational matrices [19]: given a rational n-by-n matrix M and two rational vectors $\mathbf{v}, \mathbf{w} \in \mathbb{Q}^n$, determine if there exists $k \in \mathbb{N}$ such that $A^k\mathbf{v} = \mathbf{w}$. In fact, they coincide when M is restricted to *permutation matrices* [6], i.e., 0-1 matrices with precisely one column for each row with entry 1. To see this, given a permutation π on $[n]$, simply take an n-by-n 0-1 matrix $A = (A[i, j])_{1 \leq i,j \leq n}$ such that $A[i, j] = 1$ iff $\pi(j) = i$. The reverse direction is similar. That OP over cyclic groups is in P follows from Kannan and Lipton's celebrated result [19] that OP over rational matrices is in P.

Contributions. In this paper, we provide an algorithm for the orbit problem over cyclic groups that is simpler than Kannan-Lipton's algorithm [19] and moreover runs in linear-time in the standard RAM model. To this end, we provide a linear-time reduction to the problem of solvability of systems of linear congruence equations. The reduction is simple though it exploits subtle connections to the string searching problem and number-theoretic results like the Erdös-Graham Lemma [15] concerning solutions of Diophantine equations.

As for the solvability of systems of linear congruence equations, we start off with an algorithm that runs in linear-time assuming constant-time integer arithmetic operations. However, when we measure the number of bit operations (i.e. bit complexity model), it turns out that the algorithm runs in time cubic in the number of equations in the systems. To address this issue, we restrict the problem to input instances provided by our reduction from the orbit problem. We offer two solutions. Firstly, we show that the average-case complexity of the algorithm under the bit complexity model is $O(\log^5 n)$, which is sublinear. Secondly, we provide another algorithm that uses at most linearly many bit operations *in the worst case* (though on average it is worse than the first algorithm).

Organisation. Section 2 contains definitions and basic concepts. We provide our first algorithm for solving systems of linear congruence equations in Section 3 (Algorithm 1), while we provide our linear-time reduction from the orbit problem to equations solving in Section 4 (Algorithm 2). Thus far, we assume that arithmetic operations take constant time. We deal with the issue of bit complexity in Section 5. We conclude with future work in Section 6.

2 Preliminaries

General Notations: We use log (resp. ln) to denote the logarithm function in base 2 (resp. natural logarithm). We use the standard interval notations to denote a subset of integers within that interval. For example, $[i, j)$ denotes the set $\{k \in \mathbb{Z} : i \leq k < j\}$. Likewise, for each positive integer n, we use $[n]$ to denote the set $\{1, \ldots, n\}$. We shall also extend arithmetic operations to sets of numbers in the usual way: whenever $S_1, S_2 \subseteq \mathbb{Z}$, we define $S_1 + S_2 := \{s_1 + s_2 : s_1 \in S_1, s_2 \in S_2\}$ and $S_1 S_2 := \{s_1 \times s_2 : s_1 \in S_1, s_2 \in S_2\}$. In the context of arithmetic over $2^{\mathbb{Z}}$, we will treat a number $n \in \mathbb{N}$ as the singleton set $\{n\}$. That way, for $a, b \in \mathbb{N}$, the notation $a + b\mathbb{Z}$ refers to the *arithmetic progression* $\{a + bc : c \in \mathbb{Z}\}$, where a (resp. b) is called the *offset* (resp. *period*) of the arithmetic progression. Likewise, for a subset $S \subseteq \mathbb{N}$, we use $\gcd(S)$ to denote the greatest common divisor of S.

We will use standard notations from formal language theory. Let Γ be an *alphabet* whose elements are called *letters*. A word (or a string) w over Γ is a finite sequence of elements from Γ. We use Γ^* to denote the set of all words over Γ. The length of w is denoted by $|w|$. Given a word $w = a_1 \ldots a_n$, the notation $w[i, j]$ denotes the subword $a_i \ldots a_j$. For a sequence $\sigma = i_1, \ldots, i_k \in [n]^*$ of *distinct* indices of w, we write $w[\sigma]$ to denote the word $a_{i_1} \ldots a_{i_k}$. We also define $\mathrm{RS}(w)$ to be $a_n a_1 a_2 \ldots a_{n-1}$, i.e., the word w cyclically right-shifted.

Number Theory: We will use the following basic result (cf. [11]).

Proposition 1 (Chinese Remainder Theorem). *Let $n_1, \ldots, n_k$ be pairwise relatively prime positive integers, and $n = \prod_{i=1}^{k} n_i$. The ring $\mathbb{Z}_n$ and the direct product of rings $\mathbb{Z}_{n_1} \times \cdots \times \mathbb{Z}_{n_k}$ are isomorphic under the function $\sigma : \mathbb{Z} \to \mathbb{Z}_{n_1} \times \cdots \times \mathbb{Z}_{n_k}$ with $\sigma(x) := (x \mod n_1, \ldots, x \mod n_k)$ for each $x \in \mathbb{Z}$.*

Groups: We briefly recall basic concepts from group theory and permutation groups (cf. see [7]). A *group* G is a pair $(S, \cdot)$, where S is a set and $\cdot : (S \times S) \to S$ is a binary operator satisfying: (i) associativity (i.e. $g_1 \cdot (g_2 \cdot g_3) = (g_1 \cdot g_2) \cdot g_3$), (ii) the existence of a (unique) identity element $e \in S$ such that $g \cdot e = e \cdot g = g$ for all $g \in S$, and (iii) closure under inverse (i.e. for each $g \in G$, there exists $g^{-1} \in G$ such that $g \cdot g^{-1} = g^{-1} \cdot g = e$). When it is clear from the context, we will write $g \cdot g'$ as gg'. The *order* $\text{ord}(G)$ of the group G is defined to be $|S|$. This paper concerns only finite groups, i.e., groups G with $\text{ord}(G) = |S| \in \mathbb{N}$. For each $n \in \mathbb{N}$, we define g^n by induction: (i) $g^0 = e$, and (ii) $g^n = g^{n-1} \cdot g$. The *order* $\text{ord}(g)$ of $g \in G$ is the least positive integer n such that $g^n = e$.

A *subgroup* H of $G = (S, \cdot)$ (denoted as $H \leq G$) is any group $(S', \cdot_H)$ such that $S' \subseteq S$ and $\cdot_H$ and $\cdot$ agree on S'. Given any subset $X \subseteq S$, the subgroup of G *generated* by X is defined to be the subgroup $\langle X \rangle := (S', \cdot_h)$ of G each of whose elements can be expressed as a finite product of elements of X and their inverses. If $H = \langle X \rangle$, then X is said to *generate* H. A *cyclic group* is a group generated by a singleton set $X = \{g\}$.

An *action* of a group $G = (S, \cdot)$ on a set Y is a function $\times : S \times Y \to Y$ such that for all $g, h \in S$ and $y \in Y$: (1) $(gh) \times y = g \times (h \times y)$, and (2) $e \times y = y$. The (G-)orbit containing y, denoted Gy, is the subset $\{g \times y : g \in G\}$ of Y. The action $\times$ partitions the set Y into G-orbits. When the meaning is clear, we shall omit mention of the operator $\times$, e.g, condition (2) above becomes $ey = y$.

Permutation Groups. A *permutation* on $[n]$ is any bijection $\pi : [n] \to [n]$. The set of all permutations on $[n]$ forms the *(nth) full symmetry group* S_n under functional composition. We shall use the notation Id to denote the identity element of each S_n. A word $w = a_0 \dots a_{k-1} \in [n]^*$ containing distinct elements of $[n]$ (i.e. $a_i \neq a_j$ if $i \neq j$) can be used to denote the permutation that maps $a_i \mapsto a_{i+1 \bmod k}$ for each $i \in [0, k)$ and fixes other elements of $[n]$. In this case, w is called a *cycle*, which we will often write in the standard notation $(a_0, \dots, a_{k-1})$ so as to avoid confusion. Observe that w and $\text{RS}(w)$ represent the same cycle c. We will however fix a particular ordering to represent c (e.g. the word provided as input to the orbit problem). For this reason, if $\mathbf{v} \in \Gamma^n$ for some alphabet Γ, the notation $\mathbf{v}[c]$ is well-defined (see General Notations above), which means projections of $\mathbf{v}$ onto elements with indices in c, e.g., if $\mathbf{v} = (1, 1, 1, 0)$ and $c = (1, 4, 2)$, then $\mathbf{v}[c] = (1, 0, 1)$. Any permutation can be written as a composition of disjoint cycles [7]. Each subgroup $G = (S, \cdot)$ of S_n acts on the set Γ^n (over any finite alphabet Γ) under the group action of permuting indices, i.e., for each $\pi \in S$ and $\mathbf{v} = (a_1, \dots, a_n) \in \Gamma^n$, we define $\pi\mathbf{v} := (a_{\pi(1)}, \dots, a_{\pi(n)})$.

Complexity Analysis: We will assume that permutations will be given in the input as a composition of disjoint cycles. It is easy to see that permutations can be converted back and forth in linear time from such representations and the representations of permutations as functions. The size $\|n\|$ of a number $n \in \mathbb{N}$ is defined to be the length of the binary representation of n, which is $\lfloor \log n \rfloor + 1$. The size $\|c\|$ of a cycle $c = (a_1, \dots, a_k)$ on $[n]$ is defined to be $\sum_{i=1}^{k} \|a_i\|$ (in contrast, the length $|c|$ of c is k). For a permutation $\pi = c_1 \cdots c_m$ where each c_i is a cycle, the size $\|\pi\|$ of π is defined to be $\sum_{i=1}^{m} \|c_i\|$. We will use standard asymptotic notations from analysis of algorithms (big-O and little-o), cf. [11]. We also use the standard $\sim$ notation: $f(n) \sim g(n)$ iff $\lim_{n\to\infty} f(n)/g(n) = 1$. We will use the standard RAM model that is commonly used when analysing the complexity of algorithms (cf. [11]). In Sections 3 and 4, we will assume that integer arithmetic takes constant time. Later in Section 5, we will use the *bit complexity model* (cf. [11]), wherein the running time is measured in the number of bit operations.

3 Solving a System of Modular Arithmetic Equations

Recall that a linear congruence equation is a relation of the form $x \equiv a \pmod{b}$, where $a, b \in \mathbb{N}$, whose solution set is denoted by $[\![x \equiv a \pmod{b}]\!] = a + b\mathbb{Z}$. A system of linear congruence equations is a relation of the form $\bigwedge_{i=1}^{m} x \equiv a_i \pmod{b_i}$. The set of solutions $x \in \mathbb{Z}$ to this system is denoted by $[\![\bigwedge_{i=1}^{m} x \equiv a_i \pmod{b_i}]\!]$, which equals $\bigcap_{i=1}^{m} [\![x \equiv a_i \pmod{b_i}]\!]$. The system is *soluble* / *solvable* if the solution set is nonempty. We use FALSE to denote $x \equiv 0 \pmod{2} \wedge x \equiv 1 \pmod{2}$, which is not solvable. The following proposition provides a fast symbolic method for computing solutions to systems of linear congruences.

Proposition 2. *For any solvable system of linear congruence equations* $\varphi(x) := \bigwedge_{i=1}^{m} x \equiv a_i \pmod{b_i}$, *we have* $[\![\varphi(x)]\!] = [\![x \equiv a \pmod{b}]\!]$ *for some* $a, b \in \mathbb{Z}$. *Furthermore, there exists an algorithm which computes* a, b *in linear time.*

This proposition is in fact a rather easy corollary of the following result in algorithmic number theory about solving more general linear congruence equations of the form $ax \equiv b \pmod{n}$.

Lemma 1 (Linear Congruence Theorem; see [11, Chapter 31.4]). *The equation* $ax \equiv b \pmod{n}$ *is solvable for the unknown* x *iff* $d|b$, *where* $d = \gcd(a, n)$. *Furthermore, if it is solvable, then the set of solutions equals* $x_0 + (n/d)\mathbb{Z}$, *for some* $x_0 \in [0, n/d)$ *that can be computed in time* $O(\log n)$.

This algorithm made use of the Extended Euclidean algorithm, which explains the $O(\log n)$ time complexity (see [11]). Algorithm 1 witnesses the linear-time algorithm claimed in Proposition 2. The algorithm sequentially goes through each equation $x \equiv a_i \pmod{b_i}$, while keeping the solution to the subsystem $\bigwedge_{i=1}^{j} x \equiv a_i \pmod{b_i}$ at jth iteration as an arithmetic progression $a + b\mathbb{Z}$, for some $a, b \in \mathbb{Z}$. Before we go through any equation, the set of solutions to the empty system of equations is $a + b\mathbb{Z}$ with $a = 0$ and $b = 1$. At the jth iteration, we assume that $[\![\bigwedge_{i=1}^{j-1} x \equiv a_i \pmod{b_i}]\!] = a + b\mathbb{Z}$ for some $a, b \in \mathbb{Z}$. We replace x in the equation $x \equiv a_j \pmod{b_j}$ by $a + by$ for an unknown y, which results in the new equation $\varphi(x) := by \equiv a_i - a \pmod{b_i}$. Lemma 1 gives an answer to $[\![\varphi]\!]$ as either $\emptyset$ or $a' + b'\mathbb{Z}$, for some $a' \in [0, b_i)$ and $b' \in [1, b_i]$. We substitute this solution set back to x, which gives $[\![\bigwedge_{i=1}^{j} x \equiv a_i \pmod{b_i}]\!] = (a'b + a) + bb'\mathbb{Z}$, which justifies the assignments $a := a'b + a$ and $b := bb'$.

As for the time complexity of the algorithm, at jth iteration the algorithm invokes the algorithm from Lemma 1, which runs in time $O(\log b_j)$. Therefore,

Algorithm 1. Solving a system of modular arithmetic equations

Input: A system of modular arithmetic equations $\bigwedge_{i=1}^{m} x \equiv a_i \pmod{b_i}$
Output: Solution set $[\![\bigwedge_{i=1}^{m} x \equiv a_i \pmod{b_i}]\!]$ as $\emptyset$ or an arithmetic progression $a + b\mathbb{Z}$.

```
a := 0; b := 1;
for i = 1,...,m do
    φ(y) := by ≡ a_i − a (mod b_i);
    Apply algorithm from Lemma 1 on φ returning either ∅ or a′ + b′ℤ for ⟦φ⟧;
    if ⟦φ⟧ = ∅ then return NO else a := a′b + a; b := bb′ end if
end for
return a + bℤ;
```

the total running time of our algorithm is $O(\sum_{j=1}^{m} \log b_j)$, i.e., linear in the size $\sum_{j=1}^{m}(\log a_j + \log b_j)$ of the input.

Remark 1. The number of bits that is used to maintain a and b in the worst case is linear in the size $\sum_{j=1}^{m}(\log a_j + \log b_j)$ of the input. This justifies treating a single arithmetic operation as a constant-time operation. We will address the issue of bit complexity in Section 5.

4 Reducing to Solving a System of Linear Congruence Equations

In this section, we prove the main result of the paper.

Theorem 1. *There is a linear-time algorithm for solving the orbit problem when the acting group is cyclic.*

This algorithm is a linear-time reduction from the orbit problem over cyclic groups to solving a system of linear congruence equations, which will allow us to use results from the previous section.

Before we proceed to the algorithm, the following proposition shows why the naive algorithm that checks whether $g^i(\mathbf{v}) = \mathbf{w}$, for a given permutation $g \in S_n$ and for each $i \in [0, \mathrm{ord}(g))$, actually runs in exponential time.

Proposition 3. *There exists a sequence $\{G_i\}_{i=1}^{\infty}$ of cyclic groups $G_i = \langle g_i \rangle$ such that ord(g_i) is exponential in the size $\|g_i\|$ of the permutation g_i.*

Proof. Let p_n denote the nth prime. The *Prime Number Theorem* states that $p_n \sim n \log n$ (cf. [17]). For each $i \in \mathbb{Z}_{>0}$, we define a cycle c_i of length p_i by induction on i. For $i = 1$, let $c_1 = (1, 2)$. Suppose that $c_{i-1} = (j, \ldots, k)$. In this case, we define c_i to be the cycle $(k+1, \ldots, k+p_i)$. To define the sequence $\{g_i\}_{i=1}^{\infty}$ of permutations, simply let $g_i = \Pi_{j=1}^{i} c_i$. For example, we have $g_3 = (1,2)(3,4,5)(6,7,8,9,10)$. Since c_i's are disjoint, the order $\mathrm{ord}(g_i)$ of g_i is the smallest positive integer k such that $c_j^k = \mathrm{Id}$ for all $j \in [i]$. If S_j denotes the set of integers k satisfying $c_j^k = \mathrm{Id}$, then $\mathrm{ord}(g_i)$ is precisely the smallest positive integer in the set $\bigcap_{j=1}^{i} S_j$. It is easy to see that $S_j = p_j\mathbb{Z}$, which is the set of solutions to the linear congruence equation $x \equiv 0 \pmod{p_j}$. Therefore, by the

Chinese Remainder Theorem (cf. Propositon 1), the set $\bigcap_{j=1}^{i} S_j$ coincides with the arithmetic progression $t_i\mathbb{Z}$ with $t_i := \prod_{j=1}^{i} p_j$. This implies that $\mathrm{ord}(g_i) = t_i$. Now the number t_i is also known as the *ith primorial number* [1] with $t_i \sim e^{(1+o(1))i \log i}$, which is a corollary of the Prime Number Theorem. On the other hand, the size of g_i is $\sum(i) := \sum_{j=1}^{i} p_i$, which is known to be $\sim \frac{1}{2}i^2 \ln i$ (cf. [4]). Therefore, $\mathrm{ord}(g_i)$ is exponential in $\|g_i\|$ as desired. □

Algorithm 2. Reduction to system of modular arithmetic equations

---

Input: A permutation $g = c_1 \cdots c_m \in S_n$, a finite alphabet Γ, and $\mathbf{v}, \mathbf{w} \in \Gamma^n$.
Output: A system of modular arithmetic equations, which is satisfiable iff $\exists i \in \mathbb{N}$: $g^i(\mathbf{v}) = \mathbf{w}$.

```
// First solve for each individual cycle
for all i = 1,...,m do
    Compute the length |c_i| of the cycle c_i;
    Compute an ordered list S'_i ⊆ [0,|c_i|) of numbers r with c_i^r(v[c_i]) = w[c_i];
    if S'_i = ∅ then return FALSE end if
    if |S'_i| = 1 then let a_i be the member of S_i; b_i := |c_i|; end if
    if |S'_i| > 1 then a_i := min(S'_i); a'_i := min(S'_i \ {a_i}); b_i := a'_i − a_i; end if
end for
// Now for each i ∈ [1,m] we have a modular arithmetic equation x ≡ a_i (mod b_i)
return YES iff there exists x ∈ N satisfying ⋀_{i=1}^m x ≡ a_i (mod b_i)
```

Our linear-time reduction that witnesses Theorem 1 is given in Algorithm 2. In this algorithm, the acting group is $G = \langle g \rangle$ with $g \in S_n$, expressed as a composition of disjoint cycles in a standard way, say, $g = c_1 c_2 \cdots c_m$ where each c_i is a cycle. Also part of the input is two strings $\mathbf{v} = v_1 \ldots v_n, \mathbf{w} = w_1 \ldots w_n \in \Gamma^n$ over a finite alphabet Γ. The orbit problem is to check whether $f(v) = w$ for some $f \in G$, i.e., $f = g^r$ for some $r \in \mathbb{N}$. Since c_i's are pairwise disjoint cycles, the question reduces to checking if there exists $r \in \mathbb{N}$ such that

$$\forall i \in [1,m] : (c_i^r \mathbf{v})[c_i] = \mathbf{w}[c_i] \qquad (*)$$

In other words, for each $i \in [1,m]$, applying the action c_i^r to $\mathbf{v}$ gives us $\mathbf{w}$ when restricted to the indices in c_i. Essentially, Algorithm 2 sequentially goes through each cycle c_i and computes the set S_i of solutions r to $(c_i^r\mathbf{v})[c_i] = \mathbf{w}[c_i]$ as the set of solutions to the linear congruence equation $x \equiv a_i \pmod{b_i}$. Therefore, the set of solutions to (*) is precisely the set of solutions to the system of congruence equations $\bigwedge_{i=1}^{m} x \equiv a_i \pmod{b_i}$. In the following, we will provide the details of each individual step of Algorithm 2. We will also use the following running example to illustrate the algorithm: $c = (6,5,7,3,2,1)$, $\mathbf{v} = \underline{010}0\underline{011}11$, and $\mathbf{w} = \underline{101}1\underline{100}01$, where the positions in $\mathbf{v}$ and $\mathbf{w}$ that are modified by c are underlined.

Step 1: Computing the length of cycles. This is the same as how to compute the length of a list. Therefore, computing the length $|c_i|$ can be done in time $O(\|c_i\|)$.

Step 2: Computing representatives $S'_i \subseteq [0,|c_i|)$ for S_i. During this step, we collect a subset of numbers $h \in [0,|c_i|)$ such that $c_i^h(\mathbf{v}[c_i]) = \mathbf{w}[c_i]$. A quadratic algorithm for this is easy to come up with: sequentially go through $h \in [0, |c_i|)$ while computing the current c_i^h, and save h if $c_i^h(\mathbf{v}[c_i]) = \mathbf{w}[c_i]$ holds. One way to obtain a linear-time algorithm is to reduce our problem to the *string searching problem*: given a "text" $T \in \Sigma^*$ (over some finite alphabet Σ) and a "pattern" $P \in \Sigma^*$, find all positions i in T such that $T[i, i+|P|] = P$. This problem is solvable in linear-time by Knuth-Morris-Pratt (KMP) algorithm (e.g. see [11]).

We now show how to reduce our problem to the string searching problem in linear time. Suppose that $c := c_i = (j_1, \ldots, j_k)$. We have $\mathbf{v}[c] = v_{j_1} \ldots v_{j_k}$ and $\mathbf{w}[c] = w_{j_1} \ldots w_{j_k}$.

Lemma 2. $(cv)[c] = \mathrm{RS}(v[c])$.

In other words, if $\textsc{Dom}(c) = \{j_1, \ldots, j_k\}$, the effect of c on $\mathbf{v}$ when restricted to $\textsc{Dom}(c)$ coincides with applying a cyclical right shift on the string $[c]$. Following our running example, it is easy to check that $[c] = 101010$ and $(c\mathbf{v})[c] = \mathrm{RS}(\mathbf{v}[c]) = 010101$.

Proof (of Lemma 2). Let $\mathbf{u} = u_1 \ldots u_k := (c\mathbf{v})[c]$ and $\mathbf{u} = u'_1 \ldots u'_k := \mathrm{RS}(\mathbf{v}[c])$. It suffices to show that $u_t = u'_t$ for all $t \in \mathbb{Z}_k$. By definition of RS, it follows that $u'_t = v_{j_{t-1}}$. Now suppose that $\mathbf{v}' = v'_1 \ldots v'_n := c\mathbf{v}$. Then

$$v'_j := \begin{cases} v_j & \text{if } j \notin \textsc{Dom}(c) \\ v_{j'} & \text{if } j \in \textsc{Dom}(c) \text{ and, for some } t \in \mathbb{Z}_k,\ j = j_{t+1} \text{ and } j' = j_t. \end{cases}$$

So, we have $u_t = ((c\mathbf{v})[c])[t] = (\mathbf{v}'[c])[t] = v'_{j_t} = v_{j_{t-1}}$. This proves that $u_t = u'_t$. □

Lemma 3. *For each* $r \in \mathbb{N}$, *we have* $(c^r v)[c] = \mathrm{RS}^r(v[c])$.

Lemma 3 can easily be proven by induction using Lemma 2 (see full version). Lemma 3 implies that the set $S := S_i \subseteq \mathbb{N}$ of solutions r to the equation $(c_i^r \mathbf{v})[c_i] = \mathbf{w}[c_i]$ is a finite union of arithmetic progressions of the form $a + k\mathbb{Z}$, where $k = |c_i|$ and $a \in [0, k)$. This is simply because $\mathrm{RS}^{r+k}(\mathbf{v}[c_i]) = \mathrm{RS}^r(\mathbf{v}[c_i])$. We will finitely represent S by the offsets a's and the unique period k in these arithmetic progressions.

We now show how to compute the offsets for S in linear time by a linear-time reduction to the string searching problem. Define the text $T := \mathbf{v}[c]\mathbf{v}[c]$ and the pattern $P := \mathbf{w}[c]$. Observe that, for each $r \in [0, k)$, P is matched at position r in T iff $\mathrm{RS}^{r-1}(\mathbf{v}[c]) = \mathbf{w}[c]$. Therefore, after running the KMP algorithm with the solution set S', the offsets for S will be $\{r - 1 : r \in S'\}$. Solvability for each individual equation amounts to checking that, for each cycle c_i, the set S_i of solutions for the corresponding equation is nonempty.

Example 1. Continuing with our running example, it follows that $T = \mathbf{v}[c]\mathbf{v}[c] = 101010101010$ and $P = \mathbf{w}[c] = 010101$. We see that P matches T at positions $S' = \{2, 4, 6\}$. This implies that the set S of solutions $r \in \mathbb{Z}$ to the equation $(c^r\mathbf{v})[c] = \mathbf{w}[c]$ is $(1+6\mathbb{Z}) \cup (3+6\mathbb{Z}) \cup (5+6\mathbb{Z})$. ■

Observe that, for each c_i, this step takes time $O(\|c_i\|)$. Therefore, going through all the c_i's, this step takes time $\sum_{i=1}^{m} O(\|c_i\|) = O(\sum_{i=1}^{m} \|c_i\|) = O(\|g\|)$, i.e., linear in input size.

Step 3: Representing S_i as a single arithmetic progression. In the previous step, we have computed the representatives for S_i in $[0, |c_i|)$. This only shows that S_i is a finite union of arithmetic progressions, which cannot in general be expressed as the set of solutions to a linear congruence equation. In this step, we show that S_i can be represented as a single arithmetic progression and furthermore justify why the last three lines in Algorithm 2 computes S_i.

Lemma 4 (Normal Form). *For each $i = 1, \ldots, m$, either $S_i = \emptyset$ or $S_i = a_i + b_i\mathbb{Z}$ for some $a_i, b_i \in [0, |c_i|)$ where b_i divides $|c_i|$. In the case when $|S_i'| > 1$, we have $a_i = p_1$ and $b_i = p_2 - p_1$, where $p_1 < p_2$ are the smallest numbers in S_i'. Furthermore, we may compute the pair (a_i, b_i) of numbers in time $O(\|c_i\|)$.*

To prove this lemma, we will use the following number-theoretic result by Erdös and Graham [15]. [Also see the formulation in [8, 22], in which the result was applied in automata theory.]

Proposition 4. *Let $0 < p_1 < \ldots < p_s \leq k$ be natural numbers. Then, the set $X := \{\sum_{i=1}^{s} p_i x_i : x_1, \ldots, x_s \in \mathbb{N}\} \subseteq \mathbb{N}$ coincides with the set $S \cup (a + b\mathbb{N})$, where $S \subseteq \mathbb{N}$ contains no numbers bigger than k^2, and a is the least integer bigger than k^2 that is a multiple of $b := \gcd(p_1, \ldots, p_s)$.*

Proof (of Lemma 4). We use the shorthand S (resp. c) for S_i (resp. c_i). From Step 2, we know that S is a union of arithmetic progressions $\bigcup_{j=1}^{s} (p_j + k\mathbb{Z})$, for some $p_j \in [0, k)$ and $k = |c|$. Without loss of generality, we assume that $p_1 < \cdots < p_s$. If $s \in \{0, 1\}$, then we are done. Suppose now that $s > 1$. Let $\mathbf{v}[c] = d_1 \ldots d_k$ and $\mathbf{w}[c] = d_1' \ldots d_k'$. In this case, thanks to Lemma 3, it is the case that for each $j \in [1, s]$ and $l \in [1, k]$, we have $d_{l + p_j \bmod k} = d_l'$. Let $\Delta := \{p_{h'} - p_h : \forall h < h' \in [1, s]\} \cup \{k\}$ be the set of all differences in the offsets of the arithmetic progressions union the set $\{k\}$ containing the common period. By transitivity of '=', it follows that $d_{l \bmod k} = d_{l+\delta \bmod k}$ for each $l \in [0, k)$ and $\delta \in \Delta$. Again, by transitivity of '=', it follows that $d_{l \bmod k} = d_{l+\sigma \bmod k}$ for each $l \in [0, k)$ and each number σ in the set $X := \{(\sum_{i=1}^{s} p_i x_i) + k x_{s+1} : x_1, \ldots, x_{s+1} \in \mathbb{N}\}$. By Proposition 4, we have $X = S \cup (a + b\mathbb{N})$ where $S \subseteq [0, k^2]$ and a is the least integer bigger than k^2 that is a multiple of $b := \gcd(\Delta)$. Observe also that b divides all numbers in S and so we have $d_l = d_{l'}$ for each $l, l' \in [0, k)$ with $l \equiv l' \pmod{b}$. In other words, we have $\mathbf{v} = \underbrace{\mathbf{v}' \ldots \mathbf{v}'}_{k/b \text{ times}}$, where $\mathbf{v}' = d_1 \ldots d_b$. Since $\mathrm{RS}^{p_1}(\mathbf{v}[c]) = \mathbf{w}[c]$, it follows that, for each $q \in \mathbb{N}$, $\mathrm{RS}^{p_1 + bq}(\mathbf{v}[c]) = \mathrm{RS}^{p_1}(\mathrm{RS}^{bq}(\mathbf{v}[c])) = \mathrm{RS}^{p_1}(\mathbf{v}[c]) = \mathbf{w}[c]$. Therefore, we have $S \subseteq p_1 + b\mathbb{N}$. On the other hand, since b divides k and each number in $\{p_j - p_1 : j \in [2, s]\}$, we also have $S \supseteq p_1 + b\mathbb{N}$. This gives us $S = p_1 + b\mathbb{N}$.

From Step 2, we have computed the set $S' := S \cap [0, |c|)$. If $S' = \emptyset$, we also knew that $S_i = \emptyset$. If $S' = \{p\}$ is a singleton, we have $S = p + k\mathbb{Z}$. If $|S'| > 1$, we find the two smallest numbers $p_1 < p_2$ in S'. It follows that $S = p_1 + (p_2 - p_1)\mathbb{Z}$.

Observe that this takes time $O(\|c\|)$. [In fact, it is only linear in the size of the two smallest numbers since we ignore the rest of the members of S'.] □

Example 2. Continuing with our running example, we have $S = (1+6\mathbb{Z}) \cup (3+6\mathbb{Z}) \cup (5+6\mathbb{Z}) = 1+2\mathbb{Z}$. ■

The last three lines in Algorithm 2 runs in constant time since determining whether $|S_i| = 0$, $|S_i| = 1$, or $|S_i| > 1$ requires the algorithm to explore only a constant number of elements in S_i.

Summing Up. To sum up, the time spent computing the linear congruence equation $x \equiv a_i \pmod{b_i}$ for each $i \in [1, m]$ is $O(\|c_i\|)$. Therefore, our reduction runs in time $O(\sum_{i=1}^{m} \|c_i\|) = O(\|g\|)$, which is linear in input size. Therefore, invoking Proposition 2 on the resulting system of linear congruence equations, we obtain the set of solutions to (*) in linear time.

Example 3. Let us continue with our running example. Let

$$g_1 := c(4,8) = (6,5,7,3,2,1)(4,8), \quad g_2 := c(4,8,9) = (6,5,7,3,2,1)(4,8,9).$$

Then, running Algorithm 2 on g_1 yields the system $x \equiv 1 \pmod 2 \wedge x \equiv 1 \pmod 2$, which is equivalent to $x \equiv 1 \pmod 2$. Running Algorithm 2 on g_2 yields the system $x \equiv 1 \pmod 2 \wedge x \equiv 1 \pmod 3$. Both systems are solvable. ■

Remark 2. At this stage, the reader might wonder whether the Normal Form Lemma (cf. Lemma 4) is necessary. For example, without this lemma one could directly convert the orbit problem over cyclic groups into satisfiability of positive boolean formulas (i.e. involving both disjunctions and conjunctions) where each proposition is interpreted as a linear congruence equation. [This can be construed as adding the power of disjunction to systems of linear congruence equations.] Unfortunately, it is not difficult to show that the resulting satisfiability problem is NP-complete using the techniques of Gödel numbering (cf. [16, 21]).

5 Making Do with Linearly Many Bit Operations

Thus far, we have assumed that arithmetic operations take constant time. In this section, since Algorithm 1 makes a substantial use of basic arithmetic operations, we will revisit this assumption. It turns out that, although our reduction (Algorithm 2) to solving a system of linear congruence equations runs in linear time in the bit complexity model, the algorithm for solving the system of equations (Algorithm 1) uses at least a cubic number of arithmetic operations. The main results in this section are two-fold: (1) on inputs given by our reduction, Algorithm 1 runs in sublinear time (more precisely, $O(\log^5 n)$) *on average* in the bit complexity model, and (2) there exists another algorithm for solving a system of linear congruence equations (with numbers in the input represented in unary) that runs in linear time in the bit complexity model in the worst case.

We begin with two lemmas that provide the running time of Algorithm 2 and Algorithm 1 in the bit complexity model.

Lemma 5. *Algorithm 2 runs in linear time in the bit complexity model.*

Proof. On ith iteration, the number $|c_i|$ is stored in binary counter and can be computed by counting upwards from 0 and incrementing by 1 as we go through the elements in c_i. Although a single increment by 1 might take $O(|c_i|)$ bit operations in the worst case (since we have to propagate the carry bit), it is known (e.g. see [11, Chapter 17, p. 454]) that the entire sequence of operations actually takes time $O(|c_i|)$. Finally, since addition and substraction of two numbers can easily be performed in $O(\beta)$ time on numbers that use at most β bits, the operation $b_i := a_i' - a_i$ on the last line of the iteration takes at most $O(\log |c_i|)$ time. Therefore, accounting for all the cycles, the algorithm takes $\sum_{i=1}^{m} O(\|c_i\|) = O(\sum_{i=1}^{m} \|c_i\|) = O(\|g\|)$, which is linear in the input size. □

Lemma 6. *On an input* $\bigwedge_{i=1}^{m} x \equiv a_i \pmod{b_i}$ *with* $N = \max\{b_i : i \in [1, m]\}$, *Algorithm 1 uses at most* $m \log N$ *bits to store any numeric variables. Furthermore, the algorithm runs in time* $O(m^3 \log^2 N)$ *in the bit complexity model.*

Proof. On ith iteration, the number of bits used to store a and b grow by at most $\log b_i$. On the other hand, the invariant that $a', b' \in [0, b_i)$ is always maintained on the ith iteration and so they only need at most $\log N$ bits to represent throughout the algorithm. Hence, the algorithm uses $M = O(m \log N)$ bits to store a, b, a', and b'. Extended Euclidean Algorithm runs in time $O(M^2)$ on inputs where each number uses at most M bits (cf. [11, Problem 31-2]), which also bounds the time it takes on each iteration. Therefore, the algorithm takes at most $O(mM^2) = O(m^3 \log^2 N)$ in the bit complexity model. □

We now provide an average case analysis of the running time of Algorithm 1 on system of linear congruence equations given by our reduction. The input to the orbit problem over cyclic groups includes a permutation $g \in S_n$ and two vectors $\mathbf{v}, \mathbf{w} \in \Gamma^n$. We briefly recall the setting of average-case analysis (cf. [20]). Let Π_N be the set of all inputs to the algorithm of size N. Likewise, let Σ_N be the sum of the *costs* (i.e. running time) of the algorithm on *all* inputs of size N. Hence, if $\Pi_{N,k}$ is the cost of the algorithm on input of size N, then $\Sigma_N = \sum_k k\Pi_{N,k}$. The *average case complexity of the algorithm* is defined to be Σ_N / Π_N.

Theorem 2. *The expected running time of Algorithm 1 in the bit complexity model on inputs provided by Algorithm 2 is* $O(\log^5 n)$.

Proof. The size of a single permutation $g \in S_n$ is $O(n)$ and additionally $\Pi_n = |S_n| = n!$. Suppose that g has k cycles (say, $g = c_1 \cdots c_k$). Then, Algorithm 2 produces a system of equations $\bigwedge_{i=1}^{k} x \equiv a_i \pmod{b_i}$, where $a_i, b_i \in [0, |c_i|)$. By Lemma 6, Algorithm 1 takes $O(k^3 \log^2 n)$ time in the bit complexity model, since $N := \max[b_i : i \subset [1, m]] \leq n$. In addition, the number of permutations in S_n with k cycles is precisely the definition of the *unsigned Stirling number of the first kind* $\begin{bmatrix} n \\ k \end{bmatrix}$. Therefore, we have $\Sigma_n = O\left(\sum_{k=1}^{n} (k^3 \log^2 n) \begin{bmatrix} n \\ k \end{bmatrix}\right)$ $= O\left(\log^2 n \sum_{k=1}^{n} k^3 \begin{bmatrix} n \\ k \end{bmatrix}\right)$. Therefore, it suffices to show that $\frac{1}{n!} \sum_{k=1}^{n} k^3 \begin{bmatrix} n \\ k \end{bmatrix} \sim c \log^3 n$ for a constant c. The proof can be found in the full version. □

Finally, we will now give our final main result of this section.

Theorem 3. *There exists a linear-time algorithm in the bit complexity model for solving a system of linear congruence equations when the input numbers are represented in unary.*

We now provide an algorithm that witnesses the above theorem. Let $\bigwedge_{i=1}^{m} x \equiv a_i \pmod{b_i}$ be the given system of equations. With unary representation of numbers, the size N_i of the equation $x \equiv a_i \pmod{b_i}$ is $a_i + b_i$. We use n to denote the total number of bits in the system of equations. Initially, we compute a binary representation of all the numbers a_i's, b_i's, and n as in the proof of Lemma 5, which takes linear time. Next we factorise all the numbers b_i into a product of distinct prime powers $p_{ji1}^{e_{i1}} \cdots p_{jit_i}^{e_{it_i}}$, where p_j stands for the jth prime and all e_{ij}'s are positive integers. This can be done in time $O(\sqrt{N_i}\log^2 N_i)$. To obtain this time bound, we can use any *unconditional*[1] deterministic factorisation methods like Strassen's algorithm, whose complexity was shown in [5] (cf. also see [12]) to be $O(f(N^{1/4}\log N))$ for factoring a number N, where $f(M)$ is the number of bit operations required to multiply two numbers with M bits. The standard (high-school) multiplication algorithm runs in quadratic time giving us $f(M) = O(M^2)$, which suffices for our purposes. This shows that Strassen's algorithm runs in time $O(N^{1/2}\log^2 N)$. [In practice, do factoring using the general number field sieve (cf. [11]), which performs extremely well in practice, though its complexity requires some unproven number-theoretic assumptions.]

Next, following Chinese Remainder Theorem (CRT), we compute $z_{ij} := a_i \bmod p_{ij}^{e_{ij}}$ for each $j \in [1, t_i]$. Let us analyse the time complexity for performing this. Each z_{ij} can be computed by a standard algorithm (e.g. see [11]) in time quadratic in the number of bits used to represent a_i and $p_{ij}^{e_{ij}}$. Since each of these numbers use at most $\log N_i$ bits, each z_i can be computed in time $O(\log^2 N_i)$, which is $o(N_i)$. In addition, since $e_{ij} > 1$ for each $j \in [1, t_i]$, it follows that $t_i = O(\log N_i)$. This means that the total time it takes to compute $\{z_{ij} : j \in [1, t_i]\}$ is $O(\log^3 N_i)$, which is also $o(N_i)$. So, computing this for all $i \in [1, m]$ takes time $O(\sum_{i=1}^{m} \log^3 N_i)$, which is at most linear in the input size.

In summary, for each $i \in [1, m]$, we obtained the following system of equations, which is equivalent to $x \equiv a_i \pmod{b_i}$ by CRT:

$$x \equiv z_{i1} \pmod{p_{i1}^{e_{i1}}} \quad \wedge \quad \cdots\cdots \quad \wedge \quad x \equiv z_{it_i} \pmod{p_{it_i}^{e_{it_i}}} \qquad (E_i)$$

The final step is to determine if there exists a number $x \in \mathbb{N}$ that satisfies *each* (E_i), for all $i \in [1, m]$. Loosely, we will go through all the equations and makes sure that there is no conflict between any two equations whose periods are powers of the same prime number, i.e., $x \equiv a \pmod{b}$ and $x \equiv a' \pmod{b'}$ such that $b = p^i$ and $b' = p^{i'}$ for some prime p and $i, i' \in \mathbb{Z}_{>0}$. In order to achieve this in linear-time in the bit complexity model, one has to store these equations in the memory (in the form of lookup tables) and carefully perform the lookup operations while looking for a conflict. To this end, we first compute $p_{\max} = \max\{p_{ij} : i \in [1, m], j \in [1, t_i]\}$ and $e_{\max} = \max\{e_{ij} : i \in [1, m], j \in [1, t_j]\}$.

[1] This means that the bound does not depend on any number-theoretic assumptions.

Lemma 7. *$p_{\max}$ and $e_{\max}$ can be computed using $O(n)$ many bit operations.*

Proof. The algorithm for computing $p_{\max}$ and $e_{\max}$ is a slight modification of the standard algorithm that computes the maximum number in a list, which sequentially goes through the list $n_1, \ldots, n_m$ while keeping the maximum number $n_{\max}$ in the sublist explored so far. To ensure linear-time complexity, we have to make sure that when comparing the values of n_i and $n_{\max}$, we explore at most n_i bits of $n_{\max}$ (since $n_{\max}$ is possibly much larger than n_i). This is easily achievable by assuming binary representation of these numbers *without redundant leading 0s*, e.g., the number 5 will be represented as 101, not 0101 or 00000101. That way, we will only need to inspect $\log(n_i)$ bits from $n_{\max}$ on the ith iteration, which will give a total running time of $O(\sum_{i=1}^{m} \log(n_i))$, which is linear in input size. □

Next, keep one 1-dimensional array A and one 2-dimensional array B:

$$A[1, \ldots, p_{\max}] \qquad B[1, \ldots, p_{\max}][1, \ldots, e_{\max}].$$

$A[k]$ and $B[k][e]$ will not be defined when k is not a prime number. We will use $A[k]$ as a flag indicating whether some equation of the form $x \equiv z \pmod{k^e}$ has been visited, in which case $A[k]$ will contain (z, e). In this case, we will use $B[k][e']$ (with $e' \leq e$) to store the value of $z \mod k^{e'}$.

We now elaborate how A and B are used when iterating over the equations in the system. Sequentially go through each system (E_i) of equations. For each $i \in [1, m]$, sequentially go through each equation $x \equiv z_{ij} \pmod{p_{ij}^{e_{ij}}}$, for each $j \in [1, t_i]$, and check if $A[p_{ij}]$ is defined. If it is not defined, set $A[p_{ij}] := (z_{ij}, e_{ij})$ and compute $B[p_{ij}][l] = z_{ij} \mod p^l$ for each $l \in [1, e_{ij}]$. If it is defined (say, $A[p_{ij}] = (z, e)$), then we analyse the constraints $x \equiv z \pmod{p_{ij}^e}$ and $x \equiv z_{ij} \pmod{p_{ij}^{e_{ij}}}$ simultaneously. We compare e and e_{ij} resulting in three cases:

Case 1. $e = e_{ij}$. In this case, make sure that $z = z_{ij}$ otherwise the two equations (and, hence, the entire system) cannot be satisfied simultaneously.

Case 2. $e < e_{ij}$. In this case, make sure that $z_{ij} \equiv z \pmod{p_{ij}^e}$ (otherwise, unsatisfiable) and assign $A[p_{ij}] := (z_{ij}, e_{ij})$. For each $l \in [1, e_{ij}]$, update $B[p_{ij}][l] := z_{ij} \mod p_{ij}^l$.

Case 3. $e > e_{ij}$. In this case, make sure that $z_{ij} \equiv z \pmod{p_{ij}^{e_{ij}}}$ (otherwise, unsatisfiable).

We now analyse the running time of this final step (i.e. when scanning through the subsystem (E_i)). To this end, we measure the time it takes to process each equation $x \equiv z_{ij} \pmod{p_{ij}^{e_{ij}}}$. There are two cases, which we will analyse in turn.

(Case I): when $A[p_{ij}]$ is not defined. In this case, setting $A[p_{ij}]$ takes constant time, while setting $B[p_{ij}][l]$ for all $l \in [1, e_{ij}]$ takes $O(e_{ij} \times (\log z_{ij} + \log p_{ij}^{e_{ij}})^2)$ since computing $a \mod b$ can be done in time quadratic in $\log(a) + \log(b)$. Since $e_{ij} \leq \log N_i$ and $z_{ij}, p_{ij} \leq N_i$, this expression can be simplified to $O(\log N_i \times \log^2(z_{ij} N_i p_{ij})) = O(\log^3 N_i)$.

(Case II): when $A[p_{ij}]$ is already defined, e.g., $A[p_{ij}] = (z, e)$. In this case, we will compare the values of e and e_{ij}. To ensure linear-time complexity, we will make sure that at most $\log(e_{ij})$ bits from e are read by using the trick from the proof of Lemma 7. For Case 1, we will need extra $O(\log z_{ij}) = O(\log N_i)$ time steps. For Case 2, we have $0 \leq z \leq p^{e_{ij}}$ and computing $z_{ij} \mod p_{ij}^e$ can be done in time $O(\log^2 N_i)$ as before. Updating $B[p_{ij}][l]$ for all $l \in [1, e_{ij}]$ takes $O(\log^3 N_i)$ as in the previous paragraph. For Case 3, since $e > e_{ij}$, we may access the value of $z \mod p_{ij}^{e_{ij}}$ from $B[p_{ij}][e_{ij}]$ in constant time and compare this with the value of z_{ij}. Since $z \in [0, p_{ij}^{e_{ij}})$, this takes time $O(\log N_i)$.

In summary, either case takes time at most $O(\log^3 N_i)$. Therefore, accounting for the entire subsystem (E_i), the algorithm incurs $O(\sum_{j=1}^{t_i} \log^3 N_i) = O(\log^4 N_i)$ time steps. Hence, accounting for *all* of the subsystems E_i $(i \in [1, m])$ the algorithm takes time $O(\sum_{i=1}^{m} \log^4 N_i)$, which is linear in the size of the input. This completes the proof of Theorem 3.

Remark 3. The purpose of the 2-dimensional array B above is to avoid superlinear time complexity for Case 3. We can imagine a system of linear equations $\bigwedge_{i=1}^{m} x \equiv a_i \pmod{b_i}$, where a_1 and b_1 are substantially larger than the other a_i's and b_i's $(i \in [2, m])$. In this case, without the lookup table B, checking whether $a_i \equiv a_1 \pmod{b_i}$ in Case 3 will require the algorithm to inspect the entire value of a_1, which prevents us from bounding the time complexity in terms of a_i and will yield a superlinear time complexity for our algorithm.

6 Future Work

We mention several future research avenues. Firstly, can we extend polynomial-solvability to any fixed number $k \in \mathbb{Z}_{>0}$ of group generators? The polynomial-time reduction in [10] from the graph isomorphism problem to the orbit problem requires an unbounded number of generators. In addition, the generalisation of the orbit problem over rational matrices to any fixed number k of matrices viewed as generators of (semi)groups is undecidable even when $k = 3, 4$, though results on polynomial-time solvability (hence, decidability) exist when the matrices commute (see [2] and references therein). So, polynomial-time solvability does not follow from the corresponding problem over matrices. The second problem concerns the constructive orbit problem over cyclic groups. Due to the lack of a target configuration $\mathbf{w} \in \Gamma^n$, our technique does not seem to apply directly in this case. In particular, we cannot simply use $\mathbf{w} \in \Gamma^n$ that is derived from the input configuration $\mathbf{v} \in \Gamma^n$ by separately finding the lexicographically minimum parts for each cycle in the given permutation, since this might render the system of equations insoluble.

Acknowledgment. We thank the anonymous referees for their helpful feedback. Part of the work was done when Lin was at Oxford supported by EPSRC (H026878). Zhou was supported by ARC (FT110100629).

References

1. Primorial Numbers (The On-Line Encyclopedia of Integer Sequences), http://oeis.org/A002110
2. Babai, L., Beals, R., Cai, J.-Y., Ivanyos, G., Luks, E.M.: Multiplicative equations over commuting matrices. In: SODA, pp. 498–507 (1996)
3. Babai, L., Luks, E.M.: Canonical labeling of graphs. In: STOC, pp. 171–183 (1983)
4. Bach, E., Shallit, J.: Algorithmic Number Theory. Foundations of Computing, vol. 1. MIT Press (1996)
5. Bostan, A., Gaudry, P., Schost, E.: Linear Recurrences with Polynomial Coefficients and Application to Integer Factorization and Cartier-Manin Operator. SIAM J. Comput. 36(6), 1777–1806 (2007)
6. Brualdi, R.A.: Combinatorial matrix classes. Encyclopedia of Mathematics and Its Applications, vol. 108. Cambridge University Press (2006)
7. Cameron, P.J.: Permutation Groups. London Mathematical Society Student Texts. Cambridge University Press (1999)
8. Chrobak, M.: Finite automata and unary languages. Theor. Comput. Sci. 47(3), 149–158 (1986)
9. Clarke, E.M., Emerson, E.A., Jha, S., Sistla, A.P.: Symmetry reductions in model checking. In: Hu, A.J., Vardi, M.Y. (eds.) CAV 1998. LNCS, vol. 1427, pp. 147–158. Springer, Heidelberg (1998)
10. Clarke, E.M., Jha, S., Enders, R., Filkorn, T.: Exploiting symmetry in temporal logic model checking. Formal Methods in System Design 9(1/2), 77–104 (1996)
11. Cormen, T.H., Leiserson, C.E., Rivest, R.L., Stein, C.: Introduction to Algorithms, 3rd edn. The MIT Press (2009)
12. Costa, E., Harvey, D.: Faster deterministic integer factorization. CoRR, abs/1201.2116 (2012)
13. Donaldson, A.F., Miller, A.: On the constructive orbit problem. Ann. Math. Artif. Intell. 57(1), 1–35 (2009)
14. Emerson, E.A., Sistla, A.P.: Symmetry and model checking. Formal Methods in System Design 9(1/2), 105–131 (1996)
15. Erdös, P., Graham, R.L.: On a linear diophantine problem of Frobenius. Acta Arithm. 21, 399–408 (1972)
16. Göller, S., Mayr, R., To, A.W.: On the computational complexity of verifying one-counter processes. In: LICS, pp. 235–244 (2009)
17. Hardy, G.H., Wright, E.M.: An Introduction to The Theory of Numbers, 6th edn. OUP Oxford (2008)
18. Ip, C.N., Dill, D.L.: Better verification through symmetry. Formal Methods in System Design 9(1/2), 41–75 (1996)
19. Kannan, R., Lipton, R.J.: Polynomial-time algorithm for the orbit problem. J. ACM 33(4), 808–821 (1986)
20. Sedgewick, R., Flajolet, P.: An Introduction to the Analysis of Algorithms, 2nd edn. Addison-Wesley Professional (2013)
21. Stockmeyer, L.J., Meyer, A.R.: Word problems requiring exponential time: Preliminary report. In: STOC, pp. 1–9 (1973)
22. To, A.W.: Unary finite automata vs. arithmetic progressions. Inf. Process. Lett. 109(17), 1010–1014 (2009)
23. Wahl, T., Donaldson, A.F.: Replication and abstraction: Symmetry in automated formal verification. Symmetry 2, 799–847 (2010)

A Nearly Optimal Upper Bound for the Self-Stabilization Time in Herman's Algorithm

Yuan Feng[1,2] and Lijun Zhang[3]

[1] Centre for Quantum Computation and Intelligent Systems, University of Technology Sydney, Australia
[2] AMSS-UTS Joint Research Laboratory for Quantum Computation, Chinese Academy of Sciences, Beijing, China
[3] State Key Laboratory of Computer Science, Institute of Software, Chinese Academy of Sciences, Beijing, China

Abstract. Self-stabilization algorithms are very important in designing fault-tolerant distributed systems. In this paper we consider Herman's self-stabilization algorithm and study its expected self-stabilization time. McIver and Morgan have conjectured the optimal upper bound being $0.148N^2$, where N denotes the number of processors. We present an elementary proof showing a bound of $0.167N^2$, a sharp improvement compared with the best known bound $0.521N^2$. Our proof is inspired by McIver and Morgan's approach: we find a nearly optimal closed form of the expected stabilization time for any initial configuration, and apply the Lagrange multipliers method to give an upper bound of it.

1 Introduction

In [2], Dijkstra proposed the influential notion of self-stabilization algorithms for designing fault-tolerant distributed systems. A distributed system is self-stabilizable if it will always reach *legitimate* configurations, no matter where the system starts. The system thus can recover from any transient error such as local corrupted states. The concept has many applications in the network protocol, and thus received much attention. See for example [14,3] for surveys on this topic.

Dijkstra assumed that all participating processors are identical except for a single processor which is necessary for breaking the symmetry. It is already shown by Dijkstra in 1974 that no deterministic scheduler exists which guarantees self-stabilization if all processors are identical. On the other side, Herman proposed a randomized program in [7] to break the symmetry: he proposed a self-stabilizing mutual exclusion algorithm, today known as Herman's algorithm, which stabilizes within finite steps with probability 1.

The protocol is designed for a *token ring* of N, N is odd, synchronous processors. Each processor may or may not have a token, and in a legitimate configuration only a single token exists. For any finite N, the protocol can be viewed as a finite state Markov chain with a single bottom strongly connected component (SCC) consisting of all legitimate configurations. So a legitimate configuration

P. Baldan and D. Gorla (Eds.): CONCUR 2014, LNCS 8704, pp. 342–356, 2014.

is reached with probability 1, regardless of the initial configuration. Hence, Herman's protocol is *self-stabilizing.*

Another important performance measure in designing self-stabilization protocols is the stabilization time which is the expected time until a legitimate configuration is reached. In Herman's original work [7], an upper bound $O(N^2\lceil\log N\rceil)$ for stabilization time has been established, while in 2005, several groups of researchers [6,12,13] gave an upper bound of $O(N^2)$, independently. Moreover, McIver and Morgan [12] proved that the stabilization time is actually $\Theta(N^2)$, meaning that the lower bound and upper bound coincide. They also provided a *precise* expected stabilization time for configurations with exactly three tokens.

One may expect that the story should end here from the viewpoint of complexity theory, as we already have the asymptotically tight bound for the stabilization time. However, McIver and Morgan [12] conjectured that the optimal upper bound for general configurations is $\frac{4}{27}N^2 \approx 0.148N^2$, which is obtained by equidistant three token configurations. This conjecture, simple and elegant, is indeed very difficult to prove. In recent years, it has attracted much attention to improve the bound towards this conjecture: Kiefer *et al.* [9] proved a bound of $0.64N^2$, and the authors of this paper further improved it to $0.521N^2$ [5], by simply exploiting the precise solution for the three token configurations derived in [12].

In this paper, we follow this research line by proving an upper bound of $\frac{1}{6}N^2$, approximately $0.167N^2$, for arbitrary configurations. Our bound is very close to the conjectured optimal bound, with a gap of $0.019N^2$. It is worth noting that our approach is completely elementary: for each initial configuration, we found a closed-form upper bound for the expected stabilization time, inspired by the three token formula given by McIver and Morgan. This bound is a function of the gap vector of the initial configuration, thus a multivariate function. Our result then follows by obtaining the maximum of the upper bounds over all initial configurations, using the Lagrange multipliers method.

Note that systems of interacting and annihilating particles, either on a circle or on a line, are heavily studied in areas including physics, combinatorics and neural networks [11]. Most of them focus on exploring the precise solutions, for example Balding [1] gives generating functions for the number of remaining particles at time t, and this results is transferred in [9] to Herman's setting. However, such expressions are in general very complex and difficult to analyze, see [1,4,9]. In contrast, our proof in this paper exploits mostly elementary concepts, and it is much simpler than previous techniques for analyzing Herman's algorithm [6,9]. Because of this, we are optimistic that our approach might provide alternative ways to improve worst-case analysis of such particle systems.

Related Work. In [9], an asynchronous variant of Herman's protocol is studied as well. Recently, [8] has studied the distribution of the self-stabilization time and shown that for an arbitrary t the probability of stabilization within time t is minimized under this configuration with $M = 3$. On the practical side, using the probabilistic model checker PRISM [10], McIver and Morgan's conjecture is validated for all rings with the size $N \leq 21$ that can be exhaustively analyzed.

2 Preliminaries

We assume to have N processors numbered from 0 to $N-1$, clockwise, with N odd, organized in a ring topology. Each processor may or may not have a token. A configuration with $0 < M \leq N$ tokens, M is odd, is a strictly increasing mapping $z : \{0, \ldots, M-1\} \to \{0, \ldots, N-1\}$ such that $z(0) < \cdots < z(M-1)$. For all $i \in \{0, \ldots, M-1\}$, the processor $z(i)$ has a token. We fix the ring size N throughout this paper.

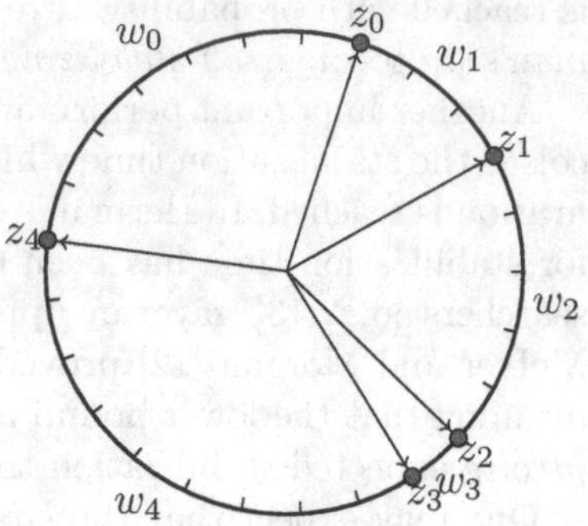

Fig. 1. A configuration with $M = 5$, $N = 25$

Herman's protocol [7] works as follows: in each time step, each processor with a token either passes its token to its clockwise neighbor with probability $\frac{1}{2}$, or keeps it with probability $\frac{1}{2}$. If a processor keeps its token and receives another one from its counterclockwise neighbor, then both of those tokens are annihilated. We refer to configurations with only one token as *legitimate* configurations. The protocol can also be viewed as a finite state Markov chain. It is easy to see that in this Markov chain there is a single bottom SCC consisting of all legitimate configurations. Thus this SCC is reached with probability 1, regardless of the initial configuration. It implies then that Herman's protocol is *self-stabilizing.*

Let S_M be the set of configurations with the number of tokens not exceeding M. Let $P_M : S_M \times S_M \to [0, 1]$ be the probabilistic transition matrix between configurations in S_M, and $\mathbb{E}_M : S_M \to [0, \infty)$ the function of expected stabilization time. The following lemma from [12], slightly modified with respect to our notations, is crucial for our discussion.

Lemma 1. *[12, Lemma 5] Let $M \geq 1$ and $v : S_M \to [0, \infty)$ be a mapping such that $v(z) = 0$ whenever $z \in S_1$ is a legitimate configuration. Suppose $(P_M \cdot v)(z) \leq v(z) - 1$ for any non-legitimate configuration z, where $P_M \cdot v$ is the mapping from S_M to $[0, \infty)$ such that*

$$(P_M \cdot v)(z) = \sum_{y \in S_M} P_M(z, y) v(y).$$

Then $\mathbb{E}_M(z) \leq v(z)$ for all $z \in S_M$.

Employing Lemma 1, McIver and Morgan were able to find a closed form for $\mathbb{E}_M$ when $M = 3$. To present their result, we need a further definition.

Definition 1 (Gap Vector). *Let $M \geq 3$ and $z \in S_M \backslash S_{M-2}$, i.e., it has exactly M tokens. We define the associated* gap vector $w = \langle w_0, w_1, \ldots, w_{M-1} \rangle$ *of z, where w_i is the gap between the tokens $z(i-1)$ and $z(i)$ defined by $w_i := z(i) - z(i-1)$ for $i = 1, \ldots, M-1$, and $w_0 = N - \sum_{i=1}^{M-1} w_i$. We denote by G_M, $M \geq 3$, the set of gap vectors corresponding to configurations from S_M, and set $G_1 = \{\langle N \rangle\}$.*

Obviously, configurations with the same gap vector have the same expected stabilization time. In other words, the value $\mathbb{E}_M(z)$ depends only on the gap vector w associated with z.

Lemma 2. *[12, Lemma 7] For any $z \in S_3$, let $w = \langle w_0, w_1, w_2 \rangle$ be the gap vector of z. Then $\mathbb{E}_3(z) = 4w_0w_1w_2/N$.*

In this paper, we will further dig the potential of Lemma 1 to give a (nearly optimal) bound on $\mathbb{E}_M$ for the general case $M \geq 3$.

3 Our Main Result

To simplify notations, we sometimes extend gap vectors, which have finite dimension, to infinite ones by appending 0 entries. That is, we let $w_i = 0$ for all $i \geq M$ if w is a gap vector of dimension M. The following definition is crucial.

Definition 2. *Let $G = \bigcup_{M=1, M \text{ is odd}}^{N} G_M$ and $F : G \to [0, \infty)$ be a mapping defined by*

$$F(\langle w_0, w_1, \cdots, w_{M-1} \rangle) = \sum_{i=0}^{\infty} w_i \cdot \left[\sum_{j=0}^{\infty} w_{i+2j+1} \cdot \left(\sum_{k=0}^{\infty} w_{i+2j+2k+2} \right) \right]. \quad (1)$$

With this definition, we can now state the main result of this paper.

Theorem 1. *For any $z \in S_M$ with the associated gap vector w,*

$$\mathbb{E}_M(z) \leq \frac{4}{N} F(w). \quad (2)$$

We can further apply the Lagrange multipliers method to compute the maximal value of $\mathbb{E}_M(z)$ for each $M \leq N$, which provides a better upper bound $\frac{1}{6}N^2 = 0.167N^2$, over the previous known bound $0.521N^2$ [5], of the expected self-stabilization time for arbitrary initial configurations (cf. Theorem 2).

The proof of Theorem 1 will be presented in the next section. But first, we apply it for some small values of M.

- $M = 3$. Then $F(\langle w_0, w_1, w_2 \rangle) = w_0w_1w_2$, and Eqn.(2) agrees with the precise bound in Lemma 2.
- $M = 5$. Then $F(w)$ equals the sum of all the products of three *neighboring gaps*:

$$F(\langle w_0, w_1, w_2, w_3, w_4 \rangle) = w_0w_1w_2 + w_1w_2w_3 + w_2w_3w_4 + w_3w_4w_0 + w_4w_0w_1 \quad (3)$$

- $M = 7$. In this case, $F(w)$ is already involved. It contains the sum of all the products of three neighboring gaps, and in addition it contains products of gaps of the form $w_iw_{i+3}w_{i+4}$. Here if we assume all arithmetic operations over the index set $\{0, \ldots, M-1\}$ are understood as modulo 7, then it can be written as:

$$F(\langle w_0, w_1, w_2, w_3, w_4, w_5, w_6 \rangle) = \sum_{i=0}^{6} w_i w_{i+1} w_{i+2} + \sum_{i=0}^{6} w_i w_{i+3} w_{i+4} \,.$$

– The explicit expressions for $M > 7$ are even more involved. It is still the sum of some products of three (not necessarily neighboring) gaps, but the pattern becomes more and more complicated. For example, products of the form $w_i w_{i+\frac{N}{3}} w_{i+\frac{2N}{3}}$ will be needed for those N which are multiples of 3.

To prove the main theorem, we first need to introduce some notation.

Definition 3. *For any configuration $z \in S_M$, we denote by $O(z)$ the bag of next-step configurations obtained from z; that is, $O(z) = \{y \in S_M : P_M(z, y) > 0\}$. Let $O_g(z)$ be the bag of gap vectors for $O(z)$; that is*

$$O_g(z) = \{w : w \text{ is the gap vector for some } y \in O(z)\}.$$

Here by bag we mean a multiset where an element can appear more than once. For simplicity, we use the set notation $\{\cdot\}$ to denote bags as well.

Actually, $O_g(z)$ is almost an ordinary set except that the gap vector associated to z occurs twice, one corresponding to the case where all tokens move, and the other where no token moves.

Note that in our setting, for each $z \in S_M \backslash S_{M-2}$, $M \geq 3$, and $y \in O(z)$, the probability $P_M(z, y)$ is always $\frac{1}{2^M}$. Let F_M^g be the function obtained by composing F with the gap function, restricting on the set of M-token configurations; that is, for any $z \in S_M \backslash S_{M-2}$, $F_M^g(z) = F(w)$ where w is the gap vector of z. Then

$$(P_M \cdot \frac{4}{N} F_M^g)(z) = \frac{4}{2^M N} \sum_{y \in O(z)} F_M^g(y) = \frac{4}{2^M N} \sum_{v \in O_g(z)} F(v).$$

The proof of our main theorem will exploit the definition of F to derive a closed form for the sum $\sum_{v \in O_g(z)} F(v)$, which is the most challenging part. With that we will be able to show

$$(P_M \cdot \frac{4}{N} F_M^g)(z) \leq \frac{4}{N} F_M^g(z) - 1$$

for all non-legitimate configuration z, and the main theorem follows from Lemma 1.

4 Proof of the Main Theorem

4.1 The 5-Token Case

To illustrate our basic ideas, let us first consider the case of 5 tokens. The function F is given in Eqn.(3), which has obviously the following properties:

- F is *rotationally symmetric*, i.e., $F(\langle w_0, \ldots, w_4 \rangle) = F(\langle w_1, w_2, w_3, w_4, w_0 \rangle)$.
- F is in *harmony* for smaller $M < 5$, i.e., assuming $w_1 = 0$,

$$F(\langle w_0, w_1, w_2, w_3, w_4 \rangle) = F(\langle w_0 + w_2, w_3, w_4 \rangle).$$

  Thus, we can freely use the 5-token formula for all 3-token configurations as well, and we will not distinguish a 5-dimensional integer vector with some of the elements being 0 with the 3-token or 1-token configuration it really represents.

These two properties will be extended for arbitrary M, and they will be exploited to prove our main theorem.

We define the one-step *gap increment vectors* for a 5-token configuration as follows.

1. Let $\Delta_1 = \langle 1, -1, 0, 0, 0 \rangle$, which corresponds to the first token passing while the others remaining. Obviously, the cases where a single token passes while the others remain can be obtained by applying Per^i to Δ_1^T, where $i \in \{0, 1, 2, 3, 4\}$ and
$$Per = \begin{pmatrix} 0&0&0&0&1 \\ 1&0&0&0&0 \\ 0&1&0&0&0 \\ 0&0&1&0&0 \\ 0&0&0&1&0 \end{pmatrix}$$
is the basic cyclic permutation matrix.
2. Let $\Delta_{2,1} = \langle 1, 0, -1, 0, 0 \rangle$, corresponding to the first two tokens passing while the others remaining, and $\Delta_{2,2} = \langle 1, -1, 1, -1, 0 \rangle$, corresponding to the first and the third tokens passing while the others remaining. Other cases where exactly 2 tokens passing can be obtained by applying the cyclic permutation matrices to either $\Delta_{2,1}$ or $\Delta_{2,2}$.
3. Let $\Delta_0 = \langle 0, 0, 0, 0, 0 \rangle$, corresponding to the case that no token, or all, moves.

Observe that the case of exactly 3 tokens passing is equivalent to exactly 2 passing, but in the opposite direction. Similar correspondence holds for exactly 1 or 4 tokens passing. Thus all the possible outcomes of a single step starting from a non-legitimate configuration $z \in S_5$ with the gap vector $w = (w_0, \cdots, w_4)$ constitute the set

$$O_g(z) = \{w \pm \Delta_0, w \pm Per^i \Delta_1^T, w \pm Per^i \Delta_{2,1}^T, w \pm Per^i \Delta_{2,2}^T : i = 0, 1, 2, 3, 4\}$$

where each element occurs with probability 1/32 (here we recall $O_g(z)$ is a bag, and $w + \Delta_0 = w - \Delta_0$). Since $F(v)$ is in harmony, in case some gaps in $v \in O_g(z)$ are equal to 0, which corresponds to a 3 or 1 token configuration, we can still use the 5-token formula.

To calculate the value $\sum_{v \in O_g(z)} F(v)$, we let

$$\square_1^i := F(w + Per^i \Delta_1^T) + F(w - Per^i \Delta_1^T)$$

for $i = 0, 1, 2, 3, 4$, and $\square_{2,1}^i$ and $\square_{2,2}^i$ be defined similarly. Note that

$$(w_0 + 1)(w_1 - 1)w_2 + (w_0 - 1)(w_1 + 1)w_2 = 2w_0w_1w_2 - 2w_2.$$

We have $\square_1^0 = 2F(w) - 2w_2 - 2w_4$. Moreover, as $F(w)$ is rotationally symmetric, and $\sum_{i=0}^4 w_i = N$, we derive $\sum_{i=0}^4 \square_1^i = 10F(w) - 4N$. In a similar way, we have $\square_{2,1}^0 = 2F(w) - 2w_1$ and $\sum_{i=0}^4 \square_{2,1}^i = 10F(w) - 2N$. The case for $\Delta_{2,2}$ is slightly complicated: the sum $\square_{2,2}^0$ can be first simplified to

$$\begin{gathered}(w_1 - 1)(w_2 + 1)(w_0 + w_3) + (w_2 + 1)(w_3 - 1)w_4 + (w_3 - 1)w_4(w_0 + 1)\\ + \quad w_4(w_0 + 1)(w_1 - 1) \quad + \quad w_4(w_0 - 1)(w_1 + 1)\\ (w_1 + 1)(w_2 - 1)(w_0 + w_3) + (w_2 - 1)(w_3 + 1)w_4 + (w_3 + 1)w_4(w_0 - 1)\end{gathered}$$

Thus $\square_{2,2}^0 = 2F(w) - 2(w_0 + w_3) - 6w_4$, and $\sum_{i=0}^4 \square_{2,2}^i = 10F(w) - 10N$. Finally, noting $F(w + \Delta_0) = F(w - \Delta_0) = F(w)$, we have $\sum_{v \in O_g(z)} F(v) = 32F(w) - 16N$. Thus

$$(P_5 \cdot \frac{4}{N} F_5^g)(z) = \frac{4}{32N}(32F(w) - 16N) = \frac{4}{N}F(w) - 2 \leq \frac{4}{N}F_5^g(z) - 1,$$

and Lemma 1 implies $\mathbb{E}_5(z) \leq \frac{4}{N} \cdot F_5^g(z)$. Using Lagrange multipliers method (cf. Theorem 2), we have then $\mathbb{E}_5(z) \leq \frac{4}{N} \cdot \frac{1}{25}N^3 = \frac{4}{25}N^2 = 0.16N^2$.

4.2 Properties of the Function F

For $M = 5$, we have seen that F is rotationally symmetric and in harmony for smaller values of M. Below we generalize these two properties for arbitrary M.

Lemma 3. *[Rotational Symmetricity] The function F is rotationally symmetric. That is, for any odd number $M \geq 3$,*

$$F(\langle w_0, w_1, \cdots, w_{M-1}\rangle) = F(\langle w_1, \cdots, w_{M-1}, w_0\rangle).$$

Proof. Let $w = \langle w_0, w_1, \cdots, w_{M-1}\rangle$ and $w' = \langle w_1, w_2, \cdots, w_{M-1}, w_0\rangle$. We need to prove $F(w) = F(w')$. Note that by Eqn.(1),

$$\begin{aligned} F(w) &= \sum_{i=0}^{M-3} w_i \sum_{j=0}^{\infty} w_{i+2j+1} \sum_{k=0}^{\infty} w_{i+2j+2k+2} \\ &= \sum_{i=0}^{M-3} w_i \sum_{j=0}^{\lfloor (M-3-i)/2 \rfloor} w_{i+2j+1} \sum_{k=0}^{\lfloor (M-3-i-2j)/2 \rfloor} w_{i+2j+2k+2}. \end{aligned}$$

The proof idea is to divide the sum above into two parts, for even and odd index i, respectively. Then we can see the relation of $F(w)$ and $F(w')$ by shifting the indices. For this purpose, we denote by

$$\Sigma_1(w) := \sum_{n=1}^{(M-3)/2} w_{2n-1} \sum_{j=0}^{(M-3-2n)/2} w_{2n+2j} \sum_{k=0}^{(M-3-2n-2j)/2} w_{2n+2j+2k+1} \quad (4)$$

$$\Sigma_2(w) := \sum_{n=0}^{(M-3)/2} w_{2n} \sum_{j=0}^{(M-3-2n)/2} w_{2n+2j+1} \sum_{k=0}^{(M-3-2n-2j)/2} w_{2n+2j+2k+2}. \quad (5)$$

Then $F(w) = \Sigma_1(w) + \Sigma_2(w)$. Note that $M-1$ is an even number, and w'_i equals w_{i+1} if $i < M-1$, and equals w_0 if $i = M-1$. For the gap vector w', we calculate that

$$\Sigma_1(w') = \sum_{n=1}^{(M-3)/2} w_{2n} \sum_{j=0}^{(M-3-2n)/2} w_{2n+2j+1} \sum_{k=0}^{(M-3-2n-2j)/2} w_{2n+2j+2k+2}$$
$$= \Sigma_2(w) - w_0 \sum_{j=0}^{(M-3)/2} w_{2j+1} \sum_{k=0}^{(M-3-2j)/2} w_{2j+2k+2}.$$

The most involved part is the sum $\Sigma_2(w')$. Note $k = (M-3-2n-2j)/2$ implies $w'_{2n+2j+2k+2} = w'_{M-1}$. Isolating the term of w'_{M-1} from the last part of $\Sigma_2(w')$, we derive:

$$\Sigma_2(w') = \sum_{n=0}^{(M-5)/2} w'_{2n} \sum_{j=0}^{(M-5-2n)/2} w'_{2n+2j+1} \sum_{k=0}^{(M-5-2n-2j)/2} w'_{2n+2j+2k+2}$$
$$+ \sum_{n=0}^{(M-3)/2} w'_{2n} \sum_{j=0}^{(M-3-2n)/2} w'_{2n+2j+1} \cdot w'_{M-1}.$$

Some subtle simplifications have been used above: the case $n = (M-3)/2$ implies $(M-3-2n)/2 = 0$ and $(M-3-2n-2j)/2 = 0$ as well, thus the corresponding term $w'_{M-3}w'_{M-2}w'_{M-1}$ appears in the sum in the last line. Similar with the case $j = (M-3-2n)/2$. Now we can further rewrite $\Sigma_2(w')$ by:

$$\Sigma_2(w') = \sum_{n=1}^{(M-3)/2} w_{2n-1} \sum_{j=0}^{(M-3-2n)/2} w_{2n+2j} \sum_{k=0}^{(M-3-2n-2j)/2} w_{2n+2j+2k+1}$$
$$+ w_0 \sum_{n=0}^{(M-3)/2} w_{2n+1} \sum_{j=0}^{(M-3-2n)/2} w_{2n+2j+2}$$
$$= \Sigma_1(w) + w_0 \sum_{j=0}^{(M-3)/2} w_{2j+1} \sum_{k=0}^{(M-3-2j)/2} w_{2j+2k+2}.$$

Thus we have $F(w') = \Sigma_1(w') + \Sigma_2(w') = \Sigma_1(w) + \Sigma_2(w) = F(w)$. □

Remark 1. We could also define the function F in Definition 2 in a rotationally symmetric way directly by, say, letting the arithmetic operations over indices be modulo M. This would save our efforts to prove Lemma 3. However, we decided to adopt the current definition for the following two reasons:

1. This definition makes the proof of Lemma 4 easier to follow;
2. The generating set $C(M)$ of the gap increment vectors in the next section is constructed inductively (Proposition 1), which is in harmony with the current definition of F, and makes the proof of the main theorem easy to follow as well.

The following lemma shows that the definition of F is in harmony for arbitrary M.

Lemma 4. *For any odd number $M \geq 3$, if $w_1 = 0$ then*

$$F(\langle w_0, w_1, w_2, \cdots, w_{M-1}\rangle) = F(\langle w_0 + w_2, w_3, \cdots, w_{M-1}\rangle).$$

Proof. The equality is obtained by directly expanding both sides according to Eqn.(1), by noting that $w_1 = 0$:

$$\begin{aligned}
F(\langle w_0, w_1, w_2, \cdots, w_{M-1}\rangle) &= \sum_{i=0}^{\infty} w_i \cdot \left[\sum_{j=0}^{\infty} w_{i+2j+1} \cdot \left(\sum_{k=0}^{\infty} w_{i+2j+2k+2}\right)\right] \\
= w_0 \cdot \left[\sum_{j=0}^{\infty} w_{2j+1} \cdot \left(\sum_{k=0}^{\infty} w_{2j+2k+2}\right)\right] &+ w_2 \cdot \left[\sum_{j=0}^{\infty} w_{2j+3} \cdot \left(\sum_{k=0}^{\infty} w_{2j+2k+4}\right)\right] \\
&+ \sum_{i=3}^{\infty} w_i \cdot \left[\sum_{j=0}^{\infty} w_{i+2j+1} \cdot \left(\sum_{k=0}^{\infty} w_{i+2j+2k+2}\right)\right] \\
= (w_0 + w_2) \cdot \left[\sum_{j=0}^{\infty} w_{2j+3} \cdot \left(\sum_{k=0}^{\infty} w_{2j+2k+4}\right)\right] \\
&+ \sum_{i=3}^{\infty} w_i \cdot \left[\sum_{j=0}^{\infty} w_{i+2j+1} \cdot \left(\sum_{k=0}^{\infty} w_{i+2j+2k+2}\right)\right] \\
= F(\langle w_0 + w_2, w_3, \cdots, w_{M-1}\rangle).
\end{aligned}$$

□

As the function F is rotationally symmetric, the above lemma indeed shows that *any* 0 entry in the gap vectors can be absorbed, without affecting the value of the F function.

4.3 Gap Increment Vector

In this section, we characterize the vectors in $O_g(z)$ with the help of gap increment vectors.

Definition 4 (Gap Increment Vector). *Let z be a configuration with w its associated gap vector. The vectors $\Delta := w' - w$, where $w' \in O_g(z)$, are called the* gap increment vector *for z.*

Moreover, as seen in the 5-token case, the set of gap increment vectors consists of pairs of *symmetric* ones:

Lemma 5. *For any gap increment vector Δ for z, both $w + \Delta$ and $w - \Delta$ are in $O_g(z)$.*

Proof. By definition, $w' := w + \Delta \in O_g(z)$. The gap vector w' is obtained from w by moving a set A of tokens forward. By symmetry, the vector $w - \Delta$ is obtained if all tokens in A stay, but other tokens move forward. □

Let $C(M)$ be a subset of gap increment vectors for M tokens such that for each non-legitimate $z \in S_M \backslash S_{M-2}$,

$$O_g(z) = \{w \pm \Delta : \Delta \in C(M)\}.$$

Without loss of generality, we assume every vector in $C(M)$ has the first entry being either 0 or 1. We would like to construct $C(M)$ in an inductive way.

When $M = 1$, obviously $C(M) = \{\langle 0 \rangle\}$. Let $z \in S_M \backslash S_{M-2}$ be a configuration with $M \geq 3$ tokens, and $w = \langle w_0, w_1, \cdots, w_{M-1} \rangle$ the associated gap vector. We first ignore the first two tokens and consider the $M - 2$ token configuration z' with gap vector $w' = \langle w_0 + w_1 + w_2, w_3, \cdots, w_{M-1} \rangle$. For each $v' \in O_g(z')$ with $v' = w' + \Delta'$ and $\Delta' \in C(M-2)$, we need to consider two cases:

1. $v'_0 = w'_0$. That is, the first gap of w' does not change. Come back to the original vector w. There are four gap vectors $v \in O_g(z)$ corresponding to this case: (i) $v_i = w_i$ for each $i = 0, 1, 2$; (ii) $v_0 = w_0$, $v_1 = w_1 + 1$, and $v_2 = w_2 - 1$; (iii) $v_0 = w_0 + 1$, $v_1 = w_1 - 1$, and $v_2 = w_2$; (iv) $v_0 = w_0 + 1$, $v_1 = w_1$, and $v_2 = w_2 - 1$. That is, corresponding to each increment vector $\Delta' \in C(M-2)$ with $\Delta'_0 = 0$, there are four increment vectors $\Delta \in C(M)$ obtained from Δ' by replacing Δ'_0 with the three-element vectors $\langle 0, 0, 0 \rangle$, $\langle 0, 1, -1 \rangle$, $\langle 1, -1, 0 \rangle$, and $\langle 1, 0, -1 \rangle$, respectively.
2. $v'_0 = w'_0 + 1$. That is, the first gap of w' increases by 1. Similar to the first case, we have for each increment vector $\Delta' \in C(M-2)$ with $\Delta'_0 = 1$, there are four increment vectors $\Delta \in C(M)$ obtained from Δ' by replacing Δ'_0 by the three-element vectors $\langle 0, 0, 1 \rangle$, $\langle 0, 1, 0 \rangle$, $\langle 1, -1, 1 \rangle$, and $\langle 1, 0, 0 \rangle$, respectively.

The items 1 and 2 above actually give us an inductive way to construct $C(M)$, $M \geq 3$, from $C(M-2)$:

Proposition 1. *Let $C(M)$ be defined above. Then $C(1) = \{\langle 0 \rangle\}$, and for any odd number $M \geq 3$,*

$$C(M) = A^\frown C^0(M-2) \cup B^\frown C^1(M-2)$$

where the operation $^\frown$ means the element-wise concatenation of vectors,

$$C^i(M-2) = \{\langle \Delta_1, \ldots, \Delta_{M-3} \rangle : \langle i, \Delta_1, \ldots, \Delta_{M-3} \rangle \in C(M-2)\}$$

for $i = 0, 1$, and

$$A := \{\langle 0,0,0 \rangle, \langle 0,1,-1 \rangle, \langle 1,-1,0 \rangle, \langle 1,0,-1 \rangle\}$$
$$B := \{\langle 0,0,1 \rangle, \langle 0,1,0 \rangle, \langle 1,-1,1 \rangle, \langle 1,0,0 \rangle\}.$$

For example, applying the above proposition, we have $C(3) = A$, and $C(5)$ is the union of the following two sets:

$$A^\frown C^0(3) = \left\{\begin{array}{l} \langle 0, \ \ 0, \ \ 0, 0, \ \ 0\rangle, \\ \langle 0, \ \ 1, -1, 0, \ \ 0\rangle, \\ \langle 1, -1, \ \ 0, 0, \ \ 0\rangle, \\ \langle 1, \ \ 0, -1, 0, \ \ 0\rangle, \\ \langle 0, \ \ 0, \ \ 0, 1, -1\rangle, \\ \langle 0, \ \ 1, -1, 1, -1\rangle, \\ \langle 1, -1, \ \ 0, 1, -1\rangle, \\ \langle 1, \ \ 0, -1, 1, \ -1\rangle \end{array}\right\}; \ B^\frown C^1(3) = \left\{\begin{array}{l} \langle 0, \ \ 0, 1, -1, \ \ 0\rangle, \\ \langle 0, \ \ 1, 0, -1, \ \ 0\rangle, \\ \langle 1, -1, 1, -1, \ \ 0\rangle, \\ \langle 1, \ \ 0, 0, -1, \ \ 0\rangle, \\ \langle 0, \ \ 0, 1, \ \ 0, -1\rangle, \\ \langle 0, \ \ 1, 0, \ \ 0, -1\rangle, \\ \langle 1, -1, 1, \ \ 0, -1\rangle, \\ \langle 1, \ \ 0, 0, \ \ 0, -1\rangle \end{array}\right\}.$$

Obviously, the cardinality of $C(M)$ is 2^{M-1}.

4.4 Properties of Gap Increment Vectors

As for the gap vectors, in the following, when the index exceeds $M-1$, we always assume 0 entries for the gap increment vectors. That is, we let $w_i = 0$ and $\Delta_i = 0$ for all $i \geq M$ if $w = (w_0, \cdots, w_{M-1})$ and $\Delta = (\Delta_0, \cdots, \Delta_{M-1})$. The following two lemmas state properties about sums of increment vectors, that will be used to simplify the sum $\sum_{v \in O_g(z)} F(v)$ later.

Lemma 6. *For any odd number $M \geq 3$,*

$$\sum_{\Delta \in C(M)} \Delta_1 \sum_{k=0}^{\infty} \Delta_{2k+2} = -2^{M-3}. \tag{6}$$

Proof. The lemma is proved by dividing the sum according to the recursive definition of the gap increment vector. Precisely, $\sum_{\Delta \in C(M)} \Delta_1 \sum_{k=0}^{\infty} \Delta_{2k+2}$ equals

$$\begin{aligned} &\sum_{\Delta' \in C^0(M-2)} 1 \cdot \left(\sum_{k=0}^{\infty} \Delta'_{2k+1} - 1 \right) + \sum_{\Delta' \in C^0(M-2)} (-1) \cdot \sum_{k=0}^{\infty} \Delta'_{2k+1} \\ &+ \sum_{\Delta' \in C^1(M-2)} (-1) \cdot \left(\sum_{k=0}^{\infty} \Delta'_{2k+1} + 1 \right) + \sum_{\Delta' \in C^1(M-2)} \sum_{k=0}^{\infty} \Delta'_{2k+1} \\ &= -|C^0(M-2)| - |C^1(M-2)| \\ &= -|C(M-2)| = -2^{M-3}. \end{aligned}$$

□

Lemma 7. *For any odd number $M \geq 1$,*

$$\sum_{\Delta \in C(M)} \sum_{j=0}^{\infty} \sum_{k=0}^{\infty} \Delta_{2j+1} \Delta_{2j+2k+2} = -(M-1)2^{M-4}. \tag{7}$$

Proof. Let $T(M)$ be the LHS of Eqn.(7).We prove by induction that $T(M) = -(M-1)2^{M-4}$. The result is obvious for $M = 1$. Suppose now that Eqn.(7) holds for $M-2$, $M \geq 3$. Then we have from Lemma 6 that

$$\begin{aligned} T(M) &= \sum_{\Delta \in C(M)} \Delta_1 \sum_{k=0}^{\infty} \Delta_{2k+2} + \sum_{\Delta \in C(M)} \sum_{j=1}^{\infty} \sum_{k=0}^{\infty} \Delta_{2j+1} \Delta_{2j+2k+2} \\ &= -2^{M-3} + 4 \cdot \sum_{\Delta \in C(M-2)} \sum_{j=0}^{\infty} \sum_{k=0}^{\infty} \Delta_{2j+1} \Delta_{2j+2k+2} \\ &= -2^{M-3} - 4(M-3)2^{M-6} = -(M-1)2^{M-4}. \end{aligned}$$

□

4.5 Proof of the Main Theorem

We are now ready to prove the main theorem. First we give a closed form for the sum $\sum_{v \in O_g(z)} F(v)$.

Lemma 8. *For any non-legitimate configuration* $z \in S_M \backslash S_{M-2}$ *with gap vector* w,

$$\sum_{v \in O_g(z)} F(v) = 2^M F(w) - (M-1)2^{M-3} N.$$

Proof. First note that

$$\begin{aligned} \sum_{v \in O_g(z)} F(v) &= \sum_{\Delta \in C(M)} [F(w+\Delta) + F(w-\Delta)] \\ &= \sum_{\Delta \in C(M)} \sum_{i=0}^{M-3} \sum_{j=0}^{\infty} \sum_{k=0}^{\infty} [(w_i + \Delta_i)(w_{i+2j+1} + \Delta_{i+2j+1})(w_{i+2j+2k+2} + \Delta_{i+2j+2k+2}) \\ &\quad + (w_i - \Delta_i)(w_{i+2j+1} - \Delta_{i+2j+1})(w_{i+2j+2k+2} - \Delta_{i+2j+2k+2})]. \end{aligned}$$

On the other hand, a simple calculation shows that for any a, b, c and x, y, z,

$$(a+x)(b+y)(c+z) + (a-x)(b-y)(c-z) = 2abc + 2xyc + 2xzb + 2yza$$

Thus we have

$$\sum_{v \in O_g(z)} F(v) = \sum_{\Delta \in C(M)} 2F(w) + \sum_{i=0}^{M-1} A_{w_i} w_i$$

where A_{w_i} is the coefficient of w_i. Using Lemma 7 we compute the coefficient A_{w_0} of w_0 as

$$A_{w_0} = \sum_{\Delta \in C(M)} 2 \cdot \sum_{j=0}^{\infty} \sum_{k=0}^{\infty} \Delta_{2j+1} \Delta_{2j+2k+2} = -(M-1)2^{M-3}.$$

As the function F is rotationally symmetric, we derive that

$$\sum_{v \in O_g(z)} F(v) = \sum_{\Delta \in C(M)} 2F(w) - (M-1)2^{M-3} \sum_{i=0}^{M-1} w_i$$
$$= 2^M F(w) - (M-1)2^{M-3} N.$$

□

Proof of the Main Theorem. From Lemma 8, we have that for any non-legitimate configuration $z \in S_M \backslash S_{M-2}$ with gap vector w,

$$(P_M \cdot \frac{4}{N} F^g_M)(z) = \frac{4}{2^M N} \sum_{v \in O_g(z)} F(v) = \frac{4}{N} F(w) - \frac{M-1}{2} \leq \frac{4}{N} F^g_M(z) - 1. \tag{8}$$

Thus, Lemma 1 implies that $\mathbb{E}_M(z) \leq \frac{4}{N} F^g_M(z) = \frac{4}{N} F(w)$. □

5 A Nearly Optimal Upper Bound

In our main theorem, we derived an upper bound for the stabilization time $\mathbb{E}_M(z)$, which is given in terms of the function $F(w)$. Furthermore, using the method of Lagrange multipliers, we can derive a nearly optimal upper bound which is independent of the initial configurations.

Theorem 2. *1. For all N and odd number $3 \leq M \leq N$, we have*

$$\max_{z \in S_M} \mathbb{E}_M(z) \leq \frac{N^2}{6} \cdot \left(1 - \frac{1}{M^2}\right).$$

2. For all N and for all initial configurations, we have $\mathbb{E}\mathbf{T} \leq \frac{1}{6} N^2$.

Proof. Item 2 is direct from Item 1. For Item 1, it suffices to show that for any $z \in S_M$ with gap vector w,

$$F(w) \leq u(M) := \frac{N^3}{24} \cdot \left(1 - \frac{1}{M^2}\right).$$

First, we use the method of Lagrange multipliers to find the critical point of $F(w)$ with the constraints $w_i \geq 0$ for each i, and $\sum_{i=0}^{M-1} w_i = N$. Here we do not require values of w_i being integers any more; they can be any nonnegative real numbers. Let

$$f(w) = F(w) + \lambda \left(\sum_{i=0}^{M-1} w_i - N \right).$$

We calculate the gradient equations for w_0 and w_2 as

$$\frac{\partial f}{\partial w_0} = \sum_{j=0}^{\infty} w_{2j+1} \sum_{k=0}^{\infty} w_{2j+2k+2} + \lambda$$

$$\frac{\partial f}{\partial w_2} = \sum_{j=0}^{\infty} w_{2j+3} \sum_{k=0}^{\infty} w_{2j+2k+4} + w_0 w_1 + w_1 \sum_{k=0}^{\infty} w_{2k+3} + \lambda.$$

By letting $\frac{\partial f}{\partial w_0} = \frac{\partial f}{\partial w_2} = 0$ and noting that $\sum_{i=0}^{M-1} w_i = N$, we derive directly:

$$w_2 + w_4 + \cdots + w_{M-1} = \frac{N - w_1}{2} \tag{9}$$

$$w_1 + w_3 + \cdots + w_{M-2} = \frac{N + w_1}{2} - w_0. \tag{10}$$

Since F is rotationally symmetric, we can derive from Eqn.(10) that

$$w_2 + w_4 + \cdots + w_{M-1} = \frac{N + w_2}{2} - w_1. \tag{11}$$

Thus $w_1 = w_2$ from Eqs.(9) and (11). By the rotational symmetry of F again, we have $w_0 = w_1 = \cdots = w_{M-1} = N/M$. Denote by w^* this (unique) critical point. Then $F(w^*) = u(M) = \frac{N^3}{24} \cdot (1 - \frac{1}{M^2})$ from Eqs.(4) and (5).

On the other hand, note that $F(w)$ is a continous multivariate function and

$$R(M) := \{w \in \mathbf{R}^M \mid w_i \geq 0, \sum_{i=0}^{M-1} w_i = N\}$$

is a compact set. It follows that $F(w)$ has a global maximum in $R(M)$. For any $w' \in R(M)$ which achieves this global maximum, if w' is an interior point of $R(M)$, then it must be a critical point. Thus $w^* = w'$, and as a result, $F(w^*) = u(M)$ is the global maximum of $F(w)$ in $R(M)$ (and so in $G(M)$). Then the theorem follows.

We now argue that w' is indeed an interior point of $R(M)$. Otherwise, w' must have some zero elements. By deleting all zero elements from w', we get a vector w'' which lies in the interior of $R(M')$ for some $M' < M$. Thus $F(w'') = F(w')$ is the global maximum of $F(w)$ in $R(M')$, so w'' is a critical point, and $F(w'') = u(M')$. From the fact that $u(M)$ is a strictly increasing function, we have

$$F(w') = F(w'') = u(M') < u(M),$$

contraditing the assumption that w' achieves the global maximum of F in $R(M)$. □

6 Conclusion and Future Work

It is conjectured that $\frac{4}{27}N^2$ is the tight upper bound of Herman's self-stabilization algorithm. Our paper provides a bound $\frac{1}{6}N^2$, which is very close to the conjectured bound. This gap, which is approximately $0.019N^2$, arises from the strict inequality in Eqn.(8) for $M \geq 5$. To make the inequality tighter, and derive a better bound is one of our further works. Our technique takes large advantage of the uniform distribution of the next-step configurations. This is not true for the asynchronous variant of Herman's protocol [9], as well as for the asymmetric case for token passing. The generalization to these cases will be our future work.

Finally, as Herman's protocol is very similar to systems of interacting and annihilating particles proposed and studied in physics, combinatorics, and neural networks, we are also interested in exploiting the possibility of extending our elementary methodology for Herman's protocol to providing approximate upper bound for the worst-case analysis of such particle systems.

Acknowledgement. Yuan Feng was partially supported by Australian Research Council (Grant Nos. DP130102764 and FT100100218). Lijun Zhang is the corresponding author (zhanglj@ios.ac.cn), and has received support from the National Natural Science Foundation of China (NSFC) under grant No. 61361136002, and the CAS/SAFEA International Partnership Program for Creative Research Teams.

References

1. Balding, D.: Diffusion-reaction in one dimension. J. Appl. Prob. 25, 733–743 (1988)
2. Dijkstra, E.: Self-stabilizing systems in spite of distributed control. Communications of the ACM 17(11), 643–644 (1974)
3. Dolev, S.: Self-Stabilization. MIT Press (2000)
4. Feller, W.: An introduction to probability theory and its applications, vol. 1. John Wiley & Sons (1968)
5. Feng, Y., Zhang, L.: A tighter bound for the self-stabilization time in Herman's algorithm. Inf. Process. Lett. 113(13), 486–488 (2013)
6. Fribourg, L., Messika, S., Picaronny, C.: Coupling and self-stabilization. Distributed Computing 18(3), 221–232 (2006)
7. Herman, T.: Probabilistic self-stabilization. Information Processing Letters 35(2), 63–67 (1990), Report at `ftp://ftp.math.uiowa.edu/pub/selfstab/H90.html`
8. Kiefer, S., Murawski, A.S., Ouaknine, J., Wachter, B., Worrell, J.: Three tokens in Herman's algorithm. Formal Asp. Comput. 24(4-6), 671–678 (2012)
9. Kiefer, S., Murawski, A.S., Ouaknine, J., Worrell, J., Zhang, L.: On Stabilization in Herman's Algorithm. In: Aceto, L., Henzinger, M., Sgall, J. (eds.) ICALP 2011, Part II. LNCS, vol. 6756, pp. 466–477. Springer, Heidelberg (2011)
10. Kwiatkowska, M.Z., Norman, G., Parker, D.: Probabilistic verification of Herman's self-stabilisation algorithm. Formal Asp. Comput. 24(4-6), 661–670 (2012)
11. Liggett, T.: Interacting particle systems. Springer (2005)
12. McIver, A., Morgan, C.: An elementary proof that Herman's ring is $\Theta(N^2)$. Inf. Process. Lett. 94(2), 79–84 (2005)
13. Nakata, T.: On the expected time for Herman's probabilistic self-stabilizing algorithm. Theoretical Computer Science 349(3), 475–483 (2005)
14. Schneider, M.: Self-stabilization. ACM Comput. Surv. 25(1), 45–67 (1993)

Bounds on Mobility

Reiner Hüchting[1], Rupak Majumdar[2], and Roland Meyer[1]

[1] University of Kaiserslautern
[2] MPI-SWS

Abstract. We study natural semantic fragments of the π-calculus: depth-bounded processes (there is a bound on the longest communication path), breadth-bounded processes (there is a bound on the number of parallel processes sharing a name), and name-bounded processes (there is a bound on the number of shared names). We give a complete characterization of the decidability frontier for checking if a π-calculus process in one subclass belongs to another. Our main construction is a general acceleration scheme for π-calculus processes. Based on this acceleration, we define a Karp and Miller (KM) tree construction for the depth-bounded π-calculus. The KM tree can be used to decide if a depth-bounded process is name-bounded, if a depth-bounded process is breadth-bounded by a constant k, and if a name-bounded process is additionally breadth-bounded. Moreover, we give a procedure that decides whether an arbitrary process is bounded in depth by a given k.

We complement our positive results with undecidability results for the remaining cases. While depth- and name-boundedness are known to be Σ_1-complete, we show that breadth-boundedness is Σ_2-complete, and checking if a process has a breadth bound at most k is Π_1-complete, even when the input process is promised to be breadth-bounded.

1 Introduction

The π-calculus is an expressive formalism for modelling and reasoning about concurrent systems. The full π-calculus is Turing-complete. From a verification perspective, much research has therefore focused on defining semantic fragments which have decidable analysis questions but retain enough expressiveness to capture practical systems. π-calculus processes model communication between components along channels. Natural restrictions on the use of these channels give rise to natural semantic fragments, like bounding the depth of communication (the longest communication chain between processes), bounding the degree of sharing (the number of processes sharing a channel), or bounding the number of channels used concurrently. These restrictions model natural resource constraints in implementations, and indeed have all been studied previously: bounds on depth lead to *depth-bounded* processes [11], bounds on sharing lead to *breadth-bounded* processes [12], and bounds on the number of concurrent channels lead to *name-bounded* processes [9]. While these bounds are defined by induction on the syntax, the restricted classes they define are *semantic*: for example, a process P is breadth-bounded if there is a bound $k \geq 0$ such that in every process reachable

P. Baldan and D. Gorla (Eds.): CONCUR 2014, LNCS 8704, pp. 357–371, 2014.

from P, every channel is shared by at most k processes. The semantic fragments are still very expressive. For example, name-bounded processes can simulate Petri nets, and depth- or breadth-bounded processes can simulate extensions of Petri nets with reset operations. At the same time, they enable algorithmic verification, e.g. coverability is decidable for depth-bounded processes [18].

Little is known about the relation between semantic fragments, beyond the fact that name-bounded processes are also depth-bounded, and the incomparability of the depth- and breadth-bounded fragments. In particular, the *classification problem* —given a process from one (sub)class, does it also belong to another?— has not been studied. By an analogue of Rice's theorem, checking if a π-calculus process belongs to a semantic fragment is likely to be undecidable. However, the status of natural classification questions, such as whether a given depth-bounded process is actually name-bounded, remains open.

The classification problem has various applications in verification. As standalone analysis, classification can judge the resource requirements of a system. For example, to check whether a system is implementable on a given platform, one may check whether it is depth-, breadth-, or name-bounded by a suitable constant k. In turn, a heap-manipulating program that is shown to violate name boundedness may have a memory leak. Within a verification effort, classification serves as a type check that precedes the actual analysis. If the check establishes certain bounds, then the following verification may employ specialized algorithms that make use of this knowledge. For example, to check coverability for general depth-bounded systems, there is only a forward procedure based on iterative refinement. However, if the system is additionally known to be breadth- or name-bounded then more efficient acceleration schemes apply. Similarly, if a bound on the depth is known, then one can use a backwards search.

In this paper, we study and completely characterize the decidability frontier for classification of π-calculus processes into the depth, breadth, and name-bounded semantic fragments, as well as their "k-restricted" versions consisting of all processes where the bound k is given explicitly. Our main construction is a general acceleration scheme for π-calculus processes, and a π-calculus analogue of the Karp-Miller construction for vector addition systems. We characterize the limits of the acceleration for depth-bounded systems and show that —using suitable break conditions— the Karp-Miller construction can decide classification questions such as "given depth-bounded P, is P name-bounded?" and "given depth-bounded P and $k \geq 0$, is P breadth-bounded by k?".

Figure 1 shows a summary of the classification problems, and highlights the results of this paper that complete the picture. The question "given a general process P and $k \geq 0$, is P depth-bounded by k?" was also open (and conjectured to be undecidable). We prove it decidable by showing the existence of a small witness in case the depth bound k is violated. Then we reduce to the coverability problem for depth-bounded processes, which is known to be decidable [18].

We complement our decidability results with undecidability results in the remaining cases. We show that checking if a breadth-bounded process is also depth- or name-bounded is Σ_1-complete, and that checking if

	$\in$ Breadth	$\in$ Breadth(k)	$\in$ Depth	$\in$ Depth(k)	$\in$ Name	$\in$ Name(k)
Proc	Σ_2	Π_1	(Σ_1)	$\surd$	(Σ_1)	$\surd$ [9]
Breadth		Π_1	(Σ_1)	$(\surd)$	Σ_1	$(\surd)$
Depth	Σ_1	$\surd$		$(\surd)$	$\surd$	$(\surd)$
Name	$\surd$	$(\surd)$		$(\surd)$		$(\surd)$

$\surd$: decidable Σ_i/Π_i : undecidable $(-)$: follows from another result
shaded box : results in this paper empty box : trivial

Fig. 1. Decidability frontier for the classification problem

a breadth-bounded process has breadth-bound at most a given k is already Π_1-complete. Additionally, while checking if a process is depth- or name-bounded is only Σ_1-complete, checking breadth boundedness is actually Σ_2-complete. Our lower bounds follow from simulations of Turing machines by depth- or breadth-bounded processes.

Related Work. The Karp-Miller construction, originally developed for Petri nets [10], is generalized to WSTS in [4,5]. The use of Karp-Miller procedures in model checking is assessed e.g. in [3]. In [6,7,8], Finkel and Goubault-Larrecq provide a general algorithmic concept that abstracts from the actual tree structure. They devise a class of systems for which this type of algorithm is guaranteed to terminate. One can prove that our domain of limits is the ideal completion of the partially ordered set of processes in the sense of [6]. Wies et al. make a similar observation in [18]. These results make the general acceleration-based algorithms developed in [7,8] applicable to depth-bounded processes. In contrast to [7,8], we explicitly construct a Karp-Miller tree in order to decide properties like name and breadth boundedness. Recently, a specific acceleration scheme was developed for name-bounded processes [9]. Our acceleration scheme applies to any process, not restricted to depth-bounded systems and subsumes the results in [9].

The key observation in our acceleration is that repeating transition sequences leads to a cyclic behaviour in the use of restricted names. A similar observation is made for ν-APNs (an extension of Petri nets using names as tokens) in [14], where it is used to establish decidability of so-called width boundedness. The property asks for a bound on the number of names that holds in all reachable ν-APN markings and is related to name boundedness. However, our decision procedure handles the more general depth-bounded systems. The notion of depth applies to other concurrency models with name creation (e.g. ν-MSR in [15]), but we are not aware of any results for the classification problem.

2 The π-Calculus

We recall the basics on π-calculus [13,16], a formalism to encode computation using processes that exchange messages over channels. Messages and channels are represented uniformly by *names* from a countable set $\mathcal{N}$. Processes communicate by synchronising on *prefixes* π that *send* $(\overline{x}\langle y\rangle)$ or *receive* $(x(y))$ message y on channel x. Using these communication primitives, we construct processes using choice ($+$), parallel composition ($\mid$), restriction (νa), and calls ($K\lfloor \tilde{a} \rfloor$). Process identifiers K in calls are associated with defining equations $K(\tilde{x}) := P$ where P is a process and $\tilde{x}$ is a vector of distinct names. Each process depends on finitely many defining equations. Formally, *π-calculus processes* P, Q, R are defined by

$$M ::= \mathbf{0} \;\mid\; \pi.P \;\mid\; M_1 + M_2 \qquad\qquad P ::= M \;\mid\; K\lfloor \tilde{a} \rfloor \;\mid\; P_1 \mid P_2 \;\mid\; \nu a.P\ .$$

We write Proc for the set of all processes. Processes M and $K\lfloor \tilde{a} \rfloor$ are called *sequential*, and we use S to indicate that a process is sequential. We denote parallel compositions by products ($\prod$), and use P^k for the k-fold parallel composition of P. Multiple restrictions $\nu a_1 \ldots \nu a_k.P$ are written as a vector $\nu\tilde{a}.P$.

A name a that is neither bound by a restriction νa nor by an input prefix $x(a)$ is called *free*. Names bound by ν are called *restricted*. A restricted name νa is *active* if it is not covered by a prefix. For instance, in the process $\nu a.a(y).\nu b.P\lfloor y \rfloor$, names a and b are restricted, a is active, b is not active, and y is bound but not restricted. We denote the set of free names in P by $fn(P)$ and the active restricted names by $arn(P)$. Since we will be able to α-convert bound names, we assume active restrictions to be unique within a process, and to be disjoint from the free names. By $P\{\tilde{a}/\tilde{x}\}$ we mean the substitution of the free names $\tilde{x}$ in P by $\tilde{a}$. We only apply substitutions that do not clash with the restricted names.

The π-calculus semantics relies on the *structural congruence* relation $\equiv$, the smallest relation that allows for α-conversion of bound names, where $+$ and $\mid$ are associative and commutative with neutral element $\mathbf{0}$, and that satisfies

$$\nu a.\mathbf{0} \equiv \mathbf{0} \qquad \nu a.\nu b.P \equiv \nu b.\nu a.P \qquad \nu a.(P \mid Q) \equiv P \mid \nu a.Q \quad \text{if } a \notin fn(P)\ .$$

The behaviour of processes is given by the *reaction relation*, the smallest relation closed under $\mid$, ν, and structural congruence, and that satisfies the rules

$$x(z).P + M \mid \overline{x}\langle y \rangle.Q + N \to P\{y/z\} \mid Q \qquad \text{and} \qquad K\lfloor \tilde{a} \rfloor \to P\{\tilde{a}/\tilde{x}\}$$

with $K(\tilde{x}) := P$ a defining equation. Up to $\equiv$, processes have only finitely many successors. The set of all processes reachable from P is the *reachability set* $\mathcal{R}(P)$.

We define the *embedding* ordering $\preceq$ on processes to satisfy $\nu\tilde{a}.P \preceq \nu\tilde{a}.(P \mid Q)$ and to be closed under structural congruence. We use $P{\downarrow} := \{Q \in \mathsf{Proc} \mid Q \preceq P\}$ to denote the downward closure of P wrt. embedding.

Our development relies on two normal forms for processes. A process $\nu\tilde{a}.P$ with $P = S_1 \mid \ldots \mid S_n$ and $\tilde{a} \subseteq fn(P)$ is in *standard form* [13], which can be obtained by maximising the scope of restricted names. Similarly, a process is rewritten to *restricted form* by minimising the scope of restricted names [12]. We write $P \equiv F_1 \mid \ldots \mid F_n$ where each F_i is sequential or of the form $\nu a.P'$ with P' again in restricted form and $a \in fn(F_i)$ for all i. The F_i are called *fragments*.

The restricted form of a process P can be used to determine its *depth*, defined as the minimal nesting depth of restrictions in its congruence class: $|P|_{\mathcal{D}} := min\{nest_\nu(Q) \mid P \equiv Q \text{ in restricted form}\}$ with $nest_\nu$ defined inductively as $nest_\nu(S) := 0$, $nest_\nu(P_1 \mid P_2) := max\{nest_\nu(P_1), nest_\nu(P_2)\}$, and $nest_\nu(\nu a.P) := 1 + nest_\nu(P)$. Similarly, the breadth $|P|_{\mathcal{B}}$ is the maximal number of sequential processes composed in parallel underneath an active restricted name.

We define the following subclasses of the set Proc of π-calculus processes. The class $\mathsf{Depth}(k)$ contains all processes P that are *bounded in depth* by k. Formally, for all $Q \in \mathcal{R}(P)$ we have $|Q|_{\mathcal{D}} \leq k$. Then the class Depth of all depth-bounded processes is the union $\cup_{k\geq 0}\mathsf{Depth}(k)$. The classes $\mathsf{Breadth}(k)$ and Breadth are defined analogously. Finally, $\mathsf{Name}(k)$ contains all processes P so that every process in $\mathcal{R}(P)$ has at most k active restricted names, and $\mathsf{Name} := \cup_{k\geq 0}\mathsf{Name}(k)$. Clearly, $\mathsf{Name}(k) \subseteq \mathsf{Depth}(k)$ for each $k \geq 0$ and so $\mathsf{Name} \subseteq \mathsf{Depth}$. It is well known that Depth and Breadth are incomparable.

Our results rely on the fact that depth-, breadth-, and name-bounded processes can be represented in a finite way:

Lemma 1. *Let $P \in \mathsf{Name}(k)$. There is a finite set of sequential processes $\{S_1, \ldots S_n\}$ so that every $Q \in \mathcal{R}(P)$ satisfies*

$$Q \equiv \nu a_1 \ldots a_k.(S_1^{m_i} \mid \ldots \mid S_n^{m_n}) \quad \textit{with} \quad m_1, \ldots, m_n \in \mathbb{N}\ .$$

Similarly, given a depth-bounded process $P \in \mathsf{Depth}$, all reachable $Q \in \mathcal{R}(P)$ can be written using finitely many sequential processes [11]. Indeed, with depth bound k, we rewrite Q to restricted form with at most k nested restrictions. Then we choose a distinguished name a_k for each nesting level:

$$Q \equiv \nu a_1.(\nu a_2.(\ldots) \mid \ldots \mid \nu a_2(\ldots))\ .$$

After this modification, the sequential processes take the form $D\sigma$ with finitely many D and finitely many substitutions σ. For $\mathsf{Depth} \cap \mathsf{Breadth}$, we obtain a finite representation using fragments [12].

Lemma 2. *Let $P \in \mathsf{Depth} \cap \mathsf{Breadth}$. There is a finite set of fragments $\{F_1, \ldots F_n\}$ so that every $Q \in \mathcal{R}(P)$ satisfies*

$$Q \equiv F_1^{m_i} \mid \ldots \mid F_n^{m_n} \quad \textit{with} \quad m_1, \ldots, m_n \in \mathbb{N}\ .$$

Decision Problems. We study decision problems of the form $(\mathcal{C}_1, \mathcal{C}_2)$ for classes of processes $\mathcal{C}_1$ and $\mathcal{C}_2$, asking for a process from class $\mathcal{C}_1$, is it also in $\mathcal{C}_2$.

3 Karp and Miller for Bounded Depth

We now describe an adaptation of the Karp-Miller algorithm for Petri nets [10] to the π-calculus. The Karp-Miller algorithm determines a finite representation of (the downward closure of) a system's reachability set. It unwinds the state space until it detects a transition sequence between comparable states. The key ingredient is then an acceleration theorem that characterizes the states reachable with arbitrary repetitions of this sequence in a symbolic way. When transferring the idea to depth-bounded processes, the unwinding finds $Q_1 \rightarrow^* Q_2$ with $Q_1 \prec Q_2$. By monotonicity, this can be repeated as

$$Q_1 \rightarrow^* Q_2 \rightarrow^* Q_3 \rightarrow^* \cdots \quad \text{with} \quad Q_1 \prec Q_2 \prec Q_3 \prec \cdots$$

Our main result is an acceleration theorem that characterizes $\{Q_i \mid i \in \mathbb{N}\}$. Acceleration for the π-calculus is more involved than for Petri nets, because we have to take the identities of restricted names into account. Consider the reaction

$$\nu a.\nu b.K_1\lfloor a,b\rfloor \to \nu a.\nu b.\left(K_1\lfloor a,b\rfloor \mid K_2\lfloor b\rfloor\right)$$

with $K_1(x,y) := \nu z.(K_1\lfloor z,x\rfloor \mid K_2\lfloor x\rfloor)$ and $K_2(x) := \mathbf{0}$. At first glance, k repetitions of this reaction should lead to processes $\nu a.\nu b.\left(K_1\lfloor a,b\rfloor \mid (K_2\lfloor b\rfloor)^k\right)$. However, due to α-conversion, a and b in source and target process are different. Since a has been renamed to b and a fresh a has been created, repetitions will lead to terms $\nu c.K_2\lfloor c\rfloor$. To see this, we repeat the reaction without renaming:

$$\begin{aligned}\nu a.\nu b.K_1\lfloor a,b\rfloor &\to \nu c.\nu a.(K_1\lfloor c,a\rfloor \mid K_2\lfloor a\rfloor)\\ &\to \nu d.\nu c.(K_1\lfloor d,c\rfloor \mid K_2\lfloor c\rfloor \mid \nu a.K_2\lfloor a\rfloor)\ .\end{aligned}$$

The example illustrates that we have to track the identity of restricted names over transitions. We have to determine if a restricted name is *stable* in the sense that it remains in the original process, or it is *fragile*, i.e. it is eventually forgotten and moves to the accelerated part. The following subsection develops a suitable notion of identity relations, afterwards we turn to the actual acceleration.

3.1 Identity Relations

To track the identity of active restricted names over transitions, we extend the reaction relation. Recall that we assume active restrictions to be unique within processes. With each reaction $P \to Q$, we associate an *identity relation* $I \subseteq (\mathcal{N} \cup \{\star\}) \times \mathcal{N}$ relating the active restrictions in P and Q. Formally, I is the smallest set that satisfies the following conditions. If reaction $P \to Q$ α-converts $a \in arn(P)$ into $b \in arn(Q)$, then we have $(a,b) \in I$. Moreover, if νc becomes active in Q, we have $(\star, c) \in I$. We write $P \to_I Q$ for these *extended reactions* in $\mathsf{Proc} \times \mathbb{P}((\mathcal{N} \cup \{\star\}) \times \mathcal{N}) \times \mathsf{Proc}$. The definition generalizes to sequences of extended reactions by composing the identity relations. Such a sequence is *faithful* if it does not use α-conversion. Formally, $I \cap (\mathcal{N} \times \mathcal{N})$ is the identity.

For the purpose of acceleration, we focus on the special case that I acts on a finite subset $\tilde{a} \subseteq \mathcal{N}$ of names, $I \subseteq (\{\star\} \cup \tilde{a}) \times \tilde{a}$. In this case, I induces a partition $\tilde{a} = \tilde{f} \uplus \tilde{s}$ as follows. The set $\tilde{f}$ of *fragile names* contains all names that are recreated in repeated applications of I. Technically, these names form a least fixed point. If $(\star, a) \in I$, then $a \in \tilde{f}$. If $a \in \tilde{f}$ and $(a,b) \in I$, then $b \in \tilde{f}$. The remaining names $\tilde{s} := \tilde{a} \setminus \tilde{f}$ are called *stable*. We use σ_I for $I \cap (\tilde{s} \times \tilde{s})$.

Lemma 3. *$\sigma_I : \tilde{s} \to \tilde{s}$ is a bijection.*

Since σ_I is a bijection, repeated applications of σ_I become periodic. The *period* of σ_I is the smallest $p \geq 1$ so that $\sigma_I^p = id$. Now $\sigma_I^x = \sigma_I^{x \bmod p}$ for all $x \in \mathbb{N}$.

We call an identity relation I *forgetful* if it immediately recreates all fragile names, as opposed to needing several applications to recreate them. Formally, I is forgetful if the fragile names are precisely the names $(\star, a)$ in I. We also call an extended reaction sequence $P \rightarrow_I^* Q$ forgetful if I is. To give an example, consider $\tilde{a} = a.b.c.d$ and $I = \{(\star, a), (a, b), (d, c), (c, d)\}$. Then $\tilde{f} = a.b$, $\tilde{s} = c.d$, $\sigma_I(c) = d$, and $\sigma_I(d) = c$. The period of σ_I is $p = 2$. Relation I is not forgetful, but $I^2 = \{(\star, a), (\star, b), (c, c), (d, d)\}$ is.

3.2 Acceleration for π-Calculus

We make precise what it means to repeat a reaction sequence. Embedding $P \preceq Q_1$ ensures $P \equiv \nu\tilde{a}.P'$ and $Q_1 \equiv \nu\tilde{a}.(P' \mid Q')$. Hence, Q_1 contains all the sequential processes of P. So if $P \rightarrow^* Q_1$ then also $Q_1 \rightarrow^* Q_2$ with reactions between the same sequential processes. We use the syntax $P \xrightarrow{\rho} Q_1$ and $Q_1 \xrightarrow{\rho} Q_2$ to indicate that the sequences rely on the same reactions ρ.

Our main result characterizes the shape of Q_k with $P \xrightarrow{\rho^k} Q_k$. The idea is to compose additional copies of Q' in parallel with the repeating process P'. These copies keep track of (i) the fragile names forgotten by P' and (ii) repeated applications of σ_I to the stable names. We state the precise result for forgetful sequences. The general case is more involved, but does not add ideas.

Theorem 1 (Acceleration). *Let $\nu\tilde{s}.\nu\tilde{f}.P \xrightarrow{\rho'}_{I'} \nu\tilde{s}'.\nu\tilde{f}'.(P' \mid \nu\tilde{f}.Q)$ be a faithful sequence that, using α-conversion, gives rise to the forgetful sequence*

$$\nu\tilde{s}.\nu\tilde{f}.P \xrightarrow{\rho}_I \nu\tilde{s}.\nu\tilde{f}.(P \mid \nu\tilde{f}_1.Q\{\tilde{f}_1/\tilde{f}\}\{\tilde{f}/\tilde{f}'\}\sigma_I) .$$

Then for all $k \in \mathbb{N}$ we have

$$\nu\tilde{s}.\nu\tilde{f}.P \xrightarrow{\rho^k} \nu\tilde{s}.\nu\tilde{f}.(P \mid \nu\tilde{f}_k.(Q_k \mid \ldots \nu\tilde{f}_2.(Q_2 \mid \nu\tilde{f}_1.Q_1)\ldots))$$

where $Q_i := Q\{\tilde{f}_i/\tilde{f}\}\{\tilde{f}_{i+1}/\tilde{f}'\}\sigma_I^{k-i+1}$ and $\tilde{f}_{k+1} := \tilde{f}$.

Proof. We use an induction on the number of repetitions of ρ. In the base case, the term $\nu\tilde{f}_k.(\ldots)$ is missing. Now assume ρ^k behaves as required. We can repeat ρ once more since embedding is a simulation [11]. We use a faithful repetition that avoids α-conversion and instantiates restricted names to fresh names:

$$\begin{aligned}\nu\tilde{s}.\nu\tilde{f}.P &\xrightarrow{\rho^k} \nu\tilde{s}.\nu\tilde{f}.(P \mid \nu\tilde{f}_k.(Q_k \mid \ldots \nu\tilde{f}_2.(Q_2 \mid \nu\tilde{f}_1.Q_1)\ldots))\\ &\xrightarrow{\rho'}_{I'} \nu\tilde{s}'.\nu\tilde{f}.(\nu\tilde{f}'.(P' \mid Q) \mid \nu\tilde{f}_k.(Q_k \mid \ldots \nu\tilde{f}_2.(Q_2 \mid \nu\tilde{f}_1.Q_1)\ldots)) .\end{aligned}$$

Since the names $\tilde{f}'$ are fresh and do not occur free in $\nu\tilde{f}_k.(\ldots)$, we can extend the scope of $\nu\tilde{f}'$. Process P forgets the names $\tilde{f}$ in the reaction sequence ρ. After swapping $\nu\tilde{f}$ and $\nu\tilde{f}'$, we can therefore restrict the scope of $\nu\tilde{f}$:

$$\begin{aligned}&\nu\tilde{s}'.\nu\tilde{f}.[\nu\tilde{f}'.(P' \mid Q) \mid \nu\tilde{f}_k.(Q_k \mid \ldots \nu\tilde{f}_2.(Q_2 \mid \nu\tilde{f}_1.Q_1)\ldots)]\\ \equiv\ &\nu\tilde{s}'.\nu\tilde{f}.\nu\tilde{f}'.(P' \mid Q \mid \nu\tilde{f}_k.(Q_k \mid \ldots \nu\tilde{f}_2.(Q_2 \mid \nu\tilde{f}_1.Q_1)\ldots))\\ \equiv\ &\nu\tilde{s}'.\nu\tilde{f}'.\nu\tilde{f}.(P' \mid Q \mid \nu\tilde{f}_k.(Q_k \mid \ldots \nu\tilde{f}_2.(Q_2 \mid \nu\tilde{f}_1.Q_1)\ldots))\\ =\ &\nu\tilde{s}'.\nu\tilde{f}'.(P' \mid \nu\tilde{f}.[Q \mid \nu\tilde{f}_k.(Q_k \mid \ldots \nu\tilde{f}_2.(Q_2 \mid \nu\tilde{f}_1.Q_1)\ldots)]) .\end{aligned}$$

We rename $\tilde{f}$ to fresh names $\tilde{f}_{k+1}$ and afterwards $\tilde{f}'$ to $\tilde{f}$. The last step is to rename $\tilde{s}'$ to $\tilde{s}$ using σ_I. The resulting processes have the required form. □

Depth boundedness of $\nu\tilde{s}.\nu\tilde{f}.P$ allows us to draw further conclusions about the shape of the processes resulting from acceleration. We now characterize this shape. We start with the assumption that $\nu\tilde{s}.\nu\tilde{f}.P$ is even name-bounded. This case explains well the finiteness obtained from periodicity of σ_I. Moreover, it illustrates the shape of limit processes in the decision procedure for name boundedness of depth-bounded processes. Under the assumption of name boundedness, the accelerated processes Q_i cannot contain the fragile names $\tilde{f}_i$ and $\tilde{f}_{i+1}$. As a consequence, these processes can be written as $Q\sigma^i$.

Lemma 4. *Under the conditions in Theorem 1 and the additional assumption that $\nu\tilde{s}.\nu\tilde{f}.P$ is name-bounded, acceleration yields*

$$\nu\tilde{s}.\nu\tilde{f}.P \xrightarrow{\rho^k} \nu\tilde{s}.(\nu\tilde{f}.P \mid \prod_{i=1}^{p}(Q\sigma_I^i)^{\lfloor \frac{k}{p} \rfloor + z_i}) .$$

Here, p is the period of σ_I and $z_i = 1$ if $i \leq k \bmod p$ and $z_i = 0$ otherwise.

Proof. We skip the substitutions $\{\tilde{f}_i/\tilde{f}\}\{\tilde{f}_{i+1}/\tilde{f}'\}$ and remove the corresponding restrictions from Q_i. This leaves us with an i-fold application of σ_I:

$$\nu\tilde{s}.\nu\tilde{f}.(P \mid \nu\tilde{f}_k.(Q_k \mid \ldots \nu\tilde{f}_2.(Q_2 \mid \nu\tilde{f}_1.Q_1)\ldots)) \equiv \nu\tilde{s}.(\nu\tilde{f}.P \mid \prod_{i=1}^{k} Q\sigma_I^i) .$$

The claim then follows from $\sigma^x = \sigma^{x \bmod p}$ for all $x \in \mathbb{N}$ and the fact that $k = \lfloor \frac{k}{p} \rfloor p + r$ where $r = k \bmod p$. □

In case $\nu\tilde{s}.\nu\tilde{f}.P$ is depth-bounded, the processes Q_i in

$$\nu\tilde{s}.\nu\tilde{f}.(P \mid \nu\tilde{f}_k.(Q_k \mid \ldots \nu\tilde{f}_2.(Q_2 \mid \nu\tilde{f}_1.Q_1)\ldots))$$

cannot connect all fragile names $\tilde{f}_{k+1}$ to $\tilde{f}_1$. Instead, we now argue that process $\nu\tilde{f}_k.(\ldots)$ falls apart into a composition of fragments that stem from a finite set.

We first note that Q induces a relation $D \subseteq \tilde{f} \times \tilde{f}$ among the fragile names. Intuitively, $(f_1, f_2) \in D$ indicates that name f_2 from some execution of ρ is connected with name f_1 from the next execution via a fragment in Q. For the formal definition, recall that $Q_i = Q\{\tilde{f}_i/\tilde{f}\}\{\tilde{f}_{i+1}/\tilde{f}'\}\sigma_I^{k-i+1}$. Substitution σ_I is a bijection among the stable names. When studying the relation among the fragile names, we can drop it. Moreover, $\{\tilde{f}_i/\tilde{f}\}$ and $\{\tilde{f}_{i+1}/\tilde{f}'\}$ are bijective renamings of $\tilde{f}$ that we also avoid. To define $D \subseteq \tilde{f} \times \tilde{f}$, we turn the remaining process Q into restricted form. The restricted form is a parallel composition $G_1 \mid \ldots \mid G_n$ of fragments. Relation $D \subseteq \tilde{f} \times \tilde{f}$ is defined to contain the pair (f_1, f_2) if there is a fragment G_i that has $f_1' \in \tilde{f}'$ and $f_2 \in \tilde{f}$ as free names. To give an example, consider $Q = G_1 \mid G_2$ where $G_1 = \nu a.(K_1\lfloor f_1', a \rfloor \mid K_2\lfloor a, f_2 \rfloor)$ and $G_2 = K_3\lfloor f_2', f_3 \rfloor$. Then we have $D = \{(f_1, f_2), (f_2, f_3)\}$.

If the process of interest is depth-bounded, relation D is guaranteed to vanish. Otherwise, the connected fragments G_i would witness unbounded depth.

Lemma 5. *If $\nu\tilde{s}.\nu\tilde{f}.P \in$ Depth, then there is $b \leq |\tilde{f}|$ so that $D^b = \emptyset$.*

The lemma shows that fragments in $Q_{i+(b-2)}$ may only share fragile names with fragments in $Q_{i+(b-3)}$, which only share names with fragments in $Q_{i+(b-4)}$ up to Q_i. Combined with periodicity of σ_I, it follows that process $\nu\tilde{f_k}.(\ldots)$ falls apart into fragments from a finite set $\{F_1, \ldots, F_t\}$. In the example, we have $b = 3$ as $D^3 = \emptyset$. So fragments in Q_{i+1} may only share fragile names with fragments in Q_i, but these fragments will not share fragile names with fragments in Q_{i-1}. Hence, we may observe $\nu f_{1,i+2}.f_{2,i+1}.f_{3,i}.(G_{1,i+1} \mid G_{2,i})$ but $G_{2,i} = K_3\lfloor f_{2,i+1}, f_{3,i}\rfloor$ will not share $f_{3,i}$ with another fragment.

Lemma 6. *If $\nu\tilde{s}.\nu\tilde{f}.P$ is depth-bounded, then acceleration yields*

$$\nu\tilde{s}.\nu\tilde{f}.P \xrightarrow{\rho^k} \nu\tilde{s}.(\nu\tilde{f}.(P \mid R_k) \mid \prod_{i=1}^{t} F_i^{n_i})$$

where the R_k are periodic, $t \in \mathbb{N}$ is the number of fragments, and $n_1, \ldots n_t \in \mathbb{N}$ grow unboundedly with k.

To actually compute the fragments $F_1, \ldots, F_t$, we iterate ρ until the fragments repeat, due to periodicity of σ_I and modulo α-conversion of fragile names. The time it takes until repetition depends on the period of σ_I and $|\tilde{f}|$. However, termination does not rely on the precise depth bound.

In the decision procedure for restricted breadth boundedness, we apply Lemma 6 in a setting where also the breadth is bounded. In this case, the stable names are guaranteed not to occur in the accelerated fragments. We derive

$$\nu\tilde{s}.\nu\tilde{f}.P \xrightarrow{\rho^k} \nu\tilde{f}.(\nu\tilde{s}.P \mid R_k) \mid \prod_{i=1}^{t} F_i^{n_i} \enspace .$$

3.3 Karp and Miller Trees for Bounded Depth

We now apply these insights about acceleration to devise a Karp and Miller algorithm for depth-bounded processes. Our goal is to separate soundness and completeness of the construction from termination. We will guarantee that the tree represents precisely the downward closure of the reachability set, but the tree computation may not terminate. Termination is studied independently in the next section where we draw conclusions about decidability. In Section 5, we show that in general the coverability set is not computable for depth-bounded systems. Hence, no Karp-Miller algorithm for depth-bounded processes can be both, sound and complete as well as terminating.

To represent the downward closure of the state space, we use *limit processes*:

$$L ::= S \mid L_1 \mid L_2 \mid \nu a.L \mid L^\omega \enspace .$$

Structural congruence is extended to limits by

$$\mathbf{0}^\omega \equiv \mathbf{0} \quad L^\omega \equiv L \mid L^\omega \quad (L^\omega)^\omega \equiv L^\omega \quad (L_1 \mid L_2)^\omega \equiv L_1^\omega \mid L_2^\omega \quad L^\omega \mid L^\omega = L^\omega \enspace .$$

Algorithm 1. Generic Karp & Miller Tree Construction

$V := \{\mathsf{root} : P\}$; $\quad \rightsquigarrow := \emptyset$; $\quad work := \mathsf{root} : P$;
while *work* not empty **do**
  pop $\mathsf{n}_1 : L_1$ from *work*;
  Break condition;
  for all $L_1 \rightarrow_I L_2$ up to $\equiv$ **do**
    if there is $\mathsf{n} : L_p \rightsquigarrow^*_{I_1} \mathsf{n}_1 : L_1$ since last acceleration so that
      $L_p \equiv \nu\tilde{s}.\tilde{f}.L$ and $L_2 \equiv \nu\tilde{s}.\tilde{f}.(L \mid Q)$ **then**
      $L_2 :=$ **Accelerate**(L_p, L_2);
    let n_2 fresh; $\quad V := V \cup \{\mathsf{n}_2 : L_2\}$; $\quad \rightsquigarrow := \rightsquigarrow \cup \{\mathsf{n}_1 : L_1 \rightsquigarrow_I \mathsf{n}_2 : L_2\}$;
    $work := work \cdot (\mathsf{n}_2 : L_2)$ provided L_2 does not occur on the path;
return $(V, \rightsquigarrow, \mathsf{root} : P)$;

With this, the reaction relation and the embedding order carry over to limit processes. To generalize the extended reaction relation to limits, we do not consider a restricted name as active if it is covered by an ω. If a reaction takes a restriction $\nu b.L\{b/a\}$ out of a term $(\nu a.L)^\omega$, we add $(\star, b)$ to the identity relation.

Our Karp-Miller construction is stated as Algorithm 1. Understand **break condition** as a no-op for the moment. The command will be instantiated in Section 4. The algorithm constructs a tree $\mathrm{KM}(P) := (V, \rightsquigarrow, \mathsf{root} : P)$ where the nodes are labelled by limit processes, starting from the root that is labelled by the given process P. To construct the tree, the algorithm maintains a worklist of nodes that have not yet been processed. As long as the worklist is not empty, the algorithm pops nodes $\mathsf{n}_1 : L_1$ and computes all successors $L_1 \rightarrow_I L_2$. In contrast to a standard state space computation, the algorithm does not immediately add L_2 to the tree, but it checks for a predecessor L_p that is strictly smaller in the following sense: $L_p \equiv \nu\tilde{s}.\tilde{f}.L$ and $L_2 \equiv \nu\tilde{s}.\tilde{f}.(L \mid Q)$. If such an L_p exists, the algorithm accelerates subterms in L_2 to ω and adds the result to the tree.

We use an acceleration scheme that is flat in the following sense. We only look for predecessors L_p starting from the last accelerated process on the path from the root. This ensures that the Karp-Miller transitions between L_p and L_2 are actually reactions. As a consequence, the additional process Q in L_2 can be assumed to be concrete because the reaction relation does not introduce ω.

To define the acceleration, assume $\nu\tilde{s}.\tilde{f}.L \xrightarrow{\rho}_I \nu\tilde{s}.\tilde{f}.(L \mid Q)$. We observe that Theorem 1 carries over to limits. As the input process is depth-bounded, Lemma 6 shows that $k \in \mathbb{N}$ repetitions of the reaction sequence yield

$$\nu\tilde{s}.\nu\tilde{f}.L \xrightarrow{\rho^k} \nu\tilde{s}.(\nu\tilde{f}.(L \mid R_k) \mid \prod_{i=1}^{t} F_i^{n_i})$$

where the R_k are periodic and $n_1, \ldots, n_t \in \mathbb{N}$ grow with k. We set

$$\mathbf{Accelerate}(\nu\tilde{s}.\tilde{f}.L, \nu\tilde{s}.\tilde{f}.(L \mid Q)) := \nu\tilde{s}.(\nu\tilde{f}.(L \mid R_1) \mid \prod_{i=1}^{t} F_i^{\omega}) \ .$$

The following theorem states that the limit processes in $\mathrm{KM}(P)$ are a sound and complete representation of the state space.

Theorem 2. *Let* $P \in \mathsf{Depth}$. *Then* $\mathcal{R}(P){\downarrow} = \mathrm{KM}(P){\downarrow} \cap \mathsf{Proc}$.

Completeness means every $Q \in \mathcal{R}(P)$ satisfies $Q \in \mathrm{KM}(P){\downarrow}$. This holds because acceleration only returns a larger process and embedding is a simulation. Soundness states that every process $Q \preceq L \in \mathrm{KM}(P)$ is covered by a reachable process, $Q \preceq R \in \mathcal{R}(P)$. The acceleration results show that for every limit $L \in \mathrm{KM}(P)$ and $k \in \mathbb{N}$ there is a process $R_{L,k} \in \mathcal{R}(P)$ that coincides with L in the concrete part and yields more than k parallel compositions of the ω terms.

4 Decidability Results

4.1 Given $P \in$ Depth, is $P \in$ Name?

To decide name boundedness, we instantiate **break condition** as follows:

if there is a predecessor $\mathsf{n} : L_p \rightsquigarrow^* \mathsf{n}_1 : L_1$ so that $\mathsf{n} : L_p$ is not inside an acceleration, $L_p \prec L_1$, and $|arn(L_p)| < |arn(L_1)|$ **then**
return not name-bounded;

That predecessor $\mathsf{n} : L_p$ is not inside an acceleration, means there is no pair of nodes $\mathsf{n}_x : L_x \rightsquigarrow^+ \mathsf{n} : L_p \rightsquigarrow^+ \mathsf{n}_y : L_y$ that have been used to accelerate L_y. The condition guarantees that the sequence $\mathsf{n} : L_p \rightsquigarrow^* \mathsf{n}_1 : L_1$ can be accelerated although it may be nested. The trick is that nested sequences inside have all the processes required to pump.

We say that Algorithm 1 *succeeds*, if it returns the tree (Line 10) and it *breaks*, if it enters the break condition (Line 4). The next observation shows that acceleration never introduces ω over an active restriction.

Lemma 7. *If* $\mathrm{KM}(P)$ *accelerates an active restriction, it breaks.*

Proof. Let L_1 be the first limit on a path with some $(\nu a.(\ldots))^\omega$. The acceleration is due to some $\mathsf{n} : L_p \rightsquigarrow^* \mathsf{n}_1 : L_1$ that occurs after the last acceleration and satisfies $L_p \prec L_1$. As L_1 is minimal, $|arn(L_p)| \in \mathbb{N}$ and $|arn(L_1)| := \omega$. □

We now show Algorithm 1 is partially correct. If $\mathrm{KM}(P)$ succeeds, it will not accelerate active restrictions by Lemma 7. Moreover, the constructed tree is finite and the maximum $max\{|arn(L)| \mid L \in \mathrm{KM}(P)\}$ is a natural number and an optimal name bound. If $\mathrm{KM}(P)$ breaks, we obtain name unboundedness with the above argumentation on nested acceleration and Lemma 6.

Lemma 8. *If* $\mathrm{KM}(P)$ *succeeds,* $P \in \mathsf{Name}$. *If* $\mathrm{KM}(P)$ *breaks,* $P \notin \mathsf{Name}$.

It remains to show that Algorithm 1 is guaranteed to terminate. First assume P is name-bounded. By Lemma 8, the algorithm does not break. To show that the tree construction succeeds, we apply König's lemma and argue that the paths in the tree are finite. Towards a contradiction, assume there was an infinite path

$$\mathsf{root} : P - \mathsf{n}_0 : L_0 \rightsquigarrow \mathsf{n}_1 : L_1 \rightsquigarrow \mathsf{n}_2 : L_2 \rightsquigarrow \ldots$$

Since the Karp-Miller construction is sound (Theorem 2) and $P \in \mathsf{Name}$, not only the reachable processes but also the limits L_i take the shape from Lemma 1:

$$L_i \equiv \nu a_1 \dots a_k.(S_1^{m_i} \mid \dots \mid S_n^{m_n})$$

where $n \in \mathbb{N}$ is fixed and potentially some m_i are accelerated to ω. We can understand the limits as vectors in $(\mathbb{N} \cup \{\omega\})^n$. The set of vectors $(\mathbb{N} \cup \{\omega\})^n$ with the usual ordering on $\mathbb{N} \cup \{\omega\}$ (applied component-wise) is well-quasi-ordered. Hence, the infinite path contains an infinite non-decreasing subsequence

$$L_{i_1} \preceq L_{i_2} \preceq \dots .$$

This sequence is strict, as Algorithm 1 never adds a process to the worklist that already occurs on the path. Since there are only n sequential processes and ωs are never removed, the number of S^ω stabilizes in some node in the path, say $\mathsf{n}_{i_k} : L_{i_k}$. Then there are no more accelerations from L_{i_k}, and $L_{i_{k+1}} \succ L_{i_k}$ can be accelerated. A contradiction to the fact that no ω is added to $L_{i_{k+1}}$.

Lemma 9. *If* $P \in \mathsf{Name}$, *then* KM(P) *succeeds.*

Now assume that $P \notin \mathsf{Name}$. By Lemma 8, the algorithm does not succeed. To show that it always breaks, assume this is not the case. With Lemma 7, we never accelerate an active restriction. Since Algorithm 1 is complete and $P \notin \mathsf{Name}$, for every $k \in \mathbb{N}$ the tree contains a limit L_k with more than k active restrictions. Thus, the tree is infinite and by König's lemma contains an infinite path

$$\mathsf{root} : P = \mathsf{n}_0 : L_0 \rightsquigarrow \mathsf{n}_1 : L_1 \rightsquigarrow \mathsf{n}_2 : L_2 \rightsquigarrow \dots .$$

We isolate the subsequence $L_{i_1}, L_{i_2}, \dots$ of limits that (i) form the lhs of an acceleration, like $\mathsf{n}_x : L_x$ above, or (ii) that lie strictly between accelerations.

We now show that the limits computed without a break form a wqo with $\preceq$. Since we never accelerate a restriction, the limits L_i are term trees over sequential processes S and S^ω. Due to soundness, the limits are bounded in depth by some $k \in \mathbb{N}$. It remains to show that S stems from a finite set. With the depth bound and the observation following Lemma 1, we can assume that the limits are flat processes with at most k nested restrictions. They can be written as

$$L_i \equiv \nu a_1.(\nu a_2.(\dots) \mid \dots \mid \nu a_2(\dots)) .$$

This in turn means $S \equiv D\sigma$ with $\sigma : \mathit{fn}(D) \to \mathit{fn}(P) \cup \{a_1, \dots, a_k\}$, where D is a syntactic subterm of P and σ is taken from a finite set of substitutions.

As the limits on the path are wqo, there is an infinite non-decreasing subsequence $L_{j_1} \preceq L_{j_2} \preceq \dots$ of the sequence $L_{i_1}, L_{i_2}, \dots$ we just considered. Again, Algorithm 1 does not add copies and the sequence is strict. Either the number of active restricted names in the limits along this path is bounded, or it grows indefinitely. If it is bounded, this contradicts infinity of the path as in Lemma 9. Otherwise, there are $L_{j_k} \prec L_{j_l}$ with $|\mathit{arn}(L_{j_k})| < |\mathit{arn}(L_{j_l})|$ and the break condition applies.

Lemma 10. *If* $P \notin \mathsf{Name}$*, then* KM(P) *breaks.*

Theorem 3. *Given* $P \in \mathsf{Depth}$*, it is decidable if* $P \in \mathsf{Name}$.

Name-bounded processes have finite Karp-Miller trees by Lemma 9. We can now decide breadth boundedness for $P \in \mathsf{Name}$ by checking if the tree contains a limit $\nu\tilde{a}.(S_1^{m_1} \mid \ldots \mid S_n^{m_n})$ where $a \in fn(S_i)$ for some $a \in \tilde{a}$ and S_i with $m_i = \omega$.

Corollary 1. *Given* $P \in \mathsf{Name}$*, it is decidable if* $P \in \mathsf{Breadth}$.

4.2 Given $P \in$ Depth and k, is $P \in$ Breadth(k)?

To derive a decision procedure, we use the following **break condition**:

if $|L_1|_{\mathcal{B}} > k$ **then return** not breadth-bounded by k;

Lemma 11. *If* KM(P) *succeeds,* $P \in \mathsf{Breadth}(k)$*. If* KM$(P)$ *breaks,* $P \notin \mathsf{Breadth}(k)$.

We now show that Algorithm 1 is guaranteed to terminate. If $P \in \mathsf{Breadth}(k)$, by Lemma 11, the break condition will not apply. By soundness of the construction, limits are composed of finitely many fragments (Lemma 2). Then an infinite path yields a contradiction to the acceleration behaviour as for name boundedness. If $P \notin \mathsf{Breadth}(k)$, by Lemma 11, the tree construction will not succeed. We show that we enter the break condition. Since $P \notin \mathsf{Breadth}(k)$, there is a process $Q \in \mathcal{R}(P)$ with $|Q|_{\mathcal{B}} > k$. As Algorithm 1 is complete, we eventually find a limit L with $Q \preceq L$ and $k < |Q|_{\mathcal{B}} \leq |L|_{\mathcal{B}}$. The break condition applies for L.

Lemma 12. *If* $P \in \mathsf{Breadth}(k)$*, then* KM$(P)$ *succeeds. If* $P \notin \mathsf{Breadth}(k)$*, then* KM$(P)$ *breaks.*

Theorem 4. *Given* $P \in \mathsf{Depth}$ *and* $k \in \mathbb{N}$*, it is decidable if* $P \in \mathsf{Breadth}(k)$.

Note that we cannot semi-decide breadth unboundedness. We show in Section 5 that breadth boundedness, despite the seeming similarity with name boundedness, is undecidable for depth-bounded processes.

4.3 Given $P \in$ Proc and k, is $P \in$ Depth(k)?

By definition, $P \in \mathsf{Depth}(k)$ iff all processes in $\mathcal{R}(P)\!\downarrow$ are in $\mathsf{Depth}(k)$.

If $\mathcal{R}(P)\!\downarrow$ is not in $\mathsf{Depth}(k)$, we claim that there is some process Q in $\mathcal{R}(P)\!\downarrow$ for which $k < |Q|_{\mathcal{D}} \leq |P| + 2k$. Let $P \rightarrow^* Q' \rightarrow Q$, such that (1) $|Q|_{\mathcal{D}} > k$, and (2) all processes in the path $P \rightarrow^* Q'$ have depth less than or equal to k. So Q is a minimal process whose depth exceeds k. Note that Q need not be unique, we arbitrarily pick a minimal process. Consider the step $Q' \rightarrow Q$. By case analysis on the possible steps, Q' either unfolds a recursive call and Q has its depth bounded by $|P| + |Q'|_{\mathcal{D}}$, or Q results from a communication and has fragments whose depth is bounded above by $|P| + 2|Q'|_{\mathcal{D}}$. In both cases, $|P|$ takes care of active restrictions that are freed in the reaction. As $|Q'|_{\mathcal{D}} \leq k$ by assumption, we have that there is a process Q in $\mathcal{R}(P)\downarrow$ whose depth is at most $|P| + 2k$.

Now, consider the set $\mathsf{Proc}_w := \{Q \in \mathsf{Proc} \mid k < |Q|_{\mathcal{D}} \leq |P| + 2k\}$. We work in the well-quasi-order of $|P| + 2k$ depth-bounded processes with the ordering $\preceq$. If $\mathcal{R}(P)\downarrow$ intersects this set, a minimal element of this set must be covered by process P. We enumerate the finitely many minimal elements of this set, and for each minimal element Q, we check if $\mathcal{R}(P)\downarrow$ intersects Q. This problem is a coverability question. Since the coverability problem is decidable for depth-bounded systems [18], we get a decision procedure to check if $P \in \mathsf{Depth}(k)$.

Theorem 5. *Given $P \in \mathsf{Proc}$ and $k \in \mathbb{N}$, it is decidable if $P \in \mathsf{Depth}(k)$.*

5 Undecidability Results

We complement our decidability results with undecidability results for the remaining questions. Recall that a problem L is in the class Σ_1 if it is semi-decidable, and in Σ_2 if it is semi-decidable using a Σ_1 oracle. A problem is Π_1 if it is the complement of a Σ_1 problem, and L is $\mathcal{C}$-complete if it is in the class $\mathcal{C}$ and there is a recursive reduction from every $L' \in \mathcal{C}$ to L. The halting problem is Σ_1-complete, the emptiness problem is Π_1-complete, and the finiteness problem is Σ_2-complete [17]. Boundedness for reset Petri nets is also Σ_1-complete [1,2].

Theorem 6 (Undecidability results).

1. *Given $P \in \mathsf{Proc}$, checking if $P \in \mathsf{Name}$ or $P \in \mathsf{Depth}$ are both Σ_1-complete.*
2. *Given $P \in \mathsf{Proc}$ and $k \in \mathbb{N}$, checking if $P \in \mathsf{Breadth}(k)$ is Π_1-complete.*
3. *Given $P \in \mathsf{Proc}$, checking if $P \in \mathsf{Breadth}$ is Σ_2-complete.*
4. *Given $P \in \mathsf{Depth}$, checking if $P \in \mathsf{Breadth}$ is Σ_1-complete.*

Given a process P, since checking $P \in \mathsf{Depth}(k)$ and $P \in \mathsf{Name}(k)$ is decidable, $P \in \mathsf{Depth}$ and $P \in \mathsf{Name}$ are both Σ_1. For the problem $P \in \mathsf{Breadth}(k)$ to be in Π_1, we semi-decide the complement. We enumerate processes until we find $Q \in \mathcal{R}(P)$ with $|Q|_{\mathcal{B}} > k$. With an oracle for this, we can semi-decide $P \in \mathsf{Breadth}$ by enumerating the possible bounds k and querying the oracle.

We show hardness for these problems. For a Turing machine M with input x, we construct a process $P(M, x)$ that simulates M and in each step increases the number of names and the depth. If M halts on x, then $P(M, x)$ is name-, and also depth-bounded. If M does not halt on x, then $P(M, x)$ is not depth-bounded.

For problems related to breadth boundedness, we show that the computation of a Turing machine M simultaneously on all possible inputs can be simulated by a process $P(M)$ in constant breadth. We increment the breadth of $P(M)$ every time M accepts an input. Restricted breadth boundedness is Π_1-hard, because we can reduce emptiness of M to this question. Breadth boundedness is Σ_2-hard, since the simulating process has bounded breadth iff $L(M)$ is finite.

To show that boundedness in breadth is Σ_1-complete even for $P \in \mathsf{Depth}$, we reduce the boundedness problem for reset Petri nets to breadth boundedness. We simulate net N with a depth-bounded $P(N)$. Tokens are send actions, consumed and produced when transitions fire. To reset a place, we create a fresh name for its tokens. The breadth of $P(N)$ is related to the token count in N, and $P(N)$ is bounded in breadth if and only if N is bounded.

This shows that the coverability set is not computable for $P \in \mathsf{Depth}$. The coverability set of $P(N)$ would allow to decide boundedness of the reset net N.

References

1. Dufourd, C., Finkel, A., Schnoebelen, P.: Reset nets between decidability and undecidability. In: Larsen, K.G., Skyum, S., Winskel, G. (eds.) ICALP 1998. LNCS, vol. 1443, pp. 103–115. Springer, Heidelberg (1998)
2. Dufourd, C., Jančar, P., Schnoebelen, P.: Boundedness of reset P/T nets. In: Wiedermann, J., Van Emde Boas, P., Nielsen, M. (eds.) ICALP 1999. LNCS, vol. 1644, pp. 301–310. Springer, Heidelberg (1999)
3. Emerson, E.A., Namjoshi, K.S.: On model checking for non-deterministic infinite-state systems. In: LICS, pp. 70–80 (June 1998)
4. Finkel, A.: A generalization of the procedure of Karp and Miller to well structured transition systems. In: Ottmann, T. (ed.) ICALP 1987. LNCS, vol. 267, pp. 499–508. Springer, Heidelberg (1987)
5. Finkel, A.: Reduction and covering of infinite reachability trees. Inf. Comp. 89(2), 144–179 (1990)
6. Finkel, A., Goubault-Larrecq, J.: Forward analysis for WSTS, part I: Completions. In: STACS, pp. 433–444 (2009)
7. Finkel, A., Goubault-Larrecq, J.: Forward analysis for WSTS, part II: Complete WSTS. In: Albers, S., Marchetti-Spaccamela, A., Matias, Y., Nikoletseas, S., Thomas, W. (eds.) ICALP 2009, Part II. LNCS, vol. 5556, pp. 188–199. Springer, Heidelberg (2009)
8. Finkel, A., Goubault-Larrecq, J.: The theory of WSTS: The case of complete WSTS. In: Haddad, S., Pomello, L. (eds.) PETRI NETS 2012. LNCS, vol. 7347, pp. 3–31. Springer, Heidelberg (2012)
9. Hüchting, R., Majumdar, R., Meyer, R.: A theory of name boundedness. In: D'Argenio, P.R., Melgratti, H. (eds.) CONCUR 2013 – Concurrency Theory. LNCS, vol. 8052, pp. 182–196. Springer, Heidelberg (2013)
10. Karp, R.M., Miller, R.E.: Parallel program schemata. J. Comput. Syst. Sci. 3(2), 147–195 (1969)
11. Meyer, R.: On boundedness in depth in the π-calculus. In: Ausiello, G., Karhumäki, J., Mauri, G., Ong, L. (eds.) IFIP TCS. IFIP, vol. 273, pp. 477–489. Springer, Heidelberg (2008)
12. Meyer, R.: A theory of structural stationarity in the π-calculus. Acta Inf. 46(2), 87–137 (2009)
13. Milner, R.: Communicating and Mobile Systems: the π-Calculus. CUP (1999)
14. Rosa-Velardo, F., de Frutos-Escrig, D.: Forward analysis for petri nets with name creation. In: Lilius, J., Penczek, W. (eds.) PETRI NETS 2010. LNCS, vol. 6128, pp. 185–205. Springer, Heidelberg (2010)
15. Rosa-Velardo, F., Martos-Salgado, M.: Multiset rewriting for the verification of depth-bounded processes with name binding. Inf. Comput. 215, 68–87 (2012)
16. Sangiorgi, D., Walker, D.: The π-calculus: a Theory of Mobile Processes. CUP (2001)
17. Soare, R.I.: Recursively enumerable sets and degrees. Springer (1980)
18. Wies, T., Zufferey, D., Henzinger, T.A.: Forward analysis of depth-bounded processes. In: Ong, L. (ed.) FOSSACS 2010. LNCS, vol. 6014, pp. 94–108. Springer, Heidelberg (2010)

Typing Messages for Free in Security Protocols: The Case of Equivalence Properties*

Rémy Chrétien[1,2], Véronique Cortier[1], and Stéphanie Delaune[2]

[1] LORIA, INRIA Nancy - Grand-Est, Nancy, France
[2] LSV, ENS Cachan & CNRS, Cachan, France

Abstract. Privacy properties such as untraceability, vote secrecy, or anonymity are typically expressed as behavioural equivalence in a process algebra that models security protocols. In this paper, we study how to decide one particular relation, namely trace equivalence, for an unbounded number of sessions.

Our first main contribution is to reduce the search space for attacks. Specifically, we show that if there is an attack then there is one that is well-typed. Our result holds for a large class of typing systems and a large class of determinate security protocols. Assuming finitely many nonces and keys, we can derive from this result that trace equivalence is decidable for an unbounded number of sessions for a class of tagged protocols, yielding one of the first decidability results for the unbounded case. As an intermediate result, we also provide a novel decision procedure in the case of a bounded number of sessions.

1 Introduction

Privacy properties such as untraceability, vote secrecy, or anonymity are typically expressed as behavioural equivalence (*e.g.* [9,5]). For example, the anonymity of Bob is typically expressed by the fact that an adversary should not distinguish between the situation where Bob is present and the situation where Alice is present. Formally, the behaviour of a protocol can be modelled through a process algebra such as CSP or the pi calculus, enriched with terms to represent cryptographic messages. Then indistinguishability can be modelled through various behavioural equivalences. We focus here on trace equivalence, denoted $\approx$. Checking for privacy then amounts into checking for trace equivalence between processes, which is of course undecidable in general. Even in the case of a bounded number of sessions, there are few decidability results and the associated decision procedures are complex [6,22,11]. In this paper, we study trace equivalence in the case of an unbounded number of sessions.

Our Contribution. Our first main contribution is a simplification result, that reduces the search space for attacks: if there is an attack, then there exists a well-typed attack. More formally, we show that if there is a witness (*i.e.* a trace) that $P \not\approx Q$ then there exists a witness which is well-typed w.r.t. P or Q, provided that P and Q are determinate

* The research leading to these results has received funding from the European Research Council under the European Union's Seventh Framework Programme (FP7/2007-2013) / ERC grant agreement n° 258865, project ProSecure, and the ANR project JCJC VIP n° 11 JS02 006 01.

P. Baldan and D. Gorla (Eds.): CONCUR 2014, LNCS 8704, pp. 372–386, 2014.

processes (intuitively, messages that are outputted are completely determined by the interactions of the protocol with the environment, *i.e.* the attacker). This typing result holds for an unbounded number of sessions and an unbounded number of nonces, that is, it holds even if P and Q contain arbitrary replications and NEW operations. It holds for any typing system provided that any two unifiable encrypted subterms of P (or Q) are of the same type. It is then up to the user to adjust the typing system such that this hypothesis holds for the protocols under consideration. For simplicity, we prove this typing result for the case of symmetric encryption and concatenation but we believe that our result could be extended to the other standard cryptographic primitives.

The finer the typing system is, the more our typing result restricts the attack search. In general, our typing result does not yield directly a decidability result since even the simple property of reachability is undecidable for an unbounded number of sessions and arbitrary nonces, even if the messages are of bounded size (*e.g.* [3]). Indeed, our typing system ensures the existence of a well-typed attack (if any) but the number of well-typed traces may remain infinite. To obtain decidability, we further assume a finite number of terms of each type (*i.e.* in particular a finite number of nonces). Decidability of trace equivalence then follows from our main typing result, for a class of *simple* protocols where each subprocess uses a distinct channel (intuitively, a session identifiers).

As an application, we consider the class of tagged protocols introduced by Blanchet and Podelski [8]. An easy way to achieve this in practice by labelling encryption and is actually a good protocol design principle [2,19]. We show that tagged protocols induce a typing system for which trace equivalence is decidable, for simple protocols and for an unbounded number of sessions (but a fixed number of nonces).

Interestingly, the proof of our main typing result involves providing a new decision procedure for trace equivalence in the case of a bounded number of sessions. This is a key intermediate result of our proof. Trace equivalence was already shown to be decidable for a bounded number of sessions (*e.g.* [22,11]) but we propose a novel decision procedure that further provides a *well-typed* witness whenever the two processes are not in trace equivalence. Compared to existing procedures (*e.g.* [22]), we show that it is only necessary to consider unification between encrypted terms. We believe that this new procedure is of independent interest since it reduces the number of traces (executions) that need to be considered. Our result could therefore be used to speed up equivalence checkers like SPEC [22]. Detailed proofs of our results can be found in [14].

Related Work. Formal methods have been very successful for the analysis of security protocols and many decision procedures and tools (*e.g.* [21,20,16]) have been proposed. However, most of these results focus on reachability properties such as confidentiality or authentication. Much fewer results exist for behavioural equivalences. Based on a procedure proposed by Baudet [6], a first decidability result has been proposed for determinate process without else branches, and for equational theories that capture most standard primitives [12]. Then Tiu and Dawson [22] have designed and implemented a procedure for open bisimulation, a notion of equivalence stronger than the standard notion of trace equivalence. Cheval *et al* [11] have proposed and implemented a procedure for processes with else branches and standard primitives. The tool AkisS [10] is also dedicated to trace equivalence but is not guaranteed to terminate. However, all these results focus on a bounded number of sessions. An exception is the tool ProVerif which can handle observational equivalence for an unbounded number of sessions [7]. It actually reasons on a stronger notion of equivalence (which may turn to be too strong in practice) and is again not guaranteed to terminate.

To our knowledge, the only decidability result for an unbounded number of sessions is [13]. It is shown that trace equivalence can be reduced to the equality of languages of pushdown automata. A key hypothesis for reducing to pushdown automata is that protocol rules have at most one variable, that is, at any execution step, any participant knows already every component of the message he received except for at most one component (*e.g.* a nonce received from another participant). Moreover variables shall not occur in key position, *i.e.* agents may not use received keys for encryption. This strongly limits the class of protocols that can be considered and the approach is strictly bound to this "one-variable" hypothesis. In contrast, we can consider here a much wider class of protocols, provided that they are tagged (which is easy to implement).

Our proof technique is inspired from the approach developed by Arapinis *et al* [4] for bounding the size of messages of an attack for the reachability case. Specifically, they show for some class of tagged protocols, that whenever there is an attack, there is a well-typed attack (for a particular typing system). We somehow extend their approach to trace equivalence and more general typing systems.

2 Model for Security Protocols

Security protocols are modelled through a process algebra inspired from [1] that manipulates terms.

2.1 Syntax

Term algebra. We assume an infinite set $\mathcal{N}$ of *names*, which are used to represent keys and nonces, and two infinite disjoint sets of *variables* $\mathcal{X}$ and $\mathcal{W}$. The variables in $\mathcal{W}$ intuitively refer to variables used to store messages learnt by the attacker. We assume a signature $\mathcal{F}$, *i.e.* a set of function symbols together with their arity. We consider:

$$\Sigma_{\mathsf{c}} = \{\mathsf{enc}, \langle\ \rangle\},\ \Sigma_{\mathsf{d}} = \{\mathsf{dec}, \mathsf{proj}_1, \mathsf{proj}_2\},\ \text{and } \Sigma = \Sigma_{\mathsf{c}} \cup \Sigma_{\mathsf{d}}.$$

The symbols dec and enc of arity 2 represent symmetric decryption/encryption. Pairing is modelled using a symbol of arity 2, denoted $\langle\ \rangle$, and projection functions are denoted proj_1 and proj_2. We further assume an infinite set of *constant symbols* Σ_0 to represent atomic data known to the attacker. The symbols in Σ_{c} are constructors whereas those in Σ_{d} are destructors. Both represent functions available to the attacker.

Given a set of A of atoms (*i.e.* names, variables, and constants), and a signature $\mathcal{F} \in \{\Sigma_{\mathsf{c}}, \Sigma_{\mathsf{d}}, \Sigma\}$, we denote by $\mathcal{T}(\mathcal{F}, \mathsf{A})$ the set of terms built from symbols in $\mathcal{F}$, and atoms in A. The subset of $\mathcal{T}(\Sigma_{\mathsf{c}}, \mathsf{A})$ which only contains terms with atoms as a second argument of the symbol enc, is denoted $\mathcal{T}_0(\Sigma_{\mathsf{c}}, \mathsf{A})$. Terms in $\mathcal{T}_0(\Sigma_{\mathsf{c}}, \Sigma_0 \cup \mathcal{N})$ are called *messages*. An attacker builds his own messages by applying functions to terms he already knows. Formally, a computation done by the attacker is modelled by a term, called a *recipe*, built on the signature Σ using (public) constants in Σ_0 as well as variables in $\mathcal{W}$, *i.e.* a term $R \in \mathcal{T}(\Sigma, \Sigma_0 \cup \mathcal{W})$. Note that such a term does not contain any name.

We denote $vars(u)$ the set of variables that occur in u. The application of a substitution σ to a term u is written $u\sigma$, and we denote $dom(\sigma)$ its *domain*. Two terms u_1 and u_2 are *unifiable* when there exists σ such that $u_1\sigma = u_2\sigma$.

The relations between encryption/decryption and pairing/projections are represented through the three following rewriting rules, yielding a convergent rewrite system:

$$\mathsf{dec}(\mathsf{enc}(x,y),y) \to x, \text{ and } \mathsf{proj}_i(\langle x_1, x_2\rangle) \to x_i \text{ with } i \in \{1,2\}.$$

Given $u \in \mathcal{T}(\Sigma, \Sigma_0 \cup \mathcal{N} \cup \mathcal{X})$, we denote by $u\!\downarrow$ its *normal form*. We refer the reader to [18] for the precise definitions of rewriting systems, convergence, and normal forms.

Example 1. Let $s, k \in \mathcal{N}$, and $u = \mathsf{enc}(s,k)$. The term $\mathsf{dec}(u,k)$ models the application of the decryption algorithm on u using k. We have that $\mathsf{dec}(u,k)\!\downarrow = s$.

Process algebra. Let Ch be an infinite set of *channels*. We consider processes built using the following grammar where $u \in \mathcal{T}(\Sigma_c, \Sigma_0 \cup \mathcal{N} \cup \mathcal{X})$, $n \in \mathcal{N}$, and $c, c' \in Ch$:

$$P, Q := 0 \mid \mathsf{in}(c,u).P \mid \mathsf{out}(c,u).P \mid (P \mid Q) \mid !P \mid \mathsf{new}\, n.P \mid \mathsf{new}\, c'.\mathsf{out}(c,c').P$$

The process 0 does nothing. The process "$\mathsf{in}(c,u).P$" expects a message m of the form u on channel c and then behaves like $P\sigma$ where σ is a substitution such that $m = u\sigma$. The process "$\mathsf{out}(c,u).P$" emits u on channel c, and then behaves like P. The variables that occur in u are instantiated when the evaluation takes place. The process $P \mid Q$ runs P and Q in parallel. The process $!P$ executes P some arbitrary number of times. The name restriction "$\mathsf{new}\ n$" is used to model the creation in a process of a fresh random number (*e.g.*, a nonce or a key) whereas channel generation "$\mathsf{new}\, c'.\mathsf{out}(c,c').P$" is used to model the creation of a new channel name that shall immediately be made public. Note that we consider only public channels. It is still useful to generate fresh (public) channel names to let the attacker identify the different sessions of a protocol (as it is often the case in practice through sessions identifiers).

We assume that names are implicitly freshly generated, thus $\mathsf{new}\, k.\mathsf{out}(c,k)$ and $\mathsf{out}(c,k)$ have exactly the same behaviour. The construction "new" becomes important in the presence of replication to distinguish whether some value k is generated at each session, *e.g.* in $!(\mathsf{new}\, k.\mathsf{out}(c,k))$ or not, *e.g.* in $\mathsf{new}\, k.(!\mathsf{out}(c,k))$.

For the sake of clarity, we may omit the null process. We also assume that processes are *name and variable distinct*, *i.e.* any name and variable is at most bound once. For example, in the process $\mathsf{in}(c,x).\mathsf{in}(c,x)$ the variable x is bound once and thus the process is name and variable distinct. By contrast, in $\mathsf{in}(c,x) \mid \mathsf{in}(c,x)$, one occurrence of the variable x would need to be renamed. We write $fv(P)$ for the set of *free variables* that occur in P, *i.e.* the set of variables that are not in the scope of an input.

We assume $Ch = Ch_0 \uplus Ch^{\mathsf{fresh}}$ where Ch_0 and Ch^{fresh} are two infinite and disjoint sets of channels. Intuitively, channels of Ch^{fresh}, denoted $ch_1, \ldots, ch_i, \ldots$ will be used in the semantics to *instantiate* the channels generated during the execution of a protocol. They shall not be part of its specification.

Definition 1. *A protocol P is a process such that P is ground,* i.e. *$fv(P) = \emptyset$; P is name and variable distinct; and P does not use channel names from Ch^{fresh}.*

Example 2. The Otway-Rees protocol [15] is a key distribution protocol using symmetric encryption and a trusted server. It can be described informally as follows:

1. $A \rightarrow B: \ M, A, B, \{N_a, M, A, B\}_{K_{as}}$
2. $B \rightarrow S: \ M, A, B, \{N_a, M, A, B\}_{K_{as}}, \{N_b, M, A, B\}_{K_{bs}}$
3. $S \rightarrow B: \ M, \{N_a, K_{ab}\}_{K_{as}}, \{N_b, K_{ab}\}_{K_{bs}}$
4. $B \rightarrow A: \ M, \{N_a, K_{ab}\}_{K_{as}}$

where $\{m\}_k$ denotes the symmetric encryption of a message m with key k, A and B are agents trying to authenticate each other, S is a trusted server, K_{as} (resp. K_{bs}) is a long term key shared between A and S (resp. B and S), N_a and N_b are nonces generated by A and B, K_{ab} is a session key generated by S, and M is a session identifier.

We propose a modelling of the Otway-Rees protocol in our formalism. We use restricted channels to model the use of unique session identifiers used along an execution of the protocol. Below, k_{as}, k_{bs}, m, n_a, n_b, k_{ab} are names, whereas a and b are constants from Σ_0. We denote by $\langle x_1, \ldots, x_{n-1}, x_n \rangle$ the term $\langle x_1, \langle \ldots \langle x_{n-1}, x_n \rangle \rangle \rangle$.

$$P_{\mathsf{OR}} = ! \,\mathsf{new}\ c_1.\mathsf{out}(c_A, c_1).P_A \mid ! \,\mathsf{new}\ c_2.\mathsf{out}(c_B, c_2).P_B \mid ! \,\mathsf{new}\ c_3.\mathsf{out}(c_S, c_3).P_S$$

where the processes P_A, P_B are given below, and P_S can be defined in a similar way.

$$\begin{aligned} P_A = {} & \mathsf{new}\ m.\mathsf{new}\ n_a.\ \mathsf{out}(c_1, \langle m, \mathsf{a}, \mathsf{b}, \mathsf{enc}(\langle n_a, m, \mathsf{a}, \mathsf{b}\rangle, k_{as})\rangle). \\ & \mathsf{in}(c_1, \langle m, \mathsf{enc}(\langle n_a, x_{ab}\rangle, k_{as})\rangle); \end{aligned}$$

$$\begin{aligned} P_B = {} & \mathsf{in}(c_2, \langle y_m, \mathsf{a}, \mathsf{b}, y_{as}\rangle).\mathsf{new}\ n_b.\mathsf{out}(c_2, \langle y_m, \mathsf{a}, \mathsf{b}, y_{as}, \mathsf{enc}(\langle n_b, y_m, \mathsf{a}, \mathsf{b}\rangle, k_{bs})\rangle). \\ & \mathsf{in}(c_2, \langle y_m, z_{as}, \mathsf{enc}(\langle n_b, y_{ab}\rangle, k_{bs})\rangle).\mathsf{out}(c_2, \langle y_m, z_{as}\rangle) \end{aligned}$$

2.2 Semantics

The operational semantics of a process is defined using a relation over configurations. A *configuration* is a pair $(\mathcal{P}; \phi)$ where:

- $\mathcal{P}$ is a multiset of ground processes.
- $\phi = \{\mathsf{w}_1 \triangleright m_1, \ldots, \mathsf{w}_n \triangleright m_n\}$ is a *frame*, *i.e.* a substitution where $\mathsf{w}_1, \ldots, \mathsf{w}_n$ are variables in $\mathcal{W}$, and $m_1, \ldots, m_n$ are messages, *i.e.* terms in $\mathcal{T}_0(\Sigma_{\mathsf{c}}, \Sigma_0 \cup \mathcal{N})$.

We often write P instead of $(\{P\}; \emptyset)$, and $P \cup \mathcal{P}$ or $P \mid \mathcal{P}$ instead of $\{P\} \cup \mathcal{P}$. The terms in ϕ represent the messages that are known by the attacker. The operational semantics of a process is induced by the relation $\xrightarrow{\alpha}$ over configurations defined below.

$(\mathsf{in}(c,u).P \cup \mathcal{P}; \phi) \xrightarrow{\mathsf{in}(c,R)} (P\sigma \cup \mathcal{P}; \phi)$ where R is a recipe such that $R\phi\!\downarrow$ is a message and $R\phi\!\downarrow = u\sigma$ for some σ with $dom(\sigma) = vars(u)$

$(\mathsf{out}(c,u).P \cup \mathcal{P}; \phi) \xrightarrow{\mathsf{out}(c,\mathsf{w}_{i+1})} (P \cup \mathcal{P}; \phi \cup \{\mathsf{w}_{i+1} \triangleright u\})$ where u is a message and i is the number of elements in ϕ

$(\mathsf{new}\ c'.\mathsf{out}(c,c').P \cup \mathcal{P}; \phi) \xrightarrow{\mathsf{out}(c,ch_i)} (P\{^{ch_i}/_{c'}\} \cup \mathcal{P}; \phi)$ where ch_i is the "next" fresh channel name available in $\mathcal{C}h^{\mathsf{fresh}}$

$(\mathsf{new}\ n.P \cup \mathcal{P}; \phi) \xrightarrow{\tau} (P\{^{n'}/_{n}\} \cup \mathcal{P}; \phi)$ where n' is a fresh name in $\mathcal{N}$

$(!P \cup \mathcal{P}; \phi) \xrightarrow{\tau} (P \cup !P \cup \mathcal{P}; \phi)$

The first rule allows the attacker to send to some process a term built from publicly available terms and symbols. The second rule corresponds to the output of a term by some process: the corresponding term is added to the frame of the current configuration, which means that the attacker can now access the sent term. Note that the term is outputted provided that it is a message. In case the evaluation of the term yields an encryption with a non atomic key, the evaluation fails and there is no output. The third rule corresponds to the special case of an output of a freshly generated channel name. In such a case, the channel is not added to the frame but it is implicitly assumed known to the attacker, as all the channel names. These three rules are the only observable actions. The two remaining rules are quite standard and are unobservable (τ action) from the point of view of the attacker. The relation $\xrightarrow{\alpha_1 \dots \alpha_n}$ between configurations (where $\alpha_1 \dots \alpha_n$ is a sequence of actions) is defined as the transitive closure of $\xrightarrow{\alpha}$.

Given a sequence of observable actions tr, we write $K \stackrel{\mathsf{tr}}{\Longrightarrow} K'$ when there exists a sequence $\alpha_1 \dots \alpha_n$ such that $K \xrightarrow{\alpha_1 \dots \alpha_n} K'$ and tr is obtained from $\alpha_1 \dots \alpha_n$ by erasing all occurrences of τ. For every protocol P, we define its *set of traces* as follows:

$$\mathsf{trace}(P) = \{(\mathsf{tr}, \phi) \mid P \stackrel{\mathsf{tr}}{\Longrightarrow} (\mathcal{P}; \phi) \text{ for some configuration } (\mathcal{P}; \phi)\}.$$

Note that, by definition of $\mathsf{trace}(P)$, $\mathsf{tr}\phi\!\downarrow$ only contains terms from $\mathcal{T}_0(\Sigma_{\mathsf{c}}, \Sigma_0 \cup \mathcal{N})$.

Example 3. Consider the following sequence tr:

$$\mathsf{tr} = \mathsf{out}(c_A, ch_1).\mathsf{out}(c_B, ch_2).\mathsf{out}(ch_1, \mathsf{w}_1).\mathsf{in}(ch_2, \mathsf{w}_1).$$
$$\mathsf{out}(ch_2, \mathsf{w}_2).\mathsf{in}(ch_2, R_0).\mathsf{out}(ch_2, \mathsf{w}_3).\mathsf{in}(ch_1, \mathsf{w}_3)$$

where $R_0 = \langle \mathsf{proj}_{1/5}(\mathsf{w}_2), \mathsf{proj}_{4/5}(\mathsf{w}_2), \mathsf{proj}_{5/5}(\mathsf{w}_2)\rangle$, and $\mathsf{proj}_{i/5}$ is used as a shortcut to extract the ith component of a 5-uplet. Actually such a sequence of actions allows one to reach the following frame with $t_{\mathsf{enc}} = \mathsf{enc}(\langle n_a, m, \mathsf{a}, \mathsf{b}\rangle, k_{as})$:

$$\phi = \{\mathsf{w}_1 \triangleright \langle m, \mathsf{a}, \mathsf{b}, t_{\mathsf{enc}}\rangle, \mathsf{w}_2 \triangleright \langle m, \mathsf{a}, \mathsf{b}, t_{\mathsf{enc}}, \mathsf{enc}(\langle n_b, m, \mathsf{a}, \mathsf{b}\rangle, k_{bs})\rangle, \mathsf{w}_3 \triangleright \langle m, t_{\mathsf{enc}}\rangle\}.$$

We have that $(\mathsf{tr}, \phi) \subset \mathsf{trace}(P_{\mathsf{OR}})$. The first five actions actually correspond to a normal execution of the protocol. Then, the agent who plays P_B will accept in input the message built using R_0, *i.e.* $u = \langle m, \mathsf{enc}(\langle n_a, m, \mathsf{a}, \mathsf{b}\rangle, k_{as}), \mathsf{enc}(\langle n_b, m, \mathsf{a}, \mathsf{b}\rangle, k_{bs})\rangle$. Indeed, this message has the expected form. At this stage, the agent who plays P_B is waiting for a message of the form: $u_0 = \langle m, z_{as}, \mathsf{enc}(\langle n_b, y_{ab}\rangle, k_{bs})\rangle$. The substitution $\sigma = \{z_{as} \triangleright t_{\mathsf{enc}},\ y_{ab} \triangleright \langle m, \mathsf{a}, \mathsf{b}\rangle\}$ is such that $u = u_0\sigma$. Once this input has been done, a message is outputted (action $\mathsf{out}(ch_3, \mathsf{w}_3)$) and given in input to P_A (action $\mathsf{in}(ch_1, \mathsf{w}_3)$).

Note that, at the end of the execution, A and B share a key but it is not the expected one, *i.e.* one freshly generated by the trusted server, but $\langle m, \mathsf{a}, \mathsf{b}\rangle$.

2.3 Trace Equivalence

Intuitively, two protocols are equivalent if they cannot be distinguished by any attacker. Trace equivalence can be used to formalise many interesting security properties, in particular privacy-type properties, such as those studied for instance in [9]. We first introduce a notion of intruder's knowledge well-suited to cryptographic primitives for which the success of decrypting is visible.

Definition 2. *Two frames ϕ_1 and ϕ_2 are* statically equivalent, *$\phi_1 \sim \phi_2$, when we have that $dom(\phi_1) = dom(\phi_2)$, and:*

- *for any recipe R, $R\phi_1\!\downarrow \in \mathcal{T}_0(\Sigma_c, \Sigma_0 \cup \mathcal{N})$ iff $R\phi_2\!\downarrow \in \mathcal{T}_0(\Sigma_c, \Sigma_0 \cup \mathcal{N})$; and*
- *for all recipes R_1 and R_2 such that $R_1\phi_1\!\downarrow, R_2\phi_1\!\downarrow \in \mathcal{T}_0(\Sigma_c, \Sigma_0 \cup \mathcal{N})$, we have that $R_1\phi_1\!\downarrow = R_2\phi_1\!\downarrow$ iff $R_1\phi_2\!\downarrow = R_2\phi_2\!\downarrow$.*

Intuitively, two frames are equivalent if an attacker cannot see the difference between the two situations they represent. If some computation fails in ϕ_1 for some recipe R, *i.e.* $R\phi_1\!\downarrow$ is not a message, it should fail in ϕ_2 as well. Moreover, ϕ_1 and ϕ_2 should satisfy the same equalities. In other words, the ability of the attacker to distinguish whether a recipe R produces a message, or whether two recipes R_1, R_2 produce the same message should not depend on the frame.

Example 4. Consider $\phi_1 = \phi \cup \{\mathsf{w}_4 \rhd \langle m, \mathsf{a}, \mathsf{b}\rangle\}$, and $\phi_2 = \phi \cup \{\mathsf{w}_4 \rhd n\}$ where n is a name. Let $R = \mathsf{proj}_1(\mathsf{w}_4)$. We have that $R\phi_1\!\downarrow = m \in \mathcal{T}_0(\Sigma_c, \Sigma_0 \cup \mathcal{N})$, but $R\phi_2\!\downarrow = \mathsf{proj}_1(n) \notin \mathcal{T}_0(\Sigma_c, \Sigma_0 \cup \mathcal{N})$, hence $\phi_1 \not\sim \phi_2$. This non static equivalence can also be established considering the recipes $R_1 = \langle \mathsf{proj}_1(\mathsf{w}_3), \mathsf{a}, \mathsf{b}\rangle$ and $R_2 = \mathsf{w}_4$. We have that $R_1\phi_1\!\downarrow, R_2\phi_1\!\downarrow \in \mathcal{T}_0(\Sigma_c, \Sigma_0 \cup \mathcal{N})$, and $R_1\phi_1\!\downarrow = R_2\phi_1\!\downarrow$ whereas $R_1\phi_2\!\downarrow \neq R_2\phi_2\!\downarrow$.

Intuitively, two protocols are *trace equivalent* if, however they behave, the resulting sequences of messages observed by the attacker are in static equivalence.

Definition 3. *A protocol P is* trace included *in a protocol Q, written $P \sqsubseteq Q$, if for every $(\mathsf{tr}, \phi) \in \mathsf{trace}(P)$, there exists $(\mathsf{tr}', \phi') \in \mathsf{trace}(Q)$ such that $\mathsf{tr} = \mathsf{tr}'$ and $\phi \sim \phi'$. The protocols P and Q are* trace equivalent, *written $P \approx Q$, if $P \sqsubseteq Q$ and $Q \sqsubseteq P$.*

As illustrated by the following example, restricting messages to only contain atoms in key position also provides the adversary with more comparison power when variables occurred in key position in the protocol.

Example 5. Let $n, k \in \mathcal{N}$ and consider the protocol $P = \mathsf{in}(c, x).\mathsf{out}(c, \mathsf{enc}(n, k))$ as well as the protocol $Q = \mathsf{in}(c, x).\mathsf{out}(c, \mathsf{enc}(\mathsf{enc}(n, x), k))$. An attacker may distinguish between P and Q by sending a non atomic data and observing whether the process can emit. Q will not be able to emit since its first encryption will fail. This attack would not have been detected if arbitrary terms were allowed in key position.

In what follows, we consider *determinate* protocols as defined in [10], *i.e.*, we consider protocols in which the attacker knowledge is completely determined (up to static equivalence) by its past interaction with the protocol participants.

Definition 4. *A protocol P is* determinate *if for any tr, and for any $(\mathcal{P}_1, \phi_1), (\mathcal{P}_2, \phi_2)$ such that $P \stackrel{\mathsf{tr}}{\Longrightarrow} (\mathcal{P}_1, \phi_1)$, and $P \stackrel{\mathsf{tr}}{\Longrightarrow} (\mathcal{P}_2, \phi_2)$, we have that $\phi_1 \sim \phi_2$.*

Assume given two determinate protocols P and Q such that $P \not\sqsubseteq Q$. A *witness of non-inclusion* is a trace tr for which there exists ϕ such that $(\mathsf{tr}, \phi) \in \mathsf{trace}(P)$ and:

- either there does not exist ϕ' such that $(\mathsf{tr}, \phi') \in \mathsf{trace}(Q)$,
- or such a ϕ' exists and $\phi \not\sim \phi'$.

A *witness of non-equivalence* for determinate protocols P and Q is a trace tr that is a witness for $P \not\sqsubseteq Q$ or $Q \not\sqsubseteq P$. Note that when a protocol P is determinate, once the sequence tr is fixed, all the frames reachable through tr are actually in static equivalence, which ensures the unicity of ϕ', if it exists, up-to static equivalence.

Example 6. We wish to check strong secrecy of the exchanged key received by the agent A for the Otway-Rees protocol. A way of doing so is to check that $P^1_{\mathsf{OR}} \approx P^2_{\mathsf{OR}}$ where the two protocols are defined as follows:

- P^1_{OR} is as P_{OR} but we add the instruction $\mathsf{out}(c_1, x_{ab})$ at the end of the process P_A;
- P^2_{OR} is as P_{OR} but we add the instruction new $n.\mathsf{out}(c_1, n)$ at the end of P_A.

The idea is to check whether an attacker can see the difference between the session key obtained by A and a fresh nonce.

As already suggested by the scenario described in Example 3, the secrecy (and so the strong secrecy) of the key received by A is not preserved. More precisely, consider the sequence $\mathsf{tr}' = \mathsf{tr}.\mathsf{out}(ch_1, \mathsf{w}_4)$ where tr is as in Example 3. In particular, $(\mathsf{tr}', \phi_1) \in \mathsf{trace}(P^1_{\mathsf{OR}})$ and $(\mathsf{tr}', \phi_2) \in \mathsf{trace}(P^2_{\mathsf{OR}})$ with $\phi_1 = \phi \cup \{\mathsf{w}_4 \rhd \langle m, \mathsf{a}, \mathsf{b}\rangle\}$ and $\phi_2 = \phi \cup \{\mathsf{w}_4 \rhd n\}$. As described in Example 4, $\phi_1 \not\sim \phi_2$ and thus tr′ is a witness of non-equivalence for P^1_{OR} and P^2_{OR}. This witness is actually a variant of a known attack on the Otway-Rees protocol [15].

3 Existence of a Well-Typed Witness of Non-equivalence

In this section, we present our first main contribution: a simplification result that reduces the search space for attacks. Roughly, when looking for an attack, we can restrict ourselves to consider well-typed traces. This results holds for a general class of typing systems and as soon as the protocols under study are determinate and type-compliant. We first explain these hypotheses and then we state our general simplification result (see Theorem 1). The proof of this simplification result involves to provide a novel decision procedure for trace equivalence in the case of a bounded number of sessions. The novelty of this decision procedure, in comparison to the existing ones, is to provide a well-typed witness whenever the two processes are not in trace equivalence. This key intermediate result is stated in Proposition 1.

3.1 Typing System

Our simplification result holds for a general class of typing systems: we simply require that types are preserved by unification and application of substitutions. These operations are indeed routinely used in decision procedures.

Definition 5. *A typing system is a pair $(\mathcal{T}, \delta)$ where $\mathcal{T}$ is a set of elements called* types, *and δ is a function mapping terms $t \in \mathcal{T}(\Sigma_c, \Sigma_0 \cup \mathcal{N} \cup \mathcal{X})$ to types τ in $\mathcal{T}$ such that:*

- *if t is a term of type τ and σ is a* well-typed *substitution,* i.e. *every variable of its domain has the same type as its image, then $t\sigma$ is of type τ,*

– *for any terms t and t' with the same type,* i.e. $\delta(t) = \delta(t')$ *and which are unifiable, their most general unifier ($mgu(t,t')$) is well-typed.*

We further assume the existence of an infinite number of constants in Σ_0 (resp. variables in $\mathcal{X}$, names in $\mathcal{N}$) of any type.

A straightforward typing system is when all terms are of a unique type, say Msg. Of course, our typing result would then be useless to reduce the search space for attacks. Which typing system shall be used typically depends on the protocols under study. We present in Section 5 a typing system that allows us to reduce the search space (and then derive decidability) for a large subclass of (tagged) protocols.

3.2 Well-Typed Trace

Whether or not a trace is well-typed is defined w.r.t. the set of *symbolic traces* of a protocol. Formally, we define $\xrightarrow{\mathsf{tr}_s}_s$ to be the transitive closure of the relation $\xrightarrow{\alpha_s}_s$ defined between processes as follows:

$$\mathsf{in}(c,u).P \cup \mathcal{P} \xrightarrow{\mathsf{in}(c,u)}_s P \cup \mathcal{P} \qquad !P \cup \mathcal{P} \xrightarrow{\tau}_s P' \cup !P \cup \mathcal{P}$$
$$\mathsf{out}(c,u).P \cup \mathcal{P} \xrightarrow{\mathsf{out}(c,u)}_s P \cup \mathcal{P} \qquad \mathsf{new}\, n.P \cup \mathcal{P} \xrightarrow{\tau}_s P\{^{n'}/_{n}\} \cup \mathcal{P}$$
$$\mathsf{new}\, c'.\mathsf{out}(c,c').P \cup \mathcal{P} \xrightarrow{\mathsf{out}(c,ch_i)}_s P\{^{ch_i}/_{c'}\} \cup \mathcal{P}$$

where P' is equal to P up to renaming of variables that do not occur yet in the trace with fresh ones (of the same type), n' is a fresh name (of the same type as n), and ch_i is the "next" fresh channel name available in $\mathcal{C}h^{\mathsf{fresh}}$.

Then, the set of *symbolic traces* $\mathsf{trace}_s(P)$ of a protocol P is defined as follows:

$$\mathsf{trace}_s(P) = \{\mathsf{tr}_s \mid P \xrightarrow{\mathsf{tr}_s}_s Q \text{ for some } Q\}.$$

Intuitively, the symbolic traces are simply all possible traces before instantiation of the variables, with some renaming to avoid unwanted captures.

Example 7. Let $P_1 = \mathsf{in}(c,x).!\mathsf{new}\, k.\, \mathsf{in}(c, \mathsf{enc}(\langle x,y\rangle,k))$. We have that:

$$\mathsf{tr}_s = \mathsf{in}(c,x).\mathsf{in}(c,\mathsf{enc}(\langle x,y_1\rangle,k_1)).\mathsf{in}(c,\mathsf{enc}(\langle x,y_2\rangle),k_2) \in \mathsf{trace}_s(P_1)$$

Indeed, the variable x is bound before replication.

As stated in the lemma below, any concrete trace is the instance of a symbolic trace.

Lemma 1. *Let P be a protocol and* $(\mathsf{tr},\phi) \in \mathsf{trace}(P)$. *We have that* $\mathsf{tr}\phi{\downarrow} = \mathsf{tr}_s\sigma$ *for some* $\mathsf{tr}_s \in \mathsf{trace}_s(P)$ *and some substitution σ.*

A well-typed trace is simply a trace that is well-typed w.r.t. one of the symbolic traces. Since keys are atomic, some executions may fail when a protocol is about to output a term that contains an encryption with a non atomic key. To detect these behaviours, we need to consider slightly ill-typed traces. Formally, we consider a special constant $\omega \in \Sigma_0$. Its usefulness is illustrated in Example 8.

Definition 6. *A* first-order trace *of P is a sequence* $\mathsf{tr} = \mathsf{tr}_s\sigma$ *where* $\mathsf{tr}_s \in \mathsf{trace}_s(P)$ *and σ is a substitution such that for any* $\mathsf{io}(c,u)$ *that occurs in* tr_s *with* $\mathsf{io} \in \{\mathsf{in},\mathsf{out}\}$ *and u not a channel, then* $u\sigma \in \mathcal{T}_0(\Sigma_c, \Sigma_0 \cup \mathcal{N} \cup \mathcal{X})$. *The trace* tr *is said to be:*

- well-typed *w.r.t. a typing system* $(\mathcal{T}, \delta)$ *if there exists such a* σ *that is well-typed;*
- pseudo-well-typed *w.r.t. a typing system* $(\mathcal{T}, \delta)$ *if there exists such* σ, *as well as* $\mathsf{c}_0 \in \Sigma_0$ *and* σ' *such that* $\sigma = \sigma'\{^{\langle \omega,\omega\rangle}/_{\mathsf{c}_0}\}$ *with* σ' *well-typed.*

Then a trace $(\mathsf{tr}, \phi) \in \mathsf{trace}(P)$ *is* well-typed *(resp.* pseudo-well-typed*) if* $\mathsf{tr}\phi\!\downarrow$ *is well-typed (resp.* pseudo-well-typed*).*

Note that Lemma 1 ensures that $\mathsf{tr}\phi\!\downarrow$ is a first-order trace of P, and a well-typed trace is also pseudo-well-typed.

Example 8. Going back to Example 5, let $\mathsf{tr} = \mathsf{in}(c, \langle \omega, \omega\rangle).\mathsf{out}(c, \mathsf{w}_1)$. We have that $(\mathsf{tr}, \{\mathsf{w}_1 \triangleright \mathsf{enc}(n,k)\}) \in \mathsf{trace}(P)$ while there exists no frame ψ such that $(\mathsf{tr}, \psi) \in \mathsf{trace}(Q)$. Consider the typing system $(\mathcal{T}, \delta)$ such that $\delta(t) = \mathsf{atom}$ for any atom or variable t and $\delta(t) = \neg\mathsf{atom}$ if t is not an atom. We can see there exists no well-typed witness of $P \not\approx Q$ (while P and Q are type-compliant as defined in Definition 7). However, the witness $(\mathsf{tr}, \{\mathsf{w}_1 \triangleright \mathsf{enc}(n,k)\})$ of $P \not\sqsubseteq Q$ is pseudo-well-typed (note that $\langle \omega, \omega\rangle$ occurs in tr). Intuitively, pseudo-well-typed traces harness the ability for the attacker to use the protocol as an oracle to test if some terms (when used in a key position) are atomic.

3.3 Type Compliance

Our main assumption on the typing of protocols is that any two unifiable encrypted subterms are of the same type. The goal of this part is to state this hypothesis formally.

Due to the presence of replication, we need to consider two copies of protocols in order to consider different instances of the variables. Given a protocol P with replication, we define its 2-unfolding $\mathsf{unfold}^2(P)$ to be the protocol such that every occurrence of a process $!R$ in P is replaced by $R \mid R$, and some α-renaming is performed on one copy to ensure names and variables distinctness of the resulting process. Note that if P is a protocol that does not contain any replication, we have that $\mathsf{unfold}^2(P) = P$.

Example 9. Let P_1 be the protocol defined in Example 7. We have that:
$\mathsf{unfold}^2(P_1) = \mathsf{in}(c,x).(\mathsf{new}\ k_1.\mathsf{in}(c, \mathsf{enc}(\langle x, y_1\rangle, k_1)) \mid \mathsf{new}\ k_2.\mathsf{in}(c, \mathsf{enc}(\langle x, y_2\rangle, k_2)))$

We write $St(t)$ for the set of *(syntactic) subterms* of a term t, and $ESt(t)$ the set of its *encrypted subterms, i.e.* $ESt(t) = \{u \in St(t) \mid u \text{ is of the form } \mathsf{enc}(u_1, u_2)\}$. We extend this notion to sets/sequences of terms, and to protocols as expected.

Definition 7. *A protocol* P *is* type-compliant *w.r.t. a typing system* $(\mathcal{T}, \delta)$ *if for every* $t, t' \in ESt(\mathsf{unfold}^2(P))$ *we have that:* t *and* t' *unifiable implies that* $\delta(t) = \delta(t')$.

3.4 Main Result

We are now ready to state our first main contribution: if there is an attack, then there is a pseudo-well-typed attack. This result holds for protocols with replications and nonces.

Theorem 1. *Let* P *and* Q *be two determinate protocols type-compliant w.r.t.* $(\mathcal{T}_1, \delta_1)$ *and* $(\mathcal{T}_2, \delta_2)$ *respectively. We have that* $P \not\approx Q$ *if, and only if, there exists a witness of non-equivalence* tr *such that:*

– *either* $(\mathsf{tr}, \phi) \in \mathsf{trace}(P)$ *for some* ϕ *and* (tr, ϕ) *is pseudo-well-typed w.r.t.* $(\mathcal{T}_1, \delta_1)$*;*
– *or* $(\mathsf{tr}, \psi) \in \mathsf{trace}(Q)$ *for some* ψ *and* (tr, ψ) *is pseudo-well-typed w.r.t.* $(\mathcal{T}_2, \delta_2)$*.*

The key step for proving Theorem 1 is to provide a decision procedure, in the bounded case (*i.e.* processes without replication), that returns a pseudo-well-typed witness of non-equivalence.

Proposition 1. *Let* P *and* Q *be two determinate protocols without replication. There exists an algorithm that decides whether* $P \approx Q$ *and if not, returns a witness* tr *of non-equivalence. Moreover, if* P *and* Q *are type-compliant w.r.t.* $(\mathcal{T}_1, \delta_1)$ *and* $(\mathcal{T}_2, \delta_2)$ *respectively, the witness* tr *of non-equivalence returned by the algorithm is such that:*

– *either* $(\mathsf{tr}, \phi) \in \mathsf{trace}(P)$ *for some* ϕ *and* (tr, ϕ) *is pseudo-well-typed w.r.t.* $(\mathcal{T}_1, \delta_1)$*;*
– *or* $(\mathsf{tr}, \psi) \in \mathsf{trace}(Q)$ *for some* ψ *and* (tr, ψ) *is pseudo-well-typed w.r.t.* $(\mathcal{T}_2, \delta_2)$*.*

The main idea is to assume given a decision procedure (for a bounded number of sessions) for reachability properties such as those proposed in [20,16,23] and to built on top of it a decision procedure for trace equivalence. Our procedure is carefully design to only allow unification between encrypted subterms. To achieve this,

1. we use as a reachability blackbox one that satisfies this requirement. Most of the existing algorithms (*e.g.* [20,16,23]) were not designed with such a goal in mind. However, in the case of the algorithm given in [16], it has already been shown how it can be turned into one that satisfies this requirement [17].
2. we design carefully the remaining of our algorithm to only consider unification between encrypted subterms.

This design allows us to provide a pseudo-well-typed witness when the protocols under study are type-compliant and not trace equivalent.

Then, relying on Proposition 1, the proof of Theorem 1 is almost immediate. Indeed, whenever two determinate type-compliant protocols P and Q are not in trace equivalence, there exists a witness of non-inclusion for $P \sqsubseteq Q$ (or $Q \sqsubseteq P$) for a bounded version of P and Q (unfolding the replications).

4 Decidability Result

Now, assuming finitely many terms of each type, and in particular finitely many nonces, we obtain a new decidability result for trace equivalence, for an unbounded number of sessions. Compared to [13], we no longer need to restrict the number of variables per transition (to one), we allow variables in key positions, and we are more flexible in the control-flow of the program (we may have arbitrary sequences of in and out actions).

4.1 Simple Processes

To establish decidability, we consider the class of simple protocols as given in [12] but we do not allow name restriction. Intuitively, simple protocols are protocols such that each copy of a replicated process has its own channel. This reflects the fact that due to IP addresses and sessions identifiers, an attacker can identify which process and which session he is sending messages to (or receiving messages from).

Definition 8. *A simple protocol P is a protocol of the form $P_U \mid P_B$ where:*

- $P_U = !\mathsf{new}\ c'_1.\mathsf{out}(c_1, c'_1).B_1 \mid \ldots \mid !\mathsf{new}\ c'_m.\mathsf{out}(c_m, c'_m).B_m$; *and*
- $P_B = B_{m+1} \mid \cdots \mid B_{m+n}$.

Each B_i with $1 \leq i \leq m$ (resp. $m < i \leq m+n$) is a ground process on channel c'_i (resp. c_i) built using the following grammar:

$B := 0 \mid \mathsf{in}(c'_i, u).B \mid \mathsf{out}(c'_i, u).B$ *where* $u \in \mathcal{T}_0(\Sigma_c, \Sigma_0 \cup \mathcal{N} \cup \mathcal{X})$.

Moreover, we assume that $c_1, \ldots, c_n, c_{n+1}, \ldots, c_{n+m}$ are pairwise distinct.

Example 10. The protocol presented in Example 2 is not simple yet: we need to consider only finitely many nonces. To achieve this, we may remove all the instructions "new n" with $n \in \mathcal{N}$ that occur in the process. Note that removing for instance "new n_a" from the process P_A means that n_a is still modelled as a name, and thus it is unknown to the attacker. However, we do not assume anymore that a fresh nonce is generated at each session.

Simple protocols form a large class of protocols that are determinate: the attacker knows exactly who is sending a message or from whom he is receiving a message. Given a simple protocol P and a sequence of observable actions tr, there is a unique configuration $(\mathcal{P}; \phi)$ (up to some internal reduction steps) such that $P \stackrel{\mathsf{tr}}{\Longrightarrow} (\mathcal{P}; \phi)$.

Lemma 2. *A simple protocol is determinate.*

4.2 Main Result

Our decidability result relies on the assumption that there are finitely many terms of each type (of the protocol), once the number of constants is bound for each type.

Formally, we say that a typing system $(\mathcal{T}, \delta)$ is *finite* if, for any set $\mathsf{A} \subseteq \mathcal{N} \cup \Sigma_0$ such that there is a finite number of names/constants of each type, then there are finitely many terms of each type, that is, for any $\tau \in \mathcal{T}$, the following set is finite and computable:

$$\{t \in \mathcal{T}(\Sigma_c, \mathsf{A}) \mid \delta(t) = \tau\}.$$

Theorem 2. *The problem of deciding whether two simple protocols P and Q, type-compliant w.r.t. some finite typing systems $(\mathcal{T}_1, \delta_1)$ and $(\mathcal{T}_2, \delta_2)$ are trace equivalent* (i.e. $P \approx Q$) *is decidable.*

Proof. (Sketch) Since simple protocols are determinate (see Lemma 2), we obtain, thanks to our typing result (Theorem 1), the existence of well-typed witness of non-equivalence when such a witness exists. We further show that we can bound the number of useful constants in the witness trace. We then derive from the finiteness of the typing system that the witness trace uses finitely many distinct terms. Therefore, after some point, the trace only reproduces already existing transitions. Using the form of simple protocols, we can then show how to shorten the length of the witness trace. □

5 Application: Tagged Protocols

In this section, we instantiate our general results (Theorems 1 and 2) by exhibiting a class of protocols that is type-compliant for rather fine-grained typing systems. We consider tagged protocols, for a notion of tagging similar to one introduced by Blanchet [8].

Assume given a protocol P and an unfolding P' of it (remember that when computing $\mathsf{unfold}^2(P)$ names and variables are renamed to avoid clashes). Let u be a term in $\mathcal{T}(\Sigma_c, \Sigma_P \cup \mathcal{N}'_P \cup \mathcal{X}'_P)$ where Σ_P, $\mathcal{N}'_P$, $\mathcal{X}'_P$ are the constants, names, and variables occurring in P', we denote by $\overline{u}$ the transformation that replaces any name and variable occurring in u by its representative in $\mathcal{N}_P$ and $\mathcal{X}_P$ where $\mathcal{N}_P$ and $\mathcal{X}_P$ are the names and variables occurring in P.

Definition 9. *A protocol P is* tagged *if there exists a substitution σ_P such that for any $s_1, s_2 \in ESt(\mathsf{unfold}^2(P))$ with s_1 and s_2 unifiable, we have that $\overline{s_1\sigma_P} = \overline{s_2\sigma_P}$.*

Tagging can easily be enforced by labelling encrypted terms, as proposed in [8].

Definition 10. *A protocol P is* strongly tagged *if:*

1. *any term in $ESt(P)$ is of the form $\mathsf{enc}(\langle \mathsf{c}, m\rangle, k)$ for some $\mathsf{c} \in \Sigma_0$; and*
2. *there exists σ_P such that for any $s, t \in ESt(P)$ with $s = \mathsf{enc}(\langle \mathsf{c}_0, s_1\rangle, s_2)$ and $t = \mathsf{enc}(\langle \mathsf{c}_0, t_1\rangle, t_2)$ for some $\mathsf{c}_0 \in \Sigma_0$, we have that $s\sigma_P = t\sigma_P$.*

The second condition requires that there is a a substitution that unifies any two tagged terms unless their tags differ. This condition is easy to achieve for executable protocols. More precisely, assume a protocol admits an execution where each protocol step (in and out) is executed once (*i.e.* there is one honest execution). This protocol can be easily strongly tagged by adding a distinct tag in each encrypted term.

Lemma 3. *Let P be a protocol. If P is strongly tagged then P is tagged.*

Example 11. In our modelling of the Otway-Rees protocol, the protocols P^1_{OR} and P^2_{OR} (as described in Example 6) are not tagged. For instance, consider the terms $s_1 = \mathsf{enc}(\langle n_a, m, \mathsf{a}, \mathsf{b}\rangle, k_{as})$ and $s_2 = \mathsf{enc}(\langle n_a, x_{ab}\rangle, k_{as})$. Both are encrypted subterms of P_A (and thus of $\mathsf{unfold}^2(P^1_{\mathsf{OR}})$ and $\mathsf{unfold}^2(P^2_{\mathsf{OR}})$) and s_1 and s_2 are unifiable. Now, let $s_3 = \mathsf{enc}(\langle z_a, k_{ab}\rangle, k_{as})$. Actually, s_3 is an encrypted subterm of P_S which is unifiable with s_2. However, there exists no substitution σ such that $\overline{s_1\sigma} = \overline{s_2\sigma} = \overline{s_3\sigma}$.

We can consider a tagged, and safer, version of the Otway-Rees protocol by introducing 4 different tags, denoted 1,2,3 and 4, that are modelled using constants from Σ_0.

$$P'_{\mathsf{OR}} = !\ \mathsf{new}\ c_1.\mathsf{out}(c_A, c_1).P'_A \mid !\ \mathsf{new}\ c_2.\mathsf{out}(c_B, c_2).P'_B \mid !\ \mathsf{new}\ c_3.\mathsf{out}(c_S, c_3).P'_S$$

$$\begin{aligned}
P'_A = {} & \mathsf{new}\ m.\mathsf{new}\ n_a.\ \mathsf{out}(c_1, \langle m, \mathsf{a}, \mathsf{b}, \mathsf{enc}(\langle 1, n_a, m, \mathsf{a}, \mathsf{b}\rangle, k_{as})\rangle).\\
& \mathsf{in}(c_1, \langle m, \mathsf{enc}(\langle 2, n_a, x_{ab}\rangle, k_{as})\rangle)\\
P'_B = {} & \mathsf{in}(c_2, \langle y_m, \mathsf{a}, \mathsf{b}, y_{as}\rangle).\\
& \mathsf{new}\ n_b.\ \mathsf{out}(c_2, \langle y_m, \mathsf{a}, \mathsf{b}, y_{as}, \mathsf{enc}(\langle 3, n_b, y_m, \mathsf{a}, \mathsf{b}\rangle, k_{bs})\rangle).\\
& \mathsf{in}(c_2, \langle y_m, z_{as}, \mathsf{enc}(\langle 4, n_b, y_{ab}\rangle, k_{bs})\rangle).\mathsf{out}(c_2, \langle y_m, z_{as}\rangle)\\
P'_S = {} & \mathsf{in}(c_3, \langle z_m, \mathsf{a}, \mathsf{b}, \mathsf{enc}(\langle 1, z_a, z_m, \mathsf{a}, \mathsf{b}\rangle, k_{as}), \mathsf{enc}(\langle 3, z_b, z_m, \mathsf{a}, \mathsf{b}\rangle, k_{bs})\rangle).\\
& \mathsf{new}\ k_{ab}.\ \mathsf{out}(c_3, \langle z_m, \mathsf{enc}(\langle 2, z_a, k_{ab}\rangle, k_{as}), \mathsf{enc}(\langle 4, z_b, k_{ab}\rangle, k_{bs})\rangle)
\end{aligned}$$

and P'^1_{OR} and P'^2_{OR} are defined similarly as P^1_{OR} and P^2_{OR} relying on P'_{OR} instead of P_{OR}. Note that tr′ is no longer a witness of $P'^1_{\mathsf{OR}} \not\approx P'^2_{\mathsf{OR}}$ as the attack has been removed by this tagging scheme. We can show that P'_{OR} is strongly tagged: consider the natural execution of P'_{OR}, matching inputs and outputs as intended. From this execution we can define:

$$\sigma_P = \{x_{ab} \triangleright k_{ab},\ y_m \triangleright m,\ y_{as} \triangleright \mathsf{enc}(\langle 1, n_a, m, \mathsf{a}, \mathsf{b}\rangle, k_{as}),\ z_{as} \triangleright \mathsf{enc}(\langle 2, n_a, k_{ab}\rangle, k_{as}),\ z_m \triangleright m,\ z_a \triangleright n_a,\ z_b \triangleright n_b\}.$$

It is then easy to check that for any two terms s_1 and s_2 that are unifiable, their instances by σ_P are actually identical.

For any tagged protocol, we can infer a finite typing system, and show the type-compliance of the tagged protocol w.r.t. this typing system. Thus, relying on Theorem 2, we derive the following decidability result for simple and tagged protocols.

Corollary 1. *The problem of deciding whether two simple and tagged protocols P and Q are trace equivalent (i.e. $P \approx Q$) is decidable.*

Proof. (Sketch) The first step of the proof consists in associating to a tagged protocol P, a typing system $(\mathcal{T}_P, \delta_P)$ such that P is type-compliant w.r.t. $(\mathcal{T}_P, \delta_P)$. Intuitively, $(\mathcal{T}_P, \delta_P)$ is simply induced by σ_P, the substitution ensuring the tagged condition in Definition 9. For example, the type of a closed term t is t itself while the type of a variable x in P is simply $x\sigma_P$. This definition is then propagated to any term. With such typing systems, we can show that the size of a term (*i.e.* number of function symbols) is smaller than the size "indicated" by its type (*i.e.* the size of the type, viewed as a term). Thus the typing system $(\mathcal{T}_P, \delta_P)$ is finite. We then conclude by applying Theorem 2. □

Example 12. Consider the protocols $\overline{P'^1_{\mathsf{OR}}}$ and $\overline{P'^2_{\mathsf{OR}}}$ obtained from P'^1_{OR} and P'^2_{OR} by removing the instructions corresponding to a name restriction. These protocols are still strongly tagged and are now simple. Thus, our algorithm can be used to check whether these two protocols are in trace equivalence or not. This equivalence actually models a notion of strong secrecy of the key received by A. Since we have bounded the number of nonces, this equivalence does not require that the key is renewed at each session but it requires the key to be indistinguishable from a (private) name, n in our setting.

6 Conclusion

Decidability results for unbounded nonces are rare and complex, even in the reachability case. One of the only results has been established by Ramanujam and Suresh [21], assuming a particular tagging scheme (which itself involves nonces). We plan to explore whether our typing result could be applied to the tagging scheme defined in [21], to derive decidability of trace equivalence in the presence of nonces.

Our main typing result relies on the design of a new procedure in the case of a bounded number of sessions, that preserves typing. Specifically, we show that it is sufficient to consider only unification between encrypted (sub)terms. We think that this result can be applied to existing decision procedures (in particular SPEC [22] and also APTE [11], with some more work) to speed up their corresponding tools. As future work, we plan to implement this optimisation and measure its benefit.

References

1. Abadi, M., Fournet, C.: Mobile values, new names, and secure communication. In: 28th Symposium on Principles of Programming Languages (POPL 2001). ACM Press (2001)
2. Abadi, M., Needham, R.M.: Prudent engineering practice for cryptographic protocols. IEEE Trans. Software Eng. 22(1), 6–15 (1996)
3. Amadio, R.M., Charatonik, W.: On name generation and set-based analysis in the Dolev-Yao model. In: Brim, L., Jančar, P., Křetínský, M., Kučera, A. (eds.) CONCUR 2002. LNCS, vol. 2421, pp. 499–514. Springer, Heidelberg (2002)
4. Arapinis, M., Duflot, M.: Bounding messages for free in security protocols. In: Arvind, V., Prasad, S. (eds.) FSTTCS 2007. LNCS, vol. 4855, pp. 376–387. Springer, Heidelberg (2007)
5. Backes, M., Hritcu, C., Maffei, M.: Automated verification of remote electronic voting protocols in the applied pi-calculus. In: 21st IEEE Computer Security Foundations Symposium (CSF 2008), pp. 195–209. IEEE Computer Society (2008)
6. Baudet, M.: Deciding security of protocols against off-line guessing attacks. In: 12th ACM Conference on Computer and Communications Security (CCS 2005). ACM Press (2005)
7. Blanchet, B., Abadi, M., Fournet, C.: Automated Verification of Selected Equivalences for Security Protocols. In: 20th Symposium on Logic in Computer Science (2005)
8. Blanchet, B., Podelski, A.: Verification of cryptographic protocols: Tagging enforces termination. In: Gordon, A.D. (ed.) FOSSACS 2003. LNCS, vol. 2620, pp. 136–152. Springer, Heidelberg (2003)
9. Bruso, M., Chatzikokolakis, K., den Hartog, J.: Formal verification of privacy for RFID systems. In: 23rd Computer Security Foundations Symposium, CSF 2010 (2010)
10. Chadha, R., Ciobâcă, Ş., Kremer, S.: Automated verification of equivalence properties of cryptographic protocols. In: Seidl, H. (ed.) ESOP 2012. LNCS, vol. 7211, pp. 108–127. Springer, Heidelberg (2012)
11. Cheval, V., Comon-Lundh, H., Delaune, S.: Trace equivalence decision: Negative tests and non-determinism. In: 18th ACM Conference on Computer and Communications Security
12. Cheval, V., Cortier, V., Delaune, S.: Deciding equivalence-based properties using constraint solving. Theoretical Computer Science 492, 1–39 (2013)
13. Chrétien, R., Cortier, V., Delaune, S.: From security protocols to pushdown automata. In: Fomin, F.V., Freivalds, R., Kwiatkowska, M., Peleg, D. (eds.) ICALP 2013, Part II. LNCS, vol. 7966, pp. 137–149. Springer, Heidelberg (2013)
14. Chrétien, R., Cortier, V., Delaune, S.: Typing messages for free in security protocols: the case of equivalence properties. Technical Report 8546, Inria (June 2014)
15. Clark, J., Jacob, J.: A survey of authentication protocol literature: Version 1.0 (1997)
16. Comon-Lundh, H., Cortier, V., Zalinescu, E.: Deciding security properties for cryptographic protocols. Application to key cycles. ACM Transactions on Computational Logic (TOCL) 11(4) (2010)
17. Cortier, V., Delaune, S.: Safely composing security protocols. Formal Methods in System Design 34(1), 1–36 (2009)
18. Dershowitz, N., Jouannaud, J.-P.: Rewrite systems. In: van Leeuwen, J. (ed.) Handbook of Theoretical Computer Science, Elsevier (1990)
19. Guttman, J.D., Thayer, F.J.: Protocol independence through disjoint encryption. In: 13th Computer Security Foundations Workshop (CSFW 2000). IEEE Comp. Soc. Press (2000)
20. Millen, J., Shmatikov, V.: Constraint solving for bounded-process cryptographic protocol analysis. In: 8th ACM Conference on Computer and Communications Security (2001)
21. Ramanujam, R., Suresh, S.P.: Tagging makes secrecy decidable with unbounded nonces as well. In: Pandya, P.K., Radhakrishnan, J. (eds.) FSTTCS 2003. LNCS, vol. 2914, pp. 363–374. Springer, Heidelberg (2003)
22. Tiu, A., Dawson, J.E.: Automating open bisimulation checking for the spi calculus. In: 23rd IEEE Computer Security Foundations Symposium (CSF 2010), pp. 307–321 (2010)
23. Tiu, A., Goré, R., Dawson, J.E.: A proof theoretic analysis of intruder theories. Logical Methods in Computer Science 6(3) (2010)

Using Higher-Order Contracts to Model Session Types★ (Extended Abstract)

Giovanni Bernardi[1] and Matthew Hennessy[2]

[1] IMDEA Software Institute, Madrid, Spain
bernargi@tcd.ie
[2] School of Computer Science and Statistics, University of Dublin, Trinity College, Ireland
matthew.hennessy@cs.tcd.ie

Abstract. Session types are used to describe and structure interactions between independent processes in distributed systems. Higher-order types are needed in order to properly structure delegation of responsibility between processes. In this paper we show that higher-order web-service contracts can be used to provide a fully-abstract model of recursive higher-order session types. The model is set-theoretic, in the sense that the denotation of a contract is given by the set of contracts with which it complies; we use a novel notion of *peer* compliance. A crucial step in the proof of full-abstraction is showing that every contract has a non-empty denotation.

1 Introduction

The purpose of this paper is to show that recursive higher-order session types [15], [11] can be given a behavioural interpretation using web-service contracts [19], which is fully-abstract with respect to the Gay & Hole subtyping [13]. Higher-order session types are necessary to handle *session delegation*, and in turn this calls for the development of a novel form of *peer compliance* between higher-order contracts. Our model interprets a higher-order session type as the set of session types, again higher-order, with which it *complies*. This is formalised by viewing session types as *contracts* [19] and using a notion of compliance, which we call *peer* compliance. The completeness of the model relies on showing that every type has at least one other type with which it complies. We prove this using the recently suggested *type complement* [6]. We also believe that this type complement captures the intuition of complementary behaviour more faithfully than the standard notion of type duality from [15]; in the full report [5] we show that type-checking systems for session types, such as in [23], can be improved by using *type complement* rather than *type duality*.

Session types: The interactions between processes in a complex distributed system often follow a pre-ordained pattern. Session types [21,15] have been proposed as a mechanism for concisely describing and structuring these interactions. As a simple example consider a system consisting of two entities

$$(\nu s)\,(\texttt{urls}?(x^+ : S).\textbf{store} \parallel \texttt{urls}![\,s^+\,].\textbf{cstmr})$$

★ Research supported by SFI project SFI 06 IN.1 1898, and FCT project PTDC/EIA-CCO/122547/2010.

P. Baldan and D. Gorla (Eds.): CONCUR 2014, LNCS 8704, pp. 387–401, 2014.

which first exchange a new private communication channel or *session*, s, over the public address of the store `urls`; using the conventions of [13] the customer sends to the store one endpoint of this private session, namely s^+, and keeps the other endpoint s^- for itself. The session type S determines the nature of the subsequent interaction allowed between the two entities; as an example S could be

$$?[\,\mathtt{Id}\,];\&\langle\, \mathtt{l}_1\colon\ ![\,\mathtt{Addr}\,];?[\,\mathtt{Int}\,];T, \mathtt{l}_2\colon\ ![\,\mathtt{Addr}\,];?[\,\mathtt{Int}\,];T\,\rangle \qquad (1)$$

where `Int`, `Addr`, `Id` are some base types of integers, addresses and credentials respectively, and $\&\langle\, \mathtt{l}_1\colon S_1, \ldots, \mathtt{l}_n\colon S_n\,\rangle$ is a *branch* type which accepts a choice between interaction on any of the predefined labels $\mathtt{l}_i$, followed by the interaction described by the residual type S_i. Thus (1) above dictates that **store** offers a sequence of four interactions on its end s^+ of the session, namely (i) reception of credential, (ii) acceptance of a choice among two commodities labelled by $\mathtt{l}_1$, and $\mathtt{l}_2$, (iii) followed by the receipt of an address, and (iv) the transmission of a price, of type `Int`; subsequent behaviour is determined by the type T.

The behaviour of **cstmr** on the other end of the session, s^-, is required to match the behaviour described by S, thus satisfying a session type which is intuitively dual to S. For example, the dual to (1) above is

$$![\,\mathtt{Id}\,];\oplus\langle\, \mathtt{l}_1\colon\ ?[\,\mathtt{Addr}\,];![\,\mathtt{Int}\,];T', \mathtt{l}_2\colon\ ?[\,\mathtt{Addr}\,];![\,\mathtt{Int}\,];T'\,\rangle \qquad (2)$$

under the assumption that T' is the dual of T. Intuitively, input is dual to output and the dual to a branch type is a *choice* type $\oplus\langle\, \mathtt{l}_1\colon S_1, \ldots\ \mathtt{l}_k\colon S_k\,\rangle$, which allows the process executing the role described by the type to choose one among the labels $\mathtt{l}_i$. These two principles lead to a general definition of the *dual* of a session type T, denoted $\overline{T}$ in [21,15].

In order to allow flexibility to the processes fulfilling the roles described by these types a subtyping relation between session types, $T \preccurlyeq S$, is essential; see [13] for a description of the crucial role played by subtyping. Intuitively $T \preccurlyeq S$ means that any process or component fulfilling the role dictated by the session type S may be used where one is required to fulfil the role dictated by T. Thus subtyping gives an intuitive *comparative semantics* to session types. In Definition 1 of Section 2 we slightly generalise the standard definition of [13], so as to account also for base types such as `Int` and `Bool`.

Recursive types are necessary in order to handle sessions which may allow interactions between their endpoints to go on indefinitely.

Example 1. [An ever-lasting session]

$$\begin{aligned} D_s &= X(y) := y \triangleright \{\,\mathtt{plus}\colon y?(x)\ \mathtt{in}\ y?(z)\ \mathtt{in}\ y![\,x+z\,].X[\,y\,]\ \square \\ &\qquad\qquad\qquad\quad \mathtt{pos}\colon y?(z)\ \mathtt{in}\ x![\,z>0\,].X[\,y\,]\,\} \\ D_c &= Y(x) := x \triangleleft \{\,\mathtt{pos}\colon x![\,\mathit{random}()\,].x?(z)\ \mathtt{in}\ Y[\,x\,]\,\} \\ P\ &= (\nu\kappa)(\mathtt{def}\ D_s\ \mathtt{in}\ \mathtt{def}\ D_c\ \mathtt{in}\ X[\,\kappa^+\,] \parallel Y[\,\kappa^-\,]) \end{aligned}$$

The peer $X[\,\kappa^+\,]$, defined by instantiating D_s, accepts over κ^+ the invocation of one of the two methods `plus` and `pos`, reads the actual parameters, sends the result of the chosen method and starts again. The peer $Y[\,\kappa^-\,]$, defined by instantiating D_c, invokes via its endpoint κ^- the method pos, sends a random number, reads the result of the invoked method, and also starts again. The composition of these two peers in P results in a never ending session in which interaction occurs between the two peers forever. Note that the definition of D_s is a recursive version of the math server of [13]. □

The type T that describes the behaviour of D_s on an endpoint k^+ is naturally expressed using recursion:

$$\mu X.\&\langle \texttt{plus}:\ ?[\texttt{Int}];?[\texttt{Int}];![\texttt{Int}];X, \texttt{pos}:\ ?[\texttt{Int}];![\texttt{Int}];X\rangle$$

Contracts: Web services [19,9] are distributed components which may be combined and extended to offer services to clients. These services are advertised using *contracts*, which are high-level descriptions of the expected behaviour of services. These contracts come equipped with a *sub-contract* relation $\texttt{cnt}_1 \sqsubseteq \texttt{cnt}_2$; intuitively this means that the contract $\texttt{cnt}_2$, or rather a service offering the behaviour described by this contract, may be used as a service which is required to provide the contract $\texttt{cnt}_1$; these abstract contracts are reminiscent of process calculi as CCS and CSP [18,14].

Contracts are very similar, at least syntactically, to sessions types; for example (2) above can very easily be read as the following process description from CCS, !(Id). ($?\texttt{l}_1$.?Addr.!Int.cnt′ + $?\texttt{l}_2$.?Addr.!Int.cnt′). In fact in Section 3 we give the obvious translation $\mathcal{M}$ from the language of session types to that of contracts; however we continue to use the two distinct languages in order to emphasise the intended use of terms. Then if we provide a behavioural theory of contracts it should be possible to explain how session types determine process behaviour via this mapping $\mathcal{M}$, at least along individual sessions. Indeed steps in this direction have already been made in [1,4] restricting session types to the first-order ones, that is types that cannot express session delegation. But, as we will now explain, the use of delegation in session types requires the use of higher-order types, and in turn higher-order contracts, for which suitable behavioural theories are lacking.

Session delegation: Consider the following system where the customer **cstmr** is replaced by **girlf** and there are now four components:

$$(\nu s)(\nu p)(\nu b)(\texttt{urls}![s^+].\texttt{urlb}![p^+].\texttt{urlb}![b^+].\textbf{girlf} \parallel \texttt{urls}?(s^+:S).\textbf{store} \parallel \texttt{urls}?(p^+:S_p).\textbf{bank} \parallel \texttt{urlbf}?(b^+:S_b).\textbf{boyf})$$

Three private sessions s, p, b are created and the positive endpoints are distributed to the **store**, **bank**, and **boyf** respectively. One possible script for the new customer **girlf** is as follows:

(i) send credential to **store**: send id on session s^-
(ii) **delegate** choice of commodity to **boyf**: send session b^- on session s^-
(iii) await **delegation** from **boyf** to arrange payment: receive session s^- back on session b^-.

Thus the session type S_b at which the boyfriend uses the session end b^+ must countenance both the reception and transmission of session ends, rather than simply data. In this case we can take S_b to be the higher-order session type $?[T_1];![T_2];\textsc{end}$, where in turn T_1 is the session type $\oplus\langle\, \mathtt{l}_1 : \; ?[\,\mathtt{Addr}\,]; \text{END}\,\rangle$ and T_2 must allow **girlf** to arrange payment through the **bank**. This in turn means that T_2 is a higher-order session type as payment will involve the transmission of the payment session p.

The combination of delegation and recursion leads to processes with complicated behaviour which in turn puts further strain on the system of session types.

Example 2. [Everlasting generation of finite sessions]
We use the syntax of [23]. Consider the process $P = (\nu\kappa_{\mathbf{o}})(\mathtt{def}\ D\ \mathtt{in}\ X[\,\kappa^{+}_{\mathbf{o}}, \kappa^{-}_{\mathbf{o}}\,])$, where $D := X(x,y) = (\nu\kappa_{\mathbf{f}})(\mathtt{throw}\ x[\,\kappa^{+}_{\mathbf{f}}\,]; \mathbf{0} \parallel \mathtt{catch}\ y(z)\ \mathtt{in} X[\,z, \kappa^{-}_{\mathbf{f}}\,])$. Intuitively, at each iteration the code $X[\,\kappa^{+}_{\mathbf{o}}, \kappa^{-}_{\mathbf{f}}\,]$ has the two endpoints of a pre-existing session, $\kappa_{\mathbf{o}}$, delegates over the endpoint $\kappa^{+}_{\mathbf{o}}$ the endpoint $\kappa^{+}_{\mathbf{f}}$, and then recursively repeats the loop using $\kappa_{\mathbf{f}}$ as pre-existing session.

According to the reduction semantics in [23] the execution of P will never give rise to a communication error or a deadlock. But the endpoint $\kappa^{+}_{\mathbf{f}}$ can only be assigned a session type of the form $\mu X.\, ![\,X\,]; \text{END}$. Such types are forbidden in [2] but they are allowed in the typing systems of [15,13,23,22]. □

If session types are to be explained behaviourally via the translation $\mathcal{M}$ into contracts, the target language of contracts needs to be higher-order. For instance, the type $?[\,T_1\,]; ![\,T_2\,]; \text{END}$ is mapped by $\mathcal{M}$ to the contract $?(\,!\mathtt{l}_1.?(\mathtt{Addr}).\,1\,).?(\,\mathtt{cnt}_2\,).\,1$, where $\mathtt{cnt}_2 = \mathcal{M}(T_2)$. This in turn means that we require a behavioural theory of higher-order contracts. This is the topic of the current paper. In particular we develop a novel sub-contract preorder, which we refer to as the *peer* sub-contract preorder $\sqsubseteq$ with the property that, for all session types,

$$S \preccurlyeq T \text{ if and only if } \mathcal{M}(S) \sqsubseteq \mathcal{M}(T) \tag{3}$$

On the left hand-side we have the subtyping preorder between session types, which determines when processes with session type T can play the role required by type S; on the right-hand side we have a behaviourally determined sub-contract preorder between the interpretation of the types as higher-order contracts. This behavioural preorder is defined in terms of a novel definition of *peer compliance* between these contracts.

In the remainder of this Introduction we briefly outline how the *peer* sub-contract preorder is defined. Intuitively $\sigma_1 \sqsubseteq \sigma_2$, where σ_i are contracts, if every contract ρ which *complies* with σ_1 also *complies* with σ_2. In turn the intuition behind *compliance* is as follows. We say that a contract ρ complies with contract σ, written $\rho \dashv_{\text{p2p}} \sigma$, if any pair of processes in the source language p, q which guarantee the contracts ρ, σ respectively, can interact indefinitely to their mutual satisfaction; in particular if no further interaction is possible between them, individually they both have reached *successful* or *happy* states. We call this concept *mutual* or *peer* compliance, as both participants are required to attain a *happy* state simultaneously. This is in contrast to [9,19,4] where an asymmetric compliance is used, in which only one participant, the client, is required to reach a *happy* state.

In this paper, rather than discussing processes in the source language, how they can interact and how they guarantee contracts, we mimic the interaction between processes using a symbolic semantics between contracts. We define judgements of the form

$$\rho \parallel \sigma \xrightarrow{\tau} \rho' \parallel \sigma' \tag{4}$$

meaning that if p, q, from the source language, guarantee the contracts ρ, σ respectively, then they can interact and evolve to processes p', q' which guarantee the residual contracts ρ', σ' respectively.

For example we will have the judgement $!\texttt{Int}.\rho' \parallel ?\texttt{Real}.\sigma' \xrightarrow{\tau} \rho' \parallel \sigma'$. On the left-hand side of the parallel constructor $\parallel$ we have a contract guaranteed by a process that supplies an `Int`; on the right-hand side there is a contract guaranteed by a process which will accept a datum that can be used as a real. Since we are assuming that integers can be interpreted as reals, that is $\texttt{Int} \preccurlyeq_b \texttt{Real}$, we know that an interaction described by the judgement above takes place.

However it is unclear when an interaction of the form

$$!(\sigma_1).\rho' \parallel ?(\sigma_2).\sigma' \xrightarrow{\tau} \rho' \parallel \sigma' \tag{5}$$

should take place. Here on the left is a contract satisfied by a process which provides a session endpoint that satisfies the contract σ_1; on the right is a contract satisfied by a process that accepts any session endpoint which guarantees the contract σ_2. Intuitively the interaction should be allowed if σ_1 is a sub-contract of σ_2, that is $\sigma_2 \lessapprox \sigma$. However the whole purpose of defining the judgements (4) above is in order to define the preorder $\lessapprox$; there is a circularity in our arguments.

We break this circularity by supposing a predefined sub-contract preorder $\mathcal{B}$ and allowing the interaction (5) whenever $\sigma_1 \;\; \mathcal{B} \;\; \sigma_2$. More generally we develop a parametrised theory, with interaction judgements of the form $\rho \parallel \sigma \xrightarrow{\tau}_{\mathcal{B}} \rho' \parallel \sigma'$ leading to a parametrised peer-compliance relation $\sigma \dashv^{\mathcal{B}}_{\text{p2p}} \rho$ which in turn leads to a parametrised sub-contract preorder $\rho_1 \sqsubseteq^{\mathcal{B}} \rho_2$. We then prove the main result of the paper, (3) above, by showing:

> There exists some preorder $\mathcal{B}_0$ over higher-order contracts such that $S \preccurlyeq T$ if and only if $\mathcal{M}(S) \sqsubseteq^{\mathcal{B}_0} \mathcal{M}(T)$

This particular preorder $\mathcal{B}_0$, which we construct and in (3) above has been referred to as $\lessapprox$, has a natural behavioural interpretation. It satisfies the behavioural equation

$$\sigma_1 \; \mathcal{B}_0 \; \sigma_2 \text{ if and only if } \sigma_1 \sqsubseteq^{\mathcal{B}_0} \sigma_2 \tag{6}$$

Moreover it is the largest preorder between higher-order contracts which satisfies (6).

The proof of (6) depends on an alternative syntactic characterisation of the set-based preorders $\sqsubseteq^{\mathcal{B}}$ which in turn relies crucially on a natural property of the peer-compliance relations:

> For every contract σ there exists a complementary contract, $\mathsf{cplmt}(\sigma)$, which complies with it, $\sigma \dashv^{\mathcal{B}}_{\text{p2p}} \mathsf{cplmt}(\sigma)$. $(\star)$

In view of the natural correspondence between contracts and session types there is a natural candidate for complementary contracts. Intuitively the dual of a type $\overline{T}$ is designed to capture the complementary behaviour expressed by the type T. Moreover the duality function on session types discussed on page 388 immediately extends to contracts; specifically we can define $\overline{\sigma}$ to be $\mathcal{M}(\overline{\mathcal{M}^{-1}(\sigma)})$.

However, somewhat surprisingly, there are contracts σ which do not comply with their duals, $\sigma \not\dashv^{\mathcal{B}}_{\text{p2P}} \overline{\sigma}$; see Example 4. However ($\star$) above can be established by using instead a different notion of dual, first proposed in [6] for typing *copyless message-passing processes*. We also believe that this alternative notion, which in this paper we call *complement*, captures the intuitive notion of complementary behaviour more faithfully than the standard duality.

Paper structure: In Section 2 we recall the standard theory of recursive higher-order session types, while Section 3 introduces higher-order contracts and our novel parametrised peer sub-contract preorder $\sqsubseteq^{\mathcal{B}}$. Although the definition of this preorder is set-theoretic, it can be characterised using only the syntactic form of contracts; this stems from the very restricted form that our higher-order contracts can take. This is also discussed in Section 3. Using this syntactic characterisation we develop enough properties of the preorders $\sqsubseteq^{\mathcal{B}}$ to ensure the existence of the particular preorder $\mathcal{B}_0$ alluded to in (6) above; this is the topic of Section 4. The complementation operator on contracts from [6], $\mathsf{cplmt}(\sigma)$, alluded to above is also defined and discussed in Section 4. Related work is then discussed in Section 5.

All the proofs and the technical details are omitted from this extended abstract, and can be found in the companion report [5].

2 Session Types

Here we recall, using the notation from [13], the standard theory of subtyping for recursive session types. The grammar for the language L_{STyp} of session type terms is given by the following grammar, which uses a collection of unspecified base types $\mathtt{BT}$, of which we enumerate a sample.

$$
\begin{aligned}
S, T \; &::= \text{END} \mid X \mid ?[M]; S \mid ![M]; S \mid \mu X.S \mid \&\langle \mathtt{l}_1 \colon S_1, \ldots, \mathtt{l}_n \colon S_n \rangle \\
&\quad\; \oplus\langle \mathtt{l}_1 \colon S_1, \ldots, \mathtt{l}_n \colon S_n \rangle \\
M, N \; &::= S \mid \mathtt{t} \\
\mathtt{t} \; &::= \mathtt{Id}, \mathtt{Addr}, \mathtt{Int}, \mathtt{Real}, \ldots
\end{aligned}
$$

In the grammar above we assume $n \geq 1$; moreover we use a denumerable set of labels, $\mathtt{L} = \{\mathtt{l}_1, \mathtt{l}_2, \mathtt{l}_3, \ldots\}$, in the *branch* and *choice* constructs. Recall from the Introduction that $\&\langle \mathtt{l}_1 \colon S_1, \ldots, \mathtt{l}_n \colon S_n \rangle$ offers different possible behaviours based on a set of labels $\{\mathtt{l}_1, \mathtt{l}_2, \mathtt{l}_3, \ldots \mathtt{l}_n\}$ while $\oplus\langle \mathtt{l}_1 \colon S_1, \ldots, \mathtt{l}_n \colon S_n \rangle$ takes a choice of behaviours; in both constructs the labels used are assumed to be distinct.

We use STyp to denote the set of session type terms in L_{STyp} which are *closed* and *guarded*; both these concepts have standard definitions, which may be found in [5, Appendix A]. We refer to the terms in STyp as session types. For instance $\mu X.X$ and $\&\langle \mathtt{tea} \colon \mu X.X \rangle$ are not in STyp.

Subtyping is defined coinductively and uses some unspecified subtyping preorder $\preccurlyeq_{\mathsf{b}}$ between base types, a typical example being $\mathtt{Int} \preccurlyeq_{\mathsf{b}} \mathtt{Real}$, meaning that an integer may be supplied where a real number is required. Recursive types are handled by a standard function $\mathsf{unfold}(T)$ which unfolds all the first-level occurrences of $\mu X.-$ in the (guarded) type T. The formal definition of unfold in turn depends on the definition of substitution $T\{S/X\}$, the syntactic substitution of the term S for all free occurrences of X in T. The details may be found in [5].

Definition 1. *[Subtyping]*
Let $\mathcal{F}\preccurlyeq : \mathcal{P}(\mathsf{STyp}^2) \longrightarrow \mathcal{P}(\mathsf{STyp}^2)$ *be the functional defined so that* $(T,U) \in \mathcal{F}\preccurlyeq(\mathcal{R})$ *whenever one of the following holds:*

(i) *if* $\mathsf{unfold}(T) = \textsc{end}$ *then* $\mathsf{unfold}(U) = \textsc{end}$
(ii) *if* $\mathsf{unfold}(T) = ?[\,\mathtt{t}_1\,];S_1$ *then* $\mathsf{unfold}(U) = ?[\,\mathtt{t}_2\,];S_2$ *and* $S_1\,\mathcal{R}\,S_2$ *and* $\mathtt{t}_1 \preccurlyeq_{\mathsf{b}} \mathtt{t}_2$
(iii) *if* $\mathsf{unfold}(T) = ![\,\mathtt{t}_1\,];S_1$ *then* $\mathsf{unfold}(U) = ![\,\mathtt{t}_2\,];S_2$ *and* $S_1\,\mathcal{R}\,S_2$ *and* $\mathtt{t}_2 \preccurlyeq_{\mathsf{b}} \mathtt{t}_1$
(iv) *if* $\mathsf{unfold}(T) = ![\,T_1\,];S_1$ *then* $\mathsf{unfold}(U) = ![\,T_2\,];S_2$ *and* $S_1\,\mathcal{R}\,S_2$ *and* $T_2\,\mathcal{R}\,T_1$
(v) *if* $\mathsf{unfold}(T) = ?[\,T_1\,];S_1$ *then* $\mathsf{unfold}(U) = ?[\,T_2\,];S_2$ *and* $S_1\,\mathcal{R}\,S_2$ *and* $T_1\,\mathcal{R}\,T_2$
(vi) *if* $\mathsf{unfold}(T) = \&\langle\,\mathtt{l}_1 : S_1, \ldots l_m : S_m\,\rangle$ *then* $\mathsf{unfold}(U) = \&\langle\,\mathtt{l}_1 : S'_1, \ldots, \mathtt{l}_n : S'_n\,\rangle$ *where* $m \le n$ *and* $S_i\,\mathcal{R}\,S'_i$ *for all* $i \in [1,\ldots,m]$
(vii) *if* $\mathsf{unfold}(T) = \oplus\langle\,\mathtt{l}_1 : S_1, \ldots l_m : S_m\,\rangle$ *then* $\mathsf{unfold}(U) = \oplus\langle\,\mathtt{l}_1 : S'_1, \ldots, \mathtt{l}_n : S'_n\,\rangle$ *where* $n \le m$ *and* $S_i\,\mathcal{R}\,S'_i$ *for all* $i \in [1,\ldots,n]$

If $\mathcal{R} \subseteq \mathcal{F}\preccurlyeq(\mathcal{R})$, *then we say that* R *is a* type simulation. *Standard arguments ensure that there exists the greatest solution of the equation* $\mathcal{R} = \mathcal{F}\preccurlyeq(\mathcal{R})$*; we call this solution the* subtyping, *and we denote it* $\preccurlyeq$. □

Intuitively $S \preccurlyeq T$ means that processes adhering to the role dictated by T may be used where processes following the role dictated by S are required. Our aim is to formalise this intuition by proving that the higher-order contracts determined by these types, respectively $\mathcal{M}(S)$ and $\mathcal{M}(T)$ are related behaviourally, using our notion of *peer* compliance.

3 Higher-Order Contracts

Here first we define higher-order session contracts and explain the set-based subcontract preorder on them; this uses the notion of *peer compliance* between them. Afterwards we characterise up-to a parameter $\mathcal{B}$ this set-based preorder. We do so by comparing the purely syntactic structure of contracts.

The grammar for the language of contract terms L_{SCts} is:

$$\rho,\sigma ::= \mathsf{1} \mid ?\mathtt{t}.\sigma \mid !\mathtt{t}.\sigma \mid !(\sigma).\sigma \mid ?(\sigma).\sigma \mid x \mid \mu x.\sigma \mid \sum_{i\in I} ?\mathtt{l}_\mathtt{i}.\sigma_i \mid \bigoplus_{i\in I} !\mathtt{l}_\mathtt{i}.\sigma_i$$

where we assume the labels $\mathtt{l}_i$s to be pairwise distinct and the set I to be non-empty. We use SCts to denote the set of terms which are guarded and closed. These will be referred to as higher-order session contracts, or simply contracts. When I is a singleton set $\{k\}$, we write $!\mathtt{l}_\mathtt{k}.\sigma_k$ and $?\mathtt{l}_\mathtt{k}.\sigma_k$ in place of $\bigoplus_{i\in I} !\mathtt{l}_\mathtt{i}.\sigma_i$ and $\sum_{i\in I} ?\mathtt{l}_\mathtt{i}.\sigma_i$.

The operational meaning of contracts is given by interpreting them as processes from a simple process calculus. To this end let Act, ranged over by λ, be the union of three sets, namely $\{\,?\mathtt{l}, !\mathtt{l} \mid \mathtt{l} \in \mathtt{L}\,\}$, $\{\,?\mathtt{t}, !\mathtt{t} \mid \mathtt{t} \in \mathtt{BT}\,\}$, and $\{\,?(\sigma), !(\sigma) \mid \sigma \in \mathsf{SCts}\,\}$. We use Act_τ to denote the set $\mathsf{Act} \cup \{\,\tau\,\}$ to emphasise that the special symbol τ is not in Act.

We define judgements of the form $\sigma_1 \xrightarrow{\mu} \sigma_2$, where $\mu \in \mathsf{Act}_\tau$ and $\sigma_1, \sigma_2 \in \mathsf{SCts}$, by using the following (standard) axioms, where $|\,I\,|$ is the cardinality of I,

$$\frac{}{\lambda.\sigma \xrightarrow{\lambda} \sigma}\ \lambda \in \mathsf{Act} \qquad \frac{}{\mu x.\sigma \xrightarrow{\tau} \sigma\{\mu x.\sigma/x\}}$$

$$\frac{}{\bigoplus_{i \in I} !\mathtt{l}_\mathtt{i}.\sigma_i \xrightarrow{\tau} !\mathtt{l}_\mathtt{i}.\sigma_i}\ |\,I\,| > 1 \qquad \frac{}{\sum_{i \in I} ?\mathtt{l}_\mathtt{i}.\sigma_i \xrightarrow{?\mathtt{l}_\mathtt{i}} \sigma_i}$$

We also have the special judgement $\mathsf{1} \xrightarrow{\checkmark}$, which formalises operationally that $\mathsf{1}$ is the *satisfied* contract. Although terms like $!\mathtt{l}.\sigma$ stand actually for singleton internal sums, we infer their semantics by using the rule for prefixes; for example $!\mathtt{l}.\sigma \xrightarrow{!\mathtt{l}} \sigma$.

In order to define the *peer compliance* between two contracts ρ, σ, we also need to say when two processes p, q satisfying these contracts can interact. This is formalised indirectly as a relation of the form $\rho \parallel \sigma \xrightarrow{\tau}_{\mathcal{B}} \rho' \parallel \sigma'$ which, as explained in the Introduction, is designed to capture the intuition that if processes p, q satisfy the contracts ρ, σ respectively, then they can interact and their residuals will satisfy the residual contracts ρ', σ' respectively.

The relation $\longrightarrow_{\mathcal{B}}$ is determined by the following inference rules:

$$\frac{\rho \xrightarrow{\tau} \rho'}{\rho \parallel \sigma \xrightarrow{\tau}_{\mathcal{B}} \rho' \parallel \sigma} \qquad \frac{\sigma \xrightarrow{\tau} \sigma'}{\rho \parallel \sigma \xrightarrow{\tau}_{\mathcal{B}} \rho \parallel \sigma'} \qquad \frac{\rho \xrightarrow{\lambda_1} \rho' \quad \sigma \xrightarrow{\lambda_2} \sigma'}{\rho \parallel \sigma \xrightarrow{\tau}_{\mathcal{B}} \rho' \parallel \sigma'}\ \lambda_1 \bowtie_{\mathcal{B}} \lambda_2$$

This reduction relation is parametrised on a relation $\sigma_1\ \mathcal{B}\ \sigma_2$ between contracts, which determines when the contract σ_1 can be accepted when σ_2 is required. Using such a $\mathcal{B}$ we define an *interaction* relation between contracts as follows:

$$\lambda_1 \bowtie_{\mathcal{B}} \lambda_2 = \begin{cases} \lambda_1 = !\mathtt{l},\ \lambda_2 = ?\mathtt{l} \\ \lambda_1 = ?\mathtt{l},\ \lambda_2 = !\mathtt{l} \\ \lambda_1 = !\mathtt{t}_1,\ \lambda_2 = ?\mathtt{t}_2 & \mathtt{t}_1 \preccurlyeq_{\mathsf{b}} \mathtt{t}_2 \\ \lambda_1 = ?\mathtt{t}_1,\ \lambda_2 = !\mathtt{t}_2 & \mathtt{t}_2 \preccurlyeq_{\mathsf{b}} \mathtt{t}_1 \\ \lambda_1 = !(\sigma_1),\ \lambda_2 = ?(\sigma_2) & \sigma_1\ \mathcal{B}\ \sigma_2 \\ \lambda_1 = ?(\sigma_1),\ \lambda_2 = !(\sigma_2) & \sigma_2\ \mathcal{B}\ \sigma_1 \end{cases}$$

Essentially the relation $\bowtie_{\mathcal{B}}$ treats $\mathcal{B}$ as a subtyping on contracts.

Definition 2. *[$\mathcal{B}$-Peer compliance]*
Let $C^{\mathrm{P2P}} : \mathcal{P}(\mathsf{SCts}^2) \times \mathcal{P}(\mathsf{SCts}^2) \longrightarrow \mathcal{P}(\mathsf{SCts}^2)$ be the rule functional defined so that $(\rho, \sigma) \in C^{\mathrm{P2P}}(\mathcal{R}, \mathcal{B})$ whenever both the following conditions hold:

(i) if $\rho \parallel \sigma \not\xrightarrow{\tau}_{\mathcal{B}}$ then $\rho \xrightarrow{\checkmark}$ and $\sigma \xrightarrow{\checkmark}$
(ii) if $\rho \parallel \sigma \xrightarrow{\tau}_{\mathcal{B}} \rho' \parallel \sigma'$ then $\rho'\ \mathcal{R}\ \sigma'$

*If $\mathcal{R} \subseteq C^{\mathrm{P2P}}(\mathcal{R}, \mathcal{B})$, then we say that $\mathcal{R}$ is a $\mathcal{B}$-*coinductive peer compliance. *Fix a $\mathcal{B}$. Standard arguments ensure that there exists the greatest solution of the equation $X = C^{\mathrm{P2P}}(X, \mathcal{B})$; we call this solution the $\mathcal{B}$-*peer compliance, *and we denote it $\dashv^{\mathcal{B}}_{\mathrm{P2P}}$.* □

The intuition here is that if $\rho \dashv^{\mathcal{B}}_{\mathrm{p2p}} \sigma$ then processes satisfying these contracts can interact safely; the co-inductive nature of the definition even allows this interaction to continue forever. But if a point is reached where no further interaction is allowed condition (i) means that both participants must be *happy* simultaneously; that is they must be able to perform the success action $\checkmark$.

Definition 3. *[$\mathcal{B}$-peer subcontract preorder]*
For $\sigma_1, \sigma_2 \in \mathsf{SCts}$ *let* $\sigma_1 \sqsubseteq^{\mathcal{B}} \sigma_2$ *whenever* $\rho \dashv^{\mathcal{B}}_{\mathrm{p2p}} \sigma_1$ *implies* $\rho \dashv^{\mathcal{B}}_{\mathrm{p2p}} \sigma_2$*, for every* $\rho \in \mathsf{SCts}$. □

The parametrised peer subcontract preorder $\sigma_1 \sqsubseteq^{\mathcal{B}} \sigma_2$ is set based, and quantifies over all peers in $\mathcal{B}$-compliance with σ_1. However, because of the restricted syntax of higher-order contracts, it turns out that $\sqsubseteq^{\mathcal{B}}$ can be characterised by the syntactic structure of σ_1 and σ_2, at least for behavioural preorders $\mathcal{B}$ which satisfy certain minimal conditions.

Definition 4. *[$\mathcal{B}$-syntactic peer preorder]*
Let $\mathcal{S} : \mathcal{P}(\mathsf{SCts}^2)\times\mathcal{P}(\mathsf{SCts}^2) \longrightarrow \mathcal{P}(\mathsf{SCts}^2)$ *be the functional defined so that* $(\sigma_1, \sigma_2) \in \mathcal{S}(\mathcal{R}, \mathcal{B})$ *whenever one of the following holds:*

(i) if $\mathsf{unfold}(\sigma_1) = \mathit{1}$ *then* $\mathsf{unfold}(\sigma_2) = \mathit{1}$
(ii) if $\mathsf{unfold}(\sigma_1) = ?\mathtt{t}_1.\sigma_1'$ *then* $\mathsf{unfold}(\sigma_2) = ?\mathtt{t}_2.\sigma_2'$ *and* $\sigma_1' \mathrel{\mathcal{R}} \sigma_2'$ *and* $\mathtt{t}_1 \preccurlyeq_{\mathtt{b}} \mathtt{t}_2$
(iii) if $\mathsf{unfold}(\sigma_1) = !\mathtt{t}_1.\sigma_1'$ *then* $\mathsf{unfold}(\sigma_2) = !\mathtt{t}_2.\sigma_2'$ *and* $\sigma_1' \mathrel{\mathcal{R}} \sigma_2'$ *and* $\mathtt{t}_2 \preccurlyeq_{\mathtt{b}} \mathtt{t}_1$
(iv) if $\mathsf{unfold}(\sigma_1) = !(\sigma_1'').\sigma_1'$ *then* $\mathsf{unfold}(\sigma_2) = !(\sigma_2'').\sigma_2'$ *and* $\sigma_1' \mathrel{\mathcal{R}} \sigma_2'$ *and* $\sigma_2'' \mathrel{\mathcal{B}} \sigma_1''$
(v) if $\mathsf{unfold}(\sigma_1) = ?(\sigma_1'').\sigma_1'$ *then* $\mathsf{unfold}(\sigma_2) = ?(\sigma_2'').\sigma_2'$ *and* $\sigma_1' \mathrel{\mathcal{R}} \sigma_2'$ *and* $\sigma_1'' \mathrel{\mathcal{B}} \sigma_2''$
(vi) if $\mathsf{unfold}(\sigma_1) = \sum_{i\in I} ?\mathtt{l}_i.\sigma_i^1$ *then* $\mathsf{unfold}(\sigma_2) = \sum_{j\in J} ?\mathtt{l}_j.\sigma_j^2$ *where* $I \subseteq J$ *and* $\sigma_i^1 \mathrel{\mathcal{R}} \sigma_i^2$ *for all* $i \in I$
(vii) if $\mathsf{unfold}(\sigma_1) = \bigoplus_{i\in I} !\mathtt{l}_i.\sigma_i^1$ *then* $\mathsf{unfold}(\sigma_2) = \bigoplus_{j\in J} !\mathtt{l}_j.\sigma_j^2$ *where* $J \subseteq I$ *and* $\sigma_j^1 \mathrel{\mathcal{R}} \sigma_j^2$ *for all* $j \in J$

Fix a $\mathcal{B}$*. Since* $\mathcal{S}$ *is monotone ([5, Lemma 3.3]), standard arguments ensure that there exists the greatest solution of the equation* $X = \mathcal{S}(X, \mathcal{B})$*; we call this solution the* $\mathcal{B}$-syntactic peer preorder*, and we denote it by* $\preccurlyeq^{\mathcal{B}}$. □

Our intention is to show that the set-theoretic relation $\sigma_1 \sqsubseteq^{\mathcal{B}} \sigma_2$ coincides with the more amenable syntactically defined relation $\sigma_1 \preccurlyeq^{\mathcal{B}} \sigma_2$, provided $\mathcal{B}$ satisfies some simple properties. In one direction the proof follows directly from the definitions of the relations at issue. In the other we need a non-trivial property of session contracts that we relegate to Section 4; see Theorem 4.

Theorem 1. *Let* $\mathcal{B}$ *be a transitive relation on session contracts. Then* $\sigma_1 \preccurlyeq^{\mathcal{B}} \sigma_2$ *implies* $\sigma_1 \sqsubseteq^{\mathcal{B}} \sigma_2$.

Example 3. Here we show that Theorem 1 requires the relation $\mathcal{B}$ to be transitive. Let $\mathcal{B}$ be $\{(1, \sigma), (\sigma, !\mathtt{l}.1)\}$, where σ is the contract $!\mathtt{l}.!\mathtt{l}.1$; this is obviously not transitive. We show that $\preccurlyeq^{\mathcal{B}} \not\subseteq \sqsubseteq^{\mathcal{B}}$. First, the relation $\mathcal{R} = \{(\sigma_1, \sigma_2), (1, 1)\}$, where $\sigma_1 = !(\sigma).1$, $\sigma_2 = !(1).1$, is a prefixed point of $\mathcal{S}$, and thus $\sigma_1 \preccurlyeq^{\mathcal{B}} \sigma_2$. Now let $\rho = ?(!\mathtt{l}.1).1$. The reason why $\sigma_1 \not\sqsubseteq^{\mathcal{B}} \sigma_2$ is that $\rho \dashv^{\mathcal{B}}_{\mathrm{p2p}} \sigma_1$, because $\{(\rho, \sigma_1), (1, 1)\}$ is a $\mathcal{B}$-coinductive compliance, while $\rho \not\dashv^{\mathcal{B}}_{\mathrm{p2p}} \sigma_2$ because of the computation $\rho \parallel \sigma_2 \stackrel{\tau}{\not\longrightarrow}_{\mathcal{B}}$. □

Theorem 2. *For every preorder on session contracts $\mathcal{B}$, $\sigma_1 \sqsubseteq^{\mathcal{B}} \sigma_2$ implies $\sigma_1 \preccurlyeq^{\mathcal{B}} \sigma_2$.*

Proof (Outline). The argument is by case analysis on $\mathsf{unfold}(\sigma_1)$. For instance, consider the case in which $\mathsf{unfold}(\sigma_1) = !(\mathtt{t}).\sigma'_1$. Thanks to Theorem 4 there exists a ρ' such that $\rho' \dashv^{\mathcal{B}}_{\text{p2P}} \sigma'_1$. It follows that $?(\mathtt{t}).\rho' \dashv^{\mathcal{B}}_{\text{p2P}} \mathsf{unfold}(\sigma_1)$, and so $\rho \dashv^{\mathcal{B}}_{\text{p2P}} \mathsf{unfold}(\sigma_2)$. This is enough to show the properties of $\mathsf{unfold}(\sigma_2)$ required by the definition of $\preccurlyeq^{\mathcal{B}}$.

Corollary 1. *For any preorder $\mathcal{B}$ over session contracts, $\sigma_1 \preccurlyeq^{\mathcal{B}} \sigma_2$ if and only if $\sigma_1 \sqsubseteq^{\mathcal{B}} \sigma_2$.*

4 Modelling Session Types

Session types and contracts, formalisms developed independently, are nevertheless just syntactic variations of each other:

$$\begin{aligned}
&\mathcal{M}(\text{END}) = 1, \quad \mathcal{M}(X) = x, \quad \mathcal{M}(\mu X.S) = \mu x.\mathcal{M}(S), && \mathcal{M}(![T];S') = !(\mathcal{M}(T)).\mathcal{M}(S'),\\
&\mathcal{M}(\&\langle \mathtt{l}_1 : S_1, \ldots, \mathtt{l}_n : S_n\rangle) = \textstyle\sum_{i\in[1;n]} ?\mathtt{l}_i.\mathcal{M}(S_i), && \mathcal{M}(![\mathtt{t}];S') = !\mathtt{t}.\mathcal{M}(S'),\\
&\mathcal{M}(\oplus\langle \mathtt{l}_1 : S_1, \ldots, \mathtt{l}_n : S_n\rangle) = \textstyle\bigoplus_{i\in[1;n]} !\mathtt{l}_i.\mathcal{M}(S_i), && \mathcal{M}(?[T];S') = ?(\mathcal{M}(T)).\mathcal{M}(S'),\\
& && \mathcal{M}(?[\mathtt{t}];S') = ?\mathtt{t}.\mathcal{M}(S')
\end{aligned}$$

Our aim is to show that the subtyping relation between session types, $S \preccurlyeq T$, can be modelled precisely by the set-based contract preorder, $\mathcal{M}(S) \sqsubseteq^{\mathcal{B}} \mathcal{M}(T)$, for a particular choice of $\mathcal{B}$. In order to determine this $\mathcal{B}$ we need to develop some properties of functionals over contracts. Let $\mathcal{P}re$ denote the collection of preorders over the set of contracts SCts; ordered set-theoretically this is a complete lattice [5, Lemma 4.2]. Let $\mathcal{F} : \mathcal{P}re \longrightarrow \mathcal{P}re$ be defined by letting $\mathcal{F}(\mathcal{B})$ be the preorder $\sqsubseteq^{\mathcal{B}}$. By Corollary 1 we know that $\mathcal{F}(\mathcal{B}) = \preccurlyeq^{\mathcal{B}}$, and therefore $\mathcal{F}(\mathcal{B}) = \nu X.\mathcal{S}(X, \mathcal{B})$, from Definition 4. Since $\mathcal{S}$ is monotone in its second parameter ([5, Lemma 3.3 (b)]), the endofunction $\mathcal{F}$ over the complete lattice $\mathcal{P}re$ is monotone. The Knaster-Tarski theorem now ensures that $\mathcal{F}$ has fixed points, in particular a maximal one.

Definition 5. *[Peer subcontract preorder]*
Let $\precsim$ denote $\nu X.\mathcal{F}(X)$, the greatest fixed point of the function $\mathcal{F}$. We refer to $\precsim$ as the Peer subcontract preorder. □

The proof that $\precsim$ provides a fully-abstract model of subtyping $\preccurlyeq$ on session types, relies on a syntactic characterisation of $\precsim$, stated in the next lemma, and it implies a result on the decidability of $\precsim$.

Lemma 1. $\precsim = \nu X.\mathcal{S}(X, X)$.

Theorem 3. *[Full-abstraction]*
For every $T, S \in \mathsf{STyp}$, $S \preccurlyeq T$ if and only if $\mathcal{M}(S) \precsim \mathcal{M}(T)$.

Proof (Outline). The subtyping $\preccurlyeq$ is the greatest fixed point of $\mathcal{F}\preccurlyeq$ by definition, $\precsim$ is the greatest fixed point of $\mathcal{S}$ because of Lemma 1, and $\mathcal{M}$ provides a bijection from prefixed points of $\mathcal{F}\preccurlyeq$ to prefixed points $\mathcal{S}$ ([5, Lemma 4.8, Lemma 4.9]). This is why full-abstraction is true.

Proposition 1. *If $\preccurlyeq_{\mathrm{b}}$ is decidable, then the relation $\sqsubseteq$ is decidable.*

Theorem 3 depends on Corollary 1, which depends on Theorem 2. In turn this theorem relies on the existence for every session contract σ of a "complementary" session contract $\mathsf{cplmt}(\sigma)$ that is in $\mathcal{B}$-peer compliance with σ, at least for $\mathcal{B}$s that satisfy certain minimal conditions. To construct $\mathsf{cplmt}(\sigma)$, the well-known *syntactic duality* of session types is an obvious candidate. This is defined inductively as follows [15]:

$$\overline{\mathrm{END}} = \mathrm{END}, \quad \overline{X} = X, \quad \overline{\mu X.S} = \mu X.\overline{S}, \quad \overline{?[M];S} = ![M];\overline{S}, \quad \overline{![M];S} = ?[M];\overline{S},$$
$$\overline{\&\langle \mathtt{l}_1 : S_1, \ldots, \mathtt{l}_n : S_n \rangle} = \oplus\langle \mathtt{l}_1 : \overline{S_1}, \ldots, \mathtt{l}_n : \overline{S_n} \rangle,$$
$$\overline{\oplus\langle \mathtt{l}_1 : S_1, \ldots, \mathtt{l}_n : S_n \rangle} = \&\langle \mathtt{l}_1 : \overline{S_1}, \ldots, \mathtt{l}_n : \overline{S_n} \rangle$$

This operator can also be applied to contracts in the obvious manner, using the injection $\mathcal{M}(-)$.

Example 4. In general it is not true that a contract σ complies with its dual $\overline{\sigma}$. To prove this, we say that the relation $\mathcal{B}$ is *reasonable* whenever $\sigma_1 \mathrel{\mathcal{B}} \sigma_2$ implies the following conditions:

i) $\mathsf{unfold}(\sigma_1) \mathrel{\mathcal{B}} \mathsf{unfold}(\sigma_2)$
ii) $\sigma_1 \xrightarrow{\lambda_1}$ and $\sigma_2 \xrightarrow{\lambda_2}$ imply that λ_1 and λ_2 are *both* input actions or *both* output actions.

If $\mathcal{B}$ is reasonable then we can find a contract σ such that $\sigma \not\dashv^{\mathcal{B}}_{\mathrm{p2p}} \overline{\sigma}$. For example take σ to be $\mu x.?(x).\mathsf{1}$; here $\overline{\sigma}$ is $\mu x.!(x).\mathsf{1}$. The behaviour of these contracts is $\sigma \xrightarrow{\tau} ?(\sigma).\mathsf{1} \xrightarrow{?(\sigma)} \mathsf{1} \xrightarrow{\checkmark}$, and $\overline{\sigma} \xrightarrow{\tau} !(\overline{\sigma}).\mathsf{1} \xrightarrow{!(\overline{\sigma})} \mathsf{1} \xrightarrow{\checkmark}$. If $\mathcal{B}$ is reasonable, then the pair $(!(\overline{\sigma}), ?(\sigma))$ is not in $\mathcal{B}$, and so σ and $\overline{\sigma}$ are not in $\mathcal{B}$-mutual compliance.

Since $\mathsf{unfold}(\sigma)$ performs inputs, while $\mathsf{unfold}(\overline{\sigma})$ performs outputs, and $\mathcal{B}$ is a reasonable relation, condition ii) above ensures that $(\,\mathsf{unfold}(\overline{\sigma}), \mathsf{unfold}(\sigma)\,) \notin \mathcal{B}$, so condition i) implies that $(\overline{\sigma}, \sigma) \notin \mathcal{B}$. This implies that $!(\overline{\sigma}) \not\preccurlyeq_{\mathcal{B}} ?(\sigma)$, and so $\sigma \parallel \overline{\sigma} \xrightarrow{\tau}_{\mathcal{B}} \mathsf{unfold}(\sigma) \parallel \mathsf{unfold}(\overline{\sigma}) \not\xrightarrow{\tau}_{\mathcal{B}}$. But this means that $\sigma \not\dashv^{\mathcal{B}}_{\mathrm{p2p}} \overline{\sigma}$ because neither $\mathsf{unfold}(\sigma)$ nor $\mathsf{unfold}(\overline{\sigma})$ perform $\checkmark$. □

In view of the previous example, we introduce a function to syntactically manipulate session contracts, whereby the result of manipulating ρ is a session contract in mutual compliance with ρ, at least for preorders $\mathcal{B}$s. In view of the encoding $\mathcal{M}$, this syntactic transformation applies equally well to session types.

Definition 6 ([Complement] [6]).
Let $\mathsf{cplmt} : L_{\mathsf{SCts}} \to L_{\mathsf{SCts}}$ *be defined inductively as follows,*

$$\mathsf{cplmt}(\mathsf{1}) = \mathsf{1}, \quad \mathsf{cplmt}(x) = x, \quad \mathsf{cplmt}(\mu x.\sigma) = \mu x.\mathsf{cplmt}(\sigma\lfloor^{\mu x.\sigma}/_{x}\rfloor),$$
$$\mathsf{cplmt}(!(\sigma'').\sigma') = ?(\sigma'').\mathsf{cplmt}(\sigma'), \quad \mathsf{cplmt}(?(\sigma'').\sigma') = !(\sigma'').\mathsf{cplmt}(\sigma'),$$
$$\mathsf{cplmt}(\textstyle\sum_{i\in I} ?\mathtt{l}_i.\sigma_i) = \textstyle\bigoplus_{i\in I} !\mathtt{l}_i.\mathsf{cplmt}(\sigma_i), \quad \mathsf{cplmt}(\textstyle\bigoplus_{i\in I} !\mathtt{l}_i.\sigma_i) = \textstyle\sum_{i\in I} ?\mathtt{l}_i.\mathsf{cplmt}(\sigma_i)$$

We say that $\mathsf{cplmt}(\sigma)$ *is the* complement *of* σ. □

In this definition the application of $\lfloor^{\sigma}/_{x}\rfloor$ to σ' stands for the substitution of σ in place of x in the message fields that appear in σ'; this is called *inner substitution* in [6]. The formal definition for our contracts is in [5, Appendix A].

Example 5. Suppose that $\sigma = \mu x.?(x).x$, then $\mathsf{cplmt}(\sigma) = \mu x.!(\sigma).x$. Observe that, intuitively, the input of σ depends on σ itself. The application of cplmt results in a contract which does not show that dependency, in that the output of $\mathsf{cplmt}(\sigma)$ does not depend on $\mathsf{cplmt}(\sigma)$.

Let us check a more involved example. We show how cplmt acts on session contracts. Let $\sigma = \mu x.\mu y.!(y)!(x).y$, and $\sigma' = \mu y.!(y)!(\sigma).y$. By definition,

$$\begin{aligned}\mathsf{cplmt}(\sigma) &= \mu x.\mathsf{cplmt}((\mu y.!(y).!(x).y)\lfloor^{\sigma}/_{x}\rfloor)\\ &= \mu x.\mathsf{cplmt}(\sigma')\\ &= \mu x.\mu y.\mathsf{cplmt}((!(y).!(\sigma).y)\lfloor^{\sigma'}/_{y}\rfloor)\\ &= \mu x.\mu y.\mathsf{cplmt}(!(\sigma').!(\sigma).y)\\ &= \mu x.\mu y.?(\sigma').?(\sigma).y\end{aligned}$$

Here again note that the contacts used in the input fields of $\mathsf{cplmt}(\sigma)$ are not defined in terms of $\mathsf{cplmt}(\sigma)$. □

In the previous example the contract σ and its complement are syntactically quite different objects, the complement being syntactically more complicated than σ,

$$\mathsf{cplmt}(\sigma) = \mu x.\mu y.?(\mu y.!(y)!(\sigma).y).?(\mu x.\mu y.!(y)!(x).y).y$$

What matters, though, are the behaviours of σ and of its complement. Those two behaviours are in $\mathcal{B}$-mutual compliance for every preorder $\mathcal{B}$. What was just argued for σ and its complement is true for every contract; the proof of it uses the commutativity of unfold and cplmt.

Proposition 2. *[Unfolding and complement commute]*
For every contract σ, $\mathsf{cplmt}(\mathsf{unfold}(\sigma)) = \mathsf{unfold}(\mathsf{cplmt}(\sigma))$.

Theorem 4. *For every preorder on contracts $\mathcal{B}$, $\rho \dashv^{\mathcal{B}}_{\text{P2P}} \mathsf{cplmt}(\rho)$ for every session contract ρ.*

In the full version of the paper [5] we argue that the notion of *complement* of a session type, Definition 6, in addition to being indispensable in the proof of Theorem 4, can also have a significant impact on type-checking systems for session types. For example the program P in Example 2 from the Introduction cannot be typed using the type-checking rules from [23]; the difficulty is the use of the duality operator $\overline{T}$ in the rule [CRes] on page 14. The bulk of the argument is that the dual of $\overline{\mu X.\,![\,X\,];\text{END}}$, that is $\mu X.\,?[\,X\,];\text{END}$, is not equivalent to $?[\,\mu X.\,![\,X\,];\text{END}\,];\text{END}$, and this hinders the necessary application of [CRes]. However we exhibit a type inference if instead $\mathsf{cplmt}(T)$ were used: the complement of $\mu X.\,![\,X\,];\text{END}$, namely $\mu X.\,?[\,\mu X.\,![\,X\,];\text{END}\,];\text{END}$, is equivalent to $?[\,\mu X.\,![\,X\,];\text{END}\,];\text{END}$, and this allows us to apply rule [CRes].

5 Related Work

In this paper we proposed a new behavioural model for recursive higher-order session types [15], which is fully-abstract with respect to the subtyping relation [13]. The denotation of a type consists of the set of higher-order contracts with which it complies, when it in turn is viewed as a contract. We use a novel notion of compliance, called *peer compliance*, which is also parametrised with a particular decidable relation $\mathcal{B}_0$, used for comparing higher-order contracts which are supplied by one peer in order to satisfy the higher-order contract required by it's partner. Moreover this relation $\mathcal{B}_0$ is the maximal solution to a natural behavioural equation over contracts.

Contracts for Web-Service: First-order contracts for web-services and an operationally defined contract compliance have been proposed first in [16], where the compliance is defined in terms of the LTS of contracts, and then, in the style of testing theory [10], the sub-contract preorder is defined using the compliance. All the subsequent works - including this paper - adhere to that style.

The most recent accounts of first-order contracts for web-services are [19,9]. A striking difference between the two papers is the treatment of infinite behaviours. In [19] infinite behaviours are expressed by recursive contracts, whereas in [9] there is no recursive construct, $\mu X.-$, and the theory accounts for infinite behaviours by using a *coinductively* defined language. Our treatment of infinite behaviours follows the lines of [19].

Session Types: Recursive higher-order session types appeared first in [15], where also the definition of type duality that we reported in Section 4 has been proposed. The authors of [15] argue in favour of program abstractions, that help programmers structure the interaction of processes around sessions. The proposed result is that a "typable program never reduces into an error" (see Theorem 5.4 (3) of [15]). In [23, pag. 86, paragraph 4], though, it is shown that that result is not true, that is the type system of [15] does not satisfy type-safety. The authors of [23] amend the type system of [15], thereby achieving type-safety (see Theorem 3.4 of [23]).

Subtyping for recursive higher-order session types has been introduced in [13], along with a *coinductive* definition of the duality. In addition to the standard type-safety result (Theorem 2), the authors show also a type-checking algorithm which they prove sound (Theorem 5) wrt the type system. The proof of completeness, though, relies on a relation between the inductive and the coinductive dualities (Proposition 5 there) which in general is false; a counter example is provided by the session type $\mu X.\,![X];\textsc{end}$. The consequence is that there is the possibility that the algorithm of [13], if employed in more general settings, may reject programs which are well-typed.

An alternative "fair" subtyping has been proposed recently in [20]. There session types are higher-order and recursive, their operational semantics is defined by parametrising the interactions of session types on pre-subtyping relations, and the fair subtyping is defined as a greatest fixed point (Definition 2.1). In our development we adopted the same technique as [20]. However, our aim was to model the standard subtyping of [13], while Padovani focuses on the properties of his new fair subtyping.

Models of Gay & Hole Subtyping: The first attempt to model the Gay & Hole subtyping of [13] in terms of a compliance preorder appeared in [17]. For a comparison of that research and our work the reader is referred to [4]. The authors of [1] have shown the first sound model of this subtyping restricted to first-order session types, by using a subset of contracts for web-services, a mutual compliance, called *orthogonality*, and the preorder generated by it. The $\mathcal{B}$-peer compliances we used in this work generalises to parametrised LTS the orthogonality of [1].

Following the approach of [1], in [4] we have shown a fully-abstract model of the subtyping for first-order session types, but using the standard asymmetric compliance and an intersection of the obvious server and client preorders. An alternative definition of the model proposed in [4] can be found in [3, Chapter 5], where the must testing of [10] is used in place of the compliance.

Semantic Subtyping: We view our main result as a behavioural or *semantic* interpretation of Gay & Hole subtyping. There is an alternative well-developed approach to semantic theories of types and subtyping [12] in which the denotation of a type is given by the set of values which inhabit it, and subtyping is simply subset inclusion. This apparent simplicity is tempered by the fact that for non-trivial languages, such as the pi-calculus [7], there is a circularity in the constructions due to the fact that determining which terms are values depends in turn on the set of types. This circularity is broken using a technique called *bootstrapping* or *stratification*, essentially an inductive approach. The research using this approach which is closest to our results on Gay & Hole subtyping may be found in [8]; this contains a treatment of a very general language of session types, an extension of Gay & Hole types. But there are essential differences. The most important is that their model does not yield a semantic theory of Gay & Hole subtyping. Their subtyping relation, $\leq$, is defined via an LTS generated by considering the transmission of values rather than session types; effectively subtyping is not allowed on messages. The resulting subtyping is very different than our focus of concern, the Gay & Hole subtyping relation $\preccurlyeq$. For example the preorder $\leq$ has bottom elements, in contrast to $\preccurlyeq$, and $?[\texttt{Int}];\textsc{end} \preccurlyeq ?[\texttt{Real}];\textsc{end}$ whereas $?[\texttt{Int}];\textsc{end} \not\leq ?[\texttt{Real}];\textsc{end}$. The particular use of *stratification* (Theorem 2.6) is also complex, and rules out the use of session types such as $\mu X. ![X];\textsc{end}$. Finally they use as types infinite regular trees whereas we prefer to work directly with recursive terms, as proposed in [13]; for example this allows us to discuss the inadequacies of the type-checking rules of [23].

Nevertheless the extended language of sessions types of [8] is of considerable significance. It would be interesting to see if it can be interpreted behaviourally using our co-inductive approach, particularly endowed with a larger subtyping preorder more akin to the standard Gay & Hole relation [13].

Acknowledgements. The authors would like to thank the reviewers, and reviewers of a previous version of this paper, for their insightful comments and questions.

References

1. Barbanera, F., de'Liguoro, U.: Two notions of sub-behaviour for session-based client/server systems. In: Kutsia, T., Schreiner, W., Fernández, M. (eds.) PPDP, pp. 155–164. ACM (2010)
2. Barbanera, F., de'Liguoro, U.: Sub-behaviour relations for session-based client/server systems (2013) (submitted for publication)
3. Bernardi, G.: Behavioural Equivalences for Web Services. Ph.D. thesis, Trinity College Dublin (2013), `https://software.imdea.org/~giovanni.bernardi`

4. Bernardi, G., Hennessy, M.: Modelling session types using contracts. In: Ossowski, S., Lecca, P. (eds.) SAC, pp. 1941–1946. ACM (2012)
5. Bernardi, G., Hennessy, M.: Using higher-order contracts to model session types. CoRR abs/1310.6176 (2013)
6. Bono, V., Padovani, L.: Typing copyless message passing. Logical Methods in Computer Science 8(1) (2012)
7. Castagna, G., De Nicola, R., Varacca, D.: Semantic subtyping for the pi-calculus. Theor. Comput. Sci. 398(1-3), 217–242 (2008)
8. Castagna, G., Dezani-Ciancaglini, M., Giachino, E., Padovani, L.: Foundations of session types. In: Porto, A., López-Fraguas, F.J. (eds.) PPDP, pp. 219–230. ACM (2009)
9. Castagna, G., Gesbert, N., Padovani, L.: A theory of contracts for web services. ACM Trans. Program. Lang. Syst. 31(5), 1–61 (2009), Supersedes the article in POPL 2008
10. De Nicola, R., Hennessy, M.: Testing equivalences for processes. Theoretical Computer Science 34, 83–133 (1984)
11. Dezani-Ciancaglini, M., de'Liguoro, U.: Sessions and session types: An overview. In: Laneve, C., Su, J. (eds.) WS-FM 2009. LNCS, vol. 6194, pp. 1–28. Springer, Heidelberg (2010)
12. Frisch, A., Castagna, G., Benzaken, V.: Semantic subtyping: Dealing set-theoretically with function, union, intersection, and negation types. J. ACM 55(4), 19:1–19:64 (2008), `http://doi.acm.org/10.1145/1391289.1391293`
13. Gay, S.J., Hole, M.: Subtyping for session types in the pi calculus. Acta Inf. 42(2-3), 191–225 (2005)
14. Hoare, C.A.R.: Communicating sequential processes. Prentice-Hall (1985)
15. Honda, K., Vasconcelos, V.T., Kubo, M.: Language primitives and type discipline for structured communication-based programming. In: Hankin, C. (ed.) ESOP 1998. LNCS, vol. 1381, pp. 122–138. Springer, Heidelberg (1998)
16. Laneve, C., Padovani, L.: The *must* preorder revisited. In: Caires, L., Vasconcelos, V.T. (eds.) CONCUR 2007. LNCS, vol. 4703, pp. 212–225. Springer, Heidelberg (2007)
17. Laneve, C., Padovani, L.: The pairing of contracts and session types. In: Degano, P., De Nicola, R., Meseguer, J. (eds.) Concurrency, Graphs and Models. LNCS, vol. 5065, pp. 681–700. Springer, Heidelberg (2008)
18. Milner, R.: Communication and concurrency. PHI Series in computer science. Prentice Hall (1989)
19. Padovani, L.: Contract-based discovery of web services modulo simple orchestrators. Theor. Comput. Sci. 411(37), 3328–3347 (2010)
20. Padovani, L.: Fair Subtyping for Multi-Party Session Types. In: De Meuter, W., Roman, G.-C. (eds.) COORDINATION 2011. LNCS, vol. 6721, pp. 127–141. Springer, Heidelberg (2011)
21. Takeuchi, K., Honda, K., Kubo, M.: An interaction-based language and its typing system. In: Halatsis, C., Maritsas, D., Philokyprou, G., Theodoridis, S. (eds.) PARLE 1994. LNCS, vol. 817, pp. 398–413. Springer, Heidelberg (1994)
22. Vasconcelos, V.T.: Fundamentals of session types. Inf. Comput. 217, 52–70 (2012)
23. Yoshida, N., Vasconcelos, V.T.: Language primitives and type discipline for structured communication-based programming revisited: Two systems for higher-order session communication. Electr. Notes Theor. Comput. Sci. 171(4), 73–93 (2007)

A Semantic Deconstruction of Session Types*

Massimo Bartoletti[1], Alceste Scalas[1], and Roberto Zunino[2]

[1] Università degli Studi di Cagliari, Italy
{bart, alceste.scalas}@unica.it
[2] Università degli Studi di Trento, Italy
roberto.zunino@unitn.it

Abstract. We investigate the semantic foundations of session types, by revisiting them in the abstract setting of labelled transition systems. The crucial insight is a simulation relation which generalises the usual syntax-directed notions of typing and subtyping, and encompasses both synchronous and asynchronous binary session types. This allows us to extend the session types theory to some common programming patterns which are not typically considered in the session types literature.

1 Introduction

Session typing is a well-established approach to the problem of correctly designing distributed applications [20,21,28]. In a nutshell, the application designer specifies the overall communication behaviour through a *choreography*, which enjoys some correctness properties (e.g. safety and progress). The overall application is the result of the composition of a set of *processes*, which are distributed over the network and interact through *sessions*. To ensure the correctness of this composition, the choreography is projected into a set of *session types*, which abstract the end-point communication behaviour of processes: if each process is type-checked against its session type, the composition of services preserves the properties enjoyed by the choreography.

The usual technical tool used to prove the correctness of a behavioural type system is *subject reduction*. Say P is a process, and T is a session type. Roughly, subject reduction guarantees that, if we have a typing judgement $\vdash P : T$, then whenever P takes a computation step $P \xrightarrow{\ell} P'$, also the type can take a similar step, i.e. there exists some T' such that $T \xrightarrow{\ell} T'$ and $\vdash P' : T'$.

This relation between processes and types somehow resembles the *simulation* relation in labelled transition systems (LTSs): a state T simulates a state P iff, whenever $P \xrightarrow{\ell} P'$, then $T \xrightarrow{\ell} T'$, for some T' which still simulates P'. This seems to suggest that $\vdash P : T$ is rooted in some kind of "process-type simulation". To elaborate further on this insight, consider a session type $T = !\mathsf{a} \oplus !\mathsf{b}$, which models an *internal* choice between two outputs.

* Work partially supported by: Aut. Reg. Sardinia (L.R.7/07 *TRICS*, P.I.A.2010 *Social Glue*), MIUR PRIN 2010-11 *Security Horizons*, EU COST Action IC1201 *BETTY*.

P. Baldan and D. Gorla (Eds.): CONCUR 2014, LNCS 8704, pp. 402–418, 2014.

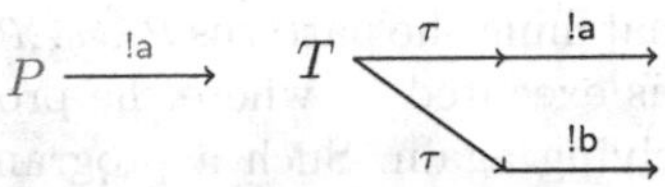

We can refine this session type as the process $P = !\mathsf{a}$ which just wants to output $!\mathsf{a}$. Intuitively, the process P respects the type T, because any client who can handle both choices in T will interact correctly with P. Now, let us consider the LTSs of P and T (on the left): we can observe that P is (weakly) simulated by T, in symbols $P \precapprox T$, because each move of P is matched by a move of T.

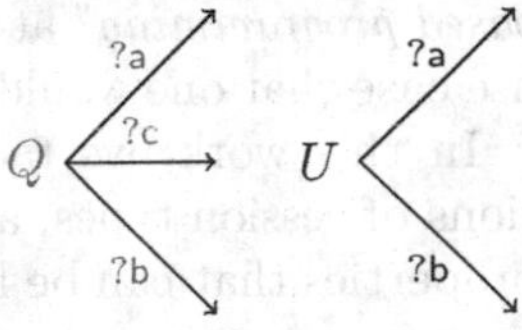

Let us now consider the type $U = ?\mathsf{a} \,\&\, ?\mathsf{b}$, which models an *external* choice between two inputs, and let $Q = ?\mathsf{a} + ?\mathsf{b} + ?\mathsf{c}$ (where $+$ is the standard CCS choice operator) which allows for an additional input $?\mathsf{c}$. Again, Q respects U: any client compatible with U will not exploit the additional choice, and will interact correctly with Q. But let us look at the LTSs of Q and U (on the right): differently from the previous case, now we have that Q is *not* weakly simulated by U (whereas the converse $U \precapprox Q$ holds). This shows that the weak simulation relation does not faithfully capture the notion of session typing: indeed, the previous examples suggest that a hypothetical "process-type simulation" should treat input and output capabilities differently: intuitively, it should be *covariant* w.r.t. outputs and *contravariant* w.r.t. inputs.

A similar kind of co/contra-variance arises when dealing with *subtyping*. The intuition is that if a session type T is subtype of U, and we have two processes P, Q such that $\vdash P : T$ and $\vdash Q : U$, then P can safely "replace" Q: i.e., each process that interacts correctly with Q will also interact correctly with P. Again, the session subtyping relations (e.g. [18]) are *covariant w.r.t. outputs and contravariant w.r.t. inputs*; moreover, they are *coinductive*. This suggests a link between the subtyping relation and our hypothetical "process-type simulation".

Several papers have studied session typing relations (e.g. [7,8,11,19,20,21,23]) and subtype preorders (e.g. [1,2,9,10,12,15,18]). Despite the variety of aims and results, all these works share a common approach: fix some syntax for types and/or processes, and then characterise typing/subtyping through *syntax-driven* definitions, usually in the form of a type system, or coinductive definitions (for subtyping). This seems in slight contrast with a common principle in concurrency theory: keeping syntax separated from semantics. Indeed, behavioural equivalences (e.g. (bi)simulation, testing, *etc.*) are typically defined over arbitrary LTSs, and then applied to calculi by providing the latter with an LTS semantics [27].

Another drawback of these syntax-driven approaches is that they do not usually consider some common programming patterns for interactive applications. For example, let us think about a server waiting for client's input: typically, the server must handle the case where such inputs do not arrive. This can be achieved via signals/exceptions handling, or other programming language constructs. In Erlang , for instance, one can write:

```
receive P1 -> Body1 ...
        Pk -> Bodyk
after   10 -> BodyT
```

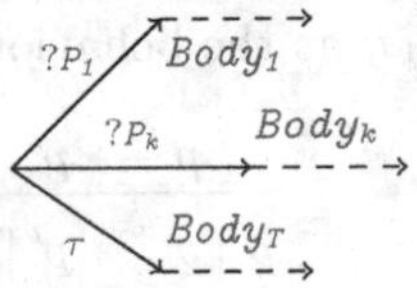

This causes `receive` to be aborted if no messages matching the patterns $P_1, \ldots, P_k$ arrive within 10 milliseconds; in this case, $Body_T$ is executed — where the program may e.g. do internal actions and start receiving again. Such a program blurs the distinction between internal/external choices: intuitively, its LTS (on the right) has a state with external inputs $?P_1, \ldots, ?P_k$ and an internal τ-move abstracting the timeout. This eludes the notion of *"structured communication-based programming"* at the roots of the session types approach [19,20]; yet, it is a use case that one would like to somehow typecheck to ensure correct interaction.

In this work, we tackle these problems by revisiting the semantic foundations of session types, aiming for behavioural, syntax-independent relations and properties that can be later applied to specific process calculi and programming languages.

Contributions. We study a behavioural theory of session types, aimed at unifying the notions of typing and subtyping, including both synchronous/asynchronous semantics. We start in §2 by setting our framework, and giving a running example. In §3 we define *I/O compliance* as a notion of correct interaction between behaviours, stricter than progress, albeit coinciding with it on synchronous session types (Theorem 1). In §4 we introduce the *I/O simulation* $\ddot{\leqslant}$ between behaviours, which is an I/O compliance-preserving preorder (Theorems 3 and 4), is a Gay-Hole subtyping relation [18] (Theorem 5), and is preserved when passing from synchronous to asynchronous session types semantics (Theorem 6). In §5 we show that $\ddot{\leqslant}$ induces syntax-driven type systems, which guarantee correct interaction (Theorem 8). Due to space constraints, the proofs of all our statements, more examples and discussion are available in [5].

2 Behaviours

In this section we exploit the semantic model of labelled transition systems (LTSs) to provide a unifying ground for the notions developed later. We consider LTSs where labels are partitioned into internal, input, and output actions, and we call *behaviours* the states of such LTSs. Then, we exploit this model to embed three calculi for concurrency: binary session types with synchronous or asynchronous semantics, and asynchronous CCS. We will sometimes use these calculi to write examples and to discuss related work, but all the main technical notions and results do apply to the general class of behaviours.

We consider an LTS $(\mathcal{U}, \mathsf{A}_\tau, \{\xrightarrow{\ell_\tau} \mid \ell_\tau \in \mathsf{A}_\tau\})$, where $\mathcal{U} = \{p, q, \ldots\}$ is a set of *behaviours*, A_τ is a set of *labels*, and $\xrightarrow{\ell_\tau} \subseteq \mathcal{U} \times \mathcal{U}$ is a *transition relation*. A_τ is partitioned into *input actions* $\mathsf{A}^? = \{?\mathsf{a}, ?\mathsf{b}, \ldots\}$, *output actions* $\mathsf{A}^! = \{!\mathsf{a}, !\mathsf{b}, \ldots\}$, and the *internal action* τ. We use an involution $\mathrm{co}(\cdot)$ such that $\mathrm{co}(?\mathsf{a}) = !\mathsf{a}$ and $\mathrm{co}(!\mathsf{a}) = ?\mathsf{a}$. We let $\ell, \ell', \ldots$ range over $\mathsf{A} = \mathsf{A}^? \cup \mathsf{A}^!$. For a set $L \subseteq \mathsf{A}$, we define $L^? = L \cap \mathsf{A}^?$ and $L^! = L \cap \mathsf{A}^!$. For all $p, q \in \mathcal{U}$, we define the *parallel composition* $p \parallel q$ as the behaviour whose transitions are given by the (standard) rules:

$$\frac{p \xrightarrow{\ell_\tau} p'}{p \parallel q \xrightarrow{\ell_\tau} p' \parallel q} \qquad \frac{q \xrightarrow{\ell_\tau} q'}{p \parallel q \xrightarrow{\ell_\tau} p \parallel q'} \qquad \frac{p \xrightarrow{\ell} p' \quad q \xrightarrow{\mathrm{co}(\ell)} q'}{p \parallel q \xrightarrow{\tau} p' \parallel q'}$$

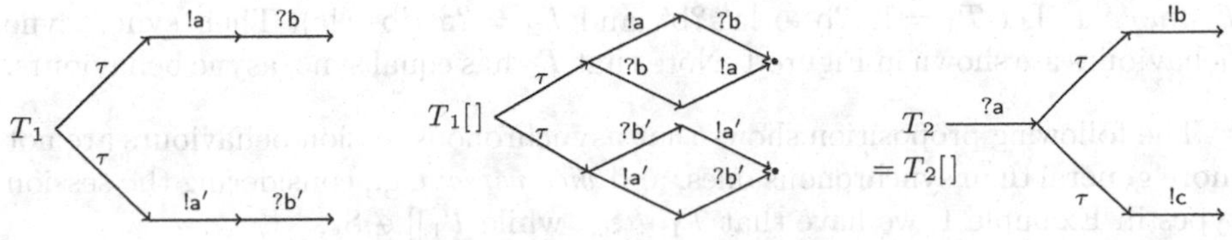

Fig. 1. Three session behaviours

We define the relation $\Rightarrow$ as the reflexive and transitive closure of $\xrightarrow{\tau}$, and $\xRightarrow{\ell_\tau}$ as $\Rightarrow\xrightarrow{\ell_\tau}\Rightarrow$. We write $p\xrightarrow{\ell_\tau}$ when $\exists p' \,.\, p \xrightarrow{\ell_\tau} p'$; we write $p\rightarrow$ when $\exists \ell_\tau \,.\, p\xrightarrow{\ell_\tau}$. We write $\mathbf{0}$ to denote any p such that $p\not\rightarrow$. We define the *weak barbs of* p as $p\Downarrow \;=\; \{\ell \mid p\xRightarrow{\ell}\}$. Hereafter, we shall consider two behaviours equal iff their transition graphs are isomorphic (i.e. equal up-to node renaming).

Session types. A session type is an abstraction of the behaviour of a process interacting with its environment. Here, we use a simple version of session types by slightly adapting those studied in [1]. Session types comprise external choice (&) among inputs (?a), internal choice ($\oplus$) among outputs (!a), and recursion. Empty choices (of any kind) represent successful termination.

Definition 1 (Session types). Session types *are terms with the syntax:*

$$T \;::=\; \&_{i\in I} ?\mathsf{a}_i.T_i \;\mid\; \bigoplus_{i\in I} !\mathsf{a}_i.T_i \;\mid\; \mathrm{rec}_X\, T \;\mid\; X$$

where (i) the set I is finite, (ii) the actions in internal/external choices are pairwise distinct, and (iii) recursion is guarded. We write $\mathbf{0}$ for the empty choice.

We present two semantics for session types: one *synchronous* (Def. 2) and one *asynchronous* (Def. 3). In both, an internal choice first commits to one of the branches $!\mathsf{a}.T$, before enabling !a. An external choice enables all its actions.

Definition 2 (Synchronous session behaviours). *We denote with $\mathcal{S}_s$ the set of behaviours of the form T (up-to unfolding), with transitions given by the rules:*

$$\&_{i\in I} ?\mathsf{a}_i.T_i \xrightarrow{?\mathsf{a}_j} T_j \;\; (j \in I) \qquad \bigoplus_{i\in I} T_i \xrightarrow{\tau} T_j \;\; (j \in I,\ |I| > 1) \qquad !\mathsf{a}.T \xrightarrow{!\mathsf{a}} T$$

For the asynchronous semantics, we consider behaviours of the form $T[\sigma]$ where σ is a sequence of output actions, modelling an unbounded *buffer*.

Definition 3 (Asynchronous session behaviours). *We denote with $\mathcal{S}_a$ the set of behaviours of the form $T[\sigma]$ (up-to unfolding), with transition rules:*

$$\Big(\bigoplus_{i\in I} !\mathsf{a}_i.T_i\Big)[\sigma] \xrightarrow{\tau} T_j\,[\sigma.!\mathsf{a}_j] \;\; (j \in I) \qquad T\,[!\mathsf{a}.\sigma] \xrightarrow{!\mathsf{a}} T\,[\sigma] \qquad \Big(\&_{i\in I} ?\mathsf{a}_i.T_i\Big)[\sigma] \xrightarrow{?\mathsf{a}_j} T_j\,[\sigma] \;\; (j \in I)$$

The async rule for $\oplus$ adds the selected output to the end of the buffer, with a τ-move. The 2nd rule says that an output !a at the head of the buffer is consumed with a !a-transition. The async rule for & is similar to the sync one.

Example 1. Let $T_1 = \mathsf{!a.?b} \oplus \mathsf{!a'.?b'}$, and $T_2 = \mathsf{?a.(!b} \oplus \mathsf{!c)}$. Their sync/async behaviours are shown in Figure 1. Note that T_2 has equal sync/async behaviours.

The following proposition shows that asynchronous session behaviours are not more general than synchronous ones, and *vice versa*: e.g., considering the session types in Example 1, we have that $T_1 \notin \mathcal{S}_a$, while $T_1[] \notin \mathcal{S}_s$.

Proposition 1. $\mathcal{S}_a \not\subseteq\not\supseteq \mathcal{S}_s$.

Definition 4 (CCS). CCS terms *have the following syntax:*

$$P, Q \ ::= \ \mathbf{0} \ \mid \ \ell_\tau.P \ \mid \ P + Q \ \mid \ P \mid Q \ \mid \ X \ \mid \ \mu_X P$$

where $+$ is non-deterministic choice, $\mid$ is parallel composition, and recursion $\mu_X P$ is guarded. Like async session behaviours, async CCS semantics use a buffer $[\sigma]$.

Definition 5 (Async CCS semantics). *We denote with* $\mathcal{P}_a$ *the set of behaviours of the form* $P[\sigma]$ *(up-to unfolding), with transitions given by the following rules (the symmetric ones for* $\mid$ *and* $+$ *are omitted):*

$$\frac{\ell_\tau \in \{\tau\} \cup \mathsf{A}^?}{\ell_\tau.P\,[\sigma] \xrightarrow{\ell_\tau} P[\sigma]} \qquad \mathsf{!a}.P\,[\sigma] \xrightarrow{\tau} P[\sigma.\mathsf{!a}] \qquad \frac{P[\sigma] \xrightarrow{\ell_\tau} P'[\sigma']}{(P+Q)[\sigma] \xrightarrow{\ell_\tau} P'[\sigma']} \qquad \frac{P[\sigma] \xrightarrow{\ell_\tau} P'[\sigma']}{(P \mid Q)[\sigma] \xrightarrow{\ell_\tau} (P' \mid Q)[\sigma']}$$

$$P\,[\mathsf{!a}.\sigma] \xrightarrow{\mathsf{!a}} P[\sigma]$$

As in async session behaviours, an output $\mathsf{!a}$ is first added at the end of the buffer, and can only be consumed from its head. Note that a behaviour *cannot* consume its own buffer: $\mid$ just allows for interleaving. Synchronization is obtained with $P[\sigma] \parallel Q[\sigma']$, i.e. using the parallel composition of LTS states: this allows P's input actions to consume Q's output buffer, and *vice versa*.

Example 2. The behaviour of the async process $\mathsf{!a}.\tau[]$ is shown as p_1 in Figure 2.

Definition 6. *We define an* encoding $[\![\,]\!]$ of session type terms into async CCS*:*

$$[\![\textstyle\bigoplus_I \mathsf{!a}_i.T_i]\!] = \sum_I \mathsf{!a}_i.[\![T_i]\!] \qquad [\![\&_I \mathsf{?a}_i.T_i]\!] = \sum_I \mathsf{?a}_i.[\![c_i]\!] \qquad [\![\mathrm{rec}_X\, T]\!] = \mu_X [\![T]\!] \qquad [\![X]\!] = X$$

By Lemma 1, an *async* session type and its encoding in async CCS are equivalent.

Lemma 1. $T[] = [\![T]\!][]$.

Proposition 2 relates async CCS behaviours with session behaviours.

Proposition 2. $\mathcal{S}_a \subsetneq \mathcal{P}_a \not\subseteq\not\supseteq \mathcal{S}_s$.

An example. The following types model a bartender (B) and a client Alice (A):

$$U_{\mathsf{B}} = \mathrm{rec}_X\,(\mathsf{?aCoffee.!coffee}.X \ \& \ \mathsf{?aBeer.(!beer}.X \oplus \mathsf{!no}.X) \ \& \ \mathsf{?pay})$$
$$T_{\mathsf{A}} = \mathsf{!aCoffee.?coffee.!pay} \ \oplus \ \mathsf{!aBeer.(?beer.!pay} \,\&\, \mathsf{?no.!pay})$$

The bartender presents an external choice &, allowing a customer to order either coffee or beer, or to eventually pay; in the first case, he will serve the coffee and then recursively wait for more orders; in the second case, he uses the internal choice $\oplus$ to decide whether to serve the beer or not — and then waits for more orders; in the third case, after the due amount (possibly 0) is paid, the interaction ends. Alice internally chooses between coffee or beer; in the first case, she waits to get the coffee and then pays; in the second case, she lets the bartender choose between serving the beer, or saying no — and in both cases, she will check out.

Intuitively, U_{B} and T_{A} are compliant, and the following processes type-check:

$$Q_{\mathsf{B}} = \mu_Y(\mathsf{?aCoffee.!coffee.}Y + \mathsf{?aBeer}\,.\,(\mathsf{!beer.}Y + \mathsf{!no.}Y) + \mathsf{?pay})$$
$$P_{\mathsf{A}} = \mathsf{!aCoffee.?coffee.!pay} + \mathsf{!aBeer}\,.\,(\mathsf{?beer.!pay} + \mathsf{?no.!pay})$$

From typing and compliance, we can deduce that $P_{\mathsf{A}}[] \parallel Q_{\mathsf{B}}[]$ synchronize and reach the successful state $\mathbf{0}[] \parallel \mathbf{0}[]$, where they agree in stopping their interaction.

Alice may also implement a *subtype* of T_{A} only asking for coffee: $T'_{\mathsf{A}} = \mathsf{!aCoffee.?coffee.!pay}$, with a corresponding process $P'_{\mathsf{A}} = \mathsf{!aCoffee.?coffee.!pay}$. Note however that the subtyping step is not necessary: P'_{A} has also type T_{A}.

So far, the structures of A's and B's processes match the structure of their types. This is a common situation in the session types literature: processes are usually written using calculi inheriting the structured communication approach pioneered by Honda *et al.* [19,20], thus reflecting the internal/external choices of types. However, in some cases things may be more complex. The bartender might have other incumbencies, and may need to stop selling beer after a certain hour:

$$Q''_{\mathsf{B}} = \mu_Y\big((\mathsf{?aCoffee.!coffee.}Y + \mathsf{?aBeer}\,.\,(\mathsf{!beer.}Y + \mathsf{!no.}Y) + \mathsf{?pay})$$
$$+\ \tau.\mu_Z(\mathsf{?aCoffee.!coffee.}Z + \mathsf{?aBeer.!no.}Z + \mathsf{?pay})\big)$$

This reminds us of the small Erlang code sample given in §1: the τ branch represents the bartender's decision to stop waiting for customer orders, perform some internal duties (e.g. clean up the bar) and then serve again — this time, refusing to sell beer. Intuitively, we would like Q''_{B} to still have the type U_{B}, since compliant customer processes (e.g. Alice's one) will still be able to interact (either before or after the τ). A process like Q''_{B}, however, is usually impossible to write (and type) using classical session calculi: their grammar does not offer a τ prefix, since it would allow for processes where the distinction between internal/external choices is blurred (contrary to the expected program structure).

Let us consider another scenario: Alice is late for work. But she realises that the bartender-customer system is *asynchronous*: the counter is a bidirectional *buffer* where drinks and money can be placed. Thus, she tries to save time by implementing the following type and process:

$$T''_{\mathsf{A}} = \mathsf{!aCoffee.!pay.?coffee} \qquad P''_{\mathsf{A}} = \mathsf{!aCoffee.(?coffee} \mid \mathsf{!pay)}$$

i.e., in her type she plans to order a coffee, put her money on the counter while B prepares her drink, and take it as soon as it is ready; in her process, she orders a coffee, and tries to grab the coffee with one hand, while putting the money on the counter with the other. P''_{A} represents an optimised program exploiting buffered communication, thus diverging from the syntactic structure of T''_{A}. Therefore, is T''_{A} a type for P''_{A}? Is T''_{A} compliant with U_{B}, and will P''_{A} interact smoothly with Q_{B} and Q''_{B}? We shall answer these questions later on in §5.

3 I/O Compliance

We now address the problem of defining a relation between behaviours to guarantee that, when combined together, they interact in a "correct" manner. Many different notions of correctness have been considered to this purpose in the literature, both for the binary [12,14,1,2] and for the multi-party settings [9,10,3,17].

We start by considering the classical, trace-based notion of compliance of [14,1], where correctness is interpreted as *progress* of the interaction. In Definition 7 we say that a behaviour p has progress with q (in symbols, $p \dashv q$) iff, whenever a τ-computation of the system $p \parallel q$ is stuck, then p has reached the final (success) state $\mathbf{0}$. Note that this notion is *asymmetric*, in the sense that p is allowed to terminate the interaction without the permission of q. This is intended to model the asymmetry between the role of a client p and that of a server q, as in [1].

Definition 7 (Progress). *We write* $p \dashv q$ *iff* $p \parallel q \Rightarrow p' \parallel q' \not\rightarrow$ *implies* $p' = \mathbf{0}$. *We write* $p \perp q$ *when* $p \dashv q$ *and* $p \vdash q$.

The following proposition states that, for session types, progress with the synchronous semantics implies progress with the async semantics. As we shall see, the main relations introduced in the rest of the paper will be preserved when passing from the synchronous to the asynchronous semantics of session types.

Proposition 3. *If* $T \dashv U$, *then* $T[] \dashv U[]$.

Example 3. We have the following relations:

$$\mathtt{!a.?b} \perp \mathtt{?a.!b} \qquad \mathtt{!a.?b} \not\dashv\vdash \mathtt{?a} \qquad \mathrm{rec}_X\, \mathtt{!a}.X \not\dashv\vdash \mathtt{?a} \qquad (\mathrm{rec}_X\, \mathtt{!a}.X)[] \perp \mathtt{?b}[]$$

$$\mathtt{!a.?b} \not\dashv\nvdash \mathtt{!b.?a} \qquad (\mathtt{!a.?b})[] \perp (\mathtt{!b.?a})[] \qquad \mathrm{rec}_X\, \mathtt{!a}.X \perp \mathrm{rec}_Y\, \mathtt{?a}.Y \qquad (\mathrm{rec}_X\, \mathtt{?a}.X)[] \not\dashv\vdash \mathtt{!b}[]$$

The progress-based notion of correctness above also relates behaviours that allow arguably incorrect interactions. For instance, $(\mathrm{rec}_X\, \mathtt{!a}.X)[] \dashv \mathtt{?b}[]$ holds, because they produce an infinite τ-trace, even if they cannot synchronise. Ideally, we would like our notion of correct interaction to be stricter, avoiding "vacuous" progress where the client p exposes I/O capabilities, but the server q cannot interact, and $p \parallel q$ merely advances via internal τ-transitions (without synchronisations). We introduce a notion of compliance enjoying such a property on general behaviours (recall from §2 that $p{\Downarrow}^{!} = (p{\Downarrow})^{!} = p{\Downarrow} \cap \mathsf{A}^{!}$):

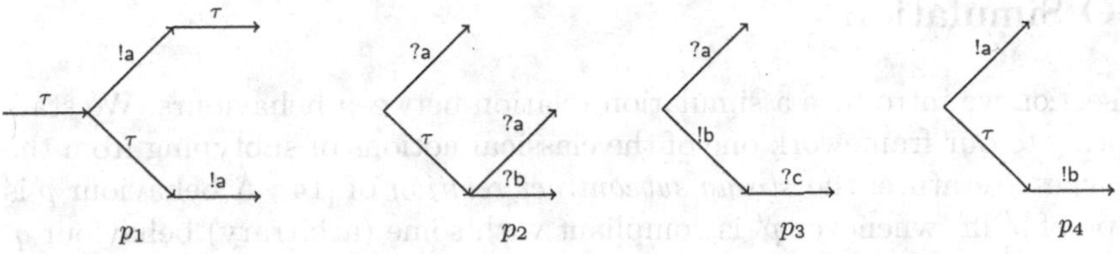

Fig. 2. Four behaviours which are not session behaviours

Definition 8. $\mathcal{R}$ *is an* I/O compliance relation *iff, when* $p\,\mathcal{R}\,q$:

a) $p{\Downarrow}^! \subseteq \mathrm{co}(q{\Downarrow}^?) \;\wedge\; \Big((p{\Downarrow}^! = \emptyset \wedge p{\Downarrow}^? \neq \emptyset) \implies \emptyset \neq q{\Downarrow}^! \subseteq \mathrm{co}(p{\Downarrow}^?)\Big)$;

b) $p \xrightarrow{\ell} p' \wedge q \xrightarrow{\mathrm{co}(\ell)} q' \implies p'\,\mathcal{R}\,q'$;

c) $p \xrightarrow{\tau} p' \implies p'\,\mathcal{R}\,q$;

d) $q \xrightarrow{\tau} q' \implies p\,\mathcal{R}\,q'$.

We write $\ddot{\lhd}$ *for the largest I/O compliance relation, and* $\ddot{\bowtie}$ *for the largest symmetric I/O compliance relation. When* $p \ddot{\bowtie} q$*, then we say that* p *and* q *are* I/O compliant.

Definition 8 can be interpreted with the game-theoretic metaphor. Let p and q be two players. Item *a*) has two conditions: by the leftmost constraint, if p wants to do some output (possibly after some τ-moves), then q must match it with its inputs; by the rightmost constraint, if p is *not* going to output, but wants to do some input, then q must be ready (possibly after some τ-moves) to do some output, and q cannot have outputs other than those accepted by p. I/O compliance must be preserved if p and q synchronise or do internal moves (items *b*), *c*), *d*)).

Lemma 2. $\ddot{\bowtie} \;=\; \ddot{\rhd} \cap \ddot{\lhd}$.

Example 4. Consider the behaviours in Figure 2. We have that $p_1 \ddot{\bowtie} p_2$, $p_2 \ddot{\bowtie} p_4$, $p_1 \ddot{\lhd} p_3$, and $p_2 \ddot{\lhd} p_3$, while all the other pairs of behaviours are not compliant.

Theorem 1 relates I/O compliance with Def. 7. If two behaviours are compliant, then they enjoy progress. The *vice versa* is not true: e.g., $(\mathrm{rec}_X\, !\mathsf{a}.X)[] \not\ddot{\lhd}$ $?\mathsf{b}[]$, coherently with our *desideratum* that correct interactions must not progress vacuously. $\ddot{\lhd}$ can relate async session behaviours which intuitively interact correctly, e.g. $(!\mathsf{a}.?\mathsf{b})[] \ddot{\bowtie} (!\mathsf{b}.?\mathsf{a})[]$. Still, $\ddot{\lhd}$ and $\dashv$ coincide in $\mathcal{S}_s$.

Theorem 1. *If* $p \ddot{\lhd} q$ *then* $p \dashv q$*. Also, if* $p, q \in \mathcal{S}_s$ *then* $p \dashv q$ *implies* $p \ddot{\lhd} q$.

Example 5. Recall the example in §2. In the sync case, $U_{\mathsf{B}} \perp T_{\mathsf{A}}$, $U_{\mathsf{B}} \ddot{\bowtie} T_{\mathsf{A}}$, $U_{\mathsf{B}} \perp T'_{\mathsf{A}}$ and $U_{\mathsf{B}} \ddot{\bowtie} T'_{\mathsf{A}}$. The same holds for their async versions. When Alice is late for work, for the *sync* types $U_{\mathsf{B}} \not\perp T''_{\mathsf{A}}$ and $U_{\mathsf{B}} \not\ddot{\bowtie} T''_{\mathsf{A}}$, due to the wrong order of Alice's actions. In the *async* case, instead, $U_{\mathsf{B}}[] \perp T''_{\mathsf{A}}[]$ and $U_{\mathsf{B}}[] \ddot{\bowtie} T''_{\mathsf{A}}[]$.

Proposition 4 says that $\ddot{\lhd}$ is preserved when passing from sync to async session behaviour. It refines Proposition 3, that deals with the weaker notion of progress.

Proposition 4. *If* $T \ddot{\lhd} U$*, then* $T[] \ddot{\lhd} U[]$.

4 I/O Simulation

In this section we introduce a simulation relation between behaviours. We start by adapting to our framework one of the classical notions of subtyping from the session types literature: the *strong subcontract relation* of [14]. A behaviour p is a subtype of p' iff, whenever p' is compliant with some (arbitrary) behaviour q, then p is compliant with q[1]. Thus, p can transparently replace p', in all contexts.

Definition 9 (Subtype). *$\sqsubseteq$ is the largest relation s.t. $p \sqsubseteq q$ implies $\forall r \,.\, q \ddot{\bowtie} r \implies p \ddot{\bowtie} r$. We write $p \sqsubseteq_{\mathbb{R}} q$ to restrict r to the set $\mathbb{R}$ (i.e., $\forall r \in \mathbb{R} \ldots$).*

Despite its elegance and generality, Def. 9 cannot be directly exploited to establish whether two behaviours are related, due to the universal quantification over all contexts. For session types, alternative characterisations of $\sqsubseteq$ have been defined, usually in the form of a syntax-driven coinductive relation [14,1]. This approach amounts to restricting p, q and r in Def. 9 to a process calculus with specific syntax and transition rules — e.g., $p, q, r \in \mathcal{S}_s$. In our semantic framework, however, behaviours are not syntax. We shall extend these characterisations from session behaviours to arbitrary ones, without resorting to a universal quantification over contexts. To do that, we define an *I/O simulation* relation on behaviours, denoted by $\ddot{\leqslant}$. We show that it is a preorder (Theorem 3), and it preserves I/O compliance (Theorem 4). $\ddot{\leqslant}$ is equivalent to the subtype relation on sync session behaviours (Theorem 5), albeit stricter on arbitrary behaviours.

Let $\mathbb{Q}$ be a set of behaviours. We write $q \Rrightarrow \mathbb{Q}$ iff $\emptyset \neq \mathbb{Q} \subseteq \{q' \mid q \Rightarrow q'\}$. By extension, we write $\mathbb{Q} \stackrel{\ell_\tau}{\Longrightarrow} q''$ iff $\exists q' \in \mathbb{Q} \,.\, q' \stackrel{\ell_\tau}{\Longrightarrow} q''$, and similarly for $\mathbb{Q} \Rightarrow q''$. We write $\mathbb{Q}{\Downarrow}$ for $\bigcup_{q' \in \mathbb{Q}} q'{\Downarrow}$, and similarly for $\mathbb{Q}{\Downarrow}^{?}$ and $\mathbb{Q}{\Downarrow}^{!}$.

Definition 10 (I/O simulation). *$\ddot{\mathcal{R}}$ is a* I/O simulation relation *iff, whenever $p \,\ddot{\mathcal{R}}\, q$, then $\exists \mathbb{Q}$ (called* predictive set*) such that $q \Rrightarrow \mathbb{Q}$, and:*

a) $p{\Downarrow}^{!} = \emptyset \implies \mathbb{Q}{\Downarrow}^{!} = \emptyset$;
b) $\mathbb{Q}{\Downarrow}^{?} \subseteq p{\Downarrow}^{?} \wedge (\mathbb{Q}{\Downarrow}^{?} = \emptyset \implies p{\Downarrow}^{?} = \emptyset)$;
c) $p \xrightarrow{\tau} p' \implies \exists q' \,.\, \mathbb{Q} \Rightarrow q' \wedge p' \,\ddot{\mathcal{R}}\, q'$;
d) $p \xrightarrow{!\mathsf{a}} p' \implies \exists q' \,.\, \mathbb{Q} \stackrel{!\mathsf{a}}{\Longrightarrow} q' \wedge p' \,\ddot{\mathcal{R}}\, q'$;
e) $p \xrightarrow{?\mathsf{a}} p' \wedge \mathbb{Q} \stackrel{?\mathsf{a}}{\Longrightarrow} \implies \exists q' \,.\, \mathbb{Q} \stackrel{?\mathsf{a}}{\Longrightarrow} q' \wedge p' \,\ddot{\mathcal{R}}\, q'$.

We write $\ddot{\leqslant}$ for the largest I/O simulation, $\ddot{\approx}$ for the largest symmetric I/O simulation, and $\ddot{=}$ for $\ddot{\leqslant} \cap \ddot{\geqslant}$.

Definition 10 can be explained in terms of a sort of simulation game between players p and q. At the first step, q predicts a suitable choice of its internal moves, via a set $\mathbb{Q}$ of states reachable from q. The outputs of $\mathbb{Q}$ must include those of p (item *d*)), and the inputs of $\mathbb{Q}$ must be included in those of p (item *b*)). Moreover, if p has no outputs, then also $\mathbb{Q}$ cannot have outputs, and if $\mathbb{Q}$ has no inputs, then also p cannot have inputs (items *a*),*b*)). Intuitively, these constraints reflect the

[1] In this paper the direction of $\sqsubseteq$ is opposite w.r.t. the subcontract relation in [14]. Moreover, we require I/O compliance in each context, while [14] only requires progress.

Table 1. Example of I/O simulation

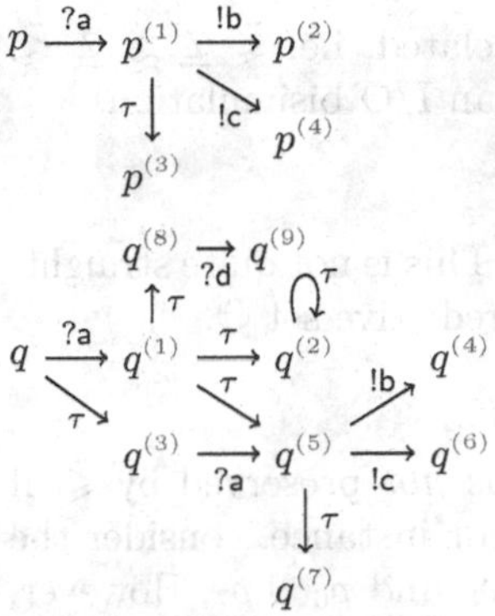

Relation	Pred. set
(p, q)	$\{q\}$
$(p^{(1)}, q^{(1)})$	$\{q^{(2)}, q^{(5)}\}$
$(p^{(1)}, q^{(5)})$	$\{q^{(5)}\}$
$(p^{(2)}, q^{(4)})$	$\{q^{(4)}\}$
$(p^{(3)}, q^{(2)})$	$\{q^{(2)}\}$
$(p^{(3)}, q^{(7)})$	$\{q^{(7)}\}$
$(p^{(4)}, q^{(6)})$	$\{q^{(6)}\}$

p is in relation with $q, q^{(3)}$, matching their ?a. Then, $p^{(1)}$ wants either to perform !b, !c or quit interacting: this behaviour is matched by $q^{(1)}$ and $q^{(5)}$; if $p^{(1)}$ follows its τ-branch to $p^{(3)}$, the latter is related with $q^{(2)}$ and $q^{(7)}$. Notice that $q^{(2)}$, does not stop, but enters in a τ-loop. Also notice that $p^{(1)}$'s predictive set has 2 elements: it cannot be $\{q^{(1)}\}$, because otherwise $p^{(1)}$ would not match ?d, reachable via $q^{(8)}$.

usual subtyping in session types: inputs (external choices) can be enlarged (if not empty), while outputs (internal choices) can be narrowed (but not emptied). The requirements above must be preserved by the moves of p. τ-moves and outputs of p must be (weakly) simulated by some process in $\mathbb{Q}$ (items *c*)–*d*)). The same holds for inputs (item *e*)), but only moves shared by p and $\mathbb{Q}$ are considered.

Example 6. Detailed examples of $\ddot{\leqslant}$ are shown in Table 1, and in [5].

Example 7. Consider Figure 3. To assess $p \ddot{\leqslant} q$, we choose a predictive set $\mathbb{Q}$ that mandates the inputs of p, and includes its outputs (note that p has an additional input ?c′ *not* offered by $\mathbb{Q}$). The same happens with the predictive set $\mathbb{R}$, assessing $q \ddot{\leqslant} r$ — but $\mathbb{R}$ must be chosen carefully: it *must* include the lower τ-branch of r, matching the branch of q with a τ-loop and no further I/O; however, it *must not* include the upper τ-branch of r, which requires ?d (not matched by q). Note that $\mathbb{R}$ and the small set inside are predictive sets for $p \ddot{\leqslant} r$.

We now study some properties of $\ddot{\leqslant}$. Lemma 3 ensures that Def. 10 is well-formed.

Lemma 3. *Let $\ddot{\mathbb{R}}$ be a set of I/O simulations. Then, $\bigcup \ddot{\mathbb{R}}$ is an I/O simulation.*

The following result relates I/O simulation with weak moves. When $p \ddot{\leqslant} q$, the relation $\ddot{\leqslant}$ is preserved by forward τ-moves of p and backward τ-moves of q.

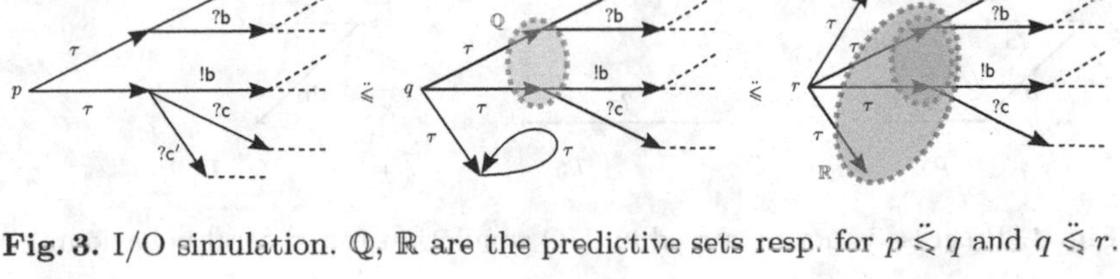

Fig. 3. I/O simulation. $\mathbb{Q}$, $\mathbb{R}$ are the predictive sets resp. for $p \ddot{\leqslant} q$ and $q \ddot{\leqslant} r$.

Lemma 4. *If $p \mathrel{\ddot{\leqslant}} q$, with $p \Rightarrow p'$ and $q' \Rightarrow q$, then $p' \mathrel{\ddot{\leqslant}} q'$.*

Weak simulation ($\precsim$) and I/O simulations are unrelated, i.e. $\mathrel{\ddot{\leqslant}} \not\subseteq \precsim \not\subseteq \mathrel{\ddot{\leqslant}}$. However, weak *bisimulation* ($\approx$) is strictly stronger than I/O bisimulation.

Theorem 2. $\approx \subsetneq \mathrel{\ddot{\approx}}$

By Theorem 3, $\mathrel{\ddot{\leqslant}}$ is a preorder, as the subtype relation. This is not quite straightforward, due to the existential quantification on the predictive set $\mathbb{Q}$.

Theorem 3. *$(\mathcal{U}, \mathrel{\ddot{\leqslant}})$ is a preorder.*

Quite surprisingly, on general behaviours *progress* is *not* preserved by $\mathrel{\ddot{\leqslant}}$: if $p \mathrel{\ddot{\leqslant}} q \dashv r$, then it is not always the case that $p \dashv r$. For instance, consider the behaviours in Figure 4. It is easy to check that $p_5 \mathrel{\ddot{\leqslant}} p_6$ and $p_6 \dashv p_7$. However, $p_5 \not\dashv p_7$: indeed, if p_7 chooses the branch !b, then p_5 is stuck waiting for ?c.

Theorem 4 is one of our main results: it states that $\mathrel{\ddot{\leqslant}}$ preserves (symmetric/asymmetric) *I/O compliance*. This is a further motivation for using $\mathrel{\ddot{\bowtie}}$ instead of $\perp$, when dealing with behaviours where these two notions do not coincide. In the example above, p_7 is not a sync session behaviour: were all behaviours in Figure 4 elements of $\mathbb{S}_s$, we would also have preserved progress (by Theorem 1).

Theorem 4. $p \mathrel{\ddot{\leqslant}} q \circ r \implies p \circ r$, *for* $\circ \in \{\mathrel{\ddot{\rhd}}, \mathrel{\ddot{\bowtie}}, \mathrel{\ddot{\lhd}}\}$.

I/O simulation can be seen as a subtyping relation on general behaviours, that is $p \mathrel{\ddot{\leqslant}} q$ allows p to be always used in place of q. For instance, assume that p is an asynchronous CCS process typed with a session type q, which in turn complies with the session type r. Then, Theorem 4 states that I/O compliance is preserved by $\mathrel{\ddot{\leqslant}}$, i.e. p is also I/O compliant with r, notwithstanding with the fact that p and r are specified in different calculi (actually, our statement is even more general, as it applies to *arbitrary* behaviours). Summing up, the process p will interact correctly with any process with type r (Theorem 8).

Theorem 5 below states that I/O simulation is stricter than Definition 9. However, the two notions coincide on synchronous session behaviours. Hence, $\mathrel{\ddot{\leqslant}}$ can be interpreted as a subtyping relation in $\mathbb{S}_s$, according to [18].

Theorem 5. *If $p \mathrel{\ddot{\leqslant}} q$, then $p \sqsubseteq q$. Also, if $p, q \in \mathbb{S}_s$, then $p \sqsubseteq_{\mathbb{S}_s} q \implies p \mathrel{\ddot{\leqslant}} q$.*

Theorem 6 generalises Propositions 3 and 4, extending to I/O simulation the set of properties preserved when passing from a sync to an async semantics.

Theorem 6. *If $T \circ T'$, then $T[] \circ T'[]$, for $\circ \in \{\mathrel{\ddot{\leqslant}}, \vdash, \perp, \dashv, \mathrel{\ddot{\rhd}}, \mathrel{\ddot{\bowtie}}, \mathrel{\ddot{\lhd}}\}$.*

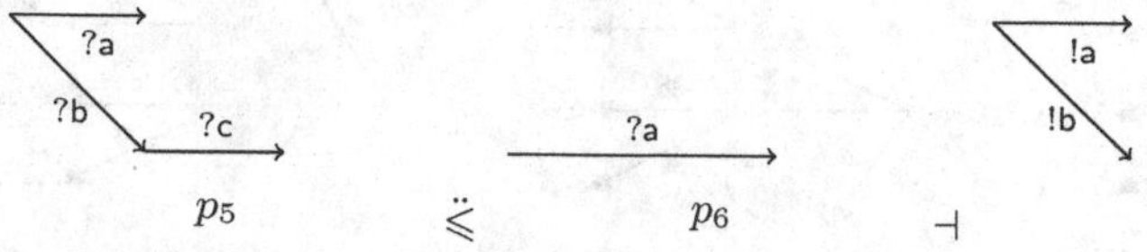

Fig. 4. Progress is not preserved by I/O simulation (on general behaviours)

$$\begin{array}{rcl} P &=& P+\mathbf{0} \\ P+Q &=& Q+P \\ P+(Q+R) &=& (P+Q)+R \\ P &=& P+P \end{array} \qquad \begin{array}{rcl} P &=& P\,|\,\mathbf{0} \\ P\,|\,Q &=& Q\,|\,P \\ P\,|\,(Q\,|\,R) &=& (P\,|\,Q)\,|\,R \\ \mu_X P &=& P[^{\mu_X P}/X] \end{array} \qquad \begin{array}{c} P \ddot{\leqslant} P \\ \mathsf{!a}\,|\,\mathsf{?b} \ddot{\leqslant} \mathsf{!a.?b} \end{array} \qquad \begin{array}{c} \dfrac{P \ddot{\leqslant} Q}{\mathsf{!a.?b}.P \ddot{\leqslant} \mathsf{!a}\,|\,\mathsf{?b}.Q}\,[|^0] \\ \dfrac{\forall i \in I\,.\,P_i \ddot{\leqslant} Q}{\sum_{i\in I} \tau.P_i \ddot{\leqslant} Q}\,[\tau] \end{array}$$

$$\frac{\forall i \in I\,.\,P_i \ddot{\leqslant} Q_i}{\sum_{i\in I} \ell_{\tau i}.P_i \ddot{\leqslant} \sum_{i\in I} \ell_{\tau i}.Q_i}\,[\ell_\tau] \qquad \frac{Q \doteq \sum_{i\in I} \mathsf{!a_i}.Q_i \quad \forall i \in I\,.\,P_i \ddot{\leqslant} Q_i \quad I \neq \emptyset}{\sum_{i\in I} \mathsf{!a_i}.P_i \ddot{\leqslant} Q + \mathsf{!b}.Q'}\,[\textsc{Int}] \qquad \frac{P \ddot{\leqslant} Q \quad R \ddot{\leqslant} S \quad \mathsf{ins}(P)\cup\mathsf{ins}(Q) = \emptyset}{P\,|\,R \ddot{\leqslant} Q\,|\,S}\,[|]$$

$$\frac{Q \doteq \sum_{i\in I} \mathsf{?a_i}.Q_i \quad \forall i \in I\,.\,P_i \ddot{\leqslant} Q_i \quad \emptyset \neq I \subseteq J \quad \forall k \in K\,.\,P_k \ddot{\leqslant} Q \quad \forall j \in J\backslash I\,.\,\mathsf{a}_j \notin \{\mathsf{a}_i\}_{i\in I}}{\sum_{j\in J} (\mathsf{?a_j}.P_j) + \sum_{k\in K} \tau.P_k \ddot{\leqslant} Q}\,[\textsc{Ext}]$$

Fig. 5. Axioms for $\ddot{\leqslant}$ in $\mathcal{P}_a^-$. ins(P) gives the set of inputs in P's body.

5 Session Types without Types

Our treatment so far does not depend on a syntactic representation of behaviours in $\mathcal{U}$. In the resulting unifying view, there are no inherent distinctions between processes and types: they are just states of an LTS. This allows us to define relations between objects which morally belong to different realms: e.g. $p \ddot{\leqslant} q$ may relate, say, an async CCS process with a (sync or async) session type.

The price for this generalisation is (seemingly) the loss of a useful feature: using syntax-based reasoning to check whether a process has a certain type, without having to deal with the semantic level. In this section, we show how this possibility can be restored in four steps: *(i)* choosing a process language and a type language (with their corresponding semantics); *(ii)* encoding the former in the latter; *(iii)* devising a sound set of axioms for $\ddot{\leqslant}$; and *(iv)* using these axioms to induce syntax-based typing rules that imply (i.e., safely approximate) $\ddot{\leqslant}$.

In this section we give a proof-of-concept of this methodology for the case of async CCS ($\mathcal{P}_a$) for processes, and async session behaviours ($\mathcal{S}_a$) as types. The encoding from types to processes for step *(ii)* is the one given in Definition 6. Proceeding to step *(iii)*, we now introduce a set of $\ddot{\leqslant}$-based relations for $\mathcal{P}_a$. We shall sometimes omit generic buffers $[\sigma]$ appearing in processes.

Lemma 5 ("Axioms" for $\ddot{\leqslant}$). *The relations in Figure 5 hold.*

The axioms in Lemma 5 are mostly straightforward. [Int] and [Ext] (with $K = \emptyset$) model the typical session typing rules for internal/external choices (resp. with outputs and inputs), allowing to add inputs and remove outputs according to $\ddot{\leqslant}$. [Ext] with $K \neq \emptyset$ handles an external choice that is interrupted (with τ-moves) and later reprised (i.e., a simple case of Erlang-style `receive...after...` behaviour, seen in §2). [|] allows the parallel composition of behaviours, provided that they cannot interfere badly (i.e., compete on the same inputs) along their reductions.

To ease the presentation, we focus on a fragment of async CCS (called $\mathcal{P}_a^-$) where *(a)* choices are guarded, *(b)* $|$ cannot appear within recursion, and *(c)* in $P \mid Q$, either P's or Q's body cannot contain inputs. Conditions *(b)* and *(c)* globally enforce the premises of [|], allowing us to simplify Def. 11 below.

Definition 11. *Let Γ be a mapping from recursion variables to pairs of $\mathcal{P}_a^-$ terms. We define $\ddot{\preccurlyeq}_\Gamma$ as the least relation between $\mathcal{P}_a^-$ terms closed under the rules obtained by replacing $\ddot{\leqslant}$ with $\ddot{\preccurlyeq}_\Gamma$ in Figure 5 — plus the following:*

$$\frac{\Gamma(X) = (P, Q)}{X \ddot{\preccurlyeq}_\Gamma Q}\ \text{[S-VAR]} \qquad \frac{P \ddot{\preccurlyeq}_{\Gamma, X:(\mu_X P, Q)} Q}{\mu_X P \ddot{\preccurlyeq}_\Gamma Q}\ \text{[S-REC]}$$

We treat the =-based relations in Figure 5 as structural congruence rules.

The rules in Def. 11 are straightforward: [S-VAR] enriches the environment by "guessing" that $P \ddot{\leqslant} Q$; [S-REC] consumes such a guess, introducing recursion.

Theorem 7 states that $(\mathcal{P}_a^-, \ddot{\preccurlyeq})$ is a preorder stricter than $(\mathcal{P}_a^-, \ddot{\leqslant})$, and it is preserved by all the operators of $\mathcal{P}_a^-$, that is ., +, and |. This enables us to use the syntactic rules $\ddot{\preccurlyeq}_\Gamma$ as a basis for a type system for $\mathcal{P}_a^-$ (as we will see in Def. 12).

Theorem 7. *$\ddot{\preccurlyeq}$ is a precongruence for $\mathcal{P}_a^-$, and $P \ddot{\preccurlyeq} Q \implies P[\sigma] \ddot{\leqslant} Q[\sigma]$.*

A non-obvious aspect of Definition 11 and Theorem 7 is that, by requiring guarded choices in $\mathcal{P}_a^-$, $\ddot{\preccurlyeq}$ is preserved by + (rule [ℓ_τ]). This is *not* directly matched by a corresponding property for $\ddot{\leqslant}$ in $\mathcal{P}_a$ without guarded choices, i.e. $P \ddot{\leqslant} Q \implies P + R \ddot{\leqslant} Q + R$. Indeed, the latter implication is false in general, because $\tau.?\mathsf{a}.P \ddot{\leqslant} ?\mathsf{a}.P$, but $?\mathsf{b} + \tau.?\mathsf{a}.P \not\ddot{\leqslant} ?\mathsf{b} + ?\mathsf{a}.P$. A similar argument holds for |, when arbitrary terms are put in parallel. This shows that $\ddot{\leqslant}$ is *not* a precongruence for $\mathcal{P}_a$, and gives reason for having $\ddot{\preccurlyeq}$ *stricter* than $\ddot{\leqslant}$.

We can now define a *syntax-directed* typing judgement relating $\mathcal{P}_a^-$ processes with session types. To this purpose, we exploit the encoding in Definition 6.

Definition 12. *We write $\Gamma \vdash P : T$ iff $P \ddot{\preccurlyeq}_\Gamma [\![T]\!]$.*

Theorem 8 states the correctness of our "typing" discipline. Suppose you have a process P with type T, and a process Q with type U. If T and U are I/O compliant, then we have that P and Q are I/O compliant, too. Thus, by Theorem 1 we have that the behaviour $P[] \parallel Q[]$ enjoys progress.

Theorem 8. *If $\vdash P : T$, $\vdash Q : U$ with $T[] \circ U[]$, then $P[] \circ Q[]$, for $\circ \in \{\ddot{\rhd}, \ddot{\bowtie}, \ddot{\lhd}\}$.*

Proof. From Def. 12 we have $P \ddot{\preccurlyeq} [\![T]\!]$; by Lemma 1, Def. 11 and Theorem 7 it follows $P[] \ddot{\leqslant} T[]$. Similarly, $Q[] \ddot{\leqslant} U[]$. Since $P \ddot{\leqslant} T[] \circ U[]$, by Theorem 4 it follows $P[] \circ U[]$. Since $Q[] \ddot{\leqslant} U[] \circ P[]$, then by Theorem 4 we conclude $Q[] \circ P[]$.

Note that $\vdash P : T$ and $\vdash Q : U$ can be inferred by a syntax-driven analysis, by the rules in Definition 11. If T and U are interpreted as *synchronous* session types, than we can use syntax-driven techniques (e.g. those in [1]) to deduce $T \circ U$ in the synchronous case; then, by Theorem 6, this result is lifted "for free" to the async case. We stress that the above result is obtained just by exploiting the properties of I/O simulation, without explicitly proving subject reduction.

Example 8. From §2, recall Alice's type T''_{A} and process P''_{A} when she is late for work. We have the following type encoding in $\mathcal{P}_a^-$: $P_{T''_{\mathsf{A}}} = [\![T''_{\mathsf{A}}]\!] = !\mathsf{aCoffee}.!\mathsf{pay}.?\mathsf{coffee}$. Then, by Definition 11, we can derive $\vdash P''_{\mathsf{A}} : T''_{\mathsf{A}}$.

Let us now consider the bartender processes and types in §2. Since in Example 5 we determined that $U_{\mathsf{B}} \ddot{\bowtie} T''_{\mathsf{A}}$, by Theorem 4 we have $Q_{\mathsf{B}} \ddot{\bowtie} P''_{\mathsf{A}}$. Also, since in Example 9 we show that $\vdash Q''_{\mathsf{B}} : U_{\mathsf{B}}$, by Theorem 8 we have $Q''_{\mathsf{B}} \ddot{\bowtie} P''_{\mathsf{A}}$.

Example 9. From §2, recall Q''_{B}, U_{B}. We can derive $\vdash Q''_{\mathsf{B}} : U_{\mathsf{B}}$.

The previous examples show that our syntax-driven rules allow to type an Erlang-style `receive...after...` behaviour, featured in the bartender process.

6 Conclusions and Related Work

We have revisited the theory of session types from a purely semantic perspective. We have defined a preorder $\ddot{\leqslant}$ between generic behaviours, which unifies the notions of typing and subtyping for session types, as well as their synchronous and asynchronous interpretations. In this work we mostly focused on behaviours arising from session types and async CCS; however, it seems that our framework can be easily exploited to analyse the properties of other behaviours populating $\mathcal{U}$ — e.g. the LTS semantics of other process calculi and programming languages.

Session types were introduced by Honda *et al.* in [19,28,20], as a type system for communication channels in a variant of the π-calculus. The resulting concept of *structured communication-based programming* has been the cornerstone of the subsequent research. In [23], session types are coupled with a "featherweight" Erlang-like language that, however, omits the problematic `receive...after...` construct described in §1. While adapting the type system of [23] to cope with such construct should be feasible, our approach allows the construction of the type system (in our case, the rules for $\ddot{\preccurlyeq}$) to be driven by an explicit underlying semantic notion (the I/O simulation). In particular, we think that the syntax-based reasoning in §5 can be extended to deal with other language constructs, beyond the Erlang-style `receive...after...` (which is treated in §5).

Some recent results extend the session types discipline to the multiparty case, starting from [21]. We expect that our approach can be extended to this setting: some insights come from the streamlined approach of [13], where the authors *"take a step back ... defining global descriptions whose restrictions are semantically justified"*. The plan is to extend the $\ddot{\leqslant}$ relation to capture the *role* of each type/process, and then to produce the syntax-based typing rules via (partial) axiomatisation for a given calculus. We also plan to address the orthogonal problem of multiple *interleaved* sessions. Two starting points are [4,26], which both introduce type systems for ensuring liveness in this setting.

[18] studies subtyping for (dyadic) session types. This topic is reprised in [1,2] with different notions of client-server compliance (e.g., allowing the client to terminate interaction or to skip messages). We took inspiration from these works, aiming at a framework general enough to replicate their notions and results.

Asynchronous dyadic session types have been addressed in [24], where type equivalence up-to buffering was defined over traces, and then approximated via syntax-based rules. A notion of compliance among services with buffers has been studied in [10], which extends [9] (albeit the setting is quite different from session types). Also [25] addresses the problem of defining compliance between service contracts. In their *weak compliance* relation, finite-state orchestrators can resolve external choices or rearrange messages in order to guarantee progress. Weak compliance is unrelated to our I/O compliance: on the one hand, the latter cannot rearrange messages; on the other hand, I/O compliance has no fixed bound on the size of the buffers. For instance, let $!a^m$ be a sequence of m !a; the async behaviours $!a.?b.!a^2.?b^2 \cdots !a^n.?b^n \cdots$ and $!b.?a.!b^2.?a^2 \cdots !b^n.?a^n \cdots$ are I/O compliant, but they are not weakly compliant, as orchestrators must have a finite rank. In [22] a bisimulation is defined to relate processes communicating via unbounded buffers. The aim of Theorem 6 is to provide for a unifying approach to these issues, by tranferring properties from the sync to the async setting.

Several works denote the successful termination of a behaviour with a specific transition label (e.g. ✓) and/or a specific state (e.g. **1** or End). In this paper, we consider two behaviours to be I/O compliant when they synchronise until the client (in the asymmetric case) simply stops interacting. It is easy to extend our framework with a success label/state, thus allowing e.g. to study a testing theory (as in [6]). For simplicity, we chose not to include it in the present work.

Our approach shares some common ground with [14,12]: the inspiration to [16] for the (synchronous) session types semantics, the idea of representing processes and contracts/types in the same LTS, thus allowing for easy reasoning about their progress/compliance properties, and the will to overcome the rigid internal/external choices dichotomy required by session types. In [14], it is assumed that some type system can abstract processes P, Q (expressed in *any* calculus) into contracts. This type system must be *"consistent"* and *"informative"*, by preserving some essential properties like e.g. visible actions and internal nondeterminism. A result in [14] is that if the abstractions of P, Q are (strongly) compliant, then P, Q will be (strongly) compliant as well. We believe that the concept of consistent/informative abstraction could be adapted to our framework: it would allow, e.g., to abstract rich process calculi (e.g. with value passing and delegation) into an LTS populated with I/O sorts (like the one adopted in this work). Beyond these general ideas, the technical developments are different: in the strong subcontract relation of [14] there is no input/output distinction, and some desirable subtypings do not hold, e.g. $?a \,\&\, ?b \not\sqsubseteq ?a$. These are restored through a "weak" subcontract relation, exploiting *filters* to suitably resolve external non-determinism. A challenging task would be that of using filters to enforce the I/O co/contra-variance typical of session types (and embodied in $\ddot{\leqslant}$).

References

1. Barbanera, F., de' Liguoro, U.: Two Notions of Sub-behaviour for Session-based Client/Server Systems. In: PPDP. ACM SIGPLAN. ACM (2010)

2. Barbanera, F., de' Liguoro, U.: Loosening the notions of compliance and sub-behaviour in client/server systems. In: ICE (2014)
3. Bartoletti, M., Cimoli, T., Zunino, R.: A theory of agreements and protection. In: Basin, D., Mitchell, J.C. (eds.) POST 2013. LNCS, vol. 7796, pp. 186–205. Springer, Heidelberg (2013)
4. Bartoletti, M., Scalas, A., Tuosto, E., Zunino, R.: Honesty by typing. In: Beyer, D., Boreale, M. (eds.) FMOODS/FORTE 2013. LNCS, vol. 7892, pp. 305–320. Springer, Heidelberg (2013)
5. Bartoletti, M., Scalas, A., Zunino, R.: A semantic deconstruction of session types. Tech. rep. (2014), `http://tcs.unica.it/publications`
6. Bernardi, G., Hennessy, M.: Compliance and testing preorders differ. In: SEFM Workshops (2013)
7. Bettini, L., Coppo, M., D'Antoni, L., De Luca, M., Dezani-Ciancaglini, M., Yoshida, N.: Global progress in dynamically interleaved multiparty sessions. In: van Breugel, F., Chechik, M. (eds.) CONCUR 2008. LNCS, vol. 5201, pp. 418–433. Springer, Heidelberg (2008)
8. Bocchi, L., Honda, K., Tuosto, E., Yoshida, N.: A theory of design-by-contract for distributed multiparty interactions. In: Gastin, P., Laroussinie, F. (eds.) CONCUR 2010. LNCS, vol. 6269, pp. 162–176. Springer, Heidelberg (2010)
9. Bravetti, M., Zavattaro, G.: Towards a unifying theory for choreography conformance and contract compliance. In: Lumpe, M., Vanderperren, W. (eds.) SC 2007. LNCS, vol. 4829, pp. 34–50. Springer, Heidelberg (2007)
10. Bravetti, M., Zavattaro, G.: Contract compliance and choreography conformance in the presence of message queues. In: Bruni, R., Wolf, K. (eds.) WS-FM 2008. LNCS, vol. 5387, pp. 37–54. Springer, Heidelberg (2009)
11. Caires, L., Vieira, H.T.: Conversation types. Theor. Comput. Sci. 411(51-52) (2010)
12. Carpineti, S., Castagna, G., Laneve, C., Padovani, L.: A formal account of contracts for Web Services. In: Bravetti, M., Núñez, M., Zavattaro, G. (eds.) WS-FM 2006. LNCS, vol. 4184, pp. 148–162. Springer, Heidelberg (2006)
13. Castagna, G., Dezani-Ciancaglini, M., Padovani, L.: On global types and multiparty session. Logical Methods in Computer Science 8(1) (2012)
14. Castagna, G., Gesbert, N., Padovani, L.: A theory of contracts for Web services. ACM TOPLAS 31(5) (2009)
15. Castagna, G., Padovani, L.: Contracts for mobile processes. In: Bravetti, M., Zavattaro, G. (eds.) CONCUR 2009. LNCS, vol. 5710, pp. 211–228. Springer, Heidelberg (2009)
16. De Nicola, R., Hennessy, M.: CCS without tau's. In: TAPSOFT, vol. 1 (1987)
17. Deniélou, P.-M., Yoshida, N.: Multiparty compatibility in communicating automata: Characterisation and synthesis of global session types. In: Fomin, F.V., Freivalds, R., Kwiatkowska, M., Peleg, D. (eds.) ICALP 2013, Part II. LNCS, vol. 7966, pp. 174–186. Springer, Heidelberg (2013)
18. Gay, S., Hole, M.: Subtyping for session types in the Pi calculus. Acta Inf. 42(2) (2005)
19. Honda, K.: Types for dyadic interaction. In: Best, E. (ed.) CONCUR 1993. LNCS, vol. 715, pp. 509–523. Springer, Heidelberg (1993)
20. Honda, K., Vasconcelos, V.T., Kubo, M.: Language primitives and type discipline for structured communication-based programming. In: Hankin, C. (ed.) ESOP 1998. LNCS, vol. 1381, pp. 122–138. Springer, Heidelberg (1998)
21. Honda, K., Yoshida, N., Carbone, M.: Multiparty asynchronous session types. In: POPL (2008)

22. Kouzapas, D., Yoshida, N., Honda, K.: On asynchronous session semantics. In: Bruni, R., Dingel, J. (eds.) FMOODS/FORTE 2011. LNCS, vol. 6722, pp. 228–243. Springer, Heidelberg (2011)
23. Mostrous, D., Vasconcelos, V.T.: Session typing for a featherweight Erlang. In: De Meuter, W., Roman, G.-C. (eds.) COORDINATION 2011. LNCS, vol. 6721, pp. 95–109. Springer, Heidelberg (2011)
24. Neubauer, M., Thiemann, P.: Session types for asynchronous communication. Universität Freiburg (2004)
25. Padovani, L.: Contract-based discovery of web services modulo simple orchestrators. Theor. Comput. Sci. 411(37) (2010)
26. Padovani, L., Vasconcelos, V.T., Vieira, H.T.: Typing liveness in multiparty communicating systems. In: Kühn, E., Pugliese, R. (eds.) COORDINATION 2014. LNCS, vol. 8459, pp. 147–162. Springer, Heidelberg (2014)
27. Sangiorgi, D.: An introduction to bisimulation and coinduction. Cambridge University Press, Cambridge (2012)
28. Takeuchi, K., Honda, K., Kubo, M.: An interaction-based language and its typing system. In: Halatsis, C., Philokyprou, G., Maritsas, D., Theodoridis, S. (eds.) PARLE 1994. LNCS, vol. 817, pp. 398–413. Springer, Heidelberg (1994)

Timed Multiparty Session Types*

Laura Bocchi, Weizhen Yang, and Nobuko Yoshida

Imperial College London, London, UK

Abstract. We propose a typing theory, based on multiparty session types, for modular verification of real-time choreographic interactions. To model real-time implementations, we introduce a simple calculus with delays and a decidable static proof system. The proof system ensures type safety and time-error freedom, namely processes respect the prescribed timing and causalities between interactions. A decidable condition on timed global types guarantees time-progress for validated processes with delays, and gives a sound and complete characterisation of a new class of CTAs with general topologies that enjoys progress and liveness.

1 Introduction

Communicating timed automata (CTAs) [14] extend the theory of timed automata [3] to enable precise specification and verification of real-time distributed protocols. A CTA consists of a finite number of timed automata synchronising over the elapsing of time and exchanging messages over unbound channels. In spite of its simplicity, the combination of timed automata [3] and communicating automata (CAs) [8] can represent many different temporal aspects from a local viewpoint. On the other hand, the model is known to be computationally hard, and it is difficult to directly link its idealised semantics to implementations of programming languages and distributed systems.

On a parallel line of research, *multiparty session types* (MPSTs) [13,6] have been proposed to describe communication protocols among two or more participants from a global viewpoint. Global types are projected to local types, against which programs can be type-checked and verified to behave correctly without deadlocks. This framework is applied in industry projects [19] and to the governance of large cyberinfrastructures [17] via the Scribble project (a MPST-based tool chain) [20].

From the theoretical side, in the untimed setting recent work brings CAs into choreographic frameworks, by seeking a correspondence with projected local types [11]. We proceed along these lines by applying the idealised mathematical semantics of CTAs to the design of MPSTs with clocks, clock constraints, and resets, in order to fill the gap between the abstract specification by CTAs and the verification of real-time programs. Surprisingly, since MPSTs inherently capture relative temporal constraints by imposing an order on the communications, they enable effective verification without limitations on topology or buffer-boundedness, unlike existing work on CTAs.

We organise our results in two parts. First we show that although time annotations increase the expressive power of global types, time-error freedom is guaranteed without additional time analysis of the types. In § 3 we give the semantics of timed global

* This work has been partially sponsored by EPSRC EP/K034413/1 and EP/K011715/1. We thank Viviana Bono and Mariangiola Dezani-Ciancaglini for their insightful comments.

P. Baldan and D. Gorla (Eds.): CONCUR 2014, LNCS 8704, pp. 419–434, 2014.

types (*TG*s) and prove soundness and completeness of the projection onto timed local types (*TL*s) (Theorem 3). In § 4 we give a simple π-calculus for programs (running as *processes*) with delays that can be used to synchronise the communications in a session. A compositional proof system enables modular verification of time-error freedom (Theorem 7): if all programs in a system are validated, then the global conversation will respect the prescribed timing and causalities between interactions. In the second part we investigate the conditions for an advanced property – time-progress – ensuring that if a process deadlocks, then its untimed counter-part would also deadlock (i.e., deadlock is not caused by time constraints). The fact that untimed processes in single sessions are deadlock-free [13] yields progress for timed processes. Time-progress is related to two delicate issues: (1) some time constraints in a *TG* may be unsatisfiable and (2) there may exist some distributed implementation of the *TG* which deadlocks. We give two sufficient conditions on *TG*s (§5) to prevent (1) and (2): *feasibility* (for each partial execution allowed by a *TG* there is a correct complete one) and *wait-freedom* (if all senders respect their time constraints, then no receiver has to wait for a message). Feasibility and wait-freedom are decidable (Proposition 8), and if we start from feasible and wait-free *TG*s, then the proof system given in part one guarantees time-progress for processes (Theorem 11). We give a sound and complete characterisation (Theorem 13) yielding a new class of CTAs which enjoys progress and liveness (Theorem 14). Conclusion and related work are in § 6. Full definitions can be found in the technical report [22].

2 Running Example: A Use-Case of a Distributed Timed Protocol

The motivating scenario developed with our partner, the Ocean Observatories Initiative (OOI) [17], is directed towards deploying a network of sensors and ocean instruments used/controlled remotely via service agents. In many OOI use-cases requests are augmented with deadlines and services are scheduled to execute at certain time intervals. These temporal requirements can be expressed by combining global protocol descriptions from MPSTs and time from CTAs. We show a protocol to calculate the average water temperature via sensor sampling. The protocol involves three participants: a master M that initiates the sampling, a worker/sensor W with fixed response time w, and an aggregator A for accumulating the data; their time constraints are expressed using clocks $x_{\mathtt{M}}$, $x_{\mathtt{W}}$, and $x_{\mathtt{A}}$, initially set to 0. Each clocks can be reset many times. Delays l (average latency of the network) and w (sampling time) are expressed in milliseconds. As in [14] (synchronous semantics) time elapses at the same pace for all the parts of the system.

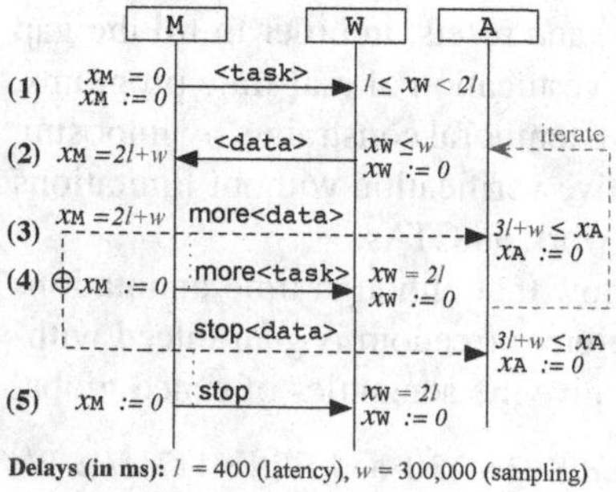

Delays (in ms): l = 400 (latency), w = 300,000 (sampling)

(1): M sends W a message of type `task` and resets $x_{\mathtt{M}}$ ($x_{\mathtt{M}}$:=0). After at least l and at most $2l$, W receives the message.

(2): W completes the task and replies to M with the temperature (of type `data`) at any time satisfying $x_{\mathtt{W}} \leq w$ and resets $x_{\mathtt{W}}$. M receives the message at time $x_{\mathtt{M}} = 2l + w$.

(3): M immediately sends A a message of type `data` with either label more (the sampling continues for another iteration) or stop (enough results have been aggregated).

(4): If more was selected then M sends W a new task with label more, resets $x_{\mathtt{M}}$ and another iteration is executed.

(5): If stop was selected, M notifies W and the session ends.

3 Timed Multiparty Session Types

Global types [6,13] are specifications of the interactions (causalities and carried data types) of multiparty sessions. A global type can be automatically projected onto a set of local types describing the session from the perspective of each single participant and used for local verification of processes. We extend global and local types with constraints on clocks, yielding *timed global types* (*TG*s) and *local session types* (*TL*s).

We use some definitions from timed automata (see [3, § 3.3], [14, § 2]): let X be a set of *clocks* ranging over $x_1,\ldots,x_n$ and taking values in $\mathbb{R}^{\geq 0}$. A *clock assignment* $\nu : X \mapsto \mathbb{R}^{\geq 0}$ returns the time of the clocks in X. We write $\nu + t$ for the assignment mapping all $x \in X$ to $\nu(x)+t$. We write ν_0 for the initial assignment mapping all clocks to 0. The set $\Phi(X)$ of *clock constraints* over X is:

$$\delta ::= \texttt{true} \mid x > c \mid x = c \mid \neg\delta \mid \delta_1 \wedge \delta_2$$

where c is a bound time constant taking values in $\mathbb{Q}^{\geq 0}$ (we derive $\texttt{false}, <, \leq, \geq, \vee$ in the standard way). The set of free clocks in δ, written $fn(\delta)$, is defined inductively as: $fn(\texttt{true}) = \emptyset, fn(x>c) = fn(x=c) = \{x\}, fn(\neg\delta) = fn(\delta)$, and $fn(\delta_1 \wedge \delta_2) = fn(\delta_1) \cup fn(\delta_2)$. We write $\delta(\vec{x})$ if $fn(\delta) = \vec{x}$ and let $\nu \models \delta$ denote that δ is satisfied by ν. A *reset* λ over X is a subset of X. When λ is $\emptyset$ then clocks are not reset, otherwise the assignment for each $x \in \lambda$ is set to 0. We write $[\lambda \mapsto 0]\nu$ for the clock assignment that is like ν except 0 is assigned to all clocks in λ.

Participants ($\mathtt{p},\mathtt{q},\mathtt{p}_1,\ldots \in \mathbb{N}$) interact via point-to-point asynchronous message passing. An interaction consists of a send action and a receive action, each annotated with a clock constraint and a (possibly empty) reset. The clock constraint specifies when that action can be executed and the reset specifies which clocks must be set to 0.

Syntax. The syntax for sorts S, timed global types G, and timed local types T is:

$$S ::= \texttt{bool} \mid \texttt{nat} \mid \ldots \mid G \mid (T,\delta)$$
$$G ::= \mathtt{p} \to \mathtt{q} : \{l_i\langle S_i\rangle\{A_i\}.G_i\}_{i\in I} \mid \mu\mathbf{t}.G \mid \mathbf{t} \mid \mathtt{end} \qquad A ::= \{\delta_{\mathtt{O}},\lambda_{\mathtt{O}},\delta_{\mathtt{I}},\lambda_{\mathtt{I}}\}$$
$$T ::= \mathtt{p}\oplus\{l_i : \langle S_i\rangle\{B_i\}.T_i\}_{i\in I} \mid \mathtt{p}\&\{l_i : \langle S_i\rangle\{B_i\}.T_i\}_{i\in I} \mid \mu\mathbf{t}.T \mid \mathbf{t} \mid \mathtt{end} \qquad B ::= \{\delta,\lambda\}$$

The sorts S include base types (`bool`, `nat`, etc.), G for shared name passing (used for the initiation of sessions of type G, cf. § 4), and (T,δ) for session delegation. Sort (T,δ) allows a participant involved in a session to delegate the remaining behaviour T; upon delegation the sender will no longer participate in the delegated session and receiver will execute the protocol described by T under any clock assignment satisfying δ. G and T in sorts do not include free type variables.

In G, type $\mathtt{p} \to \mathtt{q} : \{l_i\langle S_i\rangle\{A_i\}.G_i\}_{i\in I}$ models an *interaction*: $\mathtt{p}$ chooses a branch $i \in I$, where I is a finite set of indices, and sends $\mathtt{q}$ the branching label l_i along with a message of sort S_i. The session then continues as prescribed by G_i. Each branch is annotated with a *time assertion* $A_i = \{\delta_{\mathtt{O}i},\lambda_{\mathtt{O}i},\delta_{\mathtt{I}i},\lambda_{\mathtt{I}i}\}$, where $\delta_{\mathtt{O}i}$ and $\lambda_{\mathtt{O}i}$ are the clock constraint and reset for the output action, and $\delta_{\mathtt{I}i}$ and $\lambda_{\mathtt{I}i}$ are for the input action. We will write $\mathtt{p} \to \mathtt{q} : \langle S\rangle\{A\}.G'$ for interactions with one branch. *Recursive type* $\mu\mathbf{t}.G$ associates a *type variable* $\mathbf{t}$ to a recursion body G; we assume that type variables are guarded in the standard way and $\mathtt{end}$ occurs at least once in G (this is a common assumption e.g., [9]). We denote by $\mathcal{P}(G)$ the set of participants of G and write $G' \in G$ when G' appears in G.

As in [14] we assume that the sets of clocks 'owned' (i.e., that can be read and reset) by different participants in a *TG* are pair-wise disjoint, and that the clock constraint and reset of an action performed by a participant are defined only over the clocks owned by that participant. The example below violates this assumption.

$$G_1 = \mathtt{p} \to \mathtt{q} : \langle \mathtt{int} \rangle \{x_\mathtt{p} < 10, x_\mathtt{p}, x_\mathtt{p} < 20, x_\mathtt{p}\}$$

since both the constraints of the (send) action of p and of the (receive) action of q are defined over $x_\mathtt{p}$, and $x_\mathtt{p}$ can be owned by either p or q (similarly for the resets $\{x_\mathtt{p}\}$). Formally, we require that for all G there exists a partition $\{X(\mathtt{p}, G)\}_{\mathtt{p} \in \mathcal{P}(G)}$ of X such that $\mathtt{p} \to \mathtt{q} : \{l_i \langle S_i \rangle \{\delta_{\mathtt{O}i}, \lambda_{\mathtt{O}i}, \delta_{\mathtt{I}i}, \lambda_{\mathtt{I}i}\}.G_i\}_{i \in I} \in G$ implies $fn(\delta_{\mathtt{O}i}), \lambda_{\mathtt{O}i} \subseteq X(\mathtt{p}, G)$ and $fn(\delta_{\mathtt{I}i}), \lambda_{\mathtt{I}i} \subseteq X(\mathtt{q}, G)$ for all $i \in I$.

In T, interactions are modelled from a participant's viewpoint either as *selection* types $\mathtt{p} \oplus \{l_i : \langle S_i \rangle \{B_i\}.T_i\}_{i \in I}$ or *branching* types $\mathtt{p} \& \{l_i : \langle S_i \rangle \{B_i\}.T_i\}_{i \in I}$. We denote the *projection* of G on $\mathtt{p} \in \mathcal{P}(G)$ by $G \downarrow_\mathtt{p}$; the definition is standard except that each $\{\delta_{\mathtt{O}i}, \lambda_{\mathtt{O}i}, \delta_{\mathtt{I}i}, \lambda_{\mathtt{I}i}\}$ is projected on the sender (resp. receiver) by keeping only the output part $\{\delta_{\mathtt{O}i}, \lambda_{\mathtt{O}i}\}$ (resp. the input part $\{\delta_{\mathtt{I}i}, \lambda_{\mathtt{I}i}\}$), e.g., if $G = \mathtt{p} \to \mathtt{q} : \{l_i \langle S_i \rangle \{B_i, B'_i\}.G_i\}_{i \in I}$ then $G \downarrow_\mathtt{p} = \mathtt{q} \oplus \{l_i : \langle S_i \rangle \{B_i\}.G_i \downarrow_\mathtt{p}\}_{i \in I}$ and $G \downarrow_\mathtt{q} = \mathtt{p} \& \{l_i : \langle S_i \rangle \{B'_i\}.G_i \downarrow_\mathtt{q}\}_{i \in I}$.

Example 1 (Temperature calculation). We show below the global timed type G for the protocol in § 2 and its projection $G \downarrow_\mathtt{M}$ onto M. We write _ for empty resets.

$$\begin{array}{lll|l}
G & = & \mathtt{M} \to \mathtt{W} : \langle \mathtt{task} \rangle \{B^1_\mathtt{O}, B^1_\mathtt{I}\}.\mu \mathbf{t}.G' & B^1_\mathtt{O} = \{x_\mathtt{M} = 0, x_\mathtt{M}\} \\
G' & = & \mathtt{W} \to \mathtt{M} : \langle \mathtt{data} \rangle \{B^2_\mathtt{O}, B^2_\mathtt{I}\}. & B^1_\mathtt{I} = \{l \le x_\mathtt{W} < 2l, _\} \\
 & & \mathtt{M} \to \mathtt{A} : \{\mathsf{more}\langle \mathtt{data} \rangle \{B^3_\mathtt{O}, B^3_\mathtt{I}\}.\ \mathtt{M} \to \mathtt{W} : \mathsf{more}\langle \mathtt{task} \rangle \{B^4_\mathtt{O}, B^4_\mathtt{I}\}.\mathbf{t}, & B^2_\mathtt{O} = \{x_\mathtt{W} \le w, x_\mathtt{W}\} \\
 & & \quad \mathsf{stop}\langle \mathtt{data} \rangle \{B^3_\mathtt{O}, B^3_\mathtt{I}\}.\ \mathtt{M} \to \mathtt{W} : \mathsf{stop}\langle\rangle \{B^4_\mathtt{O}, B^4_\mathtt{I}\}.\mathtt{end}\} & B^2_\mathtt{I} = \{x_\mathtt{M} = 2l + w, _\} \\
G \downarrow_\mathtt{M} & = & \mathtt{W} \oplus \langle \mathtt{task} \rangle \{B^1_\mathtt{O}\}. & B^3_\mathtt{O} = \{x_\mathtt{M} = 2l + w, _\} \\
 & & \mu \mathbf{t}.\ \mathtt{W} \& \langle \mathtt{data} \rangle \{B^2_\mathtt{I}\}. & B^3_\mathtt{I} = \{3l + w \le x_\mathtt{A}, x_\mathtt{A}\} \\
 & & \mathtt{A} \oplus \{\mathsf{more} : \langle \mathtt{data} \rangle \{B^3_\mathtt{O}\}.\mathtt{W} \oplus \mathsf{more} : \langle \mathtt{task} \rangle \{B^4_\mathtt{O}\}.\mathbf{t}, & B^4_\mathtt{O} = \{x_\mathtt{M} = 2l + w, x_\mathtt{M}\} \\
 & & \quad \mathsf{stop} : \langle \mathtt{data} \rangle \{B^3_\mathtt{O}\}.\mathtt{W} \oplus \mathsf{stop} : \langle\rangle \{B^4_\mathtt{O}\}.\mathtt{end}\} & B^4_\mathtt{I} = \{x_\mathtt{W} = 2l, x_\mathtt{W}\}
\end{array}$$

Remark 1 (On the importance of resets). Resets in timed global types play an important role to model the same notion of time as the one supported by CTAs, yielding a more direct comparison between types and CTAs. Resets give a concise representation of several scenarios, e.g., when time constraints must be repeatedly satisfied for an unbounded number of times. This is clear from Example 1: the repetition of the same scenario across recursion instances (one for each sampling task) is modelled by resetting all clocks before starting a new recursion instance (e.g., $B^3_\mathtt{I}$, $B^4_\mathtt{O}$ and $B^4_\mathtt{I}$ on the second line of G' in Example 1).

Semantics of Timed Global Types. The LTS for *TG*s is defined over states of the form (ν, G) and labels $\ell ::= \mathtt{pq}!l\langle S \rangle \mid \mathtt{pq}?l\langle S \rangle \mid t$ where $\mathtt{pq}!l\langle S \rangle$ is a send action (i.e., p sends $l\langle S \rangle$ to q), $\mathtt{pq}?l\langle S \rangle$ is the dual receive action, and $t \in \mathbb{R}^{\ge 0}$ is a time action modelling time passing. We denote the set of labels by $\mathcal{L}$ and let $subj(\mathtt{pq}!l\langle S \rangle) = \mathtt{p}$, $subj(\mathtt{qp}?l\langle S \rangle) = \mathtt{p}$ and $subj(t) = \emptyset$.

We extend the syntax of G with $\mathtt{p} \rightsquigarrow \mathtt{q} : l\langle S \rangle \{A\}.G$ to describe the state in which message $l\langle S \rangle$ has been sent by p but not yet received by q (as in [11, § 2]). The separation of send and receive actions is used to model the asynchronous behaviour in distributed systems, as illustrated by the following example.

$$\mathtt{p} \to \mathtt{q} : \langle \mathtt{int} \rangle \{x_\mathtt{p} < 10,\ \_, x_\mathtt{q} \geq 10,\ \_\}.\mathtt{p} \to \mathtt{r} : \langle \mathtt{int} \rangle \{x_\mathtt{p} \geq 10,\ \_, \mathtt{true},\ \_\}$$
$$\xrightarrow{\mathtt{pq!}\langle \mathtt{int} \rangle} \mathtt{p} \rightsquigarrow \mathtt{q} : \langle \mathtt{int} \rangle \{x_\mathtt{p} < 10,\ \_, x_\mathtt{q} > 20,\ \_\}.\mathtt{p} \to \mathtt{r} : \langle \mathtt{int} \rangle \{x_\mathtt{p} < 10,\ \_, \mathtt{true},\ \_\}$$
$$\xrightarrow{\mathtt{pr!}\langle \mathtt{int} \rangle} \mathtt{p} \rightsquigarrow \mathtt{q} : \langle \mathtt{int} \rangle \{x_\mathtt{p} < 10,\ \_, x_\mathtt{q} > 20,\ \_\}.\mathtt{p} \rightsquigarrow \mathtt{r} : \langle \mathtt{int} \rangle \{x_\mathtt{p} \geq 10,\ \_, \mathtt{true},\ \_\}$$

After the first action $\mathtt{pq!}\langle \mathtt{int} \rangle$ the *TG* above can reduce by one of the following actions: a send $\mathtt{pr!}\langle \mathtt{int} \rangle$ (as illustrated), a receive of q, or a time step. By using intermediate states, a send action and its corresponding receive action (e.g., $\mathtt{pq!}\langle \mathtt{int} \rangle$ and $\mathtt{pq?}\langle \mathtt{int} \rangle$) are separate, hence could be interleaved with other actions, as well as occur at different times. This fine-grained semantics corresponds to local type semantics where asynchrony is modelled as message exchange through channels (see Theorem 3).

*TG*s are used as a model of the correct behaviour for distributed implementations in § 4. Therefore their semantics should only include desirable executions. We need to take special care in the definition of the semantics of time actions: if an action is ready to be executed and the associated constraint has an upper bound, then the semantics should prevent time steps that are too big and would make that clock constraint unsatisfiable. For instance in $\mathtt{p} \to \mathtt{q} : \langle \mathtt{int} \rangle \{x_\mathtt{p} \leq 20,\ \_, \mathtt{true},\ \_\}$ (assuming $x_\mathtt{p} = 0$) the LTS should allow, before the send action of p occurs, only time steps that preserve $x_\mathtt{p} \leq 20$.

More generally, we need to ensure that time actions do not invalidate the constraint of any action that is ready to be executed, or *ready action*. A ready action is an action that has no causal relationship with other actions that occur earlier, syntactically. A *TG* may have more than one ready action, as shown by the following example.

$$\mathtt{p} \to \mathtt{q} : \langle \mathtt{int} \rangle \{x_\mathtt{p} \leq 20,\ \_, \mathtt{true},\ \_\}.\mathtt{k} \to \mathtt{r} : \langle \mathtt{int} \rangle \{x_\mathtt{k} < 10,\ \_, x_\mathtt{r} = 10,\ \_\}$$

The *TG* above has two ready actions, namely the send actions of p and of k which can happen in any order due to asynchrony (i.e., an order cannot be enforced without extra communications between p and k). In this case a desirable semantics should prevent the elapsing of time intervals that would invalidate either $\{x_\mathtt{p} \leq 20\}$ or $\{x_\mathtt{k} < 10\}$.

Below, function $\mathtt{rdy}(G,D)$ returns the set, for each ready actions in G, of elements of the form $\{\delta_i\}_{i \in I}$ which are the constraints of the branches of that ready action. D is a set of participants, initially empty, used to keep track of the causal dependencies between actions. We write $\mathtt{rdy}(G)$ for $\mathtt{rdy}(G,\emptyset)$.

$$(1)\ \begin{array}{l}\mathtt{rdy}(\mathtt{p} \to \mathtt{q} : \{l_i \langle S_i \rangle \{A_i\}.G_i\}_{i \in I}, D) \\ \quad (\text{with } A_i = \{\delta_{\mathrm{O}i}, \lambda_{\mathrm{O}i}, \delta_{\mathrm{I}i}, \lambda_{\mathrm{I}i}\})\end{array} = \begin{cases} \{\{\delta_{\mathrm{O}i}\}_{i \in I}\} \cup_{i \in I} \mathtt{rdy}(G_i, D \cup \{\mathtt{p},\mathtt{q}\}) & \text{if } \mathtt{p} \notin D \\ \bigcup_{i \in I} \mathtt{rdy}(G_i, D \cup \{\mathtt{p},\mathtt{q}\}) & \text{otherwise} \end{cases}$$

$$(2)\ \mathtt{rdy}(\mathtt{p} \rightsquigarrow \mathtt{q} : l \langle S \rangle \{\delta_\mathrm{O}, \lambda_\mathrm{O}, \delta_\mathrm{I}, \lambda_\mathrm{I}\}.G, D) = \begin{cases} \{\{\delta_\mathrm{I}\}\} \cup \mathtt{rdy}(G, D \cup \{\mathtt{q}\}) & \text{if } \mathtt{q} \notin D \\ \mathtt{rdy}(G, D \cup \{\mathtt{q}\}) & \text{otherwise} \end{cases}$$

$$(3)\ \mathtt{rdy}(\mu \mathtt{t}.G, D) = \mathtt{rdy}(G, D) \qquad (4)\ \mathtt{rdy}(\mathtt{t}, D) = \mathtt{rdy}(\mathtt{end}, D) = \emptyset$$

In (1) the send action of p is ready, hence the singleton including the constraints $\{\delta_{\mathrm{O}i}\}_{i \in I}$ are added to the solution and each G_i is recursively checked. Any action in G_i involving p or q is not ready. Adding $\{\mathtt{p},\mathtt{q}\}$ to D ensures that the constraints of actions that causally depend from the first interaction are not included in the solution. (2) is similar.

Definition 2 (Satisfiability of ready actions). *We write* $\nu \models^* \mathtt{rdy}(G)$ *when the constraints of all ready actions of G are eventually satisfiable under* ν. *Formally,* $\nu \models^* \mathtt{rdy}(G)$ *iff* $\forall \{\{\delta_i\}_{i \in I}\} \in \mathtt{rdy}(G)\, \exists t \geq 0,\ j \in I.\ \nu + t \models \delta_j$.

$$\frac{j \in I \quad A_j = \{\delta_O, \lambda_O, \delta_I, \lambda_I\} \quad \nu \models \delta_O \quad \nu' = [\lambda_O \mapsto 0]\nu}{(\nu, p \to q : \{l_i\langle S_i\rangle\{A_i\}.G_i\}_{i\in I}) \xrightarrow{pq!l_j\langle S_j\rangle} (\nu', p \rightsquigarrow q : l_j\langle S_j\rangle\{A_j\}.G_j)} \lfloor\text{SELECT}\rfloor$$

$$\frac{\nu \models \delta_I \quad \nu' = [\lambda_I \mapsto 0]\nu}{(\nu, p \rightsquigarrow q : l\langle S\rangle\{\delta_O, \lambda_O, \delta_I, \lambda_I\}.G) \xrightarrow{pq?l\langle S\rangle} (\nu', G)} \quad \frac{(\nu, G[\mu t.G/t]) \xrightarrow{\ell} (\nu', G')}{(\nu, \mu t.G) \xrightarrow{\ell} (\nu', G')} \lfloor\text{BRANCH}\rfloor/\lfloor\text{REC}\rfloor$$

$$\frac{\forall k \in I \quad (\nu, G_k) \xrightarrow{\ell} (\nu', G'_k) \quad p, q \notin subj(\ell) \quad \ell \neq t}{(\nu, p \to q : \{l_i\langle S_i\rangle\{A_i\}.G_i\}_{i\in I}) \xrightarrow{\ell} (\nu', p \to q : \{l_i\langle S_i\rangle\{A_i\}.G'_i\}_{i\in I})} \lfloor\text{ASYNC1}\rfloor$$

$$\frac{(\nu, G) \xrightarrow{\ell} (\nu', G') \quad q \notin subj(\ell)}{(\nu, p \rightsquigarrow q : l\langle S\rangle\{A\}.G) \xrightarrow{\ell} (\nu', p \rightsquigarrow q : l\langle S\rangle\{A\}.G')} \quad \frac{\nu + t \models^* \mathtt{rdy}(G)}{(\nu, G) \xrightarrow{t} (\nu + t, G)} \lfloor\text{ASYNC2}\rfloor/\lfloor\text{TIME}\rfloor$$

Fig. 1. Labelled transitions for timed global types

The transition rules for *TG*s are given in Figure 1. We assume the execution always begins with initial assignment ν_0. Rule $\lfloor$SELECT$\rfloor$ models selection as usual, except that the clock constraint of the selected branch j is checked against the current assignment (i.e., $\nu \models \delta_O$) which is updated with reset λ_O. Rules $\lfloor$ASYNC1$\rfloor$ and $\lfloor$ASYNC2$\rfloor$ model interactions that appear later (syntactically), but are not causally dependent on the first interaction. Rule $\lfloor$TIME$\rfloor$ models time passing by incrementing all clocks; the clause in the premise prevents time steps that would make the clock constraints of some ready action unsatisfiable. By Definition 2, $\nu + t \models^* \mathtt{rdy}(G)$ requires the satisfiability of the constraints of *some* of the branches of (each ready action of) G, while some other branches may become unsatisfiable. In this way, the semantics of *TG*s specifies the full range of correct behaviours. For instance in $p \to q : \{l_1 : \{x_p < c, \_, \mathtt{true}, \_\},\ l_2 : \{x_p > c, \_, \mathtt{true}, \_\}\}$ one can, in some executions, let time pass until $x_p > c$ so that l_2 can be chosen. $\lfloor$TIME$\rfloor$ can always be applied to $(\nu, \mathtt{end})$ since $\nu + t \models^* \mathtt{rdy}(\mathtt{end})$ for all t.

Semantics for Timed Local Types. The LTS for *TLs* is defined over states (ν, T), labels $\mathcal{L}$ and is generated by the following rules:

$$(\nu, q\oplus\{l_i : \langle S_i\rangle\{B_i\}.T_i\}_{i\in I}) \xrightarrow{pq!l_j\langle S_j\rangle} (\nu', T_j) \quad (j \in I \ B_j = \{\delta, \lambda\} \ \nu \models \delta \ \nu' = [\lambda \mapsto 0]\nu) \ \lfloor\text{LSEL}\rfloor$$
$$(\nu, q\&\{l_i : \langle S_i\rangle\{B_i\}.T_i\}_{i\in I}) \xrightarrow{qp?l_j\langle S_j\rangle} (\nu', T_j) \quad (j \in I \ B_j = \{\delta, \lambda\} \ \nu \models \delta \ \nu' = [\lambda \mapsto 0]\nu) \ \lfloor\text{LBRA}\rfloor$$
$$(\nu, T[\mu t.T/t]) \xrightarrow{\ell} (\nu', T') \ \textit{implies} \ (\nu, \mu t.T) \xrightarrow{\ell} (\nu', T') \quad \lfloor\text{LREC}\rfloor$$
$$\nu + t \models^* \mathtt{rdy}(T) \ \textit{implies} \ (\nu, T) \xrightarrow{t} (\nu', T) \quad \lfloor\text{LTIME}\rfloor$$

Rule $\lfloor$LSEL$\rfloor$ is for send actions and its dual $\lfloor$LBRA$\rfloor$ for receive actions. In rule $\lfloor$LTIME$\rfloor$ for time passing, the constraints of the ready action of T must be satisfiable after t in ν. Note that T always has only one ready action. The definitions of $\mathtt{rdy}(T)$ and $\nu + t \models^* \mathtt{rdy}(T)$ are the obvious extensions of the definitions we have given for *TG*s.

Given a set of participants $\{1, \ldots, n\}$ we define configurations $(T_1, \ldots, T_n, \vec{w})$ where $\vec{w} ::= \{w_{ij}\}_{i\neq j\in\{1,\ldots,n\}}$ are unidirectional, possibly empty (denoted by ε), unbounded channels with elements of the form $l\langle S\rangle$. The LTS of $(T_1, \ldots, T_n, \vec{w})$ is defined as follows, with ν being the overriding union (i.e., $\oplus_{i\in\{1,\ldots,n\}}\nu_i$) of the clock assignments ν_i

of the participants. $(\nu, (T_1, \ldots, T_n, \vec{w})) \xrightarrow{\ell} (\nu', (T_1', \ldots, T_n', \vec{w}'))$ iff:

(1) $\ell = \mathtt{pq}!l\langle S\rangle \Rightarrow (\nu_\mathtt{p}, T_\mathtt{p}) \xrightarrow{\ell} (\nu_\mathtt{p}', T_\mathtt{p}') \wedge w_{\mathtt{pq}}' = w_{\mathtt{pq}} \cdot l\langle S\rangle \wedge (ij \neq \mathtt{pq} \Rightarrow w_{ij} = w_{ij}' \wedge T_i = T_i')$
(2) $\ell = \mathtt{pq}?l\langle S\rangle \Rightarrow (\nu_\mathtt{q}, T_\mathtt{q}) \xrightarrow{\ell} (\nu_\mathtt{q}', T_\mathtt{q}') \wedge l\langle S\rangle \cdot w_{\mathtt{pq}}' = w_{\mathtt{pq}} \wedge (ij \neq \mathtt{pq} \Rightarrow w_{ij} = w_{ij}' \wedge T_j = T_j')$
(3) $\ell = t \Rightarrow \forall i \neq j \in \{1, \ldots, n\}.(\nu_i, T_i) \xrightarrow{\ell} (\nu_i + t, T_i) \wedge w_{ij} = w_{ij}'$
with $\mathtt{p}, \mathtt{q}, i, j \in \{1, \ldots, n\}$.

We write $TR(G)$ for the set of visible traces obtained by reducing G under the initial assignment ν_0. Similarly for $TR(T_1, \ldots, T_n, \vec{\varepsilon})$. We denote trace equivalence by $\approx$.

Theorem 3 (Soundness and completeness of projection). *Let G be a timed global type and $\{T_1, \ldots, T_n\} = \{G \downarrow_\mathtt{p}\}_{\mathtt{p} \in \mathcal{P}(G)}$ be the set of its projections, then $G \approx (T_1, \ldots, T_n, \vec{\varepsilon})$.*

4 Multiparty Session Processes with Delays

We model processes using a timed extension of the asynchronous session calculus [6]. The syntax of the session calculus with delays is presented below.

$P ::= \overline{u}[\mathtt{n}](y).P$	Request	$\mid\ (\nu a)P$	Hide Shared	
$\mid\ u[\mathtt{i}](y).P$	Accept	$\mid\ (\nu s)P$	Hide Session	
$\mid\ c[\mathtt{p}] \triangleleft l\langle e\rangle; P$	Select	$\mid\ s : h$	Queue	
$\mid\ c[\mathtt{p}] \triangleright \{l_i(z_i).P_i\}_{i \in I}$	Branching	$h ::= \emptyset \mid h \cdot (\mathtt{p}, \mathtt{q}, m)$		(queue content)
$\mid\ \mathtt{delay}(t).P$	Delay	$m ::= l\langle v\rangle \mid (s[\mathtt{p}], \nu)$		(messages)
$\mid\ \mathtt{if}\ e\ \mathtt{then}\ P\ \mathtt{else}\ Q$	Conditional	$c ::= s[\mathtt{p}] \mid y$		(session names)
$\mid\ P \mid Q$	Parallel	$u ::= a \mid z$		(shared names)
$\mid\ \mathbf{0}$	Inaction	$e ::= v \mid \neg e \mid e' \,\mathrm{op}\, e'$		(expressions)
$\mid\ \mu X.P$	Recursion	$v ::= c \mid u \mid \mathtt{true} \mid \ldots$		(values)
$\mid\ X$	Variable			

$\overline{u}[\mathtt{n}](y).P$ sends, along u, a request to start a new session y with participants $\mathtt{1}, \ldots, \mathtt{n}$, where it participates as $\mathtt{1}$ and continues as P. Its dual $u[\mathtt{i}](y).P$ engages in a new session as participant $\mathtt{i}$. Select $c[\mathtt{p}] \triangleleft l\langle e\rangle; P$ sends message $l\langle e\rangle$ to participant $\mathtt{p}$ in session c and continues as P. Branching is dual. Request and accept bind y in P, and branching binds z_i in P_i. We introduce a new primitive $\mathtt{delay}(t).P$ that executes P after waiting exactly t units of time. Note that t is a constant (as in [5,16]). The other processes are standard. We often omit inaction $\mathbf{0}$, and the label in a singleton selection or branching, and denote with $\mathtt{fn}(P)$ the set of free variables and names of P.

We define *programs* as processes that have not yet engaged in any session, namely that have no queues, no session name hiding, and no free session names/variables.

Structural equivalence for processes is the least equivalence relation satisfying the standard rules from [6] – we recall below (first row) those for queues – plus the following rules for delays:

$$(\nu s)s : \emptyset \equiv \mathbf{0} \quad s : h \cdot (\mathtt{p}, \mathtt{q}, m) \cdot (\mathtt{p}', \mathtt{q}', m') \cdot h' \equiv s : h \cdot (\mathtt{p}', \mathtt{q}', m') \cdot (\mathtt{p}, \mathtt{q}, m) \cdot h' \quad \text{if } \mathtt{p} \neq \mathtt{p}' \text{ or } \mathtt{q} \neq \mathtt{q}'$$
$$\mathtt{delay}(t + t').P \equiv \mathtt{delay}(t).\mathtt{delay}(t').P \qquad \mathtt{delay}(0).P \equiv P$$
$$\mathtt{delay}(t).(\nu a)P \equiv (\nu a)\mathtt{delay}(t).P \qquad \mathtt{delay}(t).(P \mid Q) \equiv \mathtt{delay}(t).P \mid \mathtt{delay}(t).Q$$

In the first row: $(\nu s)s : \emptyset \equiv \mathbf{0}$ removes queues of ended sessions, the second rule permutes causally unrelated messages. In the second row: the first rule breaks delays into

$$
\begin{array}{rcll}
\overline{a}[\mathtt{n}](y).P_1 \mid \prod_{i\in\{2,..,n\}} a[\mathtt{i}](y).P_i & \longrightarrow & (\nu s)(\prod_{i\in\{1,..,n\}} P_i[s[\mathtt{i}]/y] \mid s:\emptyset)\ (s \notin \mathtt{fn}(P_i)) & \lfloor\text{LINK}\rfloor \\
s[\mathtt{p}][\mathtt{q}] \lhd l\langle e\rangle;P \mid s:h & \longrightarrow & P \mid s:h\cdot(\mathtt{p},\mathtt{q},l\langle v\rangle) \quad (e\downarrow v) & \lfloor\text{SEL}\rfloor \\
s[\mathtt{p}][\mathtt{q}] \rhd \{l_i(z_i).P_i\}_{i\in J} \mid s:(\mathtt{p},\mathtt{q},l_j\langle v\rangle)\cdot h & \longrightarrow & P_j[v/z_j] \mid s:h \quad (j\in J) & \lfloor\text{BRA}\rfloor \\
\mathtt{delay}(t).P \mid \prod_{j\in J} s_j:h_j & \longrightarrow & P \mid \prod_{j\in J} s_j:h_j & \lfloor\text{DELAY}\rfloor \\
P \longrightarrow P' \text{ (not by } \lfloor\text{DELAY}\rfloor) & \textit{imply} & P \mid Q \longrightarrow P' \mid Q & \lfloor\text{COM}\rfloor \\
\mathtt{if}\ e\ \mathtt{then}\ P\ \mathtt{else}\ Q \longrightarrow P\ \ (e\downarrow\mathtt{true}) & & \mathtt{if}\ e\ \mathtt{then}\ P\ \mathtt{else}\ Q \longrightarrow Q\ \ (e\downarrow\mathtt{false}) & \lfloor\text{IFT/IFF}\rfloor \\
P\equiv P'\ \ P'\longrightarrow Q'\ \ Q\equiv Q'\ \textit{imply}\ P\longrightarrow Q & & P\longrightarrow P'\ \textit{imply}\ (\nu n)P \longrightarrow (\nu n)P' & \lfloor\text{STR/HIDE}\rfloor
\end{array}
$$

Fig. 2. Reduction for processes

smaller intervals, and $\mathtt{delay}(0).P \equiv P$ allows time to pass for idle processes. The rules in the third row distribute delays in hiding and parallel processes.

The reduction rules are given in Figure 2. In $\lfloor\text{SEL}\rfloor$ we write $e \downarrow v$ when expression e evaluates to value v. Rule $\lfloor\text{DELAY}\rfloor$ models time passing for P. By combining $\lfloor\text{DELAY}\rfloor$ with rule $\mathtt{delay}(t).(P \mid Q) \equiv \mathtt{delay}(t).P \mid \mathtt{delay}(t).Q$ we allow a delay to elapse simultaneously for parallel processes. The queues in parallel with P always allow time passing, unlike other kinds of processes (as shown in rule $\lfloor\text{COM}\rfloor$ which models the synchronous semantics of time in [14]). Rule $\lfloor\text{COM}\rfloor$ enables part of the system to reduce as long as the reduction does not involve $\lfloor\text{DELAY}\rfloor$ on P. If P reduces by $\lfloor\text{DELAY}\rfloor$ then also all other parallel processes must make the same time step, i.e. the whole system must move by $\lfloor\text{DELAY}\rfloor$. The other rules are standard (n stands for s or a in $\lfloor\text{HIDE}\rfloor$).

Example 4 (Temperature calculation). Process $P_{\mathtt{M}}$ is a possible implementation of participant M of the protocol in Example 1, e.g., $G\downarrow_{\mathtt{M}}$. Assuming that at least one task is needed in each session, we let $\mathtt{task}()$ be a local function returning the next task and $\mathtt{more\_tasks}()$ return $\mathtt{true}$ when more tasks have to be submitted and $\mathtt{false}$ otherwise.

$$
\begin{array}{l}
P_{\mathtt{M}} = s[\mathtt{M}][\mathtt{W}] \lhd \langle\mathtt{task}()\rangle; \mu X.\mathtt{delay}(2l+w).\ \ s[\mathtt{M}][\mathtt{W}] \rhd (y); \mathtt{if\ more\_tasks}() \\
\quad \mathtt{then}\ s[\mathtt{M}][\mathtt{A}] \lhd \mathsf{more}\langle y\rangle; s[\mathtt{M}][\mathtt{W}] \lhd \mathsf{more}\langle\mathtt{task}()\rangle; X\ \mathtt{else}\ s[\mathtt{M}][\mathtt{A}] \lhd \mathsf{stop}\langle y\rangle; s[\mathtt{M}][\mathtt{W}] \lhd \mathsf{stop}\langle\rangle; \mathtt{end}
\end{array}
$$

Proof rules. We validate programs against specifications based on *TL*s, using judgements of the form $\Gamma \vdash P \rhd \Delta$ and $\Gamma \vdash e : S$ defined on the following environments:

$$
\Gamma ::= \emptyset \mid \Gamma, u:S \mid \Gamma, X:\Delta \qquad\qquad \Delta ::= \emptyset \mid \Delta, c:(\nu, T)
$$

The *type environment* Γ maps shared variables/names to sorts and process variables to their types, and the *session environment* Δ holds information on the ongoing sessions, e.g., $\Delta(s[\mathtt{p}]) = (\nu, T)$ when the process being validated is acting as p in session s specified by T; ν is a virtual clock assignment built during the validation (virtual in the sense that it mimics the clock assignment associated to T by the LTS) .

Resets can generate infinite time scenarios in recursive protocols. To ensure sound typing we introduce a condition, *infinite satisfiability*, that guarantees a regularity across different instances of a recursion.

Definition 5 (Infinitely satisfiable). *G is infinitely satisfiable if either: (1) constraints in recursion bodies have no equalities nor upper bounds (i.e., $x < c$ or $x \leq c$) and no resets occur, or (2) all participants reset at each iteration.*

In the rest of this section we assume that *TG*s are infinitely satisfiable. As usual (e.g., [13]), in the validation of P we check $\Gamma \vdash P' \rhd \Delta$ where P' is obtained by unfolding once all

$$\lfloor\text{VREQ}\rfloor \frac{\Gamma,u:G\vdash P\triangleright\Delta,y[1]:(\nu_0,G\downarrow_1) \quad dom(\nu_0)=\{x_1\}}{\Gamma,u:G\vdash \overline{u}[\mathtt{n}](y).P\triangleright\Delta} \qquad \lfloor\text{VACC}\rfloor \frac{\Gamma,u:G\vdash P\triangleright\Delta,y[\mathtt{i}]:(\nu_0,G\downarrow_\mathtt{i}) \quad dom(\nu_0)=\{x_1\} \quad \mathtt{i}\neq 1}{\Gamma,u:G\vdash u[\mathtt{i}](y).P\triangleright\Delta}$$

$$\lfloor\text{VBRA}\rfloor \frac{\forall i\in I \quad \nu\models\delta_i \quad \begin{cases}\Gamma,z_i:S_i\vdash P_i\triangleright\Delta,c:([\lambda_i\mapsto 0]\nu,T_i) & (S_i\neq(T_d,\delta_d))\\ \Gamma\vdash P_i\triangleright\Delta,c:([\lambda_i\mapsto 0]\nu,T_i),z_i:(\nu_d,T_d) \quad \nu_d\models\delta_d & (S_i=(T_d,\delta_d))\end{cases}}{\Gamma\vdash c[\mathtt{p}]\triangleright\{l_i(z_i).P_i\}_{i\in I}\triangleright\Delta,c:(\nu,\mathtt{p}\&\{l_i:\langle S_i\rangle\{\delta_i,\lambda_i\}.T_i\}_{i\in I})}$$

$$\lfloor\text{VSEL}\rfloor \frac{j\in I \quad \Gamma\vdash e:S_j \quad \nu\models\delta_j \quad \Gamma\vdash P\triangleright\Delta,c:([\lambda_j\mapsto 0]\nu,T_j) \quad (S_j\neq(T_d,\delta_d))}{\Gamma\vdash c[\mathtt{p}]\triangleleft l_j\langle e\rangle;P\triangleright\Delta,c:(\nu,\mathtt{p}\oplus\{l_i:\langle S_i\rangle\{\delta_i,\lambda_i\}.T_i\}_{i\in I})}$$

$$\lfloor\text{VDEL}\rfloor \frac{j\in I \quad \Gamma\vdash e:S_j \quad \nu\models\delta_j \quad \nu_d\models\delta_d \quad \Gamma\vdash P\triangleright\Delta,c:([\lambda_j\mapsto 0]\nu,T_j) \quad (S_j=(T_d,\delta_d))}{\Gamma\vdash c[\mathtt{p}]\triangleleft l_j\langle e\rangle;P\triangleright\Delta,c:(\nu,\mathtt{p}\oplus\{l_i:\langle S_i\rangle\{\delta_i,\lambda_i\}.T_i\}_{i\in I}),c':(\nu_d,T_d)}$$

$$\lfloor\text{VPAR}\rfloor \frac{dom(\Delta_1)\cap dom(\Delta_2)=\emptyset \quad \Gamma\vdash P_i\triangleright\Delta_i \quad i\in\{1,2\}}{\Gamma\vdash P_1\mid P_2\triangleright\Delta_1,\Delta_2} \qquad \lfloor\text{VCOND}\rfloor \frac{\Gamma\vdash e:\mathtt{bool} \quad \Gamma\vdash P_i\triangleright\Delta \quad i\in\{1,2\}}{\Gamma\vdash \mathtt{if}\ e\ \mathtt{then}\ P_1\ \mathtt{else}\ P_2\triangleright\Delta}$$

$$\lfloor\text{VTIME}\rfloor \frac{\Gamma\vdash P\triangleright\{c_i:(\nu_i+t,T_i)\}_{i\in I}}{\Gamma\vdash \mathtt{delay}(t).P\triangleright\{c_i:(\nu_i,T_i)\}_{i\in I}} \qquad \lfloor\text{VEND}\rfloor \frac{\forall c\in dom(\Delta) \quad \Delta(c)=(\nu,\mathtt{end})}{\Gamma\vdash \mathbf{0}\triangleright\Delta}$$

$$\lfloor\text{VDEF}\rfloor \frac{\Gamma,X:\Delta\vdash P\triangleright\Delta}{\Gamma\vdash \mu X.P\triangleright\Delta} \qquad \lfloor\text{VCALL}\rfloor \frac{\forall c\in dom(\Delta') \quad \Delta'(c)=(\nu,\mathtt{end})}{\Gamma,X:\Delta\vdash X\triangleright\Delta,\Delta'}$$

Fig. 3. Proof rules for programs

recursions $\mu X.P''$ occurring in P. This ensures that both the first instance of a recursion and the successive ones (all similar by infinite satisfiability) satisfy the specification.

We show in Figure 3 selected proof rules for programs. Rule $\lfloor$VREQ$\rfloor$ for session request adds a new instance of session for participant 1 to Δ in the premise. The newly instantiated session is associated with an initial assignment ν_0 for the clock of participant 1. Rule $\lfloor$VACC$\rfloor$ for session accept is similar but initiates a new session for participant i with $\mathtt{i}>1$. Rule $\lfloor$VBRA$\rfloor$ is for branching processes. For all $i\in I$, δ_i must hold under ν and the virtual clock assignments used to validate P_i is reset according to λ_i. If the received message is a session (i.e., $S_i=(T_d,\delta_d)$) a new assignment $z_i:(\nu_d,T_d)$ is added to Δ in the premise. This can be any assignment such that $\nu_d\models\delta_d$. Rule $\lfloor$VSEL$\rfloor$ for selection processes checks the constraint δ_j of the selected branch j against ν. In the premise, ν is reset as prescribed by λ_j. Rule $\lfloor$VDEL$\rfloor$ for delegation requires δ_d to be satisfied under ν_d (of the delegated session) which is removed from the premise. Rule $\lfloor$VTIME$\rfloor$ increments the clock assignments of all sessions in Δ. Rule $\lfloor$VEND$\rfloor$ validates $\mathbf{0}$ if there are no more actions prescribed by Δ. Rule $\lfloor$VDEF$\rfloor$ extends Γ with the assignment for process variable X. Rules $\lfloor$VPAR$\rfloor$ and $\lfloor$VCOND$\rfloor$ are standard. Rule $\lfloor$VCALL$\rfloor$ validates, as usual, recursive call X against $\Gamma(X)$ (and possibly some terminated sessions Δ').

Theorem 6 (Type preservation). *If* $\Gamma\vdash P\triangleright\emptyset$ *and* $P\longrightarrow P'$, *then* $\Gamma\vdash P'\triangleright\emptyset$.

In the above theorem, P is a process reduced from a program (hence Δ is $\emptyset$). A standard corollary of type preservation is error freedom. An error state is reached when a process performs an action at a time that violates the constraints prescribed by its type. To formulate this property, we extend the syntax of processes as follows: selection and branching are annotated with clock constraints and resets (i.e., $c[\mathtt{p}]\triangleleft l\langle e\rangle\{\delta,\lambda\};P$ and $c[\mathtt{p}]\triangleright\{l_i(z_i)\{\delta_i,\lambda_i\}.P_i\}_{i\in I}$); two new processes, `error` and clock process $(s[\mathtt{p}],\nu)$, are introduced. Process `error` denotes a state in which a violation has occurred, and

$(s[\mathtt{p}],\nu)$ associates a clock assignment ν to ongoing session $s[\mathtt{p}]$. The reduction rules for processes are extended as shown below.

$$\frac{\forall i \in \{1,..,n\} \quad s \notin \mathtt{fn}(P_i)}{\overline{a}[\mathtt{n}](y).P_1 \mid \prod_{i\in\{2,..,n\}} a[\mathtt{i}](y).P_i \longrightarrow (\nu s)(\prod_{i\in\{1,..,n\}}(P_i[s[\mathtt{i}]/y] \mid (s[\mathtt{i}],\nu_0)) \mid s:\emptyset)} \lfloor \textsc{Link} \rfloor$$

$$\mathtt{delay}(t).P \mid \prod_{j\in J}(s_j : h_j \mid \prod_{k\in K_j}(s_j[\mathtt{p}_k],\nu_k)) \longrightarrow P \mid \prod_{j\in J}(s_j : h_j \mid \prod_{k\in K_j}(s_j[\mathtt{p}_k],\nu_k + t)) \; \lfloor \textsc{Delay} \rfloor$$

$$\frac{e \downarrow v \qquad \nu' = [\lambda \mapsto 0]\nu \qquad \delta \models \nu}{s[\mathtt{p}][\mathtt{q}] \triangleleft \{\delta,\lambda\} l\langle e\rangle;P \mid s:h \mid (s[\mathtt{p}],\nu) \longrightarrow P \mid s:h\cdot(\mathtt{p},\mathtt{q},l\langle v\rangle) \mid (s[\mathtt{p}],\nu')} \lfloor \textsc{Sel} \rfloor$$

$$\frac{\neg\delta \models \nu}{s[\mathtt{p}][\mathtt{q}] \triangleleft \{\delta,\lambda\} l\langle e\rangle;P \mid s:h \mid (s[\mathtt{p}],\nu) \longrightarrow \mathtt{error} \mid s:h \mid (s[\mathtt{p}],\nu)} \lfloor \textsc{ESel} \rfloor$$

$\lfloor \textsc{Link} \rfloor$ introduces a clock process $(s[\mathtt{i}],\nu_0)$ with initial assignment for each participant $\mathtt{i}$ in the new session; $\lfloor \textsc{Delay} \rfloor$ increments all clock assignments; $\lfloor \textsc{Sel} \rfloor$ checks clock constraints against clock assignments and appropriately resets (the rule for branching is extended similarly); $\lfloor \textsc{ESel} \rfloor$ is an additional rule which moves to `error` when a process tries to perform a send action at a time that does not satisfy the constraint (a similar rule is added for violating receive actions). Note that $\lfloor \textsc{Sel} \rfloor$ only resets the clocks associated to participant $\mathtt{p}$ in session s and never affects clocks of other participants and sessions. The proof rules are adapted straightforwardly, with `error` not validated against any Δ.

Theorem 7 (Time-error freedom). *If* $\Gamma \vdash P \triangleright \Delta$*, and* $P \rightarrow^* P'$ *then* $P' \not\equiv \mathtt{error}$.

5 Time-Progress of Timed Processes and CTAs

This section studies a subclass of timed global types characterised by two properties, *feasibility* and *wait-freedom* and states their decidability; it then shows that these are sufficient conditions for progress of validated processes and CTAs.

Feasibility. A *TG* G is *feasible* iff $(\nu_0, G_0) \longrightarrow^* (\nu, G)$ implies $(\nu, G) \longrightarrow^* (\nu', \mathtt{end})$ for some ν'. Intuitively, G_0 is feasible if every partial execution can be extended to a terminated session. Not all *TGs* are feasible. The specified protocol may get stuck because a constraint is unsatisfiable, for example it is `false`, or the restrictions posed by previously occurred constraints are too strong. We give below a few examples of non-feasible (1,5) and feasible (2,3,4,6) global types:

1. $\mathtt{p} \rightarrow \mathtt{q} : \langle \mathtt{int} \rangle \{x_\mathtt{p} > 3,\ \_, x_\mathtt{q} = 4,\ \_\}$
2. $\mathtt{p} \rightarrow \mathtt{q} : \langle \mathtt{int} \rangle \{x_\mathtt{p} > 3 \wedge x_\mathtt{p} \leq 4,\ \_, x_\mathtt{q} = 4,\ \_\}$ 3. $\mathtt{p} \rightarrow \mathtt{q} : \langle \mathtt{int} \rangle \{x_\mathtt{p} > 3,\ \_, x_\mathtt{q} \geq 4,\ \_\}$
4. $\mathtt{q} \rightarrow \mathtt{r} : \{l_1 : \{x_\mathtt{q} > 3,\ \_, \mathtt{true},\ \_\},\ l_2 : \{x_\mathtt{q} < 3,\ \_, \mathtt{true},\ \_\}\}$
5. $\mu t.\mathtt{p} \rightarrow \mathtt{q} : \langle \mathtt{int} \rangle \{x_\mathtt{p} < 1, x_\mathtt{p}, x_\mathtt{q} = 2, x_\mathtt{q}\}.\mathtt{p} \rightarrow \mathtt{r} : \langle \mathtt{int} \rangle \{x_\mathtt{p} < 5,\ \_, \mathtt{true}, x_\mathtt{r}\}.\mathtt{t}$
6. $\mu t.\mathtt{p} \rightarrow \mathtt{q} : \langle \mathtt{int} \rangle \{x_\mathtt{p} < 1, x_\mathtt{p}, x_\mathtt{q} = 2, x_\mathtt{q}\}.\mathtt{p} \rightarrow \mathtt{r} : \langle \mathtt{int} \rangle \{x_\mathtt{p} < 1,\ \_, \mathtt{true}, x_\mathtt{r}\}.\mathtt{t}$

In (1) if $\mathtt{p}$ sends $\langle \mathtt{int} \rangle$ at time 5, which satisfies $x_\mathtt{p} > 3$, then there exists no $x_\mathtt{q}$ satisfying $x_\mathtt{q} = 4$ (considering that $x_\mathtt{q}$ must be greater than or equal to 5 to respect the global flowing of time); (2) amends (1) by restricting the earlier constraint; (3) amends (1) by relaxing the unsatisfiable constraint. In branching and selection at least one constraint associated to the branches must be satisfiable, e.g., we accept (4). In recursive *TGs*, a constraint may become unsatisfiable by constraints that occur after, syntactically, in the same recursion body. In the second iteration of (5) $x_\mathtt{q} = 2$ is made unsatisfiable by $x_\mathtt{p} < 5$ occurring in the first iteration (e.g., $\mathtt{p}$ may send $\mathtt{q}$ the message when $x_\mathtt{q} > 2$); in (6) this problem is solved by restricting the second constraint on $x_\mathtt{p}$.

Wait-freedom. In distributed implementations, a party can send a message at any time satisfying the constraint. Another party can choose to execute the corresponding receive action at any specific time satisfying the constraint without knowing when the message has been or will be sent. If the constraints in a *TG* allow a receive action before the corresponding send, a complete correct execution of the protocol may not be possible at run-time. We show below a process $P \mid Q$ whose correct execution cannot complete despite $P \mid Q$ is the well-typed implementation of a feasible *TG*.

$$
\begin{aligned}
G \; &= \mathtt{p} \to \mathtt{q} : \langle \mathtt{int} \rangle \{x_\mathtt{p} < 3 \vee x_\mathtt{p} > 3,\ \_, x_\mathtt{q} < 3 \vee x_\mathtt{q} > 3,\ \_\}.\\
&\quad \mathtt{q} \to \mathtt{p} : \{l_1 : \{x_\mathtt{q} > 3,\ \_, x_\mathtt{p} > 3,\ \_\},\ l_2 : \{x_\mathtt{q} < 3,\ \_, x_\mathtt{p} < 3,\ \_\}\}\\
G \downarrow_\mathtt{p} &= \mathtt{q} \oplus \langle \mathtt{int} \rangle \{x_\mathtt{p} < 3 \vee x_\mathtt{p} > 3,\ \_\}.\mathtt{q}\&\{l_1 : \{x_\mathtt{p} > 3,\ \_\},\ l_2 : \{x_\mathtt{p} < 3,\ \_\}\}\\
G \downarrow_\mathtt{q} &= \mathtt{p}\&\langle \mathtt{int} \rangle \{x_\mathtt{q} < 3 \vee x_\mathtt{q} > 3,\ \_\}.\mathtt{p} \oplus \{l_1 : \{x_\mathtt{q} > 3,\ \_\},\ l_2 : \{x_\mathtt{q} < 3,\ \_\}\}\\
P \; &= \mathtt{delay}(6).s[\mathtt{p}][\mathtt{q}] \triangleleft \langle 10 \rangle ; s[\mathtt{q}][\mathtt{p}] \triangleright \{l_1.0,\ l_2.0\} \quad Q = s[\mathtt{p}][\mathtt{q}] \triangleright (x).s[\mathtt{q}][\mathtt{p}] \triangleleft l_2 \langle\rangle ; 0
\end{aligned}
$$

P implements $G \downarrow_\mathtt{p}$: it waits 6 units of time, then sends q a message and waits for the reply. Q implements $G \downarrow_\mathtt{q}$: it receives a message from p and then selects label l_2; both interactions occur at time 0 which satisfies the clock constraints of $G \downarrow_\mathtt{q}$. By Theorem 7, since $\emptyset \vdash P \mid Q \triangleright s[\mathtt{p}] : (\nu_0, G \downarrow_\mathtt{p}), s[\mathtt{q}] : (\nu_0, G \downarrow_\mathtt{q})$, no violating interactions will occur in $P \mid Q$. However $P \mid Q$ cannot make any step and the session it stuck. This exhibits an intrinsic problem of G, which allows participants to have incompatible views of the timings of action (unlike error transitions in § 4 which, instead, represent constraints violations).

To prevent scenarios as the one above, we introduce a condition on *TG*s called wait-freedom, ensuring that in all the distributed implementations of a *TG*, a receiver checking the queue at any prescribed time never has to wait for a message. Formally (and using $\supset$ for logic implication): G_0 is *wait-free* iff $(\nu_0, G_0) \xrightarrow{}{}^{*} \xrightarrow{\mathtt{pq}!l\langle S \rangle} (\nu, G)$ and $\mathtt{p} \rightsquigarrow \mathtt{q} : l\langle S \rangle \{\delta_\mathtt{O}, \lambda_\mathtt{O}, \delta_\mathtt{I}, \lambda_\mathtt{I}\}.G' \in G$ imply $\delta_\mathtt{I} \supset \nu(x) \le x$ for all $x \in fn(\delta_\mathtt{I})$.

Decidability. If G is infinitely satisfiable (as also assumed by the typing in § 4), then there exists a terminating algorithm for checking that it is feasible and wait-free. The algorithm is based on a direct acyclic graph annotated with clock constraints and resets, and whose edges model the causal dependencies between actions in (the one-time unfolding of) G. The algorithm yields Proposition 8.

Proposition 8 (Decidability) *Feasibility and wait-freedom of infinitely satisfiable TGs are decidable.*

Time-progress for processes. We study the conditions under which a validated program P is guaranteed to proceed until the completion of all activities of the protocols it implements, assuming progress of its untimed counterpart (i.e., $\mathtt{erase}(P)$). The *erasure* $\mathtt{erase}(P)$ of a timed processes P is defined inductively by removing the delays in P (i.e., $\mathtt{erase}(\mathtt{delay}(t).P') = \mathtt{erase}(P')$), while leaving unchanged the untimed parts (e.g., $\mathtt{erase}(\overline{u}[\mathrm{n}](y).P') = \overline{u}[\mathrm{n}](y).\mathtt{erase}(P')$); the other rules are homomorphic.

Proposition 9 (Conformance) *If* $P \longrightarrow P'$, *then* $\mathtt{erase}(P) \longrightarrow^{*} \mathtt{erase}(P')$.

Processes implementing multiple sessions may get stuck because of a bad timing of their attempts to initiate new sessions. Consider $P = \mathtt{delay}(5).\overline{a}[2](v).P_1 \mid a[2](y).P_2$;

$\mathtt{erase}(P)$ can immediately start the session, whereas P is stuck. Namely, the delay of 5 time units introduces a deadlock in a process that would otherwise progress. This scenario is ruled out by requiring processes to only initiate sessions before any delay occurs, namely we assume processes to be *session delay*. All examples we have examined in practice (e.g., OOI use cases [17]) conform session delay.

Definition 10 (Session delay). *P is session delay if for each process occurring in P of the form* $\mathtt{delay}(t).P'$ *(with $t > 0$), there are no session request and session accept in P'.*

We show that feasibility and wait-freedom, by regulating the exchange of messages *within* established sessions, are sufficient conditions for progress of session delay processes. We say that P is a *deadlock process* if $P \longrightarrow^* P'$ where $P' \not\longrightarrow$ and $P' \not\equiv \mathbf{0}$, and that Γ is feasible (resp. wait-free) if $\Gamma(u)$ is feasible (resp. wait-free) for all $u \in dom(\Gamma)$.

Theorem 11 (Timed progress in interleaved sessions). *Let Γ be a* feasible *and* wait-free *mapping, $\Gamma \vdash P_0 \triangleright \emptyset$, and $P_0 \longrightarrow^+ P$. If P_0 is session delay,* $\mathtt{erase}(P)$ *is not a deadlock process and if* $\mathtt{erase}(P) \longrightarrow$ *then $P \longrightarrow$.*

Several typing systems guarantee deadlock-freedom, e.g. [6]. We use one instance from [13] where a single session ensures deadlock-freedom. We characterise processes implementing single sessions, or *simple*, as follows: P is simple if $P_0 \longrightarrow^* P$ for some program P_0 such that $a : G \vdash P_0 \triangleright \emptyset$, and $P_0 = \overline{a}[\mathtt{n}](y).P_1 \mid \prod_{i\in\{2,..,n\}} a[\mathtt{i}](y).P_i$ where $P_1,\ldots,P_n$ do not contain any name hiding, request/accept, and session receive/delegate.

Corollary 12 (Time progress in single sessions). *Let G be* feasible *and* wait-free, *and P be a* simple process *with $a : G \vdash P \triangleright \emptyset$. If* $\mathtt{erase}(P) \longrightarrow$, *then there exist P' and P'' such that* $\mathtt{erase}(P) \longrightarrow P'$, $P \longrightarrow^+ P''$ *and* $\mathtt{erase}(P'') = P'$.

Progress for CTAs. Our *TG*s (§ 3) are a natural extension of global types with timed notions from CTAs. This paragraph clarifies the relationships between *TG*s and CTAs. We describe the exact subset of CTAs that corresponds to *TG*s. We also give the conditions for progress and liveness that characterise a new class of CTAs.

We first recall some definitions from [3,14]. A *timed automaton* is a tuple $\mathcal{A} = (Q, q_0, \mathcal{A}ct, X, E, F)$ such that Q are the states, $q_0 \in Q$ is the initial state, $\mathcal{A}ct$ is the alphabet, X are the clocks, and $E \subseteq (Q \times Q \times \mathcal{A}ct \times 2^X \times \Phi(X))$ are the transitions, where 2^X are the resets, $\Phi(X)$ the clock constraints, and F the final states. A *network of CTAs* is a tuple $C = (\mathcal{A}_1, \ldots, \mathcal{A}_n, \vec{w})$ where $\vec{w} = \{w_{ij}\}_{i \neq j \in \{1,..,n\}}$ are unidirectional unbounded channels. The LTS for CTAs is defined on states $s = ((q_1, \nu_1), \ldots, (q_n, \nu_n), \vec{w})$ and labels $\mathcal{L}$ and is similar to the semantics of configurations except that each $\mathcal{A}_i$ can make a time step even if it violates a constraint. For instance, assume that $\mathcal{A}_1$ can only perform transition $(q_1, q_1', ij!l\langle S\rangle, \emptyset, x_i \leq 10)$ from a non-final state q_1, and that $\nu_1 = 10$, then the semantics in [14] would allow a time transition with label 10. However, after such transition $\mathcal{A}_1$ would be stuck in a non-final state and the corresponding trace would not be accepted by the semantics of [14].

In order to establish a natural correspondence between *TG*s and CTAs we introduce an additional condition on the semantics of C, similar to the constraint on ready actions in the LTS for TG (rule $\lfloor \text{TIME} \rfloor$ in § 3). We say that a time transition with label t is *specified* if $\forall i \in \{1,..,n\}$, $\nu_i + t \models^* \texttt{rdy}(q_i)$ where $\texttt{rdy}(q_i)$ is the set $\{\delta_j\}_{j\in J}$ of constraints of the outgoing actions from q_i. We say that a semantics is specified if it only allows specified time transitions. With a specified semantics, $\mathcal{A}_1$ from the example above could not make any time transition before action $ij!l\langle S\rangle$ occurs.

The correspondence between TGs and CTAs is given as a sound and complete encoding. The *encoding* from T into $\mathcal{A}$, denoted by $\mathcal{A}(T)$, follows exactly the definition in [11, § 2], but adds clock constraints and resets to the corresponding edges, and sets the final states to $\{\texttt{end}\}$. The encoding of a set of TLs $\{T_i\}_{i\in I}$ into a network of CTAs, written $\mathcal{A}(\{T_i\}_{i\in I})$, is the tuple $(\mathcal{A}(T_1),\dots,\mathcal{A}(T_n), \vec{\varepsilon})$. Let G have projections $\{T_i\}_{i\in I}$, we write $\mathcal{A}(G)$ for as $(\mathcal{A}(T_1),\dots,\mathcal{A}(T_n), \vec{\varepsilon})$.

Before stating soundness and completeness we recall, and adapt to the timed setting, two conditions from [11]: the basic property (timed automata have the same shape as TLs) and multiparty compatibility (timed automata perform the same actions as a set of projected TG). More precisely: C is *basic* when all its timed automata are deterministic, and the outgoing actions from each (q_i,C_i) are all sending or all receiving actions, and all to/from the same co-party. A state s is *stable* when all its channels are empty. C is *multiparty compatible* when in all its reachable stable states, all possible (input/output) action of each timed automaton can be matched with a corresponding complementary (output/input) actions of the rest of the system after some 1-bounded executions (i.e., executions where the size of each buffer contains at most 1 message).[1]

A *session CTA* is a basic and multiparty compatible CTA with specified semantics.

Theorem 13 (Soundness and completeness). (1) *Let G be a (projectable) TG then $\mathcal{A}(G)$ is basic and multiparty compatible. Furthermore with a specified semantics $G \approx \mathcal{A}(G)$.* **(2)** *If C is a session CTA then there exists G such that $C \approx \mathcal{A}(G)$.*

Our characterisation does not directly yield transparency of properties, differently from the untimed setting [11] and similarly to timed processes (§ 4). In fact, a session CTA itself does not satisfy progress. In the following we give the conditions that guarantee progress and liveness of CTAs. Let $s = ((q_1,\nu_1),\dots,(q_n,\nu_n),\vec{w})$ be a reachable state of C: s is a *deadlock state* if (i) $\vec{w} = \vec{\varepsilon}$, (ii) for all $i \in \{1,\dots,n\}$, (q_i,ν_i) does not have outgoing send actions, and (iii) for some $i \in \{1,\dots,n\}$, (q_i,ν_i) has incoming receiving action; s satisfies *progress* if for all s' reachable from s: (1) s' is not a deadlock state, and (2) $\forall t \in \mathbb{N}$, $((q_1,\nu_1+t),\dots,(q_n,\nu_n+t),\vec{w})$ is reachable from s in C. We say C satisfies *liveness* if for every reachable state s in C, $s \longrightarrow^* s'$ with s' final.

Progress entails deadlock freedom (1) and, in addition, requires (2) that it is always possible to let time to diverge; namely the only possible way forward cannot be by actions occurring at increasingly short intervals of time (i.e., Zeno runs).[2]

We write $TR(C)$ for the set of visible traces that can be obtained by reducing C. We extend to CTAs the trace equivalence $\approx$ defined in § 3.

Theorem 14 (Progress and liveness for CTAs). *If C is a session CTA and there exists a* feasible *G s.t. $C \approx \mathcal{A}(G)$, then C satisfies progress and liveness.*

[1] Note that multiparty compatibility allows scenarios with unbounded channels e.g., the channel from p to q in $\mu \mathbf{t}.\mathtt{p} \to \mathtt{q} : l\langle S\rangle\{A\}.\mathbf{t}$.

[2] The time divergence condition is common in timed setting and is called time-progress in [3].

6 Conclusion and Related Work

We design choreographic timed specifications based on the semantics of CTAs and MPSTs, and attest our theory in the π-calculus. The table below recalls the results for the untimed setting we build upon (first row), and summarises our results: a decidable proof system for π-calculus processes ensuring time-error freedom and a sound and complete characterisation of CTAs (second row), and two decidable conditions ensuring progress of processes (third row). These conditions also characterise a new class of CTAs, without restrictions on the topologies, that satisfy progress and liveness. We verified the practicability of our approach in an implementation of a timed conversation API for Python. The prototype [1] is being integrated into the OOI infrastructure [17].

*TG*s	π-calculus	session CTAs
untimed	type safety, error-freedom, progress [13]	Sound, complete characterisation, progress [11]
timed	type safety (Theorem 6) error-freedom (Theorem 7)	Sound, complete characterisation, (Theorem 13)
feasible, wait-free	progress (Theorems 11 and 12)	progress (Theorem 14)

Challenges of Extending MPSTs with Time. The extension of the semantics of types with time is delicate as it may introduce unwanted executions (as discussed in § 3). To capture only the correct executions (corresponding to accepted traces in timed automata) we have introduced a new condition on time reductions of *TG*s and *TL*s: *satisfiability of ready actions* (e.g., $\lfloor$TIME$\rfloor$ in Figure 1). Our main challenge was extending the progress properties of untimed types [6] and CAs [11] to timed interactions. We introduced two additional necessary conditions for the timed setting, feasibility and wait-freedom, whose decidability is non trivial, and their application to time-progress is new. The theory of assertion-enhanced MPSTs [7] (which do not include progress) could not be applied to the timed scenario due to resets and the need to ensure consistency with respect to absolute time flowing.

Reachability and verification. In our work, if a CTA derives from a feasible *TG* then error and deadlock states will not be reached. Decidability of reachability for CTAs has been proven for specific topologies: those of the form $(\mathcal{A}_1, \mathcal{A}_2, w_{1,2})$ [14] and polyforests [10]. A related approach [2] extends MSCs with timed events and provides verification method that is decidable when the topology is a single strongly connected component, which ensures that channels have an upper bound. Our results do not depend on the topology nor require a limitation of the buffer size (e.g., the example in § 2 is not a polyforest and the buffer of `A` is unlimited). On the other hand, our approach relies on the additional restrictions induced by the conversation structure of *TG*s.

Feasibility. Feasibility was introduced in a different context (i.e., defining a not too stringent notion of fairness) in [4]. This paper gives a concrete definition in the context of real-time interactions, and states its decidability for infinitely satisfiable *TG*s. [21] gives an algorithm to check deadlock freedom for timed automata. The algorithm, based on syntactic conditions on the states relying on invariant annotations, is not directly applicable to check feasibility e.g., on the timed automaton derived from a *TG*.

Calculi with Time. Recent work proposes calculi with time, for example: [18] includes time constraints inspired by timed automata into the π-calculus, [5,16] add timeouts, [12] analyses the active times of processes, and [15] for service-oriented systems. The aim of our work is different from the work above: we use timed specifications *as types* to check time properties of the interactions, rather than enriching the π-calculus syntax with time primitives and reason on examples using timed LTS (or check channels linearity as [5]). Our aim is to define a static checker for time-error freedom and progress on the basis of a semantics guided by timed automata. With this respect, our calculus is a small syntactic extension from the π-calculus and is simpler than the above calculi.

References

1. Timed conversation API for Python, `http://www.doc.ic.ac.uk/~lbocchi/TimeApp.html`
2. Akshay, S., Gastin, P., Mukund, M., Kumar, K.N.: Model checking time-constrained scenario-based specifications. In: FSTTCS. LIPIcs, vol. 8, pp. 204–215 (2010)
3. Alur, R., Dill, D.L.: A theory of timed automata. TCS 126, 183–235 (1994)
4. Apt, K.R., Francez, N., Katz, S.: Appraising fairness in distributed languages. In: POPL, pp. 189–198. ACM (1987)
5. Berger, M., Yoshida, N.: Timed, distributed, probabilistic, typed processes. In: Shao, Z. (ed.) APLAS 2007. LNCS, vol. 4807, pp. 158–174. Springer, Heidelberg (2007)
6. Bettini, L., Coppo, M., D'Antoni, L., De Luca, M., Dezani-Ciancaglini, M., Yoshida, N.: Global progress in dynamically interleaved multiparty sessions. In: van Breugel, F., Chechik, M. (eds.) CONCUR 2008. LNCS, vol. 5201, pp. 418–433. Springer, Heidelberg (2008)
7. Bocchi, L., Honda, K., Tuosto, E., Yoshida, N.: A theory of design-by-contract for distributed multiparty interactions. In: Gastin, P., Laroussinie, F. (eds.) CONCUR 2010. LNCS, vol. 6269, pp. 162–176. Springer, Heidelberg (2010)
8. Brand, D., Zafiropulo, P.: On communicating finite-state machines. J. ACM 30, 323–342 (1983)
9. Castagna, G., Dezani-Ciancaglini, M., Padovani, L.: On global types and multi-party session. Logical Methods in Computer Science 8(1) (2012)
10. Clemente, L., Herbreteau, F., Stainer, A., Sutre, G.: Reachability of communicating timed processes. In: Pfenning, F. (ed.) FOSSACS 2013. LNCS, vol. 7794, pp. 81–96. Springer, Heidelberg (2013)
11. Deniélou, P.-M., Yoshida, N.: Multiparty compatibility in communicating automata: Characterisation and synthesis of global session types. In: Fomin, F.V., Freivalds, R., Kwiatkowska, M., Peleg, D. (eds.) ICALP 2013, Part II. LNCS, vol. 7966, pp. 174–186. Springer, Heidelberg (2013)
12. Fischer, M., et al.: A new time extension to π-calculus based on time consuming transition semantics. In: Languages for System Specification, pp. 271–283 (2004)
13. Honda, K., Yoshida, N., Carbone, M.: Multiparty Asynchronous Session Types. In: POPL, pp. 273–284. ACM (2008)
14. Krčál, P., Yi, W.: Communicating timed automata: The more synchronous, the more difficult to verify. In: Ball, T., Jones, R.B. (eds.) CAV 2006. LNCS, vol. 4144, pp. 249–262. Springer, Heidelberg (2006)
15. Lapadula, A., Pugliese, R., Tiezzi, F.: C**[Equation image]**WS: A timed service-oriented calculus. In: Jones, C.B., Liu, Z., Woodcock, J. (eds.) ICTAC 2007. LNCS, vol. 4711, pp. 275–290. Springer, Heidelberg (2007)

16. López, H.A., Pérez, J.A.: Time and exceptional behavior in multiparty structured interactions. In: Carbone, M., Petit, J.-M. (eds.) WS-FM 2011. LNCS, vol. 7176, pp. 48–63. Springer, Heidelberg (2012)
17. Ocean Observatories Initiative (OOI)., `http://oceanobservatories.org/`
18. Saeedloei, N., Gupta, G.: Timed π-calculus. In: TGC. LNCS (2013) (to appear)
19. Savara JBoss Project, `http://www.jboss.org/savara`
20. Scribble Project homepage, `http://www.scribble.org`
21. Tripakis, S.: Verifying progress in timed systems. In: Katoen, J.-P. (ed.) AMAST-ARTS 1999, ARTS 1999, and AMAST-WS 1999. LNCS, vol. 1601, pp. 299–314. Springer, Heidelberg (1999)
22. Technical Report, Department of Computing, Imperial College London (May 2014) (March 2014)

A Categorical Semantics of Signal Flow Graphs

Filippo Bonchi[1], Paweł Sobociński[2], and Fabio Zanasi[1]

[1] ENS de Lyon, Université de Lyon, CNRS, INRIA, France
[2] ECS, University of Southampton, UK

Abstract. We introduce $\mathbb{IH}$, a sound and complete graphical theory of vector subspaces over the field of polynomial fractions, with relational composition. The theory is constructed in modular fashion, using Lack's approach to composing PROPs with distributive laws.

We then view string diagrams of $\mathbb{IH}$ as generalised stream circuits by using a formal Laurent series semantics. We characterize the subtheory where circuits adhere to the classical notion of signal flow graphs, and illustrate the use of the graphical calculus on several examples.

1 Introduction

We introduce a graphical calculus of string diagrams, which we call *circuits*, consisting of the following constants, sequential ; and parallel $\oplus$ composition.

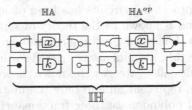

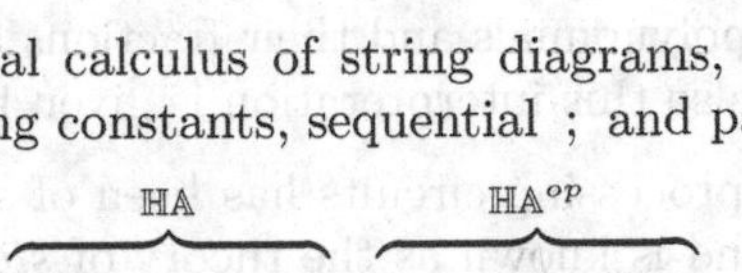

These circuits can be given a *stream semantics.* The intuition is that wires carry elements of a field k that enter and exit through boundary ports. In particular, for circuits built from components in the leftmost three columns, which we hereafter refer to as being in $\mathbb{HA}$, the signal enters from the left and exits from the right boundary. Computation in the circuit proceeds synchronously according to a global "clock", where at each iteration fresh elements are processed from input streams on the left and emitted as elements of output streams on the right.

Intuitively, 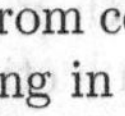 is a *copier*, duplicating its input signal; its counit 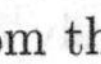 accepts any signal and discards it, producing no output; 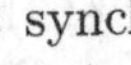 is an *adder* that takes two inputs and emits their sum, and its unit 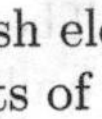 constantly outputs the signal 0; 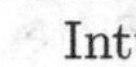 is a *delay*, or 1-place buffer that initially holds the 0 value. Finally, 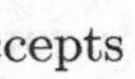 is an *amplifier*, multiplying its input by the scalar $k \in \mathsf{k}$. For circuits resulting from the other three columns, $\mathbb{HA}^{op}$, the signal flows on the opposite direction: from right to left. The behaviour is symmetric. Formally, the stream semantics of circuits in $\mathbb{HA}$ and $\mathbb{HA}^{op}$ consists of linear transformations of streams.

P. Baldan and D. Gorla (Eds.): CONCUR 2014, LNCS 8704, pp. 435–450, 2014.

Circuits in $\mathbb{IH}$ built out of all the constants above do not, in general, yield functional behaviour. Signals no longer flow in a fixed direction, analogously to how in electrical circuits physical wires are not directed. Indeed, the semantics of circuits in $\mathbb{IH}$ are not linear maps, but rather subspaces, i.e., linear relations. Passing from functions to relations gives meaning to circuits that contain feedbacks. We must also use an extended notion of streams, *Laurent series*, typical in algebraic approaches [4] to signal processing—roughly speaking, these streams are allowed to start in the past. Notably, while matrices denoted by circuits in $\mathbb{HA}$ or $\mathbb{HA}^{op}$ only contain streams with a finite number of non-zero values, the subspaces denoted by $\mathbb{IH}$ are, in general, generated by vectors of streams with infinitely many non-zero values. An example is the Fibonacci circuit (Example 3).

We characterise the stream semantics via both a universal property and an intuitive inductive definition. Furthermore, we provide a sound and complete axiomatization for proving semantic equivalence of circuits.

In order to do that, we consider another canonical semantics for circuits, prior to the stream semantics. We show (Proposition 1) that $\mathbb{HA}$ is the theory of $\mathsf{k}[x]$-matrices, where $\mathsf{k}[x]$ is the ring of polynomials with coefficients from field k. A modular construction [8] that generalises our earlier contribution [7] allows us to conclude (Theorem 1) that $\mathbb{IH}$ is the relational theory of vector subspaces over the field of fractions of $\mathsf{k}[x]$. Then, the passage to the stream semantics simply consists in interpreting polynomials and their fractions as streams. Using again the construction in [8], also this interpretation is given by a universal property.

The study of stream processing circuits has been of significant interest since at least the 1950s [16] and is known as the theory of *signal flow graphs* (SFG). Traditionally only SFGs that yield functional behaviours on ordinary streams are considered: to ensure this, circuits are restricted so that every feedback loop passes through at least one delay gate. A well-known theorem (see e.g. [14]) states that circuits in this form represent all and only the matrices with entries from $\mathsf{k}\langle x\rangle$, the ring of *rational* polynomials: those fractions where the constant term in the denominator is non-zero. A novel proof of this result has been recently given by Rutten in [18] by using coinductive and coalgebraic techniques.

Signal flow graphs are first class objects of our theory—they are a certain inductively defined family $\mathbb{SF}$ of circuits in $\mathbb{IH}$. Using its inductive definition, we can give another proof of the aforementioned theorem: $\mathbb{SF}$ is the theory of matrices over $\mathsf{k}\langle x\rangle$ (Theorem 2). The main advantage of our approach is that, by virtue of our full abstraction result, we are able to use graphical equational reasoning on signal flow graphs *directly* without translations to intermediate linear-algebraic or coalgebraic syntax.

The definition of SFGs given in this paper is very close in spirit to their classical interpretations as graphical structures, which are typically defined in an informal and intuitive fashion, before they are translated to a more formal syntax and abandoned as objects of study. Our main departure from circuit orthodoxy is that we dispense with all notions of input, output and direction of wires. Indeed, guided by the mathematics of circuits, we must consider all of these as derivative notions. By doing so, we are close in spirit to Willems' *behavioural approach* in control theory [22]. Our approach—using string diagrams, which originated in the study of free monoidal categories [20], in order to capture physical systems—can also be considered as a contribution to *network theory* [2]. Similar ideas lie behind Span(Graph) [12], the algebra of Petri nets with boundaries [10,21] and several algebras of connectors [1]. Independently, Baez and Erbele proposed an equivalent presentation of relational subspaces in their technical report [3], which appeared after the submission of this paper.

There are also close connections to recent work on graphical languages for quantum information. In [7] we used a similar modular construction to characterise the free model of the undirected phase-free version of the ZX-calculus [11]. That construction and the construction of $\mathbb{IH}$ are both instances of a more general result [8] that we sketch in this paper. Indeed, $\mathbb{IH}$ can itself be considered as a flavour of directed ZX, albeit with a very different semantic interpretation.

Structure of the Paper. In §2 we recall the required categorical notions. In §3 we show that $\mathbb{HA}$ is the graphical theory of $\mathsf{k}[x]$-matrices. In §4 we give a modular account of $\mathbb{IH}$ and show that it is the theory of relational vector subspaces over the field of fractions of $\mathsf{k}[x]$. In §5 we focus on the stream semantics and in §6 we identify an important subclass of $\mathbb{IH}$: the theory of signal flow graphs.

2 Background

$\mathbb{C}[a,b]$ is the set of arrows from a to b in a small category $\mathbb{C}$, composition of $f\colon a \to b$, $g\colon b \to c$ is denoted by $f\,;g\colon a \to c$. For $\mathbb{C}$ symmetric monoidal, $\oplus$ is the monoidal product and $\sigma_{X,Y}\colon X \oplus Y \to Y \oplus X$ the symmetry for $X, Y \in \mathbb{C}$. Given $\mathcal{F}\colon \mathbb{C}_1 \to \mathbb{C}_2$, $\mathcal{F}^{op}\colon \mathbb{C}_1^{op} \to \mathbb{C}_2^{op}$ is the induced functor on the opposite categories of $\mathbb{C}_1, \mathbb{C}_2$. If $\mathbb{C}$ has pullbacks, its span bicategory has the objects of $\mathbb{C}$ as 0-cells, spans of arrows of $\mathbb{C}$ as 1-cells and span morphisms as 2-cells. We denote with $\mathsf{Span}(\mathbb{C})$ the (ordinary) category obtained by identifying the isomorphic 1-cells and forgetting the 2-cells. Dually, if $\mathbb{C}$ has pushouts, $\mathsf{Cospan}(\mathbb{C})$ is the category obtained from the bicategory of cospans.

2.1 PROPs

A (one sorted) symmetric monoidal theory (SMT) is given by a pair (Σ, E) where Σ is the *signature*: a set of *operations* $o\colon n \to m$ with *arity* n and *coarity* m. The set of Σ-*terms* is obtained by composing operations, the identity $id_1 : 1 \to 1$ and symmetry $\sigma_{1,1} : 2 \to 2$ with ; and $\oplus$: given Σ-terms $t\colon k \to l$, $u\colon l \to m$, $v\colon m \to n$, we construct Σ-terms $t\,;u\colon k \to m$ and $t \oplus v\colon k + n \to l + n$. The elements of the set E of *equations* are pairs of Σ-terms $(l, l'.k \to l)$.

To study SMTs we use PROPs [13,15] (product and permutation categories). A PROP is a strict symmetric monoidal category with objects natural numbers, where $\oplus$ on objects is addition. Morphisms between PROPs are strict symmetric monoidal functors that act as identity on objects: PROPs and their morphisms form the category **PROP**. Given an SMT (Σ, E), one (freely) obtains a PROP where arrows $k \to l$ are Σ-terms $k \to l$ modulo the laws of symmetric monoidal categories and equations $t = t'$ where $(t, t') \in E$. There is a graphical representation of terms as string diagrams (see [20]): we call these diagrams *circuits*.

For example, let (Σ_M, E_M) be the SMT of commutative monoids. Σ_M contains two operation symbols: the multiplication — which we depict as a circuit : $2 \to 1$ — and the unit, represented as : $0 \to 1$. Graphically, the generation of Σ_M-terms amounts to "tiling" and together with the circuit  (representing $\sigma_{1,1} : 2 \to 2$) and (representing $id_1 : 1 \to 1$). Equations E_M assert associativity (A3), commutativity (A2) and identity (A1).

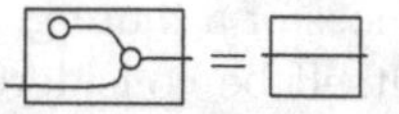 (A1) 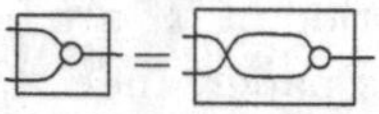 (A2) 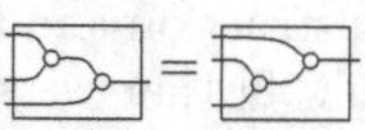 (A3)

We call $\mathbb{M}^w$ the PROP freely generated by (Σ_M, E_M).[1] A useful observation is that to give a circuit $c \in \mathbb{M}^w[n, m]$ is to give the graph of a function of type $\{0, \ldots, n-1\} \to \{0, \ldots, m-1\}$. For instance, $\oplus$: $2 \to 2$ describes the function $f : \{0, 1\} \to \{0, 1\}$ mapping both elements to 0. This yields an iso $\mathcal{S}_{\mathbb{M}^w} : \mathbb{M}^w \to \mathbb{F}$, where $\mathbb{F}$ is the PROP with arrows $n \to m$ functions $\{0, \ldots, n-1\} \to \{0, \ldots, m-1\}$. Intuitively, $\mathbb{F}$ is a "concrete" representation of the theory of commutative monoids and thus we refer to the morphism $\mathcal{S}_{\mathbb{M}^w}$ as the *denotational semantics* of $\mathbb{M}^w$.

For later reference, we introduce two more examples of free PROPs. First, let $\mathbb{K}[\mathbb{X}]$ be the PROP freely generated by the signature consisting of p for each $p \in \mathsf{k}[x]$ and the following equations, where p_1, p_2 range over $\mathsf{k}[x]$.

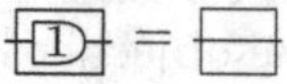 (A4) 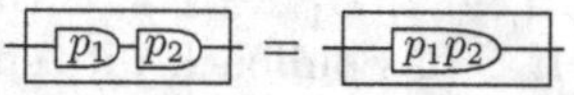 (A5)

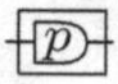

Next, let $\mathbb{C}^b$ be the PROP of (black) cocommutative comonoids, freely generated by the signature consisting of circuits , and the following equations.

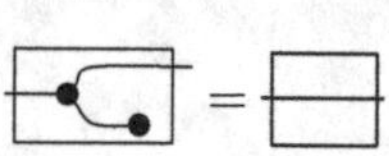 (A6) 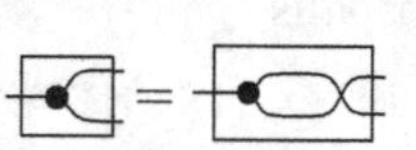 (A7) 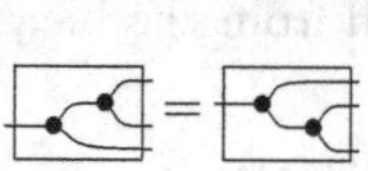 (A8)

Modulo the white vs. black colouring, the circuits of $\mathbb{C}^b$ can be seen as those of $\mathbb{M}^w$ "reflected about the y-axis". This observation yields that $\mathbb{C}^b \cong \mathbb{M}^{w\,op}$. More generally, for $\mathbb{T}$ a free PROP, $\mathbb{T}^{op}$ can be presented by operations and equations which are those of $\mathbb{T}$ reflected about the y-axis.

PROPs are also objects of a certain coslice category. First, a PRO is a strict monoidal category with objects the naturals and tensor product on objects addition. Morphisms of PROs are strict identity-on-objects monoidal functors. There is a PRO of particular interest: the PRO of permutations $\mathbb{P}$, where $\mathbb{P}[k, l]$ is empty if $k \neq l$ and otherwise consists of the permutations on the set with k elements. Now PROPs are objects of the coslice $\mathbb{P}/\mathbf{PRO}$, where $\mathbf{PRO}$ is the category of

[1] The notation w emphasizes the white colouring of circuits in Σ_M — later on, we will use the black coloring for another copy of the same PROP.

PROs. Morphisms of PROPs are thus simply morphisms of PROs that preserve the permutation structure. Working in the coslice is quite intuitive: e.g. $\mathbb{P}$ is the initial PROP and to compute the coproduct $\mathbb{T}+\mathbb{S}$ in **PROP** one must identify the permutation structures. When $\mathbb{T}$ and $\mathbb{S}$ are generated by (Σ_C, E_C) and (Σ_D, E_D) respectively, it follows that $\mathbb{T}+\mathbb{S}$ is generated by $(\Sigma_C+\Sigma_D, E_C+E_D)$.

2.2 Composing PROPs

In [13] Lack showed that co/commutative bialgebras and separable Frobenius algebras stem from different ways of "composing" $\mathbb{C}^b$ and $\mathbb{M}^w$. Just as small categories are monads in $\mathsf{Span}(\mathbf{Set})$, a PROP is a monad in a certain bicategory, and PROPs $\mathbb{T}_1$ and $\mathbb{T}_2$ can be composed via distributive law $\lambda\colon \mathbb{T}_2 \,;\mathbb{T}_1 \to \mathbb{T}_1 \,; \mathbb{T}_2$. A key observation is that the graph of λ can be seen as a set of equations. Thus, if $\mathbb{T}_1$ and $\mathbb{T}_2$ are freely generated PROPs, then so is $\mathbb{T}_1 \,; \mathbb{T}_2$.

As an example, we show how composing $\mathbb{C}^b$ and $\mathbb{M}^w$ yields the PROP of co/commutative bialgebras. First observe that circuits of $\mathbb{C}^b$ yield arrows of $\mathbb{F}^{op}$, because $\mathbb{C}^b \cong \mathbb{M}^{w\,op} \cong \mathbb{F}^{op}$. Then a distributive law $\lambda\colon \mathbb{M}^w \,; \mathbb{C}^b \Rightarrow \mathbb{C}^b \,; \mathbb{M}^w$ has type $\mathbb{F} \,; \mathbb{F}^{op} \Rightarrow \mathbb{F}^{op} \,; \mathbb{F}$, that is, it maps a pair $p \in \mathbb{F}[n,z]$, $q \in \mathbb{F}^{op}[z,m]$ to a pair $f \in \mathbb{F}^{op}[n,z]$, $g \in \mathbb{F}[z,m]$. This amounts to saying that λ maps *cospans* $n \xrightarrow{p} z \xleftarrow{q} m$ into *spans* $n \xleftarrow{f} r \xrightarrow{g} m$ in $\mathbb{F}$. Defining $n \xleftarrow{f} r \xrightarrow{g} m$ as the pullback of $n \xrightarrow{p} z \xleftarrow{q} m$ makes λ a distributive law [13]. The resulting PROP $\mathbb{C}^b \,; \mathbb{M}^w$ can be presented by operations — the ones of $\mathbb{C}^b + \mathbb{M}^w$ — and equations — the ones of $\mathbb{C}^b + \mathbb{M}^w$ together with those given by the graph of λ. By definition of λ, one can read them (in $\mathbb{C}^b + \mathbb{M}^w$) out of the pullback squares in $\mathbb{F}$. For instance:

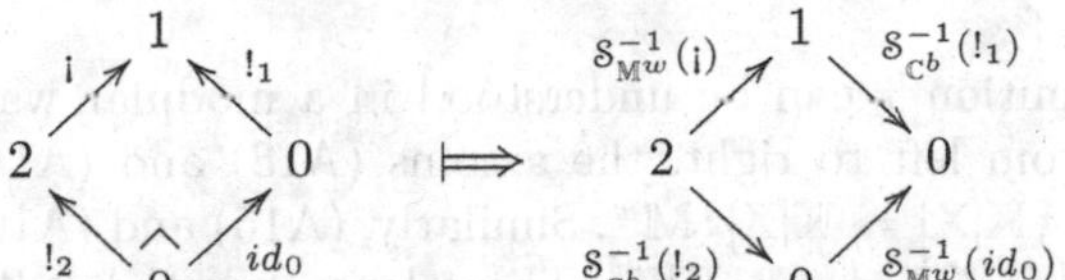

yields 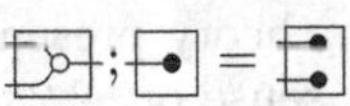 $; id_0$

where $¡\colon 2 \to 1$ and $!_n\colon 0 \to n$ are given, respectively, by finality of 1 and initiality of 0 in $\mathbb{F}$, and $\mathcal{S}_{\mathbb{C}^b}$ is the isomorphism $\mathbb{C}^b \cong \mathbb{F}^{op}$. In fact, all the equations can be derived from (those of $\mathbb{C}^b + \mathbb{M}^w$ and) just four pullbacks (*cf.* [13]) that yield:

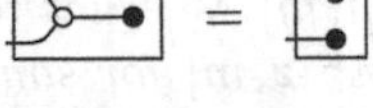 (A9) 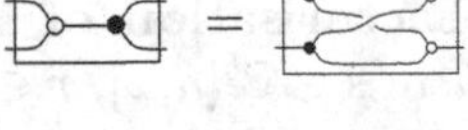 (A10)

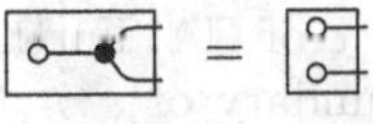 (A11) 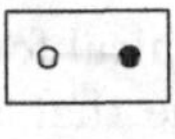 $= id_0$ (A12)

Therefore $\mathbb{C}^b \,; \mathbb{M}^w$ is the free PROP of (black-white) co/commutative bialgebras, obtained as the quotient of $\mathbb{C}^b + \mathbb{M}^w$ by (A9)-(A12). As another perspective on the same result, one can say that the PROP of co/commutative bialgebras is the theory of $\mathsf{Span}(\mathbb{F}) \cong \mathbb{F}^{op} \,; \mathbb{F}$ and each circuit $c\colon n \to m$ of this PROP can be factorised as $c = c_1 \,; c_2$, where $c_1 \in \mathbb{C}^b[n,z]$ and $c_2 \in \mathbb{M}^w[z,m]$ for some z.

3 The Theory of $\mathsf{k}[x]$ Matrices

In this section we introduce the PROP $\mathbb{HA}$ of $\mathsf{k}[x]$-Hopf Algebras and show that it is isomorphic to the category of matrices over $\mathsf{k}[x]$.

Definition 1. *The PROP* $\mathbb{HA}$ *is the quotient of* $\mathbb{C}^b + \mathbb{K}[\mathbb{X}] + \mathbb{M}^w$ *by the equations* (A9), (A11), (A10), (A12) *and the following, where* $p, p_1, p_2 \in \mathsf{k}[x]$.

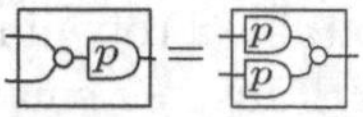 (A13) 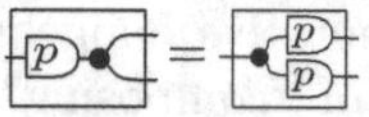 (A15) 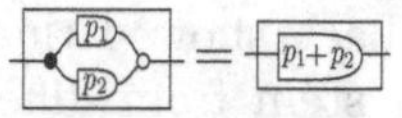 (A17)

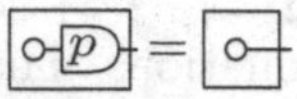 (A14) 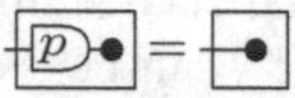 (A16) 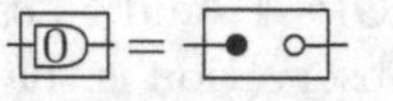 (A18)

Remark 1. $\mathbb{HA}$ is a Hopf algebra with antipode 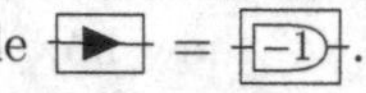. Indeed it inherits the bialgebra structure of $\mathbb{C}^b ; \mathbb{M}^w$ and (Hopf) holds by (A4), (A17) and (A18):

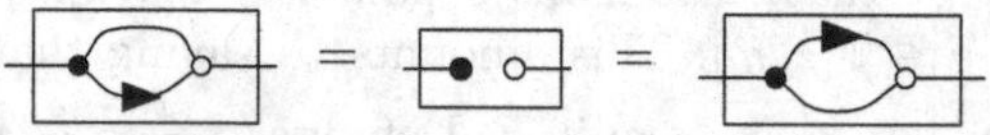 (Hopf)

Remark 2. There is an important "operational" intuition associated with circuits in $\mathbb{HA}$. First, the ports on the left are inputs, the ports on the right are outputs. The circuit constants behave as described in the Introduction, but now wires carry signals which are elements of $\mathsf{k}[x]$, rather than streams.

The theory presented in Definition 1 can be understood in a modular way, in the sense of §2.2. Reading from left to right, the axioms (A13) and (A14) present a distributive law $\sigma\colon \mathbb{M}^w ; \mathbb{K}[\mathbb{X}] \Rightarrow \mathbb{K}[\mathbb{X}] ; \mathbb{M}^w$. Similarly, (A15) and (A16) present a distributive law $\tau\colon \mathbb{K}[\mathbb{X}] ; \mathbb{C}^b \Rightarrow \mathbb{C}^b ; \mathbb{K}[\mathbb{X}]$. These laws, together with $\lambda\colon \mathbb{M}^w ; \mathbb{C}^b \Rightarrow \mathbb{C}^b ; \mathbb{M}^w$ which is presented by (A9), (A11), (A10), (A12) (*cf.* §2.2), yield the composite $\mathbb{C}^b ; \mathbb{K}[\mathbb{X}] ; \mathbb{M}^w$. We refer to [8, §3] for proofs and further details. Now, $\mathbb{HA}$ is the quotient of $\mathbb{C}^b ; \mathbb{K}[\mathbb{X}] ; \mathbb{M}^w$ by (A18) and (A17). As a consequence, it inherits the factorisation property of $\mathbb{C}^b ; \mathbb{K}[\mathbb{X}] ; \mathbb{M}^w$.

Lemma 1 (Factorisation of $\mathbb{HA}$). *Any* $c \in \mathbb{HA}[n, m]$ *is equal to* $s ; r ; t \in \mathbb{HA}[n, m]$*, where* $s \in \mathbb{C}^b[n, z]$*,* $r \in \mathbb{K}[\mathbb{X}][z, z]$ *and* $t \in \mathbb{M}^w[z, m]$ *for some* $z \in \mathbb{N}$.

Lemma 1 fixes a canonical form $s ; r ; t$ for any circuit c of $\mathbb{HA}$. Furthermore, by (A17), we can assume that any port on the left boundary of $s ; r ; t$ has at most one connection with any port on the right boundary, and by (A4),(A5) we know that any such connection passes through exactly one circuit of shape 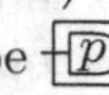. We say that a factorised circuit $s ; r ; t$ satisfying this additional requirements is in *matrix form.* Circuits in matrix form have an intuitive representation as $\mathsf{k}[x]$-matrices, as illustrated in the following example.

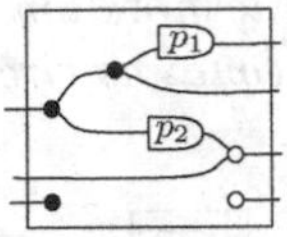

$$\begin{pmatrix} p_1 & 0 & 0 \\ 1 & 0 & 0 \\ p_2 & 1 & 0 \\ 0 & 0 & 0 \end{pmatrix}$$

Example 1. Consider the circuit $t \in \mathbb{HA}[3,4]$ and its representation as a 4×3 matrix M (on the right). For each boundary of t, the ports are enumerated from top to bottom, starting from 1. Then the entry $M_{i,j}$ has value $p \in \mathsf{k}[x]$ if, reading the circuit from the left to the right, one finds a path connecting the jth port on the left to the ith port on the right passing through a circuit 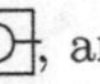, and 0 otherwise.

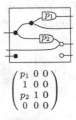

We now make the matrix semantics of circuits in $\mathbb{HA}$ formal. For this purpose, let $\mathsf{Mat}\,\mathsf{k}[x]$ be the PROP with arrows $n \to m$ the $m \times n$ $\mathsf{k}[x]$-matrices, where ; is matrix multiplication and $A \oplus B$ is defined as the matrix $\begin{pmatrix} A & 0 \\ 0 & B \end{pmatrix}$. The symmetries are the rearrangements of the rows of the identity matrix.

Definition 2. *The PROP morphism* $\mathcal{S}_{\mathbb{HA}} \colon \mathbb{HA} \to \mathsf{Mat}\,\mathsf{k}[x]$ *is defined inductively:*

 $\mapsto\ !$  $\mapsto\ ¡$ 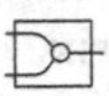 $\mapsto (1\ 1)$ 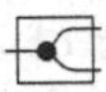 $\mapsto \begin{pmatrix} 1 \\ 1 \end{pmatrix}$ 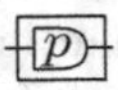 $\mapsto\ (p)$

$$s \oplus t \mapsto \mathcal{S}_{\mathbb{HA}}(s) \oplus \mathcal{S}_{\mathbb{HA}}(t) \qquad s\,;t \mapsto \mathcal{S}_{\mathbb{HA}}(s)\,;\mathcal{S}_{\mathbb{HA}}(t)$$

where $! \colon 0 \to 1$ *and* $¡ \colon 1 \to 0$ *are given by initiality and finality of* 0 *in* $\mathsf{Mat}\,\mathsf{k}[x]$*. It can be checked that* $\mathcal{S}_{\mathbb{HA}}$ *is well defined, as it respects the equations of* $\mathbb{HA}$*.*

Proposition 1. $\mathcal{S}_{\mathbb{HA}} : \mathbb{HA} \to \mathsf{Mat}\,\mathsf{k}[x]$ *is an isomorphism of PROPs.*

Proof. Since the two categories have the same objects, it suffices to prove that $\mathcal{S}_{\mathbb{HA}}$ is full and faithful. For this purpose, observe that, for a circuit c in matrix form, the matrix $\mathcal{S}_{\mathbb{HA}}(c)$ can be computed as described in Example 1. Since by Lemma 1 any circuit is equivalent to one of this shape, fullness and faithfulness follows by checking that the encoding of Example 1 is a 1-1 correspondence between matrices and circuits of $\mathbb{HA}$ in matrix form. □

4 The Theory of Relational $\mathsf{k}(x)$ Subspaces

Let $\mathsf{k}(x)$ denote the field of fractions of $\mathsf{k}[x]$. In this section we introduce the PROP $\mathbb{IH}$, whose axioms describe the interaction of two $\mathsf{k}[x]$-Hopf algebras, and we show that it is isomorphic to the PROP of $\mathsf{k}(x)$-vector subspaces.

Definition 3. *The PROP* $\mathbb{IH}$ *is the quotient of* $\mathbb{HA} + \mathbb{HA}^{op}$ *by the following equations, where* p *ranges over* $\mathsf{k}[x] \setminus \{0\}$.

The notation  *indicates both the antipodes*  *and* *: indeed, they are equal as circuits of* $\mathbb{IH}$ *by virtue of* (S5).

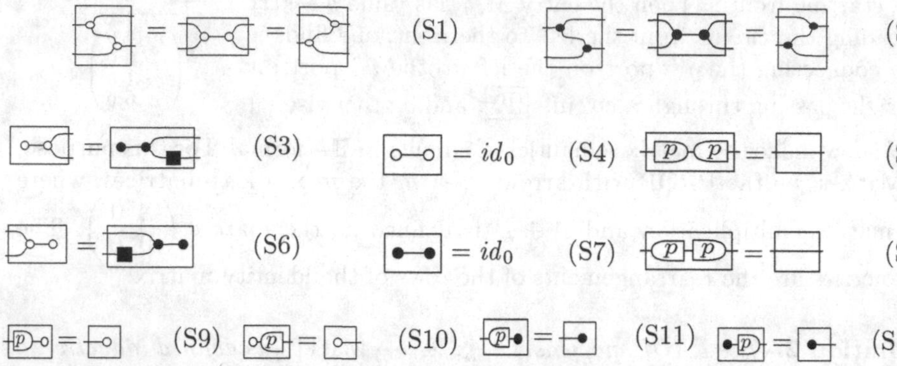

(S1)
(S2)
(S3)
= id_0
(S4)
(S5)
(S6)
= id_0
(S7)
(S8)
(S9)
(S10)
(S11)
(S12)

We now consider the task of giving a semantics to circuits of $\mathbb{IH}$. Recall that the semantics of a circuit of $\mathbb{HA}$ is a matrix, or in other words, a linear transformation. Indeed, as explained in Remark 2, circuits in $\mathbb{HA}$ can be read from left to right: ports on the left are inputs and ports on the right are outputs.

These traditional mores fail for circuits in $\mathbb{IH}$. Consider 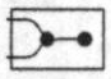 : $2 \to 0$: the component  accepts an arbitrary signal while 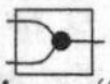 ensures that the signal is equal on the two ports. In other words, the circuit is a "bent identity wire" whose behaviour is relational: the two ports on the left are neither inputs nor outputs in any traditional sense. Indeed, only some circuits of $\mathbb{IH}$ have a functional interpretation. We now introduce the semantic domain of interest for $\mathbb{IH}$.

Definition 4. *Let* $\mathbb{SV}_{\mathsf{k}(x)}$ *be the following PROP:*

- *arrows* $n \to m$ *are subspaces of* $\mathsf{k}(x)^n \times \mathsf{k}(x)^m$ *(as a* $\mathsf{k}(x)$*-vector space).*
- *composition is relational: for subspaces* $G = \{(u,v) \mid u \in \mathsf{k}(x)^n, v \in \mathsf{k}(x)^z\}$ *and* $H = \{(v,w) \mid v \in \mathsf{k}(x)^z, w \in \mathsf{k}(x)^m\}$*, their composition is the subspace* $\{(u,w) \mid \exists v.(u,v) \in G \wedge (v,w) \in H\}$.
- *The tensor product* $\oplus$ *on arrows is given by direct sum of spaces.*
- *The symmetries* $n \to n$ *are induced by bijections of finite sets,* $\rho : n \to n$ *is associated with the subspace generated by* $\{(1_i, 1_{\rho i})\}_{i<n}$ *where* 1_k *is the binary* n*-vector with* 1 *at the* $k{+}1$*th coordinate and* 0*s elsewhere. For instance* $\sigma_{1,1} : 2 \to 2$ *is the subspace generated by* $\{(\begin{pmatrix}1\\0\end{pmatrix},\begin{pmatrix}0\\1\end{pmatrix}),(\begin{pmatrix}0\\1\end{pmatrix},\begin{pmatrix}1\\0\end{pmatrix})\}$.

Definition 5. *Let* $[v_1,\ldots,v_n]$ *denote the space generated by the vectors* $v_1 \ldots v_n$*. The PROP morphism* $\mathcal{S}_{\mathbb{IH}} : \mathbb{IH} \to \mathbb{SV}_{\mathsf{k}(x)}$ *is inductively defined on circuits* c *of* $\mathbb{IH}$ *as follows. For the operations of* $\mathbb{HA}$[2]*:*

[2] Here and in Definition 6, () denotes the only element of the space with dimension 0.

$$\boxed{-\!\!\!\text{C}} \longmapsto [(1, \begin{pmatrix} 1 \\ 1 \end{pmatrix})] \qquad\qquad \boxed{\text{D}\!-} \longmapsto [(\begin{pmatrix} 0 \\ 1 \end{pmatrix}, 1), (\begin{pmatrix} 1 \\ 0 \end{pmatrix}, 1)]$$

$$\boxed{-\bullet} \longmapsto [(1, ())] \qquad \boxed{\circ\!-} \longmapsto [((), 0)] \qquad \boxed{-\!\!\boxed{x}\!-} \longmapsto [(1, p)]$$

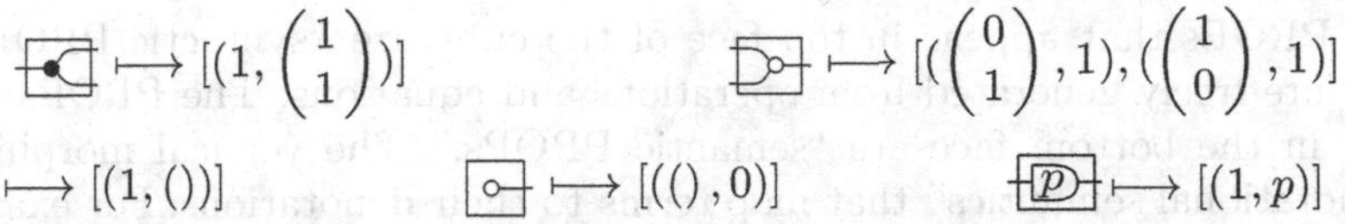

The semantics of an operation c in HA^{op} *is symmetric, e.g.* $\boxed{\bullet\!-}$ *is mapped to* $[((), 1)]$. *For the composite circuits, we define* $c_1 \oplus c_2 \mapsto \mathcal{S}_{\text{IH}}(c_1) \oplus \mathcal{S}_{\text{IH}}(c_2)$ *and* $c_1; c_2 \mapsto \mathcal{S}_{\text{IH}}(c_1); \mathcal{S}_{\text{IH}}(c_2)$. *The PROP morphism is well-defined since all the equations of* IH *are sound w.r.t.* $\mathcal{S}_{\text{IH}}$.

The circuit $\boxed{\text{D}\!-\!\bullet}$ discussed above is mapped to $\{((p,p), ()) \mid p \in \mathsf{k}(x)\} \subseteq \mathsf{k}(x)^2 \times \mathsf{k}(x)^0$. There are similar circuits in $\text{IH}[2n, 0]$ for arbitrary n.

$$\epsilon_0 := id_0 \qquad \epsilon_1 := \boxed{\text{D}\!-\!\bullet} \qquad \epsilon_2 := \boxed{} \qquad \epsilon_3 := \boxed{} \quad \cdots$$

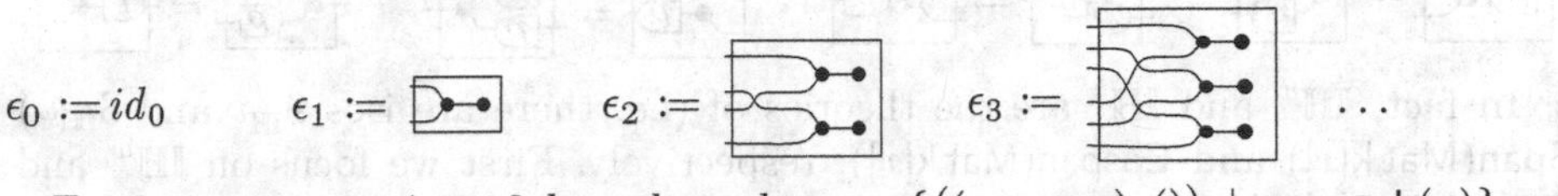

For instance, $\epsilon_2 \colon 4 \to 0$ has the subspace $\{((p, q, p, q), ()) \mid p, q \in \mathsf{k}(x)\}$ as semantics. One can define circuits from 0 to $2n$ symmetrically, starting from $\eta_2 := \boxed{\bullet\!-\!\text{C}} : 0 \to 2$. As shown in [8, §5], the ηs and the ϵs form a (self-dual) compact closed structure on the category IH. This yields a contravariant endofunctor $(\cdot)^\star$ on IH (*cf.* [19, Rmk 2.1]): for $c \colon n \to m$ a circuit, $c^\star \colon m \to n$ is defined as on the right, where $\boxed{\bullet\!-\!\text{C}}$ is notation for η_n, $\boxed{\text{D}\!-\!\bullet}$ for ϵ_n and $\boxed{n}$ for the circuit id_n.

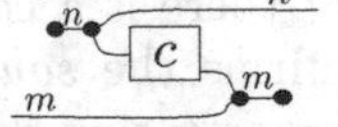

For the sequel, we also fix $\boxed{n\!-\!\boxed{x}\!-\!n}$ for the n-fold tensor product of $\boxed{-\boxed{x}\!-}$. Using the equational theory of IH, one can show (see [8, §5]) that $c^\star$ is just "c reflected about the y-axis": for example, $\boxed{\text{D}\!-}^\star = \boxed{-\text{C}}$ and $\boxed{-\boxed{x}\!-}^\star = \boxed{-\boxed{x}}$.

4.1 Soundness and Completeness of IH: The Cube Construction

In this subsection we sketch the proof of the following result, which states that the axioms of IH (Definition 3) characterise the PROP $\mathbb{SV}_{\mathsf{k}(x)}$. The details are in [8, §6-10].

Theorem 1. $\mathcal{S}_{\text{IH}} \colon \text{IH} \to \mathbb{SV}_{\mathsf{k}(x)}$ *is an isomorphism of PROPs.*

The proof is interesting in its own right because it is a *modular* account of the theory of IH. Its components are summarised by the cube diagram ($\mathcal{D}$) below.

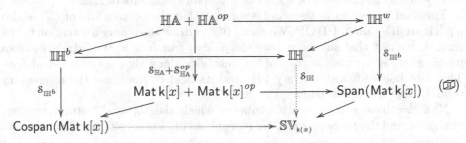

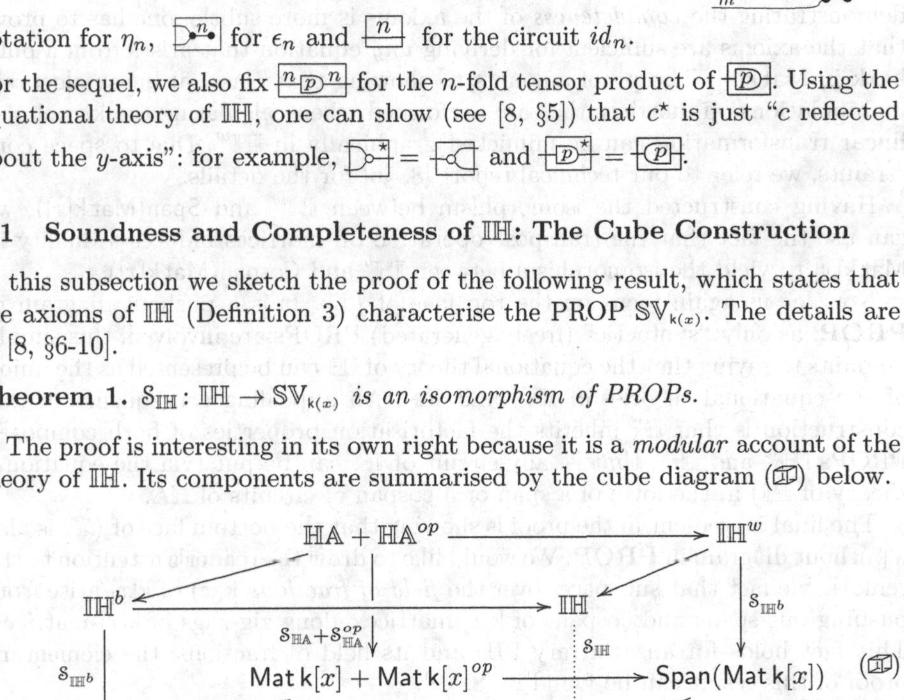

The PROPs that appear in top face of the cube are "syntactic PROPs", i.e., they are freely generated from operations and equations. The PROPs that appear in the bottom face are "semantic PROPs." The vertical morphisms are "denotational semantics" that map terms to their denotations. For example, as we showed in §3, $\mathbb{HA}$ is the theory of matrices with entries from the polynomial ring $\mathsf{k}[x]$, i.e. there is an isomorphism of PROPs $\mathcal{S}_{\mathbb{HA}} \colon \mathbb{HA} \to \mathsf{Mat}\,\mathsf{k}[x]$.

The theory $\mathbb{IH}^w$ has the presentation of $\mathbb{IH}$ (Definition 3) but with the two leftmost axioms below replacing (S7), (S8), (S11) and (S12). Dually, $\mathbb{IH}^b$ is $\mathbb{IH}$ without (S4), (S5), (S9) and (S10), and the addition of the two rightmost axioms below (the four of them are derivable in $\mathbb{IH}$, see [8]).

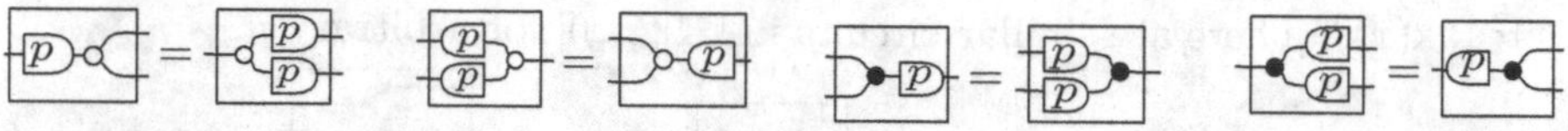

In fact, $\mathbb{IH}^w$ and $\mathbb{IH}^b$ are the theories of (i.e. there are isos $\mathcal{S}_{\mathbb{IH}^b}$ and $\mathcal{S}_{\mathbb{IH}^w}$) $\mathsf{Span}(\mathsf{Mat}\,\mathsf{k}[x])$ and $\mathsf{Cospan}(\mathsf{Mat}\,\mathsf{k}[x])$, respectively. First we focus on $\mathbb{IH}^w$ and $\mathsf{Span}(\mathsf{Mat}\,\mathsf{k}[x])$. Note that pullbacks in $\mathsf{Mat}\,\mathsf{k}[x]$ exist and are computed as in the category of sets, since $\mathsf{k}[x]$ is a principal ideal domain (PID).

The pullback construction gives a distributive law of PROPs in the sense of Lack [13] and, as we explained in §2.2, pullbacks can be understood as "adding new equations" to the theory $\mathbb{HA} + \mathbb{HA}^{op}$. Indeed, for each of the axioms of $\mathbb{IH}^w$ there is a corresponding "witnessing" pullback in $\mathsf{Mat}\,\mathsf{k}[x]$: this argument confirms the *soundness* of the theory of $\mathbb{IH}^w$ for $\mathsf{Span}(\mathsf{Mat}\,\mathsf{k}[x])$. The task of demonstrating the *completeness* of the axioms is more subtle: one has to prove that the axioms are sufficient for deriving *any* equation that arises from a pullback in $\mathsf{Mat}\,\mathsf{k}[x]$. The proof amounts to showing that classical linear algebraic manipulations on matrices that are performed when calculating the kernel of a linear transformation can be mimicked graphically in $\mathbb{IH}^w$. Due to space constraints, we refer to our technical report [8, §6] for the details.

Having constructed the isomorphism between $\mathbb{IH}^w$ and $\mathsf{Span}(\mathsf{Mat}\,\mathsf{k}[x])$, we can use the fact that the transpose operation on matrices induces a duality in $\mathsf{Mat}\,\mathsf{k}[x]$ to yield the isomorphism between $\mathbb{IH}^b$ and $\mathsf{Cospan}(\mathsf{Mat}\,\mathsf{k}[x])$.

Now let us again focus on the top face of (⧉). It is a pushout diagram in **PROP**: as only "syntactic" (freely generated) PROPs are involved, this simply amounts to saying that the equational theory of $\mathbb{IH}$ can be presented as the union of the equational theories of $\mathbb{IH}^w$ and $\mathbb{IH}^b$. An appealing consequence of this construction is that $\mathbb{IH}$ inherits the factorisation properties of both composed PROPs $\mathbb{IH}^w$ and $\mathbb{IH}^b$, that is, any circuit of $\mathbb{IH}$ can be put (via the equational theory of $\mathbb{IH}$) in the form of a span or a cospan of circuits of $\mathbb{HA}$.

The final ingredient in the proof is showing that the bottom face of (⧉) is also a pushout diagram in **PROP**. We would like to draw the reader's attention to the remarkable fact that subspaces over the *field of fractions* $\mathsf{k}(x)$ of $\mathsf{k}[x]$ arise from pushing out spans and cospans of $\mathsf{k}[x]$-matrices along zig-zags of $\mathsf{k}[x]$-matrices. This fact holds for an arbitrary PID and its field of fractions: the elementary proof of this result can be found in [8, §9].

Now, we have a commutative cube in which the top and bottom face are pushouts, and the three rear vertical morphisms are isomorphisms. The universal property of pushouts now ensures that the unique morphism $\mathcal{S}_{\mathbb{IH}} : \mathbb{IH} \to \mathbb{SV}_{\mathsf{k}(x)}$ is an isomorphism: it is easy to verify that it is the morphism of Definition 5.

5 Stream Semantics

With simple extensions of the semantics morphisms, we can interpret circuits of $\mathbb{HA}$ and $\mathbb{IH}$ in terms of streams. First we need to recall some useful notions.

A formal Laurent series (fls) is a function $\sigma\colon \mathbb{Z} \to \mathsf{k}$ for which there exists $i \in \mathbb{Z}$ such that $\sigma(j) = 0$ for all $j < i$. The *degree* of σ is the smallest $d \in \mathbb{Z}$ such that $\sigma(d) \neq 0$. We write σ as $\dots, \sigma(-1), \underline{\sigma(0)}, \sigma(1), \dots$ with position 0 underlined, or as formal sum $\sum_{i=d}^{\infty} \sigma(i)x^i$. With the latter notation, we define the sum and product of two fls $\sigma = \sum_{i=d}^{\infty} \sigma(i)x^i$ and $\tau = \sum_{i=e}^{\infty} \tau(i)x^i$ as

$$\sigma + \tau = \sum_{i=min(d,e)}^{\infty} (\sigma(i) + \tau(i))x^i \qquad \sigma \cdot \tau = \sum_{i=d+e}^{\infty} \Big(\sum_{k+j=i} \sigma(j) \cdot \tau(k) \Big) x^i \tag{1}$$

The units for $+$ and $\cdot$ are $\dots 0, \underline{0}, 0 \dots$ and $\dots 0, \underline{1}, 0 \dots$. Fls form a field $\mathsf{k}((x))$, where the inverse σ^{-1} of fls σ with degree d is given as follows.

$$\sigma^{-1}(i) = \begin{cases} 0 & \text{if } i < -d \\ \sigma(d)^{-1} & \text{if } i = -d \\ \frac{\sum_{i=1}^{n} \left(\sigma(d+i) \cdot \sigma^{-1}(-d+n-i) \right)}{-\sigma(d)} & \text{if } i = -s + n \text{ for } n > 0 \end{cases} \tag{2}$$

A *formal power series* (fps) is a fls with degree $d \geq 0$. By (1), fps are closed under $+$ and $\cdot$, but not under inverse: it is immediate by (2) that σ^{-1} is a fps iff σ has degree $d = 0$. Therefore fps form a ring which we denote by $\mathsf{k}[[x]]$.

We will refer to both fps and fls as *streams*. Indeed, fls can be thought of as sequences with an infinite future, but a finite past. Just as a polynomial p can be seen as a fraction $\frac{p}{1}$, an fps σ can be interpreted as the fls $\dots, 0, \underline{\sigma(0)}, \sigma(1), \sigma(2), \dots$. A polynomial $p_0 + p_1 x + \dots + p_n x^n$ can also be regarded as the fps $\sum_{i=0}^{\infty} p_i x^i$ with $p_i = 0$ for all $i > n$. Similarly, fractions can be regarded as fls: we define $\tilde{\cdot}\colon \mathsf{k}(x) \to \mathsf{k}((x))$ as the unique field morphism mapping $k \in \mathsf{k}$ into the stream $\dots 0, \underline{k}, 0 \dots$ and the indeterminate x into $\dots, 0, \underline{0}, 1, 0, \dots$

Differently from polynomials, fractions can denote streams with possibly infinitely many non-zero values. For instance, (1) and (2) imply that $\frac{x}{1-x-x^2}$ is the Fibonacci series $\dots, 0, \underline{0}, 1, 1, 2, 3, \dots$. Moreover, while polynomials can be interpreted as fps, fractions need the full generality of fls: $\frac{1}{x}$ denotes $\dots 0, 0, 1, \underline{0}, 0, \dots$

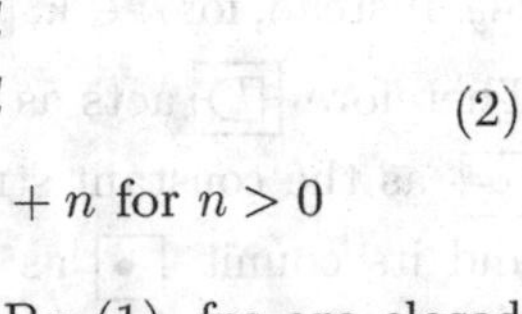

These are all ring morphisms and are illustrated by the commutative diagram on the right. At the center there is $\mathsf{k}\langle x \rangle$, the ring of *rationals*, i.e, fractions of polynomials $\frac{k_0 + k_1 x + k_2 x^2 \cdots + k_n x^n}{l_0 + l_1 x + l_2 x^2 \cdots + l_n x^n}$ where $l_0 \neq 0$. Differently from fractions, rationals denote only fps — in other words, bona fide streams that do not start "in the past". Indeed, since $l_0 \neq 0$, the inverse of $l_0 + l_1 x + l_2 x^2 \cdots + l_n x^n$ is, by (2), a fps. The streams denoted by $\mathsf{k}\langle x\rangle$ are well known in literature under the name of *rational streams* [6]. Hereafter, we will use polynomials and fractions to denote the corresponding streams. Moreover, $\mathsf{Mat}\,\mathsf{k}[[x]]$ and $\mathsf{Mat}\,\mathsf{k}\langle x\rangle$ denote the PROPs of matrices over $\mathsf{k}[[x]]$ and $\mathsf{k}\langle x\rangle$ defined analogously to $\mathsf{Mat}\,\mathsf{k}[x]$. Similarly, $\mathbb{SV}_{\mathsf{k}((x))}$ is the PROP of $\mathsf{k}((x))$-vector subspaces defined like $\mathbb{SV}_{\mathsf{k}(x)}$.

5.1 A Stream Semantics of $\mathbb{HA}$

The semantics $\mathcal{S}_{\mathbb{HA}} \colon \mathbb{HA} \to \mathsf{Mat}\,\mathsf{k}[x]$ of Definition 2 allows us to regard the circuits in $\mathbb{HA}$ as *stream* transformers. Indeed, the interpretation of a polynomial in $\mathsf{k}[x]$ as a fps in $\mathsf{k}[[x]]$ can be pointwise extended to a faithful PROP morphism $\hat{\cdot} \colon \mathsf{Mat}\,\mathsf{k}[x] \to \mathsf{Mat}\,\mathsf{k}[[x]]$. By taking $[\![\cdot]\!]_{\mathbb{HA}} = \mathcal{S}_{\mathbb{HA}} ; \hat{\cdot}$, the semantics $[\![c]\!]_{\mathbb{HA}}$ of a circuit $c \in \mathbb{HA}[n,m]$ consists of a linear map of type $\mathsf{k}[[x]]^n \to \mathsf{k}[[x]]^m$.

Remark 3. Recall the operational intuition for circuits in $\mathbb{HA}$ given in Remark 2. This intuition extends to the stream semantics, but rather than carrying elements of $\mathsf{k}[x]$ along the wires, the circuits now carry individual elements of a k-stream, processing one after the other. Inputs arrive on the left and outputs are emitted on the right. For instance, $[\![\;\;]\!]_{\mathbb{HA}} = (\,x\,)$ maps every stream $\sigma \in \mathsf{k}[[x]]$ into the stream $\sigma \cdot x$ which, by (1), is just $\underline{0}, \sigma(0), \sigma(1), \sigma(2), \ldots$ Thus behaves as a *delay*. Instead, for $k \in \mathsf{k}$, $[\![\;\;]\!]_{\mathbb{HA}} = (\,k\,)$ maps σ into $\sigma \cdot k = \underline{k\sigma(0)}, k\sigma(1), k\sigma(2), \ldots$ Therefore acts as an *amplifier*. Also behaves as an *adder* and its unit as the constant stream $\underline{0}, 0, 0 \ldots$. The comultiplication acts as *copier* and its counit as the trivial transformer taking any stream in input and giving no output.

One can readily check that this interpretation coincides with the semantics given in [18, §4.1]. Our approach has the advantage of making the circuits representation formal and allowing for equational reasoning, as shown for instance in Example 2 below. Indeed, since $[\![\cdot]\!]_{\mathbb{HA}} \colon \mathbb{HA} \to \mathsf{Mat}\,\mathsf{k}[[x]]$ is faithful, the axiomatization of $\mathbb{HA}$ is sound and complete.

Example 2. Consider the following derivation in the equational theory of $\mathbb{HA}$, where (A15) is used at each step.

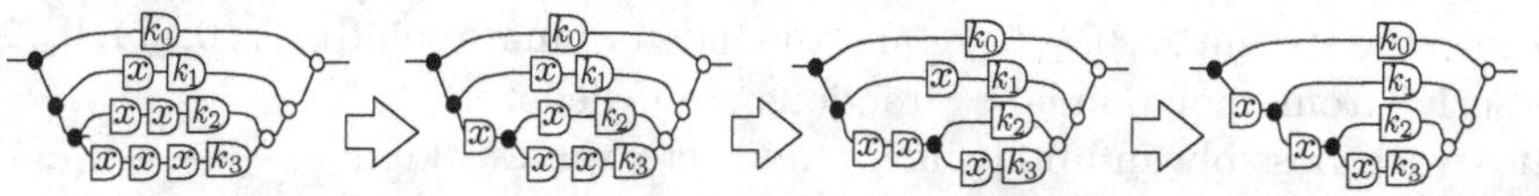
(3)

Using the stream interpretation of Remark 3, the circuits above are readily seen to implement the polynomial stream function $f \colon \sigma \mapsto \sigma \cdot p$ where $p = \underline{k_0}, k_1, k_2, k_3, 0 \ldots$ Then (3) yields a procedure that reduces the total number of delays appearing in the circuit (*cf.* [18, Prop. 4.12]). The equational theory of $\mathbb{HA}$ allows us to verify that the circuits in (3) really implement f:

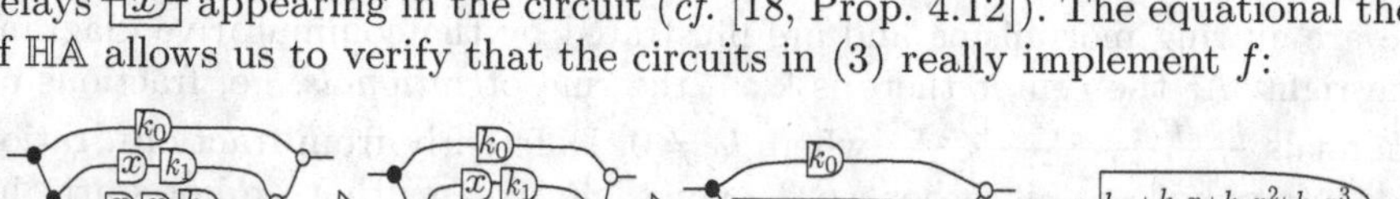
(4)

By iteratively applying (A5) and (A17), the circuit is show to be equal to $\boxed{p}$, which clearly has the 1×1 matrix (p) as semantics and thus implements f. Reading (4) in the converse direction, it yields a decomposition of $\boxed{p}$ into an equivalent circuit with only "basic gates": amplifiers $\boxed{k}$ (for $k \in \mathsf{k}$) and delays $\boxed{x}$ — in fact, such a decomposition is possible for arbitrary circuits of $\mathbb{HA}$.

5.2 A Stream Semantics of $\mathbb{IH}$

In order to give the stream interpretation of $\mathbb{IH}$, we construct the following cube, the topmost face of which coincides with the bottom face of (☐).

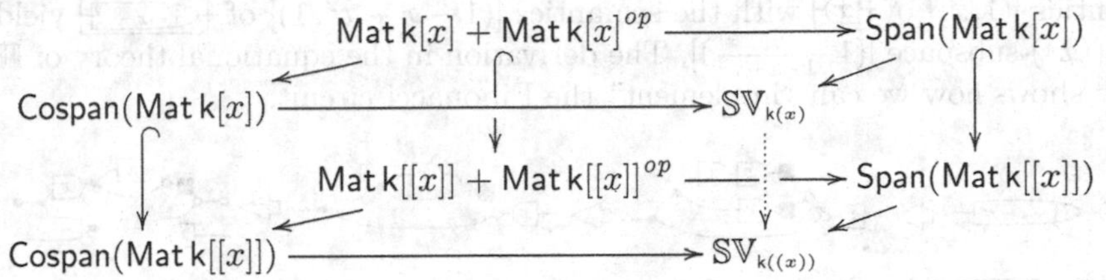

Mat k[x] + Mat k[x]^op
Span(Mat k[x])
Cospan(Mat k[x])
SV_k(x)
Mat k[[x]] + Mat k[[x]]^op
Span(Mat k[[x]])
Cospan(Mat k[[x]])
SV_k((x))

(5)

The bottom face commutes and is a pushout for the same reasons as the top face (*cf.* §4.1), because $\mathsf{k}[[x]]$ is a PID and $\mathsf{k}((x))$ is its field of fractions. The rear map is $\hat{\cdot} + \hat{\cdot}^{op} \colon \mathsf{Mat}\,\mathsf{k}[x] + \mathsf{Mat}\,\mathsf{k}[x]^{op} \to \mathsf{Mat}\,\mathsf{k}[[x]] + \mathsf{Mat}\,\mathsf{k}[[x]]^{op}$. Since $\hat{\cdot}$ preserves pullbacks, we can define the righmost vertical morphism as mapping a span $n \xleftarrow{V} z \xrightarrow{W} m$ into $n \xleftarrow{\hat{V}} z \xrightarrow{\hat{W}} m$. The leftmost vertical map is defined analogously.

One can readily check that all these morphisms are faithful and that the rear faces commute. Since the top face is a pushout, the universal property induces the faithful morphism $[\tilde{\cdot}] \colon \mathbb{SV}_{\mathsf{k}(x)} \to \mathbb{SV}_{\mathsf{k}((x))}$. This can be concretely defined by observing that $\tilde{\cdot} \colon \mathsf{k}(x) \to \mathsf{k}((x))$ can be pointwise extended to matrices and sets of vectors. For a subspace H in $\mathbb{SV}_{\mathsf{k}(x)}$, $[\tilde{H}]$ is the space in $\mathbb{SV}_{\mathsf{k}((x))}$ generated by the set of vectors $\tilde{H}$. Note that the composition of $\mathcal{S}_{\mathbb{IH}}$ (Definition 5) with $[\tilde{\cdot}]$, that we call the stream semantics of $\mathbb{IH}$, is also induced by the universal property of the topmost face of (☐).

Definition 6. *The stream semantics of $\mathbb{IH}$ is the morphism $[\![\cdot]\!]_{\mathbb{IH}} \colon \mathbb{IH} \to \mathbb{SV}_{\mathsf{k}((x))}$ defined as $\mathcal{S}_{\mathbb{IH}} \,;\, [\tilde{\cdot}]$. It can be presented as follows. For the operations of $\mathbb{HA}$:*

$$\boxed{\bullet\!\!<} \longmapsto \{(\sigma, \begin{pmatrix}\sigma\\ \sigma\end{pmatrix}) \mid \sigma \in \mathsf{k}((x))\} \qquad \boxed{>\!\!\circ} \longmapsto \{(\begin{pmatrix}\sigma\\ \tau\end{pmatrix}, \sigma+\tau) \mid \sigma, \tau \in \mathsf{k}((x))\}$$

$$\boxed{\bullet} \longmapsto \{(\sigma, ()) \mid \sigma \in \mathsf{k}((x))\} \quad \boxed{\circ} \longmapsto \{((), 0)\} \quad \boxed{p} \longmapsto \{(\sigma, \sigma \cdot p) \mid \sigma \in \mathsf{k}((x))\}$$

where 0 and p denote streams. The semantics of an operation c of $\mathbb{HA}^{op}$ is the reverse relations of $[\![c^\star]\!]_{\mathbb{IH}}$. For composite circuits, we let $c_1 \oplus c_2 \mapsto [\![c_1]\!]_{\mathbb{HA}} \oplus [\![c_2]\!]_{\mathbb{HA}}$ and $c_1 \,;\, c_2 \mapsto [\![c_1]\!]_{\mathbb{HA}} \,;\, [\![c_2]\!]_{\mathbb{HA}}$.

Since $[\tilde{\cdot}]$ is faithful, by Theorem 1, also $[\![\cdot]\!]_{\mathbb{IH}}$ is faithful.

Corollary 1 (Completeness). *For all $c_1, c_2 \in \mathbb{IH}$, $c_1 = c_2$ iff $[\![c_1]\!]_{\mathbb{IH}} = [\![c_2]\!]_{\mathbb{IH}}$.*

6 The Theory of Signal Flow Graphs

In this section we introduce an inductively defined class of circuits $\mathbb{SF}$ of $\mathbb{IH}$ that we call signal flow graphs and show that it is the theory of $\mathsf{k}\langle x\rangle$-matrices. The definition is close in spirit to the classical variations found in the literature sans inputs, outputs and directions of wires.

We start with a motivating example of a circuit not in $\mathbb{HA}$ that nevertheless gives functional behaviour on $\mathsf{k}[[x]]$.

Example 3. The rational $\frac{x}{1-x-x^2}$ denoting the Fibonacci sequence can be succintly represented in $\mathbb{IH}$ as the circuit 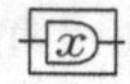 ; 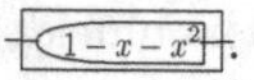. Indeed, composing the semantics $[(1,x)]$ of 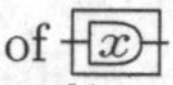 with the semantics $[(1-x-x^2,1)]$ of 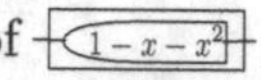 yields the $\mathsf{k}((x))$-subspace $[(1, \frac{x}{1-x-x^2})]$. The derivation in the equational theory of $\mathbb{IH}$ below shows how we can "implement" the Fibonacci circuit.

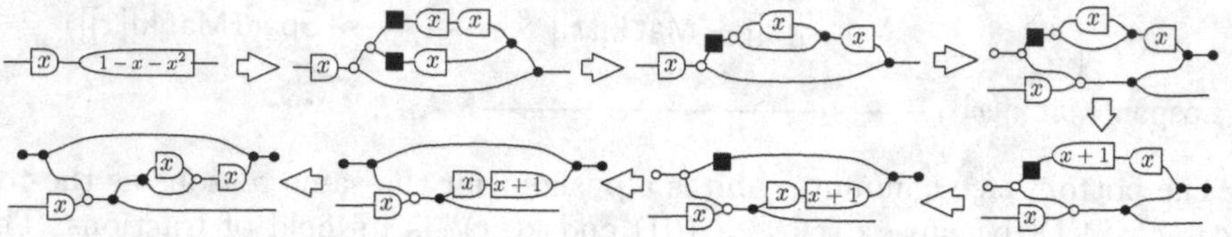

First 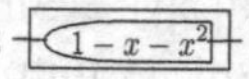 is decomposed (using (A17)^{op} from $\mathbb{HA}^{op}$) and then the circuit is "deformed" in a suitable way by essentially using the Frobenius axioms (S2)-(S1) and the compact closed structure of $\mathbb{IH}$. The resulting circuit exhibits a feedback structure. Indeed, using the intuitive operational descriptions of Remark 3 and the behaviour of 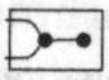, 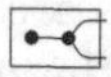 as "bent identity wires" that merely forward signals from one port to the other, the operational behaviour of the final circuit in the derivation can be "read off" the final circuit, with inputs entering on the left and outputs emitted on the right. In particular, the reader will verify that inputing the stream $\dots, 0, \underline{1}, 0, \dots$ yields the Fibonacci sequence as output. Note that the notions of "input", "output" and directionality of wires are entirely derivative.

The Fibonacci circuit belongs to the class of circuits $\mathbb{SF}$. To define it, we first introduce a particular trace structure [20, §5.1] on $\mathbb{IH}$. It is not the canonical trace induced by the compact closed structure, but rather a "guarded" version.

Definition 7. *For $n, m, z \in \mathbb{N}$, $c \in \mathbb{IH}[z+n, z+m]$, the z-feedback $\mathsf{Tr}^z(c) \in \mathbb{IH}[n,m]$ is the circuit below, for which we use the indicated shorthand notation:*

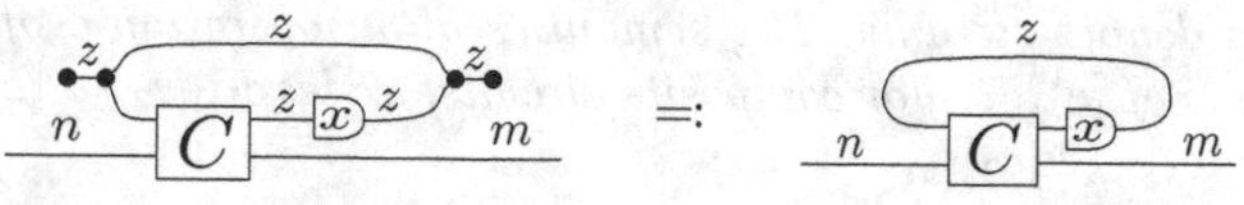

It can be verified that $\mathsf{Tr}^{(\cdot)}$ actually defines a trace on $\mathbb{IH}$. We have now all the ingredients to define the theory of signal flow graphs.

Definition 8. *Let $\mathbb{SF}$ be the following, inductively defined set of circuits.*

- *If $c \in \mathbb{HA}[n, m]$ then c is in $\mathbb{SF}$.*
- *If $c \in \mathbb{SF}[z + n, z + m]$, then $\mathsf{Tr}^z(c) \in \mathbb{IH}[n, m]$ is in $\mathbb{SF}$.*

Circuits in $\mathbb{SF}$ inherit the equational theory of $\mathbb{IH}$, that is, we say that $c = c'$ as circuits in $\mathbb{SF}$ exactly when $c = c'$ in $\mathbb{IH}$.

One can check that $\mathbb{SF}$ is a sub PROP of $\mathbb{IH}$, namely the smallest one containing $\mathbb{HA}$ and closed under the trace. This is essential for proving that $\mathbb{SF}$ is the theory of $\mathsf{k}\langle x\rangle$-matrices.

Theorem 2. *There is an isomorphism of PROPs between $\mathbb{SF}$ and $\mathsf{Mat}\,\mathsf{k}\langle x\rangle$.*

Proof. Hereafter, we sketch one direction of the isomorphism, namely from $\mathsf{Mat}\,\mathsf{k}\langle x\rangle$ to $\mathbb{SF}$. The main insight is that any (1×1 matrix with a) rational of the form $1/(k + xp)$, with $k \neq 0$ and $p \in \mathsf{k}[x]$, corresponds to a circuit in $\mathbb{SF}$, as witnessed by the following derivation in $\mathbb{IH}$:

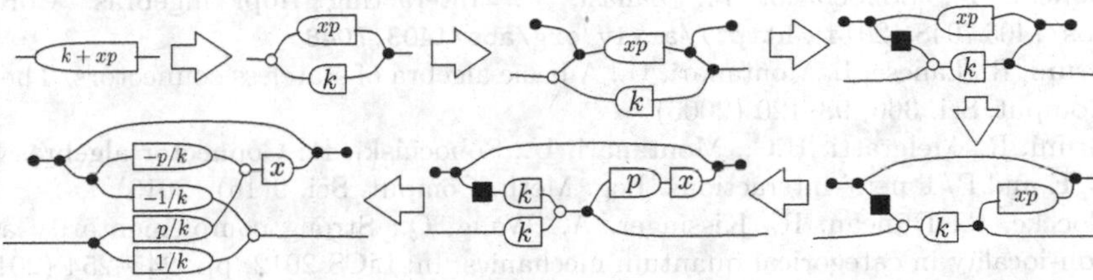

Then, every matrix in $\mathsf{Mat}\,\mathsf{k}\langle x\rangle$ is obtained by composing these circuits with those in $\mathbb{HA}$. Such composition is still in $\mathbb{SF}$, since $\mathbb{SF}$ is a PROP. □

7 Conclusions

We introduced $\mathbb{IH}$, a graphical calculus of streams transformers equipped with a compositional semantics in terms of subspaces and a sound and complete axiomatisation. We have identified a fragment of $\mathbb{IH}$ characterising signal flow graphs, which are *functional* streams transformers. To best of our knowledge, the axioms of $\mathbb{IH}$ provides the first sound and complete axiomatisation of signal flow graphs seen as syntactic entities. Indeed, previous results either restrict the class of systems (for instance [17] only considers the "closed" ones) or exploit an intermediate (co)algebraic syntax (see e.g. [5] and the references therein).

However, our interest in $\mathbb{IH}$ is not restricted to signal flow graphs: the circuits of $\mathbb{IH}$ are streams transformers which are, in general, *relational* rather than functional. Such relational behaviour emerges naturally when studying different sorts of networks [1,2,4,9–12,21].

Acknowledgements. The first and the third author acknowledge support from the ANR project 12IS02001 PACE.

References

1. Arbab, F.: Reo: a channel-based coordination model for component composition. Mathematical Structures in Computer Science 14, 329–366 (2004)
2. Baez, J.C.: Network theory (2014), `http://math.ucr.edu/home/baez/networks/`
3. Baez, J.C., Erbele, J.: Categories in control. CoRR, abs/1405.6881 (2014), `http://arxiv.org/abs/1405.6881`
4. Barnabei, M., Guerrini, C., Montefusco, L.: Some algebraic aspects of signal processing. Linear Algebra and its Applications 284(1-3), 3–17 (1998)
5. Basold, H., Bonsangue, M., Hansen, H.H., Rutten, J.: (co)algebraic characterizations of signal flow graphs. In: van Breugel, F., Kashefi, E., Palamidessi, C., Rutten, J. (eds.) Panangaden Festschrift. LNCS, vol. 8464, pp. 124–145. Springer, Heidelberg (2014)
6. Berstel, J., Reutenauer, C.: Rational series and their languages. EATCS Monographs on Theoretical Computer Science 12 (1988)
7. Bonchi, F., Sobociński, P., Zanasi, F.: Interacting bialgebras are Frobenius. In: Muscholl, A. (ed.) FOSSACS 2014. LNCS, vol. 8412, pp. 351–365. Springer, Heidelberg (2014)
8. Bonchi, F., Sobociński, P., Zanasi, F.: Interacting Hopf algebras. CoRR, abs/1403.7048 (2014), `http://arxiv.org/abs/1403.7048`
9. Bruni, R., Lanese, I., Montanari, U.: A basic algebra of stateless connectors. Theor. Comput. Sci. 366, 98–120 (2006)
10. Bruni, R., Melgratti, H.C., Montanari, U., Sobociński, P.: Connector algebras for C/E and P/T nets' interactions. Log. Meth. Comput. Sci. 9(16) (2013)
11. Coecke, B., Duncan, R., Kissinger, A., Wang, Q.: Strong complementarity and non-locality in categorical quantum mechanics. In: LiCS 2012, pp. 245–254 (2012)
12. Katis, P., Sabadini, N., Walters, R.F.C.: Span(Graph): an algebra of transition systems. In: Johnson, M. (ed.) AMAST 1997. LNCS, vol. 1349, pp. 322–336. Springer, Heidelberg (1997)
13. Lack, S.: Composing PROPs. Theor. App. Categories 13(9), 147–163 (2004)
14. Lahti, B.P.: Signal Processing and Linear Systems. Oxford University Press (1998)
15. Mac Lane, S.: Categorical algebra. Bull. Amer. Math. Soc. 71, 40–106 (1965)
16. Mason, S.J.: Feedback Theory: I. Some Properties of Signal Flow Graphs. Massachusetts Institute of Technology, Research Laboratory of Electronics (1953)
17. Milius, S.: A sound and complete calculus for finite stream circuits. In: LICS (2010)
18. Rutten, J.J.M.M.: A tutorial on coinductive stream calculus and signal flow graphs. Theor. Comput. Sci. 343(3), 443–481 (2005)
19. Selinger, P.: Dagger compact closed categories and completely positive maps (extended abstract). Electr. Notes Theor. Comput. Sci. 170, 139–163 (2007)
20. Selinger, P.: A survey of graphical languages for monoidal categories. arXiv:0908.3347v1 [math.CT] (2009)
21. Sobociński, P.: Representations of petri net interactions. In: Gastin, P., Laroussinie, F. (eds.) CONCUR 2010. LNCS, vol. 6269, pp. 554–568. Springer, Heidelberg (2010)
22. Willems, J.C.: The behavioural approach to open and interconnected systems. IEEE Contr. Syst. Mag. 27, 46–99 (2007)

Generic Forward and Backward Simulations III: Quantitative Simulations by Matrices

Natsuki Urabe and Ichiro Hasuo

University of Tokyo, Japan

Abstract. We introduce notions of simulation between semiring-weighted automata as models of quantitative systems. Our simulations are instances of the categorical/coalgebraic notions previously studied by Hasuo—hence soundness wrt. language inclusion comes for free—but are concretely presented as matrices that are subject to linear inequality constraints. Pervasiveness of these formalisms allows us to exploit existing algorithms in: searching for a simulation, and hence verifying quantitative correctness that is formulated as language inclusion. Transformations of automata that aid search for simulations are introduced, too. This verification workflow is implemented for the plus-times and max-plus semirings.

1 Introduction

Quantitative aspects of various systems are more and more emphasized in recent verification scenarios. Probabilities in randomized or fuzzy systems are a classic example; utility in economics and game theory is another. Furthermore, now that many computer systems are integrated into physical ambience—realizing so-called *cyber-physical systems*—physical quantities like energy consumption are necessarily taken into account.

Semiring-weighted Automata. It is standard in the concurrency community to model such quantitative systems by state-transition systems in which *weights* are assigned to their states and/or transitions. The semantics of such systems varies, however, depending on the interpretation of weights. If they are probabilities, they are accumulated by $\times$ along a path and summed across different paths; if weights are (worst-case) costs, they are summed up along a path and we would take max across different paths.

The algebraic structure of *semirings* then arises as a uniform mathematical language for different notions of "weight," as is widely acknowledged in the community. The subject of the current study is state-based systems with labeled transitions, in which each transition is assigned a weight from a prescribed semiring $\mathcal{S}$. We shall call them *$\mathcal{S}$-weighted automata*; and we are more specifically interested in the (weighted, finite) *language inclusion* problem and a *simulation-based* approach to it.

Language Inclusion. Let $\mathcal{A}$ be an $\mathcal{S}$-weighted automaton with labels from an alphabet Σ. It assigns to each word $w \in \Sigma^*$ a weight taken from $\mathcal{S}$—this is much like a (purely) probabilistic automaton assigns a probability to each word. Let us denote this function by $L(\mathcal{A}) \colon \Sigma^* \to \mathcal{S}$ and call it the *(weighted) language* of $\mathcal{A}$ by analogy with classic automata theory. The *language inclusion* problem $L(\mathcal{A}) \sqsubseteq L(\mathcal{B})$ asks if: $L(\mathcal{A})(w) \sqsubseteq L(\mathcal{B})(w)$ for each word $w \in \Sigma^*$, where $\sqsubseteq$ is a natural order on the semiring $\mathcal{S}$.

It is not hard to see that language inclusion $L(\mathcal{A}) \sqsubseteq L(\mathcal{B})$ has numerous applications in verification. In a typical scenario, one of $\mathcal{A}$ and $\mathcal{B}$ is a model of a *system* and

P. Baldan and D. Gorla (Eds.): CONCUR 2014, LNCS 8704, pp. 451–466, 2014.

the other expresses *specification*; and $L(\mathcal{A}) \sqsubseteq L(\mathcal{B})$ gives the definition of "the system meeting the specification." We shall further present three concrete examples. 1) $\mathcal{S}$ represents probabilities; $\mathcal{A}$ models a system; and $\mathcal{B}$ expresses the specification that certain bad behaviors—identified with words—occur with a certain probability. Then $L(\mathcal{A}) \sqsubseteq L(\mathcal{B})$ is a *safety* statement: each bad behavior occurs in $\mathcal{A}$ at most as likely as in $\mathcal{B}$. 2) $\mathcal{S}$ represents profit, $\mathcal{A}$ is a specification and $\mathcal{B}$ is a system. Then $L(\mathcal{A}) \sqsubseteq L(\mathcal{B})$ guarantees the minimal profit yielded by the system $\mathcal{B}$. 3) There are other properties reduced to language inclusion in a less trivial manner. An example is *probable innocence* [26], a quantitative notion of anonymity. See [15].

Simulation. Direct check of language inclusion is simply infeasible because there are infinitely many words $w \in \Sigma^*$. One finitary proof method—well-known for nondeterministic (i.e. possibilistic) systems—is by *(forward or backward) simulations*, whose systematic study is initiated in [23]. In the nondeterministic setting, a simulation R is a relation between states of $\mathcal{A}$ and $\mathcal{B}$ that witnesses "local language inclusion"; moreover, from the coinductive way in which it is defined, a simulation persistently witnesses local language inclusion—ultimately yielding (global) language inclusion. This property—existence of a simulation implies language inclusion—is called *soundness*.

Contribution: Weighted Forward/Backward Simulations by Matrices. In this paper we extend this simulation approach to language inclusion [23] to the quantitative setting of semiring-weighted automata. Our notions of (forward and backward) weighted simulation are not given by relations, but by *matrices* with entries from a semiring $\mathcal{S}$.

Use of matrices in automata theory is classic—in fact our framework instantiates to that in [23] when we take as $\mathcal{S}$ the Boolean semiring. This is not how we arrived here; conversely, the current results are obtained as instances of a more general theory of *coalgebraic simulations* [11, 12, 14]. There various systems are identified with a categorical construct of *coalgebras* in a Kleisli category; and fwd./bwd. simulations are characterized as lax/oplax morphisms between coalgebras. A generic soundness result (with respect to language/trace inclusion) is also proved in the general categorical terms.

This paper is devoted to concrete presentations of these categorical notions by matrices, and to their application to actual verification of quantitative systems. Presentation by matrices turns out to be an advantage: a simulation is now a matrix X that satisfies certain *linear inequalities*; and existence of such X—i.e. feasibility of linear inequalities—is so common a problem in many fields that there is a large body of existing work that is waiting to be applied. For example *linear programming (LP)* can be exploited for the plus-times semiring for probabilities; and there are algorithms proposed for other semirings such as the max-plus (tropical) one.

Our (mostly semiring-independent) workflow is as follows. A verification goal is formulated as language inclusion $L(\mathcal{A}) \sqsubseteq L(\mathcal{B})$, which we aim to establish by finding a fwd. or bwd. simulation from $\mathcal{A}$ to $\mathcal{B}$. Soundness of simulations follows from the general result in [11]. A simulation we seek for is a matrix subject to certain linear inequalities, existence of which is checked by various algorithms that exist for different semirings. We implemented this workflow for the plus-times and max-plus semirings.

This simulation-based method is sound but not necessarily complete with respect to language inclusion. Hence we introduce transformations of weighted automata—called *(forward/backward) partial execution*—that potentially create matrix simulations.

Organization of the Paper. In §2 we define semiring-weighted automata, characterize them in coalgebraic terms and recap the coalgebraic theory in [11]. These are combined to yield the notion of simulation matrix in §3. In §4 partial execution transformations of automata are described and proved correct. The framework obtained so far is applied to the plus-times and max-plus semirings, in §5 and §6, respectively. There our proof-of-concept implementations and relationship to other known simulation notions are discussed, too. The appendices referred to in the paper are found at the first author's webpage; so is the code of our implementation.

2 Preliminaries

2.1 Semiring-Weighted Automata

The notion of semiring-weighted automaton is parametrized by a semiring $\mathcal{S}$. For our purpose of applying coalgebraic theory in [11, 14], we impose the following properties.

Definition 2.1. A *commutative cppo-semiring* is a tuple $\mathcal{S} = (S, +_\mathcal{S}, 0_\mathcal{S}, \times_\mathcal{S}, 1_\mathcal{S}, \sqsubseteq)$ that satisfies the following conditions.

- $(S, +_\mathcal{S}, 0_\mathcal{S}, \times_\mathcal{S}, 1_\mathcal{S})$ is a semiring in which $\times_\mathcal{S}$, in addition to $+_\mathcal{S}$, is commutative.
- A relation $\sqsubseteq$ is a partial order on S and $(S, \sqsubseteq)$ is *ω-complete*, i.e. an increasing chain $s_0 \sqsubseteq s_1 \sqsubseteq \cdots$ has a supremum.
- Any element $s \in S$ is *positive* in the sense that $0_\mathcal{S} \sqsubseteq s$.
- Addition $+_\mathcal{S}$ and multiplication $\times_\mathcal{S}$ are monotone with respect to $\sqsubseteq$.

It follows from positivity and ω-completeness that countable sum can be straightforwardly defined in a comm. cppo-semiring $\mathcal{S}$. We will use this fact throughout the paper.

Example 2.2 (Semirings $\mathcal{S}_{+,\times}, \mathcal{S}_{\max,+}, \mathcal{B}$). The *plus-times semiring* $\mathcal{S}_{+,\times} = ([0, \infty], +, 0, \times, 1, \leq)$ is a comm. cppo-semiring, where $+$ and $\times$ are usual addition and multiplication of real numbers. This is the semiring that we will use for modeling probabilistic branching. Specifically, probabilities of successive transitions are accumulated using $\times$, and those of different branches are combined with $+$.

The *max-plus semiring* $\mathcal{S}_{\max,+} = ([-\infty, \infty], \max, -\infty, +, 0, \leq)$—also sometimes called the *tropical semiring* [24]—is also a comm. cppo-semiring. Here a number $r \in [-\infty, \infty]$ can be understood as (best-case) *profit*: they are summed up along a path, and an optimal one (max) is chosen among different branches. Another possible understanding of r is as (worst-case) *cost*. The unit for the semiring addition max is given by $-\infty$; since it must also be a zero element of the semiring multiplication $+$, we define $(-\infty) + \infty = -\infty$. In the two examples $\mathcal{S}_{+,\times}$ and $\mathcal{S}_{\max,+}$ we added ∞ so that they become ω-complete. Finally, the *Boolean semiring* $\mathcal{B} = (\{0, 1\}, \vee, 0, \wedge, 1, \leq)$ is an example that is qualitative rather than quantitative.

Definition 2.3 ($\mathcal{S}$-weighted automaton, weighted language). Let $\mathcal{S} = (S, +_\mathcal{S}, 0_\mathcal{S}, \times_\mathcal{S}, 1_\mathcal{S}, \sqsubseteq)$ be a comm. cppo-semiring. An *$\mathcal{S}$-weighted automaton* $\mathcal{A} = (Q, \Sigma, M, \alpha, \beta)$ consists of a countable state space Q, a countable alphabet Σ, transition matrices $M(a) \in S^{Q\times Q}$ for all $a \in \Sigma$, an initial row vector $\alpha \in S^Q$ and a final column vector $\beta \in S^Q$.

Let $x, y \in Q$ and $a \in \Sigma$. We write α_x and β_x for the x-th entry of α and β, respectively, and $M(a)_{x,y}$ for the (x, y)-entry of the matrix $M(a)$. Note that these entries are all elements of the semiring $\mathcal{S}$.

An $\mathcal{S}$-weighted automaton $\mathcal{A} = (Q, \Sigma, M, \alpha, \beta)$ yields a *weighted language* $L(\mathcal{A})\colon \Sigma^* \to S$. It is given by the following multiplication of matrices and vectors.

$$L(\mathcal{A})(w) := \alpha \cdot M(a_1) \cdot \cdots \cdot M(a_k) \cdot \beta \qquad \text{for each } w = a_1 \cdots a_k \in \Sigma^*. \tag{1}$$

We require a state space Q to be at most countably infinite. This is so that the matrix multiplications in (1)—by addition and multiplication of $\mathcal{S}$—are well-defined. Recall that $\mathcal{S}$ has countable sum given by supremums of suitable ω-chains.

Our interest is in establishing language inclusion between two weighted automata.

Definition 2.4 (language inclusion). We write $L(\mathcal{A}) \sqsubseteq L(\mathcal{B})$ if, for each $w \in \Sigma^*$, $L(\mathcal{A})(w) \sqsubseteq L(\mathcal{B})(w)$. The last $\sqsubseteq$ is the order of $\mathcal{S}$.

2.2 Coalgebraic Modeling of Semiring-Weighted Automata

Here we characterize semiring-weighted automata as instances of a generic coalgebraic model of branching systems—so-called (T, F)-*systems* with parameters T, F [11, 14].

Definition 2.5 ((T, F)-system). Let T be a monad and F be a functor, both on the category **Sets** of sets and functions. A (T, F)-*system* is a triple

$$\mathcal{X} = \big(\, X,\ s\colon \{\bullet\} \to TX,\ c\colon X \to TFX \,\big)$$

of a set X (the *state space*), and functions s (the *initial states*) and c (the *dynamics*).

This modeling is coalgebraic [17] in the sense that c is so-called a TF-coalgebra. In the definition we have two parameters T and F. Let us forget about their categorical structures (a *monad* or a *functor*) for a moment and think of them simply as constructions on sets. Intuitively speaking, T specifies what kind of *branching* the systems in question exhibit; and F specifies a type of *linear-time behaviors*. Here are some examples; in the example $F = 1 + \Sigma \times (\_)$ the only element of 1 is denoted by ✓ (i.e. $1 = \{✓\}$).

T	"branching"	F	"linear-time behavior"
$\mathcal{P}$	non-deterministic	$1 + \Sigma \times (\_)$	$\to$✓ or $\xrightarrow{a}$ (where $a \in \Sigma$)
$\mathcal{D}$	probabilistic	$(\Sigma + (\_))^*$	words over terminals ($a \in \Sigma$)
$\mathcal{M}_{\mathcal{S}}$	$\mathcal{S}$-weighted		& nonterminals, suited for CFG [13]

The above examples of a monad T—the *powerset monad* $\mathcal{P}$, the *subdistribution monad* $\mathcal{D}$, and the $\mathcal{S}$-*multiset monad* $\mathcal{M}_{\mathcal{S}}$ for $\mathcal{S}$—are described as follows.

$$\begin{array}{ll} \mathcal{P}X = \{X' \mid X' \subseteq X\} & \mathcal{D}X = \{f : X \to [0,1] \mid \sum_{x\in X} f(x) \le 1\} \\ \mathcal{M}_{\mathcal{S}}X = \{f : X \to S \mid \mathrm{supp}(f) \text{ is countable}\} & \end{array} \tag{2}$$

Here $\mathrm{supp}(f) = \{x \in X \mid f(x) \neq 0_{\mathcal{S}}\}$. Countable support in $\mathcal{M}_{\mathcal{S}}$ is a technical requirement so that composition $\odot$ of Kleisli arrows is well-defined (Def. 2.7).

A (T, F)-system is a state-based system with T-branching and F-linear-time behaviors. For example, when $T = \mathcal{P}$ and $F = 1 + \Sigma \times (\_)$, $s\colon \{\bullet\} \to \mathcal{P}X$ represents initial states and $c\colon X \to \mathcal{P}(1 + \Sigma \times X)$ represents one-step transitions—$✓ \in c(x)$ means x

is accepting ($x \to \checkmark$), and $(a, x') \in c(x)$ means there is a transition $x \xrightarrow{a} x'$. Overall, a $(\mathcal{P}, 1 + \Sigma \times (\_))$-system is nothing but a nondeterministic automaton.

Analogously we obtain the following, by the definition of $\mathcal{M_S}$ in (2).

Proposition 2.6 (weighted automata as (T, F)-systems). Let $\mathcal{S}$ be a comm. cppo-semiring. There is a bijective correspondence between: 1) $\mathcal{S}$-weighted automata (Def. 2.3); and 2) $\big(\mathcal{M_S}, 1 + \Sigma \times (\_)\big)$-systems whose state spaces are countable. Concretely, an $\mathcal{S}$-weighted automaton $\mathcal{A} = (Q, \Sigma, M, \alpha, \beta)$ gives rise to an $\big(\mathcal{M_S}, 1 + \Sigma \times (\_)\big)$-system $\mathcal{X_A} = (Q, s_\mathcal{A}, c_\mathcal{A})$ defined as follows. $s_\mathcal{A} \colon \{\bullet\} \to \mathcal{M_S}Q$ is given by $s_\mathcal{A}(\bullet)(x) = \alpha_x$; and $c_\mathcal{A} \colon Q \to \mathcal{M_S}(1 + \Sigma \times Q)$ is given by $c_\mathcal{A}(x)(\checkmark) = \beta_x$ and $c_\mathcal{A}(x)(a, y) = M(a)_{x,y}$. □

2.3 Coalgebraic Theory of Traces and Simulations

We review the theory of traces and simulations in [11, 14] that is based on (T, F)-systems. In presentation we restrict to $T = \mathcal{M_S}$ and $F = 1 + \Sigma \times (\_)$ for simplicity.

Kleisli Arrows. One notable success of coalgebra was a uniform characterization, in terms of the same categorical diagram, of *bisimulations* for various kinds of systems (nondeterministic, probabilistic, etc.) [17]. This works quite well for branching-time process semantics. For linear-time semantics—i.e. trace semantics—it is noticed in [25] that so-called a *Kleisli category*, in place of the category **Sets**, gives a suitable base category for coalgebraic treatment. This idea—replacing functions $X \to Y$ with *Kleisli arrows* $X \twoheadrightarrow Y$ and drawing the same diagrams—led to the development in [11, 12, 14] of an extensive theory of traces and simulations. The notion of Kleisli arrow is parametrized by a monad T: a T-Kleisli arrow $X \twoheadrightarrow_T Y$ (or simply $X \twoheadrightarrow Y$) is defined to be a function $X \to TY$, hence represents a "T-branching function from X to Y."

We restrict to $T = \mathcal{M_S}$ for simplicity of presentation. An $\mathcal{M_S}$-Kleisli arrow $f \colon X \twoheadrightarrow Y$ below is "an $\mathcal{S}$-weighted function from X to Y." In particular, for each $x \in X$ and $y \in Y$ it assigns a *weight* $f(x)(y) \in \mathcal{S}$.

Definition 2.7 (Kleisli arrow). Let X, Y be sets. An *$\mathcal{M_S}$-Kleisli arrow* (or simply a *Kleisli arrow*) from X to Y, denoted by $X \twoheadrightarrow Y$, is a function from X to $\mathcal{M_S}Y$.

We list some special Kleisli arrows: η_X, $g \odot f$ and Jf.

- For each set X, the *unit arrow* $\eta_X : X \twoheadrightarrow X$ is given by: $\eta(x)(x) = 1_\mathcal{S}$; and $\eta(x)(x') = 0_\mathcal{S}$ for $x' \neq x$. Here $0_\mathcal{S}$ and $1_\mathcal{S}$ are units in the semiring $\mathcal{S}$.
- For consecutive Kleisli arrows $f : X \twoheadrightarrow Y$ and $g : Y \twoheadrightarrow Z$, their *composition* $g \odot f : X \twoheadrightarrow Z$ is given as follows: $(g \odot f)(x)(z) := \sum_{y \in \mathrm{supp}(f(x))} f(x)(y) \times_\mathcal{S} g(y)(z)$. Since $\mathrm{supp}(f(x))$ is countable, the above sum in a cppo-semiring $\mathcal{S}$ is well-defined.
- For a (usual) function $f : X \to Y$, its *lifting* to a Kleisli arrow $Jf : X \twoheadrightarrow Y$ is given by $Jf = \eta_Y \circ f$. Here we identified $\eta_Y : Y \twoheadrightarrow Y$ with a function $\eta_Y : Y \to \mathcal{M_S}Y$.

Categorically: the first two (η and $\odot$) organize Kleisli arrows as the *Kleisli category* $\mathcal{K}\ell(\mathcal{M_S})$; and the third gives a functor $J \colon \mathbf{Sets} \to \mathcal{K}\ell(\mathcal{M_S})$ that is identity on objects.

In Prop. 2.6 we characterized an $\mathcal{S}$-weighted automaton $\mathcal{A}$ in coalgebraic terms. Using Kleisli arrows it is presented as a triple

$$\mathcal{X_A} = \big(\, Q,\ s_\mathcal{A} \colon \{\bullet\} \twoheadrightarrow Q,\ c_\mathcal{A} \colon Q \twoheadrightarrow 1 + \Sigma \times Q \,\big) . \tag{3}$$

Generic Trace Semantics. In [14], for monads T with a suitable order, a final coalgebra in $\mathcal{K}\ell(T)$ is identified. It (somehow interestingly) coincides with an initial algebra in **Sets**. Moreover, the universality of this final coalgebra is shown to capture natural notions of (finite) *trace semantics* for a variety of branching systems—i.e. for different T and F. What is important for the current work is the fact that the weighted language $L(A)$ in (1) is an instance of this generic trace semantics, as we will show in Thm. 2.10.

We shall state the results in [14] on coalgebraic traces, restricting again to $T = \mathcal{M}_{\mathcal{S}}$ and $F = 1 + \Sigma \times (\_)$ for simplicity. In the diagram (4) on the right, composition of Kleisli arrows are given by $\odot$ in Def. 2.7; J on the right is the lifting in Def. 2.7; and nil and cons are the obvious constructors of words in Σ^*. The top arrow $1+\Sigma\times(\mathsf{tr}(c))$ is the functor $1+\Sigma\times(\_)$ on **Sets**, lifted to the Kleisli category $\mathcal{K}\ell(\mathcal{M}_{\mathcal{S}})$, and applied to the Kleisli arrow $\mathsf{tr}(c)$; its concrete description is found in Def. A.4.

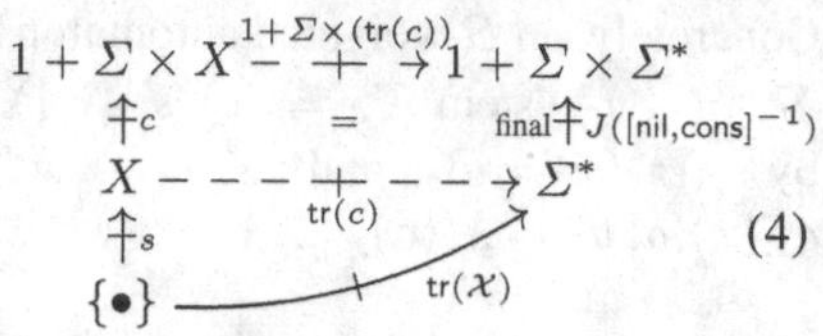

(4)

Theorem 2.8 (final coalgebra in $\mathcal{K}\ell(\mathcal{M}_{\mathcal{S}})$). Given any set X and any Kleisli arrow $c\colon X \nrightarrow 1 + \Sigma \times X$, there exists a unique Kleisli arrow $\mathsf{tr}(c)$ that makes the top square in the diagram (4) commute. □

Definition 2.9 ($\mathsf{tr}(\mathcal{X})$). Given an $(\mathcal{M}_{\mathcal{S}}, 1 + \Sigma \times (\_))$-system $\mathcal{X} = (X, s, c)$ (this is on the left in the diagram (4)), its component c induces an arrow $\mathsf{tr}(c)\colon X \nrightarrow \Sigma^*$ by Thm. 2.8. We define $\mathsf{tr}(\mathcal{X})$ to be the composite $\mathsf{tr}(c) \odot s$ (the bottom triangle in the diagram (4)), and call it the *trace semantics* of $\mathcal{X}$.

Theorem 2.10 (weighted language as trace semantics). Let $\mathcal{A}$ be an $\mathcal{S}$-weighted automaton. For $\mathcal{X}_{\mathcal{A}} = (Q, s_{\mathcal{A}}, c_{\mathcal{A}})$ induced by $\mathcal{A}$ in (3), its trace semantics $\mathsf{tr}(\mathcal{X}_{\mathcal{A}})\colon \{\bullet\} \nrightarrow \Sigma^*$—identified with a function $\{\bullet\} \to \mathcal{M}_{\mathcal{S}}\Sigma^*$, hence with a function $\Sigma^* \to \mathcal{S}$—coincides with the weighted language $L(\mathcal{A})\colon \Sigma^* \to \mathcal{S}$ in (1). □

In the last theorem we need that Σ^* is countable; this is why we assumed that Σ is countable in Def. 2.3. Henceforth we do not distinguish $L(\mathcal{A})$ and $\mathsf{tr}(\mathcal{X}_{\mathcal{A}})\colon \{\bullet\} \nrightarrow \Sigma^*$.

Forward and Backward *Kleisli* Simulations. In [11], the classic results in [23] on forward and backward simulations—for (nondeterministic) labeled transition systems—are generalized to (T, F)-systems. Specifically, fwd./bwd. simulations are characterized as *lax/oplax coalgebra homomorphisms* in a Kleisli category; and *soundness*—their existence witnesses trace inclusion—is proved once for all in a general categorical setting.

As before, we present those notions and results in [11] restricting to $T = \mathcal{M}_{\mathcal{S}}$ and $F = 1+\Sigma\times(\_)$. If $T = \mathcal{P}$ and $F = 1+\Sigma\times(\_)$ they instantiate to the results in [23].

Definition 2.11 (Kleisli simulation). Let $\mathcal{X} = (X, s, c)$ and $\mathcal{Y} = (Y, t, d)$ be $(\mathcal{M}_{\mathcal{S}}, 1+\Sigma \times (\_))$-systems (cf. Def. 2.5, Prop. 2.6 and (3)).

1. A *forward (Kleisli) simulation* from $\mathcal{X}$ to $\mathcal{Y}$ is a Kleisli arrow $f : Y \nrightarrow X$ such that $s \sqsubseteq f \odot t$ and $c \odot f \sqsubseteq (1 + \Sigma \times f) \odot d$. See Fig. 1.
2. A *backward simulation* from $\mathcal{X}$ to $\mathcal{Y}$ is a Kleisli arrow $b : X \nrightarrow Y$ such that $s \odot b \sqsubseteq t$ and $(1 + \Sigma \times b) \odot c \sqsubseteq d \odot b$.
3. A *forward-backward simulation* from $\mathcal{X}$ to $\mathcal{Y}$ consists of: a (T, F)-system $\mathcal{Z}$; a fwd. simulation f from $\mathcal{X}$ to $\mathcal{Z}$; and a bwd. simulation b from $\mathcal{Z}$ to $\mathcal{Y}$.

4. A *backward-forward simulation* from $\mathcal{X}$ to $\mathcal{Y}$ consists of: a (T, F)-system $\mathcal{Z}$; a bwd. simulation b from $\mathcal{X}$ to $\mathcal{Z}$; and a fwd. simulation f from $\mathcal{Z}$ to $\mathcal{Y}$.

$$
\begin{array}{c|c|c|c}
\begin{array}{c}
FX \xleftarrow{Ff} FY \\
{\scriptstyle c}\uparrow\ \sqsubseteq\ \uparrow{\scriptstyle d} \\
X \longleftarrow\!\!+\!f\!\text{-} Y \\
{\scriptstyle s}\nwarrow\ \sqsubseteq\ \nearrow{\scriptstyle t} \\
\{\bullet\}
\end{array}
&
\begin{array}{c}
FX \xrightarrow{Fb} FY \\
{\scriptstyle c}\uparrow\ \sqsubseteq\ \uparrow{\scriptstyle d} \\
X \text{-}b\!\!+\!\!\longrightarrow Y \\
{\scriptstyle s}\nwarrow\ \sqsubseteq\ \nearrow{\scriptstyle t} \\
\{\bullet\}
\end{array}
&
\begin{array}{c}
FX \xleftarrow{Ff} FZ \xrightarrow{Fb} FY \\
{\scriptstyle c}\uparrow\ \sqsubseteq\ \uparrow{\scriptstyle e}\ \sqsubseteq\ \uparrow{\scriptstyle d} \\
X \longleftarrow\!\!+\!f\!\text{-} Z \text{-}b\!\!+\!\!\longrightarrow Y \\
{\scriptstyle s}\nwarrow\ \sqsubseteq\ \uparrow{\scriptstyle u}\ \sqsubseteq\ \nearrow{\scriptstyle t} \\
\{\bullet\}
\end{array}
&
\begin{array}{c}
FX \xrightarrow{Fb} FZ \xleftarrow{Ff} FY \\
{\scriptstyle c}\uparrow\ \sqsubseteq\ \uparrow{\scriptstyle e}\ \sqsubseteq\ \uparrow{\scriptstyle d} \\
X \text{-}b\!\!+\!\!\longrightarrow Z \longleftarrow\!\!+\!f\!\text{-} Y \\
{\scriptstyle s}\nwarrow\ \sqsubseteq\ \uparrow{\scriptstyle u}\ \sqsubseteq\ \nearrow{\scriptstyle t} \\
\{\bullet\}
\end{array}
\\
\text{fwd. sim.} & \text{bwd. sim.} & \text{fwd.-bwd. sim.} & \text{bwd.-fwd. sim.}
\end{array}
$$

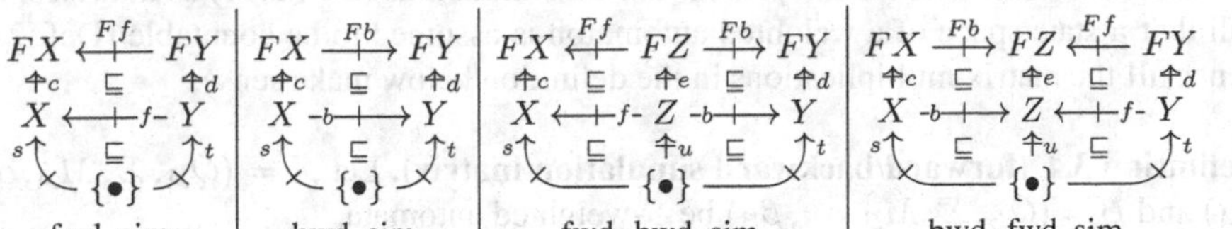

Fig. 1. Kleisli simulations (here $F = 1 + \Sigma \times (\_)$)

We write $\mathcal{X} \sqsubseteq_{\mathbf{F}} \mathcal{Y}$, $\mathcal{X} \sqsubseteq_{\mathbf{B}} \mathcal{Y}$, $\mathcal{X} \sqsubseteq_{\mathbf{FB}} \mathcal{Y}$ or $\mathcal{X} \sqsubseteq_{\mathbf{FB}} \mathcal{Y}$ if there exists a forward, backward, forward-backward, or backward-forward simulation, respectively.

(Generic) soundness is proved using the maximality of $\mathrm{tr}(c)$ in (4) among (op)lax coalgebra homomorphisms, arguing in the language of enriched category theory [11].

Theorem 2.12 (soundness). Let $\mathcal{X}$ and $\mathcal{Y}$ be $(\mathcal{M}_{\mathcal{S}}, 1 + \Sigma \times (\_))$-systems. Each of the following yields $\mathrm{tr}(\mathcal{X}) \sqsubseteq \mathrm{tr}(\mathcal{Y})$: $\{\bullet\} \twoheadrightarrow \Sigma^*$ (cf. Def. 2.9).

1. $\mathcal{X} \sqsubseteq_{\mathbf{F}} \mathcal{Y}$ 2. $\mathcal{X} \sqsubseteq_{\mathbf{B}} \mathcal{Y}$ 3. $\mathcal{X} \sqsubseteq_{\mathbf{FB}} \mathcal{Y}$ 4. $\mathcal{X} \sqsubseteq_{\mathbf{BF}} \mathcal{Y}$ □

Theorem 2.13 (completeness). The converse of soundness holds for backward-forward simulations. That is: $\mathrm{tr}(\mathcal{X}) \sqsubseteq \mathrm{tr}(\mathcal{Y})$ implies $\mathcal{X} \sqsubseteq_{\mathbf{BF}} \mathcal{Y}$.   □

3   Simulation Matrices for Semiring-Weighted Automata

In this section we fix parameters $T = \mathcal{M}_{\mathcal{S}}$ and $F = 1 + \Sigma \times (\_)$ in the generic theory in §2.3 and rephrase the coalgebraic framework in terms of matrices (whose entries are taken from $\mathcal{S}$). Specifically: Kleisli arrows become matrices; and Kleisli simulations become matrices subject to certain linear inequalities. Such matrix representations ease implementation, a feature we will exploit in later sections.

Recall that a Kleisli arrow $A \twoheadrightarrow B$ is a function $A \to \mathcal{M}_{\mathcal{S}}B$ (Def. 2.7).

Definition 3.1 (matrix representation M_f). Given a Kleisli arrow $f \colon A \twoheadrightarrow B$, its *matrix representation* $M_f \in S^{A \times B}$ is given by $(M_f)_{x,y} = f(x)(y)$.

In what follows we shall use the notations f and M_f interchangeably.

Lemma 3.2. Let $f, f' \colon A \twoheadrightarrow B$ and $g \colon B \twoheadrightarrow C$ be Kleisli arrows. We have $f \sqsubseteq f'$ if and only if $M_f \sqsubseteq M_{f'}$. Here the former $\sqsubseteq$ is between $\mathcal{M}_{\mathcal{S}}$-Kleisli arrows, and the latter order $\sqsubseteq$ is between matrices, defined entrywise. Moreover $M_{g \odot f} = M_f M_g$, computed by matrix multiplication.   □

The correspondence from $\Lambda \xrightarrow{f} B$ to $1 \mid \Sigma \times \Lambda \xrightarrow{1 + \Sigma \times f} 1 \mid \Sigma \times B$—used in (4) and in Fig. 1—can be described using matrices, too. Details are in Appendix A.2.

Lemma 3.3. Let $f \colon A \twoheadrightarrow B$ be a Kleisli arrow and M_f be its matrix representation. Then the matrix representation $M_{1+\Sigma \times f}$ is given by $I_1 \oplus (I_\Sigma \otimes M_f) \in S^{(1+\Sigma \times A) \times (1+\Sigma \times B)}$, where $\oplus$ and $\otimes$ denote *coproduct* and *the Kronecker product* of matrices: $X \oplus Y = \left(\begin{smallmatrix} X & O \\ O & Y \end{smallmatrix} \right)$ and $(x_{i,j})_{i,j} \otimes Y = (x_{i,j} Y)_{i,j}$.   □

This description of M_{Ff} generalizes from $F = 1 + \Sigma \times (\_)$ to any polynomial functor F, inductively on the construction of F. In this paper the generality is not needed.

Using Lem. 3.2–3.3, we can present Kleisli simulations (Def. 2.11) as matrices. Recall that a state space of a weighted automaton is assumed to be countable (Def. 2.3); hence all the matrix multiplications in the definition below make sense.

Definition 3.4 (forward/backward simulation matrix). Let $\mathcal{A} = (Q_{\mathcal{A}}, \Sigma, M_{\mathcal{A}}, \alpha_{\mathcal{A}}, \beta_{\mathcal{A}})$ and $\mathcal{B} = (Q_{\mathcal{B}}, \Sigma, M_{\mathcal{B}}, \alpha_{\mathcal{B}}, \beta_{\mathcal{B}})$ be $\mathcal{S}$-weighted automata.

- A matrix $X \in \mathcal{S}^{Q_{\mathcal{B}} \times Q_{\mathcal{A}}}$ is a *forward simulation matrix* from $\mathcal{A}$ to $\mathcal{B}$ if $\alpha_{\mathcal{A}} \sqsubseteq \alpha_{\mathcal{B}} X \ , \quad X \cdot M_{\mathcal{A}}(a) \sqsubseteq M_{\mathcal{B}}(a) \cdot X \quad (\forall a \in \Sigma) \ , \quad \text{and} \quad X \beta_{\mathcal{A}} \sqsubseteq \beta_{\mathcal{B}}$.
- A matrix $X \in \mathcal{S}^{Q_{\mathcal{A}} \times Q_{\mathcal{B}}}$ is a *backward simulation matrix* from $\mathcal{A}$ to $\mathcal{B}$ if $\alpha_{\mathcal{A}} X \sqsubseteq \alpha_{\mathcal{B}} \ , \quad M_{\mathcal{A}}(a) \cdot X \sqsubseteq X \cdot M_{\mathcal{B}}(a) \quad (\forall a \in \Sigma) \ , \quad \text{and} \quad \beta_{\mathcal{A}} \sqsubseteq X \beta_{\mathcal{B}}$.

The requirements on X are obtained by first translating Fig. 1 into matrices, and then breaking them up into smaller matrices using Lem. 3.3. It is notable that the requirements are given in the form of *linear inequalities*, a format often used in constraint solvers. Solving them is a topic of extensive research efforts that include [2, 6]. This fact becomes an advantage in implementing search algorithms, as we see later.

We also note that *forward* and *backward* simulation matrices have different dimensions. This difference comes from the different directions of arrows in Fig. 1.

Theorem 3.5. Let $\mathcal{A}$ and $\mathcal{B}$ be $\mathcal{S}$-weighted automata. There is a bijective correspondence between: 1) forward simulation matrices from $\mathcal{A}$ to $\mathcal{B}$; and 2) forward Kleisli simulations from $\mathcal{X}_{\mathcal{A}}$ to $\mathcal{X}_{\mathcal{B}}$. The same holds for the backward variants. □

In what follows we write $\sqsubseteq_{\mathbf{F}}, \sqsubseteq_{\mathbf{B}}$ also between $\mathcal{S}$-weighted automata. Thm. 3.5 yields: $\mathcal{A} \sqsubseteq_{\mathbf{F}} \mathcal{B}$ if and only if there is a forward simulation matrix.

Here is our core result; the rest of the paper is devoted to its application.

Corollary 3.6 (soundness of simulation matrices). Let $\mathcal{A}$ and $\mathcal{B}$ be $\mathcal{S}$-weighted automata. Existence of a forward (or backward) simulation matrix from $\mathcal{A}$ to $\mathcal{B}$—i.e. $\mathcal{A} \sqsubseteq_{\mathbf{F}} \mathcal{B}$ or $\mathcal{A} \sqsubseteq_{\mathbf{B}} \mathcal{B}$—witnesses language inclusion $L(\mathcal{A}) \sqsubseteq L(\mathcal{B})$.

Proof.
$$\begin{aligned} &\exists\,(\text{fwd./bwd. simulation matrix from } \mathcal{A} \text{ to } \mathcal{B}) \\ \overset{\text{Thm. 3.5}}{\Longleftrightarrow}\quad &\exists\,(\text{fwd./bwd. Kleisli simulation from } \mathcal{X}_{\mathcal{A}} \text{ to } \mathcal{X}_{\mathcal{B}}) \\ \overset{\text{Thm. 2.12}}{\Longrightarrow}\quad &\mathsf{tr}(\mathcal{X}_{\mathcal{A}}) \sqsubseteq \mathsf{tr}(\mathcal{X}_{\mathcal{B}}) \quad \overset{\text{Thm. 2.10}}{\Longleftrightarrow} \quad L(\mathcal{A}) \sqsubseteq L(\mathcal{B}) \ . \end{aligned}$$
□

It is classic to represent nondeterministic automata by Boolean matrices. This corresponds to the special case $\mathcal{S} = \mathcal{B}$ (the Boolean semiring) of the current framework; and a simulation matrix becomes the same thing as a (relational) simulation in [23].

Remark 3.7. The *opposite* of an $\mathcal{S}$-weighted automaton $\mathcal{A}=(Q, \Sigma, M, \alpha, \beta)$—obtained by reversing transitions and swapping initial/final states—can be naturally defined by matrix transpose, that is, ${}^t\!\mathcal{A} := (Q, \Sigma, {}^tM, {}^t\beta, {}^t\alpha)$. It is easy to see that: if X is a fwd. simulation matrix from $\mathcal{A}$ to $\mathcal{B}$, then ${}^tX$ is a bwd. simulation matrix from ${}^t\!\mathcal{A}$ to ${}^t\mathcal{B}$.

4 Forward and Backward Partial Execution

We have four different notions of simulation (Def. 2.11): fwd., bwd., fwd.-bwd., and bwd.-fwd. Our view on these is as (possibly finitary) witnesses of language inclusion.

The combined ones (fwd.-bwd. and bwd.-fwd.) subsume the one-direction ones (fwd. and bwd.)—simply take the identity arrow as one of the two simulations required. Moreover, bwd.-fwd. is complete (Thm. 2.13). Despite these theoretical advantages, the combined simulations are generally harder to find: in addition to two simulations, we have to find an intermediate system too ($\mathcal{Z}$ in Def. 2.11). Furthermore, since language inclusion for finite $\mathcal{S}_{+,\times}$-weighted automata—models of probabilistic systems—is known to be undecidable [5], existence of a bwd.-fwd. simulation is undecidable too.

Therefore in what follows we focus on the one-directional (i.e. fwd. or bwd.) simulations as proof methods for language inclusion. They have convenient matrix presentations, too, as we saw in §3. However fwd. or bwd. simulations are not necessarily complete, by a counterexample (Example A.1) or by complexity arguments (§5.1).

In this section we introduce for semiring-weighted automata their transformations—called *forward* and *backward partial execution*—that increase the number of fwd./bwd. simulation matrices. We also prove some correctness results.

Definition 4.1 (FPE, BPE). *Forward partial execution (FPE)* is a transformation of a weighted automaton that "replaces some states with their forward one-step behaviors." Concretely, given an $\mathcal{S}$-weighted automaton $\mathcal{A} = (Q, \Sigma, M, \alpha, \beta)$ and a parameter $P \subseteq Q$, the resulting automaton $\mathcal{A}_{\mathsf{FPE},P} = (Q', \Sigma, M', \alpha', \beta')$ has a state space

$$Q' = \{\checkmark \mid \exists x \in P.\, \beta_x \neq 0_{\mathcal{S}}\} + \{(a,y) \mid \exists x \in P.\, M(a)_{x,y} \neq 0_{\mathcal{S}}\} + (Q \setminus P) \ , \quad (5)$$

replacing each $x \in P$ with its forward one-step behaviors—$\checkmark$ or (a, y)—as new states. The other data M', α', β' are suitably defined following the above intuition; see Appendix A.3. Possible patterns of transformations are illustrated in Fig. 2.

Backward partial execution (BPE) in contrast "replaces states in a parameter $P \subseteq Q$ with their backward one-step behaviors." For the same $\mathcal{A}$ as above, the resulting automaton $\mathcal{A}_{\mathsf{BPE},P} = (Q', \Sigma, M', \alpha', \beta')$ has a state space

$$Q' = \{\bullet \mid \exists x \in P.\, \alpha_x \neq 0_{\mathcal{S}}\} + \{(a,y) \mid \exists x \in P.\, M(a)_{y,x} \neq 0_{\mathcal{S}}\} + (Q \setminus P) \ ,$$

replacing each $p \in P$ with its backward one-step behaviors—(a, y) with $y \xrightarrow{a} x$, and $\bullet$ if x is initial—as new states. M', α', β' are defined in Appendix A.3. See also Fig. 2.

Roughly speaking, FPE replaces a *concrete* state $p \in P$ with an *abstract* state, such as (a, q) in Q' of (5) that is thought of as a description "a state that makes an a-transition to q." The idea comes from *partial evaluation* of a program; hence the name.

The use of FPE/BPE is as follows: we aim to establish $L(\mathcal{A}) \sqsubseteq L(\mathcal{B})$; depending on whether we search for a forward or backward simulation matrix, we apply one of FPE and BPE to each of $\mathcal{A}$ and $\mathcal{B}$, according to the above table.

We shall now state correctness properties of this strategy; proofs are in Appendix A.4. *Soundness* means that discovery of a simulation after transformation indeed witnesses

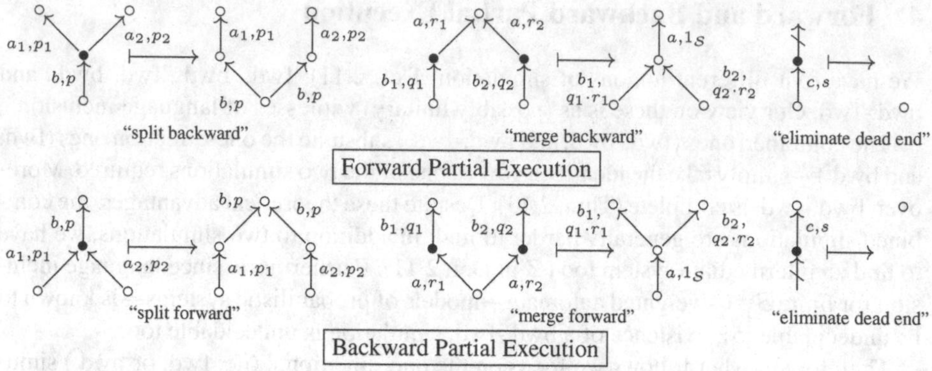

Fig. 2. Fwd./bwd. partial execution (FPE, BPE), pictorially. Black nodes need to be in P.

the language inclusion for the *original* automata. The second property—we call it *adequacy*—states that simulations that are already there are preserved by partial execution.

Theorem 4.2 (soundness of FPE/BPE). Let P, P' be arbitrary subsets. Each of the following implies $L(\mathcal{A}) \sqsubseteq L(\mathcal{B})$: 1. $\mathcal{A}_{\mathsf{FPE},P} \sqsubseteq_{\mathbf{F}} \mathcal{B}_{\mathsf{BPE},P'}$ 2. $\mathcal{A}_{\mathsf{BPE},P} \sqsubseteq_{\mathbf{B}} \mathcal{B}_{\mathsf{FPE},P'}$ □

Theorem 4.3 (adequacy of FPE/BPE). Let P, P' be arbitrary subsets. We have:
1. $\mathcal{A} \sqsubseteq_{\mathbf{F}} \mathcal{B} \Rightarrow \mathcal{A}_{\mathsf{FPE},P} \sqsubseteq_{\mathbf{F}} \mathcal{B}_{\mathsf{BPE},P'}$ 2. $\mathcal{A} \sqsubseteq_{\mathbf{B}} \mathcal{B} \Rightarrow \mathcal{A}_{\mathsf{BPE},P} \sqsubseteq_{\mathbf{B}} \mathcal{B}_{\mathsf{FPE},P'}$. □

We also show that a bigger parameter P yields a greater number of simulations. In implementation, however, a bigger P generally gives us a bigger state space which slows down search for a simulation, resulting in a trade-off situation.

Proposition 4.4 (monotonicity). Assume $P_1 \subseteq P_1'$ and $P_2 \subseteq P_2'$. We have:
1. $\mathcal{A}_{\mathsf{FPE},P_1} \sqsubseteq_{\mathbf{F}} \mathcal{B}_{\mathsf{BPE},P_2} \Rightarrow \mathcal{A}_{\mathsf{FPE},P_1'} \sqsubseteq_{\mathbf{F}} \mathcal{B}_{\mathsf{BPE},P_2'}$,
2. $\mathcal{A}_{\mathsf{BPE},P_1} \sqsubseteq_{\mathbf{B}} \mathcal{B}_{\mathsf{FPE},P_2} \Rightarrow \mathcal{A}_{\mathsf{BPE},P_1'} \sqsubseteq_{\mathbf{B}} \mathcal{B}_{\mathsf{FPE},P_2'}$. □

In fact we have a coalgebraic characterization of FPE, too, as a partial application of the functor $1 + \Sigma \times (\_)$. This characterization generalizes to a large class of (T, F)-systems, and the above correctness results can be proved generally by categorical arguments. See Appendix A.5 for details. Capturing BPE categorically is still open—it seems that BPE exists somewhat coincidentally, for the specific functor $F = 1+\Sigma\times(\_)$ for which an opposite automaton is canonically defined (cf. Rem. 3.7).

For $\mathcal{S} = \mathcal{S}_{+,\times}$ or $\mathcal{S}_{\max,+}$, we can easily see that the complement problem of language inclusion between finite $\mathcal{S}$-weighted automata is semi-decidable. Since language inclusion itself is undecidable [5, 22], language inclusion is not semidecidable either. Because existence of a simulation matrix is decidable, it can be the case that however many times we apply FPE or BPE, simulation matrices do not exist while language inclusion holds. A concrete example is found in Example A.2.

5 Simulation Matrices for Probabilistic Systems by $\mathcal{S} = \mathcal{S}_{+,\times}$

In §5 we focus on $\mathcal{S}_{+,\times}$-weighted automata which we identify as (purely) probabilistic automata (cf. Example 2.2). In §5.1 our method by simulation matrices is compared with other notions of probabilistic simulation; in §5.2 we discuss our implementation.

5.1 Other Simulation Notions for Probabilistic Systems

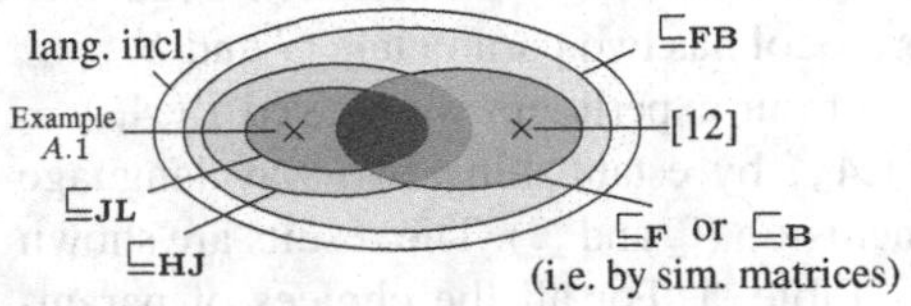

Various simulation notions have been introduced for probabilistic systems, either as a behavioral order by itself or as a proof method for language inclusion. Jonsson and Larsen's one [18] (denoted by $\sqsubseteq_{\mathbf{JL}}$) is well-known; it is shown in [12] to be a special case of Hughes and Jacobs' coalgebraic notion of simulation [16] ($\sqsubseteq_{\mathbf{HJ}}$), which in turn is a special case of fwd.-bwd. (Kleisli) simulation ($\sqsubseteq_{\mathbf{FB}}$, Def. 2.11). Comparison of all these notions (observed in [12]) is as depicted above; it follows from Thm. 2.12 that all these simulation notions are sound with respect to language inclusion.

We note that language inclusion between finite $\mathcal{S}_{+,\times}$-weighted automata is undecidable [5] while language equivalence can be determined in polynomial time [19]. The former result can account for the fact that there seem to be not many proof methods for probabilistic/quantitative language inclusion. For example, *probabilistic simulation* in [3] is possibilistic simulation between systems with both probabilistic and nondeterministic choice and not a quantitative notion like in the current study.

We also note that given finite-state $\mathcal{S}_{+,\times}$-weighted automata $\mathcal{A}$ and $\mathcal{B}$, if $\mathcal{A} \sqsubseteq_{\mathbf{F}} \mathcal{B}$ or not is decidable: existence of a solution X of the linear constraints in Def. 3.4 can be reduced to linear programming (LP) problems, and the latter are known to be decidable. The same applies to $\sqsubseteq_{\mathbf{B}}$ too.

Probabilistic systems are commonly modeled using the monad $\mathcal{D}$ (see (2))—with an explicit *normalization* condition $\sum_x d(x) \leq 1$—instead of $\mathcal{M}_{\mathcal{S}_{+,\times}}$. However there is no need to impose normalization on simulations: sometimes only "non-normalized" simulation matrices are found (Example A.3) and they are still sound.

5.2 Implementation, Experiments and Discussions

Our implementation consists of two components: $+\times$-sim and $+\times$-PE.

The program $+\times$-sim (implemented in C++) computes if a fwd. or bwd. simulation matrix X between $\mathcal{S}_{+,\times}$-weighted automata exists, and returns X if it does exist. It first combines the constraints in Def. 3.4 into a single linear inequality $A\boldsymbol{x} \leq \boldsymbol{b}$ and solves it with a linear programming solver *glpk* [1]. We note that the matrix A is sparse, having $n + anm + m$ rows, nm columns and at most $2nm + a(n^2m + nm^2)$ non-zero entries.

The program $+\times$-PE (implemented in OCaml) takes an automaton $\mathcal{A}$ and $d \in \mathbb{N}$ as input, and returns $\mathcal{A}_{\mathsf{FPE},P}$ (or $\mathcal{A}_{\mathsf{BPE},P}$, by choice) where P is chosen, by heuristics, to be $P = \{x \mid x \overbrace{\to \cdots \to}^{d} \checkmark\}$ (or $P = \{x \mid \bullet \overbrace{\to \cdots \to}^{d} x\}$, respectively).

The two programs are alternately applied to given automata, for $d = 1, 2, \ldots$, each time incrementing the parameter d for $+\times$-PE. The experiments were on an Ubuntu Linux laptop with a Core i5 2.53 GHz processor (4 cores) and 4 GB RAM.

Grades Protocol. The *grades protocol* is introduced in [19] and is used there as a benchmark: the protocol and its specification are expressed as probabilistic programs P and S; they are then translated into (purely) probabilistic automata $\mathcal{A}_{\mathsf{P}}$ and $\mathcal{A}_{\mathsf{S}}$ by a game semantics-based tool APEX [20]. By establishing $L(\mathcal{A}_{\mathsf{P}}) = L(\mathcal{A}_{\mathsf{S}})$, the protocol is shown to exhibit the same behaviors as the specification—hence is verified. The protocol has two parameters G and S.

In our experiment we proved $L(\mathcal{A}_{\mathsf{P}}) = L(\mathcal{A}_{\mathsf{S}})$ by establishing two-way language inclusion ($\sqsubseteq$ and $\sqsupseteq$). The results are shown in Table. 1. For all the choices of parameters G and S, our program $+\times$-sim was able to establish, without applying $+\times$-PE: $\mathcal{A}_{\mathsf{P}} \sqsubseteq_{\mathbf{F}} \mathcal{A}_{\mathsf{S}}$ (but not $\sqsubseteq_{\mathbf{B}}$) for the $\sqsubseteq$ direction; and $\mathcal{A}_{\mathsf{P}} \sqsupseteq_{\mathbf{B}} \mathcal{A}_{\mathsf{S}}$ (but not $\sqsupseteq_{\mathbf{F}}$) for the $\sqsupseteq$ direction. In the table, #st. and #tr. denote the numbers of states and transitions, respectively, and $|\Sigma|$ is the size of the alphabet. All these numbers are determined by APEX .

Table 1. Results for the grades protocol [19]

param. G	S	$\mathcal{A}_{\mathsf{P}}$ #st.	#tr.	$\mathcal{A}_{\mathsf{S}}$ #st.	#tr.	$\lvert\Sigma\rvert$	direction, fwd./bwd.	time (sec)	space (GB)
2	8	578	1522	130	642	11	$\mathcal{A}_{\mathsf{P}} \sqsubseteq_{\mathbf{F}} \mathcal{A}_{\mathsf{S}}$	1.77	1.21
							$\mathcal{A}_{\mathsf{P}} \sqsupseteq_{\mathbf{B}} \mathcal{A}_{\mathsf{S}}$	1.72	1.22
2	10	1102	2982	202	1202	13	$\mathcal{A}_{\mathsf{P}} \sqsubseteq_{\mathbf{F}} \mathcal{A}_{\mathsf{S}}$	9.42	4.05
							$\mathcal{A}_{\mathsf{P}} \sqsupseteq_{\mathbf{B}} \mathcal{A}_{\mathsf{S}}$	9.25	4.09
2	12	1874	5162	290	2018	15	$\mathcal{A}_{\mathsf{P}} \sqsubseteq_{\mathbf{F}} \mathcal{A}_{\mathsf{S}}$	38.60	11.51
							$\mathcal{A}_{\mathsf{P}} \sqsupseteq_{\mathbf{B}} \mathcal{A}_{\mathsf{S}}$	38.34	11.63
3	8	1923	7107	243	2163	20	$\mathcal{A}_{\mathsf{P}} \sqsubseteq_{\mathbf{F}} \mathcal{A}_{\mathsf{S}}$	44.43	12.26
							$\mathcal{A}_{\mathsf{P}} \sqsupseteq_{\mathbf{B}} \mathcal{A}_{\mathsf{S}}$	44.11	12.64
4	6	1636	7468	196	1924	23	$\mathcal{A}_{\mathsf{P}} \sqsubseteq_{\mathbf{F}} \mathcal{A}_{\mathsf{S}}$	30.28	10.39
							$\mathcal{A}_{\mathsf{P}} \sqsupseteq_{\mathbf{B}} \mathcal{A}_{\mathsf{S}}$	29.94	10.49

The table indicates that space is a bigger problem for our approach than time. In [19] four algorithms for checking language *equivalence* between $\mathcal{S}_{+,\times}$-weighted automata are implemented and compared: two are deterministic [9,27] and the other two are randomized [19]. These algorithms can process bigger problem instances (e.g. $G = 2, S = 100$ in ca. 10 sec) and, in comparison, the results in Table 1 are far from impressive. Note however that our algorithm is for language *inclusion*—an undecidable problem, unlike language *equivalence* that is in **P**, see §5.1—and hence is more general.

Crowds Protocol. Our second experiment calls for checking language *inclusion*, making the algorithms studied in [19] unapplicable. We verified some instances of the *Crowds protocol* [26] against a quantitative anonymity specification called *probable innocence* [21]. We used a general trace-based verification method in [15] for probable innocence: language inclusion $L(\mathcal{A}_{\mathsf{P}}) \sqsubseteq L(\mathcal{A}_{\mathsf{S}})$, from the model $\mathcal{A}_{\mathsf{P}}$ of a protocol in question to $\mathcal{A}_{\mathsf{P}}$'s suitable modification $\mathcal{A}_{\mathsf{S}}$, guarantees probable innocence.

Table 2. Results for the Crowds protocol

param. n	c	p_f	$\mathcal{A}_{\mathsf{P}}$ #st.	#tr.	$\mathcal{A}_{\mathsf{S}}$ #st.	#tr.	$\lvert\Sigma\rvert$	direction fwd./bwd.	time (sec)	space (GB)	d
5	1	$\frac{9}{10}$	7	44	7	56	18	$\mathcal{A}_{\mathsf{P}} \sqsubseteq_{\mathbf{F}} \mathcal{A}_{\mathsf{S}}$	52.48	0.01	2
								$\mathcal{A}_{\mathsf{P}} \sqsubseteq_{\mathbf{B}} \mathcal{A}_{\mathsf{S}}$	0.01	0.01	2
7	1	$\frac{3}{4}$	9	88	9	118	26	$\mathcal{A}_{\mathsf{P}} \sqsubseteq_{\mathbf{F}} \mathcal{A}_{\mathsf{S}}$	0.15	0.03	2
								$\mathcal{A}_{\mathsf{P}} \sqsubseteq_{\mathbf{B}} \mathcal{A}_{\mathsf{S}}$	0.02	0.01	2
10	2	$\frac{4}{5}$	12	224	12	280	54	$\mathcal{A}_{\mathsf{P}} \sqsubseteq_{\mathbf{F}} \mathcal{A}_{\mathsf{S}}$	802.47	0.35	2
								$\mathcal{A}_{\mathsf{P}} \sqsubseteq_{\mathbf{B}} \mathcal{A}_{\mathsf{S}}$	0.05	0.03	2
20	6	$\frac{4}{5}$	22	1514	22	1696	238	$\mathcal{A}_{\mathsf{P}} \sqsubseteq_{\mathbf{F}} \mathcal{A}_{\mathsf{S}}$	T/O		2
								$\mathcal{A}_{\mathsf{P}} \sqsubseteq_{\mathbf{B}} \mathcal{A}_{\mathsf{S}}$	1.32	0.78	2
30	6	$\frac{4}{5}$	32	4732	32	5112	550	$\mathcal{A}_{\mathsf{P}} \sqsubseteq_{\mathbf{F}} \mathcal{A}_{\mathsf{S}}$	S/F		
								$\mathcal{A}_{\mathsf{P}} \sqsubseteq_{\mathbf{B}} \mathcal{A}_{\mathsf{S}}$	11.84	5.99	2

The Crowds protocol has parameters n, c and p_f. In fact, for this specific protocol, a sufficient condition for probable innocence is known [26] (namely $n \geq \frac{p_f}{p_f - 1/2}(c+1)$); we used parameters that satisfy this condition. We implemented a small program that takes a choice of n, c, p_f and generates an automaton $\mathcal{A}_{\mathsf{P}}$; it is then passed to another program that generates $\mathcal{A}_{\mathsf{S}}$.

The results are in Table. 2. For each problem instance we tried both $\sqsubseteq_{\mathbf{F}}$ and $\sqsubseteq_{\mathbf{B}}$. The last column shows the final value of the parameter d for $+\times$-PE—i.e. how many times partial execution (§4) was applied.

The entry "S/F" designates that $+\times$-PE was killed because of segmentation fault caused by an oversized automaton. "T/O" means that alternate application of $+\times$-sim and $+\times$-PE did not terminate within a time limit (one hour).

We observe that backward simulation matrices were much faster to be found than forward ones. This seems to result from the shapes of the automata for this specific problem; after all it is an advantage of our fwd./bwd. approach that we can try two different directions and use the faster one. Space consumption seems again serious.

6 Simulation Matrices for $\mathcal{S}_{\max,+}$-Weighted Automata

In §6 we discuss $\mathcal{S}_{\max,+}$-weighted automata, in which weights are understood as (best-case) profit or (worst-case) cost (see Example 2.2). Such automata are studied in [7] (called Sum-*automata* there). In fact we observe that their notion of simulation—formulated in game-theoretic terms and hence called *G-simulation* here—coincides with fwd. simulation matrix. This observation is in §6.1; in §6.2 our implementation is presented.

6.1 G-Simulation by Forward Simulation Matrices

In this section we restrict to finite-state automata, in which case we can also dispose of the weight ∞. What we shall call *G-simulation* is introduced in [7], and its soundness with respect to weighted languages over *infinite-length words* $\Sigma^\omega \to [-\infty, \infty)$ is proved there. Here we adapt their definition to the current setting of finite-length words.

Definition 6.1 ($\sqsubseteq_{\mathbf{G}}$)**.** Let $\mathcal{A} = (Q_\mathcal{A}, \Sigma, M_\mathcal{A}, \alpha_\mathcal{A}, \beta_\mathcal{A})$ and $\mathcal{B} = (Q_\mathcal{B}, \Sigma, M_\mathcal{B}, \alpha_\mathcal{B}, \beta_\mathcal{B})$ be finite-state $\mathcal{S}_{\max,+}$-weighted automata. A *finite simulation game* from $\mathcal{A}$ to $\mathcal{B}$ is played by two players called *Challenger* and *Simulator*: a *strategy for Challenger* is a pair $(\rho_1 : 1 \to Q_\mathcal{A}, \tau_1 : (Q_\mathcal{A} \times Q_\mathcal{B})^+ \to 1 + \Sigma \times Q_\mathcal{A})$ of functions; a *strategy for Simulator* is a pair $(\rho_2 : Q_\mathcal{A} \to Q_\mathcal{B}, \tau_2 : (Q_\mathcal{A} \times Q_\mathcal{B})^+ \times \Sigma \times Q_\mathcal{A} \to Q_\mathcal{B})$.

A pair $(p_0 a_1 \ldots a_n p_n, q_0 a_1 \ldots a_n q_n)$ of runs on $\mathcal{A}$ and $\mathcal{B}$ is called the *outcome of strategies* (ρ_1, τ_1) *and* (ρ_2, τ_2) if:

- $\rho_1(\bullet) = p_0$, $\rho_2(p_0) = q_0$ and $\tau_1((p_0, q_0) \ldots (p_n, q_n)) = \checkmark$.
- $\tau_1((p_0, q_0) \ldots (p_i, q_i)) = (a_{i+1}, p_{i+1})$ and $\tau_2((p_0, q_0) \ldots (p_i, q_i), (a_{i+1}, p_{i+1})) = q_{i+1}$, for each $i \in [0, n-1]$.

A strategy (ρ_1, τ_1) for Challenger is *winning* if for any strategy (ρ_2, τ_2) for Simulator, their outcome (r_1, r_2) satisfies $L(\mathcal{A})(r_1) > L(\mathcal{B})(r_2)$. Here the weight $L(\mathcal{A})(r)$ of a run r is defined in the obvious way.

Finally, we write $\mathcal{A} \sqsubseteq_{\mathbf{G}} \mathcal{B}$ if there is no winning strategy for Challenger.

Theorem 6.2. Let $\mathcal{A}$ and $\mathcal{B}$ be finite-state $\mathcal{S}_{\max,+}$-weighted automata. Assume that $\mathcal{A}$ has no trap states, that is, every state has a path to $\checkmark$ whose weight is not $-\infty$. Then, $\mathcal{A} \sqsubseteq_{\mathbf{F}} \mathcal{B}$ if and only if $\mathcal{A} \sqsubseteq_{\mathbf{G}} \mathcal{B}$. □

The extra assumption can be easily enforced by eliminating trap states through backward reachability check. This does not change the (finite) weighted language.

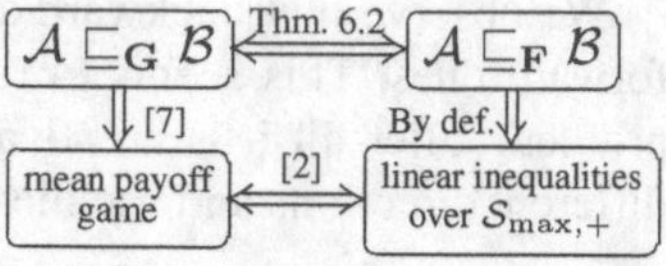

Now the situation is as shown on the right. It follows from [7] that the question if $\mathcal{A} \sqsubseteq_{\mathbf{G}} \mathcal{B}$ holds or not is reduced to a *mean payoff game* [10], whose decision problem is in $\mathbf{NP} \cap \mathbf{co\text{-}NP}$ and has a pseudo polynomial-time algorithm [28]. Moreover it is known that the decision problem of mean payoff games is equivalent to the feasibility problem of linear inequalities over $\mathcal{S}_{\max,+}$ [2]. For the latter an algorithm is proposed in [6] that is shown in [4] to be superpolynomial.

Similarly to $\mathcal{S}_{+,\times}$-weighted automata, language inclusion for $\mathcal{S}_{\max,+}$-weighted automata is undecidable [22]. We note that, by Thm. 6.2, applying FPE or BPE (§4) increases the likelihood of $\sqsubseteq_{\mathbf{G}}$ (cf. Thm. 4.3). Moreover, by exploiting symmetry of fwd. and bwd. simulation matrices (Rem. 3.7) we can define "backward G-simulation."

6.2 Implementation, Experiments and Discussions

We implemented two programs: max+-sim and max+-PE. We have seen that finding simulation matrices can be reduced to some problems that have known algorithms. Since we did not find actual software available, we implemented (in C++) the algorithm in [6] as the program max+-sim. It transforms the constraints in Def. 3.4 into an inequality $Ax \le Bx$, which in turn is made into a linear equality $A'x' = B'x'$ by adding slack variables. The last equality is solved by the algorithm in [6]. max+-PE is as in §5. It simply uses the whole state space as the parameter P.

Experiments were done on an Ubuntu Linux laptop with a Core 2 duo processor (1.40 GHz, 2 cores) and 2 GB RAM. There we faced a difficulty of finding a benchmark example: although small examples are not hard to come up with by human efforts, we could not find a good example that has parameters (like G, S in Table 1) and allows for experiments with problem instances of a varying size.

We therefore ran max+-sim for: 1) the problem if $\mathcal{A} \sqsubseteq_{\mathbf{F}} \mathcal{A}$ for randomly generated $\mathcal{A}$; and 2) the problem if $\mathcal{A} \sqsubseteq_{\mathbf{F}} \mathcal{B}$ for randomly generated $\mathcal{A}, \mathcal{B}$, and measured time and memory consumption. Although the answers are known by construction (positive for the former, and almost surely negative for the latter), actual calculation via linear inequality constraints gives us an idea about resource consumption of our simulation-based method when it is applied to real-world problems.

The outcome is as shown in Fig. 3. The parameter p is the probability with which an a-transition exists given a source state, a target state, and a character $a \in \Sigma$. Its weight is chosen from $\{0, 1, \ldots, 16\}$ subject to the uniform distribution. "Same" means checking $\mathcal{A} \sqsubseteq_{\mathbf{F}} \mathcal{A}$ and "difference" means checking $\mathcal{A} \sqsubseteq_{\mathbf{F}} \mathcal{B}$ (see above). The two problem settings resulted in comparable performance.

We observe that space consumption is not so big a problem as in the $\mathcal{S}_{+,\times}$ case (§5.2). Somehow unexpectedly, there is no big performance gap between the sparse case ($p = 0.1$) and the dense case ($p = 0.9$); in fact the sparse case consumes slightly more memory. Consumption of both time and space grows faster than linearly, which poses a question about the scalability of our approach. That said, our current implementation of the algorithm in [6] leaves a lot of room for further optimization: one possibility is use of dynamic programming (DP). After all, it is an advantage of our approach that

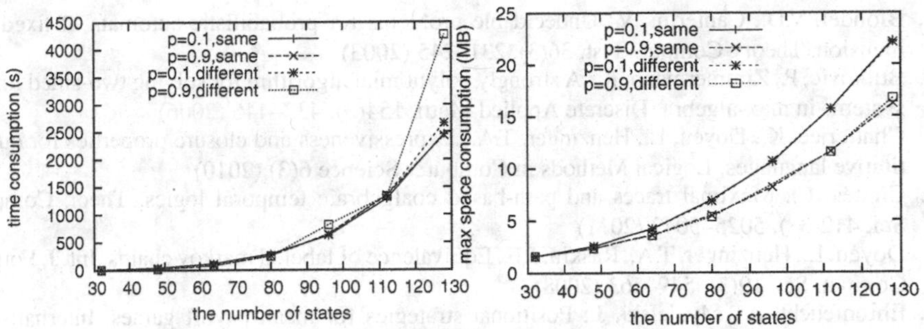

Fig. 3. Time and max space consumption for max+-sim

a simulation problem is reduced to linear inequality constraints, a subject of extensive research efforts (cf. §5.1 and §6.1).

7 Conclusions and Future Work

We introduced simulation matrices for weighted automata. While they are instances of (categorical) Kleisli simulations, their concrete presentation by matrices and linear inequalities yields concrete algorithms for simulation-based quantitative verification.

There are some directions in which the current matrix-based simulation framework can be further generalized. Generalizing F from $1 + \Sigma \times (\_)$ to any polynomial functor is mostly straightforward, as we noted after Lem. 3.3. We have not done that mainly for the space reason. Generalizing T from $\mathcal{M}_S$ (for semiring-weighted branching) to others seems more challenging. For example, in [7] weights can be given from an algebraic structure $\mathcal{S}$ in which an additive unit $0_\mathcal{S}$ does not satisfy $0_\mathcal{S} \times_\mathcal{S} x = 0_\mathcal{S}$. In this case operations like matrix multiplication becomes hard to define.

Another direction is to incorporate infinite traces, which is done in [7]. In fact our current coalgebraic backend [11] fails to do so; the work [8] will be useful here. Finally, further optimization of our implementation is obvious future work.

Acknowledgments. Thanks are due to Shota Nakagawa and the anonymous referees for useful discussions and comments. The authors are supported by Grants-in-Aid for Young Scientists (A) No. 24680001, JSPS.

References

1. The GNU linear programming kit. `http://www.gnu.org/software/glpk`
2. Akian, M., Gaubert, S., Guterman, A.E.: Tropical polyhedra are equivalent to mean payoff games. International Journal of Algebra and Computation 22(1) (2012)
3. Baier, C., Hermanns, H., Katoen, J.P.: Probabilistic weak simulation is decidable in polynomial time. Inf. Process. Lett. 89(3), 123–130 (2004)
4. Bzem, M., Nieuwenhuis, R., Rodríguez-Carbonell, E.: Exponential behaviour of the butkovic-zimmermann algorithm for solving two-sided linear systems in max-algebra. Discrete Applied Mathematics 156(18), 3506–3509 (2008)

5. Blondel, V.D., Canterini, V.: Undecidable problems for probabilistic automata of fixed dimension. Theory Comput. Syst. 36(3), 231–245 (2003)
6. Butkovic, P., Zimmermann, K.: A strongly polynomial algorithm for solving two-sided linear systems in max-algebra. Discrete Applied Math. 154(3), 437–446 (2006)
7. Chatterjee, K., Doyen, L., Henzinger, T.A.: Expressiveness and closure properties for quantitative languages. Logical Methods in Computer Science 6(3) (2010)
8. Cîrstea, C.: Maximal traces and path-based coalgebraic temporal logics. Theor. Comput. Sci. 412(38), 5025–5042 (2011)
9. Doyen, L., Henzinger, T.A., Raskin, J.F.: Equivalence of labeled markov chains. Int. J. Found. Comput. Sci. 19(3), 549–563 (2008)
10. Ehrenfeucht, A., Mycielski, J.: Positional strategies for mean payoff games. International Journal of Game Theory 8(2), 109–113 (1979)
11. Hasuo, I.: Generic forward and backward simulations. In: Baier, C., Hermanns, H. (eds.) CONCUR 2006. LNCS, vol. 4137, pp. 406–420. Springer, Heidelberg (2006)
12. Hasuo, I.: Generic forward and backward simulations II: Probabilistic simulation. In: Gastin, P., Laroussinie, F. (eds.) CONCUR 2010. LNCS, vol. 6269, pp. 447–461. Springer, Heidelberg (2010)
13. Hasuo, I., Jacobs, B.: Context-free languages via coalgebraic trace semantics. In: Fiadeiro, J.L., Harman, N.A., Roggenbach, M., Rutten, J. (eds.) CALCO 2005. LNCS, vol. 3629, pp. 213–231. Springer, Heidelberg (2005)
14. Hasuo, I., Jacobs, B., Sokolova, A.: Generic trace semantics via coinduction. Logical Methods in Computer Science 3(4) (2007)
15. Hasuo, I., Kawabe, Y., Sakurada, H.: Probabilistic anonymity via coalgebraic simulations. Theor. Comput. Sci. 411(22-24), 2239–2259 (2010)
16. Hughes, J., Jacobs, B.: Simulations in coalgebra. Theor. Comput. Sci. 327(1-2), 71–108 (2004)
17. Jacobs, B.: Introduction to coalgebra. Towards mathematics of states and observations, Draft of a book (ver. 2.0) (2012) (available online)
18. Jonsson, B., Larsen, K.G.: Specification and refinement of probabilistic processes. In: LICS, pp. 266–277. IEEE Computer Society (1991)
19. Kiefer, S., Murawski, A.S., Ouaknine, J., Wachter, B., Worrell, J.: Language equivalence for probabilistic automata. In: Gopalakrishnan, G., Qadeer, S. (eds.) CAV 2011. LNCS, vol. 6806, pp. 526–540. Springer, Heidelberg (2011)
20. Kiefer, S., Murawski, A.S., Ouaknine, J., Wachter, B., Worrell, J.: Algorithmic probabilistic game semantics—playing games with automata. Formal Methods in System Design 43(2), 285–312 (2013)
21. Konstantinos, C., Catuscia, P.: Probable innocence revisited. Theoretical Computer Science 367(1), 123–138 (2006)
22. Krob, D.: The equality problem for rational series with multiplicities in the tropical semiring is undecidable. In: Kuich, W. (ed.) ICALP 1992. LNCS, vol. 623, pp. 101–112. Springer, Heidelberg (1992)
23. Lynch, N.A., Vaandrager, F.W.: Forward and backward simulations: I. Untimed systems. Inf. Comput. 121(2), 214–233 (1995)
24. Pin, J.E.: Tropical semirings. Idempotency (Bristol, 1994), 50–69 (1998)
25. Power, J., Turi, D.: A coalgebraic foundation for linear time semantics. Electr. Notes Theor. Comput. Sci. 29, 259–274 (1999)
26. Reiter, M.K., Rubin, A.D.: Crowds: Anonymity for web transactions. ACM Trans. Inf. Syst. Secur. 1(1), 66–92 (1998)
27. Tzeng, W.G.: A polynomial-time algorithm for the equivalence of probabilistic automata. SIAM J. Comput. 21(2), 216–227 (1992)
28. Zwick, U., Paterson, M.: The complexity of mean payoff games on graphs. Theor. Comput. Sci. 158(1&2), 343–359 (1996)

A General Framework for Well-Structured Graph Transformation Systems*

Barbara König and Jan Stückrath

Universität Duisburg-Essen, Germany
{barbara_koenig,jan.stueckrath}@uni-due.de

Abstract. Graph transformation systems (GTSs) can be seen as well-structured transition systems (WSTSs), thus obtaining decidability results for certain classes of GTSs. In earlier work it was shown that well-structuredness can be obtained using the minor ordering as a well-quasi-order. In this paper we extend this idea to obtain a general framework in which several types of GTSs can be seen as (restricted) WSTSs. We instantiate this framework with the subgraph ordering and the induced subgraph ordering and apply it to analyse a simple access rights management system.

1 Introduction

Well-structured transition systems [2,9] are one of the main sources for decidability results for infinite-state systems. They equip a state space with a quasi-order, which must be a well-quasi-order (wqo) and a simulation relation for the transition relation. If a system can be seen as a WSTS, one can decide the coverability problem, i.e., the problem of verifying whether, from a given initial state, one can reach a state that covers a final state, i.e., is larger than the final state with respect to the chosen order. Often, these given final states, and all larger states, are considered to be error states and one can hence check whether an error state is reachable. Large classes of infinite-state systems are well-structured, for instance (unbounded) Petri nets and certain lossy systems. For these classes of systems the theory provides a generic backwards reachability algorithm.

A natural specification language for concurrent, distributed systems with a variable topology are graph transformation systems [19] and they usually generate infinite state spaces. In those systems states are represented by graphs and state changes by (local) transformation rules, consisting of a left-hand and a right-hand side graph. In [11] it was shown how lossy GTSs with edge contraction rules can be viewed as WSTSs with the graph minor ordering [17,18] and the theory was applied to verify a leader election protocol and a termination detection protocol [4]. The technique works for arbitrary (hyper-)graphs, i.e. the state space is not restricted to certain types of graphs. On the other hand, in order to obtain well-structuredness, we can only allow certain rule sets, for instance one has to require an edge contraction rule for each edge label.

* Research partially supported by DFG project GaReV.

P. Baldan and D. Gorla (Eds.): CONCUR 2014, LNCS 8704, pp. 467–481, 2014.

In order to make the framework more flexible we now consider other wqos, different from the minor ordering: the subgraph ordering and the induced subgraph ordering. The subgraph ordering and a corresponding WSTS were already studied in [3], but without the backwards search algorithm. Furthermore, we already mentioned the decidability result in the case of the subgraph ordering in [4], but did not treat it in detail and did not consider it as an instance of a general framework.

In contrast to the minor ordering, the subgraph ordering is not a wqo on the set of all graphs, but only on those graphs where the length of undirected paths is bounded [6]. This results in a trade-off: while the stricter order allows us to consider all possible sets of graph transformation rules in order to obtain a decision procedure, we have to make sure to consider a system where only graphs satisfying this restriction are reachable. Even if this condition is not satisfied, the procedure can yield useful partial coverability results. Also, it often terminates without excluding graphs not satisfying the restriction (this is also the case for our running example), producing exact results. We make these considerations precise by introducing *Q-restricted* WSTSs, where the order need only be a wqo on Q. In general, one wants Q to be as large as possible to obtain stronger statements.

It turns out that the results of [11] can be transferred to this new setting. Apart from the minor ordering and the subgraph ordering, there are various other wqos that could be used [8], leading to different classes of systems and different notions of coverability. In order to avoid redoing the proofs for every case, we here introduce a general framework which works for the case where the partial order can be represented by graph morphisms, which is applicable to several important cases. Especially, we state conditions required to perform the backwards search. We show that the case of the minor ordering can be seen as a special instance of this general framework and show that the subgraph and the induced subgraph orderings are also compatible. Finally we present an implementation and give runtime results. For the proofs we refer the reader to the extended version of this paper [15].

2 Preliminaries

2.1 Well-Structured Transition Systems

We define an extension to the notion of WSTS as introduced in [2,9], a general framework for decidability results for infinite-state systems, based on well-quasi-orders.

Definition 1 (Well-quasi-order and upward closure). *A quasi-order* $\leq$ *(over a set* X*) is a* well-quasi-order (wqo) *if for any infinite sequence* $x_0, x_1, x_2, \ldots$ *of elements of* X*, there exist indices* $i < j$ *with* $x_i \leq x_j$*.*

An upward-closed set *is any set* $I \subseteq X$ *such that* $x \leq y$ *and* $x \in I$ *implies* $y \in I$*. For a subset* $Y \subseteq X$*, we define its upward closure* $\uparrow Y = \{x \in X \mid \exists y \in Y : y \leq x\}$*. Then, a* basis *of an upward-closed set* I *is a set* I_B *such that* $I = \uparrow I_B$*.*

A downward-closed set, *downward closure and a basis of a downward-closed set can be defined analogously.*

The definition of wqos gives rise to properties which are important for the correctness and termination of the backwards search algorithm presented later.

Lemma 1. *Let $\leq$ be a wqo, then the following two statements hold:*

1. *Any upward-closed set I has a finite basis.*
2. *For any infinite, increasing sequence of upward-closed sets $I_0 \subseteq I_1 \subseteq I_2 \subseteq \ldots$ there exists an index $k \in \mathbb{N}$ such that $I_i = I_{i+1}$ for all $i \geq k$.*

A Q-restricted WSTS is a transition system, equipped with a quasi-order, such that the quasi-order is a (weak) simulation relation on all states and a wqo on a restricted set of states Q.

Definition 2 (Q-restricted well-structured transition system). *Let S be a set of states and let Q be a downward closed subset of S, where membership is decidable. A* Q-restricted well-structured transition system (Q-restricted WSTS) *is a transition system $\mathcal{T} = (S, \Rightarrow, \leq)$, where the following conditions hold:*

Ordering: *$\leq$ is a quasi-order on S and a wqo on Q.*
Compatibility: *For all $s_1 \leq t_1$ and a transition $s_1 \Rightarrow s_2$, there exists a sequence $t_1 \Rightarrow^* t_2$ of transitions such that $s_2 \leq t_2$.*

$$\begin{array}{ccc} t_1 & \Longrightarrow^{*} & t_2 \\ \begin{sideset}{}{}\end{sideset}\lor\!| & & \lor\!| \\ s_1 & \Longrightarrow & s_2 \end{array}$$

The presented Q-restricted WSTS are a generalization of WSTS and are identical to the classical definition, when $Q = S$. We will show how well-known results for WSTS can be transfered to Q-restricted WSTS. For Q-restricted WSTS there are two coverability problems of interest. The *(general) coverability problem* is to decide, given two states $s, t \in S$, whether there is a sequence of transitions $s \Rightarrow s_1 \Rightarrow \ldots \Rightarrow s_n$ such that $t \leq s_n$. The *restricted coverability problem* is to decide whether there is such a sequence for two $s, t \in Q$ with $s_i \in Q$ for $1 \leq i \leq n$. Both problems are undecidable in the general case (as a result of [4] and Proposition 5) but we will show that the well-known backward search for classical WSTS can be put to good use.

Given a set $I \subseteq S$ of states we denote by *Pred*(I) the set of direct predecessors of I, i.e., *Pred*$(I) = \{s \in S \mid \exists s' \in I \colon s \Rightarrow s'\}$. Additionally, we use *Pred*$_Q(I)$ to denote the restriction *Pred*$_Q(I)$ = *Pred*$(I) \cap Q$. Furthermore, we define *Pred*$^*(I)$ as the set of all predecessors (in S) which can reach some state of I with an arbitrary number of transitions. To obtain decidability results, the sets of predecessors must be computable, i.e. a so-called effective pred-basis must exist.

Definition 3 (Effective pred-basis). *A Q-restricted WSTS has an* effective pred-basis *if there exists an algorithm accepting any state $s \in S$ and returning $pb(s)$, a finite basis of $\uparrow$Pred$(\uparrow\{s\})$. It has an* effective Q-pred-basis *if there exists an algorithm accepting any state $q \in Q$ and returning $pb_Q(q)$, a finite basis of $\uparrow$Pred$_Q(\uparrow\{q\})$.*

Whenever there exists an effective pred-basis, there also exists an effective Q-pred-basis, since we can use the downward closure of Q to prove $pb_Q(q) = pb(q) \cap Q$.

Let $(S, \Rightarrow, \leq)$ be a Q-restricted WSTS with an effective pred-basis and let $I \subseteq S$ be an upward-closed set of states with finite basis I_B. To solve the general coverability problem we compute the sequence $I_0, I_1, I_2 \ldots$ where $I_0 = I_B$ and $I_{n+1} = I_n \cup pb(I_n)$. If the sequence $\uparrow I_0 \subseteq \uparrow I_1 \subseteq \uparrow I_2 \subseteq \ldots$ becomes stationary, i.e. there is an m with $\uparrow I_m = \uparrow I_{m+1}$, then $\uparrow I_m = \uparrow Pred^*(I)$ and a state of I is coverable from a state s if and only if there exists an $s' \in I_m$ with $s' \leq s$. If $\leq$ is a wqo on S, by Lemma 1 every upward-closed set is finitely representable and every sequence becomes stationary. However, in general the sequence might not become stationary if $Q \neq S$, in which case the problem becomes semi-decidable, since termination is no longer guaranteed (although correctness is).

The restricted coverability problem can be solved in a similar way, if an effective Q-pred-basis exists. Let $I^Q \subseteq S$ be an upward closed set of states with finite basis $I^Q_B \subseteq Q$. We compute the sequence $I^Q_0, I^Q_1, I^Q_2, \ldots$ with $I^Q_0 = I^Q_B$ and $I^Q_{n+1} = I^Q_n \cup pb_Q(I^Q_n)$. Contrary to the general coverability problem, the sequence $\uparrow I^Q_0 \cap Q \subseteq \uparrow I^Q_1 \cap Q \subseteq \uparrow I^Q_2 \cap Q \subseteq \ldots$ is guaranteed to become stationary according to Lemma 1. Let again m be the first index with $\uparrow I^Q_m = \uparrow I^Q_{m+1}$, and set $\Rightarrow_Q = (\Rightarrow \cap\, Q \times Q)$. We obtain the following result, of which the classical decidability result of [9] is a special case.

Theorem 1 (Coverability problems). *Let $T = (S, \Rightarrow, \leq)$ be a Q-restricted WSTS with a decidable order $\leq$.*

(i) If T has an effective pred-basis and $S = Q$, the general and restricted coverability problems coincide and both are decidable.
(ii) If T has an effective Q-pred-basis, the restricted coverability problem is decidable if Q is closed under reachability.
(iii) If T has an effective Q-pred-basis and I^Q_m is the limit as described above, then: if $s \in \uparrow I^Q_m$, then s covers a state of I^Q in $\Rightarrow$ (general coverability). If $s \notin \uparrow I^Q_m$, then s does not cover a state of I^Q in $\Rightarrow_Q$ (no restricted coverability).
(iv) If T has an effective pred-basis and the sequence I_n becomes stationary for $n = m$, then: a state s covers a state of I if and only if $s \in \uparrow I_m$.

Thus, if T is a Q-restricted WSTS and the "error states" can be represented as an upward-closed set I, then the reachability of an error state of I can be determined as described above, depending on which of the cases of Theorem 1 applies. Note that it is not always necessary to compute the limits I_m or I^Q_m, since $\uparrow I_i \subseteq \uparrow I_m$ (and $\uparrow I^Q_i \subseteq \uparrow I^Q_m$) for any $i \in \mathbb{N}_0$. Hence, if $s \in \uparrow I_i$ (or $s \in \uparrow I^Q_i$) for some i, then we already know that s covers a state of I (or of I^Q) in $\Rightarrow$.

2.2 Graph Transformation Systems

In the following we define the basics of hypergraphs and GTSs as a special form of transition systems where the states are hypergraphs and the rewriting rules are hypergraph morphisms. We prefer hypergraphs over directed or undirected graphs since they are more convenient for system modelling.

Definition 4 (Hypergraph). *Let Λ be a finite sets of edge labels and $ar\colon \Lambda \to \mathbb{N}$ a function that assigns an arity to each label. A (Λ-)*hypergraph *is a tuple (V_G, E_G, c_G, l_G^E) where V_G is a finite set of nodes, E_G is a finite set of edges, $c_G\colon E_G \to V_G^*$ is an (ordered) connection function and $l_G^E\colon E_G \to \Lambda$ is an edge labelling function. We require that $|c_G(e)| = ar(l_G^E(e))$ for each edge $e \in E_G$.*

An edge e is called incident *to a node v (and vice versa) if v occurs in $c_G(e)$.*

From now on we will often call hypergraphs simply graphs. An (elementary) *undirected path* of length n in a hypergraph is an alternating sequence $v_0, e_1, v_1, \ldots, v_{n-1}, e_n, v_n$ of nodes and edges such that for every index $1 \le i \le n$ both nodes v_{i-1} and v_i are incident to e_i and the undirected path contains all nodes and edges at most once. Note that there is no established notion of directed paths for hypergraphs, but our definition gives rise to undirected paths in the setting of directed graphs (which are a special form of hypergraphs).

Definition 5 (Partial hypergraph morphism). *Let G, G' be (Λ-)hypergraphs. A* partial hypergraph morphism *(or simply* morphism*) $\varphi\colon G \rightharpoonup G'$ consists of a pair of partial functions $(\varphi_V : V_G \rightharpoonup V_{G'}, \varphi_E : E_G \rightharpoonup E_{G'})$ such that for every $e \in E_G$ it holds that $l_G(e) = l_{G'}(\varphi_E(e))$ and $\varphi_V(c_G(e)) = c_{G'}(\varphi_E(e))$ whenever $\varphi_E(e)$ is defined. Furthermore if a morphism is defined on an edge, it must be defined on all nodes incident to it.* Total morphisms *are denoted by an arrow of the form $\to$.*

For simplicity we will drop the subscripts and write φ instead of φ_V and φ_E. We call two graphs G_1, G_2 isomorphic if there exists a total bijective morphism $\varphi : G_1 \to G_2$.

Graph rewriting relies on the notion of *pushouts.* It is known that pushouts of partial graph morphisms always exist and are unique up to isomorphism. Intuitively, for morphisms $\varphi : G_0 \rightharpoonup G_1$, $\psi : G_0 \rightharpoonup G_2$, the pushout is obtained by gluing the two graphs G_1, G_2 over the common interface G_0 and by deleting all elements which are undefined under φ or ψ (for a formal definition see [19]).

We will take pushouts mainly in the situation described in Definition 6 below, where r (the rule) is partial and connects the left-hand side L and the right-hand side R. It is applied to a graph G via a total match m. In order to ensure that the resulting morphism m' (the co-match of the right-hand side in the resulting graph) is also total, we have to require a match m to be *conflict-free* wrt. r, i.e., if there are two elements x, y of L with $m(x) = m(y)$ either $r(x), r(y)$ are both defined or both undefined. Here we consider a graph rewriting approach called the *single-pushout approach (SPO)* [7], since it relies on one pushout square, and restrict to conflict-free matches.

Definition 6 (Graph rewriting). *A* rewriting rule *is a partial morphism $r\colon L \rightharpoonup R$, where L is called left-hand and R right-hand side. A* match *(of r) is a total morphism $m\colon L \to G$, conflict-free wrt. r. Given a rule and a match, a* rewriting step *or rule application is given by a pushout diagram as shown below, resulting in the graph* H.

A graph transformation system (GTS) *is a finite set of rules* $\mathcal{R}$. *Given a fixed set of graphs* $\mathcal{G}$, *a* graph transition system *on* $\mathcal{G}$ *generated by a graph transformation system* $\mathcal{R}$ *is represented by a tuple* $(\mathcal{G}, \Rightarrow)$ *where* $\mathcal{G}$ *is the set of states and* $G \Rightarrow G'$ *if and only if* $G, G' \in \mathcal{G}$ *and* G *can be rewritten to* G' *using a rule of* $\mathcal{R}$.

$$\begin{array}{ccc} L & \xrightarrow{r} & R \\ {\scriptstyle m}\downarrow & & \downarrow{\scriptstyle m'} \\ G & \longrightarrow & H \end{array}$$

Later we will have to apply rules backwards, which means that it is necessary to compute so-called pushout complements, i.e., given r and m' above, we want to obtain G (such that m is total and conflict-free). How this computation can be performed in general is described in [10]. Note that pushout complements are not unique and possibly do not exist for arbitrary morphisms. For two partial morphisms the number of pushout complements may be infinite.

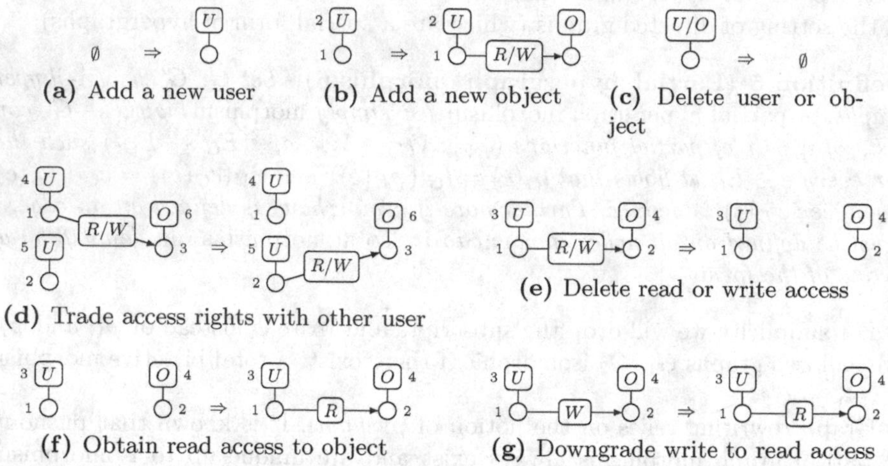
(a) Add a new user
(b) Add a new object
(c) Delete user or object
(d) Trade access rights with other user
(e) Delete read or write access
(f) Obtain read access to object
(g) Downgrade write to read access

Fig. 1. A GTS modelling a multi-user system

Example 1. To illustrate graph rewriting we model a multi-user system as a GTS (see Figure 1) inspired by [13]. A graph contains user nodes, indicated by unary U-edges, and object nodes, indicated by unary O-edges. Users can have read (R) or write (W) access rights regarding objects indicated by a (directed) edge. Note that binary edges are depicted by arrows, the numbers describe the rule morphisms and labels of the form R/W represent two rules, one with R-edges and one with W-edges.

The users and objects can be manipulated by rules for adding new users (Fig. 1a), adding new objects with read or write access associated with a user (Fig. 1b) and deleting users or objects (Fig. 1c). Both read and write access can be traded between users (Fig. 1d) or dropped (Fig. 1e). Additionally users can downgrade their write access to a read access (Fig. 1g) and obtain read access of arbitrary objects (Fig. 1f).

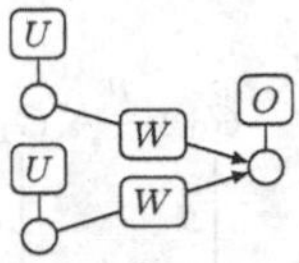

Fig. 2. An undesired state in the multi-user system

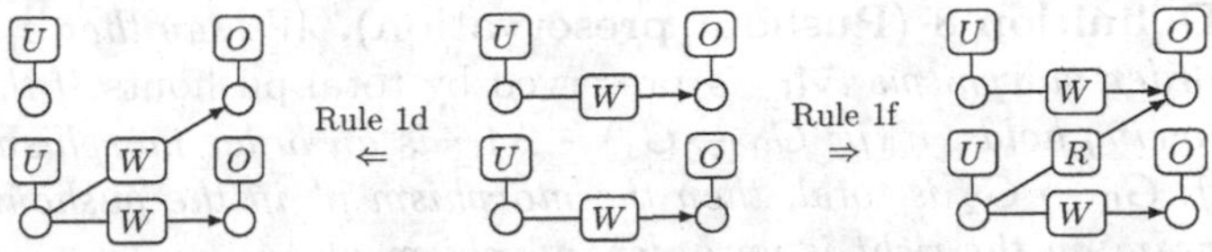

Fig. 3. Example of two rule applications

In a multi-user system there can be arbitrary many users with read access to an object, but at most one user may have write access. This means especially that any configuration of the system containing the graph depicted in Figure 2 is erroneous.

An application of the Rules 1d and 1f is shown in Figure 3. In general, nodes and edges on which the rule morphism r is undefined are deleted and nodes and edges of the right-hand side are added if they have no preimage under r. In the case of non-injective rule morphisms, nodes or edges with the same image are merged. Finally, node deletion results in the deletion of all incident edges (which would otherwise be left dangling). For instance, if Rule 1c is applied, all read/write access edges attached to the single deleted node will be deleted as well.

3 GTS as WSTS: A General Framework

In this section we state some sufficient conditions such that the coverability problems for $\mathcal{Q}$-restricted well-structured GTS can be solved in the sense of Theorem 1 (in the following we use $\mathcal{Q}$ to emphasize that $\mathcal{Q}$ is a set of graphs). We will also give an appropriate backward algorithm. The basic idea is to represent the wqo by a given class of morphisms.

Definition 7 (Representable by morphisms). *Let $\sqsubseteq$ be a quasi-order that satisfies $G_1 \sqsubseteq G_2$, $G_2 \sqsubseteq G_1$ for two graphs G_1, G_2 if and only if G_1, G_2 are isomorphic, i.e., $\sqsubseteq$ is anti-symmetric up to isomorphism.*

We call $\sqsubseteq$ representable by morphisms *if there is a class of (partial) morphisms $\mathcal{M}_\sqsubseteq$ such that for two graphs G, G' it holds that $G' \sqsubseteq G$ if and only if there is a morphism $(\mu\colon G \rightarrowtail G') \in \mathcal{M}_\sqsubseteq$. Furthermore, for $(\mu_1 : G_1 \rightarrowtail G_2), (\mu_2 : G_2 \rightarrowtail G_3) \in \mathcal{M}_\sqsubseteq$ it holds that $\mu_2 \circ \mu_1 \in \mathcal{M}_\sqsubseteq$, i.e., $\mathcal{M}_\sqsubseteq$ is closed under composition. We call such morphisms μ* order morphisms.

The intuition behind an order morphism is the following: whenever there is an order morphism from G to G', we usually assume that G' is the smaller graph that can be obtained from G by some form of node deletion, edge deletion or edge contraction. For any graphs G (which represent all larger graphs) we can now compose rules and order morphisms to simulate a co-match of a rule to some graph larger than G. However, for this construction to yield correct results, the order morphisms have to satisfy the following two properties.

Definition 8 (Pushout preservation). *We say that a set of order morphisms* $\mathcal{M}_{\sqsubseteq}$ *is* preserved by total pushouts *if the following holds: if* $(\mu\colon G_0 \rightarrowtail G_1) \in \mathcal{M}_{\sqsubseteq}$ *is an order morphism and* $g\colon G_0 \to G_2$ *is total, then the morphism* μ' *in the pushout diagram on the right is an order morphism of* $\mathcal{M}_{\sqsubseteq}$.

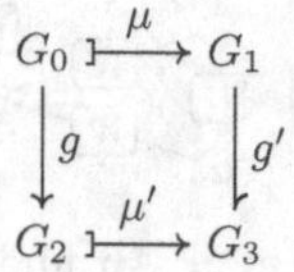

The next property is needed to ensure that every graph G, which is rewritten to a graph H larger than S, is represented by a graph G' obtained by a backward rewriting step from S, i.e. the backward step need not be applied to H.

Definition 9 (Pushout closure). *Let* $m\colon L \to G$ *be total and conflict-free wrt.* $r\colon L \rightharpoonup R$. *A set of order morphisms is called* pushout closed *if the following holds: if the diagram below on the left is a pushout and* $\mu\colon H \rightarrowtail S$ *an order morphism, then there exist graphs* R' *and* G' *and order morphisms* $\mu_R\colon R \rightarrowtail R'$, $\mu_G\colon G \rightarrowtail G'$, *such that:*

1. *the diagram below on the right commutes and the outer square is a pushout.*
2. *the morphisms* $\mu_G \circ m\colon L \to G'$ *and* $n\colon R' \to S$ *are total and* $\mu_G \circ m$ *is conflict-free wrt.* r.

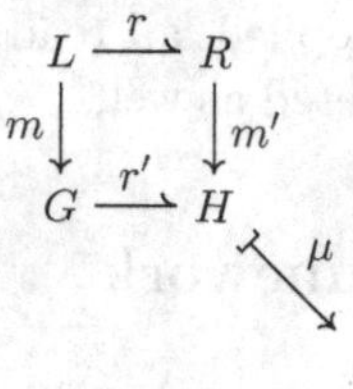

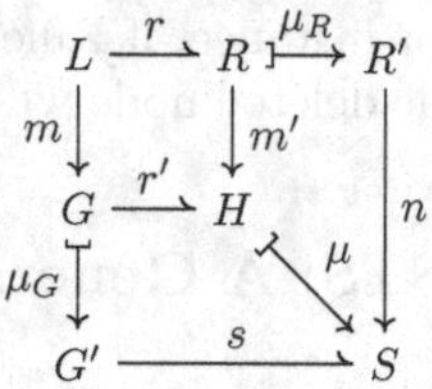

We now present a generic backward algorithm for (partially) solving both coverability problems. The procedure has two variants, which both require a GTS, an order and a set of final graphs to generate a set of minimal representatives of graphs covering a final graph. The first variant computes the sequence $I_n^{\mathcal{Q}}$ and restricts the set of graphs to ensure termination. It can be used for cases (i), (ii) and (iii) of Theorem 1, while the second variant computes I_n (without restriction) and can be used for cases (i) and (iv).

Procedure 1 (Computation of the ($\mathcal{Q}$-)pred-basis).
Input: A set $\mathcal{R}$ of graph transformation rules, a quasi-order $\sqsubseteq$ on all graphs which is a wqo on a downward-closed set $\mathcal{Q}$ and a finite set of final graphs $\mathcal{F}$, satisfying:

- The transition system generated by the rule set $\mathcal{R}$ is a $\mathcal{Q}$-restricted WSTS with respect to the order $\sqsubseteq$.
- The order $\sqsubseteq$ is representable by a class of morphisms $\mathcal{M}_{\sqsubseteq}$ (Definition 7) and this class satisfies Definitions 8 and 9.
- *Variant 1.* The set of minimal pushout complements restricted to $\mathcal{Q}$ with respect to $\sqsubseteq$ is computable, for all pairs of rules and co-matches (it is automatically finite).
  Variant 2. The set of minimal pushout complements with respect to $\sqsubseteq$ is finite and computable, for all pairs of rules and co-matches.

Preparation: Generate a new rule set $\mathcal{R}'$ from $\mathcal{R}$ in the following way: for every rule $(r : L \rightharpoonup R) \in \mathcal{R}$ and every order morphism $\mu\colon R \rightarrowtail \overline{R}$ add the rule $\mu \circ r$ to $\mathcal{R}'$. (Note that it is sufficient to take a representative $\overline{R}$ for each of the finitely many isomorphism classes, resulting in a finite set $\mathcal{R}'$.) Start with the working set $\mathcal{W} = \mathcal{F}$ and apply the first backward step.
Backward Step: Perform backward steps until the sequence of working sets $\mathcal{W}$ becomes stationary. The following substeps are performed in one backward step for each rule $(r : L \rightharpoonup R) \in \mathcal{R}'$:

1. For a graph $G \in \mathcal{W}$ compute all total morphisms $m' : R \to G$ (co-matches of R in G).
2. *Variant 1.* For each such morphism m' calculate the set $\mathcal{G}_{poc}$ of minimal pushout complement objects of m' with rule r, which are also elements of $\mathcal{Q}$. *Variant 2.* Same as Variant 1, but calculate *all* minimal pushout complements, without the restriction to $\mathcal{Q}$.
3. Add all remaining graphs in $\mathcal{G}_{poc}$ to $\mathcal{W}$ and minimize $\mathcal{W}$ by removing all graphs G' for which there is a graph $G'' \in \mathcal{W}$ with $G' \neq G''$ and $G'' \sqsubseteq G'$.

Result: The resulting set $\mathcal{W}$ contains minimal representatives of graphs from which a final state is coverable (cf. Theorem 1).

The reason for composing rule morphisms with order morphisms when doing the backwards step is the following: the graph G, for which we perform the step, might not contain a right-hand side R in its entirety. However, G can represent graphs that do contain R and hence we have to compute the effect of applying the rule backwards to all graphs represented by G. Instead of enumerating all these graphs (which are infinitely many), we simulate this effect by looking for matches of right-hand sides modulo order morphisms. We show that the procedure is correct by proving the following lemma.

Proposition 1. *Let $pb_1()$ and $pb_2()$ be a single backward step of Procedure 1 for Variant 1 and 2 respectively. For each graph S, $pb_1(S)$ is an effective $\mathcal{Q}$-pred-basis and $pb_2(S)$ is an effective pred-basis.*

4 Well-Quasi Orders for Graph Transformation Systems

4.1 Minor Ordering

We first instantiate the general framework with the minor ordering, which was already considered in [11]. The minor ordering is a well-known order on graphs, which is defined as follows: a graph G is a minor of G' whenever G can be obtained from G' by a series of node deletions, edge deletions and edge contractions, i.e. deleting an edge and merging its incident nodes according to an arbitrary partition. Robertson and Seymour showed in a seminal result that the minor ordering is a wqo on the set of all graphs [17], even for hypergraphs [18], thus case (i) of Theorem 1 applies. In [11,12] we showed that the conditions for WSTS are satisfied for a restricted set of GTS by introducing minor morphisms and proving a result analogous to Proposition 1, but only for this specific case. The resulting algorithm is a special case of both variants of Procedure 1.

Proposition 2 ([11]). *The coverability problem is decidable for every GTS if the minor ordering is used and the rule set contains edge contraction rules for each edge label.*

4.2 Subgraph Ordering

In this paper we will show that the subgraph ordering and the induced subgraph ordering satisfy the conditions of Procedure 1 for a restricted set of graphs and are therefore also compatible with our framework. For the subgraph ordering we already stated a related result (but for injective instead of conflict-free matches) in [4], but did not yet instantiate a general framework.

Definition 10 (Subgraph). *Let G_1, G_2 be graphs. G_1 is a subgraph of G_2 (written $G_1 \subseteq G_2$) if G_1 can be obtained from G_2 by a sequence of deletions of edges and isolated nodes. We call a partial morphism $\mu\colon G \rightarrowtail S$ a* subgraph morphism *if and only if it is injective on all elements on which it is defined and surjective.*

It can be shown that the subgraph ordering is representable by subgraph morphisms, which satisfy the necessary properties. Using a result from Ding [6] we can show that the set $\mathcal{G}_k$ of hypergraphs where the length of every undirected path is bounded by k, is well-quasi-ordered by the subgraph relation. A similar result was shown by Meyer for depth-bounded systems in [16]. Note that we bound undirected path lengths instead of directed path lengths. For the class of graphs with bounded directed paths there exists a sequence of graphs violating the wqo property (a sequence of circles of increasing length, where the edge directions alternate along the circle).

Since every GTS satisfies the compatibility condition of Definition 2 naturally, we obtain the following result.

Proposition 3 (WSTS wrt. the subgraph ordering). *Let k be a natural number. Every graph transformation system forms a $\mathcal{G}_k$-restricted WSTS with the subgraph ordering.*

The set of minimal pushout complements (not just restricted to $\mathcal{G}_k$) is always finite and can be computed as in the minor case.

Proposition 4. *Every $\mathcal{G}_k$-restricted well-structured GTS with the subgraph order has an effective pred-basis and the (decidability) results of Theorem 1 apply.*

By a simple reduction from the reachability problem for two counter machines, we can show that the restricted coverability problem is undecidable in the general case. Although we cannot directly simulate the zero test, i.e. negative application conditions are not possible, we can make sure that the rules simulating the zero test are applied correctly if and only if the bound k was not exceeded.

Proposition 5. *Let $k > 2$ be a natural number. The restricted coverability problem for $\mathcal{G}_k$-restricted well-structured GTS with the subgraph ordering is undecidable.*

Example 2. Now assume that an error graph is given and that a graph exhibits an error if and only if it contains the error graph as a subgraph. Then we can use Proposition 4 to calculate all graphs which lead to some error configuration.

For instance, let a multi-user system as described in Example 1 be given. Normally we have to choose a bound on the undirected path length to guarantee termination, but in this example Variant 2 of Procedure 1 terminates and we can solve coverability on the unrestricted transition system (see Theorem 1(iv)). The graph in Figure 2 represents the error in the system and by applying Procedure 1 we obtain a set of four graphs (one of which is the error graph itself), fully characterizing all predecessor graphs. We can observe that the error can only be reached from graphs already containing two W-edges going to a single object node. Hence, the error is not produced by the given rule set if we start with the empty graph and thus the system is correct.

Interestingly the backward search finds the leftmost graph below due to the depicted sequence of rule applications, which leads directly to the error graph. Thus, the error can occur even if a single user has two write access rights to an object, because of access right trading.

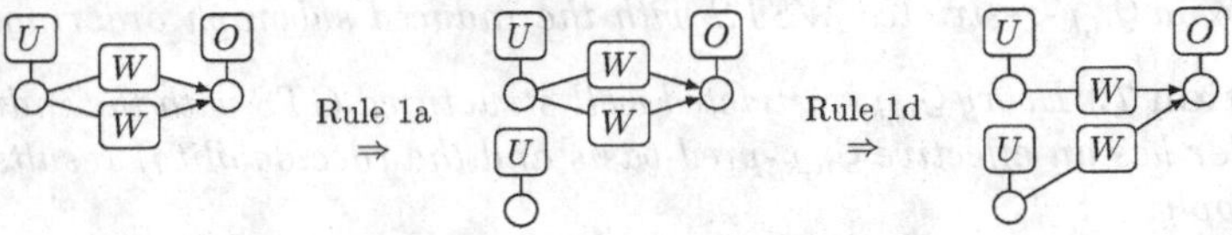

The other two graphs computed are shown below and represent states with "broken" structure (a node cannot be a user *and* an object). The left graph for instance can be rewritten to a graph larger than the left graph above, by a non-injective match of the rule in Figure 1d mapping both nodes 2 and 3 to the right node.

4.3 Induced Subgraph Ordering

As for the subgraph ordering in Section 4.2 our backward algorithm can also be applied to the induced subgraph ordering, where a graph G is considered as an induced subgraph of G' if every edge in G' connecting only nodes also present in G, is contained in G as well. Unfortunately, this ordering is not a wqo even when bounding the longest undirected path in a graph, such that we also have to bound the multiplicity of edges between two nodes. Note that this restriction is implicitly done in [6] since Ding uses simple graphs.

Furthermore, since we do not know whether the induced subgraph ordering can be extended to a wqo on (a class of) hypergraphs, we here use only *directed graphs*, where each edge is connected to a sequence of exactly two nodes. For many applications directed graphs are sufficient for modelling, also for our examples, since unary hyperedges can simply be represented by loops.

At first, this order seems unnecessary, since it is stricter than the subgraph ordering and is a wqo on a more restricted set of graphs. On the other hand, it allows us to specify error graphs more precisely, since a graph $G \in \mathcal{F}$ does not represent graphs with additional edges between nodes of G. Furthermore one could equip the rules with a limited form of negative application conditions, still retaining the compatibility condition of Definition 2.

Definition 11 (Induced subgraph). *Let G_1, G_2 be graphs. G_1 is an induced subgraph of G_2 (written $G_1 \trianglelefteq G_2$) if G_1 can be obtained from G_2 by deleting a subset of the nodes and all incident edges. We call a partial morphism $\mu\colon G \rightarrowtail S$ an* induced subgraph morphism *if and only if it is injective for all elements on which is defined, surjective, and if it is undefined on an edge e, it is undefined on at least one node incident to e.*

Proposition 6 (WSTS wrt. the induced subgraph ordering). *Let n, k be natural numbers and let $\mathcal{G}_{n,k}$ be a set of directed, edge-labelled graphs, where the longest undirected path is bounded by n and every two nodes are connected by at most k parallel edges with the same label (bounded edge multiplicity). Every GTS forms a $\mathcal{G}_{n,k}$-restricted WSTS with the induced subgraph ordering.*

Proposition 7. *Every $\mathcal{G}_{n,k}$-restricted well-structured GTS with the induced subgraph order has an effective $\mathcal{G}_{n,k}$-pred-basis and the (decidability) results of Theorem 1 apply.*

The computation of minimal pushout complements in this case is considerably more complex, since extra edges have to be added, but we also obtain additional expressiveness. In general GTS with negative application conditions do not satisfy the compatibility condition with respect to the subgraph relation, but we show in the following example, that it may still be satisfied with respect to the induced subgraph relation.

Example 3. Let the following simple rule be given, where the negative application condition is indicated by the dashed edge, i.e. the rule is applicable if and only if there is a matching only for the solid part of the left-hand side and this matching cannot be extended to match also the dashed part.

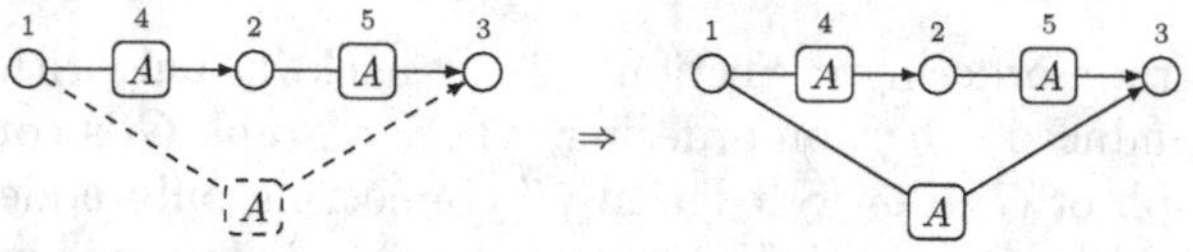

Applied to a graph containing only A-edges, this rule calculates the transitive closure and will terminate at some point. This GTS satisfies the compatibility condition wrt. the induced subgraph ordering, since for instance a directed path of length two (the left-hand side) does not represent graphs where there is an edge from the first to the last node of the graph. Therefore we can use the induced subgraph ordering and our procedure to show that a graph containing two parallel A-edges can only be reached from graphs already containing two parallel A-edges.

The principle described in the example can be extended to all negative application conditions which forbid the existence of edges but not of nodes. This is the case, because if there is no edge between two nodes of a graph, there is also no edge between these two nodes in any larger graph. Hence if there is no mapping from the negative application condition into the smaller graph, there can also be none into the larger graph. Graphs violating the negative application condition are simply not represented by the smaller graph. Hence, all graph transformation rules with such negative application conditions satisfy the compatibility condition wrt. the induced subgraph ordering. The backward step has to be modified in this case by dropping all obtained graphs which do not satisfy one of the negative application conditions.

4.4 Implementation

We implemented Procedure 1 with support for the minor ordering as well as the subgraph ordering in the tool UNCOVER. The tool is written in C++ and designed in a modular way for easy extension with further orders. The sole optimization currently implemented is the omission of all rules that are also order morphisms. It can be shown that the backward application of such rules produces only graphs which are already represented.

Table 1 shows the runtime results of different case studies, namely a leader election protocol and a termination detection protocol (in an incorrect as well as a correct version), using the minor ordering, and the access rights management protocol described in Figure 1 as well as a public-private server protocol, using the subgraph order. It shows for each case the restricted graph set $\mathcal{Q}$, the variant of the procedure used (for the minor ordering they coincide), the runtime and the number of minimal graphs representing all predecessors of error graphs.

Table 1. Runtime result for different case studies

case study	wqo	graph set $\mathcal{Q}$	variant	time	#(error graphs)
Leader election	minor	all graphs	1 / 2	3s	38
Termination detection (faulty)	minor	all graphs	1 / 2	7s	69
Termination detection (correct)	minor	all graphs	1 / 2	2s	101
Rights management	subgraph	all graphs	2	1s	4
Public-private server ($l = 5$)	subgraph	path ≤ 5	1	1s	14
Public-private server ($l = 6$)	subgraph	path ≤ 6	1	16s	16

5 Conclusion

We have presented a general framework for viewing GTSs as restricted WSTSs. We showed that the work in [11] for the minor ordering can be seen as an instance of this framework and we presented two additional instantiations, based on the subgraph ordering and the induced subgraph ordering. Furthermore we presented the management of read and write access rights as an example and discussed our implementation with very encouraging runtime results.

Currently we are working on an extension of the presented framework with rules, which can uniformly change the entire neighbourhood of nodes. In this case the computed set of predecessor graphs will be an over-approximation. More extensions are possible (possibly introducing over-approximations) and we especially plan to further investigate the integration of rules with negative application conditions as for the induced subgraph ordering. In [14] we introduced an extension with negative application conditions for the minor ordering, but still, the interplay of the well-quasi-order and conditions has to be better understood. Naturally, we plan to look for additional orders, for instance the induced minor and topological minor orderings [8] in order to see whether they can be integrated into this framework and to study application scenarios.

Related work. Related to our work is [3], where the authors use the subgraph ordering and a forward search to prove fair termination for depth-bounded systems. In [1] another wqo for well-structuring graph rewriting is considered, however only for graphs where every node has out-degree 1. It would be interesting to see whether this wqo can be integrated into our general framework. The work in [5] uses the induced subgraph ordering to verify broadcast protocols. There the rules are different from our setting: a left-hand side consists of a node and its entire neighbourhood of arbitrary size. Finally [20] uses a backwards search on graph patterns in order to verify an ad-hoc routing protocol, but not in the setting of WSTSs.

Acknowledgements. We would like to thank Roland Meyer, for giving us the idea to consider the subgraph ordering on graphs, and Giorgio Delzanno for several interesting discussions on wqos and WSTSs.

References

1. Abdulla, P.A., Bouajjani, A., Cederberg, J., Haziza, F., Rezine, A.: Monotonic abstraction for programs with dynamic memory heaps. In: Gupta, A., Malik, S. (eds.) CAV 2008. LNCS, vol. 5123, pp. 341–354. Springer, Heidelberg (2008)
2. Abdulla, P.A., Čerāns, K., Jonsson, B., Tsay, Y.: General decidability theorems for infinite-state systems. In: Proc. of LICS 1996, pp. 313–321. IEEE (1996)
3. Bansal, K., Koskinen, E., Wies, T., Zufferey, D.: Structural counter abstraction. In: Piterman, N., Smolka, S.A. (eds.) TACAS 2013. LNCS, vol. 7795, pp. 62–77. Springer, Heidelberg (2013)
4. Bertrand, N., Delzanno, G., König, B., Sangnier, A., Stückrath, J.: On the decidability status of reachability and coverability in graph transformation systems. In: Proc. of RTA 2012. LIPIcs, vol. 15, pp. 101–116. Schloss Dagstuhl, Leibniz Center for Informatics (2012)
5. Delzanno, G., Sangnier, A., Zavattaro, G.: Parameterized verification of ad hoc networks. In: Gastin, P., Laroussinie, F. (eds.) CONCUR 2010. LNCS, vol. 6269, pp. 313–327. Springer, Heidelberg (2010)

6. Ding, G.: Subgraphs and well-quasi-ordering. Jornal of Graph Theory 16, 489–502 (1992)
7. Ehrig, H., Heckel, R., Korff, M., Löwe, M., Ribeiro, L., Wagner, A., Corradini, A.: Algebraic approaches to graph transformation—part II: Single pushout approach and comparison with double pushout approach. In: Rozenberg, G. (ed.) Handbook of Graph Grammars and Computing by Graph Transformation. Foundations, vol. 1, ch. 4. World Scientific (1997)
8. Fellows, M.R., Hermelin, D., Rosamond, F.A.: Well-quasi-orders in subclasses of bounded treewidth graphs. In: Chen, J., Fomin, F.V. (eds.) IWPEC 2009. LNCS, vol. 5917, pp. 149–160. Springer, Heidelberg (2009)
9. Finkel, A., Schnoebelen, P.: Well-structured transition systems everywhere? Theoretical Computer Science 256(1-2), 63–92 (2001)
10. Heumüller, M., Joshi, S., König, B., Stückrath, J.: Construction of pushout complements in the category of hypergraphs. In: Proc. of GCM 2010 (Workshop on Graph Computation Models) (2010)
11. Joshi, S., König, B.: Applying the graph minor theorem to the verification of graph transformation systems. In: Gupta, A., Malik, S. (eds.) CAV 2008. LNCS, vol. 5123, pp. 214–226. Springer, Heidelberg (2008)
12. Joshi, S., König, B.: Applying the graph minor theorem to the verification of graph transformation systems. Technical Report 2012-01, Abteilung für Informatik und Angewandte Kognitionswissenschaft, Universität Duisburg-Essen (2012)
13. Koch, M., Mancini, L.V., Parisi-Presicce, F.: Decidability of safety in graph-based models for access control. In: Gollmann, D., Karjoth, G., Waidner, M. (eds.) ESORICS 2002. LNCS, vol. 2502, pp. 229–243. Springer, Heidelberg (2002)
14. König, B., Stückrath, J.: Well-structured graph transformation systems with negative application conditions. In: Ehrig, H., Engels, G., Kreowski, H.-J., Rozenberg, G. (eds.) ICGT 2012. LNCS, vol. 7562, pp. 81–95. Springer, Heidelberg (2012)
15. König, B., Stückrath, J.: A general framework for well-structured graph transformation systems, arXiv:1406.4782 (2014)
16. Meyer, R.: Structural Stationarity in the π-Calculus. PhD thesis, Carl-von-Ossietzky-Universität Oldenburg (2009)
17. Robertson, N., Seymour, P.: Graph minors. XX. Wagner's conjecture. Journal of Combinatorial Theory, Series B 92(2), 325–357 (2004)
18. Robertson, N., Seymour, P.: Graph minors XXIII. Nash-Williams' immersion conjecture. Journal of Combinatorial Theory Series B 100, 181–205 (2010)
19. Rozenberg, G. (ed.): Handbook of Graph Grammars and Computing by Graph Transformation. Foundations, vol. 1. World Scientific (1997)
20. Saksena, M., Wibling, O., Jonsson, B.: Graph grammar modeling and verification of ad hoc routing protocols. In: Ramakrishnan, C.R., Rehof, J. (eds.) TACAS 2008. LNCS, vol. 4963, pp. 18–32. Springer, Heidelberg (2008)

(Un)decidable Problems about Reachability of Quantum Systems

Yangjia Li and Mingsheng Ying

Tsinghua University, China and University of Technology, Sydney, Australia
liyj04@mails.tsinghua.edu.cn, Mingsheng.Ying@uts.edu.au,
yingmsh@tsinghua.edu.au

Abstract. We study the reachability problem of a quantum system modeled by a quantum automaton, namely, a set of processes each of which is formalized as a quantum unitary transformation. The reachable sets are chosen to be boolean combinations of (closed) subspaces of the state Hilbert space of the quantum system. Four different reachability properties are considered: eventually reachable, globally reachable, ultimately forever reachable, and infinitely often reachable. The main result of this paper is that all of the four reachability properties are undecidable in general; however, the last three become decidable if the reachable sets are boolean combinations without negation.

Keywords: Quantum reachability, verification, Skolem's problem, 2-counter Minsky machine, undecidability.

1 Introduction

Recently, verification of quantum systems has simultaneously emerged as an important problem from several very different fields. First, it was identified by leading physicists as one of the key steps in the simulation of many-body quantum systems [9]. Secondly, verification techniques for quantum protocols [11,3] become indispensable as quantum cryptography is being commercialised. Thirdly, verification of quantum programs [26,27] will certainly attract more and more attention, in particular after the announcement of several scalable quantum programming languages like Quipper [12].

Reachability is a fundamental issue in the verification and model-checking of both classical and probabilistic systems because a large class of verification problems can be reduced to reachability analysis [4]. Reachability of quantum systems also started to receive attention in recent years. For example, Eisert, Müller and Gogolin's notion of quantum measurement occurrence in physics [10] is essentially the reachability of null state; a certain reachability problem [25] lies at the heart of quantum control theory since the controllability of a quantum mechanical system requires that all states are reachable by choosing the Hamiltonian of the system [1]. Reachability of concurrent quantum systems modelled by quantum automata, or more generally by quantum Markov chains, was studied by the authors [28] with an application in termination analysis of (concurrent) quantum programs [16,29].

This paper is a continuation of our previous work [28,16,29], where only reachability to a single (closed) subspace of the state Hilbert space of a quantum system

P. Baldan and D. Gorla (Eds.): CONCUR 2014, LNCS 8704, pp. 482–496, 2014.

was considered. In this paper, we consider a class of much more general reachability properties; that is, we use subspaces of the state space as the basic properties (atomic propositions) about the quantum system, and then reachability properties can be defined as certain temporal logical formulas over general properties, which are formalized as boolean combinations of the subspaces. The reason for using boolean combinations rather than orthomodular lattice-theoretic combinations in the Birkhoff-von Neumann quantum logic [6] is that in applications these reachability properties will be used as a high-level specification language where boolean connectives are suitable; for example, when a physicist says that a particle will eventually enter region A or region B, the word "or" here is usually meant to be the boolean "or" but not the orthomodular "or" (see Example 1). The reachability properties that we are concerned with are:

- eventually reachability denoted by the temporal logic formula $\mathbf{F}f$;
- globally reachability denoted by $\mathbf{G}f$;
- ultimately forever reachability denoted by $\mathbf{U}f$;
- infinitely often reachability denoted by $\mathbf{I}f$,

where f is a boolean combination of the subspaces of the state Hilbert space.

We use quantum automata [14] as a formal model for (concurrent) quantum systems. Then the reachability problem can be described as: decide whether or not all the execution paths of a quantum automaton satisfy $\mathbf{F}f$, $\mathbf{G}f$, $\mathbf{U}f$, or $\mathbf{I}f$. There are two reasons for adopting this model. First, unitary operations of quantum automata are widely used as the mathematical formalism of the dynamics of closed physical systems, e.g., quantum circuits. Second, without probabilistic choices (which occur in other operations such as quantum measurements and super-operators) it can be seen more clear that the reachability problem for quantum systems is essentially more difficult than that for classical systems. In fact, we note that reachability analysis is challenging in the quantum scenario, since the state space is a continuum where some techniques that have been successfully used in the classical case will become ineffective.

1.1 Contributions of the Paper

- We prove undecidability of the above reachability problem, even with f in a very simple form containing the boolean negation. Undecidability of $\mathbf{G}f$ (globally reachable), $\mathbf{U}f$ (ultimately forever reachable) and $\mathbf{I}f$ (infinitely often reachable) comes from a straightforward reduction from the emptiness problem for quantum automata [7]. However, undecidability of $\mathbf{F}f$ (eventually reachable) requires a careful reduction from the halting problem for 2-counter Minsky machines [18]. In particular, a novel strategy is introduced in this reduction to simulate a (possibly irreversible) classical computation using a quantum automaton which is definitely reversible. These undecidability results present an impressive difference between quantum systems and classical systems because the reachability properties considered in this paper are decidable for classical systems.
- We prove decidability of the reachability problem for $\mathbf{G}f$, $\mathbf{U}f$, and $\mathbf{I}f$ with f being positive; that is, containing no negation. A key strategy in proving this decidability is to characterize how a set of states can be reached infinitely often in execution

paths of a quantum automaton. For the special case where the quantum automaton has only a single unitary operator and f is an atomic proposition, it is shown based on the Skolem-Mahler-Lech Theorem [24,17,15] that states are reached periodically, and thus the execution can be represented by a cycle graph. In general, we show that this execution graph becomes a general directed graph representing a reversible DFA (deterministic finite automaton), which can be inductively constructed.

1.2 Organization of the Paper

The main results are stated in Sec. 2 after introducing several basic definitions. In Sec. 3 we give a brief discussion about Skolem's problem and relate it to a special case of the quantum reachability problem. Then in Sec. 4, the undecidability of $\mathbf{G}f$, $\mathbf{U}f$ and $\mathbf{I}f$ is immediately proved from this connection. We also prove the undecidability of $\mathbf{F}f$ in this section using 2-counter Minsky machines. The proofs of decidable results about $\mathbf{G}f$, $\mathbf{U}f$ and $\mathbf{I}f$ for positive f and related algorithms are presented in Sec. 5. A brief conclusion is drawn in Sec. 6.

2 Basic Definitions and Main Results

2.1 A Propositional Logic for Quantum Systems

We first introduce a propositional logical language to describe boolean combinations of the subspaces of a Hilbert space. Let $\mathcal{H}$ be the state Hilbert space of a quantum system. A basic property of the system can be described by a (closed) subspace V of $\mathcal{H}$. In quantum mechanics, to check whether or not this property is satisfied, a binary (yes-no) measurement $\{P_V, P_{V^\perp}\}$ would be performed on the system's current state $|\psi\rangle$, where P_V and $P_{V^\perp}$ are the projection on V and its ortho-complement $V^\perp$, respectively. The measurement outcome is generally nondeterministic: V is considered as being satisfied in $|\psi\rangle$ with probability $\langle\psi|P_V|\psi\rangle$, and it is not satisfied with probability $\langle\psi|P_{V^\perp}|\psi\rangle = 1 - \langle\psi|P_V|\psi\rangle$. A quantitative satisfaction relation can be defined by setting a threshold $\lambda \in [0, 1]$ to the probability of satisfaction:

$$V \text{ is } (\lambda, \rhd) - \text{satisfied in } |\psi\rangle \text{ if } \langle\psi|P_V|\psi\rangle \rhd \lambda$$

where $\rhd \in \{<, \leq, >, \geq\}$. In this paper, we only consider the *qualitative* satisfaction, namely, the $(\lambda, \rhd)-$satisfaction with the threshold λ being 0 or 1. Obviously, we have:

- $V = \{|\psi\rangle \in \mathcal{H} \mid V \text{ is } (1, \geq) - \text{satisfied in } |\psi\rangle\}$;
- $V^\perp = \{|\psi\rangle \in \mathcal{H} \mid V \text{ is } (0, \leq) - \text{satisfied in } |\psi\rangle\}$.

Thus, it is reasonable to choose the set of atomic propositions to be

$$AP = \{V \mid V \text{ is a (closed) subspace of } \mathcal{H}\}.$$

Furthermore, we define a (classical) propositional logic over AP so that we can talk about, for example, that "the current state of the quantum system is in subspace U, or in V but not in W". The logical formulas are generated from AP by using boolean connectives $\neg$, $\wedge$ and $\vee$, and their semantics are inductively defined as follows: for any state $|\psi\rangle \in \mathcal{H}$,

- If $f \in AP$, then $|\psi\rangle \models f$ if $|\psi\rangle \in f$;
- $|\psi\rangle \models \neg f$ if $|\psi\rangle \models f$ does not hold;
- $|\psi\rangle \models f_1 \wedge f_2$ if $|\psi\rangle \models f_1$ and $|\psi\rangle \models f_2$;
- $|\psi\rangle \models f_1 \vee f_2$ if $|\psi\rangle \models f_1$ or $|\psi\rangle \models f_2$.

For a logical formula f, we write $\|f\|$ for the set of states that satisfy f. In general, $\|f\|$ might not be a subspace of $\mathcal{H}$. For example, for a subspace V of $\mathcal{H}$, we have:

- $\|\neg V\| = \{|\psi\rangle \in \mathcal{H} \mid V \text{ is } (1, <) - \text{satisfied in } |\psi\rangle\}$;
- $\|\neg(V^\perp)\| = \{|\psi\rangle \in \mathcal{H} \mid V \text{ is } (0, >) - \text{satisfied in } |\psi\rangle\}$.

It is clear that these classical connectives are different from their quantum counterparts interpreted as the operations in the orthomodular lattice of (closed) subspaces of $\mathcal{H}$ [6].

2.2 Reachability of Quantum Automata

Definition 1. *A quantum automaton is a 4−tuple* $\mathcal{A}=(\mathcal{H}, Act, \{U_\alpha \mid \alpha \in Act\}, \mathcal{H}_{ini})$, *where*

1. *$\mathcal{H}$ is the state Hilbert space;*
2. *Act is a finite set of actions of processes;*
3. *for each process action $\alpha \in Act$, U_α is a unitary operator in $\mathcal{H}$;*
4. *$\mathcal{H}_{ini} \subseteq \mathcal{H}$ is the subspace of initial states.*

We say that automaton $\mathcal{A}$ is finite-dimensional if its state space $\mathcal{H}$ is finite-dimensional. Throughout this paper, we only consider finite-dimensional quantum automata.

A path of $\mathcal{A}$ is generated by actions of processes, starting in an initial state:

$$p = |\psi_0\rangle \overset{U_{\alpha_0}}{\to} |\psi_1\rangle \overset{U_{\alpha_1}}{\to} |\psi_2\rangle \overset{U_{\alpha_2}}{\to} \cdots,$$

where $|\psi_0\rangle \in \mathcal{H}_{ini}$, $\alpha_n \in Act$, and $|\psi_{n+1}\rangle = U_{\alpha_n}|\psi_n\rangle$, for all $n \geq 0$. A schedule of processes is formalized as an infinite sequence $\alpha_0\alpha_1\alpha_2 \cdots \in Act^\omega$. For a given initial state $|\psi_0\rangle$ and a schedule $w \in Act^\omega$, we write the corresponding path as $p = p(|\psi_0\rangle, w)$. We further write $\sigma(p) = |\psi_0\rangle|\psi_1\rangle|\psi_2\rangle \cdots$ for the sequence of states in p. Sometimes, we simply call $\sigma(p)$ a path of $\mathcal{A}$.

Now let f be a logical formula defined in the above subsection representing a boolean combination of the subspaces of the state Hilbert space, and let

$$\sigma = |\psi_0\rangle|\psi_1\rangle|\psi_2\rangle \cdots$$

be an infinite sequence of states in $\mathcal{H}$. Formally, we define:

- (Eventually reachable): $\sigma \models \mathbf{F}f$ if $\exists i \geq 0. |\psi_i\rangle \models f$;
- (Globally reachable): $\sigma \models \mathbf{G}f$ if $\forall i \geq 0. |\psi_i\rangle \models f$;
- (Ultimately forever reachable): $\sigma \models \mathbf{U}f$ if $\overset{\infty}{\forall} i \geq 0. |\psi_i\rangle \models f$;
- (Infinitely often reachable): $\sigma \models \mathbf{I}f$ if $\overset{\infty}{\exists} i \geq 0. |\psi_i\rangle \models f$.

Here, $\overset{\infty}{\forall} i \geq 0$ means "$\exists j \geq 0, \forall i \geq j$", and $\overset{\infty}{\exists} i \geq 0$ means "$\forall j \geq 0, \exists i \geq j$". These reachability properties can be directly applied to quantum automata.

Definition 2. *Let $\mathcal{A}$ be a quantum automaton. Then for $\Delta \in \{\boldsymbol{F}, \boldsymbol{G}, \boldsymbol{U}, \boldsymbol{I}\}$, we define:*

$$\mathcal{A} \models \Delta f \text{ if } \sigma(p) \models \Delta f \text{ for all paths } p \text{ in } \mathcal{A}.$$

The reachability of a quantum automaton $\mathcal{A}$ can be stated in a different way. For any action string $s = \alpha_0\alpha_1 \cdots \alpha_n \in Act^*$, we write $U_s = U_{\alpha_n} \cdots U_{\alpha_1} U_{\alpha_0}$. If $U_s|\psi_0\rangle \models f$ for some initial state $|\psi_0\rangle \in \mathcal{H}_{ini}$, we say that s is accepted by $\mathcal{A}$ with f. The set of all accepted action strings is called the language accepted by $\mathcal{A}$ with f, and denoted by $\mathcal{L}(\mathcal{A}, f)$. For any $S \subseteq Act^*$, we write

$$\omega(S) = \{w = \alpha_0\alpha_1\alpha_2 \cdots \in Act^\omega \mid \overset{\infty}{\exists} n \geq 0,\ \alpha_0\alpha_1 \cdots \alpha_n \in S\},$$

and say that $S \subseteq Act^*$ satisfies the *liveness* (or *deadness*) property, if $\omega(S) = Act^\omega$ (or $\omega(S) = \emptyset$).

Lemma 1. *Let $\mathcal{A}$ be a quantum automaton with* $\dim \mathcal{H}_{ini} = 1$. *Then:*

1. *$\mathcal{A} \models \boldsymbol{F} f$ iff $Act^\omega = \mathcal{L}(\mathcal{A}, f) \cdot Act^\omega$;*
2. *$\mathcal{A} \models \boldsymbol{I} f$ iff $\mathcal{L}(\mathcal{A}, f)$ satisfies the liveness;*
3. *$\mathcal{A} \models \boldsymbol{G} f$ iff $\mathcal{L}(\mathcal{A}, f) = Act^*$ (i.e. $\mathcal{L}(\mathcal{A}, \neg f) = \emptyset$);*
4. *$\mathcal{A} \models \boldsymbol{U} f$ iff $Act^* - \mathcal{L}(\mathcal{A}, f)$ (i.e. $\mathcal{L}(\mathcal{A}, \neg f)$) satisfies the deadness.*

Here, $X \cdot Y$ in clause 1 is the concatenation of X and Y.

2.3 An Illustrative Example

Example 1. *Consider a quantum walk on a quadrilateral with the state Hilbert space $\mathcal{H}_4 = \text{span}\{|0\rangle, |1\rangle, |2\rangle, |3\rangle\}$. Its behaviour is described as follows:*

1. *Initialize the system in state $|0\rangle$.*
2. *Perform a measurement $\{P_{yes}, P_{no}\}$, where $P_{yes} = |2\rangle\langle 2|$, $P_{no} = I_4 - |2\rangle\langle 2|$. Here, I_4 is the 4×4 unit matrix. If the outcome is "yes", then the walk terminates; otherwise execute step 3.*
3. *Nondeterministically choose one of the two unitary operators:*

$$W_\pm = \frac{1}{\sqrt{3}} \begin{pmatrix} 1 & 1 & 0 & \mp 1 \\ \pm 1 & \mp 1 & \pm 1 & 0 \\ 0 & 1 & 1 & \pm 1 \\ 1 & 0 & -1 & \pm 1 \end{pmatrix}$$

and apply it. Then go to step 2.

It was proved in [16] that this walk terminates with a probability less than 1 if and only if a diverging state (i.e. a state with terminating probability 0) can be reached, and the set of diverging states is $PD_1 \cup PD_2$, where

$$PD_1 = \text{span}\{|0\rangle, (|1\rangle - |3\rangle)/\sqrt{2}\},$$
$$PD_2 = \text{span}\{|0\rangle, (|1\rangle + |3\rangle)/\sqrt{2}\}.$$

So, termination of the walk can be expressed as a reachability property $\mathcal{A} \models \mathbf{G}\neg(PD_1 \vee PD_2)$. Here, "$\vee$" is obviously boolean disjunction rather than the disjunction in Birkhoff-von Neumann quantum logic.

2.4 Main Theorems

Now we are ready to present the main problem considered in this paper. For $\Delta \in \{\mathbf{F}, \mathbf{G}, \mathbf{U}, \mathbf{I}\}$, the decision problem for the Δ−reachability is defined as follows:

Problem 1. *Given a finite-dimensional quantum automaton $\mathcal{A}$ and a logical formula f representing a boolean combination of the subspaces of the state Hilbert space of $\mathcal{A}$, decide whether or not $\mathcal{A} \models \Delta f$.*

For the algorithmic purpose, it is reasonable to make the convention: we identify a subspace of $\mathcal{H}$ with the projection operator on it, and assume that all the projection operators and unitary operators in automaton $\mathcal{A}$ and formula f are represented by complex matrices in a fixed orthonormal basis. Furthermore, we assume that all complex numbers have rational real and imaginary parts.

The main results of this paper can be stated as the following two theorems:

Theorem 1. *(Undecidability) For $\Delta \in \{\boldsymbol{F}, \boldsymbol{G}, \boldsymbol{U}, \boldsymbol{I}\}$, the problem whether or not $\mathcal{A} \models \Delta f$ is undecidable.*

Theorem 2. *(Decidability) For $\Delta \in \{\boldsymbol{G}, \boldsymbol{U}, \boldsymbol{I}\}$, if f contains no negation, then the problem whether or not $\mathcal{A} \models \Delta f$ is decidable.*

3 Relating Quantum Reachability to Skolem's Problem

3.1 Skolem's Problem for Linear Recurrence Sequences

For convenience of the reader, we first recall several results about Skolem's problem. A linear recurrence sequence is a sequence $\{a_n\}_{n=0}^{\infty}$ satisfying a linear recurrence relation given as follows:

$$a_{n+d} = c_{d-1}a_{n+d-1} + c_{d-2}a_{n+d-2} + \cdots + c_0 a_n, \tag{1}$$

for all $n \geq 0$, where $d, c_0, c_1, \cdots, c_{d-1}$ are all fixed constants with $c_0 \neq 0$. d is called the order of this relation. Let

$$Z = \{n \in \mathbb{N} \mid a_n = 0\} \tag{2}$$

be the set of indices of null elements of the sequence $\{a_n\}_{n=0}^{\infty}$. The problem of charac-terising Z was first studied by Skolem [24] in 1934, and his result was generalised by Mahler [17] and Lech [15].

Theorem 3 (Skolem-Mahler-Lech). *In a field of characteristic 0, for any linear recurrence sequence $\{a_n\}_{n=0}^{\infty}$, the set Z of indices of its null elements is semi-linear; that is, it is the union of a finite set and finitely many arithmetic progressions.*

The above problem was further considered in terms of decidability. The problem of deciding whether or not Z is infinite was solved by Berstel and Mignotte [5] who found an algorithm for generating all arithmetic progressions used in Theorem 3. The problem of deciding the finiteness of the complement of Z was studied by Salomaa and Soittola [23]. Their results are summarised as the following:

Theorem 4 (Berstel-Mignotte-Salomaa-Soittola). *For linear recurrence sequences* $\{a_n\}_{n=0}^{\infty}$, *it is decidable whether or not*

1. Z *is infinite;*
2. $Z = \mathbb{N}$;
3. Z *contains all except finitely many natural numbers.*

The following emptiness problem dual to item 2 in Theorem 4 was also considered in the literature, but it is still open; for details, we refer to [13,22].

Problem 2. *Given a linear recurrence sequence* $\{a_n\}_{n=0}^{\infty}$, *decide whether or not* Z *is empty.*

3.2 Skolem's Problem in Matrix Form

In this subsection, we show a useful connection between the quantum reachability problem and Skolem's problem. The linear recurrence relation Eq. (1) can be written in a matrix form:

$$a_n = u^T M^n v, \tag{3}$$

where M is the $d \times d$ matrix

$$\begin{bmatrix} c_{d-1} & c_{d-2} & \cdots & c_1 & c_0 \\ 1 & 0 & \cdots & 0 & 0 \\ 0 & 1 & \cdots & 0 & 0 \\ \vdots & \vdots & \ddots & \vdots & \vdots \\ 0 & 0 & \cdots & 1 & 0 \end{bmatrix},$$

$u = [1, 0, \cdots, 0]^T$ and $v = [a_{d-1}, a_{d-2}, \cdots, a_0]^T$ are $d-$dimensional column vectors, and ^{T} stands for transpose. On the other hand, if $\{a_n\}_{n=0}^{\infty}$ is of form Eq. (3) for general u,v and M with dimension d, then the minimal polynomial $g(x)$ of M is of order at most d, $g(M) = 0$, and a linear recurrence relation of order no greater than d is satisfied by $\{a_n\}_{n=0}^{\infty}$. Therefore, Skolem's problem can be equivalently considered in the matrix form Eq. (3).

Let us consider Problem 1 in a special case: (1) $|Act| = 1$, i.e., there is only one unitary operator U_α of $\mathcal{A}$, (2) $f = V$ is a subspace of $\mathcal{H}$, and (3) $\dim \mathcal{H}_{ini} = \dim V^\perp = 1$. Let $|\psi_0\rangle \in \mathcal{H}_{ini}$ and $|\varphi\rangle \in V^\perp$. Then we have $\mathcal{L}(\mathcal{A}, f) = \{n \in \mathbb{N} \mid \langle\varphi|U_\alpha^n|\psi_0\rangle = 0\}$. It is actually the set Z in Eq. (2) if we think of U_α, $|\varphi\rangle$ and $|\psi_0\rangle$ as M, u, and v in Eq. (3). From Lemma 1, the emptiness of Z (Problem 2), and the properties 1, 2 and 3 of Z in Theorem 4 are equivalent to $\mathcal{A} \models \mathbf{F}V$, $\mathcal{A} \models \mathbf{I}V$, $\mathcal{A} \models \mathbf{G}V$, and $\mathcal{A} \models \mathbf{U}V$, respectively. From this point of view, our decidability for a general f (Theorem 2) is somewhat a generalization of the decidable results (Theorem 4) of Skolem's problem where f is taken to be a subspace.

Now we consider an undecidable result relevant to Skolem's problem. Instead of $\{M^n \mid n \in \mathbb{N}\}$ in Eq. (3), there is a semi-group generated by a finite number of matrices $M_1, M_2, \cdots, M_k$, written as $\langle M_1, M_2, \cdots, M_k\rangle$. Then the emptiness problem can be generalised as follows:

Problem 3. *Provided $d \times d$ matrices $M_1, M_2, \cdots, M_k$ and d-dimensional vectors u and v, decide whether or not $\exists M \in \langle M_1, M_2, \cdots, M_k\rangle$ s.t. $u^T M v = 0$.*

The above problem was proved to be undecidable in [20] and [8], through a reduction from the Post's Correspondence Problem (PCP) [21]. Similar to the discussion in last subsection, we can choose M_i as unitary operators and u, v as quantum states, and then the emptiness of $\mathcal{L}(\mathcal{A}, f)$ for $f = V$ and $\dim \mathcal{H}_{ini} = \dim V^{\perp} = 1$ but with $|Act| > 1$ being allowed can be regarded as a special case of Problem 3. In fact, this problem was also proved to be undecidable by Blondel et. al. [7].

Theorem 5 (Blondel-Jeandel-Koiran-Portier). *It is undecidable whether or not $\mathcal{L}(\mathcal{A}, V)$ is empty, given a quantum automaton $\mathcal{A}$ and a subspace V with $\dim \mathcal{H}_{ini} = \dim V^{\perp} = 1$.*

4 Undecidable Results

4.1 Undecidability of $\mathcal{A} \models \mathbf{G}f$, $\mathcal{A} \models \mathbf{U}f$ and $\mathcal{A} \models \mathbf{I}f$

We can use the Theorem 5 to prove the Theorem 1 for $\Delta \in \{\mathbf{G}, \mathbf{U}, \mathbf{I}\}$. We first prove undecidability of $\mathcal{A} \models \mathbf{G}f$. Let automaton $\mathcal{A}$ be the same as in Theorem 5, but put $f = \neg V$ (not V). Then according to clause 3 of Lemma 1, $\mathcal{A} \models \mathbf{G}f$ is equivalent to the emptiness of $\mathcal{L}(\mathcal{A}, \neg(\neg V)) = \mathcal{L}(\mathcal{A}, V)$. The undecidability follows immediately from Theorem 5. To prove undecidability of $\mathcal{A} \models \mathbf{U}f$ and $\mathcal{A} \models \mathbf{I}f$, we slightly modify each quantum automaton $\mathcal{A} = (\mathcal{H}, Act, \{U_\alpha \mid \alpha \in Act\}, \mathcal{H}_{ini})$ by adding a silent action τ. Assume that $\tau \notin Act$ and $U_\tau = I$ (the identity operator in $\mathcal{H}$). Put $\mathcal{A}' = (\mathcal{H}, Act \cup \{\tau\}, \{U_\alpha \mid \alpha \in Act \cup \{\tau\}\}, \mathcal{H}_{ini})$. Then we claim:

$$\mathcal{A} \models \mathbf{G}f \text{ iff } \mathcal{A}' \models \mathbf{U}f \text{ iff } \mathcal{A}' \models \mathbf{I}f. \quad (4)$$

In fact, it is obvious that $\mathcal{A} \models \mathbf{G}f \Rightarrow \mathcal{A}' \models \mathbf{U}f \Rightarrow \mathcal{A}' \models \mathbf{I}f$ because U_τ is silent. Conversely, if $\mathcal{A} \not\models \mathbf{G}f$, then there exists $s = \alpha_0\alpha_1 \cdots \alpha_n \in Act^*$ such that $U_s|\psi_0\rangle \not\models f$. We consider the infinite sequence of actions $w = s\tau^\omega \in (Act \cup \{\tau\})^\omega$. It is clear that $\sigma(p(|\psi_0\rangle, w)) \not\models \mathbf{U}f$ and $\sigma(p(|\psi_0\rangle, w)) \not\models \mathbf{I}f$, and so $\mathcal{A}' \not\models \mathbf{U}f$ and $\mathcal{A}' \not\models \mathbf{I}f$. Finally, undecidability of $\mathcal{A} \models \mathbf{U}f$ and $\mathcal{A} \models \mathbf{I}f$ follows immediately from Eq. (4) and undecidability of $\mathcal{A} \models \mathbf{G}f$. Remarkably, the simple form of $f = \neg V$ is sufficient for undecidability.

4.2 undecidability of $\mathcal{A} \models \mathbf{F}f$

We separately prove undecidability of $\mathcal{A} \models \mathbf{F}f$. Our strategy is a reduction from the halting problem for 2-counter Minsky machines to reachability of quantum automata.

A 2-counter Minsky machine [18] is a program $\mathcal{M}$ consisting of two variables (counters) a and b of natural numbers $\mathbb{N}$, and a finite set of instructions, labeled by $l_0, l_1, \cdots, l_m$. This program starts at l_0 and halts at l_m. Each of instructions $l_0, l_1, \cdots, l_{m-1}$ is one of the following two types:

increment $l_i : c \leftarrow c+1$; goto l_j;
test-and-decrement l_i : if $c = 0$ then goto l_{j_1};
else $c \leftarrow c-1$; goto l_{j_2};

where $c \in \{a, b\}$ is one of the counters. The halting problem is as follows: given a 2-counter Minsky machine $\mathcal{M}$ together with the initial values of a and b, decide whether the computation of $\mathcal{M}$ will terminate or not. This problem is known to be undecidable.

For convenience of relating $\mathcal{M}$ to a quantum automaton, we slightly modify the definition of $\mathcal{M}$ without changing its termination:

1. Without loss of generality, we assume the initial values of a and b to be both 0. This can be done because any value can be achieved from zero by adding some instructions of increment at the beginning.
2. For each instruction l_i of test-and-decrement of c, we rewrite it as
$$\begin{aligned} l_i &: \text{ if } c = 0 \text{ then goto } l_i'; \text{ else goto } l_i''; \\ l_i' &: \text{ goto } l_{j_1}; \\ l_i'' &: c \leftarrow c-1; \text{ goto } l_{j_2}; \end{aligned} \tag{5}$$
where l_i' and l_i'' are new instructions. For $c \in \{a, b\}$, we write L_{1c} for the set of all instructions of increment of c; and we write L_{2c}, L_{2c}' and L_{2c}'' for the set of instructions l_i, the set of instructions l_i' and the set of instructions l_i'' given in Eq. (5), respectively. Now the set of all instructions of $\mathcal{M}$ becomes
$$L = L_{1a} \cup L_{1b} \cup L_{2a} \cup L_{2b} \cup L_{2a}' \cup L_{2b}' \cup L_{2a}'' \cup L_{2b}'' \cup \{l_m\}.$$
3. We rewrite l_m as
$$l_m : \text{ goto } l_m;$$
and we define that $\mathcal{M}$ terminates if l_m is reachable during the computation.

Obviously, the halting problem is also undecidable for 2-counter Minsky machines defined in this way.

We encode 2-counter Minsky machines into quantum automata so that undecidability of $\mathcal{A} \models \mathbf{F}f$ is derived from the undecidability of halting problem. More precisely, for any given 2-counter Minsky machine $\mathcal{M}$, we will construct a quantum automaton $\mathcal{A}$ and find two subspaces V and W of $\mathcal{H}$ such that

$$\mathcal{M} \text{ terminates} \Leftrightarrow \mathcal{A} \models \mathbf{F}(V \wedge \neg W). \tag{6}$$

The procedure of this construction is as follows:

1. A state of $\mathcal{M}$ is of form (a, b, x), where $a, b \in \mathbb{N}$ are the values of the two counters, and $x \in L$ is the instruction to be executed immediately. We use quantum states $|\phi_n\rangle$ and $|l\rangle$ to encode nature numbers n and instructions l, respectively. Then the corresponding quantum state in $\mathcal{A}$ is chosen as the product state $|\psi\rangle = |\phi_a\rangle|\phi_b\rangle|l\rangle$.

2. The computation of $\mathcal{M}$ is represented by the sequence of its states:

$$\sigma_{\mathcal{M}} = (a_0, b_0, x_0)(a_1, b_1, x_1)(a_2, b_2, x_2)\cdots, \tag{7}$$

where $(a_0, b_0, x_0) = (0, 0, l_0)$ is the initial state and $(a_{i+1}, b_{i+1}, x_{i+1})$ is the successor of (a_i, b_i, x_i) for all $i \geq 0$. We construct unitary operators of $\mathcal{A}$ to encode the transitions from a state to its successor. Then by successively taking the corresponding unitary operators, the quantum computation

$$\sigma_0 = |\psi_0\rangle|\psi_1\rangle\cdots, \ \forall i \geq 0 \ |\psi_i\rangle = |\phi_{a_i}\rangle|\phi_{b_i}\rangle|x_i\rangle \tag{8}$$

is achieved in $\mathcal{A}$ to encode $\sigma_{\mathcal{M}}$.

3. Note that $\sigma_{\mathcal{M}}$ is terminating if states at l_m are reachable. Then from the correspondence between $\sigma_{\mathcal{M}}$ and σ_0, termination of $\mathcal{M}$ is encoded as a reachability property $\sigma_0 \models \mathbf{F}V_0$, where V_0 is the subspace of terminating states, namely,

$$|\phi_a\rangle|\phi_b\rangle|l\rangle \in V_0 \Leftrightarrow l = l_m.$$

4. Besides σ_0, infinitely many computation paths are achievable in $\mathcal{A}$. So there is still a gap between reachability of σ_0 and that of $\mathcal{A}$. We construct two subspaces V' and W such that $\sigma \models \mathbf{F}(V' \wedge \neg W)$ for all paths σ of $\mathcal{A}$ except σ_0. We put $V = V' + V_0$, then it can be proved that

$$\mathcal{A} \models \mathbf{F}(V \wedge \neg W) \Leftrightarrow \sigma_0 \models \mathbf{F}(V \wedge \neg W) \Leftrightarrow \sigma_0 \models \mathbf{F}V_0,$$

and thus Eq. (6) is obtained from this equivalence.

5 Decidable Results

We prove Theorem 2 in this section. We write f in the disjunctive normal form. As it contains no negation, for each conjunctive clause f_i of f, $\|f_i\|$ is a subspace of $\mathcal{H}$. We write $V_i = \|f_i\| \in AP$, then f can be equivalently written as $f = \bigvee_{i=1}^m V_i$ and $\|f\| = \bigcup_{i=1}^m V_i$ is a union of finitely many subspaces of the state Hilbert space $\mathcal{H}$ of quantum automaton $\mathcal{A}$.

To decide whether or not $\mathcal{A} \models \Delta f$, we need to compute the set of all predecessor states with respect to a reachability property. Formally, for any given quantum automaton $\mathcal{A} = (\mathcal{H}, Act, \{U_\alpha \mid \alpha \in Act\}, \mathcal{H}_{ini})$ and any $|\psi\rangle \in \mathcal{H}$, we consider the automaton $\mathcal{A}(\psi) = (\mathcal{H}, Act, \{U_\alpha \mid \alpha \in Act\}, \text{span}\{|\psi\rangle\})$ for the paths starting in $|\psi\rangle$. Then for any $\Delta \in \{\mathbf{G}, \mathbf{U}, \mathbf{I}\}$, $|\psi\rangle$ is called a $(\Delta, f)-$predecessor state if $\mathcal{A}(\psi) \models \Delta f$, and we write the set of all predecessor states as

$$Y(\mathcal{A}, \Delta, f) = \{|\psi\rangle \in \mathcal{H} \mid \mathcal{A}(\psi) \models \Delta f\}.$$

Then $\mathcal{A} \models \Delta f$ can be decided by checking whether or not $\mathcal{H}_{ini} \subseteq Y(\mathcal{A}, \Delta, f)$.

5.1 Decidability of $\mathcal{A} \models \mathbf{I}f$ for Single Unitary Operator

We will prove the decidability of $\mathcal{A} \models \mathbf{I}f$ by constructing the set $Y(\mathcal{A}, \mathbf{I}, f)$. In this subsection, we do this for a special case in which $|Act| = 1$ and $m = 1$, i.e., $\mathcal{A}$ contains only a single unitary operator, and $f = V$ is a subspace. It should be pointed out that the result for this special case was proved in [5] as the decidability of finiteness Skolem's problem in the single matrix form. Here, we present our new proof as it would be useful for us to obtain a general result for finitely many unitary operators in next subsection. For convenience, we simply write Y for $Y(\mathcal{A}, \mathbf{I}, f)$ in these two subsections.

Let $Act = \{\alpha\}$, and the string α^n is simply represented by n. By an algorithm, we show that Y is a union of finitely many subspaces $Y_0, Y_1, \cdots, Y_{p-1}$ which forms a cycle graph under the unitary transformation, namely $Y_{r+1} = U_\alpha Y_r$ for all $0 \leq r \leq p-2$ and $Y_0 = U_\alpha Y_{p-1}$. Then Y can be written as $Y = \bigcup_{r=0}^{p-1} U_\alpha^r Y_0$ and $Y_0 = U_\alpha^p Y_0$. The following lemma is required for proving correctness of the algorithm.

Lemma 2. *For any unitary operator U on $\mathcal{H}$, a positive integer p can be found such that for any subspace K of $\mathcal{H}$, $U^p K = K$ provided $U^n K = K$ for some positive integer n. We call this integer p the period of U.*

Now Y can be computed by Algorithm 1. Step 1 can be done as shown in Lemma 2.

Algorithm 1.

---

1. Compute the period p of U_α;
2. Compute the maximal subspace K of V such that $U_\alpha^p K = K$;
3. Return $Y = \bigcup_{r=0}^{p-1} U_\alpha^r K$.

Step 2 can be done as follows: initially put $K_0 = V$, repeatedly compute $K_{n+1} = K_n \cap U_\alpha^p K_n$ until $K_{n+1} = K_n$, and then $K = K_n$. Sometimes, we write K as $K(U_\alpha, V)$ to show dependence of K on U_α and V. Correctness of this algorithm is presented as a lemma:

Lemma 3. *If integer p and subspace K are obtained by Algorithm 1, then*

$$Y = \bigcup_{r=0}^{p-1} U_\alpha^r K.$$

5.2 Decidability of $\mathcal{A} \models \mathbf{I}f$ for General Case

Now, we construct $Y = Y(\mathcal{A}, \mathbf{I}, f)$ for a general input: $\mathcal{A}$ and $f = \bigvee_{i=1}^m V_i$. Like the case of a single unitary operator, we can prove that Y is a union of finitely many subspaces. The result can be specifically described as follows:

Lemma 4. *Let $X = \{Y_1, Y_2, \cdots, Y_q\}$ be a set of subspaces of $\mathcal{H}$ satisfying the following three conditions:*

1. *For any* Y_i *and* $\alpha \in Act$, *there exists* Y_j *such that* $U_\alpha Y_i = Y_j$. *In other words, under the unitary transformations, these subspaces form a more general directed graph than a simple cycle graph in the case of single unitary operator.*
2. *For any simple loop (namely* $Y_{r_i} \neq Y_{r_j}$ *for different* i *and* j *in the loop)*

$$Y_{r_0} \overset{U_{\alpha_0}}{\rightarrow} Y_{r_1} \overset{U_{\alpha_1}}{\rightarrow} \cdots \overset{U_{\alpha_{k-2}}}{\rightarrow} Y_{r_{k-1}} \overset{U_{\alpha_{k-1}}}{\rightarrow} Y_{r_0},$$

there exists some $i \in \{0, 1, \cdots, k-1\}$ *and* $j \in \{1, 2, \cdots, m\}$ *such that* $Y_{r_i} \subseteq V_j$.
3. $Y \subseteq Y_1 \cup Y_2 \cup \cdots \cup Y_q$.

Then $Y = Y_1 \cup Y_2 \cup \cdots \cup Y_q$.

Therefore, to construct Y we only need to find an algorithm for constructing a set of subspaces $X = \{Y_1, Y_2, \cdots, Y_q\}$ satisfying the three conditions of Lemma 4. To this end, we invoke a lemma which is proved in [16]:

Lemma 5. *Suppose that* X_k *is the union of a finite number of subspaces of* $\mathcal{H}$ *for all* $k \geq 0$. *If* $X_0 \supseteq X_1 \supseteq \cdots \supseteq X_k \supseteq \cdots$, *then there exists* $n \geq 0$ *such that* $X_k = X_n$ *for all* $k \geq n$.

Now the set X can be computed by Algorithm 2. Step 2 is the key step in the algorithm, in which X can be replaced by a "smaller" one X' if it is not available. Due to Lemma 5, this step can only be executed a finite number of times and thus an output X satisfying condition 1 and condition 2 of Lemma 4 should be returned by the algorithm. We also note that condition 3 of Lemma 4 is always satisfied by X during the execution. So this output is just what we need.

Algorithm 2.

1. Initially put $X \leftarrow \{\mathcal{H}\}$ then jump to step 2;
2. If X satisfics condition 1 and condition 2 of Lemma 4, then return X; otherwise construct a new set X' of subspaces of $\mathcal{H}$ satisfying $Y \subseteq \cup X' \subset \cup X$, and put $X \leftarrow X'$, then repeat step 2. Here notation "$\subset$" is for "proper subset".

Now we give a detailed description of step 2. It can be properly formalized as a lemma:

Lemma 6. *Given a set* $X = \{Y_1, Y_2, \cdots, Y_q\}$ *of subspaces in which any two subspaces* Y_i *and* Y_j *do not include each other, if* X *satisfies condition 3 but does not satisfy condition 1 or condition 2 of Lemma 4, then we can algorithmically find some* $Y_i \in X$ *and its proper subspaces* $W_1, W_2, \cdots, W_l$, *such that*

$$Y \cap Y_i \subseteq W_1 \cup W_2 \cup \cdots \cup W_l. \tag{9}$$

From this lemma, we can construct X' for any given X as follows. First, we eliminate all such Y_i from X that $Y_i \subset Y_j$ for some $Y_j \in X$. Then from Lemma 6 we can find some $Y_i \in X$ and its subspaces $W_1, W_2, \cdots, W_l$ satisfying Eq. (9). We put $X' = X \cup \{W_k \mid 1 \leq k \leq l\} \backslash \{Y_i\}$, and then $\cup X' \subset \cup X$. As $Y \subseteq \cup X$, we also have $Y \subseteq \cup X'$ from Eq. (9).

5.3 Decidability of $\mathcal{A} \models \mathbf{G}f$ and $\mathcal{A} \models \mathbf{U}f$

We now prove Theorem 2 for $\Delta \in \{\mathbf{G}, \mathbf{U}\}$. We first prove the decidability of $\mathcal{A} \models \mathbf{G}f$ by computing $Y = Y(\mathcal{A}, \mathbf{G}, f)$. According to clause 3 in Lemma 1, we have

$$\begin{aligned} Y &= \{|\psi\rangle \in \mathcal{H} \mid \mathcal{L}(\mathcal{A}(\psi), f) = Act^*\} \\ &= \{|\psi\rangle \in \mathcal{H} \mid U_s|\psi\rangle \in \|f\|, \forall s \in Act^*\}. \end{aligned} \tag{10}$$

Then we obtain $\forall \alpha \in Act,\ U_\alpha Y \subseteq Y \subseteq \|f\|$. In fact, Y can be computed by Algorithm 3, and thus Y is the maximal one of sets satisfying $\forall \alpha \in Act,\ U_\alpha Y = Y \subseteq \|f\|$.

Algorithm 3.

1. $Y \leftarrow V_1 \cup V_2 \cup \cdots \cup V_m$;
2. If $U_\alpha Y \neq Y$, for some $\alpha \in Act$, then $Y \leftarrow U_\alpha^{-1} Y \cap Y$; otherwise return Y.

Correctness of Algorithm 3: We write $Y_0, Y_1, \cdots$ for the instances of Y during the execution of the algorithm. Then $Y_0 = V_1 \cup V_2 \cup \cdots \cup V_m$ and $Y_{n+1} = U_\alpha^{-1} Y_n \cap Y_n$ for some $\alpha \in Act$. It can be proved by induction on n that each Y_n is a union of finitely many subspaces of $\mathcal{H}$. Note that $Y_0 \supset Y_1 \supset Y_2 \supset \cdots$ is a descending chain. According to Lemma 5, this chain would terminates at some n, and the algorithm output is Y_n. We have $U_\alpha Y_n = Y_n$ for all $\alpha \in Act$. Now we prove $Y_n = Y$. First, since $Y \subseteq \|f\| = Y_0$ and $Y \subseteq U_\alpha^{-1} Y$ for all $\alpha \in Act$, it can be proved by induction on k that $Y \subseteq Y_k$ for all k, and particularly, $Y \subseteq Y_n$. On the other hand, As $U_s Y_n = Y_n \subseteq \|f\|$ for all $s \in Act^*$, we have $Y_n \subseteq Y$ from the definition of Y. So $Y_n = Y$. □

Next we prove the decidability of $\mathcal{A} \models \mathbf{U}f$. Indeed, we have the following lemma from which it follows that $Y(\mathcal{A}, \mathbf{U}, f) = Y(\mathcal{A}, \mathbf{G}, f)$.

Lemma 7. *$\mathcal{A} \models \mathbf{U}f$ iff $\mathcal{H}_{ini} \subseteq Y(\mathcal{A}, \mathbf{G}, f)$.*

It actually means that the reachabililty properties $\mathbf{U}f$ and $\mathbf{G}f$ are equivalent for f being positive.

6 Conclusion

We have investigated the decision problem of quantum reachability: decide whether or not a set of quantum states is reachable by a quantum system modelled by a quantum automaton. The reachable sets considered in this paper are defined as boolean combinations of (or described by classical propositional logical formula over) the set of (closed) subspaces of the state Hilbert space of the system. Four types of reachability properties have been studied: eventually reachable, globally reachable, ultimately forever reachable, and infinitely often reachable. Our major contribution is the (un)decidable results:

- All of these four reachability properties are undecidable even for a certain class of the reachable sets which are formalized by logical formulas of a simple form;

- Whenever the reachable set is a union of finitely many subspaces, the problem is decidable for globally reachable, ultimately forever reachable and infinitely often reachable. In particular, it is decidable when the reachable set contains only finitely many quantum states.

One of our main proof techniques is to demonstrate that quantum reachability problem is a generalization of Skolem's problem for unitary matrices. The undecidable results for global reachability, ultimately forever reachability and infinitely often reachability have been derived directly by employing the undecidability of a relevant emptiness problem. Nevertheless, the celebrated Skolem-Mahler-Lech theorem has been applied to the development of algorithms showing the decidable results. Another technique we have employed is to encode a 2-counter Minsky machine using a quantum automaton. It was used to prove undecidability of the eventually reachable property. This approach is interesting, since it provides a new way to demonstrate quantum undecidability other than reduction from the PCP that has been the main technique for the same purpose in previous works.

The problem whether or not $\mathcal{A} \models \mathbf{F}f$ is decidable for $\|f\|$ being a finite union of subspaces has been left unsolved. In fact, this problem is difficult even for a very special case where $|Act| = 1$ and $\|f\|$ is a single subspace. We have shown that in this case the quantum reachability problem is actually the emptiness Skolem's problem, which is still open for $d \geq 5$ [22].

The model of concurrent quantum systems used in this paper is quantum automata. For further studies, the decidable results can considered using a more general model, namely quantum Markov chains [28] where actions can be not only unitary transformations but also super-operators.

Acknowledgments. We are very grateful to Taolue Chen for helpful discussion about 2-counter Minsky machines. Wc also thank Nengkun Yu, Yuan Feng and Runyao Duan for useful suggestions. This work was partly supported by the Australian Research Council (Grant No. DP130102764) and AMSS, Chinese Academy of Sciences.

References

1. Altafini, C., Ticozzi, F.: Modeling and control of quantum systems: An introduction. IEEE Transactions on Automatic Control 57, 1898 (2012)
2. Amano, M., Iwama, K.: Undecidability on quantum finite automata. In: Proceedings of the Thirty-First Annual ACM Symposium on Theory of Computing (STOC), pp. 368–375 (1999)
3. Ardeshir-Larijani, E., Gay, S.J., Nagarajan, R.: Equivalence checking of quantum protocols. In: Piterman, N., Smolka, S.A. (eds.) TACAS 2013. LNCS, vol. 7795, pp. 478–492. Springer, Heidelberg (2013)
4. Baier, C., Katoen, J.-P.: Principles of Model Checking. MIT Press, Cambridge (2008)
5. Berstel, J., Mignotte, M.: Deux propriétés décidables des suites récurrentes linéaires. Bull. Soc. Math. France 104, 175–184 (1976)
6. Birkhoff, G., von Neumann, J.: The Logic of Quantum Mechanics. Annals of Mathematics 37, 823–843 (1936)
7. Blondel, V.D., Jeandel, E., Koiran, P., Portier, N.: Decidable and undecidable problems about quantum automata. SIAM Journal on Computing 34, 1464–1473 (2005)

8. Cassaigne, J., Karhumäki, J.: Examples of undecidable problems for 2-generator matrix semigroups. Theoretical Computer Science 204, 29–34 (1998)
9. Cirac, J.I., Zoller, P.: Goals and opportunities in quantum simulation. Nature Physics 8, 264–266 (2012)
10. Eisert, J., Müller, M.P., Gogolin, C.: Quantum measurement occurrence is undecidable. Physcal Review Letters 108, 260501 (2012)
11. Gay, S.J., Nagarajan, R., Papanikolaou, N.: Specification and verification of quantum protocols. In: Gay, S.J., Mackie, I. (eds.) Semantic Techniques in Quantum Computation, pp. 414–472. Cambridge University Press (2010)
12. Green, A.S., Lumsdaine, P.L., Ross, N.J., Selinger, P., Valiron, B.: Quipper: A scalable quantum programming language. In: Proceedings of the 34th ACM Conference on Programming Language Design and Implementation (PLDI), pp. 333–342 (2013)
13. Halava, V., Harju, T., Hirvensalo, M., Karhumäki, J.: Skolem's Problem: On the Border between Decidability and Undecidability, Technical Report 683, Turku Centre for Computer Science (2005)
14. Kondacs, A., Watrous, J.: On the power of quantum finite state automata. In: Proc. 38th Symposium on Foundation of Computer Science (FOCS), pp. 66–75 (1997)
15. Lech, C.: A note on recurring series. Ark. Mat. 2, 417–421 (1953)
16. Li, Y.J., Yu, N.K., Ying, M.S.: Termination of nondeterministic quantum programs. Acta Informatica 51, 1–24 (2014)
17. Mahler, K.: Eine arithmetische eigenschaft der Taylor koeffizienten rationaler funktionen. In: Proc. Akad. Wet. Amsterdam, vol. 38 (1935)
18. Minsky, M.L.: Computation: finite and infinite machines. Prentice-Hall (1967)
19. Nielsen, M.A., Chuang, I.L.: Quantum Computation and Quantum Information. Cambridge University Press (2000)
20. Paz, A.: Introduction to probabilistic automata. Academic Press, New York (1971)
21. Post, E.L.: A variant of a recursively unsolvable problem. Bulletin of the American Mathematical Society 52, 264–268 (1946)
22. Ouaknine, J., Worrell, J.: Decision Problems for Linear Recurrence Sequences. In: Finkel, A., Leroux, J., Potapov, I. (eds.) RP 2012. LNCS, vol. 7550, pp. 21–28. Springer, Heidelberg (2012)
23. Salomaa, A., Soittola, M.: Automata-Theoretic Aspects of Formal Power Series. Springer (1978)
24. Skolem, T.: Ein verfahren zur behandlung gewisser exponentialer gleichungen. In: Proceedings of the 8th Congress of Scandinavian Mathematicians, Stockholm, pp. 163–188 (1934)
25. Schirmer, S.G., Solomon, A.I., Leahy, J.V.: Criteria for reachability of quantum states. Journal of Physics A: Mathematical and General 35, 8551–8562 (2002)
26. Ying, M.S.: Floyd-Hoare logic for quantum programs. ACM Transactions on Programming Languages and Systems (19) (2011)
27. Ying, M.S., Yu, N.K., Feng, Y., Duan, R.Y.: Verification of quantum programs. Science of Computer Programming 78, 1679–1700 (2013)
28. Ying, S., Feng, Y., Yu, N., Ying, M.: Reachability probabilities of quantum Markov chains. In: D'Argenio, P.R., Melgratti, H. (eds.) CONCUR 2013. LNCS, vol. 8052, pp. 334–348. Springer, Heidelberg (2013)
29. Yu, N., Ying, M.: Reachability and termination analysis of concurrent quantum programs. In: Koutny, M., Ulidowski, I. (eds.) CONCUR 2012. LNCS, vol. 7454, pp. 69–83. Springer, Heidelberg (2012)

Ordered Navigation on Multi-attributed Data Words*

Normann Decker[1], Peter Habermehl[2], Martin Leucker[1], and Daniel Thoma[1]

[1] ISP, University of Lübeck, Germany
{decker,leucker,thoma}@isp.uni-luebeck.de
[2] Univ Paris Diderot, Sorbonne Paris Cité, LIAFA, CNRS, France
peter.habermehl@liafa.univ-paris-diderot.fr

Abstract. We study temporal logics and automata on multi-attributed data words. Recently, BD-LTL was introduced as a temporal logic on data words extending LTL by navigation along positions of single data values. As allowing for navigation wrt. tuples of data values renders the logic undecidable, we introduce ND-LTL, an extension of BD-LTL by a restricted form of tuple-navigation. While complete ND-LTL is still undecidable, the two natural fragments allowing for either future or past navigation along data values are shown to be Ackermann-hard, yet decidability is obtained by reduction to nested multi-counter systems. To this end, we introduce and study nested variants of data automata as an intermediate model simplifying the constructions. To complement these results we show that imposing the same restrictions on BD-LTL yields two 2ExpSpace-complete fragments while satisfiability for the full logic is known to be as hard as reachability in Petri nets.

1 Introduction

Executions of object-oriented and concurrent systems can naturally be modeled using data words. They are composed of labels from a finite alphabet together with a data value from an infinite domain. They can, for example, be considered as an interleaving of actions of an unbounded number of objects or processes, distinguished by identifiers. Recently, several formalisms based on first-order logic [4,21] or temporal logic [10,9,8] have been proposed to specify properties over data words. Automata-based models have also been considered [18,13,6,3] including data automata (DA) [4]. Usually, in these formalisms the data values can only be compared with respect to equality. More expressive relations like ordering lead fast to undecidability. The automata/logic connection has been studied extensively. For example, the satisfiability problem of two-variable first-order logic over data words was shown decidable by a reduction to the emptiness problem of DA [4]. They consist of a finite-state letter-to-letter transducer $\mathcal{A}$ and a class automaton $\mathcal{B}$. $\mathcal{A}$ changes the labels from the finite alphabet of the input data word before the data word is projected into class strings (one for each different data value) which must all be accepted by $\mathcal{B}$. Emptiness of DA was proven decidable by a reduction to the reachability problem in multi-counter systems (Petri nets, VASS) showing a deep connection between data word formalisms and counter systems (see also [8,22,2]).

* This work is partially supported by EGIDE/DAAD-Procope (LeMon). An extended version of this article can be found on arxiv.org in CoRR, arXiv:1404.6064.

P. Baldan and D. Gorla (Eds.): CONCUR 2014, LNCS 8704, pp. 497–511, 2014.

We study *multi-attributed data words* where, instead of one data value, *several* data values are associated to a given position. This important extension allows for example modeling *nested* parameterized systems where a process has subprocesses which have subprocesses and so on. We built on the logic on multi-attributed data words *basic data LTL (BD-LTL)* [14] allowing for navigation wrt. a data value. It uses the well-known LTL with past-time operators and has additionally a class quantifier over *one data value* used to bind a current data value and restrict the evaluation of the formula to the positions where the same data value appears. Decidability of the satisfiability problem was shown using a reduction to non-emptiness of DA. Adding a class quantifier over tuples makes BD-LTL undecidable like other logics over multi-attributed data words with tuple navigation [8,2].

Contributions. We consider first two fragments of BD-LTL: the *class future fragment* BD-LTL$^+$ (past operators are disallowed for navigation wrt. a data value) and the *class past fragment* BD-LTL$^-$ (restriction of future operators). Both fragments are shown 2EXPSPACE-complete using [8] and revisiting the translation from BD-LTL to DA [14]. Instead of going to general DA we translate BD-LTL$^+$ and BD-LTL$^-$ into pDA and sDA, respectively, whose emptiness problems are in EXPSPACE. In pDA (resp. sDA) the language of the class automaton is suffix- (resp. prefix-) closed allowing to use the EXPSPACE-complete coverability problem of multi-counter systems instead of its reachability problem for which no primitive recursive algorithm is known (cf. [16]). We consider both finite and infinite word semantics of the fragments.

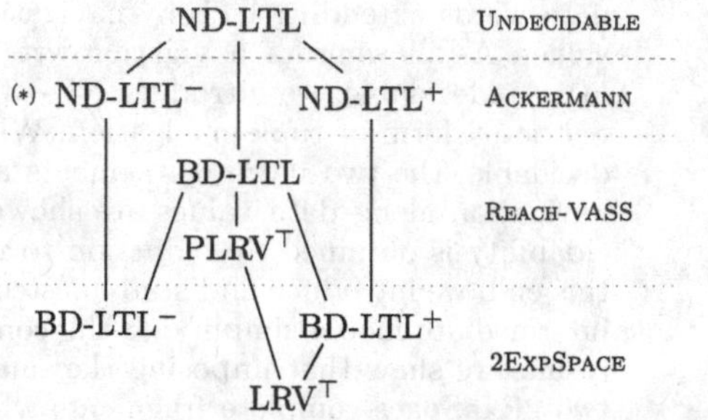

Fig. 1. Overview of the logics studied in this paper. Lines are drawn downwards to logics with lower expressiveness. The depicted complexity classes apply over finite as well as infinite words except for ND-LTL$^-$, marked by (∗), which is undecidable over infinite words.

We then define the new logic ND-LTL allowing for navigation wrt. tuples respecting a certain tree-order, i. e., there are several layers of data with nested access. For example, one can navigate on the first layer and, fixing a value, navigate on the second (see example below). Independent navigation on the whole second layer is not possible. While even with this restricted navigation ND-LTL is undecidable we obtain, as for BD-LTL, two natural fragments ND-LTL$^+$ and ND-LTL$^-$. We can prove their decidability by a translation into *nested data automata (NDA)* that we introduce as an appropriate extension of DA. k-NDA have k class automata and accept data words with k data values at each position. The i-th class automaton must accept all class strings obtained by projection of the data word using the same *first* i data values. Emptiness of k-NDA is undecidable, but shown decidable for k-sNDA (where class automata have suffix-closed languages) and k-pNDA (prefix-closed) using *nested multi-counter systems* (similar to models in [2,17]) which generalize multi-counter systems to several layers of nested counters. Their emptiness problem is undecidable, but, as they are well-structured transition systems [11,1], coverability and control state reachability are decidable. ND-LTL$^+$ and ND-LTL$^-$ are shown ACKERMANN-hard via a reduction from the control state reachability problem of reset multi-counter systems [20]. Finally, ND-LTL$^+$ is decidable over infinite words but ND-LTL$^-$ is not. Figure 1 summarizes some results.

Related Work. The logics LRV$^\top$ (based on [7]) and the more expressive LRV over multi-attributed data words studied in [8] built also on LTL and allow to state that *one* of the current data values must be seen again in the future. LRV (LRV$^\top$) can be extended to PLRV (PLRV$^\top$) with past obligations. PLRV$^\top$ is less expressive than BD-LTL[1] and we show that LRV$^\top$ is less expressive than BD-LTL$^+$. LRV (and LRV$^\top$) are 2EXPSPACE-complete like BD-LTL$^+$. We use their hardness result for our logic. The proof of the upper bound is also based on the coverability problem of multi-counter systems. However, our proof is split into smaller, structured parts. The handling of infinite word versions of our fragments is similar to theirs but we have to treat the additional problems coming from the nested data. Navigation wrt. data tuples was considered and shown undecidable but no decidable fragments were given. A logic handling data values in a very natural way is Freeze-LTL [9]. It exhibits a similar future-restriction as BD-LTL$^+$ and ND-LTL$^+$ and finite satisfiability is decidable and ACKERMANN-hard. However, satisfiability over infinite words is undecidable while it is still decidable for BD-LTL$^+$ and ND-LTL$^+$. In [2], words with nested data values were also considered. They show undecidability for the two-variable logic with two layers of nested data and the $+1$ and $<$ predicates over positions. They introduce *higher-order multi-counter automata*, a very similar model to our nested multi-counter systems. Their proof of Turing completeness could be easily adapted to nested multi-counter systems. However, the well-structuredness of the model is not exploited. If the $+1$ predicate is dropped they obtain decidability, which is orthogonal to our result as we can express the successor relation in our fragments. In [22] history register automata (HRA) have been introduced, which can easily be simulated by our pNDA. A weak variant of HRA is defined which is similar to our pDA, but only studied over finite words.

2 Preliminaries

Let $\mathbb{N} = \{0, 1, 2, \dots\}$ be the set of natural numbers and $[k] := \{1, \dots, k\}$ for $k \in \mathbb{N}, k > 0$. We denote the set of finite words over an alphabet Σ by Σ^*, the set of infinite words by Σ^ω and their union by $\Sigma^\infty = \Sigma^* \cup \Sigma^\omega$. The empty word is denoted ϵ. The *shuffle* of two words $w, w' \in \Sigma^*$ is inductively defined by $\epsilon \sqcup\!\sqcup w = w \sqcup\!\sqcup \epsilon = \{w\}$ and $aw \sqcup\!\sqcup a'w' = a(w \sqcup\!\sqcup a'w') \cup a'(aw \sqcup\!\sqcup w')$ where $a, a' \in \Sigma$. For possibly infinite words $u_1, u_2 \in \Sigma^\infty$ we let $v \in u_1 \sqcup\!\sqcup u_2$ iff for all prefixes v' of v there are prefixes u_1' and u_2' of u_1 and u_2, respectively, such that $v' \in u_1' \sqcup\!\sqcup u_2'$. The shuffle of two languages $L, L' \subseteq \Sigma^\infty$ is $L \sqcup\!\sqcup L' = \{w \sqcup\!\sqcup w' \mid w \in L, w' \in L'\}$ and $\sqcup\!\sqcup(L) = \bigcup\{M \mid M \subseteq L \sqcup\!\sqcup M\}$ denotes the infinite shuffle of a language with itself. For two sets M, N we denote by M^N the set of all mappings $f : N \to M$ from N to M. Given a partial order $(M, \sqsubseteq)$ we write $m^\downarrow = \{m' \in M \mid m' \sqsubseteq m\}$ for the downward closure of $m \in M$. We define a *tree order* $(M, \leq)$ to be a

[1] In [8] it is stated without proof that PLRV is also less expressive than BD-LTL.

partial order s. t. for all $m \in M$ its downward closure is a linear order $(m^{\downarrow}, \leq)$. Hence, we allow a tree order to contain several minimal elements (roots).

An ∞-automaton over a finite *input alphabet* Σ is a tuple $\mathcal{A} = (Q, \Sigma, \delta, I, F, B)$ where Q is a finite set of *states*, $I, F, B \subseteq Q$ are sets of *initial*, *final* and *Büchi-accepting* states, respectively, and $\delta \subseteq Q \times \Sigma \times Q$ is the *transition relation*. A *run* of $\mathcal{A}$ on a word $w_0 w_1 \ldots \in \Sigma^\infty$, $w_i \in \Sigma$ is a maximal sequence of transitions $t_0 t_1 \ldots \in \delta^\infty$ with $t_i = (q_i, w_i, q_{i+1})$ and $q_0 \in I$. It is *accepting* if it ends in a final state $q_f \in F$ or visits a Büchi-accepting state $q_b \in B$ infinitely often. $\mathcal{A}$ *accepts* w if there is an accepting run of $\mathcal{A}$ on w and the set of all accepted words is denoted $\mathcal{L}(\mathcal{A})$.

A *letter-to-letter transducer* is an ∞-automaton $\mathcal{T} = (Q, \Sigma, \Gamma, \delta, I, F, B)$ where Γ is an additional *output alphabet* and $\delta \subseteq Q \times \Sigma \times \Gamma \times Q$ is a *transition relation with output*. A word $\gamma \in \Gamma^\infty$ is an *output* of $\mathcal{T}$ if there is an accepting run of $\mathcal{T}$ labeled by γ. For $w \in \Sigma^\infty$ we denote $\mathcal{T}(w) \subseteq \Gamma^\infty$ the set of possible outputs of $\mathcal{T}$ when reading w.

Data Words and Data Languages. Let Σ be a finite alphabet, Δ an infinite set of *data values* and A a finite set of *attributes*. A *multi-attributed data word* is a finite or infinite sequence $w = w_0 w_1 \ldots \in (\Sigma \times \Delta^A)^\infty$ of pairs $w_i = (a_i, \mathbf{d}_i)$ of letters and *data valuations* $\mathbf{d}_i : A \to \Delta$. Given a valuation $\mathbf{d} \in \Delta^A$ and a set of attributes $X \subseteq A$ we denote by $\mathbf{d}|_X$ the restriction of $\mathbf{d}$ to X. We call $\mathsf{str}(w) := a_0 a_1 \ldots \in \Sigma^\infty$ the *string projection* of w. The X-*class string* of w for a data valuation $\mathbf{d} \in \Delta^X$ is the maximal projected subsequence $\mathsf{cl}(w, \mathbf{d}) := a_{i_0} a_{i_1} \ldots \in \Sigma^\infty$ of w with $0 \leq i_j \leq |w|$, $i_j < i_{j+1}$ and $\mathbf{d}_{i_j}|_X = \mathbf{d}$. We use natural numbers $1, 2, 3, \ldots$ as representatives for arbitrary data values. For a data word $w = (a_0, \mathbf{d}_0)(a_1, \mathbf{d}_1) \ldots$ we also write $\binom{a_0 a_1 \ldots}{\mathbf{d}_0 \mathbf{d}_1 \ldots}$. For $|A| = 1$ we call data words $w \in (\Sigma \times \Delta^A)^\infty$ *single-attributed*. We may then omit the functional notation and use Δ instead of Δ^A if A is not essential, e. g., writing $w \in (\Sigma \times \Delta)^\infty$.

Register Automata (RA). A *register automaton* [13] over Σ and Δ is a tuple $\mathcal{R} = (Q, \Sigma, k, \delta, I, F, B)$ where Q is a finite set of states, $I, F, B \subseteq Q$ are sets of initial, final and Büchi-accepting states, respectively, $k \geq 1$ is the number of *registers* and $\delta \subseteq Q \times 2^{[k]} \times 2^{[k]} \times \Sigma \times [k] \times Q$ is the transition relation. A *configuration* of $\mathcal{R}$ is a pair (q, v) where $q \in Q$ and $v : [k] \to \Delta \cup \{\bot\}$ is a valuation of the registers. A *run* of $\mathcal{R}$ on a single-attributed data word $w = (a_0, d_0)(a_1, d_1) \ldots \in (\Sigma \times \Delta)^\infty$ is a maximal sequence of configurations $\rho = (q_0, v_0)(q_1, v_1) \ldots$ s. t. $q_0 \in I$ and for all $0 \leq i < |w|$ there is a transition $(q_i, R_i^=, R_i^{\neq}, a_i, x_i, q_{i+1}) \in \delta$ such that $\forall_{r \in R_i^=} v_i(r) = d_i$, $\forall_{r \in R_i^{\neq}} v_i(r) \neq d_i$, $v_{i+1}(x_i) = d_i$ and $\forall_{r \neq x_i} v_{i+1}(r) = v_i(r)$. A run ρ of $\mathcal{R}$ is *accepting* if it ends in a final state $q \in F$ or it visits a Büchi-accepting state $q \in B$ infinitely often. An RA accepts a single-attributed data word w if it has an accepting run on w.

Multi-counter Systems. A *reset multi-counter system (rMCS)* is a tuple $\mathcal{M} = (Q, C, \delta, Q_0)$ where Q and C are finite sets of *(control) states* and *counters*, respectively, $Q_0 \subseteq Q$ is the set of *initial states* and, for $OP := \{\text{inc}, \text{dec}, \text{res}\}$, $\delta \subseteq Q \times OP \times C \times Q$ is the *transition relation*. A *run* of $\mathcal{M}$ is a sequence $\rho \in Q_0 \times (OP \times C \times Q)^\infty$, s. t. every subsequence (q, op, c, q') of ρ, with $q, q' \in Q$, $op \in OP$, $c \in C$, is an element of δ and counters never become negative,

i. e, there is an injection $f_\rho : \mathbb{N} \to \mathbb{N}$ that maps every position i in ρ with $(\rho_i, \rho_{i+1}) = (\text{dec}, c)$, for $c \in C$, to a position $j < i$ with $(\rho_j, \rho_{j+1}) = (\text{inc}, c)$ and $(\rho_k, \rho_{k+1}) \neq (\text{res}, c)$, for all k with $j < k < i$. An *MCS* is an rMCS where the transition relation does not use the reset operation res.

3 Local Navigation in BD-LTL

The temporal logic BD-LTL is based on LTL. Linear-time properties are formulated using temporal operators to navigate along the positions of a word. This concept is extended analogously to data words by allowing for navigation along the occurrences of a data value. While the LTL operators express properties on the global structure of the word, independent of associated data values, navigation along the class strings of a word allows for expressing a local view, e. g., modeling the behaviour of a single process.

We now recall syntax and semantics of BD-LTL [14] and define two natural fragments BD-LTL$^+$ and BD-LTL$^-$ where local navigation is restricted to future and past operators, respectively. The satisfiability problem of BD-LTL is decidable. Yet, it is known to be as hard as reachability in Petri nets [14] and we show that satisfiability in our fragments is still 2ExpSpace-hard. The next section then sharpens this result by developing a 2ExpSpace decision procedure based on restricted variants of data automata.

Let AP be a finite set of atomic propositions and A a finite set of attributes. The syntax of BD-LTL formulae consists of *position formulae* φ and *class formulae* ψ. It is defined by the following grammar where $p \in AP$, $x, y \in A$ and $r \in \mathbb{Z}$.

$$\varphi ::= p \mid \varphi \wedge \varphi \mid \neg\varphi \mid \mathrm{X}\,\varphi \mid \mathrm{Y}\,\varphi \mid \varphi\,\mathrm{U}\,\varphi \mid \varphi\,\mathrm{S}\,\varphi \mid \mathrm{C}^r_x\,\psi$$
$$\psi ::= @x \mid \psi \wedge \psi \mid \neg\psi \mid \mathrm{X}^=\,\psi \mid \mathrm{Y}^=\,\psi \mid \psi\,\mathrm{U}^=\,\psi \mid \psi\,\mathrm{S}^=\,\psi \mid \varphi$$

The semantics of BD-LTL position formulae φ is defined over models (w, i) consisting of an A-attributed data word $w = (a_0, \mathbf{d}_0)(a_1, \mathbf{d}_1)\ldots \in (\Sigma \times \Delta^A)^\infty$ over alphabet $\Sigma = 2^{AP}$ and a position $0 \leq i < |w|$. Class formulae ψ are defined over models (w, i, d) containing an additional data value $d \in \Delta$. Boolean and LTL operators are defined as usual, ignoring the data values. For the semantics of the quantifier C^r_x and class formulae ψ, let $\text{pos}_d(w) := \{i \mid 0 \leq i < |w|, \exists_{x \in A} : \mathbf{d}_i(x) = d\}$ denote the set of positions i in w where some attribute has the value $d \in \Delta$. Then,

$$\begin{array}{lll}
(w,i) & \models \mathrm{C}^r_x\,\psi & \text{if } 0 \leq i + r < |w| \text{ and } (w, i+r, \mathbf{d}_i(x)) \models \psi, \\
(w,i,d) & \models \varphi & \text{if } (w,i) \models \varphi, \\
(w,i,d) & \models @x & \text{if } \mathbf{d}_i(x) = d, \\
(w,i,d) & \models \mathrm{X}^=\,\psi & \text{if there is } j \in \text{pos}_d(w), j > i \\
 & & \quad \text{and, for the smallest such } j,\ (w,j,d) \models \psi, \\
(w,i,d) & \models \psi_1\,\mathrm{U}^=\,\psi_2 & \text{if there is } j \in \text{pos}_d(w), j \geq i \text{ s. t. } (w,j,d) \models \psi_2 \\
 & & \quad \text{and } \forall_{j' \in \text{pos}_d(w), j > j' \geq i} : (w, j', d) \models \psi_1.
\end{array}$$

The operators $\mathrm{Y}^=$ and $\mathrm{S}^=$ are furthermore defined as expected and $(w, 0) \models \varphi$ is abbreviated $w \models \varphi$. We also use the abbreviations $\top$ and $\mathrm{F}^=\,\varphi := \top\,\mathrm{U}^=\,\varphi$.

Definition 1 (BD-LTL$^\pm$). *We define the following syntactical fragments: BD-LTL without operators* $\mathrm{X}^=$ *and* $\mathrm{U}^=$ *is called BD-LTL$^-$. BD-LTL without operators* $\mathrm{Y}^=$ *and* $\mathrm{S}^=$ *is called BD-LTL$^+$.*

In [8], the *Logic of Repeating Values (LRV)* was introduced as an extension of LTL interpreted over multi-attributed data words. The additional operators are of the form $x \approx \mathrm{X}^r\, y$, $x \approx \langle\varphi?\rangle y$ and $x \not\approx \langle\varphi?\rangle y$. The former expresses that the current value of attribute x must be equal to the value of attribute y at the position r steps ahead. Similarly, the latter two express that the value of x must eventually or never, respectively, be observed as the value of y at a position where, in addition, a formula φ holds. In $\mathrm{LRV}^\top$ only $x \approx \mathrm{X}^r\, y$ and $x \approx \langle\top?\rangle y$ are allowed. $x \approx \mathrm{X}^r\, y$ and $x \approx \langle\varphi?\rangle y$ can easily be translated into BD-LTL^+: $x \approx \mathrm{X}^r\, y$ is equivalent to $\mathrm{C}_x^r\, @y$ and $x \approx \langle\varphi?\rangle y$ is equivalent to $\mathrm{C}_x^0\, \mathrm{X}^= \mathrm{F}^=(@y \wedge \varphi)$ [14]. On the contrary, LRV cannot express the operator $\mathrm{X}^=$.

Proposition 1. *BD-LTL$^+$ is strictly more expressive than LRV$^\top$.*

The satisfiability problem of $\mathrm{LRV}^\top$ (and LRV) was shown to be 2ExpSpace-hard in [8] by encoding runs of so called chain automata using exponentially many counters. The proof [8, Lemma 15] can easily be adapted to show that the variant of $\mathrm{LRV}^\top$ where past *instead* of future operators are used ($x \approx \mathrm{Y}^r\, y$, $x \approx \langle\varphi?\rangle^{-1} y$) is also 2ExpSpace-hard and as BD-LTL^- subsumes this variant we obtain a lower bound for both of our fragments.

Theorem 1 (Hardness). *The satisfiability problems of BD-LTL$^+$ and BD-LTL$^-$ are* 2ExpSpace-*hard over both, finite and infinite data words.*

4 Satisfiability of BD-LTL$^\pm$ is 2ExpSpace-Complete

This section is dedicated to an exact characterization of $\text{BD-LTL}^\pm$ satisfiability in terms of complexity. It also provides a basis for Section 6 that follows a similar structure but is technically more involved.

First, we formally define data automata and give restrictions that reflect the restrictions on our logic. They allow us to decide emptiness in ExpSpace, as opposed to full data automata for which emptiness is as hard as reachability in Petri nets [4]. Second, we briefly recall the (exponential) translation from BD-LTL to data automata [15] and show that our logical restrictions indeed carry over to the restrictions on the automata side.

4.1 ExpSpace-Variants of Data Automata

A *data automaton (DA)* is a tuple $\mathcal{D} = (\mathcal{A}, \mathcal{B})$ where the *base automaton* $\mathcal{A} = (Q, \Sigma, \Gamma, \delta_\mathcal{A}, Q_0, F_\mathcal{A}, B_\mathcal{A})$ is a letter-to-letter transducer and the *class automaton* $\mathcal{B} = (S, \Gamma, \delta_\mathcal{B}, I, F, B)$ is an ∞-automaton. A *memory function* of $\mathcal{D}$ is a mapping $f : \Delta \to S \cup \{\bot\}$ and we denote $\mathfrak{F}$ the set of all memory functions. A *configuration* of $\mathcal{D}$ is a tuple $(q, f) \in Q \times \mathfrak{F}$ consisting of a base automaton state and a memory function. A *run* of $\mathcal{D}$ on a single-attributed data word $w = (a_0, d_0)(a_1, d_1)\ldots \in (\Sigma \times \Delta)^\infty$ is a maximal sequence $\rho = (q_0, f_0)(q_1, f_1)\ldots \in (Q \times \mathfrak{F})^\infty$ such that $q_0 \in Q_0$, $\forall_{d \in \Delta} : f_0(d) = \bot$ and for all consecutive positions $i, i+1$ on ρ there is a transition $(q_i, a_i, g, q_{i+1}) \in \delta_\mathcal{A}$ of the base automaton and a transition $(s, g, s') \in \delta_\mathcal{B}$ of the class automaton such that (1) $f_{i+1}(d_i) = s'$ and (2) either $f_i(d_i) = s$, or $f_i(d_i) = \bot$ and $s \in I$, and (3) $\forall_{d' \in \Delta, d' \neq d_i} : f_i(d') = f_{i+1}(d')$.

The run ρ is *accepting* if (I) it ends in a configuration (q, f) with $q \in F_{\mathcal{A}}$ is final and $f(\Delta) \cap S \subseteq F$, or (II) there are infinitely many configurations (q, f_i) on ρ such that $q \in B_{\mathcal{A}}$ is Büchi-accepting and for each data value d occurring *last* at some position i on w the state $f_{i+1}(d) \in F$ is final and for each data value d' occurring *infinitely* often on w there are infinitely many positions j with $d_j = d'$ and $f_{j+1}(d') \in B$ is Büchi-accepting. The word w is accepted if there is an accepting run of $\mathcal{D}$.

Intuitively, the base transducer $\mathcal{A}$ reads a letter $a_i \in \Sigma$, performs a transition and outputs its label $g \in \Gamma$. The memory function maintains an *instance* of the class automaton $\mathcal{B}$ for every data value that occurred so far and spawns a new instance for a fresh data value. The (present or newly spawned) instance of $\mathcal{B}$ that corresponds to the current data value d_i, reads g and performs a step. For $\mathcal{D}$ to accept, $\mathcal{A}$ and every spawned instance of $\mathcal{B}$ needs to accept by either terminating in a final state or visiting some Büchi-accepting state infinitely often.

Definition 2 (Prefix- and suffix-closed DA). *A data automaton $\mathcal{D} = (\mathcal{A}, \mathcal{B})$ is* locally prefix-closed (pDA) *if all states of the class automaton $\mathcal{B}$ are final and Büchi-accepting. It is* locally suffix-closed (sDA) *if all states of $\mathcal{B}$ are initial.*

The construction to decide emptiness of DA given in [4] translates a DA into a multi-counter automaton (MCA) that maintains for every class automaton state the number of instances residing in it. That way, emptiness of DA reduces (for finite words) to reachability in MCA. Note that technical differences in the various notions of counter systems (e. g., MCA, MCS, VASS, Petri nets) are inessential here.

For pDA, where all class automaton states are final and Büchi-accepting, automaton instances can be dismissed in any state. The corresponding MCS thus allows for a random decrement of counters. Clearly, in such a *lossy* system the problem of reachability reduces to *coverability.* Regarding infinite words, *repeated coverability* is sufficient since every class automaton state is also Büchi-accepting. Both problems are in ExpSpace [19,12].

For an sDA we can decide if it accepts a *finite* word by reversing the automata and checking the resulting pDA for emptiness. In the rest of this section we address the remaining case of sDA emptiness wrt. *infinite* words obtaining the following result.

Theorem 2. *Emptiness of pDA and sDA over finite and infinite data words is decidable in* ExpSpace.

Let $\mathcal{D} = (\mathcal{A}, \mathcal{B})$ be an sDA with $\mathcal{A} = (Q, \Sigma, \Gamma, \delta_{\mathcal{A}}, Q_0, \emptyset, B_{\mathcal{A}})$ (we omit final states) and $\mathcal{B} = (S, \Sigma, \delta, I, F, B)$. Towards deciding emptiness of $\mathcal{D}$, we consider an accepting run ρ of $\mathcal{D}$ and separate the finite from the infinite behaviour in terms of transitions $t \in \delta$ of the class automaton: There is a position i on ρ, such that t is taken after i iff t is taken infinitely often on ρ.

The idea is now to guess (characteristics of) the configuration at this position and check that there is a finite run reaching the configuration and that starting from it there is an infinite accepting run. For the former, we construct an sDA that accepts a *finite* word iff the configuration is reachable. For the latter, we now have guaranteed infinite recurrence of all relevant transitions and can thereby reduce the problem to emptiness of an at most exponentially larger Büchi automaton.

For a set $T \subseteq \delta$ of transitions of the class automaton $\mathcal{B}$ and a state $q \in Q$ of the base automaton $\mathcal{A}$, consider the following three properties.

(A1) After taking any transition in T, $\mathcal{B}$ can eventually reach a final state from F or an accepting state from B, only by taking transitions from T.
(A2) There is a sequence $t_1 t_2 \ldots \in T^\omega$ with $t_i = (s_i, g_i, s_i')$ in which each $t \in T$ occurs infinitely often and $g_1 g_2 \ldots \in \Gamma^\omega$ is an output of $\mathcal{A}$ starting in q.
(A3) There is a reachable configuration (q, f) such that for all data values $d \in \Delta$, either (i) there is no corresponding instance of $\mathcal{B}$ ($f(d) = \bot$), or (ii) the corresponding instance of $\mathcal{B}$ is in an accepting state ($f(d) \in F$), or (iii) there is a transition $(f(d), g, s) \in T$ for some $s \in S$ and some $g \in \Gamma$.

Lemma 1. *The sDA $\mathcal{D}$ accepts an infinite data word iff there are $T \subseteq \delta$, $q \in Q$ such that the properties (A1)–(A3) hold.*

Verifying (A1) is a reachability problem in the finite graph of $\mathcal{B}$ restricted to T. Further, we can build a Büchi automaton over Γ that is non-empty iff (A2) holds: In $\mathcal{A}$, take the outputs as inputs and remove all transitions with a label not occurring on any transition in T. For each transition $(s, g, s') \in T$, intersect the automaton with the property $\mathrm{G\,F}\, g$. The size of the resulting Büchi automaton is at most $c^{(|Q|^2)}$ for a constant c. Finally, (A3) can be verified by constructing the sDA $\hat{\mathcal{D}} = (\hat{\mathcal{A}}, \hat{\mathcal{B}})$ with $\hat{\mathcal{A}} = (Q, \Sigma, \Gamma, \delta_{\mathcal{A}}, Q_0, \{q\}, \emptyset)$ and $\hat{\mathcal{B}} = (S, \Sigma, \delta, I, F \cup {}^\bullet T)$, where ${}^\bullet T := \{s \in S \mid \exists_{s' \in S, g \in \Gamma} : (s, g, s') \in T\}$, and checking it for emptiness in exponential space as above.

Lemma 2. *Given $T \subseteq \delta$ and $q \in Q$, it is decidable in* ExpSpace *if properties (A1)–(A3) are satisfied.*

To conclude, using Lemma 1 and 2 we can check the sDA $\mathcal{D}$ for emptiness wrt. infinite words by nondeterministically guessing a state $q \in Q$ and a set of transitions $T \subseteq \delta$ and verifying (A1)–(A3) in exponential space.

4.2 From BD-LTL$^\pm$ to Suffix- and Prefix-closed Data Automata

We adopt the construction given in [14] to show that a BD-LTL$^\pm$ formula can be translated into an at most exponentially larger sDA and pDA, respectively, that is nonempty iff the formula is satisfiable. First, the formula is translated into an equisatisfiable formula φ over *single*-attributed data words. Second, a data automaton $\mathcal{A}_\varphi$ is constructed that is nonempty iff φ is satisfiable.

The basic idea for eliminating multiple attributes in a formula is to encode one position of an A-attributed data word into a segment of $|A|$ positions in a single-attributed word. The temporal operators are adjusted by offsets according to the segment length. For this elimination, given in detail in [15], only additional C_x^r operators are required. A BD-LTL$^\pm$ formula hence stays in its respective fragment and grows at most polynomially in size.

The data automaton $\mathcal{A}_\varphi$ is obtained by constructing a base automaton that verifies global constraints and a class automaton that verifies local constraints (those expressed by class formulae in φ).

Let $r_{max} := \max(\{|r| \mid \mathrm{C}^r \text{ occurs in } \varphi\} \cup \{0\})$ be the largest (absolute) offset and AP_φ the set of atomic propositions used in φ. Consider new, additional propositions p_j^ψ and $=_j$ for each $-r_{\max} \leq j \leq r_{\max}$ and subformula ψ of φ.

We construct $\mathcal{A}_\varphi$ s. t. it checks on a data word that a proposition p_j^ψ holds at position i iff ψ holds at position $i+j$ and that p_0^φ holds at the very first position. It then verifies that φ holds on an input word. If φ is satisfiable then any of its models can be annotated by propositions such that it is accepted by $\mathcal{A}_\varphi$.

By straight forward equivalences we assume w. l. o. g. that all and only class operators in φ are preceded by a class quantifier C^r. The base automaton of $\mathcal{A}_\varphi$ is constructed such that it verifies the correct occurrence of propositions p_0^ψ where ψ is a position formula not starting with a C^r quantifier. When evaluating subformulae of ψ just by referring to the corresponding propositions, these are regular properties.

To capture the other cases where ψ has the form $C^r\,\psi'$ the class automaton of $\mathcal{A}_\varphi$ is constructed to verify them. As the class formulae express regular properties on the class projection they can be translated as usual. Subformulae that are position formulae are as above evaluated in terms of their corresponding propositions. If the offset r is nonzero, this shift has to be taken into account by the class automaton. This is possible since at every positions the information for r_max steps back and forth is provided by the additional propositions. While in the original construction from [14], the propositions are not needed, we require them to obtain suffix or prefix closed class automata. Also due to the possible global shift r by C^r quantifiers, the class automaton needs to know how many steps to take on the class projection in order to perform a specific number of global steps. This information can be derived from the proposition $=_j$ that are assumed to holds at position i iff position $i+j$ carries the same data value.

It is easy to construct a register automaton $\mathcal{R}$ that verify this assumption by maintaining the frame of data values. While it is known that RA can be translated into DA it is not clear how to adapt the construction from [14] to obtain sDA and pDA. The following slightly weaker lemma is, however, sufficient. We only need to verify that the data automaton constructed above accepts a word where propositions $=_j$ hold as assumed.

Lemma 3 (Simulating RA). *Given an sDA or pDA $\mathcal{D}$ and a register automaton $\mathcal{R}$, we can construct an sDA or pDA $\mathcal{D}'$, respectively, such that $\mathcal{L}(\mathcal{D}) \cap \mathcal{L}(\mathcal{R}) = \emptyset$ iff $\mathcal{L}(\mathcal{D}') = \emptyset$. $\mathcal{D}'$ is of polynomial size in the size of $\mathcal{D}$ and $\mathcal{R}$ and of exponential size in the number of registers of $\mathcal{R}$.*

Translating LTL formulae into word automata results in a state space that is at most exponential in the size of the formula and thus the construction gives an up to exponential overall blowup. Note that we assume a unary encoding for the offsets r in formulae C^r. By Theorem 2, the translation proves our completeness result for BD-LTL$^\pm$.

Theorem 3 (2ExpSpace**-completeness**)**.** *Satisfiability problems of BD-LTL$^+$ and BD-LTL$^-$ are 2ExpSpace-complete over finite and infinite data words.*

5 Ordered Navigation on Multi-attributed Data Words

As we have seen, multiple attributes do not actually enrich the models of BD-LTL. They can be eliminated due to the inability of BD-LTL to reason about their interdependencies. A natural extension is thus to allow for so-called *tuple navigation*, e. g., by adding an operator $C^r_{(x,y)}$ binding a tuple instead of single values. Class operators such as $\mathrm{X}^=$ and $\mathrm{S}^=$ then navigate along the positions of a multi-attributed data word that carry both values. Unfortunately, it is well-known that such an extension leads to undecidability. For example, LRV is known to be undecidable when being extended by tuple navigation [8]. This implies undecidability of such an extension of BD-LTL$^+$ and by similar arguments BD-LTL$^-$.

Proposition 2. *The satisfiability problem of BD-LTL$^\pm$ with tuple navigation is undecidable.*

To overcome the restrictions of BD-LTL while maintaining decidability, at least for reasonable fragments, we define the logic ND-LTL.

Definition 3 (ND-LTL). *The logic* Nested Data LTL (ND-LTL) *consists of BD-LTL formulae where the set of attributes A is enriched by a tree order relation $\leq \subseteq A \times A$. The fragments* ND-LTL$^+$ *and* ND-LTL$^-$ *are obtained by the same restrictions as for BD-LTL$^+$ and BD-LTL$^-$, respectively.*

The quantifier C^r_x in ND-LTL binds not only the value of attribute $x \in A$ but also the values of all smaller attributes. Class operators, such as $\mathrm{U}^=$, then navigate according to this tuple of values respecting, however, the attribute order in the following sense.

For an attribute $x \in A$, with downward-closure $x^\downarrow$ consisting of attributes $x_1 < x_2 < \ldots < x_n$, a mapping $\mathbf{d} \in \Delta^{x^\downarrow}$ induces a vector of data values $(\mathbf{d}(x_1), \mathbf{d}(x_2), \ldots, \mathbf{d}(x_n))$. By $\mathbf{d} \simeq \mathbf{d}'$ we denote that $\mathbf{d}$ and $\mathbf{d}'$ have the same such vector representation. Note, this can differ from the element-wise equality of the functions. Using this we define for a data word $w \in (\Sigma \times \Delta^A)^\infty$ and $\mathbf{d} \in \Delta^{x^\downarrow}$ the set $\mathrm{pos}_{\mathbf{d}}(w)$ of positions i in w where there is an attribute $y \in A$ such that $\mathbf{d} \simeq \mathbf{d}_i|_{y^\downarrow}$. ND-LTL class formulae are interpreted over models $(w, i, \mathbf{d})$ where $i \in \mathbb{N}$, $0 \leq i < |w|$, is a position in w and $\mathbf{d} \in \Delta^{x^\downarrow}$ for some $x \in A$. For position formulae φ, $x, y \in A$ and $r \in \mathbb{Z}$, we define the semantics of the C^r_x operator and class formulae ψ as follows.

$$
\begin{array}{lll}
(w,i) & \models \mathrm{C}^r_x\psi & \text{if } 0 < i+r < |w| \text{ and } (w, i+r, \mathbf{d}_i|_{x^\downarrow}) \models \psi, \\
(w,i,\mathbf{d}) & \models \varphi & \text{if } (w,i) \models \varphi, \\
(w,i,\mathbf{d}) & \models @x & \text{if } \mathbf{d}_i|_{x^\downarrow} \simeq \mathbf{d}, \\
(w,i,\mathbf{d}) & \models \mathrm{X}^= \psi & \text{if there is } j \in \mathrm{pos}_{\mathbf{d}}(w),\ j > i, \\
 & & \text{and, for the smallest such } j,\ (w,j,\mathbf{d}) \models \psi, \\
(w,i,\mathbf{d}) & \models \psi_1 \, \mathrm{U}^= \, \psi_2 & \text{if } \exists_{j \in \mathrm{pos}_{\mathbf{d}}(w), j \geq i} : (w,j,\mathbf{d}) \models \psi_2 \\
 & & \text{and } \forall_{j' \in \mathrm{pos}_{\mathbf{d}}(w), j > j' \geq i} : (w,j',\mathbf{d}) \models \psi_1.
\end{array}
$$

As before, the operators $\mathrm{Y}^=$ and $\mathrm{S}^=$ are defined as expected. The semantics of boolean and LTL operators in ND-LTL formulae remains as for BD-LTL.

Lemma 4. *For every rMCS $\mathcal{M} = (Q, C, \delta, Q_0)$, there is an ND-LTL$^-$ formula $\Phi_{\mathcal{M}}$ over the set of propositions $AP = Q \cup \{\mathrm{inc}, \mathrm{dec}, \mathrm{res}\} \cup C$ and attributes A s. t. $\Phi_{\mathcal{M}}$ is satisfiable iff there is a data word $w \in (2^{AP} \times \Delta^A)^\omega$ where $\mathsf{str}(w) = \{p_0\}\{p_1\}\ldots$ ($p_i \in AP$) and $p_0 p_1 \ldots$ is a run in $\mathcal{M}$.*

Using a pair $x_c > \hat{x}_c$ of attributes for each counter $c \in C$, a formula $\bigwedge_{c \in C} \mathrm{G}((\mathrm{res} \wedge \mathrm{X}\, c) \rightarrow \mathrm{C}^0_{\hat{x}_c} \neg \mathrm{Y}^= \top)$ can be used for specifying resets and $\bigwedge_{c \in C} \mathrm{G}((\mathrm{dec} \wedge \mathrm{X}\, c) \rightarrow$

$\mathrm{C}^0_{x_c}\,\mathrm{Y}^=(\mathrm{inc} \wedge \mathrm{X}\,c))$ assures non-negative counter values. It is clear that using a further constraint of the form $\mathrm{F}\,q$ allows for expressing control state reachability in rMCS, being ACKERMANN-hard by results on lossy channel systems in [20]. Encoding such finite runs of an rMCS *backwards*, can be done analogously within the fragment ND-LTL$^+$.

Theorem 4 (ACK-hardness). *Satisfiability of ND-LTL$^\pm$ is* ACKERMANN*-hard.*

Similarly, $\mathrm{G}\,\mathrm{F}\,q$ expresses *repeated* control state reachability in rMCS, being undecidable due to results in [5]. Further, full ND-LTL is already undecidable over *finite* words. This can be shown by considering the formula $\bigwedge_{c\in C} \mathrm{G}((\mathrm{inc} \wedge \mathrm{X}\,c) \rightarrow \mathrm{C}^0_{x_c}\,\mathrm{X}^=(\mathrm{dec} \wedge \mathrm{X}\,c))$ that ensures that for every incrementing operation, there is a following decrement on the same counter *before the next reset* on that counter. Thus, reset operations turn into *zero tests*, allowing to encode Minski machine computations where reachability is undecidable.

Theorem 5 (Undecidability). *Satisfiability of ND-LTL is undecidable over* finite and infinite *data words. Satisfiability of ND-LTL$^-$ is undecidable over* infinite *data words.*

6 Deciding Satisfiability of ND-LTL$^\pm$

Having established undecidability and hardness results for ND-LTL we finally turn to decision procedures in this section. We complete our picture by decidability results for the remaining cases of ND-LTL$^-$ over finite words and ND-LTL$^+$ over finite and infinite words. The structure follows that of Section 4 and we provide the essential ideas for lifting the constructions as well as additional arguments where needed. To capture the notion of nesting in ND-LTL we extend data automata and again provide restrictions that carry over from the logic.

We extend data automata to read multi-attributed data words by adding a class automaton for each attribute. The class automata are linearly ordered in the sense that the i-th class automaton reads refinements (subwords) of the input of the $(i-1)$-th class automaton. That way they express a linear order on the attributes which is, however, sufficient since we later show that ND-LTL formulae over a tree order can be translated into formulae over a linear order. For that reason, we only consider attribute sets $[k] = \{1, \ldots, k\}$ for $k \in \mathbb{N}$.

Definition 4 (Nested data automaton). *A* k-nested data automaton *(k-NDA) is a $(k+1)$-tuple $\mathcal{D} = (\mathcal{A}, \mathcal{B}_1, \ldots, \mathcal{B}_k)$ where $(\mathcal{A}, \mathcal{B}_i)$ is a data automaton for each $i \in [k]$. $\mathcal{D}$ is called* locally prefix-closed (pNDA) *if each $(\mathcal{A}, \mathcal{B}_i)$ is a pDA and it is called* locally suffix-closed (sNDA) *if each $(\mathcal{A}, \mathcal{B}_i)$ is an sDA.*

Let $\mathcal{D} = (\mathcal{A}, \mathcal{B}_1, \ldots, \mathcal{B}_k)$ be a k-NDA with $\mathcal{A} = (Q, \Sigma, \Gamma, \delta_\mathcal{A}, Q_0, F_\mathcal{A}, B_\mathcal{A})$ and $\mathcal{B}_i = (S_i, \Gamma, \delta_i, I_i, F_i, B_i)$. A *configuration* of $\mathcal{D}$ is a tuple $c = (q, f_1, \ldots, f_k) \in Q \times \mathfrak{F}_1 \times \ldots \times \mathfrak{F}_k$ where $\mathfrak{F}_i$ is the set of *memory functions* $f : \Delta^{[i]} \rightarrow S_i \cup \{\perp\}$ (partially) mapping i-tuples of data values to states.

A *run* of $\mathcal{D}$ on an $[k]$-attributed data word $w = (a_0, \mathbf{d}_0)(a_1, \mathbf{d}_1) \ldots \in (\Sigma \times \Delta^{[k]})^\infty$ is a maximal sequence $\rho = (q_0, f_{1,0}, \ldots, f_{k,0})(q_1, f_{1,1}, \ldots, f_{k,1}) \ldots$ of configurations where $q_0 \in Q_0$, $f_{i,0}(\Delta^{[i]}) = \{\perp\}$ and for each consecutive positions

$n, n+1$ on ρ there is a transition $(q_n, a_n, g, q_{n+1}) \in \delta_{\mathcal{A}}$ for $g \in \Gamma$ of the base automaton and a transition $(s_i, g, s_i') \in \delta_i$ for each class automaton $\mathcal{B}_i$ such that (1) $f_{i,n+1}(\mathbf{d}_n|_{[i]}) = s_i'$ and (2) either $f_{i,n}(\mathbf{d}_n|_{[i]}) = s_i$, or $f_{i,n}(\mathbf{d}_n|_{[i]}) = \bot$ and $s_i \in I_i$, and (3) $\forall_{\mathbf{d}' \in \Delta^{[i]}, \mathbf{d}' \neq \mathbf{d}_n|_{[i]}} : f_{i,n}(\mathbf{d}') = f_{i,n+1}(\mathbf{d}')$.

A run of $\mathcal{D}$ on w is (finitely) *accepting* if it ends in a configuration $(q, f_1, \ldots f_k)$ with $q \in F_{\mathcal{A}}$ and $\forall_{i \in [k]} f_i(\Delta^{[i]}) \subseteq F_i \cup \{\bot\}$. Moreover, it is accepting if there are infinitely many configurations $(q, f_{1,n}, \cdots, f_{k,n})$ on ρ such that $q \in B_{\mathcal{A}}$ is Büchi-accepting and for each level $i \in [k]$ and each data valuation $\mathbf{d} \in \Delta^{[i]}$ there is either (I) *no* position m with $\mathbf{d}_m|_{[i]} = \mathbf{d}$, or (II) a *last* position m with $\mathbf{d}_m|_{[i]} = \mathbf{d}$ and the state $f_{i,m+1}(\mathbf{d}) \in F$ is final, or (III) there are *infinitely many* positions m where $\mathbf{d}_m|_{[i]} = \mathbf{d}$ and $f_{i,m+1}(\mathbf{d}) \in B_i$ is Büchi-accepting.

The idea of deciding emptiness of pNDA and sNDA is, again, to translate them into multi-counter systems, which this time will be nested. Similar notions of such nested systems can be found in [17,2].

Definition 5 (k-nMCS). *A* k-nested multi-counter system (k-nMCS) *is a tuple* $\mathcal{M} = (Q, \delta, I)$ *with a finite set of states* Q*, a set of* initial states $I \subseteq Q$*, and a* transition relation $\delta \subseteq (\bigcup_{i \in [k]} Q^i) \times Q^k$.

A *multiset* over a set S is a mapping $m \in \mathbb{N}^S$. For a k-nMCS $\mathcal{M} = (Q, \delta, I)$, the set of *configurations of level* i are defined inductively (from k to 0) as $C_k = Q$ and $C_{i-1} = Q \times \mathbb{N}^{C_i}$. The set of *configurations* of $\mathcal{M}$ is then $C_{\mathcal{M}} = C_0$. We can see an element of C_0 as a term constructed over unary function symbols Q, constants Q and the binary operator $+$. The terms are considered modulo associativity and commutativity of the $+$ operator which does not appear on the top level. For example $q_0(q_1(q_3(q_5+q_5+q_6)+q_3(q_6+q_6))+q_1(q_3(q_6+q_6)+q_3(q_6+q_5+q_5))+q_2(q_7(q_8))+q_2(q_7(q_8)))$ corresponds to $(q_0, \{(q_1, \{(q_3, \{q_5 : 2, q_6 : 1\}) : 2, (q_3, \{q_6 : 2\}) : 2\}) : 1, (q_2, \{(q_7, \{q_8 : 1\}) : 1\}) : 2\})$.

Now, the transition relation $\rightarrow \subseteq C_{\mathcal{M}} \times C_{\mathcal{M}}$ on configurations can be easily defined as a rewrite rule. For $((q_0, q_1, \ldots, q_i), (q_0', q_1', q_2', \ldots, q_k')) \in \delta$, we have $(q_0, X_1 + q_1(X_2 + \ldots q_i(X_{i+1}) \ldots)) \rightarrow (q_0', X_1 + q_1'(X_2 + \ldots q_i'(X_{i+1} + q_{i+1}'(q_{i+2}' \cdots q_{k-1}'(q_k')))))$ where $X_i \in \mathbb{N}^{C_i}$. As usual we denote by $\rightarrow^*$ the reflexive and transitive closure of $\rightarrow$.

A *well-quasi-ordering (WQO)* on a set C is a pre-order $\preceq$ such that, for any infinite sequence $c_0, c_1, c_2, \ldots$ there are i, j with $i < j$ and $c_i \preceq c_j$. A WQO $\preceq$ on a set C induces a WQO $\preceq_m$ on multisets over C as follows. Let $B = \{b_1, \ldots, b_n\}$ and $B' = \{b_1', \ldots, b_{n'}'\}$ two multisets over C. Then, $B \preceq_m B'$ iff there is an injection h from $[n]$ to $[n']$ with $b_i \preceq b_{h(i)}'$. Let $\preceq_k$ be the WQO $=$ (equality relation) on the set of states Q of the k-nMCS $\mathcal{M}$. We iterate the construction and obtain a WQO $\preceq_1$ on $C_{\mathcal{M}}$. It can be easily seen that the transition relation $\rightarrow$ of k-nMCS is *monotonic* wrt. $\preceq_1$, i. e., if $c_1 \preceq_1 c_2$ and $c_1 \rightarrow c_3$ then $c_2 \rightarrow c_4$ for some c_4 with $c_2 \preceq_1 c_4$. A k-nMCS is hence a *well-structured transition system* [1] and we obtain the following lemma.

Lemma 5 (Coverability). *Let* $\mathcal{M} = (Q, \delta, I)$ *be a* k*-nMCS,* $c \in C_{\mathcal{M}}$ *a configuration and* $q \in Q$ *a state. The* coverability problem *of checking if there is a configuration* $c' \in C_{\mathcal{M}}$ *with* $c \preceq c'$ *such that* $(q, \emptyset) \rightarrow^* c'$*, is decidable.*

Given a k-NDA $\mathcal{D} = (\mathcal{A}, \mathcal{B}_1, \ldots, \mathcal{B}_k)$ where $\mathcal{A} = (Q_0, \Sigma, \Gamma, \delta_0, I_0, F_0, \emptyset)$ and $\mathcal{B}_i = (Q_i, \Gamma, \delta_i, I_i, F_i, \emptyset)$ for $i \in [k]$ without Büchi-accepting states, we can construct a k-nMCS $\mathcal{M}_{\mathcal{D}} = (\bigcup_{i=0}^{k} Q_i, \delta, I_0)$ where δ mimics the possible transitions of $\mathcal{A}$ and the $\mathcal{B}_i$. Then $\mathcal{D}$ is nonempty iff a configuration can be reached in $\mathcal{M}_{\mathcal{D}}$ containing only states from $F := \bigcup_{i=0}^{k} F_i$ If all all class automata states are final ($\mathcal{D}$ is a pNDA) this reduces to *coverability* in $\mathcal{M}_{\mathcal{D}}$. The case where $\mathcal{D}$ is an sNDA can be reduced to the case of pNDA as above by reversing the base and the class automata. As is shown later, emptiness of pNDA *with* Büchi-accepting is undecidable. The remaining case of sNDA is addressed in the rest of this section. We present conditions that are necessary and sufficient for for an sNDA to nonempty and sketch how they can be checked.

Let $\mathcal{D} = (\mathcal{A}, \mathcal{B}_1, \ldots, \mathcal{B}_k)$ be a k-sNDA where $\mathcal{A} = (Q, \Sigma, \Gamma, \delta_{\mathcal{A}}, I_{\mathcal{A}}, \emptyset, B_{\mathcal{A}})$ and $\mathcal{B}_i = (S_i, \Gamma, \delta_i, S_i, F_i, B_i)$. For a configuration $c = (q, f_1, \ldots, f_k)$ of $\mathcal{D}$, a data valuation $\mathbf{d} \in \Delta^{[1]}$ with $f_1(\mathbf{d}) \neq \bot$ corresponds to an "active" instance of the class automaton $\mathcal{B}_1$. Consider the set $m := \{\mathbf{d}' \in \Delta^{[i]} \mid i \in [k], f_i(\mathbf{d}') \neq \bot, \mathbf{d}'(1) = \mathbf{d}(1)\}$ of data valuations *depending* on $\mathbf{d}$. It is prefix-closed wrt. the linear order on $[k]$ and can hence be considered as a *tree* with root $\mathbf{d}$ (level 1). Define a labeling $s : m \to \bigcup_{i \in [k]} S_i$ attaching to each node $\mathbf{d}' \in \Delta^{[i]}$ (level i) in m the current state of the corresponding class automaton instance, i. e., $s(\mathbf{d}') := f_i(\mathbf{d}')$, and repeatedly delete all leaf nodes of m that are final states. Let M_c be the (finite) set of all such labeled trees (m, s) for a configuration c.

As done similar in Section 4, we characterize a configuration that splits the finite from the infinite behaviour on an accepting run of $\mathcal{D}$. For a set of transitions $T \subseteq \delta_1$ of $\mathcal{B}_1$, a state $q \in Q$ of $\mathcal{A}$ and a finite set M of finite trees labeled by states from $S_1 \cup \ldots \cup S_k$, consider the following properties.

(B1) For all $t_1 \in T$ there is a sequence $t_1 t_2 \ldots \in T^{\infty}$, $t_i = (s_i, g_i, s_i')$, inducing an accepting run of $\mathcal{B}_1$ and $g_1 g_2 \ldots \in ⧢(\mathcal{L}(\mathcal{B}_2) \cap ⧢(\ldots \cap ⧢(\mathcal{L}(\mathcal{B}_{k-1}) \cap ⧢\mathcal{L}(\mathcal{B}_k))\ldots))$.

(B2) There is a sequence $t_1 t_2 \ldots \in T^{\omega}$ with $t_i = (s_i, g_i, s_i')$ in which each $t \in T$ occurs infinitely often and $g_1 g_2 \ldots \in \Gamma^{\omega}$ is an output of $\mathcal{A}$ starting in q.

(B3) There is a reachable configuration $c = (q, f_1, \ldots, f_k)$ with $M = M_c$ such that for all $i \in [k]$ and all $\mathbf{d} \in \Delta^{[i]}$ either (i) $f_i(\mathbf{d}) = \bot$ (there is no corresponding instance), or (ii) $\forall_{\mathbf{d}' \in \Delta^{[k]} \text{ s.t. } \mathbf{d}'|_{[i]} = \mathbf{d}} \forall_{j \geq i} : f_j(\mathbf{d}'|_{[j]}) \in F_j$ (the corresponding instance and all instances depending on it are in a final state), or (iii) $\exists_{g \in \Gamma, s' \in S} : (f_1(\mathbf{d}|_{[1]}), g, s') \in T$ (there is a transition applicable to the corresponding instance of $\mathcal{B}_1$).

(B4) For each tree $(m, s) \in M$ there is a second labeling $\gamma : m \to \Gamma^{\infty}$ such that, for the root $r \in m$, the label $\gamma(r)$ is accepted by $\mathcal{B}_1$ restricted to T when starting in state $s(r) \in S_1$ and for all nodes $v \in m$ on a level $i > 1$ (i) $\gamma(v)$ is accepted by $\mathcal{B}_i$ starting in state $s(v) \in S_i$ and (ii) $\gamma(v)$ must be a shuffle of the labels of the direct children of v and a (possibly infinite) number of words from the shuffle set $⧢(\mathcal{L}(\mathcal{B}_{i+1}) \ldots \cap ⧢(\mathcal{L}(\mathcal{B}_{k-1}) \cap ⧢\mathcal{L}(\mathcal{B}_k))\ldots)$.

Lemma 6. *The sNDA $\mathcal{D}$ accepts an infinite data word iff there are $T \subseteq \delta_1$, $q \in Q$ and a set M of finite trees labeled by states from $S_1 \cup \ldots \cup S_k$ s. t. properties (B1)–(B4) hold.*

Based on these conditions we obtain a decision procedure by nondeterministically guessing the state q and the set of transitions T. We verify (B2) as above by constructing and analyzing a Büchi automaton. Then we compute compute a set of candidates for M that satisfy (B3) using the following idea. Construct a k-sNDA $\tilde{\mathcal{D}} = (\tilde{\mathcal{A}}, \tilde{\mathcal{B}}_1, \ldots, \tilde{\mathcal{B}}_k)$, without Büchi-accepting states, from $\mathcal{D}$ by taking q as only final state in $\tilde{\mathcal{A}}$. In each step, $\tilde{\mathcal{A}}$ guesses whether the currently active instance of $\tilde{\mathcal{B}}_1$ performs its last step entering a source state s of some transition $(s, g, s') \in T$. In that case it marks the current output by some flag. $\tilde{B}_1$ simulates $\mathcal{B}_1$ and verifies that $\tilde{\mathcal{A}}$ guessed correctly. Each other class automaton $\tilde{\mathcal{B}}_i$ ($i > 1$) simulates $\mathcal{B}_i$. Upon reading the flag it moves to a final copy of the state they would have moved to otherwise. The configurations in which $\tilde{\mathcal{D}}$ can accept are exactly those configurations reachable by $\mathcal{D}$ that satisfy (B3). We apply the standard saturation algorithm for well-structured transition systems where constraints are propagated from the target control state backwards along the edges of the nMCS. After its termination, the algorithm computed the minimal preconditions for reaching a the target state. On a reversed sNDA, this can be understood as a forward propagation computing minimal post-conditions. In this case the target state is q and the minimal post conditions characterize the minimal configurations $(q, f_1, \ldots, f_k)$ that can be reached. Here, minimal means with the smallest number of instances of some class automaton. The post-conditions hence give us all minimal sets M when reaching q. These are the (finitely many) candidates for (B4) since if none of those satisfies the properties any larger one will not either. For testing the candidates M to comply (B4) and T to satisfy (B1), the essential idea is to let the shuffle requirements be checked by a $(k-1)$-sNDA built by modifying the components of $\mathcal{D}$. Such an automaton is constructed for each $(m, s) \in M$ and each $t \in T$, respectively, and can, by induction, be checked for emptiness.

Theorem 6. *Emptiness of sNDA is decidable over finite and infinite data words. Emptiness of pNDA is decidable over finite data words.*

From ND-LTL to NDA. The translation from ND-LTL$^{\pm}$ to sNDA and pNDA, respectively, follows closely the one for BD-LTL in Section 4.2. For an ND-LTL formula over arbitrarily ordered attributes, a word has at every position a tree of attributes with n maximal paths of length of at most k. The first step is to translate this formula to a formula over the linearly order set of attributes $[k]$ and encode each position of such a word by a segment of length n, where each position within a segment corresponds to a maximal path in the tree order $(A, \leq)$. This step is crucial as NDA only navigate according to linearly ordered attributes. For translating the obtained formula φ into an NDA, the set AP_{φ} of atomic propositions used by φ is extended by propositions p_j^{ψ} and $=_j^x$ for each $-r_{\max} \leq j \leq r_{\max}$ and subformula ψ and attribute x of φ, where r_{max} denotes the largest (absolute) value used by C_x^r operators. As before positional formulae can be checked by the base automaton. Class formulae of the form $\mathrm{C}_x^r\,\psi$ can be handled by the local automaton corresponding to attribute x. Propositions $=_j^x$ are checked separately for each attribute x by adapting Lemma 3.

Now, together with Theorem 6, we obtain a decision procedure for ND-LTL$^{\pm}$.

Theorem 7. *Satisfiability of ND-LTL^{+} is decidable over* finite and infinite *data words. Satisfiability of ND-LTL^{-} is decidable over* finite *data words.*

Corollary 1. *Emptiness of pNDA wrt. infinite data words is undecidable.*

References

1. Abdulla, P.A.: Well (and better) quasi-ordered transition systems. Bulletin of Symbolic Logic 16(4), 457–515 (2010)
2. Björklund, H., Bojańczyk, M.: Shuffle expressions and words with nested data. In: Kučera, L., Kučera, A. (eds.) MFCS 2007. LNCS, vol. 4708, pp. 750–761. Springer, Heidelberg (2007)
3. Björklund, H., Schwentick, T.: On notions of regularity for data languages. Theor. Comput. Sci. 411(4-5), 702–715 (2010)
4. Bojanczyk, M., David, C., Muscholl, A., Schwentick, T., Segoufin, L.: Two-variable logic on data words. ACM Trans. Comput. Log. 12(4), 27 (2011)
5. Bouajjani, A., Mayr, R.: Model checking lossy vector addition systems. In: Meinel, C., Tison, S. (eds.) STACS 1999. LNCS, vol. 1563, pp. 323–333. Springer, Heidelberg (1999)
6. Bouyer, P., Petit, A., Thérien, D.: An algebraic approach to data languages and timed languages. Inf. Comput. 182(2), 137–162 (2003)
7. Demri, S., D'Souza, D., Gascon, R.: Temporal logics of repeating values. J. Log. Comput. 22(5), 1059–1096 (2012)
8. Demri, S., Figueira, D., Praveen, M.: Reasoning about data repetitions with counter systems. In: LICS, pp. 33–42. IEEE Computer Society (2013)
9. Demri, S., Lazic, R.: LTL with the freeze quantifier and register automata. ACM Trans. Comput. Log. 10(3) (2009)
10. Demri, S., Lazic, R., Nowak, D.: On the freeze quantifier in constraint LTL: Decidability and complexity. Inf. Comput. 205(1), 2–24 (2007)
11. Finkel, A., Schnoebelen, P.: Well-structured transition systems everywhere! Theor. Comput. Sci. 256(1-2), 63–92 (2001)
12. Habermehl, P.: On the complexity of the linear-time μ-calculus for Petri nets. In: Azéma, P., Balbo, G. (eds.) ICATPN 1997. LNCS, vol. 1248, pp. 102–116. Springer, Heidelberg (1997)
13. Kaminski, M., Francez, N.: Finite-memory automata. Theor. Comput. Sci. 134(2), 329–363 (1994)
14. Kara, A., Schwentick, T., Zeume, T.: Temporal logics on words with multiple data values. In: Lodaya, K., Mahajan, M. (eds.) FSTTCS. LIPIcs, vol. 8, pp. 481–492 (2010)
15. Kara, A., Schwentick, T., Zeume, T.: Temporal logics on words with multiple data values. CoRR abs/1010.1139 (2010)
16. Leroux, J.: Vector addition system reachability problem: A short self-contained proof. In: Ball, T., Sagiv, M. (eds.) POPL, pp. 307–316. ACM (2011)
17. Lomazova, I.A., Schnoebelen, P.: Some decidability results for nested Petri nets. In: Bjorner, D., Broy, M., Zamulin, A.V. (eds.) PSI 1999. LNCS, vol. 1755, pp. 208–220. Springer, Heidelberg (2000)
18. Neven, F., Schwentick, T., Vianu, V.: Finite state machines for strings over infinite alphabets. ACM Trans. Comput. Log. 5(3), 403–435 (2004)
19. Rackoff, C.: The covering and boundedness problems for vector addition systems. Theor. Comput. Sci. 6, 223–231 (1978)
20. Schnoebelen, P.: Revisiting Ackermann-hardness for lossy counter machines and reset Petri nets. In: Hlinený, P., Kucera, A. (eds.) MFCS 2010. LNCS, vol. 6281, pp. 616–628. Springer, Heidelberg (2010)
21. Schwentick, T., Zeume, T.: Two-variable logic with two order relations. Logical Methods in Computer Science 8(1) (2012)
22. Tzevelekos, N., Grigore, R.: History-register automata. In: Pfenning, F. (ed.) FOSSACS 2013. LNCS, vol. 7794, pp. 17–33. Springer, Heidelberg (2013)

Verification for Timed Automata Extended with Unbounded Discrete Data Structures

Karin Quaas*

Universität Leipzig, Germany

Abstract. We study decidability of verification problems for timed automata extended with unbounded discrete data structures. More detailed, we extend timed automata with a pushdown stack. In this way, we obtain a strong model that may for instance be used to model real-time programs with procedure calls. It is long known that the reachability problem for this model is decidable. The goal of this paper is to identify subclasses of timed pushdown automata for which the language inclusion problem and related problems are decidable.

1 Introduction

Timed automata were introduced by Alur and Dill [4], and have since then become a popular standard formalism to model real-time systems. An undeniable reason for the success of timed automata is the PSPACE decidability of the *language emptiness problem* [4]. A major drawback of timed automata is the undecidability [4] of the *language inclusion problem*: given two timed automata $\mathcal{A}$ and $\mathcal{B}$, does $L(\mathcal{A}) \subseteq L(\mathcal{B})$ hold? The undecidability of this problem prohibits the usage of automated verification algorithms for analysing timed automata, where $\mathcal{B}$ can be seen as the specification that is supposed to be satisfied by the system modelled by $\mathcal{A}$. However, if $\mathcal{B}$ is restricted to have at most one clock, then the language inclusion problem over finite timed words is decidable (albeit with non-primitive recursive complexity) [31]. Another milestone in the success story of timed automata is the decidability of the *model checking problem* for timed automata and Metric Temporal Logic (MTL, for short) over finite timed words [32].

Timed automata can express many interesting time-related properties, and even with the restriction to a single clock, they allow one to model a large class of systems, including, for example, the internet protocol TCP [30]. If we want to reason about real-time programs with procedure calls, or about the number of events occurring in computations of real-time systems, we have to extend the model of timed automata with some unbounded discrete data structure. In 1994, Bouajjani et al. [7] extended timed automata with discrete counters and a pushdown stack and proved that the satisfiability of reachability properties for several subclasses of this model is decidable. Nine years later, it was shown that

* The author is supported by *DFG*, project QU 316/1-1.

P. Baldan and D. Gorla (Eds.): CONCUR 2014, LNCS 8704, pp. 512–526, 2014.

the binary reachability relation for *timed pushdown systems* is decidable [14]. Decidability of the reachability problem was also proved for several classes of *timed counter systems* [8], mainly by simple extensions of the classical region-graph construction [4]. The language inclusion problem, however, is to the best of our knowledge only considered in [18] for the class of timed pushdown systems. In [18] it is stated that the language inclusion problem is decidable if $\mathcal{A}$ is a timed pushdown automaton, and $\mathcal{B}$ is a one-clock timed automaton. The proof is based on an extension of the proof for the decidability of the language inclusion problem for the case that $\mathcal{A}$ is a timed automaton without pushdown stack [31]. Unfortunately, and as is well known, the proof in [18] is not correct.

In this paper, we prove that different to what is claimed in [18], the language inclusion problem for the case that $\mathcal{A}$ is a pushdown timed automaton and $\mathcal{B}$ is a one-clock timed automaton is undecidable. This is even the case if $\mathcal{A}$ is a deterministic instance of a very restricted subclass of timed pushdown automata called *timed visibly one-counter nets*. On the other hand, we prove that the language inclusion problem is decidable if $\mathcal{A}$ is a timed automaton and $\mathcal{B}$ is a timed automaton extended with a finite set of counters that can be incremented and decremented, and which we call *timed counter nets*. As a special case, we obtain the decidability of the *universality problem* for timed counter nets: given a timed automaton $\mathcal{A}$ with input alphabet Σ, does $L(\mathcal{A})$ accept the set of all timed words over Σ? Finally, we give the precise decidability border for the universality problem by proving that the universality problem is undecidable for the class of *timed visibly one-counter automata*. We remark that all results apply to extensions of timed automata over *finite* timed words.

2 Extensions of Timed Automata with Discrete Data Structure

We use $\mathbb{Z}$, $\mathbb{N}$ and $\mathbb{R}_{\geq 0}$ to denote the integers, the non-negative integers and the non-negative reals, respectively.

We use Σ to denote a finite alphabet. A *timed word* over Σ is a non-empty finite sequence $(a_1, t_1) \dots (a_k, t_n) \in (\Sigma \times \mathbb{R}_{\geq 0})^+$ such that the sequence $t_1, \dots, t_n$ of timestamps is non-decreasing. We say that a timed word is *strictly monotonic* if $t_{i-1} < t_i$ for every $i \in \{2, \dots, n\}$. We use $T\Sigma^+$ to denote the set of finite timed words over Σ. A set $L \subseteq T\Sigma^+$ is called a *timed language*.

Let $\mathcal{X}$ be a finite set of *clock variables* ranging over $\mathbb{R}_{\geq 0}$. We define *clock constraints* ϕ over $\mathcal{X}$ to be conjunctions of formulas of the form $x \sim c$, where $x \in \mathcal{X}$, $c \in \mathbb{N}$, and $\sim \in \{<, \leq, =, \geq, >\}$. We use $\Phi(\mathcal{X})$ to denote the set of all clock constraints over $\mathcal{X}$. A *clock valuation* is a mapping from $\mathcal{X}$ to $\mathbb{R}_{\geq 0}$. A clock valuation ν satisfies a clock constraint ϕ, written $\nu \models \phi$, if ϕ evaluates to true according to the values given by ν. For $\delta \in \mathbb{R}_{\geq 0}$ and $\lambda \subseteq \mathcal{X}$, we define $\nu + \delta$ to be $(\nu + \delta)(x) = \nu(x) + \delta$ for each $x \in \mathcal{X}$, and we define $\nu[\lambda := 0]$ by $(\nu[\lambda := 0])(x) = 0$ if $x \in \lambda$, and $(\nu[\lambda := 0])(x) = \nu(x)$ otherwise.

Let Γ be a finite stack alphabet. We use Γ^* to denote the set of finite words over Γ, including the empty word denoted by ε. We define a finite set Op of *stack operations* by $\mathsf{Op} := \{pop(a), push(a) \mid a \in \Gamma\} \cup \{noop, empty?\}$.

A *timed pushdown automaton* is a tuple $\mathcal{A} = (\Sigma, \Gamma, \mathcal{L}, \mathcal{L}_0, \mathcal{L}_f, \mathcal{X}, E)$, where

- $\mathcal{L}$ is a finite set of *locations*,
- $\mathcal{L}_0 \subseteq \mathcal{L}$ is the set of initial locations,
- $\mathcal{L}_f \subseteq \mathcal{L}$ is the set of accepting locations,
- $E \subseteq \mathcal{L} \times \Sigma \times \Phi(\mathcal{X}) \times \mathsf{Op} \times 2^{\mathcal{X}} \times \mathcal{L}$ is a finite set of edges.

A *state* of $\mathcal{A}$ is a triple (l, ν, u), where $l \in \mathcal{L}$ is the current location, the clock valuation ν represents the current values of the clocks, and $u \in \Gamma^*$ represents the current stack content, where the top-most symbol of the stack is the left-most symbol in the word u, and the empty word ε represents the empty stack. We use $\mathcal{G}^{\mathcal{A}}$ to denote the set of all states of $\mathcal{A}$. A timed pushdown automaton $\mathcal{A}$ induces a transition relation $\Rightarrow_{\mathcal{A}}$ on $(\mathcal{G}^{\mathcal{A}} \times \mathbb{R}_{\geq 0} \times \Sigma \times \mathcal{G}^{\mathcal{A}})$ as follows: $\langle (l, \nu, u), \delta, a, (l', \nu', u') \rangle \in \Rightarrow_{\mathcal{A}}$, if, and only if, there exists some edge $(l, a, \phi, \mathsf{op}, \lambda, l') \in E$ such that $(\nu + \delta) \models \phi$, $\nu' = (\nu + \delta)[\lambda := 0]$, and (i) if $\mathsf{op} = \mathit{pop}(a)$ for some $a \in \Gamma$, then $u = a \cdot u'$; (ii) if $\mathsf{op} = \mathit{push}(a)$ for some $a \in \Gamma$, then $u' = a \cdot u$; (iii) if $\mathsf{op} = \mathit{empty?}$, then $u = u' = \varepsilon$; (iv) if $\mathsf{op} = \mathit{noop}$, then $u' = u$. A *run* of $\mathcal{A}$ is a finite sequence $\prod_{1 \leq i \leq n} \langle (l_{i-1}, \nu_{i-1}, u_{i-1}), \delta_i, a_i, (l_i, \nu_i, u_i) \rangle$ such that $\langle (l_{i-1}, \nu_{i-1}, u_{i-1}), \delta_i, a_i, (l_i, \nu_i, u_i) \rangle \in \Rightarrow_{\mathcal{A}}$ for every $i \in \{1, \ldots, n\}$. A run is called *successful* if $l_0 \in \mathcal{L}_0$, $\nu_0(x) = 0$ for every $x \in \mathcal{X}$, $u_0 = \varepsilon$, and $l_n \in \mathcal{L}_f$. With a run we associate the timed word $(a_1, \delta_1)(a_2, \delta_1 + \delta_2) \ldots (a_n, \Sigma_{1 \leq i \leq n} \delta_i)$. The language accepted by a timed automaton, denoted by $L(\mathcal{A})$, is defined to be the set of timed words $w \in T\Sigma^+$ for which there exists a successful run of $\mathcal{A}$ that w is associated with.

Next we define some subclasses of timed pushdown automata; see Fig. 1 for a graphical overview. We start with timed extensions of *one-counter automata* [16,28] and *one-counter nets* [24,1]. A *timed one-counter automaton* is a timed pushdown automaton where the stack alphabet is a singleton. By writing *push* and *pop* we mean that we increment and decrement the counter, respectively, whereas *empty?* corresponds to a zero test. A *timed one-counter net* is a timed one-counter automaton without zero tests, *i.e.*, the *empty?* operation is not allowed. We remark that for both classes, the execution of an edge of the form $(l, a, \phi, \mathit{pop}, \lambda, l')$ is *blocked* if the stack is empty. Next, we consider the timed extension of an interesting subclass of pushdown automata called *visibly pushdown automata* [5]. A *timed visibly pushdown automaton* is a timed pushdown automaton for which the input alphabet Σ can be partitioned into three pairwise disjoint sets $\Sigma = \Sigma_{\mathsf{int}} \cup \Sigma_{\mathsf{call}} \cup \Sigma_{\mathsf{ret}}$ of *internal*, *call*, and *return* input symbols, respectively, and such that for every edge $(l, a, \phi, \mathsf{op}, \lambda, l')$ the following conditions are satisfied:

- $a \in \Sigma_{\mathsf{int}}$ if, and only if, $\mathsf{op} = \mathit{noop}$,
- $a \in \Sigma_{\mathsf{call}}$, if, and only if, $\mathsf{op} = \mathit{push}(b)$ for some $b \in \Gamma$,
- $a \in \Sigma_{\mathsf{ret}}$ if, and only if, $\mathsf{op} = \mathit{empty?}$ or $\mathsf{op} = \mathit{pop}(b)$ for some $b \in \Gamma$.

A *timed visibly one-counter automaton* (timed visibly one-counter net, respectively) is a timed one-counter automaton (timed one-counter net, respectively) that is also a timed visibly pushdown automaton. We say that a timed visibly one-counter net with no clocks is *deterministic* if for all $e = (l, a, \mathtt{true}, \mathsf{op}', \emptyset, l')$, $e' = (l, a, \mathtt{true}, \mathsf{op}'', \emptyset, l'') \in E$ with $e \neq e'$ we have either $\mathsf{op}' = pop$ and $\mathsf{op}'' = empty?$, or $\mathsf{op}' = empty?$ and $\mathsf{op}'' = pop$.

Finally, we define the class of *timed counter nets*, which generalizes timed one-counter nets, but is not a subclass of timed pushdown automata. A timed counter net of dimension n is a tuple $\mathcal{A} = (\Sigma, n, \mathcal{L}, \mathcal{L}_0, \mathcal{L}_f, \mathcal{X}, E)$, where $\mathcal{L}, \mathcal{L}_0, \mathcal{L}_f$ are the sets of locations, initial locations and accepting locations, respectively, and $E \subseteq \mathcal{L} \times \Sigma \times \Phi(\mathcal{X}) \times \{0, 1, -1\}^n \times 2^{\mathcal{X}} \times \mathcal{L}$ is a finite set of edges. A state of a timed counter net is a triple $(l, \nu, \boldsymbol{v})$, where $l \in \mathcal{L}$, ν is a clock valuation, and $\boldsymbol{v} \in \mathbb{N}^n$ is a vector representing the current values of the counters. We define $\langle (l, \nu, \boldsymbol{v}), \delta, a, (l', \nu', \boldsymbol{v}') \rangle \in \Rightarrow_{\mathcal{A}}$ if, and only if, there exists some edge $(l, a, \phi, \boldsymbol{c}, \lambda, l') \in E$ such that $(\nu + \delta) \models \phi$, $\nu' = (\nu + \delta)[\lambda := 0]$, and $\boldsymbol{v}' = \boldsymbol{v} + \boldsymbol{c}$, where vector addition is defined pointwise. Note that, similar to pop operations on an empty stack, transitions which result in the negative value of one of the counters are *blocked*. The notions of *runs*, *successful runs*, *associated timed words* and *the language accepted by* $\mathcal{A}$, are defined analogously to the corresponding definitions for timed pushdown automata.

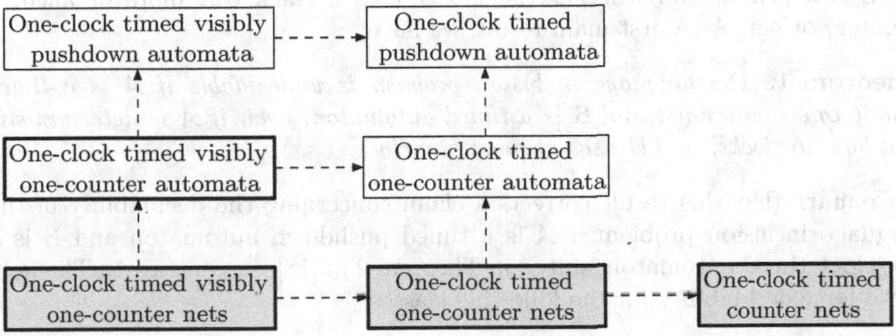

Fig. 1. Extensions of one-clock timed automata with discrete data structures. The subclass relation is represented by dashed arrows. The language emptiness problem is decidable for all classes. The classes in grey boxes have a decidable universality problem, the classes in white boxes have an undecidable universality problem, where the corresponding results for classes in boxes with bold line are new and presented in this paper.

3 Main Results

In this section, we present the main results of the paper. We are interested in the language inclusion problem $L(\mathcal{A}) \subseteq L(\mathcal{B})$, where $\mathcal{A}$ and $\mathcal{B}$ are extensions of timed automata with discrete data structure. Recall that according to standard notation in the field of verification, in this problem formulation $\mathcal{B}$ is seen as the *specification*, and $\mathcal{A}$ is the system that should satisfy this specification, *i.e.*, $\mathcal{A}$ should be a *model* of $\mathcal{B}$. As a special case of this problem, we consider the universality problem, *i.e.*, the question whether $L(\mathcal{A}) = T\Sigma^+$ for a given automaton

$\mathcal{A}$. In general, the two problems are undecidable for timed pushdown automata. This follows on the one hand from the undecidability of the universality problem for timed automata [4], and on the other hand from the undecidability of the universality problem for pushdown automata. In fact, it is long known that the universality problem is undecidable already for non-deterministic one-counter automata [22,26].

However, there are interesting decidability results for subclasses of timed pushdown automata: The language inclusion problem is decidable if $\mathcal{A}$ is a timed automaton, and $\mathcal{B}$ is a timed automaton with at most one clock [31]. As a special case, the universality problem for timed automata is decidable if only one clock is used. The language inclusion problem is also decidable if $\mathcal{A}$ is a one-counter net and $\mathcal{B}$ is a finite automaton, and if $\mathcal{A}$ is a finite automaton and $\mathcal{B}$ is a one-counter net [27]. The universality problem for non-deterministic one-counter nets has recently been proved to have non-primitive recursive complexity [25]. Further we know that the universality and language inclusion problems are decidable if $\mathcal{A}$ and $\mathcal{B}$ are visibly pushdown automata [6].

Hence it is interesting to consider the two problems for the corresponding subclasses of timed pushdown automata. It turns out that the decidability status changes depending on whether the *model* uses a stack (or, more detailed: a counter) or not. As a first main result, we have:

Theorem 1. *The language inclusion problem is undecidable if $\mathcal{A}$ is a timed visibly one-counter net and $\mathcal{B}$ is a timed automaton, even if $\mathcal{A}$ is deterministic and has no clocks, and $\mathcal{B}$ uses at most one clock.*

We remark that this result corrects a claim concerning the decidability of the language inclusion problem if $\mathcal{A}$ is a timed pushdown automaton and $\mathcal{B}$ is a one-clock timed automaton, stated in Theorem 2 in [18]. In contrast to Theorem 1, we have decidability for the following classes:

Theorem 2. *The language inclusion problem is decidable with non-primitive recursive complexity if $\mathcal{A}$ is a timed automaton and $\mathcal{B}$ is a one-clock timed counter net.*

As a special case of this result (and with the lower bound implied by the corresponding result for one-clock timed automata [2]), we obtain:

Corollary 1. *The universality problem for one-clock timed counter nets is decidable with non-primitive recursive complexity.*

The next two sections are devoted to the proofs of Theorems 1 and 2. We will also give some interesting consequences of these results respectively of their proofs. Amongst others, we prove the undecidability of model checking problem for timed visibly one-counter nets and MTL over finite timed words. After this, in Sect. 5, we will prove the following theorem:

Theorem 3. *The universality problem for one-clock timed visibly one-counter automata is undecidable.*

This is in contrast to the decidability of the universality problem for the two underlying models of one-clock timed automata [31] and visibly one-counter automata, which form a subclass of visibly pushdown automata [6]. We also want to point out that this result is stronger than a previous result on the undecidability of the universality problem for one-clock timed visibly pushdown automata (Theorem 3 in [18]), and our proof closes a gap in the proof of Theorem 3 in [18]. Further, we can infer from Corollary 1 and Theorem 3 the exact decidability border for the universality problem of timed pushdown automata, which lies between timed visibly one-counter nets and timed visibly one-counter automata.

4 Undecidability Results

In this section, we prove Theorem 1. The proof is a reduction of an undecidable problem for *channel machines.*

4.1 Channel Machines

Let A be a finite alphabet. We define the order $\leq$ over the set of finite words over A by $a_1a_2\ldots a_m \leq b_1b_2\ldots b_n$ if there exists a strictly increasing function $f:\{1,\ldots,m\}\to\{1,\ldots,n\}$ such that $a_i = b_{f(i)}$ for every $i\in\{1,\ldots,m\}$.

A *channel machine* consists of a finite-state automaton acting on an unbounded fifo channel. Formally, a channel machine is a tuple $\mathcal{C} = (S, s_I, M, \Delta)$, where

- S is a finite set of *control states,*
- $s_I \in S$ is the initial control state,
- M is a finite set of *messages,*
- $\Delta \subseteq S \times L \times S$ is the transition relation over the label set $L = \{!m, ?m \mid m \in M\} \cup \{\mathit{empty}?\}$.

Here, $!m$ corresponds to a *send* operation, $?m$ corresponds to a *read* operation, and *empty*? is a test which returns `true` if and only if the channel is empty. Without loss of generality, we assume that s_I does not have any incoming transitions, *i.e.*, $(s,l,s')\in\Delta$ implies $s'\neq s_I$. Further, we assume that $(s_I,l,s')\in\Delta$ implies $l = \mathit{empty}?$. A *configuration* of $\mathcal{C}$ is a pair (s,x), where $s\in S$ is the control state and $x\in M^*$ represents the contents of the channel. We use $\mathcal{H}^\mathcal{C}$ to denote the set of all configurations of $\mathcal{C}$. The rules in Δ induce a transition relation $\to_\mathcal{C}$ on $(\mathcal{H}^\mathcal{C}\times L\times\mathcal{H}^\mathcal{C})$ as follows:

- $\langle(s,x),!m,(s',x')\rangle \in \to_\mathcal{C}$ if, and only if, there exists some transition $(s,!m,s')\in\Delta$ and $x' = x\cdot m$, *i.e.*, m is added to the tail of the channel.
- $\langle(s,x),?m,(s',x')\rangle \in \to_\mathcal{C}$ if, and only if, there exists some transition $(s,?m,s')\in\Delta$ and $x = m\cdot x'$, *i.e.*, m is removed from the head of the channel.
- $\langle(s,x),\mathit{empty}?,(s',x')\rangle \in \to_\mathcal{C}$ if, and only if, there exists some transition $(s,\varepsilon,s')\in\Delta$ and $x=\varepsilon$, *i.e.*, the channel is empty, and $x' = x$.

Next, we define a second transition relation $\rightsquigarrow_{\mathcal{C}}$ on $(\mathcal{H}^{\mathcal{C}} \times L \times \mathcal{H}^{\mathcal{C}})$. The relation $\rightsquigarrow_{\mathcal{C}}$ is a superset of $\rightarrow_{\mathcal{C}}$. It contains some additional transitions which result from *insertion errors.* We define $\langle(s, x_1), l, (s, x_1')\rangle \in \rightsquigarrow_{\mathcal{C}}$, if, and only if, there exist $x, x' \in M^*$ such that $\langle(s, x), l, (s', x')\rangle \in \rightarrow_{\mathcal{C}}$, $x_1 \leq x$, and $x' \leq x_1'$. A *computation* of $\mathcal{C}$ is a finite sequence $\prod_{1 \leq i \leq k} \langle(s_{i-1}, x_{i-1}), l_i, (s_i, x_i)\rangle$ such that $\langle(s_{i-1}, x_{i-1}), l_i, (s_i, x_i)\rangle \in \rightsquigarrow_{\mathcal{C}}$ for every $i \in \{1, \ldots, k\}$. We say that a computation is *error-free* if for all $i \in \{1, \ldots, k\}$ we have $\langle(s_{i-1}, x_{i-1}), l_i, (s_i, x_i)\rangle \in \rightarrow_{\mathcal{C}}$. Otherwise, we say that the computation is *faulty.*

The proof of Theorem 1 is a reduction from the following undecidable [12,2] control state reachability problem: given a channel machine $\mathcal{C}$ with control states S and $s_F \in S$, does there exist an error-free computation of $\mathcal{C}$ from (s_I, ε) to (s_F, x) for some $x \in M^*$? We remark that the analogous problem for faulty computations is decidable [33]. The idea of our reduction is as follows: Given a channel machine $\mathcal{C}$, we define a timed language $L(\mathcal{C})$ consisting of all timed words that encode potentially faulty computations of $\mathcal{C}$ that start in (s_I, ε) and end in (s_F, x) for some $x \in M^*$. Then we define a timed visibly one-counter net $\mathcal{A}$ such that $L(\mathcal{A}) \cap L(\mathcal{C})$ contains exactly *error-free* encodings of such computations. In other words, we use $\mathcal{A}$ to exclude the encodings of faulty computations from $L(\mathcal{C})$, obtaining undecidability of the non-emptiness problem for $L(\mathcal{A}) \cap L(\mathcal{C})$. Finally, we define a one-clock timed automaton $\mathcal{B}$ that accepts the complement of $L(\mathcal{C})$; hence the problem of deciding whether $L(\mathcal{A}) \not\subseteq L(\mathcal{B})$ is undecidable.

4.2 Encoding Faulty Computations

For the remainder of Section 4, let $\mathcal{C} = (S, s_I, M, \Delta)$ be a channel machine and let $s_F \in S$. Define $\Sigma_{\mathsf{int}} := (S \backslash \{s_I\}) \cup M \cup L \cup \{\#\}$, $\Sigma_{\mathsf{call}} := \{s_I, +\}$, and $\Sigma_{\mathsf{ret}} := \{-, \star\}$, where $+, -, \#$ and $\star$ are fresh symbols that do not occur in $S \cup M \cup L$. We define a timed language $L(\mathcal{C})$ over $\Sigma = \Sigma_{\mathsf{int}} \cup \Sigma_{\mathsf{call}} \cup \Sigma_{\mathsf{ret}}$ that consists of all timed words that encode computations of $\mathcal{C}$ from (s_I, ε) to (s_F, x) for some $x \in M^*$. The definition of $L(\mathcal{C})$ follows the ideas presented in [32].

Let $\gamma = \prod_{1 \leq i \leq k} \langle(s_{i-1}, x_{i-1}), l_i, (s_i, x_i)\rangle$ be a computation of $\mathcal{C}$ with $s_0 = s_I$, $x_0 = \varepsilon$, and $s_k = s_F$. For each $i \in \{0, \ldots, k\}$, the configuration (s_i, x_i) is encoded by a timed word of duration one. This timed word starts with the symbol s_i at some time t_i. If the content of the channel x_i is of the form $m_1 m_2 \ldots m_j$, then s_i is followed by the symbols $m_1, m_2, \ldots, m_j$ in this order. The timestamps of these symbols must be in the interval $(t_i, t_i + 1)$. Due to the denseness of the time domain, one can indeed store the channel content in one time unit without any upper bound on j. For encoding the computation, we glue together the encodings of the single configurations as follows: Every control state symbol s_{i-1} is followed by l_i after exactly one time unit, and by s_i after exactly two time units. For every message symbol m between s_{i-1} and l_i, there is a copy of m after two time units, unless it is removed from the channel by a read operation. There are no symbols between l_i and s_i, and the encoding of the last configuration ends with $\star$ one time unit after s_k. Note: we *do not* require that for every message symbol m between s_{i-1} and l_i there is a copy of m two time units *before.* It is the absence of exactly this condition that causes $L(\mathcal{C})$ to contain encodings of *faulty* computations.

For our reduction to work, we change the idea from [32] in some details. As mentioned above, we will later define a timed visibly one-counter net $\mathcal{A}$ to exclude faulty computations from $L(\mathcal{C})$. The idea is to let $\mathcal{A}$ *guess* the maximum number n of messages occurring between control state symbols and the following label symbol. While reading the encoding of the first configuration of a computation, it increments the counter n times. The automaton does not do any operations on the counter until it reads the encoding of the last configuration, where it decrements the counter whenever it reads symbols occurring between the control state and the label symbol. Since it can decrement the counter at most n times, it can only accept encodings of error-free computations. We define a timed language

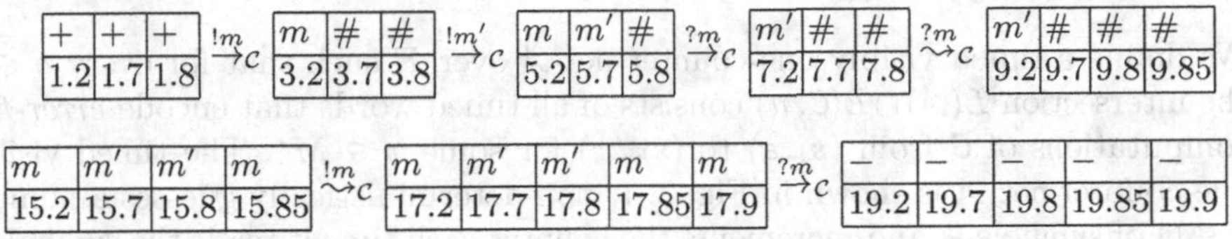

Fig. 2. Examples for the encoding of the channel content for $n = 3$

$L(\mathcal{C}, n)$ for every $n \in \mathbb{N}$. For illustrating the definition, we use the examples in Fig. 2. Since *push*, *pop* and *noop* require symbols from the call, internal, and return input alphabet, respectively, we use three fresh extra symbols $+, \#$, and $-$. In the encoding of the initial configuration, we use n occurrences of the call symbol $+$ as placeholder for message symbols, see the first timed word in Fig. 2(a). In the encoding of all following configurations except for the last one we use the internal symbol $\#$ as placeholder. In the encoding of the last configuration, we use the return symbol $-$ as placeholder, see the last timed word in Fig. 2(b). Every time some $!m$ operation occurs, all symbols in the encoding of the current configuration have a copy after exactly two time units, except for the first free placeholder symbol, which is replaced by m in the encoding of the next configuration, see for instance the first two transitions in Fig. 2(a). Every time an error-free $?m$ operation occurs, the first message symbol in the encoding of the current configuration (which should be m) is replaced by a new placeholder symbol at the end of the encoding of the next configuration, and the timestamps of the other symbols are shifted one position, see the third transition in Fig. 2(a). A faulty read operation due to an insertion error is encoded by the insertion of a new placeholder symbol at the end of the encoding of the next configuration, see the last transition in Fig. 2(a). We also insert a new symbol if a faulty send operation due to a too small choice of n is happening (first transition in Fig. 2(b)). Note, however, that this transition is not faulty due to an insertion error of the channel machine.

Let $w \in L(\mathcal{C}, n)$ for some $n \in \mathbb{N}$. We use $\max(w)$ to denote the maximum number of symbols in $M \cup \{\#, +, -\}$ that occur in w between a control state symbol and a symbol in $L \cup \{\star\}$. Let $\gamma = \prod_{1 \leq i \leq k} \langle (s_{i-1}, x_{i-1}), l_i, (s_i, x_i) \rangle$ be a computation of $\mathcal{C}$. We use $\max(\gamma)$ to denote the maximum length of the channel content occurring in γ, formally: $\max(\gamma) := \max\{|x_i| \mid 0 \leq x_i \leq k\}$.

Lemma 1. *For each error-free computation γ of $\mathcal{C}$ from (s_I, ε) to (s_F, x) for some $x \in M^*$, there exists some timed word $w \in L(\mathcal{C}, \max(\gamma))$ such that $\max(w) = \max(\gamma)$.*

Lemma 2. *For each $n \in \mathbb{N}$ and $w \in L(\mathcal{C}, n)$ with $\max(w) = n$, there exists some error-free computation γ of $\mathcal{C}$ from (s_I, ε) to (s_F, x) for some $x \in M^*$ with $\max(\gamma) \leq n$.*

4.3 Excluding Faulty Computations

We define a timed visibly one-counter net $\mathcal{A}$ over Σ such that for every $n \in \mathbb{N}$ the intersection $L(\mathcal{A}) \cap L(\mathcal{C}, n)$ consists of all timed words that encode *error-free* computations of $\mathcal{C}$ from (s_I, ε) to (s_F, x) for some $x \in M^*$. The timed visibly one-counter net $\mathcal{A}$ is shown in Fig. 3. It non-deterministically guesses a number $n \in \mathbb{N}$ of symbols $+$ and increments the counter each time it reads the symbol $+$. When $\mathcal{A}$ leaves l_1, the value of the counter is $n+1$. After that, the counter value is not changed until the state symbol s_F is read. Then, while reading symbols in $\{-, \star\}$, the counter value is decremented. Note that $\mathcal{A}$ can reach the final location l_4 only if the number of the occurrences of symbol $-$ between s_F and $\star$ is *at most* n. Note that $\mathcal{A}$ does not use any clock, and it is deterministic.

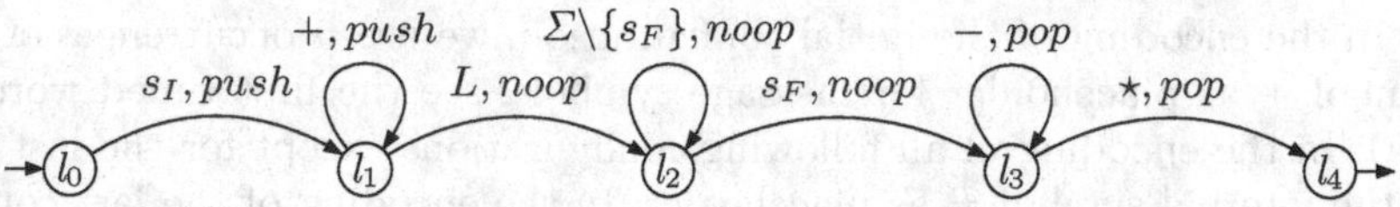

Fig. 3. The deterministic timed visibly one-counter net $\mathcal{A}$ for excluding insertion errors

Lemma 3. *$\mathcal{C}$ has an error-free computation from (s_I, ε) to (s_F, x) for some $x \in M^*$, if, and only if, there exists $n \in \mathbb{N}$ such that $L(\mathcal{C}, n) \cap L(\mathcal{A}) \neq \emptyset$.*

We finally define $L(\mathcal{C}) := \bigcup_{n \in \mathbb{N}} L(\mathcal{C}, n)$.

Corollary 2. *There exists some error-free computation of $\mathcal{C}$ from (s_I, ε) to (s_F, x) for some $x \in M^*$ if, and only if, $L(\mathcal{A}) \cap L(\mathcal{C}) \neq \emptyset$.*

4.4 The Reduction

Finally, we define a one-clock timed automaton $\mathcal{B}$ such that $L(\mathcal{B}) = T\Sigma^+ \backslash L(\mathcal{C})$. The construction of $\mathcal{B}$ follows the same ideas as, *eg.*, in [3]: $\mathcal{B}$ is the union of several one-clock timed automata, each of them violating one of the conditions of the definition of $L(\mathcal{C})$. For instance, the timed automaton in Fig. 4 accepts the set of timed words over Σ violating the condition that for every control state

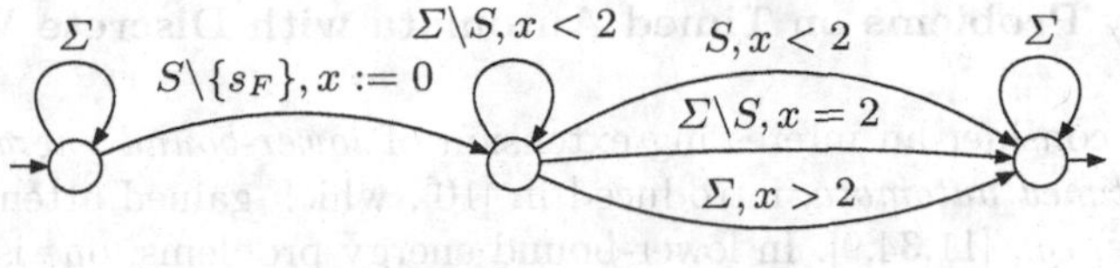

Fig. 4. A timed automaton violating condition 2 of $L(\mathcal{C})$

symbol different from s_F there must be some control state symbol after two time units.

By Corollary 2, there exists some error-free computation of $\mathcal{C}$ from (s_I, ε) to (s_F, x) for some $x \in M^*$ if, and only if, $L(\mathcal{A}) \cap L(\mathcal{C}) \neq \emptyset$. The latter is equivalent to $L(\mathcal{A}) \not\subseteq L(\mathcal{B})$. Hence, the language inclusion problem is undecidable. □

4.5 Undecidability of the Model Checking Problem for MTL

The proof idea of Theorem 1 can be used to show the undecidability of the following model checking problem: given a timed visibly one-counter net $\mathcal{A}$, and an MTL formula φ, is every $w \in L(\mathcal{A})$ a model of φ? Recall that this problem is decidable for the class of timed automata [32]. We prove that adding a visibly counter without zero test already makes the problem undecidable. Recall that MTL only allows to express restrictions on time, and it does not allow for any restrictions on the values of the counters. In fact, it is known that as soon as we add to MTL the capability for expressing restrictions on the values of a counter that can be incremented and decremented, model checking is undecidable [35]. The proof of the following theorem is based on the fact that - like one-clock timed automata - MTL can encode computations of channel machines with insertion errors [32].

Theorem 4. *The model checking problem for timed visibly one-counter nets and MTL is undecidable, even if the timed visibly one-counter net does not use any clocks and is deterministic.*

We would like to remark that the proof of Theorem 4 shares some similarities with the proof of the undecidability of model checking one-counter machines (*i.e.*, one-counter automata without input alphabet) and Freeze LTL with one register ($\mathrm{LTL}_1^{\downarrow}$, for short) [16]. In [15], it is proved that $\mathrm{LTL}_1^{\downarrow}$ can encode computations of *counter automata with incrementing errors*. Similar to the situation for MTL and channel machines, $\mathrm{LTL}_1^{\downarrow}$ can however not encode *error-free* computations of counter automata. In [16], a one-counter machine is used to repair this incapability, resulting in the undecidability of the model checking problem. The one-counter machine in [16] does not use zero tests; however, we point out that in contrast to our visibly timed one-counter net the one-counter machine in [16] is *non-deterministic*. Indeed, model checking *deterministic* one-counter machines and $\mathrm{LTL}_1^{\downarrow}$ is decidable [16].

4.6 Energy Problems on Timed Automata with Discrete Weights

Next we will consider an interesting extension of *lower-bound energy problems on weighted timed automata*, introduced in [10], which gained attention in the last years, see, *eg.*, [11,34,9]. In lower-bound energy problems, one is interested whether in a given automaton with some weight variable whose value can be increased and decreased, there exists a successful run in which all accumulated weight values are never below zero. Similar problems have also been considered for untimed settings, *eg.*, [29,19,20,13].

A *timed automaton with discrete weights* (dWTA, for short) is syntactically the same as a timed one-counter net. In the semantical graph induced by a dWTA, however, we allow the value of the counter (or, the *weight variable*) to become negative. Hence the value of the weight variable does not influence the behaviour of the dWTA, because, different to timed one-counter nets, transitions that result in negative values are not blocked. We remark that for the simple reasons that the value of the weight variable does not influence the behaviour of dWTA and MTL does not restrict the values of the weight variable, the model checking problem for dWTA and MTL is decidable, using the same algorithm as for timed automata [32]. We define the *energy model checking problem* for dWTA and MTL as follows: given a dWTA $\mathcal{A}$ and an MTL formula φ, does there exist some accepting run ρ of $\mathcal{A}$ such that the value of the weight variable is always non-negative, and the timed word w associated with ρ satisfies φ? For the special case $\varphi = \texttt{true}$, the problem is decidable in polynomial time for one-clock dWTA [10].

Theorem 5. *The energy problem for dWTA and MTL is undecidable, even if the dWTA uses no clocks.*

5 Decidability Result

In this section, we shortly explain the proof idea for the decidability of the language inclusion problem if $\mathcal{A}$ is a timed automaton, and $\mathcal{B}$ is a one-clock timed counter net. The proof is a generalization of the proof for the case where both $\mathcal{A}$ and $\mathcal{B}$ are timed automata [31]. The idea is to reduce the language inclusion problem to a reachability problem on an infinite graph constructed from the joint state space of $\mathcal{A}$ and $\mathcal{B}$. The decidability of the reachability problem on our infinite graph is implied by the fact that the graph is a downward-compatible well-structured transition system [21]. For taking into account the additional information on the values of the counters, we have to define a new well-quasi-order on the state space of the graph. This new well-quasi-order is based on the product of the equality order $=$ on a finite alphabet and the pointwise order $\leq^n$ on $\mathbb{N}^n$ (where n is the number of counters of $\mathcal{B}$), which is a well-quasi-order by Dickson's Lemma [17]. We use several applications of Higman's Lemma [23] to prove that our quasi-order is a well-quasi-order.

6 The Universality Problem for Visibly One-Counter Automata

We prove that allowing the counter in a one-clock timed visibly one-counter net to be tested for zero, results in the undecidability of the universality problem. The undecidability of the universality problem for the more general class of one-clock visibly pushdown automata was already stated in Theorem 3 in [18]. The proof in [18] is a reduction of the halting problem for two-counter machines. Given a two-counter machine $\mathcal{M}$, one can define a timed language $L(\mathcal{M})$ that consists of all timed words encoding a halting computation of $\mathcal{M}$. Then a timed visibly pushdown automaton $\mathcal{A}$ is defined that accepts the complement of $L(\mathcal{M})$. Altogether, $L(\mathcal{A}) = T\Sigma^+$ if, and only if, $\mathcal{M}$ does not have a halting computation. The definition of $L(\mathcal{M})$ is similar to the definition of $L(\mathcal{C})$ in the proof of Theorem 1. Recall that in the definition of $L(\mathcal{C})$ we did not include a condition that requires every symbol to have a matching symbol two time units *before*, and, as we mentioned, this is the reason for $L(\mathcal{C})$ to contain timed words encoding *faulty* computations of $\mathcal{C}$. However, in the definition of $L(\mathcal{M})$ in [18], such a "backward-looking" condition is used. In the proof in [18], it is unfortunately not clear how the one-clock timed visibly pushdown automaton $\mathcal{A}$ can detect violations of this condition[1].

Here, we give a complete proof for the subclass of timed visibly one-counter automata. Like the proof of Theorem 1, the proof is a reduction of the control state reachability problem for channel machines. We however remark that one can similarly use a reduction of the halting problem for two-counter machines.

Proof of Theorem 3. Let $\mathcal{C} = (S, s_I, M, \Delta)$ be a channel machine, and let $s_F \in S$. Define Σ in the same way as in the proof of Theorem 1. For every $n \in \mathbb{N}$, we define a timed language $L_{\mathsf{ef}}(\mathcal{C}, n)$ that consists of all timed words over Σ that encode *error-free* computations of $\mathcal{C}$ from (s_I, ε) to (s_F, x) for some $x \in M^*$. Formally, $L_{\mathsf{ef}}(\mathcal{C}, n)$ is defined using the same conditions as the ones for $L(\mathcal{C}, n)$ in the proof of Theorem 1 plus an additional condition that requires every symbol in the encoding of a configuration to have a matching symbol two time units before. As we have mentioned in the proof of Theorem 1, this excludes encodings of faulty computations. We thus have for $L_{\mathsf{ef}}(\mathcal{C}) = \bigcup_{n \geq 1} L_{\mathsf{ef}}(\mathcal{C}, n)$:

Lemma 4. *There exists some error-free computation of* $\mathcal{C}$ *from* (s_I, ε) *to* (s_F, x) *for some* $x \in M^*$*, if, and only if,* $L_{\mathsf{ef}}(\mathcal{C}) \neq \emptyset$*.*

Next, we define a timed visibly one-counter automaton with a single clock such that $L(\mathcal{A}) = T\Sigma^+ \setminus L_{\mathsf{ef}}(\mathcal{C})$. Hence, by the preceding lemma, $L(\mathcal{A}) \neq T\Sigma^+$ if, and only if, there exists some error-free computation of $\mathcal{C}$ from (s_I, ε) to (s_F, x) for some $x \in M^*$.

[1] More detailed, it is not clear how to construct one-clock timed automata $N_{\neg f_r \leftarrow f_c}$ and $N_{\neg g_r \leftarrow g_c}$ mentioned on p. 10 in [18]. Recall that in the proof for undecidability of the universality problem for timed automata with two or more clocks, it is exactly this backward-looking condition that requires *two* clocks [3].

$\mathcal{A}$ is the union of several one-clock timed automata and one timed visibly one-counter automaton with no clocks. We already know from the proof of Theorem 1, that violations of conditions of $L(\mathcal{C}, n)$ can be detected by one-clock timed automata. For detecting violations of the additional condition, we use the visibly one-counter automaton shown in Fig. 5. The automaton non-deterministically guesses the maximum number n of occurrences of the symbol $+$. When leaving l_1, the value of the counter is $n+1$. The final location l_4, however, can only be reached while reading $-$ or $\star$ if the value of the counter is zero. This means that there must be some symbol for which there is no matching symbol two time units before. □

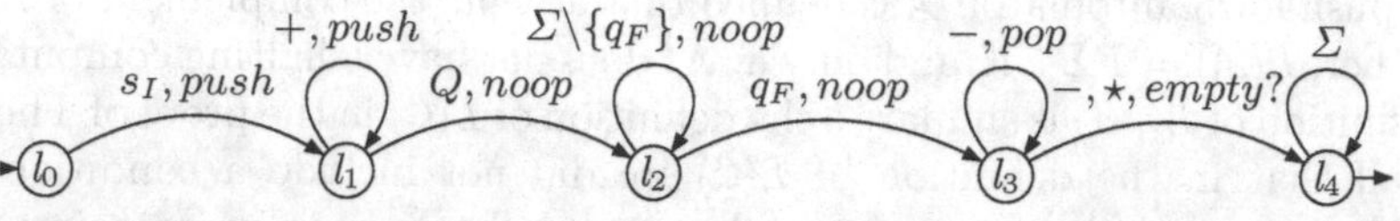

Fig. 5. The timed visibly one-counter automaton for recognizing timed words violating the additional "backwards-looking" condition of $L_{\text{ef}}(\mathcal{C})$

7 Conclusion and Open Problems

The main conclusion of this paper is that even for very weak extensions of timed automata with counters it is impossible to automatically verify whether a given specification is satisfied. On the other hand, we may use one-clock timed counter nets as specifications to verify timed automata. This increases so far known possibilities for the verification of timed automata: For instance, the timed language $L = \{(a^m b^n, \bar{\tau}) \mid m \geq n\}$ can be accepted by a timed one-counter net without any clocks, but not by a timed automaton.

An interesting problem is to figure out a (decidable) extension of LTL that is capable of expressing properties referring to both time and discrete data structures.

We remark that all our results hold for automata defined over *finite* timed words. We cannot expect the decidability of, *eg.*, the universality problem for one-clock timed counter nets over infinite timed words, as the same problem is already undecidable for the subclass of one-clock timed automata [2].

Acknowledgements. I would like to thank Michael Emmi and Rupak Majumdar for helpful discussions on their work on timed pushdown automata. I further would like to thank James Worrell very much for pointing me to MTL's capability of encoding faulty computations of channel machines.

References

1. Abdulla, P.A., Cerans, K.: Simulation is decidable for one-counter nets (extended abstract). In: Sangiorgi, D., de Simone, R. (eds.) CONCUR 1998. LNCS, vol. 1466, pp. 253–268. Springer, Heidelberg (1998)

2. Abdulla, P.A., Deneux, J., Ouaknine, J., Quaas, K., Worrell, J.: Universality analysis for one-clock timed automata. Fundam. Inform. 89(4), 419–450 (2008)
3. Adams, S., Ouaknine, J., Worrell, J.B.: Undecidability of universality for timed automata with minimal resources. In: Raskin, J.-F., Thiagarajan, P.S. (eds.) FORMATS 2007. LNCS, vol. 4763, pp. 25–37. Springer, Heidelberg (2007)
4. Alur, R., Dill, D.L.: A theory of timed automata. Theoretical Computer Science 126(2), 183–235 (1994)
5. Alur, R., Madhusudan, P.: Decision problems for timed automata: A survey. In: Bernardo, M., Corradini, F. (eds.) SFM-RT 2004. LNCS, vol. 3185, pp. 1–24. Springer, Heidelberg (2004)
6. Alur, R., Madhusudan, P.: Visibly pushdown languages. In: Babai, L. (ed.) STOC, pp. 202–211. ACM (2004)
7. Bouajjani, A., Echahed, R., Robbana, R.: On the automatic verification of systems with continuous variables and unbounded discrete data structures. In: Antsaklis, P., Kohn, W., Nerode, A., Sastry, S. (eds.) Hybrid Systems II. LNCS, vol. 999, pp. 64–85. Springer, Heidelberg (1995)
8. Bouchy, F., Finkel, A., Sangnier, A.: Reachability in timed counter systems. Electr. Notes Theor. Comput. Sci. 239, 167–178 (2009)
9. Bouyer, P., Fahrenberg, U., Larsen, K.G., Markey, N.: Timed automata with observers under energy constraints. In: Johansson, K.H., Yi, W. (eds.) HSCC, pp. 61–70. ACM (2010)
10. Bouyer, P., Fahrenberg, U., Larsen, K.G., Markey, N., Srba, J.: Infinite runs in weighted timed automata with energy constraints. In: Cassez, F., Jard, C. (eds.) FORMATS 2008. LNCS, vol. 5215, pp. 33–47. Springer, Heidelberg (2008)
11. Bouyer, P., Larsen, K.G., Markey, N.: Lower-bound-constrained runs in weighted timed automata. Perform. Eval. 73, 91–109 (2014)
12. Brand, D., Zafiropulo, P.: On communicating finite-state machines. J. ACM 30(2), 323–342 (1983)
13. Brázdil, T., Jančar, P., Kučera, A.: Reachability games on extended vector addition systems with states. In: Abramsky, S., Gavoille, C., Kirchner, C., Meyer auf der Heide, F., Spirakis, P.G. (eds.) ICALP 2010. LNCS, vol. 6199, pp. 478–489. Springer, Heidelberg (2010)
14. Dang, Z.: Pushdown timed automata: a binary reachability characterization and safety verification. Theor. Comput. Sci. 302(1-3), 93–121 (2003)
15. Demri, S., Lazić, R.: LTL with the freeze quantifier and register automata. ACM Trans. Comput. Log. 10(3) (2009)
16. Demri, S., Lazić, R.S., Sangnier, A.: Model checking Freeze LTL over one-counter automata. In: Amadio, R.M. (ed.) FOSSACS 2008. LNCS, vol. 4962, pp. 490–504. Springer, Heidelberg (2008)
17. Dickson, L.E.: Finiteness of the odd perfect and primitive abundant numbers with n distinct prime factors. Amer. J. Math. 35, 413–422 (1913)
18. Emmi, M., Majumdar, R.: Decision problems for the verification of real-time software. In: Hespanha, J.P., Tiwari, A. (eds.) HSCC 2006. LNCS, vol. 3927, pp. 200–211. Springer, Heidelberg (2006)
19. Ésik, Z., Fahrenberg, U., Legay, A., Quaas, K.: Kleene algebras and semimodules for energy problems. In: Van Hung, D., Ogawa, M. (eds.) ATVA 2013. LNCS, vol. 8172, pp. 102–117. Springer, Heidelberg (2013)
20. Fahrenberg, U., Juhl, L., Larsen, K.G., Srba, J.: Energy games in multiweighted automata. In: Cerone, A., Pihlajasaari, P. (eds.) ICTAC 2011. LNCS, vol. 6916, pp. 95–115. Springer, Heidelberg (2011)

21. Finkel, A., Schnoebelen, P.: Well-structured transition systems everywhere! Theor. Comput. Sci. 256(1-2), 63–92 (2001)
22. Greibach, S.A.: An infinite hierarchy of context-free languages. J. ACM 16(1), 91–106 (1969)
23. Higman, G.: Ordering by divisibility in abstract algebras. Proceedings of the London Mathematical Society 2, 236–366 (1952)
24. Hofman, P., Lasota, S., Mayr, R., Totzke, P.: Simulation over one-counter nets is PSPACE-complete. In: Seth, A., Vishnoi, N.K. (eds.) FSTTCS. LIPIcs, vol. 24, pp. 515–526. Schloss Dagstuhl - Leibniz-Zentrum fuer Informatik (2013)
25. Hofman, P., Totzke, P.: Trace inclusion for one-counter nets revisited. CoRR, abs/1404.5157 (2014)
26. Ibarra, O.H.: Restricted one-counter machines with undecidable universe problems. Mathematical Systems Theory 13, 181–186 (1979)
27. Jančar, P., Esparza, J., Moller, F.: Petri nets and regular processes. J. Comput. Syst. Sci. 59(3), 476–503 (1999)
28. Jančar, P., Kucera, A., Moller, F., Sawa, Z.: DP lower bounds for equivalence-checking and model-checking of one-counter automata. Inf. Comput. 188(1), 1–19 (2004)
29. Juhl, L., Guldstrand Larsen, K., Raskin, J.-F.: Optimal bounds for multiweighted and parametrised energy games. In: Liu, Z., Woodcock, J., Zhu, H. (eds.) He Festschrift. LNCS, vol. 8051, pp. 244–255. Springer, Heidelberg (2013)
30. Information Sciences Institute of the University of Southern California. Transmission Control Protocoll (DARPA Internet Program Protocol Specification) (1981), `http://www.faqs.org/rfcs/rfc793.html`
31. Ouaknine, J., Worrell, J.: On the language inclusion problem for timed automata: Closing a decidability gap. In: LICS, pp. 54–63. IEEE Computer Society (2004)
32. Ouaknine, J., Worrell, J.: On the decidability of metric temporal logic. In: LICS, pp. 188–197. IEEE Computer Society (2005)
33. Ouaknine, J., Worrell, J.B.: On metric temporal logic and faulty turing machines. In: Aceto, L., Ingólfsdóttir, A. (eds.) FOSSACS 2006. LNCS, vol. 3921, pp. 217–230. Springer, Heidelberg (2006)
34. Quaas, K.: On the interval-bound problem for weighted timed automata. In: Dediu, A.-H., Inenaga, S., Martín-Vide, C. (eds.) LATA 2011. LNCS, vol. 6638, pp. 452–464. Springer, Heidelberg (2011)
35. Quaas, K.: Model checking metric temporal logic over automata with one counter. In: Dediu, A.-H., Martín-Vide, C., Truthe, B. (eds.) LATA 2013. LNCS, vol. 7810, pp. 468–479. Springer, Heidelberg (2013)

Reducing Clocks in Timed Automata while Preserving Bisimulation

Shibashis Guha*, Chinmay Narayan, and S. Arun-Kumar

Indian Institute of Technology Delhi, New Delhi, India
{shibashis,chinmay,sak}@cse.iitd.ac.in

Abstract. Model checking timed automata becomes increasingly complex with the increase in the number of clocks. Hence it is desirable that one constructs an automaton with the minimum number of clocks possible. The problem of checking whether there exists a timed automaton with a smaller number of clocks such that the timed language accepted by the original automaton is preserved is known to be undecidable. In this paper, we give a construction, which for any given timed automaton produces a timed bisimilar automaton with the least number of clocks. Further, we show that such an automaton with the minimum possible number of clocks can be constructed in time that is doubly exponential in the number of clocks of the original automaton.

1 Introduction

Timed automata [3] is a formalism for modelling and analyzing real time systems. The complexity of model checking is dependent on the number of clocks of the timed automaton(TA) [3,2]. Many model checking and reachability problems use a region graph or a zone graph for the timed automaton whose sizes are exponential in the number of clocks. Hence it is desirable to construct a timed automaton with the minimum number of clocks that preserves some property of interest. It is known that given a timed automaton, checking whether there exists another timed automaton accepting the same timed language as the original one but with a smaller number of clocks is undecidable [12]. In this paper, we show that checking the existence of a timed automaton with a smaller number of clocks that is timed bisimilar to the original timed automaton is however decidable. Our method is constructive and we provide a 2-EXPTIME algorithm to construct the timed bisimilar automaton with the least possible number of clocks. We also note that if the constructed automaton has a smaller number of clocks, then it implies that there exists an automaton with a smaller number of clocks accepting the same timed language.

Related Work: In [9], an algorithm has been provided to reduce the number of clocks of a given timed automaton and produce a new timed automaton that is timed bisimilar to the original one. The algorithm detects a set of *active clocks*

* The research of Shibashis Guha was supported by Microsoft Corporation and Microsoft Research India under the Microsoft Research India PhD Fellowship Award.

P. Baldan and D. Gorla (Eds.): CONCUR 2014, LNCS 8704, pp. 527–543, 2014.

at every location and partitions these active clocks into classes such that all the clocks belonging to a class in the partition always have the same value. However, this may not result in the minimum possible number of clocks since the algorithm works on the timed automaton directly rather than on its semantics. Thus if a constraint associated with clock x implies a constraint associated with clock y, and both of them appear on an edge, then the constraint with clock y can be eliminated. However, the algorithm of [9] does not capture such implication. Also by considering constraints on more than one outgoing edge from a location, e.g. $l_0 \xrightarrow{a,x\leq 3,\emptyset} l_1$ and $l_0 \xrightarrow{a,x>3,\emptyset} l_2$ collectively, we may sometimes eliminate the constraints that may remove some clock. This too has not been accounted for by the algorithm of [9].

In [19], it has been shown that no algorithm can decide the minimality of the number of clocks while preserving the timed language and for the non-minimal case find a timed language equivalent automaton with fewer clocks. Also for a given timed automaton, the problem of finding whether there exists another TA with fewer clocks accepting the same timed language is undecidable [12].

Another result appearing in [16] which uses the region-graph construction is the following. A (C, M)-automaton is one with C clocks and M is the largest integer appearing in the timed automaton. Given a timed automaton A, a set of clocks C and an integer M, checking the existence of a (C, M)-automaton that is timed bisimilar to A is shown to be decidable in [16]. The method in [16] constructs a logical formula called the *characteristic formula* and checks whether there exists a (C, M)-automaton that satisfies it. Further, it is shown that a pair of automata satisfying the same characteristic formula are timed bisimilar. The problem that we solve in the current paper was in fact left open in [16].

The rest of the paper is organized as follows: in Section 2, we describe timed automata and introduce several concepts that will be used in the paper. We also describe the construction of the zone graph used in reducing the number of clocks. In Section 3, we discuss our approach in detail along with a few examples. Section 4 is the conclusion.

2 Timed Automata

Formally, a *timed automaton* (TA) [3] is defined as a tuple $A = (L, Act, l_0, C, E)$ where L is a finite set of locations, Act is a finite set of visible actions, $l_0 \in L$ is the initial location, C is a finite set of clocks and $E \subseteq L \times \mathcal{B}(C) \times Act \times 2^C \times L$ is a finite set of *edges.* The set of constraints or guards on the edges, denoted $\mathcal{B}(C)$, is given by the grammar $g ::= \; x \bowtie k \mid g \wedge g$, where $k \in \mathbb{N}$ and $x \in C$ and $\bowtie \in \{\leq, <, =, >, \geq\}$. Given two locations l, l', a transition from l to l' is of the form (l, g, a, R, l') i.e. a transition from l to l' on action a is possible if the constraints specified by g are satisfied; $R \subseteq C$ is a set of clocks which are reset to zero during the transition.

The semantics of a timed automaton(TA) is described by a *timed labelled transition system* (TLTS) [1]. The timed labelled transition system $T(A)$ generated by A is defined as $T(A) = (Q, Lab, Q_0, \{\xrightarrow{\alpha} \mid \alpha \in Lab\})$, where $Q =$

$\{(l,v) \mid l \in L, v \in \mathbb{R}_{\geq 0}{}^{|C|}\}$ is the set of *states*, each of which is of the form (l,v), where l is a location of the timed automaton and v is a valuation in $|C|$ dimensional real space where each clock is mapped to a unique dimension in this space; $Lab = Act \cup \mathbb{R}_{\geq 0}$ is the set of labels. Let v_0 denote the valuation such that $v_0(x) = 0$ for all $x \in C$. $Q_0 = (l_0, v_0)$ is the initial state of $T(A)$. A transition may occur in one of the following ways:
(i) *Delay transitions* : $(l,v) \xrightarrow{d} (l, v+d)$. Here, $d \in \mathbb{R}_{\geq 0}$ and $v+d$ is the valuation in which the value of every clock is incremented by d.
(ii) *Discrete transitions* : $(l,v) \xrightarrow{a} (l',v')$ if for an edge $e = (l,g,a,R,l') \in E$, $v \models g, v' = v_{[R \leftarrow \bar{0}]}$, where $v_{[R \leftarrow \bar{0}]}$ denotes that every clock in R has been reset to 0, while the remaining clocks are unchanged. From a state (l,v), if $v \models g$, then there exists an a-transition to a state (l',v'); after this, the clocks in R are reset while those in $C \backslash R$ remain unchanged.

For simplicity, we do not consider annotating locations with clock constraints (known as *invariant conditions* [15]). Our results extend in a straightforward manner to timed automata with invariant conditions. In Section 3, we provide the modifications to our method for dealing with location invariants. We now define various concepts that will be used in the rest of the paper.

Definition 1. *Let $A = (L, Act, l_0, E, C)$ be a timed automaton, and $T(A)$ be the TLTS corresponding to A.*

1. ***Timed trace:*** *A sequence of delays and visible actions $d_1a_1d_2a_2 \ldots d_na_n$ is called a timed trace iff there is a sequence of transitions $p_0 \xrightarrow{d_1} p_1 \xrightarrow{a_1} p_1' \xrightarrow{d_2} p_2 \xrightarrow{a_2} p_2' \cdots \xrightarrow{d_n} p_n \xrightarrow{a_n} p'$ in $T(A)$, with p_0 being the initial state of the timed automaton.*
2. ***Zone:*** *A zone Z is a set of valuations $\{v \in \mathbb{R}_{\geq 0}^{|C|} \mid v \models \beta\}$, where β is of the form $\beta ::= \ x \bowtie k \mid x - y \bowtie k \mid \beta \wedge \beta$, k is an integer, $x, y \in C$ and $\bowtie \in \{\leq, <, =, >, \geq\}$. $Z \uparrow$ denotes the future of the zone Z. $Z \uparrow = \{v + d \mid v \in Z, d \geq 0\}$ is the set of all valuations reachable from Z by time elapse. A zone is a convex set of clock valuations.*
3. ***Pre-stability:*** *A zone Z_1 of location l_1 is pre-stable with respect to another zone Z_2 of location l_2 if $Z_1 \subseteq preds(Z_2)$ or $Z_1 \cap preds(Z_2) = \emptyset$ where $preds(Z) \stackrel{def}{=} \{v \in \mathbb{R}_{\geq 0}^{|C|} \mid \exists v' \in Z, \exists l, l' \in L, \exists \alpha \in Lab$ such that $(l,v) \xrightarrow{\alpha} (l',v')\}$. Here $l = l'$ if α is a delay action.*
4. ***Canonical decomposition:*** *Let $g = \bigwedge_{i=1}^{n} \gamma_i \in \mathcal{B}(C)$, where each γ_i is an* elementary constraint *of the form $x_i \bowtie k_i$, such that $x_i \in C$ and k_i is a non-negative integer. A canonical decomposition of a zone Z with respect to g is obtained by splitting Z into a set of zones $Z_1, \ldots, Z_m$ such that for each $1 \leq j \leq m$, and $1 \leq i \leq n$, either $\forall v \in Z_j, v \models \gamma_i$ or $\forall v \in Z_j$, $v \not\models \gamma_i$. For example, consider the zone $Z = x \geq 0 \wedge y \geq 0$ and the guard $x \leq 2 \wedge y > 1$. Z is split with respect to $x \leq 2$, and then with respect to $y > 1$, hence into four zones : $x \leq 2 \wedge y \leq 1$, $x > 2 \wedge y \leq 1$, $x \leq 2 \wedge y > 1$ and $x > 2 \wedge y > 1$. An elementary constraint $x_i \bowtie k_i$ induces the hyperplane $x_i = k_i$ in a zone graph of the timed automaton.*

5. ***Zone graph:*** *Given a timed automaton* $A = (L, Act, l_0, C, E)$, *a* zone graph $\mathcal{G}_A$ *of* A *is a transition system* $(S, s_0, Lep, \rightarrow)$, *that is a finite representation of* $T(A)$. *Here* $Lep = Act \cup \{\varepsilon\}$. $S \subseteq L \times \mathcal{Z}$ *is the set of nodes of* $\mathcal{G}_A$, $\mathcal{Z}$ *being the set of zones. The node* $s_0 = (l_0, Z_0)$ *is the initial node such that* $v_0 \in Z_0$. $(l_i, Z) \xrightarrow{a} (l_j, Z')$ *iff* $l_i \xrightarrow{g,a,R} l_j$ *in* A *and* $Z' \subseteq ([Z \cap g]_{R \leftarrow \bar{0}})\uparrow$ *obtained after canonical decomposition of* $([Z \cap g]_{R \leftarrow \bar{0}})$. *For any* Z *and* Z', $(l_i, Z) \xrightarrow{\varepsilon} (l_i, Z')$ *iff there exists a delay* d *and a valuation* v *such that* $v \in Z$, $v + d \in Z'$ *and* $(l_i, v) \xrightarrow{d} (l_i, v + d)$ *is in* $T(A)$. *Here the zone* Z' *is called a* delay successor *of zone* Z, *while* Z *is called the* delay predecessor *of* Z'. *The relation* ε *is reflexive and transitive and so is the* delay successor *relation. We denote the set of delay successor zones of* Z *with* $ds(Z)$. *A zone* $Z' \neq Z$ *is called the* immediate delay successor *of a zone* Z *iff* $Z' \in ds(Z)$ *and* $\forall Z'' \in ds(Z) : Z'' \neq Z$ *and* $Z'' \neq Z'$, $Z'' \in ds(Z')$. *We call a zone* Z *corresponding to a location to be a* base zone *if* Z *does not have a delay predecessor other than itself.*
6. *A hyperplane* $x = k$ *is said to* bound *a zone* Z *from above if* $\exists v \in Z\, [\, \forall d \in \mathbb{R}_{\geq 0}\, [\, (v + d)(x) \succ k \iff (v + d)(x) \notin Z\,]\,]$, *where* $\succ \in \{>, \geq\}$. *A zone, in general, can be bounded above by several hyperplanes. A hyperplane* $x = k$ *is said to* bound *a zone* Z *fully from above if* $\forall v \in Z\, [\, \forall d \in \mathbb{R}_{\geq 0}\, [\, (v + d)(x) \succ k \iff (v + d)(x) \notin Z\,]\,]$. *Analogously, we can also say that a hyperplane* $x = k$ *bounds a zone from below if* $\exists v \in Z\, [\, \forall d \in \mathbb{R}_{\geq 0}\, [\, (v - d)(x) \prec k \iff (v - d)(x) \notin Z]]$, *where* $\prec \in \{<, \leq\}$. *We can also define a hyperplane bounding a zone fully from below in a similar manner.*
   When not specified otherwise, in this paper, a hyperplane bounding a zone implies that it bounds the zone from above. A zone Z *is bounded above if it has an immediate delay successor zone.*

We create a zone graph such that for any location l, the zones Z and Z' of any two nodes (l, Z) and (l, Z') in the zone graph are disjoint and all zones of the zone graph are pre-stable. This zone graph is constructed in two phases in time exponential in the number of the clocks. The first phase performs a forward analysis of the timed automaton while the second phase ensures pre-stability in the zone graph. The forward analysis may cause a zone graph to become infinite [4]. Several kinds of abstractions have been proposed in the literature [8,4,5] to make the zone graph finite. We use *location dependent maximal constants* abstraction [4] in our construction. In phase 2 of the zone graph creation, the zones are further split to ensure that the resultant zone graph is pre-stable. The following lemma states an important property of the zone graph which will further be used for clock reduction.

Lemma 1. *Pre-stability ensures that if the zone* Z *in any node* (l, Z) *in the zone graph is bounded above, then it is bounded fully from above by a hyperplane* $x = h$, *where* $x \in C$ *and* $h \in \mathbb{N}$.

Some approaches for preserving convexity and implementing pre-stability have been discussed in [20]. As an example, consider the timed automaton in Figure 1. The pre-stable zones of location l_1 are shown in the right side of the

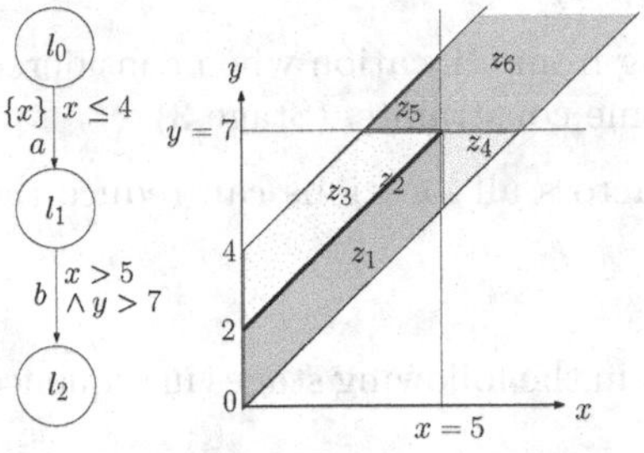

Fig. 1. A timed automaton and the zones for location l_1

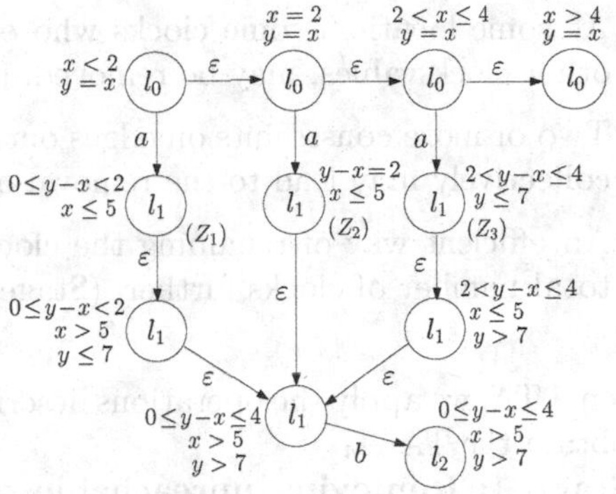

Fig. 2. Zone Graph for the TA in Figure 1

figure. In this paper, from now on, unless stated otherwise, *zone graph* will refer to this form of the pre-stable zone graph that is described above. An algorithmic procedure for the construction of the zone graph is given in [13].

A relation $\mathcal{R} \subseteq Q \times Q$ is a *timed simulation* relation if the following conditions hold for any two timed states $(p, q) \in \mathcal{R}$.
$\forall\, a \in Act,\, p \xrightarrow{a} p' \implies \exists q' : q \xrightarrow{a} q'$ and $(p', q') \in \mathcal{R}$ and
$\forall d \in \mathbb{R}_{\geq 0},\, p \xrightarrow{d} p' \implies \exists q' : q \xrightarrow{d} q'$ and $(p', q') \in \mathcal{R}$.
A *timed bisimulation* relation is a symmetric timed simulation. Two timed automata are timed bisimilar if and only if their initial states are timed bisimilar. Using product construction on region graphs, timed bisimilarity for timed automata was shown to be decidable in EXPTIME [7].

3 Clock Reduction

Unlike the method described in [9], which works on the syntactic structure of the timed automaton, we use a semantic representation, the zone graph described in Section 2 to capture the behaviour of the timed automaton. This helps us to reduce the number of clocks in a more effective way. For a given TA A, we first describe a sequence of stages to construct a TA A_4 that is timed bisimilar to A. Later we prove the minimality in terms of the number of clocks for the TA A_4. The operations involved in our procedure use a *difference bound matrix* (DBM) [6,10] representation of the zones. A DBM for a set $C = \{x_1, x_2, \ldots, x_n\}$ of n clocks is an $(n+1)$ square matrix M where an extra variable x_0 is introduced such that the value of x_0 is always 0. An element M_{ij} is of the form $(m_{ij}, \prec)$ where $\prec \in \{<, \leq\}$ such that $x_i - x_j \prec m_{ij}$. The following are important considerations in reducing the number of clocks.

- There may be some clock constraints on an edge of the TA that are never enabled. Such edges and constraints may be removed. (Stage 1)
- Splitting some locations may lead to a reduction in the number of the clocks. (Stage 2)

- At some location, some clocks whose values may be expressed in terms of other clock values, may be removed. (Stage 2)
- Two or more constraints on edges outgoing from a location when considered collectively may lead to the removal of some constraints. (Stage 3)
- An efficient way of renaming the clocks across all locations can reduce the total number of clocks further. (Stage 4)

Given a TA, we apply the operations described in the following stages in sequence to obtain the TA A_4.

Stage 1: Removing unreachable edges and associated constraints: This stage involves creating the pre-stable zone graph of the given timed automaton, as described in Section 2. The edges and their associated constraints that are never enabled in an actual transition are removed while creating the zone graph. Suppose there is an edge $l_i \xrightarrow{g,a,R} l_j$ in A but in the zone graph, a corresponding transition of the form $(l_i, Z) \xrightarrow{a} (l_j, Z')$ does not exist. This implies that the transition $l_i \xrightarrow{g,a,R} l_j$ is never enabled and hence is removed from the timed automaton. Since the edges that do not affect any transition get removed during this stage, we have the following lemma trivially.

Lemma 2. *The operations in stage 1 produce a timed automaton A_1 that is timed bisimilar to the original TA A.*

The time required in this stage is proportional to the size of the zone graph and hence *exponential* in the number of clocks of the timed automaton.

Stage 2: Splitting locations and removing constraints not affecting transitions: Locations may also require to be split in order to reduce the number of clocks of a timed automaton. Let us consider the example of the timed automaton in Figure 1 and its zone graph in Figure 2. There are three base zones corresponding to location l_1 in the zone graph, i.e. $Z_1 = \{0 \le y - x < 2, x \le 5\}$, $Z_2 = \{y - x = 2, x \le 5\}$ and $Z_3 = \{2 < y - x \le 4, y \le 7\}$. This stage splits l_1 into three locations l_{1_1}, l_{1_2} and l_{1_3} (one for each of the base zones Z_1, Z_2 and Z_3) as shown in Figure 3(a). While the original automaton, in Figure 1, contains two elementary constraints on the edge between l_1 and l_2, the modified automaton, in Figure 3(a), contains only one of these two elementary constraints on the outgoing edges from each of l_{1_1}, l_{1_2} and l_{1_3} to l_2. Subsequent stages modify it further to generate an automaton using a single clock as in Figure 3(b).

Splitting ensures that only those constraints, that are relevant for every valuation in the base zone of a newly created location, appear on the edges originating from that location. Since the clocks can be reused while describing the behaviours from each of the individual locations created after the split, this may lead to a reduction in the number of clocks.

We describe a formal procedure for splitting a location into multiple locations in Algorithm 1.

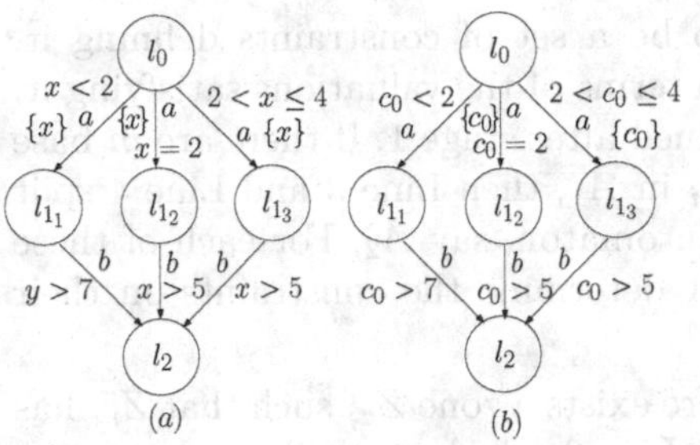

Fig. 3. Splitting locations of the TA in Figure 1

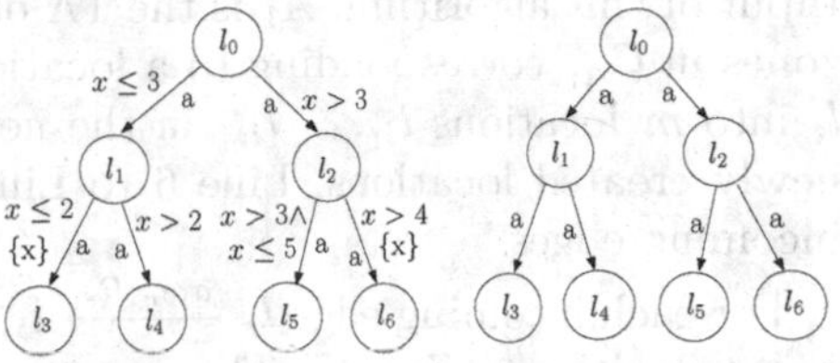

Fig. 4. The two timed automata are timed bisimilar

Algorithm 1. Algorithm for splitting locations
Input: Timed automaton A_1 obtained after stage 1
Output: Modified TA A_2 after applying stage 2 splitting procedures

---

$A_2 := A_1$ ▷ A_1 is the TA obtained from A after the first stage
for each location l_i in A_1 **do** ▷ i is the index of the location
  Split l_i into m locations $l_{i_1}, \ldots, l_{i_m}$ in A_2 ▷ Let m be the number of base zones of l_i
  Remove location l_i and all incoming and outgoing edges to and from l_i from A_2
  for each j in 1 to m **do**
    for each incoming edge $l_r \xrightarrow{a, g_r, R_r} l_i$ in A_1 **do**
      ▷ Split the constraints on the incoming edges to l_i for the newly created locations
      $Z'_{i_j} := Z_{i_j} \uparrow \cap \mathbb{R}^{|C|}_{\geq 0}{}_{[R_r \leftarrow \bar{0}]}$ ▷ Let Z_{i_j} be the base zone corresponding to l_{i_j}
      ▷ Let Z_{r_j} is a zone of location l_r from which there is an a transition to Z_{i_j}
      Let g_{r_j} be the weakest formula such that
      $g_r \wedge free(Z'_{i_j}, R_r) \wedge Z_{r_j} \Rightarrow g_{r_j}$ and g_{r_j} has a subset of the clocks used in g_r.
      Create an edge $l_r \xrightarrow{a, g_{r_j}, R_r} l_{i_j}$ in A_2
    end for
    for each outgoing edge $l_i \xrightarrow{a, g_i, R_i} l_r$ in A_1 **do**
      if $Z_{i_j} \uparrow \cap g_i = \emptyset$ **then**
        Do not create this edge from l_{i_j} to l_r in A_2 since it is never going to be enabled for any valuation of Z_{i_j};
      else
        Let S_r be the set of elementary constraints in g_i
        loop
          if $\exists s' \in S_r,\ s.t.\ Z_{i_j} \uparrow \wedge (\bigwedge_{s \in S_r \setminus \{s'\}} s) \Rightarrow Z_{i_j} \uparrow \wedge s'$ **then**
            $S_r = S_r \setminus \{s'\}$
          else
            Create an edge $l_{i_j} \xrightarrow{a, g_{i'}, R_i} l_r$ in A_2, where $g_{i'} = \bigwedge_{s \in S_r}$
            Break;
          end if
        end loop
      end if
    end for
  end for
end for

Note that a zone can be considered to be a set of constraints defining it. Similarly a guard can also be considered in terms of the valuations satisfying it. Input of this algorithm, A_1 is the TA obtained after stage 1. If there are m base zones in $\mathcal{G}_{A_1}$ corresponding to a location l_i in A_1, then Line 3 and Line 4 split l_i into m locations $l_{i_1}, \cdots l_{i_m}$ in the new automaton, say A_2. For each of these newly created locations, Line 6 to Line 11 determine the constraints on their incoming edges.

For each incoming edge $l_r \xrightarrow{a,g_r,R_r} l_i$, there exists a zone Z_{r_j} such that Z_{r_j} has an a transition to Z_{i_j}, the jth base zone of l_i. Line 8 calculates the lower bounding hyperplane of Z_{i_j} by resetting the clocks R_r in the intersection of $Z_{i_j} \uparrow$ with $\mathbb{R}^{|C|}_{\geq 0}$. In Line 11, $free(Z'_{i_j}, R_r)$ represents a zone that becomes the same as Z'_{i_j} after resetting the clocks in R_r. Further, g_{r_j} is calculated as the weakest guard that simultaneously satisfies the constraints g_r, Z_{r_j} and $free(Z'_{i_j}, R_r)$ and has the same set of clocks as in g_r. For our running example, if we consider $Z_{i_j} = Z_1$ then we have $Z'_{i_j} = Z_{i_j} \uparrow \cap \mathbb{R}^2_{[x \leftarrow \bar{0}]} = \{x = 0, y < 2\}$, $Z_{r_j} = \{x = y, x < 2\}$ and $free(Z'_{i_j}, \{x\}) = \{x \geq 0, y < 2\}$. We can see that $x < 2$ is the weakest formula such that $x \leq 4 \wedge x \geq 0 \wedge y < 2 \wedge x = y \wedge x < 2 \Rightarrow x < 2$ holds and hence $g_{r_j} = \{x < 2\}$.

Loop from Line 14 to Line 28 determines the constraints on the outgoing edges from these new locations. Line 15 checks if the zone $Z_{i_j} \uparrow$ has any valuation that satisfies the guard g_i on an outgoing edge from location l_i. If no satisfying valuation exists then this transition will never be enabled from l_{i_j} and hence this edge is not added in A_2. Loop from Line 19 to Line 26 checks if some elementary constraints of the guard are implied by other elementary constraints of the same guard. If it happens then we can remove those elementary constraints from the guard that are implied by the other elementary constraints.

For our running example, the modified automaton of Figure 3(a) does not contain the constraint $x > 5$ on the edge from l_{l_1} to l_2 even though it was present on the edge from l_1 to l_2. The reason being that the future of the zone of l_{l_1} (that is $0 \leq y - x < 2$) along with the constraint $y > 7$ implies $x > 5$ hence we do not need to put $x > 5$ explicitly on the outgoing edge from l_{1_1} to l_2. Such removal of elementary constraints helps future stages to reduce the number of clocks. The maximum number of locations produced in the timed automaton as a result of the split is bounded by the number of zones in the zone graph. This is exponential in the number of clocks of the original TA A. However, we note that the base zones of a location l in the original TA are distributed across multiple locations as a result of the split of l and no new valuations are created. This gives us the following lemma.

Lemma 3. *The splitting procedure described in this stage does not increase the number of clocks in A_2, but the number of locations in A_2 may become exponential in the number of clocks of the given TA A. However, there is no addition of new valuations to the underlying state space of the original TA A and corresponding to every state (l, v) of a location l in the original TA A, exactly one state (l_i, v) is created in the modified TA A_2, where l_i is one of the newly created locations as a result of splitting l.*

Splitting locations and removing constraints as described above do not alter the behaviour of the timed automaton that leads us to the following lemma.

Lemma 4. *The operations in stage 2 produce a timed automaton A_2 that is timed bisimilar to the TA A_1 obtained at the end of stage 1.*

The number of locations after the split can become exponential in the number of the clocks. The constraints on the incoming edges of l are also split appropriately into constraints on the incoming edges of the newly created locations. Hence this stage too runs in time that is *exponential* in the number of the clocks of the timed automaton.

Stage 3: Removing constraints by considering multiple edges with the same action: We consider the example in Figure 4. Note that the constraints $x \leq 3$ and $x > 3$ on the edges from l_0 to l_1 and from l_0 to l_2 respectively could as well be merged together to produce a constraint without any clock.

For every action a enabled at any location l, this stage checks whether a guard enabling that action at l can be merged with another guard enabling the same action at that location such that timed bisimilarity is preserved. The transformation made in this stage has been formally described in Algorithm 2. The input to this algorithm is the TA obtained after stage 2, say A_2. For each location l_i, the algorithm does the following: for every action $a \in Act$, it determines the zones of l_i from which action a is enabled. We call this set $\mathcal{Z}_{ia}$. Zone graph construction and splitting of locations in stage 2 ensures that all zones in $\mathcal{Z}_{ia}$ form a linear chain connected by ε edges as shown in Figure 5. We use $\lessdot$ to capture this total ordering relation. Let us use ordered indexed variable $1, \ldots, m$ to name the zones in this total order, i.e. $Z_{i_1} \lessdot \cdots \lessdot Z_{i_k} \lessdot Z_{i_{k+1}} \cdots \lessdot Z_{i_m}$. Lemma 1 ensures that for each Z_{i_k}, $k \geq 1$, that is bounded above, there exists a hyperplane that bounds the zone fully from above and similarly, for each Z_{i_k}, $k > 1$ there exists a hyperplane that bounds the zone fully from below. For a zone Z, let $LB(Z)$ and $UB(Z)$ denote these lower and upper bounding hyperplanes of Z respectively. Further, $UB(Z)$ is ∞ if Z is not bounded from above.

Let $\Gamma_{(l_i,a)} = \{g \mid l_i \xrightarrow{a,g,R} l' \in E_{A_2}\}$ be the set of guards on the outgoing edges from l_i in A_2 which are labelled with a. For any $g \in \Gamma_{(i,a)}$, let us define the following;

- $\mathrm{Strt}(g)_{(l_i,a)} = Z \in \mathcal{Z}_{ia}$ is the zone in $\mathcal{Z}_{ia}$ which is bounded from below by the same constraints as the lower bound of the constraints in g.
- $\mathrm{End}(g)_{(l_i,a)} = Z \in \mathcal{Z}_{ia}$ is the zone in $\mathcal{Z}_{ia}$ which is bounded from above by the same constraints as the upper bound of the constraints in g.
- $\mathrm{Ran}(g)_{(l_i,a)} = \{Z \in \mathcal{Z}_{ia} \mid \mathrm{Strt}(g)_{(l_i,a)} \lessdot Z \wedge Z \lessdot \mathrm{End}(g)_{(l_i,a)}\} \cup \{\mathrm{Strt}(g)_{(l_i,a)}\} \cup \{\mathrm{End}(g)_{(l_i,a)}\}$ is the set of zones ordered by $\lessdot$ relation in between $\mathrm{Strt}(g)_{(l_i,a)}$ and $\mathrm{End}(g)_{(l_i,a)}$.

In Algorithm 2, we use a rather informal notation $g := [C1, C2]$ to denote that $C1$ and $C2$ are the constraints defining the lower and the upper bounds of g respectively. If g does not have any constraint defining the upper bound then $C_2 = \infty$. We define a total order $\lll$ on $\Gamma_{(i,a)}$ such that for any $g, g' \in \Gamma_{(i,a)}$,

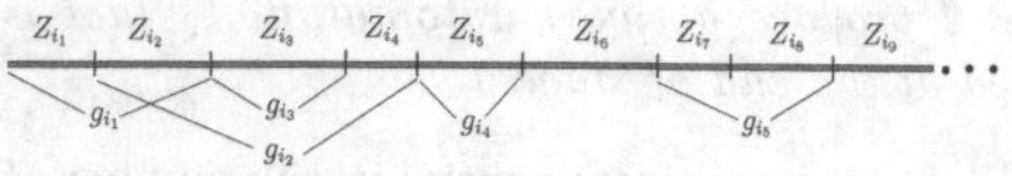

Fig. 5. Merging constraints in stage 3

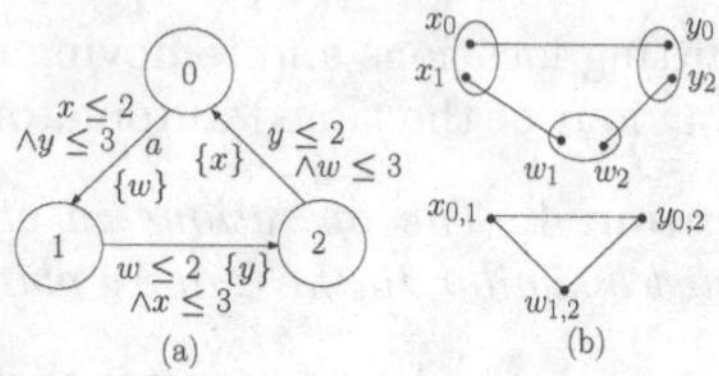

Fig. 6. Colouring clock graph

$g \lll g'$ iff $\exists Z \in \text{Ran}(g)_{(l_i,a)}$ such that $Z < Z'$ for all $Z' \in \text{Ran}(g')_{(l_i,a)}$. Similar to the zones let us use ordered indexed variable $g_{i_1}, \ldots, g_{i_p}$ to denote $g_{i_1} \lll \cdots g_{i_k} \lll g_{i_{k+1}} \cdots \lll g_{i_p}$. One such total order on guards is shown in Figure 5. The loop from Line 5 to Line 38 in Algorithm 2 traverses the elements of $\Gamma_{(i,a)}$ in this total order with the help of a variable *next* initialized to 2. In every iteration of this loop the invariant $g_{curr} \lll g_{i_{next}}$ holds. Three possibilities exist based on whether the set union of zones corresponding to these guards is (i) not convex (ii) convex but non-overlapping, or (iii) convex as well as overlapping.

If the union is non-convex then both g_{curr} and index are changed in Line 7 to pick the next ordered pair in this order. For cases (ii) and (iii), new guards are created by merging corresponding zones as long as the modified automaton preserves timed bisimilarity. If timed bisimilarity is preserved then the modified automaton A' is set as the current automaton which is A_3 (Line 13 and Line 29) and *next* is incremented to process the next guard. Otherwise the guard g_{curr} is set to $g_{i_{next}}$ and *next* is incremented by 1 (Line 15 and Line 36). Therefore the only difference in these two cases is in creating the new guard.

For case (ii), convex but non-overlapping zones, a new guard is created from the lower bound of $\text{Strt}(g_{curr})_{(l_i,a)}$ and the upper bound of $\text{End}(g_{i_{next}})_{(l_i,a)}$. For case (iii), there are three possibilities of combining guards, mentioned in Line 20, Line 22 and Line 24. The first possibility is the same as in case (ii). The second and the third possibilities are replacing the upper bound of g_{curr} with the lower bound of $\text{Strt}(g_{i_{next}})_{(l_i,a)}$ and the lower bound of $g_{i_{next}}$ with the upper bound of $\text{End}(g_{curr})_{(l_i,a)}$ respectively.

A zone graph captures the behaviour of the timed automaton and hence timed bisimilarity between two TAs can be checked using their zone graphs [21,13]. This is why we create the pre-stable zone graph as described in Section 2 as it enables one to directly check timed bisimilarity on this zone graph [13].

Lemma 5. *The operations in stage 3 produce a timed automaton A_3 that is timed bisimilar to the TA A_2 obtained at the end of stage 2.*

As mentioned above, in this stage, while merging the constraints, timed bisimilarity is checked and the number of bisimulation checks is bounded by the number of zones in the zone graph of the TA obtained after stage 2. From Lemma 3, in stage 2, no new valuations are added to the underlying state space of the original TA A, and corresponding to every valuation (l, v), exactly one valuation (l_i, v) is created, where l and l_i are as described in Lemma 3. Thus the number of zones

---

Algorithm 2. Algorithm for stage 3
Input: Timed automaton A_2 obtained after stage 2
Output: Modified TA A_3 after applying stage 3 procedures

1: $A_3 := A_2$ ▷ A_2 is the TA obtained from A after the first two stages
2: **for each** location l_i in A_2 **do**   ▷ i is the index of the location, the set of locations do not change in this stage
3:     **for each** $a \in sort(l_i)$ **do**   ▷ $sort(l_i)$ is the set of actions in l_i that can be performed from l_i
4:        $g_{curr} := g_{i_1}$, $next := 2$
5:        **while** $next < |\Gamma_{(i,a)}| - 1$ **do**
6:           **if** $\text{Ran}(g_{curr})_{(l_i,a)} \cup \text{Ran}(g_{i_{next}})_{(l_i,a)}$ is not convex **then**
7:             $g_{curr} := g_{i_{next}}$, $next := next + 1$
8:           **else if** $\text{Ran}(g_{i_{next}})_{(l_i,a)} \cap \text{Ran}(g_{curr})_{(l_i,a)} = \emptyset$ **then** ▷ non-overlapping but contiguous
9:             $g'_{curr} := [LB(\text{Strt}(g_{curr})_{(l_i,a)}), UB(\text{End}(g_{i_{next}})_{(l_i,a)})]$
10:             $g'_{i_{next}} := g'_{curr}$
11:             Let A' be the TA obtained by replacing all occurrences of g_{curr} and $g_{i_{next}}$ with g'_{curr} and $g'_{i_{next}}$ respectively in A_3
12:             **if** A' is timed bisimilar to A_3 **then**
13:                $A_3 := A'$, $g_{curr} := g'_{i_{next}}$, $next := next + 1$
14:             **else**
15:                $g_{curr} := g_{i_{next}}$, $next := next + 1$
16:             **end if**
17:           **else**   ▷ $\text{Ran}(g_{curr})_{(l_i,a)}$ and $\text{Ran}(g_{i_{next}})_{(l_i,a)}$ have overlapping zones
18:             ▷ There are three ways to combine g_{curr} and $g_{i_{next}}$, and
19:             ▷ Resultant new guards should be checked for timed bisimilarity in the following order
20:             (i). $g'_{curr} := [LB(\text{Strt}(g_{curr})_{(l_i,a)}), UB(\text{End}(g_{i_{next}})_{(l_i,a)})]$,
21:             $g'_{i_{next}} := g'_{curr}$
22:             (ii). $g'_{curr} := [LB(\text{Strt}(g_{curr})_{(l_i,a)}), LB(\text{Strt}(g_{i_{next}})_{(l_i,a)})]$,
23:             $g'_{i_{next}} := [LB(\text{Strt}(g_{i_{next}})_{(l_i,a)}), UB(\text{End}(g_{i_{next}})_{(l_i,a)})]$
24:             (iii). $g'_{curr} := [LB(\text{Strt}(g_{curr})_{(l_i,a)}), UB(\text{End}(g_{curr})_{(l_i,a)})]$,
25:             $g'_{i_{next}} := [UB(\text{End}(g_{curr})_{(l_i,a)}), UB(\text{End}(g_{i_{next}})_{(l_i,a)})]$
26:             **while** $1 \le i \le 3$ **do**   ▷ Corresponding to the three cases above
27:                Let A' be the TA obtained by replacing all occurrences of g_{curr} and $g_{i_{next}}$ with the ith g'_{curr} and $g'_{i_{next}}$ respectively in A_3
28:                **if** A' is timed bisimilar to A_3 **then**
29:                   $A_3 := A'$, $g_{curr} := g'_{i_{next}}$, $next := next + 1$
30:                   **Break**
31:                **else**
32:                   $i := i + 1$
33:                **end if**
34:             **end while**
35:             **if** $i = 4$ **then**   ▷ Bisimilarity could not be preserved in any of these three cases
36:                $g_{curr} := g_{i_{next}}$, $next := next + 1$
37:             **end if**
38:           **end if**
39:        **end while**
40:     **end for**
41: **end for**

---

in the zone graph of A_2 is still exponential in the number of cocks of the original TA A. Checking timed bisimilarity is done in EXPTIME [7,17]. A zone graph is constructed prior to every bisimulation check and the construction is done in EXPTIME. Hence this entire stage runs in EXPTIME.

Stage 4: Finding Active clocks, clock replacement and renaming: Given a location l, an iterative method for finding the set of active clocks at l,denoted $act(l)$, is given in [9]. The method has been modified and stated below for the case where clock assignments of the form $x := y, x, y \in C$ are disallowed.

Determining active clocks : For a location l, let $clk(l)$ be the set of clocks that appear on the constraints in the outgoing edges of l. Let $\rho : (2^C \times E) \rightarrow 2^C$ be a partial function such that for an edge $e = l \xrightarrow{g,a,R} l'$, $\rho(act(l'), e)$ gives the set of active clocks of l' that are not reset along e. For all $l \in L$, $act(l)$ is the limit of the convergent sequence $act_0(l) \subseteq act_1(l) \ldots$ such that $act_0(l) := clk(l)$ and $act_{i+1}(l) := act_i(l) \cup \bigcup_{e=(l,g,a,R,l') \in E} \rho(act_i(l'), e)$.

Removing redundant resets : Once we find the active clocks of a location l, we remove all resets of clock x on the incoming edges of l if $x \notin act(l)$.

Partitioning active clocks : Using the DBM representation of the zones, one can determine from the set of active clocks in every location whether some of the clocks in the timed automaton can be expressed in terms of other clocks and thus be removed. Any $x, y \in act(l)$ belong to an equivalence class iff the same relation of the form $x - y = k$, for some fixed integer k is maintained between these clocks across all zones of l. This is checked using the DBM of the zones of l. In this case either x can be replaced by $y + k$ or y can be replaced by $x - k$. Let π_l be the partition induced by this equivalence relation.

We note that the size of the largest partition does not give the minimum number of clocks required to represent a TA while preserving timed bisimulation. An example is shown in Figure 6(a). Though the automaton in the figure has two active clocks partitioned into two different classes in every location, a timed bisimilar TA cannot be constructed with only two clocks. Assigning the minimum number of clocks to represent the timed automaton so that timed bisimilarity is preserved can be reduced to the problem of finding the chromatic number of a graph as described below.

Clock graph colouring and clock renaming : A clock graph, G_{A_3}, for the timed automaton A_3 is constructed in the following way. This graph contains a vertex for each class in the partition π_l, for every location l. Let V_l be the set of vertices corresponding to the classes of π_l. For each pair of distinct vertices $r_1, r_2 \in V_l$, an edge between r_1 and r_2 exists denoting that r_1 and r_2 cannot be assigned the same colour. This is because two classes in the partition π_l cannot be represented using the same clock.

Moreover, if at least one clock, say c, is common in two classes corresponding to two different locations without any intervening reset of c then only one vertex represents these two classes. For example, in Figure 6, clock x is active in both locations 0 and 1. $\{x\}$ forms a class in the partition of the active clocks for each of locations 0 and 1. Thus we create vertices x_0 and x_1 corresponding to these two classes. However, since there is no intervening reset of clock x between locations 0 and 1, the vertices x_0 and x_1 are merged together and the resultant graph is termed the *clock graph*. Thus after merging some classes into one class, the resultant class can have active clocks corresponding to multiple locations. For a class $\mathcal{T}$, let $loc(\mathcal{T})$ represent the set of locations whose active clocks are members of $\mathcal{T}$.

Finding the minimum number of clocks to represent the TA A_3 is thus equivalent to colouring its clock graph with the minimum number of colours so that no two adjacent vertices have the same colour. The number of colours gives the minimum number of clocks required to represent the TA. If a colour c is assigned to a vertex r, then all the clocks in the class corresponding to r, say $\mathcal{T}$, are renamed c. The value of c can be chosen to be equal to some clock in $\mathcal{T}$ that is considered to be the *representative clock* for that class. The constraints involving the rest of these clocks in $\mathcal{T}$ are adjusted appropriately and any resets of the clocks, different from the representative clock, present on the incoming edges to l such that $l \in loc(\mathcal{T})$ are also removed.

For example, suppose vertex r corresponds to a class $\mathcal{T}$ having clocks x, y and z such that the valuations of the clocks are related as : $x - y = k_1$ and $y - z = k_2$. If colour c is assigned to vertex r, then the clocks x, y and z in class $\mathcal{T}$ are replaced with c. If the value of clock c is chosen to be the same as clock y, then every occurrence of x in $\mathcal{T}$ is replaced with $y + k_1$, while every occurrence of z in $\mathcal{T}$ is replaced with $y - k_2$ in the constraints involving x and z. The corresponding resets of clocks x and z are also removed.

In Figure 6(a), a TA with three locations is shown. In locations 0, 1 and 2, the sets of active clocks are $\{x, y\}$, $\{w, x\}$ and $\{w, y\}$ respectively. At every location, in this example, each of the active clocks itself makes a class of the partition. Since there are six classes in total, we draw initially six vertices. As mentioned earlier, the vertices x_0 and x_1 are merged into a single vertex. Similarly w_1, w_2 and y_0, y_2 are also merged. We call the resultant vertices $x_{0,1}$, $w_{1,2}$ and $y_{0,2}$. Adding the edges as described previously, we get the clock graph which is a triangle as shown in Figure 6(b). Thus the chromatic number of this graph is 3 which translates to the number of clocks obtained through the operations in stage 4. Since this stage consists of finding the active clocks and renaming them, we have the following lemma.

Lemma 6. *The operations in stage 4 produce a timed automaton A_4 that is timed bisimilar to the TA A_3 obtained after stage 3.*

We look at the complexity of the operations in this stage. The sequence of computation of active clocks converges within n iterations and every iteration runs in time $O(|E|)$, where there are n locations and $|E|$ edges respectively in the timed automaton after the first three stages. This is due to the fact that in iteration i, for some location l, its active clocks are updated so as to include those active clocks of the locations l' such that there exists a path of length at most i between l and l' and these clocks are not reset along this path. In each iteration, each edge is traversed once for updating the set of active clocks of the locations. Thus the complexity of finding active clocks is $O(n \times |E|)$.

Partitioning the active clocks of each of the locations too requires traversing the zone graph and checking the clock relations from the DBM of the zones. This can be done in time equal to the order of the size of the zone graph times the size of DBM which is in EXPTIME.

Finally, determining the chromatic number of a graph is possible in time exponential in the number of the vertices of the graph [18,11]. Since the number of locations after the splitting operation in stage 2 is exponential in the number of clocks in A, renaming the clocks using the clock graph runs in time doubly exponential in the number of the clocks of the original timed automaton A. Thus we have the following theorem.

Theorem 1. *The stages mentioned above run in 2-EXPTIME.*

In the presence of an invariant condition, considering an edge $l \xrightarrow{g,a,R} l'$, a zone Z' of l' is initially created such that $Z' = (Z \cap g)_{[R \leftarrow \bar{0}]}\uparrow \cap I(l')$, if $(Z \cap g)_{[R \leftarrow \bar{0}]}\uparrow \cap I(l') \neq \emptyset$, where $I(l')$ is the invariant on location l'. The edge can be removed from the timed automaton if for all zones Z of l, $(Z \cap g)_{[R \leftarrow \bar{0}]} \uparrow \cap I(l') = \emptyset$. The invariant condition on location l' can be entirely removed if for every incoming edge $l \xrightarrow{g,a,R} l'$ and for each zone Z of l, $(Z \cap g)_{[R \leftarrow \bar{0}]} \uparrow \cap I(l') = (Z \cap g)_{[R \leftarrow \bar{0}]} \uparrow$ holds. Considering the clock relations in the zones of a location, an elementary constraint in the invariant too is removed if it is implied by the rest of the elementary constraints in the invariant. In stage 4, $clk(l)$ becomes the union of the set of clocks appearing in the constraints on the outgoing edges from l and the set of clocks appearing in $I(l)$.

Proof of Minimality of Clocks: Let A_4 be the TA obtained from a TA A through the four stages described earlier. We can show that for each location l in A_4, for every clock $x \in act(l)$, there exists at least one constraint involving clock x which is indispensable for any TA that preserves timed bisimilarity.

Lemma 7. *In the TA A_4, for each location l and clock $x \in act(l)$, there exists at least one constraint involving x on some outgoing edge of l or on the outgoing edge of another location l' reachable from l such that there is at least one path from l to l' without any intervening reset of x. Moreover the TA obtained by removing the constraint is not timed bisimilar to the given timed automaton A.*

Proof sketch : We prove this lemma by induction on the structure of the timed automaton A_4. Let us suppose that we have edges in the timed automaton A_4 from location l to locations $l_1, \ldots, l_m$. By induction hypothesis, the lemma holds for $l_1, \ldots, l_m$. Note that $act(l) = clk(l) \cup (\bigcup_{i=1,\, e=(l,g,a,R_i,l_i) \in E}^{m} (act(l_i) \setminus R_i))$. Consider a clock $c_0 \in act(l)$. If $c_0 \in act(l_i)$, for some $i \in \{1, \ldots, m\}$, and c_0 is not reset on the edge between l and l_i, then from the induction hypothesis, the lemma holds for l trivially. Otherwise, if for each $i \in \{1, \ldots, m\}$, such that $c_0 \in act(l_i)$ implies c_0 is reset on the edge from l to l_i, we can show that there exists a constraint on an outgoing edge from l involving c_0 that cannot be removed while preserving timed bisimilarity. □

Definition 2. ***Minimal bisimilar TA***: *For a given timed automaton D, a minimal bisimilar TA is one that is timed bisimilar to D and has the minimum number of clocks possible.*

Fact 1. *For every TA D, there exists a minimal bisimilar TA.*

One can show that the clocks of a TA that is minimal bisimilar to A_4 can replace the clocks of A_4 which gives us the following lemma.

Lemma 8. *The timed automaton A_4 has the same number of clocks as a minimal bisimilar TA for A.*

Proof sketch : We consider a minimal bisimilar TA D_1 for A and apply the operations in the four stages on it to produce a TA D. Since D_1 is already minimal, D has the same number of clocks as D_1. The transformations ensure that in the resultant zone graph of D, for each bounded zone in every location, there exists a hyperplane that fully bounds it. Similarly in the zone graph of A_4 too, each bounded zone of every location is fully bounded by a hyperplane.

We use the zone graphs of A_4 and D for mapping the clocks of A_4 to the clocks of D. From Lemma 7, for a location l_{A_4} of A_4, corresponding to every clock $x \in act(l_{A_4})$, there is a hyperplane of the form $x = k$ corresponding to a constraint $x \bowtie k$ bounding a zone that cannot be removed if timed bisimulation has to be preserved. Since A_4 and D are timed bisimilar, there is a corresponding hyperplane, say $y = k'$, induced by a constraint $y \bowtie k'$ which too cannot be removed from the zone graph of D while preserving timed bisimulation. Clock x in A_4 can thus be renamed y. The clock renaming in stage 4 ensures that, any further renaming cannot reduce the number of clocks in A_4. Hence if the clocks of A_4 can be replaced with the clocks of D, we have $|C_{A_4}| \leq |C_D|$, thus giving $|C_{A_4}| = |C_D|$ since D is a minimal bisimilar TA for A. □

Theorem 2. *There exists an algorithm to construct a TA A_4 that is timed bisimilar to a given TA A such that among all the timed automata that are timed bisimilar to A, A_4 has the minimum number of clocks. Further the algorithm runs in time that is doubly exponential in the number of clocks of A.*

4 Conclusion

In this paper, we have described an algorithm, which given a timed automaton A, produces another timed automaton A_4 with the smallest number of clocks that is timed bisimilar to A. It also follows trivially that A_4 accepts the same timed language as A. The problem solved in this paper was left open in [16]

For reducing the number of clocks of the timed automaton, we rely on a semantic representation of the timed automaton rather than its syntactic form as in [9]. This helps us to reason about the behaviour of the timed automaton more effectively. Besides, the zone graph we use in our approach is usually much smaller in size than the region graph and its size is independent of the constants used in the timed automaton. There is an exponential increase in the number of locations while producing the TA with the minimal number of clocks. However, there is no addition of new valuations to the underlying state space of the timed automaton since the splitting of a location l, described in stage 2, involves distributing the zones of l across the locations l is split into.

References

1. Aceto, L., Ingólfsdóttir, A., Larsen, K.G., Srba, J.: Reactive Systems: Modelling, Specification and Verification. Cambridge University Press (2007)
2. Alur, R., Courcoubetis, C., Dill, D.: Model-checking in dense real-time. Information and Computation 104, 2–34 (1993)
3. Alur, R., Dill, D.L.: A theory of timed automata. Theoretical Computer Science 126, 183–235 (1994)
4. Behrmann, G., Bouyer, P., Fleury, E., Larsen, K.G.: Static guard analysis in timed automata verification. In: Garavel, H., Hatcliff, J. (eds.) TACAS 2003. LNCS, vol. 2619, pp. 254–270. Springer, Heidelberg (2003)
5. Behrmann, G., Bouyer, P., Larsen, K.G., Pelanek, R.: Lower and upper bounds in zone-based abstractions of timed automata. International Journal on Software Tools for Technology Transfer 8, 204–215 (2006)
6. Berthomieu, B., Menasche, M.: An enumerative approach for analyzing time petri nets. In: IFIP Congress, pp. 41–46 (1983)
7. Čerāns, K.: Decidability of bisimulation equivalences for parallel timer processes. In: Probst, D.K., von Bochmann, G. (eds.) CAV 1992. LNCS, vol. 663, pp. 302–315. Springer, Heidelberg (1993)
8. Daws, C., Tripakis, S.: Model checking of real-time reachability properties using abstractions. In: Steffen, B. (ed.) TACAS 1998. LNCS, vol. 1384, pp. 313–329. Springer, Heidelberg (1998)
9. Daws, C., Yovine, S.: Reducing the number of clock variables of timed automata. In: IEEE Proc. RTSS 1996, pp. 73–81. IEEE Computer Society Press (1996)
10. Dill, D.L.: Timing assumptions and verification of finite-state concurrent systems. In: Sifakis, J. (ed.) CAV 1989. LNCS, vol. 407, pp. 197–212. Springer, Heidelberg (1990)
11. Eppstein, D.: Small maximal independent sets and faster exact graph coloring. Journal of Graph Algorithms and Applications 7, 131–140 (2003)
12. Finkel, O.: Undecidable problems about timed automata. In: Asarin, E., Bouyer, P. (eds.) FORMATS 2006. LNCS, vol. 4202, pp. 187–199. Springer, Heidelberg (2006)
13. Guha, S., Krishna, S.N., Narayan, C., Arun-Kumar, S.: A unifying approach to decide relations for timed automata and their game characterization. In: Express/SOS, pp. 47–62 (2013)
14. Henzinger, T.A., Nicollin, X., Sifakis, J., Yovine, S.: Symbolic Model Checking for Real-time Systems. In: Proceedings of the Seventh Annual Symposium on Logic in Computer Science (LICS 1992), Santa Cruz, California, USA, June 22-25, pp. 394–406. IEEE Computer Society (1992)
15. Henzinger, T.A., Nicollin, X., Sifakis, J., Yovine, S.: Symbolic model checking for real-time systems. In: The Proceedings of Logic in Computer Science, LICS (1992)
16. Laroussinie, F., Larsen, K.G., Weise, C.: From timed automata to logic – and back. In: Wiedermann, J., Hájek, P. (eds.) MFCS 1995. LNCS, vol. 969, pp. 529–539. Springer, Heidelberg (1995)

17. Laroussinie, F., Schnoebelen, P.: The state explosion problem from trace to bisimulation equivalence. In: Tiuryn, J. (ed.) FOSSACS 2000. LNCS, vol. 1784, pp. 192–207. Springer, Heidelberg (2000)
18. Lawler, E.L.: A note on the complexity of the chromatic number problem. Information Processing Letters 5, 66–67 (1976)
19. Tripakis, S.: Folk theorems on the determinization and minimization of timed automata. Information Processing Letters 99, 222–226 (2006)
20. Tripakis, S., Yovine, S.: Analysis of timed systems using time-abstracting bisimulations. Formal Methods in System Design 18, 25–68 (2001)
21. Weise, C., Lenzkes, D.: Efficient scaling-invariant checking of timed bisimulation. In: Reischuk, R., Morvan, M. (eds.) STACS 1997. LNCS, vol. 1200, pp. 177–188. Springer, Heidelberg (1997)

Qualitative Concurrent Parity Games: Bounded Rationality*

Krishnendu Chatterjee

IST Austria

Abstract. We study two-player *concurrent* games on finite-state graphs played for an infinite number of rounds, where in each round, the two players (player 1 and player 2) choose their moves independently and simultaneously; the current state and the two moves determine the successor state. The objectives are ω-regular winning conditions specified as *parity* objectives. We consider the *qualitative analysis* problems: the computation of the *almost-sure* and *limit-sure* winning set of states, where player 1 can ensure to win with probability 1 and with probability arbitrarily close to 1, respectively. In general the almost-sure and limit-sure winning strategies require both *infinite-memory* as well as *infinite-precision* (to describe probabilities). While the qualitative analysis problem for concurrent parity games with infinite-memory, infinite-precision randomized strategies was studied before, we study the *bounded-rationality* problem for qualitative analysis of concurrent parity games, where the strategy set for player 1 is restricted to bounded-resource strategies. In terms of precision, strategies can be deterministic, uniform, finite-precision, or infinite-precision; and in terms of memory, strategies can be memoryless, finite-memory, or infinite-memory. We present a precise and complete characterization of the qualitative winning sets for all combinations of classes of strategies. In particular, we show that uniform memoryless strategies are as powerful as finite-precision infinite-memory strategies, and infinite-precision memoryless strategies are as powerful as infinite-precision finite-memory strategies. We show that the winning sets can be computed in $\mathcal{O}(n^{2d+3})$ time, where n is the size of the game structure and $2d$ is the number of priorities (or colors), and our algorithms are symbolic. The membership problem of whether a state belongs to a winning set can be decided in NP $\cap$ coNP. Our symbolic algorithms are based on a characterization of the winning sets as μ-calculus formulas, however, our μ-calculus formulas are crucially different from the ones for concurrent parity games (without bounded rationality); and our memoryless witness strategy constructions are significantly different from the infinite-memory witness strategy constructions for concurrent parity games.

1 Introduction

In this work we consider the qualitative analysis (computation of almost-sure and limit-sure winning sets) for concurrent parity games. In prior works [15,8] the qualitative

* The research was partly supported by FWF Grant No P 23499-N23, FWF NFN Grant No S11407-N23 (RiSE), ERC Start grant (279307: Graph Games), and Microsoft faculty fellows award.

P. Baldan and D. Gorla (Eds.): CONCUR 2014, LNCS 8704, pp. 544–559, 2014.

analysis for concurrent parity games have been studied for the general class of infinite-memory, infinite-precision randomized strategies. In this work, we study the *bounded rationality* problem where the resources of the strategies are limited, and establish a precise and complete characterization of the qualitative analysis of concurrent parity games for combinations of resource-limited strategies. We start with the basic background of concurrent parity games and qualitative analysis.

Concurrent Parity Games. A two-player (player 1 and player 2) concurrent game is played on a finite-state graph for an infinite number of rounds, where in each round, the players independently choose moves, and the current state and the two chosen moves determine the successor state. In *deterministic* concurrent games, the successor state is unique; in *probabilistic* concurrent games, the successor state is given by a probability distribution. The outcome of the game (or a *play*) is an infinite sequence of states. These games were introduced by Shapley [35], and have been one of the most fundamental and well studied game models in stochastic graph games. We consider ω-regular objectives; where given an ω-regular set Φ of plays, player 1 wins if the outcome of the game lies in Φ. Otherwise, player 2 wins, i.e., the game is zero-sum. Such games occur in the synthesis and verification of reactive systems [13,33,31], and ω-regular objectives (that generalizes regular languages to infinite words) provide a robust specification language that can express all specifications (e.g. safety, liveness, fairness) that arise in the analysis of reactive systems ([1,19,2]). Concurrency in moves is necessary for modeling the synchronous interaction of components [17]. Parity objectives can express all ω-regular conditions, and we consider concurrent parity games.

Qualitative and Quantitative Analysis. The player-1 *value* $v_1(s)$ of the game at a state s is the limit probability with which player 1 can guarantee Φ against all strategies of player 2. The player-2 *value* $v_2(s)$ is analogously the limit probability with which player 2 can ensure that the outcome of the game lies outside Φ. The *qualitative* analysis of games asks for the computation of the set of *almost-sure* winning states where player 1 can ensure Φ with probability 1, and the set of *limit-sure* winning states where player 1 can ensure Φ with probability arbitrarily close to 1 (states with value 1); and the *quantitative* analysis asks for precise computation of values. Concurrent (probabilistic) parity games are determined [30], i.e., for each state s we have $v_1(s) + v_2(s) = 1$. The qualitative analysis for concurrent parity games was studied in [15,8] and the quantitative analysis in [18,7,3].

Difference of Turn-Based and Concurrent Games. Traditionally, the special case of *turn-based* games has received most attention. In turn-based games, in each round, only one of the two players has a choice of moves. In turn-based deterministic games, all values are 0 or 1 and can be computed using combinatorial algorithms [36,34,28]; in turn-based probabilistic games, values can be computed by iterative approximation [11,14]. Concurrent games significantly differ from turn-based games in requirement of strategies to play optimally. A *pure* strategy must, in each round, choose a move based on the current state and the history (i.e., past state sequence) of the game, whereas, a *randomized* strategy in each round chooses a probability distribution over moves (rather than a single move). In contrast to turn-based deterministic and probabilistic games with parity objectives, where pure memoryless (history-independent) optimal strategies exist [21,38,20,12,3], in concurrent games, both randomization and infinite-memory are required for limit-sure winning [15] (also see [23] for results on pushdown concurrent games, [27,16,6] on complexity of strategies required in concurrent reachability games, and [24,26] on complexity of related concurrent game problems).

Bounded Rationality. The qualitative analysis for concurrent parity games with infinite-memory, infinite-precision randomized strategies was studied in [15,8]. The strategies for qualitative analysis for concurrent games require two different types of infinite resource: (a) infinite-memory, and (b) infinite-precision in describing the probabilities in the randomized strategies; (see example in [15] that limit-sure winning in concurrent Büchi games require both infinite-memory and infinite-precision). In many applications, such as synthesis of reactive systems, infinite-memory and infinite-precision strategies are not implementable in practice. Thus though the theoretical solution of infinite-memory and infinite-precision strategies was established in [15], the strategies obtained are not realizable in practice, and the theory to obtain implementable strategies in such games has not been studied before. In this work we consider the *bounded rationality* problem for qualitative analysis of concurrent parity games, where player 1 (that represents the controller) can play strategies with bounded resource. To the best of our knowledge this is the first work that considers the bounded rationality problem for concurrent ω-regular graph games. The motivation is clear as controllers obtained from infinite-memory and infinite-precision strategies are not implementable.

Strategy Classification. In terms of precision, strategies can be classified as pure, uniformly random, bounded-finite-precision, finite-precision, and infinite-precision (in increasing order of precision to describe probabilities of a randomized strategy). In terms of memory, strategies can be classified as memoryless, finite-memory and infinite-memory. In [15] the almost-sure and limit-sure winning characterization under infinite-memory, infinite-precision strategies were presented. In this work, we present (i) a complete and precise characterization of the qualitative winning sets for bounded resource strategies, (ii) symbolic algorithms to compute the winning sets, and (iii) complexity results to determine whether a given state belongs to a qualitative winning set.

Our Results. Our contributions are summarized below.

1. We show that pure memoryless strategies are as powerful as pure infinite-memory strategies in concurrent games (Proposition 1). This result is straight-forward, obtained by a simple reduction to turn-based probabilistic games.
2. Uniform memoryless strategies are more powerful than pure infinite-memory strategies (the fact that randomization is more powerful than pure strategies follows from the classical matching pennies game), and we show that uniform memoryless strategies are as powerful as finite-precision infinite-memory strategies (Proposition 2). Thus our results show that if player 1 has only finite-precision strategies, then no memory is required and uniform randomization is sufficient. Hence very simple (uniform memoryless) controllers can be obtained for the entire class of finite-precision infinite-memory controllers. The result is obtained by a reduction to turn-based stochastic games, and the main technical contribution is the characterization of the winning sets for uniform memoryless strategies by a μ-calculus formula. The μ-calculus formula not only gives a symbolic algorithm, but is also in the heart of other proofs of the paper.

3. In case of bounded-finite-precision strategies, the almost-sure and limit-sure winning sets coincide (Proposition 2). For almost-sure winning, uniform memoryless strategies are also as powerful as infinite-precision finite-memory strategies (Proposition 3). In contrast infinite-memory infinite-precision strategies are more powerful than uniform memoryless strategies for almost-sure winning. For limit-sure winning, we show that infinite-precision memoryless strategies are more powerful than bounded-finite-precision infinite-memory strategies, and infinite-precision memoryless strategies are as powerful as infinite-precision finite-memory strategies (Proposition 4). Our results show that if infinite-memory is not available, then no memory is required (memoryless strategies are as powerful as finite-memory strategies). The result is obtained by using the μ-calculus formula for the uniform memoryless case: we show that a μ-calculus formula that combines the μ-calculus formula for almost-sure winning for uniform memoryless strategies and limit-sure winning for reachability with memoryless strategies exactly characterizes the limit-sure winning for parity objectives for memoryless strategies. The fact that we show that in concurrent parity games, though infinite-memory strategies are necessary, memoryless strategies are as powerful as finite-memory strategies, is in contrast with many other examples of graph games which require infinite-memory. For example, in perfect-information multi-dimensional (such as multi-dimensional mean-payoff) games [9] as well as partial-observation stochastic parity games [10,5] infinite-memory strategies are necessary and finite-memory strategies are strictly more powerful than memoryless strategies.
4. As a consequence of the characterization of the winning sets as μ-calculus formulas we obtain symbolic algorithms to compute the winning sets. We show that the winning sets can be computed in $\mathcal{O}(n^{2d+3})$ time, where n is the size of the game structure and $2d$ is the number of priorities (or colors), and our algorithms are symbolic.
5. The decision problem of whether a state belongs to a winning set lies in NP $\cap$ coNP.

In short, our results show that if infinite-memory is not available, then memory is useless, and if infinite-precision is not available, then uniform memoryless strategies are sufficient. Let P, U, bFP, FP, IP denote pure, uniform, bounded-finite-precision with bound b, finite-precision, and infinite-precision strategies, respectively, and M, FM, IM denote memoryless, finite-memory, and infinite-memory strategies, respectively. For $A \in \{P, U, bFP, FP, IP\}$ and $B \in \{M, FM, IM\}$, let $Almost_1(A, B, \Phi)$ denote the almost-sure winning set under player-1 strategies that are restricted to be both A and B for a parity objective Φ (and similar notation for $Limit_1(A, B, \Phi)$). Our results can be summarized by the following equalities and strict inclusion:

$$\begin{aligned} & Almost_1(P,M,\Phi) = Almost_1(P,IM,\Phi) = Limit_1(P,IM,\Phi) \subsetneq Almost_1(U,M,\Phi) \\ = & Almost_1(FP,IM,\Phi) = \bigcup_{b>0} Limit_1(bFP,IM,\Phi) = Almost_1(IP,FM,\Phi) \subsetneq Almost_1(IP,IM,\Phi); \end{aligned} \tag{1}$$

and

$$\begin{aligned} \bigcup_{b>0} Limit_1(bFP, IM, \Phi) & \subsetneq Limit_1(IP, M, \Phi) = Limit_1(IP, FM, \Phi) \\ = & \; Limit_1(FP, M, \Phi) = Limit_1(FP, IM, \Phi) \subsetneq Limit_1(IP, IM, \Phi). \end{aligned} \tag{2}$$

Comparison with Turn-Based Games and [8]. Our μ-calculus formulas and the correctness proofs are non-trivial generalizations of both the result of [22] for turn-based deterministic parity games and the result of [16] for concurrent reachability games. Our algorithms, that are obtained by characterization of the winning sets as μ-calculus formulas, are considerably more involved than those for turn-based games. Our proof structure of using μ-calculus formulas to characterize the winning sets, though similar to [8], has several new aspects. In contrast to the proof of [8] that constructs witness infinite-memory strategies for both players from the μ-calculus formulas, our proof constructs memoryless witness strategies for player 1 from our new μ-calculus formulas, and furthermore, we show that in the complement set of the μ-calculus formulas for every finite-memory strategy for player 1 there is a witness memoryless spoiling strategy of the opponent. Thus the witness strategy constructions are different from [8]. Since our μ-calculus formulas and the predecessor operators are different from [8] the proofs of the complementations of the μ-calculus formulas are also different. Moreover [8] only concerns limit-sure winning and not almost-sure winning. Note that in [15] both almost-sure and limit-sure winning was considered, but as shown in [8] the predecessor operators suggested for limit-sure winning (which was a nested stacked predecessor operator) in [15] require modification for correctness proof, and similar modification is also required for almost-sure winning. Thus some results from [15] related to almost-sure winning require a careful proof.

Techniques. One of the key difficulty in concurrent parity games is that the recursive characterization of turn-based games completely fail for concurrent games. All results for concurrent parity games [16,15,18,8] rely on μ-calculus formulas. A μ-calculus formula is a succinct description of a nested iterative algorithm, and thus provides a very general technique. The key challenge and ingenuity is always to come up with the appropriate μ-calculus formula with the right predecessor operators (i.e., the right algorithm), establish duality (complementation of the formulas), and then construct from μ-calculus formulas the witness strategies in concurrent games (i.e., the correctness proof). Our results are also based on μ-calculus formula characterization (nested iterative algorithms), however, the predecessor operators and construction of witness strategies (the heart of the proofs) are quite different from the previous results.

2 Definitions

In this section we define game structures, strategies, objectives, and winning modes.

Probability Distributions. For a finite set A, a *probability distribution* on A is a function $\delta\colon A \mapsto [0,1]$ such that $\sum_{a\in A} \delta(a) = 1$. We denote the set of probability distributions on A by $\mathcal{D}(A)$. Given $\delta \in \mathcal{D}(A)$, we denote by $\mathrm{Supp}(\delta) = \{x \in A \mid \delta(x) > 0\}$ the *support* of δ.

Concurrent Game Structures. A (two-player) *concurrent stochastic game structure* $\mathcal{G} = \langle S, A, \Gamma_1, \Gamma_2, \delta\rangle$ consists of the following components.

- A finite state space S and a finite set A of moves (or actions).
- Two move assignments $\Gamma_1, \Gamma_2 : S \mapsto 2^A \setminus \emptyset$. For $i \in \{1,2\}$, assignment Γ_i associates with each state $s \in S$ the nonempty set $\Gamma_i(s) \subseteq A$ of moves available to player i at state s. For technical convenience, we assume that $\Gamma_i(s) \cap \Gamma_j(t) = \emptyset$ unless $i = j$ and $s = t$, for all $i, j \in \{1,2\}$ and $s, t \in S$. If this assumption is not met, then the moves can be trivially renamed to satisfy it.

– A probabilistic transition function $\delta: S \times A \times A \mapsto \mathcal{D}(S)$, which associates with every state $s \in S$ and moves $a_1 \in \Gamma_1(s)$ and $a_2 \in \Gamma_2(s)$ a probability distribution $\delta(s, a_1, a_2) \in \mathcal{D}(S)$ for the successor state.

Plays. At every state $s \in S$, player 1 chooses a move $a_1 \in \Gamma_1(s)$, and simultaneously and independently player 2 chooses a move $a_2 \in \Gamma_2(s)$. The game then proceeds to the successor state t with probability $\delta(s, a_1, a_2)(t)$, for all $t \in S$. For all states $s \in S$ and moves $a_1 \in \Gamma_1(s)$ and $a_2 \in \Gamma_2(s)$, we indicate by $Dest(s, a_1, a_2) = \mathrm{Supp}(\delta(s, a_1, a_2))$ the set of possible successors of s given moves a_1, a_2. A *path* or a *play* is an infinite sequence $\omega = \langle s_0, s_1, s_2, \ldots \rangle$ of states such that for all $k \geq 0$, there exists $a_1^k \in \Gamma_1(s_k)$ and $a_2^k \in \Gamma_2(s_k)$ such that $s_{k+1} \in Dest(s_k, a_1^k, a_2^k)$. We denote by Ω the set of all paths. For a play $\omega = \langle s_0, s_1, s_2, \ldots \rangle \in \Omega$, we define $Inf(\omega) = \{s \in S \mid s_k = s \text{ for infinitely many } k \geq 0\}$ to be the set of states that occur infinitely often in ω.

Size of a Game. The *size* of a concurrent game is the sum of the size of the state space and the number of the entries of the transition function, i.e., $|S| + \sum_{s \in S, a \in \Gamma_1(s), b \in \Gamma_2(s)} |Dest(s, a, b)|$.

Turn-Based Stochastic Games and MDPs. A game structure $\mathcal{G}$ is *turn-based stochastic* if at every state at most one player can choose among multiple moves; that is, for every state $s \in S$ there exists at most one $i \in \{1, 2\}$ with $|\Gamma_i(s)| > 1$. A game structure is a player-2 *Markov decision process* if for all $s \in S$ we have $|\Gamma_1(s)| = 1$, i.e., only player-2 has choice of actions in the game.

Equivalent Game Structures. Given two game structures $\mathcal{G}_1 = \langle S, A, \Gamma_1, \Gamma_2, \delta_1 \rangle$ and $\mathcal{G}_2 = \langle S, A, \Gamma_1, \Gamma_2, \delta_2 \rangle$ on the same state and action space, with a possibly different transition function, we say that $\mathcal{G}_1$ is equivalent to $\mathcal{G}_2$ (denoted $\mathcal{G}_1 \equiv \mathcal{G}_2$) if for all $s \in S$ and all $a_1 \in \Gamma_1(s)$ and $a_2 \in \Gamma_2(s)$ we have $\mathrm{Supp}(\delta_1(s, a_1, a_2)) = \mathrm{Supp}(\delta_2(s, a_1, a_2))$.

Strategies. A *strategy* for a player is a recipe that describes how to extend a play. Formally, a strategy for player $i \in \{1, 2\}$ is a mapping $\pi_i: S^+ \mapsto \mathcal{D}(A)$ that associates with every nonempty finite sequence $x \in S^+$ of states, representing the past history of the game, a probability distribution $\pi_i(x)$ used to select the next move. The strategy π_i can prescribe only moves that are available to player i; i.e., for all sequences $x \in S^*$ and states $s \in S$, we require that $\mathrm{Supp}(\pi_i(x \cdot s)) \subseteq \Gamma_i(s)$. We denote by Π_i the set of all strategies for player $i \in \{1, 2\}$. Given a state $s \in S$ and two strategies $\pi_1 \in \Pi_1$ and $\pi_2 \in \Pi_2$, the probabilities of events are uniquely defined [37], where an *event* $\mathcal{A} \subseteq \Omega$ is a measurable set of paths. For an event $\mathcal{A} \subseteq \Omega$, we denote by $\Pr_s^{\pi_1, \pi_2}(\mathcal{A})$ the probability that a path belongs to $\mathcal{A}$ when the game starts from s and the players use the strategies π_1 and π_2.

Classification of Strategies. We classify strategies according to their use of *randomization* and *memory*. We first present the classification according to randomization.

1. *(Pure).* A strategy π is *pure (deterministic)* if for all $x \in S^+$ there exists $a \in A$ such that $\pi(x)(a) = 1$. Thus, deterministic strategies are equivalent to functions $S^+ \mapsto A$.
2. *(Uniform).* A strategy π is *uniform* if for all $x \in S^+$ we have $\pi(x)$ is uniform over its support, i.e., for all $a \in \mathrm{Supp}(\pi(x))$ we have $\pi(x)(a) = \frac{1}{|\mathrm{Supp}(\pi(x))|}$.
3. *(Finite-precision).* A strategy π is *finite-precision* if there exists a bound $b \in \mathbb{N}$ such that for all $x \in S^+$ and all actions a we have $\pi(x)(a) = \frac{i}{j}$, where $i, j \in \mathbb{N}$ and

$0 \leq i \leq j \leq b$ and $j > 0$, i.e., the probability of an action played by the strategy is a multiple of some $\frac{1}{j}$, with $j \in \mathbb{N}$ such that $j \leq b$.

We denote by $\Pi_i^P, \Pi_i^U, \Pi_i^{FP}$ and Π_i^{IP} the set of pure (deterministic), uniform, finite-precision, and infinite-precision (or general) strategies for player i, respectively. Observe that we have the following strict inclusion in general: $\Pi_i^P \subsetneq \Pi_i^U \subsetneq \Pi_i^{FP} \subsetneq \Pi_i^{IP}$.

1. *(Finite-memory)*. Strategies in general are *history-dependent* and can be represented as follows: let $\mathcal{M}$ be a set called *memory* to remember the history of plays (the set $\mathcal{M}$ can be infinite in general). A strategy with memory can be described as a pair of functions: (a) a *memory update* function $\pi_u : S \times \mathcal{M} \mapsto \mathcal{M}$, that given the memory $\mathcal{M}$ with the information about the history and the current state updates the memory; and (b) a *next move* function $\pi_n : S \times \mathcal{M} \mapsto \mathcal{D}(A)$ that given the memory and the current state specifies the next move of the player. A strategy is *finite-memory* if the memory $\mathcal{M}$ is finite.
2. *(Memoryless)*. A *memoryless* strategy is independent of the history of play and only depends on the current state. Formally, for a memoryless strategy π we have $\pi(x \cdot s) = \pi(s)$ for all $s \in S$ and all $x \in S^*$. Thus memoryless strategies are equivalent to functions $S \mapsto \mathcal{D}(A)$.

We denote by Π_i^M, Π_i^{FM} and Π_i^{IM} the set of memoryless, finite-memory, and infinite-memory (or general) strategies for player i, respectively. Observe that we have the following strict inclusion in general: $\Pi_i^M \subsetneq \Pi_i^{FM} \subsetneq \Pi_i^{IM}$.

Objectives. We specify objectives for the players by providing the set of *winning plays* $\Phi \subseteq \Omega$ for each player. In this paper we study only zero-sum games [32,25], where the objectives of the two players are complementary. A general class of objectives are the Borel objectives [29]. A *Borel objective* $\Phi \subseteq S^\omega$ is a Borel set in the Cantor topology on S^ω. In this paper we consider *ω-regular objectives* [36], which lie in the first $2^1\!/_2$ levels of the Borel hierarchy (i.e., in the intersection of Σ_3 and Π_3). We will consider the following ω-regular objectives.

- *Reachability and safety objectives*. Given a set $T \subseteq S$ of "target" states, the reachability objective requires that some state of T be visited. The set of winning plays is thus $\text{Reach}(T) = \{\omega = \langle s_0, s_1, s_2, \ldots \rangle \in \Omega \mid \exists k \geq 0.\ s_k \in T\}$. Given a set $F \subseteq S$, the dual safety objective is defined as $\text{Safe}(F) = \{\omega = \langle s_0, s_1, s_2, \ldots \rangle \in \Omega \mid \forall k \geq 0.\ s_k \in F\}$.
- *Büchi and co-Büchi objectives*. Given a set $B \subseteq S$ of "Büchi" states, the Büchi objective requires that B is visited infinitely often. Formally, the set of winning plays is $\text{Büchi}(B) = \{\omega \in \Omega \mid \mathit{Inf}(\omega) \cap B \neq \emptyset\}$. Given $C \subseteq S$, the co-Büchi objective requires that all states visited infinitely often are in C. Formally, the set of winning plays is $\text{co-Büchi}(C) = \{\omega \in \Omega \mid \mathit{Inf}(\omega) \subseteq C\}$.
- *Parity objectives*. For $c, d \in \mathbb{N}$, we let $[c..d] = \{c, c+1, \ldots, d\}$. Let $p : S \mapsto [0..d]$ be a function that assigns a *priority* $p(s)$ to every state $s \in S$, where $d \in \mathbb{N}$. The *Even parity objective* requires that the maximum priority visited infinitely often is even. Formally, the set of winning plays is defined as $\text{Parity}(p) = \{\omega \in \Omega \mid \max(p(\mathit{Inf}(\omega))) \text{ is even}\}$. The dual *Odd parity objective* is defined as $\text{coParity}(p) = \{\omega \in \Omega \mid \max(p(\mathit{Inf}(\omega))) \text{ is odd}\}$. Büchi and co-Büchi objectives are simpler and special cases of parity objectives with two priorities.

Given a set $U \subseteq S$ we use LTL notations $\Box U, \Diamond U, \Box\Diamond U$ and $\Diamond\Box U$ to denote $\text{Safe}(U)$, $\text{Reach}(U)$, $\text{Büchi}(U)$ and co-$\text{Büchi}(U)$, respectively.

Winning Modes. Given an initial state $s \in S$, an objective Φ, and a class Π_1^C of strategies we consider the following *winning modes* for player 1:

- *(Almost).* We say that player 1 *wins almost surely* with the class Π_1^C if the player has a strategy in Π_1^C to win with probability 1, i.e., formally $\exists \pi_1 \in \Pi_1^C . \forall \pi_2 \in \Pi_2 . \Pr_s^{\pi_1,\pi_2}(\Phi) = 1$.
- *(Limit).* We say that player 1 *wins limit surely* with the class Π_1^C if the player can ensure to win with probability arbitrarily close to 1 with Π_1^C, i.e., for all $\varepsilon > 0$ there is a strategy for player 1 in Π_1^C that ensures to win with probability at least $1-\varepsilon$. Formally we have $\sup_{\pi_1 \in \Pi_1^C} \inf_{\pi_2 \in \Pi_2} \Pr_s^{\pi_1,\pi_2}(\Phi) = 1$.

We abbreviate the winning modes by *Almost* and *Limit*, respectively. We call these winning modes the *qualitative* winning modes. Given a game structure G, for $C_1 \in \{P, U, FP, IP\}$ and $C_2 \in \{M, FM, IM\}$ we denote by $Almost_1^G(C_1, C_2, \Phi)$ (resp. $Limit_1^G(C_1, C_2, \Phi)$) the set of almost-sure (resp. limit-sure) winning states for player 1 in G when the strategy set for player 1 is restricted to $\Pi_1^{C_1} \cap \Pi_1^{C_2}$. If the game structure G is clear from the context we omit the superscript G.

Mu-Calculus, Complementation, and Levels. Consider a mu-calculus expression $\Psi = \mu X . \psi(X)$ over a finite set S, where $\psi : 2^S \mapsto 2^S$ is monotonic. The least fixpoint $\Psi = \mu X . \psi(X)$ is equal to the limit $\lim_{k\to\infty} X_k$, where $X_0 = \emptyset$, and $X_{k+1} = \psi(X_k)$. For every state $s \in \Psi$, we define the *level* $k \geq 0$ of s to be the integer such that $s \notin X_k$ and $s \in X_{k+1}$. The greatest fixpoint $\Psi = \nu X . \psi(X)$ is equal to the limit $\lim_{k\to\infty} X_k$, where $X_0 = S$, and $X_{k+1} = \psi(X_k)$. For every state $s \notin \Psi$, we define the *level* $k \geq 0$ of s to be the integer such that $s \in X_k$ and $s \notin X_{k+1}$. The *height* of a mu-calculus expression $\lambda X . \psi(X)$, where $\lambda \in \{\mu, \nu\}$, is the least integer h such that $X_h = \lim_{k\to\infty} X_k$. An expression of height h can be computed in $h+1$ iterations. Given a mu-calculus expression $\Psi = \lambda X . \psi(X)$, where $\lambda \in \{\mu, \nu\}$, the complement $\neg\Psi = S \setminus \Psi$ of λ is given by $\overline{\lambda} X . \neg\psi(\neg X)$, where $\overline{\lambda} = \mu$ if $\lambda = \nu$, and $\overline{\lambda} = \nu$ if $\lambda = \mu$.

Mu-Calculus Formulas and Algorithms. As descrived above that μ-calculus formulas $\Psi = \mu X . \psi(X)$ (resp. $\Psi = \nu X . \psi(X)$) represent an iterative algorithm that successively iterates $\psi(X_k)$ till the least (resp. greatest) fixpoint is reached. Thus in general, a μ-calculus formulas with nested μ and ν operators represents a nested iterative algorithm. Intuitively, a μ-calculus formula is a succinct representation of a nested iterative algorithm.

Distributions and One-Step Transitions. Given a state $s \in S$, we denote by $\chi_1^s = \mathcal{D}(\Gamma_1(s))$ and $\chi_2^s = \mathcal{D}(\Gamma_2(s))$ the sets of probability distributions over the moves at s available to player 1 and 2, respectively. Moreover, for $s \in S$, $X \subseteq S$, $\xi_1 \in \chi_1^s$, and $\xi_2 \in \chi_2^s$ we denote by $P_s^{\xi_1,\xi_2}(X) = \sum_{a\in\Gamma_1(s)} \sum_{b\in\Gamma_2(s)} \sum_{t\in X} \xi_1(a) \cdot \xi_2(b) \cdot \delta(s, a, b)(t)$ the one-step probability of a transition into X when players 1 and 2 play at s with distributions ξ_1 and ξ_2, respectively.

Theorem 1 *The following assertions hold: (1)* [12] *For all turn-based stochastic game structures G with a parity objective Φ we have $Almost_1(P, M, \Phi) = Almost_1(IP, IM, \Phi) = Limit_1(P, M, \Phi) = Limit_1(IP, IM, \Phi)$. (2)* [15] *Let G_1 and G_2 be two equivalent game structures with a parity objective Φ, then we have*

(1). $Almost_1^{G_1}(IP, IM, \Phi) = Almost_1^{G_2}(IP, IM, \Phi)$; and (2). $Limit_1^{G_1}(IP, IM, \Phi) = Limit_1^{G_2}(IP, IM, \Phi)$.

3 Finite-Precision Strategies

In this section we present our results for pure, uniform and finite-precision strategies. We start with the characterization for pure strategies.

Pure Strategies. The following result shows that for pure strategies, memoryless strategies are as strong as infinite-memory strategies, and the almost-sure and limit-sure sets coincide.

Proposition 1 *Given a concurrent game structure G and a parity objective Φ we have $Almost_1^G(P, M, \Phi) = Almost_1^G(P, IM, \Phi) = Limit_1^G(P, IM, \Phi)$.*

Algorithm and Complexity. In the proof of the above proposition we establish a linear reduction to turn-based stochastic games. Thus the set $Almost_1(P, M, \Phi)$ can be computed using the algorithms for turn-based stochastic parity games (such as [12]). We have the following results.

Theorem 2 *Given a concurrent game structure G, a parity objective Φ, and a state s, whether $s \in Almost_1(P, IM, \Phi) = Limit_1(P, IM, \Phi)$ can be decided in NP $\cap$ coNP.*

Uniform and Finite-Precision Strategies. We will present the characterization for uniform and finite-precision strategies. We will also present symbolic algorithms (via μ-calculus formula characterization) to compute the winning sets, which is the main result of this section. First note that it follows from the classical matching penny game example that $Almost_1(P, M, \Phi) \subsetneq Almost_1(U, M, \Phi)$ We show that uniform memoryless strategies are as powerful as finite-precision infinite-memory strategies and the almost-sure and limit-sure sets coincide for finite-precision strategies.

Proposition 2 *Given a concurrent game structure G and a parity objective Φ we have $Almost_1^G(U, M, \Phi) = Almost_1^G(FP, IM, \Phi) = \bigcup_{b>0} Limit_1^G(bFP, IM, \Phi)$.*

Computation of $Almost_1(U, M, \Phi)$. It follows from Proposition 2 that the computation of $Almost_1(U, M, \Phi)$ can be achieved by a reduction to a turn-based stochastic game. We now present the main technical result of this subsection which presents a symbolic algorithm to compute $Almost_1(U, M, \Phi)$. The symbolic algorithm developed in this section is crucial for analysis of infinite-precision finite-memory strategies, where the reduction to a turn-based stochastic game cannot be applied. The symbolic algorithm is obtained via a μ-calculus formula characterization and we now introduce the predecessor operators for the μ-calculus formula.

Basic Predecessor Operators. The *predecessor* operators Pre_1 (pre) and Apre_1 (almost-pre) are defined for all $s \in S$ and $X, Y \subseteq S$ by:

$$\mathrm{Pre}_1(X) = \{s \in S \mid \exists \xi_1 \in \chi_1^s \,.\, \forall \xi_2 \in \chi_2^s \,.\, P_s^{\xi_1,\xi_2}(X) = 1\};$$
$$\mathrm{Apre}_1(Y, X) = \{s \in S \mid \exists \xi_1 \in \chi_1^s \,.\, \forall \xi_2 \in \chi_1^s \,.\, P_s^{\xi_1,\xi_2}(Y) = 1 \wedge P_s^{\xi_1,\xi_2}(X) > 0\}.$$

Intuitively, the $\mathrm{Pre}_1(X)$ is the set of states such that player 1 can ensure that the next state is in X with probability 1, and $\mathrm{Apre}_1(Y,X)$ is the set of states such that player 1 can ensure that the next state is in Y with probability 1 and in X with positive probability.

Principle of General Predecessor Operators. While the operators Apre and Pre suffice for solving Büchi games, for solving general parity games, we require predecessor operators that are best understood as the combination of the basic predecessor operators. We use the operators $\uplus$ and $\Cap$ to combine predecessor operators; the operators $\uplus$ and $\Cap$ are different from the usual union $\cup$ and intersection $\cap$. Roughly, let α and β be two set of states for two predecessor operators, then the set $\alpha \uplus \beta$ requires that the distributions of player 1 satisfy the disjunction of the conditions stipulated by α and β. We first introduce the operator Apre$\uplus$Pre. For all $s \in S$ and $X_1, Y_0, Y_1 \subseteq S$, we define

$$\mathrm{Apre}_1(Y_1, X_1) \uplus \mathrm{Pre}_1(Y_0) =$$
$$\left\{ s \in S \mid \exists \xi_1 \in \chi_1^s . \forall \xi_2 \in \chi_2^s . \left[\begin{array}{c} (P_s^{\xi_1,\xi_2}(X_1) > 0 \wedge P_s^{\xi_1,\xi_2}(Y_1) = 1) \\ \bigvee \\ P_s^{\xi_1,\xi_2}(Y_0) = 1 \end{array} \right] \right\}.$$

Note that the above formula corresponds to a disjunction of the predicates for Apre_1 and Pre_1. However, it is important to note that the distributions ξ_1 for player 1 to satisfy (ξ_2 for player 2 to falsify) the predicate must be *the same*. In other words, $\mathrm{Apre}_1(Y_1, X_1) \uplus \mathrm{Pre}_1(Y_0)$ is *not* equivalent to $\mathrm{Apre}_1(Y_1, X_1) \cup \mathrm{Pre}_1(Y_0)$.

General Predecessor Operators. We first introduce two predecessor operators as follows:

$$\mathrm{APreOdd}_1(i, Y_n, X_n, \ldots, Y_{n-i}, X_{n-i})$$
$$= \mathrm{Apre}_1(Y_n, X_n) \uplus \mathrm{Apre}_1(Y_{n-1}, X_{n-1}) \uplus \cdots \uplus \mathrm{Apre}_1(Y_{n-i}, X_{n-i});$$

$$\mathrm{APreEven}_1(i, Y_n, X_n, \ldots, Y_{n-i}, X_{n-i}, Y_{n-i-1})$$
$$= \mathrm{Apre}_1(Y_n, X_n) \uplus \mathrm{Apre}_1(Y_{n-1}, X_{n-1}) \uplus \cdots \uplus \mathrm{Apre}_1(Y_{n-i}, X_{n-i}) \uplus \mathrm{Pre}_1(Y_{n-i-1}).$$

We show Theorem 3, and the proof methodology is as follows: (a) we define appropriate predecessor operators and prove duality with respect to APreOdd and APreEven; (b) we show that for the set characterized by the μ-calculus formula of Theorem 3 there is a witness uniform memoryless strategy to ensure the parity objective with probability 1; (c) we show that in the complement μ-calculus formula with the dual predecessor operators, for every memoryless strategy for player 1, there is a counter-strategy for player 2 to ensure that the complement of the parity objective is satisfied with positive probability. Establishing the duality of the predecessor operators, μ-calculus complementation, and establishing equivalence of the μ-calculus formulas and almost-sure winning for uniform memoryless strategies by producing appropriate witness strategies from μ-calculus formula is the heart of the proof. The μ-calculus formulas and the correctness proof are non-trivial generalizations of the result of [22] for turn-based deterministic parity games (e.g., the μ-calculus formula of [22] only requires the Pre_1 operator and does not deal with concurrency), and the long and subtle proof is given in complete details in [4]. The proof structure is similar to the results of [8] (journal version of [15]) but with many important differences (e.g., the witness strategy we construct is a uniform memoryless strategy, as compared to an infinite-memory strategy in [8]). Also note that in the result below we only focus on uniform memoryless strategies, and the equivalence to finite-precision strategies and uniform memoryless strategies has been shown in Proposition 2.

Theorem 3 *For all concurrent game structures $\mathcal{G}$ over state space S, for all parity objectives $Parity(p)$ for player 1, with $p : S \mapsto [1..2n]$ the following assertions hold.*

1. *We have $Almost_1(U, M, Parity(p)) = W$, where W is defined as follows*

$$\nu Y_{n-1}.\mu X_{n-1}\dots\nu Y_0.\mu X_0.\left[\begin{array}{c} B_{2n}\cap Pre_1(Y_{n-1}) \\ \cup \\ B_{2n-1}\cap APreOdd_1(0,Y_{n-1},X_{n-1}) \\ \cup \\ B_{2n-2}\cap APreEven_1(0,Y_{n-1},X_{n-1},Y_{n-2}) \\ \cup \\ \vdots \\ B_2\cap APreEven_1(n-2,Y_{n-1},X_{n-1},\dots,X_1,Y_0) \\ \cup \\ B_1\cap APreOdd_1(n-1,Y_{n-1},X_{n-1},\dots,Y_0,X_0) \end{array}\right]$$

   and $B_i = p^{-1}(i)$ is the set of states with priority i, for $i \in [1..2n]$.
2. *The set $Almost_1(U, M, Parity(p))$ can be computed symbolically using the above expression in time $\mathcal{O}(|S|^{2n+1} \cdot \sum_{s\in S} 2^{|\Gamma_1(s)\cup\Gamma_2(s)|})$; and for $s \in S$ whether $s \in Almost_1(U, M, Parity(p))$ can be decided in NP $\cap$ coNP.*

Intuitively, the μ-calculus formula of Theorem 3 is obtained from the μ-calculus formula for turn-based deterministic parity games of [22] by replacing the predecessor operator (Pre_1) of the μ-calculus formula of [22] appropriately with our new predecessor operators ($APreEven_1$ and $APreOdd_1$) (see detailed description in [4]). Thus from the understanding of the μ-calculus formula of [22] for turn-based deterministic games and our predecessor operators, the μ-calculus formula of Theorem 3 can be understood. The NP $\cap$ coNP bound follows directly from the μ-calculus expressions as the players can guess the *ranking function* (see [4] for the standard definition of ranking function, also see [22]) of the μ-calculus formula and the support of the uniform distribution at every state to witness that the predecessor operator is satisfied, and the guess can be verified in polynomial time. Observe that the computation through μ-calculus formulas is symbolic and more efficient than enumeration over the set of all uniform memoryless strategies of size $O(\prod_{s\in S}|\Gamma_1(s)\cup\Gamma_2(s)|)$ (for example, with constant action size and constant d, the μ-calculus formula is polynomial, whereas enumeration of strategies is exponential). The μ-calculus formulas of [22] can be obtained from the μ-calculus formula of Theorem 3 by replacing all predecessor operators with the Pre_1 predecessor operator (details in [4]).

Proposition 3 *Given a concurrent game structure G and a parity objective Φ we have $Almost_1(IP, FM, \Phi) = Almost_1(U, FM, \Phi) = Almost_1(U, M, \Phi)$.*

For almost-sure winning uniform memoryless strategies are as powerful as finite-precision infinite-memory strategies (Proposition 2) as well as infinite-precision finite-memory strategies (Proposition 3). In contrast, infinite-precision infinite-memory strategies are more powerful than uniform memoryless strategies, i.e., $\mathit{Almost}_1(U, M, \Phi) \subsetneq \mathit{Almost}_1(IP, IM, \Phi)$ (for details see [4]). The propositions and examples of this section establish all the results for equalities and inequalities (1) of Section 1. The fact that $\mathit{Limit}_1(IP, FM, \Phi) \subsetneq \mathit{Limit}_1(IP, IM, \Phi)$ was shown in [15]. The fact that we have $\bigcup_{b>0} \mathit{Limit}_1(bFP, IM, \Phi) = \mathit{Almost}_1(U, M, \Phi)$, and the result of [16] that for reachability objectives memoryless limit-sure winning strategies exist and limit-sure winning is different from almost-sure winning established that $\bigcup_{b>0} \mathit{Limit}_1(bFP, IM, \Phi) \subsetneq \mathit{Limit}_1(IP, M, \Phi)$. Thus we have all the results of (1) and (2), other than $\mathit{Limit}_1(IP, M, \Phi) = \mathit{Limit}_1(IP, FM, \Phi) = \mathit{Limit}_1(FP, M, \Phi) = \mathit{Limit}_1(FP, IM, \Phi)$ and we establish this in the next section.

4 Infinite-Precision Strategies

The results of the previous section already characterize that for almost-sure winning infinite-precision finite-memory strategies are no more powerful than uniform memoryless strategies. In this section we characterize the limit-sure winning for infinite-precision finite-memory strategies. We require a new predecessor operator, Lpre (limit-pre). For $s \in S$ and $X, Y \subseteq S$, the two-argument predecessor operator Lpre is defined as follows:

$$\mathrm{Lpre}_1(Y, X) = \{ s \in S \mid \forall \alpha > 0. \exists \xi_1 \in \chi_1^s . \forall \xi_2 \in \chi_2^s . [P_s^{\xi_1,\xi_2}(X) > \alpha \cdot P_s^{\xi_1,\xi_2}(\neg Y)] \}.$$

The operator $\mathrm{Lpre}_1(Y, X)$ is the set of states such that player 1 can choose distributions to ensure that the probability to progress to X can be made arbitrarily large as compared to the probability of escape from Y. In other words, the probability to progress to X divided by the sum of the probability to progress to X and to escape Y can be made arbitrarily close to 1 (in the limit 1).

Limit-Sure Winning for Memoryless Strategies. The results of [16] show that for reachability objectives, memoryless strategies suffice for limit-sure winning, and the limit-sure winning set can be characterized by a μ-calculus formula with Lpre_1. We now show with an example that limit-sure winning for Büchi objectives with memoryless strategies is not simply limit-sure reachability to the set of almost-sure winning states. Consider the game shown in Fig 1 with actions $\{a, b\}$ for player 1 and $\{c, d, e\}$ for player 2 at s_0. States s_1, s_2 are absorbing, and the unique successor of s_3 is s_0. The Büchi objective is to visit $\{s_1, s_3\}$ infinitely often. The only almost-sure winning state is s_1. The state s_0 is not almost-sure winning because at s_0 if player 1 plays b with positive probability the counter move is d, otherwise the counter move is c. Hence either s_2 is reached with positive probability or s_0 is never left. Moreover, player 1 cannot limit-sure reach the state s_1 from s_0, as the move e ensures that s_1 is never reached. Thus in this game the limit-sure reach to the almost-sure winning set is only state s_1. We now show that for all ε, there is a memoryless strategy to ensure the Büchi objective with probability at least $1 - \varepsilon$ from s_0. At s_0 the memoryless strategy plays a with probability $1 - \varepsilon$ and b with probability ε. Fixing the strategy for player 1 we

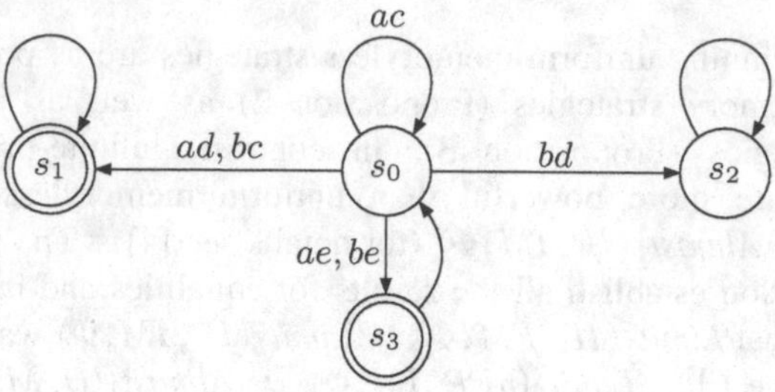

Fig. 1. A Büchi game

obtain an MDP for player 2, and in the MDP player 2 has an optimal pure memoryless strategy. If player 2 plays the pure memoryless strategy e, then s_3 is visited infinitely often with probability 1; if player 2 plays the pure memoryless strategy c, then s_1 is reached with probability 1; and if player 2 plays the pure memoryless strategy d, then s_1 is reached with probability $1 - \varepsilon$. Thus for all $\varepsilon > 0$, player 1 can win from s_0 and s_3 with probability at least $1 - \varepsilon$ with a memoryless strategy.

Limit-Winning Set for Büchi Objectives. We first present the characterization of the set of limit-sure winning states for concurrent Büchi games from [15] for infinite-memory and infinite-precision strategies. The limit-sure winning set is characterized by the following formula: $\nu Y_0.\mu X_0.[(B \cap \mathrm{Pre}_1(Y_0)) \cup (\neg B \cap \mathrm{Lpre}_1(Y_0, X_0))]$. Our characterization of the limit-sure winning set for memoryless infinite-precision strategies would be obtained as follows: we will obtain sequence of sets of states $X_0 \subseteq X_1 \subseteq \ldots \subseteq X_k$ such that from each X_i for all $\varepsilon > 0$ there is a memoryless strategy to ensure that $\Diamond X_{i-1} \cup (\Box\Diamond B \cap \Box(X_i \setminus X_{i-1}))$ is satisfied with probability at least $1 - \varepsilon$. We consider the following μ-calculus formula:

$$\nu Y_1.\mu X_1.\nu Y_0.\mu X_0.\begin{bmatrix} (B \cap \mathrm{Pre}_1(Y_0) \mathbin{\uplus\!\!\!\!\!\ast} \mathrm{Lpre}_1(Y_1, X_1)) \cup \\ (\neg B \cap \mathrm{Apre}_1(Y_0, X_0) \mathbin{\uplus\!\!\!\!\!\ast} \mathrm{Lpre}_1(Y_1, X_1)) \end{bmatrix}$$

Let Y^* be the fixpoint, and since it is a fixpoint we have

$$Y^* = \mu X_1.\nu Y_0.\mu X_0.\begin{bmatrix} (B \cap \mathrm{Pre}_1(Y_0) \mathbin{\uplus\!\!\!\!\!\ast} \mathrm{Lpre}_1(Y^*, X_1)) \cup \\ (\neg B \cap \mathrm{Apre}_1(Y_0, X_0) \mathbin{\uplus\!\!\!\!\!\ast} \mathrm{Lpre}_1(Y^*, X_1)) \end{bmatrix}$$

Thus Y^* is computed as least fixpoint as sequence of sets $X_0 \subseteq X_1 \ldots \subseteq X_k$, and X_{i+1} is obtained from X_i as

$$\nu Y_0.\mu X_0.\begin{bmatrix} (B \cap \mathrm{Pre}_1(Y_0) \mathbin{\uplus\!\!\!\!\!\ast} \mathrm{Lpre}_1(Y^*, X_i)) \cup \\ (\neg B \cap \mathrm{Apre}_1(Y_0, X_0) \mathbin{\uplus\!\!\!\!\!\ast} \mathrm{Lpre}_1(Y^*, X_i)) \end{bmatrix}$$

The $\mathrm{Lpre}_i(Y^*, X_i)$ is similar to limit-sure reachability to X_i, and once we rule out $\mathrm{Lpre}_1(Y^*, X_i)$, the formula simplifies to the almost-sure winning under memoryless strategies. In other words, from each X_{i+1} player 1 can ensure with a memoryless strategy that either (i) X_i is reached with limit probability 1 or (ii) the game stays in $X_{i+1} \setminus X_i$ and the Büchi objective is satisfied with probability 1. It follows that $Y^* \subseteq \mathit{Limit}_1(\mathit{IP}, M, \Box\Diamond B)$. In other words the Lpre_1 operator needs to be combined with the $\mathrm{APreOdd}_1$ and $\mathrm{APreEven}_1$ operators with $\mathbin{\uplus\!\!\!\!\!\ast}$ to characterize the limit-sure winning set for memoryless strategies. We will show that in the complement set there exists constant

$\eta > 0$ such that for all finite-memory infinite-precision strategies for player 1 there is a counter strategy to ensure the complementary objective with probability at least $\eta > 0$. The heart of the proof is again in establishing duality of predecessor operators, complement of μ-calculus formulas, and producing appropriate witness strategies from the μ-calculus formulas (the detailed proof is in [4]).

The General Principle. The general principle to obtain the μ-calculus formula for limit-sure winning for memoryless infinite-precision strategies is as follows: we consider the μ-calculus formula for the almost-sure winning for uniform memoryless strategies, then add a $\nu Y_{n+1} \mu X_{n+1}$ quantifier and add the $\text{Lpre}_1(Y_{n+1}, X_{n+1}) \uplus$ to every predecessor operator. Intuitively, when we replace Y_{n+1} by the fixpoint Y^*, then we obtain a sequence X_i of sets of states for the least fixpoint computation of X_{n+1}, such that from X_{i+1} either X_i is reached with limit probability 1 (by the $\text{Lpre}_1(Y^*, X_{n+1})$ operator), or the game stays in $X_{i+1} \setminus X_i$ and then the parity objective is satisfied with probability 1 by a memoryless strategy. The μ-calculus formula and the correctness proof are non-trivial generalizations of both the result of [22] (for turn-based deterministic parity games which uses only Pre_1 operator and does not deal with concurrency) and the result of [16] for concurrent reachability games (that uses only Lpre_1 operator and has only one quantifier alternation of $\nu\mu$ as compared to the general nested fixpoint characterization required for parity objectives). We have the following result.

Theorem 4 *For all concurrent game structures $\mathcal{G}$ over state space S, for all parity objectives $\Phi = Parity(p)$ for player 1, with $p : S \mapsto [1..2n]$, the following assertions hold.*

1. *We have $Limit_1(IP, M, \Phi) = Limit_1(IP, FM, \Phi)$, and $Limit_1(IP, FM, \Phi) = W$, where W is defined as follows*

$$\nu Y_n.\mu X_n.\nu Y_{n-1}.\mu X_{n-1}.\cdots \nu Y_1.\mu X_1.\nu Y_0.\mu X_0.$$
$$\left[\begin{array}{c} B_{2n} \cap Pre_1(Y_{n-1}) \uplus Lpre_1(Y_n, X_n) \\ \cup \\ B_{2n-1} \cap APreOdd_1(0, Y_{n-1}, X_{n-1}) \uplus Lpre_1(Y_n, X_n) \\ \cup \\ B_{2n-2} \cap APreEven_1(0, Y_{n-1}, X_{n-1}, Y_{n-2}) \uplus Lpre_1(Y_n, X_n) \\ \cup \\ B_{2n-3} \cap APreOdd_1(1, Y_{n-1}, X_{n-1}, Y_{n-2}, X_{n-2}) \uplus Lpre_1(Y_n, X_n) \\ \cup \\ \vdots \\ B_2 \cap APreEven_1(n-2, Y_{n-1}, X_{n-1}, \ldots, Y_1, X_1, Y_0) \uplus Lpre_1(Y_n, X_n) \\ \cup \\ B_1 \cap APreOdd_1(n-1, Y_{n-1}, X_{n-1}, \ldots, Y_0, X_0) \uplus Lpre_1(Y_n, X_n) \end{array}\right]$$

   and $B_i = p^{-1}(i)$ is the set of states with priority i, for $i \in [1..2n]$.
2. *The set $Limit_1(IP, FM, \Phi)$ can be computed symbolically using the above μ-calculus expression in time $\mathcal{O}(|S|^{2n+2} \cdot \sum_{s \in S} 2^{|\Gamma_1(s) \cup \Gamma_2(s)|})$; and for $s \in S$ whether $s \in Limit_1(IP, FM, \Phi)$ can be decided in NP $\cap$ coNP.*

Proposition 4 *Given a concurrent game structure G and a parity objective Φ we have $Limit_1(IP, M, \Phi) = Limit_1(FP, M, \Phi) = Limit_1(FP, FM, \Phi) = Limit_1(FP, IM, \Phi)$*

Construction of Infinite-Precision Strategies. Note that for infinite-precision strategies we are interested in the limit-sure winning set, i.e., for every $\varepsilon > 0$, there is a strategy to win with probability $1-\varepsilon$, but not necessarily a strategy to win with probability 1. The proof of Theorem 4 constructs for every $\varepsilon > 0$ a memoryless strategy that ensures winning with probability at least $1-\varepsilon$.

Independence from Precise Probabilities. The computation of all the predecessor operators only depends on the supports of the transition function, and not on the precise transition probabilities.

Theorem 5 *Let $\mathcal{G}_1 = (S, A, \Gamma_1, \Gamma_2, \delta_1)$ and $\mathcal{G}_2 = (S, A, \Gamma_1, \Gamma_2, \delta_2)$ be two concurrent game structures that are equivalent, i.e., $\mathcal{G}_1 \equiv \mathcal{G}_2$. Then for all parity objectives Φ, for all $C_1 \in \{P, U, FP, IP\}$ and $C_2 \in \{M, FM, IM\}$ we have (a) $Almost_1^{\mathcal{G}_1}(C_1, C_2, \Phi) = Almost_1^{\mathcal{G}_2}(C_1, C_2, \Phi)$; and (b) $Limit_1^{\mathcal{G}_1}(C_1, C_2, \Phi) = Limit_1^{\mathcal{G}_2}(C_1, C_2, \Phi)$.*

All cases of the above theorem, other than when $C_1 = IP$ and $C_2 = IM$ follow from our results, and the result for $C_1 = IP$ and $C_2 = IM$ follows from the results of [15].

Concluding Remarks. In this work we studied the bounded rationality problem for qualitative analysis in concurrent parity games, and presented a precise characterization. The theory of bounded rationality for quantitative analysis is future work, where we believe our results will be helpful.

References

1. Abadi, M., Lamport, L., Wolper, P.: Realizable and unrealizable specifications of reactive systems. In: Ausiello, G., Dezani-Ciancaglini, M., Rocca, S.R.D. (eds.) ICALP 1989. LNCS, vol. 372, pp. 1–17. Springer, Heidelberg (1989)
2. Alur, R., Henzinger, T.A., Kupferman, O.: Alternating-time temporal logic. In: FOCS 1997, pp. 100–109. IEEE (1997)
3. Chatterjee, K.: Stochastic omega-Regular Games. PhD thesis, UC Berkeley (2007)
4. Chatterjee, K.: Bounded rationality in concurrent parity games. CoRR, abs/1107.2146 (2011)
5. Chatterjee, K., Chmelik, M., Tracol, M.: What is decidable about partially observable Markov decision processes with omega-regular objectives. In: CSL (2013)
6. Chatterjee, K., de Alfaro, L., Henzinger, T.A.: Strategy improvement in concurrent reachability games. In: QEST 2006. IEEE (2006)
7. Chatterjee, K., de Alfaro, L., Henzinger, T.A.: The complexity of quantitative concurrent parity games. In: SODA 2006. IEEE (2006)
8. Chatterjee, K., de Alfaro, L., Henzinger, T.A.: Qualitative concurrent parity games. ACM ToCL (2011)
9. Chatterjee, K., Doyen, L., Henzinger, T.A., Raskin, J.-F.: Generalized mean-payoff and energy games. In: FSTTCS, pp. 505–516 (2010)
10. Chatterjee, K., Doyen, L., Nain, S., Vardi, M.Y.: The complexity of partial-observation stochastic parity games with finite-memory strategies. In: Muscholl, A. (ed.) FOSSACS 2014. LNCS, vol. 8412, pp. 242–257. Springer, Heidelberg (2014)
11. Chatterjee, K., Henzinger, T.A.: Strategy improvement and randomized subexponential algorithms for stochastic parity games. In: Durand, B., Thomas, W. (eds.) STACS 2006. LNCS, vol. 3884, pp. 512–523. Springer, Heidelberg (2006)

12. Chatterjee, K., Jurdziński, M., Henzinger, T.A.: Simple stochastic parity games. In: Baaz, M., Makowsky, J.A. (eds.) CSL 2003. LNCS, vol. 2803, pp. 100–113. Springer, Heidelberg (2003)
13. Church, A.: Logic, arithmetic, and automata. In: Proc. of Int. Cong. of Math., pp. 23–35 (1962)
14. Condon, A.: The complexity of stochastic games. I&C 96, 203–224 (1992)
15. de Alfaro, L., Henzinger, T.A.: Concurrent ω-regular games. In: LICS, pp. 141–154 (2000)
16. de Alfaro, L., Henzinger, T.A., Kupferman, O.: Concurrent reachability games. TCS 386(3), 188–217 (2007)
17. de Alfaro, L., Henzinger, T.A., Mang, F.Y.C.: The control of synchronous systems. In: Palamidessi, C. (ed.) CONCUR 2000. LNCS, vol. 1877, pp. 458–473. Springer, Heidelberg (2000)
18. de Alfaro, L., Majumdar, R.: Quantitative solution of omega-regular games. Journal of Computer and System Sciences 68, 374–397 (2004)
19. Dill, D.L.: Trace Theory for Automatic Hierarchical Verification of Speed-independent Circuits. The MIT Press (1989)
20. Dziembowski, S., Jurdzinski, M., Walukiewicz, I.: How much memory is needed to win infinite games? In: LICS 1997, pp. 99–110. IEEE (1997)
21. Emerson, E.A., Jutla, C.: The complexity of tree automata and logics of programs. In: FOCS 1988, pp. 328–337. IEEE (1988)
22. Emerson, E.A., Jutla, C.: Tree automata, mu-calculus and determinacy. In: FOCS, pp. 368–377. IEEE (1991)
23. Etessami, K., Yannakakis, M.: Recursive concurrent stochastic games. In: Bugliesi, M., Preneel, B., Sassone, V., Wegener, I. (eds.) ICALP 2006. LNCS, vol. 4052, pp. 324–335. Springer, Heidelberg (2006)
24. Etessami, K., Yannakakis, M.: On the complexity of Nash equilibria and other fixed points. SIAM J. Comput. 39(6), 2531–2597 (2010)
25. Filar, J., Vrieze, K.: Competitive Markov Decision Processes. Springer (1997)
26. Hansen, K.A., Koucký, M., Lauritzen, N., Miltersen, P.B., Tsigaridas, E.P.: Exact algorithms for solving stochastic games (extended abstract). In: STOC, pp. 205–214 (2011)
27. Hansen, K.A., Koucký, M., Miltersen, P.B.: Winning concurrent reachability games requires doubly-exponential patience. In: LICS, pp. 332–341 (2009)
28. Jurdziński, M., Paterson, M., Zwick, U.: A deterministic subexponential algorithm for solving parity games. In: SODA 2006, pp. 117–123. ACM-SIAM (2006)
29. Kechris, A.: Classical Descriptive Set Theory. Springer (1995)
30. Martin, D.A.: The determinacy of Blackwell games. J. of Symbolic Logic 63(4), 1565–1581 (1998)
31. Pnueli, A., Rosner, R.: On the synthesis of a reactive module. In: POPL 1989, pp. 179–190. ACM Press (1989)
32. Raghavan, T.E.S., Filar, J.A.: Algorithms for stochastic games — a survey. ZOR — Methods and Models of Op. Res. 35, 437–472 (1991)
33. Ramadge, P.J., Wonham, W.M.: Supervisory control of a class of discrete-event processes. SIAM Journal of Control and Optimization 25(1), 206–230 (1987)
34. Schewe, S.: Solving parity games in big steps. In: Arvind, V., Prasad, S. (eds.) FSTTCS 2007. LNCS, vol. 4855, pp. 449–460. Springer, Heidelberg (2007)
35. Shapley, L.S.: Stochastic games. Proc. Nat. Acad. Sci. USA 39, 1095–1100 (1953)
36. Thomas, W.: Automata on infinite objects. In: Handbook of Theoretical Computer Science, vol. B, ch. 4, pp. 135–191. Elsevier Science Publishers, Amsterdam (1990)
37. Vardi, M.Y.: Automatic verification of probabilistic concurrent finite-state systems. In: FOCS 1985, pp. 327–338. IEEE (1985)
38. Zielonka, W.: Infinite games on finitely coloured graphs with applications to automata on infinite trees. In: TCS, vol. 200(1-2), pp. 135–183 (1998)

Adding Negative Prices to Priced Timed Games*

Thomas Brihaye[1], Gilles Geeraerts[2,**], Shankara Narayanan Krishna[3], Lakshmi Manasa[3], Benjamin Monmege[2], and Ashutosh Trivedi[3]

[1] Université de Mons, Belgium
thomas.brihaye@umons.ac.be
[2] Université libre de Bruxelles, Belgium
{gigeerae,benjamin.monmege}@ulb.ac.be
[3] IIT Bombay, India
{krishnas,manasa,trivedi}@cse.iitb.ac.in

Abstract. Priced timed games (PTGs) are two-player zero-sum games played on the infinite graph of configurations of priced timed automata where two players take turns to choose transitions in order to optimize cost to reach target states. Bouyer et al. and Alur, Bernadsky, and Madhusudan independently proposed algorithms to solve PTGs with nonnegative prices under certain divergence restriction over prices. Brihaye, Bruyère, and Raskin later provided a justification for such a restriction by showing the undecidability of the optimal strategy synthesis problem in the absence of this divergence restriction. This problem for PTGs with one clock has long been conjectured to be in polynomial time, however the current best known algorithm, by Hansen, Ibsen-Jensen, and Miltersen, is exponential. We extend this picture by studying PTGs with both negative and positive prices. We refine the undecidability results for optimal strategy synthesis problem, and show undecidability for several variants of optimal reachability cost objectives including reachability cost, time-bounded reachability cost, and repeated reachability cost objectives. We also identify a subclass with bi-valued price-rates and give a pseudo-polynomial (polynomial when prices are nonnegative) algorithm to partially answer the conjecture on the complexity of one-clock PTGs.

1 Introduction

Timed automata [2] equip finite automata with a finite number of real-valued variables—aptly called clocks—that evolve with a uniform rate. The syntax of timed automata also permits specifying *transition guards* and *location (state) invariants* using the constraints over clock valuations, and resetting the clocks as a means to remember the time since the execution of a transition. Timed automata is a well-established formalism to specify time-critical properties of real-time systems. Priced timed automata [3,4] (PTAs) extend timed automata with

* The research leading to these results has received funding from the European Union Seventh Framework Programme (FP7/2007-2013) under Grant Agreement n601148 (CASSTING).

** Supported by a 'Crédit aux chercheurs' number 1808881 of the F.R.S./FNRS.

P. Baldan and D. Gorla (Eds.): CONCUR 2014, LNCS 8704, pp. 560–575, 2014.

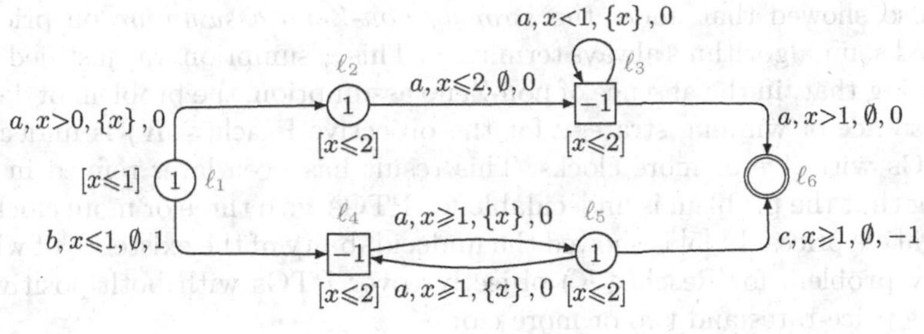

Fig. 1. A price timed game arena with one clock

price information by augmenting locations with price-rates and transitions with discrete prices. The natural reachability-cost optimization problem for PTAs is known to be decidable with the same complexity [6] as the reachability problem (PSPACE-complete), and forms the backbone of many applications of timed automata including scheduling and planning.

Priced timed games (PTGs) extend the reachability-cost optimization problem to the setting of competitive optimization problem, and form the basis of optimal controller synthesis [19] for real-time systems. We study turn-based variant of these games where the game arena is a PTA with a partition of the locations between two players Player 1 and Player 2. A play of such a game begins with a token in an initial location, and at every step the player controlling the current location proposes a valid timed move, i.e., a time delay and a discrete transition, and the state of the system is modified accordingly. The play stops if the token reaches a location from a distinguished set of *target locations*, and the payoff of the play is equal to the cost accumulated before reaching the target location. If the token never reaches a target location then the game continues forever, and the payoff in this case is $+\infty$ irrespective of actual cost of the infinite play. We characterize a PTG according to the objectives of Player 1. Since we study zero-sum games, the objective of Player 2 is also implicitly defined. We study PTGs with the following objectives: (*i*) *Constrained-price reachability* objective $\mathsf{Reach}(\bowtie K)$ is to achieve a payoff C of the play such that $C \bowtie K$ where $\bowtie \in \{\leqslant, <, =, >, \geqslant\}$ and $K \in \mathbb{N}$; (*ii*) *Bounded-time reachability* objective $\mathsf{TBReach}(K, T)$ is to keep the payoff of the play less than K while keeping the total time elapsed within T units; and (*iii*) *Repeated reachability* objective $\mathsf{RReach}(\eta)$ is to visit target infinitely often with a payoff in the interval $[-\eta, \eta]$.

An example of PTG with clock variable x and six locations is given in Fig. 1. We depict Player 1 locations as circles and Player 2 locations as boxes. The numbers inside locations denote their price-rates, while the clock constraints next to a location depicts its invariant. We denote a transition, as usual, by an arrow between two location annotated by a tuple a, g, r, c where a is the label, g is the guard, r is the clocks reset set, and c is the cost of the transition.

Related Work. PTGs with constrained-price reachability objective $\mathsf{Reach}(\leqslant K)$ were independently introduced in [9] and [1], with semi-algorithms to decide the existence of winning strategy for Player 1 in PTGs with nonnegative prices.

They also showed that under the *strongly non-Zeno assumption* on prices the proposed semi-algorithms always terminate. This assumption was justified in [11] by showing that, in the absence of non-Zeno assumption, the problem of deciding the existence of winning strategy for the objective $\mathsf{Reach}(\leqslant K)$ is undecidable for PTGs with five or more clocks. This result has been later refined in [7] by showing that the problem is undecidable for PTGs with three or more clocks and nonnegative prices. In [5] is showed the undecidability of the existence of winning strategy problem for $\mathsf{Reach}(\leqslant K)$ objective over PTGs with both positive and negative price-rates and two or more clocks.

On a positive side, the existence of winning strategy for $\mathsf{Reach}(\leqslant K)$ problem for PTGs with one clock when the price-rates are restricted to values 0 and $d \in \mathbb{N}$ has been shown decidable in [11], by proving that the semi-algorithms in [9,1] always terminate. However, the authors did not provide any complexity analysis of their algorithm. One-clock PTGs with nonnegative prices are reconsidered in [10], and a 3-EXPTIME algorithm is given to solve the problem, while the best known lower bound is PTIME. A tighter analysis of the problem is presented in [20] that lowered the known complexity of this problem to EXPTIME, namely $2^{O(n^2+m)}$ where n is the number of locations and m is the number of transitions. A significant improvement over the complexity $(m12^n n^{O(1)})$ was given in [15] by improving the analysis of the semi-algorithms by [9,1].

Contributions. We consider PTGs with both negative and positive prices. We show that deciding the existence of a winning strategy for reachability objective $\mathsf{Reach}(\bowtie K)$ is undecidable for PTGs with two or more clocks. In [18], a theory of time-bounded verification has been proposed, arguing that restriction to bounded-time domain reclaims the decidability of several key verification problems. As an example, we cite [12] where authors recovered the decidability of the reachability problem for hybrid automata under time-bounded restriction. We begin studying PTGs with bounded reachability objective $\mathsf{TBReach}(K, T)$ hoping that the problem may be decidable due to time-bounded restriction. However, we answer this question negatively by showing undecidability of the existence of winning strategy problem for PTGs with six or more clocks. We also show the undecidability for the corresponding problem for repeated reachability objective $\mathsf{RReach}(\eta)$ for PTGs with three or more clocks.

On the positive side, we introduce a previously unexplored subclass of one-clock PTGs, called one-clock bi-valued priced timed games (1BPTGs), where the price-rates of locations are taken from a set of two integers from $\{-d, 0, d\}$ (with d any positive integer). None of the previously cited algorithms can be applied in this case since we do not assume non-Zenoness of prices and consider both positive and negative prices. After showing a determinacy result for 1BPTGs, we proceed to give a pseudo-polynomial time algorithm to compute the value and ε-optimal strategy for both players with $\mathsf{Reach}(\leqslant K)$ objective. The complexity drops to polynomial for 1BPTGs if the price-rates are non-negative integers. This gives a polynomial time algorithm for the one-clock PTG problem studied in [11]. Due to lack of space, full proofs of the results are given in [13].

2 Reachability-Cost Games on Priced Game Graphs

PTGs can be considered as a succinct representation of some games on uncountable state space characterized by the configuration graph of timed automata.

We begin by introducing the concepts and notations related to such more general game arenas that we call priced game graphs.

Definition 1. *A* priced game graph *is a tuple* $\mathcal{G} = (V, A, E, \pi, V_f)$ *where:*
- $V = V_1 \uplus V_2$ *is the set of vertices partitioned into the sets* V_1 *and* V_2*;*
- A *is a set of labels called actions;*
- $E : V \times A \to V$ *is the edge function defining the set of labeled edges;*
- $\pi : V \times A \to \mathbb{R}$ *is the price function that assigns prices to edges; and*
- $V_f \subseteq V$ *is the set of target vertices.*

We call a game graph finite *if both* V *and* A *are finite and with rational prices.*

A reachability-cost game begins with a token placed on some initial vertex v_0. At each round, the player who controls the current vertex v chooses an action $a \in A$ and the token is moved to the vertex $E(v, a)$. The two players continue moving the token in this fashion, and give rise to an infinite sequence of vertices and actions called a play of the game. Formally, a finite play r is a finite sequence of vertices and actions $\langle v_0, a_0, v_1, a_1, \ldots, a_{n-1}, v_n\rangle$ where for each $0 \leqslant i < n$ we have that $v_{i+1} = E(v_i, a_i)$; we write $\mathsf{Last}(r)$ for the last vertex of a finite play, here $\mathsf{Last}(r) = v_n$. An infinite play is defined analogously. We write $\mathsf{FPlay}_\mathcal{G}$ ($\mathsf{FPlay}_\mathcal{G}(v)$) for the set of finite plays (starting from the vertex v) of the game graph $\mathcal{G}$. We often omit the subscript when the game arena is clear form the context. We similarly define Play and $\mathsf{Play}(v)$ for the set of infinite plays. For all $k \geqslant 0$, we let $r[k]$ be the prefix $\langle v_0, a_0, \ldots, a_{k-1}, v_k\rangle$ of r, and we denote by $\mathsf{Cost}(r[k]) = \sum_{i=0}^{k-1} \pi(v_i, a_i)$ its cost. We write $\mathsf{Stop}(r)$ for the index of the first target vertex in r, i.e., $\mathsf{Stop}(r) = \inf\{k : v_k \in F\}$. We define the cost of an infinite run $r = \langle v_0, a_1, v_1, \ldots\rangle$ as $\mathsf{Cost}(r) = +\infty$ if $\mathsf{Stop}(r) = \infty$ and $\mathsf{Cost}(r) = \mathsf{Cost}(r[\mathsf{Stop}(r)])$, otherwise.

A strategy for a Player i (for $i \in \{1, 2\}$) is a partial function $\sigma : \mathsf{FPlay} \to A$ that is defined for a run $r = \langle v_0, a_0, v_1, \ldots, a_{n-1}, v_n\rangle$ if $v_n \in V_i$ and is such that $E(v_n, \sigma(r))$ is defined, i.e., there is a $\sigma(r)$-labeled outgoing transition from v_n. We denote by $\mathsf{Strat}_i(\mathcal{G})$ (or Strat_i when the game arena is clear) the set of strategies for Player i. Given a strategy profile $(\sigma_1, \sigma_2) \in \mathsf{Strat}_1 \times \mathsf{Strat}_2$ for both players, and an initial vertex $v \in V$, the unique infinite play $\mathsf{Play}(v, \sigma_1, \sigma_2) = \langle v_0, a_0, v_1, \ldots v_k, a_k, v_{k+1}, \ldots\rangle$ is such that for all $k \geqslant 0$ if $v_k \in V_i$, for $i = 1, 2$, then $a_{k+1} = \sigma_i(r[k])$ and $v_{k+1} = E(v_k, a_{k+1})$. A strategy σ is said to be *memoryless* (or *positional*) if, for all finite plays $r, r' \in \mathsf{FPlay}$ with $\mathsf{Last}(r) = \mathsf{Last}(r')$ we have that $\sigma(r) = \sigma(r')$. Similarly, *finite-memory strategies* can be defined as implementable with Moore machines, see [14] for a formal definition.

We consider optimal reachability-cost games on priced game graphs, where the goal of Player 1 is to minimize the reachability-cost, while the goal of Player 2 is the opposite. The standard concepts of upper value and lower value of the optimal reachability-cost game are defined in straightforward manner. Formally, the upper-value $\overline{\mathsf{Val}}_\mathcal{G}(v)$ and lower value $\underline{\mathsf{Val}}_\mathcal{G}(v)$ of a game starting from a vertex v is defined as $\overline{\mathsf{Val}}_{\mathcal{G}}(v) = \inf_{\sigma_1 \in \mathsf{Strat}_1} \sup_{\sigma_2 \in \mathsf{Strat}_2} \mathsf{Cost}(\mathsf{Play}(v, \sigma_1, \sigma_2))$ and $\underline{\mathsf{Val}}_{\mathcal{G}}(v) = \sup_{\sigma_2 \in \mathsf{Strat}_2} \inf_{\sigma_1 \in \mathsf{Strat}_1} \mathsf{Cost}(\mathsf{Play}(v, \sigma_1, \sigma_2))$. It is easy to see that $\underline{\mathsf{Val}}_{\mathcal{G}}(v) \leqslant \overline{\mathsf{Val}}_{\mathcal{G}}(v)$ for every vertex v. We say that a game is *determined* if the lower and the upper values match for every vertex v, and in this case, we say that the optimal value of the game exists and we let $\mathsf{Val}_{\mathcal{G}}(v) = \underline{\mathsf{Val}}_{\mathcal{G}}(v) = \overline{\mathsf{Val}}_{\mathcal{G}}(v)$. The determinacy of these games follow from Martin's determinacy theorem, and an alternative proof is given in [14].

In the following, we write $\mathsf{Cost}(v, \sigma_1)$ for the value of the strategy σ_1 of Player 1 from vertex v, i.e., $\mathsf{Cost}(v, \sigma_1) = \sup_{\sigma_2 \in \mathsf{Strat}_2} \mathsf{Cost}(\mathsf{Play}(v, \sigma_1, \sigma_2))$. A strategy σ_1^* of Player 1 is said to be optimal from v if $\mathsf{Cost}(v, \sigma_1^*) = \overline{\mathsf{Val}}_{\mathcal{G}}(v)$. Optimal strategies do not always exist, hence we also define ε-optimal strategies. For $\varepsilon > 0$, a strategy σ_1 is an ε-optimal strategy if for all vertex $v \in V$, $\mathsf{Cost}(v, \sigma_1) \leqslant \overline{\mathsf{Val}}_{\mathcal{G}}(v) + \varepsilon$. In this paper we exploit the following result from [14].

Theorem 1 ([14]). *Let $\mathcal{G}$ be a finite priced game graph.*

1. *Deciding $\mathsf{Val}_{\mathcal{G}}(v) = +\infty$ is in Polynomial Time.*
2. *Deciding $\mathsf{Val}_{\mathcal{G}}(v) = -\infty$ is in NP $\cap$ co-NP, can be achieved in pseudo-polynomial time[1] and is as hard as solving mean-payoff games [21].*
3. *Given $-\infty < \mathsf{Val}_{\mathcal{G}}(v) < +\infty$ for every vertex v, optimal strategies exist for both players. In particular, Player 2 has optimal memoryless strategies, while Player 1 has optimal finite-memory strategies. Moreover, the values $\mathsf{Val}_{\mathcal{G}}(v)$, as well as optimal strategies, can be computed in pseudo-polynomial time.*

It must be noticed that, in the presence of negative costs, and even when every vertex v has a finite value $\mathsf{Val}_{\mathcal{G}}(v) \in \mathbb{R}$, memoryless optimal strategies may not exist for Player 1, as pointed out in [14, Example 1].

3 Priced Timed Games

In order to formally introduce priced timed games, we need to define the concepts of clocks, clock valuations, constraints, and zones. Let $\mathcal{X}$ be a finite set of real-valued variables called *clocks*. A clock valuation on $\mathcal{X}$ is a function $\nu \colon \mathcal{X} \to \mathbb{R}_{\geqslant 0}$ and we write $V(\mathcal{X})$ for the set of clock valuations. Abusing notation, we also treat a valuation ν as a point in $\mathbb{R}^{|\mathcal{X}|}$. If $\nu \in V(\mathcal{X})$ and $t \in \mathbb{R}_{\geqslant 0}$ then we write $\nu + t$ for the clock valuation defined by $(\nu + t)(c) = \nu(c) + t$ for all $c \in \mathcal{X}$. For $C \subseteq \mathcal{X}$, we write $\nu[C := 0]$ for the valuation where $\nu[C := 0](c)$ equals 0 if $c \in C$ and $\nu(c)$ otherwise. A clock constraint over $\mathcal{X}$ is a conjunction of simple constraints of the form $c \bowtie i$ or $c - c' \bowtie i$, where $c, c' \in \mathcal{X}$, $i \in \mathbb{N}$ and $\bowtie \in \{<, >, =, \leqslant, \geqslant\}$. A clock zone is a finite set of clock constraints that defines a convex set of clock valuations. We write $Z(\mathcal{X})$ for the set of clock zones over the set of clocks $\mathcal{X}$.

Definition 2. *A priced timed game is a tuple $\mathcal{A} = (L, \mathcal{X}, \mathrm{Inv}, \Sigma, \delta, \omega, L_f)$ where:*

- *$L = L_1 \uplus L_2$ is a finite set of locations, partitioned into the sets L_1 and L_2;*
- *$\mathcal{X}$ is a finite set of clocks;*

[1] Polynomial time if the prices are encoded in unary.

– $\mathrm{Inv}\colon L \to Z(\mathcal{X})$ *associates an invariant to each location;*
– Σ *is a finite set of labels;*
– $\delta\colon L \times \Sigma \to Z(\mathcal{X}) \times 2^{\mathcal{X}} \times L$ *is a transition function that maps a location* $\ell \in L$ *and label* $a \in \Sigma$ *to a clock zone* $\zeta \in Z(\mathcal{X})$ *representing the guard on the transition, a set of clocks* $R \subseteq \mathcal{X}$ *to be reset and successor location* $\ell' \in L$;
– $\omega\colon L \cup \Sigma \to \mathbb{Z}$ *is the price function; and*
– *and* $L_f \subseteq L$ *is the set of target locations.*

A configuration of a PTG is a tuple $(\ell, \nu) \in L \times V$ where ℓ is a location, ν is a clock valuation and $\nu \in \mathrm{Inv}(\ell)$. A timed action is a tuple $\tau = (t, a) \in \mathbb{R}_{\geqslant 0} \times \Sigma$ where t is a time delay and a is a label. In the following, for a timed move $\tau = (t, a) \in \mathbb{R}_{\geqslant 0} \times \Sigma$, we let $\mathsf{del}(\tau) = t$ be the delay part and $\mathsf{lab}(\tau) = a$ be the label part. The semantics of a PTG is given as an infinite priced game graph.

Definition 3 (Semantics). *The semantics of a PTG* $\mathcal{A} = (L, \mathcal{X}, \mathrm{Inv}, \Sigma, \delta, \omega, L_f)$ *is given as a priced game graph* $[\![\mathcal{A}]\!] = (S, \Gamma, \Delta, \kappa, S_f)$ *where*
– $S = \{(\ell, \nu) \in L \times V \mid \nu \in \mathrm{Inv}(\ell)\}$ *is the set of configurations of the PTG;*
– $\Gamma = \mathbb{R}_{\geqslant 0} \times \Sigma$ *is the set of timed moves;*
– $\Delta\colon S \times \Gamma \to S$ *is the transition function defined by* $(\ell', \nu') = \Delta((\ell, \nu), (t, a))$ *if* $\delta(\ell, a) = (\zeta, R, \ell')$ *such that* $\nu + t \in \zeta$, $\nu + t' \in \mathrm{Inv}(\ell)$ *for all* $0 \leqslant t' \leqslant t$, *and* $\nu' = (\nu + t)[R := 0]$;
– $\kappa : S \times \Gamma \to \mathbb{R}$ *is such that* $\kappa((\ell, \nu), (t, a)) = \omega(\ell) \times t + \omega(a)$; *and*
– $S_f \subseteq S$ *is such that* $(\ell, \nu) \in S_f$ *iff* $\ell \in L_f$.

The concepts of a play, its cost, and strategies of players for a PTG $\mathcal{A}$ is defined via corresponding objects for its semantic priced game graph $[\![\mathcal{A}]\!]$. In the previous section we introduced games with reachability-cost objective for priced game graphs. We also study the following winning objectives for Player 1 in the context of priced timed games; the objective for Player 2 is the opposite.

1. **Constrained-price reachability.** The constrained-price reachability objective $\mathsf{Reach}(\leqslant K)$ is to keep the payoff within a given bound $K \in \mathbb{N}$. Objectives $\mathsf{Reach}(\bowtie K)$ for constrains $\bowtie \in \{<, =, >, \geqslant\}$ are defined analogously.
2. **Bounded-time reachability.** Given constants $K, T \in \mathbb{N}$, the bounded-time reachability objective $\mathsf{TBReach}(K, T)$ is to keep the payoff of the play less than or equal to K while keeping the total time elapsed within T units.
3. **Repeated reachability.** For this objective, we consider slightly different semantics of the game where the play continues forever, and the repeated reachability objective $\mathsf{RReach}(\eta)$, $\eta \in \mathbb{R}_{\geqslant 0}$ is to visit target locations infinitely often each time with a payoff in a given interval $[-\eta, \eta]$.

In Section 4, we sketch the proof of the following negative result regarding the decidability of PTGs with these objectives. This result is particularly surprising for bounded-time reachability objective, since bounded-time restriction has been shown to recover decidability in many related problems [18,12].

Theorem 2. *Let* $\mathcal{A}$ *be a priced timed game arena. The decision problems corresponding to the existence of winning strategy for following objectives are undecidable:*

1. $\mathsf{Reach}(\bowtie K)$ *objective for PTGs with two or more clocks and arbitrary prices;*
2. $\mathsf{TBReach}(K,T)$ *objective for PTGs with five or more clocks; and prices 0,1;*
3. $\mathsf{RReach}(\eta)$ *objective for PTGs with three or more clocks and arbitrary prices.*

To recover decidability, we consider a subclass of one-clock PTGs. In this subclass, the set of clocks $\mathcal{X}$ is a singleton $\{x\}$, and price-rates of the locations come from a doubleton set $\{p^-, p^+\}$ with $p^- < p^+$ two distinct elements of $\{-1, 0, 1\}$ (no condition is made on the prices $\omega(a)$ of labels $a \in \Sigma$). We call these restricted games *one-clock bi-valued priced timed games*, abbreviated as $1\mathrm{PTG}(p^-, p^+)$, or 1BPTG if p^- and p^+ do not matter. All our results may easily be extended to the case where p^- and p^+ are taken from the set $\{-d, 0, d\}$ with $d \in \mathbb{N}$. We devote Section 5 to the proof of the following decidability results.

Theorem 3. *We have the following results:*
1. 1BPTG*s are determined.*
2. *The value of a* 1BPTG *can be computed in pseudo-polynomial time.*
3. *Given that a* 1BPTG *has a finite value, an* ε*-optimal strategy for Player 1 can be computed in pseudo-polynomial time.*
4. *Aforementioned complexities drop to polynomial time for* $1\mathrm{PTG}(0,1)$ *with prices of labels taken from* $\mathbb{N}$.

4 Undecidability Results

In this section we provide a proof sketch of our undecidability result (Theorem 2) by reducing the halting problem for two counter machines (see [17]) to the existence of a winning strategy for Player 1 for the desired objective. For all the three objectives, given a two counter machine, we construct a PTG $\mathcal{A}$ whose building blocks are the modules for instructions. In these reductions the objective of Player 1 is linked to a faithful simulation of various increment, decrement, and zero-test instructions of the machine by choosing appropriate delays to adjust the clocks to reflect changes in counter values. The goal of Player 2 is then to verify the simulation performed by Player 1. Proofs of correctness of the reductions, as well as more details can be found in the appendix.

Constrained-Price Reachability Objectives $\mathsf{Reach}(\bowtie K)$**.** The result in the case $\mathsf{Reach}(\leqslant K)$ is a consequence of the result in [5]. Undecidability for other comparison operators $\bowtie$ is a new contribution. We only consider the objective $\mathsf{Reach}(=1)$ in this section, since proofs for other constraints are similar. Our reduction uses a PTG with two clocks x_1 and x_2, arbitrary price-rates for locations and no prices for labels. Each counter machine instruction (increment, decrement, and test for zero value) is specified using a PTG module. The main invariant in our reduction is that upon entry into a module, we have that $x_1 = \frac{1}{5^{c_1}7^{c_2}}$ and $x_2 = 0$ where c_1 (respectively, c_2) is the value of counter C_1 (respectively, C_2). We outline the simulation of a decrement instruction for counter C_1 in Fig. 2. Let us denote by $x_{old} = \frac{1}{5^{c_1}7^{c_2}}$ the value of x_1 while entering the

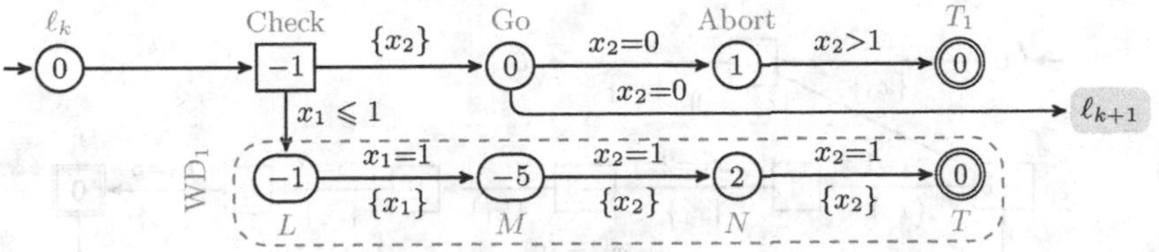

Fig. 2. Decrement module for the objection Reach(=1)

module. At the location ℓ_{k+1} of the module, $x_1 = x_{new}$ should be $5x_{old}$ to correctly decrement counter C_1. At location ℓ_k, Player 1 spends a non-deterministic amount of time $t_k = x_{new} - x_{old}$ such that $x_{new} = 5x_{old} + \varepsilon$. To correctly decrement C_1, ε should be 0, and t_k must be $\frac{4}{5^{c_1}7^{c_2}}$. At location Check, Player 2 could choose to go to Go (in order to continue the simulation of the machine) or go to the widget WD_1, if he suspects that $\varepsilon \neq 0$. If Player 2 spends time $t > 0$ in the location Check before proceeding to Go, then Player 1 can enter the location Abort (to abort the simulation), rather than going to ℓ_{k+1}. Player 1 spends $1+t$ time in location Abort and reaches a target T_1 with cost 1 (and thus achieve his objective). However, if $t = 0$ then entering location Abort will make the cost to be greater than 1 (which is losing for Player 1). If Player 2 decides to enter widget WD_1, then the cost upon reaching the target in the widget WD_1 is $1 + \varepsilon$ which is 1 iff $\varepsilon = 0$.

Bounded-Time Reachability Objective. We sketch the reduction for objective $\mathsf{TBReach}(K, T)$. Our reduction uses a PTG with price-rates 0 or 1 on locations, and zero prices on labels, along with five clocks x_1, x_2, z, a, b. On entry into a module for the $(k+1)$th instruction, we always have one of the two clocks x_1, x_2 with value $\frac{1}{2^{k+c_1}3^{k+c_2}}$ and other is 0. Clock z keeps track of the total time elapsed during simulation of an instruction: we always have $z = 1 - \frac{1}{2^k}$ at the end of simulating kth instruction. Thus, time $\frac{1}{2}$ is spent simulating the first instruction, $\frac{1}{4}$ for the second instruction and so on, so that the total time spent in simulating the main modules is less than 1. The main challenge here is to ensure that only a bounded time is spent along the entire simulation, along with updating the counter values correctly. Clocks a, b are used for rough work. For instance, if the $(k+1)$th instruction ℓ_{k+1} is an increment of C_1, and we have $x_1 = \frac{1}{2^{k+c_1}3^{k+c_2}}$, while $a = b = x_2 = 0$, and $z = 1 - \frac{1}{2^k}$, then at the end of the module simulating ℓ_{k+1}, we want $x_2 = \frac{1}{2^{k+1+c_1+1}3^{k+1+c_2}}$ and $x_1 = 0$ and $z = 1 - \frac{1}{2^{k+1}}$.

Repeated Reachability Objective. Finally, we consider the repeated reachability objective $\mathsf{RReach}(\eta)$. Our reduction uses a PTG with 3 clocks, and arbitrary price-rates, but zero prices for labels. On entry into a module, we have $x_1 = \frac{1}{5^{c_1}7^{c_2}}$, $x_2 = 0$ and $x_3 = 0$, where c_1, c_2 are the values of C_1 and C_2. Fig. 3 shows module to simulate decrement C_1. Location ℓ_k is entered with $x_1 = \frac{1}{5^{c_1}7^{c_2}}$, $x_2 = 0$ and $x_3 = 0$. To correctly decrement C_1, Player 1 should choose a delay of $\frac{4}{5^{c_1}7^{c_2}}$ in location ℓ_k. At location Check, no time can elapse

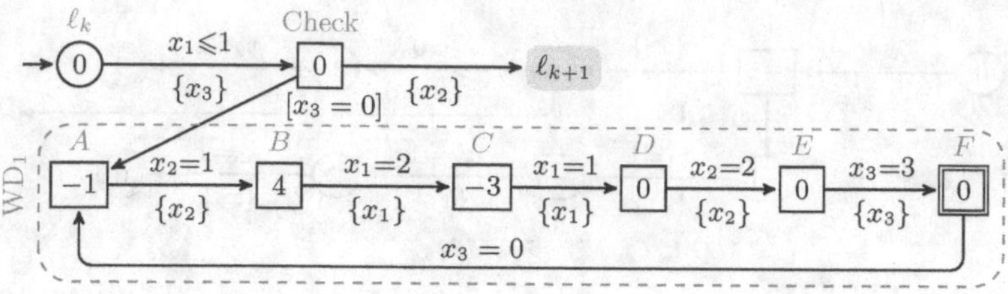

Fig. 3. Decrement module for Repeated reachability objective

because of the invariant. If Player 1 makes an error, and delays $\frac{4}{5^{c_1}7^{c_2}} + \varepsilon$ at ℓ_k ($\varepsilon \neq 0$) then Player 2 can jump in widget WD$_1$. The cost of going from location A to F is ε; each time we come back to A, the clock values with which A was entered are restored. Clearly, if $\varepsilon \neq 0$, Player 2 can incur a cost that is not in $[-\eta, \eta]$ by taking the loop from A to F a large number of times.

5 One-Clock Bi-Valued Priced Timed Games

This section is devoted to the proof of Theorem 3. First of all, let us assume that all 1BPTGs $\mathcal{A}$ we consider are *bounded*, i.e., that there is a global invariant in every location, of the form $x \leqslant M_K$ (where M_K denotes the greatest constant appearing in the clock guards and invariants of $\mathcal{A}$). This restriction comes w.l.o.g since every 1BPTG arena can be made bounded with a polynomial algorithm.[2]

Our proof of Theorem 3 is based on an extension of the classical notion of regions in timed automata, in the spirit of the regions introduced to define the corner point abstraction [8]. Indeed, to take the price into account, ε-optimal strategies do not take uniform decisions on the classical regions. That is why we need to subdivide each classical region into three parts: two small parts around the corners of the region (that we will call *borders* in the following, considering our one-clock setting), and a big part in-between. We will show that considering only strategies that never jump into those big parts is sufficient (Lemma 1). Lemma 2, later, shows a stronger result that one can restrict attention to strategies that play closer and closer to the borders of the regions as time elapses. Finally, we combine these results to show that a finite abstraction of 1BPTGs is sufficient to compute the value as well as ε-optimal strategies (Lemma 3). This not only yields the desired result, but also provides us further insight into the shape of ε-optimal strategies for both players.

5.1 Reduction to η-Region-Uniform Strategies

Since we only consider one-clock PTGs, we need not consider the standard Alur-Dill regional equivalence relation. Instead, we consider special region equivalence relation characterized by the intervals with constants appearing in guards

[2] By introducing auxiliary states in order to reset the clock x at every time unit once its value goes beyond M_k. The polynomial complexity holds only for one-clock PTGs.

and invariants of $\mathcal{A}$ inspired by Laroussinie, Markey, and Schnoebelen construction [16]. Let $0{=}M_0{<}M_1{<}\cdots{<}M_K$ be the integers appearing in guards and invariants of $\mathcal{A}$. We say that two valuations $\nu, \nu' \in \mathbb{R}_{\geqslant 0}$ are region-equivalent (or lie in the same region), and we write $\nu \sim \nu'$, if for every $k \in \{0, \dots, K\}$, $\nu \leqslant M_k$ iff $\nu' \leqslant M_k$, and $\nu \geqslant M_k$ iff $\nu' \geqslant M_k$. We define the set of regions to be the set of equivalence classes of $\sim$. We extend the equivalence relation $\sim$ from valuations to configurations in a straightforward manner. We also generalize the regional equivalence relation to the plays. For two (finite or infinite) plays $r = \langle(\ell_0, \nu_0), (t_0, a_0), \dots\rangle$ and $r' = \langle(\ell'_0, \nu'_0), (t'_0, a'_0), \dots\rangle$ we say that $r \sim r'$ if the lengths of r and r' are equal, and they define sequences of regional equivalent states (i.e., $(\ell_i, \nu_i) \sim (\ell'_i, \nu'_i)$ for all $i \geqslant 0$) and follow equivalent timed actions (i.e., $a_i = a'_i$ and $\nu_i + t_i \sim \nu'_i + t'_i$ for all $i \geqslant 0$). We also consider a refinement of region equivalence relation that we call the η-region equivalence relation, and we write $\sim_\eta$, for a given $\eta \in (0, \frac{1}{3})$. Intuitively, $\nu \sim_\eta \nu'$ if both valuations are close or far from any borders of the regions, with respect to the distance η.

Definition 4 (η-regions). *For valuations $\nu, \nu' \in \mathbb{R}_{\geqslant 0}$ we say that $\nu \sim_\eta \nu'$ if $\nu \sim \nu'$ and for every $k \in \{0, \dots, K-1\}$, $|\nu - M_k| \leqslant \eta$ iff $|\nu' - M_k| \leqslant \eta$, and $\nu \geqslant M_K - \eta$ iff $\nu' \geqslant M_K - \eta$. We assume the natural order $\preceq$ over η-regions by their lower bounds. We call η-regions the equivalence classes of $\sim_\eta$. We also extend the relation $\sim_\eta$ to configurations and runs.*

For instance, if $M_1 = 2$ and $M_2 = 3$, the set of η-regions is given by $\{\{0\}, (0, \eta], (\eta, 2-\eta), [2-\eta, 2), \{2\}, (2, 2+\eta], (2+\eta, 3-\eta), [3-\eta, 3), \{3\}, (3, +\infty)\}$. We next introduce the strategies of a restricted shape with the properties that they depend only on the η-region abstraction of runs; their decision is uniform over each η-region; and they play η-close to the borders of the regions.

Definition 5 (η-region uniform strategies). *Let $\eta \in (0, \frac{1}{3})$ be a constant. A strategy $\sigma \in \mathsf{Strat}_1 \cup \mathsf{Strat}_2$ is said to be* η-region-uniform *if*

- *for all finite run $r \sim_\eta r'$ ending respectively in (ℓ, ν) and (ℓ, ν') (in particular $\nu \sim_\eta \nu'$) we have $\nu + \mathsf{del}(\sigma(r)) \sim_\eta \nu' + \mathsf{del}(\sigma(r'))$ and $\mathsf{lab}(\sigma(r)) = \mathsf{lab}(\sigma(r'))$;*
- *for every finite run r ending in (ℓ, ν), if $\nu + \mathsf{del}(\sigma(r)) \in (M_k, M_{k+1})$, we have $\nu + \mathsf{del}(\sigma(r)) \in (M_k, M_k + \eta] \cup [M_{k+1} - \eta, M_{k+1})$.*

We write UStrat_1^η and UStrat_2^η for the set of η-region-uniform strategies for Players 1 and 2. We also define upper-value $\overline{\mathsf{UVal}}^\eta(s)$ when both players are restricted to use only η-region-uniform strategies. Formally,

$$\overline{\mathsf{UVal}}^\eta(s) = \inf_{\sigma_1 \in \mathsf{UStrat}_1^\eta} \sup_{\sigma_2 \in \mathsf{UStrat}_2^\eta} \mathsf{Cost}(\mathsf{Play}(s, \sigma_1, \sigma_2)), \text{ for all } s \in S.$$

Example 1. Consider PTG $\mathcal{A}_1$ shown in Fig. 4 (that is not a 1BPTG since there are three distinct price-rates). A strategy of Player 2 is entirely described by the time spent in the initial location with initial valuation 0. For example, Player 2 can choose to delay $1/2$ time units before jumping in the next location. Indeed, the lower and upper value of the game is $-\frac{1}{2}$. However, this strategy is not η-region-uniform. Instead, an η-region-uniform strategy will delay t time units with $t \in [0, \eta] \cup [1 - \eta, 1]$. Hence, the upper value when players can only use η-region-uniform strategies is equal to -1.

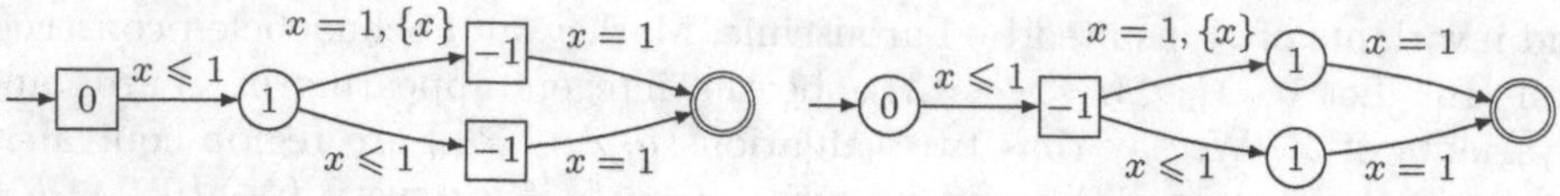

Fig. 4. The value in the left-side one-clock PTG $\mathcal{A}_1$ with price-rates in $\{-1, 0, 1\}$ is $-\frac{1}{2}$, while the value in the right-side PTG $\mathcal{A}_2$ is $\frac{1}{2}$

Contrary to this example, the next lemma shows that, in 1BPTGs, the upper value of the game increases when we restrict ourselves to η-region-uniform strategies. Intuitively, every cost that Player 2 can secure with general strategies, it can also secure it with η-region-uniform strategies against η-region-uniform strategies of Player 1.

Lemma 1. $\overline{\mathsf{Val}}(s) \leqslant \overline{\mathsf{UVal}}^{\eta}(s)$, *for every* 1BPTG $\mathcal{A}$, $s \in S$ *and* $\eta \in (0, \frac{1}{3})$.

5.2 Reduction to η-Convergent Strategies

A similar result concerning the lower values of the games can be shown in case of η-region-uniform strategies. In subsequent proofs, we need a stronger result to avoid situations detailed in Example 2, where player 2 needs infinite precision to play incrementally closer to borders (as well as an infinite memory). For this reason, we restrict the shape of strategies to force them to play at distance $\frac{\eta}{2^n}$ of borders when playing the nth round of the game. The slight asymmetry in the definitions for the two players is exploited in proving subsequent results.

Definition 6 (η-convergent strategies). *Let* $\eta \in (0, \frac{1}{3})$ *be a constant. A strategy* $\sigma \in \mathsf{Strat}_1 \cup \mathsf{Strat}_2$ *is said to be* η-convergent *if* σ *is* η*-region-uniform and for all finite run* r *of length* n *ending in* (ℓ, ν):

- *if* $\sigma \in \mathsf{Strat}_1$, *there exists* k *such that either* $|\nu + \mathsf{del}(\sigma(r)) - M_k| \leqslant \frac{\eta}{2^{n+1}}$, *or* $\mathsf{del}(\sigma(r)) = 0$ *and* $\nu \in (M_k + \frac{\eta}{2^{n+1}}, M_k + \eta]$;
- *if* $\sigma \in \mathsf{Strat}_2$, *there exists* k *such that either* $\nu + \mathsf{del}(\sigma(r)) \in \{M_k + \frac{\eta}{2^{n+1}}\} \cup [M_k - \frac{\eta}{2^{n+1}}, M_k)$, *or* $\mathsf{del}(\sigma(r)) = 0$ *and* $\nu \in (M_k + \frac{\eta}{2^{n+1}}, M_k + \eta]$.

We let CStrat_1^{η} *and* CStrat_2^{η} *be respectively the set of* η*-convergent strategies for Player 1 and Player 2, and we define, for every configuration* $s \in S$, $\underline{\mathsf{CVal}}^{\eta}(s) = \sup_{\sigma_2 \in \mathsf{CStrat}_2^{\eta}} \inf_{\sigma_1 \in \mathsf{CStrat}_1^{\eta}} \mathsf{Cost}(\mathsf{Play}(s, \sigma_1, \sigma_2))$.

Example 2. Consider the 1BPTG $\mathcal{A}_3$ composed of a vertex per player, on top of the target vertex. In its vertex, having price-rate 0, Player 1 must choose between going to the target vertex, or going to the vertex of Player 2 by resetting clock x. In its vertex, having price-rate -1, Player 2 must go back to the vertex of Player 1, with a guard $x > 0$: hence, Player 2 would like to exit as soon as possible, but because of the guard, he must spend some time before exiting. If Player 2 plays according to a finite-memory strategy, there must be a bound ε such that Player 2 always stays in his state for a duration bounded from below by ε, and Player 1 can exploit it by letting the game continue for an arbitrarily long time to achieve an arbitrarily small payoff. On the other hand, if Player 2 plays an infinite-memory η-convergent strategy by staying in his location for a duration $\varepsilon/2^n$ in his n-th visit to its location, Player 2 ensures a payoff $-\varepsilon$ for an arbitrarily small $\varepsilon > 0$, resulting in the value 0 of the game.

It is clear from the previous example that Player 2 needs infinite-memory strategies to optimize his objective. The following lemma formalizes our intuition that the lower value of the game decreases when we restrict ourselves to η-convergent strategies. Intuitively, every cost that Player 1 can secure with general strategies, it can also secure it with η-convergent strategies against an η-convergent strategy of Player 2.

Lemma 2. $\underline{\mathsf{CVal}}^\eta(s) \leqslant \underline{\mathsf{Val}}(s)$, *for every* 1BPTG $\mathcal{A}$, $s \in S$ *and* $\eta \in (0, \frac{1}{3})$.

Observe that this lemma fails to hold when location price-rates can take more than two values as exemplified by arena $\mathcal{A}_2$ in Fig. 4. It shows a game with three distinct prices with lower and upper value equal to 1/2. However, when restricted to η-convergent strategies, the lower value equals 1.

Our next goal is to find a common bound being both a lower bound on $\underline{\mathsf{CVal}}^\eta(s)$ and an upper bound on $\overline{\mathsf{UVal}}^\eta(s)$ by studying the value of a reachability-cost game on a finitary abstraction of 1BPTGs.

5.3 Finite Abstraction of 1BPTGs

We now construct a finite price game graph $\tilde{\mathcal{A}}$ from any 1BPTG $\mathcal{A}$, as a finite abstraction of the infinite weighted game $[\![\mathcal{A}]\!]$, based on η-regions. Since we have learned that η-region-uniform strategies suffice, we limit ourselves to playing at a distance at most η from the borders of regions. Observe that only η-regions close to the borders are of interest, and moreover η-regions after the maximal constant M_K are not useful since $\mathcal{A}$ is bounded. Let $\mathcal{I}^\eta_\mathcal{A}$ be the set of remaining "useful" η-regions. For example, if constant appearing in the PTG are $M_1 = 2$ and $M_2 = 3$, we have $\mathcal{I}^\eta_\mathcal{A} = \{\{0\}, (0, \eta], [2-\eta, 2), \{2\}, (2, 2+\eta], [3-\eta, 3), \{3\}\}$. We next define the *delay* between two such η-regions $I \preceq J$, denoted by $d(I, J)$, as the closest integer of $q' - q$, where q (respectively, q') is the lower bound of interval I (respectively, J). For example, $d((2, 2+\eta], [3-\eta, 3)) = 1$ and $d(\{0\}, [2-\eta, 2)) = 2$.

Definition 7. *For every* 1BPTG $\mathcal{A}$ *we define its border abstraction as a finite priced game graph* $\tilde{\mathcal{A}} = (V = V_1 \uplus V_2, A, E, \pi, V_f)$ *where:*
- $V_i = \{(\ell, I) \mid \ell \in L_i, I \in \mathcal{I}^\eta_\mathcal{A}, I \subseteq \mathrm{Inv}(\ell)\}$ *for* $i \in \{1, 2\}$;
- $A = \mathcal{I}^\eta_\mathcal{A} \times \Sigma$;
- E *is the set of tuples* $((\ell, I), (J, a), (\ell', J'))$ *such that* $I \preceq J$ *and for all* $I \preceq K \preceq J$ *we have* $K \subseteq \mathrm{Inv}(\ell)$ *and* $J \subseteq \zeta$ *and* $J' = J[R := 0]$ *with* $(\zeta, R, \ell') = \delta(\ell, a)\}$;
- $\pi((\ell, I), (J, a), (\ell', J')) = \omega(\ell) \times d(I, J) + \omega(a)$; *and*
- $V_f = \{(\ell, I) \mid \ell \in L_f, I \in \mathcal{I}^\eta_\mathcal{A}\}$.

In a border abstraction game $\tilde{\mathcal{A}}$, the meaning of action (J, a) is that the player wants to let time elapse until it reaches the η-region J, then playing label a. It simulates any timed move (t, a) with t any delay reaching a point in J.

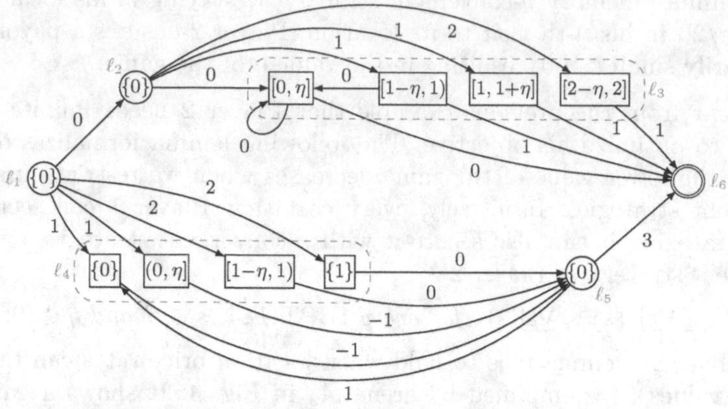

Fig. 5. Finite weighted game associated with the 1BPTG of Fig. 1

Example 3. Consider the border abstraction of the 1BPTG of Fig. 1 shown in Fig. 5. Observe that we depict only a succinct representation of the real abstraction, since we only show the reachable part of the game from $(\ell_1, 0)$, and we have removed multiple edges (introduced due to label hiding) and kept only the most useful ones for the corresponding player. For example, consider the location $(\ell_5, \{0\})$. There are edges labelled by (J, a) for every interval $J \in \mathcal{I}^\eta_{\mathcal{A}}$, all directed to $(\ell_4, \{0\})$ due to a reset being performed there. We only show the best possible edge—the one with lowest price—since location ℓ_5 belongs to Player 1, who seeks to minimise cost. Each vertex contains the η-region it represents. Thanks to Theorem 1, it is possible to compute the optimal value as well as optimal strategies for both players. Here, the value of state $(\ell_1, 0)$ is 1, and an optimal strategy for Player 1 is to follow action $(\{0\}, a)$ (i.e., jump to ℓ_2 immediately), and then action $(\{1\}, a)$ (i.e., to delay 1 time unit, before jumping in ℓ_3).

Lemma 3. *Let $\mathcal{A}$ be a* 1BPTG *and $\tilde{\mathcal{A}}$ be its border abstraction. Suppose that for all $0 \leqslant k \leqslant K$ and $\ell \in L$ we have that $\mathsf{Val}_{\tilde{\mathcal{A}}}((\ell, \{M_k\}))$ is finite. Then, for all $\varepsilon > 0$, there is $\eta > 0$ s.t.* $\overline{\mathsf{UVal}}^\eta_{\mathcal{A}}((\ell, M_k)) - \varepsilon \leqslant \mathsf{Val}_{\tilde{\mathcal{A}}}((\ell, \{M_k\})) \leqslant \underline{\mathsf{CVal}}^\eta_{\mathcal{A}}((\ell, M_k)) + \varepsilon$.

Combining this result with Theorem 1 we obtain the following.

Corollary 1. 1BPTG*s are determined and we can compute their values in pseudo-polynomial time. Moreover, in case the values are finite, ε-optimal strategies exist for both players: Player 2 may require infinite memory strategies, whereas finite memory is sufficient for Player 1. Finally, ε-optimal strategies can also be computed in pseudo-polynomial time.*

Proof. In case of infinite values $\mathsf{Val}_{\tilde{\mathcal{A}}}((\ell, \{M_k\}))$, we can show directly that $\overline{\mathsf{Val}}_{\mathcal{A}}((\ell, M_k)) = \mathsf{Val}_{\tilde{\mathcal{A}}}((\ell, \{M_k\})) = \underline{\mathsf{Val}}_{\mathcal{A}}((\ell, M_k))$. Otherwise, let $\varepsilon > 0$. By Lemma 3, we know that there exists $\eta > 0$ such that for every location $\ell \in L$ and integer $0 \leqslant k \leqslant K$:

$$\overline{\mathsf{UVal}}^\eta_{\mathcal{A}}((\ell, M_k)) - \varepsilon \leqslant \mathsf{Val}_{\tilde{\mathcal{A}}}((\ell, \{M_k\})) \leqslant \underline{\mathsf{CVal}}^\eta_{\mathcal{A}}((\ell, M_k)) + \varepsilon.$$

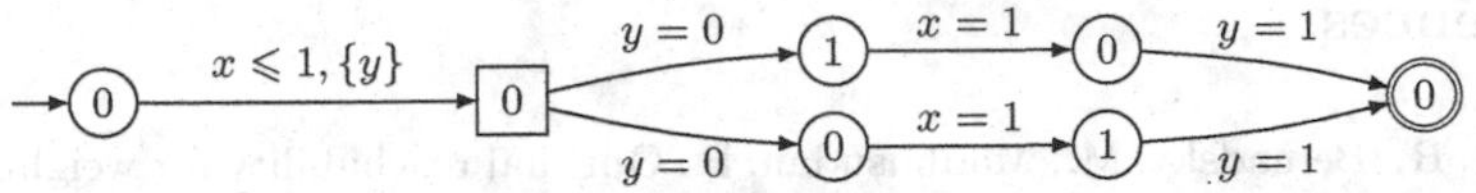

Fig. 6. A two-clock PTG with prices of locations in $\{0, +1\}$ and value 1/2

Moreover Lemma 1 and 2 show that:

$$\underline{\mathsf{CVal}}^{\eta}((\ell, M_k)) \leqslant \underline{\mathsf{Val}}((\ell, M_k)) \leqslant \overline{\mathsf{Val}}((\ell, M_k)) \leqslant \overline{\mathsf{UVal}}^{\eta}((\ell, M_k)) .$$

Both inequalities combined permit to obtain

$$\mathsf{Val}_{\tilde{\mathcal{A}}}((\ell, \{M_k\})) - \varepsilon \leqslant \underline{\mathsf{Val}}((\ell, M_k)) \leqslant \overline{\mathsf{Val}}((\ell, M_k)) \leqslant \mathsf{Val}_{\tilde{\mathcal{A}}}((\ell, \{M_k\})) + \varepsilon .$$

Taking the limit when ε tends to 0, we obtain that $\underline{\mathsf{Val}}((\ell, M_k)) = \overline{\mathsf{Val}}((\ell, M_k)) = \mathsf{Val}_{\tilde{\mathcal{A}}}((\ell, \{M_k\}))$. Therefore, 1BPTG are determined. Moreover, in case of finite values, the proof of Lemma 3 permits to construct ε-optimal η-region-uniform strategies σ_1^* (with finite memory) and σ_2^* (which is moreover η-convergent). □

In the case of 1BPTGs, the finite values are integers. This property fails if we allow more than one clock, as shows Fig. 6 with a two-clock PTG with price-rates in $\{0, 1\}$ and optimal value $\frac{1}{2}$. It also fails if we allow more than two price-rates as was shown in Fig. 4. However for 1PTG(0, 1) with prices of labels in $\mathbb{N}$, the value of the game is necessarily nonnegative disallowing the case $-\infty$. The case $+\infty$ can be detected in polynomial time. If the value is not $+\infty$, the exact computation in the finite abstraction $\tilde{\mathcal{A}}$ can be performed in polynomial time (see [14] or [15]), resulting in a polynomial algorithm for PTGs. The sketch of Theorem 3 is now complete. Notice that our proof shows that optimal value functions (as defined in [10,20,15]) of such games have a polynomial number of line segments, and hence algorithms presented in [10,20,15] are indeed polynomial time.

6 Conclusion

We revisited games with reachability objective on PTGs with both positive and negative price-rates. We showed undecidability of all classes of constrained-price reachability objectives with two or more clocks. We also observed that adding bounded-time restriction does not recover decidability, even with nonnegative prices. We also partially answer the question regarding polynomial-time algorithm for one-clock PTGs by showing that for a bi-valued variant the problem is in pseudo-polynomial time. However, the existence of a polynomial-time algorithm for multi-priced one-clock PTGs with nonnegative price-rates, and the existence of algorithm for computing ε-optimal strategies for PTGs with arbitrary number of clocks remain open problems.

References

1. Alur, R., Bernadsky, M., Madhusudan, P.: Optimal reachability for weighted timed games. In: Díaz, J., Karhumäki, J., Lepistö, A., Sannella, D. (eds.) ICALP 2004. LNCS, vol. 3142, pp. 122–133. Springer, Heidelberg (2004)
2. Alur, R., Dill, D.L.: A theory of timed automata. TCS 126(2), 183–235 (1994)
3. Alur, R., La Torre, S., Pappas, G.J.: Optimal paths in weighted timed automata. TCS 318(3), 297–322 (2004)
4. Behrmann, G., Fehnker, A., Hune, T., Larsen, K.G., Pettersson, P., Romijn, J.M.T., Vaandrager, F.W.: Minimum-cost reachability for priced timed automata. In: Di Benedetto, M.D., Sangiovanni-Vincentelli, A.L. (eds.) HSCC 2001. LNCS, vol. 2034, pp. 147–161. Springer, Heidelberg (2001)
5. Berendsen, J., Chen, T., Jansen, D.N.: Undecidability of cost-bounded reachability in priced probabilistic timed automata. In: Chen, J., Cooper, S.B. (eds.) TAMC 2009. LNCS, vol. 5532, pp. 128–137. Springer, Heidelberg (2009)
6. Bouyer, P., Brihaye, T., Bruyère, V., Raskin, J.-F.: On the optimal reachability problem of weighted timed automata. FMSD 31(2), 135–175 (2007)
7. Bouyer, P., Brihaye, T., Markey, N.: Improved undecidability results on weighted timed automata. IPL 98(5), 188–194 (2006)
8. Bouyer, P., Brinksma, E., Larsen, K.G.: Optimal infinite scheduling for multi-priced timed automata. FMSD 32(1), 3–23 (2008)
9. Bouyer, P., Cassez, F., Fleury, E., Larsen, K.G.: Optimal strategies in priced timed game automata. In: Lodaya, K., Mahajan, M. (eds.) FSTTCS 2004. LNCS, vol. 3328, pp. 148–160. Springer, Heidelberg (2004)
10. Bouyer, P., Larsen, K.G., Markey, N., Rasmussen, J.I.: Almost optimal strategies in one-clock priced timed games. In: Arun-Kumar, S., Garg, N. (eds.) FSTTCS 2006. LNCS, vol. 4337, pp. 345–356. Springer, Heidelberg (2006)
11. Brihaye, T., Bruyère, V., Raskin, J.-F.: On optimal timed strategies. In: Pettersson, P., Yi, W. (eds.) FORMATS 2005. LNCS, vol. 3829, pp. 49–64. Springer, Heidelberg (2005)
12. Brihaye, T., Doyen, L., Geeraerts, G., Ouaknine, J., Raskin, J.-F., Worrell, J.: Time-bounded reachability for monotonic hybrid automata: Complexity and fixed points. In: Van Hung, D., Ogawa, M. (eds.) ATVA 2013. LNCS, vol. 8172, pp. 55–70. Springer, Heidelberg (2013)
13. Brihaye, T., Geeraerts, G., Krishna, S.N., Manasa, L., Monmege, B., Trivedi, A.: Reachability-cost games with negative weights. Technical report (2014), `http://arxiv.org/abs/1404.5894`
14. Brihaye, T., Geeraerts, G., Monmege, B.: Reachability-cost games with negative weights. Technical report (2014), `http://www.ulb.ac.be/di/verif/monmege/download/priced-games.pdf`
15. Hansen, T.D., Ibsen-Jensen, R., Miltersen, P.B.: A faster algorithm for solving one-clock priced timed games. In: D'Argenio, P.R., Melgratti, H. (eds.) CONCUR 2013 – Concurrency Theory. LNCS, vol. 8052, pp. 531–545. Springer, Heidelberg (2013)
16. Laroussinie, F., Markey, N., Schnoebelen, P.: Model checking timed automata with one or two clocks. In: Gardner, P., Yoshida, N. (eds.) CONCUR 2004. LNCS, vol. 3170, pp. 387–401. Springer, Heidelberg (2004)
17. Minsky, M.L.: Computation: Finite and infinite machines. Prentice-Hall, Inc. (1967)

18. Ouaknine, J., Worrell, J.: Towards a theory of time-bounded verification. In: Abramsky, S., Gavoille, C., Kirchner, C., Meyer auf der Heide, F., Spirakis, P.G. (eds.) ICALP 2010. LNCS, vol. 6199, pp. 22–37. Springer, Heidelberg (2010)
19. Ramadge, P.J., Wonham, W.M.: The control of discrete event systems. IEEE 77, 81–98 (1989)
20. Rutkowski, M.: Two-player reachability-price games on single-clock timed automata. In: QAPL 2011. EPTCS, vol. 57, pp. 31–46 (2011)
21. Zwick, U., Paterson, M.: The complexity of mean payoff games on graphs. TCS 158, 343–359 (1996)

Tight Game Abstractions of Probabilistic Automata*

Falak Sher Vira and Joost-Pieter Katoen

Software Modeling and Verification
RWTH Aachen University
D-52056 Aachen, Germany

Abstract. We present a new game-based abstraction technique for probabilistic automata (PA). The key idea is to use *distribution*-based abstraction – preserving novel distribution-based (alternating) simulation relations – rather than classical *state*-based abstraction. These abstractions yield (simple) probabilistic game automata (PGA), turn-based 2 player stochastic games in which moves of both players – as opposed to classical stochastic games – yield distributions over states. As distribution-based (alternating) simulation relations are pre-congruences for composite PGA, abstraction can be done compositionally. Our abstraction yields tighter upper and lower bounds on (extremal) reachability probabilities than state-based abstraction. This shows the potential superiority over state-based abstraction of PA and Markov decision processes.

1 Introduction

Probabilistic automata [1] (PA) extend labelled transition systems by allowing targets of transitions to be distributions over states rather than simply states. As transitions emanating from a state can be equally labelled, PA slightly generalize Markov decision processes (MDPs). This enables a natural way of putting PA in parallel. Due to the presence of non-determinism and discrete probabilistic branching, PA are convenient for modelling randomized distributed algorithms and security protocols. They are also popular semantic models for probabilistic process algebras and form the backbone of the PIOA language.

To combat the well-known state space explosion problem, abstractions of PA that go beyond bisimulation have received quite some attention. Whereas abstract PA [2,3,4] build upon concepts from modal transition systems and constraint functions, [5] uses three-valued abstraction yielding interval Markov chains, while [6] aggregates MDPs by separating the non-determinism in the MDPs from that introduced by abstraction. This naturally yields turn-based stochastic 2-player games [7,8], where one player controls the non-determinism in the MDPs, whereas the other is in charge of the non-determinism from the abstraction. Game-based MDPs abstraction yields upper and lower bounds

* This research is supported by the EU FP7 SENSATION Project and the EU Marie-Curie Project MEALS.

P. Baldan and D. Gorla (Eds.): CONCUR 2014, LNCS 8704, pp. 576–591, 2014.

on reachability probabilities, and significantly improves standard MDPs model checking as evidently shown by several case studies [6]. Besides, this game-based abstraction is optimal in the sense of abstract interpretation [9].

Although the aforementioned abstraction techniques are different in nature, they have in common that the abstraction is state-based. That is to say, abstract models simulate concrete models in a step-wise manner [10]. The key idea in this paper is to treat distributions rather than states as first-class citizens, and relax state-based simulation to *distribution*-based simulations. Our abstractions yield (simple) probabilistic game automata (PGA), turn-based 2-player stochastic games in which moves of both players – as opposed to classical stochastic games (SGs) [7,8] – yield distributions over states. The new abstraction technique yields tighter upper and lower bounds on (extremal) reachability probabilities than state-based abstraction. This shows the potential superiority over *state*-based game-based MDP abstraction [6], and puts the optimality result of [9] in perspective.

Abstract models are probabilistic game automata (PGA), in fact simplified versions of the games in [11]. These games feature action-labelled transitions, in which every player non-deterministically makes a move and randomly picks the next state. We define two distribution-based pre-orders between abstract and concrete PGA: *simulation* and *alternating simulation* relations. Simulation relations are of interest when both players have identical objectives, whereas alternating simulation relations are useful for competitive objectives. Both relations are shown to be pre-congruences w.r.t. parallel composition of (a class of) PGA, enabling *compositional* abstraction of P(G)A. The pre-orders are the key to distribution-based abstraction, a technique distinguishing the non-deterministic behaviour of concrete distributions from that of the distributions induced by the abstraction. This enables merging concrete distributions having similar behaviour in the abstraction.

Put in a nutshell, the major contributions of this paper are: (1) a distribution-based abstraction framework of PA using a slight generalisation of stochastic games, (2) elementary results for distribution-based (alternating) simulation relations such as congruence properties and comparison to state-based simulation, and (3) distribution-based abstraction yields tighter bounds for extremal reachability probabilities than state-based abstraction.

Organization. Section 2 sets the ground for this paper and introduces SGs and PGA. Section 3 and 4 present (alternating) simulation relations for PGA. Section 5 treats two abstraction techniques. Section 6 presents game composition and the fact that abstraction is compositional. Section 7 presents results on bounding extremal reachability probabilities for PA. Section 8 discusses the special case of MDPs abstraction and compares to [6] while Section 9 concludes.

2 Preliminaries

Distributions. A *distribution* μ is a function on a countable set S iff $\mu : S \to [0,1]$ and $0 < \sum_{s\in S} \mu(s) \leq 1$; its support set is given as $\mathrm{Supp}(\mu) = \{s \in S \mid$

$\mu(s) > 0\}$; and its mass w.r.t. set $S' \subseteq S$ is given as $\mu(S') = \sum_{s \in S'} \mu(s)$. Let $|\mu| = \mu(S)$ denote the size of the distribution μ; μ is a *full distribution* iff $|\mu| = 1$, otherwise, it is a sub-distribution. Let $\text{Dist}(S)$ denote the set of distributions over S. Let $\iota_s \in \text{Dist}(S)$ denote the *Dirac* distribution for $s \in S$, i.e, $\iota_s(s) = 1$. A distribution μ'' can be split into sub-distributions μ and μ', say, represented as $\mu'' = \mu \oplus \mu'$, iff $\mu''(s) = \mu(s) + \mu'(s)$ for $s \in S$. Since $\oplus$ is associative and commutative, we use the notation $\bigoplus$ for finite sums. A distribution is sometimes represented as $\mu = [\![\mu(s)s \mid s \in \text{Supp}(\mu)]\!]$, where $[\![$ and $]\!]$ differentiate a set of probabilities from an ordinary set. For $0 \leq c \leq 1$, $c{\cdot}\mu$ denotes the distribution defined by: $(c{\cdot}\mu)(s) = c \cdot \mu(s)$. For a distribution μ, the conditional distribution w.r.t. a set $A \subseteq \text{Supp}(\mu)$ is given as: $\mu_{\downarrow_A}(s) = \frac{\mu(s)}{\mu(A)}$ for $s \in A$, and $\mu_{\downarrow_A}(s) = 0$ if $s \notin A$; if $A = \text{Supp}(\mu)$, we omit A and simply write $\mu_{\downarrow}$.

Probability Measures and Spaces. Let Ω be a non-empty set and $\mathcal{F} \subseteq 2^{\Omega}$. $\mathcal{F}$ is a σ-field on Ω iff: (1) $\emptyset \in \mathcal{F}$; (2) $A \in \mathcal{F} \Rightarrow \Omega \backslash A \in \mathcal{F}$; (3) $A_1, A_2, A_3, ... \in \mathcal{F} \Rightarrow \bigcup_{i \geq 1} A_i \in \mathcal{F}$. The elements of $\mathcal{F}$ are *measurable sets* and $(\Omega, \mathcal{F})$ is a *measurable space*. A function $\Pr : \mathcal{F} \to [0,1]$ is a *probability measure* on $(\Omega, \mathcal{F})$ iff $\Pr(\Omega) = 1$ and if $A_1, A_2, ...$ are disjoint elements in $\mathcal{F}$, then $\Pr(\bigcup_i A_i) = \sum_i \Pr(A_i)$. $(\Omega, \mathcal{F}, \Pr)$ is called a *measurable space*. For any $\mathcal{A} \subseteq \mathcal{F}$, there exists a unique smallest σ-field that contains $\mathcal{A}$ [12]; and given that $\mathcal{A}$ satisfies certain conditions [12], a *probability measure* defined on $\mathcal{A}$ can be uniquely extended to the σ-field containing $\mathcal{A}$.

Probabilistic Automata (PA). PA [1] is an extension of labelled transition systems (LTS) in which the target of any action-labelled transition is a distribution over states instead of a single state. Let UAct be a countable universe actions including the internal action τ. Formally,

Definition 1. *A* Probabilistic Automaton *is a tuple $\mathcal{M} = (S, A, \Delta, s_0)$ where S is a non-empty, countable set of states with initial state $s_0 \in S$; $A \subseteq$ UAct; and $\Delta \subseteq S \times A \times \text{Dist}(S)$ is a set of transitions.*

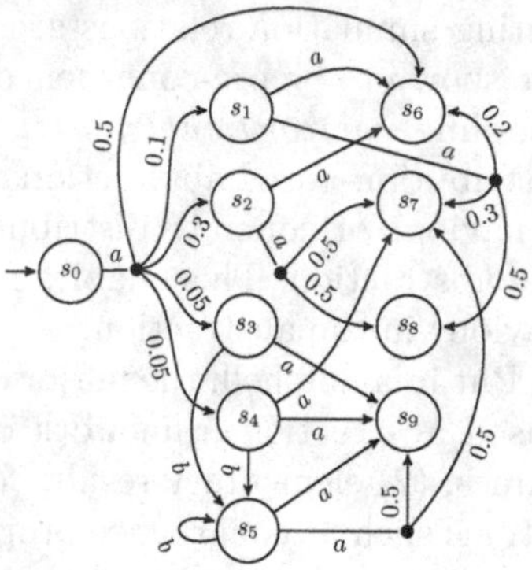

Fig. 1. A PA

In the sequel, $\mathcal{M} = (S, A, \Delta, s_0)$ is assumed to be a finitely branching PA.

Stochastic Games (SGs). A 2-player SG [7,8] is a game of chance played between two players, say, player 1 and player 2. The game arena is a bipartite graph – having, say, S_1 and S_2 as sets of vertices – in which each player owns a specific set of vertices; say, the players 1 and 2 own S_1 and S_2 respectively. The game is started by player 1 and evolves in a turn-based fashion. Starting from the initial state in S_1, player 1 non-deterministically chooses an action-distribution pair. Based on the selected distribution a state in S_2, say s_2, is randomly selected and control is passed to player 2. Player 2 non-deterministically selects an enabled action in s_2, uniquely picks a successor of s_2 and passes control back to player 1. This goes on until some goal is achieved either by player 1 or player 2.

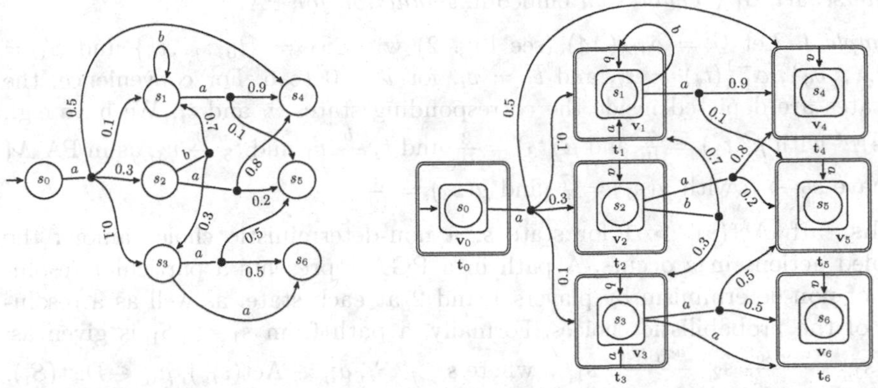

Fig. 2. A PA $\mathcal{M}$ (left) and its embedding $\mathcal{G} = \alpha_{\text{PA}}(\mathcal{M})$ (right)

Definition 2. *A* Stochastic Game *is a tuple* $\mathcal{G} = (S, \{S_1, S_2\}, A, \Delta, s_0)$ *where* S *is a non-empty, countable set of states, partitioned into* S_1 *and* S_2*, with* $s_0 \in S_1$*;* $A \subseteq \text{UAct}$*; and* $\Delta \subseteq (S_1 \times A \times \text{Dist}(S_2)) \cup (S_2 \times A \times \text{D}(S_1))$ *is a set of probabilistic transitions where* $\text{D}(S_1) \subseteq \text{Dist}(S_1)$ *is the set of Dirac distributions over* S_1.

We often denote $(s, a, \mu) \in \Delta$ by $s \xrightarrow{a} \mu$ and $\text{Act}(s)$ as the set of enabled actions from state s, i.e., $\text{Act}(s) = \{a \in A \mid \exists \mu \in \text{Dist}(S) : s \xrightarrow{a} \mu\}$. We assume that a game is started by player 1 and $|\text{Act}(s)| > 0$ for $s \in S$. Note that PA are SGs in which $\forall s \in S_2, a, b \in A : (s \xrightarrow{a} \mu \wedge s \xrightarrow{b} \nu)$ implies $\mu = \nu$ and $|\text{Supp}(\mu)| = 1$.

Simple Probabilistic Game Automata (PGA). In SGs, player 1 moves yield distributions over states, while player 2 moves yield states. In PGA, player 2 moves also yield distributions over states.

Definition 3. *A* Simple Probabilistic Game Automaton *is a tuple* $\mathcal{G} = (S, \{S_1, S_2\}, A, \Delta, s_0)$ *where* S, S_1, S_2, A *and* s_0 *are as for* SGs, *and* $\Delta \subseteq S_{1+x} \times A \times \text{Dist}(S_{2-x})$ *where* x *is a bit.*

Our PGA are simplified versions of the probabilistic game automata in [11]. SGs are a subclass of PGA in which $\text{Dist}(S_1)$ is a set of Dirac distributions. In the sequel, let $\mathcal{G} = (S, \{S_1, S_2\}, A, \Delta, s_0)$ be a finitely branching PGA. For depicting PGA we represent states in S_1 and S_2 as rectangles and double rectangles respectively. In case of PA, states are circles. In the following, we show how a PA can be embedded into a PGA. For a state $s \in S$, let $\overline{s}$ be a copy of s.

Definition 4. *For* PA $\mathcal{M}$*, the bijective* embedding *function* $\alpha : S \to S'_2$ *induces the* PGA $\alpha(\mathcal{M}) = \mathcal{G}' = (S', \{S'_1, S'_2\}, A', \Delta', s'_0)$ *where* $A' = A$, $S'_1 = \{\overline{s'} \mid s' \in S'_2\}$ *—* S'_1 *is a copy of* S'_2 *—,* $s'_0 = \alpha(s_0)$ *and for every* $s' \in S'_2$*:*

1. $\overline{s'} \xrightarrow{a} \mu'$ *iff* $\alpha^{-1}(s') \xrightarrow{a} \mu$ *and* $\mu'(u') = \mu(\alpha^{-1}(u'))$ *for all* $u' \in S'_2$*,*
2. $s' \xrightarrow{a} \iota_{\overline{s'}}$ *iff* $\alpha^{-1}(s') \in \text{Supp}(\mu)$ *for some* $u \in S$ *such that* $(u, a, \mu) \in \Delta$ *in* $\mathcal{M}$.

In the sequel, α_{PA} *denotes an* embedding *function for* PA.

Example 1. Let $\mathcal{G} = \alpha_{\mathrm{PA}}(\mathcal{M})$ (see Fig. 2) with $S_2 = \{t_0, \ldots, t_6\}$ and $S_1 = \{v_0, \ldots, v_6\}$, $\alpha_{\mathrm{PA}}^{-1}(t_i) = s_i$, and $\overline{t_i} = v_i$, for $i = 0$ to 6. For convenience, the s_i states are depicted inside the corresponding states v_i and t_i. We have e.g., $v_2 \xrightarrow{b} \mu'$ with $\mu'(t_1) = \frac{7}{10}$ and $\mu'(t_3) = \frac{3}{10}$ and $t_1 \xrightarrow{b} v_1$ and $t_3 \xrightarrow{b} v_3$, as in PA $\mathcal{M}$ we have $s_2 \xrightarrow{b} \mu$ with $\mu(s_1) = \frac{7}{10}$ and $\mu(s_3) = \frac{3}{10}$.

Paths. If $|\mathrm{Act}(s)| > 1$ for state s, a non-deterministic choice among the enabled actions in s occurs. A path in a PGA represents a particular resolution of non-determinism by players 1 and 2 at each state, as well as a resolution of the probabilistic choices. Formally, a path from $s_{1_0} \in S_1$ is given as: $\pi = s_{1_0} \xrightarrow{a_{1_0}, \mu_{1_0}} s_{2_0} \xrightarrow{a_{2_0}, \mu_{2_0}} s_{1_1} \ldots$ where $s_{i_k} \in S_i$, $a_{i_k} \in \mathrm{Act}(s_{i_k})$, $\mu_{i_k} \in \mathrm{Dist}(S_i)$, $\mu_{1_k}(s_{2_k}) > 0$ and $\mu_{2_k}(s_{1_{k+1}}) > 0$ for all $i \in \{1, 2\}$ and $k \geq 0$; if $k < k'$ for some $k' \in \mathbb{N}^+$, then π is called *finite* path, otherwise *infinite* path. For a finite path π_{fin}, let $\mathrm{last}_i(\pi_{\mathrm{fin}})$ denote the last state in S_i for $i \in \{1, 2\}$ in π_{fin}. Let $\mathrm{Path}_{\mathrm{fin}}(\mathcal{G})$ and $\mathrm{Path}_{\mathrm{inf}}(\mathcal{G})$ denote the set of finite and infinite paths in a PGA $\mathcal{G}$ respectively, and $\mathrm{Paths}(\mathcal{G}) = \mathrm{Path}_{\mathrm{fin}}(\mathcal{G}) \cup \mathrm{Path}_{\mathrm{inf}}(\mathcal{G})$.

Schedulers. In order to analyse reachability properties on $\mathcal{G}$, we resolve non-determinism at all game states by means of a *scheduler* (also known as *policy, strategy or adversary*). Let κ_i be the scheduler for S_i, $i \in \{1, 2\}$. We consider *deterministic memoryless* (DM) schedulers as they suffice for reachability probabilities on PGA [11]. DM-schedulers select an action-distribution pair only on the basis of the current state. More specifically, for bit x, a *deterministic* scheduler $\kappa_{(1+x)}$ maps a finite path π_{fin} to a pair in $\mathrm{Act}(last_{(1+x)}(\pi_{\mathrm{fin}})) \times \mathrm{Dist}(S_{(2-x)})$; and a *memoryless* scheduler $\kappa_{(1+x)}$ assures that for finite paths π_{fin} and π'_{fin}, $last_{(1+x)}(\pi_{\mathrm{fin}}) = last_{(1+x)}(\pi'_{\mathrm{fin}})$ implies $\kappa_{(1+x)}(\pi_{\mathrm{fin}}) = \kappa_{(1+x)}(\pi'_{\mathrm{fin}})$.

A path π under a pair of DM-schedulers (κ_1, κ_2) is of the form $\pi = s_{1_0} \xrightarrow{a_{1_0}, \mu_{1_0}} s_{2_0} \xrightarrow{a_{2_0}, \mu_{2_0}} s_{1_1} \ldots$ where $\kappa_i(s_{i_k}) = (a_{i_k}, \mu_{i_k})$ for $i \in \{1, 2\}$ and $k \geq 0$. Let $\mathrm{Paths}_{\kappa_2}^{\kappa_1}(\mathcal{G})$ be the set of paths of PGA $\mathcal{G}$ under DM-schedulers (κ_1, κ_2). The DM-schedulers (κ_1, κ_2) on PGA $\mathcal{G}$ induce a countably infinite Markov chain. This allows us to construct a measurable space $(\mathrm{Paths}_{\kappa_2}^{\kappa_1}(\mathcal{G}), \mathcal{F}_{\kappa_2}^{\kappa_1}, \mathrm{Pr}_{\kappa_2}^{\kappa_1})$ over the (infinite) paths of $\mathcal{G}$ under (κ_1, κ_2). The problem of computing reachability probabilities on $\mathcal{G}$ reduces to a *stochastic shortest path problem* [13,14] (for details, see Section 7). As reachability analysis is performed on closed versions of systems, we define a function that yields closed versions of PGA.

Definition 5. *For* PGA $\mathcal{G}$, *let* PGA $\tau(\mathcal{G}) = \mathcal{G}' = (S', \{S'_1, S'_2\}, A', \Delta', s'_0)$ *with* $S' = S$, $s'_0 = s_0$, $A' = \{\tau\}$ *and* $\Delta' = \{(s, \tau, \mu) \mid (s, a, \mu) \in \Delta\}$.

Combined Hyper Transitions. We adapt hyper and combined transitions – convex combinations of sets of transitions – for PA [1,15] to PGA which are later on used in definitions.

Definition 6. *For* PGA $\mathcal{G}$ *with* $s \in S$ *and* $\mu \in \mathrm{Dist}(S)$, *we write:*

- $\mu \xrightarrow{a} \eta$ *is a* hyper transition *iff* $\eta = \bigoplus\{\mu(s) \cdot \rho \mid \exists s \in \mathrm{Supp}(\mu) : s \xrightarrow{a} \rho\}$. *Let* $\Delta(\mu, a) = \{\eta \mid \exists \eta \in \mathrm{Dist}(S) : \mu \xrightarrow{a} \eta\}$.

- $s \xrightarrow{a}_c \eta$ *is a* combined transition *iff there is a finite indexed set* $\{(c_i, \eta_i)\}_{i \in I}$ *such that* $s \xrightarrow{a} \eta_i$ *and* $c_i \in \mathbb{R}_{\geq 0}$ *for all* $i \in I$, $\sum_{i \in I} c_i = 1$ *and* $\eta = \bigoplus_{i \in I} c_i \cdot \eta_i$.
- $\mu \xrightarrow{a}_c \eta$ *is a* combined hyper transition *iff* $\eta = \bigoplus\{\mu(s) \cdot \rho \mid \exists s \in \mathrm{Supp}(\mu) : s \xrightarrow{a}_c \rho\}$.

Simulation. The notion of simulation for probabilistic processes [10] is a pre-order on the state space requiring that whenever state u simulates state s, then u can mimic the stepwise behaviour of s but may have more behaviour. This notion can be lifted to distributions over states using weight functions [10]:

Definition 7. *Let S be a finite, non-empty set of states, and let $\mu, \mu' \in \mathrm{Dist}(S)$. For $R \subseteq S \times S$, μ is* simulated *by μ' w.r.t. R, denoted $\mu R \mu'$, iff there exists a weight function $\delta : S \times S \to [0,1]$ such that for all $u, v \in S$:(1) $\delta(u,v) > 0 \Rightarrow uRv$, (2) $\sum_{s \in S} \delta(u,s) = \mu(u)$ and (3) $\sum_{s \in S} \delta(s,v) = \mu'(v)$.*

We now recall Segala's probabilistic simulation relation [1] for PA.

Definition 8. *$R \subseteq S \times S$ is a* simulation *relation for* PA $\mathcal{M}$ *iff for every sRs', $s \xrightarrow{a} \mu$ implies $s' \xrightarrow{a}_c \mu'$ with $\mu R \mu'$. Let $\prec_{\mathrm{pa}}$ be the largest simulation relation.*

We can lift $\prec_{\mathrm{pa}}$ to PA in the usual way: $\mathcal{M} \prec_{\mathrm{pa}} \mathcal{M}'$ for PA $\mathcal{M}$ and $\mathcal{M}'$, with initial states s_0 and s_0', iff $s_0 \prec_{\mathrm{pa}} s_0'$ in the disjoint union of $\mathcal{M}$ and $\mathcal{M}'$. In the sequel, we will adopt this convention to all simulation relations.

3 Simulation Relations on Stochastic Games

Simulation relations are typically defined over the states of models; however, in the probabilistic setting, coarser relations have been considered over the distributions over states [1,16,17]. We define simulation relations for PGA, that are state-based as well as distribution-based, and prove them to be pre-orders. Later on, these relations form the basis to compare a PGA with its abstraction.

Definition 9. *$R \subseteq \bigcup_{j \in \{1,2\}} S_j \times S_j$ is a* state-based simulation (SBS) *relation on* PGA $\mathcal{G}$ *iff for every sRs', $s \xrightarrow{a} \mu$ implies $s' \xrightarrow{a}_c \mu'$ with $\mu R \mu'$. Let $\prec_{\mathrm{sb}}$ be the largest* SBS *relation.*

This asserts that for sRs', an a-transition from s implies a combined a-transition from s' such that the resulting distributions are related as by Def. 7. It is not difficult to show that $\prec_{\mathrm{sb}}$ is a preorder. Moreover, $\prec_{\mathrm{sb}} = \prec_{\mathrm{pa}}$ for PA.

Definition 10. *$R \subseteq \bigcup_{j \in \{1,2\}} \mathrm{Dist}(S_j) \times \mathrm{Dist}(S_j)$ is a* distribution-based simulation (DBS) *relation on* PGA $\mathcal{G}$ *iff for every $\mu R \mu'$, (1) $\mu = \bigoplus_{s' \in \mathrm{Supp}(\mu')} \mu_{s'} : \mu'(s') = |\mu_{s'}|$ and $\mu_{s'\downarrow} R \iota_{s'}$, (2) $\mu \xrightarrow{a} \rho$ implies $\mu' \xrightarrow{a}_c \rho'$ such that $|\rho| \leq |\rho'|$ and $\rho_{\downarrow} R \rho'_{\downarrow}$. Let $\prec_{\mathrm{db}}$ be the largest* DBS *relation. We write $s \prec_{\mathrm{db}} s'$ iff $\iota_s \prec_{\mathrm{db}} \iota_{s'}$.*

By constraint (1), μ splits into sub-distributions as per the support of μ', i.e., for every $s' \in \mathrm{Supp}(\mu')$, there exists a sub-distribution μ_s' of μ such that the conditional distribution of μ_s' is related to $\iota_{s'}$. By constraint (2), an a-transition from μ to *some* ρ implies a combined a-transition from μ' to ρ' such that the mass of ρ' is at least that of ρ and their conditional distributions are related.

Example 2. In Fig. 3, $\mu = [\![0.3s_3, 0.3s_4, 0.4s_5]\!] \prec_{\text{db}} \iota_{s_0}$ as $R = \{(\iota_{s_1}, \iota_{s_1}), (\iota_{s_2}, \iota_{s_2}), ([\![0.3s_3, 0.3s_4, 0.4s_5]\!], \iota_{s_0}), ([\![0.5s_1, 0.5s_2]\!], [\![0.5s_1, 0.5s_2]\!])\}$ is a DBS relation. Let us check the conditions of Def. 10 for μ and ι_{s_0}. The constraint (1) trivially holds for μ and ι_{s_0}. For the a-transition from μ to $\rho = [\![0.3s_1, 0.3s_2]\!]$, there is an a-transition from ι_{s_0} to $\rho' = [\![0.5s_1, 0.5s_2]\!]$ such that $|\rho| \leq |\rho'|$ and $\rho_{\downarrow} R \rho'$. The same holds for the b-transitions from μ and ι_{s_0}, thus fulfilling constraint (2). Note that no SBS relation exists associating s_0 with any other state in Fig. 3.

4 Alternating Simulation Relations

To compare two-player stochastic games with competitive objectives (e.g., if player 1 maximises the probability to reach a certain goal state, her opponent (player 2) will try to minimize this quantity), we use *alternating* simulation relations. Our state-based alternating simulation relations are inspired by the notions of alternating simulation [18] and strong probabilistic game simulation [19].

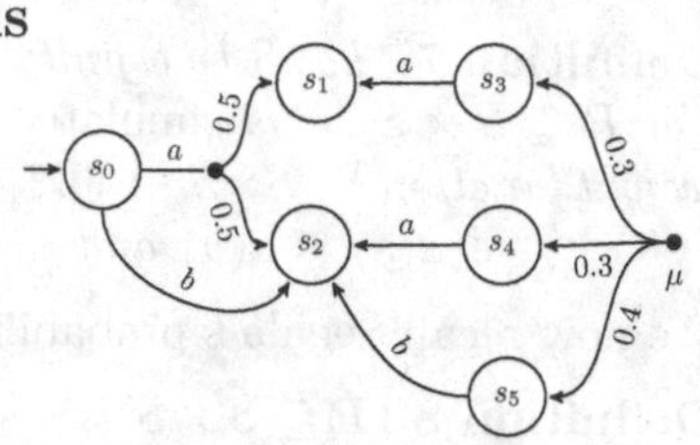

Fig. 3. $[\![0.5s_3, 0.5s_4]\!] \prec_{\text{db}} \iota_{s_0}$ but $s_i \not\prec_{\text{sb}} s_0$ for $i \in \{3, 4\}$

Definition 11. *$R \subseteq \bigcup_{j\in\{1,2\}} S_j \times S_j$ is a* state-based alternating simulation (SBAS) *relation for* PGA $\mathcal{G}$ *iff for every sRs' the following holds: (1) if $s, s' \in S_1$, then $s' \xrightarrow{a} \mu'$ implies $s \xrightarrow{a}_{\text{c}} \mu$ such that $\mu R \mu'$, (2) if $s, s' \in S_2$, then $s \xrightarrow{a} \mu$ implies $s' \xrightarrow{a}_{\text{c}} \mu'$ such that $\mu R \mu'$. Let $\preccurlyeq_{\text{sb}}$ be the largest* SBAS *relation.*

Intuitively, in case of player 1 states, the behaviour of s' is simulated by that of s; whereas in case of player 2 states, it is the other way round. The first constraint asserts that if $s, s' \in S_1$, then an a-transition from s' implies a combined a-transition from s and the resulting distributions are related with each other by Def. 7. The second constraint asserts that if $s, s' \in S_2$, the similar conditions as in (1) hold for every transition from s. It is easy to show that $\preccurlyeq_{\text{sb}}$ is a preorder.

Remark 1. The strong probabilistic game simulation relation in [19] [Def. 6.10] is obtained by merging Def. 9 and 11 and lifting them to player 2 states.

Definition 12. *$R \subseteq \bigcup_{j\in\{1,2\}} \text{Dist}(S_j) \times \text{Dist}(S_j)$ is a* distribution-based alternating simulation (DBAS) *relation for* PGA $\mathcal{G}$ *iff for every $\mu R \mu'$: (1) $\mu = \bigoplus_{s'\in\text{Supp}(\mu')} \mu_{s'} : |\mu_{s'}| = \mu'(s')$ and $\mu_{s'\downarrow} R \iota_{s'}$, (2) if $\mu, \mu' \in \text{Dist}(S_1)$, $\mu' \xrightarrow{a} \rho'$ implies $\mu \xrightarrow{a}_{\text{c}} \rho$ such that $|\rho| \geq |\rho'|$ and $\rho_{\downarrow} R \rho'_{\downarrow}$, (3) if $\mu, \mu' \in \text{Dist}(S_2)$, $\mu \xrightarrow{a} \rho$ implies $\mu' \xrightarrow{a}_{\text{c}} \rho'$ such that $|\rho| \leq |\rho'|$ and $\rho_{\downarrow} R \rho'_{\downarrow}$. Let $\preccurlyeq_{\text{db}}$ be the largest* DBAS *relation. We write $s \preccurlyeq_{\text{db}} s'$ iff $\iota_s \preccurlyeq_{\text{db}} \iota_{s'}$.*

The constraint (1) is the same as in Def. 10. By constraint (2), if $\mu, \mu' \in \text{Dist}(S_1)$, then an a-transition from μ' to ρ' implies a combined a-transition from μ to ρ such that the mass of ρ is at least the mass of ρ' and the conditional distribution of ρ is in relation R with that of ρ'. And by constraint (3), if $\mu, \mu' \in \text{Dist}(S_2)$, the similar conditions as in (2) hold for every transition from μ. Note that if the state space is not partitioned (as for PA), then simulation relations coincide with alternating simulation relations:

Proposition 1. $\prec_x = \preccurlyeq_x^{-1}$ *for* PA, *where* $x \in \{\text{sb}, \text{db}\}$.

Theorem 1. $\prec_{\text{db}}$ *and* $\preccurlyeq_{\text{db}}$ *are preorders.*

Although a state-based (alternating) simulation relation can be lifted from states to distributions over states (by Def. 7), an example can be constructed showing state-based (lifted to distributions over states) and distribution-based (alternating) simulation relations are not ordered in general but for closed PGA.

Proposition 2. $\prec_{\text{sb}}/\preccurlyeq_{\text{sb}}$ *and* $\prec_{\text{db}}/\preccurlyeq_{\text{db}}$ *are incomparable for* PGA; *and* $\prec_{\text{sb}}/\preccurlyeq_{\text{sb}} \subseteq \prec_{\text{db}}/\preccurlyeq_{\text{db}}$ *for closed* PGA.

At the end of this section, we highlight that if PGA are in a state-based or a distribution-based relation, their closed versions are also in that relation.

5 Game Abstraction

In this section, we show that PGA can act as appropriate abstract models for PA. We do so by considering abstractions of PGA that are embeddings of PA. Let $\mathcal{G}$ be a PGA with $S = \{S_1, S_2\}$. Intuitively, the state space S_2 of $\mathcal{G}$ is partitioned and each partition is represented by a single state in the abstract state space S_2'. This step induces a partition of S_1. We propose two different techniques for the partition of S_1: (a) S_1 states having similar behaviour under the player 2 partition S_2' are grouped (*state-based abstraction*); (b) the (sub-)distributions (over S_1) that have similar behaviour are grouped (*distribution-based abstraction*). In the sequel, we show that the latter technique is more precise as well as concise than the former one.

Let (α, γ) be an abstraction-concretization pair such that $\alpha : S \to S'$ is a surjection and $\gamma : S' \to 2^S$ is the corresponding concretization function. That is, $\alpha(s)$ is the abstract state of s whereas $\gamma(s')$ is the set of concrete states abstracted by s'. The abstraction of distribution μ is given as $\alpha(\mu)(s') = \mu(\gamma(s'))$. The functions α and γ are lifted to sets of states or sets of distributions in a point-wise manner.

Definition 13. *For* PGA $\mathcal{G}$, *the* state-based abstraction *function* $\alpha : S \to S'$ *induces the* PGA $\alpha(\mathcal{G}) = \mathcal{G}' = (S', \{S_1', S_2'\}, A', \Delta', s_0')$ *where* $A' = A$; $S_i' = \alpha(S_i)$ *for* $i \in \{1, 2\}$; $\forall u', v' \in S_1' : \Lambda'(u') = \Lambda'(v')$ *implies* $u' = v'$; $s_0' = \alpha(s_0)$ *and for every* $s' \in S'$:

1. *if* $s' \in S_1'$, *then: (a)* $s' \xrightarrow{a} \mu'$ *iff* $\forall s \in \gamma(s') : s \xrightarrow{a} \mu$ *such that* $\alpha(\mu) = \mu'$, *(b)* $\exists s \in \gamma(s') : s \xrightarrow{a} \mu$ *implies* $s' \xrightarrow{a} \mu'$ *such that* $\alpha(\mu) = \mu'$,
2. *if* $s' \in S_2'$, *then: (a)* $s' \xrightarrow{a} \mu'$ *implies* $\exists s \in \gamma(s') : s \xrightarrow{a} \mu$ *such that* $\alpha(\mu) = \mu'$, *(b)* $\exists s \in \gamma(s') : s \xrightarrow{a} \mu$ *implies* $s' \xrightarrow{a}_{\text{c}} \mu'$ *such that* $\alpha(\mu) = \mu'$.

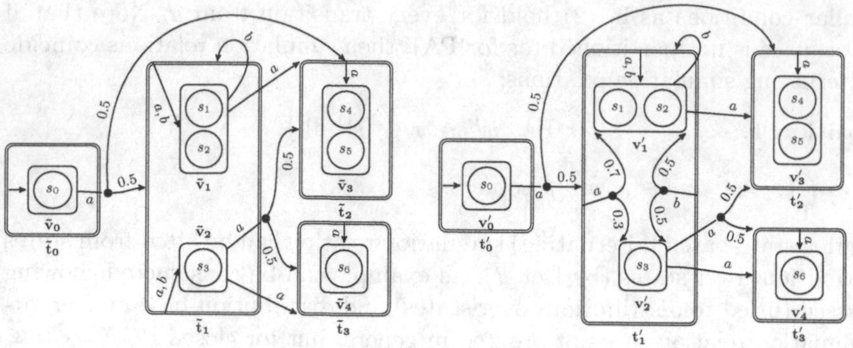

Fig. 4. For game $\mathcal{G}$ (Fig. 2 right), $\tilde{\mathcal{G}} = \alpha_{\text{sb}}(\mathcal{G})$ (left) and $\mathcal{G}' = \alpha_{\text{db}}(\mathcal{G})$ (right)

In the sequel, $(\alpha_{\text{sb}}, \gamma_{\text{sb}})$ *denotes a pair of* state-based abstraction-concretization *functions for* PGA.

By constraint (1) all player 1 states in the concrete model whose transitions become similar after abstraction — that can be found by considering their ordinary transitions instead of combined transitions — are aggregated; thus every state in S'_1 has a unique set of transitions enabled from it. Besides, (2) transitions of player 2 abstract states are derived from their concrete states, whose convex combination simulate the (abstract) transitions of concrete states.

Example 3. Let $\tilde{\mathcal{G}} = \alpha_{\text{sb}}(\mathcal{G})$ (Fig. 4 left) where $\mathcal{G}$ is the PGA of Fig. 2 (see page 579) with $\gamma_{\text{sb}}(\tilde{t_0}) = \{t_0\}$, $\gamma_{\text{sb}}(\tilde{t_1}) = \{t_1, t_2, t_3\}$, $\gamma_{\text{sb}}(\tilde{t_2}) = \{t_4, t_5\}$ and $\gamma_{\text{sb}}(\tilde{t_3}) = \{t_6\}$. Consider v_1, v_2 and v_4, v_5 in S_1; the transitions of v_1 and v_2 are the same after abstraction; therefore, they are merged into $\tilde{v}_1$. The same applies to v_4 and v_5. Now consider the transitions from $\tilde{t}_1$; for each concrete transition from t_1, t_2 and t_3, there is a corresponding abstract transition from $\tilde{t}_1$; thus, $\tilde{t}_1$ simulates (according to Def. 9) t_1, t_2 and t_3 (after abstraction).

Theorem 2. *For* PGA $\mathcal{G}$, $\mathcal{G} \prec_{\text{sb}} \alpha_{\text{sb}}(\mathcal{G})$ *and* $\mathcal{G} \preccurlyeq_{\text{sb}} \alpha_{\text{sb}}(\mathcal{G})$.

Definition 14. *For* PGA $\mathcal{G}$, *the* distribution-based abstraction *function* $\alpha : S \to S'$ *induces the* PGA $\alpha(\mathcal{G}) = \mathcal{G}' = (S', \{S'_1, S'_2\}, A', \Delta', s'_0)$ *where* $A' = A$; $S'_i = \alpha(S_i)$ *for* $i \in \{1, 2\}$; $\forall u', v' \in S'_1$: $\Delta'(u') = \Delta'(v')$ *implies* $u' = v'$; $s'_0 = \alpha(s_0)$ *and for all* $\mu' \in \text{Dist}(S')$:

1. $\forall \mu \in \gamma(\mu') : \mu = \bigoplus_{s' \in \text{Supp}(\mu')} \mu_{s'} : \mu'(s') = |\mu_{s'}| \wedge \alpha(\mu_{s'})_{\downarrow} = \iota_{s'}$,
2. *if* $\mu' \in \text{Dist}(S'_1)$, *then:*
   (a) $\mu' \xrightarrow{a} \eta'$ *iff* $\forall \mu \in \gamma(\mu') : \mu \xrightarrow{a} \eta$ *such that* $|\eta| = |\eta'|$ *and* $\alpha(\eta)_{\downarrow} = \eta'_{\downarrow}$,
   (b) $\exists \mu \in \gamma(\mu') : \mu \xrightarrow{a} \eta$ *implies* $\mu' \xrightarrow{a} \eta'$ *such that* $|\eta| = |\eta'|$ *and* $\alpha(\eta)_{\downarrow} = \eta'_{\downarrow}$,
3. *if* $\mu' \in \text{Dist}(S'_2)$, *then:*
   (a) $\mu' \xrightarrow{a} \eta'$ *implies* $\exists \mu \in \gamma(\mu') : \mu \xrightarrow{a} \eta$ *such that* $|\eta| \leq |\eta'|$ *and* $\alpha(\eta)_{\downarrow} = \eta'_{\downarrow}$,
   (b) $\exists \mu \in \gamma(\mu') : \mu \xrightarrow{a} \eta$ *implies* $\mu' \xrightarrow{a}_{\text{c}} \eta'$ *such that* $|\eta| \leq |\eta'|$ *and* $\alpha(\eta)_{\downarrow} = \eta'_{\downarrow}$.

In the sequel, $(\alpha_{\mathrm{db}}, \gamma_{\mathrm{db}})$ *denotes a pair of* distribution-based abstraction-concretization *functions for* PGA.

As in a state-based abstraction, all player 1 states in a distribution-based abstraction of a PGA have a unique set of transitions enabled from them. However, the distribution-based abstraction differs from the state-based one in several ways: (1) asserts the splitting of every concrete distribution μ of μ' into sub-distributions as per the support of μ', i.e., $\mu = \bigoplus_{s' \in \mathsf{Supp}(\mu')} \mu_{s'}$, and the conditional distribution of $\mu_{s'}$ is abstracted by $\iota_{s'}$. By (2a), when μ' is defined over S'_1, then μ' has an a-transition to *some* distribution η' if its every concrete distribution μ has an a-transition to some distribution η such that the masses of η and η' coincide and (the conditional distribution of) η is abstracted by (that of) η'; moreover, (2b) all transitions from μ are present (after abstraction) from μ'. In fact, all concrete distributions of μ' have similar behaviour after abstraction, that can be asserted by considering their ordinary transitions instead of combined transitions. By (3a), when μ' is defined over S'_2, then μ' has an a-transition to *some* distribution η' if a concrete distribution μ (of μ') has an a-transition to some distribution η such that the mass of η' is at least that of η and (the conditional distribution of) η is abstracted by (that of) η'. Moreover, (3b) the transitions of concrete distributions of μ' are simulated by the convex combination of transitions of μ'.

Example 4. Let $\mathcal{G}' = \alpha_{\mathrm{db}}(\mathcal{G})$ (Fig. 4 right) for $\mathcal{G}$ in Fig.2 with $\gamma_{\mathrm{db}}(t'_0) = \{t_0\}$, $\gamma_{\mathrm{db}}(t'_1) = \{t_1, t_2, t_3\}$, $\gamma_{\mathrm{db}}(t'_2) = \{t_4, t_5\}$ and $\gamma_{\mathrm{db}}(t'_3) = \{t_6\}$. As the abstract state space is the same as for the state-based abstraction in the previous example, the transitions of concrete states v_1 and v_2 are the same after abstraction; therefore, they are merged into v'_1. The same applies to v_4 and v_5. Now, consider the transition $v'_0 \xrightarrow{a} [\![0.5t'_1, 0.5t'_2]\!]$ and its corresponding concrete transition $v_0 \xrightarrow{a} [\![0.1t_1, 0.3t_2, 0.1t_3, 0.5t_4]\!]$; note that $[\![0.1t_1, 0.3t_2, 0.1t_3, 0.5t_4]\!]$ can be split into sub-distributions as per the support of $[\![0.5t'_1, 0.5t'_2]\!]$. Consider the abstract distribution $[\![0.5t'_1]\!]$ and its concrete distribution $[\![0.1t_1, 0.3t_2, 0.1t_3]\!]$; for $[\![0.1t_1, 0.3t_2, 0.1t_3]\!]_\downarrow \xrightarrow{a} [\![0.1v_1, 0.3v_2, 0.1v_3]\!]_\downarrow$, there is a $[\![0.5t'_1]\!]_\downarrow \xrightarrow{a}_{\mathrm{c}} [\![0.4v'_1, 0.1v'_2]\!]_\downarrow$ and $[\![0.1v_1, 0.3v_2, 0.1v_3]\!]_\downarrow \in \gamma_{\mathrm{db}}([\![0.4v'_1, 0.1v'_2]\!]_\downarrow)$; and for $[\![0.1t_1, 0.3t_2, 0.1t_3]\!]_\downarrow \xrightarrow{b} [\![0.1v_1, 0.1v_3]\!]_\downarrow$, there is a $[\![0.5t'_1]\!]_\downarrow \xrightarrow{b} [\![0.25v'_1, 0.25v'_2]\!]_\downarrow$ and $[\![0.1v_1, 0.1v_3]\!]_\downarrow \in \gamma_{\mathrm{db}}([\![0.25v'_1, 0.25v'_2]\!]_\downarrow)$. Now consider the b-transition from v'_1 to $\iota_{t'_1}$; we have two concrete b-transitions: from v_1 to ι_{t_1} and from v_2 to $[\![0.7t_1, 0.3t_3]\!]$. For $\iota_{t_1} \xrightarrow{b} \iota_{v_1}$, there is a $\iota_{t'_1} \xrightarrow{b} \iota_{v'_1}$; and for $[\![0.7t_1, 0.3t_3]\!] \xrightarrow{b} [\![0.7v_1, 0.3v_3]\!]$, there is a $\iota_{t'_1} \xrightarrow{b}_{\mathrm{c}} [\![0.7v'_1, 0.3v'_2]\!]$. Same is the case with a-transitions from ι_{t_1} and $[\![0.7t_1, 0.3t_3]\!]$.

In the previous example, only those states in S_1 whose transitions became the same after abstraction were aggregated. The next example illustrates that S_1 states having different transitions after abstraction can also be aggregated.

Example 5. For PA $\mathcal{M}$ (Fig. 1), let $\mathcal{G} = \alpha_{\mathrm{PA}}(\mathcal{M})$ be its induced game. Let $\tilde{\mathcal{G}} = \alpha_{\mathrm{sb}}(\mathcal{G})$ (Fig. 5 left) be the state-based abstract model of $\mathcal{G}$ with $\gamma_{\mathrm{sb}}(\tilde{t}_0) = \{t_0\}$,

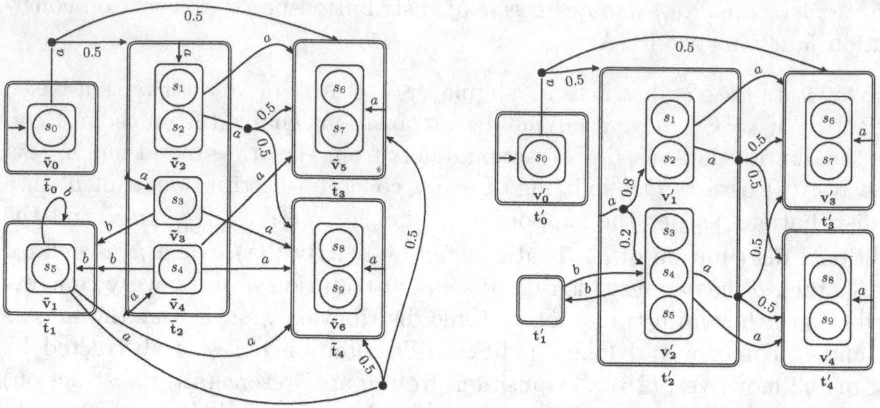

Fig. 5. For PA $\mathcal{M}$ (Fig. 1), $\tilde{\mathcal{G}} = \alpha_{\text{sb}}(\alpha_{\text{PA}}(\mathcal{M}))$ (left) and $\mathcal{G}' = \alpha_{\text{db}}(\alpha_{\text{PA}}(\mathcal{M}))$ (right). Considering each probabilistic transition as two transitions, $|\tilde{\Delta}| = 20$ and $|\tilde{S}_1| = 7$; whereas $|\Delta'| = 14$ and $|S'_1| = 5$.

$\gamma_{\text{sb}}(\tilde{t}_1) = \{t_5\}$, $\gamma_{\text{sb}}(\tilde{t}_2) = \{t_1, t_2, t_3, t_4\}$, $\gamma_{\text{sb}}(\tilde{t}_3) = \{t_6, t_7\}$ and $\gamma_{\text{sb}}(\tilde{t}_4) = \{t_8, t_9\}$. Let $\mathcal{G}' = \alpha_{\text{db}}(\mathcal{G})$ (Fig.5 right) be the distribution-based abstract model of $\mathcal{G}$ with the same partition as above. Consider the distribution $[\![0.8v'_1, 0.2v'_2]\!]$ such that $[\![0.1v_1, 0.3v_2]\!]_\downarrow$ and $[\![0.05v_3, 0.05v_4]\!]_\downarrow$ are the corresponding distributions for v'_1 and v'_2 respectively. Note that $v'_2 \xrightarrow{a} \iota_{t'_4}, [\![0.5t'_3, 0.5t'_4]\!]$ iff $[\![0.05v_3, 0.05v_4]\!]_\downarrow \xrightarrow{a} \iota_{t_9}, [\![0.5t_7, 0.5t_9]\!]$. Similarly, $v'_2 \xrightarrow{b} \iota_{t'_1}$ iff $[\![0.05v_3, 0.05v_4]\!]_\downarrow \xrightarrow{b} \iota_{t_5}$. Moreover, the concrete distribution ι_{v_5} has the same behaviour as $[\![0.05v_3, 0.05v_4]\!]_\downarrow$, therefore, v_3, v_4 and v_5 are merged into v'_2. This example shows that distribution-based abstraction induces more concise models than state-based abstraction. Note that for PGA $\mathcal{G}$, $R = \{(s, \alpha_{\text{db}}(s) \mid s \in S\}$ is not an SBS relation.

Theorem 3. *For* PGA $\mathcal{G}$, $\mathcal{G} \prec_{\text{db}} \alpha_{\text{db}}(\mathcal{G})$ *and* $\mathcal{G} \precsim_{\text{db}} \alpha_{\text{db}}(\mathcal{G})$.

Both Th. 2 and 3 are of importance when showing (in Section 7) that abstraction provides upper- and lower-bounds on extremal reachability probabilities in PGA (and thus PA).

Distribution- vs. State-Based Abstraction. Like for simulation relations, state-based abstraction is not a special case of distribution-based abstraction. We observe that for every possible partition of state space, we can have a state-based abstract model of PGA, but not a distribution-based abstract model; however, for closed versions of PGA — PGA having $A = \{\tau\}$ —, we can have state-based as well as distribution-based abstract models.

Proposition 3. $\alpha_{\text{db}}(\tau(\mathcal{G}))$ *is well-defined for* PGA $\mathcal{G}$.

By the above proposition, we mean that for every partition of state space of a closed PGA, we can construct a distribution-based abstract model, which is not the case with other PGA (not closed). However, for some PGA (not closed) we can have partitions of state space that can define distribution-based abstract models by aggregating states. Moreover, although we do not aggregate any states when the partition is S, α_{db} is defined for this partition. In the sequel, we assume that $\alpha_{\text{db}}(\mathcal{G})$ is defined for PGA $\mathcal{G}$.

Now we prove that distribution-based abstraction is more precise than state-based abstraction. In fact, when two abstract models, obtained by a state-based and a distribution-based abstraction, have the same state space; then the latter one is at least as precise as the former one. Formally,

Theorem 4. *For* PGA $\mathcal{G}$, $\alpha_{\text{sb}}(S) = \alpha_{\text{db}}(S)$ *implies* $\alpha_{\text{db}}(\mathcal{G}) \prec \alpha_{\text{sb}}(\mathcal{G})$ *where* $\prec \in \{\prec_{\text{sb}}, \preccurlyeq_{\text{sb}}\}$.

6 Composition

We define a composition operator for a class of PGA that can act as abstract models of PA. The operator is defined in a TCSP-like manner, i.e., it is parametrized by a set of actions that need to be performed simultaneously by both games; other actions occur autonomously. For distributions μ and μ', let the point-wise product $\mu \| \mu' : S \times S \to [0,1]$ be given as: $\mu\|\mu'(s,s') = \mu(s) \cdot \mu'(s')$ for $s, s' \in S$.

Definition 15. *The* parallel composition *of* PGA $\mathcal{G}$ *and* $\mathcal{G}'$ *w.r.t. synchronization set* $\bar{A} \subseteq (A \cap A') \backslash \{\tau\}$ *is given as:* $\mathcal{G} \|_{\bar{A}} \mathcal{G}' = (S \times S', \{S_1 \times S'_1, S \times S' \backslash S_1 \times S'_1\}, A \cup A', \bar{\Delta}, (s_0, s'_0))$, *where for all* $a \in A \cup A'$ *and* $(s,s') \in S \times S'$, $(s,s') \xrightarrow{a}_{\text{c}} \mu\|\mu'$ *iff one of the following holds:*

1. *if* $(s,s') \in S_1 \times S'_1$, *then (i)* $a \in \bar{A}$, $s \xrightarrow{a}_{\text{c}} \mu$ *and* $s' \xrightarrow{a}_{\text{c}} \mu'$, *or (ii)* $a \in A$, $s \xrightarrow{a}_{\text{c}} \mu$ *and* $\iota_{s'} = \mu'$, *or (iii)* $a \in A'$, $s' \xrightarrow{a}_{\text{c}} \mu'$ *and* $\iota_s = \mu$,
2. *if* $(s,s') \in S_2 \times S'_2$, *then* $a \in \bar{A}$, $s \xrightarrow{a}_{\text{c}} \mu$ *and* $s' \xrightarrow{a}_{\text{c}} \mu'$,
3. *else, (i)* $s \in S_2$, $s \xrightarrow{a}_{\text{c}} \mu$ *and* $\iota_{s'} = \mu'$, *or (ii)* $s' \in S'_2$, $s' \xrightarrow{a}_{\text{c}} \mu'$ *and* $\iota_s = \mu$.

Note that the state space of our composite game is disjointly dividable based on the actions which are enabled. Although, we allow composition of $S_1(S_2)$ states with that of $S'_2(S'_1)$ states, but only player 2 can make a move in such a state. (1) asserts that states in $S_1 \times S'_1$ can either synchronize with each other or act independently. Note that a state in $S_2 \times S'_2$ is only reached by a synchronizing action performed by players of type 1 in some $S_1 \times S'_1$ state; and (2) asserts that the next state is reached only by some synchronizing action. (3) tells that for a state in $S_{(1+x)} \times S'_{(2-x)}$, where x is a bit, no synchronization occurs and only player 2 can make a move independently. Note that such a state can only be reached by a non-synchronizing action.

Theorem 5. *For any set* $\bar{A}$ *and* $x \in \{\text{sb}, \text{db}\}$, $\prec_x$ *and* $\preccurlyeq_x$ *are pre-congruences w.r.t.* $\|_{\bar{A}}$.

Like for APA [4], our state-based and distribution-based abstractions for PGA are compositional. Intuitively, the composite PGA may be exponentially larger in size as compared to the composing ones. This problem could be avoided by applying abstraction prior to composition as illustrated by the following theorem.

Theorem 6. *For* PGA $\mathcal{G}$ *and* $\mathcal{G}'$, *synchronization set* $\bar{A}$ *and abstraction functions* α_x, α'_x; $\alpha_x(\mathcal{G})\,||_{\bar{A}}\,\alpha'_x(\mathcal{G}') = (\alpha_x \times \alpha'_x)(\mathcal{G}||_{\bar{A}}\mathcal{G}')$ *up to isomorphism, where* $x \in \{\text{sb}, \text{db}\}$ *and* $\alpha_x \times \alpha'_x$ *is defined as* $(\alpha_x \times \alpha'_x)(s, s') = (\alpha_x(s), \alpha'_x(s'))$.

7 Preservation of Reachability Probabilities

This section presents how optimal (i.e., maximal and minimal) reachability probabilities are preserved under abstraction. We first define some notations and definitions. Let $\Pr^{\kappa_1}_{\kappa_2}(T)$ be the probability of the set of paths from the initial state s_0 that reach some state in $T \subseteq S$ under schedulers (κ_1, κ_2) for PGA $\mathcal{G}$.

Definition 16. [11] *For* PGA $\mathcal{G}$, *the* optimal probabilities *of reaching* $T \subseteq S$ *for players 1 and 2 are defined as:* $\sup_{\kappa_1}\inf_{\kappa_2} \Pr^{\kappa_1}_{\kappa_2}(T)$ *and* $\inf_{\kappa_1}\sup_{\kappa_2} \Pr^{\kappa_1}_{\kappa_2}(T)$.

Intuitively, the reachability probability to a set T of target states is optimal for player 1 under scheduler κ iff for every scheduler κ_2 of player 2, $\inf_{\kappa_2} \Pr^{\kappa}_{\kappa_2}(T) = \sup_{\kappa_1}\inf_{\kappa_2} \Pr^{\kappa_1}_{\kappa_2}(T)$. Similarly, we can define optimal reachability probability for player 2. For PGA $\mathcal{G}$ and $T \subseteq S$, we write:

- $\max^{\blacktriangledown}(T) = \sup_{\kappa_1}\inf_{\kappa_2} \Pr^{\kappa_1}_{\kappa_2}(T)$ and $\max^{\blacktriangle}(T) = \sup_{\kappa_1}\sup_{\kappa_2} \Pr^{\kappa_1}_{\kappa_2}(T)$
- $\min^{\blacktriangledown}(T) = \inf_{\kappa_1}\inf_{\kappa_2} \Pr^{\kappa_1}_{\kappa_2}(T)$ and $\min^{\blacktriangle}(T) = \inf_{\kappa_1}\sup_{\kappa_2} \Pr^{\kappa_1}_{\kappa_2}(T)$.

Note that the values $\max^{\blacktriangledown}(T)$ and $\min^{\blacktriangle}(T)$ are the optimal reachability probabilities for players 1 and 2 respectively, which can be achieved by DM-schedulers [11]. The values $\max^{\blacktriangle}(T)$ and $\min^{\blacktriangledown}(T)$ – for which both players collaborate with each other – can be obtained similarly. For games with finite state spaces these values can be computed through value iteration [13,14] or by linear programming.

Let $w : S \to [0, 1]$ be a probability valuation function mapping a state s to the probability of reaching target states $T \subseteq S$ from s. The probability valuation functions $W = \{w \mid w : S \to [0, 1]\}$ form a complete lattice $(W, \leq, \bot, \top)$ with order $\leq$, bottom element $\bot \in W$ and top element $\top \in W$. We write $w \leq w'$ iff $w(s) \leq w'(s)$ for $s \in S$. $\bot(s) = 0$ and $\top(s) = 1$ for $s \in S$. Moreover, w can be lifted from states to distributions over states as $w(\mu) = \sum_{s \in S} \mu(s) \cdot w(s)$ for $\mu \in \text{Dist}(S)$.

Definition 17. *Let* PGA $\tau(\mathcal{G})$ *and* $T \subseteq S$. *For reachability goals* $\mathbf{1}, \mathbf{2} \in \{\min, \max\}$ *for players 1, 2 respectively, the* probability valuation transformer $\text{Prt}^{\mathbf{1}}_{\mathbf{2}} : W \to W$ *is defined for* $w \in W$, $s \in S$ *and* $n \in \mathbb{N}$ *as:*

$$(\text{Prt}^{\mathbf{1}}_{\mathbf{2}})^n(w)(s) = \begin{cases} 1 & s \in T,\, n \geq 0 \\ 0 & n = 0,\, s \notin T \\ \mathbf{1}\{w(\mu) \mid s \xrightarrow{\tau} \mu\} & s \in S_1,\, n > 0 \\ \mathbf{2}\{w(\mu) \mid s \xrightarrow{\tau} \mu\} & s \in S_2,\, n > 0 \end{cases}$$

For $n > 0$, when $s \in S_1$, then for the next iteration the reachability probability from s is the optimal value of the set $\{w(\mu) \mid s \xrightarrow{\tau} \mu\}$ w.r.t. objective **1**; whereas when $s \in S_2$, it is w.r.t. objective **2**. Note that $\mathrm{Prt}_{\mathbf{2}}^{\mathbf{1}}$ is a monotonic function over W and, by Tarski's theorem [20], has a least and a greatest fixpoint. This definition provides the basis to compute reachability probabilities. A similar function has been defined in [11].

The next theorem shows that simulation/alternating simulation relations between PGA provide bounds on their reachability probabilities when players collaborate/compete with each other. In fact, simulation relations between PGA bound $\max^{\blacktriangle}$ and $\min^{\blacktriangledown}$ values, and alternating simulation relations $\max^{\blacktriangledown}$ and $\min^{\blacktriangle}$ values.

Theorem 7. *For $x \in \{\mathrm{sb}, \mathrm{db}\}$, let PGA $\mathcal{G}$ and $\mathcal{G}'$ with $\mathcal{G} \prec_x \mathcal{G}'$ and $\mathcal{G} \preccurlyeq_x \mathcal{G}'$. Let $T \subseteq S$ such that $T' = \{s' \in S' \mid \exists s \in T : s \prec_x s'\}$ and $T'' = \{s' \in S' \mid \exists s \in T : s \preccurlyeq_x s'\}$, then: (1) $\min^{\blacktriangledown}(T') \leq \min^{\blacktriangledown}(T)$ and $\max^{\blacktriangle}(T) \leq \max^{\blacktriangle}(T')$, (2) $\min^{\blacktriangle}(T) \leq \min^{\blacktriangle}(T'')$ and $\max^{\blacktriangledown}(T'') \leq \max^{\blacktriangledown}(T)$*

As abstractions of PGA preserve simulation and alternating simulation relations, their optimal probabilities are bounded by their abstract models. This is laid down in the following corollary, a direct consequence of Th. 2, 3 and 7:

Corollary 1. *Let $\mathcal{G} = \alpha_{\mathrm{PA}}(\mathcal{M})$ for PA $\mathcal{M}$, and $x \in \{\mathrm{sb}, \mathrm{db}\}$ with $\mathcal{G}' = \alpha_x(\mathcal{G})$. Let $T \subseteq S_2$ such that $T' = \alpha_x(T)$. Then $\min^{\blacktriangledown}(T') \leq \min(T) \leq \min^{\blacktriangle}(T')$ and $\max^{\blacktriangledown}(T') \leq \max(T) \leq \max^{\blacktriangle}(T')$.*

Note that for every $s \in T$, we have $s \prec_x \alpha_x(s)$ and $s \preccurlyeq_x \alpha_x(s)$ for $x \in \{\mathrm{sb}, \mathrm{db}\}$. Moreover, the target states are only player 2 states as they represent the partitions of the concrete states of PA. Next, as one of the main results of this work, we show that distribution-based abstraction of PA is more precise than state-based abstraction. This result is a direct consequence Th. 4 and 7.

Corollary 2. *Let $\mathcal{G} = \alpha_{\mathrm{PA}}(\mathcal{M})$ for PA $\mathcal{M}$, $\mathcal{G}_{\mathrm{sb}} = \alpha_{\mathrm{sb}}(\mathcal{G})$ and $\mathcal{G}_{\mathrm{db}} = \alpha_{\mathrm{db}}(\mathcal{G})$ with $\alpha_{\mathrm{sb}}(S) = \alpha_{\mathrm{db}}(S)$. Let $T \subseteq S_2$ such that $T_{\mathrm{sb}} = \alpha_{\mathrm{sb}}(T)$ and $T_{\mathrm{db}} = \alpha_{\mathrm{db}}(T)$. Then $\min(T) \leq \min^{\blacktriangle}(T_{\mathrm{db}}) \leq \min^{\blacktriangle}(T_{\mathrm{sb}})$ and $\max^{\blacktriangledown}(T_{\mathrm{sb}}) \leq \max^{\blacktriangledown}(T_{\mathrm{db}}) \leq \max(T)$.*

Example 6. The minimum probability in PA $\mathcal{M}$ (Fig. 2) to reach state s_6 is 0.05. By Corollary 1, this probability lies in $[0, 0.25]$ for $\alpha_{\mathrm{sb}}(\mathcal{G}) = \tilde{\mathcal{G}}$ (Fig. 4 left). Instead, $\alpha_{\mathrm{db}}(\mathcal{G}) = \mathcal{G}'$ (Fig. 4 right) yields $[0, 0.125]$.

8 Distribution-Based Game Abstraction of MDP

In [6], abstract models of Markov decision processes (MDP) are given as stochastic games (SG). These abstractions coincide with our state-based abstractions. The abstract models of MDP induced by our *distribution-based abstraction* are PGA that generalize SG. By Th. 4, our distribution-based abstraction induces more precise abstract models than state-based abstraction. This shows the superiority of our distribution-based abstraction technique over [6]. The following corollary follows from Def. 5 and Th. 4. It asserts that our distribution-based abstraction induces more precise abstractions than [6].

Corollary 3. *For* PA $\mathcal{M}'$, *let* $\mathcal{G} = \alpha_{\mathrm{PA}}(\mathcal{M}')$. *If* $\alpha_{\mathrm{sb}}(S) = \alpha_{\mathrm{db}}(S)$, *then:* $\alpha_{\mathrm{db}}(\tau(\mathcal{G})) \prec_{\mathrm{sb}} \alpha_{\mathrm{sb}}(\tau(\mathcal{G}))$ *and* $\alpha_{\mathrm{db}}(\tau(\mathcal{G})) \preccurlyeq_{\mathrm{sb}} \alpha_{\mathrm{sb}}(\tau(\mathcal{G}))$.

One may argue that although PGA-based abstract models of MDP are at least as precise as SG-based ones this comes at the expense of larger games, — e.g. more space is required to store the target distributions of player 2 transitions. The following example shows that for abstracting $\mathcal{G}$ — an embedding on MDP — with $\alpha_{\mathrm{db}}(S_2) = \alpha_{\mathrm{sb}}(S_2)$, $\alpha_{\mathrm{db}}(\mathcal{G})$ is at least as precise as $\alpha_{\mathrm{sb}}(\mathcal{G})$ and $|\alpha_{\mathrm{db}}(S)| \leq |\alpha_{\mathrm{sb}}(S)|$. (Recall that the same partition of player 2 states does not imply the same partition for player 1 states, as shown in Example 5).

Example 7. The maximum probability in PA $\mathcal{M}$ (Fig. 1) to reach states $\{s_8, s_9\}$ equals 0.3. By Corollary 1, this probability lies in $[0.25, 0.5]$ for the state-based abstraction $\tilde{\mathcal{G}}$ (Fig. 5 left). Instead, distribution-based abstraction $\mathcal{G}'$ (Fig. 5 right) yields $[0.3, 0.3]$. Moreover, ignoring player 2 transitions — such that the successor states from player 2 states are decided non-deterministically as in [6] — yields $[0.25, 0.5]$ in $\tilde{\mathcal{G}}$ and $\mathcal{G}'$. However, in terms of number of transitions and states, the size of $\mathcal{G}'$ is smaller than $\tilde{\mathcal{G}}$ (see Fig. 5).

As a side result of our achievement, we put the result of [9][Th. 2] in perspective: game-based abstraction is the optimal *state-based* abstraction, but not the optimal abstraction preserving reachability probabilities.

9 Conclusion

We gave two abstraction techniques — state-based and distribution-based — for PA, and presented PGA as abstract models for PA. We defined a composition operator for a class of PGA that act as abstract models for PA; and gave two notions of simulation and alternating simulation relations for PGA that are precongruences w.r.t. composition. Our distribution-based abstraction is more precise as well as concise than the one in [19]. Future work includes the application of this work to practical case studies, and the extension of abstraction-refinement framework, in [19], for PA.

Acknowledgement. We thank the reviewers for the constructive feedback, and helping us improve the quality of the paper.

References

1. Segala, R.: Modeling and Verification of Randomized Distributed Real-Time Systems. PhD thesis, Massachusetts Institute of Technology (1995)
2. Delahaye, B., Katoen, J.-P., Larsen, K.G., Legay, A., Pedersen, M.L., Sher, F., Wasowski, A.: Abstract probabilistic automata. Information and Computation 232, 66–116 (2013)
3. Delahaye, B., Katoen, J.-P., Larsen, K.G., Legay, A., Pedersen, M.L., Sher, F., Wąsowski, A.: Abstract probabilistic automata. In: Jhala, R., Schmidt, D. (eds.) VMCAI 2011. LNCS, vol. 6538, pp. 324–339. Springer, Heidelberg (2011)

4. Sher, F., Katoen, J.-P.: Compositional abstraction techniques for probabilistic automata. In: Baeten, J.C.M., Ball, T., de Boer, F.S. (eds.) TCS 2012. LNCS, vol. 7604, pp. 325–341. Springer, Heidelberg (2012)
5. Katoen, J.-P., Klink, D., Leucker, M., Wolf, V.: Three-valued abstraction for probabilistic systems. J. Log. Algebr. Program. 81(4), 356–389 (2012)
6. Kattenbelt, M., Kwiatkowska, M.Z., Norman, G., Parker, D.: A game-based abstraction-refinement framework for Markov decision processes. Formal Methods in System Design 36(3), 246–280 (2010)
7. Shapley, L.S.: Stochastic games. Proceedings of the National Academy of Sciences of the United States of America 39(10), 1095–1100 (1953)
8. Condon, A.: The complexity of stochastic games. Information and Computation 96, 203–224 (1992)
9. Wachter, B., Zhang, L.: Best probabilistic transformers. In: Barthe, G., Hermenegildo, M. (eds.) VMCAI 2010. LNCS, vol. 5944, pp. 362–379. Springer, Heidelberg (2010)
10. Jonsson, B., Larsen, K.G.: Specification and refinement of probabilistic processes. In: LICS, pp. 266–277. IEEE CS Press (1991)
11. Condon, A., Ladner, R.E.: Probabilistic game automata. Journal of Computer and System Sciences 36(3), 452–489 (1988)
12. Ash, R.B., Doléans-Dade, C.A.: Probability & Measure Theory, 2nd edn. Academic Press (2000)
13. Bertsekas, D.P., Tsitsiklis, J.N.: An analysis of stochastic shortest path problems. Mathematics of Operations Research 16, 580–595 (1991)
14. de Alfaro, L.: Computing minimum and maximum reachability times in probabilistic systems. In: Baeten, J.C.M., Mauw, S. (eds.) CONCUR 1999. LNCS, vol. 1664, pp. 66–81. Springer, Heidelberg (1999)
15. Lynch, N.A., Segala, R., Vaandrager, F.W.: Observing branching structure through probabilistic contexts. SIAM J. Comput. 37(4), 977–1013 (2007)
16. Eisentraut, C., Hermanns, H., Zhang, L.: On probabilistic automata in continuous time. In: LICS, pp. 342–351. IEEE CS Press (2010)
17. Doyen, L., Henzinger, T.A., Raskin, J.F.: Equivalence of labeled Markov chains. Int. J. Found. Comput. Sci. 19(3), 549–563 (2008)
18. Alur, R., Henzinger, T.A., Kupferman, O., Vardi, M.Y.: Alternating refinement relations. In: Sangiorgi, D., de Simone, R. (eds.) CONCUR 1998. LNCS, vol. 1466, pp. 163–178. Springer, Heidelberg (1998)
19. Kattenbelt, M.: Automated Quantitative Software Verification. PhD thesis, University of Oxford (2010)
20. Tarski, A., et al.: A lattice-theoretical fixpoint theorem and its applications. Pacific Journal of Mathematics 5(2), 285–309 (1955)

Author Index